Birkhäuser

Science Networks. Historical Studies

Volume 66

This series is devoted to historical studies in the fields of mathematics, physics, and astronomy, including their applications. It is mainly composed of monographs, although it may occasionally include critical editions of primary sources and biographies of important actors. It aims to attract a multiple audience including scientists, historians, philosophers, and graduate students throughout the academic world. Submitted projects are examined by an international and multidisciplinary committee of experts. The publication language is preferentially English, and the diffusion is international.

Holger Becker

A Historical Dictionary of Mathematical Terms

Volume I: A–I

Birkhäuser

Holger Becker
IBS IT & Business School Oldenburg
Oldenburg, Germany

ISSN 1421-6329 ISSN 2296-6080 (electronic)
Science Networks. Historical Studies
ISBN 978-3-032-02872-3 ISBN 978-3-032-02873-0 (eBook)
https://doi.org/10.1007/978-3-032-02873-0

Mathematics Subject Classification (2020): 01-02, 01-XX, 01AXX, 01-00, 01-11

This book is published under the imprint Birkhäuser, www.birkhauser-science.com by the registered company Springer Nature Switzerland AG
The registered company address is: Gewerbestrasse 11, 6330 Cham, Switzerland

To my parents

Preface

If the dream is won
Though everything is lost
We will pay the price
But we will not count the cost
– Neil Peart / John Barth

This book is meant to fill a gap in the literature: there is no *one* reference work where detailed and reliable information on the history of mathematical terms can be looked up. Instead, such information is scattered in various kinds of secondary sources such as general historical dictionaries, specialized historical dictionaries as well as books and articles on the history of mathematics. Additionally, the quality of the secondary literature varies from excellent to misleading, incomplete, or just wrong. The history of some terms cannot be looked up anywhere.

I became dissatisfied with this situation while I was working on my PhD thesis, which dealt with mathematical terminology in the 19th century. I needed to know the precise origins of certain terms like *group*, *isomorphism*, *field* – these are comparatively easy to trace. But other terms, for example *skew field*, are more difficult. At the time (2003), the first reference I had was from the OED, which cited a source from 1965. When reading various books and journals, I easily found earlier usages of the English term, although it soon became clear that the term was in fact a German creation (*Schiefkörper*), first used in print by Emil Artin in 1927, who explicitly attributed the term to B. L. van der Waerden. Further questions came up: why *schief* 'slanted, not straight'? I found the answer, eventually, in 19th-century sources in which the French word *gauche*, literally meaning 'left' but also 'skew', was used with reference to determinants and matrices satisfying $a_{ij} = -a_{ji}$.

Eventually, I decided to compile a historical dictionary of mathematical terms myself. I wanted a book that covered the core terminology, discussing in detail the origins and histories of these terms, based on the primary sources, and giving citations from the original authors as well as historical illustrations. Whenever available, I also wanted this book to contain comments made by those who coined a new word as well as comments by those who first accepted or refected them. For many terms, various statements of approval have been found, others provoked (sometimes very strong) statements of rejection, and with some terms opinions are divided. While such statements often indicate the success or non-success of a terminological suggestion, sometimes even the strongest disapproval does not impede the success of a new creation: for example, *eigenvalue* has been called “a barbarism without sense or convenience, and equally repellent to German and to English ears”.

All in all, it took me about 20 years (with some interruptions) to do all the research necessary for this book, write it, and also typeset it. During this process, many challenges came up, and I also made various unexpected findings. And with all that said, I am now most pleased to present my dictionary to the public and hope readers will enjoy consulting it. It is for all those who are interested in the history of mathematical terms.

Oldenburg in Oldenburg, Germany *Holger Becker*
May 2025

Acknowledgements

First of all, I would like to thank Winfried Boeder (1937-2023) for his longtime support, beginning with my studies in the early 1990s. Apart from sparking my interest in historical semantics, he supervised my PhD thesis and afterwards often discussed the progress on the dictionary with me. He sacrificed a lot of time helping me understand many passages in Latin and Greek, and often translated and analyzed them for me. The book would not be what it is without his support.

Erhard Scholz read and commented on various entries. I have happily adopted his suggestions for improving the book. I also wish to thank the anonymous reviewers. Their comments on numerous entries as well as their general remarks on the dictionary helped me revise and rethink a number of entries. At their advice I also deleted some unnecessary entries, wrote some new ones, and extended various entries considerably.

I would also like to thank Berni Metcalfe of the National Library of Ireland, who supplied me with some rare texts I wanted to be included in the article for *inversion* as used in geometry. The librarians at the Landesbibliothek Oldenburg and at the library of the Carl von Ossietzky University in Oldenburg were very helpful with the many interlibrary loans I made over the years.

I am also very grateful to Jessica Sänger for valuable advice on some technical matters, and to Jonathan Blaney for constructive comments on the preface, the introduction, and other things.

I thank Frida Trotter from Birkhäuser for valuable assistance and support during this project. I am very grateful to her as well as to the editors, Olivier Darrigol and Peter Ullrich, for inclusion of my book in *Science Networks*.

I would finally like to thank my wife and "the community" (you know who you are) for their enduring support while I was writing this book. They were convinced that I was engaged in a never-ending project and could not believe it when I eventually proclaimed that I actually finished my book.

Last, I thank my parents for always believing in me and for their unconditional support throughout the years.

Contents

Introduction

In this section, I would like to explain some general aspects concerning the dictionary, in particular, its overall structure, its coverage, the structure of the entries, and the information contained in them. The section will close with a brief discussion of how the dictionary was made.

The general structure of the dictionary is straightforward. Apart from the usual front matter, there is an index (in both volumes) just before the actual dictionary. This index contains a list of all entries (there are about 1,150) and also refers to words that are discussed in the dictionary but do not have an entry of their own. The index is therefore more comprehensive than an ordinary table of contents would have been. The dictionary itself is arranged alphabetically. Some entries serve only as cross-references to other entries; for example, *parabola* does not have entry of its own and thus points to the relevant entry *conic section.*

In the electronic version, the cross-references and the entries in the index can be clicked and the corresponding entry will be opened. This works within each volume, but not across the two volumes.

In this dictionary, an attempt has been made to cover many commonly used mathematical terms from a large variety of mathematical fields. These include geometry (including differential, algebraic, and analytic geometry), trigonometry, arithmetic and number theory, algebra (including linear algebra), calculus and mathematical analysis, mathematical logic and set theory, topology, combinatorics, graph theory, functional analysis, probability and measure theory, category theory, and, to a lesser extent, statistics and stochastics.

Decisions have to be made in the making of a dictionary, the most fundamental one concerning the issue of which terms to include, and which ones to exclude. I wanted the dictionary to contain entries for the most important concepts and objects as well as for the major branches of mathematics. To achieve this, I proceeded by comparing the entries in Weisstein's *Encyclopedia of Mathematics* with various standard textbooks in the mathematical branches covered in the dictionary. On the basis of this comparison, a word list containing the terms to be included was established.

Terms that seemed to be too specialized were excluded as well as certain other kinds of terms; in particular, numerals, names of conjectures, theorems, problems, equations, etc. are not covered in the dictionary (the only exception is the entry for *triangle inequality*).

The terms and meanings discussed in the dictionary usually still exist and are in common use, another decision deliberately made. There are very few exceptions to this rule. Examples are the entries *indivisible* and *fluxion*, now obsolete terms which did, however, play an important role in the development of the calculus.

A general caveat about the selection of terms is in order here: there is no "objective" way of deciding which terms to include. Criteria like "the most important" terms, or possibly "the most central" or "essential" ones, cannot be defined in a satisfactory way, and they yield no clearcut ways to decide whether a specific term should be included or not. Ultimately, I made the decisions that I made, and I hope they are reasonable and acceptable. But I am also aware that dictionary makers often face criticism for the choices they make.

Let me now move on to the entries themselves.

Usually, each entry covers one term. However, it often seemed to make sense to group certain words together as, for example, in the entries *elliptic function* and *elliptic integral*, *cohomology* and related *"co*-terms*"*, and *countable* and *uncountable*. The combination of what might have been several entries into just one entry is meant to reduce overlaps and repetitions.

The entries themselves discuss the origin and development of a given term. Every entry has the following goals, where possible:

1. To say explicitly and precisely who used the term first or at least when it came up,
2. To illustrate first uses and discuss historical comments on the term,
3. To indicate possible earlier occurrences of the concept, especially if in these cases there are paraphrases which may be helpful to understand the terminology later chosen,
4. To explain why a particular name was chosen,
5. To elucidate the development of the term, especially in terms of semantic changes and loans,
6. To name synonyms and competing terms,
7. To give precise references for the original texts discussed,
8. To indicate relevant secondary literature.

Not all goals could always be attained. For example, the histories of some terms are so complex that whole books could be and have been written about them; this concerns words like *analysis* or *algebra*. In such cases, simplifications have been necessary, but at least an outline of major shifts of meaning has been given and references to more detailed secondary sources have been indicated.

Another example: while I have made great efforts to trace the origins of the terms covered in the book, there are some words that appear to defy such efforts. This resistance applies to various words used in the calculus, for instance. A different, rather notorious example is *golden section*: the concept is ancient, but *golden* in this sense is not, despite many mistaken claims to the contrary. Until recently, the earliest known usages of this term were from the 1830s. In 2019, Roger Herz-Fischler presented a usage from 1789, and I added one from 1717. In this book, I am happy to present still earlier usages from 1707 and 1710, but if these are really the earliest ones remains an open question.

The structure of the entries varies to some extent. Entries discussing classical terms such as *circle* usually first deal with the Greek terminology, then move on to classical or medieval Latin, and possibly to other languages such as Sanskrit and classical Arabic, and eventually treat the term in "modern" languages such as English, German, and French.

Many entries start with a definition of the concept in question, often quoted from the original literature. In the case of very long, highly technical, or somewhat cumbersome definitions requiring multiple explanations, paraphrases outlining the gist of the concept have been given. For example, in the entry *game* a relatively short paraphrase is cited, since the axiomatic definition given by von Neumann runs to over a complete page.

Some entries have a complex structure; this typically happens when a word changed its meaning several times. The entries for *algebra* or *statistics*, for example, have for this reason been divided into several sections. For some particularly polysemous words such as *conjugate* or *modulus*, additional overviews are given in introductory entries.

I have always tried to find interesting passages that could be cited directly. Direct quotations from primary sources have, in particular, been included for definitions and often for terminological comments. Quotations from languages other than English have been translated. The original text is usually given in a footnote (sometimes in shortened form), but for some languages such as Greek, Arabic, and Russian, the original has not been cited. The translations are mine unless otherwise indicated.

Sources are usually quoted by page numbers, but some books are quoted by paragraphs, allowing easier access to the cited passage if there are various editions. If quoting in terms of page numbers is impossible (which is often the case with older texts), the folio or structural part is given. Classical works such as those by Plato or Aristotle are cited in the traditional way; e. g. Plato (*Timaeus* (88b). In the case of alphabetically arranged reference works such as the OED or Harris' *Lexicon technicum*, I have just cited the name of the work without indicating the specific entry when it is clear from the context which entry is meant. Unless otherwise indicated, the emphasis in the quotations is always the original one. A typographical convention: in the text, I use italics when a word is referred to (except in the quotations from original works), and single quotation marks to indicate meanings, as in the sentence: The word *folium* means 'leaf'.

Various entries contain illustrations found in the primary sources. Most of the figures are taken directly from these sources, but some have been redrawn as closely as possible.

At the end of most entries there is a note pointing out relevant secondary literature. "Relevant" means that the references deal with the history of the concept under discussion (or just with certain aspects of the history) or that they discuss the history of the terminology itself. Some of the literature cited in the notes is actually quite old and should be read with appropriate caution; I still found those texts useful and worth pointing them out. Let me explain the reasons for this by means of some examples. In the entries of particular curves, for instance, I have often cited works by Loria and Teixeira, which date from the early 20th century. While their books are not always reliable because in some cases they miss relevant authors or contain mistakes, I still found them useful because they offer very precise references to the primary literature, which is often not the case with later literature on curves. This precision made it easy to verify their claims, and it also helped me find competing terminology which I might otherwise not have found. A second example is Tropfke, whose books published between the 1920s and 1940 are often cited in entries about elementary mathematical terms. Tropfke also worked with original sources, and also cited very precisely, sometimes even giving line numbers. He worked very systematically and was meticulous in tracing the history of terms and even today, many of his assertions for first uses have not been been antedated. A third example is the classic work by von Braunmühl on the history of trigonometry, dating from the early 20th century. His two-volume book is still useful in many respects, while in others it is outdated. His work is now supplemented by numerous articles on the history of trigonometry as well as by the books by Van Brummelen. As a final example, I have sometimes cited articles from the *Encyklopädie der mathematischen Wissenschaften mit Einschluss ihrer Anwendungen* and its French counterpart. Again, these articles often rather comprehensively cover the topics they deal with. With all of these sources, as indeed for all secondary sources on first uses, I never just relied on them but always checked the primary sources.

In conclusion, I cited these sources when I found them useful for my work and, potentially, also for the work of others. But caution is in order, since parts of them may be outdated, or they may contain anachronisms or be unreliable in some places.

Apart from these general remarks, let me also point out that in various entries such as those for *commutator*, *hyperbolic spiral*, *dual space*, and *polytope*, I cited some secondary sources that were useful in preparing the respective entry but I also pointed out some deficiencies that I saw in them.

Regarding the notes, I should add that, occasionally, additional notes have been included; these cover certain special aspects that may be of interest. For example, the entry *sine* contains a note on the history of the discovery of the word's etymology, and the entry *slope* discusses a hypothesis concerning the use of m as in $y = mx + n$.

Each entry has a separate reference section listing the primary and secondary literature cited in the entry. This does increase the size of the book, but it makes looking up the references very convenient, and this user-friendliness made me opt for this approach. A single bibliography at the end of the book would have been very long and, in my opinion, tedious to consult. In the reference sections, very long book titles have been shortened.

Note that when reference to Euclid's *Elements* is made, which is regularly the case in connection with geometrical and arithmetical terms, the text is from Heiberg's standard edition. For reasons of space, this edition has not been listed in the respective bibliographies. All other editions and translations of the *Elements* referred to *are* listed in the bibliographies.

A technical point: in the cases of surnames having a prefix such as *von*, the respective item in the bibliography will be sorted under the surname. For example, van der Waerden will be found under W, not under V.

An additional point on the bibliography and the sources: for many mathematicians, collected papers are available. In this dictionary, however, reference has almost always been made to the original publications. The reason is that the standards for collected papers have not always been as high as they are nowadays, and there are cases where editorial changes have been made without indicating them. A case in point are the collected papers of Weierstraß, which contain unmarked redactions made by Weierstraß himself; see, for example, the entry *circle of convergence*. Collected papers *have* been referred to for editorial comments and unpublished material such as manuscripts or correspondence.

To conclude this introduction, I would now like to give an overview of how this dictionary was made.

As there are hardly any precursors for historical dictionaries of special languages, the methodology used has mostly been developed in analogy to methods long known and applied in general historical dictionaries like the OED. In essence, this dictionary is based on extensive reading of and excerpting from primary and secondary literature as well as the equally extensive use of electronic searches.

After establishing a list of entries to be included in the dictionary as described above, the second step involved the identification of major texts, usually books and articles, from all the subject fields the dictionary was supposed to cover. This identification was an ongoing process. Initially, I just produced a list of major sources in algebra, topology, etc., but the list became more refined in the course of time: for example, I produced lists of 12th- and 13th-century mathematical texts in Latin (often translations from Arabic and Greek), lists of early mathematical texts in the vernaculars, or, to give a rather different example, a list of 19th- and early 20th-century texts on graph theory. Additionally, I made a list of the major mathematical journals.

The next step consisted in taking excerpts. Note that at the outset of the project (2005), electronic searches were only possible to a limited extent, and these excerpts were made in the traditional way. That means, I created a database and went through the assembled sources. During this process, notable usages of the terms to be included were keyed into the database. Such notable usages were those where a definition was given, where interesting comments on a term were made (for example, historical comments or judgements on how appropriate a term was considered to be), or usages which seemed to be interesting for some other reason.

The three steps should not be thought of as a linear process, with one step beginning after the previous step was finished. The word list has been modified several times (although only to a limited extent), and the corpus of texts to be processed grew constantly, as did the database with notable usages.

The procedure described above in connection with primary sources was also carried out in connection with secondary sources.

Note that while the text corpus as described above contains a very large number of items, it can in no way be said to be "comprehensive". One major reason is that there are still many sources that have never been published in print and only exist in manuscripts. This gap especially concerns medieval mathematical texts in Latin or Arabic as well as 13th- to 15th-century mathematical sources in modern languages, even though the number of edited sources is continually growing. There seems to be quite a high potential for improvement in these fields. Unpublished manuscripts have not generally been consulted in the making of this dictionary, but a few exceptions have been made: some manuscripts have been cited when they contained clear examples of early uses of arithmetical or geometrical terms, for example, and I also consulted some transcripts of lectures held by Weierstraß, Klein, and others. The cited manuscripts are all publicly available online.

The possibilities of doing electronic searches increased in the course of time, and extensive searches have been made in the preparations for all entries. The older, "hand-made" material still proved to be useful because electronic searches depend on the degree that optical character recognition works accurately, and the accuracy various considerably. Especially with older texts, it may be rather low.

An illustrative rather than exhaustive list of electronic research possibilities is, in no particular order:

- `https://books.google.com`
- `https://www.english-corpora.org/eebo/`
- `http://dml.mathdoc.fr`
- `https://www.jstor.org`
- `https://link.springer.com`
- `https://projecteuclid.org/`
- `http://www.ams.org/epubsearch/`
- `https://www.jstage.jst.go.jp`
- `http://www.numdam.org`
- Corpus corporum `https://www.mlat.uzh.ch`
- Aristoteles latinus database `https://www.brepols.net/series/ald-o`

The scope of these sites varies considerably. For example, Google Books includes books on practically all subject matters in various languages, Early English Books Online (EEBO) focuses on early printed texts in English, while on NUMDAM (NUMérisation de Documents Anciens Mathématiques) mathematical texts are available, mostly, but not solely, in French. The last two sites on the list were sometimes useful in tracing Latin terms.

Apart from searching my database, doing electronic searches and, of course, reading the relevant secondary sources on the history of mathematics, I usually also consulted available reference works (where appropriate), such as the ones found in the list of abbreviations. Reference works available online (such as the OED) have last been consulted in late 2024; antedatings to the OED (unless otherwise indicated, I refer to the revised entries in OED3) have been indicated in the respective entries. I am also planning to inform the editors of the OED about all the antedatings that I have found.

I also consulted various additional reference works, editions of mathematical texts that contain useful glossaries, and articles that discuss mathematical terminology; when used, these have, of course, been cited in the respective bibliographies.

A final note: while this book has been proofread many times, it seems unlikely, considering its considerable length and the wealth of material used, that it is free of errors. If you find any errors, or if you have any suggestions for additions or modifications, feel free to let me know at BeckerHDMT@gmail.com.

List of Figures

List of Tables

List of Abbreviations

DEAF	Dictionnaire Étymologique de l'Ancien Français
DFSM	Dictionnaire du Français Scientifique Médiéval
DMF	Dictionnaire du Moyen Français
DWB	Deutsches Wörterbuch
GDLI	Grande dizionario della lingua italiana
LS	Lewis - Short, A Latin Dictionary
LSJ	Liddell - Scott - Jones, A Greek-English Lexicon
MED	Middle English Dictionary
OED	Oxford English Dictionary
OLD	Oxford Latin Dictionary
TLL	Thesaurus Linguae Latinae
TLF	Trésor de la Langue Française
TLIO	Tesoro della Lingua Italiana delle Origini

Index

Page numbers in bold refer to entries. The index contains all headwords of the dictionary entries as well as some notable usages of certain terms. Roman numerals indicate the volume, Arabic ones the corresponding page.

The Dictionary

A

Abelian equation and *Abelian group*

See *commutative.*

Abelian function, *Abelian integral*, and *Abelian differential*

In the introduction of Abel (1828),[1] an addition theorem is stated; an informal paraphrase of this theorem is given by Kleiman (2004, p. 308): "the integrals of arbitrary algebraic differentials[2] form a very large class of transcendental functions with an addition formula; if an arbitrary sum of functions arising from the same differential cannot be given by a single function of the same sort, as is the case for elliptic functions, at least the sum can be given by a sum of a specific number of them, plus certain algebraic and transcendental terms". Legendre (1828, p. 191) applied this theorem to hyperelliptic integrals and used the term *ultra-elliptic* to refer to the more general integrals (see Kleiman 2004, p. 308).

In turn, Legendre's book was reviewed by Jacobi (1832, p. 415), who suggested *Abelian transcendents* (*Abelsche Transcendenten*) for these integrals. He wrote:

> To the transcendental functions $\int \frac{f(x)\,dx}{\sqrt{X}}$, where X exceeds the fourth degree, Legendre gives the name *hyperelliptical* (*ultra-elliptiques*). We wish to call them *Abelian transcendents*, for it was Abel who first introduced them into analysis and who first made evident their great importance by his far-reaching theorem. For this theorem, as the most fitting monument to this extraordinary genius, the name *Abelian theorem* would be very appropriate. For we happily agree with the author that it carries the full imprint of the depth of his ideas. Since it enunciates in a simple manner, without the vast setup of mathematical formalism, the deepest and most comprehensive mathematical thought, we consider it to be the greatest

[1] A fuller article from the time (whose publication was significantly delayed, however) is Abel (1841). [2] The term *integrals of arbitrary algebraic differentials* was used by Abel (1828, p. 313) for Abelian integrals.

H. Becker, *A Historical Dictionary of Mathematical Terms*, Science Networks. Historical Studies 66, https://doi.org/10.1007/978-3-032-02873-0_1

> mathematical discovery of our time although only future–perhaps distant–hard work will be able to reveal its whole importance.[1]

However, the terminology was soon changed to *Abelian integral*. In print, the term seems to have been first used (in Latin) by Richelot (1834, p. 182). It appears to be due to Jacobi as well, who used it, for example, in a paper written, according to Koenigsberger (1904, p. 153), in the early 1830s but only published posthumously (Jacobi 1858).

Not long afterwards, the terminology is also found in other languages. Ellis (1847, p. 74), for example, used English *Abelian integral* and attributed it to Jacobi.

In the following years, Abelian integrals attracted considerable attention. Two important publications were Weierstraß (1856) and Riemann (1857), both entitled "Theory of Abelian Functions". As explained by Markushevich (1996, p. 249), they used the term in different ways: "The fact is that Weierstrass [...] defined Abelian functions to be the single-valued analytic functions of several variables to which his solution of the Jacobi problem led. Riemann, however, in accordance with the tradition of the time (which also originated with Jacobi), understood Abelian functions to be simply Abelian integrals".

Towards the end of the 19th century, *Abelian differential* also came into use. Its precise origin is obscure,[2] but a clear usage is found in Jordan (1894, p. 538): "Let us consider an Abelian differential $R(x, y)dx$, where R designates a rational function and y is linked to x by an algebraic equation $f(x, y) = 0$".[3] In the 20th century, new definitions were brought up, for example by Wang (1954, p. 776): "An abelian differential of the first kind is, by definition, a closed holomorphic 1-form".

See also *hyperelliptic*.

References

Abel, Niels Henrik (1828). "Remarques sur quelques propriétés générales d'une certaine sorte de fonctions transcendantes". *Journal für die reine und angewandte Mathematik* 3, pp. 313–323.

Abel, Niels Henrik (1841). "Mémoire sur une propriété générale d'une classe très étendue de fonctions transcendantes". *Mémoires présentés par divers savants à l'Académie Royale des Sciences de l'Institut de France* 7, pp. 176–264.

Ellis, Robert Leslie (1847). "Report on the Recent Progress of Analysis (Theory of the Comparison of Transcendentals)". *Report of the British Association for the Advancement of Science* (1846), pp. 34–90.

Jacobi, Carl Gustav Jacob (1832). "Legendre, *Théorie des fonctions elliptiques, troisième supplément* [review]". *Journal für die reine und angewandte Mathematik* 8, pp. 413–417.

[1] "Legendre giebt den Transcendenten $\int \frac{f(x)dx}{\sqrt{X}}$, wenn X den vierten Grad übersteigt, den Namen der *hyperelliptischen* (*ultra-elliptiques*). Wir würden sie die *Abelschen Transcendenten* nennen, da Abel zuerst sie in die Analysis eingeführt und durch ein umfassendes Theorem ihre grosse Bedeutung nachgewiesen hat. Diesem Theoreme selbst dürfte wohl vorzugsweise, als dem schönsten Monumente dieses ausserordentlichen Geistes, der Name des *Abelschen Theorems* zukommen. Denn gern stimmen wir dem Verfasser bei, dass es das ganze Gepräge seiner Gedankentiefe trägt. Wir halten es, wie es in einfacher Gestalt ohne Apparat von Calcul den tiefsten und umfassendsten mathematischen Gedanken ausspricht, für die grösste mathematische Entdeckung unserer Zeit, obgleich erst eine künftige, vielleicht späte, grosse Arbeit ihre ganze Bedeutung aufweisen kann". Translation from Scriba (1973, p. 53) (slightly adapted). [2] Koenigsberger (1879, p. 337) already seems to have used the term. [3] "Considérons la différentielle abélienne $R(x, y)dx$, R désignant une fonction rationnelle et y étant lié à x par une équation algébrique $f(x, y) = 0$".

Jacobi, Carl Gustav Jacob (1858). "Über die Substitution $(ax^2 + 2bx + c)y^2 + 2(a'x^2 + 2b'x + c')y + a''x^2 + 2b''x + c'' = 0$ und über die Reduction der *Abel*schen Integrale erster Ordnung in die Normalform". *Journal für die reine und angewandte Mathematik* 55, pp. 1–14.
Jordan, Camille (1894). *Cours d'analyse*. Tome deuxème. Calcul intégral. Paris.
Kleiman, Steven L. (2004). "What is Abel's Theorem Anyway?" In: *The Legacy of Niels Henrik Abel*. The Abel Bicentennial, Oslo. Ed. by Olav Arnfinn Laudal and Ragni Piene. Berlin et al., pp. 395–440.
Koenigsberger, Leo (1879). "Ueber eine Beziehung der complexen Multiplication der elliptischen Integrale zur Reduction gewisser Klassen Abelscher Integrale auf elliptische". *Journal für die reine und angewandte Mathematik* 86, pp. 317–352.
Koenigsberger, Leo (1904). *Carl Gustav Jacob Jacobi*. Leipzig.
Legendre, Adrien-Marie (1828). *Traité des fonctions elliptiques et des intégrales eulériennes*. Tome troisième. Paris.
Markushevich, Alexei Ivanovich (1996). "Analytic Function Theory". In: *Mathematics of the 19th Century*. Geometry. Analytic Function Theory. Ed. by Andrey Nikolaevich Kolmogorov and Adolph Andrei Pavlovich Yushkevich. Basel, pp. 119–272.
Richelot, Friedrich Julius (1834). "De integralibus Abelianis primi ordinis commentatio prima". *Journal für die reine und angewandte Mathematik* 12, pp. 181–233.
Riemann, Bernhard (1857). "Theorie der Abel'schen Functionen". *Journal für die reine und angewandte Mathematik* 54, pp. 115–155.
Scriba, Christoph J. (1973). "Jacobi, Carl Gustav Jacob". In: *Dictionary of Scientific Biography*. Ed. by Charles Gillispie. Vol. 7. New York, pp. 50–55.
Wang, Hsien-Chung (1954). "Complex Parallisable Manifolds". *Proceedings of the American Mathematical Society* 5, pp. 771–776.
Weierstraß, Karl (1856). "Theorie der *Abel*'schen Functionen". *Journal für die reine und angewandte Mathematik* 52, pp. 285–379.

Abelian variety and *Jacobian variety*

According to Kleiman (2004, p. 333), the now standard definition of an Abelian variety as a "compact algebraic group variety: not only do its points form a group, but the operations of adding and of inverting are given by polynomial maps" was first given by Weil (1948, p. 28).

As of around 1920, *Abelian variety* begins to be commonly used in various languages. A definition is found in Lefschetz (1921, p. 364):

> An *Abelian* variety of genus p, V_p, is a variety whose non-homogeneous point coördinates are equal to $2p$-ply periodic meromorphic functions of p arguments $u_1, u_3, \dots u_p$, or whose homogeneous point coördinates are proportional to theta's of the same order and continuous characteristic. The variety is algebraic (Weierstrass) and of dimensionality p. When the periods are those of a set of independent integrals of the first kind of a curve of genus p, V_p is called a *Jacobi* variety.

Lefschetz (1919, p. 104) had actually used the term before. He referred to Scorza (1916) for some terms (although not specifically for *Abelian variety*), and indeed, on p. 269 of that paper, Italian *varietà abeliane* appears, as it did in Scorza (1911, p. 363). This is, however, not the first occurrence of the term. In the *Giornale di matematiche di Battaglini* 48 (1910), p. 175, there is a summary of a lecture given by Severi, and *varietà abeliane* as well as *varietà di Jacobi* are on the list of topics.

On the broadness of the notion of Abelian variety in early texts, Kleiman (2004, p. 334) remarks that "it included any variety whose functions form a subfield of finite index in a given field of Abelian functions, in other words, a variety that is parameterized on a classical open set by Abelian functions (see Scorza 1916, pp. 330f.). For example, an Abelian surface is just a hyperelliptic surface (see Scorza 1916, pp. 330f.). The index was termed the 'rank' by Enriques and Severi in 1909 according to Scorza (1916); however, on p. 269, he said he would use the term 'Abelian variety' only when the rank is 1".[1] Kleiman (2004, p. 333) also explains that the name is due to the following "important result: an Abelian variety is parameterized globally by Abelian functions with common periods".

The term *Jacobian variety* has a similar history. A modern definition was given by Weil (1948, p. 68) but, as seen in the quotation by Lefschetz above, the term had already been in use around 1920. Its first English occurrence seems to be Lefschetz (1919, p. 106). The corresponding German appeared in Severi (1921, p. 271).[2] In print, *varietà di Jacobi* already appeared in Severi (1911, p. 541) and in the course descriptions from 1910 cited above. It thus also seems to go back to Severi.

References

Kleiman, Steven L. (2004). "What is Abel's Theorem Anyway?" In: *The Legacy of Niels Henrik Abel*. The Abel Bicentennial, Oslo. Ed. by Olav Arnfinn Laudal and Ragni Piene. Berlin et al., pp. 395–440.

Lefschetz, Solomon (1919). "On the Real Folds of Abelian Varieties". *Proceedings of the National Academy of Sciences* 5, pp. 103–106.

Lefschetz, Solomon (1921). "On Certain Numerical Invariants of Algebraic Varieties with Application to Abelian Varieties". *Transactions of the American Mathematical Society* 22, pp. 327–406.

Scorza, Gaetano (1911). "Sopra una classe di varietà algebriche a tre dimensioni con un gruppo ∞^2 di trasformazioni birazionali in sè". *Atti della Reale Accademia dei Lincei. Rendiconti* 20, pp. 361–368.

Scorza, Gaetano (1916). "Intorno alla teoria generale delle matrici di Riemann e ad alcune sue applicazioni". *Rendiconti del Circolo Matematico di Palermo* 41, pp. 263–379.

Severi, Francesco (1908). *Lezioni di geometria algebrica*. Padova.

Severi, Francesco (1911). "Sulle superficie e varietà algebriche irregolari di genere geometrico nullo". *Atti della Reale Accademia dei Lincei. Rendiconti* 20, pp. 537–546.

Severi, Francesco (1921). *Vorlesungen über algebraische Geometrie*. Geometrie auf einer Kurve. Riemannsche Flächen. Abelsche Integrale. Berechtigte deutsche Übersetzung von Eugen Löffler. Leipzig and Berlin.

Weil, André (1948). *Variétés abéliennes et courbes algébriques*. Paris.

abscissa

The abscissa is the horizontal coordinate of a point in a plane Cartesian coordinate system, but the origin of the term goes back to conic sections. Apollonius, in his treatise on conics, used the verb *ἀποτέμνειν* 'to cut off' to indicate that one can take a determined segment on a line of indefinite length. The verb is used, for example, in prop. 20 of book I: "If in a parabola two straight lines are dropped ordinatewise to the diameter, the squares on them will be to each other as the straight lines cut off by them on the diameter beginning from the vertex are to each other" (Apollonius of Perga 1891, p. 72).[3] The "straight lines cut off" are just the abscissas.

---

[1] The quotation from Kleiman is word by word, but his citation style has been adapted. [2] The book is a translation of Severi (1908), in which the term is not found. As stated in the preface of the translation, the text was updated by Severi and contained various additions. [3] Translation from Apollonius of Perga (2013, p. 38).

In an important Latin edition of Apollonius' *Conics*, Commandino translated the verb *ἀποτέμνειν* by Latin *abscindere* (Apollonius of Perga 1566, fol. 19[r]), which also means 'to cut off'. Commandino does not use the noun *abscissa,* which is derived from *abscindere,* in the sense under discussion, however.[1]

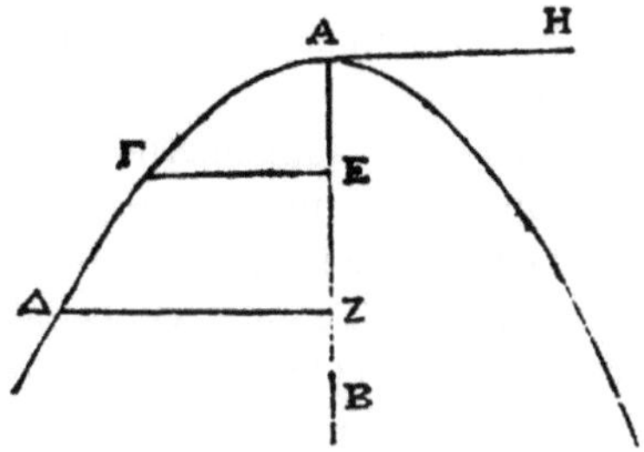

Figure 1: Abscissas. From Halley (1710, p. 46). The ordinates are ΔZ and ΓE, and the corresponding abscissas are AZ and AE.

In connection with systems of coordinates, *abscissa* was used by Mengoli (1659, p. 367);[2] in the current sense, however, the term became known through Leibniz, who began using it in 1673 for the vertical rather than the horizonal coordinate (see, for example, Leibniz 2008, pp. 492f.).[3] In print, it is found, for example, in Leibniz (1684, p. 467) and in Leibniz (1692, p. 170); in the latter article, *coordinate* is also used. Leibniz has been said to have established *abscissa* as a mathematical term (Struik 1969, p. 272), and indeed after the publications of Leibniz the term was adopted by others such as Craig (1685, p. 4).

French *abscisse* in Leibniz' sense is found in Catelan (1694, p. 147); de l'Hôpital (1696, p. 1) used French *coupée* (from *couper* 'to cut') for the abscissa, a literal translation of the Latin term. In German, the term was first used by Wolff (1710, p. 121), who also included an entry for the word in his mathematical dictionary (Wolff 1716, p. 3). The OED cites [anon.] (1694, p. 114) as the first use of *abscissa* in English; the form *absciss* was used slightly earlier, in 1692, by Newton (Turnbull 1961, p. 225).

In a letter to Leibniz from 1698, Wallis mentioned some alternative terms: "intercepted diameter or versed sine (which you call abscissa)" (Leibniz 2011, p. 838).[4] The first of these appears to be due to Mydorge (1631, p. 3), who used *intercepta diameter*. This term was in common use and appeared, for example, in Wallis (1655, p. 27); Mersenne (1636, Traitez de la voix et des chants, p. 40) translated it by *diametre entrecoupee*. The second term mentioned by Wallis was taken over from trigonometry,

[1] He does use the word in the general sense 'segment', for example in Apollonius of Perga (1566, fol. 98[v]). See Bortolotti (1923, p. xxix). [2] For an account of Mengoli's system of coordinates, see Massa Esteve (2006, pp. 94f.). [3] Tou (2021) remarks: "In more modern language, we would say that Leibniz's abscissa is a distance measured along the vertical coordinate axis, and the ordinate is the length of the perpendicular connecting the curve to the vertical coordinate axis". [4] "intercepta-Diameter vel Sinus versus (quam tu Abscissam vocas)".

where the versed sine was commonly called *sagitta*; in the sense of 'abscissa', it was used by Dechales (1674, p. 793).

Further alternative terms were suggested by de la Hire (1679, p. 216), who used *tige*, lit. 'trunk', for the abscissa, and *rameaux*, lit. 'branches', for the ordinates. He later abandoned these terms in favour of the new ones (de la Hire 1713, p. 9). An earlier attempt at unifying the terminology is found in Pascal (1658, p. 18), who suggested *ordonnée à l'axe* 'ordinate to the axis' (abscissa) and *ordonnée à la base* 'ordinate to the base' (ordinate).

Note. It is commonly stated on various internet pages that it may have been Stefano degli Angeli who coined the term in 1659. This statement appears to go back to Cantor (1892, p. 821), who cited degli Angeli (1659) as the earliest use of the word *abscissa* that he was aware of, a statement he later repeated (Cantor 1913, p. 898) and that was adopted by Cajori (1894, p. 185), who referred to Cantor and wrote: "The term *abscissa* occurs for the first time in a Latin work of 1659, written by Stefano degli Angeli". Tropfke (1903, p. 426) protested against the attribution to degli Angeli, pointing out the word was not a term at all with degli Angeli (it occurs very rarely in degli Angeli's book, and is also not defined); similar criticism was made by Eneström (1909-10, pp. 175f.). Also, Cajori (1919, p. 175) later revised his opinion, stating that *abscissa* as a technical term came up in the 18th century and that "in the more general sense of a 'distance' it was used earlier"; he cited Cavalieri and degli Angeli as examples.

Note. For a historical overview of the terminology, see Tropfke (1924, pp. 116–118) and Dingeldey and Fabry (1915, pp. 1f.)

References

[anon.] (1694). "Tractatus Mathematicus de Figuram Curvilinearum [review]". *Philosophical Transactions of the Royal Society of London* 18, pp. 113–114.

degli Angeli, Stefano (1659). *Miscellaneum hyperbolicum, et parabolicum*. Venezia.

Apollonius of Perga (1566). *Conicorum libri quattuor*. Ed. by Federico Commandino. Bologna.

Apollonius of Perga (1891). *Apollonii Pergaei quae graece exstant cum commentariis antiquis*. Vol. I. Ed. by J.L. Heiberg. Leipzig.

Apollonius of Perga (2013). *Conics*. Books I-IV. Books I-III translated by R. Catesby Taliaferro. Book IV translated by Michael N. Fried. Santa Fe.

Bortolotti, Ettore (1923). *Lezioni di geometria analitica*. Volume primo. Bologna.

Cajori, Florian (1894). *A History of Mathematics*. New York.

Cajori, Florian (1919). *A History of Mathematics*. London and New York.

Cantor, Moritz (1892). *Vorlesungen über Geschichte der Mathematik*. Zweiter Band. Von 1200 bis 1668. Leipzig.

Cantor, Moritz (1913). *Vorlesungen über Geschichte der Mathematik*. Zweiter Band. Von 1759 bis 1799. Leipzig.

Catelan, François (1694). "Difficulté sur la solution d'un problème de M. Bernoulli". *Journal des savants* (1694), pp. 146–150.

Craig, John (1685). *Methodus Figurarum Lineis Rectis & Curvis Comprehensarum Quadraturas Determinandi*. London.

Dechales, Claude François Milliet (1674). *Cursus seu mundus mathematicus*. Tomus tertius. Lyon.

Dingeldey, Friedrich and Eugène Fabry (1915). "Coniques". *Encyclopédie des sciences mathématiques pures et appliquées* III.17, pp. 1–162.

Eneström, Gustaf (1909-10). "Kleine Bemerkungen zur letzten Auflage von Cantors ‚Vorlesungen über Geschichte der Mathematik'". *Bibliotheca Mathematica* 10, pp. 164–179.

Halley, Edmund (1710). *Apollonii Pergaei Conicorum libri octo, et Sereni Antissensis De sectione cylindri & coni libri duo*. Oxford.
de la Hire, Philippe (1679). *Nouveaux elemens des sections coniques*. Paris.
de la Hire, Philippe (1713). "Remarques sur la construction des Lieux géométriques et des Equations". *Mémoires de l'Académie royale des sciences* (1710), pp. 7–45.
de l'Hôpital, Guillaume François Antoine (1696). *Analyse des infiniments petits, pour l'intelligence des lignes courbes*. Paris.
Leibniz, Gottfried Wilhelm (1684). "Nova Methodus pro maximis et minimis, itemque tangentibus, quae nec fractas nec irrationales quantitates moratur, et singulare pro illis calculi genu". *Acta Eruditorum* (Oct. 1684), pp. 467–473.
Leibniz, Gottfried Wilhelm (1692). "De linea ex lineis numero infinitis ordinatim ductis inter se concurrentibus formata [...]" *Acta Eruditorum* (Apr. 1692), pp. 168–171.
Leibniz, Gottfried Wilhelm (2008). *Sämtliche Schriften und Briefe*. Siebente Reihe. Mathematische Schriften. Vierter Band. 1670-1673. Infinitesimalmathematik. Berlin.
Leibniz, Gottfried Wilhelm (2011). *Sämtliche Schriften und Briefe*. Dritte Reihe. Mathematischer, naturwissenschaftlicher und technischer Briefwechsel. Siebenter Band. Juli 1696-Dezember 1698. Hannover.
Massa Esteve, Maria Rosa (2006). "Algebra and geometry in Pietro Mengoli (1625–1686)". *Historia Mathematica* 33, pp. 82–112.
Mengoli, Pietro (1659). *Geometriae speciosae elementa*. Bologna.
Mersenne, Marin (1636). *Harmonie universelle, contenant la theorie et la pratique de la musique*. Paris.
Mydorge, Claude (1631). *Prodromi catoptricorum et dioptricorum*. Libri primus et secundus. Paris.
Pascal, Blaise (1658). *Lettre de A. Dettonville à Monsieur de Carcavy*. Paris.
Struik, Dirk Jan, ed. (1969). *A Source Book in Mathematics, 1200-1800*. Harvard and London.
Tou, Erik R. (2021). "Math Origins: The Language of Change". *Convergence* (May 2021). URL: `https://old.maa.org/press/periodicals/convergence/math-origins-the-language-of-change`.
Tropfke, Johannes (1903). *Geschichte der Elementar-Mathematik in systematischer Darstellung*. Zweiter Band. Leipzig.
Tropfke, Johannes (1924). *Geschichte der Elementar-Mathematik*. Sechster Band. Analysis. Analytische Geometrie. Berlin and Leipzig.
Turnbull, H.W., ed. (1961). *The Correspondence of Isaac Newton*. Volume III. 1688–1694. Cambridge.
Wallis, John (1655). *De sectionibus conicis*. Oxford.
Wolff, Christian (1710). *Der Anfangs-Gründe aller Mathematischen Wissenschaften Letzter Theil*. Halle.
Wolff, Christian (1716). *Mathematisches Lexicon*. Leipzig.

absolute and *absolute value*

The origin of the word *absolute* goes back to Latin *absolutus*, from the verb *absolvo* 'to loosen from'. In the 17th century, the word is found in various mathematical texts. This is reflected in Harris' *Lexicon technicum*, for example, in which an *absolute number* in an equation is defined as one "which always possesseth one entire part of side of the Equation, and is always a known Quantity" (Harris 1704, s. v. absolute).[1] The reason for calling such a quantity *absolute* seems to be that such quantities are "loosened" in that they stand alone on one side of the equation and that as such they are not linked to other quantities by the + or − signs. Such a definition of the term is found in Prestet (1689, p. 60), for example. He spoke of *grandeur absoluë* 'absolute

[1] Harris also points out that Viète had called this *Homogeneum Comparationis*. The term is similarly defined in Chambers' *Cyclopaedia*, the *Encyclopédie*, and in Hutton's *Mathematical Dictionary*.

quantity' and also mentioned the synonym *grandeur incomplêxe* 'incomplex quantity'. The latter designation was preferred by many writers, for example by Guisnée (1705, p. iij). Schott (1661, p. 528) also used *numeri absoluti sive simplices*; in other words, he used *simple* instead of *incomplex*.

The modern meaning as found in *absolute value* seems to have developed in the second half of the 18th century, particularly in connection with inequalities. At that time, the phrase "abstraction being made of the sign" ("abstraction faite du signe") is found in various papers. d'Alembert (1768, p. 171) uses it,[1] and so does Lagrange in many of his papers, for example in Lagrange (1769, p. 177): "when we say that α shall be $< \frac{A}{2}$, we understand that α and A are taken positively, even though they can be positive or negative; so that in the comparison of the numbers α and A, only their absolute value needs to be considered".[2]

Around 1800, the phrase "abstraction faite du signe" or variants of it are used to explain *absolute* or *absolutely*. For example, Lagrange (1797, p. 153) uses it to explain "taken absolutely" ("pris absolument"), Carnot (1801, p. 2) explains the term *absolute value* (*valeur absolue*) by the same phrase, and Gauß (1801, § 1) states that "[t]he modulus must obviously be taken *absolutely*, i. e. without sign".[3]

By 1800, the term is thus quite commonly used, at least in French texts. It is soon afterwards found in other languages, for example in English in Lacroix (1816, p. 612). An alternative came up in connection with complex numbers, when Argand (1814-15, p. 208) suggested French *module* (see *modulus* (complex number)), a term also used by Cauchy (1821, p. 183). However, Cauchy (1821, p. 2) preferred *valeur numérique* 'numerical value' with respect to quantities in general. In Germany, Weierstraß (1856, p. 289) suggested *absoluter Betrag* (*Betrag* lit. 'amount') or *absoluter Wert* (*Wert* 'value') as a replacement for Argand's *module*, because the latter already had a number of meanings, for example with respect to elliptic and Abelian functions. Weierstraß' use appears to have had a significant influence since, overall, *absolute value*, which Kronecker (1901, p. 84) called "very advantageous" ("sehr vorteilhaft"), has become standard today.[4]

See also *norm*.

Note. The notion is already found in Leibniz (1710, p. 156), who used the word *moles*, lit. 'a shapeless, huge, heavy mass, huge bulk' and also 'greatness, might, power, strength, great quantity, heap' (LSJ): "$a - b$ signifies the difference between a and b when a is the greater, $b - a$ when b is the greater, and this absolute value

---

[1] On d'Alembert's use of the phrase, Grabiner (2005, pp. 60f.) comments: "though he [d'Alembert] did not have a notation for absolute value, he did have the idea, consistently using the expression 'abstraction being made of the sign'". [2] "quand nous disons que α doit être $< \frac{A}{2}$, nous entendons que α, & A soient pris positivement, quoiqu'ils puissent être d'ailleurs positifs, ou négatifs; de sorte qu'on ne doit avoir égard dans cette comparaison des nombres α, & A, qu'à leur valeur absolue". [3] "Modulus manifesto semper *absolute* i. e. sine omni signo est sumendus". Translation from Gauß (1986, § 1). [4] Bölling (1994, pp. 60f.) has pointed out that Weierstraß redacted some of the articles in his collected papers without indicating the alterations. This applies to the term *absolute value* as well as to the symbol $||$. In particular, his paper on the theory of power series from 1841, but first published as Weierstraß (1894), cannot be used as evidence that Weierstraß used the term and symbol in 1841.

[*moles*] may however be called itself $a - b$, by understanding that the absolute value of $+2$ and 2, for example, is the same, namely, $+2$".[5]

Note. For the history of the notion, see Gagatsis and Thomaidis (1995) and Gagatsis and Panaoura (2014). Grabiner (2005, esp. pp. 61-67) discusses the use of absolute values in d'Alembert and Lagrange. See also Schubring (2006), where many historical comments on the term *absolute* can be found.

References

d'Alembert, Jean Le Rond (1768). "Réflexions sur les Suites & sur les Racines Imaginaires". In: *Opuscules mathématiques*. Tome V. Paris, pp. 171–215.

Argand, Jean-Robert (1814-15). "Réflexions sur la nouvelle théorie des imaginaires, suivies d'une application à la démonstration d'un théorème d'analise". *Annales de mathématiques pures et appliquées* 5, pp. 197–209.

Bölling, Reinhard (1994). "Karl Weierstrass – Stationen eines Lebens". *Jahresbericht der Deutschen Mathematiker-Vereinigung* 96, pp. 56–75.

Cajori, Florian (1928). *A History of Mathematical Notations*. Volume I. Notations in elementary mathematics. Chicago.

Carnot, L.N.M. (1801). *De la corrélation des figures de géométrie*. Paris.

Cauchy, Augustin-Louis (1821). *Cours d'analyse de l'École Royale Polytechnique*. I.re partie. Analyse algébrique. Paris.

Gagatsis, Athanasios and Areti Panaoura (2014). "A multidimensional approach to explore the understanding of the notion of absolute value". *International Journal of Mathematical Education in Science and Technology* 45, pp. 159–173.

Gagatsis, Athanasios and Jannis Thomaidis (1995). "Eine Studie zur historischen Entwicklung und didaktischen Transposition des Begriffs 'absoluter Betrag'". *Journal für Mathematik-Didaktik* 16, pp. 3–46.

Gauß, Carl Friedrich (1801). *Disquisitiones arithmeticae*. Leipzig.

Gauß, Carl Friedrich (1986). *Disquisitiones Arithmeticae*. English Edition. Translated by A.A. Clarke. Revised by W.C. Waterhouse. New York et al.

Grabiner, Judith V. (2005). *The Origins of Cauchy's Rigorous Calculus*. Mineola, N.Y.

Guisnée, Nicolas (1705). *Application de l'algebre a la geometrie*. Paris.

Harris, John (1704). *Lexicon Technicum*. London.

Kronecker, Leopold (1901). *Vorlesungen über Zahlentheorie*. Erster Band. Ed. by Kurt Hensel. Leipzig.

Lacroix, Sylvestre François (1816). *An Elementary Treatise on the Differential and Integral Calculus*. Translated from the French. Cambridge.

Lagrange, Joseph-Louis (1769). "Sur la solution des problèmes indéterminés du second degré". *Nouveaux Mémoires de l'Académie Royale des Sciences et Belles-Lettres de Berlin* (1767), pp. 165–310.

Lagrange, Joseph-Louis (1797). *Théorie des fonctions analytiques*. Paris.

Leibniz, Gottfried Wilhelm (1710). "Monitum de characteribus algebraicis". *Miscellanea Berolinensia* (1710), pp. 155–160.

Prestet, Jean (1689). *Nouveaux elemens des mathematiques*. Second volume. Paris.

Schott, Caspar (1661). *Cursus Mathematicus, Sive Absoluta Omnium Mathematicarum Disciplinarum Encyclopaedia, In Libros XXVIII. digesta*. Würzburg.

Schubring, Gert (2006). *Conflicts Between Generalization, Rigor, and Intuition*. Number Concepts Underlying the Development of Analysis in 17th-19th Century France and Germany. New York.

Weierstraß, Karl (1856). "Theorie der *Abel*'schen Functionen". *Journal für die reine und angewandte Mathematik* 52, pp. 285–379.

---

[5] "differentia inter a et b, significat $a - b$, si a sit majus, et $b - a$ si b sit majus, quod etiam appellari potest moles ipsius $a - b$, intelligendo (exempli causa) ipsius $+2$ & ipsius -2 molem esse eandem, nempe $+2$". Translation from Cajori (1928, p. 223).

Weierstraß, Karl (1894). "Zur Theorie der Potenzreihen". In: *Mathematische Werke*. Erster Band. Berlin, pp. 67–74.

absolute continuity

The term *absolutely continuous* was introduced by Vitali in the early 20th century.[1] Vitali (1904-05, p. 1021) proceeded from a real function F on an interval (a, b); (α, β) is an interval contained in (a, b). He then called $F(\beta) - F(\alpha)$ the *increment* of $F(x)$ in (α, β) and the sum of such increments over a group of distinct (α, β)-intervals, if finite and determinate, the *increment of* $F(x)$ *in this group*.[1] On this basis, Vitali then defined $F(x)$ to be *absolutely continuous* (*assolutamente continua*) if for every $\sigma > 0$ there is a $\mu > 0$ such that the absolute value of the increment of $F(x)$ over every group of intervals of sum less than μ is less than σ.[2]

The term was soon adopted, for example by Lebesgue (1910, p. 365) and de la Vallée Poussin (1915, p. 75); the English term appears in Porter (1915, p. 109). In all of these texts, the term is attributed to Vitali.

Note. For the history of the concept, see Hawkins (1979, pp. 77, 142–154) and Biacino (2020).

References

D'Arcais, Francesco (1899). *Corso di calcolo infinitesimale*. Seconda edizione con aggiunte e modificazioni. Volume I. Padova.
Arzelà, Cesare (1883). "Un' osservazione intorno alle serie di funzioni". *Rendiconto delle sessioni dell'Accademia delle scienze dell'Istituto di Bologna* (1882-83), pp. 142–159.
Biacino, Loredana (2020). "Development of the Theory of the Functions of Real Variables in the First Decades of the Twentieth Century". *Transversal* 9, pp. 1–38.
Hawkins, Thomas (1979). *Lebesgue's Theory of Integration*. Its Origins and Development. Second Edition. Providence, R.I.
Lebesgue, Henri (1910). "Sur l'intégration des fonctions discontinues". *Annales scientifiques de l'École Normale Supérieure* 27, pp. 361–450.
Porter, M.B. (1915). "Concerning absolutely continuous functions". *Bulletin of the American Mathematical Society* 22, pp. 109–111.
de la Vallée Poussin, Charles Jean (1915). "Sur l'intégrale de Lebesgue". *Transactions of the American Mathematical Society* 16, pp. 435–501.
Vitali, Giuseppe (1904-05). "Sulle funzioni integrali". *Atti dell'Accademia delle Scienze di Torino* 40, pp. 1021–1034.

absolute convergence

Like the term *unconditional convergence*, *absolute convergence* also appears to be due to Weierstraß. As Viertel (2014, p. 124) points out, the first known definition is from lectures from 1863: "it shall further be assumed that there exists a positive whole number r such that $\sum_n \frac{1}{a_n^\lambda}$ converges absolutely for $\lambda \geq r$, i. e. for such

[1] Note that before Vitali the term was sometimes used in a different sense, for example by Arzelà (1883) and D'Arcais (1899). [1] "incremento di $F(x)$ in un gruppo di intervalli parziali di (a, b) distinti". [2] "Se per ogni numero $\sigma > 0$ existe un numero $\mu > 0$ tale che sia minore di σ il modulo dell'incremento di $F(x)$ in ogni gruppo di ampiezza minore di μ di intervalli parziali di (a, b) distinti, si dirà che $F(x)$ è *assolutamente continua*".

values of λ the series $\sum_n \frac{1}{A_n^\lambda}$ is convergent" (with $A_n = |a_n|$) (Weierstraß 1915, pp. 104f.).[3]

In print, the term is first found around 1870, for example in Heine (1870, p. 358) and du Bois-Reymond (1870, p. 1), who use *absolutely* and *unconditionally convergent* synonymously.[1] Hoüel (1871, p. 111) uses French *absolument convergente*,[2] Smith (1874, p. 156) English *absolutely convergent.*

Some further aspects of the concept are discussed in the entry *unconditional convergence.*

Note. For the history of the concept, see Sørensen (2010, esp. pp. 55-64), Viertel (2014, esp. pp. 25f., 45–48, 122–124), and Remmert (1991, pp, 29f.).

References

du Bois-Reymond, Paul (1870). *Antrittsprogramm enthaltend neue Lehrsätze über die Summen unendlicher Reihen.* Berlin.

Darboux, Gaston (1875). "Mémoire sur les fonctions discontinues". *Annales scientifiques de l'École Normale Supérieure* 4, pp. 57–112.

Heine, Eduard (1870). "Ueber trigonometrische Reihen". *Journal für die reine und angewandte Mathematik* 71, pp. 353–365.

Hoüel, Jules (1870). "Hankel, Hermann: Untersuchungen über die unendlich oft oscillirenden und unstetigen Functionen". *Bulletin des sciences mathématiques et astronomiques* 1, pp. 117–124.

Hoüel, Jules (1871). *Cours de calcul infinitésimal.* Première partie. Paris.

Remmert, Reinhold (1991). *Theory of Complex Functions.* Translated by Robert B. Burckel. New York.

Smith, H.J.S. (1874). "On the Higher Singularities of Plane Curves". *Proceedings of the London Mathematical Society* 6, pp. 153–182.

Sørensen, Henrik Kragh (2010). "Throwing Some Light on the Vast Darkness that is Analysis: Niels Henrik Abel's Critical Revision and the Concept of Absolute Convergence". *Centaurus* 52, pp. 38–72.

Viertel, Klaus (2014). *Geschichte der gleichmäßigen Konvergenz.* Ursprünge und Entwicklungen des Begriffs in der Analysis des 19. Jahrhunderts. Wiesbaden.

Weierstraß, Karl (1915). *Mathematische Werke.* Fünfter Band. Vorlesungen über die Theorie der elliptischen Funktionen. Berlin.

abundant number

See *perfect number.*

accumulation point

See *limit point.*

acute

The word *acute* comes from Latin *acutus* 'sharp, pointed', which is derived from a verb meaning 'to sharpen'. As a mathematical term, it is found in Latin texts at least since the first century; with respect to angles, the term was first used by Pliny the Elder (Ayuso García 2008, p. 387). The land surveyor Balbus defined: "An acute

[3] "ausserdem [soll] aber angenommen werden, es existiere eine ganze positive Zahl r von der Beschaffenheit, dass $\sum_n \frac{1}{a_n^\lambda}$ für $\lambda \geq r$ absolut convergirt, dass mithin für solche Werthe von λ die Reihe $\sum_n \frac{1}{A_n^\lambda}$ convergent ist". Weierstraß (1915, p. 105) also states that absolute convergence implies unconditional convergence. [1] The synonymous usage is often found in the 19th century; for example, Darboux (1875, p. 77) defined an absolutely convergent series as one in which the sum is independent of the order of its terms. [2] In Hoüel (1870, p. 122), *absolument convergente* is used to translate *unbedingt konvergent* 'unconditionally convergent'.

angle is more compressed than a right angle" (Blume, Lachmann, and Rudorff 1848, p. 101).[3]

The Latin term is a translation of the Greek term ὀξύς (lit. 'sharp' and 'sour, acid'), used by Euclid and some writers before him (Mugler 1958-59, pp. 308f.). Euclid (*Elements*, book I, def. 12) defines: "An acute angle is an angle less than a right angle".[1] The term also appears in def. 21: "an acute-angled triangle [is] that which has its three angles acute".[2]

In modern languages, the term can be found from the 14th century onwards. For example, the German term *spitzer Winkel* ('pointed angle'), which is still standard, was already used around 1400 in the *Geometria Culmensis* (Mendthal 1886, p. 17). Widmann (1489, fol. C 3^{v}) called such angles *sharpened* (*gescherfft*), and Schmid (1539, p. 9) called them *narrow* or *pointed* (*eng, spitzig*).

French *aigu* was first used in the 14th century by Oresme (DMF), who said that "the pyramid has the sharpest angles of all the regular figures" (Oresme 1968, p. 650).[3] In the first printed book on geometry in French, de Bouelles (1511, fol. ii^{v}) defined the term *angle aigu* as an angle that is smaller than a right one. The Italian term *acuto* is also first found in the 14th century (TLIO).

In English, *acute angle* first appears in print (OED) in Billingsley's translation of Euclid's *Elements* (Euclid 1570, fol. 3^{r}); L. Digges and T. Digges (1571, fol. B ij^{r}), whose book may have been written around 1560, use both *acute* and *sharp*. The latter term is earlier found in Benese (c1537, fol. B iii^{v}) and Recorde (1551, fol. A ij^{r}).

In short, the Greek model, which is based the term on the meanings 'sharp' and 'pointed', was followed by Latin and various modern languages, with some attempts to establish indigenous terms such as *sharp*.

Note. A brief historical overview of the terminology is found in Tropfke (1940, pp. 66f.). For the Latin term, see Ayuso García (2008, pp. 387f.) and Guillaumin (2020, p. 21); for the Greek, see Mugler (1958-59, pp. 308f.), and for the Arabic, see Rashed (2017, p. 167).

References

Ayuso García, Manuel (2008). "La terminología latina de la geometría en Marciano Capela". PhD thesis. Madrid: Universidad Nacional de Educación a Distancia.

Benese, Richard (c1537). *This boke sheweth the maner of measurynge of all maner of lande*. London.

Blume, Friedrich, Karl Lachmann, and Adolf Rudorff, eds. (1848). *Die Schriften der römischen Feldmesser*. Erster Band. Texte und Zeichnungen. Berlin.

de Bouelles, Charles (1511). *Geometrie en francoys*. Paris.

Campbell, Brian (2000). *The Writings of the Roman Land Surveyors*. Introduction, Text, Translation and Commentary. London.

Digges, Leonard and Thomas Digges (1571). *A geometrical practise, named Pantometria diuided into three bookes, Longimetra, Planimetra, and Stereometria*. London.

Euclid (1570). *The Elements of Geometrie of the most auncient Philosopher Euclide of Megara*. Faithfully (now first) translated into the Englishe toung, by H. Billingsley. London.

Euclid (1956). *The Thirteen Books of Euclid's Elements*. Volume I. Introduction and Books I, II. Ed. by Thomas L. Heath. New York.

[3] "acutus angulus est conpressior recto". Translation from Campbell (2000, p. 209). [1] Translation from Euclid (1956, p. 153). [2] Translation from Euclid (1956, p. 154). [3] "de toutes les figures regulieres elle [la pyramide] a les angles plus acuz". Translation from Oresme (1968, p. 650).

Guillaumin, Jean-Yves (2020). *Dictionnaire de la terminologie latine ancienne de l'arithmétique et de la géométrie*. Paris.
Mendthal, Hans (1886). *Geometria culmensis*. Ein agronomischer Tractat aus der Zeit des Hochmeisters Conrad von Jungingen (1393-1407). Leipzig.
Mugler, Charles (1958-59). *Dictionnaire historique de la terminologie géométrique des Grecs*. Paris.
Oresme, Nicole (1968). *Le livre du ciel et du monde*. Edited by Albert D. Menut and Alexander J. Denomy, translated with an introduction by Albert D. Menut. Madison.
Rashed, Roshdi, ed. (2017). *Lexique historique de la langue scientifique arabe*. Hildesheim.
Recorde, Robert (1551). *The patheway to knowledg*. London.
Schmid, Wolffgang (1539). *Das erst buch der Geometria*. Nürnberg.
Tropfke, Johannes (1940). *Geschichte der Elementar-Mathematik*. Vierter Band. Ebene Geometrie. Dritte, verbesserte und vermehrte Auflage besorgt von Dr. Kurt Vogel. Berlin and Leipzig.
Widmann, Johannes (1489). *Behende vnd hubsche Rechenung auff allen kauffmanschafft*. Leipzig.

adèle

See *idèle*.

addition

For the four elementary operations of arithmetic, there is generally a large variety of terms. For example, Tropfke (1980, pp. 191f.) and Riley (2000, pp. 7–14) cite more than ten different ways of expressing 'to add' and 'addition' in Greek. Some of these are based on the verb *τιθέναι* 'to put, place'; Riley (2000, p. 9) cites *προστιθέναι* as a "customary word" used, for example, by Plato.[1] It was also used by Euclid in common notion 2: "If equals be added [*προστεθῇ*] to equals, the wholes are equal".[2]

The variety in Latin is similar; according to Tropfke (1980, p. 192), Boethius, for example, used at least seven different ways of expressing 'to add' and 'addition', including *addere* and *additio*. This variety remained in existence after Boethius, but after some 13th century publications, *addere* and *additio* gradually were more commonly used. In particular, they appeared in Sacrobosco's highly popular *De arte numerandi*: "Additio est numeri vel numerorum ad numerum aggregatio, ut videatur summa excrescens" (Pedersen 1983, p. 178).[3] The text was translated into English in the 15th century, where a very literal translation was given: "Addicioun is of nombre, other of nombres, unto nombre or to nombres aggregacioun, that me may see that that is come therof as excressent" ([anon.] 1922a, p. 35).[4]

The earliest use (OED) of the English term is found in the *Crafte of Nombrynge*, from about 1425: "Addicion is a castyng togedur of twoo nomburys into one nombre" ([anon.] 1922b, p. 8). The earliest French algorism (13th century) has the verb *assambler* (Mortet 1908/09, p. 61); the noun *addicion* and the verb *ajoster* (mod. *ajouter*) are also found in the 13th century (Karpinski and Waters 1928, p. 62). Italian *giùngere* was used in 1307 by Jacopo da Firenze (Høyrup 2007, p. 195).[5] The German verb *addieren* is found in an algorism from about 1445.[6]

Note. Overviews of the terminology in various languages are found in Tropfke (1980, pp. 191–194) and Smith (1958, pp. 88–90). For the Greek terms, see also

---

[1] On this verb, see also Mugler (1958-59, pp. 367f.). [2] Translation from Euclid (1956, p. 223).
[3] Another popular text in which *addere* and *addicio* were used is the "Carmen de Algorismo" from about 1220 (de Villa Dei 1922, p. 73). [4] *excressent* here means 'result in an increase'. Note that the word *sum* long meant 'result' and was not limited to addition. [5] This may be the first use in Italian; the form *aggiùngere* appears slightly earlier (TLIO). [6] For the German terminology, see Tropfke (1980, p. 193) and Schirmer (1912, pp. 2f.).

Riley (2000, pp. 7–14). An overview of the Latin arithmetical terminology from the 12th and 13th centuries is found in Allard (1990, pp. 152–180); for some earlier examples, see also Guillaumin (2020, pp. 22, 24, 167) and Schmidt (1914, p. 153). The Arabic terminology is discussed by Rashed (2017, pp. 148–150, 459–465).

References

[anon.] (1922a). "The Art of Nombryng". In: *The Earliest Arithmetics in English*. Ed. by Robert Steele. London, pp. 33–51.

[anon.] (1922b). "The Craft of Nombrynge". In: *The Earliest Arithmetics in English*. Ed. by Robert Steele. London, pp. 3–32.

Allard, André (1990). "La formation du vocabulaire latin de l'arithmétique médiévale". In: *Méthodes et instruments du travail intellectuel au moyen âge*. Études sur le vocabulaire. Ed. by Olga Weijers. Turnhout, pp. 137–181.

Euclid (1956). *The Thirteen Books of Euclid's Elements*. Volume I. Introduction and Books I, II. Ed. by Thomas L. Heath. New York.

Guillaumin, Jean-Yves (2020). *Dictionnaire de la terminologie latine ancienne de l'arithmétique et de la géométrie*. Paris.

Høyrup, Jens (2007). *Jacopo da Firenze's Tractatus Algorismi and Early Italian Abbacus Culture*. Basel.

Karpinski, Louis Charles and E.G.R. Waters (1928). "A Thirteenth Century Algorism in French Verse". *Isis* 11, pp. 45–84.

Mortet, Victor (1908/09). "Le plus ancien traité français d'algorisme". *Bibliotheca Mathematica* 9, pp. 55–64.

Mugler, Charles (1958-59). *Dictionnaire historique de la terminologie géométrique des Grecs*. Paris.

Pedersen, Fritz S., ed. (1983). *Petri Philomenae de Dacia et Petri de Sancto Audomaro*. Opera quadrivalia. Pars I. Opera Petri Philomenae. København.

Rashed, Roshdi, ed. (2017). *Lexique historique de la langue scientifique arabe*. Hildesheim.

Riley, Marc (2000). "Terminology for Arithmetic Operations in Ancient Greek". URL: `http://www.csus.edu/indiv/r/rileymt/` (retrieved Sept. 22, 2017).

Schirmer, Alfred (1912). *Der Wortschatz der Mathematik nach Alter und Herkunft untersucht*. Straßburg.

Schmidt, Max Carl Paul (1914). *Zur Entstehung und Terminologie der elementaren Mathematik*. Zweite verbesserte und stark vermehrte Auflage. Leipzig.

Smith, David Eugene (1958). *History of Mathematics*. Volume II. Special Topics of Elementary Mathematics. New York.

Tropfke, Johannes (1980). *Geschichte der Elementarmathematik*. Band 1: Arithmetik und Algebra. Vollständig neu bearbeitet von Kurt Vogel, Karin Reich, Helmuth Gericke. Berlin and New York.

de Villa Dei, Alexander (1922). "Carmen de Algorismo". In: *The Earliest Arithmetics in English*. Ed. by Robert Steele. London, pp. 72–80.

addition (extensions)

In the 19th century, *addition* begins to be used in new ways. One the one hand, it is found in connection with new kinds of numbers such as complex and hypercomplex numbers; on the other hand, it is applied to various objects other than numbers, for example algebraic structures. In the following, some examples of such extended uses will be given.

For example, Wessel (1799, p. 473) used Danish *adderer* and *sum* in his discussion of the geometrical representation of complex numbers: "Two straight lines are added together, when one joins them together so that one begins where the other one ends, and next one draws a straight line from the first to the last point of the joined lines, and takes this to be their sum".[1] The term *addition* was also used by Argand (1806, p. 26) and others who dealt with the geometrical representation of complex numbers; Warren (1828, p. 1), for example, defined: "The *sum* of two quantities is the diagonal of the parallelogram whose sides are the two quantities".

In connection with quaternions, the term was used by Hamilton (1844, p. 425): "Quaternions are added or subtracted by adding or subtracting their constituents", and Cayley (1858, p. 19) gave the usual rule for the addition of matrices and also remarked that "[a] matrix is not altered by the addition or subtraction of the matrix zero"; he also discussed the additive inverse, which he called *opposite*.

The concept of a direct sum was used by Scheffers (1891, p. 323):

> The composition of several systems A, B, C, ... to just one system S is very similar to the addition in elementary mathematics, since here as there the commutative and the associative law hold [...] This entitles us to symbolically call the composition of several systems A, B, C, ... to just one system S the *addition*: $S = A + B + C + ...$[1]

References

Argand, Jean-Robert (1806). *Essai sur une manière de représenter les quantités imaginaires dans les constructions géométriques*. Paris.

Cayley, Arthur (1858). "A Memoir on the Theory of Matrices". *Philosophical Transactions of the Royal Society of London* 148, pp. 17–37.

Hamilton, William Rowan (1844). "On a new Species of Imaginary Quantities connected with a theory of Quaternions". *Proceedings of the Royal Irish Academy* 2, pp. 424–434.

Scheffers, Georg (1891). "Zurückführung complexer Zahlensysteme auf typische Formen". *Mathematische Annalen* 39, pp. 293–390.

Warren, John (1828). *A Treatise on the Geometrical Representation of the Square Roots of Negative Quantities*. Cambridge.

Wessel, Caspar (1799). "Om Directionens analytiske Betegning, et Forsøg, anvendt fornemmelig til plane og sphæriske Polygoners Opløsning". *Nye Samling af det Kongelige Danske Videnskabernes Selskabs Skrifter* 5, pp. 469–518.

Wessel, Caspar (1999). "On the Analytical Representation of Direction. An Attempt Applied Chiefly to Solving Plane and Spherical Polygons. Translated from Danish by Flemming Damhus". In: *Caspar Wessel: On the Analytical Representation of Direction*. Ed. by Bodil Branner and Jesper Lützen. Copenhagen, pp. 101–143.

adjacency matrix

As part of the studies in the newly founded discipline of sociometry, in the 1940s various authors such as Forsyth and Katz (1946) suggested using matrices. Such square matrices, called *sociomatrices* by Moreno (1946, p. 348), were constructed thus: considering a group of n persons, the element (i, j) is 1 if i and j stand in a given relation such as communication or power. Otherwise, the value is 0. Harary and Ross (1957a, p. 205) called such as matrix a *group matrix* (where *group* is to be understood in a sociological sense); an abstract of this paper (Harary and Ross 1957b,

---

[1] "To rette Linier adderes, naar man først føier dem sammen, saaledes at den ene begynder, hvor den anden slipper, derefter drager fra de sammenføiedes første til sidste Punct en ret Linie, og antager saa denne for de sammenføiedes Sum". Translation from Wessel (1999, p. 105). [1] "Die Zusammensetzung mehrerer Systeme A, B, C, ... zu einem einzigen S hat grosse Aehnlichkeit mit der Addition der elementaren Mathematik. Denn wie dort gilt hier das commutative und das associative Gesetz [...] Dies berechtigt uns, die Zusammensetzung mehrerer Systeme A, B, C, ... zu einem einzigen S symbolisch als *Addition* zu bezeichnen: $S = A + B + C + ...$".

p. 535) has *adjacency matrix* and this seems to be the earliest attestation. The notion thus originated in sociometry and received a designation appropriate to this science. Harary and Ross introduced *adjacency matrix* in this context, too, but they must already have interpreted the notion in graph-theoretical terms, renaming it because in a purely graph-theoretical context a component (i, j) of an adjacency matrix is 1 if the vertices i and j are adjacent, i. e. joined by an edge.[1]

The concept is already found in the 19th century: as pointed out by Gropp (2004, p. 222), it was used by Brunel (1894, p. ix).

References

Brunel, Georges (1894). "Sur quelques problèmes d'*Analysis Situs*". *Mémoires de la Société des sciences physiques et naturelles de Bordeaux* 4, pp. ix–xii.

Forsyth, Elaine and Leo Katz (1946). "A Matrix Approach to the Analysis of Sociometric Data: Preliminary Report". *Sociometry* 9 (4), pp. 340–347.

Gropp, Harald (2004). "'Réseaux réguliers' or regular graphs—Georges Brunel as a French pioneer in graph theory". *Discrete Mathematics* 276, pp. 219–227.

Harary, Frank (1956). "On the number of dissimilar line-subgraphs of a given graph". *Pacific Journal of Mathematics* 6, pp. 57–64.

Harary, Frank and Ian C. Ross (1957a). "A Procedure for Clique Detection Using the Group Matrix". *Sociometry* 20, pp. 205–215.

Harary, Frank and Ian C. Ross (1957b). "A Procedure for Clique Detection Using the Group Matrix [abstract]". *The American Mathematical Monthly* 64, pp. 534–535.

Moreno, Jacob Levy (1946). "Sociogram and Sociomatrix: A Note to the Paper by Forsyth and Katz". *Sociometry* 9, pp. 348–349.

adjoin (algebra)

In algebra, the term was suggested by Galois, who also gave a definition in his famous "First Memoir": "When we thus agree to regard certain quantities as known, we shall say that we *adjoin* them to the equation which it is required to solve. We shall say that these quantities are *adjoined* to the equation" (Galois 2011, p. 108).[2]

In print, the terminology is found in Galois (1832, p. 568) and in the first edition of Galois' works published by Liouville (Galois 1846, p. 418). When Galois' works gradually became known, the terminology was taken over, for example by Betti (1852, p. 87), Allégret (1856, p. 860), and Jordan (1865, p. 770). The German term was used by Dedekind in the 1850s (Scharlau 1982, p. 55), while for English term the OED cites Klein's *Lectures on the Ikosahedron*: "We then *adjoin*, as Galois expresses it, these $R_1, R_2, \ldots$ and accordingly widen the *rationality domain*, to use the language of *Herr Kronecker*" (Klein 1888, p. 93). With Klein, as with Kronecker (1882, p. 6), the term is used with reference to a rationality domain[3], not, as with Galois, with reference to equations.

References

Allégret, Alexandre (1856). "Théorèmes nouveaux relatifs à l'algèbre et à la théorie des nombres". *Comptes rendus de l'Académie des Sciences de Paris* 43, pp. 860–863.

---

[1] The adjective *adjacent* is earlier than *adjacency matrix* and appears, for example, in Harary (1956, p. 58). [2] "Lorsque nous conviendrons de regarder ainsi comme connues de certaines quantités, nous dirons que nous les adjoignons à l'équation qu'il s'agit de résoudre. Nous dirons que ces quantités sont adjointes à l'équation". Translation from Galois (2011, p. 109). [3] For this term, see *field*.

Betti, Enrico (1852). “Sulla risoluzione delle equazioni algebriche”. *Annali di scienze matematiche e fisiche* 3, pp. 49–115.
Galois, Évariste (1832). “Travaux mathématiques d’Évariste Galois”. *Revue encyclopédique* 55, pp. 566–576.
Galois, Évariste (1846). “Œuvres mathématiques d’Évariste Galois [Ed. by Joseph Liouville]”. *Journal de mathématiques pures et appliquées* 11, pp. 381–444.
Galois, Évariste (2011). “The First Memoir”. In: *The Mathematical Writings of Évariste Galois*. Ed. by Peter M. Neumann. Zürich, pp. 105–168.
Jordan, Camille (1865). “Commentaire sur la mémoire de Galois”. *Comptes rendus de l’Académie des sciences* 60, pp. 770–774.
Klein, Felix (1888). *Lectures on the Ikosahedron and the Solution of Equations of the Fifth Degree*. Translated by George Gavin Morrice. London.
Kronecker, Leopold (1882). *Grundzüge einer arithmetischen Theorie der algebraischen Grössen*. Berlin.
Scharlau, Winfried (1982). “Unveröffentlichte algebraische Arbeiten Richard Dedekinds aus seiner Göttinger Zeit 1855–1858”. *Archive for History of Exact Sciences* 27, pp. 335–367.

adjoint (linear algebra)

The term *adjoint* (Latin *adiuncta*) was introduced by Gauß (1801, § 267) in connection with ternary forms:

> [...] we will also designate such a form by
>
> $$\begin{pmatrix} a, & a', & a'' \\ b, & b', & b'' \end{pmatrix}$$
>
> By letting
>
> $$b^2 - a'a'' = A, b'^2 - aa'' = A', b''^2 - aa' = A''$$
>
> $$ab - b'b'' = B, a'b' - bb'' = B', a''b'' - bb' = B'',$$
>
> we will get another form
>
> $$\begin{pmatrix} A, & A', & A'' \\ B, & B', & B'' \end{pmatrix} \dots F$$
>
> which we will call the *adjoint* of the form
>
> $$\begin{pmatrix} a, & a', & a'' \\ b, & b', & b'' \end{pmatrix} \dots f.$$[1]

This terminology was also known to Cauchy (1815, p. 64), who referred to Gauß, and went on to define a *système adjoint* (p. 65). Hawkins (2013, p. 89) comments on Cauchy’s definition: “In present-day terms, Cauchy’s adjoint B is just the matrix of cofactors of A [...] Nowadays, the adjoint (or adjunct) of A is the transpose of Cauchy’s adjoint”. Cauchy’s terminology appears in various 19th-century textbooks such as Baltzer (1857, p. 25).[2] By the 20th century, the definition had changed, as illustrated by Bôcher (1907, p. 30): “By the adjoint A of a matrix a is understood another matrix of the same order in which the element in the *i*th row and *j*th column is

[1] “Ponendo $bb - a'a'' = A$, $b'b' - aa'' = A'$, $b''b'' - aa' = A''$, $ab - b'b'' = B$, $a'b' - bb'' = B'$, $a''b'' - bb' = B''$, oritur alia forma $\left(\begin{smallmatrix} A, & A', & A'' \\ B, & B', & B'' \end{smallmatrix}\right) \dots F$, quam *formae* $\left(\begin{smallmatrix} a, & a', & a'' \\ b, & b', & b'' \end{smallmatrix}\right) \dots f$, *adiunctam* dicemus”. Translation from Gauß (1986, § 267). [2] Salmon (1859, pp. 15f.) used *reciprocal* (of a determinant) instead.

the cofactor of the element in the jth row and ith column of a".[3] The same definition is found in Veblen and Young (1910, p. 269); however, see also the later statement by Aitken (1944, p. 52): "In most books on determinants an untransposed adjugate is used, but for consistency and unification of theory it is preferable to transpose". Aitken himself used *adjugate* but names *adjoint* as an alternative; Price (1947, p. 77) additionally remarks: "Aitken, along with many writers in the British Isles, uses *adjugate* where *adjoint* is commonly used in America".[1]

The term *adjoint* (here German *adjungiert*) was also used in a different sense; Born and Jordan (1930, p. 30) defined it as the matrix that results from taking the transposed matrix and then replacing each component by its complex conjugate. A common alternative is *conjugate transpose*, which was used by Moore (1922, p. 38), for example.

Note. For a detailed discussion of Cauchy's adjoint and its relation to Gauß' term, see Hawkins (2013, pp. 88–93).

References

Aitken, A.C. (1944). *Determinants and Matrices*. Edinburgh.

Baltzer, Richard (1857). *Theorie und Anwendung der Determinanten*. Leipzig.

Bôcher, Maxime (1907). *Introduction to Higher Algebra*. New York.

Born, Max and Pascual Jordan (1930). *Elementare Quantenmechanik*. Zweiter Band der Vorlesungen über Atommechanik. Berlin.

Cauchy, Augustin-Louis (1815). "Sur les fonctions qui ne peuvent obtenir que deux valeurs égales et de signes contraires par suite des transpositions opérées entre les variables qu'elles renferment". *Journal de l'École polytechnique* 10, cahier 17, pp. 29–112.

Dodgson, C.L. (1867). "Condensation of Determinants, Being a New and Brief Method for Computing their Arithmetical Values". *Proceedings of the Royal Society of London*.

Gauß, Carl Friedrich (1801). *Disquisitiones arithmeticae*. Leipzig.

Gauß, Carl Friedrich (1986). *Disquisitiones Arithmeticae*. English Edition. Translated by A.A. Clarke. Revised by W.C. Waterhouse. New York et al.

Hawkins, Thomas (2013). *The Mathematics of Frobenius in Context*. A Journey Through 18th to 20th Century Mathematics. New York.

Moore, E.H. (1922). "On power series in general analysis". *Mathematische Annalen* 86, pp. 30–39.

Price, G.B. (1947). "Some Identities in the Theory of Determinants". *The American Mathematical Monthly* 54, pp. 75–90.

Salmon, George (1859). *Lessons Introductory to the Modern Higher Algebra*. Dublin.

Veblen, Oswald and John Wesley Young (1910). *Projective Geometry*. Volume I. Boston.

adjoint curve

Cayley (1875, p. 359), summarizing some terms introduced in Brill and Noether (1874), gave the following definition of *adjoint curve*:

> I recall some of the fundamental notions: we have a basis-curve which to fix the ideas may be taken to be of the order $n, = p + 1$, with $\frac{1}{2}p(p-3)$ dps [double points], and therefore of the "Geschlecht" or deficiency p; any curve of order $n-3, = p-2$ passing thro' the $\frac{1}{2}p(p-3)$ *dps* is said to be an adjoint curve.

---

[3] Bôcher notes in a footnote: "Notice the interchange of rows and columns here, which in the case of adjoint determinants, being immaterial and sometimes inconvenient, was not made". [1] The term *adjugate* was originally introduced (OED) by Dodgson (1867, p. 152) in the following sense: "if there be a square block, and if a new block be formed, such that each of its terms is the determinant of the complemental minor of the corresponding term of the first block, the second block is said to be *adjugate* to the first".

As Cayley (who first used the English term) and also Clebsch (1876, p. 429) rightly pointed out, the term goes back to Brill and Noether (1874, pp. 271f.), who used German *adjungirte Curve*. French *courbe adjointe* appears in Clebsch (1880, p. 135).

It should be noted that the restriction to curves of order $n - 3$ made by Cayley was not made by Brill and Noether. They also remarked (Brill and Noether 1874, p. 272) that in an earlier publication (Brill and Noether 1873, p. 118), they had generally called adjoint curves *curves of character* φ, while for those of order $n - 3$ they had used *curve* φ. The designation *special* for the latter case is found in Salmon (1873, p. 169); the passage in question is a note by Fiedler and is not part of Salmon's original text.

Note. Remarks on the history of the concept can be found in Kleiman (2004, esp. pp. 410-412) and Gray (1998).

References

Brill, Alexander and Max Noether (1873). "Ueber die algebraischen Functionen und ihre Anwendung in der Geometrie". *Nachrichten von der Königl. Gesellschaft der Wissenschaften und der Georg-Augusts-Universität zu Göttingen* (1873), pp. 116–132.

Brill, Alexander and Max Noether (1874). "Ueber die algebraischen Functionen und ihre Anwendung in der Geometrie". *Mathematische Annalen* 7, pp. 269–310.

Cayley, Arthur (1875). "On the group of Points G_4^1 on a sextic curve with five double points". *Mathematische Annalen* 8, pp. 359–362.

Clebsch, Alfred (1876). *Vorlesungen über Geometrie*. Bearbeitet und herausgegeben von Ferdinand Lindemann. Erster Band. Geometrie der Ebene. Leipzig.

Clebsch, Alfred (1880). *Leçons sur la géométrie*. Tome deuxième. Paris.

Gray, Jeremy (1998). "The Riemann-Roch theorem and geometry, 1854-1914". *Proceedings of the International Congress of Mathematicians. Berlin 1998* 3, pp. 811–822.

Kleiman, Steven L. (2004). "What is Abel's Theorem Anyway?" In: *The Legacy of Niels Henrik Abel*. The Abel Bicentennial, Oslo. Ed. by Olav Arnfinn Laudal and Ragni Piene. Berlin et al., pp. 395–440.

Salmon, George (1873). *Analytische Geometrie der höheren ebenen Curven*. Deutsch bearbeitet von Wilhelm Fiedler. Leipzig.

adjoint differential equation

Adjoint differential equations were already used by Lagrange (1766, p. 186) but not given a name until the second half of the 19th century.[1] The following definition was given by Craig (1889, p. 471), who preferred *associate equation*, however:

> The differential equation being given in the form
>
> $$\frac{d^n y}{dx^n} + R_2 \frac{d^{n-2} y}{dx^{n-2}} + R_3 \frac{d^{n-3} y}{dx^{n-3}} + \cdots + R_n y = 0,$$
>
> its Lagrangian associate is
>
> $$\frac{d^n v}{dx^n} + \frac{d^{n-2}}{dx^{n-2}}(vR_2) - \frac{d^{n-3}}{dx^{n-3}}(vR_3) + \cdots + (-1)^n vR_n = 0.$$

Craig (ibid.) notes that earlier in his book he used *adjunct equation*, and that when he wrote the earlier parts of his books, he was not yet aware of Forsyth (1888,

[1] For Lagrange's use of the concept, see Gray (2021, pp. 14–16).

p. 425), who used *adjoint equation.*[2] Forsyth, in turn, may have taken over the term from Halphen (1884, p. 125), whom he cites earlier on in his paper (Forsyth 1888, p. 384). The actual originator of the term, however, is Fuchs (1873, p. 183), who used German *adjungirte Differentialgleichung.*[1]

References

Craig, Thomas (1889). *A Treatise on Linear Differential Equations*. Volume I. New York.
Forsyth, Andrew Russell (1888). "Invariants, Covariants, and Quotient-Derivatives Associated with Linear Differential Equations". *Philosophical Transactions of the Royal Society of London* 179, pp. 377–489.
Fuchs, Lazarus (1873). "Ueber Relationen, welche für die zwischen je zwei singulären Punkten erstreckten Integrale der Lösungen linearer Differentialgleichungen stattfinden". *Journal für die reine und angewandte Mathematik* 76, pp. 177–213.
Gray, Jeremy (2021). *Change and Variations*. A History of Differential Equations to 1900. Cham.
Halphen, Georges Henri (1884). "Mémoire sur la réduction des équations différentielles linéaires aux formes intégrables". *Mémoires présentés par divers savants á l'Académie des sciences de l'Institut de France* 28, pp. 1–307.
Ince, Edward L. (1927). *Ordinary Differential Equations*. London.
Kline, Morris (1972). *Mathematical Thought From Ancient To Modern Times*. Volume 2. New York and Oxford.
Lagrange, Joseph-Louis (1766). "Solution de différens problêmes de calcul intégral". *Mémoires de l'Académie royale des Sciences de Turin* (1762-65), pp. 179–380.

adjoint functor

The term *adjoint functor* was suggested by Kan (1958, p. 294), who defined:

> Let $\mathcal{X}$ and $\mathcal{Z}$ be categories and let $S : \mathcal{X} \to \mathcal{Z}$ and $T : \mathcal{Z} \to \mathcal{X}$ be covariant functors. Then S is called a *left adjoint* of T and T a *right adjoint* of S if there exists a natural equivalence
>
> $$\alpha : H(S(\mathcal{X}), \mathcal{Z}) \to H(\mathcal{X}, T(\mathcal{Z})).$$
>
> An important property of adjoint functors is that each determines the other up to a unique natural equivalence.

The term earlier appeared in Kan (1957, p. 450), who referred to his forthcoming paper from 1958. Kan's term spread quickly and was soon translated; for example, Shih (1958-59, p. 5) uses French *foncteurs adjoints*.

Note. For the history of the concept, see Krömer (2007, pp. 202–206).

References

Kan, Daniel M. (1957). "On c.s.s. Complexes". *American Journal of Mathematics* 79, pp. 449–476.
Kan, Daniel M. (1958). "Adjoint Functors". *Transactions of the American Mathematical Society* 87, pp. 294–329.
Krömer, Ralf (2007). *Tool and Object*. A History and Philosophy of Category Theory. Basel.
Shih, Weishu (1958-59). "Ensembles simpliciaux et opérations cohomologiques". *Séminaire Henri Cartan* 11, exposé 7, pp. 1–10.

---

[2] Craig's precise words are: "When Chapter I was written [...] I had not seen Forsyth's memoir, and had not been able to find an adopted English term for Lagrange's '*équation adjointe*,' so I used the word *adjunct*, suggested by the German '*adjungirte*,' and not unlike the French '*adjointe*.' It seems better now, however, to employ the word *associate*, or, when speaking simply of Lagrange's '*équation adjointe*,' the word *adjoint*". [1] This was pointed out by Ince (1927, p. 124) and Kline (1972, p. 487), for example.

adjoint operator

The concept of adjoint operator was used by Riesz (1910, p. 479); he used the term *Transponierte* (transposed) of a transformation. Banach and Schauder extended the concept to Banach spaces, and Banach (1929, p. 235) suggested the term *opération adjointe*. Schauder (1930, p. 184), citing Banach, preferred *konjugierte* or *transponierte Funktionaloperation*, and Banach (1932, p. 100) slightly later also used *opération associée* and *opération conjuguée*. Nevertheless, his *opération adjointe*, the basis of *adjoint operator*, became the commonly used term.

Note. For the history of the terminology, see Pietsch (2007, p. 35).

References

Banach, Stefan (1929). "Sur les fonctionnelles linéaires II". *Studia Mathematica* 1, pp. 223–239.
Banach, Stefan (1932). *Théorie des opérations linéaires*. Warszawa.
Pietsch, Albrecht (2007). *History of Banach Spaces and Linear Operators*. Boston, Mass. et al.
Riesz, Frigyes (1910). "Untersuchungen über Systeme integrierbarer Funktionen". *Mathematische Annalen* 69, pp. 449–497.
Schauder, Juliusz (1930). "Über lineare, vollstetige Funktionaloperationen". *Studia Mathematica* 2, pp. 183–196.

affine

This word comes from Latin *affinis* (from *ad finis*) 'bordering upon; related, akin', where *ad* means 'to' and *finis* means 'end, border'. In mathematics, the notion appears in Clairaut (1733, p. 486) under the name *courbes de même espèce* ('curves of the same species'), while the term under discussion was first used by Euler (1748, § 442), who defined *affine curves* (Latin *curvas affines*): "Just as in similar curves the homologous abscissas and ordinates increase and decrease in the same ratio, so if abscissas follow one ratio and the ordinates follow a different ratio, the curves are no longer similar. Nevertheless curves with this property do have some relationship to each other, and we say that the two curves are affine".[1] According to Rosenfeld (1988, p. 144), "[b]y introducing this term Euler apparently wished to emphasize that 'affine curves' are less closely related than similar curves and far less closely related than 'similar and equal'—that is, congruent—curves".

Until the end of the 18th century, the term remains rare, but it is discussed by Kästner (1794, p. 248), who also considers and ultimately dismisses some potential German equivalents like *Verwandtschaft*, lit. 'kinship, relatedness, relationship'. Möbius (1827, p. 195) retained the expression when he defined the more general *affine figures* (German *affine Figuren*). For affine mappings or transformations he used German *Affinität*.[2] Möbius discussed various properties of affine transformations, two of which were later used to define them: they preserve lines as well as parallelism (Möbius 1827, pp. 196f.). Definitions of the explicit terms *affine mapping* or *affine transformation* are of later date; they are found, for example, in Study (1885, p. 9) and Klein (1896, p. 53),

[1] "Quemadmodum in Curvis similibus Abscissae et Applicatae homologae in eadem ratione sive augentur sive diminuuntur; ita, si Abscissae aliam sequantur rationem, aliam vero Applicatae, Curvae non amplius orientur similes. Verum tamen, quia Curvae hoc modo ortae inter se quandam Affinitatem tenent, has Curvas *affines* vocabimus". Translation from Euler (1990, p. 278). [2] He had already done so in Möbius (1823, p. 58).

After Möbius used the term, it spread and is found in Magnus (1833, p. 51), Bellavitis (1835, p. 251), and Plücker (1835, p. 58), for example. The English terminology is of later date; the OED cites Page (1894-95, p. 65) as the first use of *affine*, while *affinity* is of earlier date and already found in Brande and Cox (1865, p. 487).[3] French *affinité* is found in Euler (1797, § 442) and in Cauchy's review of Möbius's book (Cauchy 1828, p. 79).

Note. For the history of the term, see Rosenfeld (1988, pp. 143f.) and Papperitz (1910, p. 570). Early historical comments are found in Möbius (1827, p. x) and Chasles (1837, pp. 218, 563).

References

Bellavitis, Giusto (1835). "Saggio di applicazioni di un nuovo methodo di geometria analytica (Calcolo delle equipollenze)". *Annali delle Scienze del Regno Lombardo-Veneto* 5, pp. 244–259.

Brande, William Thomas and George William Cox (1865). *A Dictionary of Science, Literature, & Art*. Vol. I. London.

Cauchy, Augustin-Louis (1828). "Der barycentrische Calcul [review]". *Bulletin des sciences mathématiques, astronomiques, physiques et chimiques* 9, pp. 77–80.

Chasles, Michel (1837). *Aperçu historique sur l'origine et le développement des méthodes en géométrie*. Bruxelles.

Clairaut, Alexis Claude (1733). "Sur les courbes que l'on forme en coupant une surface courbe quelconque, par un plan donné de position". *Mémoires de l'Académie royale des sciences* (1731), pp. 483–493.

Euler, Leonhard (1748). *Introductio in analysin infinitorum*. Tomus secundus. Lausanne.

Euler, Leonhard (1797). *Introduction à l'analyse infinitésimale*. Traduite du latin en français, avec des Notes & des Éclaircissements, par J.B.Labey. Tome Second. Paris.

Euler, Leonhard (1990). *Introduction to Analysis of the Infinite*. Book II. Translated by John D. Blanton. New York.

Kästner, Abraham Gotthelf (1794). *Anfangsgründe der Analysis endlicher Grössen*. Dritte, stark vermehrte Auflage. Göttingen.

Klein, Felix (1896). *Ausgewählte Kapitel der Zahlentheorie I*. Vorlesung, gehalten im Wintersemester 1895/96. Ausgearbeitet von A. Sommerfeld. Göttingen.

Magnus, Ludwig Immanuel (1833). *Sammlung von Aufgaben und Lehrsätzen aus der analytischen Geometrie*. Berlin.

Möbius, August Ferdinand (1823). *Beobachtungen auf der Universitäts-Sternwarte zu Leipzig*. Leipzig.

Möbius, August Ferdinand (1827). *Der barycentrische Calcul*. Leipzig.

Page, James M. (1894-95). "Transformation Groups Applied to Ordinary Differential Equations". *Annals of Mathematics* 9, pp. 59–69.

Papperitz, Erwin (1910). "Darstellende Geometrie". *Encyklopädie der mathematischen Wissenschaften mit Einschluss ihrer Anwendungen* III AB 6, pp. 517–595.

Plücker, Julius (1835). *System der analytischen Geometrie*. Berlin.

Rosenfeld, Boris A. (1988). *A History of Non-Euclidean Geometry*. Evolution of the Concept of a Geometric Space. Translated by Abe Shenitzer. New York.

Study, Eduard (1885). *Über die Massbestimmung extensiver Grössen*. Wien.

affine geometry

In lectures given in 1895 and 1896, Klein suggested a classification of geometry into elementary geometry (*Elementargeometrie*), affine geometry (*affine Geometrie*), and projective geometry (*projective Geometrie*). He defined the new term *affine geometry*

[3] This is a dictionary and the word is found in the entry *collinear*, which was written by T. A. Hirst.

as follows: “Affine geometry is only interested in those properties of figures that are not destroyed by affine operations” (Klein 1896, p. 53).[1]

The term appears soon afterwards in some books on geometry. In particular, Vahlen (1905, ch. iv) included a whole chapter entitled “Affine Geometrie”, and Heffter and Koehler (1905, p. iii) also took over the term and explicitly referred (p. 16) to Klein’s lectures for the classification.

Klein’s terminology was translated into English and French in various reviews of Vahlen (1905) and Heffter and Koehler (1905); see Young (1906, p. 272), Veblen (1906, p. 505), and Bosmans (1906, p. 607).

When Klein’s collected papers were first published, Klein noted that he did not consider affine geometry in his “Erlanger Programm” (Klein 1872) but that he “first did this in detail in 1895-96 in [his] [...] lectures about number theory and then later on always included it in the beginning” (Klein 1921, pp. 320f.).[1]

References

Bosmans, Henri (1906). “*Lehrbuch der analytischen Geometrie* ... von L. Heffter und C. Koehler [review]”. *Revue des questions scientifiques* 9, pp. 607–608.

Heffter, Lothar and Carl Koehler (1905). *Lehrbuch der analytischen Geometrie*. Erster Band. Geometrie in den Grundgebilden erster Stufe und in der Ebene. Leipzig and Berlin.

Klein, Felix (1872). *Vergleichende Betrachtungen über neuere geometrische Forschungen*. Programm zum Eintritt in die philosophische Facultät und den Senat der k. Friedrich-Alexanders-Universität zu Erlangen. Erlangen.

Klein, Felix (1896). *Ausgewählte Kapitel der Zahlentheorie I*. Vorlesung, gehalten im Wintersemester 1895/96. Ausgearbeitet von A. Sommerfeld. Göttingen.

Klein, Felix (1921). *Gesammelte mathematische Abhandlungen*. Erster Band. Herausgegeben von R. Fricke und A. Ostrowski. Berlin.

Tobies, Renate (1996). “The Reception of H. Grassmann’s Mathematical Achievements by A. Clebsch and His School”. In: *Hermann Günther Graßmann (1809-1877)*. Visionary Mathematician, Scientist and Neohumanist Scholar. Ed. by Gert Schubring. Dordrecht, pp. 117–130.

Vahlen, Karl Theodor (1905). *Abstrakte Geometrie*. Untersuchungen über die Grundlagen der Euklidischen und Nicht-Euklidischen Geometrie. Leipzig.

Veblen, Oswald (1906). “*Abstrakte Geometrie*. By Karl Theodor Vahlen [review]”. *Bulletin of the American Mathematical Society* 12, pp. 505–506.

Young, William Henry (1906). “*Abstrakte Geometrie* by Karl Theodor Vahlen [review]”. *The Mathematical Gazette* 3, pp. 271–272.

affine space

In his book *Space–Time–Matter*, Weyl (1918, p. 16) presented a system of axioms for vector spaces and affine spaces; for affine spaces, he gave the two axioms:

1. Every pair of points A and B determines a vector $\mathfrak{a}$; expressed symbolically $\overrightarrow{AB} = \mathfrak{a}$. If A is any point and $\mathfrak{a}$ any vector, there is one and only one point B for which $\overrightarrow{AB} = \mathfrak{a}$.
2. If $\overrightarrow{AB} = \mathfrak{a}$, $\overrightarrow{BC} = \mathfrak{b}$, then $\overrightarrow{AC} = \mathfrak{a} + \mathfrak{b}$.[2]

---

[1] “Die affine Geometrie interessiert sich nur für solche Eigenschaften der Figuren, welche durch affine Operationen nicht zerstört werden”. [1] Translation from Tobies (1996, p. 122). [2] “1. Je zwei Punkte A und B bestimmen einen Vektor $\mathfrak{a}$; in Zeichen $\overrightarrow{AB} = \mathfrak{a}$. Ist A irgend ein Punkt, $\mathfrak{a}$ irgend ein Vektor, so gibt es einen und nur einen Punkt B, für welchen $\overrightarrow{AB} = \mathfrak{a}$ ist. 2. Ist $\overrightarrow{AB} = \mathfrak{a}$, $\overrightarrow{BC} = \mathfrak{b}$, so ist $\overrightarrow{AC} = \mathfrak{a} + \mathfrak{b}$”. Translation from Weyl (1922a, p. 18).

Weyl did not explicitly use *affine space* (*affiner Raum*) at that time; the term does appear, however, in the revised edition published a year later (Weyl 1919, p. 105). The English and French terms first appear in translations of Weyl's book (1922, p. 118, 1922, p. 102) but they were also adopted by other writers such as Cartan (1923, pp. 325, 359) and Fréchet (1925, p. 60, 1926, pp. 27–36).

The concept has been redefined various times, for example by Schreier and Sperner (1931, p. 9): "The totality of all n-tuples of real numbers is called *n-dimensional affine space*".[1] Zariski (1939, p. 250) used any algebraically closed ground field of characteristic zero instead of the real numbers.

Note. For Weyl's introduction of the terminology, see Rosenfeld (1988, pp. 262–264).

References

Cartan, Élie (1923). "Sur les variétés à connexion affine (première partie)". *Annales scientifiques de l'École Normale Supérieure* 40, pp. 325–412.

Fréchet, Maurice (1925). "Les espaces vectoriels abstraits". *Bulletin of the Calcutta Mathematical Society* 16, pp. 51–62.

Fréchet, Maurice (1926). "Les espaces abstraits topologiquement affines". *Acta Mathematica* 47, pp. 25–52.

Rosenfeld, Boris A. (1988). *A History of Non-Euclidean Geometry*. Evolution of the Concept of a Geometric Space. Translated by Abe Shenitzer. New York.

Schreier, Otto and Emanuel Sperner (1931). *Einführung in die Analytische Geometrie und Algebra*. Erster Band. Leipzig and Berlin.

Schreier, Otto and Emanuel Sperner (1951). *Introduction to Modern Algebra and Matrix Theory*. New York.

Weyl, Hermann (1918). *Raum · Zeit · Materie*. Vorlesungen über allgemeine Relativitätstheorie. Berlin.

Weyl, Hermann (1919). *Raum · Zeit · Materie*. Vorlesungen über allgemeine Relativitätstheorie. Dritte, umgearbeitete Auflage. Berlin.

Weyl, Hermann (1922a). *Space–Time–Matter*. Translated from the German by Henry L. Rose. London.

Weyl, Hermann (1922b). *Temps, espace, matière*. Leçons sur la théorie de la relativité générale. Traduites sur la quatrième édition allemande par Gustave Juvet et Robert Leroy. Paris.

Zariski, Oscar (1939). "Some Results in the Arithmetic Theory of Algebraic Varieties". *American Journal of Mathematics* 61, pp. 249–294.

algebra

Like most of the names for mathematical branches, *algebra* has a rather complex history. In the following, the origin of the terminology including its spread to Latin and modern languages will first be discussed. Afterwards, some further developments are presented, including synonymous or at least overlapping terms and some variations in the understanding of the concepts. This leads to some discussion of the concept in the 19th and 20th centuries.

1. Origin and spread

The name *algebra* goes back to Arabic *al-jabr*, which word was part of *al-jabr waʾl-muqābala*, lit. 'restoration and confrontation', a name for algebra frequently found in medieval texts. Originally, *al-jabr* was used when a subtracted quantity was moved to the other side of the equation; Oaks and Alkhateeb (2007, p. 57) explain that "it

[1] "Die Gesamtheit aller n-Tupel reeller Zahlen heißt der *affine Raum von n Dimensionen*". Translation from Schreier and Sperner (1951, p. 5).

takes an understanding of medieval subtraction to explain why it [*al-jabr*] was used. For this one needs to see $A - B$ as a diminished A in need of restoration". The second term *al-muqābala* was used when like terms on opposite sides of an equation are combined. For example, in $50 + X = 29 + 10x$ the numbers 50 and 29 are combined to yield the simplified equation $21 + X = 10x$ (Oaks and Alkhateeb 2007, p. 52).

The combined phrase *al-jabr wa'l-muqābala* "was used to mean 'applying *al-jabr* and/or *al-muqābala* to simplify the equation,' as well as being the name of the art of algebra. When used in the former sense, it was sometimes shortened to just *al-muqābala*, and in the latter sense to *al-jabr*" (Oaks and Alkhateeb 2007, p. 53). Oaks (2024, p. 129) characterizes algebra in the following terms: "In contrast to the range of meanings the word 'algebra' assumes today, medieval Arabic algebra was a specific technique of numerical problem-solving, with its own technical vocabulary, methods, and concepts. It was not used to state identities like $(x + y)^2 = x^2 + y^2 + 2xy$, formulas like $A = \pi r^2$, curves like $y = 3x^2 + 2$, or physical laws like $d = \frac{1}{2}gt^2$. Apart from rare exceptions of its application in proofs, algebra was used only to find unknown numbers".[1]

al-Khwārizmī's *Brief Book on Calculation by Algebra* was very influential in spreading algebra in Europe.[2] In the 12th century, it was translated into Latin, and it is through these translations that the name *algebra* became known in Europe. In 1145, Robert of Chester used the transliterated *algebre et almuchabolae* in his translation (Hughes 1989, p. 29). Gerard of Cremona wrote *algebra et almuchabala* (Hughes 1986, p. 233) and the *Liber Mahameleth* has *gebra et muchabala* (Sesiano 2014, p. 194). As in the Arabic texts, the full phrase was often shortened to *algebra*.

Additionally, *al-jabr* was often translated by *restauratio* and *al-muqābala* by *oppositio*; the corresponding verbs *restaurare* and *opponere* also appear. This terminology was used by Robert of Chester (Hughes 1989, p. 31) and by Gerard of Cremona (Hughes 1986, p. 218), for example, and it also appears in the *Dixit Algorismi*, a translation of al-Khwārizmī's treatise on Hindu-Arabic numerals (Allard 1992, p. 1).

The term is commonly found in the spelling *algeber*,[3] which is due to a mistaken attribution of the field to a certain Geber. The 15th-century abacist Raffaello Canacci wrote: "The role of algebra, which rule Guglielmo a Lunis[4] has translated from the Arabic into our language. And the said Guglielmo and others say it was composed by an Arabic master of truly great insight, even though some others say it was one whose name is Geber" (Procissi 1954, p. 302).[5] This attribution is also made in various books, for example in Stifel (1544, fol. 226^{v}) and Ramus (1586, p. 322). Others, such as Wallis (1685, p. 5), protested: "this learning [algebra] was pursued by *Arabic* authors [...] From them it had the name of *Algebra*; not (as some would have it) from *Geber*, whom they conjecture (without any good ground that I know of) to have been its first inventer; but [...] from its *Arabic* name, *Al-gjábr W'al-mokábala*".[6]

---

[1] Various medieval definitions and characterizations of algebra are discussed by Oaks (2024, pp. 134f.). [2] For a discussion of its influence, see Oaks (2024). [3] This spelling is still sometimes used in the 18th century, for example by Kaschube (1717, p. 537). [4] On this elusive person and the question whether he translated al-Khwārizmī, see Kaunzner (1985, pp. 10–12) and Hughes (1986, pp. 22–25). [5] Translation from Høyrup (2019, p. 336). [6] For a discussion of various attributions to Geber, see Cifoletti (1996).

In German, the term is first found in a manuscript on algebra from 1461 (Gerhardt 1871, p. 142); in print, it was used by Widmann (1489, fol. a2^{r}).[7] For French, the DMF cites an isolated attestation from 1376 as a first use; in print, the term was used by de La Roche (1520, fol. 42^{r}). According to the OED, the English term was first used by Recorde (1551, fol. a iij^{v}), where "the rule of Algeber" is found; Billingsley also mentions the term in his translation of Euclid's *Elements*: "That more secret and subtill part of Arithmetike, commonly called Algebra" (Euclid 1570, fol. 228^{v}).

2. Further developments

In her discussion of "Narratives of Algebra in Early Printed European Texts", Stedall (2012, p. 217) remarks that "[t]he range of techniques and subject matter covered by the word 'algebra' has shifted and changed ever since the first printed algebra texts established algebra as an important part of the mathematical knowledge of western Europe. Such changes have continued right up to the present day, but even during the first half of the sixteenth century we see both a single object with a variety of names, and a single name with evolving meanings".

Some examples for the variety of names are: *greater art* (*arte magiore*) (Pacioli 1494, fol. 67^{r}),[1] *Ars Magna* (Cardano 1545), *Ars analytica* (Viète 1591), and *coss*,[2] to name just a few.

Algebra has also been defined in various ways. For example, Ramus (1560, fol. 2^{r}) called it "that part of arithmetic that sets up its own proper numbers from continuously proportional figures",[3] where figures are unknowns and their powers. Balam (1653) regarded algebra as the "doctrine of Composing, Inferring, and Resolving an Equation", as the title of his book indicates.[4]

A broadened view of algebra is found in the 18th century, for example in Euler's *Algebra*: "From this it appears all magnitudes may be expressed by numbers; and that the foundation of all the mathematical sciences must be laid in a complete treatise on the science of numbers; and in an accurate examination of the different possible methods of calculation. This fundamental part of mathematics is called analysis [Analytic], or algebra" (Euler 1771, p. 5).[5]

---

[7] For the German term, see Strauß et al. (1995, pp. 360f.). [1] Pacioli opposed this to *arte menore* (*lesser art*), commercial arithmetic. [2] *coss* is based on Italian *cosa* (lit. 'thing', term for the unknown); *regola della cosa* as a vernacular name for algebra is found in various 14th-century Italian manuscripts (Franci and Toti Rigatelli 1988, pp. 12–15). In German texts, the word *coss* was used; see Schirmer (1912, pp. 13f.) for various attestations. [3] "Algebra est pars Arithmeticae, quae e figuratis continue proportionalibus numerationem quandam propriam instituit". Translation from Van Egmond (1988, p. 142). [4] See Stedall (2012) for various early discussions of algebra in print; see also Cifoletti (1996), whose article focuses on early discussions of the history of algebra. As to algebra as the theory of equations, Stedall (2011, p. vii) remarks: "In the sixteenth century there had been no general 'theory of equations', only a collection of piecemeal methods and results. By the eighteenth century, however, and in particular during the 1760s, it could finally be said that a theory was beginning to emerge". [5] "Hieraus ist klar, daß sich alle Größen, durch Zahlen ausdrücken lassen, und also der Grund aller mathematischen Wissenschaften darinn gesetzt werden muß, daß man die Lehre von den Zahlen, und alle Rechnungsarten, so dabey vorkommen können, genau in Erwägung ziehe, und vollständig abhandele. Dieser Grundtheil der Mathematic wird, die Analytic oder Algebra genennet". Translation from Euler (1821, p. 2). Some comments on Euler's definition can be found in Katz and Parshall (2014, p. 6) and Tropfke (1980, pp. 5f.). Schubring (2006, p. 258) remarks that "Euler was the first to consistently present algebra as a science of numbers and to conceptually separate numbers clearly from quantities in algebra".

Euler's statement also shows a development that goes back to Viète: "After Viète's introduction of the analytical use of his new algebra in geometry and number theory, the term *analysis* was soon used as practically synonymous with the term *algebra*, both indicating letter-algebra. Later, when special symbols for objects or processes involving limits, indivisibles, or infinitesimals were introduced, it became customary to denote letter-algebra combined with these new symbols as *analysis infinitorum* ('analysis of infinite quantities') and, in contrast, letter-algebra itself as *analysis finitorum* ('analysis of finite quantities'). The 'infinite quantities' included infinitely small ones" (Bos 2001, p. 10). This aspect is discussed in more detail in the entry *analysis*.

3. The 19th century

On the one hand, the subject matter of algebra was considerably changed and extended in the 19th century. On the other hand, definitions of algebra remained essentially the same. These two aspects shall be discussed in this section.

As to the first point, one important development is that the concept of "composition" was extended to objects other than numbers. This is seen in the works of Gauß, Cauchy, and Galois, for example; see *product* (extensions). In England, composition was considered from an abstract viewpoint, and it is here that, acccording to Bourbaki (1994, p. 52), "the clearest progress towards abstraction is made". One major proponent of what was called *symbolical algebra* was Peacock, who defined it as the "science of symbols and their combinations constructed upon its own rules, which may be applied to arithmetic and to all other sciences by interpretation" (Peacock 1834, pp. 194f.). As Koppelman (1971, p. 216) remarks, "the generality of this statement is misleading, because he [Peacock] insisted that the laws of arithmetic must always be valid". In particular, this means that, for example, non-commutative systems would not be considered. A more general approach was suggested by Gregory (1840), who removed the dependence on arithmetic.[1]

Overall, numerous new objects such as vectors, quaternions, matrices, or hypercomplex numbers as well as new subject fields such as linear algebra or algebraic number theory were created in this period. Importantly, various algebraic structures such as groups or fields gradually arose. For these matters, see the respective entries in this book.

As to the second point mentioned above, in the 19th century, algebra was still often defined along traditional lines. Serret (1849, p. 1), for example, gave the classical definition: "Algebra is, strictly speaking, the *analysis of equations*".[2] Similar remarks apply to "the last important textbook on algebra published in the 19th century" (Corry 2005, p. 690), namely Weber's *Lehrbuch der Algebra* (Weber 1895-96). As explained by Corry (2001, p. 170), in the book algebra is still seen "as the discipline of polynomial equations and polynomial forms. Abstract concepts such as groups, in so far as they appear, are subordinate to the main classical tasks of algebra". Weber's book "incorporates an entire body of new individual ideas and techniques developed along the 19th century, and in doing so it provides a full picture of what algebraic

[1] For detailed discussions of Gregory's work, see Verburgt (2018) and Allaire and Bradley (2002), for example. For the origins of symbolical algebra, see also Durand-Richard (1990, 2011). [2] "L'Algèbre est, à proprement parler, l'*Analyse des équations*".

knowledge looked like at the time. In spite of the knowledge it adds over books like Serret's or Jordan's, the picture of algebra it presents does not differ essentially from theirs. On the other hand, in spite of including a great deal of material that would eventually be incorporated as the basis of van der Waerden's presentation, it does not envisage the kind of fundamental change in conception that *Moderne Algebra* intended to imply" (Corry 2005, p. 691). This modern understanding of algebra will be dealt with in the following last section of this entry.

4. Abstract algebra

Steinitz' article entitled "Algebraic Theory of Fields" (Steinitz 1910) has been described as "a fundamental work which may be considered as the origin of today's concept of algebra" (Bourbaki 1994, p. 83). In his article, Steinitz (1910, p. 167) began by remarking that he was using *field* as abstractly and generally as Weber (1893) did. However, Steinitz also indicated an important difference: "Whereas Weber's aim was a general treatment of Galois theory, independent of the numerical meaning of the elements, for us it is the concept of field which represents the focus of interest".[1]

Steinitz' abstract point of view was adopted by various mathematicians, and the abstract approach spread in the 1920s. Emmy Noether's article "Abstract Structure of the Ideal Theory in algebraic Number and Function Fields" (Noether 1927) is a major example. It is at this time that the term *abstract algebra* came up. Rella (1929, p. A18), for example, used it for the "theory of abstractly defined fields, integral domains, and rings".[2] The term was also used by Hasse (1927, p. 81), and Rella's statement summarizes Hasse's point of view.[3]

The definition given by Mac Lane (1939, p. 18) also reflects the developments of the 19th and 20th centuries: "Algebra tends to the study of the explicit structure of postulationally defined systems closed with respect to one or more rational operations". A commonly found definition was given by Bourbaki (1942, p. ii), who defined *algebraic structure* and added that "it is the study of these structures which constitutes Algebra".[4] A major landmark of algebra in this sense was van der Waerden's classic *Modern Algebra* (van der Waerden 1930-31). According to Schlote (2005, p. 901), it "was the first textbook of algebra systematically to present, on an abstract set-theoretic basis using axiomatic methods, the various areas of this subject as a homogeneous whole. Thereby it made a decisive contribution to the dissemination and fulfillment of new structural conceptions in algebra and in the whole of mathematics".

Note. For historical overviews of the terminology, see Smith (1958, pp. 386–393) and Tropfke (1980, pp. 1–7). The Arabic terminology is discussed in great detail by Oaks and Alkhateeb (2007); see also Heeffer (2008). Some fairly recent books on the history of algebra include Katz and Parshall (2014) and Alten et al. (2014); Gray

[1] "Während aber bei *Weber* das Ziel eine allgemeine, von der Zahlenbedeutung der Elemente unabhängige, Behandlung der *Galois*schen Theorie ist, steht für uns der Körperbegriff selbst im Mittelpunkt des Interesses". Translation from Corry (2004, p. 193). [2] "Theorie der abstrakt definierten Körper, Integritätsbereiche und Ringe". [3] For Hasse's definition of algebra, see Hasse (1926, pp. 46–50) and Hasse (1927, pp. 79–81); for historical comments on Hasse's understanding of algebra, see Corry (2004, pp. 57–60). [4] "c'est l'étude de ces structures qui constitue l'Algèbre". – See also *structure*.

(2018) and Corry (2004) focus on the history of abstract algebra. For Galois' work and its spread, see the detailed discussion by Ehrhardt (2012).

References

Allaire, Patricia R. and Robert E. Bradley (2002). "Symbolical Algebra as a Foundation for Calculus: D.F. Gregory's Contribution". *Historia Mathematica* 29, pp. 395–426.

Allard, André, ed. (1992). *Muḥammad ibn Mūsā al-Khwārizmī: Le Calcul Indien (Algorismus).* Histoire des textes, édition critique, traduction et commentaire des plus anciennes versions latines remaniées du XIIe siècle. Paris.

Alten, Heinz-Wilhelm et al. (2014). *4000 Jahre Algebra.* Geschichte – Kulturen – Menschen. Zweite, aktualisierte und ergänzte Auflage. Berlin and Heidelberg.

Balam, Richard (1653). *Algebra: Or, The doctrine of Composing, Inferring, and Resolving an Equation.* London.

Bos, Henk J.M. (2001). *Redefining Geometrical Exactness.* Descartes' Transformation of the Early Modern Concept of Construction. New York.

Bourbaki, Nicolas (1942). *Algèbre.* Chapitre I. Structures algébriques. Paris.

Bourbaki, Nicolas (1994). *Elements of the History of Mathematics.* Translated from the French by John Meldrum. Berlin.

Cardano, Gerolamo (1545). *Artis magnae sive de regulis algebraicis liber unus.* Nürnberg.

Cifoletti, Giovanna (1996). "The creation of the history of Algebra in the sixteenth century". In: *Mathematical Europe.* History, Myth, Identity. Ed. by Catherine Goldstein, Jeremy Gray, and Jim Ritter. Paris, pp. 123–144.

Corry, Leo (2001). "Mathematical Structures from Hilbert to Bourbaki: The Evolution of an Image of Mathematics". In: *Changing Images in Mathematics.* From the French Revolution to the New Millennium. Ed. by Umberto Bottazzini and Amy Dahan Dalmedico. London and New York, pp. 167–185.

Corry, Leo (2004). *Modern Algebra and the Rise of Mathematical Structures.* Second revised edition. Basel.

Corry, Leo (2005). "Heinrich Weber, *Lehrbuch der Algebra* (1895–1896)". In: *Landmark Writings in Western Mathematics 1640-1940.* Ed. by Yvor Grattan-Guinness. Amsterdam et al., pp. 690–699.

Durand-Richard, Marie-José (1990). "Genèse de l'Algèbre symbolique en Angleterre: une influence possible de J. Locke". *Revue d'histoire des sciences* 43, pp. 129–180.

Durand-Richard, Marie-José (2011). "Peacock's Arithmetic: An Attempt to Reconcile Empiricism to Universality". *Indian Journal of History of Science* 46, pp. 251–311.

Van Egmond, Warren (1988). "How algebra came to France". In: *Mathematics from Manuscript to Print.* 1300–1600. Ed. by Cynthia Hay. Oxford, pp. 127–144.

Ehrhardt, Caroline (2012). *Itinéraire d'un texte mathématique.* Les réélaborations d'un mémoire d'Évariste Galois au XIXe siècle. Paris.

Euclid (1570). *The Elements of Geometrie of the most auncient Philosopher Euclide of Megara.* Faithfully (now first) translated into the Englishe toung, by H. Billingsley. London.

Euler, Leonhard (1771). *Vollständige Anleitung zur Algebra.* Erster Theil. St. Petersburg.

Euler, Leonhard (1821). *An Introduction to the Elements of Algebra.* Ed. by John Farrar. Cambridge, N.Eng.

Franci, Raffaella and Laura Toti Rigatelli (1988). "Fourteenth-Century Italian Algebra". In: *Mathematics from Manuscript to Print.* 1300–1600. Ed. by Cynthia Hay. Oxford, pp. 11–29.

Gerhardt, C.I. (1871). "Zur Geschichte der Algebra in Deutschland. Zweiter Theil". *Monatsberichte der Königlich Preußischen Akademie der Wissenschaften zu Berlin* (1870), pp. 141–153.

Gray, Jeremy (2018). *A History of Abstract Algebra.* From Algebraic Equations to Modern Algebra. Oxford.

Gregory, Duncan Farquharson (1840). "On the Real Nature of Symbolical Algebra". *Transactions of the Royal Society of Edinburgh* 14, pp. 208–216.

Hasse, Helmut (1926). *Höhere Algebra.* I. Lineare Gleichungen. Berlin and Leipzig.

Hasse, Helmut (1927). *Höhere Algebra.* II. Gleichungen höheren Grades. Berlin and Leipzig.

Heeffer, Albrecht (2008). "A Conceptual Analysis of Early Arabic Algebra". In: *The Unity of Science in the Arabic Tradition*. Science, Logic, Epistemology and their Interactions. Ed. by Shahid Rahman, Tony Street, and Hassan Tahiri. Berlin and Heidelberg, pp. 89–128.

Høyrup, Jens (2019). *Selected Essays on Pre- and Early Modern Mathematical Practice*. Cham.

Hughes, Barnabas (1986). "Gerard of Cremona's Translation of al-Khwārizmī's al-Jabr. A Critical Edition". *Mediaeval Studies* 48, pp. 211–263.

Hughes, Barnabas, ed. (1989). *Robert of Chester's Translation of al-Khwārizmī's al-Jabr*. A New Critical Edition. Stuttgart.

Kaschube, Johann Wenceslaus (1717). *Cursus mathematicus, oder deutlicher Begrief der mathematischen Wissenschaften*. Jena.

Katz, Victor J. and Karen Hunger Parshall (2014). *Taming the Unknown*. A History of Algebra from Antiquity to the Early Twentieth Century. Princeton.

Kaunzner, Wolfgang (1985). "Über eine frühe lateinische Bearbeitung der Algebra al-Khwārizmīs in MS Lyell der Bodleian Library Oxford". *Archive for History of Exact Sciences* 32, pp. 1–16.

Koppelman, Elaine (1971). "The Calculus of Operations and the Rise of Abstract Algebra". *Archive for History of Exact Sciences* 8, pp. 155–242.

Mac Lane, Saunders (1939). "Some Recent Advances in Algebra". *The American Mathematical Monthly* 46, pp. 3–19.

Noether, Emmy (1927). "Abstrakter Aufbau der Idealtheorie in algebraischen Zahl- und Funktionenkörpern". *Mathematische Annalen* 96, pp. 26–61.

Oaks, Jeffrey A. (2024). "The Legacy of Al-Khwārazmī in History of Algebra". In: *Research in History and Philosophy of Mathematics*. The CSHPM 2022 Volume. Ed. by Maria Zack and David Waszek. Cham, pp. 127–181.

Oaks, Jeffrey A. and Haitham M. Alkhateeb (2007). "Simplifying equations in Arabic algebra". *Historia Mathematica* 34, pp. 45–61.

Pacioli, Luca (1494). *Summa de Arithmetica Geometria Proportioni et Proportionalita*. Venezia.

Peacock, George (1834). "Report on the Recent Progress and Present State of certain Branches of Analysis". *Report of the British Association for the Advancement of Science* (1833), pp. 185–352.

Procissi, Angiolo (1954). "I 'Ragionamenti d'algebra' di R. Canacci". *Bollettino dell'Unione Matematica Italiana Serie* 9, pp. 300–326.

Ramus, Petrus (1560). *Algebra*. Paris.

Ramus, Petrus (1586). *Arithmetices libri duo, et algebrae totidem*. Ed. by Lazarus Schöner. Frankfurt.

Recorde, Robert (1551). *The patheway to knowledg*. London.

Rella, Tonio (1929). "Höhere Algebra II, Gleichungen höheren Grades [review]". *Monatshefte für Mathematik und Physik* 36, A17–A18.

de La Roche, Estienne (1520). *Larismethique nouellement composee*. Lyon.

Schirmer, Alfred (1912). *Der Wortschatz der Mathematik nach Alter und Herkunft untersucht*. Straßburg.

Schlote, Karl-Heinz (2005). "B.L. van der Waerden, *Moderne Algebra*, first edition (1930–1931)". In: *Landmark Writings in Western Mathematics 1640-1940*. Ed. by Yvor Grattan-Guinness. Amsterdam et al., pp. 901–916.

Schubring, Gert (2006). *Conflicts Between Generalization, Rigor, and Intuition*. Number Concepts Underlying the Development of Analysis in 17th-19th Century France and Germany. New York.

Serret, Joseph-Alfred (1849). *Cours d'algèbre supérieure*. Paris.

Sesiano, Jacques, ed. (2014). *Liber Mahameleth*. Cham.

Smith, David Eugene (1958). *History of Mathematics*. Volume II. Special Topics of Elementary Mathematics. New York.

Stedall, Jacqueline A. (2011). *From Cardano's Great Art to Lagrange's Reflections*. Filling a Gap in the History of Algebra. Zürich.

Stedall, Jacqueline A. (2012). "Narratives of Algebra in Early Printed European Texts". In: *Pluralité de l'algèbre à la Renaissance*. Ed. by Sabine Rommevaux, Maryvonne Spiesser, and Maria Rosa Massa Esteve. Paris, pp. 217–236.

Steinitz, Ernst (1910). "Algebraische Theorie der Körper". *Journal für die reine und angewandte Mathematik* 137, pp. 167–309.

Stifel, Michael (1544). *Arithmetica integra*. Nürnberg.
Strauß, Gerhard et al., eds. (1995). *Deutsches Fremdwörterbuch*. 2. Auflage. Band 1: a-Präfix – Antike. Berlin and New York.
Tropfke, Johannes (1980). *Geschichte der Elementarmathematik*. Band 1: Arithmetik und Algebra. Vollständig neu bearbeitet von Kurt Vogel, Karin Reich, Helmuth Gericke. Berlin and New York.
Verburgt, Lukas M. (2018). "Duncan F. Gregory and Robert Leslie Ellis: second-generation reformers of British mathematics". *Intellectual History Review* 28, pp. 369–397.
Viète, François (1591). *In artem analyticam isagoge*. Tours.
van der Waerden, Bartel Leendert (1930-31). *Moderne Algebra*. 2 vols. Berlin.
Wallis, John (1685). *A Treatise of Algebra, both Historical and Practical*. London.
Weber, Heinrich (1893). "Die allgemeinen Grundlagen der Galois'schen Gleichungstheorie". *Mathematische Annalen* 43, pp. 521–549.
Weber, Heinrich (1895-96). *Lehrbuch der Algebra*. Braunschweig.
Widmann, Johannes (1489). *Behende vnd hubsche Rechenung auff allen kauffmanschafft*. Leipzig.

algebra (structure)

It was Benjamin Peirce who first used *algebra* not just for the discipline, but also for a mathematical structure. His paper "Linear associative algebra" was circulated in lithographed form in 1870 (Peirce 1870), and published in 1881 (Peirce 1881). In this text the word appears in the form *an algebra* (e. g., p. 98) but also in the plural (e. g., p. 99). Wedderburn (1908, p. 78) proposed a simplification of the terminology: "At Professor Dickson's suggestion I have used the word *algebra* as equivalent to Peirce's *linear associative algebra* which is too long for convenient use".

An interesting comment on the term was made by Bochner (1974, p. 830): "AN algebra was, around 1920, still totally unfamiliar to a student on the Continent [...] In fact, AN algebra, as a term and object, is in my present-day view, the most outstanding 19th century 'native' American contribution to present-day mathematics"; on Peirce's article cited above, he adds: "[this article] not only speaks of *an* algebra, but it also uses the plural 'algebras,' and this probably for the very first time anywhere in the English language" (ibid.). In his classic book on algebra, van der Waerden (1931, p. 149) defines a *hypercomplex system* (*hyperkomplexes System*) but also points out the recency of the synonymous *algebra*. His definition makes it clear that by his time an algebra was understood as a ring which is also a vector space over one of its subfields.[1]

The term *number system* in this context is frequently found in the German literature of the 19th century; for example, *Zahlensystem* was used by Hankel (1867, p. 36), a frequently referred to text, and by Molien (1893, p. 83).

See also *hypercomplex*.

Note. For the history of the notion, see Frei (2003), Frei (2007), and Schlote (1987). Some comments on the name can be found in Krömer (1998, pp. 91f.).

References

Bochner, Salomon (1974). "Mathematical Reflections". *The American Mathematical Monthly* 81, pp. 827–852.
Dickson, Leonard Eugene (1923). *Algebras and their Arithmetics*. Chicago.
Frei, Günther (2003). "Zur Geschichte der Arithmetik der Algebren (1843-1932)". *Elemente der Mathematik* 58, pp. 156–168.

---

[1] For a definition along these lines, see already Dickson (1923, p. 9).

Frei, Günther (2007). “Developments in the Theory of Algebras over Number Fields: A New Foundation for the Hasse Norm Residue Symbol and New Approaches to Both the Artin Reciprocity Law and Class Field Theory”. In: *Episodes in the History of Modern Algebra (1800–1950)*. Ed. by Jeremy J. Gray and Karen Hunger Parshall. Providence, R.I., pp. 117–151.
Hankel, Hermann (1867). *Vorlesungen über die complexen Zahlen und ihre Functionen*. In zwei Theilen. I. Theil: Theorie der complexen Zahlsysteme. Leipzig.
Krömer, Ralf (1998). “Zur Geschichte des axiomatischen Vektorraumbegriffs”. Diploma thesis. Universität des Saarlandes.
Molien, Theodor (1893). “Ueber Systeme höherer complexer Zahlen”. *Mathematische Annalen* 41, pp. 83–156.
Peirce, Benjamin (1870). “Linear Associative Algebra. Read before the National Academy of Sciences”. Washington, D.C. DOI: 10.3931/e-rara-21570.
Peirce, Benjamin (1881). “Linear Associative Algebra”. *American Journal of Mathematics* 4, pp. 97–229.
Schlote, Karl-Heinz (1987). “Zur Entstehung der Algebrentheorie bis zu ihrer Formulierung als abstrakte algebraische Theorie”. PhD thesis. Leipzig: Karl-Marx-Universität.
van der Waerden, Bartel Leendert (1931). *Moderne Algebra*. Zweiter Teil. Berlin.
Wedderburn, J.H.M. (1908). “On Hypercomplex Numbers”. *Proceedings of the London Mathematical Society* 6, pp. 77–118.

algebraic

The adjective *algebraic* was originally only used in the sense ‘pertaining to algebra’, where *algebra* itself has undergone various changes in meaning. While algebra is found in Latin texts in the 12th century, the adjective is of later origin. Early uses for *algebratico* are found in Pacioli (1494, e. g., fol. 138^r);[1] Cardano (1545) uses *algebraicus* in the title of his *Artis magnae sive de regulis algebraicis liber unus* (book one of the great art of algebraic rules).

The terminological distinction between *algebraic* and *transcendent* is due to Leibniz. In one of his articles, he stated that transcendental quantities “transcend all algebraic equations” (Leibniz 1686, p. 294).[2] For further usages, discussion, and references, see *transcendental*.

Notable uses in the 18th century include Euler’s distinction between algebraic and transcendental functions. He defined: “Functions are divided into algebraic and transcendental. The former are those made up from algebraic operations only, the latter are those which involve transcendental operations” (Euler 1748a, § 7).[3] Euler also applied the distinction to surfaces (Euler 1748b, appendix, ch. I.6).

Details regarding specific terms such as *algebraic number* can be found in the following entries. For the terms *algebraic extension* and *algebraically closed*, see *extension*.

References

Blåsjö, Viktor (2017). *Transcendental Curves in the Leibnizian Calculus*. Cambridge, Mass.
Cardano, Gerolamo (1545). *Artis magnae sive de regulis algebraicis liber unus*. Nürnberg.
Euler, Leonhard (1748a). *Introductio in analysin infinitorum*. Tomus primus. Lausanne.
Euler, Leonhard (1748b). *Introductio in analysin infinitorum*. Tomus secundus. Lausanne.

[1] More references to Pacioli are found in Ricci (1994, p. 22). [2] “omnem Algebraicam aequationem transcendat”. Translation from Blåsjö (2017, p. 7). [3] “Functiones dividuntur in Algebraicas & Transcendentes; illae sunt, quae componuntur per operationes algebraïcas solas, hae vero in quibus operationes transcendentes insunt”. Translation from Euler (1988, § 7), slightly modified.

Euler, Leonhard (1988). *Introduction to Analysis of the Infinite*. Book I. Translated by John D. Blanton. New York.
Leibniz, Gottfried Wilhelm (1686). "De Geometria recondita et Analysi Indivisibilium et infinitorum". *Acta Eruditorum* (June 1686), pp. 292–300.
Pacioli, Luca (1494). *Summa de Arithmetica Geometria Proportioni et Proportionalita*. Venezia.
Ricci, Laura (1994). "Il lessico matematico della Summa di Luca Pacioli". *Studi di lessicografia italiana* 12, pp. 5–71.

algebraic geometry

In early usages of *algebraic geometry* the word is used in some sense of the word *algebraic*. For example, in his book *A System of Algebraic Geometry*, Lardner (1823, p. viii) defended the title of the book: "The title 'Algebraic Geometry' has been preferred to either of the titles 'Analytic Geometry' or 'the Application of Algebra to Geometry', because the one is equivocal and the other circumlocutory".[1]

Algebraic geometry as it is known today, however, is essentially a development of the second half of the 19th century, during which time relatively concrete problems were studied, and of the 20th century, when solid foundations of the field were developed.

Towards the end of the 19th century, Segre (1891, p. 49) described the field in these terms:

> The idea of this geometry, i. e., the properties of an algebraic curve which remain invariant under a rational transformation of the curve into itself, is first found in a work of analysis, in the great memoir of Riemann on the theory of abelian functions. It is here that we first find fully developed the notion of *genus* [...] [The Riemann-Roch theorem] gave to modern geometry one of its most important and fruitful theorems. [...] And all through the geometry of algebraic curves the analytic researches on algebraic functions and their integrals have had applications of the greatest importance. It is sufficient to recall in this connection Abel's theorem.[2]

The actual term *geometria algebrica* was not used by Segre in this article yet; it appears, for example, in Beltrami (1900, p. 265), but features prominently in Severi's book *Lezioni di geometria algebrica* (Severi 1908); after that time, it gradually became more common.

When the new foundations of the field were laid, the designation was kept, as exemplified by Weil's book *Foundations of Algebraic Geometry* (Weil 1946) and Grothendieck's *Fondements de la géométrie algébrique* (Grothendieck 1962).

Note. For the history of algebraic geometry, see Dieudonné (1985), Gray (1989), and Coolidge (1940, pp. 117–317).

---

[1] Earlier uses of *algebraic geometry* are already found in the 18th century, for example in Caracciolo (1759), whose book is entitled *Geometria algebrica universa*, and Liger (1743, p. 2). [2] "Il concetto di questa geometria, cioè di proprietà dell'ente algebrico invariabili per trasformazioni razionali dell'ente stesso, si trova per la prima volta in un lavoro analitico: nella grande memorie di Riemann sulla *Teorie delle funczioni Abeliane*. È qui che viene usata la prima volta la nozione del *genere* [...] ha dato uno die più importanti e fecondi teoremi alla geometria moderna. [...] E per tutta quanta la geometria delle curve algebriche e sui loro integrali hanno avuto applicazioni importantissime: basti recordare in proposito il teorema di Abel". Translation from Segre (1904, pp. 449f.).

References

Beltrami, Eugenio (1900). "Francesco Brioschi. Commemorazione". *Rendiconti del Circolo Matematico di Palermo* 14, pp. 262–274.

Caracciolo, Giovanni Battista (1759). *Geometria algebrica universa*. Tomus primus. Roma.

Coolidge, Julian Lowell (1940). *A History of Geometrical Methods*. Oxford.

Dieudonné, Jean (1985). *History of Algebraic Geometry*. An Outline of the History and Development of Algebraic Geometry. Translated by Judith D. Sally. Monterey, California.

Gray, Jeremy (1989). "Algebraic Geometry in the late Nineteenth Century". In: *The History of Modern Mathematics*. Volume I: Ideas and Their Reception. Ed. by David E. Rowe and John McCleary. Boston, pp. 361–385.

Grothendieck, Alexander (1962). *Fondements de la géométrie algébrique*. Paris.

Lardner, Dionysius (1823). *A System of Algebraic Geometry*. Vol. I. The Geometry of Plane Curves. London.

Liger, Pierre (1743). *Dissertation sur la geometrie*. Paris.

Segre, Corrado (1891). "Su alcuni indirizzi nelle investigazioni geometriche". *Rivista di matematica* 1, pp. 42–66.

Segre, Corrado (1904). "On some tendencies in geometric investigations". *Bulletin of the American Mathematical Society* 10, pp. 442–468.

Severi, Francesco (1908). *Lezioni di geometria algebrica*. Padova.

Weil, André (1946). *Foundations of Algebraic Geometry*. New York.

algebraic integer

The usual definition of algebraic integers is due to Dedekind (1871, pp. 436f.):[1]

> The totality of all algebraic numbers also clearly constitutes a field. [...] we now define a number α to be an *algebraic integer* when it is the root of an equation with rational integer coefficients. Here we remark once and for all that by the *coefficients* of a function of degree m
>
> $$F(x) = cx^m + c_1x^{m-1} + c_2x^{m-2} + \cdots + c_m$$
>
> or of an equation $F(x) = 0$, we invariably mean the m quotients
>
> $$-\tfrac{c_1}{c}, +\tfrac{c_2}{c} \ldots (-1)^m \tfrac{c_m}{c}.$$ [2]

Dedekind here used German *ganze algebraische Zahl* and also the shorter *ganze Zahl*.

Dedekind (1894, p. 524) later remarked that he did not know whether the term had been used before his 1871 publication and indeed, earlier usages in print do not seem to exist. However, the same term also appears in Kronecker's major work "Elements of an arithmetic theory of the algebraic magnitudes" (Kronecker 1882, p. 14), in which Kronecker gave the same definition. While his work was published in the early 1880s, he had already worked on his theory in the 1850s and may well have used the term before Dedekind (Edwards 1980, p. 322).

---

[1] That Dedekind was the first to publish this general definition is generally acknowledged, for example by Narkiewicz (2004, p. 69).

[2] "Der Inbegriff aller algebraischen Zahlen bildet offenbar ebenfalls einen Körper. Wir wollen nun [...] eine Zahl α eine *ganze algebraische Zahl* nennen, wenn sie die Wurzel einer Gleichung ist, deren Coefficienten rationale ganze Zahlen sind, wobei wir ein- für allemal bemerken, dass wir unter den Coefficienten einer Function $m$ten Grades $F(x) = cx^m + c_1x^{m-1} + c_2x^{m-2} + \cdots + c_m$ oder der Gleichung $F(x) = 0$ stets die m Quotienten $-\frac{c_1}{c}, +\frac{c_2}{c} \ldots (-1)^m \frac{c_m}{c}$ verstehen". Translation from Avigad (2004, p. 25).

The French term *nombre entier algébrique* was used by Dedekind (1877, p. 2) himself; he also suggested the shorter *nombre entier*. The English terminology came up in the 1890s: Mathews (1892, p. 228) used *algebraical integer*, and Shaw (1896, p. 102) *algebraic integer*. Occasionally, the translation *integral algebraic number* was used, for example in Weber (1897, p. 182).

Note. For the history of the concept, see Edwards (1980); see also Ferreirós (2007, pp. 94–99) and Katz and Parshall (2014, pp. 394–399).

References

Avigad, Jeremy (2004). "Dedekind's 1871 version of the theory of ideals". URL: `https://kilthub.cmu.edu/articles/Dedekind_s_1871_Version_of_the_Theory_of_Ideals/6491342/1` (retrieved July 25, 2007).

Dedekind, Richard (1871). "Ueber die Composition der binären quadratischen Formen". In: *Vorlesungen über Zahlentheorie von P.G. Lejeune Dirichlet*. Herausgegeben und mit Zusätzen versehen von R. Dedekind. Zweite, umgearbeitete und vermehrte Auflage. Braunschweig, pp. 380–497.

Dedekind, Richard (1877). *Sur la théorie des nombres entiers algébriques*. Paris.

Dedekind, Richard (1894). "Ueber die Theorie der ganzen algebraischen Zahlen". In: *Vorlesungen über Zahlentheorie von P.G. Lejeune Dirichlet*. Herausgegeben und mit Zusätzen versehen von R. Dedekind. Vierte, umgearbeitete und vermehrte Auflage. Braunschweig, pp. 434–657.

Edwards, Harold M. (1980). "The Genesis of Ideal Theory". *Archive for History of Exact Sciences* 23, pp. 321–378.

Ferreirós, José (2007). *Labyrinth of Thought*. A History of Set Theory and Its Role in Modern Mathematics. Second revised edition. Basel.

Katz, Victor J. and Karen Hunger Parshall (2014). *Taming the Unknown*. A History of Algebra from Antiquity to the Early Twentieth Century. Princeton.

Kronecker, Leopold (1882). *Grundzüge einer arithmetischen Theorie der algebraischen Grössen*. Berlin.

Mathews, George Ballard (1892). *Theory of Numbers*. Part I. Cambridge.

Narkiewicz, Władysław (2004). *Elementary and Analytic Theory of Algebraic Numbers*. Third edition. Berlin.

Shaw, James Byrnie (1896). "Theory of Numbers and of Equations". *Bulletin of the American Mathematical Society* 3, pp. 97–107.

Weber, Heinrich (1897). "Transcendental numbers. An authorized translation by Professor W. W. Beman of Chapter XXV. of Vol. II., of the *Lehrbuch der Algebra*". *Bulletin of the American Mathematical Society* 3, pp. 174–195.

algebraic number

The concept of algebraic number was given a precise definition by Abel, which he presented in a manuscript probably written in 1828 but first published in 1839. In the manuscript, Abel first defined what it meant for a quantity to be expressed algebraically by other quantities, namely by a limited number of the usual algebraic operations, i. e. addition, subtraction, multiplication, division, and extraction of roots "with prime exponents" (Abel 1881, p. 224).[1] He then defined a *nombre algébrique*: "A quantity that can be expressed algebraically by unity is called an algebraic number".[2] The term becomes more common in the second half of the 19th century.

---

[1] "Une quantité y est dite pouvoir s'exprimer algébriquement par plusieurs autres quantités, lorsqu'on peut la former de ces dernières à l'aide d'un nombre limité des opérations suivantes: 1. Addition. 2. Soustraction. 3. Multiplication. 4. Division. 5. Extraction de racines avec des exposans premiers". [2] "Une quantité qui peut s'exprimer algébriquement par l'unité s'appelle un nombre algébrique".

The now usual definition was given by Cantor (1874, p. 258), for example: "By a real algebraic number one generally understands a real number ω which satisfies a non-constant equation of the form: $a_0\omega^n + a_1\omega^{n-1} + \cdots + a_n = 0$, where n, a_0, a_1, $\ldots$ a_n are integers".[3] Essentially the same definition was given by Dedekind (1879, p. 452); in a different formulation it had already appeared in Dedekind (1871, p. 427). These definitions were known earlier but not usually under the name *algebraic number*.

The English term does not seem to have come up until about 1890; it is found in Bolza (1890-91, p. 110).[1]

Note. For the history of the concept, see, for example, Ferreirós (2007, esp. pp. 94-99), Katz and Parshall (2014, pp. 394–399), and Haubrich (1992).

References

Abel, Niels Henrik (1881). *Œuvres complètes.* Nouvelle édition. Tome second. Ed. by Ludvig Sylow Sylow and Sophus Lie. Christiania.

Bolza, Oskar (1890-91). "On the Theory of Substitution-Groups and its Applications to Algebraic Equations". *American Journal of Mathematics* 13, pp. 59–144.

Cantor, Georg (1874). "Ueber eine Eigenschaft des Inbegriffs aller reellen algebraischen Zahlen". *Journal für die reine und angewandte Mathematik* 77, pp. 258–262.

Dedekind, Richard (1871). "Ueber die Composition der binären quadratischen Formen". In: *Vorlesungen über Zahlentheorie von P.G. Lejeune Dirichlet.* Herausgegeben und mit Zusätzen versehen von R. Dedekind. Zweite, umgearbeitete und vermehrte Auflage. Braunschweig, pp. 380–497.

Dedekind, Richard (1879). "Ueber die Theorie der ganzen algebraischen Zahlen". In: *Vorlesungen über Zahlentheorie von P.G. Lejeune Dirichlet.* Herausgegeben und mit Zusätzen versehen von R. Dedekind. Dritte, umgearbeitete und vermehrte Auflage. Braunschweig, pp. 434–627.

Ewald, William (1996). *From Kant to Hilbert.* A Source Book in the Foundations of Mathematics. Volume II. Oxford.

Ferreirós, José (2007). *Labyrinth of Thought.* A History of Set Theory and Its Role in Modern Mathematics. Second revised edition. Basel.

Haubrich, Ralf (1992). "Zur Entstehung der algebraischen Zahlentheorie Richard Dedekinds". PhD thesis. Georg-August-Universität Göttingen.

Katz, Victor J. and Karen Hunger Parshall (2014). *Taming the Unknown.* A History of Algebra from Antiquity to the Early Twentieth Century. Princeton.

algebraic topology

The German term *algebraische Topologie* was first introduced by Alexandroff (1932, p. 23). He justified the new expression in the following terms: "We prefer this expression to the otherwise customary term 'combinatorial' topology, since we consider much broader applications of algebraic methods and concepts than the word 'combinatorial' would include".[2]

Hopf (1933, p. 41) used the mixed *algebraisch-kombinatorische Topologie*, and the English term is then found in Lefschetz (1935, p. 220)[3] and Steenrod (1936,

---

[3] "Unter einer reellen algebraischen Zahl wird allgemein eine reelle Zahlgrösse ω verstanden, welche einer nicht identischen Gleichung von der Form genügt: $a_0\omega^n + a_1\omega^{n-1} + \cdots + a_n = 0$, wo n, a_0, a_1, $\ldots$ a_n ganze Zahlen sind". Translation from Ewald (1996, p. 840). [1] The OED cites a text from 1893 as a first use for *algebraic integer.* [2] "Wir ziehen diesen Ausdruck dem sonst üblichen Terminus ‚kombinatorische' Topologie vor, denn es handelt sich hier um eine viel weitere Anwendung der algebraischen Methoden und Grundbegriffe, als das Wort ‚Kombinatorik' es vermuten läßt". Translation from Alexandroff (1960, p. 27). [3] This article by Lefschetz is cited by the OED as the first use in English.

p. 661); French *topologie algébrique* appears in Threlfall (1936, p. 242) and van Dantzig (1936, p. 275). The term became more common after the publication of Lefschetz' book entitled *Algebraic Topology* (Lefschetz 1942).

See also *combinatorial topology*.

Note. The introduction of the term was pointed out by Volkert (2002, p. 291). For the history of the concept, see Henn and Puppe (1990) and Dieudonné (2009). James (1999) discusses the transition from combinatorial to algebraic topology.

References

Alexandroff, Pavel S. (1932). *Einfachste Grundbegriffe der Topologie*. Berlin.

Alexandroff, Pavel S. (1960). *Elementary Concepts of Topology*. Translated by Alan E. Parley. New York.

van Dantzig, David (1936). "Nombres universels ou $\nu!$-adiques avec une introduction sur l'algèbre topologique". *Annales scientifiques de l'École Normale Supérieure* 53, pp. 275–307.

Dieudonné, Jean (2009). *A History of Algebraic and Differential Topology, 1900–1960*. Boston.

Henn, Hans-Werner and Dieter Puppe (1990). "Algebraische Topologie". In: *Ein Jahrhundert Mathematik 1890 - 1990*. Festschrift zum Jubiläum der DMV. Ed. by Gerd Fischer et al. Braunschweig and Wiesbaden, pp. 673–716.

Hopf, Heinz (1933). "Die Klassen der Abbildungen der n-dimensionalen Polyeder auf die n-dimensionale Sphäre". *Commentarii mathematici Helvetici* 5, pp. 39–54.

James, Ioan Mackenzie (1999). "From Combinatorial to Algebraic Topology". In: *History of Topology*. Ed. by Ioan Mackenzie James. Amsterdam, pp. 561–573.

Lefschetz, Solomon (1935). "Application of Chain-Deformations to Critical Points and Extremals". *Proceedings of the National Academy of Sciences* 21, pp. 220–222.

Lefschetz, Solomon (1942). *Algebraic Topology*. Providence, R.I.

Steenrod, Norman (1936). "Universal Homology Groups". *American Journal of Mathematics* 58, pp. 661–701.

Threlfall, William (1936). "Quelques résultats récents de la topologie des variétés". *L'enseignement mathématique* 35, pp. 242–255.

Volkert, Klaus (2002). *Das Homöomorphismusproblem insbesondere der 3-Mannigfaltigkeiten, in der Topologie 1892-1935*. Paris.

algebraic variety and *algebraic set*

Definitions of algebraic varieties and sets came up in the late 19th century. For example, Segre (1891, p. 47) used Italian *varietà algebrica* and defined:[1] "A variety is said to be *algebraic* when its elements are the totality of those whose coordinates satisfy given algebraic equations (in which also indeterminate parameters may appear rationally)".[2] Segre used this definition on various occasions, for example in Segre (1894, p. 46), where he referred to Molk (1885) and Kronecker (1882) for the concept. The term was not used by Molk and Kronecker.

In German texts, *algebraische Mannigfaltigkeit* has long been the usual designation, used by van der Waerden (1926, p. 183), for example:

> The rigorous foundation of the theory of algebraic varieties [*algebraische Mannigfaltigkeiten*] in n-dimensional spaces can only be given in terms of ideal theory because the definition of an algebraic variety itself leads immediately to polynomial ideals. Indeed, a variety is called

[1] This is the first definition of *algebraic variety* that has been found in the preparation of this article. The expression as such can be found in earlier texts. [2] "Una varietà se dice *algebrica* quando i suoi elementi sono tutti quelli aventi le coordinate espresse da funzioni algebriche date di uni o più parametri variabili indipendenti". Translation from Segre (1904, p. 447) (adapted).

> algebraic, if it is given by algebraic equations in the n coordinates, and the lefthand sides of all equations that follow from the given ones form a polynomial ideal.[3]

Similarly, Hardcastle (1900, p. 122) used *algebraic manifold*, while *algebraic variety* became more common after Lefschetz (1916, p. 197) began using it.[4]

A change of meaning in the terminology occurred around 1950 and is reflected in Baldassarri (1956, p. 1):

> Let k be a subfield of an algebraically closed field k^* and $\boldsymbol{A}^n(k^*)$ (or $\boldsymbol{P}^n(k^*)$) the affine (or projective) n-dimensional space over k^*. A subset H of $\boldsymbol{A}^n(k^*)$ is called an algebraic set over k or a k-set if the points of H are the zeros of a set $\mathfrak{F}$ of polynomials in $k[\xi_l, \xi_2, \dots, \xi_n] = k[\xi]$, where $\xi_1, \xi_2, \dots, \xi_n$ are indeterminates [...] The k-sets of $\boldsymbol{A}^n(k^*)$ form a system closed both by intersection and union provided the latter be repeated finitely: a k-set is said irreducible or a k-variety V if it is not a proper union of two k-sets.

Baldassarri (1956, p. 16) goes on to comment on the change: "Algebraic sets, in the realm of classical geometry, were called varieties and our varieties were then characterised as irreducible". For the latter characterization, Baldassarri cites Bertini (1907, pp. 189f.) and Kronecker (1882).[1]

The new term *algebraic set* (French *ensemble algébrique*) was suggested by Samuel in his PhD-thesis from 1949 (Samuel 1951, p. 218) as a replacement of Weil's *bunch of varieties* (Weil 1946, pp. xi, 84). This change was also accepted by Weil, who, in a joint paper with Lang, said that they will follow the terminology of Weil (1946) "except that a bunch of varieties will be called an algebraic set" (Lang and Weil 1954, p. 820).

With Grothendieck's refoundation of algebraic geometry, the definition of algebraic varieties was reformulated in terms of schemes. Such a definition is found in Borel and Serre (1964-65, p. 131), for example.

Note. The history of the concept is discussed by Dieudonné (1985) and Craighero (1972).

References

Baldassarri, Mario (1956). *Algebraic Varieties*. Berlin et al.

Bertini, Eugenio (1907). *Introduzione alla geometria proiettiva degli iperspazi*. Con appendice sulle curve algebriche e loro singolarità. Pisa.

Borel, Armand and Jean-Pierre Serre (1964-65). "Théorèmes de finitude en cohomologie galoisienne". *Commentarii mathematici Helvetici* 39, pp. 111–164.

Craighero, Pier Carlo (1972). "L'evoluzione del concetto di varietà algebrica". *Cultura e scuola* 11, pp. 178–214.

Dieudonné, Jean (1985). *History of Algebraic Geometry*. An Outline of the History and Development of Algebraic Geometry. Translated by Judith D. Sally. Monterey, California.

Hardcastle, Frances (1900). "Report on the Present State of the Theory of Point-groups. Part I". *Report of the British Association for the Advancement of Science* (1900), pp. 121–131.

---

[3] "Die exakte Begründung der Theorie der algebraischen Mannigfaltigkeiten in n-dimensionalen Räumen kann nur mit den Hilfsmitteln der Idealtheorie geschehen, weil schon die Definition einer algebraischen Mannigfaltigkeit unmittelbar auf Polynomideale führt. Eine Mannigfaltigkeit heißt ja algebraisch, wenn sie durch algebraische Gleichungen in den n Koordinaten bestimmt wird, und die linken Seiten aller Geichungen, die aus diesen Gleichungen folgen, bilden ein Polynomideal". Translation from Schappacher (2007, p. 250). [4] The term is earlier found in Kasner (1905, p. 573). [1] Baldassarri (ibid.) also refers to Samuel (1955, p. 129), where a useful comparison of the terminology as used by himself, Zariski, van der Waerden, Severi, and Hodge-Pedoe is found.

Kasner, Edward (1905). "The Present Problems of Geometry". In: *Congress of Arts and Science. Universal Exposition, St. Louis, 1904. Volume I.* Ed. by Howard J. Rogers. Boston and New York, pp. 559–586.
Kronecker, Leopold (1882). *Grundzüge einer arithmetischen Theorie der algebraischen Grössen.* Berlin.
Lang, Serge and André Weil (1954). "Number of Points of Varieties in Finite Fields". *American Journal of Mathematics* 76, pp. 819–827.
Lefschetz, Solomon (1916). "The Arithmetic Genus of an Algebraic Manifold Immersed in Another". *Annals of Mathematics* 17, pp. 197–212.
Molk, Jules (1885). "Sur une notion qui comprend celle de la divisibilité et sur la théorie générale de l'élimination". *Acta Mathematica* 6, pp. 1–166.
Samuel, Pierre (1951). "La notion de multiplicité en algèbre et en géométrie algébrique". *Journal de mathématiques pures et appliquées* 30, pp. 159–274.
Samuel, Pierre (1955). *Méthodes d'algèbre abstraite en géométrie algébrique.* Berlin et al.
Schappacher, Norbert (2007). "A Historical Sketch of B.L. van der Waerden's work in algebraic geometry: 1926–1946". In: *Episodes in the History of Modern Algebra (1800–1950).* Ed. by Jeremy Gray and Karen Hunger Parshall. Providence, R.I., pp. 245–283.
Segre, Corrado (1891). "Su alcuni indirizzi nelle investigazioni geometriche". *Rivista di matematica* 1, pp. 42–66.
Segre, Corrado (1894). "Introduzione alla geometria sopra un ente algebrico semplicemente infinito". *Annali di Matematica Pura e Applicata* 22, pp. 41–142.
Segre, Corrado (1904). "On some tendencies in geometric investigations". *Bulletin of the American Mathematical Society* 10, pp. 442–468.
van der Waerden, Bartel Leendert (1926). "Zur Nullstellentheorie der Polynomideale". *Mathematische Annalen* 96, pp. 183–208.
Weil, André (1946). *Foundations of Algebraic Geometry.* New York.

algorithm

The term *algorithm* has a long and complex history which involves misinterpretations of the word with lasting effects and various semantic shifts.

The story begins with the spread of Indian numerals to the Islamic world in the ninth century. The mathematician al-Khwārizmī, often referred to as the father of algebra, described their use in a treatise later referred to as "On the calculation of the Indians". This treatise is "the oldest known Arabic work in which the Indian decimal place-value system and operations with it are described" (Folkerts 2001, p. 15).

While the original Arabic text of al-Khwarizmi's treatise is lost, it began to be translated into Latin from the early twelfth century onwards.[1] Some of the manuscripts point to the origin of the term: They start with *dixit Algorizmi* 'al-Khwarizmi said'. The name *algorismus* was then given to the Indian kind of calculation in general; the term can be found from around 1150 on and replaced *helcep sarracenicum* 'Saracen calculation' (Burnett 2006, p. 18). By the thirteenth century, the name can be found in numerous writings in many European places. Two treatises that were influential in its spreading were the *Carmen de Algorismo*[2] 'song of algorism' by Alexander de Villa Dei and the *Algorismus vulgaris*[3] 'common algorism' by Johannes de Sacrobosco.

Knowledge about the origin of the term was soon lost, however. In the *Algorismus vulgaris* just mentioned, for example, it is stated that "[a] certain philosopher named Algus wrote this brief science of numbering, for which reason it is called *Algorismus*,

[1] For an edition, see Folkerts (1997). [2] For an edition, see Halliwell (1839, pp. 73–83). On the work, see Ambrosetti (2016). [3] Editions include de Sacrobosco (1841), Curtze (1897), and de Sacrobosco (1983).

which is understood to be the art of numbering or the introductory art into number" (Curtze 1897, p. 1).[4] The name *Algus* is a corruption of *al-Khwārizmī*. A similar explanation is given in the 15th-century "The Art of Nombryng", where also some more derivations of the term are offered:

> ffirst we most se, what is the propre name therofe, and fro whens the name come: [...] The name is clepede [called] Algorisme, hade out of Algore, other of Algos, in grewe [Greek], That is clepide in englisshe art other craft, And of Rithmus that is callede nombre. So algorisme is clepede the art of nombryng, other it is had ofe en or in, and gogos that is introduccioun, and Rithmus nombre, that is to say Interduccioun of nombre. And thirdly it is hade of the name of a kyng that is clepede Algo and Rythmus; So callede Algorismus ([anon.] 1922, p. 33).

The passage suggests an association with the Greek word for 'number', *ἀριθμός*, and this association is the reason why the spelling *algorismus* was gradually changed to *algorithmus*.[1]

In sum, the term *algorismus* began to be used in the 12th century and referred to the calculation with Indian numerals, and a treatise that dealt with this topic was also called *algorismus*.

The scope of the term was soon widened, however. For example, in the middle of the 14th century Nicole Oresme wrote a treatise entitled *Algorismus proportionum*[2] 'algorism of ratios'. Also, the word was used as an alternative to *practical arithmetic*, for example by Ciruelo (1495), who wrote a *Tractatus arithmetice practice qui dicitur algorismus* 'Treatise on practical arithmetic called algorismus'.

Another extension of the scope is found in connection with the calculus. Thus, Leibniz (1684, p. 467) writes: "From knowledge of this *Algorithm*, as I call it, of this calculus, which I call *differential*, all other differential equations can be found by a similar calculation; also maxima and minima, and likewise tangents, can be found with no need to remove fractions or irrationals or other roots, which, however, had to be done according to methods previously published".[3] Referring to this passage, Wolff (1716, p. 38) included an entry for *algorithmus infinitesimalis* in his mathematical dictionary. Similar examples can be found in English texts as well. Thus, Harris (1702, p. 118) speaks of the "Algorithm or Arithmetick of Fluxions", and Ditton (1706, preface) of the "Dark Algorithm of Nothings [infinitely small quantities]".

The widening of the meaning is explicitly stated by Colson (1736, p. 151), who equates *algorithm* with *method of operation*, and confirmed in various reference works of the 18th century. For example, d'Alembert et al. (1784, p. 37) state that the

---

[4] "anc igitur scientiam numerandi compendiosam edidit philosophus nomine ALGUS, <quae> unde *algorismus* nuncupatur, vel ars numerandi, vel introductio in numerum". Translation from Grant (1974, p. 94). [1] It took a long time to relate *algorithmus* to al-Khwārizmī. Various vain attempts have been compiled by Chasles (1859); see also Cantor (1867, pp. 267f.) and Cantor (1907, pp. 714f.). The etymology here discussed was, according to Allard (1994, p. 235), first proposed by Reinaud (1849, pp. 303f.). [2] For an edition and an English translation, see Curtze (1868) and Grant (1965).

[3] "Ex cognito hoc velut *Algorithmo*, ut ita dicam, calculi hujus, quem voco *differentialem*, omnes aliae aequationes differentiales inveniri possunt per calculum communem, maximaeque et minimae, itemque tangentes haberi, itemque tangentes haberi, ita ut opus non sit tolli fractas aut irrationales aut alia vincula, quod tamen faciendum fuit secundum Methodos hactenus editas". Translation from (Stedall 2008, p. 129).

"word is taken to mean, in general, the method and notation of all types of calculation. In this sense, we say the algorithm of the integral calculus, the algorithm of the exponential calculus, the algorithm of sines, etc.".[4] Because of the generalization of the meaning, Klügel (1803, p. 69) regards the example terms given by d'Alembert as pleonastic.

Thus, by the end of the 18th century the meaning of the term has considerably widened, and in the 19th century it continued to be used in this way. This extension was at first often applied to wider fields such as the calculus, but the word was also used with respect to rather specific topics, as exemplified by Oresme's use, or in the 18th century, by Euler. For example, in his *Specimen algorithmi singularis* 'A specimen of a singular algorithm', Euler declared that there are certain quantities "of such a nature that they require a unique algorithm" (Euler 1764, p. 53) – he was talking about certain quantities connected with continued fractions. He uses the word in a similarly specific way in "On the use of a new algorithm in solving Pell's equation" (Euler 1767).[1] Another comparable example is the application of the term *algorithm* to the Euclidean method for finding the greatest common divisor (see, for example, Lejeune Dirichlet 1863, p. 8). At the end of the 19th century, Shaw (1895, p. 3) characterized mathematics as "the science of algorithms" and defined the word thus: "Any system of deduction of necessary conclusions, or transmutation of laws or ideas into other forms, is an *algorithm*. Thus quaternions is an algorithm. Graphical mechanics is an algorithm".

The modern notion, in contrast, typically includes semantic features such as finiteness, but these, while they may have been implicit, were not spelt out before the 20th century. At the same time, when the modern notion of algorithm (or rather, a notion resembling the modern one) was used, it was usually designated under a different name. For example, in the formulation of Hilbert's tenth problem, Hilbert is usually interpreted as meaning an algorithm when he used the German word *Verfahren*, lit. 'process, procedure': "Given a Diophantine equation with any number of unknown quantities and with rational integral numerical coefficients: To devise a process according to which it can be determined in a finite number of operations whether the equation is solvable in rational integers".[2] Another example is from Borel (1912, p. 162): "I intentionally leave aside the practical length of operations, which can be shorter or longer; the essential point is that each operation can be executed in finite time with a safe method that is unambiguous".[3]

---

[4] "Le même mot se prend, en général, pour désigner la méthode & la notation de toute espèce de calcul. En ce sens, on dit l'algorithme du calcul integral, l'algorithme du calcul exponentiel, l'algorithme du calcul des sinus, &c." Translation from Chabert (1999, p. 2). [1] On these uses in Euler, see Lüneburg (2008, p. 335). [2] "Eine Diophantische Gleichung mit irgend welchen Unbekannten und mit ganzen rationalen Zahlencoefficienten sei vorgelegt: man soll ein Verfahren angeben, nach welchem sich mittelst einer endlichen Anzahl von Operationen entscheiden läßt, ob die Gleichung in ganzen rationalen Zahlen lösbar ist". Translation from Hilbert (1901, p. 458).
[3] "Je laisse intentionnellement de côté le plus ou moins grande longueur pratique des opérations; l'essentiel est que chacune de ces opérations soit exécutable en un temps fini, par une méthode sûre et sans ambiguïté". Translation from Avigad and Brattka (2014, p. 5), who also stated that this passage "makes it clear that Borel had an intuitive notion of an algorithm in mind".

In the 1950s, the word was still not in regular use (Fant 1993, p. 3)[4], and there are only occasional attestations such as in Church (1936, p. 351): "It is clear that for any recursive function of positive integers there exists an algorithm using which any required particular value of the function can be effectively calculated". Similarly, Kleene (1943, p. 59) used *algorithmic* in the formulation of what was later called the *Church-Turing thesis*: "In setting up a complete algorithmic theory, what we do is to describe a procedure, performable for each set of values of the independent variables, which procedure necessarily terminates and in such manner that from the outcome we can read a definite answer, 'Yes' or 'No,' to the question, 'Is the predicate value true?'". In a later work, Kleene (1952, p. 301) stated: "The thesis [the Church-Turing thesis] and its converse provide the exact definition of the notion of a calculation (decision) procedure or algorithm". The quotation illustrates the use of the synonyms *calculation procedure* and *decision procedure*; other synonyms include *mechanical process*, used, for example, by Turing (1936, p. 262), and *effective method*, which Rosser (1939, p. 55) defined as follows: "'Effective method' is here used in the rather special sense of a method each step of which is precisely predetermined and which is certain to produce the answer in a finite number of steps".

Towards the middle of the 20th century the term gradually acquires the current meaning (Fant 1993, p. 3). In particular, Markov (1954, p. 3) gave the following characterization: "In mathematics, 'algorithm' is commonly understood to be an exact prescription, defining a computational process, leading from various initial data to the desired result".[1] As criteria, he named definiteness, generality, and conclusiveness. At the same time, a characterisation was also given by Kolmogorov (1953, p. 175). It is at this time that the term "caught on very quickly in the computing community" (Fant 1993, p. 3).

Note. For the history of the notion and terminology, see Fant (1993), Chabert (1999), Thomas (2015), Rapaport (2012), Ausiello (2013), and Lüneburg (2008, ch. 4). Hill (2016) focuses on 20th-century and current definitions.

References

[anon.] (1922). "The Art of Nombryng". In: *The Earliest Arithmetics in English*. Ed. by Robert Steele. London, pp. 33–51.

d'Alembert, Jean Le Rond et al. (1784). *Encyclopédie méthodique*. Mathématiques. Tome premier. Paris.

Allard, André (1994). "Les sources arithmétiques et le calcul indien dans le Liber abbaci". In: *Leonardo Fibonacci*. Il tempo, le opere, l'eredità scientifica. Ed. by Marcello Morelli and Marco Tangheroni. Pisa, pp. 81–96.

Ambrosetti, Nadia (2016). "Algorithmic in the 12th Century: The Carmen de Algorismo by Alexander de Villa Dei". In: *History and Philosophy of Computing*. Third International Conference, HaPoC 2015. Ed. by Fabio Gadducci and Mirko Tavosanis. Cham, pp. 71–86.

Ausiello, Giorgio (2013). "Algorithms, An Historical Perspective". In: *The Power of Algorithms*. Inspiration and Examples in Everyday Life. Ed. by Giorgio Ausiello and Rossella Petreschi. Heidelberg et al., pp. 3–26.

Avigad, Jeremy and Vasco Brattka (2014). "Computability and analysis: the legacy of Alan Turing". In: *Turing's Legacy*. Developments from Turing's Ideas in Logic. Ed. by Rod Downey. Cambridge, pp. 1–47.

---

[4] According to Fant, "[t]he term algorithm was not, apparently, a commonly used mathematical term in America or Europe before Markov, a Russian, introduced it". [1] Translation from Markov (1961, p. 1).

Borel, Émile (1912). "Le calcul des intégrales définies". *Journal de mathématiques pures et appliquées* 8, pp. 159–210.
Burnett, Charles (2006). "The Semantics of Indian Numerals in Arabic, Greek and Latin". *Journal of Indian Philosophy* 34, pp. 15–30.
Cantor, Moritz (1867). *Mathematische Beiträge zum Kulturleben der Völker*. Halle.
Cantor, Moritz (1907). *Vorlesungen über Geschichte der Mathematik*. Erster Band. Von den ältesten Zeiten bis zum Jahre 1200 n. Chr. Leipzig.
Chabert, Jean-Luc, ed. (1999). *A History of Algorithms*. From the Pebble to the Microchip. Berlin and Heidelberg.
Chasles, Michel (1859). "[Correspondance]". *Comptes rendus de l'Académie des sciences* 48, pp. 1054–1061.
Church, Alonzo (1936). "An Unsolvable Problem of Elementary Number Theory". *American Journal of Mathematics* 58, pp. 345–363.
Ciruelo, Pedro (1495). *Tractatus arithmetice practice qui dicitur algorismus*. Paris.
Colson, John (1736). "The Method of Fluxions and Infinite Series; or, A Perpetual Comment upon the foregoing Treatise". In: *The Method of Fluxions and Infinite Series*. London, pp. 141–339.
Curtze, Maximilian, ed. (1868). *Der Algorismus Proportionum des Nicole Oresme*. Berlin.
Curtze, Maximilian (1897). *Petri Philomeni de Dacia in Algorismum vulgarem Johannis de Sacrobosco commentarius*. Kopenhagen.
Ditton, Humphry (1706). *An Institution of Fluxions*. London.
Euler, Leonhard (1764). "Specimen algorithmi singularis". *Novi Commentarii academiae scientiarum Petropolitanae* 15, pp. 53–69.
Euler, Leonhard (1767). "De usu novi algorithmi in problemate Pelliano solvendo". *Novi Commentarii academiae scientiarum Petropolitanae* 11, pp. 28–66.
Fant, Karl M. (1993). "A critical review of the notion of algorithm in computer science". In: *Proceedings of the 1993 ACM conference on Computer science*. Ed. by Stan C. Kwasny and John F. Buck. New York, pp. 1–6.
Folkerts, Menso (1997). *Die älteste lateinische Schrift über das indische Rechnen nach al-Ḫwārizmī*. Edition, Übersetzung und Kommentar ... unter Mitarbeit von Paul Kunitzsch. München.
Folkerts, Menso (2001). "Early Texts on Hindu-Arabic Calculation". *Science in Context* 14, pp. 13–38.
Grant, Edward (1965). "Part I of Nicole Oresme's *Algorismus proportionum*". *Isis* 56, pp. 327–341.
Grant, Edward, ed. (1974). *A Source Book in Medieval Science*. Cambridge, Mass.
Halliwell, James Orchard, ed. (1839). *Rara Mathematica*. Or, A collection of treatises on the mathematics and subjects connected with them. London.
Harris, John (1702). *A New Short Treatise of Algebra*. London.
Hilbert, David (1901). "Mathematical problems". *Bulletin of the American Mathematical Society* 8, pp. 437–479.
Hill, Robin K. (2016). "What an Algorithm Is". *Philosophy and Technology* 29, pp. 35–59.
Kleene, Stephen C. (1943). "Recursive Predicates and Quantifiers". *Transactions of the American Mathematical Society* 53, pp. 41–73.
Kleene, Stephen C. (1952). *Introduction to Metamathematics*. Princeton.
Klügel, Georg Simon (1803). *Mathematisches Wörterbuch*. Erste Abtheilung: Die reine Mathematik. Erster Theil von A bis D. Leipzig.
Kolmogorov, Andrey Nikolaevich (1953). "О понятии алгоритма". *Uspekhi Matematicheskikh Nauk* 8, pp. 175–176.
Leibniz, Gottfried Wilhelm (1684). "Nova Methodus pro maximis et minimis, itemque tangentibus, quae nec fractas nec irrationales quantitates moratur, et singulare pro illis calculi genu". *Acta Eruditorum* (Oct. 1684), pp. 467–473.
Lejeune Dirichlet, Gustav (1863). *Vorlesungen über Zahlentheorie*. Ed. by Richard Dedekind. Braunschweig.
Lüneburg, Heinz (2008). *Von Zahlen und Größen*. Dritthalbtausend Jahre Theorie und Praxis. Band I. Basel.

Markov, Andrey A. (1954). "Теория алгорифмов". *Trudy Matematicheskogo Instituta imeni VA Steklova* 42, pp. 3–375.
Markov, Andrey A. (1961). *The Theory of Algorithms*. English Translation by Jacques J. Schorrkon, and Program for Scientific Translations staff. Jerusalem.
Rapaport, William J. (2012). "Semiotic systems, computers, and the mind: how cognition could be computing". *International Journal of Signs and Semiotic Systems* 2, pp. 32–71.
Reinaud, Joseph Toussaint (1849). *Mémoire géographique, historique et scientifique sur l'Inde*. Paris.
Rosser, J. Barkley (1939). "An Informal Exposition of Proofs of Gödel's Theorems and Church's Theorem". *Journal of Symbolic Logic* 4, pp. 53–60.
de Sacrobosco, Johannes (1841). "De arte numerandi". In: *Rara Mathematica*. Ed. by James Orchard Halliwell. London, pp. 1–26.
de Sacrobosco, Johannes (1983). "Algorismus Joannis de Sacrobosco cum glossis". In: *Petri Philomenae de Dacia et Petri de Sancto Audomaro*. Opera quadrivalia. Pars I. Opera Petri Philomenae. Ed. by Fritz S. Pedersen. København, pp. 174–201.
Shaw, James Byrnie (1895). *Mathematics*. The Science of Algorithms. Jacksonville.
Stedall, Jacqueline A. (2008). *Mathematics Emerging*. A Sourcebook 1540 – 1900. Oxford.
Thomas, Wolfgang (2015). "Algorithms: From Al-Khwarizmi to Turing and Beyond". In: *Turing's Revolution*. The Impact of His Ideas about Computability. Ed. by Giovanni Sommaruga and Thomas Strahm. Cham, pp. 29–42.
Turing, Alan (1936). "On Computable Numbers, with an Application to the Entscheidungsproblem". *Proceedings of the London Mathematical Society* 42, pp. 230–265.
Wolff, Christian (1716). *Mathematisches Lexicon*. Leipzig.

almost everywhere

The notion of "almost everywhere" means that a certain property holds over a set except for a subset of measure zero. The notion is found in Lebesgue (1901, p. 1027) and Lebesgue (1902, p. 258); in the latter text, he uses the expression "except for the points of a certain set of measure zero".[1] Soon afterwards, Lebesgue uses the term *presque partout* 'almost everywhere' in correspondence from 1904 with Maurice Fréchet, and in a letter from January 1905 he also explained: "save for a set of measure zero (almost everywhere)" (Taylor and Dugac 1981, pp. 155, 157).[2]

In print, the term first appears in 1906, for example in Lebesgue (1906, p. 7).[3] In the same year, Fatou (1906, p. 368), referring to Lebesgue, also uses the term, without definition, but then defines the notion under the name *en générale* ('in general'): "To abbreviate, we will often say: 'in general' instead of 'with the exception of a set of measure zero'" (Fatou 1906, p. 375).[4] The English term *almost everywhere* appears in Young (1913, p. 187) and Porter (1915, p. 79). German *fast überall* is found in Plancherel (1913, p. 577). Another way of expressing the property is illustrated by Pell (1911, p. 137): "As usual, we regard two functions of $\mathfrak{F}$ as equal if they differ only on a set of points of content [measure] zero. We call two such functions 'essentially equal'".[5]

---

[1] "sauf aux points d'un ensemble [...] de mesure nulle". [2] "sauf pour un ensemble de mesure nulle (presque partout)". [3] Lebesgue (1928, p. 179) erroneously states that he had already used the term in the first edition (Lebesgue 1904). [4] "Pour abréger, nous dirons souvent: «en général», au lieu de «sauf exception pour un ensemble de mesure nulle ...»". Fatou refers to Lebesgue (1905), where the term does *not* occur. Possibly Lebesgue had used the term in that talk. [5] A similar example is found in Brand (1912-13, p. 102).

Denjoy (1915, p. 162) later criticized Lebesgue's term because he regarded it as ambiguous and tried to replace it – with little success – by French *pleine épaisseur* 'whole thickness'; unsurprisingly, Lebesgue (1928, p. 179) rejected this criticism.

As Hawkins (1979, p. 59) points out, Harnack (1882, p. 263) earlier introduced an analogous notion. Harnack uses the term *in general* (German *im allgemeinen*) when a property always holds with the exception of what Harnack called a *discrete set* (*diskrete Menge*), a set of content zero. Hardy (1901, p. 70) picked up the expression and defined: "A function will be said to possess a property *in general* in an interval when it possesses it except at a closed, enumerable set of points".

Some authors such as Kowalewski (1908, p. 14) used the term *almost all* (German *fast alle*) in a more restricted sense: "When the terms of a series with a finite number of exceptions have a certain property, we will say briefly that 'almost all' terms of the sequence have this property".[1]

Note. The history of the notion is discussed in Medvedev (1991, pp. 110–120), who also discusses the prehistory, i. e. "the principle of negligibility of sets of points of various types"; see also van Dalen and Monna (1972, pp. 109f.) and Hawkins (1979, esp. pp. 59, 138).

References

Brand, Louis (1912-13). "On Infinite Systems of Linear Integral Equations". *Annals of Mathematics* 14, pp. 101–118.

van Dalen, Dirk and A.F. Monna (1972). *Sets and integration*. An outline of the development. Groningen.

Denjoy, Arnaud (1915). "Sur les fonctions dérivées sommables". *Bulletin de la Société Mathématique de France* 43, pp. 161–248.

Fatou, Pierre (1906). "Séries trigonométriques et séries de Taylor". *Acta Mathematica* 30, pp. 335–400.

Hardy, Godfrey Harold (1901). "On differentiation and integration under the integral sign". *The Quarterly Journal of Pure and Applied Mathematics* 32, pp. 66–140.

Harnack, Axel (1882). "Vereinfachung der Beweise in der Theorie der Fourierschen Reihe". *Mathematische Annalen* 19, pp. 235–279.

Hawkins, Thomas (1979). *Lebesgue's Theory of Integration*. Its Origins and Development. Second Edition. Providence, R.I.

Heine, Eduard (1870). "Ueber trigonometrische Reihen". *Journal für die reine und angewandte Mathematik* 71, pp. 353–365.

Kowalewski, Gerhard (1908). *Einführung in die Infinitesimalrechnung mit einer historischen Übersicht*. Leipzig.

Lebesgue, Henri (1901). "Sur une généralisation d'intégrale définie". *Comptes rendus de l'Académie des sciences* 132, pp. 1025–1027.

Lebesgue, Henri (1902). "Intégrale, Longueur, Aire". *Annali di Mathematica Pura ed Applicata* 7, pp. 231–359.

Lebesgue, Henri (1904). *Leçons sur l'intégration et la recherche des fonctions primitives*. Paris.

Lebesgue, Henri (1905). "Sur une condition de convergence des séries de Fourier". *Comptes rendus de l'Académie des sciences* 140, pp. 1378–1381.

Lebesgue, Henri (1906). "Sur les fonctions dérivées". *Atti della Reale Accademia dei Lincei. Rendiconti* 15, pp. 3–8.

[1] "Wenn die Glieder einer Folge mit einer endlichen Anzahl von Ausnahmen eine gewisse Eigenschaft haben, so wollen wir kurz sagen, daß 'fast alle' Glieder der Folge diese Eigenschaft haben". Heine (1870, p. 354) had earlier used *im allgemeinen* 'in general' in the same sense.

Lebesgue, Henri (1928). *Leçons sur l'intégration et la recherche des fonctions primitives.* Deuxième édition. Paris.
Medvedev, Fyodor A. (1991). *Scenes from the History of Real Functions.* Translated from the Russian by Roger Cooke. Basel.
Pell, Anna Johnson (1911). "Biorthogonal Systems of Functions". *Transactions of the American Mathematical Society* 12, pp. 135–164.
Plancherel, Michel (1913). "Zur Konvergenztheorie der Integrale $\lim_{z=\infty} \int_a^z f(x) \cos xy dx$". *Mathematische Annalen* 74, pp. 573–578.
Porter, M.B. (1915). "Vallée Poussin's Cours d'Analyse [review]". *Bulletin of the American Mathematical Society* 22, pp. 77–85.
Taylor, Angus E. and Pierre Dugac (1981). "Quatre lettres de Lebesgue à Fréchet". *Revue d'histoire des sciences* 34, pp. 149–169.
Young, William Henry (1913). "On the Formation of Usually Convergent Fourier Series". *Proceedings of the Royal Society of London* 88, pp. 178–188.

almost periodic

Almost periodic functions were introduced by Harald Bohr. In a paper published in French, he used *fonctions presque périodiques* (Bohr 1923, p. 737):

> Let $f(x) = u(x) + iv(x)$ be a complex function of the real variable x, continuous in the interval $-\infty < x < \infty$, and let ε be a positive number. The real number $\tau = \tau(f, \varepsilon)$ shall be called an *almost period* belonging to the function $f(x)$ and to the number ε if, for any x, the equality
>
> $$|f(x + \tau) - f(x)| < \varepsilon$$
>
> holds.
>
> Definition. – The function $f(x)$ is called *almost periodic*, if to each ε corresponds a length $L = L(\varepsilon)$ such that each interval $\alpha < x < \beta$ of length L contains at least one almost period $\tau = \tau(f, \varepsilon)$.[1]

He then presented the new notion more extensively in Bohr (1925a) and other papers; in Bohr (1925b, pp. 111–117) he showed that quasi-periodic functions are a special case of almost periodic functions.

The terminology was very soon in common use; the obvious English translation was used by Besicovitch (1926), for example. A generalization to arbitrary topological groups was presented by von Neumann (1934).

Note. Some remarks on the history of the concept and further references can be found in Berg (1989), for example.

References

Berg, Christian (1989). "Introduction to the Almost Periodic Functions of Bohr". *Matematisk-fysiske Meddelelser* 42, pp. 15–24.
Besicovitch, A.S. (1926). "On Generalized Almost Periodic Functions". *Proceedings of the London Mathematical Society* 25, pp. 495–512.

---

[1] "Soient $f(x) = u(x) + iv(x)$ une fonction complexe de la variable réelle x continue dans l'intervalle $-\infty < x < \infty$, et ε une quantité positive. Le nombre réel $\tau = \tau(f, \varepsilon)$ sera une *presque période* appartenant à la fonction $f(x)$ et au nombre ε, si, pour x quelconque, on a l'égalité $|f(x + \tau) - f(x)| < \varepsilon$. Définition. – La fonction $f(x)$ est dite *presque périodique*, si à tout ε correspond une longeur $L = L(\varepsilon)$ telle que tout intervalle $\alpha < x < \beta$ de longeur L contienne au moins une presque période $\tau = \tau(f, \varepsilon)$" (emphasis adapted).

Bohr, Harald (1923). “Sur les fonctions presque périodiques”. *Comptes rendus de l’Académie des sciences* 177, pp. 737–739.
Bohr, Harald (1925a). “Zur Theorie der fastperiodischen Funktionen. I. Eine Verallgemeinerung der Theorie der Fourierreihen”. *Acta Mathematica* 45, pp. 29–127.
Bohr, Harald (1925b). “Zur Theorie der fastperiodischen Funktionen. II. Zusammenhang der fastperiodischen Funktionen mit Funktionen von unendlich vielen Variabeln; gleichmässige Approximation durch trigonometrische Summen”. *Acta Mathematica* 46, pp. 101–214.
von Neumann, John (1934). “Almost Periodic Functions in a Group. I”. *Transactions of the American Mathematical Society* 36, pp. 445–492.

almost prime

The term appears to be due to Rényi, who used it in two papers written in March 1947 (Rényi 1947, p. 455, 1948, p. 57).[2] He introduced *almost prime* (Russian почти простого) for “integers which have not more prime factors than a certain constant k” (Rényi 1948, p. 57).[1] The English term was used soon afterwards by Rényi (1951, p. 69) himself.[2]

References

Erdős, Paul (1950). “On Almost Primes”. *The American Mathematical Monthly* 57, pp. 404–407.
Rényi, Alfréd (1947). “О представлении четных чисел в виде суммы простого и почти-простого числа”. *Doklady Akademii Nauk SSSR* 56, pp. 455–458.
Rényi, Alfréd (1948). “О представлении четных чисел в виде суммы простого и почти простого числа”. *Izvestiya Akademii Nauk SSSR* 12, pp. 57–78.
Rényi, Alfréd (1951). “On the large sieve of Ju V. Linnik”. *Compositio Mathematica* 8, pp. 68–75.
Rényi, Alfréd (1962). “On the representation of an even number as the sum of a prime and of an almost prime”. *American Mathematical Society Translations* 19, pp. 299–321.

almost uniform convergence

Almost uniform convergence was introduced by Weyl (1909, p. 225), who called it *wesentlich-gleichmäßig* (from *wesentlich* ‘essentially’ and *gleichmäßig* ‘uniform’):

> I say that the series
> $$u_1(x) + u_2(x) + \cdots, \tag{1}$$
> where $u_1(x)$, $u_2(x)$, ..., are any functions defined on $0 \le x \le 1$, converges *essentially uniformly* if for every positive number $\varepsilon < 1$ there exists a set U_ε of measure $1-\varepsilon$ contained in the interval $0 \ldots 1$ such that for all values of x belonging to U_ε the series (1) converges uniformly.[3]

The terminology was quickly adopted and translated, for example by Plancherel (1910, p. 292), who used French *convergence uniforme en général* and by Lauricella (1911, p. 529), who also translated *wesentlich* by *in generale*; both Plancherel and

---

[2] In the various papers referred to by Rényi, the term was not used. [1] Translation from Rényi (1962, p. 299). [2] Rényi there just used it in a footnote, in a translation of the title of one of his papers written in Russian. – Erdős (1950, p. 404) used *almost prime* in the meaning ‘pseudoprime’.
[3] “Sind $u_1(x)$, $u_2(x)$ irgendwelche für $0 \le x \le 1$ definierte Funktionen, so sage ich, die Reihe (1) $u_1(x) + u_2(x) + \cdots$ konvergiere *wesentlich-gleichmäßig*, wenn zu jeder positiven Zahl $\varepsilon < 1$ eine im Intervall $0 \ldots 1$ gelegene Menge $\mathfrak{A}_\varepsilon$ vom Maße (mesure) $1-\varepsilon$ gefunden werden kann, so daß die Reihe (1) für alle zu $\mathfrak{A}_\varepsilon$ gehörigen Werte x gleichmäßig konvergiert”. Translation from Medvedev (1991, pp. 114f.).

Lauricella quote Weyl for the term.[4] Lalesco (1912, p. 91), referring to all of these authors, suggested *convergence presque uniforme.*

This term stuck, although, as Halmos (1950, pp. 89f.) later remarked:

> the phrase "almost uniform convergence" is a somewhat confusing (but unfortunately standard) misnomer, which conflicts with the "almost everywhere" terminology. Some such phrase as "nearly uniform convergence" might come closer to suggesting the true state of affairs; as it stands, some care has to be exercised to distinguish between almost uniform convergence and almost everywhere uniform convergence.

Note. See Medvedev (1991, pp. 114f.) for Weyl's introduction of the concept.

References

Dunford, Nelson (1935). "Integration in General Analysis". *Transactions of the American Mathematical Society* 37, pp. 441–453.

Halmos, Paul R. (1950). *Measure theory.* New York.

Lalesco, Trajan (1912). *Introduction à la théorie des équations intégrales*. Paris.

Lauricella, Giuseppe (1911). "Sulla risoluzione dell'equazione integrale di 1ª specie". *Atti della Reale Accademia dei Lincei. Rendiconti* 20, pp. 528–536.

Medvedev, Fyodor A. (1991). *Scenes from the History of Real Functions.* Translated from the Russian by Roger Cooke. Basel.

Plancherel, Michel (1910). "Contribution à l'étude de la représentation d'une fonction arbitraire par des intégrales définies". *Rendiconti del Circolo Matematico di Palermo* 30, pp. 289–335.

Weyl, Hermann (1909). "Über die Konvergenz der Reihen, die nach Orthogonalfunktionen fortschreiten". *Mathematische Annalen* 67, pp. 225–245.

Young, William Henry and Grace Chisholm Young (1913). "On the theorem of Riesz-Fischer". *The Quarterly Journal of Pure and Applied Mathematics* 44, pp. 49–88.

alternating **(algebraic structure)**

In his book on the theory of hypercomplex number systems, Hankel (1867) included a chapter on what he called *alternating numbers* (*alternierende Zahlen*). He defined such numbers by assuming that the units ι_k are subject to the conditions $\iota_1\iota_1 = 0$, $\iota_2\iota_2 = 0$, ..., $\iota_n\iota_n = 0$ and $\iota_k\iota_m = -\iota_m\iota_k$. From these assumptions he deduced that for $\alpha = a_1\iota_1 + a_2\iota_2 + \cdots a_n\iota_n$ and $\beta = b_1\iota_1 + b_2\iota_2 + \cdots b_n\iota_n$ multiplication is anticommutative ($\alpha\beta = -\beta\alpha$) and that $\alpha\alpha = 0$ for all α. A system of numbers that followed these rules he called an *alternierendes Zahlensystem* (alternating number system).

In English texts, Hankel's *alternierende Zahlen* was translated as *alternate numbers,* for example by Spottiswoode (1871, p. 149); for the algebraic structure, Shaw (1907, p. 83) used *alternate algebra.* French *nombres alternés* already appears in Hoüel (1867, p. 321).

Hankel (1867, p. 140) himself gave a historical account of such numbers, citing, for example, Graßmann (1844) and de Saint-Venant (1845).

[4] Weyl's German term was also mentioned by W. H. Young and G. C. Young (1913, p. 50), who, however, did not translate it: "We have preferred not to use Weyl's nomenclature, 'wesentlich gleichmässige Konvergenz,' nor F. Reisz's [sic], 'convergence en mesure,' but rather to state in full the property under consideration". The English term does not appear to have been used before the mid-1930s; it appears in Dunford (1935, p. 442).

References

Graßmann, Hermann (1844). *Die Wissenschaft der extensiven Grösse, oder die Ausdehnungslehre, eine neue mathematische Disziplin.* Erster Theil, die lineale Ausdehnungslehre enthaltend. Leipzig.

Hankel, Hermann (1867). *Vorlesungen über die complexen Zahlen und ihre Functionen.* In zwei Theilen. I. Theil: Theorie der complexen Zahlsysteme. Leipzig.

Hoüel, Jules (1867). "Théorie élémentaire des quantités complexes. Première partie. Algèbre des quantités complexes". *Mémoires de la Société des sciences physiques et naturelles de Bordeaux* 5, pp. 1–64.

de Saint-Venant, Adhémar Barré (1845). "Mémoire sur les sommes et les différences géométriques et leur usage pour simplifier la mécanique". *Comptes rendus de l'Académie des sciences* 21, pp. 620–625.

Shaw, James Byrnie (1907). *Synopsis of Linear Associative Algebra.* A Report on Its Natural Development and Results Reached Up to the Present Time. Washington, D.C.

Spottiswoode, William (1871). "Remarks on some Recent Generalizations of Algebra". *Proceedings of the London Mathematical Society* 4, pp. 147–164.

alternating function

The term *alternating function* can be attributed to Cauchy, who use French *fonction alternée*. In his *Cours*, he defined:

> An *alternating* function of several quantities is one which changes sign, but keeps the same value next to the sign, when we interchange two of these quantities. Consequently, by a series of such exchanges, the function becomes alternatingly positive and negative (Cauchy 1821, p. 73).[1]

The term first appeared in Cauchy (1815, p. 29). The corresponding German term is found in a translation of Cauchy's book (Cauchy 1828, p. 55). Jacobi (1841, p. 365) used the Latin *functio alternans*, while English *alternating function* first appears in Cotterill (1868, p. 103). Both Jacobi and Cotterill attributed the term to Cauchy.

References

Bradley, Robert E. and C. Edward Sandifer (2009). *Cauchy's* Cours d'analyse. An Annotated Translation. Dordrecht et al.

Cauchy, Augustin-Louis (1815). "Sur les fonctions qui ne peuvent obtenir que deux valeurs égales et de signes contraires par suite des transpositions opérées entre les variables qu'elles renferment". *Journal de l'École polytechnique* 10, cahier 17, pp. 29–112.

Cauchy, Augustin-Louis (1821). *Cours d'analyse de l'École Royale Polytechnique.* I.re partie. Analyse algébrique. Paris.

Cauchy, Augustin-Louis (1828). *Lehrbuch der algebraischen Analysis.* Aus dem Französischen übersetzt von C.L.B. Huzler. Königsberg.

Cotterill, T. (1868). "Proofs of the Fundamental Formulæ of the Higher Geometry in Spaces of Two and Three Dimensions". *The Oxford, Cambridge, and Dublin Messenger of Mathematics* 4, pp. 98–105.

Jacobi, Carl Gustav Jacob (1841). "De functionibus alternantibus earumque divisione per productum e differentiis elementorum conflatum". *Journal für die reine und angewandte Mathematik* 22, pp. 360–371.

---

[1] "Une fonction *alternée* de plusieurs quantités est celle qui change de signe, mais en conservant au signe près la même valeur, lorsqu'on échange deux de ces quantités entre elles; en sorte que, par une suite de semblables échanges, la fonction devienne alternativement positive et négative". Translation from Bradley and Sandifer (2009, p. 51).

alternating group

The term *alternating group* (French *groupe alterné*) was introduced by Jordan (1870, pp. 62f.) in the following terms: "The substitutions that result from an even number of transpositions form a group [...] This group is called the *alternating group*; the functions it leaves invariable are called *alternating*".[2] Jordan had used the term earlier, without definition, in Jordan (1868, p. 836). The name can be explained by the relation to *alternating functions*, a term due to Cauchy (1815, pp. 29f.).

Note that *groupe alterne* is found in a manuscript by Galois; he does not give a definition. Since this manuscript by Galois was not published at the time, Jordan seems to have coined the term independently. For Galois and the alternating group, see Neumann (2011, pp. 296, 383f.).

Petersen (1878, p. 272) adopted the concept under the German term *alterne Gruppe*; Netto (1882, p. 33) used the now usual *alternierende Gruppe*. English *alternating group* is found in Cole (1886, p. 275).

References

Cauchy, Augustin-Louis (1815). "Sur les fonctions qui ne peuvent obtenir que deux valeurs égales et de signes contraires par suite des transpositions opérées entre les variables qu'elles renferment". *Journal de l'École polytechnique* 10, cahier 17, pp. 29–112.

Cole, Frank Nelson (1886). "A Contribution to the Theory of the General Equation of the Sixth Degree". *American Journal of Mathematics* 8, pp. 265–286.

Jordan, Camille (1868). "Théorèmes généraux sur les substitutions". *Comptes rendus de l'Académie des sciences* 66, pp. 836–837.

Jordan, Camille (1870). *Traité des substitutions et des équations algébriques*. Paris.

Netto, Eugen (1882). *Substitutionentheorie und ihre Anwendungen auf die Algebra*. Leipzig.

Neumann, Peter M. (2011). *The Mathematical Writings of Évariste Galois*. Zürich.

Petersen, Julius (1878). *Theorie der algebraischen Gleichungen*. Kopenhagen.

alternating series

The term *alternating series* cannot be attributed to a single person. It does not explicitly occur before the 19th century. For the English term, for example, the OED cites Leslie (1817, p. 201), and French *série alternée* was used and also defined by Blanchet (1838, pp. 19f.). Neither Leslie nor Blanchet would have felt their usages to be a terminological innovation, since the concept had been in use since the 17th century, frequently paraphrased using the verb *to alternate*.

An example of such a paraphrase is a paper by Hutton, entitled "A new Method for the Valuation of Numeral Infinite Series, whose Terms are alternately (+) Plus and (-) Minus" (Hutton 1786, pp. 11–37).

But similar paraphrases can already be found in the works of Leibniz and Wallis, for example. In particular, Leibniz used them in a letter to Johann Bernoulli from 1713, in which he discusses the convergence of alternating series (Gerhardt 1855-56, p. 926), or also in notes from 1672/73 (Leibniz 2003, p. 127). Earlier, Wallis (1668, p. 753) had used the formulation "alternatim negatis & affirmatis" in connection with the geometric series.

---

[2] "Les substitutions qui résultent d'un nombre pair de transpositions forment un groupe [...] Ce groupe se nomme le *groupe alterné*; les fonctions qu'il laisse invariables sont dites *alternées*".

References

Blanchet, P.H. (1838). *Compléments de mathématiques spéciales*. Paris.
Gerhardt, C.I., ed. (1855-56). *Leibnizens mathematische Schriften*. Erste Abtheilung. Band III. Halle.
Hutton, Charles (1786). *Tracts, Mathematical and Philosophical*. London.
Leibniz, Gottfried Wilhelm (2003). *Sämtliche Schriften und Briefe*. Siebente Reihe. Mathematische Schriften. Dritter Band. 1672-1676. Differenzen, Folgen, Reihen. Hannover.
Leslie, John (1817). *The Philosophy of Arithmetic*. Edinburgh.
Wallis, John (1668). "Logarithmotechnia Nicolai Mercatoris [review]". *Philosophical Transactions of the Royal Society of London* 3, pp. 753–759.

***alternative* (algebraic structure)**

This adjective has various applications in mathematics, especially in algebra.

The terminology goes back to Zorn (1930), who discussed what he called the *alternative law* of multiplication, a weakened form of the associative law: $(aa)b = a(ab)$, $(ab)a = a(ba)$, $(ba)a = b(aa)$ (Zorn 1930, p. 125).[1] Apart from the term *alternative ring*, Zorn (1930, p. 142) also coined the term *alternative field* (*Alternativkörper*).

These terms were quickly adopted by Moufang (1933, p. 209, 1934, p. 419), who attributed them to Zorn. The English terminology also seems to be due to Zorn; *alternative ring* is found in Zorn (1940, p. 57), *alternative field* in Zorn (1941, p. 678).

The reason for the name was explicitly pointed out by Smiley (1951, p. 36): "The name is derived from the fact that the *associator* $(a, b, c) = (ab)c - a(bc)$ is an alternating function of its arguments".

Note. For the history of the concept, see Cerroni (2004).

References

Cerroni, Cinzia (2004). "Non-Desarguian geometries and the foundations of geometry from David Hilbert to Ruth Moufang". *Historia Mathematica* 31, pp. 320–336.
Kirmse, Johannes (1924). "Über die Darstellbarkeit natürlicher ganzer Zahlen als Summen von acht Quadraten und über ein mit diesem Problem zusammenhängendes nichtkommutatives und nichtassoziatives Zahlensystem". *Berichte über die Verhandlungen der Sächsischen Akademie der Wissenschaften zu Leipzig (Mathematisch-physische Klasse)* 76, pp. 63–82.
Moufang, Ruth (1933). "Alternativkörper und der Satz vom vollständigen Vierseit (D_9)". *Abhandlungen aus dem Mathematischen Seminar der Universität Hamburg* 9, pp. 207–222.
Moufang, Ruth (1934). "Zur Struktur von Alternativkörpern". *Mathematische Annalen* 110, pp. 416–430.
Smiley, M.F. (1951). "Some questions concerning alternative rings". *Bulletin of the American Mathematical Society* 57, pp. 36–43.
Zorn, Max (1930). "Theorie der alternativen Ringe". *Abhandlungen aus dem Mathematischen Seminar der Universität Hamburg* 8, pp. 123–147.
Zorn, Max (1940). "Alternative rings with nilpotent elements [abstract]". *Bulletin of the American Mathematical Society* 46, p. 57.
Zorn, Max (1941). "Alternative Rings and Related Questions I: Existence of the Radical". *Annals of Mathematics* 43, pp. 676–686.

altitude

See *height*.

[1] Zorn (1930, p. 123) points out that this concept is found in Kirmse (1924).

amalgamated

The English term *amalgamated subgroup* first appears in print as *generalized free product with amalgamated subgroups* (Neumann 1948, p. 590):[1] "Denote by $\mathfrak{G}_a$ given groups, finite or infinite in number, each of which contains a subgroup $\mathfrak{U}_a$ isomorphic to a fixed group $\mathfrak{U}$ [...] the free product of the groups $\mathfrak{G}_a$ with the amalgamated subgroup $\mathfrak{U}$ [...] can be loosely described as the largest group generated by the groups $\mathfrak{G}_a$".

Neumann refers to Schreier (1927) for the concept; in Schreier's paper, the term *freies Produkt mit vereinigten Untergruppen* 'free product with united (combined) subgroups' is used. The terminology thus goes back to Schreier, with Neumann having suggested the English translation. Both the German and the English terms were quickly adopted, for example by Magnus (1931, p. 62) and by Baer (1949, p. 706), respectively.

Note. The introduction of the concept is discussed in detail by Chandler and Magnus (1982, pp. 102–112).

References

Baer, Reinhold (1949). "Free Sums of Groups and Their Generalizations. An Analysis of the Associative Law". *American Journal of Mathematics* 71, pp. 706–742.

Chandler, Bruce and Wilhelm Magnus (1982). *The History of Combinatorial Group Theory*. A Case Study in the History of Ideas. New York.

Magnus, Wilhelm (1931). "Untersuchungen über einige unendliche diskontinuierliche Gruppen". *Mathematische Annalen* 105, pp. 52–74.

Neumann, Hanna (1944). "Sub-group Structure of Free Products of Groups with an Amalgamated Subgroup". PhD thesis. University of Oxford.

Neumann, Hanna (1948). "Generalized Free Products with Amalgamated Subgroups". *American Journal of Mathematics* 70, pp. 590–625.

Schreier, Otto (1927). "Die Untergruppen der freien Gruppen". *Abhandlungen aus dem Mathematischen Seminar der Universität Hamburg* 6, pp. 161–183.

amicable numbers

The 4th-century Greek mathematician Iamblichus first mentioned amicable numbers, which are two numbers so related that the sum of the proper divisors of one number is equal to the sum of the proper divisors of the other number. Using the word *φίλοι* ('loving, friendly, pleasant'), he wrote:

> They [the Pythagoreans] call a pair of numbers "friendly," endowing them with virtues and good qualities, e. g., 284 and 200, when the parts of each generate the other. This accords with Pythagoras' definition of friendship. For when he was asked, "what is a friend?" he answered "another self." And this is clearly the case with these numbers (Iamblichus 2014, p. 102).[1]

The notion mainly spread via Arabic texts, where the terminology was taken over literally: Rashed (1994, p. 310), discussing abundant, deficient, perfect, and amicable numbers, points out that "[t]he Arabic terminology in this field leaves no doubt about the Greek origin of the term [*amicable*]". The Arabic term *al-mutaḥāba* comes from

---

[1] As the title of Neumann (1944) indicates, the term earlier appears in Neumann's PhD thesis, which has not been not consulted. [1] Translation from Cohen and Drabkin (1958, p. 13). Ore (1948, p. 97) regards the attribution to the Pythagoreans as questionable.

a verb meaning 'to love each other', and it appears in a treatise by Thābit ibn Qurra, in which he defined: "As for *amicable* numbers, they are two numbers such that if one adds all the divisors of one of the two, the sum is equal to the other number, which is the *conjugate* of the one whose divisors were added".[2] ibn Qurra also points out that neither Euclid nor Nicomachus dealt with amicable numbers.

The later Western terminology is based on translations from the Arabic,[3] but the exact origins are difficult to trace. Rashed (1994, pp. 282f.) discusses various Arabic writers dealing with the concept and asks: "Did this tradition – and this is the question here – remain unknown to European mathematicians? Though highly unlikely, such an eventuality nevertheless remains probable until clear documentary proof is discovered". In other words, it is not known whether there are any medieval Latin texts in which the concept is dealt with, and so it happens that the term *amicable* appears first in French rather than Latin.

The French term *nombres amiables* was used in the 1480s by Chuquet (1881, p. 69), who said that "among the imperfect numbers there are some which are amicable and of marvellous familiarity with each other".[1] In print, this statement is repeated in de La Roche (1520, fol. 3^{v}). The Latin term *numeri amicabiles* is then found in Cardano (1539, fol. H i^{r}) and Bongo (1585, p. 105). The term was also used by van Schooten (1657, p. 419); in various texts such as Klügel (1803, p. 247) and Hutton (1795, p. 104) the term is attributed to him.

The English term *ami(c)able* does not seem to have been regularly used before the early 18th century. An early usage in print is found in *The New World of English Words* (Phillips 1706);[2] Newton used *amicable* in notes that may go back to 1664 (Newton 1967, p. 54).[3] The German *befreundet* (an adjective derived from *Freund* 'friend') seems to be due to Klügel (1803, p. 243).

Note. For overviews of the history of the concept and terminology, see Dickson (1952, pp. 38–50) and Lee and Madachy (1972). Rashed (1994, ch. 4) discusses the concept in Arabic texts; see also Rashed (2017, pp. 501f.). For amicable numbers in Hebrew texts, see Lévy (1996).

References

[anon.] (1823). "On the knowledge of amicable numbers among the Arabians". *The Oriental Magazine, and Calcutta Review* 2, pp. 241–246.

Bongo, Pietro (1585). *Mysticae numerorum significationis liber*. Bergamo.

Cardano, Gerolamo (1539). *Practica arithmetice*. Milano.

Chuquet, Nicolas (1881). *Le Triparty en la Science des Nombres*. Ed. by Aristide Marre. Rome.

Cohen, Morris R. and I.E. Drabkin (1958). *A Source Book in Greek Science*. New York.

Dickson, Leonard Eugene (1952). *History of the Theory of Numbers*. Volume I. Divisibility and primality. New York.

Flegg, Graham, Cynthia Hay, and Barbara Moss, eds. (1985). *Nicolas Chuquet, Renaissance mathematician*. A study with extensive translation of Chuquet's mathematical manuscript completed in 1484. Dordrecht.

---

[2] Translation from Katz (2007, p. 561). For the Arabic text, see Rashed (2009, p. 91). [3] This was already pointed out in [anon.] (1823, p. 246). [1] "Entre les nombres Imparfaitz Il sen treuue deux amyables et de merueilleuse familiarite lung auec laultre". Translation from Flegg, Hay, and Moss (1985, p. 60). [2] Kersey, who revised this edition of Phillips' book, mentions Ozanam in connection with the term; Ozanam used it in his mathematical dictionary (Ozanam 1691, p. 26), for example. [3] The OED dates *amiable* to 1721 and *amicable* to 1743.

Hutton, Charles (1795). *A Philosophical and Mathematical Dictionary*. In two volumes. Vol. I. London.
Iamblichus (2014). *In Nicomachi arithmeticam*. Introduction, texte critique, traduction française et notes de commentaire. Ed. by Nicolas Vinel. Pisa and Roma.
Katz, Victor J., ed. (2007). *The Mathematics of Egypt, Mesopotamia, China, India, and Islam*. A Sourcebook. Princeton.
Klügel, Georg Simon (1803). *Mathematisches Wörterbuch*. Erste Abtheilung: Die reine Mathematik. Erster Theil von A bis D. Leipzig.
Lee, Elvin J. and Joseph S. Madachy (1972). "The History and Discovery of Amicable Numbers, I, II, III". *Journal of Recreational Mathematics* 5, pp. 77–93, 153–173, 231–249.
Lévy, Tony (1996). "L'histoire des nombres amiables: le témoignage des textes hébreux médiévaux". *Arabic Sciences and Philosophy* 6, pp. 63–87.
Newton, Isaac (1967). *The Mathematical Papers of Isaac Newton*. Volume I. 1664-1666. Ed. by D.T. Whiteside. Cambridge.
Ore, Øystein (1948). *Number Theory and its History*. New York.
Ozanam, Jacques (1691). *Dictionnaire mathematique*. Paris.
Phillips, Edward (1706). *The New World of English Words*. Revised, Corrected, Improved ... by John Kersey. London.
Rashed, Roshdi (1994). *The Development of Arabic Mathematics*. Between Arithmetic and Algebra. Translated by A. F. W. Armstrong. Dordrecht.
Rashed, Roshdi, ed. (2009). *Thābit ibn Qurra*. Science and Philosophy in Ninth-Century Baghdad. Berlin and New York.
Rashed, Roshdi, ed. (2017). *Lexique historique de la langue scientifique arabe*. Hildesheim.
de La Roche, Estienne (1520). *Larismethique nouellement composee*. Lyon.
van Schooten, Frans (1657). *Exercitationum Mathematicorum libri quinque*. Liber primus. Leiden.

amphichiral

See *chiral*.

amplitude

The Latin word *amplitudo* is derived from *amplus*, which means 'large' or 'spacious'. The noun means 'width, wide extent, breadth, size' and also has the figurative meanings 'dignity, grandeur' (LS).

An early technical meaning is recorded in Alsted (1613, p. 132), who defined the amplitude of an angle as its "size"; thus, the amplitude of a right angle is 90 degrees. When exactly the term began to be used in this meaning is hard to determine, but it was certainly common in the second half of the 16th century; Schreckenfuchs (1569, p. 147), for example, remarked that the amplitude of an angle is measured in degrees.

Another technical meaning was introduced by Galileo Galilei: "The reader is reminded that above I have called the horizontal line *cb* the 'amplitude' of the semi-parabola *ab*; the axis *ac* of this parabola, I have called its 'altitude'; but the line *ea* the fall along which determines the horizontal speed I have called the 'sublimity'" (Galilei 1638, p. 253).[1] In a French translation by Mersenne published in 1639, Mersenne used *amplitude* to translate Galilei's term (Galilei 1639, p. 237).

Hamilton (1844, p. 426) introduced the *amplitude* of a quaternion. He later remarked that he "believes that what originally led him [Hamilton] to use the terms 'modulus' and 'amplitude,' was a recollection of M. Cauchy's nomenclature respecting

[1] "Advertatur insuper, semiparabolae *ab* Amplitudinem à me vocari horizontalem *cb*; Altudinem, *ac* nempe, eiusdem parabolae axem. Lineam vero *ea*, ex cuius descensu determinatur impetus horizontalis, Sublimitatem appello". Translation from Galilei (1914, p. 260).

the usual imaginaries of algebra" (Hamilton 1846, p. 27). Others, for example Baltzer (1865, pp. 186f.), also attributed *amplitude* in the sense of 'argument' in connection with complex numbers to Cauchy. However, the term has not been found at all in Cauchy's works in this sense.

With reference to functions, the term seems to have been first used by Hankel (1870, p. 11), who defined German *Amplitude* as the difference of the maximal and the minimal value of a function. With reference to periodic functions, the term is used in a slightly different way. Corresponding definitions come up in the 20th century; a clear example is Palmer and Krathwohl (1921, p. 169): "The greatest absolute value of the ordinates of a periodic function is called the *amplitude* of the function".

Brief remarks on the origin of *amplitude* in connection with elliptic integrals can be found under *elliptic function*.

References

Alsted, Johann Heinrich (1613). *Methodus admirandorum mathematicorum*. Herborn.

Baltzer, Richard (1865). *Die Elemente der Mathematik*. Erster Band. Gemeine Arithmetik, Allgemeine Arithmetik, Algebra. Zweite verbesserte Auflage. Leipzig.

Galilei, Galileo (1638). *Discorsi e dimostrazioni matematiche, intorno à due nuove scienze*. Leiden.

Galilei, Galileo (1639). *Les nouvelles pensées de Galilée*. Trans. by Marin Mersenne. Paris.

Galilei, Galileo (1914). *Dialogues Concerning Two New Sciences*. Translated from the Italian and Latin into English by Henry Crew and Alfonso de Salvio. New York.

Hamilton, William Rowan (1844). "On a new Species of Imaginary Quantities connected with a theory of Quaternions". *Proceedings of the Royal Irish Academy* 2, pp. 424–434.

Hamilton, William Rowan (1846). "On Quaternions; or on a new System of Imaginaries in Algebra". *The London, Edinburgh and Dublin Philosophical Magazine* 29, pp. 26–31.

Hankel, Hermann (1870). *Untersuchungen über die unendlich oft oscillirenden und unstetigen Functionen*. Tübingen.

Palmer, Claude Irwin and William Charles Krathwohl (1921). *Analytic Geometry*. New York.

Schreckenfuchs, Erasmus Oswald (1569). *Commentaria in Sphaeram Ioannis de Sacrobusto*. Basel.

anallagmatic

In a paper from 1864, Moutard considered surfaces transformable into themselves by *inversion* (geometry). To designate the concept, he created a neologism: "I propose to give them the name *anallagmatic* [French *anallagmatique*] (α privative, $\alpha\lambda\lambda\alpha\sigma\sigma\omega$, I change)" (Moutard 1864, p. 307). Thus, the term expresses the defining property that the surfaces do not change under inversion. Moutard's term was quickly adopted, for example by Crofton (1867, p. 35) and Clifford (1870), although Cayley (1871, p. 188) suggested that "perhaps 'bicircular' would be a more convenient name" and Loria (1910, p. 429) found the term "a little indistinct"[1].

The term was also used in a very different sense. As explained by Rouse Ball (1914, p. 65),

> [a]nother variety of a pavement made with square tiles of two colours was invented by Sylvester, who termed it anallagmatic. In the ordinary chess-board, if any two rows or any two columns are placed in juxtaposition, cell to cell, the cells which are side by side are either all of the same colour or all of different colours. In an anallagmatic arrangement, the cells are so coloured (with two colours) that when any two columns or any two rows are placed together side by side, half the cells next to one another are of the same colour and half are of different colours.

[1] "ein wenig unbestimmt".

The term first appeared in Sylvester (1868, p. 112), who explained that he chose it because of "the relation of the lines or columns to one another remaining unaltered as we pass from one to another". He illustrated the concept with a 16×16-square (see fig. 2). The concept was quickly picked up by Lucas (1878).

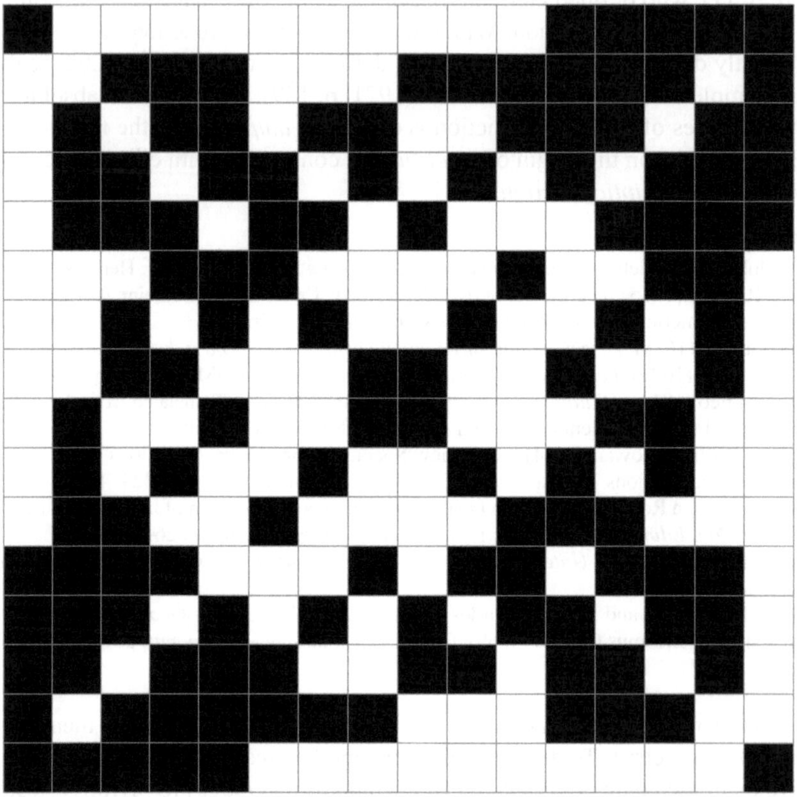

Figure 2: Anallagmatic pavement. From Sylvester (1868).

Sylvester's anallagmatic pavements also appeared in a different form in Hadamard (1893), where (implicitly) matrices are used that contain only the values $+1$ and -1 and whose rows are mutually orthogonal. Williamson (1944, p. 65) later called these *Hadamard matrices*.

Note. For historical discussions of anallagmatic pavements, see Décaillot (2002) and Anderson et al. (2007, p. 14).

References

Anderson, Ian et al. (2007). "Design Theory: Antiquity to 1950". In: *Handbook of Combinatorial Designs*. Second Edition. Ed. by Charles J. Colbourn and Dinitz Jeffrey H. Boca Raton et al., pp. 11–22.

Cayley, Arthur (1871). "Sketch of Recent Researches upon Quartic and Quintic Surfaces". *Proceedings of the London Mathematical Society* 3, pp. 186–195.

Clifford, William Kingdon (1870). "On the Umbilici of Anallagmatic Surfaces [Title only]". *Report of the British Association for the Advancement of Science* (1869), p. 9.
Crofton, M.W. (1867). "On Various Properties of Bicircular Quartics". *Proceedings of the London Mathematical Society* 2, pp. 33–45.
Décaillot, Anne-Marie (2002). "Géométrie des tissus. Mosaïques. Échiquiers. Mathématiques curieuses et utiles". *Revue d'histoire des mathématiques* 8, pp. 145–206.
Hadamard, Jacques (1893). "Résolution d'une question relative aux déterminants". *Bulletin des sciences mathématiques* 17, pp. 240–246.
Loria, Gino (1910). *Spezielle algebraische und transzendente ebene Kurven.* Theorie und Geschichte. Autorisierte, nach dem italienischen Manuskript bearbeitete deutsche Ausgabe von Fritz Schütte. Zweite Auflage. Erster Band. Leipzig.
Lucas, Édouard (1878). "Sur l'échiquier anallagmatique de M. Sylvester". *Association française pour l'avancement des sciences* 6, pp. 213–214.
Moutard, Théodore (1864). "Note sur la transformation par rayons vecteurs réciproques". *Nouvelles annales de mathématiques* 3, pp. 306–309.
Rouse Ball, Walter William (1914). *Mathematical Recreations and Essays*. Sixth edition. London.
Sylvester, James Joseph (1868). "Note on analagmatic pavements, Quest. 2552". In: *Mathematical Questions with their Solutions*. From the 'Educational Times'. Vol. X. From July to December, 1868. Ed. by W.J. Miller. London, p. 112.
Williamson, John (1944). "Hadamard's determinant theorem and the sum of four squares". *Duke Mathematical Journal* 11, pp. 65–81.

analysis

This term is highly polysemous, so much so, that Pasini (1997, p. 35) was led to say that "[t]he concept of analysis is notoriously, for its part, a polycephalous monster". In the following, the origin of the terminology and its adoption by Viète will first be discussed. Afterwards, the 17th century will be focused on, especially the use of the term by Leibniz and Newton. The following centuries will then be discussed in chronological order.

1. Origin and adoption by Viète

The Greek word *ἀνάλυσις* literally means 'loosing, releasing' (LSJ).[1] It entered into mathematics in ancient times and is, as explained by Heath (1921, p. 291), "according to the ancient view nothing more than a series of successive reductions of a theorem or a problem till it is finally reduced to a theorem or problem already known". Its use and its opposition to synthesis is best illustrated by Pappus:

> Analysis [*ἀνάλυσις*] then takes that which is sought as if it were admitted and passes from it through its successive consequences to something which is admitted as the result of synthesis: for in analysis we assume that which is sought as if it were (already) done [...], and we inquire what it is from which this results, and again what is the antecedent cause of the latter, and so on, until by so retracing our steps we come upon something already known or belonging to the class of first principles, and such a method we call analysis as being solution backwards [...] But in synthesis [*σύνθεσις*], reversing the process, we take as already done that which was last arrived at in the analysis and, by arranging in their natural order as consequences what were before antecedents, and successively connecting them one with another, we arrive finally at the construction of what was sought; and this we call synthesis (Pappus of Alexandria 1877, p. 634).[2]

The terminology was later referred to by Viète (1591, fol. 4^r):

[1] For a concise overview of the Greek term, see Mugler (1958-59, pp. 57f.). [2] Translation from Heath (1956, p. 138). For discussions of this passage, see the works by Panza cited in the note at the end of this article.

> There is a certain way of seeking the truth in mathematics, which Plato is said to have first discovered, called Analysis by Theon, and defined by him as: assuming what is sought as though given, from the consequences the truth is given. As opposed to Synthesis: assuming what is given, from the consequences one arrives at and understands what is sought. And although the ancients proposed only two kinds of analysis, zetetics and poristics, to which the definition of Theon wholly applies, I have nevertheless also added a third kind, which is called rhetics or exegetics, agreeing that it is zetetic by which are found equalities in the ratios of magnitudes, between that which is sought and those that are given. Poristic is that by which the truth of a stated theorem is examined from the equality or ratio. Exegetic is that by which the magnitude one seeks is discovered from that stated equality or ratio. And therefore the whole analytic art taking these three tasks to itself, may be defined as the doctrine of proper discovery in mathematics.[1]

Note that in this passage and elsewhere, Viète uses *analytic art* for what is otherwise referred to as *algebra*. The reason is explained by Rusnock (1997, p. 55): "In Viète's work, the link with ancient geometrical analysis is quite clear: in a very precise sense, one assumes the solution to be made, assigning letters to the knowns and the unknowns, and uses the algebraic techniques to reduce the problem to 'the given' (i. e. specific numbers, or constants/parameters) when this is possible. Thus, for example, a geometrical problem might lead to a quadratic equation like $ax^2 + bx = c$, where x, by the very writing down of the equation, is in a certain sense 'assumed as if it were known', and where a, b, c are 'given' parameters determined by the geometrical situation".[2] Stedall (2008, p. 48) also refers to this passage and comments: "It was here [...] that the word 'analysis' first entered modern European mathematics; it has since evolved through several changes of meaning but has never disappeared from the mathematical lexicon".

Breger (2016b, p. 141) remarks that, due to Viète's use of the term *analytic art*, there developed a tendency to equate *analytic* and *algebraic*. Breger (ibid.) also points out that a corresponding definition was given by Gregory (1667, p. 9): "When a quantity is composed by the addition, subtraction, multiplication, division, [or] root extraction of [some] quantities, we say that it is analytically composed".[3]

---

[1] "Est veritatis inquirendae via quaedam in mathematicis, quam Plato primus invenisse dicitur, a Theone nominata Analysis, et ab eodem definita, adsumptio quesiti tanquam concessi per consequentia ad verum concessum. Ut contrà Synthesis, Adsumptio concessi per consequentia ad quaesiti finem & comprehensionem. Et quanquam veteres duplicem tantùm proposuerunt Analysin ζητητικὴν καὶ ποριστικὴν, ad quas definitio Theonis maximè pertinet, consituti tamen etiam tertiam speciem, quae dicatur ῥητικὴ ἢ ἐξηγητική, consentaneum est, ut sit Zetetice quâ invenitur aequalitas proportione magnitudinis, de quâ quaeritur, cum iis quae data sunt. Poristice, quâ de aequalitate vel proportione ordinati Theorematis veritas examinatur. Exegetice, quâ ex ordinata aequalitate vel proportione ipsa de qua quaeritur exhibetur magnitudo. Atque adeò tota ars Analytice triplex illud sibi vendicans officium definiatur, Doctrina bene inveniendi in Mathematicis". Translation from Stedall (2008, p. 50). For a discussion and explanation of this passage, see Panza (1997, pp. 402–405). As for a possible influence on Viète, Sasaki (2003, p. 256) remarks that "the Ramist mathematician Schöner explicitly defined algebra as the analytical part of arithmetic". Schöner used *analytic* in this way in a comment on Ramus' *Algebra*; see Ramus (1586, p. 322). [2] See also Grabiner (1995, p. 86) and Russ (1980, p. 161). [3] "Quandò quantitas componitur ex quantitatum additione, subductione, multiplicationr divisione, radicum extractione; dicimus illam componi analyticè". Translation from Ferraro and Panza (2012, p. 105).

2. Analysis in the age of Leibniz and Newton and in the 18th century

The term *analysis* was frequently used by Leibniz,[4] often in the form *analysis infinitorum*: "For transcendental problems, wherever dimensions and tangents occur that have to be found by computation, there can hardly be found a calculus more useful, shorter, and more universal than my *differential calculus*, or *analysis of indivisibles and infinites*" (Leibniz 1686, p. 296).[1] The term was in common use and featured prominently in Euler's *Introductio in analysin infinitorum* (Euler 1748).

Of course, Leibniz was by no means the only person who contributed to a change of the meaning of *analysis*. As Boyer (1954, p. 456) comments on Newton: "When Newton invented the calculus, he was in large measure responsible for the extension of the word analysis to this field. In his first exposition of the new subject, based upon the use of infinite series, he wrote (in about 1669)":

> And whatever the common Analysis performs by Means of Equations of a finite Number of Terms (provided that can be done) this new method can always perform the same by Means of infinite Equations: So that I have not made any Question of giving this the Name of *Analysis* likewise. For the Reasonings in this are no less certain than in the other; nor the Equations less exact; albeit we Mortals whose reasoning Powers are confined within narrow Limits, can neither express, not so conceive all the Terms of these Equations, as to know exactly from thence the Quantities we want: Even as the surd Roots of finite Equations can neither be so exprest by Numbers, nor any analytical Contrivance, that the Quantity of any one of them can be so distinguished from all the rest, as to be understood exactly. To conclude, we may justly reckon that to belong to the *Analytic* Art, by the Help of which the Areas and Lengths &c. of Curves may be exactly and geometrically determined (when such a thing is possible) (Newton 1745, p. 340).[2]

This change in the meaning of the word was observed by Wolff (1716, pp. 51f.) in his mathematical dictionary:

> *Analysis*, the art of resolution, is a science for solving questions as yet unanswered, or for discovering unknown truths from certain known ones. As today for the most part specious arithmetic, algebra and the differential calculus of Herrn von Leibniz are used for this, it is customary for these methods to be called collectively *analysis*.[3]

Wolff also included entries for various types of analysis, including *analysis finitorum* and *analysis infinitorum*.[4] In another reference work, Chambers' *Cyclopaedia*, *analysis* was defined as "a method of solving, or resolving mathematical problems" (Chambers 1728, p. 82), but it was also seen as a branch of mathematics, divided into

---

[4] On the word *analysis* as used by Leibniz, Rabouin (2013, p. 109) states: "The meaning of the term 'analysis' in Leibniz's work is multifarious and it is doubtful that one could ever succeed in gathering this variety of meanings into a unified whole". See also Breger (2016a, pp. 167–170), Knobloch (2015, pp. 90–96), and Pasini (1997). [1] "Porro quoniam ad problemata Transcendentia, ubicunque dimensiones, tangentesque occurrunt, calculo tractanda, vix quicquam utilius, brevius, universalius fingi potest *calculo* meo *differentiali* seu *analysi indivisibilium, atque infinitorum*". Translation from Struik (1969, p. 281). [2] Newton's original text is found in Newton (1968, pp. 240, 242).

[3] "*Analysis*, die Aufflösungskunst, Ist eine Wissenschafft die verborgenen Fragen aufzulösen, oder aus einigen erkandten Wahrheiten andere noch unbekandte zu erfinden. Da man nun heute zu Tage meistens die Buchstaben-Rechenkunst, die Algebra und die Differential-Rechnung des Herrn von Leibnitz dazu brauchet; pfleget man diese Erfindungs-Künste zusammen die *Analysin* zu nennen". Translation from Rusnock (1997, pp. 66f.). [4] He also has entries for *Analysis Diophantea*, *Analysis potentiarum*, and *Analysis transcendens*.

analysis of finite quantities ("what we otherwise call specious arithmetic, or algebra") and *analysis of infinites*, "called also the new Analysis, [which] is particularly used for the Method of Fluxions, or the Differential Calculus". This view was also shared by d'Alembert (1751), whose article *analyse* for the *Encyclopédie* is, to a large extent, a translation of Chambers' articles for *analysis* and for *analytics*.[5] By the end of the 18th century, there are clear indications that analysis could be seen as a branch of mathematics in its own right; for example, the *Encyclopédie méthodique* contains an article on the word *pures* (*mathématiques pures*), in which arithmetic, algebra, analysis, and geometry are given as examples of pure mathematics (d'Alembert et al. 1785, p. 682).[1]

Breger (2016a, p. 170) summarizes the state of the terminology around the year 1800:

> By the end of the 18th or the beginning of the 19th century, *analysis* had become a washy term. There was the analysis of the finite (Viète's analysis or algebra) or the lower analysis and the analysis of the infinite or higher analysis. Parallel to this, an extra-mathematical meaning of analysis developed in the course of the 18th century; *analysis* was increasingly becoming simply the word for an investigation. Thus the higher and lower analysis might appear as a more or less complicated mathematical investigation.

As an example, Breger (ibid.) cites Fischer (1808, p. 85): "I can name not one single author in whose works I have found a clear and definite explanation of analysis".[2] Another example is Gergonne (1816-17), who also discussed the confusion in the terminology.[3] Ohm (1826, p. ix) suggested avoiding the term altogether: "Could one not banish this word *analysis*, which is used as a designation of one of the most important and most comprehensive parts of mathematics? It could be replaced by the more distinctive *calculus*".[4]

3. Analysis in the 19th century and beyond

As summarized by Gray (2015, p. 2), "[t]he 18th century calculus dealt in both single and many variable problems, it embraced differential equations, first ordinary and then partial, it had a theory of differentiation and integration, it was full of clever techniques. In Euler's hands particularly, it was a theory of functions: dependent variables were functions of independent variables". The 19th century, on the other hand, saw new developments:

> The 19th century has often been called the age of rigour. This is a correct characterization in the sense that analysis was given a foundation that we still recognize as satisfactory. The rigourisation was not just a question of clarifying a few basic concepts and changing the proofs of a few basic theorems; rather it invaded almost every part of analysis and changed it into the discipline we now learn in high schools and at universities. [...] However, it would be wrong to assume that in the 19th century the problem of rigour was considered to be the most pressing question in analysis. [...] the developments of new technical theorems of

---

[5] For a discussion of the use of *analysis* by Chambers and d'Alembert including a comparison of their articles, see Gilain (2010). [1] For a detailed discussion of the autonomisation of analysis, see Gilain (2010). [2] "Ich weiß keinen einzigen Schriftsteller zu nennen, bei welchem ich eine deutliche und bestimmte Erklärung von der Analysis gefunden hätte". Translation from Breger (2016a, p. 170). [3] See Dahan-Dalmédico (1986) for discussion of Gergonne's text. [4] "Könnte man dieses Wort Analysis, zur Bezeichnung eines der wichtigsten und umfassendsten Theile der Mathematik gebraucht, nicht verbannen? Man könnte ja das das bezeichnendere Kalkul dafür setzen".

> mathematics provided one of the main backgrounds for the growing interest in foundational questions. Fourier series were particularly important in this respect since they challenged the old ideas about the concepts of function, integral, convergence, continuity, etc., but differential equations, potential theory, elliptic functions and other areas also contributed to the rigourisation process (Lützen 2003, p. 155).[5]

A major proponent of rigour in analysis was, of course, Cauchy. In the introduction to his *Cours d'analyse*, he wrote: "As for the methods, I have sought to give them all the rigor which one demands from geometry, so that one need never rely on arguments drawn from the generality of algebra" (Cauchy 1821, p. ij).[1] In an announcement of his book, he also used the expression *analyse moderne* (Cauchy 1826, p. 74). The second major proponent of rigour was, of course, Weierstraß, whose ideas spread through his lectures. For details of the development of analysis in the 19th century, see the various entries in this book on concepts used in analysis and the calculus (and also those in topology).[2]

The 20th century saw many further advances; see Pier (2001) for an account. We shall only mention one terminological aspect here, namely the introduction of *non-standard analysis*, which term is well known to have been introduced by Robinson (1961). He gave a brief historical account himself:

> In the fall of 1960 it occurred to me that the concepts and methods of contemporary Mathematical Logic are capable of providing a suitable framework for the development of the Differential and Integral Calculus by means of infinitely small and infinitely large numbers. [...] The resulting subject was called by me Non-standard Analysis since it involves and was, in part, inspired by the so-called Non-standard models of Arithmetic whose existence was first pointed out by T. Skolem (Robinson 1966, p. vii).

Such non-standard models of arithmetic had been introduced by Skolem (1934); they can be regarded as "proper extensions of the natural number system which have, in some sense, 'the same properties' as the natural numbers" (Robinson 1967, p. 818). The name *non-standard model* was used by Henkin (1947, p. 65, 1950, p. 84).

See also *calculus*.

Note. The history of the terminology is more complex than it has been presented here. For an overview, see Boyer (1954). Panza (1997, 2007, 2012) discusses *analysis* in antiquity and the 16th and 17th centuries. Breger (2016a) focuses on the 17th century. An overview of the scope of analysis in the 20th century is given by Pier (2001, pp. 1–12), who discusses numerous 20th-century statements on analysis.

References

d'Alembert, Jean Le Rond (1751). "Analyse". *Encyclopédie, ou Dictionnaire raisonné des sciences, des arts et des métiers* 1, pp. 400–401.

d'Alembert, Jean Le Rond et al. (1785). *Encyclopédie méthodique*. Mathématiques. Tome second. Paris.

---

[5] For further discussion of rigour in analysis see, for example, Bottazzini (2001) and Archibald (2008). [1] "Quant aux méthodes, j'ai cherché à leur donner toute la rigueur qu'on exige en géométrie, de manière à ne jamais recourir aux raisons tirées de la généralité de l'algèbre". Translation from Bradley and Sandifer (2009, p. 1). [2] Two monographies that focus on the history of analysis in the 19th century are Gray (2015) and Bottazzini (1986). A shorter overview is given by Lützen (2003).

Archibald, Tom (2008). "The Development of Mathematical Rigor in Analysis". In: *The Princeton Companion to Mathematics*. Ed. by Timothy Gowers, June Barrow-Green, and Imre Leader. Princeton, pp. 117–129.

Bottazzini, Umberto (1986). *The Higher Calculus*. A History of Real and Complex Analysis from Euler to Weierstrass. Translated from the Italian by W. van Egmond. New York et al.

Bottazzini, Umberto (2001). "From Paris to Berlin: Contrasted Images of Nineteenth-Century Mathematics". In: *Changing Images in Mathematics*. From the French Revolution to the New Millennium. Ed. by Umberto Bottazzini and Amy Dahan Dalmedico. London and New York, pp. 31–47.

Boyer, Carl B. (1954). "Analysis: Notes on the Evolution of a Subject and a Name". *The Mathematics Teacher* 47, pp. 450–462.

Bradley, Robert E. and C. Edward Sandifer (2009). *Cauchy's* Cours d'analyse. An Annotated Translation. Dordrecht et al.

Breger, Herbert (2016a). "Analysis as a feature of 17th century mathematics". In: *Kontinuum, Analysis, Informales – Beiträge zur Mathematik und Philosophie von Leibniz*. Ed. by Wenchao Li. Berlin and Heidelberg, pp. 159–173.

Breger, Herbert (2016b). "Analysis und Beweis". In: *Kontinuum, Analysis, Informales – Beiträge zur Mathematik und Philosophie von Leibniz*. Ed. by Wenchao Li. Berlin and Heidelberg, pp. 137–146.

Cauchy, Augustin-Louis (1821). *Cours d'analyse de l'École Royale Polytechnique*. I.$^{\text{re partie.}}$ Analyse algébrique. Paris.

Cauchy, Augustin-Louis (1826). "*Cours d'analyse de l'École Royale Polytechnique* [announcement]". *Mémoires de l'Académie des sciences de l'Institut de France* 5, pp. 74–75.

Chambers, Ephraim (1728). *Cyclopaedia, or, An universal dictionary of arts and sciences*. Volume the First. London.

Dahan-Dalmédico, Amy (1986). "Un texte de philosophie mathématique de Gergonne. Mémoire inédit déposé à l'Académie de Bordeaux". *Revue d'histoire des sciences* 39, pp. 97–126.

Euler, Leonhard (1748). *Introductio in analysin infinitorum*. Lausanne.

Ferraro, Giovanni and Marco Panza (2012). "Lagrange's theory of analytical functions and his ideal of purity of method". *Archive for History of Exact Sciences* 66, pp. 95–197.

Fischer, Ernst Gottfried (1808). *Untersuchung über den eigentlichen Sinn der höheren Analysis*. Berlin.

Gergonne, Joseph Diez (1816-17). "De l'analise et de la synthèse, dans les sciences mathématiques". *Annales de mathématiques pures et appliquées* 7, pp. 345–372.

Gilain, Christian (2010). "La place de l'analyse dans la classification des mathématiques: de l'*Encyclopédie* à la *Méthodique*". *Recherches sur Diderot et sur l'Encyclopédie* 45, pp. 109–128.

Grabiner, Judith V. (1995). "Descartes and Problem-Solving". *Mathematics Magazine* 68, pp. 83–97.

Gray, Jeremy (2015). *The Real and the Complex*. A History of Analysis in the 19th Century. Cham et al.

Gregory, James (1667). *Vera circuli et hyperbolae quadratura*. Padova.

Heath, Thomas L. (1921). *A History of Greek Mathematics*. Volume I. From Thales to Euclid. Oxford.

Heath, Thomas L. (1956). *The Thirteen Books of Euclid's Elements*. Volume I. Introduction and Books I, II. New York.

Henkin, Leon (1947). "The Completeness Of Formal Systems". PhD thesis. Princeton University.

Henkin, Leon (1950). "Completeness in the Theory of Types". *The Journal of Symbolic Logic* 15, pp. 81–91.

Knobloch, Eberhard (2015). "Analyticité, équipollence et théorie des courbes chez Leibniz". In: *G.W. Leibniz, Interrelations Between Mathematics and Philosophy*. Ed. by Norma B. Goethe, Philip Beeley, and David Rabouin. Dordrecht et al., pp. 89–110.

Leibniz, Gottfried Wilhelm (1686). "De Geometria recondita et Analysi Indivisibilium et infinitorum". *Acta Eruditorum* (June 1686), pp. 292–300.

Lützen, Jesper (2003). "The Foundation of Analysis in the 19th century". In: *A History of Analysis*. Ed. by Hans Niels Jahnke. Providence, R.I., pp. 155–195.

Mugler, Charles (1958-59). *Dictionnaire historique de la terminologie géométrique des Grecs*. Paris.
Newton, Isaac (1745). "Of Analysis of Equations of an infinite Number of Terms". In: *Two Treatises of the Quadrature of Curves and Of Analysis of Equations of an infinite Number of Terms*. Ed. by John Stewart. London, pp. 321–343.
Newton, Isaac (1968). *The Mathematical Papers of Isaac Newton*. Volume II. 1667-1670. Ed. by D.T. Whiteside. Cambridge.
Ohm, Martin (1826). *Die allgemeine Grössenlehre und die ebene Raumgrössenlehre*. Berlin.
Panza, Marco (1997). "Classical Sources for the Concepts of Analysis and Synthesis". In: *Analysis and Synthesis in Mathematics*. History and Philosophy. Ed. by Marco Panza and Michael Otte. Dordrecht, pp. 365–414.
Panza, Marco (2007). "What is new and what is old in Viète's *analysis restituta* and *algebra nova*, and where do they come from? Some reflections on the relations between algebra and analysis before Viète". *Revue d'histoire des mathématiques* 13, pp. 85–153.
Panza, Marco (2012). "From velocities to fluxions". In: *Interpreting Newton*. Critical Essays. Ed. by Andrew Janiak and Eric Schliesser. Cambridge, pp. 219–254.
Pappus of Alexandria (1877). *Pappi Alexandrini Collectionis*. Volumen II. Ed. by Friedrich Hultsch. Berlin.
Pasini, Marco (1997). "*Arcanum Artis Inveniendi*: Leibniz and Analysis". In: *Analysis and Synthesis in Mathematics*. History and Philosophy. Ed. by Marco Panza and Michael Otte. Dordrecht, pp. 35–46.
Pier, Jean-Paul (2001). *Mathematical Analysis During the 20th Century*. Oxford.
Rabouin, David (2013). "*Analytica Generalissima Humanorum Cognitionum*. Some Reflections on the Relationship between Logical and Mathematical Analysis in Leibniz". *Studia Leibnitiana* 45, pp. 109–130.
Ramus, Petrus (1586). *Arithmetices libri duo, et algebrae totidem*. Ed. by Lazarus Schöner. Frankfurt.
Robinson, Abraham (1961). "Non-standard Analysis". *Proceedings of the Royal Academy of Sciences of Amsterdam A* 64, pp. 432–440.
Robinson, Abraham (1966). *Non-standard Analysis*. Amsterdam.
Robinson, Abraham (1967). "Nonstandard arithmetic". *Bulletin of the American Mathematical Society* 73, pp. 818–843.
Rusnock, Paul (1997). "Bolzano and the Traditions of Analysis". *Grazer Philosophische Studien* 53, pp. 61–85.
Russ, Stephen Bruce (1980). "The Mathematical Works of Bernard Bolzano published between 1804 and 1817". PhD thesis. Open University.
Sasaki, Chikara (2003). *Descartes' Mathematical Thought*. Dordrecht.
Skolem, Thoralf (1934). "Über die Nicht-Charakterisierbarkeit der Zahlenreihe mittels endlich oder abzählbar unendlich vieler Aussagen mit ausschliesslich Zahlenvariablen". *Fundamenta Mathematicae* 23, pp. 150–161.
Stedall, Jacqueline A. (2008). *Mathematics Emerging*. A Sourcebook 1540 – 1900. Oxford.
Struik, Dirk Jan, ed. (1969). *A Source Book in Mathematics, 1200-1800*. Harvard and London.
Viète, François (1591). *In artem analyticam isagoge*. Tours.
Wolff, Christian (1716). *Mathematisches Lexicon*. Leipzig.

analytic

The meanings of *analytic* strongly depend on the noun *analysis*, which has been used in various different meanings over time; see *calculus* and *analysis*. In the following entries, some commonly used terms involving the word are discussed.

analytic function

The term *analytic function* became widely known through Lagrange's work *Théorie des fonctions analytiques*, which first appeared in 1797. While the term features prominently in the title, in Lagrange's book itself it is hardly ever used. Various historians such as Ferraro and Panza (2012, p. 110) have explained that the term *analytique* is simply used in the sense of d'Alembert (1751, p. 403) as that which "belongs to the analysis, or is of the same nature as analysis, or is done through analysis".[1] This explanation is indirectly supported by Lagrange (1799, p. 235) himself: "In algebra stricto sensu, one only considers primitive functions that result from ordinary algebraic operations; this is the first branch of the theory of functions. In the second branch, one considers derivative functions, and it is this branch that we simply designate with the name 'theory of analytical functions', and that encompasses all that which pertains to the new calculi".[1]

The term was in use before Lagrange, however. Yuschkevitch (1974, pp. 133–136) cites de Condorcet as having first used French *fonction analytique* in an unpublished treatise called *Traité du calcul intégral* (1778), in which he defines: "I suppose that I have a certain number of quantities $x, y, z, \ldots, F$, and that for every determinate value of $x, y, z, \ldots$, etc., F has one or more determinate values which correspond to them; I say that F is a function of x, y, z" (de Condorcet 1786, p. 2).[2] While in the definition, only *fonction* appears, the margin reads "what is understood by an analytic function". The term is found even earlier, though, for example in Monge (1770-73, p. 21), [anon.] (1777, p. 93), and de Condorcet (1765, p. 36); no definitions are given there.[3]

Lagrange (1797, p. 1) defined an (analytic) function as follows: "One calls a 'function' of one or several quantities any expression of calculation into which these quantities enter in any way whatsoever combined or not with other quantities which are regarded as having same given and invariable values, whereas the quantities of the function may receive any possible value".[4] Analytic functions in this sense included polynomials, rational functions, exponential and trigonometric functions,[5] and they were assumed to be representable as a power series (Fraser 1987, pp. 39f.).[6]

This assumed property was later used by Weierstraß to define analytic functions, an approach he followed, for example, in an early paper from 1842 that was first

---

[1] "qui appartient à l'analyse, ou qui est de la nature de l'analyse ou qui se fait par la voie de l'analyse". Translation from Ferraro and Panza (2012, p. 110). – The same explanation of the term was given by Yuschkevitch (1974, pp. 134f.) and Dugac (2003, p. 71). [1] "Dans l'Algèbre proprement dite, on ne considère que les fonctions primitives qui résultent des opérations algébriques ordinaires; c'est la première branche de la théorie des fonctions. Dans la seconde branche on considère les fonctions dérivées, et c'est cette branche que nous désignons simplement par le nom de *Théorie des fonctions analytiques*, et qui comprend tout ce qui a rapport aux nouveaux calculs". Translation from Ferraro and Panza (2012, p. 107). [2] "Je suppose que j'aie un certain nombre de quantités $x, y, z \ldots F$, & que pour chaque valeur déterminée de $x, y, z \ldots$ &c, F ait une ou plusieurs valeurs déterminées qui y répondent; je dis que F est une fonction de x, y, z". Translation from Bottazzini (1986, p. 36). [3] But Monge explains that a *fonction analytique* is "subject to the law of continuity". [4] "On appelle *fonction* d'une ou de plusieurs quantités toute expression de calcul dans laquelle ces quantités entrent d'une manière quelconque, mêlées ou non avec d'autres quantités qu'on regarde comme ayant des valeurs données et invariables, tandis que les quantités de la fonction peuvent recevoir toutes les valeurs possibles". Translation from Ferraro and Panza (2012, p. 103). – The English term *analytical function* is found in this sense in [anon.] (1799, p. 481). [5] See also Lagrange (1797, p. 28), where x^m, a^x, $\log x$, $\sin x$ and $\cos x$ are all counted among the "simple analytic functions" (*fonctions simples analytiques*). [6] Markushevich (1996, p. 119) describes this assumption in the following terms: "*Analytically representable functions are analytic functions*: this is one of the basic principles of eighteenth-century mathematical analysis [...] This principle was tacitly assumed by almost all of the mathematicians of the first quarter of the nineteenth century" (original emphasis).

published in his collected works (Weierstraß 1894, p. 81).[7] Weierstraß' approach differed considerably from Lagrange's, but he and many others kept the term.[8] Klein (1926, p. 254) later commented:

> Lagrange had taken the series as his starting point. He called a function analytic if it admits a development in power series; yet he operated with the series only in a formal way, not troubling himself with questions of convergence. [...] These developments of Lagrange still have influence today. The name 'analytic function', coined by Lagrange – I conjecture that he meant by it no more than 'functions that are useful in analysis' – is still accepted today, though its conceptual content has become quite different.[1]

See also the entries for *holomorphic*, *regular*, and *entire* functions.

Note. For the history of analytic functions, see, for example, Valiron (1948), Yuschkevitch (1976), and Monna (1972). Fraser (1987) as well as Ferraro and Panza (2012) focus on Lagrange, Manning (1975) and Dugac (1973) on Weierstraß.

References

[anon.] (1777). "[d'Alembert, *Opuscules* vol. 6 (review)]". *Histoire de l'Académie royale des sciences* (1773), pp. 87–94.

[anon.] (1799). "La Grange's *Théorie des fonctions analytiques*". *The Monthly Review* 28, pp. 481–499.

d'Alembert, Jean Le Rond (1751). "Analytique". *Encyclopédie, ou Dictionnaire raisonné des sciences, des arts et des métiers* 1, pp. 403–404.

Bottazzini, Umberto (1986). *The Higher Calculus*. A History of Real and Complex Analysis from Euler to Weierstrass. Translated from the Italian by W. van Egmond. New York et al.

de Condorcet, Jean-Antoine-Nicolas (1765). *Du calcul intégral*. Paris.

de Condorcet, Jean-Antoine-Nicolas (1786). *Traité du calcul intégral*. Paris. URL: `http://gallica.bnf.fr/ark:/12148/btv1b86262724`. [unpublished].

Dugac, Pierre (1973). "Eléments d'analyse de Karl Weierstrass". *Archive for History of Exact Sciences* 10, pp. 41–176.

Dugac, Pierre (2003). *Histoire de l'analyse*. Autour de la notion de limite et de ses voisinages. Ed. by Bernard Bru and Roger Laurent. Paris.

Ferraro, Giovanni and Marco Panza (2012). "Lagrange's theory of analytical functions and his ideal of purity of method". *Archive for History of Exact Sciences* 66, pp. 95–197.

Fraser, C.G. (1987). "Joseph Louis Lagrange's Algebraic Vision of the Calculus". *Historia Mathematica* 14, pp. 38–53.

Klein, Felix (1926). *Vorlesungen über die Entwicklung der Mathematik im 19. Jahrhundert*. Teil I. Berlin.

Klein, Felix (1979). *Development of Mathematics in the 19th Century*. Translated by M. Ackerman. Brookline, Mass.

Lagrange, Joseph-Louis (1797). *Théorie des fonctions analytiques*. Paris.

Lagrange, Joseph-Louis (1799). "Discours sur l'objet de la Théorie des Fonctions analytiques". *Journal de l'École polytechnique* 2, cahier 6, pp. 232–235.

Manning, Kenneth R. (1975). "The Emergence of the Weierstrassian Approach to Complex Analysis". *Archive for History of Exact Sciences* 14, pp. 297–383.

Markushevich, Alexei Ivanovich (1996). "Analytic Function Theory". In: *Mathematics of the 19th Century*. Geometry. Analytic Function Theory. Ed. by Andrey Nikolaevich Kolmogorov and Adolph Andrei Pavlovich Yushkevich. Basel, pp. 119–272.

---

[7] Note that when this paper was first published, redactions were made without indicating them. Weierstraß may not have used the term *analytische Funktion* at that time but he certainly did so later. For a discussion of the redactions and references, see *absolute*. [8] On the development of Weierstraß' notion of analytic function, see Manning (1975, pp. 366–373). [1] Translation from Klein (1979, p. 103).

Monge, Gaspard (1770-73). “Sur la détermination des fonctions arbitraires dans les intégrales de quelques équations aux différences partielles”. *Mélanges de Philosophie et de Mathématique de la Societé Royale de Turin* 5, pp. 16–78.
Monna, A.F. (1972). “The Concept of Function in the 19th and 20th Centuries, in Particular with Regard to the Discussions between Baire, Borel and Lebesgue”. *Archive for History of Exact Sciences* 9, pp. 57–84.
Valiron, Georges (1948). “Formation et évolution de la notion de fonction analytique d’une variable”. In: *Les grands courants de la pensée mathématique*. Ed. by François Le Lionnais. Marseille, pp. 157–172.
Weierstraß, Karl (1894). “Definition analytischer Functionen einer Veränderlichen vermittelst algebraischer Differentialgleichungen”. In: *Mathematische Werke*. Erster Band. Berlin, pp. 75–84.
Yuschkevitch, Adolph P. (1974). “La notion de fonction chez Condorcet”. In: *For Dirk Struik*. Scientific, Historical and Political Essays in Honor of Dirk J. Struik. Ed. by Robert S. Cohen, John J. Stachel, and Marx W. Wartofsky. Dordrecht, pp. 131–139.
Yuschkevitch, Adolph P. (1976). “The Concept of Function up to the Middle of the 19th Century”. *Archive for History of Exact Sciences* 16, pp. 37–85.

analytic geometry

In its modern sense, this term is first found in Lacroix’s *Traité du calcul différentiel et du calcul intégral*.[1] He wrote:

> In carefully avoiding all geometric constructions, I would have the reader realize that there exists a way of looking at geometry which one might call *analytic geometry*, and which consists in deducing the properties of extension from the smallest possible number of principles by purely analytic methods, as Lagrange has done in his mechanics with regard to the properties of equilibrium and movement (Lacroix 1797, p. xxv).[2]

As can be seen from the quotation, Lacroix was influenced by Lagrange’s *Mécanique analytique* (1788); he also mentions that, in his belief, Monge was “the first who thought of presenting under this form the application of Algebra to Geometry” (Lacroix 1797, p. xxv).[3] The choice of the adjective *analytic* is understandable because at the time it was commonly used as a synonym of *algebraic*; see *analysis*.

Before Lacroix, the term had been occasionally used in the sense of any application of calculating algebra to geometry. In this meaning the term is used in the book *Analytische Geometrie* by Schenmark (1779) and in Meinert (1790, p. 369), for example.[4] Boyer (1956, p. 155) points out an early use “analogous to that of today” in Rolle (1711, p. 419), where Rolle describes analytic geometry as falling into two classes: “in the one, geometric questions are transformed into problems of algebra, and in the other, one is occupied only with the graphical solution of these problems”.[5]

---

[1] The origin of the term has been pointed out by Tropfke (1924, p. 112) and Boyer (1956, p. 211), for example. [2] “En écartant avec soin toutes les constructions géométriques, j’ai voulu faire au Lecteur qu’il existoit une manière d’envisager la Géométrie, qu’on pourroit appeler *Géométrie analytique*, et qui consisteroit à déduire les propriétés de l’étendue du plus petit nombre possible de principes, par des méthodes purement analytiques”. Translation from Boyer (1956, p. 211). [3] “Monge est, je crois, le premier qui ait pensé à présenter sous cette forme l’application de l’Algèbre à la Géométrie”. Translation from Domingues (2008, p. 100). [4] Boyer (1956, p. 216) gives further references for texts that have *analytic* in the title. [5] “Toutes les Recherches que l’on a faites dans la Geometrie Analytique se réduisent à deux Classes. Dans l’une on se propose the transformer les Questions de cette Science en Problèmes d’Algebre; & dans l’autre on ne s’occupe que du soin de résoudre ces Problèmes”.

French *géométrie analytique* is earlier found in [anon.] (1703, p. 480), and English *analytic geometry* was used in Sturm (1700, preface, xiii), who wanted to prepare his readers for "the Analytick Geometry of the Moderns, which is the highest apex of the Mathematicks".[6] In a different sense, the term was used by d'Alembert, who associates the term with the calculus: "If one represents a point as flowing, it will trace a line; and a line which flows will engender a surface, etc. This way of considering the generation of dimensions or the properties of bodies can be the first foundation of modern geometry, i. e., of the analytical geometry which uses the differential and integral calculus" (d'Alembert 1765, p. 871).[1]

The new term spread quickly after Lacroix used it. As pointed out by Domingues (2008, p. 105), it first appeared in the title of a work in Lefrançois (1804) and then in the widely read Biot (1805).

The term has been much criticized since the adjective *analytic* suggests, at least from a 20th-century perspective, that analytical methods are used; Peschl (1954, p. 2) called the term "misleading" for this reason. A strong opponent against using *analytic geometry* in its classical sense is Dieudonné (1970, p. 140):

> It is absolutely intolerable to use analytical geometry for linear algebra with coordinates, still called analytical geometry in the elementary books. Analytical geometry in this sense never existed. There are only people who do linear algebra badly, by taking coordinates and this they call analytical geometry. Out with them! Everyone knows that analytical geometry is the theory of analytical spaces, one of the deepest and most difficult theories of all mathematics.

As Dieudonné (1985, p. vii) later points out, analytic geometry as the theory of analytic spaces is due to Serre, who suggested this use in Serre (1956, p. 1).

Note. For the history of analytic geometry, see Boyer (1956) and Domingues (2008, esp. pp. 103-105).

References

[anon.] (1703). "Remarques de M. Rolle, de l'Académie Royale des sciences, touchant le Probleme general des Tangentes [review]". *Journal des sçavans* (1703), pp. 478–480.

d'Alembert, Jean Le Rond (1765). "Point". *Encyclopédie, ou Dictionnaire raisonné des sciences, des arts et des métiers* 12, pp. 871–872.

Biot, Jean-Baptiste (1805). *Essai de géométrie analytique, appliqué aux courbes et aux surfaces du second ordre*. Seconde édition. Paris.

Boyer, Carl B. (1956). *History of Analytic Geometry*. New York.

Dieudonné, Jean (1970). "The Work of Nicholas Bourbaki". *The American Mathematical Monthly* 77, pp. 134–145.

Dieudonné, Jean (1985). *History of Algebraic Geometry*. An Outline of the History and Development of Algebraic Geometry. Translated by Judith D. Sally. Monterey, California.

Domingues, João Caramalho (2008). *Lacroix and the Calculus*. Basel.

Johnson, Dale M. (1979). "The Problem of the Invariance of Dimension in the Growth of Modern Topology, Part I". *Archive for History of Exact Sciences* 20, pp. 97–188.

Lacroix, Sylvestre François (1797). *Traité du calcul différentiel et du calcul intégral*. Tome premier. Paris.

---

[6] In the original Latin, the term used was *analysin geometricam* (Sturm 1689, preface, xiii). Thus, the English translation is a free one. [1] "Si l'on se représente qu'un *point* coule, il tracera une ligne; & une ligne qui couleroit engendreroit une surface, &c. Cette maniere de considérer la génération des dimensions ou des propriétés des corps, paroît être le premier fondement de la Géométrie moderne, c'est-à-dire, de la Géométrie analytique qui fait usage du calcul différentiel & intégral". Translation from Johnson (1979, p. 107).

Lagrange, Joseph-Louis (1788). *Méchanique analitique*. Paris.
Lefrançois, Frederic-Louis (1804). *Essais de géométrie analytique*. Seconde édition. Paris.
Meinert, Friedrich (1790). *Lehrbuch der Mathematik*. Zweiter Theil. Gemeine Geometrie und ebene Trigonometrie; gemeine Analysis oder Algebra, und analytische Geometrie. Halle.
Peschl, Ernst (1954). *Analytische Geometrie*. Bonn.
Rolle, Michel (1711). "De l'evanouissement des quantitez inconnues dans la geometrie analytique". *Mémoires de l'Académie royale des sciences* (1709), pp. 543–583.
Schenmark, Nicolaus (1779). *Analytische Geometrie*, worin nebst den ersten Gründen der Algebra, ihre Anwendung auf die Elementar-Geometrie und die Kegelschnitte enthalten. Kopenhagen.
Serre, Jean-Pierre (1956). "Géométrie algébrique et géométrie analytique". *Annales de l'Institut Fourier* 6, pp. 1–42.
Sturm, Johann Christoph (1689). *Mathesis enucleata*. Nürnberg.
Sturm, Johann Christoph (1700). *Mathesis enucleata, or, The elements of the mathematicks*. Made English by J.R.A.M. and R.S.S. London.
Tropfke, Johannes (1924). *Geschichte der Elementar-Mathematik*. Sechster Band. Analysis. Analytische Geometrie. Berlin and Leipzig.

analytic set

The concept of analytic set was first presented in two short notes by Suslin (1917) and Luzin (1917); an analytic set was there called *ensemble (A)*.[1] The concept was developed by Luzin in the 1920, and in Luzin (1925, p. 1318) and Luzin (1926, p. 250) he began using the term *ensemble analytique*. In the second of these papers, he defined: "We call *analytic set* the set of values of a countably discontinuous function".[2]

Luzin (1926, p. 250) remarks, however, that the term is due to Lebesgue: "The name *analytic set* is given to these sets following a proposition by H. Lebesgue to call *analytic sets* all those sets that can be defined by analytic equalities or inequalities".[3] The original suggestion was made in Lebesgue (1905, p. 165).

Luzin explained the term on several occasions:[4]

> The reason for this denomination is quite clear: as any function
>
> $$x = f(t)$$
>
> continuous on $\mathcal{I}_t$ (with respect to that domain) is a function of class 1 in Mr Baire's classification, there exists a series of polynomials of t with rational coefficients
>
> $$P_1(t) + P_2(t) + \cdots + P_n(t) \ldots$$

---

[1] A definition appears in Suslin (1917, p. 89). For some comments, see Uspenskii (1985, p. 124).
[2] "On appelle *ensemble analytique* l'ensemble des valeurs d'une fonction dénombrablement discontinue". [3] "Le nom *d'ensembles analytiques* est donné à ces ensembles suivant une proposition de M. Lebesgue de nommer *ensembles analytiques* tous les ensembles qui peuvent être définis par des égalités ou inégalités analytiques". The attribution to Lebesgue was repeated in Luzin (1927, p. 6).
[4] See also Luzin (1927, pp. 5f.).

convergent for any value of t in $(0, 1)$ and the sum of which is $f(t)$. Hence any linear analytic set is the set of the values taken in $(0, 1)$ by the sum of a series of polynomials with rational coefficients and is thus defined by an analytic equality (Luzin 1930, pp. 135f.).[5]

The alternative name *Suslin set* (*Suslinsche Menge*) first appears in Hausdorff (1927, p. 91); the term was suggested by Aleksandrov in conversation with Hausdorff, according to Aleksandrov's autobiography (Alexandroff 1979, p. 286). Sierpiński (1950, p. 29) later remarked that it would be more appropriate to call them *Suslin and Luzin sets*.

Note. For the history of the concept and the terminology, see Uspenskii (1985); see also Lorentz (2001) and Kanamori (2012, pp. 27f.).

References

Alexandroff, Pavel S. (1979). "Pages from an autobiography". *Russian Mathematical Surveys* 34, pp. 267–302.

Hausdorff, Felix (1927). *Mengenlehre*. Zweite, neubearbeitete Auflage. Leipzig.

Kanamori, Akihiro (2012). "Set Theory from Cantor to Cohen". In: *Handbook of the History of Logic*. Vol. 6. Sets and Extensions in the Twentieth Century. Ed. by Dov M. Gabbay, Akihiro Kanamori, and John Woods. Amsterdam, pp. 1–71.

Lebesgue, Henri (1905). "Sur les fonctions représentables analytiquement". *Journal de mathématiques pures et appliquées* 1, pp. 139–216.

Lorentz, George G. (2001). "Who Discovered Analytic Sets?" *The Mathematical Intelligencer* 23, pp. 28–32.

Luzin, Nikolai (1917). "Sur la classification de M. Baire". *Comptes rendus de l'Académie des sciences* 164, pp. 91–94.

Luzin, Nikolai (1925). "Sur un problème de M. Emile Borel et les ensembles projectifs de M. Henri Lebesgue; les ensembles analytiques". *Comptes rendus de l'Académie des sciences* 180, pp. 1318–1320.

Luzin, Nikolai (1926). "Mémoire sur les ensembles analytiques et projectifs". *Matematicheskii Sbornik* 33, pp. 237–290.

Luzin, Nikolai (1927). "Sur les ensembles analytiques". *Fundamenta Mathematicae* 10, pp. 1–95.

Luzin, Nikolai (1930). *Leçons sur les ensembles analytiques et leurs applications*. Paris.

Pier, Jean-Paul (2001). *Mathematical Analysis During the 20th Century*. Oxford.

Sierpiński, Wacław (1950). *Les ensembles projectifs et analytiques*. Paris.

Suslin, Mikhail (1917). "Sur une définition des ensembles mesurables B sans nombres transfinis". *Comptes rendus de l'Académie des Sciences de Paris* 164, pp. 88–91.

Uspenskii, V.A. (1985). "Luzin's contribution to the descriptive theory of sets and functions: concepts, problems, predictions". *Russian Mathematical Surveys* 40, pp. 97–134.

angle

Greek mathematicians used the word γωνία for plane and solid angles. This term literally means 'corner'[1] and was already used as a mathematical term before Euclid (see, e. g., Mugler 1958-59, p. 112). It has been suggested that mathematicians took over the term from architecture, where an angle was the interior corner formed by two walls (Szabó 1992, p. 317).

---

[5] "La raison de cette dénomination est très claire: comme toute fonction $x = f(t)$ continue dans $\mathfrak{I}_t$ (relativement à ce domaine) est une fonction de classe 1 de la classification de M. Baire, il existe *une série de polynômes en t à coefficients rationnels* $P_1(t) + P_2(t) + \cdots + P_n(t) \ldots$ convergente pour toute valeur de t dans $(0, 1)$ et dont la somme est $f(t)$. Donc, tout ensemble analytique linéaire est l'ensemble des valeurs que prend dans $(0, 1)$ la somme d'une série de polynômes à coefficients rationnels et, par suite, est défini au moyen d'une égalité *analytique*". Translation from Pier (2001, p. 204).

[1] The word is etymologically related to *knee*.

Euclid defined plane angles in book I, def. 8 of the *Elements*: "A plane angle is the inclination [κλίσις] to one another of two lines[2] in a plane which meet one another and do not lie in a straight line".[3] The term was certainly standard by Euclid's time. An extensive discussion of the notion was later given by Proclus; one of the definitions he cites is this: "Carpus of Antioch says the angle is a quantity, specifically, the distance between the containing lines or planes" (Proclus 1873, pp. 125f.).[4]

In Latin texts, *γωνία* is usually translated as *angulus*,[5] which also means 'angle, corner' and derives from a root meaning 'crooked, bent'. The term is found in classical Latin (TLL); an early example in a geometric context is in Balbus (Blume, Lachmann, and Rudorff 1848, p. 101), a land surveyor who lived around the turn of the first to the second century.

The Latin word is the basis for the terminology is various modern languages. Spanish *angulo*, for example, is found in the late 13th century (Bossong 1978, p. 132), and the same applies to French *angle* (Dörr 1998, pp. 92f.) and Italian *angolo* (Feola 2008, p. 144). The English term is found in the 14th century; the MED cites from a manuscript of Trevisa's translation of *De proprietatibus rerum*: "A corner hatte angulus and is þe touchinge of tweye lynes þat toucheþ eiþer oþer". Recorde (1551, fol. A ij^{r}) defined: "whan so euer any suche meetyng of lines doth happen, the place of their metyng is called an *Angle or corner*". The German term *Winkel* originally also means 'corner'; it first appears in the 14th century in the so-called *Geometria culmensis* (Mendthal 1886, p. 20).

Aspects of the problematic nature of angles are discussed in the entry *angle of contingence*; see also *curvilinear angle*.

Note. An overview of the terminology is given by Tropfke (1940, pp. 63–66). For the Greek term, see Mugler (1958-59, pp. 109–112) and Heath in Euclid (1956, pp. 176–181). For the Arabic term, see Rashed (2017, pp. 314–333); for the Latin, see Ayuso García (2008, pp. 363–370), Guillaumin (1998) and Guillaumin (2020, pp. 36f.). The history of the notion is discussed by Rashed (2015); see also Matos (1990-91), Lejbowicz (1997), Itard (1964), and Gandz (1929).

References

Ayuso García, Manuel (2008). "La terminología latina de la geometría en Marciano Capela". PhD thesis. Madrid: Universidad Nacional de Educación a Distancia.

Blume, Friedrich, Karl Lachmann, and Adolf Rudorff, eds. (1848). *Die Schriften der römischen Feldmesser*. Erster Band. Texte und Zeichnungen. Berlin.

Bossong, Georg, ed. (1978). *Los Canones de Albateni*. Herausgegeben sowie mit Einleitung, Anmerkungen und Glossar versehen. Tübingen.

Dörr, Stephen (1998). *Der älteste Astronomietraktat in französischer Sprache: L'Introductoire d'astronomie*. Edition und lexikalische Analyse. Tübingen.

Euclid (1956). *The Thirteen Books of Euclid's Elements*. Volume I. Introduction and Books I, II. Ed. by Thomas L. Heath. New York.

Feola, Francesco (2008). *Gli esordi della geometria in volgare*. Un volgarizzamento trecentesco della *Practica Geometriae* di Leonardo Pisano. Firenze.

Gandz, Solomon (1929). "The Origin of Angle-Geometry". *Isis* 12, pp. 452–481.

---

[2] Note that *line* here does not necessarily mean 'straight line'. [3] Translation from Euclid (1956, p. 153). [4] Translation from Proclus (1970, p. 101). See Proclus (1970, pp. 98–102) for Proclus' discussion. [5] Guillaumin (1998) also discusses other terms used by land surveyors.

Guillaumin, Jean-Yves (1998). "Les différents noms de l'angle chez les agrimensores latins". *Revue des Études Anciennes* 90, pp. 411–417.
Guillaumin, Jean-Yves (2020). *Dictionnaire de la terminologie latine ancienne de l'arithmétique et de la géométrie*. Paris.
Itard, Jean (1964). "Quelques remarques sur la notion d'angle et sur l'angle de contingence". In: *Mélanges Alexandre Koyré*. I. Paris, pp. 346–360.
Lejbowicz, Max (1997). "La géométrie de l'angle au XIe siècle latin et la tradition euclidienne". In: *Langages et philosophie*. Hommage à Jean Jolivet. Ed. by Alain de Libera, Abdelali Elamrani-Jamal, and Alain Galonnier. Paris, pp. 31–47.
Matos, José (1990-91). "The Historical Development of the Concept of Angle". *The Mathematics Educator* 1, pp. 4–11, 18–25.
Mendthal, Hans (1886). *Geometria culmensis*. Ein agronomischer Tractat aus der Zeit des Hochmeisters Conrad von Jungingen (1393-1407). Leipzig.
Mugler, Charles (1958-59). *Dictionnaire historique de la terminologie géométrique des Grecs*. Paris.
Proclus (1873). *In primum Euclidis elementorum librum commentarii*. Ed. by Gottfried Friedlein. Leipzig.
Proclus (1970). *Commentary on the First Book of Euclid's Elements*. Translated with Introduction and Notes by Glenn R. Morrow. Princeton.
Rashed, Roshdi (2015). *Angles et Grandeur*. D'Euclide à Kamāl al-Dīn al-Fārisī. Boston and Berlin.
Rashed, Roshdi, ed. (2017). *Lexique historique de la langue scientifique arabe*. Hildesheim.
Recorde, Robert (1551). *The patheway to knowledg*. London.
Szabó, Árpád (1992). *Das geozentrische Weltbild*. München.
Tropfke, Johannes (1940). *Geschichte der Elementar-Mathematik*. Vierter Band. Ebene Geometrie. Dritte, verbesserte und vermehrte Auflage besorgt von Dr. Kurt Vogel. Berlin and Leipzig.

angle of contingence

The discussion of angles in Euclid's *Elements* focuses almost exclusively on rectilineal angles, that is, angles contained by straight lines. An exception is book III, prop. 16, where the angle of contingence, the angle between the circumference of a circle and its tangent (see fig. 3), is used: "The straight line drawn at right angles to the diameter of a circle from its extremity will fall outside the circle, and into the space between the straight line and the circumference another straight line cannot be interposed; further the angle of the semicircle is greater, and the remaining angle less, than any acute rectilineal angle".[1] Euclid simply called it the *remaining angle* because it is complementary to the angle of the semicircle (the angle between the diameter and its circumference). This angle has been the subject of much discussion; it was debated, for example, whether a magnitude could be assigned to this angle and whether it was an angle at all (see, e. g., Kasner 1945).

Much later, Proclus designated this angle as 'horn-like' (Greek *κέρατοιδής*, from *κέρας* 'horn'): "an angle may be contained also by a straight line and a circumference, and in either of two ways: either by a straight line and a convex circumference, such as the angle in a semicircle; or by a straight line and a concave circumference, like the horned angle" (Proclus 1873, p. 127).[2]

The 12th-century translator Gerard of Cremona used the word *cornee*, from Latin *cornu* 'horn', in an explanation of such angles: "the species of angles which are made from a straight line and a circumflex line, which angles are called horned, are two"

---

[1] Translation from Euclid (1956, p. 37). [2] Translation from Proclus (1970, pp. 102f.). – For further attestations, see Mugler (1958-59, p. 249) and Vitrac (1998).

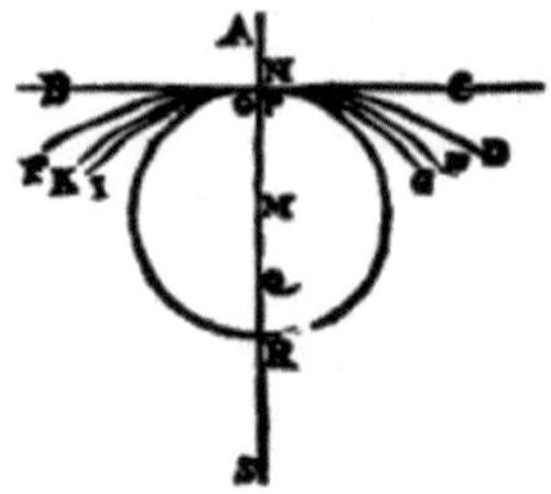

Figure 3: Angle of contingence. From Mersenne (1625, p. 866).

(Tummers 1994, p. 11).[3] Greek κέρατοιδής was sometimes literally taken over into Latin, for example by Valla (1501, book X, ch. 15) and later by Dasypodius (1573, fol. 22[r]), who both have *ceratoides*. More frequently, the Greek term was translated into Latin as *angulus cornicularis*. This term is found in a translation of Proclus' commentary on the *Elements* (Proclus 1560, p. 71). Variants include *corniculatus* (Dasypodius 1573, fol. 8[v]) and *cornutus* (Archimedes 1615, prologomena).[1]

The term *angle of contingence* (Latin *angulus contingentiae*) appears to be a creation of the 13th century.[2] It was used, for example, by Campanus of Novara (Busard 2005, p. 121), Jordanus de Nemore (Clagett 1984, p. 189), Alexander Nequam (Nequam 1988, p. 189), and William of Auxerre (William of Auxerre 1500, fol. 72[r]). The noun *contingentia* is derived from the verb *contingere* 'to touch', which was in use in the 12th century (see *tangent* (geometry)). It should also be noted that the Arabic *zāwiyat al-tamāss*, lit. 'angle of contact (tangency)' (*al-tamāss* comes from a root meaning 'to touch'), was in use earlier and can be found, for example, in al-Haytham's commentary on Euclid's postulates (Rashed 2017, p. 326).

Finally, the term *angulus contactus* is an innovation first found in the 16th century. It was used by Cardano (1554, p. 430)[3] and then by Peletier in his edition of the *Elements* (Euclid 1557, p. 74). It quickly became a frequently used alternative to *angulus contingentiae*.

The English term *angle of contingence* is first found in Euclid (1570, fol. 80[v]). In Ramus (1636, p. 195), *touch angle* is used and it is explained that the angle is so called "because it is made of a right line and periphery like unto a horn"; Hobbes (1656, epistle dedicatory) uses *angle of contact*.

In French, *angle du touchement*, from *toucher* 'to touch' is used in a translation of Cardano's *De subtilitate* (Cardano 1556, fol. 306[v]); in Euclid (1564, fol. 86[v]), the similar *angle de l'atouchement* is used. The terms *angle de contact* and *angle de*

[3] "Species quoque angulorum qui fiunt ex linea recta et circumflexa que dicuntur cornee, sunt due". Translation from Lo Bello (2003, p. 25). [1] Such a translation based on Latin is already found in the 12th century (Anaritius 1899, p. 11): "the species of angles which are made from a straight line and a circumflex line, which angles are called horned" ("Species generum angulorum, qui fiunt ex linea recta et circonflexa, qui dicuntur cornei"). Translation from Tummers (1994, p. 25). [2] Eneström (1910-11, p. 81) suspected that the term is probably older. [3] The term is not in the first edition (Cardano 1550).

contingence are found in de Gramont (1619, p. 10). Italian *angolo della contingentia* appears to have been first used by Tartaglia (Euclid 1543, fol. xliii^v).

Ryff (1547, fol. xiii^r) has German *Winckel der betreffung*; in Euclid (1562, p. 79), the frequently used *winckel der berürung* is introduced (both terms are based on the notion of 'touch').[4] *Winckel der Contingenz* is found in del Monte (1629, p. 21).

See also *curvature*.

Note. For the Greek term, see Mugler (1958-59, p. 249); for the Arabic, see Rashed (2017, pp. 326–329). For the history of the concept, see Rashed (2015, ch. 1) and Matos (1990-91). The controversy about the angle of contingence is discussed by Kasner (1945), Maierù (1990), Beeley and Scriba (2008), and by Heath in Euclid (1956, pp. 39–43), for example.

References

Anaritius (1899). *Anaritii in decem libros priores Elementorum Euclidis commentarii*. Ex interpretatione Gherardi Cremonensis. Ed. by Maximilian Curtze. Leipzig.

Archimedes (1615). *Archimedis opera quae extant*. Ed. by David Rivault. Paris.

Beeley, Philip and Christoph Scriba (2008). "Controversy and Modernity. John Wallis and the Seventeenth-Century Debate on the Nature of the Angle of Contact". *Acta Historica Leopoldina* 54, pp. 431–450.

Busard, H.L.L. (2005). *Campanus of Novara and Euclid's* Elements. Volume I. Stuttgart.

Cardano, Gerolamo (1550). *De subtilitate libri XXI*. Paris.

Cardano, Gerolamo (1554). *De subtilitate libri XXI*. Basel.

Cardano, Gerolamo (1556). *Les livres de Hierome Cardanus medecin milannois intitulés de la Subtilité, & subtiles inventions, ensemble les causes occultes, & raisons d'icelles*. Traduis de Latin en Francois, par Richard le Blanc. Paris.

Clagett, Marshall (1984). *Archimedes in the Middle Ages*. Volume V. Quasi-Archimedean Geometry in the Thirteenth Century. Philadelphia.

Dasypodius, Conrad (1573). *Lexikon seu Dictionarium mathematicum*. Strasbourg.

Eneström, Gustaf (1910-11). "Kleine Bemerkungen zur letzten Auflage von Cantors ‚Vorlesungen über Geschichte der Mathematik'". *Bibliotheca Mathematica* 11, pp. 81–90.

Euclid (1543). *Euclide Megarense philosopho*. Diligentemente reassettato, et alla integrita ridotto per ... Nicolo Tartalea. Venezia.

Euclid (1557). *In Euclidis Elementa Geometrica Demonstrationum Libri sex*. Ed. by Jacques Peletier. Lyon.

Euclid (1562). *Die Sechs Erste Bücher Euclidis*. Auß Griechischer sprach in die Teütsch gebracht ... Durch Wilhelm Holtzmann genannt Xylander. Basel.

Euclid (1564). *Les Six Premiers Livres des Elements*. Traduicts et commentez par Pierre Forcadel de Besies. Paris.

Euclid (1570). *The Elements of Geometrie of the most auncient Philosopher Euclide of Megara*. Faithfully (now first) translated into the Englishe toung, by H. Billingsley. London.

Euclid (1956). *The Thirteen Books of Euclid's Elements*. Volume II. Books III-IX. Ed. by Thomas L. Heath. New York.

de Gramont, Scipion (1619). *De la nature, qualitez et prerogatives admirables du poinct*. Paris.

Hobbes, Thomas (1656). *Six Lessons to the Professors of the Mathematiques, One of Geometry, the Other of Astronomy*. London.

Kasner, Edward (1945). "The Recent Theory of the Horn Angle". *Scripta Mathematica* 11, pp. 263–267.

Lo Bello, Anthony (2003). *Gerard of Cremona's Translation of the Commentary of Al-Nayrizi on Book I of Euclid's* Elements of Geometry. With an Introductory Account of the Twenty-Two Early Extant Arabic Manuscripts of the *Elements*. Leiden and Boston.

---

[4] The first term, *betreffung*, appears to be based on a less literal meaning of *contingere*, 'to affect, to concern'.

Maierù, Luigi (1990). "'... in Christophorum Clavium de Contactu Linearum Apologia'. Considerazioni attorno alla polemica fra Peletier e Clavio circa l'angolo di contatto (1579–1589)". *Archive for History of Exact Sciences* 41, pp. 115–137.
Matos, José (1990-91). "The Historical Development of the Concept of Angle". *The Mathematics Educator* 1, pp. 4–11, 18–25.
Mersenne, Marin (1625). *La vérité des sciences*. Paris.
del Monte, Guidobaldo (1629). *Mechanischer Kunst-Kammer Erster Theil*. Frankfurt.
Mugler, Charles (1958-59). *Dictionnaire historique de la terminologie géométrique des Grecs*. Paris.
Nequam, Alexander (1988). *Speculum Speculationum*. Edited by Rodney M. Thomson. Oxford.
Proclus (1560). *In primum Euclidis elementorum librum commentariorum libri quattuor*. Ed. by Francesco Barozzi. Padua.
Proclus (1873). *In primum Euclidis elementorum librum commentarii*. Ed. by Gottfried Friedlein. Leipzig.
Proclus (1970). *Commentary on the First Book of Euclid's Elements*. Translated with Introduction and Notes by Glenn R. Morrow. Princeton.
Ramus, Petrus (1636). *Via regia ad geometriam. The Vvay to Geometry*. Written in Latine by Peter Ramus, and now Translated and much enlarged by the Learned Mr. William Bedwell. London.
Rashed, Roshdi (2015). *Angles et Grandeur*. D'Euclide à Kamāl al-Dīn al-Fārisī. Boston and Berlin.
Rashed, Roshdi, ed. (2017). *Lexique historique de la langue scientifique arabe*. Hildesheim.
Ryff, Walther Hermann (1547). *Der furnembsten, notwendigsten, der gantzen Architectur angehörigen mathematischen und mechanischen Künst eygentlicher Bericht*. Nürnberg.
Tummers, Paul M. J. E. (1994). *Anaritius' Commentary on Euclid*. The Latin Translation, I-IV. Nijmegen.
Valla, Giorgio (1501). *De expetendis, et fugiendis rebus opus*. Venezia.
Vitrac, Bernard (1998). "L'angle corniculaire dans la tradition grecque des Éléments d'Euclide". URL: `https://hal.archives-ouvertes.fr/hal-00457940` (retrieved Jan. 23, 2015).
William of Auxerre (1500). *Summa aurea in quattuor libros sententiarum*. París.

anharmonic

See *cross ratio*.

annihilator and ***annulator***

The terms *annihilator* and *annulator* do not seem to appear until the 1930s, and they were defined in various contexts.

German *Annulator* was first used by Pontryagin (1931, p. 173), who originally defined it in connection with what he called *group pairs* (*Gruppenpaare*):

> Definition I. Two groups U and V define a *group pair* relative to M (*the modulus*)[1] when any ordered pair of elements x, y – where x is an element of U and y is an element of V – is associated with an element k of M, namely, the product of the two elements x and y:
>
> $$k = x \cdot y$$
>
> such that one always has:
>
> $$(x + x') \cdot y = x \cdot y + x' \cdot y \qquad \text{(first distributivity law)}$$
>
> and
>
> $$x \cdot (y + y') = x \cdot y + x \cdot y' \qquad \text{(second distributivity law)}$$
>
> [...] Definition II. Let A be an arbitrary subgroup of U; the totality of all elements y of V with the property that for any x in A one has:

[1] M is assumed to be a cyclic group.

$$x \cdot y = 0$$

will be called the *annihilator* [*Annulator*] of A in V, and will be denoted by (V, A).[2]

In this sense the term was also used by Komatu (1936, p. 214), who attributed it to Pontryagin.

Vandiver (1934, p. 582) used *annulator* for a semigroup S: "If there is an element O_l in S such that $O_l A = O_l$ for any A in S, then O_l is said to be a *left-annulator* of S and if $AO_r = O$ for any A, then O_r is said to be a *right-annulator* of S. Also, if an element O is both a right- and left-annulator of S, it is called an *annulator* of S". Jacobson (1937, p. 212) gave a definition of *annihilator* with respect to Lie algebras or associated algebras, while McCoy (1942, p. 287) defined for rings: "If $ab = ba = 0$, $b \neq 0$, we may say also that b is an *annihilator* of a, or that a is *annihilated* by b".

Another variant is found in Everett (1939, p. 274): "If R is a ring, the two-sided ideal N of elements z of R such that $zr = 0$ (all $r \in R$), is called the left annulling ideal of R". In this sense, *annulierendes Linksideal* had earlier been used by van der Waerden (1931, p. 126).

References

Everett, C.J. (1939). "Rings as groups with operators". *Bulletin of the American Mathematical Society* 45, pp. 274–279.

Jacobson, Nathan (1937). "Abstract Derivation and Lie Algebras". *Transactions of the American Mathematical Society* 42, pp. 206–224.

Komatu, Atuo (1936). "Über die Dualität der Überdeckungen". *Proceedings of the Imperial Academy* 12, pp. 213–214.

McCoy, Neal H. (1942). "Remarks on Divisors of Zero". *The American Mathematical Monthly* 49, pp. 286–295.

Pontryagin, Lev (1931). "Über den algebraischen Inhalt topologischer Dualitätssätze". *Mathematische Annalen* 105, pp. 165–205.

Pontryagin, Lev (2012). "On the algebraic content of topological duality theorems". Translated by D.H. Delphenich. URL: `http://www.neo-classical-physics.info/topology.html` (retrieved Dec. 30, 2019).

Vandiver, H.S. (1934). "On the Foundations of a Constructive Theory of Discrete Commutative Algebra". *Proceedings of the National Academy of Sciences* 20, pp. 579–584.

van der Waerden, Bartel Leendert (1931). *Moderne Algebra*. Zweiter Teil. Berlin.

anticommutative

An anticommutative operation $\circ$ is one for which $a \circ b = -b \circ a$ holds. This notion is originally found in connection with vector products (see *vector product*). Hankel (1867, p. 120) called an anticommutative system *alternate* or *alternating* (German *alternierend*). Sylvester finds this term "very inadvisable" since "the epithet *alternate* has no meaning except by oblique reference to alternants in the sense of determinants"; he prefers to use his own term *polar multiplication* (see Clifford 1878b, p. 127).[1] Clifford (1878a, p. 350) adopts Sylvester's terminology: "Following a suggestion of

---

[2] "[...] Definition II. Es sei A eine beliebige Untergruppe von U, die Gesamtheit aller Elemente y von V von der Eigenschaft, daß für jedes x aus A $x \cdot y = 0$ ist, heißt der Annulator von A in V und wird mit (V, A) bezeichnet". Translation from Pontryagin (2012, p. 8).

[1] Sylvester's statement appears as an editorial note to Clifford's letter.

Professor Sylvester, I call that kind of multiplication in which the sign of the product is reversed by an interchange of two adjacent factors, *polar* multiplication; because the product ab has opposite properties at its two ends, so that $ab = -ba$" (Clifford 1878a, p. 350). The term had appeared earlier in print in a note on the 1875 meeting of the British Association for the Advancement of Science: In *The Athenaeum*, Sept. 11, 1875, p. 341, it says that "Prof. Clifford astonished the Section by some remarkable applications of Grassmann's 'polar multiplication' defined by the law that *ba* is *minus ab*".

Another term that was used in this respect was *skew*. It was Gibbs (1881-84, p. 5) who called the vector product *skew product*, and he also spoke of *skew multiplication*. Knott (1893, p. 220), who criticizes the term, supposes that the name was chosen because "this product exists only when the vectors are inclined or skew towards one another" but this is explanation is unlikely to be correct because the word had earlier been used in algebraic contexts in a similar way. See *skew* (algebra).

The term *anticommutative* appears in the 20th century. It was used by McAulay (1908, p. 506) and Wilson and Lewis (1912, p. 406), for example. The German term *antikommutativ* is found in Schouten (1914, pp. 28, 97).

References

Clifford, William Kingdon (1878a). "Applications of Grassmann's Extensive Algebra". *American Journal of Mathematics* 1, pp. 350–358.
Clifford, William Kingdon (1878b). "Extract of a Letter to Mr. Sylvester from Prof. Clifford of University College, London". *American Journal of Mathematics* 1, pp. 126–128.
Gibbs, Josiah Willard (1881-84). *Elements of Vector Analysis*. New Haven.
Hankel, Hermann (1867). *Vorlesungen über die complexen Zahlen und ihre Functionen*. In zwei Theilen. I. Theil: Theorie der complexen Zahlsysteme. Leipzig.
Knott, Cargill Gilston (1893). "Recent Innovations in Vector Theory". *Proceedings of the Royal Society of Edinburgh* 19, pp. 212–237.
McAulay, Alexander (1908). "Algebra after Hamilton, or Multenions". *Proceedings of the Royal Society of Edinburgh* 28, pp. 503–585.
Schouten, Jan Arnoldus (1914). *Grundlagen der Vektor- und Affinoranalysis*. Leipzig and Berlin.
Wilson, Edwin B. and Gilbert N. Lewis (1912). "The Space-Time Manifold of Relativity. The Non-Euclidean Geometry of Mechanics and Electromagnetics". *Proceedings of the American Academy of Arts and Sciences* 48, pp. 389–507.

antilogarithm

Like the term *logarithm*, this term was introduced by Napier (1614, p. 8), where Latin *antilogarithmus* is defined as the logarithm of the cosine.[1] Napier's book was soon translated, where the definition reads: "And they are also the Logarithmes of the complements of the arches and sines towards the right hand, which we call *Antilogarithmes*" (Napier 1616, p. 11).

The term was soon adopted by Kepler (1627, preface, p. 18), for example, and also appears in Urquhart (1645, p. 91) as well as in the first English mathematical dictionary (Moxon 1679, p. 10).

References

Kepler, Johannes (1627). *Tabulae Rudolphinae*. Ulm.

[1] "Itemque bi sunt arcuum, & sinuum dextrorum Logarithmi complementorum, quos antilogarithmos appellamus".

Moxon, Joseph (1679). *Mathematicks made easie or, a compleat Mathematical Dictionary*. London.
Napier, John (1614). *Mirifici logarithmorum canonis descriptio*. Edinburgh.
Napier, John (1616). *A Description of the Admirable Table of Logarithms*. Translated by Edward Wright. London.
Urquhart, Thomas (1645). *The Trissotetras*. Or, a most exquisite table for resolving all manner of triangles, whether plaine or sphericall, rectangular or obliquangular. London.

antiparallel

This term, sometimes used in elementary mathematics, was defined by Cajori (1896, p. 262) as follows (see fig. 4): "if CD is so drawn as to make angles a and b equal, then one of the two lines AB and CD is the *anti-parallel* of the other, with reference to the angle O".

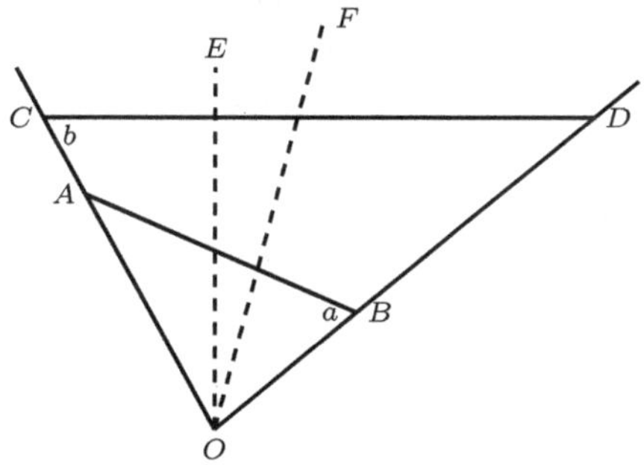

Figure 4: Antiparallel lines. From Cajori (1896, p. 262).

Cajori also makes some historical remarks, pointing out that the Latin word was used by Leibniz; it appears in Leibniz (1691, p. 279). The term is not due to Leibniz, however; it was already used in French by Arnauld (1667, p. 210), who also remarked that antiparallels have an effect contrary to those of parallels, which is why he chose the term (p. 212). The English term appears in Stone's *Mathematical Dictionary* in the entry *parallel* (Stone 1726): "Anti-parallels are those Lines that make the same Angles with two Lines cutting them; but contrary Ways".

Note. The origin of the term has been pointed out by Mackay (1883, p. 64), for example.

References

Arnauld, Antoine (1667). *Nouveaux elemens de geometrie*. Paris.
Cajori, Florian (1896). *A History of Elementary Mathematics*. New York.
Leibniz, Gottfried Wilhelm (1691). "De linea in quam flexile se pondere proprio curvat". *Acta Eruditorum* (June 1691), pp. 277–281.
Mackay, J.S. (1883). "The Triangle and its Six Scribed Circles". *Proceedings of the Edinburgh Mathematical Society* 1, pp. 4–128.
Stone, Edmund (1726). *A New Mathematical Dictionary*. London.

antisymmetric

As defined by Frobenius (1878, p. 36), "[a] substitution of the form $\pm A^{-1}A'$ is called *antisymmetric* by Mr Rosanes".[1] Frobenius refers to Rosanes (1875, p. 61), where the term was first used in this sense (in a different notation).[2]

The concept of antisymmetric relation is part of the definition of partially ordered sets, and in this sense *antisymmetric* is much better known today. When Hausdorff (1914, p. 139) introduced the term *partially ordered set*, however, he did not yet make use of the term *antisymmetric*, which does not appear to have come up until the mid-1930s. It was used by Birkhoff (1937, p. 311) and Carathéodory (1938, p. 32), neither of whom states that he is introducing a new term. It is therefore either already older or its explanation was thought to be obvious.

The term is not be confused with *asymmetrical relation*, which was introduced by Russell (1901, p. 32): "When ARB implies BRA, I call R a *symmetrical* relation; otherwise, I call R *asymmetrical*".[3] As in Höfler (1922, p. 251), the term *antisymmetrical relation* was sometimes used in Russell's sense.

References

Birkhoff, Garrett (1937). "An extended arithmetic". *Duke Mathematical Journal* 3, pp. 311–316.

Carathéodory, Constantin (1938). "Entwurf für eine Algebraisierung des Integralbegriffs". *Sitzungsberichte der mathematisch-naturwissenschaftlichen Abteilung der Bayerischen Akademie der Wissenschaften zu München* (1938), pp. 27–69.

Frobenius, Georg (1878). "Ueber lineare Substitutionen und bilineare Formen". *Journal für die reine und angewandte Mathematik* 84, pp. 1–63.

Hausdorff, Felix (1914). *Grundzüge der Mengenlehre*. Leipzig.

Hawkins, Thomas (1977). "Another look at Cayley and the theory of matrices". *Archives Internationales d'Histoire des Sciences* 27, pp. 82–112.

Höfler, Alois (1922). *Logik*. Zweite, sehr vermehrte Auflage. Wien.

Rosanes, Jacob (1875). "Ueber die Transformation einer quadratischen Form in sich selbst". *Journal für die reine und angewandte Mathematik* 80, pp. 52–72.

Russell, Bertrand (1901). "On the notion of order". *Mind* 10, pp. 30–51.

Russell, Bertrand (1903). *The Principles of Mathematics*. Vol. I. Cambridge.

arc

In classical Greek the word *περιφέρεια* was used to designate an arc of a circle. This word had another meaning: it could also designate the circumference of a circle (see *circumference*). In its sense of 'arc', *περιφέρεια* was used in the *Elements* (book I, def. 18), for example: "A semicircle is the figure contained by the diameter and the circumference cut off by it".[1] The term was already used in this sense before Euclid (Mugler 1958-59, p. 345).

The term *arc* goes back to Latin *arcus*, which means 'bow' and, derivatively, 'arch, vault, rainbow' (LSJ). Its geometrical meaning can be found in an agricultural treatise written in the 1st century by Columella (1941, p. 12): "Every piece of land is square, or long, or wedge-shaped, or triangular, or round, or else presents the form of a

---

[1] "Eine Substitution von der Gestalt $\pm A^{-1}A'$ nennt Herr Rosanes [...] *antisymmetrisch*". – A' is the conjugate form. [2] See Hawkins (1977, p. 98) for brief notes on Rosanes' introduction of the term. [3] See also Russell (1903, pp. 25, 203). [1] Translation from Euclid (1956, p. 154).

semi-circle or of the arc [*arcus*] of a circle, sometimes also of a polygon",[2] and also in an anonymous treatise on land measuring (Guillaumin 1996, pp. 208f.). The term seems to be based on its curved shape rather than on a connection with the shooting instrument. Such a connection can, however, be found in the context of trigonometry.

The "bow metaphor" essentially concerns the three trigonometric notions arc, chord, and sagitta or versed sine (see *chord* for an explanation). It is found in Sanskrit, then in Arabic, and, in the 12th century, in Latin, and the corresponding terminology can be traced to Sanskrit writings. Datta and Singh (1983, p. 40) name *cāpa* and *dhanu* as terms which both literally mean 'bow', and then, in trigonometry, 'arc'.[3]

The same metaphorical transfer can be found in the Arabic term *qaws*, which was already used by al-Khwārizmī (Gandz 1932, p. 71).[4] In Latin translations the term appears in the 12th century, for example in Plato of Tivoli's translation of Savasorda's *Liber embadorum* (Curtze 1902, pp. 12, 100) and in his translation *De motu stellarum* of al-Battānī's astronomical work (al-Battānī 1537, fol. 6^v, 14^r).[5] From then on, the term is used by various writers such as Thomas Bradwardine[1], Dominicus de Clavasio[2], and Johannes von Gmunden.[3]

In Arabic translations of the *Elements*, *περιφέρεια* was translated by *qaws* when the arc was meant (Rashed 2017, p. 676). This practice also had consequences for the Latin translations of the 12th century: while in a translation made directly from the Greek, *arcus* is *not* found (Busard 1987, p. 67), it *is* used in translations based on Arabic sources (see, e. g., Busard 1983, p. 87).

As for modern languages, German *Bogen* 'bow', still used today, is found at around 1400 in the *Geometria culmensis*, the first known geometry written in German (Mendthal 1886, p. 71). The French term *arc* is found in print in de Bouelles (1511, fol. ii^v),[4] and, in English, Recorde (1551, fol. B i^v) uses *arche lyne*: "But and if the line goe crosse the circle, and passe beside the centre, then is it called *a corde*, or *a stryng line* [...] And the compassed line that aunswereth to it, is called *an arche lyne*, or *a bowe lyne*".

Note. For the Greek term, see Mugler (1958-59, pp. 344f.), Tropfke (1940, p. 132), and Heath in Euclid (1956, p. 184). For the Arabic term, see Rashed (2017, pp. 675f.). An account of various topological definitions of the term is found in Herrlich, Hušek, and Preuß (2008).

References

al-Battānī, Muḥammad ibn Jābir (1537). *De motu stellarum*. Nürnberg.
de Bouelles, Charles (1511). *Geometrie en francoys*. Paris.
Burgess, Ebenezer (1858). "Translation of the Surya-Siddhanta, a text-book of Hindu Astronomy". *Journal of the American Oriental Society* 6, pp. 141–498.

[2] "Omnis ager aut quadratus, aut longus, aut cuneatus, aut triquetrus, aut rotundus, aut etiam semicirculi, vel arcus, nonnumquam etiam plurimum angulorum formam exhibet". Translation from Columella (1941, p. 13). [3] See also Burgess (1858, p. 431), Rodet (1879, p. 11), and Puttaswamy (2012, p. 108). [4] See also Ruska (1917, p. 106). [5] In the 12th century, the term is also found in some anonymous writings (Folkerts 1971, pp. 24f.). [1] See Molland's edition of his *Geometrica speculativa* (Molland 1989, p. 61). [2] Busard (1976, p. 538). [3] Busard (1971, p. 538). [4] The term was earlier used (DMF) by Chuquet (1979, p. 409) and Oresme (1968, p. 379). Italian *arco* was first used in the early 14th century (TLIO).

Busard, H.L.L. (1971). "Der Traktat De sinibus, chordis et arcubus von Johannes von Gmunden". *Denkschriften der Österreichischen Akademie der Wissenschaften. Mathematisch-naturwissenschaftliche Klasse* 116, pp. 73–113.

Busard, H.L.L. (1976). "The Practica Geometriae of Dominicus de Clavasio". *Archive for History of Exact Sciences* 2, pp. 520–575.

Busard, H.L.L. (1983). *The First Latin Translation of Euclid's Elements Commonly Ascribed to Adelard of Bath*. Books I-VIII and books X.36-XV.2. Toronto.

Busard, H.L.L. (1987). *The Mediaeval Latin Translation of Euclid's Elements Made Directly from the Greek*. Stuttgart.

Chuquet, Nicolas (1979). *La Géométrie*. Première géométrie algébrique en langue française (1484). Introduction, texte et notes par Hervé l'Huillier. Paris.

Columella, Lucius Junius Moderatus (1941). *On Agriculture*. With a recension of the text and an English translation by E. S. Forster and Edward H. Heffner. Vol. II. London and Cambridge, Mass.

Curtze, Maximilian (1902). "Der Liber Embadorum des Savasorda in der Übersetzung des Plato von Tivoli". In: *Urkunden zur Geschichte der Mathematik im Mittelalter und der Renaissance*. Erster Theil. Leipzig, pp. 1–183.

Datta, Bibhutibhusan and Avadhesh Narayan Singh (1983). "Hindu Trigonometry (revised by Kripa Shankar Shukla)". *Indian Journal of History of Science* 18, pp. 39–108.

Euclid (1956). *The Thirteen Books of Euclid's Elements*. Volume I. Introduction and Books I, II. Ed. by Thomas L. Heath. New York.

Folkerts, Menso (1971). "Anonyme lateinische Euklidbearbeitungen aus dem 12. Jahrhundert". *Denkschriften der Österreichischen Akademie der Wissenschaften. Mathematisch-naturwissenschaftliche Klasse* 116, pp. 5–42.

Gandz, Solomon (1932). *The Mishnat ha Middot*. The First Hebrew Geometry of about 150 C.E. and the Geometry of Muhammad Ibn Musa Al-Khowarizmi. Berlin.

Guillaumin, Jean-Yves (1996). *Balbus, Présentation systématique de toutes les figures. Podismus et textes connexes*. Extraits d'Epaphrodite et de Vitruvius Rufus; La mesure des jugères. Napoli.

Herrlich, Horst, Mirek Hušek, and Gerhard Preuß (2008). "Hausdorffs Studien über Kurven, Bögen und Peano-Kontinua". In: *Felix Hausdorff. Gesammelte Werke*. Band III. Mengenlehre (1927, 1935). Deskriptive Mengenlehre und Topologie. Ed. by Ulrich Felgner et al. Berlin and Heidelberg, pp. 798–839.

Mendthal, Hans (1886). *Geometria culmensis*. Ein agronomischer Tractat aus der Zeit des Hochmeisters Conrad von Jungingen (1393-1407). Leipzig.

Molland, George, ed. (1989). *Thomas Bradwardine, Geometria Speculativa*. Latin Text and English Translation with an Introduction and a Commentary. Stuttgart and Wiesbaden.

Mugler, Charles (1958-59). *Dictionnaire historique de la terminologie géométrique des Grecs*. Paris.

Oresme, Nicole (1968). *Le livre du ciel et du monde*. Edited by Albert D. Menut and Alexander J. Denomy, translated with an introduction by Albert D. Menut. Madison.

Puttaswamy, T.K. (2012). *Mathematical Achievements of Pre-modern Indian Mathematicians*. London.

Rashed, Roshdi, ed. (2017). *Lexique historique de la langue scientifique arabe*. Hildesheim.

Recorde, Robert (1551). *The patheway to knowledg*. London.

Rodet, Léon (1879). *Leçons de calcul d'Âryabhata*. Paris.

Ruska, Julius (1917). *Zur ältesten arabischen Algebra und Rechenkunst*. Heidelberg.

Tropfke, Johannes (1940). *Geschichte der Elementar-Mathematik*. Vierter Band. Ebene Geometrie. Dritte, verbesserte und vermehrte Auflage besorgt von Dr. Kurt Vogel. Berlin and Leipzig.

Archimedean ordered field

This terminology was introduced by Artin and Schreier (1926, p. 279):

> We now want to investigate ordered fields which satisfy either the Archimedean axiom or a certain generalization of it.

> Let G be an ordered field, and K a subfield. An element α of G is said to be "infinitely large with respect to K" if $c <| \alpha |$ for all elements c in K, or, on the other hand, "infinitely small with respect to K" if $0 <| \alpha |< c$ for all positive elements c in K. [...]
>
> We call an ordered field containing K "Archimedean with respect to K" if it contains no elements which are infinitely large (or infinitely small) with respect to K.[1]

The *Archimedean axiom* alluded to by Artin and Schreier is this: if $a > b$, then there is a multiple of b such that $nb > a$ (Stolz 1885, p. 69). Its name spread through Stolz (1883, p. 75) and was first suggested in Stolz (1881, p. 269). Stolz explicitly remarked that the axiom was known before Archimedes; this is the reason why the designation was criticized (for example, by Zeuthen 1904, p. 541). Nevertheless, the term was quickly adopted and spread, not least through Hilbert's use of it (e. g., Hilbert 1899, p. 19).

The terminology suggested by Artin and Schreier was immediately adopted, for example by Baer (1927, p. 7) and van der Waerden (1930, p. 211), who both referred to Artin and Schreier. It should also be noted that it may not have been felt to be new at the time since the epithet *(non-)Archimedean* had already been applied to number systems (*Zahlensysteme*) by Hilbert (1899, pp. 27f.) and to systems of magnitudes (*Größensysteme*) by Hahn (1907, p. 606).

References

Artin, Emil and Otto Schreier (1926). "Algebraische Konstruktion reeller Körper". *Abhandlungen aus dem Mathematischen Seminar der Universität Hamburg* 5, pp. 85–99.

Artin, Emil and Otto Schreier (2007). "The Algebraic Construction of Real Fields". In: *Exposition by Emil Artin: A Selection*. Ed. by Michael Rosen. Providence, R.I. and London, pp. 273–283.

Baer, Reinhold (1927). "Über nicht-Archimedisch geordnete Körper". *Sitzungsberichte der Heidelberger Akademie der Wissenschaften* 8, pp. 3–13.

Ehrlich, Philip (1995). "Hahn's *Über die nichtarchimedischen Grössensysteme* and the Development of the Modern Theory of Magnitudes and Numbers to Measure Them". In: *From Dedekind to Gödel*. Essays on the Development of the Foundations of Mathematics. Ed. by Jaakko Hintikka. Dordrecht, pp. 165–213.

Hahn, Hans (1907). "Über die nichtarchimedischen Grössensysteme". *Sitzungsberichte der Kaiserlichen Akademie der Wissenschaften (mathematisch-naturwissenschaftliche Klasse)* 116, pp. 601–655.

Hilbert, David (1899). *Grundlagen der Geometrie*. Leipzig.

Stolz, Otto (1881). "B. Bolzano's Bedeutung in der Geschichte der Infinitesimalrechnung". *Mathematische Annalen* 18, pp. 255–279.

Stolz, Otto (1883). "Zur Geometrie der Alten, insbesondere über ein Axiom des Archimedes". *Mathematische Annalen* 22, pp. 504–519.

Stolz, Otto (1885). *Vorlesungen über allgemeine Arithmetik*. Erster Theil: Allgemeines und Arithmetik der reellen Zahlen. Leipzig.

van der Waerden, Bartel Leendert (1930). *Moderne Algebra*. Erster Teil. Berlin.

---

[1] "Wir wollen jetzt geordnete Körper untersuchen, deren Ordnung dem Axiom des Archimedes bzw. einer gewissen Verallgemeinerung davon genügt. Sei G ein geordneter Körper, K ein Unterkörper von K. Ein Element α aus G heißt ‚in bezug auf K unendlich groß', wenn $| \alpha |> c$ für jedes Element c aus K, dagegen ‚in bezug auf K unendlich klein', wenn $0 <| \alpha |< c$ für jedes positive Element c aus K. [...] Wir nennen nun einen geordneten Körper über K ‚in bezug auf K Archimedisch', wenn er keine in bezug auf K unendlich großen (oder kleinen) Elemente enthält". Translation from Artin and Schreier (2007, p. 279). – For the origin of the terms *infinitely large* and *infinitely small* in this context, see Ehrlich (1995, pp. 174 and 197, fn. 10).

Zeuthen, H.G. (1904). "Gebrauch und Mißbrauch historischer Benennungen in der Mathematik". In: *Atti del Congresso internazionale di scienze storiche (Roma, 1-9 aprile 1903)*. Roma, pp. 536–542.

Archimedean valuation

If a valuation satisfies the ultrametric inequality $\|a+b\| \leq \max(\|a\|, \|b\|)$ (with a, b elements of a field), it is called *non-Archimedean*, otherwise *Archimedean*. In this sense, this terminology was introduced by Ostrowski (1916, p. 272), who used German *unarchimedisch* and *archimedisch*.[1] The corresponding English terms seem to have been first used by White (1924, pp. 346f.), who referred to Ostrowski.

References

Ostrowski, Alexander (1916). "Über einige Lösungen der Funktionalgleichung $\varphi(x) \cdot \varphi(y) = \varphi(xy)$". *Acta Mathematica* 41, pp. 271–284.

Roquette, Peter (2002). "History of Valuation Theory, Part I". In: *Valuation Theory and its Applications*. Volume I. Ed. by Franz-Viktor Kuhlmann, Salma Kuhlmann, and Murray Marshall. Providence, R.I., pp. 291–355.

White, Francis Puryer (1924). "Pure mathematics". *Science Progress in the Twentieth Century* 18, pp. 345–354.

arcwise connected

The term *arcwise connected* was introduced by Moore (1923, p. 294): "A point set M is said to be arc-wise connected if every two distinct points of M are the extremities of a simple continuous arc which lies wholly in M". Menger (1929, p. 210) suggested the German *bogenverknüpft* (from *verknüpfen* 'to link, connect') as a translation; French *connexité par arcs* was used by Kuratowski and Menger (1930, p. 170).

See *connected* for further historical information.

References

Kuratowski, Kazimierz and Karl Menger (1930). "Remarques sur la théorie axiomatique de la dimension". *Monatshefte für Mathematik und Physik* 37, pp. 169–174.

Menger, Karl (1929). "Über die Dimension von Punktmengen III. Zur Begründung einer axiomatischen Theorie der Dimension". *Monatshefte für Mathematik und Physik* 36, pp. 193–218.

Moore, Robert Lee (1923). "Report on continuous curves from the viewpoint of analysis situs". *Bulletin of the American Mathematical Society* 29, pp. 289–302.

area

General terms for 'area' in Greek mathematics were rarely used. One reason is that, as Tropfke (1940, p. 172) remarks, that a word like *κύκλος*, defined as a plane figure, could also stand for the area of the circle. Similarly, Mugler (1958-59, p. 172) points out that the mathematicians did not feel a need for such a general term until their geometry became more practical. Two words that were used for 'area' are *χωρίον*, which means 'space enclosed by lines, area, figure' (LSJ), attested since Aristotle (Mugler 1958-59, pp. 450f.), and *ἐμβαδόν*[1], which is of much later origin and first used regularly by Hero of Alexandria (Mugler 1958-59, p. 172), particularly in his *Metrica* (Bruins 1964, e. g., p. 68).

The latter word was taken over literally into Latin and appears in Latin as *embadon* in a treatise by the land surveyor Balbus (Blume, Lachmann, and Rudorff 1848, p. 96),

[1] The origin of the terminology has been poined out by Roquette (2002, p. 297). [1] From the verb ἐμβαίνω 'step in, upon'.

who wrote (probably around the year 100): "you take the length by the breadth; this makes an *embadon*".[2] In the second century, the land surveyor Nipsus used *embadum* (Blume, Lachmann, and Rudorff 1848, p. 297). This term remained in use for a long time; also, *embadometria* was sometimes used as a synonym for *planimetry*.

Latin *area* literally means 'ground for a house, a building-spot' (LS). In the sense under discussion, the term was used by Vitruvius (Vitruvius 1934, p. 198) and by the land surveyor Frontinus (Blume, Lachmann, and Rudorff 1848, p. 32), for example. In the Pseudo-Boethian *Geometria II*, the word *embadum* is explained by the word *area*: "embadum, id est area" (Boethius 1867, p. 404).

French *aire* is a direct successor of Latin *area* and is already found in the *Pratike de geometrie* (Victor 1979, p. 476), dating from the second half of the 13th century, and also appears in the first printed French book on geometry (de Bouelles 1511, fol. x^v); Italian *area* dates from the same time and is found (in the form *aria*) in a late 13th-century manuscript (Arrighi 1991, p. 12). The German term *Inhalt*, lit. 'content', appears at the end of the 15th century; it is used in Widmann (1489, fol. E 7^v) and Roritzer (c1487, fol. 5^r).[1] The compound *Flächeninhalt* is of later origin; it is only in the second half of the 18th century that it is used with some frequency.

In English, L. Digges (1556, e. g., fol. B iii^r) uses *content*;[2] in L. Digges and T. Digges (1571, fol. L $iiij^r$) it says: "Multiplie one of the equall sides in it selfe, the halfe of the producte is the Area or superficiall contente".[3] Billingsley also uses *area* in his translation of the *Elements*: "the area of a triangle, is that space, which is contayned within the sydes of a triangle" (Euclid 1570, fol. E i^v).

See also *volume*.

Note. A brief discussion of the terminology is found in Tropfke (1940, pp. 172f.); see also Gandz (1927). For Greek *ἐμβαδόν*, see Mugler (1958-59, p. 172); for *χωρίον*, see Mugler (1958-59, pp. 450f.) and also Ayuso García (2008, pp. 1381–1383).

References

Arrighi, Gino (1991). "Maestro umbro (sec. XIII). Amastramento de l'arte de la geometria (Cod. 2404 dalla Biblioteca Riccardiana di Firenze)". *Bollettino della deputazione di Storia patria per l'Umbria* 87, pp. 5–31.

Ayuso García, Manuel (2008). "La terminología latina de la geometría en Marciano Capela". PhD thesis. Madrid: Universidad Nacional de Educación a Distancia.

Blume, Friedrich, Karl Lachmann, and Adolf Rudorff, eds. (1848). *Die Schriften der römischen Feldmesser*. Erster Band. Texte und Zeichnungen. Berlin.

Boethius, Anicius Manlius Severinus (1867). *De institutione arithmetica libri duo, De institutione musica libri quinque*. Accedit Geometria quae fertur Boetii. Ed. by Gottfried Friedlein. Leipzig.

de Bouelles, Charles (1511). *Geometrie en francoys*. Paris.

Bruins, E.M., ed. (1964). *Codex Constantinopolitanus*. Palatii Veteris No. 1. Part Two. Greek Text. Leiden.

Campbell, Brian (2000). *The Writings of the Roman Land Surveyors*. Introduction, Text, Translation and Commentary. London.

Digges, Leonard (1556). *A boke named Tectonicon*. London.

[2] "ducis longitudinem per latitudinem: facit embadon". Translation from Campbell (2000, p. 207).

[1] For the German term, see Schirmer (1912, p. 34), Tropfke (1940, pp. 173f.), and Reiner (1961, pp. 66f.). Tropfke and Reiner present a number of alternative terms. [2] Digges uses *content* also for 'volume' (L. Digges 1556, fol. D iii*v*). [3] L. Digges and T. Digges (1571) is cited by the OED as the first use of *area* in this sense; at least parts of the text of the book may have been written in the 1560s.

Digges, Leonard and Thomas Digges (1571). *A geometrical practise, named Pantometria diuided into three bookes, Longimetra, Planimetra, and Stereometria*. London.
Euclid (1570). *The Elements of Geometrie of the most auncient Philosopher Euclide of Megara*. Faithfully (now first) translated into the Englishe toung, by H. Billingsley. London.
Gandz, Solomon (1927). "On Three Interesting Terms Relating to Area". *The American Mathematical Monthly* 33, pp. 80–86.
Mugler, Charles (1958-59). *Dictionnaire historique de la terminologie géométrique des Grecs*. Paris.
Reiner, Karl (1961). *Die Terminologie der ältesten mathematischen Werke in deutscher Sprache nach den Beständen der Bayerischen Staatsbibliothek*. München.
Roritzer, Matthäus (c1487). *Geometria deutsch*. [Nürnberg].
Schirmer, Alfred (1912). *Der Wortschatz der Mathematik nach Alter und Herkunft untersucht*. Straßburg.
Tropfke, Johannes (1940). *Geschichte der Elementar-Mathematik*. Vierter Band. Ebene Geometrie. Dritte, verbesserte und vermehrte Auflage besorgt von Dr. Kurt Vogel. Berlin and Leipzig.
Victor, Stephen K. (1979). *Practical Geometry in the High Middle Ages*. "Artis cuiuslibet consummatio" and the "Pratike de geometrie". Philadelphia.
Vitruvius (1934). *On Architecture*. In two volumes. II. Edited from the Harleian manuscript 2767 and translated into English by Frank Granger. London and New York.
Widmann, Johannes (1489). *Behende vnd hubsche Rechenung auff allen kauffmanschafft*. Leipzig.

argument

The earliest technical meaning of the term *argument*, upon which later mathematical meanings are based, is given in Moxon's mathematical dictionary in the following terms: "The Arch by which we seek another Arch, (or part of a circle) is called by Mathematicians, the Argument, in resemblance of Argumentation in Logick; for as Logical Argument leads us to knowledge of something before unknown, so this Arch notifies us to another Arch unknown, proportional" (Moxon 1679, s. v. argument).[1] In more general terms, an argument is some quantity from which another quantity may be deduced. The term was also used in mathematical or astronomical tables, in which the argument is a variable quantity from which another quantity can be calculated. This technical use has been traced back to the 12th century (von Braunmühl 1900, p. 49, fn. 1), where it appears in Adelard of Bath's translation of the astronomical tables of al-Khwārizmī (Suter 1914, p. 46).[2]

In the 19th century, *argument* begins to be used in connection with functions. For example, Jacobi (1829, p. 1) starts his treatise on elliptic functions with the following words: "The most memorable integrals which are exhibited by the formula $\int \frac{d\varphi}{\sqrt{1-k^2\sin^2\varphi}}$ and which constitute the first kind of elliptic functions, as they are called nowadays, depend on two arguments, on the amplitude φ and the modulus k".[3] The restriction to *elliptic* functions was dropped at about the same time, for example by Bartels (1837, p. 48): "If $\varphi(x)$ designates a function of any variable quantity x, which we call the *argument* of the function [...]".[4]

[1] Moxon's definition is a translation of Vitalis (1668, p. 63). [2] For further discussion, see Kunitzsch (1977, p. 33) and Bossong (1978, pp. 137f.). [3] "Integralia maxime memorabilia, quae Formula exhibentur $\int \frac{d\varphi}{\sqrt{1-k^2\sin^2\varphi}}$, et quae Functionum Ellipticarum, quae dicuntur, primam speciem constituunt, ab Argumento duplice pendent, et ab Amplitudo φ et a Modulo k". – Jacobi (1829, p. 31) also used the term in a more specialized way when he defined the argument (Latin *argumentum*) of an elliptic function. Verhulst (1841, p. 108) explicitly links the word *argument* as used in tables with this use. [4] "Wenn $\varphi(x)$ eine Function irgend einer veränderlichen Grösse x, die wir das *Argument* der Function nennen, bezeichnet, [...]". See already Bartels (1833, p. 4).

The argument of a complex number is also related to the older usage. The term was introduced in this sense by Cauchy (1838, p. 987): "In every imaginary expression, the real part and the coefficient of $\sqrt{-1}$ are [...] the respective products of a real and positive quantity, which is called the *module*, with the sine or the cosine of a certain arc or angle, which we call the *argument*".[5] Various other terms have been used in this sense. As Burkhardt (1897, p. 12) states, "there is no agreement on a definite name for the angle φ; it has been called argument, amplitude, declination, anomaly, *arcus*".[6]

References

Bartels, J.M.C. (1833). *Vorlesungen über mathematische Analysis*. Erster Band. Dorpat.
Bartels, J.M.C. (1837). *Vorlesungen über mathematische Analysis*. Ed. by F.G.W. Struve. Dorpat.
Bossong, Georg, ed. (1978). *Los Canones de Albateni*. Herausgegeben sowie mit Einleitung, Anmerkungen und Glossar versehen. Tübingen.
von Braunmühl, Anton (1900). *Vorlesungen über die Geschichte der Trigonometrie*. Erster Teil. Von den ältesten Zeiten bis zur Erfindung der Logarithmen. Leipzig.
Burkhardt, Heinrich (1897). *Funktionentheoretische Vorlesungen*. Erster Teil. Leipzig.
Cauchy, Augustin-Louis (1838). "Mémoire sur la réflexion et la réfraction de la lumière. 1re partie". *Comptes rendus de l'Académie des sciences* 7, pp. 985–992, 1044–1050.
Jacobi, Carl Gustav Jacob (1829). *Fundamenta nova theoriae functionum ellipticarum*. Königsberg.
Kunitzsch, Paul (1977). *Mittelalterliche astronomisch-astrologische Glossare mit arabischen Fachausdrücken*. München.
Moxon, Joseph (1679). *Mathematicks made easie or, a compleat Mathematical Dictionary*. London.
Study, Eduard (1898). "Theorie der gemeinen und höheren complexen Grössen". *Encyklopädie der mathematischen Wissenschaften mit Einschluss ihrer Anwendungen* I A 4, pp. 148–183.
Suter, Heinrich, ed. (1914). *Die astronomischen Tafeln des Muhammed ibn Musa al-Khwarizmi in der Bearbeitung des Maslama ibn Ahmed al-Madjriti und der latein. Übersetzung des Athelhard von Bath auf Grund der Vorarbeiten von A. Björnbo und R. Besthorn*. København.
Verhulst, Pierre-François (1841). *Traité élémentaire des fonctions elliptiques*. Bruxelles.
Vitalis, Hieronymus (1668). *Lexicon mathematicum, astronomicum et geometricum*. Paris.

arithmetic* and *number theory

The term *arithmetic* goes back to Greek *ἀριθμός* 'number'. The traditional understanding and its opposition to *logistic* is explained by Heath (1921, p. 13):

> By arithmetic Plato meant, not arithmetic in our sense, but the science which considers numbers in themselves, in other words, what we mean by the Theory of Numbers. He does not, however, ignore the art of calculation (arithmetic in our sense); he speaks of number and calculation (*ἀριθμόν καί λογισμόν*) and observes that 'the art of calculation (*λογιστική*) and arithmetic (*ἀριθμητική*) are both concerned with number'.[1]

For example, in Plato's *Philebus* (56d-57a), Socrates asks: "Are there not two kinds of arithmetic, that of the people and that of philosophers? [...] And how about the arts of reckoning [*λογιστική*] and measuring as they are used in building and in

[5] "Dans toute expression imaginaire, la partie réelle et coefficient de $\sqrt{-1}$ sont [...] les produits respectifs d'une quantité réelle et positive qu'on nomme le *module* par le sinus et le cosinus d'un certain arc ou angle que nous appellerons l'*argument*". [6] "Über einen bestimmten Namen für den Winkel φ hat man sich nicht geeinigt; man findet ihn als Argument, Amplitude, Abweichung, Anomalie, *Arcus* bezeichnet". – Study (1898, p. 54) lists the same terms but also adds *azimuth*.

[1] Heath refers to Plato's *Republic* (vii 522c, 525a, 526b). For detailed discussions of the complexities and development of the terms *arithmetic* and *logistic*, see Fowler (1999, esp. ch. 4) and Klein (1992).

trade when compared with philosophical geometry and elaborate computations – shall we speak of each of these as one or as two?", to which Protarchus answers "I should say that each of them was two".[2] The Greek philosopher Aristoxenus later mentioned that such a distinction was already made by Pythagoras: "Pythagoras seems to have valued the science of numbers most of all and to have advanced it, separating it from the merchants' business and likening all things to numbers. For number contains all else as well, and there is a ratio between all the numbers to each other" (Wehrli 1967, p. 14).[3]

Latin *arithmetica* is a borrowing from the Greek, attested in classical Latin (Guillaumin 2020, pp. 43f.). A major publication was Boethius' *Institutio arithmetica* (Boethius 1867). The explicit distinction between practical and theoretical arithmetic in Latin texts came up in the 12th century. It was used, for example, by Dominicus Gundissalinus in his translation of a text by al-Fārābī (Gundisalvo 1954, p. 85) and in the *Liber Mahameleth*:

> The study of number considered by itself is called theoretical or speculative; that involving matter, practical or active. Since it belongs to arithmetical science to deal with number from both points of view, there is accordingly a theoretical and a practical arithmetic. The kinds of practical arithmetic are numerous: there is the science of collecting numbers, that of separating them, the science of commercial transactions, the science of determining unknowns by means of numbers, and many others (Sesiano 2014, p. 1).[1]

These Latin usages have also been influenced by Arabic sources; for example, the philosopher al-Fārābī wrote in the 10th century:

> As for arithmetic, there are two sciences which are known by this name: practical arithmetic and theoretical arithmetic. Practical arithmetic inquires into numbers insofar as the numbers are fixed to things numbered, i. e., bodies and other things whose enumeration it is necessary to have, like men, horses, dinars, dirhams, or other things possessed of number. This is the arithmetic that the people use in the commercial transactions of the market and the city. Theoretical arithmetic inquires into numbers in the abstract insofar as they are separated in the mind from bodies and everything else which is numbered by them (al-Fārābī 1932, p. 33).[2]

The English term is found in the early 14th century and was originally commonly used in the form *arsmetrike*, from an erroneous *ars metrica* 'art of measure' (OED).[3] The wrong spelling was later corrected: "Bothe names are corruptly writen Arsmetrike for Arithmetyke (as the Grekes call it) and Awgrym for Algorisme (as Arabyans sounde it) whiche bothe betoken the science of nombrynge. For arithmos in greke, is called

[2] Translation from Plato (1925, pp. 361, 363). [3] Translation from Zhmud (2006, p. 218). [1] "Illa autem consideratio qua numerus per se attenditur dicitur theorica vel speculativa, qua vero in materia dicitur practica vel activa. Et quoniam artis arimethice est utroque modo de numero tractare, ideo arimethica alia est theorica, alia practica. Practice autem species multe sunt, quoniam alia est scientia coniungendi numeros, alia disiungendi, alia est scientia negotiandi, alia est scientia per numeros occulta inveniendi, et multe alie". Translation from Sesiano (2014, p. 581). [2] Translation from Levey and Petruck (1965, pp. ix–x). For practical arithmetic in Arabic texts, see also Freudenthal (2016, pp. 82–88); a detailed discussion is Rebstock (1992). [3] The incorrect spelling is found in various languages; an early example is Plato of Tivoli's 12th-century translation of the *Liber embadorum*, where, in a definition of relatively prime numbers, "Boetius in arismetricis" is referred to (Curtze 1902, p. 16). Smith (1958, p. 8) gives some more examples.

number, [and] of it cometh arithmetike, the arte of nomberynge" (Recorde 1543, fol. 6*r*). In French, the term is found in the 12th century (TLF); in German texts, the term is found from the 14th century onwards, often in Latin form (Strauß et al. 1996, pp. 217–220).

Just like in the ancient languages, *arithmetic* as used in modern languages could refer to both theoretical arithmetic (as exemplified by Boethius' *De institutione arithmetica*) and practical arithmetic (dealing with practical aspects of calculations). Spiesser (2003, pp. 9–18) discusses this aspect in detail, noting that *arithmetic* is "a generic term that encompasses everything that concerns number" (p. 12).[4] Practical arithmetic also included treatises on commercial arithmetic, a field which was "important for the growing class of merchants operating within the new mercantile economy of the fourteenth and fifteenth century" (Heeffer 2022, p. 215).[5]

Also, after the Renaissance, at different times and with different authors, the word covered different fields. As examples, Tropfke (1980, p. 5) cites, among others, the *Arithmetica infinitorum* (Wallis 1656), which contains results on infinite series and products, and the *Arithmetica universalis* (Newton 1707), which covers algebra.

In the 18th century, the word *number theory* often becomes associated with the works of Fermat, Euler, Lagrange, and some others.[1] Legendre (1797), for example, mentions these names in the preface of his *Essai sur la théorie des nombres*. At about the same time, Gauß publishes his seminal work on number theory, in which he makes some remarks about the terminology:

> what is commonly called Arithmetic hardly extends beyond the art of enumerating and calculating (i. e. expressing numbers by suitable symbols, for example by a decimal representation, and carrying out arithmetic operations). It often includes some subjects which certainly do not pertain to Arithmetic (like the theory of logarithms) and others which are common to all quantities. As a result it seems proper to call this subject Elementary Arithmetic [*arithmetica elementaris*] and to distinguish from it Higher Arithmetic [*arithmetica sublimior*] which properly includes more general inquiries concerning integers. We consider only Higher Arithmetic in the present volume (Gauß 1801, pp. vii-viii).[2]

In the course of the 19th century, the distinction between arithmetic and number theory was further developed. Crelle (1845, pp. iv-v), for example, sharply distinguished between them, using the terms *Arithmetik* and *Theorie der Zahlen*. Ambiguities remained, however:

> By "arithmetic," German authors sometimes meant the theory of natural numbers, sometimes the theory of real or even complex numbers. Indeed, with arithmetization in the late XIXth

[4] Spiesser also discusses some related (partially) synonymous terms. [5] For the origin and development of commercial arithmetic, see also Spiesser (2017) and Spiesser (2022). [1] The word was already used at a time when no such association would have been possible, for example by von Peuerbach (1511, fol. A 1^{v}) and Bodin (1581, fol. 14^{r}). [2] "Sed quum ea, quae Arithmetices nomine vulgo traduntur, vix vitra artem numerandi et calculandi (i. e. numeros per signa idonea e. g. secundum systema decadicum exhibendi, operationesque arithmeticas perficiendi) extendantur, adiectis nonnullis quae vel ad Arithmeticam omnino non pertinent (vt doctrina de logarithmis) vel saltem numeris integris non sunt propria sed ad omnes quantitates patent: e re esse videtur, duas Arithmeticae partes distinguere, illa que ad Arithmeticam elementarem referre, omnes autem disquisitiones generales de numerorum integrorum affectionibus propriis *Arithmeticae Sublimiori*, de qua sola hic sermo erit, vindicare". Translation from Gauß (1986, p. xvii). While Gauß here distinguishes between elementary and higher arithmetic, he also uses *theoria numerorum* occasionally (Gauß 1801, e. g., §56).

century, it was not uncommon to expand the meaning of "arithmetic" to include all of pure mathematics. In particular, algebra and real and complex analysis were regarded as parts of general arithmetic (Ferreirós 2007, p. 260).

An example of such a wide understanding is found in Kronecker (1887, p. 338):

Indeed, arithmetic stands to the other two mathematical disciplines, geometry and mechanics, as the whole of mathematics does to astronomy and the other natural sciences. Arithmetic too renders manifold services to geometry and mechanics, and in return receives from her sister disciplines an abundance of stimulation. The word 'arithmetic' is here not to be understood in the usual restricted sense, but rather as including all mathematical disciplines with the exception of geometry and mechanics–especially, therefore, algebra and analysis. And I also believe that we shall one day succeed in 'arithmetizing' the entire content of all these mathematical disciplines–that is, in grounding them solely on the number-concept taken in its narrowest sense [...].[1]

In the 20th century, some writers such as Serre (1970), whose book is entitled *Cours d'arithmétique*, used *arithmetic* instead of the then more common *number theory*.

Note. For the history of the terminology, see Heath (1921, pp. 13–16), Smith (1958, pp. 7f.), Tropfke (1980, pp. 1–7), and Guillaumin (2020, pp. 43f.). For the Arabic term, see Rashed (2017, pp. 189–198). Ferreirós (2007, pp. 260–263) discusses aspects of the meaning of *arithmetic* in the 19th century.

References

Bodin, Jean (1581). *Apologie de René Herpin pour la Republique de J. Bodin*. Paris.

Boethius, Anicius Manlius Severinus (1867). *De institutione arithmetica libri duo, De institutione musica libri quinque*. Accedit Geometria quae fertur Boetii. Ed. by Gottfried Friedlein. Leipzig.

Crelle, August Leopold (1845). *Encyklopädische Darstellung der Theorie der Zahlen und einiger anderer damit in Verbindung stehender analytischer Gegenstände*. Erster Band. Berlin.

Curtze, Maximilian, ed. (1902). *Urkunden zur Geschichte der Mathematik im Mittelalter und der Renaissance*. Erster Theil. Leipzig.

Ewald, William (1996). *From Kant to Hilbert*. A Source Book in the Foundations of Mathematics. Volume II. Oxford.

al-Fārābī (1932). *Catálogo de las ciencias*. Edición y traducción castellana por Ángel González Palencia. Madrid.

Ferreirós, José (2007). "*'Ο Θεὸς 'Αριθμητίζει*: The Rise of Pure Mathematics as Arithmetic with Gauss". In: *The Shaping of Arithmetic after C. F. Gauss's Disquisitiones Arithmeticae*. Ed. by Catherine Goldstein, Norbert Schappacher, and Joachim Schwermer. Berlin and Heidelberg, pp. 235–268.

Fowler, David (1999). *The Mathematics of Plato's Academy*. A New Reconstruction. Oxford.

Freudenthal, Gad (2016). "Abraham Ibn Daud, Avendauth, Dominicus Gundissalinus and Practical Mathematics in Mid-Twelfth Century Toledo". *Aleph* 16, pp. 61–106.

[1] "In der That steht die Arithmetik in ähnlicher Beziehung zu den anderen beiden mathematischen Disciplinen, der Geometrie und Mechanik, wie die gesammte Mathematik zur Astronomie und den andern Naturwissenschaften; auch die Arithmetik erweist der Geometrie und Mechanik mannigfache Dienste und empfängt dagegen von ihren Schwester-Disciplinen eine Fülle von Anregungen. Dabei ist aber das Wort ‚Arithmetik' nicht in dem üblichen beschränkten Sinne zu verstehen, sondern es sind alle mathematischen Disciplinen mit Ausnahme der Geometrie und Mechanik, also namentlich die Algebra und Analysis, mit darunter zu begreifen. Und ich glaube auch, dass es dereinst gelingen wird, den gesammten Inhalt aller dieser mathematischen Disciplinen zu ‚arithmetisiren', d h. einzig und allein auf den im engsten Sinne genommenen Zahlbegriff zu gründen [...]". Translation from Ewald (1996, p. 949).

Gauß, Carl Friedrich (1801). *Disquisitiones arithmeticae*. Leipzig.
Gauß, Carl Friedrich (1986). *Disquisitiones Arithmeticae*. English Edition. Translated by A.A. Clarke. Revised by W.C. Waterhouse. New York et al.
Guillaumin, Jean-Yves (2020). *Dictionnaire de la terminologie latine ancienne de l'arithmétique et de la géométrie*. Paris.
Gundisalvo, Domingo (1954). *De scientiis*. Ed. by Manuel Alonso Alonso. Madrid.
Heath, Thomas L. (1921). *A History of Greek Mathematics*. Volume I. From Thales to Euclid. Oxford.
Heeffer, Albrecht (2022). "Arithmetic in the Renaissance". In: *Encyclopedia of Renaissance Philosophy*. Ed. by Marco Sgarbi. Cham, pp. 215–219.
Klein, Jacob (1992). *Greek Mathematical Thought and the Origin of Algebra*. Translated by Eva Brann. New York.
Kronecker, Leopold (1887). "Über den Zahlbegriff". *Journal für die reine und angewandte Mathematik* 101, pp. 337–355.
Legendre, Adrien-Marie (1797). *Essai sur la théorie des nombres*. Paris.
Levey, Martin and Marvin Petruck (1965). *Kushyār ibn Labbān: Principles of Hindu Reckoning*. Madison and Milwaukee.
Newton, Isaac (1707). *Arithmetica universalis*. Cambridge.
von Peuerbach, Georg (1511). *Institutiones in Arithmetricam*. Wien.
Plato (1925). *Statesman. Philebus. Ion*. Translated by Harold North Fowler. London.
Rashed, Roshdi, ed. (2017). *Lexique historique de la langue scientifique arabe*. Hildesheim.
Rebstock, Ulrich (1992). *Rechnen im islamischen Orient*. Die literarischen Spuren der praktischen Rechenkunst. Darmstadt.
Recorde, Robert (1543). *The Ground of Artes*. London.
Serre, Jean-Pierre (1970). *Cours d'arithmétique*. Paris.
Sesiano, Jacques, ed. (2014). *Liber Mahameleth*. Cham.
Smith, David Eugene (1958). *History of Mathematics*. Volume II. Special Topics of Elementary Mathematics. New York.
Spiesser, Maryvonne, ed. (2003). *Une arithmétique commerciale du XVe siècle*. Le *Compendy de la praticque des nombres* de Barthélemy de Romans. Turnhout.
Spiesser, Maryvonne (2017). "La naissance d'un genre, le traité d'arithmétique commerciale (XIVe-XVe siècles)". URL: `https://images.math.cnrs.fr/la-naissance-dun-genre-le-traite-darithmetique-commerciale-xive-xvie-s/` (retrieved Sept. 11, 2024).
Spiesser, Maryvonne (2022). "L'arithmétique et les marchands. Regards croisés entre commerce et mathématiques (XVe-XVIe siècles)". *Mémoires de l'Académie des sciences, inscriptions et belles-lettres de Toulouse* 183, pp. 43–56.
Strauß, Gerhard et al., eds. (1996). *Deutsches Fremdwörterbuch*. 2. Auflage. Band 2: Antinomie – Azur. Berlin and New York.
Tropfke, Johannes (1980). *Geschichte der Elementarmathematik*. Band 1: Arithmetik und Algebra. Vollständig neu bearbeitet von Kurt Vogel, Karin Reich, Helmuth Gericke. Berlin and New York.
Wallis, John (1656). *Arithmetica infinitorum*. In: *Operum Mathematicorum*. Pars altera. Oxford, pp. 1–198.
Wehrli, Fritz (1967). *Die Schule des Aristoteles*. Texte und Kommentar. Heft II. Aristoxenos. Basel and Stuttgart.
Zhmud, Leonid (2006). *The Origin of the History of Science in Classical Antiquity*. Translated from the Russian by Alexander Chernoglazov. Berlin and New York.

arithmetic **(mean, proportion, progression)**

Greek *ἀριθμητικός* was used by Archytas (as quoted by Porphyry) in his definition of *arithmetic mean*[1]: "There are three means in music: one is the arithmetic, the second geometric and the third subcontrary [...] The mean is arithmetic, whenever three terms are in proportion by exceeding one another in the following way: by that

[1] Archytas' terms for 'mean' are discussed by Huffman (2005, pp. 177f.).

which the first exceeds the second, by this the second exceeds the third" (Huffman 2005, p. 162).[2]

The term *arithmetic proportion* (*ἀναλογία ἀριθμητική*) was used by Aristotle (*Nicomachean Ethics* V.iii.13) and by Nicomachus in his *Introduction to Arithmetic* (II, 22) (Nicomachus of Gerasa 1866, p. 122), for example. This term appeared as *proportionalitas arithmetica* in Boethius' *Institutio arithmetica* (II, 42) (Boethius 1867, p. 139); he often used *medietas geometrica* 'geometric mean' interchangeably.

The first French use of the adjective *arithmétique* occurred, according to the TLF, in Oresme's translation of Aristotle's *Ethics* (Oresme 1940, p. 288); he wrote *proporcionalité arismetrique*. Recorde (1543, fol. 97ᵛ) appears to have been the first who used *arithmeticall* in his discussion of a *progression arithmeticall*. The origin of the corresponding Latin *progressio geometrica* is quite obscure; in print, it appeared in von Peuerbach (c1495, s.v. De progressione). This text was published posthumously and so the term must have been known much earlier; it was also recorded in Reisch (1503, fol. g iijʳ).

Note. For the history of the terminology, see also Cajori (1923, 1928).

References

Boethius, Anicius Manlius Severinus (1867). *De institutione arithmetica libri duo, De institutione musica libri quinque*. Accedit Geometria quae fertur Boetii. Ed. by Gottfried Friedlein. Leipzig.

Cajori, Florian (1923). "Origin of the names arithmetical and geometrical progression and proportion". *School Science and Mathematics* 22, pp. 734–737.

Cajori, Florian (1928). "Ciruelo on the Names 'Arithmetical' and 'Geometrical' Proportions and Progressions". *Isis* 10, pp. 363–366.

Huffman, Carl A. (2005). *Archytas of Tarentum*. Pythagorean, Philosopher and Mathematician King. Cambridge.

Nicomachus of Gerasa (1866). *Nicomachi Geraseni Pythagorei Introductionis Arithmeticae Libri II*. Ed. by Richard Hoche. Leipzig.

Oresme, Nicole (1940). *Le livre de éthiques d'Aristote*. Edited by Albert D. Menut. New York.

von Peuerbach, Georg (c1495). *Algorismus*. Wien.

Recorde, Robert (1543). *The Ground of Artes*. London.

Reisch, Gregor (1503). *Margarita philosophica*. Freiburg.

articulation point

An articulation point in a graph is a vertex that, if removed, disconnects the graph. Such a node is considered as a joint: "By its articulations the graph is – visually speaking – decomposed into its joints" (König 1936, p. 225).[1]

In this sense the term was introduced (in French) by Sainte-Laguë (1923, p. 27): "We call *articulation* each point A such that the points can be partitioned into two groups such that the passage from one point of one group to a point of the other group can only be made by passing through A".[2] The word *articulation* is here used in the literal meaning 'joint'; it was adopted, for example, by König (1932, pp. 156f.), who referred to Sainte-Laguë.

---

[2] Translation from Huffman (2005, p. 163). [1] "Durch seine Articulationen wird der Graph – anschaulich ausgedrückt – ‚in seine Glieder zerlegt'". [2] "On appelle *articulation* tout sommet A tel que l'on puisse répartir les sommets en deux groupes, le passage d'un sommet d'un groupe à un sommet de l'autre ne pouvant se faire qu'en passant par A". – The term reappears in Sainte-Laguë (1926, pp. 4f.) with a slightly rephrased definition.

There are various terminological alternatives. In particular, the term *cut vertex* was suggested by Whitney (1932, p. 342) and is still in common use. Harary and Norman (1953, p. 135) used both *articulation point* and *cutpoint*. The latter term originated in the study of curves. It was used by Ayres (1929, p. 596): "The point P of the connected set M is said to be a *non-cut point* or a *cut point* according as the set $M - P$ is connected or not". As pointed out by Ayres, the use of this term in this sense is due to Moore (1923, p. 101).[3]

See also *bridge*.

Note. The origin of the term is briefly discussed by König (1936, p. 225). For the topological concept, see Willard (1970, p. 313).

References

Ayres, W.L. (1929). "Concerning the Arc-Curves and Basic Sets of a Continuous Curve, Second Paper". *The American Mathematical Monthly* 31, pp. 595–612.

Harary, Frank and Robert Z. Norman (1953). "The Dissimilarity Characteristic of Husimi Trees". *Annals of Mathematics* 58, pp. 134–141.

König, Dénes (1932). "Über trennende Knotenpunkte in Graphen (nebst Anwendungen auf Determinanten und Matrizen)". *Acta scientiarum mathematicarum* 6, pp. 155–179.

König, Dénes (1936). *Theorie der endlichen und unendlichen Graphen*. Leipzig.

Menger, Karl (1932). *Kurventheorie*. Herausgegeben unter Mitarbeit von Georg Nöbeling. Berlin and Leipzig.

Moore, Robert Lee (1923). "Concerning the Cut-Points of Continuous Curves and of Other Closed and Connected Point-sets". *Proceedings of the National Academy of Sciences* 9, pp. 101–106.

Sainte-Laguë, André (1923). "Les réseaux". *Annales de la Faculté des sciences de Toulouse* 15, pp. 27–86.

Sainte-Laguë, André (1926). *Les réseaux (ou graphes)*. Paris.

Whitney, Hassler (1932). "Non-separable and planar graphs". *Transactions of the American Mathematical Society* 34, pp. 339–363.

Willard, Stephen (1970). *General Topology*. Reading, Mass.

associate

The origin of the term *socius* (lit. 'companion') was pointed out by Gauß (1801, § 77): "When relative to the modulus p the product of two numbers a, b is congruent to unity, we will call the numbers *associates*, as Euler did".[1]

Gauß does not give a precise reference to Euler; the term is found in Euler (1783, p. 134), a text written in 1772, and in Euler (1774, p. 93), in which he defined "associated residues" (*residua sociata*), which was also picked up by Gauß (1801, § 109).

In Gauß (1807, § 77), the translation *nombres associés* was used; Lejeune Dirichlet (1828, p. 390), referring to Euler and Gauß, preferred *nombres correspondans*. The English term *associate numbers* appears to be due to Talbot (1864, p. 47).

As reflected in van der Waerden (1930, p. 63), the concept was later generalized and used in connection with rings: "Sometimes, two quantities like a and $b = a\varepsilon^{-1}$

[3] Apart from *cut point* and *cut vertex*, König (1936, p. 225) also mentions the German term *Zerlegungspunkt* (lit. 'decomposition point') as a synonym, without giving a reference. The rare term was used by Menger (1932, p. 153). [1] "Quando secundum modulum p, productum numereorum a, b unitati est congruum, numeros a, b cum ill. Euler, *socios* vocemus". Translation from Gauß (1986, § 77), slightly modified.

that differ only by a unit as a factor, are called 'associate quantities'".[2] In this sense, but only with respect to algebraic integers, the terminology was suggested by Dedekind (1877a, p. 63, 1877b, p. 8).

Note. For the history of the concept, see Dickson (1952, pp. 64–66, 73).

References

Dedekind, Richard (1877a). *Sur la théorie des nombres entiers algébriques*. Paris.
Dedekind, Richard (1877b). *Über die Anzahl der Ideal-Classen in den verschiedenen Ordnungen eines endlichen Körpers*. Braunschweig.
Dickson, Leonard Eugene (1952). *History of the Theory of Numbers*. Volume I. Divisibility and primality. New York.
Euler, Leonhard (1774). "Demonstrationes circa residua ex divisione potestatum per numeros primos resultantia". *Novi Commentarii academiae scientiarum Petropolitanae* 18, pp. 85–135.
Euler, Leonhard (1783). "Disquisitio accuratior circa residua ex divisione quadratorum altiorumque potestatum per numeros primos relicta". In: *Opuscula Analytica*. Tomus secundus. Petersburg, pp. 121–156.
Gauß, Carl Friedrich (1801). *Disquisitiones arithmeticae*. Leipzig.
Gauß, Carl Friedrich (1807). *Recherches arithmétiques*. Traduites par A.-C.-M. Poullet-Delisle. Paris.
Gauß, Carl Friedrich (1986). *Disquisitiones Arithmeticae*. English Edition. Translated by A.A. Clarke. Revised by W.C. Waterhouse. New York et al.
Lejeune Dirichlet, Gustav (1828). "Démonstrations nouvelles de quelques théorèmes relatifs aux nombres". *Journal für die reine und angewandte Mathematik* 3, pp. 390–393.
Talbot, H.F. (1864). "On the Theory of Numbers". *Transactions of the Royal Society of Edinburgh* 23, pp. 45–52.
van der Waerden, Bartel Leendert (1930). *Moderne Algebra*. Erster Teil. Berlin.

associative

The word was first used[1] in mathematics by Hamilton (1844, pp. 429f.) in his discussion of quaternions. Hamilton first discusses distributivity, then adds that the multiplication of quaternions is not commutative and finally points out that "it will be found that another important property of the old multiplication is preserved, or extended to the new, namely, that which may be called the *associative* character of the operation". The idea behind the terminology appears to be that, no matter how the factors "associate" in the multiplication, the result remains the same. Hamilton symbolically wrote $\text{Q} \cdot \text{Q}'\text{Q}'' = \text{QQ}' \cdot \text{Q}''$.

The German term was first used by Möbius (1859, p. 140) and popularized by Hankel (1867, p. 3), who both referred to Hamilton for the term. Earlier, Graßmann (1844, pp. 3–4) paraphrased the notions of associativity and commutativity, but used the term *einfach* 'simple' for compositions that have both properties. The French term appears in Hamilton (1853, p. 283).

Note. Crowe (1985, pp. 15f.) gives a concise overview of the history of the terminology. See also Miller (n.d.) for an additional statement on Hamilton's paper.

References

Crowe, Michael J. (1985). *A History of Vector Analysis*. The Evolution of the Idea of a Vectorial System. New York.

[2] "Man nennt bisweilen zwei Größen wie a und $b = a\varepsilon^{-1}$, die sich nur um eine Einheit als Faktor unterscheiden, ‚assoziierte Größen'". [1] See, e. g., Crowe (1985, p. 16) and the OED.

Graßmann, Hermann (1844). *Die Wissenschaft der extensiven Grösse, oder die Ausdehnungslehre, eine neue mathematische Disziplin.* Erster Theil, die lineale Ausdehnungslehre enthaltend. Leipzig.
Hamilton, William Rowan (1844). "On a new Species of Imaginary Quantities connected with a theory of Quaternions". *Proceedings of the Royal Irish Academy* 2, pp. 424–434.
Hamilton, William Rowan (1853). "Sur les quaternions". *Nouvelles annales de mathématiques* 12, pp. 275–283.
Hankel, Hermann (1867). *Vorlesungen über die complexen Zahlen und ihre Functionen.* In zwei Theilen. I. Theil: Theorie der complexen Zahlsysteme. Leipzig.
Miller, Jeff (n.d.). "Earliest Known Uses of Some of the Words of Mathematics". URL: `https://mathshistory.st-andrews.ac.uk/Miller/mathword/`.
Möbius, August Ferdinand (1859). "Neuer Beweis des in Hamiltons Lect. on Quaternions aufgestellten associativen Princips bei Zusammensetzung von Bögen grösster Kreise einer Kugelfläche". *Berichte über die Verhandlungen der Königlich Sächsischen Gesellschaft der Wissenschaften zu Leipzig (Mathematisch-Physische Klasse)* 11, pp. 138–149.

associator

This algebraic terminology came up in connection with the *alternative law* of multiplication, a weakened form of the associative law: $(aa)b = a(ab)$, $(ab)a = a(ba)$, $(ba)a = b(aa)$ (Zorn 1930, p. 125). See *alternative* (algebraic structure).

The term *associator* appears to be due to Zorn (1941). At the beginning of his paper, Zorn explains that his account "is phrased almost entirely in terms of 'associators' $(a, b, c) = (ab)c - a(bc)$" (Zorn 1941, p. 676). The term was soon adopted and is found in various publications on alternative algebraic structures such as Bruck (1946, p. 250).

References

Bruck, R.H. (1946). "Contributions to the Theory of Loops". *Transactions of the American Mathematical Society* 60, pp. 245–354.
Zorn, Max (1930). "Theorie der alternativen Ringe". *Abhandlungen aus dem Mathematischen Seminar der Universität Hamburg* 8, pp. 123–147.
Zorn, Max (1941). "Alternative Rings and Related Questions I: Existence of the Radical". *Annals of Mathematics* 43, pp. 676–686.

astroid

Like the astronomical *asteroid*, this term etymologically means 'having the shape of a star' (from Greek *ἀστήρ* 'star'). The term was suggested by Littrow (1838, p. 229). He originally used *Astrois* and explained the term as meaning 'star-shaped curve'.[1] Littrow also used the ending *-is* with other curves (e. g., *Cyclois* instead of *Cykloide*).

The German term did not find much acceptance in this form, and it seems to have been Jentzsch (1860) who suggested *Astroide*. In French, the term *astroïde* is found in Amstein (1877, p. 188), and the earliest English attestation seems to be Lamb (1897, p. 359), who says that "[t]his curve [which Lamb discusses under the name *four-cusped hypocycloid*] is sometimes called the 'astroid'".

Apart from the alternative *four-cusped hypocycloid*, which already appeared as *hypocycloid with four cusps* in Peacock (1820, p. 199), there were a number of competitors.[2] Bellavitis (1854, p. 82) used Italian *tetracuspide* on account of its four

[1] "sternförmige Curve". [2] See Loria (1910, pp. 265–267).

cusps, while Matthiessen (1868, p. 232) suggested German *Paracykel* and Montucci (1865, p. 441) French *cubo-cycloïde*.

Note. For the history of the concept and terminology, see Loria (1910, pp. 264–273) and Teixeira (1908, pp. 332f.).

References

Amstein, Hermann (1877). "Un exemple de représentation conforme". *Bulletin de la Société Vaudoise des Sciences Naturelles* 15, pp. 175–198.

Bellavitis, Giusto (1854). *Sposizione del metodo delle equipollenze*. Modena.

Jentzsch, Hermann (1860). "Theorie der Astroiden, einer neuen Classe von Curven". PhD thesis. Greifswald.

Lamb, Horace (1897). *An Elementary Course on the Infinitesimal Calculus*. Cambridge.

Littrow, Joseph Johann (1838). *Anfangsgründe der gesammten Mathematik*. Wien.

Loria, Gino (1910). *Spezielle algebraische und transzendente ebene Kurven*. Theorie und Geschichte. Autorisierte, nach dem italienischen Manuskript bearbeitete deutsche Ausgabe von Fritz Schütte. Zweite Auflage. Erster Band. Leipzig.

Matthiessen, Ludwig (1868). "Ueber die mechanische Construction einiger Curven, welche sich zur Auflösung des Problems von der Duplication des Würfels verwenden lassen". *Archiv der Mathematik und Physik* 48, pp. 229–235.

Montucci, Henry (1865). "Mémoire sur la résolution numérique des équations du cinquième degré, et de quelques autres équations". *Comptes rendus de l'Académie des sciences* 70, pp. 440–441.

Peacock, George (1820). *A Collection of Examples of the Applications of the Differential and Integral Calculus*. Cambridge.

Teixeira, Francisco Gomes (1908). *Traité des courbes spéciales remarquables plans et gauches*. Tome I. Coïmbra.

asymptote

Greek *ἀσύμπτωτος* literally means 'not capable of meeting'. According to Tropfke (1924, p. 145), the word reached terminological status with Apollonius. In book 2, prop. 1 of his *Conics*, Apollonius states: "If a straight line touch an hyperbola at its vertex, and from it on both sides of the diameter a straight line is cut off equal to the straight line equal in square to the fourth of the figure, then the straight lines drawn from the center of the section to the ends thus taken on the tangent will not meet the section" (Apollonius of Perga 1891, p. 192).[1] At the end of the proof, Apollonius uses *ἀσύμπτωτοι* for these asymptotes of the hyperbola.[2] The word was also used in a wider meaning, namely for any lines that do not meet; this use is also found in Apollonius (book 2, prop. 14) and Proclus: "Of these [lines] some are asymptotic, namely, those which however far extended never meet, and others that do intersect are symptotic. Of asymptotic lines some are in the same plane with one another, others not; and of the asymptotes that are in the same plane, some are always equidistant from one another, others are constantly diminishing the distance between themselves and their straight lines, like the hyperbola and the conchoid" (Proclus 1873, p. 177).[3]

In Latin, the term is found in the 13th century. At this time, William of Moerbeke translated various texts by Archimedes, in which he used the transliterated form *asymptota* (Clagett 1976, e. g. p. 244).[4] In some later texts, asymptotes were also

[1] Translation from Apollonius of Perga (2013, p. 117). [2] Archimedes did not use the term yet. He called the asymptotes the *nearest lines to the section of the obtuse-angled cone* (Archimedes 1880, p. 276); translation from Archimedes (1897, p. 100). [3] Translation from Proclus (1970, p. 113).
[4] According to Clagett (1976, p. 35), the transliteration presumably first occurs with Moerbeke.

called *non coincidentes*, for example in Valla (1501, book X, ch. lix).[5] However, with the translation of Archimedes's works by Commandinus (Apollonius of Perga 1566, fol. 43^{v}), *asymptote* gradually became a standard term. Commandinus remarks that, in Greek, the expression *ἀσυμπτώτους τῃ τομῃ* (asymptotes to the section) is used but that there is also the shorter form *ἀσυμπτώτους*, and asks for permission to take over this Greek word into Latin (Apollonius of Perga 1566, fol. 44^{r}).[6]

The French term is found in Mersenne (1624, p. 169), while the English term is first found in Hobbes (1656, p. 145), where *asymptotes* are explained as lines that "will alwayes come neerer and neerer together, but never touch one another". The German term is found in the early 18th century in Wolff (1710, p. 146), who also included the term in his mathematical dictionary (Wolff 1716, p. 213). An earlier attempt to find German terms is found in Sturm (1670, p. 204), who suggests *Unberührende*, lit. 'non-touching (line)', and later also calls asymptotes *niemals zusammenkommend*, lit. 'never coming together', because, as he says, even though they come closer and closer to the hyperbolic line they never touch it (Sturm 1670, p. 317).

Note. For the Greek term, see Mugler (1958-59, p. 86), Tropfke (1924, p. 145), and Heath in Apollonius of Perga (1896, pp. clxiii f.). For the Arabic, see Rashed (2017, pp. 246–248).

References

Apollonius of Perga (1566). *Conicorum libri quattuor*. Ed. by Federico Commandino. Bologna.

Apollonius of Perga (1891). *Apollonii Pergaei quae graece exstant cum commentariis antiquis*. Vol. I. Ed. by J.L. Heiberg. Leipzig.

Apollonius of Perga (1896). *Treatise on Conic Sections*. Edited in Modern Notation ... by T. L. Heath. Cambridge.

Apollonius of Perga (2013). *Conics*. Books I-IV. Books I-III translated by R. Catesby Taliaferro. Book IV translated by Michael N. Fried. Santa Fe.

Archimedes (1880). *Opera omnia*. Cum commentariis Eutocii. Volumen I. Ed. by J.L. Heiberg. Leipzig.

Archimedes (1897). *The Works of Archimedes*. Edited in Modern Notation with Introductory Chapters by T. L. Heath. Cambridge.

Clagett, Marshall (1976). *Archimedes in the Middle Ages*. Volume II. The Translations from the Greek by William of Moerbeke. Philadelphia.

Clagett, Marshall (1978). *Archimedes in the Middle Ages*. Volume III. The Fate of the Medieval Archimedes 1300-1565. Philadelphia.

Clagett, Marshall (1980). *Archimedes in the Middle Ages*. Volume IV. A Supplement on the Medieval Latin Traditions of Conic Sections (1150-1566). Philadelphia.

Hobbes, Thomas (1656). *Elements of Philosophy*. London.

Mersenne, Marin (1624). *L'Impiété des Deistes*. Seconde partie. Paris.

Mugler, Charles (1958-59). *Dictionnaire historique de la terminologie géométrique des Grecs*. Paris.

Proclus (1873). *In primum Euclidis elementorum librum commentarii*. Ed. by Gottfried Friedlein. Leipzig.

Proclus (1970). *Commentary on the First Book of Euclid's Elements*. Translated with Introduction and Notes by Glenn R. Morrow. Princeton.

Rashed, Roshdi, ed. (2017). *Lexique historique de la langue scientifique arabe*. Hildesheim.

Sturm, Johann Christoph (1670). *Des unvergleichlichen Archimedis Kunst-Bücher*. Nürnberg.

---

[5] For further writers who used this term, see Clagett (1978, p. 812) and Clagett (1980, p. 247).

[6] "Has Graeci *ἀσυμπτώτους τῃ τομῃ* vel simpliciter *ασυμπτώτους* appellant. quare nobis deinceps, ut uno verbo dicamus, graeca voce uti liceat".

Tropfke, Johannes (1924). *Geschichte der Elementar-Mathematik.* Sechster Band. Analysis. Analytische Geometrie. Berlin and Leipzig.
Valla, Giorgio (1501). *De expetendis, et fugiendis rebus opus.* Venezia.
Wolff, Christian (1710). *Der Anfangs-Gründe aller Mathematischen Wissenschaften Letzter Theil.* Halle.
Wolff, Christian (1716). *Mathematisches Lexicon.* Leipzig.

asymptotic convergence

See *convergence in measure.*

asymptotic series

Asymptotic series were so named by Poincaré (1886). In the first paragraph of the paper, entitled *séries asymptotiques*, he began by noting that "all mathematicans know the remarkable properties of Stirling's Series. This series $\log \Gamma(x+1) = \frac{1}{2}\log(2\pi) + (x+\frac{1}{2})\log x - x + \frac{B_1}{1\cdot 2}\frac{1}{x} - \frac{B_2}{3\cdot 4}\frac{1}{x^2} + \frac{B_3}{5\cdot 6}\frac{1}{x^3} - \cdots$ is always divergent. Nevertheless, it can legitimately be used for very large values of x" (Poincaré 1886, p. 295).[1] Poincaré then explained that the series "asymptotically represents" ("représente asymptotiquement") the function $\log \Gamma(x+1)$, "i. e., if S_n is the sum of the first terms of this series up to and including the term $\pm\frac{B_n}{2n(2n-1)}\frac{1}{x^n}$, the expression $x^{n+1}[\log \Gamma(x+1) - S_n]$ tends to 0 when x increases indefinitely" (ibid.).[2]

This prepared him for the definition of an asymptotic series:

> I say that a divergent series
>
> $$A_0 + \frac{A_1}{x} + \frac{A_2}{x^2} + \cdots \frac{A_n}{x^n} + \cdots$$
>
> where the sum of the first $n+1$ terms is S_n, asymptotically represents a function $J(x)$ if the expression $x^n(J - S_n)$ tends to 0 when x increases indefinitely (Poincaré 1886, p. 296).[3]

It did not take long for the terminology to be adopted; for example, Schlesinger (1895, p. 424) included a section on asymptotic representations in his handbook of differential equations. The English terminology is found in Maclaurin (1899, p. 66).

There was one competing term: at the same time as Poincaré, Stieltjes (1886, p. 202) also studied asymptotic series but called them *semi-convergent* (*séries semi-convergentes*). His term was not new: Legendre (1810, p. 466) already used *demi-convergente.* Knopp (1956, p. 536) explains that the term "emphasizes the fact that so far as numerical purposes are concerned they behave almost like convergent series" but that "it has not held its ground [...] The reason is that it clashes with the terminology which is customary, particularly in France, whereby our conditionally convergent series are called *semi-convergent*". Borel (1901, p. 26) and Mansion

---

[1] "Tous les géomètres connaissent les curieuses propriétés de la série de STIRLING. Cette serie $\log \Gamma(x+1) = \frac{1}{2}\log(2\pi) + (x+\frac{1}{2})\log x - x + \frac{B_1}{1\cdot 2}\frac{1}{x} - \frac{B_2}{3\cdot 4}\frac{1}{x^2} + \frac{B_3}{5\cdot 6}\frac{1}{x^3} - \cdots$ est toujours divergente. Cependant on peut en faire légitimement usage pour les valeurs très grandes de x". – B_n denotes the nth Bernoulli number, Γ the gamma function. [2] "c'est à dire que si S_n est la somme des premiers termes de cette série jusques et y compris le terme: $\pm\frac{B_n}{2n(2n-1)}\frac{1}{x^n}$, l'expression $x^{n+1}[\log \Gamma(x+1) - S_n]$ tendra vers 0 quand x croîtra indéfiniment". [3] "Je dirai qu'une série divergente $A_0 + \frac{A_1}{x} + \frac{A_2}{x^2} + \cdots \frac{A_n}{x^n} + \cdots$ où la somme des $n+1$ premiers termes est S_n, représente asymptotiquement une fonction $J(x)$ si l'expression $x^n(J - S_n)$ tend vers 0 quand x croît indéfiniment". Translation from Roy (2011, p. 480).

(1901, p. 338), for example, explicitly pointed out that indeed this was why they preferred Poincaré's term.

Note. For the history of the concept, see Kline (1972, pp. 1103–1109), Ferraro (2008, passim), and Roy (2011, ch. 24).

References

Borel, Émile (1901). *Leçons sur les séries divergentes*. Paris.

Ferraro, Giovanni (2008). *The Rise and Development of the Theory of Series up to the Early 1820s*. New York.

Kline, Morris (1972). *Mathematical Thought From Ancient To Modern Times*. Volume 3. New York and Oxford.

Knopp, Konrad (1956). *Infinite Sequences and Series*. Translated by Frederick Bagemihl. New York.

Legendre, Adrien-Marie (1810). "Recherches sur diverses sortes d'intégrales définies". *Mémoires de la classe des sciences mathématiques et physiques de l'Institut de France* (1809), pp. 416–509.

Maclaurin, R.C. (1899). "On the solutions of the equation $(V^2 + \kappa^2)\psi = 0$ in elliptic coordinates and their physical applications". *Transactions of the Cambridge Philosophical Society* 17, pp. 41–108.

Mansion, Paul (1901). "Sur quelques désignations relatives aux séries". *L'enseignement mathématique* 3, pp. 333–338.

Poincaré, Henri (1886). "Sur l'équilibre d'une masse fluide animée d'un mouvement de rotation". *Acta Mathematica* 8, pp. 295–344.

Roy, Ranjan (2011). *Sources in the Development of Mathematics*. Infinite Series and Products from the Fifteenth to the Twenty-first Century. Cambridge.

Schlesinger, Ludwig (1895). *Handbuch der Theorie der linearen Differentialgleichungen*. Erster Band. Leipzig.

Stieltjes, Thomas Joannes (1886). "Recherches sur quelques séries semi-convergentes". *Annales scientifiques de l'École Normale Supérieure* 3, pp. 201–258.

atlas and *chart*

Nowadays, manifolds are defined in terms of an *atlas*, which consists of *charts*. In this sense, the French terms *atlas* and *carte locale* were first used by Ehresmann (1943, p. 628). An early adopter was Reeb (1947, p. 1613).

Note. The origin of the terms has been pointed out by Volkert (2002, p. 97), for example.

References

Ehresmann, Charles (1943). "Sur les espaces fibrés différentiables". *Comptes rendus de l'Académie des sciences* 216, pp. 628–630.

Reeb, Georges (1947). "Variétés feuilletées, feuilles voisines". *Comptes rendus de l'Académie des sciences* 224, pp. 1613–1614.

Volkert, Klaus (2002). *Das Homöomorphismusproblem insbesondere der 3-Mannigfaltigkeiten, in der Topologie 1892-1935*. Paris.

automorphic

This term derives from Greek *αυτός* 'self' and *μορφή* 'form, shape'.

In 1858 Arthur Cayley used the term *automorphic transformation* in the context of invariant theory: "The question of the automorphic linear transformation of the function $x^2 + y^2 + z^2$, that is the transformation by linear substitutions, of this function into a function $x^2 + y^2 + z^2$ of the same form" (Cayley 1859, p. 39). The term has never been widely used, but it was adopted, for example, by Smith (1862, p. 294).

The term *automorphic function* (German *automorphe Funktion*) was introduced by Klein (1890, p. 94). Such functions can informally be understood as "functions [...] that remain invariable under certain linear substitutions of the independent variables" (Klein 1923, p. 577).[1] In the same paper, Klein (1890, p. 95) also introduced the term *automorphic form* (*automorphe Form*).

Klein's terminology was soon adopted and literally translated into other languages; for example, English *automorphic function* was already used by Burnside (1892, p. 52) and Forsyth (1893, p. 619).

The competing terminology *fonction fuchsienne* (for a large class of automorphic functions) was suggested by Poincaré (1881, p. 333) in hommage to Lazarus Fuchs.[2] Klein strongly opposed it, both in correspondence with Poincaré (Petiau 1956, p. 33) and also in print, first in a note appended to Poincaré (1882, p. 564), arguing that it was Riemann and Schwarz who provided the basic work in this field. Eventually, Poincaré admitted that Klein was right in his reservations; on June 27, 1881, he wrote: "I would have chosen a different name [for the functions] had I known of Schwarz's work, but I only knew of it from your letter after the publication of my results" and also added that it would now be too late to change this because that would offend Fuchs (Petiau 1956, p. 36).[1]

The algebraic sense is discussed under *automorphism*.

Note. The history of the concept is discussed by Lehner (1964, pp. 19–27), Kline (1972, pp. 726–730), and Gray (1999Gray1999history). Klein's own account of the prehistory of automorphic functions (Klein 1923) and Fricke (1913) are still worth reading. The dispute between Klein and Poincaré about the appropriate term has been discussed by Rowe (1992) and Gray (2013, pp. 224–246), for example. The correspondence itself has been edited by Petiau (1956). See also Brüning (1997), who discusses private notes by Klein on the topic.

References

Brüning, Reinhart (1997). "The formation of the theory of automorphic functions: Felix Klein's notes about Henri Poincaré's early publications". *Philosophia Scientiae* 2, pp. 77–89.

Burnside, William Snow (1892). "On a Class of Automorphic Functions". *Proceedings of the London Mathematical Society* 23, pp. 49–88.

Cayley, Arthur (1859). "A Memoir on the Automorphic Linear Transformation of a Bipartite Quadric Function". *Philosophical Transactions of the Royal Society of London* 148, pp. 39–46.

Forsyth, Andrew Russell (1893). *Theory of Functions of a Complex Variable*. Cambridge.

Fricke, Robert (1913). "Automorphe Funktionen mit Einschluß der elliptischen Modulfunktionen". *Encyklopädie der mathematischen Wissenschaften mit Einschluss ihrer Anwendungen* II B 4, pp. 349–470.

Gray, Jeremy (2013). *Henri Poincaré*. A Scientific Biography. Princeton.

Gray, Jeremy (1999Gray1999history). "A brief history of automorphic function theory, 1880–1930". In: *Non-Euclidean Geometry in the Theory of Automorphic Functions*. Jacques Hadamard. Ed. by Jeremy Gray and Abe Shenitzer. Providence, R.I.

---

[1] "Funktionen [...] welche bei gewissen linearen Substitutionen der unabhängigen Variablen ungeändert bleiben". [2] Poincaré had earlier asked Fuchs for permission to use this term; see Petiau (1956, p. 18). [1] "En ce qui concerne M. Fuchs et la dénomination de fonctions fuchsiennes, il est clair que j'aurais pris une autre dénomination si j'avais connu le travail de M. Schwarz". Translation from Gray (2013, p. 229).

Klein, Felix (1890). "Zur Theorie der allgemeinen Laméschen Functionen". *Nachrichten von der Königl. Gesellschaft der Wissenschaften und der Georg-Augusts-Universität zu Göttingen* (1890), pp. 85–95.
Klein, Felix (1923). "Zur Vorgeschichte der automorphen Funktionen". Dritter Band. In: *Gesammelte mathematische Abhandlungen*. Ed. by Robert Fricke, Hermann Vermeil, and Erich Bessel-Hagen. Berlin, pp. 577–586.
Kline, Morris (1972). *Mathematical Thought From Ancient To Modern Times*. Volume 2. New York and Oxford.
Lehner, Joseph (1964). *Discontinuous Groups and Automorphic Functions*. Providence, R.I.
Petiau, Gérard, ed. (1956). *Œuvres de Henri Poincaré*. Tome XI. Mémoires divers – Hommages à Henri Poincaré. Livre du Centenaire de la naissance de Henri Poincaré. Paris.
Poincaré, Henri (1881). "Sur les fonctions fuchsiennes". *Comptes rendus de l'Académie des sciences* 92, pp. 333–335.
Poincaré, Henri (1882). "Sur les Fonctions Uniformes qui se reproduisent par des Substitutions Linéaires". *Mathematische Annalen* 19, pp. 553–564.
Rowe, David E. (1992). "Klein, Mittag-Leffler, and the Klein-Poincaré Correspondence of 1881 – 1882". In: *Amphora*. Festschrift for Hans Wussing on the Occasion of his 65th Birthday. Ed. by Sergej S. Demidov et al. Basel, pp. 597–618.
Smith, H.J.S. (1862). "Report on the Theory of Numbers. Part III". *Report of the British Association for the Advancement of Science* (1861), pp. 292–340.

automorphism

Like the term *automorphic*, this word is essentially made up of the Greek words for 'same' and 'form'. As an algebraical term, *automorphism* was coined on the analogy to *homomorphism* and *isomorphism* by Frobenius (1901, p. 1324):

> Let $\mathfrak{O} = Q_0 + Q_1 + \ldots + Q_{m-1}$ be a group of order m that is permutable with the element R, and let $R^{-1}Q_\mu R = Q'_\mu$, then also $\mathfrak{O} = Q'_0 + Q'_1 + \ldots + Q'_{m-1}$. The *isomorphism of $\mathfrak{O}$ into itself* obtained in this way, or *automorphism of* $\mathfrak{O}$ for short, I designate by (R), and I say that it is *effected* by the element R.[1]

The expression *isomorphism in(to) itself* used in this definition had earlier been used by Hölder (1893, p. 313), for example. Note that at that time an "isomorphism" was not generally required to be bijective, so that *isomorphism* often means 'homomorphism' (see *isomorphism*).

In English, the term was used by Miller (1902, p. 119): "A simple isomorphism of a group G with itself has been called a holomorphism[2] or an automorphism of G"; he refers to Frobenius (1901, p. 1324) for the term. French *automorphisme* first appears in de Séguier (1904, p. 100).

Note. Brief historical notes on the term and concept can be found in Loewy (1910, p. 192) and Magnus (1939, p. 14).

References

Frobenius, Georg (1901). "Über auflösbare Gruppen. V". *Sitzungsberichte der Königlich Preußischen Akademie der Wissenschaften zu Berlin* (1901), pp. 1324–1329.

---

[1] "Ist $\mathfrak{O} = Q_0 + Q_1 + \ldots + Q_{m-1}$ eine Gruppe der Ordnung m, die mit den Elementen R vertauschbar ist, und ist $R^{-1}Q_\mu R = Q'_\mu$, so ist auch $\mathfrak{O} = Q'_0 + Q'_1 + \ldots + Q'_{m-1}$. Den so erhaltenen *Isomorphismus von $\mathfrak{O}$ in sich* oder kurz *Automorphismus* von $\mathfrak{O}$ bezeichne ich mit (R), und ich sage, er werde durch das Element R *bewirkt*". [2] For the term *holomorphism*, Miller refers to Miller (1900-01, p. 78), where the term is found but no definition is given. It seems the term is Miller's creation; Loewy (1910, p. 192), in particular, attributed it to him, citing Miller (1902).

Hölder, Otto (1893). "Die Gruppen der Ordnungen p^3, pq^2, pqr, p^4". *Mathematische Annalen* 43, pp. 301–412.

Loewy, Alfred (1910). "Algebraische Gruppentheorie". In: *Repertorium der höheren Mathematik*. Erster Band. Analysis. Ed. by Paul Epstein and Heinrich Emil Timerding. Leipzig and Berlin, pp. 168–249.

Magnus, Wilhelm (1939). "Allgemeine Gruppentheorie". *Enzyklopädie der mathematischen Wissenschaften mit Einschluss ihrer Anwendungen.* Zweite, völlig neubearbeitete Auflage I 1,9, pp. 1–51.

Miller, George Abram (1900-01). "On a Special Class of Abelian Groups". *Annals of Mathematics* 2, pp. 77–80.

Miller, George Abram (1902). "Second report on recent progress in the theory of groups of finite order". *Bulletin of the American Mathematical Society* 9, pp. 106–123.

de Séguier, Jean Armand (1904). *Théorie des groupes finis*. Éléments de la théorie des groupes abstraits. Paris.

axis

In classical Greek, *ἄξων* had the concrete sense 'axis of a chariot'. As documented by Mugler (1958-59, pp. 68f.), the term was already used by Aristotle in connection with cones, and by Autolycus in connection with spheres; Apollonius used the term for the axis of a conic section and Euclid gave definitions for the axis of a sphere (*Elements*, book XI, def. 15), the axis of a cylinder and of a cone of revolution (XI, def. 22, def. 19).

In Arabic texts, the terms *miḥwar* 'axle, axis' and *sahm* are used (the latter means 'arrow' and also translates the trigonometrical term *sagitta*). Rashed (2017, pp. 212–215) cites Hunain ibn Ishāq, who used *miḥwar* in his translation of Euclid's def. 22, for example. al-Bīrūnī (1934, p. 19), on the other hand, used *sahm* for the axis of a cone.

Latin *axis* is found, with reference to the axis of the earth, in classical Latin (LS). In the 12th- and 13th-century translations of the *Elements*, the term is used on a regular basis (see, e. g., Busard 1983, p. 300 and Busard 1987, p. 310). In some texts, particularly in some translations attributed to Gerard of Cremona, Arabic *miḥwar* appears as *meguar*; see, for example, Anaritius (1899, p. 7).[1]

In English, the geometrical term appears in the 14th century (MED); it was used by John Trevisa in his translation of *De proprietatibus rerum* by Bartholomaeus Anglicus. The first printed version of this translation has: "Axis is a certayne lyne vnderstonde that stretchyth streyghte by the myddell of a balle. Other of an other thynge fro one pole to a nother" (Bartholomaeus Anglicus 1495, book VIII, ch. vi). Recorde (1551, fol. Bij[r]) uses *axelyne* or *axtre*, Billingsley uses *axe* (Euclid 1570, fol. 316[r]), while L. Digges and T. Digges (1571, fol. Qij[v]) prefer *axis*. French *axe* is also used in a 14th-century translation of the *De proprietatibus rerum* (Bartholomaeus Anglicus 1482, book VIII, ch. vi), and the German term *Achse* is also found during that time (von Megenberg 1980, p. 6).[2]

Note. For the Greek term, see Mugler (1958-59, pp. 68f.) and Kierboe (1913-14); for the Arabic, see Rashed (2017, 212–215, pp. 389f.).

---

[1] See also Bjørnbo (1905, pp. 245f.) and Kunitzsch (1977, pp. 18f.) for further references. [2] The spelling there is *ahs*.

References

Anaritius (1899). *Anaritii in decem libros priores Elementorum Euclidis commentarii.* Ex interpretatione Gherardi Cremonensis. Ed. by Maximilian Curtze. Leipzig.
Bartholomaeus Anglicus (1482). *Le Proprietaire des choses*. Ed. by Pierre Farget. Trans. by Jean Corbichon. Lyon.
Bartholomaeus Anglicus (1495). *De proprietatibus rerum*. Trans. by John Trevisa. Westminster.
al-Bīrūnī, Abū Rayḥān Muḥammad ibn Aḥmad (1934). *The Book of Instruction in the Elements of the Art of Astrology*. Written in Ghaznah, 1029 A. D. Reproduced from Brit. Mus. MS. Or. 8349. The Translation facing the Text by R. Ramsay Wright. London.
Bjørnbo, Axel Anthon (1905). "Gerhard von Cremonas Übersetzung von Alkwarizmis Algebra und von Euklids Elementen". *Bibliotheca Mathematica* 6, pp. 239–248.
Busard, H.L.L. (1983). *The First Latin Translation of Euclid's Elements Commonly Ascribed to Adelard of Bath.* Books I-VIII and books X.36-XV.2. Toronto.
Busard, H.L.L. (1987). *The Mediaeval Latin Translation of Euclid's Elements Made Directly from the Greek.* Stuttgart.
Digges, Leonard and Thomas Digges (1571). *A geometrical practise, named Pantometria diuided into three bookes, Longimetra, Planimetra, and Stereometria.* London.
Euclid (1570). *The Elements of Geometrie of the most auncient Philosopher Euclide of Megara.* Faithfully (now first) translated into the Englishe toung, by H. Billingsley. London.
Kierboe, T. (1913-14). "Bemerkungen über die Terminologie des Archimedes". *Bibliotheca Mathematica* 14, pp. 33–40.
Kunitzsch, Paul (1977). *Mittelalterliche astronomisch-astrologische Glossare mit arabischen Fachausdrücken.* München.
von Megenberg, Konrad (1980). *Die deutsche Sphaera.* Ed. by Francis B. Brévart. Tübingen.
Mugler, Charles (1958-59). *Dictionnaire historique de la terminologie géométrique des Grecs.* Paris.
Rashed, Roshdi, ed. (2017). *Lexique historique de la langue scientifique arabe.* Hildesheim.
Recorde, Robert (1551). *The patheway to knowledg.* London.

axis of symmetry

The systematic study of symmetry is a 19th-century development. However, appeals to an axis of symmetry can be found earlier in the 18th century.

Mauduit (1757, p. 53), for example, states that "the parabola [is] a symmetric curve divided into two equal halves by the axis";[1] an analogous remark on the hyperbola is found in Mauduit (1757, p. 266). With reference to an ellipse, Sauri (1774, p. 42) comments: "This symmetry means that the parts m of the curve are situated on one side of the axis in the same way as the corresponding and opposite parts M are situated on the other side of the same axis".[2]

A spatial example dates back to 1776, when Meusnier presented a paper on the curvature of surfaces. In this paper, published in 1785, Meusnier (1785, pp. 482f.) discussed an elliptic paraboloid and called it *symétrique* with respect to an axis

---

[1] "la parabole étant une courbe simétrique & divisée en deux parties égales par l'axe". – The axis of a curve was defined by d'Alembert (1751, p. 905) without using the term *symmetry*: "the *axis* of a curve is, in general, a line drawn in the plane of this curve which divides the curve into two equal, similar, and similarly placed parts on either side of this line" ("l'*axe* d'une courbe est en général une ligne tirée dans le plan de cette courbe, & qui divise la courbe en deux parties égales, semblables, & semblablement posées de part & d'autre de cette ligne"). Translation from Hon and Goldstein (2008, p. 194). d'Alembert's definition is essentially the same as that given by Euler (1748, § 341), who used the name *orthogonal diameter* (*diameter orthogonalis*), however. [2] "Cette symmétrie fait que les parties m de la courbe sont situées d'un côté de l'axe de la même maniere que les parties correspondantes & opposées M le sont de l'autre côté du même axe".

(*axe*). Earlier, Bézout (1767, p. 416) had already discussed another solid that was "symmetric with respect to one of its axes".[3]

de Prony (1790, p. 82) considered what he called *symmetric solids* (*corps symmétriques*) "relative to a plane or an axis" ("relativement à un plan ou à un axe"). de Prony does not discuss the notion any further, but in a German translation of de Prony's book, the term is explained by the translator: "A solid is here called *symmetric* with respect to a plane, if it is cut by this plane into two equal and similar halves; but [it is called *symmetric*] with respect to an axis, if each plane lying in such an axis cuts the body into two equal and similar halves" (de Prony 1794, p. 90).[4]

The axis of symmetry as well as centres and planes of symmetry play a more central role in various 19th-century studies. The explicit *axe de symétrie* is found in Dupin (1825, p. 148). Bravais (1849, p. 141) defined it in the following terms: "I shall call *axis of symmetry* of a polyhedron a straight line AB such that when the polyhedron revolves around AB by an angle Q, the new positions of the vertices coincide with the old ones".[1]

References

d'Alembert, Jean Le Rond (1751). "Axe". *Encyclopédie, ou Dictionnaire raisonné des sciences, des arts et des métiers* 1, pp. 904–906.

Bézout, Étienne (1767). *Cours de mathématiques*. Suite de la quatrième partie. Contenant l'application des Principes généraux de la Mécanique, à différents cas de Mouvement & d'Équilibre. Paris.

Bravais, Auguste (1849). "Mémoire sur les polyèdres de forme symétrique". *Journal de mathématiques pures et appliquées* 14, pp. 141–180.

Dupin, Charles (1825). *Géometrie et méchanique des arts et métiers et des beaux-arts*. Tome deuxième. Bruxelles.

Euler, Leonhard (1748). *Introductio in analysin infinitorum*. Tomus secundus. Lausanne.

Hon, Giora and Bernard R. Goldstein (2008). *From Summetria to Symmetry*. The Making of a Revolutionary Scientific Concept. New York.

Mauduit, Antoine-René (1757). *Les elemens des sections coniques*. Paris.

Meusnier, Jean-Baptiste (1785). "Mémoire sur la courbure des surfaces". *Mémoires de mathématique et de physique, presentés à l'Académie royale des sciences, par divers sçavans* 10, pp. 477–510.

de Prony, Gaspard (1790). *Nouvelle architecture hydraulique*. Première partie. Paris.

de Prony, Gaspard (1794). *Neue Architektura hydraulika*. Erster Theil. Aus dem Französischen von Karl Christian Langsdorf. Frankfurt.

Sauri, Jean (1774). *Cours complet de mathématiques*. Tome second. Paris.

Scholz, Erhard (1989). *Symmetrie – Gruppe – Dualität*. Zur Beziehung zwischen theoretischer Mathematik und Anwendungen in Kristallographie und Baustatik des 19. Jahrhunderts. Basel.

---

[3] "symmétrique à l'égard de l'un [...] de ces axes". [4] "Symmetrisch heißt hier ein Körper in Ansehung einer Ebene, wenn er durch diese Ebene in zween gleiche und ähnliche Hälften getheilt wird; in Ansehung einer Axe aber, wenn iede durch solche Axe gelegte Ebene den Körper in zwo gleiche und ähnliche Hälften theilt". [1] "Je nommerai *axe de symétrie* d'un polyèdre, une droite AB [...] telle, qu'en faisant tourner le polyèdre d'un angle Q autour de AB, les nouveaux lieux des sommets coïncident avec les anciens". – For the work of Bravais on symmetry, see Scholz (1989, pp. 74–93).

B

Baire set

A modern definition of *Baire set* is given by Bogachev (2007, p. 12): "Another important σ-algebra on a topological space X is generated by all sets of the form $x \in X : f(x) > 0$, where f is a continuous function on X. This σ-algebra is called the *Baire σ-algebra* and is denoted by $\mathcal{B}_a(X)$. It is clear that this is the smallest σ-algebra with respect to which all continuous functions on X are measurable. The same σ-algebra is generated by the class of all bounded continuous functions. The sets in $\mathcal{B}_a(X)$ are called the *Baire sets* in the space X".

The term was explained by Halmos (1950, p. 223):

> The term 'Baire set' is suggested by the term 'Baire function' as used in analysis. If $\mathcal{B}$ is the smallest class of functions which contains all continuous functions and contains the limit of every pointwise (but not necessarily uniformly) convergent sequence of functions in it, then the functions of $\mathcal{B}$ are called the *Baire functions* on X. A necessary and sufficient condition that a set be a Baire set is that it be a Borel set and that its characteristic function be a Baire function.[1]

As pointed out by Pietsch (2007, p. 112), the term is due to Kakutani and Kodaira (1944, p. 444), who used German *Bairesche Menge*.[2]

References

Baire, René (1899). *Sur les fonctions de variables reelles*. Paris.
Bogachev, Vladimir I. (2007). *Measure Theory*. Volume II. Berlin and Heidelberg.
Halmos, Paul R. (1950). *Measure theory*. New York.
Hausdorff, Felix (1914). *Grundzüge der Mengenlehre*. Leipzig.
Kakutani, Shizuo and Kunihiko Kodaira (1944). "Über das Haarsche Mass in der lokal bikompakten Gruppe". *Proceedings of the Imperial Academy* 20, pp. 444–450.
Kodaira, Kunihiko (1941). "Über die Gruppe der messbaren Abbildungen". *Proceedings of the Imperial Academy* 17, pp. 18–23.
Pietsch, Albrecht (2007). *History of Banach Spaces and Linear Operators*. Boston, Mass. et al.

[1] The term *Baire function* seems to have been first used by Hausdorff (1914, p. 390) and de la Vallée Poussin (1916, p. 35). Baire functions and sets appear in Baire (1899). [2] Kodaira (1941, p. 21) had originally used, somewhat confusingly, *Borelsche Menge*. The definitions given by Kakutani and Kodaira and Halmos differ from Bogachev's.

H. Becker, *A Historical Dictionary of Mathematical Terms*, Science Networks. Historical Studies 66, https://doi.org/10.1007/978-3-032-02873-0_2

de la Vallée Poussin, Charles Jean (1916). *Intégrales de Lebesgue. Fonctions d'ensemble. Classes de Baire*. Leçons professées au Collège de France. Paris.

balanced (sets, graphs, matrices)

The French term *ensemble équilibré* was apparently first used by Dieudonné (1942, p. 110), who defined it in the following terms: "in a vector space E, a set K [...] is called *balanced* if, for all $x \neq 0$, there is a $\lambda \neq 0$ such that $\lambda x \in K$".[1] Bourbaki (1953, p. 5) also used the term, but with a modified definition: "Let K be non-discrete valued field, E a left vector space over K; we say that a subset M of E is balanced if, for all $x \in M$ and all $\lambda \in K$ such that $|\lambda| \leq 1$, it is true that $\lambda x \in M$".[2] The English translation *balanced* was used in this sense in the 1950s; Taylor and Halberg (1957, p. 97) is perhaps the earliest example.

The term *balanced* also appeared in graph theory. It was suggested by Harary (1954, p. 143): "An s-graph is in *balance* if all its cycles are positive". *s-graph* was Harary's shortcut for *signed graph*, which he introduced in the same place (in such graphs, all edges have a positive or a negative sign).

The concept was later transferred to hypergraphs by Berge (1969, p. 52): "A graphoid[3] $(X, \mathscr{U})$ is said to be *balanced* if every odd cycle (i. e., cycle with an odd number of edges) is such that one of its edges contains at least three of its points (or 'balanced cycle')". Shortly afterwards, Berge (1972, p. 29) also used the term in connection with matrices: "A (0, 1)-matrix is said to be *balanced* if it is the incidence matrix of a balanced hypergraph".

References

Berge, Claude (1969). "The Rank of a Family of Sets and Some Applications to Graph Theory". In: *Recent Progress in Combinatorics*. Proceedings of the Third Waterloo Conference on Combinatorics, May 1968. Ed. by William Thomas Tutte. New York, pp. 49–57.

Berge, Claude (1972). "Balanced matrices". *Mathematical Programming* 2, pp. 19–31.

Bourbaki, Nicolas (1953). *Espaces vectoriels topologiques*. Chapitre I. Espaces vectoriels topologiques sur un corps valué. Chapitre II. Ensembles convexes et espaces localement convexes. Paris.

Bourbaki, Nicolas (2003). *Topological Vector Spaces*. Chapters 1-5. Berlin and Heidelberg.

Dieudonné, Jean (1942). "La dualité dans les espaces vectoriels topologiques". *Annales scientifiques de l'École Normale Supérieure* 59, pp. 107–139.

Harary, Frank (1954). "On the notion of balance of a signed graph". *Michigan Mathematical Journal* 2, pp. 143–146.

Taylor, Angus E. and Charles J.A. Halberg (1957). "General theorems about a bounded linear operator and its conjugate". *Journal für die reine und angewandte Mathematik* 198, pp. 93–111.

ball and *disk*

The French term *boule à n dimensions* appears to have been first used by Leray (1935-36, p. 1), who defined it by the condition $x_1^2 + x_2^2 + \cdots + x_n^2 < 1$. It was immediately adopted by Marty (1935-36, p. 5) and also appeared in Dieudonné (1942,

---

[1] "dans un espace vectoriel E, un ensemble K [...] est dit *équilibré* si, pour tout $x \neq 0$, il existe $\lambda \neq 0$ tel que $\lambda x \in K$". – Strangely, Dieudonné introduced his definition by the word *rappelons* 'let us recall'. An earlier printed usage than the one presented here has not been found. [2] "Soient K un corps valué non discret, E un espace vectoriel à gauche sur K; on dit qu'une partie M de E est équilibré si, pour tout $x \in M$ et tout $\lambda \in K$ tel que $|\lambda| \leq 1$, on a $\lambda x \in M$". Translation adapted from Bourbaki (2003, p. I.6). [3] Berge's term for a hypergraph at the time.

p. 128), who explicitly used *boule fermée* for the case that "$<$" is replaced by "$\leq$". English *open* and *closed ball* appeared in Chevalley (1946, p. 58).[1]

The terminology also appeared as *boule euclidienne ouverte* and *fermée* in Bourbaki (1947, p. 38), which is worth pointing out because sometimes the term is incorrectly attributed to Bourbaki, for example by Mashaal (2006, p. 105).

Bourbaki (1947, p. 38) used *disque* on the same occasion. The members of Bourbaki would seem unlikely to have regarded this to be an innovation, since German *Kreisscheibe* (from the words for 'circle' and 'disc') had been in use much earlier. For example, in Lazzeri and Bassani (1911, p. 26) (a translation from the Italian) the term was used for a closed disk,[2] and opposed to *Kreislinie*, the circle. The translator explicitly remarked that *Kreisscheibe* translates Italian *cerchio* and that *Kreislinie* translates *circolo* or *circonferenza*, while in German *Kreis* is often used for both concepts. *Kreisscheibe* was also used in topology, for example by Kerékjártó (1923, e. g., p. 120) and Tietze and Vietoris (1928, p. 201). French *disque* was also used before Bourbaki, for example by Kuratowski (1930, p. 183).

References

Bourbaki, Nicolas (1947). *Topologie génerale*. Chapitre V. Groupes à un paramètre. Chapitre VI. Espaces numériques et espaces projectifs. Chapitre VII. Groupes additifs de R^n. Chapitre VIII. Nombres complexes. Paris.

Chevalley, Claude (1946). *Theory of Lie Groups I*. Princeton.

Dieudonné, Jean (1942). "La dualité dans les espaces vectoriels topologiques". *Annales scientifiques de l'École Normale Supérieure* 59, pp. 107–139.

Kerékjártó, Béla (1923). *Vorlesungen über Topologie*. I. Flächentopologie. Berlin.

Kuratowski, Kazimierz (1930). "Sur une propriété des continus Péaniens plans". *Fundamenta Mathematicae* 15, pp. 180–184.

Lazzeri, Giulio and Anselmo Bassani (1911). *Elemente der Geometrie*. Aus dem Italienischen übersetzt von P. Treutlein. Leipzig and Berlin.

Leray, Jean (1935-36). "Propriétés topologiques des transformations continues". *Séminaire de Mathématiques dit de Julia* 3, exposé 3, pp. 1–22.

Marty, Frédéric (1935-36). "Recouvrements. Groupe fondamental". *Séminaire de Mathématiques dit de Julia* 3, exposé 8, pp. 1–17.

Mashaal, Maurice (2006). *Bourbaki*. A Secret Society of Mathematicians. Providence.

Tietze, Heinrich and Leopold Vietoris (1928). "Beziehungen zwischen den verschiedenen Zweigen der Topologie". *Encyklopädie der mathematischen Wissenschaften mit Einschluss ihrer Anwendungen* III AB 13, pp. 141–237.

barrel* and *barrelled space

The English term *barrel* is a literal translation of French *tonneau* and was first used by Dieudonné (1953, p. 499): "In a locally convex space E, a *barrel* is a *closed, convex, symmetric*, and *absorbing* set". Dieudonné (ibid.) did not use *barrelled space* in this paper but spoke of *t-spaces* instead (the *t* standing for *tonnelé*); the term was used soon afterwards, however, for example by Warner (1956, p. 192).

Dieudonné referred to various publications by Bourbaki in his paper, and French *tonneau* and *espace tonnelé* were originally coined by Bourbaki (1950, p. 6). Beaulieu (1999, p. 228) cites the terms as an example of "mathematical terminology by which external realities gave a down-to-earth flavor to technical terms".

---

[1] The OED dates the term to 1953. [2] It is there defined kinematically in terms of a moving radius.

References

Beaulieu, Liliane (1999). "Bourbaki's Art of Memory". *Osiris* 14, pp. 219–251.
Bourbaki, Nicolas (1950). "Sur certains espaces vectoriels topologiques". *Annales de l'Institut Fourier* 2, pp. 5–16.
Dieudonné, Jean (1953). "Recent developments in the theory of locally convex vector spaces". *Bulletin of the American Mathematical Society* 59, pp. 495–512.
Warner, Seth (1956). "Inductive Limits of Normed Algebras". *Transactions of the American Mathematical Society* 82, pp. 190–216.

barycentric

The term *barycentric calculus* (*barycentrischer Calcul*) was coined by Möbius. In the preface of his book on the topic, he pointed out some works (he names Carnot and L'Huilier, for example) in which the centre of gravity was considered from a purely geometrical perspective, an approach which Möbius also followed:

> The present researches also proceed from the same elementary and purely geometrical concept of the center of gravity. What first stimulated these researches was consideration of the fruitfulness of the law, that each system of weighted points has only one center of gravity, and that thus, in whatever sequence one brings the points into connection, the result is that one and the same point must always be found. The simple technique, by means of which I was able to prove more geometrical laws, stimulated me to find a suitable algorithm for still greater simplification of such investigations (Möbius 1827, p. iv).[1]

This approach is evidently also the basis for the terminology (Möbius 1827, p. 17). The French term *calcul barycentrique* was used in Cauchy's review of Möbius' book (Cauchy 1828, p. 78); the English term is found in Salmon (1852, p. 16).

The term *barycentric coordinates* was not explicitly used by Möbius; it appears in Schläfli (1849, p. 100).

See also *centroid*.

Note. For discussion of Möbius' calculus, see Crowe (1985, pp. 48f.) and Wußing (1984, pp. 35–42).

References

Cauchy, Augustin-Louis (1828). "Der barycentrische Calcul [review]". *Bulletin des sciences mathématiques, astronomiques, physiques et chimiques* 9, pp. 77–80.
Crowe, Michael J. (1985). *A History of Vector Analysis*. The Evolution of the Idea of a Vectorial System. New York.
Möbius, August Ferdinand (1827). *Der barycentrische Calcul*. Leipzig.
Salmon, George (1852). *A Treatise on the Higher plane curves intended as a sequel to the Treatise on Conic Sections*. Dublin.
Schläfli, Ludwig (1849). "Anwendung des barycentrischen Calculs auf die Bestimmung der grössten einem Vierseit eingeschriebenen und der kleinsten einem Viereck umschriebenen Ellipse". *Archiv der Mathematik und Physik* 12, pp. 99–129.
Wußing, Hans (1984). *The Genesis of the Abstract Group Concept*. A Contribution to the History of the Origin of Abstract Group Theory. Tr. Abe Shenitzer. Cambridge, Mass.

---

[1] "Von demselben elementaren und rein geometrischen Begriffe des Schwerpunkts gehen auch die vorliegenden Untersuchungen aus. Die erste Veranlassung hierzu war die Erwägung der Fruchtbarkeit des Satzes, dass jedes System gewichtiger Punkte nur einen Schwerpunkt hat, und dass daher, in welcher Folge man auch die Punkte nach und nach in Verbindung bringt, zuletzt doch immer ein und derselbe Punkt gefunden werden muss. Die einfache Art, womit ich dadurch, mehrere geometrische Sätze zu beweisen, mich im Stande sah, bewog mich, zu noch grösserer Vereinfachung solcher Untersuchungen einen dafür passenden Algorithmus auszumitteln". Translation from Crowe (1985, p. 49).

base (geometry)

The word *base* and *basis* both go back to the same Greek word: *βάσις* in general language meant 'step, stand, that upon which one stands'. The metaphorical meaning 'fundamental part' often used today does not seem to have existed in classical Greek, but the word did have the meaning 'foundation' (LSJ). The word is found in Greek mathematical texts to designate the base of plane or solid figures such as triangles or pyramids; it was in regular use with Euclid and others before him (Mugler 1958-59, pp. 94–96). Euclid did not define the term but used it frequently in his *Elements*.

With these meanings the term also appears as Latin *basis*; Cicero is the first known classical writer who used it for the base of a triangle (LS, TLL).

French *base* is found in the *Pratike de geometrie* (Victor 1979, e. g., p. 474), which dates from the 13th century; the term was also used in the late 15th century in Chuquet's *Géométrie* (Chuquet 1979, p. 110). At about the same time, the word also appears in a German text: "Note that the lowest line is called *base*" (Widmann 1489, fol. C 6^{r}).[1] Schmid (1539, p. 5) suggested *Grundlinie* ('ground line').[2]

For the English term, the OED cites L. Digges (1556, fol. B i^{v}): "The Base of any triangle is here called that side, whiche is cut squyrewyse of the hangynge lyne"; in L. Digges and T. Digges (1571, fol. T iij^{r}), the term is also applied to solids: "Any one of the Figures wherewith these solides be enuironed, is called the base of that solide".[3] Recorde (1551, fol. A^{iiij}), with respect to plane figures, preferred to suggest an English term: "one line lyeth flatte (whiche is named the *ground line*)".

Note. For a historical overview of the terminology, see Tropfke (1940, pp. 83–87). For the Greek term, see Mugler (1958-59, pp. 94–96).

References

Chuquet, Nicolas (1979). *La Géométrie*. Première géométrie algébrique en langue française (1484). Introduction, texte et notes par Hervé l'Huillier. Paris.

Digges, Leonard (1556). *A boke named Tectonicon*. London.

Digges, Leonard and Thomas Digges (1571). *A geometrical practise, named Pantometria diuided into three bookes, Longimetra, Planimetra, and Stereometria*. London.

Mugler, Charles (1958-59). *Dictionnaire historique de la terminologie géométrique des Grecs*. Paris.

Recorde, Robert (1551). *The patheway to knowledg*. London.

Schirmer, Alfred (1912). *Der Wortschatz der Mathematik nach Alter und Herkunft untersucht*. Straßburg.

Schmid, Wolffgang (1539). *Das erst buch der Geometria*. Nürnberg.

Strauß, Gerhard et al., eds. (1997). *Deutsches Fremdwörterbuch*. 2. Auflage. Band 3: Baby–Cutter. Berlin and New York.

Tropfke, Johannes (1940). *Geschichte der Elementar-Mathematik*. Vierter Band. Ebene Geometrie. Dritte, verbesserte und vermehrte Auflage besorgt von Dr. Kurt Vogel. Berlin and Leipzig.

Victor, Stephen K. (1979). *Practical Geometry in the High Middle Ages*. "Artis cuiuslibet consummatio" and the "Pratike de geometrie". Philadelphia.

Widmann, Johannes (1489). *Behende vnd hubsche Rechenung auff allen kauffmanschafft*. Leipzig.

[1] "Und darumb merck daz die vnderste lini geheyssen wirt basis". – For the German term, see Strauß et al. (1997, p. 165) and Schirmer (1912, pp. 8f.). [2] The original spelling was *Grund lini*.

[3] L. Digges and T. Digges (1571, e. g., fol. Y iij^{v}) also use *basis*, both for triangles and solids.

base (logarithm)

The term *base* in the theory of logarithms goes back to Euler. In his book *Introductio in analysin infinitorum* he used the Latin term *basis*:

> Just as, given a number a, for any value of z, we can find the value of y, so, in turn, given a positive value for y, we would like to give a value for z, such that $a^z = y$. This value of z, insofar as it is viewed as a function of y, it is called the *Logarithm* of y. The discussion about logarithms supposes that there is some fixed constant to be substituted for a, and this number is the *base* for the logarithm" (Euler 1748, § 102).[1]

The term was soon adopted and translated into other languages. For example, Fenn (1772, p. 152) uses English *base*, the first use in English (OED). Bossut (1773, p. 432) uses the French word, and German *Basis* is found in Karsten (1768, pp. 21f.).[2]

Note. Euler's use has been pointed out by Tropfke (1980, p. 322).

References

Bossut, Charles (1773). *Traité élémentaire d'algébre*. Paris.

Euler, Leonhard (1748). *Introductio in analysin infinitorum*. Tomus primus. Lausanne.

Euler, Leonhard (1988). *Introduction to Analysis of the Infinite*. Book I. Translated by John D. Blanton. New York.

Fenn, Joseph (1772). *Second Volume of the Instructions Given in the Drawing School*. Dublin.

Karsten, Wenceslaus Johann Gustav (1768). "Abhandlung von den Logarithmen verneinter Größen. Erste Abtheilung". *Abhandlungen der Churfürstlich-Baierischen Akademie der Wissenschaften* 5, pp. 3–108.

Lorenz, Johann Friedrich (1785). *Die Elemente der Mathematik in sechs Büchern*. Erster Theil. Arithmetik, Geometrie und Analysis. Leipzig.

Tropfke, Johannes (1980). *Geschichte der Elementarmathematik*. Band 1: Arithmetik und Algebra. Vollständig neu bearbeitet von Kurt Vogel, Karin Reich, Helmuth Gericke. Berlin and New York.

Vega, Georg (1782). *Vorlesungen über die Mathematik*. Erster Band. Wien.

basis

Algebraic uses of the term *basis* in the sense of a system of independent generators date from around 1870.[3] In this meaning, the German term *Basis* was used by Dedekind in connection with complex number fields. Dedekind noted that each number ω of a field can be represented in one and only one way in the form $\omega = h_1\omega_1 + h_2\omega_2 + \ldots h_n\omega_n = \sum h_i\omega_i$ (with rational numbers h_i) and then defined: "We call the number n the *degree*, the system of n independent numbers ω_i a *basis of the field Ω*, and the n numbers h_i the *coordinates of the number ω* corresponding to this basis" (Dedekind 1871, p. 425).[4] He also defined the basis of a module (§ 161); in that definition, however, the elements of the basis were not required to be independent.[5]

---

[1] "Quemadmodum autem, dato numero a, ex quovis valore ipsius z reperiri potest valor ipsius y, ita vicissim, dato valore quocunque affirmativo ipsius y, conveniens dabitur valor ipsius z, ut sit $a^z = y$; iste autem valor ipsius z, quatenus tanquam Functio ipsius y spectatur, vocari solet *Logarithmus* ipsius y. Supponit ergo doctrina Logarithmorum numerum certum constantem loco a substituendum, qui propterea vocatur *basis* Logarithmorum". Translation from Euler (1988, § 102). [2] Vega (1782, p. 202) preferred German *Grundzahl* (from words for 'base' and 'number'), Lorenz (1785, p. 102) also names *Fundamentalgröße* 'fundamental quantity'. [3] For the general history of the term, see *base*. [4] "Wir wollen die Zahl n den *Grad*, ferner den Complex der n unabhängigen Zahlen ω_i eine *Basis des Körpers Ω*, und die n Zahlen h_i die dieser Basis entsprechenden *Coordinaten der Zahl ω* nennen". Translation from Avigad (2004, p. 13). [5] See G. H. Moore (1995, pp. 288f.).

This choice of terminology may also have influenced Frobenius and Stickelberger (1879, p. 219), when they defined the *Basis* of a group.[1] As applied to vector spaces, the term seems to go back to Dedekind and Weber (1882, p. 199).[2] In these definitions, the term *basis* is to be understood as 'system of generators'. When a basis in the current sense is meant, some further qualification was used. For example, Dedekind and Weber (1882, p. 199) called a basis a *linearly irreducible basis* (*linear irreductible Basis*).

It took some time until the terminology was generally adopted, and alternative terms are easily found. For example, Peano, who gave an axiomatic definition of vector spaces, called the elements of the basis *enti di riferimento* 'reference entities' (Peano 1888, p. 144), while Pincherle (1901, p. 8) preferred *sistema fondamentale*. English *basis-system* was used E. Moore (1896, p. 214); he referred to Frobenius and Stickelberger for the term.

In the 20th century, *basis* is used more regularly. Examples in connection with vector spaces include Weyl (1918, p. 17) and Carathéodory (1918, p. 310); both authors require the independence of the basis elements.

Earlier, Graßmann had used the term *System von Einheiten* 'system of units'. He defined: "I say that a magnitude a is *derived* from the magnitudes b, c, ... by the numbers β, γ, ... if, $a = \beta b + \gamma c + ...$, where β, γ, ... are real numbers, rational or irrational, different from zero or not. I also say that in this case a is *numerically derived* from b, c, ..." (Graßmann 1862, p. 1).[3] This notion of derivation is then used in the following definition of a basis: "I define as a *system of units* any assembly of magnitudes, standing in no numerical relation to one another, that can serve for the derivation of other magnitudes by means of arbitrary numbers" (Graßmann 1862, p. 2).[4]

See also *fundamental system* and *generating* (algebra).

Note. For the history of the concept, see Krömer (1998), Dorier (1990), and Dorier (1996).

References

Avigad, Jeremy (2004). "Dedekind's 1871 version of the theory of ideals". URL: `https://kilthub.cmu.edu/articles/Dedekind_s_1871_Version_of_the_Theory_of_Ideals/6491342/1` (retrieved July 25, 2007).

Carathéodory, Constantin (1918). *Vorlesungen über reelle Funktionen*. Leipzig.

Dedekind, Richard (1871). "Ueber die Composition der binären quadratischen Formen". In: *Vorlesungen über Zahlentheorie von P.G. Lejeune Dirichlet*. Herausgegeben und mit Zusätzen versehen von R. Dedekind. Zweite, umgearbeitete und vermehrte Auflage. Braunschweig, pp. 380–497.

Dedekind, Richard (1894). "Ueber die Theorie der ganzen algebraischen Zahlen". In: *Vorlesungen über Zahlentheorie von P.G. Lejeune Dirichlet*. Herausgegeben und mit Zusätzen versehen von R. Dedekind. Vierte, umgearbeitete und vermehrte Auflage. Braunschweig, pp. 434–657.

---

[1] For this notion, Kronecker (1871, p. 885) used the alternative *fundamental system* (German *Fundamentalsystem*). [2] The term is also found in Dedekind (1894, p. 468), where the basis is understood to be "irreducible". [3] "Ich sage, eine Grösse a sei aus den Grössen b, c, ... *abgeleitet*, wenn $a = \beta b + \gamma c + ...$, wo β, γ, ... reelle Zahlen sind, gleichviel ob rational oder irrational, ob gleich null oder verschieden von null. Auch sage ich, a sei in diesem Falle *numerisch abgeleitet* aus b, c, ...". Translation from Graßmann (2000, p. 3). [4] "Ein *System von Einheiten* nenne ich jeden Verein von Grössen, welche in keiner Zahlbeziehung zu einander stehen, und welche dazu dienen sollen, um aus ihnen durch beliebige Zahlen andere Grössen abzuleiten". Translation from Graßmann (2000, p. 3).

Dedekind, Richard and Heinrich Weber (1882). “Theorie der algebraischen Functionen einer Veränderlichen”. *Journal für die reine und angewandte Mathematik* 92, pp. 181–290.
Dorier, Jean-Luc (1990). “Analyse historique de l’émergence des concepts élémentaires d’algèbre linéaire”. Paris. URL: `https://hal.archives-ouvertes.fr/hal-02138152/` (retrieved July 9, 2020).
Dorier, Jean-Luc (1996). “Basis and dimension – from Grassmann to van der Waerden”. In: *Hermann Günther Graßmann (1809-1877)*. Visionary Mathematician, Scientist and Neohumanist Scholar. Ed. by Gert Schubring. Dordrecht, pp. 175–196.
Frobenius, Georg and Ludwig Stickelberger (1879). “Ueber Gruppen von vertauschbaren Elementen”. *Journal für die reine und angewandte Mathematik* 86, pp. 217–262.
Graßmann, Hermann (1862). *Die Ausdehnungslehre*. Vollständig und in strenger Form bearbeitet. Berlin.
Graßmann, Hermann (2000). *Extension Theory*. Translated by Lloyd C. Kannenberg. Providence, R.I.
Krömer, Ralf (1998). “Zur Geschichte des axiomatischen Vektorraumbegriffs”. Diploma thesis. Universität des Saarlandes.
Kronecker, Leopold (1871). “Auseinandersetzung einiger Eigenschaften der Klassenzahl idealer complexer Zahlen”. *Monatsberichte der Königlich Preußischen Akademie der Wissenschaften zu Berlin* (1870), pp. 881–889.
Moore, E.H. (1896). “A doubly-infinite system of simple groups”. In: *Mathematical Papers read at the International Mathematical Congress (Chicago, 1893)*. New York, pp. 208–243.
Moore, Gregory H. (1995). “The axiomatization of linear algebra: 1875-1940”. *Historia Mathematica* 22, pp. 262–303.
Peano, Guiseppe (1888). *Calcolo geometrico secondo l’Ausdehnungslehre di H. Grassmann, preceduto dalle operazioni della logica deduttiva*. Torino.
Pincherle, Salvatore (1901). *Le operazioni distributive e le loro applicazioni all’analisi*. In collaborazione con Ugo Amaldi. Bologna.
Weyl, Hermann (1918). *Raum · Zeit · Materie*. Vorlesungen über allgemeine Relativitätstheorie. Berlin.

basis (topology)

The term *basis* (of a topological space) may be due to Kuratowski (1933, p. 101), who used it for “a series R_1, R_2, ..., of open sets (non-empty) such that each open set (non-empty) is a sum of certain terms of that series”.[1] German *Basis* was also used in this sense by Hausdorff (1934, p. 279, 1935, p. 283) and by Alexandroff and Hopf (1935, p. 42). The English term *base* appears in Whyburn (1934, p. 787) and Čech (1937, p. 824); Tukey (1940, p. 27) uses *basis*. The term was also adopted by Bourbaki (1940, p. 11).

References

Alexandroff, Pavel S. and Heinz Hopf (1935). *Topologie*. Erster Band. Berlin.
Bourbaki, Nicolas (1940). *Topologie génerale*. Chapitre I. Structures topologiques. II. Structures uniformes. Paris.
Čech, Eduard (1937). “On Bicompact Spaces”. *Annals of Mathematics* 38, pp. 823–844.
Hausdorff, Felix (1934). “Über innere Abbildungen”. *Fundamenta Mathematicae* 23, pp. 279–291.
Hausdorff, Felix (1935). *Mengenlehre*. Dritte Auflage. Berlin.
Kuratowski, Kazimierz (1933). *Topologie I*. Espaces métrisables, espaces complets. Warszawa and Lwów.
Tukey, John W. (1940). *Convergence and Uniformity in Topology*. Princeton.

[1] “une suite R_1, R_2, ... d’ensembles ouverts (non vides) tels que chaque ensemble ouvert (non vide) est une somme de certains termes de cette suite”.

Whyburn, Gordon T. (1934). "*Topologie*. I. By C. Kuratowski. Warszawa-Lwow [review]". *Bulletin of the American Mathematical Society* 40, pp. 787–788.

Bernoulli number

In the *Ars conjectandi*, Bernoulli (1713, pp. 97f.) considered sums of powers of consecutive integers like $1 + 8 + 27 + 64 + 126 + 216 + 343 = 784$ and stated, without proof, the formula

$$\sum_{k=1}^{n-1} k^p = \sum_{k=0}^{p} \frac{B_k}{k!} \frac{p!}{(p+1-k)!} n^{p+1-k}.$$

This formula, here given in modern notation, contains the numbers B_k, the first values of which are $B_0 = 1$, $B_1 = -\frac{1}{2}$, $B_2 = \frac{1}{6}$, $B_3 = 0$, $B_4 = -\frac{1}{30}$.[1]

These numbers arise, as Sandifer (2007, p. 171) puts it, "in a dazzling variety of applications in analysis, numerical analysis and number theory". de Moivre (1730, Supplement, p. 19) already called them *numeros Bernoullianos*, and so did Euler (1755, p. 420). The earliest uses in modern languages are somewhat difficult to ascertain: German *Bernoullische Zahlen* appears in a review of Euler's book cited above ([anon.] 1757, p. 1150), and so does French *Nombres Bernoulliens* ([anon.] 1758, p. 143). English *numbers of Bernoulli* seems to be of much later date; it is found in Herschel (1816, p. 33).

As is well known today, the Japanese mathematician Seki discovered these numbers at about the same time as Bernoulli; for an account, see Kitagawa (2022).

Note. For the history of the concept, see Beery (2009), Sandifer (2007, pp. 171–176), and Kitagawa (2022).

References

[anon.] (1757). "[review of Euler (1755, *Institutiones calculi differentialis*]". *Göttingische Anzeigen von gelehrten Sachen* (1757), pp. 1145–1152.

[anon.] (1758). "Institutiones calculi differentialis [review]". *Bibliothèque germanique, ou Histoire littéraire de l'Allemagne, de la Suisse et des pays du Nord* 23, pt. 1, pp. 132–156, 262–285.

Beery, Janet (2009). "Sums of Powers of Positive Integers". *Convergence* (April 2009). URL: `https://old.maa.org/press/periodicals/convergence/sums-of-powers-of-positive-integers-introduction`.

Bernoulli, Jakob (1713). *Ars coniectandi*. Basel.

Euler, Leonhard (1755). *Institutiones calculi differentialis*. St. Petersburg.

Herschel, J.F.W. (1816). "On the development of exponential functions; together with several new theorems relating to finite differences". *Philosophical Transactions of the Royal Society of London* 106, part I, pp. 25–45.

Kitagawa, Tomoko L. (2022). "The Origin of the Bernoulli Numbers: Mathematics in Basel and Edo in the Early Eighteenth Century". *The Mathematical Intelligencer* 44, pp. 46–56.

de Moivre, Abraham (1730). *Miscellanea analytica de seriebus et quadraturis*. London.

Sandifer, C. Edward (2007). *How Euler Did It*. Washington, D.C.

Bessel function

The solutions of the linear differential equations of the second order $x^2y'' + xy' + (x^2 - \nu^2)y = 0$ (with ν a complex parameter) are called *Bessel functions*. This

[1] This account follows Sandifer (2007). Note, however, that Bernoulli numbers can be defined in various ways and often $B_1 = \frac{1}{2}$.

name was suggested by Schlömilch (1853, p. 482), who used German *Bessel'sche Function* and referred to Bessel (1826) for the concept.[1] Schlömilch's suggestion became better known through Schlömilch (1857), and the English translation *Bessel's function* is found soon afterwards, for example in Bache (1862, p. 264).

Bessel functions are sometimes called *cylinder functions*. This name goes back to Heine (1868, p. 128), who suggested German *Cylinderfunction* (although in the title of his paper *Fourier-Bessel function* appeared).[2] As Dutka (1995, p. 107) explains, the name arises "from their frequent occurrence in the analysis of physical problems involving circular cylinders".

Note. The early history of Bessel functions is discussed by Dutka (1995). The origins of the term and its alternative have been pointed out by Demidov, Petrova, and Simonov (1998, p. 148).

References

Bache, A.D. (1862). "Abstract of a Discussion of the Horizontal Component of the Magnetic Force, From Observations Made at the Girard College Observatory, Philadelphia, in the Years 1840-'41-'42-'43-'44-'45". *American Journal of Science* 34, pp. 261–272.

Bessel, Friedrich Wilhelm (1826). "Untersuchung des Theils der planetarischen Störungen, welcher aus der Bewegung der Sonne entsteht". *Abhandlungen der Königlichen Akademie der Wissenschaften zu Berlin* 1824, pp. 1–52.

Demidov, S.S., S.S. Petrova, and N.I. Simonov (1998). "Ordinary differential equations". In: *Mathematics of the 19th Century*. Function Theory According to Chebyshev. Ordinary Differential Equations. Calculus of Variations. Theory of Finite Differences. Ed. by Andrey Nikolaevich Kolmogorov and Adolph Andrei Pavlovich Yushkevich. Basel, pp. 83–196.

Dutka, Jacques (1995). "On the Early History of Bessel Functions". *Archive for History of Exact Sciences* 49, pp. 105–134.

Fourier, Joseph (1822). *Théorie analytique de la chaleur*. Paris.

Heine, Eduard (1868). "Die *Fourier-Bessel*sche Function". *Journal für die reine und angewandte Mathematik* 69, pp. 128–141.

Lipschitz, Rudolf (1859). "Ueber ein Integral der Differentialgleichung $\frac{\partial^2 I}{\partial x^2 I} + \frac{1}{x}\frac{\partial I}{\partial x} + I = 0$". *Journal für die reine und angewandte Mathematik* 56, pp. 189–196.

Neumann, Carl (1861). *Allgemeine Lösung des Problemes über den stationären Temperaturzustand eines homogenen Körpers*. Halle.

Schlömilch, Oskar (1853). "Integralfunctionen". *Allgemeine Encyklopädie der Wissenschaften und Künste* 30, pp. 478–483.

Schlömilch, Oskar (1857). "Ueber die Bessel'sche Funktion". *Zeitschrift für Mathematik und Physik* 2, pp. 137–165.

beta function

See *gamma function*.

bifurcation

This term was introduced by Poincaré (1885, p. 270), who used French *forme de bifurcation*. Charpentier, Ghys, and Lesne (2010, p. 9) explain the notion and the term in the following terms: in "his study of the forms of equilibrium of fluids in rotation [... Poincaré] established his theorem on the 'exchange of stability': under certain

---

[1] *Bessel'sche Function* has sometimes been incorrectly attributed to Lipschitz, for example by Neumann (1861, p. 3). Neumann refers to Lipschitz (1859), where *Bessel'sche Transcendente* appears in the page headers, but not in the text itself. *Transcendente* there means 'transcendental function'. [2] They were dealt with in Fourier (1822, ch. 6), for example.

conditions, when two solution curves cross, the solutions of one of the branches lose stability whereas those of the other gain it (he therefore called the solution at a crossing point a 'bifurcation form')".

According to Bruter (1986, p. 70), "[i]t is very likely that he [Poincaré] uses a railway image" because Poincaré (1905, p. 242), when he discusses trifurcations, explicitly explains some related terms as having been borrowed from "the art of railways".[1]

Occasionally, it is suggested that the German term *Abzweigung* appears in Jacobi (1834) (for example, by Iooss and Joseph 1980, p. 11). Jacobi did not use this term, however.

References

Bruter, Claude Paul (1986). "Bifurcation and continuity". In: *Dynamical Systems. A Renewal of Mechanism*. Centennial of George David Birkhoff. Ed. by Simon Diner, Daniel Fargue, and Georges Lochak. Singapore and Philadelphia, pp. 70–74.

Charpentier, Éric, Étienne Ghys, and Annick Lesne (2010). "Introduction". In: *The Scientific Legacy of Poincaré*. Translated by Joshua Bowman. Ed. by Éric Charpentier, Étienne Ghys, and Annick Lesne. Providence, R.I. and London, pp. 1–16.

Iooss, Gérard and Daniel D. Joseph (1980). *Elementary Stability and Bifurcation Theory*. New York.

Jacobi, Carl Gustav Jacob (1834). "Ueber die Figur des Gleichgewichts". *Annalen der Physik und Chemie* 33, pp. 229–233.

Poincaré, Henri (1885). "Sur l'équilibre d'une masse fluide animée d'un mouvement de rotation". *Acta Mathematica* 7, pp. 259–380.

Poincaré, Henri (1905). "Sur les lignes géodésiques des surfaces convexes". *Transactions of the American Mathematical Society* 6, pp. 237–274.

bijective

As is well known, the terms *bijective*, *injective*, and *surjective* as well as the corresponding nouns are all due to Bourbaki, who first uses the term *application bijective* and *bijection* in Bourbaki (1954, p. 80). Bourbaki notes that instead of saying that $f : A \to B$ is bijective, one also says that "f puts A and B in biunivocal correspondence".[2] Bourbaki also points out that *permutation* is used if the two involved sets are the same. The term *biunivocal* had been in use since about 1890; De Paolis (1890, p. 90), for example, used Italian *corrispondenza biunivoca* for a biunivocal correspondence.[3]

Another alternative Bourbaki considered appears in an undated redaction of the *Théorie des ensembles*, which shows how Bourbaki gradually approaches the now standard terminology: "One says that f is an ambijection, or an ambijective function, if f is injective and projective[4]" (Bourbaki n.d., p. 12).[5] The Latin prefix *ambi-* means 'both' and suggests that the function is both injective and surjective.

---

[1] "Ces deux expressions [*foyers en point*, *foyers en talon*] sont empruntées à l'art des chemins de fer; je me représente une trifurcation, où nous avons la voie principale représenté par la géodésique OR et deux embranchements représentés par les deux branches de la caustique; un train suivrait la ligne principale en allant de O vers R rencontrerait les aiguilles en point dans le 1er cas et en talon dans le 2^{d}". [2] "f met A et B en correspondence biunivoque". This shows how uneasy the use of *biunivocal* is because Bourbaki also cites *biunivoque* as a synonym for *injective*. [3] Another alternative term, *biuniforme*, applied to substitutions or transformations, was suggested by Picard (1887, p. 42). [4] The word *projective* here is an earlier synonym for *surjective*. [5] "On dit que f est une ambijection, ou que f est une application ambijective, si f est injective et projective".

In English, the terms *bijective* and *bijection* appear in Chevalley (1956, p. 1), and while Gandy (1959, p. 72) stated and asked "[s]tandard terms are badly needed for 'one-to-one,' 'onto,' and 'one-to-one onto'; will Bourbaki's 'injection,' 'surjection,' and 'bijection' prove acceptable?", the terms soon were generally accepted. German *bijektiv* is found in Scheja (1958, p. 11).

For the use of the concept as a one-to-one-correspondence by Steiner and Cantor, see *power* (sets). Ferreirós (2007, pp. 23f.) points out that this concept is found in various texts on projective geometry in the 1820s and 30s, for example in Steiner (1832, pp. xiv–xv): "First a straight line and a planar bundle of lines will be related to each other, so that their elements get matched, that is, so that a certain line in the bundle corresponds to each point in the straight line".[1] The verb *entsprechen* used by Steiner literally means 'to correspond'. The terminology of correspondence was systematically used by Weyr (1869, pp. vii, 6), who created terms like *ein-eindeutig* for a one-to-one correspondence (from *ein* 'one' and *deuten* 'to interpret'). For Weyr, *ein-eindeutig* meant that one element of a "formation" (*Gebilde*) corresponds to *exactly* one element of another "formation".

References

Bourbaki, Nicolas (1954). *Théorie des ensembles.* Chapitre I. Déscription de la mathématique formelle. Chapitre II. Théorie des ensembles. Paris.

Bourbaki, Nicolas (n.d.). "Rédaction n°172. Ensembles. Chapitre II, §§ 3 et 6 (état 8)". URL: `http://archives-bourbaki.ahp-numerique.fr/items/show/586` (retrieved June 9, 2019).

Chevalley, Claude (1956). *Fundamental Concepts of Algebra.* New York.

Ferreirós, José (2007). *Labyrinth of Thought.* A History of Set Theory and Its Role in Modern Mathematics. Second revised edition. Basel.

Gandy, R.O. (1959). "N. Bourbaki. *Théorie des ensembles* [review]". *The Journal of Symbolic Logic* 24, pp. 71–73.

De Paolis, Riccardo (1890). "Teoria dei gruppi geometrici e delle corrispondenze che si possono stabilire tra i loro elementi". *Memorie di matematica e di fisica della Società italiana delle Scienze* 7, pp. 1–165.

Picard, Émile (1887). "Sur une classe d'équations différentielles". *Comptes rendus de l'Académie des sciences* 104, pp. 41–43.

Scheja, Günter (1958). "Verzweigte Holomorphiehüllen". *Sitzungsberichte der Bayerischen Akademie der Wissenschaften (mathematisch-naturwissenschaftliche Klasse)* (1958), pp. 8–18.

Steiner, Jacob (1832). *Systematische Entwickelung der Abhängigkeit geometrischer Gestalten von einander.* Erster Theil. Berlin.

Weyr, Emil (1869). *Theorie der mehrdeutigen geometrischen Elementargebilde und der algebraischen Curven und Flächen als deren Erzeugnisse.* Leipzig.

bilinear

Early attestations suggest that the term first appears as German *bilinear*, which is used in several publications from 1864. Baltzer (1864, p. 164) uses *bilineare Function* in the now usual sense and refers to Jacobi (1857) for the term. Jacobi, however, does not literally speak of *bilinear* functions but uses *zweifach linear* 'doubly linear' instead (Jacobi 1857, p. 265). The term *bilinear* is also found in Christoffel (1864, p. 255), a paper also referred to by Baltzer, and it also appears in the titles of Kronecker (1867)

---

[1] "Zuerst werden eine Gerade und ein ebener Strahlbüschel aufeinander bezogen, so dass ihre Elemente gepaart sind, d. h., dass jedem Punkt der Geraden ein bestimmter Strahl des Strahlbüschels entspricht". Translation from Ferreirós (2007, p. 23).

and Christoffel (1868). The name, as is also explained by Baltzer, is due to the fact that the function $f = \sum a_{ik}x_iy_k$ is a linear function of both $x_1, ..., x_n$ and $y_1, ..., y_n$.

Note. For a historical discussion of bilinear forms, see Brechenmacher (2006, esp. pp. 205-244).

References

Baltzer, Richard (1864). *Theorie und Anwendung der Determinanten*. Zweite vermehrte Auflage. Leipzig.

Brechenmacher, Frédéric (2006). "Histoire du théorème de Jordan de la décomposition matricielle (1870-1930). Formes de représentation et méthodes de décomposition". PhD thesis. École des hautes études en sciences sociales.

Christoffel, Elwin Bruno (1864). "Verallgemeinerung einiger Theoreme des Herrn Weierstrass". *Journal für die reine und angewandte Mathematik* 63, pp. 255–272.

Christoffel, Elwin Bruno (1868). "Theorie der bilinearen Functionen". *Journal für die reine und angewandte Mathematik* 68, pp. 253–272.

Jacobi, Carl Gustav Jacob (1857). "Über eine elementare Transformation eines in Bezug auf jedes von zwei Variabeln-Systemen linearen und homogenen Ausdruckes". *Journal für die reine und angewandte Mathematik* 53, pp. 265–270.

Kronecker, Leopold (1867). "Ueber bilineare Formen". *Monatsberichte der Königlich Preußischen Akademie der Wissenschaften zu Berlin* (1866), pp. 597–612.

binary

The Latin word *binarius* means 'that contains or consists of two' and is made up of *bini* 'two by two, two to each, two each, two at a time' and the suffix *-arius*, which was used to form adjectives from nouns or numerals (LS). The word as such is found in post-classical Latin, but its use in the sense of 'binary arithmetic' dates from the 17th century.

The Spanish philosopher Juan Caramuel y Lobkowitz is generally regarded as being the first who published on binary arithmetic (see, e. g., Glaser 1981, p. 20).[1] Having raised the question whether there is one arithmetic or many (Caramuel y Lobkowitz 1670, p. xliii), he discusses a number of non-decimal systems, starting with what he called *binaria arithmetica* (Caramuel y Lobkowitz 1670, p. xlv). Caramuel also tabulated the binary numbers from 0 to 32 (see table 1).

The term became better known through Leibniz (1705, p. 89), who used French *Arithmétique binaire ou dyadique*; in the article, Leibniz also presented calculations with binary numbers.[2] The Latin terminology is then found in Wolff's mathematical dictionary (Wolff 1716, p. 169). English *binary* in this sense came up in the 1720s; it is found in Chambers' *Cyclopaedia*, which contains an entry for *binary arithmetic* (Chambers 1728, p. 103).[3]

The explicit *binary number* is of later origin; French *nombre binaire* was recorded in the *Encyclopédie, ou dictionnaire universel raisonné des connaissances humaines* ([anon.] 1775, p. 701). However, the term was not in frequent use until the second half of the 20th century.

---

[1] For the earlier history of the notion, see Glaser (1981, ch. II). Knutzen (1742) already pointed out Caramuel's use of the concept. [2] The word *dyadique* is from Greek *δυαδικός* 'of or for the number two'. Leibniz had dealt with binary arithmetic in the late 1670s; see Zacher (1973). He never (or at least not before 1711) seems to have seen Caramuel's *Mathesis biceps*, which is clear from correspondence from 1709 and 1711. See Knobloch (1973, p. 21). [3] Chambers also has an entry for *binary number* but in the unrelated sense "that compose'd of two units".

0	0	a0000	16
a	1	a000a	17
a0	2	a00a0	18
aa	3	a00aa	19
a00	4	a0a00	20
a0a	5	a0a0a	21
aa0	6	a0aa0	22
aaa	7	a0aaa	23
a000	8	aa000	24
a00a	9	aa00a	25
a0a0	10	aa0a0	26
a0aa	11	aa0aa	27
aa00	12	aaa00	28
aa0a	13	aaa0a	29
aaa0	14	aaaa0	30
aaaa	15	aaaaa	31
a0000	16	a00000	32

Table 1: Binary arithmetic introduced by Caramuel y Lobkowitz (1670, p. xlvi). He used an a instead of a 1.

Note. For the history of binary arithmetic, including the occurrence of the concept before Caramuel, see Glaser (1981); see also Shirley (1951), Zacher (1973), and Ineichen (2008).

References

[anon.] (1775). "Arithmétique". *Encyclopédie, ou dictionnaire universel raisonné des connaissances humaines. Supplément* 1, pp. 692–703.

Caramuel y Lobkowitz, Juan (1670). *Mathesis biceps, vetus et nova*. Campania.

Chambers, Ephraim (1728). *Cyclopaedia, or, An universal dictionary of arts and sciences*. Volume the First. London.

Glaser, Anton (1981). *History of Binary and Other Nondecimal Numeration*. Los Angeles.

Ineichen, Robert (2008). "Leibniz, Caramuel, Harriot und das Dualsystem". *Mitteilungen der deutschen Mathematiker-Vereinigung* 16, pp. 12–15.

Knobloch, Eberhard (1973). *Die mathematischen Studien von G.W. Leibniz zur Kombinatorik.* Auf Grund fast ausschließlich handschriftlicher Aufzeichnungen dargelegt und kommentiert. Wiesbaden.

Knutzen, Martin (1742). "Von dem Wahren Auctore der *Arithmeticae Binariae*, oder so genannten Leibnitzianischen Dyadic". *Philosophischer Bücher-Saal* 3, pp. 218–222.

Leibniz, Gottfried Wilhelm (1705). "Explication de l'Arithmétique binaire". *Mémoires de l'Académie royale des sciences* (1703), pp. 85–89.

Shirley, John W. (1951). "Binary Numeration Before Leibniz". *American Journal of Physics* 19, pp. 452–454.

Wolff, Christian (1716). *Mathematisches Lexicon*. Leipzig.

Zacher, Hans J. (1973). *Die Hauptschriften zur Dyadik von G.W. Leibniz.* Ein Beitrag zur Geschichte des binären Zahlensystems. Frankfurt.

binary* and *ternary form

In 1801, Gauß introduced the terms *binary form* and *ternary form* (Latin *forma binaria* and *forma ternaria*) (from the Latin word for 'two' and 'three'):

Thus far we have restricted our discussion to functions of the second degree with *two* unknowns and there was no need to give them a special name. But manifestly this topic is only a very special part of a more general study of *rational integral algebraic functions which are homogeneous of some degree in several unknowns*. Such functions can be conveniently divided according to their degree, as *forms of the second, third, fourth degree, etc.* and according to the number of unknowns into *binary, ternary, quaternary, etc. forms* [...]. Thus what we have simply been calling *forms* will be called *binary forms of the second degree*. But functions like

$$Ax^2 + 2Bxy + Cy^2 + 2Dxz + 2Eyz + Fz^2$$

(where A, B, C, D, E, F are integers) are called *ternary forms of the second degree* and so forth (Gauß 1801, § 266).[1]

In the French translation of Gauß' work, *forme binaire* as well as *forme trinaire* and *forme ternaire* are used (Gauß 1807, § 266). The English terminology is of later origin; an example is found in Sylvester (1851, p. 222): "I shall further call such a function binary, ternary, quaternary, and so forth, according to the number of variables contained in the functions F, G, H, &c. thus brought into coalition".

Note. For the history of the concept, see, for example, Dickson (1952).

References

Dickson, Leonard Eugene (1952). *History of the Theory of Numbers*. Volume III. Quadratic and higher forms. New York.

Gauß, Carl Friedrich (1801). *Disquisitiones arithmeticae*. Leipzig.

Gauß, Carl Friedrich (1807). *Recherches arithmétiques*. Traduites par A.-C.-M. Poullet-Delisle. Paris.

Gauß, Carl Friedrich (1986). *Disquisitiones Arithmeticae*. English Edition. Translated by A.A. Clarke. Revised by W.C. Waterhouse. New York et al.

Sylvester, James Joseph (1851). "On Extensions of the Dialytic Method of Elimination". *The London, Edinburgh, and Dublin Philosophical Magazine and Journal of Science* 2, pp. 221–230.

binomial

This term goes back to Euclid (*Elements*, book X, prop. 36), where (interpreted from a modern perspective) a sum consisting of two parts ($a + \sqrt{b}$ or $\sqrt{a} + \sqrt{b}$ with a, b rational) is called *ἐκ δύο ὀνομάτων* 'from two names'. His definition is given in geometric terms:[2] "If two rational straight lines commensurable in square only be added together, the whole is irrational; and let it be called *binomial*".[3] Euclid here uses Greek *ὄνομα* ('name') in its meaning 'term', so $a + \sqrt{b}$ is understood as

---

[1] "Hactenus disquisitionem nostram ad tales functiones secundi gradus restrinximus, quae *duas* indeterminatas implicant, neque opus fuit, denominationem specialem ipsis tribuere. Sed manifesto hoc argumentum tamquam sectionem maxime particularem disquisitionis generalissimae *de functionibus algebraicis rationalibus integris homogeneis plurium indeterminatarum et plurium dimensionum* considerare, talesque functiones secundum multitudinem dimensionum in *formas secundi, tertii, quarti gradus etc.*, secundum multitudinem indeterminatarum autem in *formas binarias, ternarias, quaternarias etc.* commode distinguere possumus. Formae itaque, hactenus simpliciter sic dictae, vocabuntur *formae binariae secundi gradus*; tales autem functiones ut $Axx + 2Bxy + Cyy + 2Dxz + 2Eyz + Fzz$) (denotantibus A, B, C, D, E, F integros datos) dicentur *formae ternariae secundi gradus* et sic porro". Translation from Gauß (1986, § 266).

[2] For accounts of how the geometric concepts translate into algebraic ones, see, e. g., Fowler (1999) and Knorr (1983). [3] Translation from Euclid (1956, p. 83).

composed of two terms. This usage of ἐκ δύο ὀνομάτων is actually older than Euclid; it was used by Theaetetus and in the pseudo-Aristotelian *De lineis insecabilibus* (Acerbi 2005, p. 331).[1]

In the 12th century, the Greek phrase was literally translated by *ex duobos nominibus* in some texts, for example in a translation of the *Elements* made directly from the Greek (Busard 1987, p. 240). At the same time, however, the term *binomium* also appears: Gerard of Cremona uses it in his translation of a commentary on the *Elements* (Anaritius 1899, p. 328). It was also used by Fibonacci in his *Liber abaci* (Fibonacci 1857, p. 357).

In the same sense, French *binome* first appears (TLF) in Peletier (1554, p. 136), and English *binomial* in Recorde (1557, fol. Pp iv^{r}), according to the OED. In German, the term first appears in Widmann (1489, fol. g 8^{v}) and Rudolff (1525, fol. F vi^{v}).[2]

In the 16th century, the concept was generalized by Tartaglia (1559, fol. 4^{r}), who called $ax^m + bx^n$ *binomio di dignità algebratice* (*dignità* is here used in the sense of 'power'). A further generalization is found in Stevin's *L'Arithmétique*, where he defines *multinomial* (*multinomie*) as a "number consisting of several quantities"[3] and uses *binomial* (*binomie*) for the special case of two quantities (Stevin 1585, p. 22). Thus, *binomial* with Stevin receives the modern meaning of an algebraic sum consisting of two terms.

Note. For the history of the concept and the terminology, see Tropfke (1933, p. 69), Netto and Le Vavasseur (1907, p. 4), and Spitzer (1941).

References

Acerbi, Fabio (2005). "A Reference to Perfect Numbers in Plato's *Theaetetus*". *Archive for History of Exact Sciences* 59, pp. 319–348.

Anaritius (1899). *Anaritii in decem libros priores Elementorum Euclidis commentarii*. Ex interpretatione Gherardi Cremonensis. Ed. by Maximilian Curtze. Leipzig.

Busard, H.L.L. (1987). *The Mediaeval Latin Translation of Euclid's Elements Made Directly from the Greek*. Stuttgart.

Euclid (1956). *The Thirteen Books of Euclid's Elements*. Volume III. Books X-XIII and Appendix. Ed. by Thomas L. Heath. New York.

Fibonacci (1857). *Scritti di Leonardo Pisano*. Pubblicati da Baldassare Boncompagni. Volume I. Roma.

Fowler, David (1999). "An Invitation to Read Book X of Euclid's *Elements*". *Historia Mathematica* 19, pp. 233–264.

Heath, Thomas L. (1949). *Mathematics in Aristotle*. Oxford.

Knorr, Wilbur R. (1983). "'La croix des mathématiciens': The Euclidean theory of irrational lines". *Bulletin of the American Mathematical Society* 9, pp. 41–69.

Netto, Eugen and Raymond Le Vavasseur (1907). "Les fonctions rationnelles". *Encyclopédie des sciences mathématiques pures et appliquées* I.9, pp. 1–232.

Peletier, Jacques (1554). *L'algebre de Jaques Peletier du Mans, departie an deus livres*. Lyon.

Recorde, Robert (1557). *The Whetstone of Witte*. London.

Rudolff, Christoph (1525). *Behend vnnd Hübsch Rechnung durch die kunstreichen regeln Algebra*. Straßburg.

Schirmer, Alfred (1912). *Der Wortschatz der Mathematik nach Alter und Herkunft untersucht*. Straßburg.

Spitzer, Leo (1941). "Binôme". *Modern Language Notes* 56, pp. 426–429.

---

[1] See also Heath (1949, pp. 255f.) and Knorr (1983, pp. 43, 63). [2] For the German term, see Schirmer (1912, p. 11). [3] "Multinomie algebraique est un nombre consistant de plusieurs diverses quantitez".

Stevin, Simon (1585). *L'Arithmetique*. Leiden.
Tartaglia, Niccolò (1559). *La sesta parte del general trattato di numeri, et misure*. Venezia.
Tropfke, Johannes (1933). *Geschichte der Elementar-Mathematik*. Zweiter Band. Allgemeine Arithmetik. Dritte, verbesserte und vermehrte Auflage. Berlin and Leipzig.
Widmann, Johannes (1489). *Behende vnd hubsche Rechenung auff allen kauffmanschafft*. Leipzig.

binomial coefficient

Binomial coefficients have been known since the Middle Ages,[1] but the current terminology only begins to develop in the 16th century, when Stifel (1544, fol. 34^r) used *numeri binomiales* for the concept.[2]

Oughtred (1631, p. 27) called binomial coefficients *unciae*, which are explained as "numerall figures" in Oughtred (1647, p. 41). The word originally meant 'one-twelth'.[3] The term is found in numerous publications and was also used by Euler (1748, §262).

The explicit *binomial coefficient* (German *Binomialcoefficient*) was regularly used by Kästner, for example in Kästner (1760, p. 385) and Kästner (1761, p. 80); it appears he was the first to do so, but he did not formally introduce the term. It was also used by other German writers such as Klügel (1770, p. 48) and Hindenburg (1779, p. 71). The English term does not seem to come up until the 19th century; an early use is Broughton (1814, p. 15).

References

Broughton, Charles (1814). *Memoir Respecting a New Theory of Numbers*. Part first. Edinburgh.
Euler, Leonhard (1748). *Introductio in analysin infinitorum*. Tomus primus. Lausanne.
Hindenburg, Carl Friedrich (1779). *Infinitinomii dignitatum exponentis indeterminati historia leges ac formulae*. Göttingen.
Kästner, Abraham Gotthelf (1760). *Anfangsgründe der Analysis endlicher Grössen*. Göttingen.
Kästner, Abraham Gotthelf (1761). *Anfangsgründe der Analysis des Unendlichen*. Göttingen.
Kline, Morris (1972). *Mathematical Thought From Ancient To Modern Times*. Volume 1. New York and Oxford.
Klügel, Georg Simon (1770). *Analytische Trigonometrie*. Braunschweig.
Knuth, Donald E. (1997). *The Art of Computer Programming*. Volume 1. Fundamental algorithms. Third Edition. Reading, Mass.
Oughtred, William (1631). *Arithmeticae in numeris et speciebus institutio*. London.
Oughtred, William (1647). *The Key of the Mathematicks*. London.
Stifel, Michael (1544). *Arithmetica integra*. Nürnberg.
Viète, François (1646). "Zeteticorum libri quinque". In: *Opera mathematica*. Ed. by Frans van Schooten. Leiden, pp. 42–81.
Viète, François (1983). *The Analytic Art*. Nine Studies in Algebra, Geometry and Trigonometry from the *Opus Restitutae Mathematicae Analyseos, seu Algebrâ Novâ*. Translated by T. Richard Witmer. New York.

binomial distribution

The precise origin of the term *binomial distribution* is somewhat obscure. The term was used by Pearson (1895, p. 351) without further explanation. The earliest textbook

---

[1] See, for example, Knuth (1997, esp. pp. 52-54). [2] Kline (1972, p. 272) incorrectly states that "Stifel introduced the term 'binomial coefficient'". [3] Oughtred may have been inspired by Viète's use of the same word for certain fractions or parts, for example in Viète (1646, p. 45). On the term, Witmer comments: "The word *uncia* [...] originally meant one-twelfth. Viète, it is clear, generalized it to mean any fraction" (Viète 1983, p. 236).

to devote a whole chapter to the binomial distribution, under that name, appears to be Yule (1911, ch. xv).

Note. For the history of the concept, see Laudański (2013, pp. 87–98).

References

Laudański, Ludomir (2013). *Between Certainty and Uncertainty*. Statistics and Probability in Five Units with Notes on Historical Origins and Illustrative Numerical Examples. Berlin and Heidelberg.

Pearson, Karl (1895). "Contributions to the Mathematical Theory of Evolution. II. Skew Variation in Homogeneous Material". *Philosophical Transactions of the Royal Society of London* 186, pp. 343–414.

Yule, George Udny (1911). *An Introduction to the Theory of Statistics*. London.

binomial series

The binomial series is closely tied to Newton, who discussed it in a letter to Henry Oldenburg from 1676.[1] The term itself was not yet used by Newton, however. In print, *binomial series* is found in Taylor (1717-19, p. 615) and Hodgson (1723, p. 670).

Note. For the history of the concept, see, for example, Edwards (1979, pp. 178–187), Guicciardini (2003, pp. 75f.), and González-Velasco (2011, pp. 122–135).

References

Edwards, C.H. (1979). *The Historical Development of the Calculus*. New York et al.

González-Velasco, Enrique A. (2011). *Journey through Mathematics*. Creative Episodes in Its History. New York.

Guicciardini, Niccolò (2003). "Newton's Method and Leibniz' Calculus". In: *A History of Analysis*. Ed. by Hans Niels Jahnke. Providence, R.I., pp. 73–103.

Hodgson, James (1723). *A System of the Mathematics*. Vol. I. London.

Taylor, Brook (1717-19). "An Attempt towards the Improvement of the Method of Approximating, in the Extraction of the Roots of Equations in Numbers". *Philosophical Transactions of the Royal Society of London* 30, pp. 610–622.

Turnbull, H. W., ed. (1960). *The Correspondence of Isaac Newton*. Volume II. 1676–1687. Cambridge.

binormal

This term was coined as French *binormale* by de Saint-Venant (1845, p. 17) for "that normal that is perpendicular to the osculating plane".[2] Saint-Venant (ibid.) points out that the binormal is normal to two consecutive elements of the curve, which accounts for the name. He adds that he also considered the term *pernormale* "because this normal is in some way more than the others" (*per-* is a prefix indicating intensity). Also, he says he would have preferred the term *principal normal*, but he avoided this as the latter already had an established mathematical meaning (see *principal normal*).

According to Reich (1973, p. 285), *binormal* did not catch on at first; for example, Serret (1853, p. 193) used *axe du plan osculateur* 'axis of the osculating plane', and Frenet (1853, p. 437) just used *axe* 'axis'. The term gradually became more popular, however, and appeared, for example, in Schell's book on space curves (Schell 1859, p. 6). The English term was already used by Price (1852, p. 448), who explained that the line "has the distinctive name *binormal*, being, as it is, perpendicular to two

---

[1] The letter is found in Turnbull (1960, pp. 20–32). For the first publications of the binomial theorem, see Turnbull (1960, pp. 153f.). [2] "celles des normales qui est perpendiculaire au plan osculateur".

consecutive elements of the curve, while all other normals are perpendicular to only one".

Note. The terminology is briefly discussed by Reich (1973, p. 285) and Delcourt (2011, p. 274).

References

Delcourt, Jean (2011). "Analyse et géométrie, histoire des courbes gauches de Clairaut à Darboux". *Archive for History of Exact Sciences* 65, pp. 229–293.

Frenet, Jean Frédéric (1853). "Sur les courbes à double courbure". *Journal de mathématiques pures et appliquées* 17, pp. 437–447.

Price, Bartholomew (1852). *A Treatise on Infinitesimal Calculus*. Vol. I. Differential Calculus. Oxford.

Reich, Karin (1973). "Die Geschichte der Differentialgeometrie von Gauss bis Riemann (1828-1868)". *Archive for History of Exact Sciences* 11, pp. 273–382.

de Saint-Venant, Adhémar Barré (1845). "Mémoire sur les lignes courbes non planes". *Journal de l'École polytechnique* 18, cahier 30, pp. 1–76.

Schell, Wilhelm (1859). *Allgemeine Theorie der Curven doppelter Krümmung in rein geometrischer Darstellung*. Leipzig.

Serret, Joseph-Alfred (1853). "Observations sur deux notes de M. Bonnet". *Comptes rendus de l'Académie des sciences* 36, pp. 432–436.

bipartite

A bipartite graph is a graph whose sets of vertices can be decomposed into two disjoint sets such that no two graph vertices within the same set are adjacent, which is equivalent to saying that all its circuits are of even length.

The notion was introduced by Dénes König, who in 1914 presented an important result on bipartite graphs at a congress and called such a graph "graph with even circuits" (*graphe à circuits pairs*); the paper was not published until 1923, however (see König 1923, p. 447).

In 1916 König published an article in Hungarian (König 1916a) which also appeared in a German translation; in that article he introduced the term *paarer Graph* for the same concept, where *paar* is used in its meaning 'even' (König 1916b, p. 454). The term *bipartite* was introduced as French *bipartie* by Sainte-Laguë (1923, p. 29). It literally means 'of two parts'.

The word *bipartite* was earlier also used in connection with curves: "We saw [...] that curves of the third degree were either altogether continuous, or bipartite, that is to say, consisting of a continuous infinite part, together with an oval or the projection of an oval" (Salmon 1852, p. 200).

Note. A brief historical account of the graph-theoretical terminology is given by König (1990, p. 297); see also Gropp (1998, esp. p. 273).

References

Gropp, Harald (1998). "On Combinatorial Papers of König and Steinitz". *Acta Applicandae Mathematica* 52, pp. 271–276.

König, Dénes (1916a). "Graphok és alkalmazásuk a determinánsok és a halmazok elméletére". *Mathematikai és természettudományi értesítő* 34, pp. 104–119.

König, Dénes (1916b). "Über Graphen und ihre Anwendung auf Determinantentheorie und Mengenlehre". *Mathematische Annalen* 77, pp. 453–465.

König, Dénes (1923). "Sur un problème de la théorie générale des ensembles et la théorie des graphes". *Revue de métaphysique et de morale* 30, pp. 443–449.

König, Dénes (1990). *Theory of Finite and Infinite Graphs*. Translated by Richard McCoart. Boston.
Sainte-Laguë, André (1923). "Les réseaux". *Annales de la Faculté des sciences de Toulouse* 15, pp. 27–86.
Salmon, George (1852). *A Treatise on the Higher plane curves intended as a sequel to the Treatise on Conic Sections*. Dublin.

birational

Birational transformations were originally conceived of as being 'doubly rational' in the sense that their inverse also preserves rationality. The term seems to appear first as *trasformazione birazionale* in two Italian papers from 1869: in Jung and Armenante (1869), the term appears only in the title, in Cremona and Casorati (1869, p. 620), it is mentioned in a definition of *genus*. In both papers, it is explicitly stated that the term translates German *eindeutige Transformation*, lit. 'unambiguous, unequivocal transformation'. That term had been used by Clebsch and Gordan (1866, ch. 3), who defined such transformations as "those in which the x's can be rationally expressed through the y's and, inversely, the y's through the x's" (p. 52).[1]

The concept had earlier been discussed by Riemann (1857, p. 133) under the name *rational substitution* (German *rationale Substitution*):

> We shall now consider *all irreducible equations between two complex variables, which can transformed into one another by rational transformations [rationale Substitutionen], as belonging to the same class*. Thus $F(s,z) = 0$ and $F_1(s_1, z_l) = 0$ belong to the same class, if rational functions of s_1 and z_l can be found which, when substituted for s and z respectively, transform the equation $F(s,z) = 0$ into the equation $F_1(s_1, z_l) = 0$; while equally s_1 and z_l are rational functions of s and z.[2]

The terminology thus varied between *rational* and *birational*; Darboux (1872, p. 223), for example, used them synonymously. Mansion (1878, p. 61) clearly stated that *birational* is the better term. In English, it came up in the 1880s; according to the *Proceedings of the London Mathematical Society* 16, p. 6, Cremona held at talk entitled "On a Birational Transformation of Space of Three Dimensions" at the annual meeting in 1884.

Note. For the history of the notion, see Coolidge (1940, pp. 195–230, 285–293), Kline (1972, pp. 932–934), Gray (1989), and Čižmár (1995).

References

Čižmár, Ján (1995). "Birationale Transformationen (ein historischer Überblick)". *Periodica Polytechnica Mechanical Engineering* 39, pp. 9–24.
Clebsch, Alfred and Paul Gordan (1866). *Theorie der Abelschen Functionen*. Leipzig.
Coolidge, Julian Lowell (1940). *A History of Geometrical Methods*. Oxford.
Cremona, Luigi and Felice Casorati (1869). "Intorno al numero de' moduli dell'equazioni e delle curve algebriche di dato genere. Osservazioni". *Reale Istituto lombardo di scienze e lettere. Rendiconti* 2, pp. 620–625.

---

[1] "diejenigen [Transformationen], in denen sich die x ebensowohl rational durch die y, als umgekehrt die y rational durch die x ausdrücken lassen". [2] "Man betrachte nun als zu Einer *Klasse* gehörend *alle irreductiblen algebraischen Gleichungen zwischen zwei veränderlichen Grössen, welche sich durch rationale Substitutionen in einander transformiren lassen*, so dass $F(s,z) = 0$ und $F_1(s_1, z_l) = 0$ zu derselben Klasse gehören, wenn sich für s und z solche rationale Functionen von s_1 und z_1 setzen lassen, dass $F(s,z) = 0$ in $F_1(s_1, z_l) = 0$ übergeht und zugleich s_1 und z_1 rationale Functionen von s und z sind". Translation from Riemann (2004, p. 111). The emphasis is original.

Darboux, Gaston (1872). "Sur une méthode nouvelle pour l'étude des courbes tracées sur les surfaces algébriques". *Bulletin des sciences mathématiques et astronomiques* 3, pp. 221–224.

Gray, Jeremy (1989). "Algebraic Geometry in the late Nineteenth Century". In: *The History of Modern Mathematics*. Volume I: Ideas and Their Reception. Ed. by David E. Rowe and John McCleary. Boston, pp. 361–385.

Jung, Giuseppe and Angelo Armenante (1869). "Sulle trasformazioni birazionali o univoche, e sulle curve normale e subnormale del genere *p*". *Giornale di matematiche* 7, pp. 235–253.

Kline, Morris (1972). *Mathematical Thought From Ancient To Modern Times*. Volume 3. New York and Oxford.

Mansion, Paul (1878). "Sur la théorie des transformations birationnelles planes en général". *Nouvelle correspondance mathématique* 1, pp. 61–69.

Riemann, Bernhard (1857). "Theorie der Abel'schen Functionen". *Journal für die reine und angewandte Mathematik* 54, pp. 115–155.

Riemann, Bernhard (2004). *Collected Papers*. Translated from the 1892 edition by Roger Baker, Charles Christenson and Henry Orde. Heber City.

bisect **and** ***bisector***

The angle bisector theorem was stated by Euclid in his *Elements* (book VI, prop. 3):

> If an angle of a triangle be bisected and the straight line cutting the angle cut the base also, the segments of the base will have the same ratio as the remaining sides of the triangle; and, if the segments of the base have the same ratio as the remaining sides of the triangle, the straight line joined from the vertex to the point of section will bisect the angle of the triangle".[1]

In this proposition and in general, to express that a figure is divided into two equal parts, Greek mathematicians used the verb *διχοτομεῖν*, which literally means 'to cut in half' (see Mugler 1958-59, p. 147 for further examples). It is also in this way that the proposition was originally translated, as exemplified by Billingsley's translation, where the same proposition begins with: "If an angle of a triangle be deuided into two equal parts [...]" (Euclid 1570, fol. 157r). Similar ways of expression can already be found in the translations of the *Elements* made in the 12th and 13th centuries.

The actual verb *bisect* (from *bi-* 'twice, two' and *secare* 'to cut') and related forms are of much later origin. The term was regularly used by Ramus (1569, p. 153), who suggested *bisectio* as a translation of *διχοτομία*.[2]

Instead of referring to the "line that cuts an angle in half", the term *bisector* (in French *bissectrice*) later began to be used. Its exact origin is unclear, but it does not seem to have come up until the 19th century. The English term is found in a collection of examination questions, and the first usage in this collection dates from 1817 ([anon.] 1823, p. 39). The French term seems to be of only slightly later origin; it is found in Bergery (1828, p. 121). In German, the term *Winkelhalbierende* (from the words for 'angle' and 'halving') also begins to be used at this time; it appears in van Swinden (1834, p. 128).[3]

References

[anon.] (1823). *Dublin Problems*. Being a Collection of Questions Proposed to the Candidates for the Gold Medal at the General Examinations, from 1816 to 1822 Inclusive. Dublin.

---

[1] Translation from Euclid (1956, pp. 195f.). [2] The word is also defined in Ramus (1567, p. 3); see also Ramus (1636, p. 50), where the translator observes: "That Bisecare, doth here, and in other places following, signifie to cutte a thing into two equall parts or portions; And so *Bisegmentum*, to be one such portion; And *Bisectio*, such a like cutting or division". [3] The text is a translation from the Dutch; in the original, no corresponding term is used.

Bergery, Claude-Lucien (1828). *Géométrie appliquée à l'industrie*. Seconde édition. Metz.
Euclid (1570). *The Elements of Geometrie of the most auncient Philosopher Euclide of Megara*. Faithfully (now first) translated into the Englishe toung, by H. Billingsley. London.
Euclid (1956). *The Thirteen Books of Euclid's Elements*. Volume II. Books III-IX. Ed. by Thomas L. Heath. New York.
Mugler, Charles (1958-59). *Dictionnaire historique de la terminologie géométrique des Grecs*. Paris.
Ramus, Petrus (1567). *Geometria*. Paris.
Ramus, Petrus (1569). *Scholarum Mathematicarum, Libri Unus et Triginta*. Basel.
Ramus, Petrus (1636). *Via regia ad geometriam. The Vvay to Geometry*. Written in Latine by Peter Ramus, and now Translated and much enlarged by the Learned Mr. William Bedwell. London.
van Swinden, Jan Hendrik (1834). *Elemente der Geometrie*. Aus dem Holländischen übersetzt und vermehrt von C.F.A. Jacobi. Jena.

block **(graph)**

As defined by Harary (1949, p. 405), a *block* of a graph is a "maximal connected subgraph containing no cut points of itself". A related notion is already found in Sainte-Laguë (1926, p. 5), who used French *membre* for particular kinds of blocks. Whitney (1932, p. 344) used the term *component*, while König (1936, p. 225) preferred German *Glied* 'member'. Harary's use of *block* in the above quotation may be the first for this term in English; he points out elsewhere (Harary and Norman 1953, p. 332) that he adopted it from König.

Harary (1963, p. 2) later also used the term *block graph*: "The *block-graph* $B(G)$ *of a given graph* is that graph whose points are the blocks B_1, B_2, ..., B_N of G and whose lines are determined by taking two points B_i and B_j as adjacent if and only if they contain a cut point of G in common. A graph is called a *block-graph* if it is the block-graph of some graph".

Note. The history of the notion and the terminology are discussed in König (1990, p. 393).

References

Harary, Frank (1949). "An Elementary Theorem on Graphs". *The American Mathematical Monthly* 66, pp. 405–407.
Harary, Frank (1963). "A characterization of block-graphs". *Canadian Mathematical Bulletin* 6, pp. 1–6.
Harary, Frank and Robert Z. Norman (1953). "The Dissimilarity Characteristic of Husimi Trees". *Annals of Mathematics* 58, pp. 134–141.
König, Dénes (1936). *Theorie der endlichen und unendlichen Graphen*. Leipzig.
König, Dénes (1990). *Theory of Finite and Infinite Graphs*. Translated by Richard McCoart. Boston.
Sainte-Laguë, André (1926). *Les réseaux (ou graphes)*. Paris.
Whitney, Hassler (1932). "Non-separable and planar graphs". *Transactions of the American Mathematical Society* 34, pp. 339–363.

Boolean algebra **and** ***Boolean ring***

Boolian algebra [sic] began to be used[1] by C. S. Peirce in the 1880s.[2] In print, it appears in Peirce (1885, e. g., p. 191); a definition was given by Peirce in his entry for

---

[1] OED2 cites Sylvester (1851, p. 192) for the first use of the adjective *Boolean*: "The Hessian, or as it ought to be termed, the first Boolian Determinant". [2] Some Internet sources, including Wikipedia, claim that Peirce first used *Boolean algebra* in a paper written around 1880 called "A Boolian Algebra with One Constant". However, the title of this paper was supplied by the editors when the paper was first published (Hartshorne and Weiss 1933, pp. 13–18), and *Boolian algebra* does not occur in the text.

the word *algebra* in the *Century Dictionary* (1889-1891): "a logical algebra, invented by the English mathematician George Boole [...], for the solution of problems in ordinary logic. It also has a connection with the theory of probabilities". With Peirce, the term appears to denote a certain approach to algebra rather than an algebraic structure.[1]

A set of postulates for a *Boolean algebra* or rather, for *Boolean algebras*, was presented by Sheffer (1913), who remarked (p. 481): "We employ the term *Boolean algebras* in its plural form for the following reasons: (1) none of the equivalent postulate-sets here referred to is in terms of its undefined entities one-valued ('categorical')–that is, each determines not a single algebra but a class of algebras; one should not speak, therefore, of '*der* identische Kalkül' (Schröder), '*the* Algebra of Symbolic Logic' (Whitehead), or '*the* algebra of logic' (Huntington) [...]".[2]

Boolean ring was suggested by Stone (1936, p. 39), for which he gave a concise definition: "A ring in which every element is idempotent, satisfying the law $aa = a$, is called a Boolean ring". The term was quickly adopted, for example by McCoy and Montgomery (1937, p. 455).

As pointed out by Miller (n.d.), an interesting terminological observation was made by Halmos (1956, p. 356):

> Terminological purists sometimes object to the Boolean use of the word "algebra." The objection is not really cogent. In the first place, the theory of Boolean algebras has not yet collided, and it is not likely to collide, with the theory of linear algebras. In the second place, a collision would not be catastrophic; a Boolean algebra is, after all, a linear algebra over the field of integers modulo 2. [...] While, to be sure, a shorter and more suggestive term than "Boolean algebra" might be desirable, the nomenclature is so thoroughly established that to change now would do more harm than good.

References

Halmos, Paul R. (1956). "The Basic Concepts of Algebraic Logic". *The American Mathematical Monthly* 63, pp. 363–387.

Hartshorne, Charles and Paul Weiss, eds. (1933). *Collected Papers of Charles Sanders Peirce*. Volume IV. Cambridge, Mass.

Kloesel, Christian J.W., ed. (1982). *Writings of Charles S. Peirce*. A Chronological Edition. Volume 4: 1879–1884. Bloomington and Indianapolis.

McCoy, Neal H. and Deane Montgomery (1937). "A representation of generalized Boolean rings". *Duke Mathematical Journal* 3, pp. 455–459.

Miller, Jeff (n.d.). "Earliest Known Uses of Some of the Words of Mathematics". URL: `https://mathshistory.st-andrews.ac.uk/Miller/mathword/`.

Peirce, Charles Sanders (1885). "On the Algebra of Logic: A Contribution to the Philosophy of Notation". *American Journal of Mathematics* 7, pp. 180–202.

Sheffer, Henry Maurice (1913). "A Set of Five Independent Postulates for Boolean Algebras, with Application to Logical Constants". *Transactions of the American Mathematical Society* 14, pp. 481–488.

---

[1] In 1883, Peirce wrote: "And when I see such writers as George Bruce Halsted speaking of the *real power* of the Boolian algebra, I laugh in my sleeve. Not but that it has its utility. Some difficult problems can be resolved by its help. I admit it is useful, just as aluminium is a useful metal, for very nice opera glasses can be made of it, though it is far from being useful like iron, zinc or tin" (Kloesel 1982, p. 509). [2] As further reasons he mentioned that "Peano's *Formulario Mathematico* and Whitehead and Russell's *Principia Mathematica* [...] have a far stronger title to the name 'algebra of logic'" and that "[t]he Algebra of Symbolic Logic, viewed as a distinct algebra, is due to Boole" (the latter statement is a quotation from Whitehead 1898, p. 115).

Stone, Marshall Harvey (1936). "The Theory of Representation for Boolean Algebras". *Transactions of the American Mathematical Society* 40, pp. 37–111.
Sylvester, James Joseph (1851). "Sketch of a memoir on elimination, transformation, and canonical forms". *The Cambridge and Dublin Mathematical Journal* 6, pp. 186–200.
Whitehead, Alfred North (1898). *A Treatise on Universal Algebra*. Volume I. Cambridge.

Borel set

A Borel set is a set in a topological space that can be constructed from open or closed sets through the operations of countable union and countable intersection. In this sense, the German term *Borelsche Menge* was coined by Hausdorff (1914, p. 305). Lebesgue (1904, p. 109) had earlier used *ensemble mesurable B*. The concept is due to Borel (1898, ch. III). Hausdorff's term eventually became standard; in English, it seems to have been first used by Blumberg (1920, p. 127).

Note that German *Borelsche Menge* was already suggested by Schoenflies (1900, pp. 109f.), but he used it for the so-called G_δ*-sets*, the countable intersection of open sets.[1]

Note. For the history of the concept and the terminology, see Kanovei and Koepke (2002, p. 775).

References

Blumberg, Henry (1920). "Hausdorff's Grundzüge der Mengenlehre [review]". *Bulletin of the American Mathematical Society* 27, pp. 116–129.
Borel, Émile (1898). *Leçons sur la théorie des fonctions*. Paris.
Hausdorff, Felix (1914). *Grundzüge der Mengenlehre*. Leipzig.
Kanovei, Vladimir and Koepke (2002). "Deskriptive Mengenlehre in Hausdorffs *Grundzügen der Mengenlehre*". In: *Felix Hausdorff. Gesammelte Werke*. Band II. Grundzüge der Mengenlehre. Ed. by Egbert Brieskorn et al. Berlin, pp. 773–787.
Lebesgue, Henri (1904). *Leçons sur l'intégration et la recherche des fonctions primitives*. Paris.
Schoenflies, Arthur (1900). "Die Entwickelung der Lehre von den Punktmannigfaltigkeiten". *Jahresbericht der Deutschen Mathematiker-Vereinigung* 8, pp. 1–250.
Young, William Henry and Grace Chisholm Young (1906). *The Theory of Sets of Points*. Cambridge.

bornological and *bornivorous*

The French terms *ensemble bornivore* and *espace bornologique*, both based on the word *borné* 'bounded', were suggested by Bourbaki (1950, p. 10):

> Given a locally convex space E, we shall say that a subset A of E is a *bornivorous* set if it is *convex, circled*, and *absorbs every bounded set* in E. Every bornivorous set generates E. Every convex and circled neighbourhood of 0 is bornivorous. We shall say that a separated locally convex space is *bornological* if, inversely, *every bornivorous set in E is a neighbourhood of* 0.[2]

The terminology was quickly adopted, for example by Takenouchi (1952, p. 59); in English, *bornological* appeared in Dieudonné (1953, p. 503) and Nachbin (1954,

---

[1] G_δ*-Menge* is due to Hausdorff (1914, p. 305) (G stands for *Gebiet* 'domain', and δ for *Durchschnitt* 'intersection'). W. H. Young and G. C. Young (1906, p. 63) earlier used *inner limiting sets*. [2] "Étant donné un espace localement convexe E, nous dirons qu'une partie A de E est un ensemble *bornivore* s'il est *convexe, cerclé*, et *absorbe tout ensemble borné* dans E. Tout ensemble bornivore engendre E. Tout voisinage convexe et cerclé de 0 est bornivore. Nous dirons qu'un espace localement convexe séparé E est *bornologique* si inversement, *tout ensemble bornivore dans E est un voisinage de* 0".

p. 471). For the concept, Bourbaki and the other sources cited refer to Mackey (1946), where it is discussed under the name *boundedly closed, relatively strong space* (pp. 520, 523).

References

Bourbaki, Nicolas (1950). "Sur certains espaces vectoriels topologiques". *Annales de l'Institut Fourier* 2, pp. 5–16.

Dieudonné, Jean (1953). "Recent developments in the theory of locally convex vector spaces". *Bulletin of the American Mathematical Society* 59, pp. 495–512.

Mackey, George W. (1946). "On Convex Topological Linear Spaces". *Transactions of the American Mathematical Society* 60, pp. 519–537.

Nachbin, Leopoldo (1954). "Topological Vector Spaces of Continuous Function". *Proceedings of the National Academy of Sciences* 40, pp. 471–474.

Takenouchi, Osamu (1952). "Sur les espaces linéaires localement convexes". *Mathematical Journal of Okayama University* 2, pp. 57–84.

boundary (geometry)

In Euclid's *Elements*, we find this definition: "A *boundary* [*ὅρος*] is that which is an extremity [*πέρας*] of anything" (*Elements*, book I, def. 13).[1] The defined term *ὅρος* literally means 'boundary, landmark' and was also used in the meanings 'definition' as well as 'term of a ratio or proportion' (LSJ). Proclus later remarked that *πέρας* is the more general term, and *ὅρος* should be applied to plane surfaces and areas, but not to lines. He also says that *ὅρος* "has been at home in geometry from the beginning, for geometry is the art by which men were accustomed to measure lands and keep their boundary marks [*ὅρους*] distinct" (Proclus 1873, p. 136).[2]

There are essentially three Latin terms that translate *ὅρος* in this context, *extremitas*, *finis*, and *terminus*. The former, for example, was used by Balbus, who gave a definition from the perspective of land surveying: "A boundary is the place up to which rights of possession are granted to everyone, or the point up to which everyone maintains his own property" (Blume, Lachmann, and Rudorff 1848, p. 98).[3] *finis* was used by Isidorus of Sevilla (Isidore of Seville 1911, book III, xii), *terminus* by Martianus Capella (*De nuptiis* 710) (Willis 1983, p. 252). The latter term was also regularly used in the 12th and 13th-century translations and redactions of the *Elements*, and its successors in the Romance languages also appear in the respective first translations. Thus, Tartaglia has *termine* (Euclid 1543, fol. ix^r), and Forcadel *terme* (Euclid 1564, fol. 3^r).

An English alternative is found in Recorde (1551, fol. A iii^r): "because you shall not muse what I dooe call *a bound*, I mean therby a generall name, betokening the beginning, end and side, of any forme [figure]". The word *boundary* in this sense was not in common use at first; it gradually spread after Simson used it in his edition of the *Elements*: "A term or boundary, is the extremity of any thing" (Euclid 1756, p. 3).

Note. For the Greek terms, see Mugler (1958-59, pp. 316f., 337f.); for the Latin ones, see Guillaumin (2020, pp. 130f., 133f., 290f.).

---

[1] Translation from Euclid (1956, p. 153). [2] Translation from Proclus (1970, p. 109). [3] "Extremitas est quo usque uni cuique possidendi ius concessum est, aut quo usque quisque suum servat". Translation from Campbell (2000, p. 209).

References

Blume, Friedrich, Karl Lachmann, and Adolf Rudorff, eds. (1848). *Die Schriften der römischen Feldmesser*. Erster Band. Texte und Zeichnungen. Berlin.
Campbell, Brian (2000). *The Writings of the Roman Land Surveyors*. Introduction, Text, Translation and Commentary. London.
Euclid (1543). *Euclide Megarense philosopho*. Diligentemente reassettato, et alla integrita ridotto per ... Nicolo Tartalea. Venezia.
Euclid (1564). *Les Six Premiers Livres des Elements*. Traduicts et commentez par Pierre Forcadel de Besies. Paris.
Euclid (1756). *The Elements of Euclid*. By Robert Simson. Glasgow.
Euclid (1956). *The Thirteen Books of Euclid's Elements*. Volume I. Introduction and Books I, II. Ed. by Thomas L. Heath. New York.
Guillaumin, Jean-Yves (2020). *Dictionnaire de la terminologie latine ancienne de l'arithmétique et de la géométrie*. Paris.
Isidore of Seville (1911). *Etymologiarum sive Originum libri XX*. Tomus I. Ed. by Wallace M. Lindsay. Oxford.
Mugler, Charles (1958-59). *Dictionnaire historique de la terminologie géométrique des Grecs*. Paris.
Proclus (1873). *In primum Euclidis elementorum librum commentarii*. Ed. by Gottfried Friedlein. Leipzig.
Proclus (1970). *Commentary on the First Book of Euclid's Elements*. Translated with Introduction and Notes by Glenn R. Morrow. Princeton.
Recorde, Robert (1551). *The patheway to knowledg*. London.
Willis, James, ed. (1983). *Martianus Capella*. Leipzig.

boundary (topology)

A definition of *boundary* in a topological setting was given by Cantor. He remarked that the points of a "continuous manifold" M_μ can be split into two classes, "*interior points* [*innere Puncte*] and *points on the boundary* [*Puncte auf der Begrenzung*]; a point p of M_μ is called *interior* point if around p as a centre a $K_{\mu-1}$ [a sphere] can be described with a radius so small that all points within an on $K_{\mu-1}$ belong to the domain M_μ; every point p of M_μ, for which such a construction is *not* possible, belongs to the *boundary* of M_μ" (Cantor 1879, p. 130).[1]

Peano (1887, p. 158) used Italian *punto limite* for this concept, an unfortunate choice because the same word was also used for limit points. Jordan (1893, p. 20) suggested French *point frontière* and *frontière*, and English *boundary point* was used by Harkness (1894, p. 138) in a review of Jordan's book. Variations of the terminology appear in various topological contexts; for example, French *frontière* was also used by Poincaré (1895, p. 6) in his famous article "Analysis situs".

---

[1] "Die Puncte einer stetigen M_μ zerfallen in zwei Classen, in sogenannte *innere Puncte* und in *Puncte auf der Begrenzung*; ein Punct p heißt *innerer* Punct, wenn sich um p als Mittelpunct eine $K_{\mu-1}$ von so kleinem r beschreiben läßt, daß alle Puncte innerhalb und auf $K_{\mu-1}$ zum Gebiete M_μ gehören; jeder Punct p von M_μ, bei welchem eine solche Construction *nicht* möglich ist, wird zur *Begrenzung* von M_μ gezählt". – The word *Begrenzung* was already used by Riemann (1851, p. 6), for example in his definition of cross-cuts, which he defines as "lines which cut through the interior from one boundary point simply (no point occurring multiply) to another boundary point" ("Linien, welche von einem Begrenzungspunkte das Innere einfach – keinen Punkt mehrfach – bis zu einem Begrenzungspunkte durchschneiden"). Translation from Riemann (2004, p. 6).

References

Cantor, Georg (1879). "Ueber einen Satz aus der Theorie der stetigen Mannigfaltigkeiten". *Nachrichten von der Königl. Gesellschaft der Wissenschaften und der Georg-Augusts-Universität zu Göttingen* (1879), pp. 127–135.

Harkness, James (1894). "Jordan's Cours d'Analyse". *Bulletin of the American Mathematical Society* 3, pp. 135–141.

Jordan, Camille (1893). *Cours d'analyse*. Tome premier. Calcul différentiel. Paris.

Peano, Guiseppe (1887). *Applicazioni geometriche del calcolo infinitesimale*. Torino.

Poincaré, Henri (1895). "Analysis situs". *Journal de l'École polytechnique* 1, pp. 1–121.

Riemann, Bernhard (1851). "Grundlagen für eine allgemeine Theorie der Functionen einer veränderlichen complexen Grösse". PhD thesis. Georg-August-Universität Göttingen.

Riemann, Bernhard (2004). *Collected Papers*. Translated from the 1892 edition by Roger Baker, Charles Christenson and Henry Orde. Heber City.

bounded function **and** ***sequence***

Boundedness of sequences was explicitly defined by Gauß in a manuscript dating from around 1800. He said that if a sequence, "however far it is continued, contains no member that is larger than a given quantity λ, then λ can be called an upper bound for the sequence (limes supra seriem, une limite en plus)" (Gauß 1917, p. 391).[1] The manuscript was written in German and Gauß used the term *obere Grenze* 'upper bound (or limit)'. Gauß there also defined the corresponding term *untere Grenze* 'lower bound' (Gauß 1917, p. 392).

The same term *obere Grenze* was also occasionally used by Weierstraß, for example in lectures from 1886 (Weierstraß 1988, p. 25); more commonly, however, Weierstraß used the same term for the limit superior.

The first printed definition of bounded functions appears to be due to Jordan. It reads: "A function $f(x, y, \ldots)$ is called *bounded* in a set E on which it is defined if the values it takes for the diverse points $(x, y, \ldots)$ of this set form a bounded set" (Jordan 1893, p. 32).[2] Jordan used French *fonction bornée*.

Schoenflies (1900, p. 179), who pointed out Jordan's authorship of the term, preferred to call such a function *endlich* 'finite'. Fouët (1902, p. 40) used *fonction limitée* instead but gave *fonction finie* and *bornée* as alternatives. The English term *bounded function* comes up in the early 20th century; it was used by Veblen and Lennes (1907, p. 47).

References

Ferraro, Giovanni (2008). *The Rise and Development of the Theory of Series up to the Early 1820s*. New York.

Fouët, Edouard-André (1902). *Leçons élémentaires sur la théorie des fonctions analytiques*. Première partie. Paris.

Gauß, Carl Friedrich (1917). "Grundbegriffe der Lehre von den Reihen". In: *Werke*. Zehnten Bandes erste Abteilung. Göttingen, pp. 390–395.

---

[1] "so weit man sie auch fortsetzt, kein Glied vorkommt, das größer als eine gebene Größe λ wäre, so kann man λ eine obere Grenze für die Reihe nennen (limes supra seriem, une limite en plus)" (note that the word *Reihe*, here used for sequences, was later reserved for series). Gauß additionally defined the limit superior and the limit inferior. The manuscript was first published by Schlesinger (1912); for discussion, see Sebestik (1992, pp. 103–105) and Ferraro (2008, p. 335). [2] "Une fonction $f(x, y, \ldots)$ est dite *bornée*, dans un ensemble E pour lequel elle est définie, si les valeurs qu'elle prend pour les divers points $(x, y, \ldots)$ de cet ensemble forment un ensemble borné".

Jordan, Camille (1893). *Cours d'analyse*. Tome premier. Calcul différentiel. Paris.
Schlesinger, Ludwig (1912). "Über Gauß' Arbeiten zur Funktionentheorie". *Nachrichten von der Königl. Gesellschaft der Wissenschaften zu Göttingen (mathematisch-physikalische Klasse)* (1912. Beiheft), pp. 1–143.
Schoenflies, Arthur (1900). "Die Entwickelung der Lehre von den Punktmannigfaltigkeiten". *Jahresbericht der Deutschen Mathematiker-Vereinigung* 8, pp. 1–250.
Sebestik, Jan (1992). *Logique et mathématique chez Bernard Bolzano*. Paris.
Veblen, Oswald and Nels Johann Lennes (1907). *Introduction to Infinitesimal Analysis*. Functions of a real variable. New York.
Weierstraß, Karl (1988). *Ausgewählte Kapitel aus der Funktionenlehre*. Vorlesung, gehalten in Berlin 1886. Ed. by Reinhard Siegmund-Schultze. Leipzig.

bounded set

Bounded sets play an important role in the calculus. In the 19th century, the concept was usually paraphrased in some way.

One example can be seen in Cauchy's explanation of a continuous function: "the function $f(x)$ is continuous with respect to x between the given limits if, between these limits, an infinitely small increment in the variable always produces an infinitely small increment in the function itself (Cauchy 1821, pp. 34f.).[1]

The concept is also found in the various statements of the so-called Bolzano-Weierstraß theorem. An early formulation by Weierstraß from 1865 reads: "If, in a bounded part [*begrenzter Teil*] of the plane, there are infinitely many points with a given property, then there is at least one point (inside that part or on its boundary) such that in every neighborhood of this point there are infinitely many points having the given property" (Weierstraß 1865/66, fol. 16^{v}).[2]

General definitions of bounded sets are not found until the end of the 19th century, however. In particular, Jordan (1892, p. 73) defined *ensemble borné* 'bounded set' in the following terms: "A set of points (x, y) shall be called *bounded* if the coordinates of all its points remain contained between some fixed numbers M and n".[3] Shortly afterwards, he also defined the terms *bounded from above* (*borné supérieurement*) and *bounded from below* (*borné inférieurement*) (Jordan 1893, pp. 22f.). A similar definition of a bounded set had earlier been given by Riquier (1890, p. 266) under the name *portion d'espace limitée*.

The terminology is found soon afterwards in various textbooks. For example, Fouët (1902, p. 26) uses Jordan's *ensemble borné* and gives *fini* and *limité* as alternatives; Tannery (1904, p. 66) defines *borné en haut* 'bounded from above' and names *borné à droite* 'bounded on the right' as an alternative. He also defines the corresponding terms *borné en bas* and *borné à gauche*.

---

1 "En d'autres termes, la fonction $f(x)$ restera continue par rapport à x centre les limites données, si, entre ces limites, un accroissement infiniment petit de la variable produit toujours un accroissement infiniment petit de la fonction elle-même". Translation from Cauchy (1821, p. 6). 2 "Wenn es innerhalb eines allseitig begrenzten Theiles der Ebene unendlich viele Punkte giebt, denen eine bestimmte Eigenschaft zukommt, so giebt es (innerhalb oder auf der Begrenzung) mindestens einen Punkt derart, daß es in jeder Umgebung von ihm unendlich viele Punkte von jener Eigenschaft giebt". Translation from Moore (2008, p. 222), who also cites this as the earliest known version of the theorem. 3 "Un ensemble de points (x, y) sera dit *borné* si les coordonnées de tous ces points restent comprises entre des nombres fixes M et n".

Carslaw (1906, p. 27), who referred to Tannery, adopted this and used *bounded on the right* and *on the left*. Veblen and Lennes (1907, p. 3) use *upper* and *lower bound*, Goursat (1904, p. 140) *upper* and *lower limit*.

See also *infimum* and *limit inferior*.

Note. Jordan's and Riquier's definitions are referred to in Pringsheim (1899, p. 45).

References

Carslaw, H.S. (1906). *Introduction to the Theory of Fourier's Series and Integrals and the Mathematical Theory of the Conduction of Heat*. London.

Cauchy, Augustin-Louis (1821). *Cours d'analyse de l'École Royale Polytechnique*. I.re partie. Analyse algébrique. Paris.

Fouët, Edouard-André (1902). *Leçons élémentaires sur la théorie des fonctions analytiques*. Première partie. Paris.

Goursat, Édouard (1904). *A Course in Mathematical Analysis*. Translated by Earle Raymond Hedrick. Vol. I. Boston.

Jordan, Camille (1892). "Remarques sur les intégrales définies". *Journal de mathématiques pures et appliquées* 8, pp. 69–99.

Jordan, Camille (1893). *Cours d'analyse*. Tome premier. Calcul différentiel. Paris.

Moore, Gregory H. (2008). "The emergence of open sets, closed sets, and limit points in analysis and topology". *Historia Mathematica* 35, pp. 220–241.

Pringsheim, Alfred (1899). "Grundlagen der allgemeinen Funktionenlehre". *Encyklopädie der mathematischen Wissenschaften mit Einschluss ihrer Anwendungen* II A 1, pp. 1–53.

Riquier, Charles (1890). "Sur les fonctions continues d'un nombre quelconque de variables et sur le principe fondamental de la théorie des équations algébriques". *Annales scientifiques de l'École Normale Supérieure* 7, pp. 265–288.

Tannery, Jules (1904). *Introduction à la théorie des fonctions d'une variable*. Tome I. Paris.

Veblen, Oswald and Nels Johann Lennes (1907). *Introduction to Infinitesimal Analysis*. Functions of a real variable. New York.

Weierstraß, Karl (1865/66). "Principien der Theorie der analytischen Functionen [lecture notes taken by Moritz Pasch at the University of Berlin]". URL: `https://nbn-resolving.de/urn:nbn:de:hebis:26-digisam-172481` (retrieved Oct. 22, 2022).

bounded variation **(calculus)**

See *variation*.

box product

In his *Elements of Vector Analysis*, Gibbs (1881-84, p. 7) considered a product consisting of both a vector and a scalar product and remarked: "It will also be seen that $\alpha \times \beta \cdot \gamma$ represents in numerical value the volume of the parallelopiped of which α, β, and γ (supposed drawn from a common origin), are the edges [...] This kind of product may be called the scalar product of the three vectors". This is how the term *scalar triple product* came about, although this exact wording was not yet used by Gibbs.

The term *box product* was later coined by Taylor (1939, p. 46): "$z \cdot (x \times y)$ is a scalar whose numerical value is the *volume* of the parallelepiped whose edges are x, y, and z. [...] Because of this important interpretation we shall call the scalar $z \cdot (x \times y)$ the 'box product'". The new term was adopted by Brand (1947, p. 42), who also names Taylor as the originator. The alternative *mixed triple product* is also occasionally used, for example by Synge and Griffith (1942, p. 248).

References

Brand, Louis (1947). *Vector and Tensor Analysis*. New York.
Gibbs, Josiah Willard (1881-84). *Elements of Vector Analysis*. New Haven.
Synge, John L. and Byron A. Griffith (1942). *Principles of Mechanics*. New York and London.
Taylor, James Henry (1939). *Vector Analysis*. With an Introduction to Tensor Analysis. New York.

box topology

The term *box* as used in topology appears to be due to Kelley (1955, p. 107): "A base for the *box* topology for the cartesian product $\times\{X_a : a \in A\}$ is the family of all sets $\times\{U_a : a \in A\}$ where U is open in X_a for each a in A. Hence the cartesian product of open sets is open relative to the box topology".

The concept goes back to Tietze (1923, p. 298) and his definition of a *Produktraum* (*product space*); see Herrlich, Hušek, and Preuß (2008, p. 780).

References

Herrlich, Horst, Mirek Hušek, and Gerhard Preuß (2008). "Hausdorffs Studien zu Fundamentalkonstruktionen der Topologie". In: *Felix Hausdorff. Gesammelte Werke*. Band III. Mengenlehre (1927, 1935). Deskriptive Mengenlehre und Topologie. Ed. by Ulrich Felgner et al. Berlin and Heidelberg, pp. 778–786.
Kelley, John L. (1955). *General Topology*. Princeton.
Tietze, Heinrich (1923). "Beiträge zur allgemeinen Topologie. I. Axiome für verschiedene Fassungen des Umgebungsbegriffs". *Mathematische Annalen* 88, pp. 290–311.

brachistochrone

In June 1696, Johann Bernoulli challenged the mathematical community with the following problem: "Let two points A and B be given in a vertical plane. Determine the path AMB of a moving body M, along which, descending by its own gravity, and starting to move at point A, it arrives at the other point B in the shortest time" (Johann Bernoulli 1696, p. 269).[1]

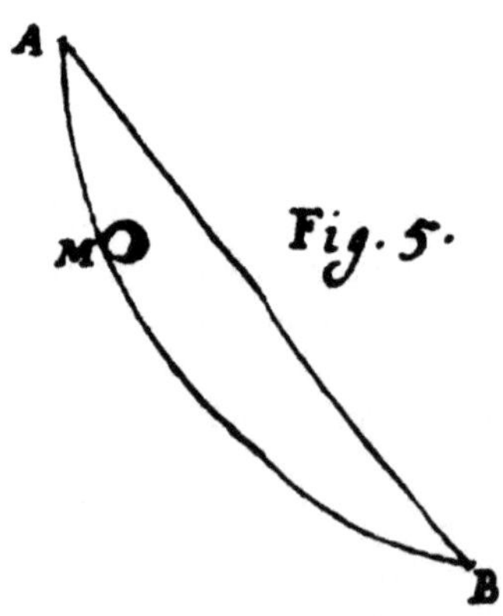

Figure 5: The brachistochrone. From Johann Bernoulli (1696, fig. 5).

[1] "Datis in plano verticali duobus punctis A & B ... assignare Mobili M, viam AMB, per quam gravitate sua descendens & moveri incipiens a puncto A, brevissimo tempore perveniat ad alterum punctum B". Translation from Engelsman (1984, p. 30).

Shortly afterwards in the same year, Bernoulli and Leibniz discussed the name of the curve. On July 21, 1696, Bernoulli wrote: "I have given the curve the name 'Brachystochrone' for the reason you see therein; but if the name 'Tachystoptote' pleases you more, I will let this be submitted in its place" (Leibniz 2011, pp. 49f.).[1] The terms under discussion express different views on the same problem: *linea brachystochronia* is based on Greek *βράχιστος* 'shortest' and *χρόνος* 'time' and thus means 'line of shortest time', whereas *linea tachystoptota* is based on Greek *τάχιστος* 'fastest' and *πίπτειν* 'to fall' and thus means 'line of swiftest fall' or 'quickest descent' (the latter term had been suggested by Leibniz in June 1695; see Leibniz 2004, p. 799). To this letter, Leibniz replied a few days later: "The name of Brachystochrone pleases me for its general significance; when it is a question of the fall of a heavy particle the name Tachystoptote possesses special significance" (Leibniz 2011, p. 75).[2]

In print, the term appears soon afterwards in Johann Bernoulli (1697, p. 206); it was taken over into other languages in unchanged form, for example in French (Johann Bernoulli 1698, p. 425). The term *curva oligochrona* (from Greek words for 'little time'), proposed by Jakob Bernoulli (1697, p. 211), did not find much appeal.[3]

Note. The history of the curve is discussed in Loria (1911, p. 87). Knobloch (2012) focuses on the brachistochrone in Leibniz' writings. For a discussion of the brachistochrone problem, Goldstine (1980, pp. 30–32) and Freguglia and Giaquinta (2016, ch. 2).

References

Bernoulli, Jakob (1697). "Solutio problematum fraternorum". *Acta Eruditorum* (May 1697), pp. 211–217.

Bernoulli, Johann (1696). "Supplementum defectus Geometriae Cartesianae circa Inventionem Locorum". *Acta Eruditorum* (June 1696), pp. 264–269.

Bernoulli, Johann (1697). "Curvatura radii in diaphanis non uniformibus". *Acta Eruditorum* (May 1697), pp. 206–211.

Bernoulli, Johann (1698). "Extrait d'une Lettre de M. Bernoulli, Professeur de Groningue, du 22 Août 1698, pour servir de Réponse à celle de son Frère Professeur à Bâle". *Journal des sçavans* (Dec. 1698), pp. 477–480.

Engelsman, Steven B. (1984). *Families of Curves and the Origins of Partial Differentiation*. Amsterdam and New York.

Freguglia, Paolo and Mariano Giaquinta (2016). *The Early Period of the Calculus of Variations*. Basel.

Goldstine, Herman H. (1980). *A History of the Calculus of Variations from the 17th through the 19th Century*. New York and Berlin.

Knobloch, Eberhard (2012). "Leibniz and the Brachistochrone". In: *Optimization Stories*. 21st International Symposium on Mathematical Programming, Berlin, August 19-24, 2012. Ed. by Martin Grötschel. Bielefeld, pp. 15–18.

Leibniz, Gottfried Wilhelm (2004). *Sämtliche Schriften und Briefe*. Dritte Reihe. Mathematischer, naturwissenschaftlicher und technischer Briefwechsel. Sechster Band. 1694-Juni 1696. Hannover.

[1] "Dederam lineae nomen *Brachistochronae* , ob rationem quam ibi videbis, sed si magis arrideat nomen Tachystoptotae, per mitto ut hoc in illius nomen ubique substituatur". Translation from Goldstine (1980, p. 31). [2] "Brachystochronae appellatio magis placet pro generali significatione; Tachystoptotae vero nomen posset speciali accommodari, cum agitur de gravis descensu seu casu". Translation from Goldstine (1980, p. 31). [3] For Jakob Bernoulli's approach to the problem, see Freguglia and Giaquinta (2016, pp. 45–49).

Leibniz, Gottfried Wilhelm (2011). *Sämtliche Schriften und Briefe*. Dritte Reihe. Mathematischer, naturwissenschaftlicher und technischer Briefwechsel. Siebenter Band. Juli 1696-Dezember 1698. Hannover.
Loria, Gino (1911). *Spezielle algebraische und transzendente ebene Kurven*. Theorie und Geschichte. Autorisierte, nach dem italienischen Manuskript bearbeitete deutsche Ausgabe von Fritz Schütte. Zweite Auflage. Zweiter Band. Leipzig.

braid

This topological term was introduced by Artin (1925-26) as German *Zopf* 'braid'. At the beginning of his paper,[1] he informally characterized braids thus: "by a braid we essentially understand an interweaving of threads, as the name already says" (Artin 1925-26, p. 47).[2] After giving a technical definition, Artin also considers the composition of braids and uses the term *braid group* (*Gruppe der Zöpfe*) in this paper (p. 55). The English term is found in Bohnenblust (1947) and in Artin's "Theory of braids" (Artin 1947, p. 102), in which he approached the subject rigourously and corrected some of his former proofs.

Although there is no direct relation to Artin's notion, it is perhaps worth pointing out that long before Artin, Vandermonde (1774, p. 566) had already used the French *tresse* 'braid' in a mathematical context:

> The worker who makes a *braid*, a *net*, or *knots*, does not conceive of them by relations of magnitude, but by those of situation: what he sees is the order in which the threads are interlaced. It would thus be useful to have a system of calculation that conforms better to the course of the worker's mind, a notation which would only represent the idea which he forms of his product, and which could suffice to reproduce a similar one for all times.[3]

Note. For the history of the concept, see Epple (1999b), Epple (1999a, esp. pp. 68-74, 139f., 183–192, 314–320), and Friedman (2019). A comparatively non-technical account of braids was given by Artin (1950).

References

Artin, Emil (1925-26). "Theorie der Zöpfe". *Abhandlungen aus dem Mathematischen Seminar der Universität Hamburg* 4, pp. 47–72.
Artin, Emil (1947). "Theory of Braids". *Annals of Mathematics* 48, pp. 101–126.
Artin, Emil (1950). "The theory of braids". *American Scientist* 38, pp. 112–129.
Bohnenblust, Frederic (1947). "The Algebraical Braid Group". *Annals of Mathematics* 48, pp. 127–136.
Epple, Moritz (1999a). *Die Entstehung der Knotentheorie*. Kontexte und Konstruktionen einer modernen mathematischen Theorie. Braunschweig.
Epple, Moritz (1999b). "Geometric Aspects in the Development of Knot Theory". In: *History of Topology*. Ed. by Ioan Mackenzie James. Amsterdam, pp. 301–357.
Friedman, Michael (2019). "Mathematical formalization and diagrammatic reasoning: the case study of the braid group between 1925 and 1950". *British Journal for the History of Mathematics* 34, pp. 43–59.

---

[1] Artin credits Schreier for "considerable support" in the writing of his paper. [2] "Dabei ist unter einem Zopf im wesentlichen ein Geflecht aus Fäden zu verstehen, wie der Name schon sagt".
[3] "L'ouvrier qui fait une *tresse*, un *réseau*, des *nœuds*, ne les conçoit pas par les rapports de grandeur, mais par ceux de situation: ce qu'il y voit, c'est l'ordre dans lequel sont entrelacés les fils. Il seroit donc utile d'avoir un système de calcul plus conforme à la marche de l'esprit de l'ouvrier, une notation qui ne représentât que l'idée qu'il se forme de son ouvrage, & qui pût suffire pour en refaire un semblable dans tous les temps". Translation from Epple (1999b, p. 311).

Vandermonde, Alexandre-Théophile (1774). “Remarques sur les problèmes de situation”. *Mémoires de l'Académie royale des sciences* 10, pp. 566–574.

branch

In his dictionary entry for *branch*, Barlow (1814) gave a common definition for the term, which is

> used to denote certain parts of a curve, which are infinitely extended without returning upon themselves; being also called infinite branches [...] In order to understand the better the nature of these branches, let x represent absciss [sic], and y the ordinate of a curve; then y being expressed in terms of x, if this latter be taken positive, y will have a certain number of values, and the same if x will be taken negative; and the curve will have as many branches, as there are real values of y corresponding to both the negative and positive values of x.

The definition was by no means new; it already appeared in Wolff (1747, pp. 340f.), who used the Latin word *crus* ‘leg’ and German *Ast* ‘branch’, and d'Alembert (1752, p. 395), where French *branche* was used. d'Alembert there also remarked that the meaning of the term was not fixed at the time. He gave, referring to Cramer (1750, p. 10), the example $y = \frac{xx}{6a} + x + \frac{5}{6}a$, which curve, he said, is ordinarily regarded as having one branch; “however”, he added, “this branch is sometimes counted as two, since it extends to infinity on the positive x side, and on the negative x side”.[1]

The terminology is found in Newton's *Enumeratio*, which text goes back to the late 1660s; in early drafts (Newton 1968, p. 94) as well as in print (Newton 1704, p. 142), Newton used the word *crus*, and distinguished between hyperbolic and parabolic branches: “I define a hyperbolic branch as one which constantly approaches some asymptote, a parabolic branch to be that which, although infinite, has no asymptote” (Newton 1704, p. 142).[2] Harris (1710, s. v. curve) translated *crus* by *leg*, Reyneau (1708, p. 571) by French *branche*.

References

d'Alembert, Jean Le Rond (1752). “Branche de courbe”. *Encyclopédie, ou Dictionnaire raisonné des sciences, des arts et des métiers* 2, p. 395.
Barlow, Peter (1814). *A New Mathematical and Philosophical Dictionary*. London.
Cramer, Gabriel (1750). *Introduction à l'analyse des lignes courbes algébriques*. Genève.
Harris, John (1710). *Lexicon Technicum*. Vol. II. London.
Newton, Isaac (1704). “Enumeratio linearum tertii ordinis”. In: *Opticks*. London, pp. 138–162.
Newton, Isaac (1860). *Enumeration of Lines of the Third Order*. Trans. by C.M.R. Talbot. London.
Newton, Isaac (1968). *The Mathematical Papers of Isaac Newton*. Volume II. 1667-1670. Ed. by D.T. Whiteside. Cambridge.
Reyneau, Charles-René (1708). *Usage de l'analyse*. Tome II. Paris.
Wolff, Christian (1747). *Vollständiges Mathematisches Lexicon*. Erster Theil. Leipzig.

branch point

Branch points play an essential role in Riemann's theory of functions. In his 1851 PhD thesis, he used *Windungspunkt* ‘winding point’ (Riemann 1851, p. 5) but also slipped into *Zweigpunkt* (*Zweig* means ‘branch, twig’) in one place (Riemann 1851, p. 12).

[1] “Cependant cette branche est quelquefois comptée pour deux, parce qu'elle s'étend à l'infini du côté des x positives, & du côté des x négatives”. [2] “Crus *Hyperbolicum* voco quod ad Asymptoton aliquam in infinitum appropinquat, *Parabolicum* quod Asymptoto destituitur”. Translation from Newton (1860, p. 10).

In 1857 he changed the terminology to *Verzweigungsstelle* (*Verzweigung* 'branching, ramification', *Stelle* 'place'): "the different prolongations of a given function in a given region of the z-plane will be called *branches* of the original function and a point around which one branch continues into another a *branch-point* of the function" (Riemann 1857a, pp. 102f.).[1] Riemann (1857b, p. 4) also uses *Verzweigungswerth* (*Wert* 'value') in the same sense, and also *Verzweigungspunkt* (1857, p. 103).[2]

Riemann's terminology was soon translated into other languages. In his translation of Riemann's thesis, Betti used *punto di giramento* 'point of gyration' (Riemann 1859, p. 292); Hoüel (1870, p. 100) used French *point de ramification*. While Smith (1874, p. 161) translated *Windungspunkt* by *spire* or *spiral point*, *branch-point* is soon found in Lodge (1876, p. 386) and Clifford (1877, p. 293).[3]

Note. For the history of the notion, see Bottazzini (1986, esp. ch. 6) and Bottazzini and Gray (2013, esp. ch. 5).

References

Bottazzini, Umberto (1986). *The Higher Calculus*. A History of Real and Complex Analysis from Euler to Weierstrass. Translated from the Italian by W. van Egmond. New York et al.

Bottazzini, Umberto and Jeremy Gray (2013). *Hidden Harmony – Geometric Fantasies*. The Rise of Complex Function Theory. New York.

Briot, Charles and Jean Claude Bouquet (1875). *Théorie des fonctions elliptiques*. Deuxième édition. Paris.

Clifford, William Kingdon (1877). "On the Canonical Form and Dissection of a Riemann's Surface". *Proceedings of the London Mathematical Society* 2, pp. 292–304.

Hoüel, Jules (1870). "Théorie élémentaire des quantités complexes. Troisième partie. Théorie des fonctions multiformes". *Mémoires de la Société des sciences physiques et naturelles de Bordeaux* 8, pp. 97–175.

Lodge, Oliver J. (1876). "On some Problems connected with the Flow of Electricity in a Plane". *The London, Edinburgh, and Dublin Philosophical Magazine* 1, pp. 373–389.

Riemann, Bernhard (1851). "Grundlagen für eine allgemeine Theorie der Functionen einer veränderlichen complexen Grösse". PhD thesis. Georg-August-Universität Göttingen.

Riemann, Bernhard (1857a). "Allgemeine Voraussetzungen und Hülfsmittel für die Untersuchung von Functionen unbeschränkt veränderlicher Größen". *Journal für die reine und angewandte Mathematik* 54, pp. 101–104.

Riemann, Bernhard (1857b). "Beiträge zur Theorie der durch die Gauss'sche Reihe $F(\alpha, \beta, \gamma, x)$ darstellbaren Functionen". *Abhandlungen der Königlichen Gesellschaft der Wissenschaften zu Göttingen* 7, pp. 3–22.

Riemann, Bernhard (1859). "Fondamenti di una teorica generale delle funzione di una variable complessa (tr. E. Betti)". *Annali di Matematica Pura e Applicata* 2, pp. 288–304, 337–356.

Riemann, Bernhard (2004). *Collected Papers*. Translated from the 1892 edition by Roger Baker, Charles Christenson and Henry Orde. Heber City.

Smith, H.J.S. (1874). "On the Higher Singularities of Plane Curves". *Proceedings of the London Mathematical Society* 6, pp. 153–182.

---

[1] "[...] sollen die verschiedenen Fortsetzungen *einer* Function für denselben Theil der z-Ebene *Zweige* dieser Function genannt werden und ein Punkt, um welchen sich ein Zweig einer Function in einen andern fortsetzt *eine Verzweigungsstelle* dieser Function". Translation from Riemann (2004, p. 80). [2] At a value where no branching takes place, Riemann (1857b, p. 4) also calls a function *monodromic*, a term he took over from Cauchy (see *monodromy*). He also transfers *monodromic* to German by creating the word *einändrig*, which does not otherwise exist in German. [3] Briot and Bouquet (1875, p. 12) called a branch point a *critical point* (*point critique*).

bridge

According to König (1936, p. 179), bridges in graphs seem to have first been considered by Tait (1880). Tait does not use a specific term, though, but says that "the points form a group connected by a *single* line only with the rest". His illustration is shown in fig. 6.

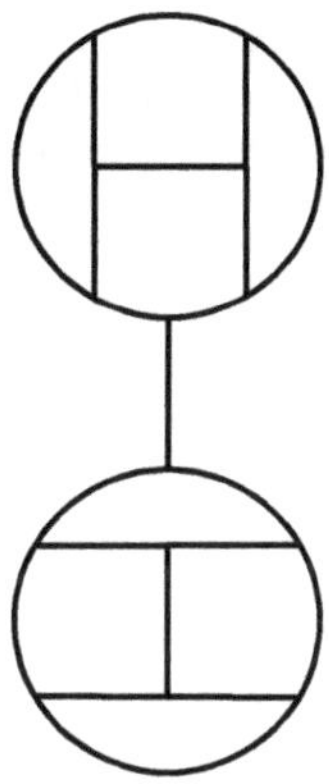

Figure 6: A bridge. From Tait (1880, plate xvi).

Lucas (1891, p. 115), referring to Tait, defines a bridge as an edge whose removal disconnects the graph. He uses French *isthme*, from Greek *ισθμός*, lit. 'neck', a word that was used in geography to designate a neck of land between two seas. Lucas adds that this definition also justifies the choice of the term. English *isthmus* appears to be a creation of the 20th century; it is found in Frink (1926, p. 492): "the stem of a leaf, or an 'isthmus'".

Apart from *isthme*, Lucas (1894, p. 135) also used *chemin isolant* 'isolating path'; evidently, he adopted this term from Fleury (1883, p. 259).

The term *bridge* (German *Brücke*) was coined by Ahrens (1897, p. 314); it also appears in the first book on graph theory by König (1936, p. 179). English *bridge* is a translation of this term; it may be due to Harary and Norman (1953, p. 8): "A *bridge* is a line of a connected graph whose removal separates the graph into two components each of which has more than one point".

Additional alternative terminology was suggested by Ore (1962, p. 81): "An edge $E = (a, b)$ is called a *separating edge* or a *bridge* or a *cut edge* in G if in the graph G_1 obtained from G by the removal of E the vertices a and b are disconnected".

See also *articulation point*.

Note. König (1936, p. 179) gives brief but precise comments on the history of the term. For an English translation, see König (1990, p. 330).

References

Ahrens, Wilhelm (1897). "Ueber das Gleichungssystem einer Kirchhoff'schen galvanischen Stromverzweigung". *Mathematische Annalen* 49, pp. 311–324.

Fleury, Henri (1883). "Deux problèmes de géométrie de situation". *Journal de mathématiques élémentaires* 2, pp. 257–261.

Frink, Orrin (1926). "A Proof of Petersen's Theorem". *Annals of Mathematics* 27, pp. 491–493.

Harary, Frank and Robert Z. Norman (1953). *Graph Theory as a Mathematical Model in Social Science*. Ann Arbor.

König, Dénes (1936). *Theorie der endlichen und unendlichen Graphen*. Leipzig.

König, Dénes (1990). *Theory of Finite and Infinite Graphs*. Translated by Richard McCoart. Boston.

Lucas, Édouard (1891). *Théorie des nombres*. Tome premier. Paris.

Lucas, Édouard (1894). *Récréations mathématiques*. IV. Paris.

Ore, Øystein (1962). *Theory of Graphs*. Providence, R.I.

Tait, Peter Guthrie (1880). "Note on a Theorem in Geometry of Position". *Transactions of the Royal Society of Edinburgh* 29, pp. 657–660.

C

cactus

The general word goes back to Greek κάκτος 'prickly plant' via Latin *cactus* 'cardoon'. As a mathematical term, it was first used by Harary and Uhlenbeck (1953, p. 315). Honouring Kodi Husimi, who used such graphs in his 1950 paper (Husimi 1950), they first defined a *Husimi tree* as "a connected graph in which no line lies on more than one cycle". They then went on to define a cactus as a Husimi tree in which every cycle is a triangle. Harary and Norman (1953, p. 134) point out that the term was chosen on account of the "pictorial appearance".

The word *cactus* had already featured earlier in mathematics: A *cactoid* was defined by Moore (1929, p. 81) for a particular kind of continuous curve: "A *cactoid* (opuntioid) is a bounded continuous curve M lying in space of three dimensions and such that (a) every non-degenerate maximal cyclic subset of M is a simple closed surface and (b) no point of M lies in a bounded complementary domain of any subcontinumn of M".

References

Harary, Frank and Robert Z. Norman (1953). "The Dissimilarity Characteristic of Husimi Trees". *Annals of Mathematics* 58, pp. 134–141.

Harary, Frank and George E. Uhlenbeck (1953). "On the Number of Husimi Trees, I". *Proceedings of the National Academy of Sciences* 39, pp. 315–322.

Husimi, Kodi (1950). "Note on Mayers' Theory of Cluster Integrals". *The Journal of Chemical Physics* 18, pp. 682–684.

Moore, Robert Lee (1929). "Concerning Upper Semi-continuous Collections". *Monatshefte für Mathematik und Physik* 36, pp. 81–88.

calculus

The Latin word *calculus* means 'small stone, pebble', and the verb *calculare* acquired the meaning 'to count' because pebbles were used for counting.[1] These words are

[1] The Roman grammarian Sextus Pompeius Festus (who probably flourished in the later 2nd century) already stated (Lindsay 1913, p. 40) that *calculus* is a diminutive of *calx* 'limestone', and this etymology of *calculus* is found in most dictionaries such as the OLD (an exception is Ernout and Meillet 2001, p. 89). This view has been questioned, however; see Loicq (1960) and André (1978, pp. 54f.). The theory is that *calculus* goes back to a pre-Indo-European root **kal*- 'stone' via reduplication (i. e., the repetition of the root).

H. Becker, *A Historical Dictionary of Mathematical Terms*, Science Networks. Historical Studies 66, https://doi.org/10.1007/978-3-032-02873-0_3

essentially loan translations of Greek words: ψῆφος means 'pebble', and derived from this is the verb ψηφίζεσθαι 'to calculate (with pebbles)'. Herodotus (*Histories* ii, 36), for example, asserted that "[i]n writing and in reckoning with pebbles the Greeks move the hand from left to right, but the Egyptians from right to left".[1] The Latin words later gave rise to uses in modern languages. In English, for example, *calculation* as well as the now obsolete *calcule* are attested both as a verb and as a noun in the 14th century; *calculus* as a method of calculating dates from the 17th century (OED).

In short, the history of the term already starts in Greek, and by the 17th century *calculus* was used in a very general sense ('calculation'). The specialization to the now usual mathematical sense is largely due to Leibniz, who used *calculus differentialis* in a manuscript from July 1676 (Leibniz 2008, p. 607). In print, this term appears in Leibniz (1684, p. 469): "this calculus [*calculus*], which I call *differential*".[2] Several years later, between 1691 and 1692, Bernoulli wrote his text on the integral calculus, entitled "Lectiones mathematicae de methodo integralium"; the first lecture is on the "nature and calculus of integrals" (Bernoulli 1742, p. 387).[3]

Thus, while early uses of *calculus* in the calculus may often be translated by *calculation*, it is through these uses that the word eventually acquired a new sense. Translations into languages others than Latin or French are found at the beginning of the 18th century. For example, the OED cites Raphson (1702, s. v. fluent) as a first use of *differential calculus*; Craig (1706, p. 261) used *integral calculus*. In Germany, Wolff (1716, pp. 283, 290) used *Differential-Rechnung* and *Integral-Rechnung* in his mathematical dictionary (*Rechnung* corresponds in meaning to *calculation*); he had already used these terms in Wolff (1710, p. 6). Some other kinds of calculi dealt with in Wolff's dictionary are *calculus astronomicus* (the calculation of the course of the sun and the stars), *calculus exponentialis*, and *calculus situs*.[4]

Newton's *De Methodis Serierum et Fluxionum* from 1671 gave rise to the commonly used term *method of fluxions*. While his book was published only in 1736 (Newton 1736), some of its contents became known much before and *method of fluxions* (or similar terms like *doctrine of fluxions*) are found in print well before the publication of Newton's book, for example in Wallis (1693, p. 396), where also *inverse method of fluxions* (for the integral calculus) is found. Early books on these topics are Hayes (1704) and Cheyne (1703).

One of the major proponents of the calculus in the 18th century was Euler. In his book on the differential calculus, he remarked: "Now, analysis of the infinite is concerned with the discovery of both differentials and integrals, and for this reason it is divided into two principal parts, one of which is called differential calculus, and the other is integral calculus. In the former, rules are given for finding differentials

---

[1] Translation from Thomas (1957, p. 35). On this passage, Rihll (1999, p. 49) comments: "The mention of pebbles in some literary contexts, such as Herodotus 2.36.2, implied the use of abacus, but the nature of such an abacus in Greek society is very uncertain". Heath (1921, p. 48) cites a number of historical sources that attest the use of pebbles for reckoning. [2] "calculi hujus, quem voco differentialem". Translation from Stedall (2008, p. 129). [3] "De Natura & Calculo Integralium". [4] For the last term, see *topology*.

of any quantity. In the latter a way of investigating integrals of given differentials is shown" (Euler 1755, § 148).[1]

For aspects of the further development, see *analysis* and the various entries in this dictionary for concepts of the calculus. For the term *calculus of variations*, see *variation* (calculus of variations). This term (as well as others) shows that, while *calculus* may often be used to mean differential calculus and integral calculus, the word is still found in its older meaning 'calculation'.

A final remark concerning the relation of *calculus* to *analysis*. An interesting aspect of the further development was pointed out by Clapham (1990, p. 5): "The term 'analysis' has also come to be used to indicate a rather more rigourous approach to the topics of calculus, and to the foundations of the real number system".

References

André, Jacques (1978). *Les mots à redoublement en latin*. Paris.

Bernoulli, Johann (1742). "Lectiones mathematicae de methodo integralium". In: *Opera omnia*. Tomus tertius. Lausanne and Genf, pp. 385–558.

Cheyne, George (1703). *Fluxionum methodus inversa*. London.

Clapham, Christopher (1990). *A Concise Dictionary of Mathematics*. Oxford and New York.

Craig, John (1706). "Of the Quadratures of Geometrically irrational Figures". *Miscellanea Curiosa* 2, pp. 251–262.

Ernout, Alfred and Alfred Meillet (2001). *Dictionnaire étymologique de la langue française*. Retirage de la 4e édition augmentée d'additions et de corrections par Jacques André. Paris.

Euler, Leonhard (1755). *Institutiones calculi differentialis*. St. Petersburg.

Euler, Leonhard (2000). *Foundations of Differential Calculus*. Translated by John D. Blanton. New York.

Hayes, Charles (1704). *A Treatise of Fluxions*. London.

Heath, Thomas L. (1921). *A History of Greek Mathematics*. Volume I. From Thales to Euclid. Oxford.

Leibniz, Gottfried Wilhelm (1684). "Nova Methodus pro maximis et minimis, itemque tangentibus, quae nec fractas nec irrationales quantitates moratur, et singulare pro illis calculi genu". *Acta Eruditorum* (Oct. 1684), pp. 467–473.

Leibniz, Gottfried Wilhelm (2008). *Sämtliche Schriften und Briefe*. Siebente Reihe. Mathematische Schriften. Fünfter Band. 1674-1676. Infinitesimalmathematik. Berlin.

Lindsay, Wallace M. (1913). *Sexti Pompei Festi*. De verborum significatu quae supersunt cum Pauli epitome. Leipzig.

Loicq, Jean (1960). "*Calcvlvs* et la formation des diminutifs en latin". *L'Antiquité Classique* 29, pp. 30–50.

Newton, Isaac (1736). *The Method of Fluxions and Infinite Series*. Trans. by John Colson. London.

Raphson, Joseph (1702). *A Mathematical Dictionary*. Abridg'd from Monsieur *Ozanam*, and Others. London.

Rihll, T.E. (1999). *Greek Science*. Oxford.

Stedall, Jacqueline A. (2008). *Mathematics Emerging*. A Sourcebook 1540 – 1900. Oxford.

Thomas, Ivor (1957). *Selections Illustrating the History of Greek Mathematics*. I. From Thales to Euclid. London and Cambridge, Mass.

Wallis, John (1693). *De Algebra tractatus*. In: *Opera Mathematica*. Volumen alterum. Oxford, pp. 1–482.

---

[1] "Analysis infinitorum igitur cum in differentialibus tum in integralibus inveniendis versatur, & hanc ob rem in duas praecipuas partes dividitur, quarum altera vocatur Calculus differentialis, altera Calculus integralis. In priori praecepta traduntur quantitatum quarumvis differentialia inveniendi; in posteriori vero via monstratur differentialium propositorum integralia investigandi". Translation from Euler (2000, § 148).

Wolff, Christian (1710). *Der Anfangs-Gründe aller Mathematischen Wissenschaften Letzter Theil.* Halle.
Wolff, Christian (1716). *Mathematisches Lexicon.* Leipzig.

canonical

The Latin word *canon* means 'marking or measuring line; hence, a rule, canon, model' (LS). The word was taken from Greek *κανών*, which literally means 'straight rod, bar, esp. to keep a thing straight'; it has various derived meanings such 'rule, model, standard' and was also used in astronomy for tables of dates (LSJ).

The term *canonical equation* (*aequatio canonica*) was first used by Harriot (1631). One example he gave was $aa + ba - ca = +bc$ (a being the unknown in Harriot's notation), a form that showed the relation between the coefficients and the roots. Harriot explained the name in the following terms:

> These two sorts of equations are regarded as canonical because, when they are applied as canons or yardsticks, the number of roots in common equations is determined [...]; hence the name 'canonical' should be given to them not from the form in which they are established but from their use as instruments in this way (Harriot 1631, p. 5).[1]

Wallis (1685, p. 129) later commented on Harriot's terminology:

> [...] which he [Harriot] calls *Canonical Equations*. And this he so calls, [...] partly because, by Vieta and others, an Equation was then reckoned to be well and regularly ordered, when the Unknown part was separated from the Known part equal to it; and its highest part (as aa) clear of any Affection. And partly, because he doth after make use of these *Canonicals*, as Paradigmes or Paterns [sic], by which he doth afterwards examine *Common Equations*, as they chance to occur".

In the 18th century, the term is found in other contexts. In particular, Hermann (1728, p. 151) applied the term to differential equations;[2] this usage was widely adopted, for example by Euler (1738, p. 158), who referred to Hermann for the term.[3] Riccati and Saladini (1765, p. 117), discussing equations for straight lines, called $my + nx + p = 0$ an *aequatio canonica*.

More uses are found in the 19th century. In 1810, Lagrange presented some differential equations which, he said, are given "in the simplest possible form";[4] he wrote them as $\frac{d\Omega}{d\alpha}dt = d\lambda$ and $\frac{d\Omega}{d\lambda}dt = -d\alpha$ (Lagrange 1810, p. 349).[5] Jacobi (1837, p. 66) later suggested the term *forme canonique* in this respect. The term was soon in common use, although it seems to have caused uncertainty: "This is the celebrated 'canonical form' of the equations of motions of a system, though why it has been so called it would be hard to say" (Thomson and Tait 1867, p. 254). A posthumously published paper suggests that for Jacobi *canonical* meant the same as *normal*: "The

---

[1] "Haec duae aequationum species pro canonicis habentur, quia per earum applicationem tanquam per canones sive regulas, radicum numerus in æquationibus communibus determinatur, [...] unde non à constitutionis forma sed ab instrumentario huiusmodi usu canonicarum nomen illis debetur". Translation from Seltman and Goulding (2007, p. 27). – For some discussion of the concept, see Seltman and Goulding (2007, p. 5) and Stedall (2003, pp. 14–16). [2] A brief discussion of Hermann's use is discussed by Cantor (1901, p. 877). [3] Euler also used *canonical* in number-theoretical contexts; for examples, see Weil (2001, pp. 155, 249). [4] "la forme la plus simple qu'il soit possible". [5] Lagrange (1815, p. 122) himself once used *équation de la forme canonique* for a different equation.

elliptic integral of the third species being reduced to its normal or canonical form [...]" (Jacobi 1882, p. 479).[6]

The term also appears in invariant theory. Sylvester (1851, p. 193) attributed it to Hermite: "I now proceed to the consideration of the more peculiar branch of my inquiry, which is as to the mode of reducing Algebraical Functions to their simplest and most symmetrical, or as my admirable friend M. Hermite well proposes to call them, their Canonical forms". As Hermite does not appear to have used this term in print before Sylvester, he may have suggested it in private correspondence.[1]

A *forme canonique* in connection with matrices, now more usually called the *Jordan normal form*, was presented by Jordan (1870, p. 126).[2] Kronecker (1875, p. 72) criticized: "In fact, the expression 'canonical form' or 'simple canonical form' [...] has no generally accepted meaning and in and of itself represents a concept devoid of objective content. [...] In reality, the so-called canonical or normal forms are usually determined merely by the tendency of the investigation and hence are only regarded as the means, not the goal of the investigation".[3]

Finally, the term *canonical basis* (in connection with modules or vector spaces) was used by Bourbaki (1947, p. 16), to whom it may be due.

References

Bourbaki, Nicolas (1947). *Algèbre*. Chapitre II. Algèbre linéaire. Paris.

Brechenmacher, Frédéric (2006). "Histoire du théorème de Jordan de la décomposition matricielle (1870-1930). Formes de représentation et méthodes de décomposition". PhD thesis. École des hautes études en sciences sociales.

Cantor, Moritz (1901). *Vorlesungen über Geschichte der Mathematik*. Dritter Band. Von 1668 bis 1758. Leipzig.

Euler, Leonhard (1738). "Quamodo data quacunque curva inveniri oporteat aliam quae cum data quodammodo iuncta ad tautochronismum producendum sit idonea". *Commentarii academiae scientiarum Petropolitanae* 5, pp. 143–159.

Harriot, Thomas (1631). *Artis Analyticae Praxis*. London.

Hawkins, Thomas (1977). "Weierstrass and the Theory of Matrices". *Archive for History of Exact Sciences* 17, pp. 119–163.

Hawkins, Thomas (2013). *The Mathematics of Frobenius in Context*. A Journey Through 18th to 20th Century Mathematics. New York.

Hermann, Jacob (1728). "De calculo integrali". *Commentarii academiae scientiarum Petropolitanae* 1, pp. 149–167.

Hermite, Charles (1854). "Sur la théorie des fonctions homogènes à deux indeterminées". *The Cambridge and Dublin Mathematical Journal* 9, pp. 172–217.

Jacobi, Carl Gustav Jacob (1837). "Note sur l'intégration des équations différentielles de la dynamique". *Comptes rendus de l'Académie des sciences* 5, pp. 61–67.

Jacobi, Carl Gustav Jacob (1882). "Nouvelle théorie de la rotation d'un corps de revolution grave suspendu en un point quelconque de son axe". In: *Gesammelte Werke*. Zweiter Band. Berlin, pp. 477–492.

---

[6] "L'intégrale elliptique de la troisième espèce étant réduite à sa forme normale ou canonique [...]".

[1] It is found in Hermite (1854, p. 179). [2] The concept was independently discovered by Weierstraß (1869). For historical accounts, see Hawkins (1977) and Brechenmacher (2006). [3] "In der That hat der Ausdruck «canonische Form» oder «einfache canonische Form» [...] keinerlei allgemein massgebende Bedeutung und bezeichnet an und für sich einen Begriff ohne jeden objectiven Inhalt. [...] In Wahrheit sind überhaupt die sogenannten canonischen oder Normalformen lediglich durch die Tendenz der Untersuchung bestimmt und daher nur als Mittel, nicht aber als Zweck der Forschung anzusehen". Translation from Hawkins (2013, p. 147).

Jordan, Camille (1870). *Traité des substitutions et des équations algébriques*. Paris.
Kronecker, Leopold (1875). "Über Schaaren von quadratischen und bilinearen Formen". *Monatsberichte der Königlich Preußischen Akademie der Wissenschaften zu Berlin* (1874), pp. 59–76.
Lagrange, Joseph-Louis (1810). "Sur la théorie de la variation des constantes arbitraires dans les problèmes de mécaniques, dans lequel on simplifie l'application des formules générales à ces problèmes". *Mémoires de la Classe des Sciences Mathématiques et Physiques de l'Institut de France* (1809), pp. 343–352.
Lagrange, Joseph-Louis (1815). *Mécanique analytique*. Nouvelle édition. Tome second. Paris.
Riccati, Vincenzo and Girolamo Saladini (1765). *Institutiones analyticae*. Tomus primus. Bologna.
Seltman, Muriel and Robert Goulding, eds. (2007). *Thomas Harriot's Artis Analyticae Praxis*. An English Translation with Commentary. New York.
Stedall, Jacqueline A. (2003). *The Greate Invention of Algebra*. Thomas Harriot's Treatise on Equations. Oxford.
Sylvester, James Joseph (1851). "Sketch of a memoir on elimination, transformation, and canonical forms". *The Cambridge and Dublin Mathematical Journal* 6, pp. 186–200.
Thomson, William and Peter Guthrie Tait (1867). *Treatise on Natural Philosophy*. Vol. I. Oxford.
Wallis, John (1685). *A Treatise of Algebra, both Historical and Practical*. London.
Weierstraß, Karl (1869). "Zur Theorie der bilinearen und quadratischen Formen". *Monatsberichte der Königlich Preußischen Akademie der Wissenschaften zu Berlin* (1868), pp. 310–338.
Weil, André (2001). *Number Theory*. An approach through history from Hammurapi to Legendre. Boston.

cardinal

Latin *cardinalis* is derived from *cardo*, which literally means 'door hinge' and which developed the sense 'that on which something depends or turns'. Both meanings are reflected in *cardinalis*, and it was the latter meaning that gave rise to the meaning 'principal, chief', on which the use of *cardinal* in its ecclesiastical sense is based (cardinals being of principal importance), and the same applies to cardinal numbers. These were first used in grammar, for example by the grammarian Priscian (5th/6th century) (LSJ).

The mathematical meaning of the term goes back to Cantor (1895, p. 481): "We call by the name 'power' [*Mächtigkeit*] or 'cardinal number' [*Cardinalzahl*] of M the general concept which, by means of our active faculty of thought, arises from the set M when we make abstraction of the nature of its various elements m and of the order in which they are given".[1]

This was not the first time Cantor used the term, though. *Cardinalzahl* already appeared in print in Cantor (1887, p. 83); in the same paper, he also considered *valence* (German *Valenz*) as an alternative. The term is also found in his correspondence from 1886 (Meschkowski and Nilson 1991, pp. 254f.).

The English term *cardinal number* is found in Tanner (1899, p. 304).

For a discussion of the synonym *Mächtigkeit*, see *power* (sets).

Note. For the history of the concept, see Deiser (2010).

---

[1] ",Mächtigkeit' oder ,Cardinalzahl' von M nennen wir den Allgemeinbegriff, welcher mit Hülfe unseres activen Denkvermögens dadurch aus der Menge M hervorgeht, dass von der Beschaffenheit ihrer verschiedenen Elemente m und von der Ordnung ihres Gegebenseins abstrahirt wird". Translation from Cantor (1915, p. 86) (slightly modified).

References

Cantor, Georg (1887). "Mittheilungen zur Lehre vom Transfiniten". *Zeitschrift für Philosophie und philosophische Kritik* 91, pp. 81–125, 252–270.

Cantor, Georg (1895). "Beiträge zur Begründung der transfiniten Mengenlehre". *Mathematische Annalen* 46, pp. 481–512.

Cantor, Georg (1915). "Contributions to the Founding of the Theory of Transfinite Numbers (First Article)". In: Jourdain, Philip E.B. *Contributions to the Founding of the Theory of Transfinite Numbers*. Chicago, pp. 85–136.

Deiser, Oliver (2010). "On the Development of the Notion of a Cardinal Number". *History and Philosophy of Logic* 31, pp. 123–143.

Meschkowski, Herbert and Wilfried Nilson, eds. (1991). *Georg Cantor*. Briefe. Berlin.

Tanner, H.W. Lloyd (1899). "Sur les fondements de la théorie des ensembles [review]". *The Mathematical Gazette* 1, pp. 304–305.

cardioid

The cardioid is a plane curve named, in Latin, by de Castillon (1741). The name *curva cardiode* literally means 'heart-shaped curve' (from Greek καρδία 'heart') and was chosen on account of the curve's seeming resemblance to a heart (de Castillon 1741, p. 779). The term spread quickly and appeared, for example, in a supplement to Chambers' *Cyclopaedia* (Chambers 1753, s. v. cardioid).

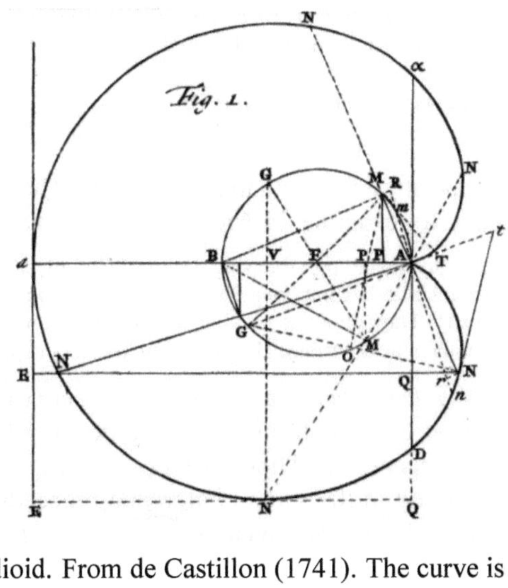

Figure 7: The cardioid. From de Castillon (1741). The curve is traced by a point on the perimeter of a circle that is rolling around a fixed circle of the same radius.

Earlier, the curve had been discussed by, for example, Ozanam (1691, pp. 102–104), who called it *cycloïde géométrique* 'geometric cycloid'. Ozanam here uses *cycloïde* in the sense of 'epicycloid' and explicitly says (p. 102) that he calls it *geometric* because the curve has an equation; in others words, he uses *geometric* in Descartes' sense (see *transcendental*). Brocard and Lemoyne (1919, p. 57) attribute the curve to the Dutch mathematician Jacobus Koërsma,[1] without giving a reference; as pointed out by Brocard (1901), the curve is indeed described in a note in the *Bibliothèque universelle et historique* (vol. 12, pp. 565-568) for the year 1689, but the publication announced therein, *Nouveaux Principes des Mathématiques*, does not seem to have been published.[1] The curve was also dealt with by Cramer (1750, p. 430), but he was not aware of Castillion's term and did not give it a name.

Note. For further historical references and discussion, see Brocard and Lemoyne (1919, pp. 83–98), Loria (1910, pp. 152–157), and [anon.] (1884).

References

[anon.] (1884). "Nachschrift der Redaktion zu vorstehendem Artikel. (Litteratur zur Kardioïde)". *Zeitschrift für mathematischen und naturwissenschaftlichen Unterricht* 15, pp. 271–273.

Brocard, Henri (1901). "[Réponse] 1171". *L'Intermédiaire des mathématiciens* 8, p. 8.

Brocard, Henri and Timoléon Lemoyne (1919). *Courbes géométriques remarquables*. (Courbes spéciales). Planes et gauches. Tome 1. Paris.

Carré, Louis (1706). "Examen d'une courbe formée par le moyen de cercle". *Mémoires de l'Académie royale des sciences* (1705), pp. 56–61.

de Castillon, Jean (1741). "de Curva Cardioïde, de Figura sua sic dicta". *Philosophical Transactions of the Royal Society of London* 41, pp. 778–781.

Chambers, Ephraim (1753). *A Supplement to Mr. Chambers's Cyclopaedia, or, Universal Dictionary of Arts and Sciences*. Vol. I. London.

Cramer, Gabriel (1750). *Introduction à l'analyse des lignes courbes algébriques*. Genève.

Loria, Gino (1910). *Spezielle algebraische und transzendente ebene Kurven*. Theorie und Geschichte. Autorisierte, nach dem italienischen Manuskript bearbeitete deutsche Ausgabe von Fritz Schütte. Zweite Auflage. Erster Band. Leipzig.

Ozanam, Jacques (1691). *Dictionnaire mathematique*. Paris.

catastrophe

The term *catastrophe* was introduced into mathematics by Thom in the second half of the 1960s.[2] He used it, for example, in Thom (1969, pp. 318f.):

> It is not too difficult a task to find all possible singularities $V(x)$ of finite codimension not exceeding four. These singularities are important, because they may appear on our space–time in a structurally stable way. They give rise to what we call the "elementary catastrophies [sic]", when we interpret them as describing dynamical fields on our space–time.

The term *generalized catastrophe* occurs in the same paper (p. 324) as well as the classification of elementary catastrophes into seven types (p. 320). In print, French *catastrophe élémentaire* appeared in Thom (1968b, p. 155), a talk held in 1966.[3] English *catastrophe* is found in the same conference proceedings (Thom 1968a, p. 34). An exposition of the theory was given by Thom in his *Stabilité structurelle et morphogénèse* (Thom 1972).

The term *catastrophe theory* came up soon after Thom's first publications on the topic; it was used by Zeeman (1971, p. 1557), for example.

Note. For the history of the concept, see Arnol'd (1994).

---

[1] Carré (1706, pp. 72f.) seems to have been the first who mentioned Koërsma in this respect.

[1] In the note, no explicit term is used for the cardioid. [2] According to Sheynin (2003, p. 349), "Nekrasov [...], in connection with the mathematical study of indeterminacies, mentioned the then not yet existing theory of catastrophes (he even used the term 'catastrophe')". Sheynin refers to Nekrasov (1916, p. 23). [3] The classification of elementary catastrophes is found on pp. 157f., the term *catastrophe généralisée* on p. 160.

References

Arnol'd, V.I. (1994). "Catastrophe Theory. Translated from the Russian by N.D. Kazarinoff". In: *Dynamical Systems V*. Ed. by V.I. Arnol'd. Berlin, pp. 207–264.

Nekrasov, P.A. (1916). *Средняя школа, математика и научная подготовка учителей*. Petrograd.

Sheynin, Oscar (2003). "Nekrasov's Work on Probability: The Background". *Archive for History of Exact Sciences* 57, pp. 337–353.

Thom, René (1968a). "Comments". In: *Towards a Theoretical Biology*. 1. Prologomena. Ed. by C.H. Waddington. Chicago, pp. 32–41.

Thom, René (1968b). "Une théorie dynamique de la morphogenèse". In: *Towards a Theoretical Biology*. 1. Prologomena. Ed. by C.H. Waddington. Chicago, pp. 152–179.

Thom, René (1969). "Topological models in biology". *Topology* 8, pp. 313–335.

Thom, René (1972). *Stabilité structurelle et morphogénèse*. Essai d'une théorie générale des modèles. New York.

Zeeman (1971). "The Geometry of Catastrophe". *Times Literary Supplement* (Dec. 10), pp. 1556–1557.

categorical (axioms)

A categorical system of axioms is one which allows, up to isomorphism, only one model. The axioms for real numbers are an important example. This notion was explicitly discussed by Huntington (1902, p. 264) under the name *sufficient* and described in the following terms: "there is essentially *only one* such assemblage possible".[1]

The term *categorical* first appears in print in a paper by Oswald Veblen. Before discussing his use of the term, however, a terminological caveat should be useful:

> reading of Veblen *1904* alone does not reveal with certainty whether by 'categorical' Veblen means "categorical" or "complete". But Huntington 1902, to which Veblen refers, (1) had already defined categoricity (denoted by the word 'sufficiency') and (2) had not defined completeness. Moreover, Huntington *1905* (p. 17 fn) attributes the idea of completeness (under the term 'categorical') to Veblen *1904*. Therefore, Veblen probably means "complete" by the word 'categorical'. Moreover, regardless of which concept is meant it is still clear that Veblen took categoricity and completeness to be coextensive (which may help to explain why it is so hard to separate the ideas in his writing). [...] The situation is further complicated by the fact that Veblen *1904*, 346 incorrectly attributes the idea of completeness to Hilbert *1899* (Corcoran 1981, p. 117).

With reference to a system of twelve axioms for geometry that he discusses in his paper, Veblen states:

> Inasmuch as the terms *point* and *order* are undefined one has a right, in thinking of the propositions, to apply the terms in connection with any class of objects of which the axioms are valid propositions. It is part of our purpose however to show that there is *essentially only one* class of which the twelve axioms are valid. In more exact language, any two classes K and K' of objects that satisfy the twelve axioms are capable of a one-to-one correspondence such that if any three elements A, B, C of K are in the order ABC, the corresponding elements of K' are also in the order ABC. Consequently any proposition which can be made in terms of points and order either is in contradiction with our axioms or is equally true of all classes that verify our axioms. The validity of any possible statement in these terms is therefore completely determined; and so any further axioms would have to be considered redundant. [...] A system of axioms such as we have described is called *categorical*, whereas

[1] This meant that any two models are isomorphic (Huntington 1902, p. 277). The term *isomorphic* was not used by Huntington.

one to which it is possible to add independent axioms [...] is called *disjunctive* (Veblen 1904, p. 346).

A more concise definition appears in Veblen (1906, p. 303): "A mathematical science is said to be determined categorically by a set of postulates or conditions if it can be proved that any two classes of objects satisfying the conditions are capable of a one-to-one correspondence preserving the relations described in the postulates". The word was also used in the same meaning by Young (1911, p. 49): "When a set of assumptions satisfies the condition that *any* two concrete applications of it are simply isomorphic, the set of assumptions is said to be *categorical*".

Veblen (1904, p. 346) remarks that the term was suggested to him by the philosopher John Dewey. W. Kneale and M. Kneale (1962, p. 388) explain why the term may have been chosen: "Probably Veblen and Dewey had in mind a contrast between categorical and disjunctive propositions; but in origin 'categorical' means 'predicative', rather than 'non-disjunctive', and apart from that the word has the disadvantage of being much over-worked in other contexts".

Veblen (1904, p. 346) also remarks that "[t]he categorical property of a system of proposition is referred to by Hilbert in his 'Axiom der Vollständigkeit" [axiom of completeness]". This axiom, added by Hilbert after the first version his *Foundations of Geometry* was published, states that "the elements of geometry form a system which not susceptible of extension, if we regard the five groups of axioms as valid" (Hilbert 1902, p. 25). It has frequently been pointed out that Veblen's equation of categoricity and completeness is questionable: "when he [Veblen] writes that 'the categorical property of a system of propositions is referred to by Hilbert in his 'Axiom der Vollständigkeit' he apparently misinterprets Hilbert's axiom, or at least describes it in a misleading way" (Awodey and Reck 2002, p. 17).[1]

Nevertheless, in various mathematical texts, the term *complete* is frequently found as an alternative to *categorical*. For example, Fraenkel (1922, p. 155) lists German *vollständig* 'complete' as a synonym, and Keyser (1922, p. 128) lists all three terms mentioned above.[2] Carnap (1927, pp. 364f.) published an article in which German *monomorph* is found instead. According to Carnap, the term was suggested to him by the philosopher Herbert Feigl.

Generally it can be said that Veblen's terminology was first accepted by American mathematicians, for example by Huntington (1906, p. 26). The French philosopher Couturat (1905, p. 169), on the other hand, found *categorical* and *disjunctive* "unclear and unhappy" ("peu claires et peu heureuses") and preferred Huntington's terminology. The German term *kategorisch* was used by Fraenkel (1912, p. 76) but gained wider currency only in the 1920s.

See also *consistent* (axioms), *complete* (axioms), and *independent* (axioms).

[1] See also Scanlan (1991, p. 994) and Mancosu, Zach, and Badesa (2009, pp. 326f.); the latter remark on this definition: "there is a certain ambiguity between defining categoricity as the property of admitting only one model (up to isomorphism) and conflating the notion with a consequence of it, namely, what we would now call semantical completeness. Veblen, however, rightly states that in the case of a categorical theory, further axioms would be redundant even if they were not deducible from the axioms by a finite number of inferences". They also note that the term *isomorphism* in this connection was not yet common at the time. [2] Fraenkel (1928, pp. 347–354) later distinguished several kinds of completeness.

Note. For the history of the concept and terminology, see Awodey and Reck (2002), Mancosu, Zach, and Badesa (2009, pp. 326–328), Moore (1988, pp. 108f.), Volkert (2015, pp. 177–180), and Corcoran (1980).

References

Awodey, Steve and Erich H. Reck (2002). "Completeness and Categoricity, Part I: Nineteenth-century Axiomatics to Twentieth-century Metalogic". *History and Philosophy of Logic* 23, pp. 1–30.

Carnap, Rudolf (1927). "Eigentliche und uneigentliche Begriffe". *Symposion* 1, pp. 355–374.

Corcoran, John (1980). "Categoricity". *History and Philosophy of Logic* 1, pp. 187–207.

Corcoran, John (1981). "From Categoricity to Completeness". *History and Philosophy of Logic* 3, pp. 113–119.

Couturat, Louis (1905). *Les principes des mathématiques*. Paris.

Fraenkel, Adolf (1912). "Axiomatische Begründung von *Hensels* p-adischen Zahlen". *Journal für die reine und angewandte Mathematik* 141, pp. 43–76.

Fraenkel, Adolf (1922). "Axiomatische Begründung der transfiniten Kardinalzahlen. I". *Mathematische Zeitschrift* 13, pp. 153–188.

Fraenkel, Adolf (1928). *Einleitung in die Mengenlehre*. Dritte umgearbeitete und stark erweiterte Auflage. Berlin.

Hilbert, David (1902). *The Foundations of Geometry*. Authorised translation by E.J. Townsend. Chicago.

Huntington, Edward V. (1902). "A Complete Set of Postulates for the Theory of Absolute Continuous Magnitude". *Transactions of the American Mathematical Society* 3, pp. 264–279.

Huntington, Edward V. (1905). "A Set of Postulates for Real Algebra". *Transactions of the American Mathematical Society* 6, pp. 17–41.

Huntington, Edward V. (1906). "The Fundamental Laws of Addition and Multiplication in Elementary Algebra". *Annals of Mathematics* 8, pp. 1–44.

Keyser, Cassius Jackson (1922). *Mathematical Philosophy*. A Study of Fate and Freedom. New York.

Kneale, William and Martha Kneale (1962). *The Development of Logic*. Oxford.

Mancosu, Paolo, Richard Zach, and Calixto Badesa (2009). "The Development of Mathematical Logic from Russell to Tarski, 1900–1935". In: *The Development of Modern Logic*. Ed. by Leila Haaparanta. Oxford, pp. 318–470.

Moore, Gregory H. (1988). "The Emergence of First-Order Logic". In: *History and Philosophy of Modern Mathematics*. Ed. by William Aspray and Philip Kitcher. Minneapolis, pp. 95–135.

Scanlan, Michael (1991). "Who were the American Postulate Theorists?" *Journal of Symbolic Logic* 56, pp. 981–1002.

Veblen, Oswald (1904). "A system of axioms for geometry". *Transactions of the American Mathematical Society* 5, pp. 343–384.

Veblen, Oswald (1906). "*The Continuum as a Type of Order: An Exposition of the Modern Theory*. By Edward V. Huntington [review]". *Bulletin of the American Mathematical Society* 12, pp. 302–305.

Volkert, Klaus, ed. (2015). *David Hilbert*. Grundlagen der Geometrie (Festschrift 1899). Berlin and Heidelberg.

Young, John Wesley (1911). *Lectures on Fundamental Concepts of Algebra and Geometry*. New York.

category

Like the term *group*, *category* had been used informally before it was given a precise mathematical meaning. Such informal senses relate to the idea of classifying objects, or categorizing them. As an example, Krömer (2007, pp. 74f.) cites a text by Weil (1940, p. 109) in which the author uses *categories of groups* (French *catégories de groupes*). Similar examples can easily be found earlier, for example in Miller (1929, p. 370).

The term was "officially" introduced by Eilenberg and Mac Lane (1945, p. 237):[1] "A *category* $\mathfrak{A} = \{A, \alpha\}$ is an aggregate of abstract elements A (for example, groups), called the *objects* of the category, and abstract elements α (for example, homomorphisms), called *mappings* of the category".

Mac Lane (1971, pp. 29f.) later explained the choice of the term and referred to "the pleasure of purloining words from the philosophers: 'Category' from Aristotle and Kant, 'Functor' from Carnap". However, Krömer (2007, p. 75) justly points out that "Mac Lane wants us to believe that they were inspired by the philosophical uses of the term [...]; I think that the truth rather is that both these philosophical definitions and their mathematical concept are inspired by the common language use of the term". As for the spreading of the terminology, Weibel (1999, p. 806) remarks that "the new language of Category Theory did not gain wide acceptance until the appearance of the books Eilenberg and Steenrod (1952) and Cartan and Eilenberg (1956) in the 1950's".

Note. For the history of the notion, see Krömer (2007) and Marquis (2009).

References

Cartan, Henri and Samuel Eilenberg (1956). *Homological Algebra*. Princeton.
Eilenberg, Samuel and Saunders Mac Lane (1943). "General theory of natural equivalences". *The Journal of Symbolic Logic* 8, p. 40.
Eilenberg, Samuel and Saunders Mac Lane (1945). "General Theory of Natural Equivalences". *Transactions of the American Mathematical Society* 58, pp. 231–294.
Eilenberg, Samuel and Norman Steenrod (1952). *Foundations of Algebraic Topology*. Princeton.
Krömer, Ralf (2007). *Tool and Object*. A History and Philosophy of Category Theory. Basel.
Mac Lane, Saunders (1971). *Categories for the Working Mathematician*. New York et al.
Marquis, Jean-Pierre (2009). *From a Geometrical Point of View*. A Study of the History and Philosophy of Category Theory. Dordrecht.
Miller, George Abram (1929). "Groups Which Admit Three-fourths Automorphisms". *Proceedings of the National Academy of Sciences* 15, pp. 369–372.
Weibel, Charles A. (1999). "History of Homological Algebra". In: *History of Topology*. Ed. by Ioan Mackenzie James. Amsterdam, pp. 797–836.
Weil, André (1940). *L'intégration dans les groupes topologiques et ses applications*. Paris.

catenary

The curve of a hanging chain was studied by various 17th-century mathematicians, notably by Galilei (1638). He said that "this chain curves in a parabolic shape"[1] (p. 146) and that "the cord thus hung, whether much or little stretched, bends in a line that is very close to parabolic" (p. 284).[2] Galilei calls the curve *catenella*, which means 'little chain'.

The curve was further investigated in the 1690s, and the term was then suggested by Huygens in a letter, dated Nov. 18, 1690, to Leibniz: "the curve of the hanging

---

[1] The term appears earlier in an abstract of that paper (Eilenberg and Mac Lane 1943, p. 40).

[1] "questa catenella si piega in figura parabolica". [2] "la corda così tesa, e poco o molto tirata, si piega in linee, le quali assai si avvicinano alle paraboliche". Translation from Galilei (1989, p. 256). Thus, contrary to a widespread belief, Galileo does not claim that the catenary *is* a parabola but that it is *similar* to it. See, e. g., Sonar (2018, pp. 328–333) and the references cited there.

chain, which may be called *catenary* [*Catenaria*)]" (Huygens 1901, p. 537).[3] The term is derived from Latin *catena* 'chain'.

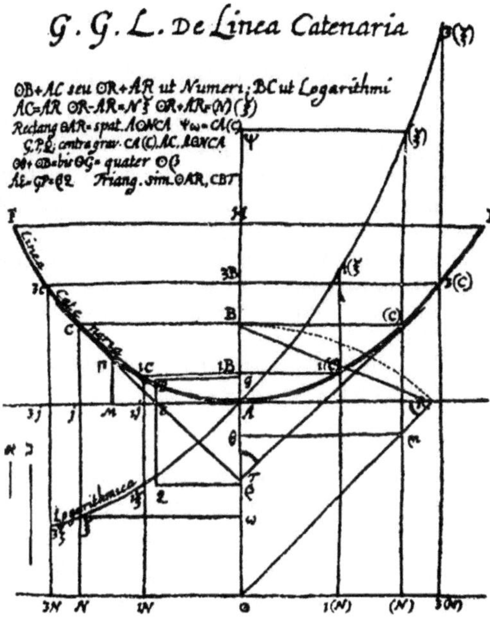

Figure 8: The catenary. From Leibniz (1691).

Various alternative terms have been used for this curve.[1] For example, a common French term is *chaînette*, lit. 'little chain', used in correspondence between Leibniz and Huygens from 1691 (see, e. g., Leibniz 2003, p. 158) and slightly later appearing in print in Leibniz (1692, p. 147). Also, *funicular (curve)* is used in Leibniz (1691, p. 277) as *linea catenaria vel Funicularis*; the term is derived from Latin *funiculus* 'rope, cord'. The German term *Kettenlinie*, lit. 'chain line', was used by Clüver (1707, p. 366) and was also included in Wolff's dictionary (Wolff 1716, p. 321); English *catenary* is first found in Harris (1704, s. v. catenaria).

Note. For the history of the notion, see Bukowski (2008), Sonar (2018, pp. 328–333), and Truesdell (1960, esp. pp. 64-88).

References

Bukowski, John (2008). "Christiaan Huygens and the Problem of the Hanging Chain". *The College Mathematics Journal* 39, pp. 2–11.

Cesàro, Ernesto (1886). "Les lignes barycentriques". *Nouvelles annales de mathématiques* 5, pp. 511–520.

Clüver, Detlev (1707). *Disquisitiones philosophicae, Oder historische Anmerckungen über die nützlichsten Sachen der Welt.* Hamburg.

[3] "la ligne de la chaine pendante, qu'on peut nommer *Catenaria*". [1] For a more general class of curves, Cesàro (1886, p. 75) suggested French *alysoïdes*. The term is based on the Greek word for 'chain'. See Loria (1911, p. 207).

Galilei, Galileo (1638). *Discorsi e dimostrazioni matematiche, intorno à due nuove scienze*. Leiden.
Galilei, Galileo (1989). *Two New Sciences Including Centers of Gravity and Force of Percussion*. Translated, with a New Introduction and Notes, by Stillman Drake. Second Edition. Toronto.
Harris, John (1704). *Lexicon Technicum*. Vol. I. London.
Huygens, Christiaan (1901). *Œuvres complètes*. Tome neuvième. Correspondance 1685–1690. La Haye.
Leibniz, Gottfried Wilhelm (1691). "De linea in quam flexile se pondere proprio curvat". *Acta Eruditorum* (June 1691), pp. 277–281.
Leibniz, Gottfried Wilhelm (1692). "De la chainette, ou solution d'un problème fameux proposé par Galilei". *Journal des sçavans* (1692), pp. 147–153.
Leibniz, Gottfried Wilhelm (2003). *Sämtliche Schriften und Briefe*. Dritte Reihe. Mathematischer, naturwissenschaftlicher und technischer Briefwechsel. Fünfter Band. 1691-1693. Hannover.
Loria, Gino (1911). *Spezielle algebraische und transzendente ebene Kurven*. Theorie und Geschichte. Autorisierte, nach dem italienischen Manuskript bearbeitete deutsche Ausgabe von Fritz Schütte. Zweite Auflage. Zweiter Band. Leipzig.
Sonar, Thomas (2018). *The History of the Priority Dispute between Newton and Leibniz*. Mathematics in History and Culture. Cham.
Truesdell, Charles (1960). *The Rational Mechanics of Flexible or Elastic Bodies, 1638-1788*. Zürich.
Wolff, Christian (1716). *Mathematisches Lexicon*. Leipzig.

catenoid

A catenoid is a particular kind of surface generated by the rotation of a catenary. Euler (1744, ch. V, § 47) discovered catenoids and some of their essential properties; in particular he noted that they are non-trivial minimal surfaces.[1] The name, though, was only introduced much later by Plateau (1858, p. 35). He used French *caténoïde*, which means 'chain-like' (from Latin *catena* 'chain') and said the name is "based on the shape of its meridian line".[2] The English term *catenoid* is found in a translation of Plateau's 1859 paper (Plateau 1865, p. 351).

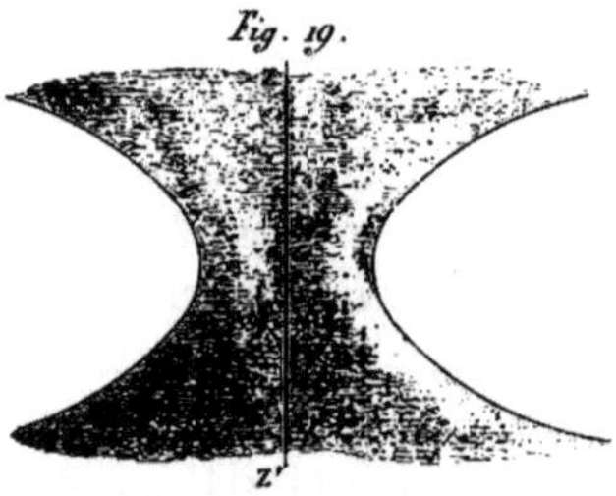

Figure 9: The catenoid as illustrated by Plateau. From Plateau (1859, following p. 52).

Discussing the same surface, Bour (1862, p. 30) suggested an alternative term: "it has seemed to me indispensable to give it [the catenoid] a name easy to pronounce and remember. One can choose that of *alysséide*, from Greek *ἁλύσιον*, 'little chain'".[3]

[1] According to Reich (2007, p. 481), this is "the first example of a minimal surface in history".
[2] "J'appelle cette figure le *caténoide*, aussi d'après la forme de sa ligne méridienne". [3] "il m'a paru indispensable de lui donner un nom facile à prononcer et à retenir. On peut choisir celui d'*alysséide*, du grec *ἁλύσιον*, chaînette".

References

Bour, Edmond (1862). "Théorie de la déformation des surfaces". *Journal de l'École polytechnique* 22, cahier 39, pp. 1–148.

Euler, Leonhard (1744). *Methodus inveniendi lineas curvas maximi minimive proprietate gaudentes, sive Solutio problematis isoperometrici latissimo sensu accepti.* Lausanne and Geneva.

Plateau, Joseph (1858). "Recherches expérimentales et théoriques sur les figures d'équilibre d'une masse liquide sans pesanteur". *Annales de chimie et de physique* 53, pp. 26–41.

Plateau, Joseph (1859). "Recherches expérimentales et théoriques sur les figures d'équilibre d'une masse liquide sans pesanteur. Quatrième série". *Mémoires de l'Académie Royale des Sciences, des Lettres et des Beaux-Arts de Belgique* 31, pp. 1–52.

Plateau, Joseph (1865). "Experimental and Theoretical Researches on the Figures of Equilibrium of a Liquid Mass Withdrawn from the Action of Gravity. Fourth Series". *Annual Report of the Board of Regents of the Smithsonian* 1864, pp. 338–369.

Reich, Karin (2007). "Euler's Contribution to Differential Geometry and its Reception". In: *Leonhard Euler*. Life, Work and Legacy. Ed. by Robert E. Bradley and C. Edward Sandifer. Amsterdam, pp. 479–502.

cathetus

The term was coined as Greek *κάθετος*, originally used as a technical term meaning 'plumb line' and derived from a verbal root meaning 'to let down'. As such, it was commonly translated by words meaning 'perpendicular'. Later, the word is commonly found, next to *base* and *hypotenuse*, for one of the sides of a triangle. The meaning 'the sides that are adjacent to the right angle' is of very late origin and dates from the 17th century.

Such plumb lines were presumably used in construction to verify that something was built vertically (Schmidt 1914, pp. 121f.). The tie to verticality also makes it plausible why in Greek mathematics there was only one cathetus in a triangle, not two. Euclid (*Elements*, book I, def. 10) used this term in his definition of 'perpendicular': "When a straight line standing on a straight line makes the adjacent angles equal to one another, each of the equal angles is right, and the straight line standing on the other is called a *perpendicular* [*κάθετος*] to that on which it stands".[1] The term was in use before Euclid (Mugler 1958-59, p. 236).

The Latin term is first attested in Vitruvius (see Schmidt 1914, p. 247) and has been in common use since. For the further development of the word in this sense, see *perpendicular* and *normal*.

The classical names of the three sides of a right-angled triangle are thus *basis*, *cathetus*, and *hypotenusa* (fig. 10). The 12th century theologian Hugh of St. Victor, for example, defined them thus: "Consider a right triangle. Its base is the lower line, set along the ground. Its perpendicular is a line at the base end, raised straight up, so that it inclines to neither side, but makes a right angle, where it meets the base. The hypotenuse is a line from the top of the perpendicular to the far and opposite end of the ground line (i. e., the base)" (Baron 1956, p. 188).[2]

---

[1] Translation from Euclid (1956, p. 153). [2] "Sit itaque triangulum orthogonium, cuius basis est linea inferius secundum planum directa iacens; kathetus autem linea a fine iacentis in directum eleuata, ita ut in neutram partem inclinetur et, ubi iacentem contingit, angulum rectum efficiat; hypotenusa uero linea a summitate katheti ad terminum iacentis linee, id est basis". Translation from Hugh of St. Victor (1991, p. 35).

According to Tropfke (1940, p. 84), this combination of the three terms has been customary from the times of Heron of Alexandria until the 18th century. The corresponding Greek terminology is found in the Pseudo-Heronian *Geometrica* (Heron of Alexandria 1912, p. 178); in Latin, is is found in an agrimensorial tract called *Podismus* (Blume, Lachmann, and Rudorff 1848, p. 297).

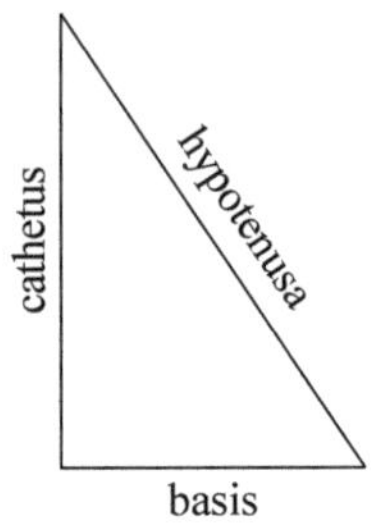

Figure 10: Classical names of the sides of a right-angled triangle

An early use in a modern language is in a Spanish text from the second half of the 13th century, where the term is used in the sense of 'perpendicular' (Bossong 1978, pp. 160f.); Italian *catheto* was used in a 14th-century translation of Fibonacci's *Practica Geometriae*, both in the sense of 'perpendicular' and for one of the sides in a right-angled triangle (Feola 2008, pp. 149f.), as in fig. 10. The German term is found in Widmann (1489, fol. C 6^v). At about the same time, Chuquet (1979, p. 110) defined: "All triangles are constituted by three straight lines of which one is taken for the base and the two others for the two hypotenuses [sic]. The line descending perpendicularly from the angle of the hypotenuses onto the base is called the *cathetus*".[1] As for English, Recorde (1551, fol. A iv^r) uses *ground line* for the base, *perpendicular or plumme line* for the cathetus, and *bias line* for the hypotenuse. According to the OED, English *cathetus* is first found in L. Digges and T. Digges (1571, fol. T ij^v), where the word is used for the height of a solid.[2]

A further change of meaning took place in the 17th century. Tropfke (1940, pp. 85–87), who systematically studied the definitions of *cathetus*, found the first "modern" definition in Jungius (1627, p. 7): "In a right-angled triangle, the side that subtends the right angle is called *hypotenuse*, the remaining sides, which are adjacent to the right angle, are called *catheti*".[3] A similar definition had earlier been given by Napier (1614, p. 22), who, however, did not use the word *cathetus* but Latin *crus*, which means 'leg': "In rectangulis crura vocamus, quae rectum angulum ambient: hypotenusam, quae subtendit". In the English translation published two

[1] "Tous triangles sont constituez par troys lignes droictes dont l'une est prise pour la base et les deux aultres pour les deux ypothenusses; et la ligne descendant perpendiculierement de l'angle des ypothenusses sus la base est dicte cathetuse". Translation from Flegg, Hay, and Moss (1985, p. 241), slightly modified. Note that Chuquet allows for *two* "hypotenuses". [2] For some uses of Latin *cathetus* in the sense of 'height', see Tropfke (1940, p. 86). [3] "In triangulo rectangulo latus angulo recto subtensum hypotenusa dicitur; reliqua recto adiacentia, dicuntur catheti".

years later, this was translated as: "In right angled triangles wee call the sides that are about the right angle, Leggs; and that which subtendeth the right angle wee call the *Hypotenuse*" (Napier 1616, p. 31). Such definitions using *leg* instead of *cathetus* are also given by Vlacq (1636, p. 5), Cavalieri (1643, p. 1), Pardies (1671, Table des mots), and by Newton in the 1680s (Newton 1971, p. 136).

Throughout the 17th century and later, however, the word *cathetus* is mostly used in the older senses; its new meaning is also not covered in dictionaries such as Moxon's *Mathematicks made easie*, where *cathetus* is defined as "[a] Greek word for a *Perpendicular*" (Moxon 1679, p. 28). In Harris' *Lexicon technicum*, however, there are two relevant entries: one for *catheti* "in a Right Angled Triangle, are the Sides including the Right Angle" and one for *cathetus*, which "signifies the Perpendicular only, the other Leg being called the base". Wolff (1716, p. 322) only includes the new meaning of the term.

See also *perpendicular* and *normal*.

Note. A historical overview of the term is found in Tropfke (1940, pp. 83–87). For the Greek term, see Mugler (1958-59, pp. 235f.) and Schmidt (1914, pp. 121f., 246–248); for the Latin, see Ayuso García (2008, pp. 527f.).

References

Ayuso García, Manuel (2008). "La terminología latina de la geometría en Marciano Capela". PhD thesis. Madrid: Universidad Nacional de Educación a Distancia.

Baron, Roger (1956). "Hugonis de Sancto Victore *Practica Geometriae*". *Osiris* 12, pp. 176–224.

Blume, Friedrich, Karl Lachmann, and Adolf Rudorff, eds. (1848). *Die Schriften der römischen Feldmesser*. Erster Band. Texte und Zeichnungen. Berlin.

Bossong, Georg, ed. (1978). *Los Canones de Albateni*. Herausgegeben sowie mit Einleitung, Anmerkungen und Glossar versehen. Tübingen.

Cavalieri, Bonaventura (1643). *Trigonometria plana et sphaerica, linearis & logarithmica*. Bologna.

Chuquet, Nicolas (1979). *La Géométrie*. Première géométrie algébrique en langue française (1484). Introduction, texte et notes par Hervé l'Huillier. Paris.

Digges, Leonard and Thomas Digges (1571). *A geometrical practise, named Pantometria diuided into three bookes, Longimetra, Planimetra, and Stereometria*. London.

Euclid (1956). *The Thirteen Books of Euclid's Elements*. Volume I. Introduction and Books I, II. Ed. by Thomas L. Heath. New York.

Feola, Francesco (2008). *Gli esordi della geometria in volgare*. Un volgarizzamento trecentesco della *Practica Geometriae* di Leonardo Pisano. Firenze.

Flegg, Graham, Cynthia Hay, and Barbara Moss, eds. (1985). *Nicolas Chuquet, Renaissance mathematician*. A study with extensive translation of Chuquet's mathematical manuscript completed in 1484. Dordrecht.

Harris, John (1704). *Lexicon Technicum*. London.

Heron of Alexandria (1912). *Heronis Alexandrini opera quae supersunt omnia*. Volumen IV. Heronis definitiones cum variis collectionibus. Heronis quae feruntur geometrica. Ed. by J.L. Heiberg. Leipzig.

Hugh of St. Victor (1991). *Practical Geometry (Practica Geometriae)*. Attributed to Hugh of St. Victor. Translated from the Latin with an Introduction, Notes and Appendices by Frederick A. Homann. Milwaukee.

Jungius, Joachim (1627). *Geometria empirica*. Rostock.

Moxon, Joseph (1679). *Mathematicks made easie or, a compleat Mathematical Dictionary*. London.

Mugler, Charles (1958-59). *Dictionnaire historique de la terminologie géométrique des Grecs*. Paris.

Napier, John (1614). *Mirifici logarithmorum canonis descriptio*. Edinburgh.

Napier, John (1616). *A Description of the Admirable Table of Logarithms*. Translated by Edward Wright. London.

Newton, Isaac (1971). *The Mathematical Papers of Isaac Newton*. Volume IV. 1674-1684. Ed. by D.T. Whiteside. Cambridge.
Pardies, Ignace Gaston (1671). *Elemens de geometrie*. Paris.
Recorde, Robert (1551). *The patheway to knowledg*. London.
Schmidt, Max Carl Paul (1914). *Zur Entstehung und Terminologie der elementaren Mathematik*. Zweite verbesserte und stark vermehrte Auflage. Leipzig.
Tropfke, Johannes (1940). *Geschichte der Elementar-Mathematik*. Vierter Band. Ebene Geometrie. Dritte, verbesserte und vermehrte Auflage besorgt von Dr. Kurt Vogel. Berlin and Leipzig.
Vlacq, Adriaan (1636). *Tables de sinus, tangentes, secantes*. Gouda.
Widmann, Johannes (1489). *Behende vnd hubsche Rechenung auff allen kauffmanschafft*. Leipzig.
Wolff, Christian (1716). *Mathematisches Lexicon*. Leipzig.

Cauchy distribution

While Poisson (1824, p. 278) already called attention to random variables with the probability density $f(x) = \frac{1}{\pi}\frac{1}{1+x^2}$, the corresponding distribution is named after Cauchy (1853). The name came up in the 1920s, when Lévy (1925, p. 179) used the name *loi de Cauchy* 'Cauchy's law' in this connection.[1]

The terminology was adopted by Lurquin (1926) and, in a review of that paper, the German terms *Cauchysche Verteilung* 'Cauchy's distribution' and *Cauchysche Verteilungsfunktion* 'Cauchy's distribution function' appear (Boehm 1926, p. 522). *Cauchy distribution* did not immediately catch on but became more common from the mid-1930s on; in English, it seems to have been first used by Koopman (1936, p. 408).

Note. For the history of the concept, see Stigler (1974) and also the article *witch*. Miller (n.d.) discusses some alternative names used in physics.

References

Boehm, C. (1926). "Lurquin, C. Sur la loi de probabilité de Cauchy [review]". *Jahrbuch über die Fortschritte der Mathematik* 52, p. 522.
Cauchy, Augustin-Louis (1853). "Mémoire sur les résultats moyens d'un très-grand nombre des observations". *Comptes rendus de l'Académie des sciences* 37, pp. 381–385.
Fischer, Hans (2011). *A History of the Central Limit Theorem*. From Classical to Modern Probability Theory. New York.
Koopman, B.O. (1936). "On Distributions Admitting a Sufficient Statistic". *Transactions of the American Mathematical Society* 39, pp. 399–409.
Lévy, Paul (1925). *Calcul des probabilités*. Paris.
Lurquin, Constant (1926). "Sur la loi de probabilité de Cauchy". *Bulletins de la Classe des sciences (Académie royale de Belgique)* 12, pp. 856–870.
Miller, Jeff (n.d.). "Earliest Known Uses of Some of the Words of Mathematics". URL: `https://mathshistory.st-andrews.ac.uk/Miller/mathword/`.
Poisson, Siméon Denis (1824). "Sur la Probabilité des résultats moyens des Observations". *Connaissance des tems pour 1827*, pp. 273–302.
Stigler, Stephen M. (1974). "Cauchy and the witch of Agnesi: An historical note on the Cauchy distribution". *Biometrika* 61, pp. 375–380.

caustic

In the 1670s, Tschirnhaus experimented with burning mirrors and observed that "the concentration of light occurs not only at the focus, but also along an 'entire curved

---

[1] According to Fischer (2011, p. 225), by a probability law of a variable X Lévy meant "the mapping $S \mapsto P(X \in S)$ for subsets S of $\mathbb{R}$, which were not further specified". See Lévy (1925, p. 136).

line, which is produced by the intersections of reflected rays'" (Joets 2008, p. 157).[2] He did not give the curve a name at that time, but in a later publication (Tschirnhaus 1690, p. 170) he introduced Latin *curva caustica*.[1] The term reflects the context in which the curve occurs, since *caustica* is based on Greek *καυστικός* 'capable of burning'.

The curve has therefore also been called *burning curve*, for example by Carré (1705, p. 70), who used French *ligne Caustique ou brûlante* (from *brûler* 'to burn'). The German term *Brennlinie* (from *brennen* 'to burn') is a similar construction, found in Wolff (1716, p. 467). The English term *caustick* is in Harris (1704).

The terminological distinction between *diacaustic* and *catacaustic* curves is due to Bernoulli (1693, p. 244), who used Latin *diacaustica* and *catacaustica.*

The name *caustic surface* seems to be due to Malus (1811, p. 241), who used French *surface caustique*. This paper was read in 1807 but not published until 1811; an earlier printed attestation is found in Malus (1807, p. 60), for example.

Note. Brief overviews of the terminology are given in Loria (1911, pp. 300–302) and Brocard and Lemoyne (1919, p. 131). Bruce, Giblin, and Gibson (1984) give an overview of the concept's history; Joets (2008) discusses the concept in Greek antiquity. On Tschirnhaus's work on caustics, information can be found in Kracht and Kreyszig (1990) and Scarpello and Scimone (2005).

References

Bernoulli, Jakob (1693). "Curvae dia-causticae, earum relatio ad evolutas". *Acta Eruditorum* (June 1693), pp. 244–256.

Brocard, Henri and Timoléon Lemoyne (1919). *Courbes géométriques remarquables*. (Courbes spéciales). Planes et gauches. Tome 1. Paris.

Bruce, J.W., P.J. Giblin, and C.G. Gibson (1984). "Caustics through the looking glass". *The Mathematical Intelligencer* 6, pp. 47–58.

Carré, Louis (1705). "Sur les caustiques". *Histoire de l'Académie royale des sciences* (1703), pp. 69–97.

Harris, John (1704). *Lexicon Technicum*. London.

Joets, Alain (2008). "Caustics in Greek antiquity". *Banach Center Publications* 82, pp. 157–161.

Kracht, Manfred and Erwin Kreyszig (1990). "E.W. von Tschirnhaus: His Role in Early Calculus and His Work and Impact on Algebra". *Historia Mathematica* 17, pp. 16–35.

Loria, Gino (1911). *Spezielle algebraische und transzendente ebene Kurven*. Theorie und Geschichte. Autorisierte, nach dem italienischen Manuskript bearbeitete deutsche Ausgabe von Fritz Schütte. Zweite Auflage. Zweiter Band. Leipzig.

Malus, Étienne-Louis (1807). "Mémoire sur l'optique". *Bulletin des sciences, par la Société philomathique de Paris* 1, pp. 59–60.

Malus, Étienne-Louis (1811). "Traité d'optique". *Mémoires présentés à l'Institut des sciences par divers savants* 2, pp. 214–302.

Scarpello, Giovanni Mingari and Aldo Scimone (2005). "The Work of Tschirnhaus, La Hire and Leibniz on Catacaustics and the Birth of the Envelopes of Lines in the 17th Century". *Archive for History of Exact Sciences* 59, pp. 223–250.

Shapiro, Alan E. (1990). "The *Optical Lectures* and the foundations of the theory of optical imagery". In: *Before Newton*. The Life and Times of Isaac Barrow. Ed. by Mordechai Feingold. Cambridge, pp. 105–178.

Tschirnhaus, Ehrenfried Walter (1682). "Inventa nova exhibita Parisiis Societati Regiae Scientiarum". *Acta Eruditorum* (Nov. 1682), pp. 364–365.

---

[2] The statement quoted by Joets is from Tschirnhaus (1682, p. 364). [1] Shapiro (1990, p. 176), Brocard and Lemoyne (1919, p. 131).

Tschirnhaus, Ehrenfried Walter (1690). "Curva Geometrica, quae seipsam sui evolutione describit". *Acta Eruditorum* (Apr. 1690), pp. 169–172.
Wolff, Christian (1716). *Mathematisches Lexicon*. Leipzig.

cell

In his series of papers on topology, Poincaré constructed manifolds from smaller building blocks. Poincaré used French *case*, which literally means 'compartment, square or rectangle delimited by a surface'. In particular, in the article "Second complement to analysis situs", he discussed his constructive approach:[1]

> Consider a closed manifold of p dimensions. We assume that this manifold is subdivided so as to form a polyhedron P of p dimensions. The elements of that polyhedron are called the a_i^p; they are separated from each other by $(p-1)$-dimensional manifolds called the a_i^{p-1}; these are separated from each other by $(p-2)$-dimensional manifolds called the a_i^{p-2}; and so on until we arrive at the vertices of the polyhedron which are called the a_i^0. [...] We sometimes denote the a_i^0 by the name *vertices* [*sommets*], the a_i^1 by the name *edges* [*arêtes*], the a_i^2 by the name *faces* [*faces*], the a_i^3 by the name *cells* [*cases*], the a_i^4 by the name *hypercells* [*hypercases*] (Poincaré 1900, pp. 278f.).[2]

The term *cell* came up a few years after Poincaré's second complement was published. It appears, for example, in a paper by Steinitz, in which the basic elements of topology are discussed. These elements are, according to Steinitz, the point, the line segment [*Strecke*], "the (simply connected) surface piece [*Flächenstück*], the (simple, sphere-like) space piece [*Raumstück*]" (Steinitz 1908, p. 30).[3] He then calls these, respectively, *cells* (*Zellen*) of 0th, 1st, 2nd, and 3rd dimension (Steinitz 1908, p. 31).[4]

With varying definitions, the term was also used by Tietze (1908, p. 21) and by Veblen and Alexander (1912-13, p. 165). The latter defined: "Now let $[P]$ be a set of objects (e. g., points or lines,–we shall always refer to these objects as points) such that there exists a one-to-one reciprocal correspondence between $[P]$ and the points interior to and on the boundary of an n-dimensional simplex. Then the points of $[P]$ which correspond to the interior of the simplex are said to constitute an n-dimensional cell E_n, and those which correspond to the boundary of the simplex are said to constitute the boundary of the cell". Veblen and Alexander (ibid.) also used n-*cell*.

Ehresmann (1934, p. 411) later redefined French *cellule*, saying that "as far as homology and homotopy are concerned, cells of a more general kind can fulfil the same role as the simplexes".[5] His approach is explained by Dieudonné (2009, p. 221): "Ehresmann was led to consider a partition of a space X into 'cells' of a more general type than those in use up to that time. For each dimension p a cell of dimension p is still homeomorphic to an open ball in R^p, but its boundary in X is not necessarily homeomorphic to S_{p-1} as for usual cells".

---

[1] In the cited passage, Poincaré recalled terminology which he originally introduced in the first complement (Poincaré 1899, p. 314). [2] Translation from Poincaré (2010, pp. 111f.). [3] "das (einfach zusammenhängende) Flächenstück, das (einfache, kugelartige) Raumstück". The terms *Flächenstück* and *Raumstück* are from Dehn and Heegaard (1907, pp. 157, 161). [4] "Zellen 0ter, 1ter, 2ter, 3ter Dimension". [5] "en ce qui concerne l'homologie et l'homotopie, des cellules d'une espèce plus générale peuvent remplir le meme rôle que les simplexes".

References

Dehn, Max and Poul Heegaard (1907). "Analysis situs". *Encyklopädie der mathematischen Wissenschaften mit Einschluss ihrer Anwendungen* III AB3, pp. 153–220.
Dieudonné, Jean (2009). *A History of Algebraic and Differential Topology, 1900–1960*. Boston.
Ehresmann, Charles (1934). "Sur la topologie de certains espaces homogènes". *Annals of Mathematics* 35, pp. 396–443.
Poincaré, Henri (1899). "Complément à l'analysis situs". *Rendiconti del Circolo Matematico di Palermo* 13, pp. 285–343.
Poincaré, Henri (1900). "Second complément à l'analysis situs". *Proceedings of the London Mathematical Society* 32, pp. 277–308.
Poincaré, Henri (2010). *Papers on Topology. Analysis Situs* and Its Five Supplements. Translated by John Stillwell. Providence, R.I.
Steinitz, Ernst (1908). "Beiträge zur Analysis situs". *Sitzungsberichte der Berliner Mathematischen Gesellschaft* 7, pp. 29–49.
Tietze, Heinrich (1908). "Über die topologischen Invarianten mehrdimensionaler Mannigfaltigkeiten". *Monatshefte für Mathematik und Physik* 19, pp. 1–118.
Veblen, Oswald and James W. Alexander (1912-13). "Manifolds of N Dimensions". *Annals of Mathematics* 14, pp. 163–178.

cell complex

Abstract cell complexes were studied by Listing (1862), Steinitz (1908), and Tucker (1933), for example, and certain geometric cell complexes were discussed by Poincaré (1895, p. 101) under the name *polyèdre*. However, while Listing introduced the word *complex*, and *cell* is found in the early 20th century, in none of these papers is the term *cell complex* actually used. The reason presumably is that the complexes considered were made up of cells, and there was no need for such a compound.

Explicitly, *cell complex* began to be used, very occasionally, in the 1920s. In particular, German *Zellenkomplex* appeared in Bilz (1923, p. 1) and van der Waerden (1930, p. 124); Kneser (1926, p. 7) used the term *cell building* (*Zellgebäude*). *Zellenkomplex* is more frequently found from the 1930s onwards; it was used in the important books by Seifert and Threlfall (1934, p. 233) and Alexandroff and Hopf (1935, p. 126).

Note. For the history of the concept and its various definitions, see Klette (2000) and Scholz (1999).

References

Alexandroff, Pavel S. and Heinz Hopf (1935). *Topologie*. Erster Band. Berlin.
Bilz, Ernst (1923). "Beitrag zu den Grundlagen der kombinatorischen Analysis situs". *Mathematische Zeitschrift* 18, pp. 1–41.
Klette, Reinhard (2000). "Cell Complexes through Time". *Proceedings of the SPIE* 4117, pp. 134–145.
Kneser, Hellmuth (1926). "Die Topologie der Mannigfaltigkeiten". *Jahresbericht der Deutschen Mathematiker-Vereinigung* 34, pp. 1–14.
Listing, Johann Benedict (1862). "Der Census räumlicher Complexe, oder Verallgemeinerung des Euler'schen Satzes von den Polyëdern". *Abhandlungen der Königlichen Gesellschaft der Wissenschaften zu Göttingen* 10, pp. 97–182.
Poincaré, Henri (1895). "Analysis situs". *Journal de l'École polytechnique* 1, pp. 1–121.
Scholz, Erhard (1999). "The Concept of Manifold, 1850-1950". In: *History of Topology*. Ed. by Ioan Mackenzie James. Amsterdam, pp. 25–64.
Seifert, Herbert and William Threlfall (1934). *Lehrbuch der Topologie*. Leipzig.

Steinitz, Ernst (1908). "Beiträge zur Analysis situs". *Sitzungsberichte der Berliner Mathematischen Gesellschaft* 7, pp. 29–49.
Tucker, Albert William (1933). "An Abstract Approach to Manifolds". *Annals of Mathematics* 34, pp. 191–243.
van der Waerden, Bartel Leendert (1930). "Kombinatorische Topologie". *Jahresbericht der Deutschen Mathematiker-Vereinigung* 39, pp. 121–139.

central projection

While the concept behind the term is much older[1], *central projection* is not found until the 18th century. Lambert (1772, p. 109) introduced it thus: "Finally, the eye was placed into the midpoint of the earth, and this method of projection, since no other name is known to me, shall be called the *central projection*".[2] The reason for choosing this term is implied by Lambert himself: the eye of the beholder is placed in the *centre* of the earth.

The term was soon adopted by Karsten (1775, p. 665), but also as French *projection centrale* by Lagrange (1781, p. 162); Lacroix (1791, p. 37) gave *perspective centrale* as an alternative. The English term first appears, according to the OED, in Cadell (1816, p. 2).

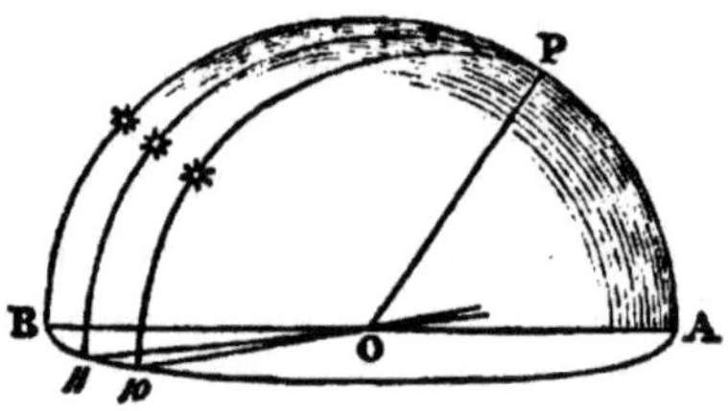

Figure 11: The central projection. From De Morgan (1836, p. 29).

An alternative name is *gnonomic projection*. Of the few authors who comment on the history of this term, Günther (1883, pp. 138f.) vaguely says that it is of French origin, and Snyder (1993, pp. 18f.) points to attestations of English *gnomic* and *gnomonic* from 1749 and 1836, respectively. While it unclear who exactly coined this term, it has already been in use in the 17th century: Latin *proiectio gnonomica* was used by Maignan (1648, p. 54), English *gnomonick projection* by Collins (1669, p. 933), and French *projection gnomonique* by Ozanam (1691, pp. 471f.). The name is based on the word *gnomon*; as explained by Snyder (1987, p. 164), "[t]he name Gnomonic is derived from the fact that the meridians radiate from the pole (or are spaced, on the equatorial aspect) just as the corresponding hour markings on a sundial for the same central latitude. The gnomon of the sundial is the elevated straightedge

[1] The belief that Thales used this projection appears to be due to d'Azevac (1863, pp. 129f.). According to Snyder (1993, p. 18), *horoscope* (lit., 'time observer') and *horologium* (sundial) used to be other designations used for this projection. [2] "Endlich setzte man das Auge in den Mittelpunct der Erde, und diese Projectionsart mag, weil mir kein anderer Name bekannt ist, die *Centralprojection* heißen". Translation based on Lambert (2011, p. 5).

pointed toward the pole and casting its shadow on the various hour markings as the sun moves across the sky".

This type of projection has been attributed to Grienberger (1612) by various authors, for example by Kästner (1798, pp. 406f.) and de Lalande (1803, p. 157); the latter additionally points out that Kircher (1645, p. 562) called Grienberger the inventor of the concept.

Note. On the history of the concept, see Kästner (1798), De Morgan (1836, pp. 50–54), Günther (1883), Keuning (1955, pp. 6f.), and Snyder (1993, pp. 18f.).

References

d'Azevac, Armand (1863). *Coup d'oeil historique sur la projection des cartes de géographie*. Paris.

Cadell, William Archibald (1816). *On the lines that divide each semidiurnal arc into six equal parts*. Edinburgh.

Collins, John (1669). "An Account Concerning the Resolution of Equations in Numbers". *Philosophical Transactions of the Royal Society of London* 4, pp. 929–934.

Grienberger, Christoph (1612). *Catalogus veteres affixarum Longitudines ac Latitudines conferens cum novis*. Imaginum cælestium Prospectiva duplex. Roma.

Günther, Siegmund (1883). "Die gnomonische Kartenprojektion". *Zeitschrift der Gesellschaft für Erdkunde zu Berlin* 18, pp. 137–149.

Karsten, Wenceslaus Johann Gustav (1775). *Lehrbegrif der gesamten Mathematik*. Der Siebende Theil. Greifswald.

Kästner, Abraham Gotthelf (1798). "Etwas zur Empfehlung der Central-Projection". *Allgemeine geographische Ephemeriden* 2, pp. 401–412.

Keuning, Johannes (1955). "The history of geographical map projections until 1600". *Imago Mundi* 12, pp. 1–24.

Kircher, Athanasius (1645). *Ars magna lucis et vmbrae in decem libros digesta*. Roma.

Lacroix, Sylvestre François (1791). "Sur une projection proposée par feu M. Lorgna". *Bulletin des sciences, par la Société philomathique de Paris* 2, pp. 37–39.

Lagrange, Joseph-Louis (1781). "Sur la Construction des Cartes Géographiques". *Nouveaux Mémoires de l'Académie Royale des Sciences et Belles-Lettres de Berlin* (1779), pp. 161–210.

de Lalande, Jérôme (1803). *Bibliographie astronomique: avec l'histoire de l'astronomie depuis 1781 jusq'à 1802*. Paris.

Lambert, Johann Heinrich (1772). "Anmerkungen und Zusätze zur Entwerfung der Land- und Himmelscharten". In: *Beyträge zum Gebrauch der Mathematik und deren Anwendung*. Dritter Theil. Berlin, pp. 105–199.

Lambert, Johann Heinrich (2011). *Notes and Comments on the Composition of Terrestrial and Celestial Maps*. Translated and Introduced by Waldo R. Tobler. Redlands.

Maignan, Emmanuel (1648). *Perspectiva horaria sive de horographia gnomonica tum theoretica tum practica*. Roma.

De Morgan, Augustus (1836). *An Explanation of the Gnomonic Projection of the Sphere*. London.

Ozanam, Jacques (1691). *Dictionnaire mathematique*. Paris.

Snyder, John P. (1987). *Map Projections—A Working Manual*. (U.S. Geological Survey Professional Paper 1395). Washington, D.C.

Snyder, John P. (1993). *Flattening the Earth*. Two Thousand Years of Map Projections. Chicago.

centralizer

The term *centralizer* is found in group theory and was introduced by Hall (1933, p. 88): "If L is any subgroup of a group G, the *centraliser* of L in G consists by definition of all those elements of G which are permutable with every element of L.

The central of G is simply the centraliser of G itself in G". The term was quickly adopted by Birkhoff (1933, p. 257).[1]

Note. Brief notes on the origin of the term can be found in Magnus (1939, p. 9).[2]

References

Birkhoff, Garrett (1933). "A theorem on transitive groups". *Proceedings of the Cambridge Philosophical Society* 29, pp. 257–259.

Hall, Philip (1933). "A Contribution to the Theory of Groups of Prime-Power Order". *Proceedings of the London Mathematical Society* 36, pp. 29–95.

Hall, Philip (1936). "On a Theorem of Frobenius". *Proceedings of the London Mathematical Society* 40, pp. 468–501.

Magnus, Wilhelm (1939). "Allgemeine Gruppentheorie". *Enzyklopädie der mathematischen Wissenschaften mit Einschluss ihrer Anwendungen.* Zweite, völlig neubearbeitete Auflage I 1,9, pp. 1–51.

centre

This term goes ultimately goes back to Greek *κέντρον* 'sharp point, prick'; it referred to the tip of the compass, which at the same time is the center of the circle. It was thus natural to call the center of the circle its *κέντρον* (Schmidt 1906, pp. 15f., 44). The term is attested in the oldest mathematical texts in Greek, for example in Autolycus' *On a moving sphere*, where it is used with reference to the sphere (Mogenet 1950, p. 196); it was also used with reference to other objects such as ellipses and ellipsoids (Mugler 1958-59, pp. 246–248). Mugler (1958-59, p. 248) notes that with Plato, the usage is not fixed: in his *Parmenides* (137e), Plato defines a circle (called "the round") thus: "The round [...] is that of which the extremes are everywhere equally distant from the center".[1] Plato does not use *κέντρον* here, but *μέσος* 'middle'.

In the *Elements* (book I, deff. 15 and 16), Euclid defines the circle and its centre (*κέντρον*) thus: "A *circle* is a plane figure contained by one line such that all the straight lines falling upon it from one point among those lying within the figure equal one another. And the point is called the *center* of the circle".[2]

The corresponding Latin term is *centrum*, which is found in late classical Latin, for example in Vitruvius (Ayuso García 2008, p. 921). While this term can be regarded as standard since that time, other words were in use. Occasionally, for example, the centre was simply called *point*; for example, Isidore of Sevilla in his *Etymologies* states that a circle "has a point in its middle, on which everything centers, which people call the geometric center, and which Latin speakers call the 'point of the circle'" (Isidore of Seville 1911, book III, xii).[3] The term *point* (*punctum*) referred to by Isidore can be found in the writings of Roman land surveyors, for example in Balbus (Blume, Lachmann, and Rudorff 1848, p. 102). The term *middle point of the circle* (*circuli media nota*) was used by Martianus Capella (Willis 1983, p. 252), for example.[4]

---

[1] Hall's paper was read in March 1932, Birkhoff's in March 1933; Birkhoff also credits Hall with the theorem he proves in his paper. [2] Magnus refers to Hall (1936, p. 470) rather than to Hall (1933).

[1] Translation from Plato (1926, p. 237). – Plato's definition has been cited as the earliest definition of a circle (Tropfke 1940, p. 141). [2] Translation from Euclid (1956, p. 153). [3] "Quarum prima circulus est figura plana, quae vocatur circumducta; cuius in medio punctus est, quo cuncta convergunt, quod centrum geometriae vocant, Latini punctum circuli nuncupant". Translation from Isidore of Seville (2006, p. 93). [4] The term *nota* is one of the Latin words for 'point'. See *point*.

French *centre* is attested in the 13th century; Dörr (1998, p. 109) cites an astronomical treatise from 1270 as the first attestation. A "definition" was given by Oresme (1968, p. 60): "a point called the center, in the middle of the circle".[5] Italian *centro* also dates from the 13th century (TLIO).

English *centre* is of slightly later date; the OED cites the late 14th-century *The Equatorie of the Planetis* as the first attestation; Recorde (1551, fol. Bi^r) states that a circle "hath in the middell of it a pricke or centre, from whiche all the lines that be drawen to the circumference[1] are equall all in length". The term *middle point* is also found in the 14th century, for example in John Trevisa's translation of the encyclopaedic *De proprietatibus rerum* (Seymour 1975, p. 1368).

The German standard term *Mittelpunkt*, lit. 'middle point', dates from 14th century and is found in the "German Sphaera" (von Megenberg 1980, p. 6), a translation of *Liber de Sphaera mundi* by Johannes de Sacrobosco.

For *centre of curvature*, see *curvature*.

Note. A brief historical overview of the terminology is found in Tropfke (1940, pp. 131f.). For the Greek term, see Mugler (1958-59, pp. 246–248) and Wersinger (2014); for the Latin, see Guillaumin (1997) and Ayuso García (2008, pp. 921–926); for the Arabic, see Rashed (2017, pp. 308–312).

References

Ayuso García, Manuel (2008). "La terminología latina de la geometría en Marciano Capela". PhD thesis. Madrid: Universidad Nacional de Educación a Distancia.

Blume, Friedrich, Karl Lachmann, and Adolf Rudorff, eds. (1848). *Die Schriften der römischen Feldmesser*. Erster Band. Texte und Zeichnungen. Berlin.

Dörr, Stephen (1998). *Der älteste Astronomietraktat in französischer Sprache: L'Introductoire d'astronomie*. Edition und lexikalische Analyse. Tübingen.

Euclid (1956). *The Thirteen Books of Euclid's Elements*. Volume I. Introduction and Books I, II. Ed. by Thomas L. Heath. New York.

Guillaumin, Jean-Yves (1997). "Les noms latins du point géométrique". In: *Atti del II Seminario Internazionale di Studi sui Lessici Tecnici Greci e Latini (Messina, 14-16 déc. 1995)*. Ed. by Paola Radici Colace. Napoli, pp. 85–106.

Isidore of Seville (1911). *Etymologiarum sive Originum libri XX*. Tomus I. Ed. by Wallace M. Lindsay. Oxford.

Isidore of Seville (2006). *The* Etymologies *of Isidore of Seville*. Ed. by Stephen A. Barney et al. Cambridge.

von Megenberg, Konrad (1980). *Die deutsche Sphaera*. Ed. by Francis B. Brévart. Tübingen.

Mogenet, Joseph (1950). *Autolycus de Pitane*. Histoire du texte suivie de l'édition critique des traités de la sphère en mouvement et des levers et couchers. Louvain.

Mugler, Charles (1958-59). *Dictionnaire historique de la terminologie géométrique des Grecs*. Paris.

Oresme, Nicole (1968). *Le livre du ciel et du monde*. Edited by Albert D. Menut and Alexander J. Denomy, translated with an introduction by Albert D. Menut. Madison.

Plato (1926). *Cratylus. Parmenides. Greater Hippias. Lesser Hippias*. Translated by H. N. Fowler. London.

Rashed, Roshdi, ed. (2017). *Lexique historique de la langue scientifique arabe*. Hildesheim.

Recorde, Robert (1551). *The patheway to knowledg*. London.

Schmidt, Max Carl Paul (1906). *Zur Entstehung und Terminologie der elementaren Mathematik*. Leipzig.

---

[5] "un point appellé centre qui est ou milieu du cercle". Translation from Oresme (1968, p. 61).

[1] Corrected from *circumfernece*.

Seymour, Maurice Charles, ed. (1975). *On the Properties of Things*. John Trevisa's Translation of Bartholomaeus Anglicus' *De Proprietatibus Rerum*. A Critical Text. Volume II. Oxford.
Tropfke, Johannes (1940). *Geschichte der Elementar-Mathematik*. Vierter Band. Ebene Geometrie. Dritte, verbesserte und vermehrte Auflage besorgt von Dr. Kurt Vogel. Berlin and Leipzig.
Wersinger, Anne Gabrielle (2014). "«Le milieu et la ronde». Contribution à l'étude des représentations géométriques et politiques du cercle et de l'harmonie chez Homère, les Pythagoriciens, Platon et Euclide". *Philosophia* 44, pp. 55–74.
Willis, James, ed. (1983). *Martianus Capella*. Leipzig.

centre (algebra)

The *centre* of a group was named *central* by de Séguier (1904, p. 57): "the set of the normal elements of *G* is a normal divisor which will be called the *central* of *G*".[1] The French term was translated as *central* by Dickson (1904, p. 162), a word which was also used in various textbooks such as Hilton (1908, p. 64). The change to *centre* seems to date from the 1920s, when Speiser (1923, p. 15) used German *Zentrum*, a term also used by van der Waerden (1930, p. 28) in his influential book *Modern Algebra*.

The term was also transferred to various other algebraic structures. For example, van der Waerden (1926, p. 719) applied it to group rings and said that the term in this sense was due to Emmy Noether; Artin (1927, p. 246) applied the term to skew fields.

Note. Brief notes on the origin of the term can be found in Magnus (1939, p. 9).

References

Artin, Emil (1927). "Über einen Satz von Herrn J. H. Maclagan Wedderburn". *Abhandlungen aus dem Mathematischen Seminar der Universität Hamburg* 5, pp. 245–250.
Dickson, Leonard Eugene (1904). "De Séguier's theory of abstract groups". *Bulletin of the American Mathematical Society* 11, pp. 159–164.
Hilton, Harold (1908). *An Introduction to the Theory of Groups of Finite Order*. Oxford.
Magnus, Wilhelm (1939). "Allgemeine Gruppentheorie". *Enzyklopädie der mathematischen Wissenschaften mit Einschluss ihrer Anwendungen*. Zweite, völlig neubearbeitete Auflage I 1,9, pp. 1–51.
de Séguier, Jean Armand (1904). *Théorie des groupes finis*. Éléments de la théorie des groupes abstraits. Paris.
Speiser, Andreas (1923). *Die Theorie der Gruppen von endlicher Ordnung*. Berlin and Heidelberg.
van der Waerden, Bartel Leendert (1926). "Über die fundamentalen Identitäten der Invariantentheorie". *Mathematische Annalen* 95, pp. 706–735.
van der Waerden, Bartel Leendert (1930). *Moderne Algebra*. Erster Teil. Berlin.

centre (curve)

The concept of a centre of a conic section was extended by Newton. In section II of his *Enumeration of Lines of the Third Order*, "The Properties of Conic Sections are analogous to those of Curves of higher orders", he first recalls the terminology in connection with conic sections:

> If parallel straight lines terminated by the curve be drawn in a conic section, the straight line bisecting two of them will bisect all the others, and is called a *diameter*, and the bisected lines are called *ordinates* to the diameter, and the intersection of all the diameters is the *centre*; the

[1] "l'ensemble des éléments normaux de *G* est un diviseur normale qui sera dit *central* de *G*".

intersection of the diameter with the curve is called the *vertex*; that diameter, whose ordinates are rectangular to it, being called the *axis* (Newton 1704, p. 140).[2]

He then proceeds (ibid.):

In like manner, in curves of the second genus[1], if any two parallel straight lines are drawn, meeting the curve in three points; the straight line which cuts these parallel lines so that the sum of the two segments meeting the curve on one side of the secant, will cut in the same ratio all lines parallel to these, provided they also meet the curve in three points; that is, so that the sum of the two parts on one side of the secant, shall equal the third part on the other side. These three parts, thus equal, may be called *ordinates*, and the secant or cutting line to which the ordinates are applied, the *diameter*; the intersection of diameter and curve, the *vertex*; and the intersection of two diameters, the *centre*. The diameter having rectangular ordinates, if any exist, may also be called an *axis*; and where all the diameters meet in a point, that point will be the *general centre*.[2]

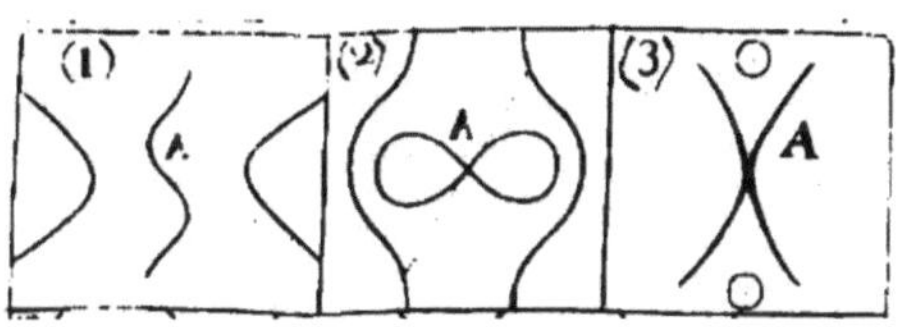

Figure 12: The centre (A) of a curve. From de Gua de Malves (1740).

The terminology was picked up, in particular, by de Gua de Malves (1740, pp. 1f.) and Cramer (1750, p. 144), who both referred to Newton.

Note. A brief historical note is found in Berzolari (1906, p. 333).

References

Berzolari, Luigi (1906). "Allgemeine Theorie der höheren ebenen algebraischen Kurven". *Encyklopädie der mathematischen Wissenschaften mit Einschluss ihrer Anwendungen* III C 4, pp. 313–456.

Cramer, Gabriel (1750). *Introduction à l'analyse des lignes courbes algébriques*. Genève.

de Gua de Malves, Jean Paul (1740). *Usages de l'analyse de Descartes pour découvrir, sans le secours du calcul différentiel, les propriétés des lignes géométriques de tous les ordres*. Paris.

[2] "Nam si rectae plures parallelae & ad conicam sectionem utrinq; terminatae ducantur, recta duas earum bisecans bisecabit alias omnes,ideoq; dicitur *Diameter* figurae & rectae bisectae dicuntur *Ordinatim applicatae* ad Diametrum, & concursus omnium Diametrorum est *Centrum* figurae, & intersectio Curvae & diametri *Vertex* nominatur, & diameter illa *Axis* est cui ordinatim applicatae insistunt ad angulos rectos". Translation from Newton (1860, p. 8) (italics added). [1] A curve of the second genus is the same as a line of the third order. [2] "Et ad eundem modum in Curvis secundi generis, si rectae duae quaevis parallelae ducantur occurrentes Curvae in tribus punctis: recta quae ita secat has parallelas ut fumma duarum partium ex uno secantis latere ad curvam terminatarum, aequetur parti tertiae ex altero latere ad curvam terminatae, eodem modo secabit omnes alias his parallelas curvaeq; in tribus punctis occurrentes rectas, hoc est, ita ut summa partium duarum ex uno ipsius latere semper aequetur parti tertiae ex altero latere. Has itaq; tres partes quae hinc inde aequantur, Ordinatim applicatas & rectam secantem cui ordinatim applicantur *Diametrum* & intersectionem diametri & *Verticem* & concursum duarum diametrorum *Centrum* nominare licet. Diameter autem ad Ordinatas rectangula si modo aliqua sit, etiam *Axis* dici potest, & ubi omnes diametri in eodem puncto concurrunt istud erit *Centrum generale*". Translation from Newton (1860, p. 8) (italics added).

Newton, Isaac (1704). “Enumeratio linearum tertii ordinis”. In: *Opticks*. London, pp. 138–162.
Newton, Isaac (1860). *Enumeration of Lines of the Third Order*. Trans. by C.M.R. Talbot. London.

centre of similitude

The term *centre of similitude* was coined as Latin *centrum simulitudinis* by Euler in 1777 (see fig. 13):

> If two similar figures are described in the same plane, and if we let the larger one be called AB and the corresponding smaller one be called ab, then there always will be, in that very plane, a certain point Γ that relates similarly to either segment, so that the triangles ΓAB and Γab are similar. Let the point be called the *center of similitude* of the two proposed similar figures (Euler 1795, p. 154).[1]

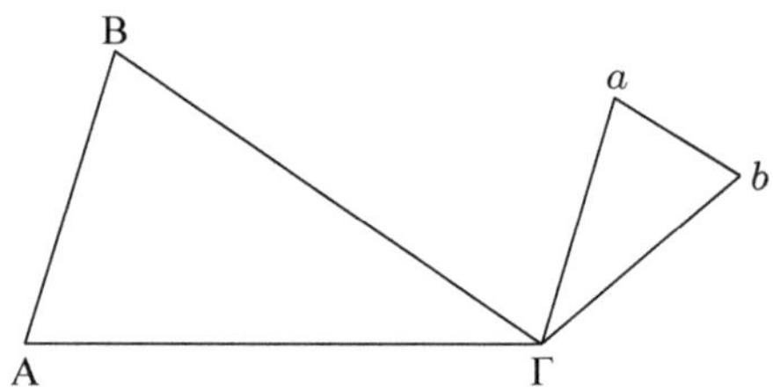

Figure 13: The centre of similitude. From Euler (1795, Tab. III).

The term was picked up by various French mathematicians such as Durrande (1815-16, p. 334) and Monge (1816, p. 4);[2] it was in fact Monge who was originally credited with the term, for example by Steiner and Gergonne (1826-27, p. 288) and Puissant (1824, p. 99). The English term first appears in Davies (1830, p. 114).[3] In German, the term *Ähnlichkeitspunkt*, lit. ‘point of similitude’, was suggested by Grunert (1833, p. 19) and adopted by Plücker (1835, p. 61).

See also *homothetic*.

Note. For the history of the concept, see Archibald (1915, 1916) and White (2007, pp. 317f.).

References

Archibald, Raymond Clare (1915). “Centers of Similitude of Circles and Certain Theorems Attributed to Monge. Were They Known to the Greeks?” *The American Mathematical Monthly* 22, pp. 6–12.
Archibald, Raymond Clare (1916). “Historical Note on Centers of Similitude of Circles”. *The American Mathematical Monthly* 23, pp. 159–161.
Davies, Thomas Stephens (1830). “Researches in the Geometry of Three Dimensions”. *New Series of The Mathematical Repository* 5, part II, pp. 111–121, 156–168.
Davies, Thomas Stephens (1835). “First Solution”. *New Series of The Mathematical Repository* 6, pp. 68–73.

[1] “Si habeantur duae figurae similes in eodem plano descriptae, quarum maioris quodpiam latus sit AB, minoris vero latus respondens ab, semper dabitur in eodem plano certum punctum Γ, quod ad utramque figuram similiter referatur, ita ut figurae ΓAB et Γab sint inter se similes, hocque punctum Γ appelletur centrum similitudinis binarium figurarum similium propositarum”. Translation from Euler (2019, p. 1). [2] Monge’s text was written in 1814. [3] Davies (1835, p. 69) later suggested *pole of similitude*.

Durrande, J.B. (1815-16). “Démonstration des deux théorèmes énoncés à la page 172 de ce volume, et de quelques autres théorèmes analogues”. *Annales de mathématiques pures et appliquées* 6, pp. 326–339.
Euler, Lconhard (1795). “De centro simulitudinis”. *Nova Acta Academiae Scientarum Imperialis Petropolitinae* 9, pp. 154–169.
Euler, Leonhard (2019). “De Centro Similitudinis (On the Center of Similitude)”. URL: `https://scholarlycommons.pacific.edu/euler-works/693/` (retrieved Dec. 27, 2019).
Grunert, Johann August (1833). *Supplemente zu Georg Simon Klügel's Wörterbuche der reinen Mathematik*. Erste Abtheilung. A bis D. Leipzig.
Monge, Gaspard (1816). “Du centre de similitude de deux courbes semblables”. *Correspondance sur l'École impériale polytechnique* 3, pp. 4–5.
Plücker, Julius (1835). *System der analytischen Geometrie*. Berlin.
Puissant, Louis (1824). *Recueil de diverses propositions de géométrie, résolues ou démontrées par l'analyse algébrique*. Troisième édition. Paris.
Steiner, Jacob and Joseph Diez Gergonne (1826-27). “Théorie générale des contacts et des intersections des cercles”. *Annales de mathématiques pures et appliquées* 17, pp. 285–315.
White, Homer S. (2007). “The Geometry of Leonhard Euler”. In: *Leonhard Euler*. Life, Work and Legacy. Ed. by Robert E. Bradley and C. Edward Sandifer. Amsterdam, pp. 303–321.

centre of symmetry

The term *centre of symmetry* is not frequently found before the 1850s. An example is Dupin (1826, pp. 96f.), where French *centre de symétrie* is applied to rectangles and rhombi (*lozanges*) and defined as the point of intersections of the two axes of symmetry of these objects. Another example is Tripon (1846, pp. 9f.), who applied it to polygons and states that symmetry can be defined with respect to a point called *centre de symétrie*. Tripon also uses the term with respect to two polygons (see fig. 14).[1]

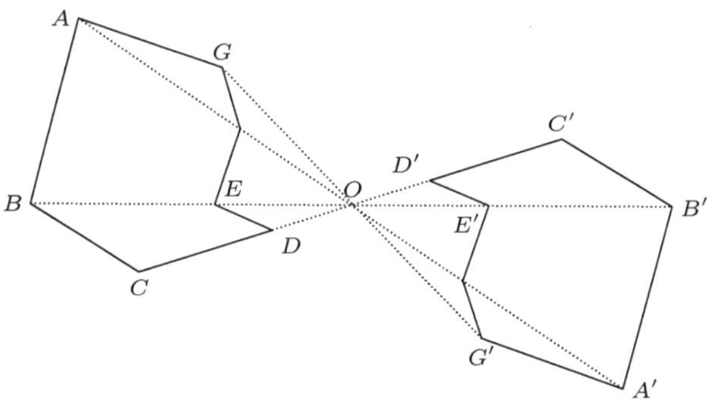

Figure 14: Symmetric polygons. From Tripon (1846, p. 10).

Bravais (1849b, p. 137) defined the term with reference to polyhedra: “I will call *pole of symmetry* of two polyhedra the point through which pass all the straight lines that pairwise join the homologous vertices of the two polyhedra. If one regards the two polyhedra P, p as forming a unique polyhedron (P, p), the same point will be

[1] A usage of the English term is in Moseley (1834, p. 38).

called the *centre of symmetry*".[2] The term was quickly adopted; for example, Möbius (1851, p. 26) literally translated it by German *Mittelpunkt*, the usual German term for the centre of a circle.

Towards the end of the 19th century, the concept of inversion, clearly present in the work of Bravais (and others), is explicitly used in definitions of the centre of symmetry; see, for example, Schoenflies (1891, p. 43): "If a figure is transformed into itself by inversion, one says that it has *symmetry with respect to a point* and calls this point the *centre of symmetry*".[1]

While, as stated before, the terminology is not frequently found before about 1850, there is an earlier usage of *centre* dating from the 18th century. This usage appears to go back de la Caille (1744), who at first defined what he calls a *symmetric polygon* (*polygone symmétrique*): "the symmetric polygons are those that are composed of parallel and equal sides" (de la Caille 1744, p. 143).[2] One property of such polygons is that when a diagonal is drawn from one angle to an opposite angle, this diagonal divides such a polygon into two equal and similar halves "because on the one side of the diagonal there are as many equal triangles that are situated in the same manner as on the other side" (de la Caille 1744, pp. 145f.).[3]

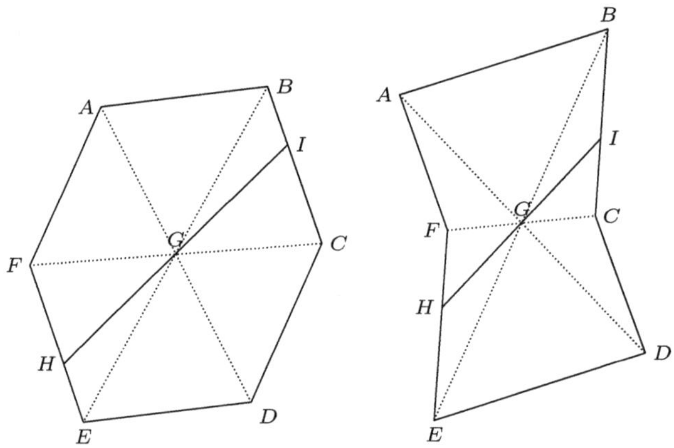

Figure 15: Symmetric polygons. From de la Caille (1744).

[2] "Je nommerai *pôle de symétrie* des deux polyèdres P, p le point par lequel passent toutes les droites joignant deux à deux les sommets homologues des deux polyèdres. Si l'on considère les deux polyèdres comme formant un polyèdre unique (P, p), ce même point sera dit *centre de symétrie* du polyèdre (P, p)". – A similar definition is found in Bravais (1849a, p. 141). [1] "Wenn eine Figur durch Inversion gegen einen Punkt in sich übergeht, so sagt man, sie besitze *Symmetrie gegen einen Punkt* und nennt diesen Punkt ihr *Symmetriecentrum*". – A similar definition was given by Hilton (1903, p. 32). [2] "Les Polygones symmétriques sont ceux qui sont composés de côtés paralléles & égaux". [3] "car il y a d'une côté de la diagonale autant de triangles égaux & posés de la même manière que de l'autre côté".

Because of this property, de la Caille (1744, p. 146) then defines: "One can therefore call the point where the diagonals meet the *centre of the symmetric polygon*".[4] This usage can be interpreted as an early example of point symmetry, but de la Caille would probably have understood it in terms of respective symmetry, for which see *symmetric*. de la Caille was by no means the only one who used *centre* in this sense; the term is also found in Martin (1781, p. 220), for example.

Note. For the history of symmetry in the 19th century, see Scholz (1989a, ch. 1); see also Scholz (1989b).

References

Bravais, Auguste (1849a). "Mémoire sur les polyèdres de forme symétrique". *Journal de mathématiques pures et appliquées* 14, pp. 141–180.

Bravais, Auguste (1849b). "Note sur les polyèdres symétriques de la géométrie". *Journal de mathématiques pures et appliquées* 14, pp. 137–140.

de la Caille, Nicolas-Louis (1744). *Leçons élémentaires de mathématiques, ou Élémens d'algèbre et de géométrie*. Paris.

Dupin, Charles (1826). *Géometrie et méchanique des arts et métiers et des beaux-arts*. Tome deuxième. Paris.

Hilton, Harold (1903). *Mathematical Crystallography and the Theory of Groups of Movements*. Oxford.

Martin, Roger (1781). *Elémens de mathématiques à l'usage des écoles de philosophie*. Toulouse.

Möbius, August Ferdinand (1851). "Ueber symmetrische Figuren". *Berichte über die Verhandlungen der Königlich Sächsischen Gesellschaft der Wissenschaften zu Leipzig (Mathematisch-Physische Klasse)* 3, pp. 19–28.

Moseley, Henry (1834). *A Treatise on Mechanics*. London.

Schoenflies, Arthur (1891). *Krystallsysteme und Krystallstructur*. Leipzig.

Scholz, Erhard (1989a). *Symmetrie – Gruppe – Dualität*. Zur Beziehung zwischen theoretischer Mathematik und Anwendungen in Kristallographie und Baustatik des 19. Jahrhunderts. Basel.

Scholz, Erhard (1989b). "The Rise of Symmetry Concepts in the Atomistic and Dynamistic Schools of Crystallography, 1815-1830". *Revue d'histoire des sciences* 42, pp. 109–122.

Tripon, Jean-Baptiste (1846). *Traité élémentaire de topographie et de lavis des plans*. Paris.

centroid

A relatively large number of terms have been suggested to designate this notion; these terms are based on certain physical or geometrical properties.

The term *centre of gravity* (Greek *κέντρον τοῦ βάρεος*, with *βάρος* meaning 'weight') is first found in the writings of Archimedes (Mugler 1958-59, p. 247). In particular, in his treatise *On the Equilibrium of Planes* Archimedes showed that the three medians meet in a point, which he called the *centre of gravity*: "the centre of gravity of any triangle is at the intersection of the lines drawn from any two angles to the middle points of the opposite sides respectively".[1] Archimedes does not define the term, but definitions are given by later writers such as Pappus, who defined it as "the point within a body which is such that, if the weight be conceived to be

---

[4] "On peut donc appeler le point où se coupent les diagonales, le centre du Polygone symmétrique, à cause de l'égalité des rayons qui vont aux angles opposées". [1] Translation from Archimedes (1897, p. 201).

suspended from the point, it will remain at rest in any position in which it is put" (Pappus of Alexandria 1878, p. 1030).[2]

The Latin term *centrum gravitatis* (from *gravis* 'heavy') does not seem to occur in classical Latin. It was used by William of Moerbeke (13th century), who translated Archimedes' treatise mentioned above and used both *centrum gravium* and *centrum gravitatis* (Clagett 1976, pp. 116, 123). An early usage in print is Valla (1501, book XIII, ch. 2).

According to the OED, English *centre of gravity* first appears in L. Digges and T. Digges (1591, p. 178): "There is in the Axis of euery Peece one onely pointe, whereupon if the Bodye of the Peece should be suspended, she shall hang in a perfect Counterpoyse, and that pointe I call the Center of her Grauity or point of Counterpoyse".[1] In German, Jacob (1565, fol. 300^v) used *Centrum der Schwäre* (from *schwer* 'heavy'), while the now standard *Schwerpunkt* first appears in Sturm (1670, p. 227) as *Schwäre-Punct*. French *centre de gravité* appears to date from the 17th century; it was used in Stevin (1634, Statique, p. 457).

Around 1740, two more terms were suggested. The first one was *centre of inertia* (Latin *centrum inertiae*), which was brought up by König (1738, p. 34).[2] The second term was *centre of masses* (French *Centre de Masses*), suggested by Bernoulli (1741, p. 75).[3] Both terms are based on certain physical properties. Euler (1765, § 286) later commented on these terms:

> The centre of mass or inertia is the same point, as that which is commonly called the centre of gravity; moreover since this point thus is essential to all bodies, in order that for these bodies it can be agreed upon on account of inertia alone, and moreover gravity must be considered as an extrinsic force acting on bodies, thus I have preferred to attribute to that the name of the centre of mass or inertia, since that is understood to be determined by inertia only.[4]

Around 1800, further terms were brought up. First, Robison (1801, p. 375) suggested *centre of position*:

> in every body, or system of bodies, there is a point such, that if a plane be passed through it *in any direction whatever*, the sum if all perpendiculars on one side of the plane is equal to the sum of all the perpendiculars on the other side. The point *P*, thus selected, may, with great propriety, be called the *centre of position* of the body or system.

Second, Carnot (1801, p. 154) used *centre of mean distances* (French *centre des moyennes distances*). Carnot's approach was purely geometrical: He considered a

---

[2] Translation from Heath (1921, p. 430). – Assis (2010, pp. 123–136) discusses various further definitions from Greek antiquity; for the concept in Archimedes, see Stein (1930). Another term that was apparently used in Greek antiquity is what has been translated as *centre of inclination*, used by Hero of Alexandria in his book *Mechanica*. This text is only extant in an Arabic translation (with the exception of some fragments). For discussion, see Heath (1921, pp. 350f.), Drachmann (1963), and Assis (2010, pp. 127–130). [1] Later in the same text (p. 192), Digges also uses "Center of Grauitie, or Counterpoyse Center". [2] Gehler (1791, p. 214) and Stäckel (1908, p. 560) cite König as the creator of the term. [3] The term is attributed to Bernoulli by various contemporaries, e. g. by d'Alembert (1743, p. 53). [4] "Centrum massae seu inertiae idem est punctum, quod vulgo centrum gravitatis vocatur; cum autem hoc punctum ita omnibus corporibus sit essentiale, ut iis ob inertiam solam conveniat, gravitas autem pro vi extrinsecus in corpora agente sit habenda, malui ei nomen centri massae seu inertiae tribuere, ut intelligatur id per solam inertiam determinari". Translation from Euler (2009, § 286).

given set of points and associated with the points a focus (*foyer*) "whose distance from an arbitrary plane (or axis) is equal to the arithmetic mean of the algebraic distances of the given points from the same plane (or axis)" (Altshiller-Court 1960, p. 35); this property explains why he chose the name. Carnot (1801, pp. 158f.) went on to point out that his centre of mean distances is the same as the centre of gravity considered in mechanics. The term is mostly used by French writers but also occurs in English texts such as [anon.] (1804, p. 552) and Gregory (1806, pp. 54f).

Finally, the term *centroid* was suggested by Davies (1845, p. 58):

> The point now found, as is well known, is that usually, but improperly, *in geometry* at least, named the centre of gravity of the triangle ABC. Carnot has proposed as a geometrical name, *le centre des moyennes distances*: but whilst this is inconveniently lengthy as a term, it is also objectionable from its implication of an erroneous theorem, when taken generally. The term *centroïd* seems to answer all purposes.

See also *barycentric*.

Note. For the history of the concept and some terminological remarks, see Altshiller-Court (1960). Assis (2010, pp. 123–136) discusses various definitions from Greek antiquity.

References

[anon.] (1804). "Carnot's Geometry of Position". *The Critical Review* 2, pp. 546–556.

d'Alembert, Jean Le Rond (1743). *Traité de dynamique*. Paris.

Altshiller-Court, Nathan (1960). "Notes on the centroid". *The Mathematics Teacher* 53, pp. 33–35.

Archimedes (1897). *The Works of Archimedes*. Edited in Modern Notation with Introductory Chapters by T. L. Heath. Cambridge.

Assis, Andre K.T. (2010). *Archimedes, the Center of Gravity, and the First Law of Mechanics*. Montreal.

Bernoulli, Daniel (1741). "Traité sur le flux et reflux de la mer". In: *Pièces qui ont remporté le Prix de l'Académie royale des Sciences, en MDCCXL*. Paris, pp. 53–191.

Carnot, L.N.M. (1801). *De la corrélation des figures de géométrie*. Paris.

Clagett, Marshall (1976). *Archimedes in the Middle Ages*. Volume II. The Translations from the Greek by William of Moerbeke. Philadelphia.

Davies, Thomas Stephens (1845). "On the algebraical analysis of porisms". *The Mathematician* 1, pp. 42–64.

Digges, Leonard and Thomas Digges (1591). *A geometrical practise, named Pantometria diuided into three bookes, Longimetra, Planimetra, and Stereometria*. London.

Drachmann, A.G. (1963). "Fragments from Archimedes in Heron's Mechanics". *Centaurus* 8, pp. 91–146.

Euler, Leonhard (1765). *Theoria motus corporum solidorum seu rigidorum*. Rostock and Greifswald.

Euler, Leonhard (2009). "Theoria Motus Corporum Solidorum Seu Rigidorum". Translated and annotated by Ian Bruce. URL: `http://www.17centurymaths.com/contents/mechanica3.html` (retrieved Jan. 22, 2016).

Gehler, Johann Samuel Traugott (1791). *Physikalisches Wörterbuch oder Versuch einer Erklärung der vornehmsten Begriffe und Kunstwörter der Naturlehre*. Vierter Theil von See bis Z. Leipzig.

Gregory, Olinthus (1806). *A Treatise of Mechanics, Theoretical, Practical, and Descriptive*. Vol. I. London.

Heath, Thomas L. (1921). *A History of Greek Mathematics*. Volume II. From Aristarchus to Diophantus. Oxford.

Jacob, Simon (1565). *Ein New und Wolgegründt Rechenbuch*. Frankfurt.

König, Samuel (1738). "De centro inertiae atque gravitatis". *Nova Acta Eruditorum* (1738), pp. 34–48.

Mugler, Charles (1958-59). *Dictionnaire historique de la terminologie géométrique des Grecs*. Paris.

Pappus of Alexandria (1878). *Pappi Alexandrini Collectionis*. Voluminis III Tomus I. Ed. by Friedrich Hultsch. Berlin.
Robison, John (1801). "Position, Centre of". *Supplement to the third edition of the Encyclopædia Britannica* II, pp. 373–375.
Stäckel, Paul (1908). "Elementare Dynamik der Punktsysteme und starren Körper". *Encyklopädie der mathematischen Wissenschaften mit Einschluss ihrer Anwendungen* IV 6, pp. 435–684.
Stein, Walter (1930). "Der Begriff des Schwerpunktes bei Archimedes". *Quellen und Studien zur Geschichte der Mathematik* B 2, pp. 221–243.
Stevin, Simon (1634). *Les Œeuvres Mathematiques*. Leiden.
Sturm, Johann Christoph (1670). *Des unvergleichlichen Archimedis Kunst-Bücher*. Nürnberg.
Valla, Giorgio (1501). *De expetendis, et fugiendis rebus opus*. Venezia.

***chain* (algebra)**

The ascending chain condition was systematically used by Emmy Noether. In her famous paper "Ideal Theory in Ring Domains", she proved what she called the "Theorem of the Finite Chain" ("Satz von der endlichen Kette"):

> If $\mathfrak{M}, \mathfrak{M}_1, \mathfrak{M}_2, \ldots, \mathfrak{M}_\nu, \ldots$ is a countably infinite system of ideals in Σ in which each ideal is divisible by the following one, then all ideals after a finite index n are identical; $\mathfrak{M}_n = \mathfrak{M}_{n+1} = \ldots$ In other words: If $\mathfrak{M}, \mathfrak{M}_1, \mathfrak{M}_2, \ldots, \mathfrak{M}_\nu, \ldots$ gives a simply ordered chain of ideals such that each ideal is a proper divisor of its immediate predecessor, then the chain terminates in a finite number of steps (Noether 1921, pp. 30f.).[1]

In a footnote, Noether pointed out that the theorem had been used in connection with modules by Dedekind (1894, p. 523), and that Dedekind had also used the term *Kette* there. Thus, both the concept and the terminology go back to Dedekind.[2] Noether (1927) later resumed the concept in her paper "Abstract Construction of Ideal Theory".

In English, the terminology appears in Ore (1933, p. 733), who presented what he called the "*condition of finite divisor chains* or simply the *chain condition*[:] *Every chain of ideals* $\mathfrak{a}_1 < \mathfrak{a}_2 < \ldots$, *each dividing the preceding, must break off after a finite number of terms*".

The counterpart of the ascending chain condition is the descending chain condition, studied by Artin (1927, p. 252). Rings in which this condition holds were later called *Artinian rings*. According to Dieudonné (1982, p. 621), this term is due to Bourbaki. In print, it first appears in Samuel (1947-48, p. 2)[3] and seems to have spread via Dieudonné (1952, p. 44), who used it at the International Congress of Mathematicians in 1950.

Note. For a discussion of Noether (1921, 1927), see Corry (2004, pp. 225–244). See also *Noetherian ring*.

References

Artin, Emil (1927). "Zur Theorie der hyperkomplexen Zahlen". *Abhandlungen aus dem Mathematischen Seminar der Universität Hamburg* 5, pp. 251–260.

[1] "Ist $\mathfrak{M}, \mathfrak{M}_1, \mathfrak{M}_2, \ldots, \mathfrak{M}_\nu, \ldots$ ein abzählbar unendliches System von Idealen in Σ, von denen jedes durch das folgende teilbar ist, so sind von einem endlichen Index n an alle Ideale identisch, $\mathfrak{M}_n = \mathfrak{M}_{n+1} = \ldots$ M.a.W.: Bildet $\mathfrak{M}, \mathfrak{M}_1, \mathfrak{M}_2, \ldots, \mathfrak{M}_\nu, \ldots$ eine einfach geordnete Kette von Idealen derart, dass jedes Ideal ein echter Teiler des unmittelbar vorangehenden ist, so bricht die Kette im Endlichen ab". Translation from Noether (2014, p. 8). [2] See also Corry (2004, pp. 114f.).
[3] Samuel was a member of Bourbaki.

Corry, Leo (2004). *Modern Algebra and the Rise of Mathematical Structures*. Second revised edition. Basel.
Dedekind, Richard (1894). “Ueber die Theorie der ganzen algebraischen Zahlen”. In: *Vorlesungen über Zahlentheorie von P.G. Lejeune Dirichlet*. Herausgegeben und mit Zusätzen versehen von R. Dedekind. Vierte, umgearbeitete und vermehrte Auflage. Braunschweig, pp. 434–657.
Dieudonné, Jean (1952). “Les idéaux minimaux dans les anneaux associatifs”. Volume II. In: *Proceedings of the International Congress of Mathematicians 1950*. Providence, R.I., pp. 44–48.
Dieudonné, Jean (1982). “The Work of Bourbaki During the Last Thirty Years”. *Notices of the American Mathematical Society* 29, pp. 618–623.
Noether, Emmy (1921). “Idealtheorie in Ringbereichen”. *Mathematische Annalen* 83, pp. 24–66.
Noether, Emmy (1927). “Abstrakter Aufbau der Idealtheorie in algebraischen Zahl- und Funktionenkörpern”. *Mathematische Annalen* 96, pp. 26–61.
Noether, Emmy (2014). “Ideal Theory in Rings”. Translated by Daniel Berlyne. URL: `https://arxiv.org/abs/1401.2577` (retrieved Apr. 2, 2021).
Ore, Øystein (1933). “Abstract ideal theory”. *Bulletin of the American Mathematical Society* 39, pp. 728–745.
Samuel, Pierre (1947-48). “Éléments de la théorie des algèbres”. *Séminaire Dubreil* 1, pp. 1–15.

chain (Markov)

According to Seneta (2006, p. 1) and other sources, Markov first dealt with chains in a paper from 1906; Markov used Russian цепь (Markov 1906, p. 149). French *chaîne* appeared in Markov (1910, p. 88), German *Kette* in Markov (1912, p. 298); both texts are translations of other papers by Markov.

Even though such early translations existed, the term was rarely used before the mid-1920s. At that time, Bernstein (1926, p. 1462, 1927, p. 40) resumed the topic and introduced the name *chaînes de A. Markoff*. As witnessed by Bru (2003, p. 140), the term was in common use after the ICM in 1928 in Bologna. This is exemplified by Hadamard’s speech at the congress; he explains (Hadamard 1931, p. 133) that Pólya drew his attention to Markov (1912) and to Bernstein (1927). The English term *Markov chain* is dated to 1938 by the OED.

Note. For the history of the concept, see, for example, Link (2006), Bru (2003), and Seneta (2006).

References

Bernstein, S.N. (1926). “Sur les sommes de quantités dépendantes”. *Bulletin de l’Académie des Sciences de l’URSS* 20, pp. 1459–1478.
Bernstein, S.N. (1927). “Sur l’extension du théorème limite du calcul des probabilités aux sommes de quantités dépendantes”. *Mathematische Annalen* 97, pp. 1–59.
Bru, Bernard (2003). “Souvenirs de Bologne”. *Journal de la société française de statistique* 144, pp. 135–226.
Hadamard, Jacques (1931). “Sur le battage des cartes et ses relations avec la mécanique statistique”. In: *Atti dei Congresso Internazionale dei Matematici*. Tomo V. Bologna, pp. 133–139.
Link, David (2006). “Chains to the West: Markov’s Theory of Connected Events and Its Transmission to Western Europe”. *Science in Context* 19, pp. 561–589.
Markov, Andrey A. (1906). “Распространение закона больших чисел на величины, зависящие друг от друга”. *Izvestiya Fiziko-matematicheskogo obschestva pri Kazanskom universitete* 15, pp. 135–156.
Markov, Andrey A. (1910). “Recherches sur un cas remarquable d’épreuves dépendantes”. *Acta Mathematica* 33, pp. 87–104.
Markov, Andrey A. (1912). “Ausdehnung der Sätze über die Grenzwerte in der Wahrscheinlichkeitsrechnung auf eine Summe verketteter Größen”. In: *Wahrscheinlichkeitsrechnung*. Nach der

zweiten Auflage des russischen Werkes übersetzt von Heinrich Liebmann. Leipzig and Berlin, pp. 272–298.
Seneta, Eugene (2006). "Markov and the creation of Markov chains". In: *MAM 2006: Markov Anniversary Meeting*. Ed. by Amy N. Langville and William J. Stewart. Raleigh, pp. 1–20.

chain (topology)

The concept of chain (more precisely of a k-chain) is used in combinatorial topology for any linear combination of k-cells (see *cell*) with regard to a presupposed ring of coefficients. Such linear combinations had already been used by Poincaré (1899) in a semi-formalized way. The terminology of *chain* was introduced by Alexander in 1926, who gave a definition in terms of formal linear combinations:[1]

> An *elementary i-chain* of a complex Φ will be defined as any symbolical expression of the form $\pm V_0 V_1 \ldots V_i$, where the marks V_0, V_1, ... V_i denote the vertices of an i-cell of Φ. [...] Now, let the elementary i-chains [...] of the complex Φ be denoted by the abridged symbols
>
> $$\pm E_s^i \qquad (s = 1, 2, \ldots, \alpha^i),$$
>
> respectively. Then, any linear combination
>
> $$K^i = x^1 E_1^i + x^2 E_2^i + \cdots + x^{\alpha^i} E_{\alpha^i}^i$$
>
> of elementary i-chains E_s^i with arbitrary integer coefficients x^s will be called an *i-chain* of Φ. [...] A *chain* K of a complex Φ will be any linear combination of elementary i-chains of Φ, whether or not the elementary i-chains are all of the same dimensionality (Alexander 1926, pp. 310f.).[2]

The term was soon adopted, for example by Hopf (1929, p. 495), who used German *Kette* and referred to Alexander for the term. Lefschetz (1930, p. 16) also attributed it to Alexander and called it "apt". French *chaîne* was used by de Rham (1931, p. 5).

References

Alexander, James W. (1920). "A Proof of Jordan's Theorem About a Simple Closed Curve". *Annals of Mathematics* 21, pp. 180–184.
Alexander, James W. (1926). "Combinatorial analysis situs". *Transactions of the American Mathematical Society* 28, pp. 301–329.
Hopf, Heinz (1929). "Über die algebraische Anzahl von Fixpunkten". *Mathematische Zeitschrift* 29, pp. 493–524.
Katz, Victor J. (1999). "Differential Forms". In: *History of Topology*. Ed. by Ioan Mackenzie James. Amsterdam, pp. 111–122.
Lefschetz, Solomon (1930). *Topology*. New York.
Poincaré, Henri (1899). "Complément à l'analysis situs". *Rendiconti del Circolo Matematico di Palermo* 13, pp. 285–343.
de Rham, Georges (1931). "Sur l'*Analysis situs* des variétés à n dimensions". *Journal de mathématiques pures et appliquées* 10, pp. 115–200.

---

[1] Alexander (1920, p. 180) had earlier used *chain* with a different sense: "A *chain* will be a generalized polygon consisting of a finite number of non-intersecting *edges* (which may be either line segments or rays), and *vertices* (the end points of the edges), where at each vertex there end an even number of edges". [2] For some comments on Alexander's definition, see Katz (1999, p. 119).

chain complex

The notion of a chain complex plays an important role in Eilenberg and Steenrod's *Foundations of Algebraic Topology*. Before giving a formal definition, they stated: "A chain complex (called a group system by W. Mayer) is a lower sequence of groups in which the composition of any two successive homomorphisms is zero" (Eilenberg and Steenrod 1952, p. 124). The term *group system* (German *Gruppensystem*) had indeed been used by Mayer (1939, p. 40) before.

The introduction of the concept is generally ascribed to Mayer (1929), for example by Dieudonné (2009, p. 39).[3] In his paper, Mayer used German *Komplexring* 'ring of complexes' (Mayer 1929, p. 2). The term is not frequently found at first, but it was referred to by Tucker (1933, p. 198) in his definition of a *complex*. The actual term *chain complex* came up shortly afterwards and was used by Lefschetz (1936, p. 716, 1938, p. 384).

Note. For the history of the notion, see Krömer (2007, pp. 51, 195f.).

References

Dieudonné, Jean (2009). *A History of Algebraic and Differential Topology, 1900–1960*. Boston.
Eilenberg, Samuel and Norman Steenrod (1952). *Foundations of Algebraic Topology*. Princeton.
Krömer, Ralf (2007). *Tool and Object*. A History and Philosophy of Category Theory. Basel.
Lefschetz, Solomon (1936). "Locally connected sets and their applications". *Matematicheskii Sbornik* 1, pp. 715–718.
Lefschetz, Solomon (1938). "On Chains of Topological Spaces". *Annals of Mathematics* 39, pp. 383–396.
Mayer, Walther (1929). "Über abstrakte Topologie". *Monatshefte für Mathematik* 36, pp. 1–42.
Mayer, Walther (1939). "Topologische Gruppensysteme". *Monatshefte für Mathematik und Physik* 47, pp. 40–86.
Poincaré, Henri (1899). "Complément à l'analysis situs". *Rendiconti del Circolo Matematico di Palermo* 13, pp. 285–343.
Tucker, Albert William (1933). "An Abstract Approach to Manifolds". *Annals of Mathematics* 34, pp. 191–243.
Weibel, Charles A. (1999). "History of Homological Algebra". In: *History of Topology*. Ed. by Ioan Mackenzie James. Amsterdam, pp. 797–836.

chaos

In print, the term first appears (OED) in May (1974, p. 645), who explicitly refers to Yorke for the term: "some of the very simplest nonlinear difference equations even for single species exhibit a spectrum of dynamical behavior which, as the intrinsic growth rate r increases, goes from a stable equilibrium point, to stable cyclic oscillations between 2 population points, to stable cycles with 4 points, then 8 points, and so on, through to a regime which can only be described as chaotic (a term coined by I. A. Yorke)". The paper referred to, Li and Yorke (1975), is entitled "Period three implies chaos".

Li and Yorke did not give a definition of *chaos*. They considered difference equations such as the logistic equation $x_{n+1} = rx_n[1 - x_n/K]$ and observed:

> We approach these equations with the viewpoint that irregularities and chaotic oscillations of complicated phenomena may sometimes be understood in terms of the simple model, even if that model is not sufficiently sophisticated to allow accurate numerical predictions. Lorenz

[3] According to Weibel (1999, p. 800), the concept first appeared in Poincaré (1899).

> [...][1] took this point of view in studying turbulent behavior in a fascinating series of papers. He showed that a certain complicated fluid flow could be modelled by such a sequence $x, F(x), F^2(x), \cdots$, which retained some of the chaotic aspects of the original flow. [...] In this paper we analyze a situation in which the sequence $\{F^n(x)\}$ is non-periodic and might be called "chaotic" (Li and Yorke 1975, pp. 985f.).

On the various concepts of chaos, Yorke (2009, p. 1232) later wrote that "the definition of chaos depends on the context; that is, it depends on what you know about a system or what you want to prove".

Note. For the history of chaos theory, see Oestreicher (2007) and Bacaër (2011, ch. 24), for example.

References

Bacaër, Nicolas (2011). *A Short History of Mathematical Population Dynamics*. London.

Li, Tien-Yien and James A. Yorke (1975). "Period Three Implies Chaos". *The American Mathematical Monthly* 82, pp. 985–992.

May, Robert M. (1974). "Biological Populations with Nonoverlapping Generations: Stable Points, Stable Cycles, and Chaos". *Science* 186, pp. 645–647.

Oestreicher, Christian (2007). "A history of chaos theory". *Dialogues in Clinical Neuroscience* 9, pp. 279–289.

Yorke, James A. (2009). "Response to David Ruelle". *Notices of the American Mathematical Society* 56, pp. 1232–1233.

character

The Latin term *character* was introduced by Gauß (1801, § 230) in his *Disquitiones arithmeticae* with respect to quadratic forms:

> Therefore all numbers that can be represented by a given primitive form F with determinant D will have a fixed relationship to the individual prime divisors of D (by which they are not divisible). And odd numbers that can be represented by F will have also a fixed relationship to the numbers 4 and 8 in certain cases [...] We will call this type of relationship to each of these numbers the *character* or the *particular character* of the form F.[1]

As an extension of this number-theoretical notion, Dedekind (1879, p. 581) defined a complex-valued homomorphism on an Abelian group and said that such a function can be called *character* (German *Charakter*):

> the function $\chi(\mathfrak{a})$ also possesses the property that it takes the same value on all ideals $\mathfrak{a}$ belonging to the same class A; this value is therefore appropriately denoted by $\chi(A)$ and is clearly always an hth root of unity. Such functions χ, which in an extended sense can be termed *characters*, always exist.[2]

---

[1] Li and Yorke here refer to various papers by Edward Lorenz. [1] "Omnes igitur numeri qui per formam primitiuam datam F determinantis D repraesentari possunt, relationem fixam habebunt ad singulos diuisores primos ipsius D (per quos quidem ipsi non sunt diuisibiles), numeri impares vero qui per F possunt repraesentari, in quibusdam casibus etiam ad numeros 4 et 8 relationem fixam habebunt [...] Talem relationem ad singulos hos numeros, *characterem* seu *characterem particularem* formae F vocabimus". Translation from Gauß (1986, § 230). – Gauß' notion is explained by Hawkins (1971, pp. 145f.), Lemmermeyer (2007, pp. 535f.), and Curtis (1999, pp. 28f.). [2] "[...] die Function $\chi(\mathfrak{a})$ ausser der Eigenschaft [...] noch die andere besitzt, für alle derselben Classe A angehörenden Ideale $\mathfrak{a}$ denselben Werth anzunehmen, welcher mithin zweckmässig durch $\chi(A)$ bezeichnet wird und offenbar immer eine h^{te} Wurzel der Einheit ist. Solche Functionen χ, die man im erweiterten Sinn *Charaktere* nennen kann, existiren immer". Translation from Hawkins (1971, p. 149).

The notion was further investigated by Weber, who used German *Gruppencharakter* 'group character' in Weber (1886, p. 200) and in his *Lehrbuch der Algebra* (Weber 1896, p. 44), and by Frobenius (1896, p. 985); the latter also credits Gauß and Dedekind for the terminology. The English term is found in Dickson (1902, p. 30).

Note. Hawkins (1971, 1974, 2013, ch. 13) analyses the history of group characters in detail.

References

Curtis, Charles W. (1999). *Pioneers of Representation Theory*. Frobenius, Burnside, Schur, and Brauer. Providence, R.I.

Dedekind, Richard (1879). "Ueber die Theorie der ganzen algebraischen Zahlen". In: *Vorlesungen über Zahlentheorie von P.G. Lejeune Dirichlet*. Herausgegeben und mit Zusätzen versehen von R. Dedekind. Dritte, umgearbeitete und vermehrte Auflage. Braunschweig, pp. 434–627.

Dickson, Leonard Eugene (1902). "An Elementary Exposition of Frobenius's Theory of Group-Characters and Group-Determinants". *Annals of Mathematics* 4, pp. 25–49.

Frobenius, Georg (1896). "Über Gruppencharaktere". *Sitzungsberichte der Königlich Preußischen Akademie der Wissenschaften zu Berlin* (1896), pp. 985–1021.

Gauß, Carl Friedrich (1801). *Disquisitiones arithmeticae*. Leipzig.

Gauß, Carl Friedrich (1986). *Disquisitiones Arithmeticae*. English Edition. Translated by A.A. Clarke. Revised by W.C. Waterhouse. New York et al.

Hawkins, Thomas (1971). "The Origins of the Theory of Group Characters". *Archive for History of Exact Sciences* 7, pp. 142–170.

Hawkins, Thomas (1974). "New Light on Frobenius' Creation of the Theory of Group Characters". *Archive for History of Exact Sciences* 12, pp. 217–243.

Hawkins, Thomas (2013). *The Mathematics of Frobenius in Context*. A Journey Through 18th to 20th Century Mathematics. New York.

Lemmermeyer, Franz (2007). "The Development of the Principal Genus Theorem". In: *The Shaping of Arithmetic after C. F. Gauss's Disquisitiones Arithmeticae*. Ed. by Catherine Goldstein, Norbert Schappacher, and Joachim Schwermer. Berlin and Heidelberg, pp. 529–561.

Weber, Heinrich (1886). "Theorie der Abel'schen Zahlkörper". *Acta Mathematica* 8, pp. 193–263.

Weber, Heinrich (1896). *Lehrbuch der Algebra*. Zweiter Band. Braunschweig.

characteristic **(algebra)**

In algebra, the characteristic of a field was introduced by Steinitz (1910, p. 168) as German *Charakteristik*: "In every field a smallest field or prime field is contained; the latter is either of the type of the field of rational numbers (characteristic 0) or of the type of the residue system of a prime number p (characteristic p)".[1]

The term was soon adopted by others such as Noether (1915, p. 127); it was applied to rings by Fraenkel (1916, p. 19). The English term is found in Gokhale (1922, p. 299).

References

Fraenkel, Adolf (1916). *Über gewisse Teilbereiche und Erweiterungen von Ringen*. Leipzig.

Gokhale, Vishnu Dattatreya (1922). "Concerning Compact Kürschak Fields". *American Journal of Mathematics* 44, pp. 297–316.

Noether, Emmy (1915). "Die allgemeinsten Bereiche aus ganzen transzendenten Zahlen". *Mathematische Annalen* 77, pp. 103–128.

Steinitz, Ernst (1910). "Algebraische Theorie der Körper". *Journal für die reine und angewandte Mathematik* 137, pp. 167–309.

---

[1] "In jedem Körper ist ein kleinster Körper oder Primkörper enthalten; derselbe ist entweder vom Typus des Körpers der rationalen Zahlen (Charakteristik 0) oder vom Typus des Restklassensystems einer Primzahl p (Charakteristik p)".

characteristic (linear algebra)

In a paper on bilinear forms, Frobenius (1878, p. 10) gave a definition of characteristic determinants that was practically identical to the modern one and coined a new term: "The determinant of $rE - A$ is an entire function of r of the nth degree and shall be called the *characteristic determinant* or *function* of A".[2] Frobenius referred to Cauchy (1840, p. 53), where the corresponding term *équation caractéristique* was introduced.[1] According to various historians such as Hawkins (2013, p. 104), the term appears there for the first time.[2]

The term *characteristic polynomial* does not appear to have required special introduction to 19th century writers; it appears at the beginning of the 20th century, for example as French *polynome caractéristique* (Autonne 1910, p. 84). Wintner (1929, p. 14) also used *secular polynomial* (German *Säkularpolynom*). For this and other related terms, see *eigenvalue*.

Note. For the history of characteristic equations, which already appear in the 18th century, and explanations of Cauchy's introduction of the term, see Kline (1972, pp. 800f.), Hawkins (2013, pp. 103–106), and Tou (2018).

References

Autonne, Léon (1910). "Sur les matrices linéaires échangeables à une matrice donnée". *Journal de l'École polytechnique* 14, pp. 83–131.

Cauchy, Augustin-Louis (1839). "Mémoire sur l'intégration des équations linéaires". *Comptes rendus de l'Académie des sciences* 8, pp. 827–830, 845–865, 889–907, 931–937.

Cauchy, Augustin-Louis (1840). "Mémoire sur l'intégration des équations linéaires". In: *Exercises d'analyse et de physique mathématique*. Tome premier. Paris, pp. 53–100.

Frobenius, Georg (1878). "Ueber lineare Substitutionen und bilineare Formen". *Journal für die reine und angewandte Mathematik* 84, pp. 1–63.

Hawkins, Thomas (2013). *The Mathematics of Frobenius in Context*. A Journey Through 18th to 20th Century Mathematics. New York.

Kline, Morris (1972). *Mathematical Thought From Ancient To Modern Times*. Volume 2. New York and Oxford.

Tou, Erik R. (2018). "Eigenvectors and Eigenvalues". *Convergence* (February 2018). URL: `https://old.maa.org/press/periodicals/convergence/math-origins-eigenvectors-and-eigenvalues`.

Wintner, Aurel (1929). *Spektraltheorie der unendlichen Matrizen*. Einführung in den analytischen Apparat der Quantenmechanik. Leipzig.

characteristic (logarithm)

The characteristic of a common logarithms is its integral part; the term was introduced by Briggs (1624, ch. 4):

> But of all logarithms, the particular ones I have adopted are these which are assigned to one and ten; certainly the nearest [whole numbers] to these must be the numbers which are continued in the ten-fold ratio, namely by a hundred, a thousand, etc, which are all described

[2] "Die Determinante von $rE - A$ ist eine ganze Function $n$ten Grades von r und soll die *characteristische Determinante* oder *Function* von A genannt werden". [1] Cauchy's definition is found on p. 57 of that paper. [2] The terminology is earlier found in Cauchy (1839, p. 846); Cauchy (1840) is a reproduction of that paper.

by zeros only after the first place to the left, which we call the characteristic, which indicates to us how many units the whole number advances beyond the place of unity.[3]

The term was quickly adopted by others, for example by Vlacq (1628, p. 4) and Guldin (1635, p. 198) .

In the posthumously published *Logarithmicall Arithmetike*, the English term *characteristic* is found and the alternative term *index* is also given: "Every Logarithme hath a figure placed afore it on the left hand called a *Characteristicke* or an *Index*" (Briggs 1631, p. 1). At the same time, the English term is also found in a slightly modified form as *character* or *characteristical note* in Norwood (1631, pp. 8f.).

Note. Brief notes on the origin of the term can be found in Tropfke (1933, p. 252) and Smith (1958, p. 514).

References

Briggs, Henry (1624). *Arithmetica logarithmica*. London.

Briggs, Henry (1631). *Logarithmicall Arithmetike*. London.

Briggs, Henry (2006). "Arithmetica logarithmica". Translated and annotated by Ian Bruce. URL: http://www.17centurymaths.com/contents/albriggs.html (retrieved May 12, 2008).

Guldin, Paul (1635). *De centro gravitatis*. Liber primus. Vienna.

Norwood, Richard (1631). *Trigonometrie, Or, the Doctrine of Triangles*. London.

Smith, David Eugene (1958). *History of Mathematics*. Volume II. Special Topics of Elementary Mathematics. New York.

Tropfke, Johannes (1933). *Geschichte der Elementar-Mathematik*. Zweiter Band. Allgemeine Arithmetik. Dritte, verbesserte und vermehrte Auflage. Berlin and Leipzig.

Vlacq, Adriaan (1628). *Arithmetica logarithmica*. Editio Secunda. Gouda.

***characteristic* (topology)**

The German term *Characteristik* was introduced by Kronecker (1870, p. 161) in connection with systems of functions. Johnson (1981, p. 136) provides an approachable discussion of Kronecker's definition as seen from a geometrical point of view:

> Suppose we have n real, differentiable functions in n-space: $x_1 = F_1(z_1, z_2, \ldots, z_n)$, $x_2 = F_2(z_1, z_2, \ldots, z_n), \ldots, x_n = F_n(z_1, z_2, \ldots, z_n)$ and an $(n-1)$-dimensional (differentiable) manifold [..]: $F_0(z_1, z_2, \ldots, z_n) = 0$, so that the functions give rise to another $(n-1)$-dimensional manifold $\Phi(x_1, x_2, \ldots, x_n) = 0$. We assume that these are both closed manifolds. If we now set all of the x's equal to zero, except two, say x_h and x_k, thereby defining a hyperplane in n-space, we obtain a closed line on the $(n-1)$-dimensional manifold $\Phi = 0$. This line winds around the origin a certain number of times in a prescribed sense and, inasmuch as Kronecker shows that this winding number (*Windungszahl*) is independent of the choice of hyperplane through the origin, it is *characteristic* for the given system of functions on the manifold $F_0 = 0$.[1]

Kronecker's algebraic concept of characteristic was also reconsidered by Dyck (1886, p. 55, 1888, p. 465), who took over the term from Kronecker and showed that "the Kronecker characteristic could be expressed in terms of his own purely topologically defined characteristic, which in fact was equivalent to the Euler characteristic

---

[3] "Logarithmorum autem omnium praecipui sunt ij qui Vnitati tribuuntur & Denario, istis vero proximi ijs numeris debentur qui in decupla ratione continuantur, nempe Centenario, Millenario, &c., qui omnes cyphris tantum describuntur praeter primam notam versus sinistram, quam Characteristicam appellare poterimus, quos nobis indicet quot notis numerus integer progreditur ultra locum Vnitatis". Translation from Briggs (2006, ch. 4).

[1] Kronecker's original definition is discussed by Siegberg (1981, pp. 126–128).

and even equal up to sign" (Scholz 1999, p. 40). According to Dombrowski (1990, p. 332), it is Dyck's work that is probably responsible for the later designation *Euler characteristic*.[2] The eponym *Euler characteristic* dates from the 1920s; it is found as German *Eulersche Charakteristik* in Hopf (1925, p. 131).[3]

See also *index* (functions), *degree* (topology), and *winding number*.

References

Bollinger, Maja (1972). "Geschichtliche Entwicklung des Homologiebegriffs". *Archive for History of Exact Sciences* 9, pp. 94–170.

Dombrowski, Peter (1990). "Differentialgeometrie". In: *Ein Jahrhundert Mathematik 1890 - 1990*. Festschrift zum Jubiläum der DMV. Ed. by Gerd Fischer et al. Braunschweig and Wiesbaden, pp. 323–360.

Dyck, Walther (1886). "Beiträge zur Analysis situs. II. Mittheilung". *Berichte über die Verhandlungen der Königlich Sächsischen Gesellschaft der Wissenschaften zu Leipzig (Mathematisch-Physische Klasse)* 38, pp. 53–69.

Dyck, Walther (1888). "Beiträge zur Analysis situs. I. Aufsatz. Ein- und zweidimensionale Mannigfaltigkeiten". *Mathematische Annalen* 32, pp. 457–512.

Hopf, Heinz (1925). "Die Curvatura integra Clifford-Kleinscher Raumformen". *Nachrichten von der Gesellschaft der Wissenschaften zu Göttingen (mathematisch-physikalische Klasse)* (1925), pp. 131–141.

Johnson, Dale M. (1981). "The Problem of the Invariance of Dimension in the Growth of Modern Topology, Part II". *Archive for History of Exact Sciences* 25, pp. 85–266.

Kronecker, Leopold (1870). "Über Systeme von Functionen mehrerer Variabeln". *Monatsberichte der Königlich Preußischen Akademie der Wissenschaften zu Berlin* (1869), pp. 159–193.

Mawhin, Jean (2000). "Poincaré's early use of *Analysis situs* in nonlinear differential equation: Variations around the theme of Kronecker's integral". *Philosophia Scientiae* 4, pp. 103–143.

Scholz, Erhard (1999). "The Concept of Manifold, 1850-1950". In: *History of Topology*. Ed. by Ioan Mackenzie James. Amsterdam, pp. 25–64.

Siegberg, Hans Willi (1981). "Some Historical Remarks Concerning Degree Theory". *The American Mathematical Monthly* 88, pp. 125–139.

Volkert, Klaus (2002). *Das Homöomorphismusproblem insbesondere der 3-Mannigfaltigkeiten, in der Topologie 1892-1935*. Paris.

characteristic function **(sets)**

The *characteristic function* of sets was introduced as French *fonction caractéristique* by de la Vallée Poussin (1915, p. 440): "Let E a measurable set (B).[1] We define a function equal to 1 in E and to 0 in CE [the complement]. We shall call it the *characteristic function in E*".[2] The term also appears in de la Vallée Poussin (1916, p. 7), where also the noun *characteristic* is used. The English term is found in Hardy (1919, p. 93), while the German term was used by Schlesinger and Plessner (1926, p. 30).[3]

The alternative term *indicator* is due, according to Feller (1966, p. 102), to Loève. It appears in Loève (1950, p. 322), where he introduces the "*indicator* of the event A, which takes values 1 and 0 respectively, according as A occurs or does not occur". He needed this alternative term because he used *characteristic function* for the integral

---

[2] See Bollinger (1972, pp. 111–116) and Volkert (2002, pp. 57–67) for an explanation of Dyck's approach. [3] For Poincaré's role in the development of the concept, see Mawhin (2000). [1] This means that E is measurable in the sense of Borel. [2] "Soit E un ensemble mesurable (B). Définissions une fonction égale à 1 dans E et à 0 dans CE. Nous l'appellerons la *fonction caractéristique de E*". – The concept is earlier found in Lebesgue (1904, p. 101). [3] Slightly earlier, Kamke (1925, p. 39) suggested *Inhaltsfunktion*, lit. 'content function'.

$f(u) = \int_{-\infty}^{+\infty} e^{iux} dF(x), -\infty < u < +\infty$. The term spread through Loève (1955, p. 9).

Note. For the history of the term, see also Miller (n.d.).

References

Feller, William (1966). *An Introduction to Probability Theory and Its Applications*. Volume II. New York.

Hardy, Godfrey Harold (1919). "Notes on some points in the integral calculus L: On the integral of Stieltjes and the formula for integration by parts". *The Messenger of Mathematics* 48, pp. 90–100.

Kamke, Erich (1925). *Das Lebesguesche Integral*. Eine Einführung in die neuere Theorie der reellen Funktionen. Leipzig.

Lebesgue, Henri (1904). *Leçons sur l'intégration et la recherche des fonctions primitives*. Paris.

Loève, Michel (1950). "Fundamental Limit Theorems of Probability Theory". *The Annals of Mathematical Statistics* 21, pp. 321–338.

Loève, Michel (1955). *Probability Theory*. Foundations – Random Sequences. New York.

Miller, Jeff (n.d.). "Earliest Known Uses of Some of the Words of Mathematics". URL: `https://mathshistory.st-andrews.ac.uk/Miller/mathword/`.

Schlesinger, Ludwig and Abraham Plessner (1926). *Lebesguesche Integrale und Fouriersche Reihen*. Berlin and Leipzig.

de la Vallée Poussin, Charles Jean (1915). "Sur l'intégrale de Lebesgue". *Transactions of the American Mathematical Society* 16, pp. 435–501.

de la Vallée Poussin, Charles Jean (1916). *Intégrales de Lebesgue. Fonctions d'ensemble. Classes de Baire*. Leçons professées au Collège de France. Paris.

characteristic function **(probability theory)**

In the now usual sense, the term *characteristic function* (French *fonction caractéristique*) was defined by Lévy (1925, p. 161): "We shall call *characteristic function*, and designate by $\varphi(t)$, the probable value of e^{itx}, i. e. $\varphi(t) = \sum \alpha_h e^{itx_h}$".[1] In a footnote, Lévy remarked that the term is due to Poincaré, who, however, used it for the probable value of e^{tx}.[2] The term was indeed first used in the second edition of the *Calcul des probabilités* (Poincaré 1912, p. 206).

The concept is older; it is found in Laplace (1810) and Cauchy (1853, p. 381), who called it *fonction auxiliaire*.

Poincaré's term was soon in common use; the English term appears to be due to Daniell (1921, p. 149). von Mises (1919, p. 26) preferred to use *complex adjuncts* (*komplexe Adjunkte*), which did not find many followers.[3]

Note. The origin of the term and the history of the concept are discussed by Fischer (2011).

References

Cauchy, Augustin-Louis (1853). "Mémoire sur les résultats moyens d'un très-grand nombre des observations". *Comptes rendus de l'Académie des sciences* 37, pp. 381–385.

Daniell, P.J. (1921). "Integral Products and Probability". *American Journal of Mathematics* 43, pp. 143–162.

Fischer, Hans (2011). *A History of the Central Limit Theorem*. From Classical to Modern Probability Theory. New York.

---

[1] "Nous appellerons *fonction caractéristique*, et désignerons par $\varphi(t)$, la valeur probable de e^{itx}, c'est-à-dire $\varphi(t) = \sum \alpha_h e^{itx_h}$". [2] The difference between Lévy's and Poincaré's definition was already pointed out in Lévy (1924, p. 56). [3] For this term, see also Siegmund-Schultze (2006, p. 439).

Laplace, Pierre-Simon (1810). "Mémoire sur les approximations des formules qui sont fonctions de très-grands nombres et sur leur application aux probabilités". *Mémoires de la classe des sciences mathématiques et physiques de l'Institut de France* 10, pp. 353–415.
Lévy, Paul (1924). "Théorie des erreurs. La loi de Gauss et les lois exceptionnelles". *Bulletin de la Société Mathématique de France* 52, pp. 49–85.
Lévy, Paul (1925). *Calcul des probabilités*. Paris.
von Mises, Richard (1919). "Fundamentalsätze der Wahrscheinlichkeitsrechnung". *Mathematische Zeitschrift* 4, pp. 1–97.
Poincaré, Henri (1912). *Calcul des probabilités*. Deuxième édition. Paris.
Siegmund-Schultze, Reinhard (2006). "Probability in 1919/20: the von Mises-Pólya-Controversy". *Archive for History of Exact Sciences* 60, pp. 431–515.

characteristic subgroup

The term *characteristic subgroup* (German *charakteristische Untergruppe*) was introduced by Frobenius (1895, p. 183) in the following terms:

> Suppose that in some way a more general group $\mathfrak{H}'$ is found, in which $\mathfrak{H}$ is contained as an invariant subgroup. If R is an element of $\mathfrak{H}'$, and if $\mathfrak{A}$ is an invariant subgroup of $\mathfrak{H}$, then $R^{-1}\mathfrak{A}R = \mathfrak{B}$ is an invariant subgroup of $R^{-1}\mathfrak{H}R = \mathfrak{H}$. It is possible that $\mathfrak{B}$ is different from $\mathfrak{A}$. But if $R^{-1}\mathfrak{A}R = \mathfrak{A}$ for each element R of each possible extended group, then $\mathfrak{A}$ shall be called a *characteristic* subgroup of $\mathfrak{H}$.[1]

Frobenius (1895, p. 184) then proved that for each isomorphism of H into itself A is transformed into itself, a property that is now usually taken as the definition of the concept.

The term was immediately adopted by Burnside (1896, p. 357), who referred to Frobenius and also included it in his book on group theory (Burnside 1897, p. 232). de Séguier (1904, p. 100) translated the term as *diviseur caractéristique* 'characteristic divisor' (*divisor* was a common synonym of *subgroup*).

References

Burnside, William Snow (1896). "On the Isomorphism of a Group with Itself". *Proceedings of the London Mathematical Society* 27, pp. 354–367.
Burnside, William Snow (1897). *Theory of Groups of Finite Order*. Cambridge.
Frobenius, Georg (1895). "Über endliche Gruppen". *Sitzungsberichte der Königlich Preußischen Akademie der Wissenschaften zu Berlin* (1895), pp. 163–194.
de Séguier, Jean Armand (1904). *Théorie des groupes finis*. Éléments de la théorie des groupes abstraits. Paris.

chart

See *atlas*.

chiral and *amphicheiral*

As Mislow (1999, p. 1) states, "[c]hirality, in its many and varied manifestations, is ubiquitous; a concept rooted in mathematics, it permeates all branches of the natural sciences". A much-quoted definition was given by Thomson (1894, p. 27, fn.):

---

[1] "Angenommen, es sei auf irgend eine Art eine allgemeinere Gruppe $\mathfrak{H}'$ gefunden, in der $\mathfrak{H}$ als invariante Untergruppe enthalten ist. Ist R ein Element von $\mathfrak{H}'$, und ist $\mathfrak{A}$ eine invariante Untergruppe von $\mathfrak{H}$, so ist $R^{-1}\mathfrak{A}R = \mathfrak{B}$ eine invariante Untergruppe von $R^{-1}\mathfrak{H}R = \mathfrak{H}$. Es ist möglich, dass $\mathfrak{B}$ von $\mathfrak{A}$ verschieden ist. Wenn aber für jedes Element R jeder möglichen erweiterten Gruppe $R^{-1}\mathfrak{A}R = \mathfrak{A}$ ist, so möge $\mathfrak{A}$ eine *characteristische* Untergruppe von $\mathfrak{H}$ genannt werden".

I call any geometrical figure or group of points, *chiral*, and say it has chirality, if its image in a plane mirror, ideally realized, cannot be brought to coincide with itself. Two equal and similar right hands are homochirally similar. Equal and similar right and left hands are heterochirally similar or 'allochirically' similar (but heterochirally is better). They are also called 'enantiomorphs' after a usage introduced, I believe, by German writers. Any chiral object and its image in a plane mirror are heterochirally similar.

The term *chiral* derives from the Greek word *χείρ*, which means 'hand'. Thomson had coined the term much earlier: in 1873, he gave a talk called "Note on homocheiral and heterocheiral similarity" (Thomson 1875), of which, however, only the title is extant. The term is also found in Thomson and Tait (1879, p. 74), for example.[2]

The term *enantiomorph*, which Thomson correctly attributes to "German writers", is made up of Greek *ἐναντίος* 'opposite' and *μορφή* 'form', and is due to Naumann (1856, p. 104).[1]

The notion had been discussed earlier, though. As Marbach (1856, pp. 451f.) observes, Pasteur (1850, p. 480) used *hémiédrie non superposable* "because their image cannot be superposed upon them, in the same way that a right hand glove does not fit the left hand".[2] And in the 18th century, Kant also discussed the phenomenon under the name *incongruent counterpart* (German *inkongruentes Gegenstück*): "An object which is completely like and similar to another, although it cannot be included exactly within the same limits, I call its *incongruent counterpart*" (cited from Kant 1807, p. 78).[3] He gave examples such as spherical triangles and body parts, especially the hands.[4]

In mathematics, the terminology is mainly known through the term *amphicheiral*, which was introduced by Tait (1879, p. 160) with respect to knots and which he defined as "right or left-handed indifferently" (Greek *ἀμφί* means 'both, on both sides'). Such knots can be deformed into their own mirror image. Tait, who was certainly aware of Thomson's terms, chose this term because "amphicheiral instruments" such as chisels or axes, as opposed to scissors, can be used with both hands. The term is soon found in other languages; for example, Dingeldey (1890, p. 9) used German *amphicheiral*.[5]

See also *symmetric*.

Note. For the history of the notion, see Mislow (1999), Gal (2007) and Cintas (2007); see also Bentley (2010). Epple (1998, pp. 359–361) focuses on the knot-theoretical concept.

References

Bentley, Ronald (2010). "Chiral: A Confusing Etymology". *Chirality* 22, pp. 1–2.

Cintas, Pedro (2007). "Tracing the origins and evolution of chirality and handedness in chemical language". *Angewandte Chemie International Edition* 46, pp. 4016–4024.

---

[2] Thomson originally used the spelling *cheiral*, which is closer to the Greek, and later changed this to *chiral*. [1] See Gal (2007) on Naumann's creation of the term. [2] "parceque leur image ne peut leur être superposée, pas plus que le gant de la main droite ne s'adapterait à la main gauche". – On Pasteur's terminology, which comes from crystallography, see Gal (2011). [3] "Ich nenne einen Körper, der einem andern völlig gleich und ähnlich ist, ob er gleich nicht in eben denselben Grenzen kann beschlossen werden, sein incongruentes Gegenstück". Translation from Kant (1968, p. 41). [4] For a discussion of Kant's notion, see, e. g., Hon and Goldstein (2008, ch. 7). On possible influences on Kant and earlier occurences of the notion, see Vaihinger (1922, pp. 530f., fn. 3), Tropfke (1923, p. 123), and Volkert (2018, ch. 1). [5] This mathematical notion had earlier been dealt with by Listing in 1849. See Little (1900, p. 771).

Dingeldey, Friedrich (1890). *Topologische Studien über die aus ringförmig geschlossenen Bändern durch gewisse Schnitte erzeugbaren Gebilde*. Leipzig.
Epple, Moritz (1998). "Topology, Matter, and Space, I: Topological Notions in 19th-Century Natural Philosophy". *Archive for History of Exact Sciences* 52, pp. 297–392.
Gal, Joseph (2007). "Carl Friedrich Naumann and the Introduction of *enantio* Terminology: A Review and Analysis on the 150th Anniversary". *Chirality* 19, pp. 89–98.
Gal, Joseph (2011). "Louis Pasteur, language, and molecular chirality. I. Background and dissymmetry". *Chirality* 23, pp. 1–16.
Hon, Giora and Bernard R. Goldstein (2008). *From Summetria to Symmetry*. The Making of a Revolutionary Scientific Concept. New York.
Kant, Immanuel (1807). "Von dem ersten Grunde des Unterschiedes der Gegenden im Raume". In: *Sammlung einiger bisher unbekannt gebliebener kleiner Schriften*. Zweyte sehr vermehrte Auflage. Königsberg, pp. 71–80.
Kant, Immanuel (1968). "On the first Grounds of the Distinction of Regions in Space". In: *Selected Pre-critical Writings and Correspondence with Beck*. Translated and introduced by G.B. Kerferd and D.E. Walford. Manchester, pp. 36–43.
Little, C.N. (1900). "Non-Alternate ± Knots". *Transactions of the Royal Society of Edinburgh* 39, pp. 771–778.
Marbach, Hermann (1856). "Ueber die Enantiomorphie und die optischen Eigenschaften von Krystallen des tesseralen Systems". *Annalen der Physik und Chemie* 9, pp. 451–466.
Mislow, Kurt (1999). "Molecular Chirality". In: *Topics in Stereochemistry*. Volume 22. Ed. by S.E. Denmark. New York, pp. 1–82.
Naumann, Carl Friedrich (1856). *Elemente der theoretischen Krystallographie*. Leipzig.
Pasteur, Louis (1850). "Nouvelles recherches sur les relations qui peuvent exister entre la forme cristalline, la composition chimique et le phénomène de la polarisation rotatoire". *Comptes rendus de l'Académie des sciences* 31, pp. 480–483.
Tait, Peter Guthrie (1879). "On Knots". *Transactions of the Royal Society of Edinburgh* 28, pp. 145–190.
Thomson, William (1875). "Note on Homocheiral and Heterocheiral Similarity [title only]". *Proceedings of the Royal Society of Edinburgh* 8, p. 70.
Thomson, William (1894). *The Molecular Tactics of a Crystal*. Oxford.
Thomson, William and Peter Guthrie Tait (1879). *Treatise on Natural Philosophy*. Vol. I. Part I. New Edition. Cambridge.
Tropfke, Johannes (1923). *Geschichte der Elementar-Mathematik*. Fünfter Band. I. Ebene Trigonometrie. II. Sphärik und sphärische Trigonometrie. Berlin and Leipzig.
Vaihinger, Hans (1922). *Kommentar zu Kants Kritik der reinen Vernunft*. Zweiter Band. Stuttgart et al.
Volkert, Klaus (2018). *In höheren Räumen*. Der Weg der Geometrie in die vierte Dimension. Berlin.

chord

Greek geometers did not have a term for a chord in a circle. Instead, they used a paraphrase which can be translated as "straight line in a circle".[1] This paraphrase is found is Euclid's *Elements*, for example in book III, def. 4: "In a circle straight lines are said *to be equally distant from the centre* when the perpendiculars drawn to them from the centre are equal".[2] It was also used by Ptolemy (Toomer 1984, p. 17).

The origin of the term *chord* is therefore not found in ancient Greek mathematics; instead, like in the case of *sine*, it is found in India. The terminology can perhaps best be understood in connection with the related terms *arc* and *sagitta* (*versed sine*). Together, they make up a metaphor, the "bow metaphor" (fig. 16).

---

[1] See Tropfke (1940, pp. 137f.), Szabó (1992, p. 347), Kiselman (2015). In particular, the word *χορδή* was not used in the sense of 'chord' (see below for its meanings). [2] Translation from Euclid (1956, p. 1).

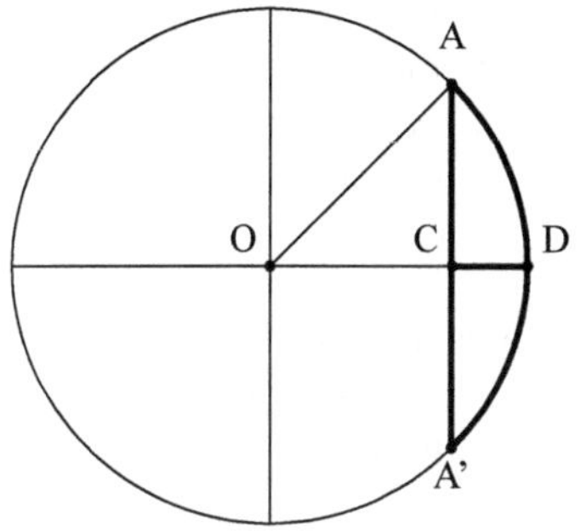

Figure 16: The bow metaphor

The bow metaphor is a transfer from archery to trigonometry. In the figure, the bold lines indicate what looks like a bow, bow-string, and arrow, and these notions gave rise to the corresponding trigonometrical terms in various languages such as Sanskrit and Arabic, and also to the term *sine*.[1]

The synonymous Sanskrit words *jyā* and *jīvā*, which literally mean 'string of a bow', were used to designate the chord (Datta and Singh 1983, p. 40). In Arabic texts, these words appear as *watar*, which has the same literal meaning.[2]

Finally, in 12th-century translations Latin *corda* 'catgut, string of an instrument' (LSJ) is found, a word borrowed from Greek *χορδή* with the same meaning. For example, Plato of Tivoli used it in his translation of al-Battānī's *De motu stellarum*, and also in his translation of the *Liber embadorum*, a text written in Hebrew.[3] Gerard of Cremona uses the term occasionally, and it is also found in a redaction of Euclid's *Elements*, ascribed to Robert of Chester.[4] From then on, the term is adopted by numerous other writers. Before the bow metaphor appeared in Latin mathematical texts, paraphrases based on the Greek model can also be found.[5]

As for modern languages, French *corde* was used by Oresme in the 14th century (Oresme 1968, p. 176); in print, it appeared in de Bouelles (1511, fol. iii^{v}). Italian *corda* is found in a late 13th-century manuscript (Arrighi 1989, p. 119).[6] German *Sehne* appeared in Schmid (1539, p. 4). The English term is first found (OED) in Recorde (1551, fol. Bi^{v}): "If the line goe crosse the circle, and passe beside the centre, then is it called a *corde*, or a *stryngline*".

Two further terms also used to designate the chord are the following. First, Latin *subtensa* or *linea subtendens* can be explained by the verb *subtendere*: a chord subtends – 'stretches under' – an arc.[7] It is not clear when exactly this was

[1] The bow metaphor has also been mentioned by Ruska (1895, p. 127), Simon (1901, p. 80), Gandz (1932, pp. 18f.), Worrell (1944, p. 99), Bossong (1978, pp. 137, 179), and Rosenfeld (1988, p. 11).
[2] The word is attested at least since al-Khwārizmī, who used the word in a geometrical treatise (Gandz 1932, p. 73). [3] See al-Battānī (1537, fol. 6^{r}) and Curtze (1902, p. 100), respectively.
[4] For Gerard of Cremona, see Anaritius (1899, pp. 127, 136), for Robert of Chester, see Busard and Folkerts (1992, p. 127). [5] For example, the Pseudo-Boethian *Geometria* contains the paraphrase 'straight line in a circle' (Boethius 1867, p. 379). [6] For further references to Italian usages, see Feola (2008, p. 153). [7] A corresponding Greek phrase can be found in Theon of Smyrna (1878, p. 190).

first used. Tropfke (1940, p. 137) cites Fibonacci, who used both *corda* and *linea subtendens* in his *Practica geometriae*, and Johannes Nemorarius, who used both *corda* and *subtensa*, so early attestations date from the early 13th century.[1] Second, Latin *inscripta* 'the inscribed (line)' was apparently first suggested by Ramus (1569, Geometria, e. g. p. 103) (see also Ramus 1567, p. 37); Magini (1592, fol. 3ʳ) points out that the term was used by Ramus, Fincke, Bressieu, "and others".[2]

Note. A brief historical discussion of the terminology is in Tropfke (1940, p. 137). For the Sanskrit term, see Datta and Singh (1983, p. 40). Various remarks on the history of the concept can be found in the works cited in the note in the entry *trigonometry*.

References

Anaritius (1899). *Anaritii in decem libros priores Elementorum Euclidis commentarii*. Ex interpretatione Gherardi Cremonensis. Ed. by Maximilian Curtze. Leipzig.

Arrighi, Gino (1989). "Maestro umbro (sec. XIII), Livero de l'abbecho (Cod. 2404 della Biblioteca Riccardiana di Firenze)". *Bollettino della deputazione di Storia patria per l'Umbria* 86, pp. 5–140.

al-Battānī, Muḥammad ibn Jābir (1537). *De motu stellarum*. Nürnberg.

Boethius, Anicius Manlius Severinus (1867). *De institutione arithmetica libri duo, De institutione musica libri quinque*. Accedit Geometria quae fertur Boetii. Ed. by Gottfried Friedlein. Leipzig.

Bossong, Georg, ed. (1978). *Los Canones de Albateni*. Herausgegeben sowie mit Einleitung, Anmerkungen und Glossar versehen. Tübingen.

de Bouelles, Charles (1511). *Geometrie en francoys*. Paris.

Busard, H.L.L. and Menso Folkerts, eds. (1992). *Robert of Chester's (?) Redaction of Euclid's* Elements, *the so-called Adelard II Version*. Basel.

Clagett, Marshall (1984). *Archimedes in the Middle Ages*. Volume V. Quasi-Archimedean Geometry in the Thirteenth Century. Philadelphia.

Curtze, Maximilian (1902). "Der Liber Embadorum des Savasorda in der Übersetzung des Plato von Tivoli". In: *Urkunden zur Geschichte der Mathematik im Mittelalter und der Renaissance*. Erster Theil. Leipzig, pp. 1–183.

Datta, Bibhutibhusan and Avadhesh Narayan Singh (1983). "Hindu Trigonometry (revised by Kripa Shankar Shukla)". *Indian Journal of History of Science* 18, pp. 39–108.

Euclid (1956). *The Thirteen Books of Euclid's Elements*. Volume II. Books III-IX. Ed. by Thomas L. Heath. New York.

Feola, Francesco (2008). *Gli esordi della geometria in volgare*. Un volgarizzamento trecentesco della *Practica Geometriae* di Leonardo Pisano. Firenze.

Fibonacci (1862). *Scritti di Leonardo Pisano*. Pubblicati da Baldassare Boncompagni. Volume II. Roma.

Fincke, Thomas (1583). *Geometriae rotundi libri XIII*. Basel.

Gandz, Solomon (1932). *The Mishnat ha Middot*. The First Hebrew Geometry of about 150 C.E. and the Geometry of Muhammad Ibn Musa Al-Khowarizmi. Berlin.

Kiselman, Christer O. (2015). "Euclid's straight lines". *Nordisk matematisk tidskrift* 60, pp. 145–169.

Magini, Giovanni Antonio (1592). *De planis triangulis*. Venice.

Oresme, Nicole (1968). *Le livre du ciel et du monde*. Edited by Albert D. Menut and Alexander J. Denomy, translated with an introduction by Albert D. Menut. Madison.

Ramus, Petrus (1567). *Geometria*. Paris.

Ramus, Petrus (1569). *Arithmeticae libri duo. Geometriae septem et viginti*. Basel.

Recorde, Robert (1551). *The patheway to knowledg*. London.

Rosenfeld, Boris A. (1988). *A History of Non-Euclidean Geometry*. Evolution of the Concept of a Geometric Space. Translated by Abe Shenitzer. New York.

---

[1] See Fibonacci (1862, e. g., pp. 85, 92) and, for Nemorarius, Clagett (1984, p. 376). [2] Tropfke (1940, p. 137) attributed the term to Fincke (1583, pp. 10f., 68).

Ruska, Julius (1895). "Zur Geschichte des Sinus". *Zeitschrift für Mathematik und Physik (hist.-lit. Abtheilung)* 40, pp. 126–128.
Schmid, Wolffgang (1539). *Das erst buch der Geometria*. Nürnberg.
Simon, Max (1901). *Euclid und die sechs planimetrischen Bücher*. Leipzig.
Theon of Smyrna (1878). *Theonis Smyrnaei philosophi Platonici expositio rerum mathematicarum ad legendum Platonem utilium*. Ed. by Eduard Hiller. Leipzig.
Szabó, Árpád (1992). *Das geozentrische Weltbild*. München.
Toomer, G.J. (1984). *Ptolemy's Almagest*. London.
Tropfke, Johannes (1940). *Geschichte der Elementar-Mathematik*. Vierter Band. Ebene Geometrie. Dritte, verbesserte und vermehrte Auflage besorgt von Dr. Kurt Vogel. Berlin and Leipzig.
Worrell, W.H. (1944). "The terminology of Arabic goniometrical works". *Journal of Near Eastern Studies* 3, pp. 91–100.

chordal graph

In a talk held in 1960, Berge (1961, p. 114) introduced two new terms: "A graph is called a *Gallai graph* if every elementary cycle of size greater than 3 has a *chord*, i. e. an edge connecting two nonconsecutive nodes of the cycle".[1] At about the same time, he also used the corresponding French terminology *corde* and *graphe Gallaïen* (Berge 1960, p. 146).

An early adopter of *chord* was Dirac (1961, p. 71), for example, who, at the same time, suggested an alternative for *Gallai graph*: "A graph in which every circuit with more than three vertices has at least one chord, is called a *rigid circuit graph*". Another proposal was made by Berge himself, namely *triangulated graph* (Berge 1965, p. 91). Both terms found their followers. A new synonym was put forward and justified by Gavril (1972, p. 180): "A graph is called chordal if every simple circuit v_1–v_2–v_3–$\cdots$ v_l–v_1 has a chord [...] Some call chordal graphs 'triangulated.' However, the term has a somewhat similar, but different meaning in the theory of planar graphs.[2] If this terminology were accepted, it would allow valid statements like: 'Some planar triangulated graphs are not triangulated'".

The concept is first found in Hajnal and Surányi (1958, p. 115).

References

Berge, Claude (1960). "Les problèmes de coloration en théorie des graphes". *Publications de l'Institut de statistique de l'Université de Paris* 9, pp. 123–160.
Berge, Claude (1961). "Färbung von Graphen, deren sämtliche bzw. deren ungerade Kreise starr sind". *Wissenschaftliche Zeitschrift der Martin-Luther-Universität Halle-Wittenberg* 10, pp. 114–115.
Berge, Claude (1965). "Graph Theory and Applications". In: *Mathematics and Social Sciences I*. Proceedings of the Seminars of Menthon-Saint-Bernard, France (1-27 July 1960) and of Gösing, Austria (3-27 July 1962). Ed. by Saul Sternberg et al. Paris and The Hague, pp. 83–97.
Berge, Claude and Jorge L. Ramírez Alfonsín (2001). "Origins and genesis". In: *Perfect Graphs*. Ed. by Bruce A. Reed and Jorge L. Ramírez Alfonsín. Chichester, pp. 1–12.
Dirac, Gabriel Andrew (1961). "On rigid circuit graphs". *Abhandlungen aus dem Mathematischen Seminar der Universität Hamburg* 25, pp. 71–76.

---

[1] "Ein Graph G werde ein *Gallaischer Graph* genannt, wenn jeder Elementarkreis, dessen Länge größer als 3 ist, eine *Sehne* besitzt, d. h. eine Kante, welche zwei nicht-konsekutive Knoten des Kreises verbindet". Translation from Berge and Ramírez Alfonsín (2001, p. 7). It is also remarked there that Gallai himself stated that the attribution to him was mistaken. [2] A planar graph is called *triangulated* when the addition of any edge results in a nonplanar graph.

Gavril, Fănică (1972). "Algorithms for Minimum Coloring, Maximum Clique, Minimum Covering by Cliques, and Maximum Independent Set of a Chordal Graph". *SIAM Journal on Computing* 1, pp. 180–187.

Hajnal, András and János Surányi (1958). "Über die Auflösung von Graphen in vollständige Teilgraphen". *Annales Universitatis Scientiarum Budapestinensis de Rolando Eötvös Nominatae* 1, pp. 113–121.

chromatic number and *chromatic polynomial*

The term *chromatic* is found in graph theory in connection with map colouring. In this context Wernicke (1904, p. 413) introduced the German term *chromatische Zahl* 'chromatic number', which he defined as "the least number of colours by which neighbouring regions can always be distinguished".[1] The term was adopted by Tietze (1905, p. 22); the English term seems to appear first in Franklin (1939, p. 206), while French *nombre chromatique* is found in Errera (1921, p. 57). Dirac (1951, p. 347) later also called a graph k-*chromatic*: "A graph has *chromatic number* k, or is k-*chromatic*, if there exists a smallest positive integer k such that the nodes of the graph can be divided into k mutually disjoint (colour) classes so that no two nodes in the same class are joined by an edge".

The chromatic polynomial of a graph counts the number of ways a graph can be coloured using no more than a given number of colours. First defined by Birkhoff (1912, p. 42) and generalized by Whitney (1932), it was not given a name until Birkhoff and Lewis published a paper entitled "Chromatic Polynomials" in 1946:

> Let P_n be a map of n regions and let $P_n(\lambda)$ denote the number of ways that P_n can be colored using some or all of λ given colors. Then it is well known that $P_n(\lambda)$ can be written as a polynomial in λ, identically zero in the case of a pseudo-map, but otherwise of the nth degree [...] These polynomials are called *chromatic polynomials* (Birkhoff and Lewis 1946, p. 360).

Note. For the history of these notions, see Biggs, Lloyd, and Wilson (1998, ch. 7).

References

Biggs, Norman L., E. Keith Lloyd, and Robin J. Wilson (1998). *Graph Theory, 1736-1936*. Oxford.

Birkhoff, George D. (1912). "A Determinant Formula for the Number of Ways of Coloring a Map". *Annals of Mathematics* 14, pp. 42–46.

Birkhoff, George D. and D. C. Lewis (1946). "Chromatic Polynomials". *Transactions of the American Mathematical Society* 60, pp. 355–451.

Dirac, Gabriel Andrew (1951). "Note on the colouring of graphs". *Mathematische Zeitschrift* 54, pp. 347–353.

Errera, Alfred (1921). "Du coloriage des cartes et de quelques questions d'*analysis situs*". PhD thesis. Université libre de Bruxelles.

Franklin, Philip (1939). "The four color problem". *Scripta Mathematica* 6, pp. 149–156, 197–210.

Tietze, Heinrich (1905). "Über das Problem der Nachbargebiete im Raum". *Monatshefte für Mathematik und Physik* 16, pp. 211–216.

Wernicke, Paul (1904). "Über den kartographischen Vierfarbensatz". *Mathematische Annalen* 58, pp. 413–426.

Whitney, Hassler (1932). "The Coloring of Graphs". *Annals of Mathematics* 33, pp. 688–718.

[1] "die *chromatische Zahl*, wie wir die geringste Anzahl von Farben bezeichnen dürfen, mit denen Nachbarländer stets unterschieden werden können".

circle

The circle counts among the oldest mathematical figures; it was already treated in Egyptian and Babylonian texts (Tropfke 1940, p. 131). In Greek mathematics, the term *κύκλος*, originally 'wheel', was used to refer to the circle since presocratic times.

In what appears to be the oldest definition of a circle (Tropfke 1940, p. 141), Plato (*Parmenides* 137e) does not use *κύκλος* but instead defines 'the round': "The round [...] is that of which the extremes are everywhere equally distant from the center".[1] Euclid, using *κύκλος*, gives a different definition as a two-dimensional figure: "A circle is a plane figure contained by one line such that all the straight lines falling upon it from one point among those lying within the figure equal one another" (*Elements*, book I, def. 15).[2]

In Latin, the term *circulus*, diminutive of *circus* 'circle', was used as a translation of the Greek *κύκλος*, as witnessed by Cicero: "There are two forms that excel all others, among solid bodies the globe (for so we may translate the Greek *sphaera*), and among plane figures the round or circle, the Greek *kyklos*".[3] In mathematical contexts, the term was used by the Roman land surveyors, for example by Balbus (Blume, Lachmann, and Rudorff 1848, p. 104) and by Censorinus (Sallmann 1983, p. 69).

The Latin term was the basis for English *circle*, French *cercle* and for German *Zirkel*, which competed for some time with the now standard *Kreis*.

The French term *cercle* is found in the 13th-century (TLF, DMF), for example in the *Pratike de geometrie* (Victor 1979, p. 481); Italian *cerchio* also dates from the late 13th (TLIO). English *circle* is found in the early 14th century (OED); the MED cites from an English translation of *De proprietatibus rerum* by Bartholomaeus Anglicus: "þe cercle is a lyne y drawe al rounde aboute fro a certein poynt to þe same poynt, and eueriche party of þat lyne is yliche fer fro þe myddel point".

The German terms *Kreis* and *Zirkel* first appear in print (as *kraysz* and *cirkel*) in a lexicographical work from about 1400, printed in 1482 ([anon.] 1482, fol. r ii^{v}). *Kreis* had earlier been used in the middle of the 14th century by Konrad von Mengenberg in his translation of Sacrobosco's *Sphaera mundi* (von Megenberg 1980, p. 6), and *Zirkel* appears in the *Geometria culmensis* from around 1400 (Mendthal 1886, p. 65).[4] Until the 19th century, *Zirkel* is more frequently used than *Kreis* (Tropfke 1940, p. 138).

Note. The history of the notion and terminology are discussed by Tropfke (1940, pp. 131, 138f., 141f.) and Heath in Euclid (1956, pp. 183–185). For the Greek term, see Mugler (1958-59, pp. 260f.); for the Latin (including synonyms), see Ayuso García (2008, pp. 833–840). The Arabic terminology is discussed by Souissi (1968, p. 164) and Rashed (2017, pp. 258–262).

References

[anon.] (1482). *Vocabularius Theutonicus*. Nürnberg.

[1] Translation from Plato (1926, p. 237). [2] Translation from Euclid (1956, p. 153). – A circle is thus defined as a plane figure. [3] "cumque duae formae praestantes sint, ex solidis globus (sic enim *σφαῖραν* interpretari placet), ex planis autem circulus aut orbis, qui *κύκλος* Graece dicitur". Translation from Cicero (1967, p. 169). [4] For the German terminology, see also Schirmer (1912, p. 40) and Reiner (1961, pp. 76–79).

Ayuso García, Manuel (2008). "La terminología latina de la geometría en Marciano Capela". PhD thesis. Madrid: Universidad Nacional de Educación a Distancia.
Blume, Friedrich, Karl Lachmann, and Adolf Rudorff, eds. (1848). *Die Schriften der römischen Feldmesser*. Erster Band. Texte und Zeichnungen. Berlin.
Cicero (1967). *De natura deorum. Academica*. With an English translation by H. Rackham. Cambridge, Mass.
Euclid (1956). *The Thirteen Books of Euclid's Elements*. Volume I. Introduction and Books I, II. Ed. by Thomas L. Heath. New York.
von Megenberg, Konrad (1980). *Die deutsche Sphaera*. Ed. by Francis B. Brévart. Tübingen.
Mendthal, Hans (1886). *Geometria culmensis*. Ein agronomischer Tractat aus der Zeit des Hochmeisters Conrad von Jungingen (1393-1407). Leipzig.
Mugler, Charles (1958-59). *Dictionnaire historique de la terminologie géométrique des Grecs*. Paris.
Plato (1926). *Cratylus. Parmenides. Greater Hippias. Lesser Hippias*. Translated by H. N. Fowler. London.
Rashed, Roshdi, ed. (2017). *Lexique historique de la langue scientifique arabe*. Hildesheim.
Reiner, Karl (1961). *Die Terminologie der ältesten mathematischen Werke in deutscher Sprache nach den Beständen der Bayerischen Staatsbibliothek*. München.
Sallmann, Klaus, ed. (1983). *Censorini de die natali liber ad Q. Caerellium*. Leipzig.
Schirmer, Alfred (1912). *Der Wortschatz der Mathematik nach Alter und Herkunft untersucht*. Straßburg.
Souissi, Mohammed (1968). *La langue des mathématiques en arabe*. Tunis.
Tropfke, Johannes (1940). *Geschichte der Elementar-Mathematik*. Vierter Band. Ebene Geometrie. Dritte, verbesserte und vermehrte Auflage besorgt von Dr. Kurt Vogel. Berlin and Leipzig.
Victor, Stephen K. (1979). *Practical Geometry in the High Middle Ages*. "Artis cuiuslibet consummatio" and the "Pratike de geometrie". Philadelphia.

circle of convergence

The French term *cercle de convergence* was first introduced by Bouquet and Briot (1853, p. 264) in a brief note. In a longer article published three years later, they explained: "the series is convergent for all values z situated in the interior of this circle, which we shall call for this reason *circle of convergence*" (Briot and Bouquet 1856, p. 95).[1] They also used *rayon de convergence* (p. 101); both terms also appear in Briot and Bouquet (1859, pp. 13, 19).

Briot and Bouquet point to Cauchy (1844, p. 388) for the concept; is is found earlier, though, for example in Cauchy (1821); for further references, see Pringsheim (1898, p. 81) and Laugwitz and Neuenschwander (1994, pp. 64, 67).

The term was quickly adopted; Marie (1860, p. 458) already states that it is in frequent use. The English terms *circle* and *radius of convergence* are found in the late 1880s; they were used by Chrystal (1889, p. 129), for example. The German terms *Convergenzkreis* and *Convergenzradius* are found in print in Fischer (1862, pp. 18, 26).

In Germany, various related terms were in use. The most common ones are *Convergenzbezirk* 'district of convergence' and *Convergenzbereich* 'domain of convergence'. Like many German terms in complex analysis, these may be due to Weierstraß; such terms are, however, difficult to trace because many of them were first used in lectures rather than printed publications.[2] Weierstraß (1880, p. 719) gave the following

[1] "la série est convergente pour toutes les valeurs de z situées dans l'intérieur de ce cercle, que nous appellerons pour cette raison *cercle de convergence*". [2] Early attestations in print are: Roch (1863, p. 186): *Convergenzbezirk*, Wernicke (1866, p. 24): *Bereich*, Thomé (1866, p. 324): *Convergenzgebiet*, Wangerin (1870, p. 86): *Convergenzbereich*.

definition: "The totality of those values of x for which the series $\sum_{\nu=0}^{\infty} f_\nu(x)$ has a finite value shall be called the domain of convergence of this series".[1]

Note on Weiterstraß. Bölling (1994, pp. 60f.) has pointed out that Weierstraß redacted some of the articles in his collected papers without indicating the alterations. These redactions do not just concern papers previously unpublished (see *absolute*). An example from Weierstraß' article on the theory of analytic faculties (1856): in the collected works, the words *absoluter Betrag* 'absolute value' and *innerhalb des Convergenzgebietes* 'within the domain of convergence' are used (Weierstraß 1894a, p. 204); in the original printed paper, however, *analytischer Modul* and *zwischen zwei Grenzen a und b* 'between two boundaries a and b' appear instead (Weierstraß 1856, p. 47). Thus, the use of *Convergenzbezirk* in a paper originally written in 1841 (Weierstraß 1894b, p. 67) is to be regarded as such a redaction and certainly not as evidence that Weierstraß already used the term in the 1840s. Instead, he may have started using the term in the late 1860s.

Note. The origin of the terminology in Briot and Bouquet is briefly pointed out by Bottazzini and Gray (2013, p. 53).

References

Bölling, Reinhard (1994). "Karl Weierstrass – Stationen eines Lebens". *Jahresbericht der Deutschen Mathematiker-Vereinigung* 96, pp. 56–75.

Bottazzini, Umberto and Jeremy Gray (2013). *Hidden Harmony – Geometric Fantasies*. The Rise of Complex Function Theory. New York.

Bouquet, Jean Claude and Charles Briot (1853). "Recherches sur les séries ordonnées suivant les puissances croissantes d'une variable imaginaire". *Comptes rendus de l'Académie des sciences* 36, p. 264.

Briot, Charles and Jean Claude Bouquet (1856). "Étude des fonctions d'une variable imaginaire (Premier Mémoire)". *Journal de l'École polytechnique* 21, cahier 36, pp. 85–131.

Briot, Charles and Jean Claude Bouquet (1859). *Théorie des fonctions doublement périodiques et, en particulier, des fonctions elliptiques*. Paris.

Cauchy, Augustin-Louis (1821). *Cours d'analyse de l'École Royale Polytechnique*. I.re partie. Analyse algébrique. Paris.

Cauchy, Augustin-Louis (1844). *Exercises d'analyse et de physique mathématique*. Tome troisième. Paris.

Chrystal, George (1889). *Algebra*. An Elementary Text Book for the Higher Classes of Secondary Schools and for Colleges. Part II. Edinburgh.

Fischer, Hermann (1862). *Briot und Bouquet's Theorie der doppelt-periodischen Functionen und insbesondere der elliptischen Transcendenten*. Halle.

Laugwitz, Detlef and Erwin Neuenschwander (1994). "Riemann and the Cauchy-Hadamard Formula for the Convergence of Power Series". *Historia Mathematica* 21, pp. 64–70.

Marie, Maximilien (1860). "Nouvelle théorie des fonctions de variables imaginaires. Troisième Partie (Suite)". *Journal de mathématiques pures et appliquées* 5, pp. 457–474.

Pringsheim, Alfred (1898). "Irrationalzahlen und Konvergenz unendlicher Prozesse". *Encyklopädie der mathematischen Wissenschaften mit Einschluss ihrer Anwendungen* I A 3, pp. 49–146.

Roch, Gustav (1863). "Ueber Functionen complexer Grössen". *Zeitschrift für Mathematik und Physik* 8, pp. 12–26, 183–203.

Thomé, Ludwig Wilhelm (1866). "Ueber die Kettenbruchentwicklung der *Gauss*schen Function $F(\alpha, 1, \gamma, x)$". *Journal für die reine und angewandte Mathematik* 66, pp. 322–343.

---

[1] "Die Gesammtheit derjenigen Werthe von x, für welche die Reihe $\sum_{\nu=0}^{\infty} f_\nu(x)$ einen endlichen Werte hat, nenne ich den Convergenzbereich dieser Reihe".

Wangerin, Albert (1870). "W. Wernicke. Ueber die Bewegung eines von einem Rotationsellipsoide angezogenen Punktes ohne Anfangsgeschwindigkeit [review]". *Die Fortschritte der Physik im Jahre 1867* 23, p. 86.
Weierstraß, Karl (1856). "Über die Theorie der analytischen Facultäten". *Journal für die reine und angewandte Mathematik* 51, pp. 1–60.
Weierstraß, Karl (1880). "Zur Functionenlehre". *Monatsberichte der Königlich Preußischen Akademie der Wissenschaften zu Berlin* (1880), pp. 719–743.
Weierstraß, Karl (1894a). *Mathematische Werke*. Erster Band. Berlin.
Weierstraß, Karl (1894b). "Zur Theorie der Potenzreihen". In: *Mathematische Werke*. Erster Band. Berlin, pp. 67–74.
Wernicke, Wilhelm (1866). "Ueber die Bewegung eines von einem Rotations-Ellipsoide angezogenen Punktes ohne Anfangsgeschwindigkeit". *Erster Jahresbericht über die Luisenstädtische Gewerbeschule in Berlin*, pp. 1–26.

circumference

The term *circumference* goes back to Greek *περιφέρεια*, from *περί* 'around' and *φέρειν* 'to carry'. This word was first used to refer to a part of a circumference, an arc, and later to the whole circumference (Tropfke 1940, p. 132). Euclid uses but never defines the term because, as Heath (1956, p. 184) states, "no explanation was needed. Though the word *περιφέρεια* does not occur in Plato, Aristotle uses it several times (1) in the general sense of *contour* without any special mathematical signification, (2) mathematically, with reference to the rainbow and the circumference, as well as an arc, of a circle. Hence Euclid was perfectly justified in employing the word in Deff. 17, 18 [of book I] and elsewhere, but leaving it undefined as being a word universally understood and not involving in itself any mathematical conception".

The Latin term *circumferentia* is a literal translation of the Greek term (from *circum* 'around' and *ferrere* 'to carry'). It first appears around the second century in the context of Roman land surveying and was used by Hyginus and Balbus (Blume, Lachmann, and Rudorff 1848, pp. 99, 189), for example, and around the year 400 by Martianus Capella (Ayuso García 2008, p. 952). It is also generally used in 12th-century translations of the *Elements* (see, e. g., Busard and Folkerts 1992, p. 114) and other texts.[1]

French *circonférence* is found in the 13th century; Dörr (1998, p. 111) presents an astronomical treatise from around 1270 as the earliest attestation. Italian *circonferenza* appeared at about the same time (TLIO). In German, the Latin term remained virtually unchanged as *circumferencia* in Widmann (1489, fol. C 4^{r});[2] in the first translation of the *Elements* into German and other texts, *Umkreis*[3] is found (Euclid 1562, p. 2), while the now standard *Umfang* does not appear until the 17th century (DWB); Beyer (1619, p. 13) is an early attestation.[4] In English, the term appears in the late 14th century (MED and OED both cite John Gower's *Confessio Amantis*).

Note. Tropfke (1940, p. 132) gives a brief overview of the Greek and Latin terms meaning 'circumference'. For the Greek term *περιφέρεια*, see Mugler (1958-59, pp. 344f.) and Heath (1956, pp. 183–185). The Latin terminology is discussed in

[1] In the first Latin translation based on a Greek original, the Greek term was Latinized as *periferia* (Busard 1987, p. 27). [2] Widmann (1489, fol. C 5^{r}) also writes *circumferencz*. [3] The spelling in the cited source is *Umbkhrais*. [4] He wrote *Umbfang*. For the German terms, see also Tropfke (1940, pp. 138f.) and Reiner (1961, pp. 78f.).

detail by Ayuso García (2008, pp. 951–959) and Guillaumin (2020, pp. 59–61), the Arabic in Rashed (2017, pp. 216-219, 257f.).

References

Ayuso García, Manuel (2008). "La terminología latina de la geometría en Marciano Capela". PhD thesis. Madrid: Universidad Nacional de Educación a Distancia.

Beyer, Johann Hartmann (1619). *Logistica Decimalis*. Frankfurt.

Blume, Friedrich, Karl Lachmann, and Adolf Rudorff, eds. (1848). *Die Schriften der römischen Feldmesser*. Erster Band. Texte und Zeichnungen. Berlin.

Busard, H.L.L. (1987). *The Mediaeval Latin Translation of Euclid's Elements Made Directly from the Greek*. Stuttgart.

Busard, H.L.L. and Menso Folkerts, eds. (1992). *Robert of Chester's (?) Redaction of Euclid's* Elements, *the so-called Adelard II Version*. Basel.

Dörr, Stephen (1998). *Der älteste Astronomietraktat in französischer Sprache: L'Introductoire d'astronomie*. Edition und lexikalische Analyse. Tübingen.

Euclid (1562). *Die Sechs Erste Bücher Euclidis*. Auß Griechischer sprach in die Teütsch gebracht ... Durch Wilhelm Holtzmann genannt Xylander. Basel.

Guillaumin, Jean-Yves (2020). *Dictionnaire de la terminologie latine ancienne de l'arithmétique et de la géométrie*. Paris.

Heath, Thomas L. (1956). *The Thirteen Books of Euclid's Elements*. Volume I. Introduction and Books I, II. New York.

Mugler, Charles (1958-59). *Dictionnaire historique de la terminologie géométrique des Grecs*. Paris.

Rashed, Roshdi, ed. (2017). *Lexique historique de la langue scientifique arabe*. Hildesheim.

Reiner, Karl (1961). *Die Terminologie der ältesten mathematischen Werke in deutscher Sprache nach den Beständen der Bayerischen Staatsbibliothek*. München.

Tropfke, Johannes (1940). *Geschichte der Elementar-Mathematik*. Vierter Band. Ebene Geometrie. Dritte, verbesserte und vermehrte Auflage besorgt von Dr. Kurt Vogel. Berlin and Leipzig.

Widmann, Johannes (1489). *Behende vnd hubsche Rechenung auff allen kauffmanschafft*. Leipzig.

cissoid

The term *cissoid* comes from *κίσσος* 'ivy' and means 'ivy-shaped'. Two different curves are designated under this name, and both these curves were identified in the 17th century: on the one hand, a curve discussed by Pappus and Proclus, on the one hand, another curve known as the cissoid of Diocles.

The ancient curve is briefly mentioned by Pappus, for example in book 4 of his *Collection* (Pappus of Alexandria 2010, p. 145), while Proclus gives some more information on it. Thus, he states that the curve was already discussed by Geminus (Proclus 1970, pp. 90f.), who flourished in the 1st century BC. He also says that the name is due to a resemblance with ivy leaves (Proclus 1970, p. 102). Most importantly, he describes the curve as "closed": "Some lines, he [Geminus] says, are finite and enclose a figure, like the circle, the perimeter of an ellipse, the cissoid, and many others" (Proclus 1970, p. 138). This is a property that the cissoid of Diocles does not have (Toomer 1976, p. 24), and that is an essential argument why the two curves must be assumed to be different. Toomer (1976, p. 24) also points out that "[w]e now know that Diocles himself did not call the curve 'cissoid'. More significantly, neither does Eutocius[1] [...] I therefore consider it in the highest degree unlikely that the name 'cissoid' was ever applied in antiquity to Diocles' curve".[2]

[1] Eutocius described Diocles' curve in his commentary on Archimedes' *On the Sphere and Cylinder*. See Archimedes (2004, pp. 279–284). [2] The nature of the curve discussed by Proclus and Pappus is discussed in detail by Knorr (1987, pp. 246–263).

Toomer (1976, p. 25) then raises the question "who was responsible for the identification of Diocles' curve and the name 'cissoid'", but at the time was unable to give an answer except to say that the identification was made by Roberval and Fermat in 1640 (see de Fermat 1894, p. 201).[1] An answer to this question had indeed already been given by Itard (1974, p. 87), who points out that the identification is probably due to David Rivault, editor of Archimedes' works. Rivault there identifies Diocles' curve and the term *cissoid* (*linea cissoidea*) (see Archimedes 1615, pp. 92, 96f.), without justifying this step.[2] This identification was soon taken over by others such as Mersenne (1625, pp. 860f.) and Aggiunti (1627, p. 32).

The Latin term *cissoid* (for the first of the two curves discussed here) appears to be due to Barozzi,[3] who used it in a translation of Proclus' commentary on Euclid's *Elements* (Proclus 1560, pp. 64, 72).

Note. For comparatively recent literature on the history of the term, see Toomer (1976, pp. 24f.) and the detailed discussion by Knorr (1987, pp. 246–263). Older relevant literature includes Loria (1910, pp. 36–46), Teixeira (1908, pp. 1–26), and Tannery (1884, pp. 109–112).

References

Aggiunti, Niccolò (1627). *Oratio de mathematicae laudibus*. Roma.

Archimedes (1615). *Archimedis opera quae extant*. Ed. by David Rivault. Paris.

Archimedes (2004). *The Works of Archimedes*. Translation and Commentary. Volume 1: The Two Books *On the Sphere and the Cylinder*. Edited and translated by Reviel Netz. Cambridge.

Clavius, Christopher (1604). *Geometria practica*. Roma.

Descartes, René (1637). *La Géométrie*. In: *Discours de la méthode*. Leiden, pp. 295–413.

de Fermat, Pierre (1894). *Œuvres*. Tome deuxième. Correspondance. Ed. by Paul Tannery and Charles Henry. Paris.

Ferroni, Pietro (1792). *De calculo integralium exercitatio mathematica*. Firenze.

Itard, Jean (1974). "Archimède, t. IV: *Commentaires d'Eutocius et fragments*, texte établi et traduit par Charles Mugler [review]". *Revue d'histoire des sciences* 27, pp. 87–88.

Knorr, Wilbur R. (1987). *The Ancient Tradition of Geometric Problems*. Boston, Mass.

Loria, Gino (1910). *Spezielle algebraische und transzendente ebene Kurven*. Theorie und Geschichte. Autorisierte, nach dem italienischen Manuskript bearbeitete deutsche Ausgabe von Fritz Schütte. Zweite Auflage. Erster Band. Leipzig.

Mersenne, Marin (1625). *La vérité des sciences*. Paris.

Pappus of Alexandria (2010). *Book 4 of the* Collection. Edited With Translation and Commentary by Heike Sefrin-Weis. London.

Proclus (1560). *Commentariorum ad universam mathematicam disciplinam Libri IIII*. A Francisco Barocio primum iam Romanae linguae venustate donati. Padua.

Proclus (1970). *Commentary on the First Book of Euclid's Elements*. Translated with Introduction and Notes by Glenn R. Morrow. Princeton.

Tannery, Paul (1884). "Notes pour l'histoire des lignes et surfaces courbes dans l'antiquité (3)". *Bulletin des sciences mathématiques et astronomiques* 8, pp. 101–112.

Teixeira, Francisco Gomes (1908). *Traité des courbes spéciales remarquables plans et gauches*. Tome I. Coïmbra.

---

[1] Toomer also points out that, for example, Descartes uses the term in his *Geométrie* (Descartes 1637, p. 317)) but notes that Descartes mentions the term out of context and therefore an identification cannot be securely assumed. [2] Earlier writers were more careful: in Valla (1501, book XIII, ch. 2) and Clavius (1604, pp. 299f.), Diocles' curve is discussed but the term *cissoid* is *not* used. Before Itard, the identification had critically been commented on by Ferroni (1792, p. 200). [3] As pointed out by Toomer (1976, p. 25).

Toomer, G.J., ed. (1976). *Diocles. On Burning Mirrors*. The Arabic Translation of the Lost Greek Original. Berlin et al.
Valla, Giorgio (1501). *De expetendis, et fugiendis rebus opus*. Venezia.

class (curve)

The term *class of a curve* was introduced by Gergonne (1827-28, p. 151), who defined it thus: "A plane curve is said to be of the mth class when m real or ideal tangents can be drawn to it from a given point in its plane".[1]

The term was quickly adopted, for example by Plücker (1830, p. 109) and Steiner (1832, p. 149), who both used German *Klasse* and referred to Gergonne for the term. The English term comes up at around 1850; it was used by Kirkman (1850, p. 74) and Salmon (1852, p. 1).

Gergonne (1827-28, p. 151) also defined the term *degree* (*degré*) in the same place: "A plane curve is said to be of the mth degree when it has m real or ideal intersections with a given line".[2] This usage was not generally adopted; Plücker (1830, p. 109), for example, used *order* (*Ordnung*) instead because *degree* was ambiguous.

Note. For the introduction of the term, see Boyer (1956, p. 251) and Gray (2011, pp. 57–61).

References

Boyer, Carl B. (1956). *History of Analytic Geometry*. New York.
Gergonne, Joseph Diez (1827-28). "Rectification de quelques théorèmes énoncés dans les Annales". *Annales de mathématiques pures et appliquées* 18, pp. 149–154.
Gray, Jeremy (2011). *Worlds Out of Nothing*. A Course in the History of Geometry in the 19th Century. London et al.
Kirkman, Thomas P. (1850). "Some hints on line-coordinates". *The Lady's and Gentleman's Diary* (1850), pp. 73–85.
Plücker, Julius (1830). "Über eine neue Art, in der analytischen Geometrie Puncte und Curven durch Gleichungen darzustellen". *Journal für die reine und angewandte Mathematik* 6, pp. 107–146.
Salmon, George (1852). *A Treatise on the Higher plane curves intended as a sequel to the Treatise on Conic Sections*. Dublin.
Steiner, Jacob (1832). *Systematische Entwickelung der Abhängigkeit geometrischer Gestalten von einander*. Erster Theil. Berlin.

class (equivalence relations)

The term *class* in the context of (particular kinds of) equivalence relations first was used by Gauß[3] in his *Disquisitiones arithmeticae*, where, on the basis of a definition of *equivalent forms*, he pointed out:

> *all the forms of this determinant can be distributed into as many* c l a s s e s *as there are periods*, that is by putting the forms which are properly equivalent to the form F into the first class, those which are properly equivalent to G in the second class, etc. In this way all

[1] "Une courbe plane est dite de m.[ième] classe, lorsqu'on peut lui mener d'un même point de son plan m tangentes réelles ou idéales". Translation from Gray (2011, p. 59). [2] "Une courbe plane est dite du m.[ième] degré, lorsqu'elle a avec une même droite m intersections réelles ou idéales". Translation from Gray (2011, p. 59). [3] The term has been attributed to Gauß by Weil (2001, p. 319), for example.

> the forms contained in the same class will be properly equivalent, and forms contained in different classes cannot be properly equivalent (Gauß 1801, § 195).[4]

Gauß (1801, § 222) refers to Lagrange (1775) and Lagrange (1777) for the notion and points out that "he [Lagrange] showed that for a given determinant we can find a finite number of forms so arranged that each form of that determinant is equivalent to one of these, and thus all forms of a given determinant can be distributed into classes". Lagrange used neither the word *équivalent* nor *classe*.[1]

The term *class* in this sense was in common use in the 19th century and adopted, for example, by Seeber (1831, p. 21) and Lejeune Dirichlet (1839, p. 335). Others also used the terms *equivalence* and *class* in connection with other concepts of equivalence. For example, Kummer (1847, p. 324) applied them to his ideal complex numbers: "Two ideal complex numbers that, multiplied by one and the same ideal number, give the same genuine complex number I call *equivalent* or say belong to the same class".[2]

However, throughout the 19th century *class* is only used in connection with various *particular* kinds of equivalence relations, since the general concept of equivalence relation only comes up in the 20th century.

The actual compound *equivalence class* essentially dates from the 1920s; Netto's (1896, p. 246) use of *Aequivalenzclasse* is exceptional. von Neumann (1926, p. 197) used *Äquivalenzklasse* for a particular equivalence relation: "An equivalence class that contains only fundamental sequences shall be called an *ideal number*".[3] The same term was also used by Levi (1929, p. 2), also for a particular equivalence relation (equivalence of configurations). The term did not require any explanation for von Neumann and Levi.

General definitions of equivalence relations were given by Hasse (1926, p. 17) and van der Waerden (1930, p. 14); both used the word *class* (*Klasse*). For the term *equivalence class* in connection with equivalence relations in general, Asghari (2019, p. 4661) refers to Tukey (1940, p. 4): "It is well known that an *equivalence relation* (that is, one which generates a reflexive, transitive and symmetric ordered system) divide[s] the set on which it is defined into mutually exclusive equivalence classes".[4]

Alternative names include German *Abstraktionsklasse* 'abstraction class' (Carnap 1928, p. 102); this term was cited by Birkhoff (1935, p. 446), who preferred *x-category*, however. The word *category* had earlier been used by Clifford (1933, p. 869): "Conjugacy as thus defined is an equivalence relation, that is, it is reflexive, symmetric, and transitive, and hence leads to a dissection of G into categories $J(a)$, where $J(a)$ is defined to be the class of elements of G conjugate to the element a". As late as 1942, Ore (1942, p. 574) suggested *block*.

---

[4] "*omnes formas huius determinantis in totidem c l a s s e s distribui posse, quot habeantur periodi,* scilicet referendo eas quae formae F proprie aequivalent in primam classem, eas quae fortune G proprie aequivalent in secundam etc. Hoc modo omnes formae in eadem classe contentae proprie aequivalentes erunt, formae vero e classibus diversis non poterunt proprie aequivalere". Translation from (Gauß 1986, § 195). [1] Translation from (Gauß 1986, § 222). [2] "Zwei ideale complexe Zahlen, welche, mit einer und derselben idealen Zahl multiplicirt, beide zu wirklichen complexen Zahlen machen, nenne ich *äquivalent* oder derselben Classe angehörig". Translation from Gray (2018, p. 328). [3] "Eine Aequivalenzklasse, die lauter Fundamentalfolgen enthält, nennen wir eine *ideale Zahl*". [4] Albert (1937, p. 5) had earlied used *classes of equivalent elements*.

Note. For the history of the concept, see Asghari (2019, esp. pp. 4660f.); many of the examples discussed in this article have also been pointed out there. For Lagrange's and Gauß' concept of equivalence, see also Bashmakova and Rudakov (2001, pp. 86–91) and Goldstein and Schappacher (2007, esp. 8-13).

References

Albert, A.A. (1937). *Modern Higher Algebra*. Chicago.

Asghari, Amir (2019). "Equivalence: an attempt at a history of the idea". *Synthese* 196, pp. 4657–4677.

Bashmakova, I.G. and E.I. Rudakov (2001). "Algebra and Algebraic Number Theory". In: *Mathematics of the 19th Century*. Mathematical Logic. Algebra. Number Theory. Probability Theory. Second revised edition. Ed. by Andrey Nikolaevich Kolmogorov and Adolph Andrei Pavlovich Yushkevich. Basel et al., pp. 35–85.

Birkhoff, Garrett (1935). "On the structure of abstract algebras". *Proceedings of the Cambridge Philosophical Society* 31, pp. 433–454.

Carnap, Rudolf (1928). *Der logische Aufbau der Welt*. Hamburg.

Clifford, A.H. (1933). "A System Arising from a Weakened Set of Group Postulates". *Annals of Mathematics* 34, pp. 865–871.

Gauß, Carl Friedrich (1801). *Disquisitiones arithmeticae*. Leipzig.

Gauß, Carl Friedrich (1986). *Disquisitiones Arithmeticae*. English Edition. Translated by A.A. Clarke. Revised by W.C. Waterhouse. New York et al.

Goldstein, Catherine and Norbert Schappacher (2007). "A Book in Search of a Discipline (1801-1860)". In: *The Shaping of Arithmetic after C. F. Gauss's Disquisitiones Arithmeticae*. Ed. by Catherine Goldstein, Norbert Schappacher, and Joachim Schwermer. Berlin and Heidelberg, pp. 3–65.

Gray, Jeremy (2018). *A History of Abstract Algebra*. From Algebraic Equations to Modern Algebra. Oxford.

Hasse, Helmut (1926). *Höhere Algebra*. I. Lineare Gleichungen. Berlin and Leipzig.

Kummer, Ernst Eduard (1847). "Zur Theorie der complexen Zahlen". *Journal für die reine und angewandte Mathematik* 35, pp. 319–326.

Lagrange, Joseph-Louis (1775). "Recherches d'arithmétique". *Nouveaux Mémoires de l'Académie Royale des Sciences et Belles-Lettres de Berlin* (1773), pp. 265–312.

Lagrange, Joseph-Louis (1777). "Suite des recherches d'arithmétique". *Nouveaux Mémoires de l'Académie Royale des Sciences et Belles-Lettres de Berlin* (1775), pp. 323–356.

Lejeune Dirichlet, Gustav (1839). "Recherches sur diverses applications de l'Analyse infinitésimale à la Théorie des Nombres". *Journal für die reine und angewandte Mathematik* 19, pp. 324–369.

Levi, Friedrich (1929). *Geometrische Konfigurationen*. Mit einer Einführung in die kombinatorische Flächentopologie. Leipzig.

Netto, Eugen (1896). "Über die arithmetisch-algebraischen Tendenzen Leopold Kronecker's". In: *Mathematical Papers read at the International Mathematical Congress (Chicago, 1893)*. New York, pp. 243–252.

von Neumann, John (1926). "Zur Prüferschen Theorie der idealen Zahlen. Erste Mitteilung". *Acta litterarum ac scientiarum Regiae Universitatis Hungaricae Francisco-Josephinae (Sectio scientiarum mathematicarum)* 2, pp. 193–227.

Ore, Øystein (1942). "Theory of equivalence relations". *Duke Mathematical Journal* 9, pp. 573–627.

Seeber, Ludwig August (1831). *Untersuchungen über die Eigenschaften der positiven ternären quadratischen Formen*. Freiburg.

Tukey, John W. (1940). *Convergence and Uniformity in Topology*. Princeton.

van der Waerden, Bartel Leendert (1930). *Moderne Algebra*. Erster Teil. Berlin.

Weil, André (2001). *Number Theory*. An approach through history from Hammurapi to Legendre. Boston.

class (set theory)

In set theory, the term was sometimes used instead of *set*, for example by Peano (1891, p. 3) and, following Peano, by Russell (1903, e. g., pp. 68f.).

A distinction between classes and sets was drawn by von Neumann (1925, p. 229); however, instead of a *class*, he originally spoke of a *domain* (German *Bereich*): "'sets' are the sets that (in the earlier terminology) are 'not too big', and 'domains' are all totalities irrespective of their 'size'".[1] By *too big* von Neumann meant, as he had already explained in a letter to Zermelo from 1923: "A set is 'too big' if and only if it is equivalent to the set of all things" (Meschkowski 1967, p. 272).[2]

Bernays (1937, p. 66) later replaced the word *domain* by *class*: "According to the leading idea of the von Neumann set theory we have to deal with two kinds of individuals, which we may distinguish as *sets* and *classes*. The distinction may be thought of this way, that a set is a multitude forming a proper thing, whereas as class is a predicate regarded only with respect to its extension".

Note. For von Neumann's introduction of the distinction between classes and sets, see Ferreirós (2007, pp. 378–382).

References

Bernays, Paul (1937). "A System of Axiomatic Set Theory–Part I". *The Journal of Symbolic Logic* 2, pp. 65–77.

Ferreirós, José (2007). *Labyrinth of Thought*. A History of Set Theory and Its Role in Modern Mathematics. Second revised edition. Basel.

van Heijenoort, Jean (2002). *From Frege to Gödel*. A Source Book in Mathematical Logic, 1879–1931. Cambridge, Mass.

Meschkowski, Herbert (1967). *Probleme des Unendlichen*. Werk und Leben Georg Cantors. Braunschweig.

von Neumann, John (1925). "Eine Axiomatisierung der Mengenlehre". *Journal für die reine und angewandte Mathematik* 154, pp. 219–240.

Peano, Guiseppe (1891). "Principii di logica matematica". *Rivista di matematica* 1, pp. 1–10.

Russell, Bertrand (1903). *The Principles of Mathematics*. Vol. I. Cambridge.

class field

The term *class field* first appears as German *Classenkörper*[3] in Weber (1891, p. 439). From today's perspective, he at first gave a restricted definition but extended the concept on several later occasions (Weber 1897, p. 98, 1908, p. 607).

The term was also used by Hilbert in his "Number report" (Hilbert 1897, p. 279), who remarks that the term is due to "the close connexion [...] between the field K and certain ideal classes of k".[4] The English term appears in Hilbert (1901, p. 459).

Frei (1989, p. 425) remarks that "[t]he modern concept of a class field is due to Teiji Takagi (1875-1960), who established in a long and fundamental paper in 1920 [(Takagi 1920)] class field theory as the theory of all abelian algebraic extensions $\boldsymbol{K}$, $\boldsymbol{K}/\boldsymbol{k}$, with respect to an algebraic number field $\boldsymbol{k}$ as a base field".

---

[1] "‚Menge' heißen (in der früheren Terminologie) die ‚nicht zu großen' Mengen, und ‚Bereiche' sind alle Gesamtheiten ohne Rücksicht auf ihre ‚Größe'". Translation based on van Heijenoort (2002, p. 403). [2] "Eine Menge ist dann und nur dann ‚zu gross', wenn sie der Menge aller Dinge aequivalent ist". Translation based on Ferreirós (2007, p. 379). [3] The current spelling is *Klassenkörper*. [4] "[...] der engen Beziehung, die [...] der Körper K zu gewissen Idealklassen des Körpers k aufweist". Translation from Hilbert (1998, p. 111).

Note. For the history of the concept, its origin in the work of Kronecker, and comments on the various definitions given by Weber and others, see Hasse (1967), Masahito (1994), Katsuya (1994), and Frei (1989).

References

Frei, Günther (1989). "Heinrich Weber and the Emergence of Class Field Theory". In: *The History of Modern Mathematics*. Volume I: Ideas and Their Reception. Ed. by David E. Rowe and John McCleary. Boston, pp. 425–450.

Hasse, Helmut (1967). "History of class field theory". In: *Algebraic Number Theory*. Ed. by J.W.S. Cassels and A. Fröhlich. London, pp. 266–279.

Hilbert, David (1897). "Die Theorie der algebraischen Zahlkörper". *Jahresbericht der Deutschen Mathematiker-Vereinigung* 4, pp. 175–546.

Hilbert, David (1901). "Mathematical problems". *Bulletin of the American Mathematical Society* 8, pp. 437–479.

Hilbert, David (1998). *The Theory of Algebraic Number Fields*. Translated from the German by Iain T. Adamson. With an introduction by Franz Lemmermeyer and Norbert Schappacher. Berlin.

Katsuya, Miyake (1994). "The Establishment of the Takagi-Artin Class Field Theory". In: *The Intersection of History and Mathematics*. Ed. by Sasaki Chikara, Sugiura Mitsuo, and Joseph W. Dauben. Basel, pp. 109–128.

Masahito, Takase (1994). "Three Aspects of the Theory of Complex Multiplication". In: *The Intersection of History and Mathematics*. Ed. by Sasaki Chikara, Sugiura Mitsuo, and Joseph W. Dauben. Basel, pp. 91–108.

Takagi, Teiji (1920). "Über eine Theorie des relativ-abelschen Zahlkörpers". *Journal College of Science, Imperial University of Tokio* 41, pp. 1–133.

Weber, Heinrich (1891). *Elliptische Functionen und algebraische Zahlen*. Braunschweig.

Weber, Heinrich (1897). "Ueber Zahlengruppen von algebraischen Körpern". *Mathematische Annalen* 48, pp. 433–473.

Weber, Heinrich (1908). *Lehrbuch der Algebra*. Zweite Auflage. Dritter Band. Braunschweig.

clique

According to Prell (2012, p. 155), the first sociological studies on cliques date from the 1940s; she cites Warner (1941, p. 792), Luce and Perry (1949, p. 97), and Festinger (1949, p. 156) as studies in which the terminology was used. The latter defines a clique as "more than two individuals, all of whom choose each other mutually".[1] Luce and Perry (1949, p. 97) define: "A subset of a group forms a *clique* provided that it consists of three or more members each in the symmetric relation to each other member of the subset, and provided further that there can be found no element outside the subset that is in the symmetric relation to each of the elements of the subset".

In a slightly rephrased form, the term also appears in Harary and Ross (1957, p. 205): "A *clique* [...] is a maximal subgroup of at least three members in which each member is in the relation to every other member". They refer to Festinger as cited above. The term also appears in Harary (1957, p. 906), a paper written in 1955. The condition that a clique induces a complete subgraph is recognizable in all three definitions.

The term *clique graph* appears in the late 1960s; it is found in Hamelink (1968, p. 192), who defined: "A *clique graph* of G is a graph F and a 1-1, onto mapping α from $V(F)$ to the set of cliques of G which preserves incidence".

Note. An overview of definitions of *clique* is found in Harary (1959, p. 391).

---

[1] Festinger (ibid.) calls this "an extreme type of clique".

References

Festinger, Leon (1949). "The Analysis of Sociograms using Matrix Algebra". *Human Relations* 2, pp. 153–158.
Hamelink, Ronald C. (1968). "A Partial Characterization of Clique Graphs". *Journal of Combinatorial Theory* 5, pp. 192–197.
Harary, Frank (1957). "The number of dissimilar supergraphs of a linear graph". *Pacific Journal of Mathematics* 7, pp. 903–911.
Harary, Frank (1959). "Graph Theoretic Methods in the Management Sciences". *Management Science* 5, pp. 387–403.
Harary, Frank and Ian C. Ross (1957). "A Procedure for Clique Detection Using the Group Matrix". *Sociometry* 20, pp. 205–215.
Luce, R. Duncan and Albert D. Perry (1949). "A method of matrix analysis of group structure". *Psychometrika* 14, pp. 95–116.
Prell, Christina (2012). *Social Network Analysis*. London.
Warner, W. Lloyd (1941). "Social Anthropology and the Modern Community". *American Journal of Sociology* 46, pp. 785–796.

closed (algebra)

In algebra, a closed operation is one which yields a result that belongs to the same set. This notion was already used by Galois: "if in such a group one has substitutions S and T, one is sure to have the substitution ST" (Galois 2011, p. 114).[1] Cayley (1854, p. 41) expressed the notion in similar terms in his definition of a group: "A set of symbols $1, \alpha, \beta, \ldots$, all of them different, and such that the product of any two of them (no matter in what order) or the product of any one of them into itself, belongs to the set, is said to be a *group*".

From around 1870, in German mathematical texts the words *abgeschlossen* and *geschlossen*, both of which mean 'closed', begin to be used by several mathematicians, for example by Klein and Lie (1871, p. 56): "*Two arbitrary transformations of the system produce, when applied one after another, independently of their order, the same new transformation. This new transformation is itself a transformation of the system.* In view of the first property the transformations of the system are called *permutable*, with reference to the second the system is called *closed*".[2] In a footnote they add: "The term 'a closed system of transformations' therefore corresponds exactly to what is called 'a group of substitutions' in the theory of substitutions".[3]

In the same year, Dedekind (1871, p. 424) used the term in connection with fields: "By a *field* we will mean every infinite system of real or complex numbers so closed

[1] "si dans un pareil groupe on a les substitutions S et T, on est sûr d'avoir la substitution ST". Translation from Galois (2011, p. 115). [2] "*Zwei beliebige Transformationen des Systems geben, hintereinander angewandt, unabhängig von ihrer Reihenfolge, dieselbe neue Transformation. Diese neue Transformation ist selbst eine Transformation des Systems.* Mit Rücksicht auf die erste Eigenschaft heißen die Transformationen des Systems *vertauschbar*, mit Bezug auf die zweite heißt das System *geschlossen*". – On this passage, see Wußing (1984, p. 185) and Hawkins (2000, pp. 17f.). [3] "Der Ausdruck 'ein geschlossenes System von Transformationen' entspricht also ganz dem, was man in der Theorie der Substitutionen als 'eine Gruppe von Substitutionen' zu bezeichnen pflegt".

in itself and perfect that addition, subtraction, multiplication, and division of any two of these numbers again yields a number of the system".[4]

The English term is a translation of the German; it is found from around 1880 on, e. g. in Story (1880, p. 367).

Note. For the history and terminology, see Becker (2005, pp. 181–184).

References

Becker, Holger (2005). "Semantische und lexikalische Aspekte der mathematischen Fachsprache des 19. Jahrhunderts". PhD thesis. Carl von Ossietzky Universität Oldenburg. URL: `http://oops.uni-oldenburg.de/59/`.

Cayley, Arthur (1854). "On the theory of groups, as depending on the symbolic equation $\theta^n = 1$". *The London, Edinburgh, and Dublin Philosophical Magazine and Journal of Science* 7, pp. 40–47, 408–409.

Dedekind, Richard (1871). "Ueber die Komposition der binären quadratischen Formen". In: *Vorlesungen über Zahlentheorie von P.G. Lejeune Dirichlet.* Herausgegeben und mit Zusätzen versehen von R. Dedekind. Zweite, umgearbeitete und vermehrte Auflage. Braunschweig, pp. 380–497.

Galois, Évariste (2011). "The First Memoir". In: *The Mathematical Writings of Évariste Galois.* Ed. by Peter M. Neumann. Zürich, pp. 105–168.

Hawkins, Thomas (2000). *Emergence of the Theory of Lie Groups.* An Essay in the History of Mathematics 1869-1926. New York.

Klein, Felix and Sophus Lie (1871). "Ueber diejenigen ebenen Curven, welche durch ein geschlossenes System von einfach unendlich vielen vertauschbaren Transformationen in sich übergehen". *Mathematische Annalen* 4, pp. 50–84.

Kleiner, Israel (2007). *A History of Abstract Algebra.* Basel.

Story, William E. (1880). "On the Theory of Rational Derivation on a Cubic Curve". *American Journal of Mathematics* 3, pp. 356–387.

Wußing, Hans (1984). *The Genesis of the Abstract Group Concept.* A Contribution to the History of the Origin of Abstract Group Theory. Tr. Abe Shenitzer. Cambridge, Mass.

closed set

The term *closed set* was coined by Cantor. He used German *abgeschlossene Menge*, of which the English term is a literal translation, and gave the following definition: "If a point set P is such that its derivative $P^{(1)}$ is contained in P as a divisor [...] we shall call P a *closed* set" (Cantor 1884b, p. 470).[1] The *derivative* used by Cantor is the set of the limit points of a set (see *derived* (further uses)), and thus the definition means that a closed set contains all its limit points.

French *ensemble fermé* appeared in Cantor (1884a, p. 388), while the English term was used in Hardy (1901, p. 70): "A set of points is *closed*, if it includes its first derived set; *perfect* if the two are identical". Note, however, that *closed* was already used in a related sense by Harkness and Morley (1898, p. 124): "The system of all the points in the circle (c, R) will be called the *open* region (c, R); the system of points in and on the circle will be called the *closed* region (c, R)".

Note. A detailed account of the history of closed sets and related concepts is found in Moore (2008).

---

[4] "Unter einem *Körper* wollen wir jedes System von unendlich vielen reellen oder komplexen Zahlen verstehen, welches in sich so abgeschlossen und vollständig ist, daß die Addition, Subtraktion, Multiplikation und Division von je zwei Zahlen immer wieder eine Zahl desselben Systems hervorbringt". Translation from Kleiner (2007, p. 66).

[1] "Wenn eine Punktmenge P so beschaffen ist, dass ihre Ableitung $P^{(1)}$ in ihr als Divisor enthalten ist [...] so wollen wir P eine *abgeschlossene* Menge nennen".

References

Cantor, Georg (1884a). "De la puissance des ensembles parfaits de points". *Acta Mathematica* 4, pp. 381–392.
Cantor, Georg (1884b). "Ueber unendliche, lineare Punktmannichfaltigkeiten. 6". *Mathematische Annalen* 23, pp. 453–488.
Hardy, Godfrey Harold (1901). "On differentiation and integration under the integral sign". *The Quarterly Journal of Pure and Applied Mathematics* 32, pp. 66–140.
Harkness, James and Frank Morley (1898). *Introduction to the Theory of Analytic Functions*. London.
Moore, Gregory H. (2008). "The emergence of open sets, closed sets, and limit points in analysis and topology". *Historia Mathematica* 35, pp. 220–241.

closure

The closure of a set first appears[1] in Cantor (1883), who does not use a special term for it. Such a term *is* found in Pierpont (1905, p. 423): "If $\mathfrak{A}$ does not contain all of its limiting points, these may be adjoined to $\mathfrak{A}$. The resulting aggregate is called the *completed aggregate*".[2] Note that *aggregate* is here used in the sense of 'set'.

Carathéodory (1918, p. 57) then coined the German term *abgeschlossene Hülle*, lit. 'closed hull', which is still in use.[3] The name was soon adopted (and also attributed to Carathéodory) by Bohr (1919, p. 146) and Hahn (1921, p. 70), for example. Fréchet (1921, p. 367) translated the word by French *fermeture*, from *fermer* 'to close'; de Possel (1935-36, pp. 3f.) suggested *adhérence* as an alternative because it seemed to him "more convenient" than the usual *fermeture*. English *closure* appears to have been introduced by Lefschetz (1930, p. 271): "Finally we shall find it convenient to introduce, as now quite customary, the *closure* $\bar{A}$, of a point-set A, or sum of A plus all its limit points".

Note. The literature on the history of the concept is scarce; see Zoretti and Rosenthal (1924, pp. 865f.) for brief historical comments on the terminology.

References

Bohr, Harald (1919). "Zur Theorie der allgemeinen Dirichletschen Reihen". *Mathematische Annalen* 79, pp. 136–156.
Cantor, Georg (1883). "Ueber unendliche, lineare Punktmannichfaltigkeiten. 4". *Mathematische Annalen* 21, pp. 51–58.
Cantor, Georg (1984). *Über unendliche, lineare Punktmannigfaltigkeiten*. Arbeiten zur Mengenlehre aus den Jahren 1872–1884. Herausgegeben und kommentiert von G. Asser. Leipzig.
Carathéodory, Constantin (1918). *Vorlesungen über reelle Funktionen*. Leipzig.
Fréchet, Maurice (1921). "Sur les ensembles abstraits". *Annales scientifiques de l'École Normale Supérieure* 38, pp. 341–388.
Hahn, Hans (1921). *Theorie der reellen Funktionen*. Berlin.
Lefschetz, Solomon (1930). "On Transformations of Closed Sets". *Annals of Mathematics* 31, pp. 271–280.
Pierpont, James (1905). *Lectures on the Theory of Functions of Real Variables*. Volume I. Boston.
de Possel, René (1935-36). "Espaces topologiques". *Séminaire de Mathématiques dit de Julia* 3, exposé 1, pp. 1–18.
Zoretti, Ludovic and Arthur Rosenthal (1924). "Die Punktmengen". *Encyklopädie der mathematischen Wissenschaften mit Einschluss ihrer Anwendungen* II C 9a, pp. 852–1030.

---

[1] See the editorial comments in Cantor (1984, p. 165). [2] The term is also in Pierpont (1905, p. 522). This usage was pointed out by Zoretti and Rosenthal (1924, pp. 865f.). [3] The name may be based on an analogy to *convex hull* (German *konvexe Hülle*), which was introduced in 1916. See *convex*.

clothoid

The clothoid is a curve whose curvature increases linearly with its arc length. Although it was already studied in the 17th century, it did not receive its name until 1886. The term was coined by Cesàro (1886, p. 512): "The clothoid [*clothoïde*] is a plane curve whose curvature varies in proportion to its arc".[4] The name derives from a Greek verb meaning 'to twist by spinning', and was presumably chosen because the curve consists of two symmetric spirals (Loria 1911, p. 71).

The curve has also been named after various mathematicians. Two examples are the terms *Euler spiral* (Euler 1744) and *Cornu spiral* (Cornu 1874).

Note. For the history of the concept, see Loria (1911, pp. 70–73), Teixeira (1909, pp. 102–107) and, more recently, Levien (2008).

References

Cesàro, Ernesto (1886). "Les lignes barycentriques". *Nouvelles annales de mathématiques* 5, pp. 511–520.

Cornu, Alfred (1874). "Études sur la diffraction; méthode géométrique pour la discussion des problèmes de diffraction". *Comptes rendus de l'Académie des sciences* 78, pp. 113–117.

Euler, Leonhard (1744). *Methodus inveniendi lineas curvas maximi minimive proprietate gaudentes, sive Solutio problematis isoperometrici latissimo sensu accepti.* Lausanne and Geneva.

Levien, Raph (2008). *The Euler spiral: a mathematical history.* Tech. rep. EECS Department, University of California, Berkeley. URL: `http://www.eecs.berkeley.edu/Pubs/TechRpts/2008/EECS-2008-111.html` (retrieved Nov. 29, 2017).

Loria, Gino (1911). *Spezielle algebraische und transzendente ebene Kurven.* Theorie und Geschichte. Autorisierte, nach dem italienischen Manuskript bearbeitete deutsche Ausgabe von Fritz Schütte. Zweite Auflage. Zweiter Band. Leipzig.

Teixeira, Francisco Gomes (1909). *Traité des courbes spéciales remarquables plans et gauches.* Tome II. Coïmbra.

cobordant

The term *cobordant* is due to Thom. In his paper from 1954, he set out the foundations of the theory of cobordism, using the following definition: "Two oriented compact manifolds V and V' of the same dimension k are called *cobordant* (notation: $V \simeq V'$), if the manifold $V' \cup (-V)$, which is the disjoint union of V' and V, the latter taken with the opposite orientation, is *null-cobordant*" (Thom 1954, p. 64).[1] The terminology appeared earlier in Thom (1953, p. 1733).

While Dold (1956, p. 25), for example, translated the term by *äquivalent*, *cobordant* was quickly in common in use; at the ICM in 1958, Thom received a Fields Medal for his work.

Note. Remarks on the history of the concept can be found in May (1999).

References

Dold, Albrecht (1956). "Erzeugende der Thomschen Algebra $\mathfrak{N}$". *Mathematische Zeitschrift* 65, pp. 25–35.

May, J.P. (1999). "Stable Algebraic Topology, 1945-1966". In: *History of Topology*. Ed. by Ioan Mackenzie James. Amsterdam, pp. 665–723.

---

[4] "La clothoïde est une ligne plane, dont la courbure varie proportionnelement à l'arc". [1] "Deux variétés compactes V, V', de même dimension k, *orientées*, seront dites *cobordantes* (notation: $V \simeq V'$), si la variété $V' - V$), réunion de la variété V' et de la variété V dont on a renversé l'orientation, est une *variété-bord*". Translation from Thom (2007, p. 182).

Thom, René (1953). "Variétés différentiables cobordantes". *Comptes rendus de l'Académie des sciences* 236, pp. 1733–1735.
Thom, René (1954). "Quelques propriétés globales des variétés différentiables". *Commentarii Mathematici Helvetici* 28, pp. 17–86.
Thom, René (2007). "Some General Properties of Differentiable Manifolds". In: *Topological Library. Part 1: Cobordisms and Their Applications*. Translated by V.O. Manturov. Ed. by S.P. Novikov and I.A. Taimanov. New Jersey, pp. 131–209.

cochleoid

There are three similar names of curves derived from the word *κόχλος* 'shell-fish with a spiral shell, land snail' (LSJ): First, *cochlias* for the cylindrical helix; second, *cochloid* as an early alternative for *conchoid*, and third, *cochleoid* (also *cochlioid*) for a curve defined in polar coordinates by $\rho = \alpha\frac{\sin(\theta)}{\theta}$.

The first term (Greek *κοχλίας*) was mentioned by Proclus, who discusses "the cylindrical helix, which is traced by a point moving uniformly along a straight line that is moving round the surface of a cylinder. This moving point generates a helix any part of which coincides homoeomerously[1] with any other, as Apollonius has shown in his treatise *On the Cochlias*" (Proclus 1873, p. 105).[2] Latin *cochlea* was later used by Wallis (1670, p. 637), for example. He explained it on the basis of its similarity to a shellfish and also gave English *screw* and French *vis* as synonyms. In Wallis (1685, p. 103), he described it as a "spiral about a cylinder".

The second term is found in book 4 of Pappus' *Collection*, where it refers to the conchoid of Nicomedes (Pappus of Alexandria 2010, p. 127). This usage is rare and was soon corrected. See *conchoid*.

The third term first appeared in two related publications from 1884. The Dutch engineer Falkenburg published an article in which he mentions that he "has discovered a line, which shows particularly remarkable properties and is characterized by a snail-like shape. He[3] therefore called it *cochleoid*, from *cochlea* 'snail shell'" (Falkenburg 1884, p. 259).[4] The term is also found in Benthem (1884, p. 76), where also the corresponding Dutch term *slakkenlijn*, lit. 'snail line', appears. Benthem refers to Falkenburg and also says that Falkenburg pointed out the curve to him some time ago and "*we* called it snail line [emphasis added]".[5] The term may thus have been a joint creation.

The curve had previously been studied by various mathematicians; John Perks appears to be the first to have mentioned it. See Wallis, Gregory, and Caswell (1699), who reported on Perks' discovery, and the relevant article (Perks 1702).[6]

Note. For the history of the cochleoid, see Teixeira (1909, pp. 96–101), Loria (1911, pp. 26–36), and Woods (1924).

References

Benthem, A. (1884). "De slakkenlijn of cochleoïde". *Nieuw Archief voor Wiskunde* 10, pp. 76–80.

[1] As explained by Knorr (1987, p. 296), *homoeomerous* means that "any part of it [the cochlias] can be superimposed over any part uniformly". [2] Translation from Proclus (1970, p. 85). [3] *He* refers to Falkenburg, who otherwise refers to himself as "the author". [4] "eine Linie entdeckt, welche besonders merkwürdige Eigenschaften zeigt, und sich durch eine schneckenartige Gestalt characterisiert. Er hat sie daher *Cochleoïde* genannt, von *cochlea* = Schneckenhaus". [5] "die wij Slakkenlijn [...] noemden". [6] Perks' article was published anonymously; Wölffing (1900, p. 98) showed that Perks was the author. On Perks, see also Pedersen (1963).

Falkenburg, C. (1884). "Die Cochleoide". *Archiv der Mathematik und Physik* 70, pp. 259–267.
Knorr, Wilbur R. (1987). *The Ancient Tradition of Geometric Problems*. Boston, Mass.
Loria, Gino (1911). *Spezielle algebraische und transzendente ebene Kurven*. Theorie und Geschichte. Autorisierte, nach dem italienischen Manuskript bearbeitete deutsche Ausgabe von Fritz Schütte. Zweite Auflage. Zweiter Band. Leipzig.
Pappus of Alexandria (2010). *Book 4 of the* Collection. Edited With Translation and Commentary by Heike Sefrin-Weis. London.
Pedersen, Olaf (1963). "Master John Perks and his Mechanical Curves". *Centaurus* 8, pp. 1–18.
Perks, John (1702). "The construction of a Quadratrix to the Circle, being the Curv described by its Equable Evolution". *Philosophical Transactions of the Royal Society of London* 22, pp. 446–446.
Proclus (1873). *In primum Euclidis elementorum librum commentarii*. Ed. by Gottfried Friedlein. Leipzig.
Proclus (1970). *Commentary on the First Book of Euclid's Elements*. Translated with Introduction and Notes by Glenn R. Morrow. Princeton.
Teixeira, Francisco Gomes (1909). *Traité des courbes spéciales remarquables plans et gauches*. Tome II. Coïmbra.
Wallis, John (1670). *Mechanica*. Sive, De Motu, Tractatus Geometricus. London.
Wallis, John (1685). *A Treatise of Angular Sections*. In: *A Treatise of Algebra, both Historical and Practical*. London, 1–68 (appendix).
Wallis, John, David Gregory, and John Caswell (1699). "A Letter of Dr. Wallis to Dr. Sloan, concerning the Quadrature of the Parts of the Lunula of Hippocrates Chius, performed by Mr. John Perks; with the further Improvements of the same, by Dr. David Gregory, and Mr. John Caswell". *Philosophical Transactions of the Royal Society of London* 21, pp. 411–418.
Wölffing, Ernst (1900). "Bibliografia della cocleoide". *Bollettino di bibliografia e storia delle scienze matematiche* 3, pp. 97–99.
Woods, Roscoe (1924). "The Cochlioid". *The American Mathematical Monthly* 31, pp. 222–227.

codomain

The term *codomain* designates the target set of a function. Originally, however, it was used in a more general sense (with respect to relations), as can be illustrated by Keyser (1909, p. 956), who appears to have been the first who used the English term: "The writing xRy means to assert that x has the relation R to y, so that a relation has *sense* or direction; the symbols ρ and $\breve{\rho}$, called respectively the *domain* and the *codomain* of R, denote respectively the classes of terms that may stand before R and after R".

The term is not due to Keyser, though; French *domaine* and *codomaine* were used in this sense by Couturat (1905, pp. 28f.); Russell (1903, p. 97), to whom the terminology goes back, used *domain* and *converse domain*.[1] To explain the term, Couturat (1905, p. 29) explicitly points out that the codomain is the domain of the converse relation.

Banach (1922, p. 133) used the French terms *domaine* and *contre-domaine* when he defined an operator (*opération*): "An operation [operator] is a univocal relation yRx, that is, yRx and zRx entail $y = z$ for all x, y, z. Each relation involves a domain (the reserve of the y) and a *counter-domain* (the reserve of the x) or *field*".[2] The term *contre-domaine* never reached widespread use with reference to functions, although

[1] In an earlier publication, Russell (1900-01, p. 116) uses *domain* bot not yet *converse domain*.
[2] "L'*opération* c'est une relation univoque yRx c'est-à-dire, telle que yRx et zRx entraîne $y = z$ pout tout x, y, z. Chaque relation comporte un *domain* (c'est la réserve des y) et un *contre-domaine* (la réserve des x) ou *champ*".

it reappeared in Banach's influential *Théorie des opérations linéaires* (Banach 1932, p. 16).

A clear distinction between the *range* (image set) and the *codomain* (target set) was suggested by Birkhoff and Mac Lane (1953, p. 119). For a transformation $\phi : S \to T$, they defined: "The set S is called the *domain* of ϕ, and T its *codomain*. Note that the set $S\phi$ of all images under ϕ of elements in S, which is usually called the *range* of $S\phi$, may comprise only part of the codomain T".[1]

See also *range* and *image*.

References

Banach, Stefan (1922). "Sur les opérations dans les ensembles abstraits et leur application aux équations intégrales". *Fundamenta Mathematicae* 3, pp. 133–181.
Banach, Stefan (1932). *Théorie des opérations linéaires*. Warszawa.
Birkhoff, Garrett and Saunders Mac Lane (1953). *A Survey of Modern Algebra*. Revised edition. New York.
Couturat, Louis (1905). *Les principes des mathématiques*. Paris.
Keyser, Cassius Jackson (1909). "The Thesis of Modern Logistic". *Science* 30, pp. 949–963.
Russell, Bertrand (1903). *The Principles of Mathematics*. Vol. I. Cambridge.
Russell, Bertrand (1900-01). "Sur la logique des relations avec des applications à la théorie des séries". *Rivista di matematica* 7, pp. 115–148.

codimension

The French term *codimension* was introduced by Bourbaki (1947, p. 37): "A subspace V of a vector space E is said to be *of finite codimension* in E if the quotient space E/V is of finite dimension. The dimension of E/V is then called the *codimension* of V in E".[2]

The term was in common use soon afterwards; for example, the English term was used by Artin (1953, p. 476), the German one by Schöneborn (1952, p. 203). It was also soon applied in other contexts; for example, Cohen (1952, p. 60) defined the term for compact spaces.

References

Artin, Emil (1953). "*Éléments de mathématique*, by N. Bourbaki [review]". *Bulletin of the American Mathematical Society* 59, pp. 474–479.
Bourbaki, Nicolas (1947). *Algèbre*. Chapitre II. Algèbre linéaire. Paris.
Cohen, Haskell (1952). "Codimension. I". *Bulletin of the American Mathematical Society* 58, p. 60.
Schöneborn, Heinz (1952). "Über Linearformenmoduln unendlichen Ranges. II". *Journal für die reine und angewandte Mathematik* 189, pp. 193–203.

coefficient

Literally, *coefficient* means 'cooperating to produce a result' (OED). As a mathematical term, Latin *coefficiens* was first used by Viète. In print, it appears in his *Introduction to the Analytic Art* (Viète 1591, fol. 5^r, 8^v). According to Oaks (2018, p. 276), this is the first time Viète uses the term in connection with algebra.

---

[1] *codomain* was not used in the first edition of the *Survey*. [2] "On dit qu'un sous-espace V d'un espace vectoriel E est *de codimension finie* dans E si l'espace quotient E/V est de dimension finie. La dimension de E/V est alors appelée la *codimension* de V dans E" (emphasis added).

The French term was used in Viète (1630, p. 9) and Hume (1636, p. 213).[3] It was also adopted by Oughtred (1631, p. 50); the English term is in Oughtred (1647, p. 121). Rahn (1659, p. 1) translated the term literally into German *Mitwirker*[4], where *mit* 'with' corresponds to the prefix *co-*, and *wirken* means 'to have an effect'.

See also *binomial coefficient*.

Note. For brief discussions of the terminology, see Tropfke (1937, pp. 29f.) and Smith (1958, p. 393).[1]

References

Girard, Albert (1629). *Invention nouvelle en algèbre*. Amsterdam.

Hume, James (1636). *Algebre de Viete, d'une methode nouvelle, claire, et facile*. Paris.

Oaks, Jeffrey A. (2018). "François Viète's revolution in algebra". *Archive for History of Exact Sciences* 72, pp. 245–302.

Oughtred, William (1631). *Arithmeticae in numeris et speciebus institutio*. London.

Oughtred, William (1647). *The Key of the Mathematicks*. London.

Rahn, Johann Heinrich (1659). *Teutsche Algebra*. Zürich.

Smith, David Eugene (1958). *History of Mathematics*. Volume II. Special Topics of Elementary Mathematics. New York.

Tropfke, Johannes (1937). *Geschichte der Elementar-Mathematik*. Dritter Band. Proportionen, Gleichungen. Dritte, verbesserte und vermehrte Auflage. Berlin and Leipzig.

Viète, François (1591). *In artem analyticam isagoge*. Tours.

Viète, François (1630). *L'algèbre nouvelle*. Traduicte en françois par A. Vasset. Paris.

cofinal **and** ***coinitial***

These terms are due to Hausdorff, who introduced them in 1906. For an ordered set M, he defined the German terms *confinal* and *coinitial*: "M is said to be *cofinal* with its subset A if no element of M follows A" and "M is said to be *coinitial* with its subset A if no element of M precedes A" (Hausdorff 1906, p. 124).[2] As noted by the editors of Hausdorff's works (Hausdorff 2002, pp. 593f.), the concept appears at the same time in Hessenberg (1906, p. 559), where the term *kernel* (*Kern*) is used.

Hausdorff used the terms on various occasions, including his classic work on set theory (Hausdorff 1914, p. 86). The English terms appears in a review of that book (Blumberg 1920, p. 121).

Note. For the origin of the concept, see the editorial comments in Hausdorff (2002, pp. 593f.).

References

Blumberg, Henry (1920). "Hausdorff's Grundzüge der Mengenlehre [review]". *Bulletin of the American Mathematical Society* 27, pp. 116–129.

Hausdorff, Felix (1906). "Untersuchungen über Ordnungstypen". *Berichte über die Verhandlungen der Königlich Sächsischen Gesellschaft der Wissenschaften zu Leipzig (Mathematisch-Physische Klasse)* 58, pp. 106–169.

Hausdorff, Felix (1914). *Grundzüge der Mengenlehre*. Leipzig.

Hausdorff, Felix (2002). *Gesammelte Werke*. Band II. Grundzüge der Mengenlehre. Ed. by Egbert Brieskorn et al. Berlin and Heidelberg.

---

[3] Girard (1629, fol. A^{v}) used *coefficient* for the multiplier of a product, *efficient* the multiplicand.

[4] He wrote *Mittwürker*. [1] Note, however, that Tropfke cites Viète's *Ad logisticem speciosam notae priores* as the first use but mistakenly claims this book was first published in 1591, whereas in fact it was not published until 1631. [2] "M heißt mit A *confinal*, wenn kein Element von M auf A folgt [...] M heißt mit A *coinitial*, wenn kein Element von M dem A vorangeht". Translation from Plotkin (2005, p. 58).

Hessenberg, Gerhard (1906). *Grundbegriffe der Mengenlehre*. Göttingen.
Plotkin, Jacob M., ed. (2005). *Hausdorff on Ordered Sets*. Providence, R.I.

cogredient and *contragredient*

The terms *cogredient* and *contragredient* were originally used by Sylvester in invariant theory. Sylvester (1853, p. 543) defined: "A system of variables is cogredient to another system when it is subject to undergo simultaneously therewith linear substitutions of a like kind, and contragredient when it is subject to undergo linear substitutions simultaneously therewith but of a contrary kind". He first suggested these terms in Sylvester (1852, p. 53). The terminology was adopted by Salmon (1859, p. 56), for example, who gave the following definition: "Two sets of variables $x, y, z; x', y', z'$ are said to be *cogredient* when it is understood that they are always to be transformed by the same linear substitutions, that is to say, that if we substitute for x, $\alpha_1 X + \beta_1 Y + \gamma_1 Z$, we are at the same time to substitute for x', $\alpha_1 X' + \beta_1 Y' + \gamma_1 Z'$". The corresponding definition for *contragredient* is found in Salmon (1859, p. 64). Sylvester never explained the terms, but *cogredient* (which contains the prefix *co-* 'together' and the word *gradi* 'to step') expresses just what it defines, and *contragredient* is a natural counterpart.

The terms were also used in connection with matrices and forms; they were introduced by Frobenius (1878, p. 19). Frobenius proceeded from an equation $PAQ = B$ with $\det(P)$ and $\det(Q)$ both $\neq 0$, and called P and Q "*substitutions* by which A goes over into B". These "substitutions" were called *cogredient* if $P = Q'$ (Q' being the transpose of Q) and *contragredient* if $P = Q^{-1}$. Cogredient matrices were later called *congruent*, for example by MacDuffee (1933, p. 51).

The terms were also applied to automorphisms; see *inner automorphism*.

References

Frobenius, Georg (1878). "Ueber lineare Substitutionen und bilineare Formen". *Journal für die reine und angewandte Mathematik* 84, pp. 1–63.
MacDuffee, Cyrus Colton (1933). *The Theory of Matrices*. Berlin.
Salmon, George (1859). *Lessons Introductory to the Modern Higher Algebra*. Dublin.
Sylvester, James Joseph (1852). "On the principles of the calculus of forms". *The Cambridge and Dublin Mathematical Journal* 7, pp. 52–97, 179–214.
Sylvester, James Joseph (1853). "On a Theory of the Syzygetic relations of two rational integral functions, comprising an application to the Theory of Sturm's Functions, and that of the greatest Algebraical Common Measure". *Philosophical Transactions of the Royal Society of London* 143, pp. 407–548.

cohomology and related "*co*-terms"

As remarked by Whitney (1988, p. 109), the Moscow 1935 conference on topology "marked the true birth of cohomology theory". In this context, the relevant talks were given by Alexander and Kolmogorov (Alexander 1936, Kolmogorov 1936).[1]

The "*co*-terminology" is largely due to Whitney. It appears in print in 1937, when Whitney (1937c, p. 788) defined *coboundary*, *cocycle*, *cohomologous* and *cohomology group*:

[1] Alexander (1936) just contains the title of the talk. For a discussion of the conference and these and further relevant papers by Alexander and Kolmogorov, see Massey (1999, pp. 583–587).

> Let ${}^{r}\partial_{ij}$ be the incidence number of the cells σ_i^r and σ_j^{r-1}. Define the *boundary* and *coboundary* of the chain $A^r = \sum \alpha_i \sigma_i^r$ by $\partial A^r = \sum^r \partial_i^j \alpha_i \sigma_j^{r-1}$, $\delta A^r = \sum^{r+1} \partial_j^i \alpha_i \sigma_j^{r+1}$. Then A^r is a *cycle* or *cocycle* if $\partial A^r = 0$ or $\delta A^r = 0$; A^r is *homologous* to B^r, $A^r \sim B^r$, if $A^r - B^r$ is a boundary; similarly, for *cohomologous*, for which we use the symbol $\backsim$. If we identify homologous r-cycles, we obtain the rth *homology group*; similarly for the rth *cohomology group*.[2]

The term *cochain* is not found in these papers, since "he [Whitney] was dealing only with finite complexes, [and] identified chains and cochains" (Massey 1999, p. 588). It appears that *cochain* was first used by Tucker (1939, p. 646).

At the same time, Whitney (1937a, p. 35, fn. 1) remarked on the terminology:

> The relation of "coboundary" is becoming of increasing importance. Other terms have been used: dual boundary, upper boundary, inverse boundary, boundary in the dual subdivision, derived (of a function). The present term (accepted by E. Čech) offers definite advantages. In differential geometry, the (exact alternating) covariant tensors and the contravariant differentials play the same rôle as cocycles and (contra)cycles in the combinatorial theory; in fact, they may be obtained directly by a passage to limit. Moreover, the prefix co is very convenient to handle.

Note. Detailed studies of the history of cohomology were given by Massey (1999) and Basbois (2009); see also Mac Lane (1978) and Steenrod (1952, p. 48).

References

Alexander, James W. (1936). "On the ring of a complex and the combinatory theory of integration". *Matematicheskii Sbornik* 43, p. 617.

Basbois, Nicolas (2009). "La naissance de la cohomologie des groupes". PhD thesis. Université Nice Sophia Antipolis.

Kolmogorov, Andrey Nikolaevich (1936). "Homologiering des Komplexes und des lokal-bikompakten Raumes". *Matematicheskii Sbornik* 43, pp. 701–706.

Mac Lane, Saunders (1978). "Origins of the cohomology of groups". *L'Enseignement mathématique* 24, pp. 1–29.

Massey, William S. (1999). "Origins of the Cohomology of Groups". In: *History of Topology*. Ed. by Ioan Mackenzie James. Amsterdam, pp. 579–603.

Steenrod, Norman (1952). *Foundations of Algebraic Topology*. Princeton.

Tucker, Albert William (1939). "The Algebraic Structure of Complexes". *Proceedings of the National Academy of Sciences* 25, pp. 643–647.

Whitney, Hassler (1937a). "On matrices of integers and combinatorial topology". *Duke Mathematical Journal* 3, pp. 35–45.

Whitney, Hassler (1937b). "On Products in a Complex". *Proceedings of the National Academy of Sciences* 23, pp. 285–291.

Whitney, Hassler (1937c). "Topological properties of differentiable manifolds". *Bulletin of the American Mathematical Society* 43, pp. 785–805.

Whitney, Hassler (1938). "On Products in a Complex". *Annals of Mathematics* 39, pp. 397–432.

Whitney, Hassler (1988). "Moscow 1935: Topology Moving Toward America". In: *A Century of Mathematics in America*. Part I. Ed. by Peter Duren. Providence, R.I., pp. 97–117.

cokernel

The French term *conoyau* was introduced by Cartan and Serre (1952, p. 394): "The *cokernel* of a homomorphism $A \to B$ is the quotient of B by the image of A".[1] It

---

[2] The terms are also found in Whitney (1937b); the full version of this short paper is Whitney (1938). [1] "Le *conoyau* d'un homomorphisme $A \to B$ est le quotient de B par l'image de A".

also appears in Cartan (1953, p. 44), for example, and was also soon translated by English *cokernel*, for example by Buchsbaum (1955, p. 1). Buchsbaum also points out that the concept is found in Mac Lane (1950, p. 490), who remarked that "[t]hese two statements can be so reformulated as to be strictly dual in our sense (the notion of 'kernel' must be replaced by a suitable notion of 'coimage,' [...])".

References

Buchsbaum, David A. (1955). "Exact Categories and Duality". *Transactions of the American Mathematical Society* 80, pp. 1–34.

Cartan, Henri (1953). "Variétés analytiques complexes et cohomologie". In: *Colloque sur les fonctions de plusieurs variables tenu à Bruxelles du 11 au 14 mars 1953*. Liège, pp. 41–55.

Cartan, Henri and Jean-Pierre Serre (1952). "Espaces fibrés et groupes d'homotopie. II. Applications". *Comptes rendus de l'Académie des sciences* 234, pp. 393–395.

Mac Lane, Saunders (1950). "Duality for groups". *Bulletin of the American Mathematical Society* 56, pp. 485–516.

collinear and *collineation*

The term *collineation* was first introduced in Möbius' *Barycentric Calculus*, in which Möbius discussed general geometric transformations under the name *Verwandtschaft*, lit. 'relationship', and described as the essential property of a collineation that to three points on a line in one figure correspond three points on a line in another figure.[1] This also justifies the name, since 'lying on one line' is exactly what *collinear* means (Möbius 1827, p. xii).

Möbius defines collineations in these terms:

> Thus, the essence of this new relationship consists in that, with two plane or solid spaces, to each point of the first space corresponds a point in the other space in such a way that, when in the first space an arbitrary line is drawn, from all points that are hit by this line (*collineantur*), the corresponding points in the other space can also be joined by a line. For this reason, this relationship has been called the relationship of *collineation*. Figures between which it holds are called *collinearly related*, or simply *collinear figures* (Möbius 1827, p. 302).[2]

In the preface to his book, Möbius (1827, p. xii) explicitly credited the philologist Benjamin Gotthold Weiske for the term, who advised Möbius on terminological issues.

The corresponding French term is found in Cauchy (1828, p. 78) and Magnus (1832, pp. 56f.); the English term is in Rutherford (1847, p. 123).

Magnus (1832, p. 56) also introduced the terms *centre of collineation* (*centre de collinéation*) and *axis of collineation* (*axe de collinéation*); the corresponding German terms are found in Magnus (1833, pp. 43f.). Rutherford (1847, pp. 126f.) translated these as *collinear centre* and *collinear axis*; Smith (1869, p. 246) has

---

[1] Collineations were earlier considered by Möbius in a little known text (Möbius 1823, p. 61); he there called them *Verwandtschaft nach der geraden Linie*, which may be translated by *relationship by the straight line*. [2] "Das Wesen dieser neuen Verwandtschaft besteht also darin, dass bei zwei ebenen oder körperlichen Räumen, jedem Punkte des einen Raums ein Punkt in dem anderen Raume dergestalt entspricht, dass, wenn man in dem einen Raume eine beliebige Gerade zieht, von allen Punkten, welche von dieser Geraden getroffen werden (*collineantur*), die ensprechenden Punkte in dem anderen Raume gleichfalls durch eine Gerade verbunden werden können. Es ist deshalb diese Verwandtschaft die Verwandtschaft der *Collineation* genannt worden. Figuren, zwischen denen sie statt findet, heissen *collinear verwandt*, oder schlechthin *collineare Figuren*".

centre of collineation. Magnus (1837, p. vi) is also responsible for the term *central collineation*.

See also *homography*, a term widely used synonymously.

Note. For the introduction of the term and Möbius' work, see Wußing (1984, pp. 35–42) and Rosenfeld (1988, pp. 149–151).

References

Cauchy, Augustin-Louis (1828). "Der barycentrische Calcul [review]". *Bulletin des sciences mathématiques, astronomiques, physiques et chimiques* 9, pp. 77–80.

Magnus, Ludwig Immanuel (1832). "Nouvelle méthode pour decouvrir des théorèmes de géometrie". *Journal für die reine und angewandte Mathematik* 8, pp. 51–63.

Magnus, Ludwig Immanuel (1833). *Sammlung von Aufgaben und Lehrsätzen aus der analytischen Geometrie*. Berlin.

Magnus, Ludwig Immanuel (1837). *Sammlung von Aufgaben und Lehrsätzen aus der analytischen Geometrie des Raumes*. Erste Abtheilung. Berlin.

Möbius, August Ferdinand (1823). *Beobachtungen auf der Universitäts-Sternwarte zu Leipzig*. Leipzig.

Möbius, August Ferdinand (1827). *Der barycentrische Calcul*. Leipzig.

Rosenfeld, Boris A. (1988). *A History of Non-Euclidean Geometry*. Evolution of the Concept of a Geometric Space. Translated by Abe Shenitzer. New York.

Rutherford, William (1847). "On the theory of collineation". *The Mathematician* 2, pp. 123–128, 189–198.

Smith, H.J.S. (1869). "On the Focal Properties of Homographic Figures". *Proceedings of the London Mathematical Society* 2, pp. 196–248.

Wußing, Hans (1984). *The Genesis of the Abstract Group Concept*. A Contribution to the History of the Origin of Abstract Group Theory. Tr. Abe Shenitzer. Cambridge, Mass.

colourable

The famous Four-Colour Conjecture, which states that any map in the plane is 4-colourable, was first mentioned in print by Augustus De Morgan:

> When a person colours a map, say the countries in a kingdom, it is clear he must have so many different colours that every pair of countries which have some common boundary line–not a mere meeting of two corners–must have different colours. Now, it must have been always known to map-colourers that *four* different colours are enough (De Morgan 1860, p. 502).[1]

The adjective *colourable* in this context does not seem to have come up until the 1920s. It is found in Franklin (1922, p. 226), who used it thus: "An example is also given of a map of 42 regions which, although colorable in four colors, satisfies all the conditions which have been derived for an irreducible map".

Note. For this history of the problem, see Soifer (2009, pp. 147–162), Biggs, Lloyd, and R. J. Wilson (1998, ch. 7), and Mayer (1982).

References

Biggs, Norman L., E. Keith Lloyd, and Robin J. Wilson (1998). *Graph Theory, 1736-1936*. Oxford.

Franklin, Philip (1922). "The Four Color Problem". *American Journal of Mathematics* 44, pp. 225–236.

---

[1] This passage was pointed out by J. Wilson (1976). It is now known that De Morgan considered this conjecture as early as 1852. For discussion, see Soifer (2009, pp. 147–162), where also De Morgan's letter to Hamilton from October 23, 1852, is reproduced. The letter contains the first known statement of the conjecture.

Mayer, Jean (1982). "Le théorème des quatre couleurs: notice historique et aperçu technique". *Cahiers du séminaire d'histoire des mathématiques* 3, pp. 43–62.
De Morgan, Augustus (1860). "The Philosophy of Discovery [review]". *The Athenaeum* 1694, pp. 501–503.
Soifer, Alexander (2009). *The Mathematical Coloring Book.* Mathematics of Coloring and the Colorful Life of Its Creators. New York.
Wilson, John (1976). "New light on the origin of the four-color conjecture". *Historia Mathematica* 3, pp. 329–330.

combination

In his famous *Ars conjectandi*, Bernoulli (1713, p. 82) gave the following definition of *combinatio*: "*Combinations* of things are collections, such that from a given multitude of things some are removed and joined together, no attention being paid to their order or position".[1] In addition to defining the term, Bernoulli also went on to discuss some alternatives:

> Some call these collections *combinations*, *com3nations*, *com4nations*, etc., all of which are usually described by the single word *combinations*, even though this word in its stricter sense seems to connote only those collections that join things together two at a time. For this reason, others prefer to use the more general terms *complications* [*complicationes*] or *complexions* [*complexiones*]. Others more appropriately call them *choices* [*electiones*], so that they may also understand the selections of things in which single things are taken by themselves or even in which nothing at all is taken (Bernoulli 1713, p. 82).[2]

The formulation "in its stricter sense" points to the etymology of the term, a point which was explained by Wallis (1685, p. 113):

> If, by *Combination*, we understand the taking of Two or more (but not of One, or None;) then, out of the Number of cases before found, we must abate so many as is the Number of things exposed, and one more. For, of those, so many as is the Number of things exposed, answers to the cases of 1. And one more, answers to the case of taking None: But all the rest are Combinations in that sense. For though *Combination* (as coming from *Bini*) in its proper signification extend onely to the taking of *Couples*, (or Two's;) yet in common acceptation the word is now used of greater Numbers. And, in *English* also, we scruple not to say, that Three, or Four, (or more than so) are *Coupled* together; that is connected.

The extended sense of *combination* for any number of objects was not new when Wallis wrote this; Mersenne (1635, p. 118), for example, had already made a similar comment. The word was used in this extended sense by Cardano (1539, ch. 51) and Buckley (1567, fol. B ij^{v}); according to Haller (1999), Cardano may be responsible for this change.[3]

---

[1] "Combinationes rerum sunt Conjunctiones, juxta quas exta datâ rerum multitudine nonnullae eximuntur, interque se conjunguntur nullo ordinis situsve ipsarum respectu habito". Translation from Bernoulli (2006, p. 201). [2] "Conjunctiones ipsas nonnulli vocant Combinationes, Con3nationes, Con4nationes, &c. quas omnes vulgò unâ voce *Combinationum* complecti solent, tametsi haec vox strictiori significatu propriè non nisi illas conjunctiones indigitare videatur, quibus res binae invicem junguntur. Quamobrem alii generaliori voce *Complicationum* vel *Complexionum* uti malunt: alii magis appositè *Electiones* vocant, ut & illae subintelligi possint rerum acceptiones, quibus res singulæ seorsim sumuntur, aut quibus etiam nulla planè sumitur". Translation from Bernoulli (2006, p. 201). [3] Knobloch (2013, p. 127) additionally remarks: "It is worth noting that, for Cardano, a combination involved at least two elements: for him, selections of one or no elements were not combinations".

The alternatives Bernoulli mentioned are indeed found in various earlier texts.

For example, *electio* was used by van Schooten (1657, p. 373); Strode (1678, p. 1) attributed it to him: "By Election of Quantities, I mean (as *Francis Schoten* in his *Miscellanies*) any number of Quantities being given, how many several ways I may take them without having respect to their places; as a, b, c, may be taken seven ways; as, a, b, c, ab, ac, bc, and abc".[4]

The word *complexio* (lit. 'folding together') was used by Leibniz (1666, p. 4); for k-combinations, he wrote *com2natio* (*combinatio*), *con3natio* (*conternatio*), *con4natio* (*conquaternatio*), etc.[1] As noted by Knobloch (1974, p. 412), *conternatio*, *conquaternatio*, etc. are already found in Mersenne (1635, p. 135).

complicatio essentially has the same literal meaning as *complexio*, and may for that reason have been listed by Bernoulli. Harriot used the word in notes (Maclean 2005, p. 77), but does not appear to have used the word in print.

Another term, not mentioned[2] by Bernoulli, is *conjunctio*. It was used by Clavius (1585, pp. 34f.) and, following him, Mersenne (1625, e. g., p. 535) and Hérigone (1634, p. 119). Cardano (1570, p. 185) preferred *conjugatio*.

In conclusion, in the 17th century, variation did not only exist in the sense that several words were used for the same concept, but also in the sense that words like *combination* were used in different senses.[3] After the publication of Bernoulli's influential *Ars conjectandi*, this variation declined.

Note. For overviews of the history of the concept, see Knobloch (1973, pp. 1–23) and Tropfke (1924, pp. 63–70).

References

Bernoulli, Jakob (1713). *Ars coniectandi*. Basel.

Bernoulli, Jakob (2006). *The Art of Conjecturing, together with Letter to a Friend on Sets in Court Tennis*. Translated with an introduction and notes by Edith Dudley Sylla. Baltimore.

Buckley, William (1567). *Arithmetica memorativa*. London.

Cardano, Gerolamo (1539). *Practica arithmetice*. Milano.

Cardano, Gerolamo (1570). *Opus novum de proportionibus numerorum, ponderum, sonorum, aliarumque rerum mensurandum*. Basel.

Clavius, Christopher (1585). *In Sphaeram Ioannis de Sacro Bosco commentarius*. Roma.

Haller, Rudolf (1999). "Combinatio: nicht Buckley, sondern Cardano - und die $(2^n - 1)$-Regel". *Praxis der Mathematik* 41, pp. 59–61.

Hérigone, Pierre (1634). *Cursus mathematicus*. Tomus secundus. Paris.

Knobloch, Eberhard (1973). *Die mathematischen Studien von G.W. Leibniz zur Kombinatorik*. Auf Grund fast ausschließlich handschriftlicher Aufzeichnungen dargelegt und kommentiert. Wiesbaden.

Knobloch, Eberhard (1974). "The Mathematical Studies of G.W. Leibniz on Combinatorics". *Historia Mathematica* 1, pp. 409–430.

Knobloch, Eberhard (2013). "Renaissance combinatorics". In: *Combinatorics*. Ancient and Modern. Ed. by Robin Wilson and John J. Watkins. Oxford, pp. 123–145.

Knuth, Donald E. (2013). "Two thousand years of combinatorics". In: *Combinatorics*. Ancient and Modern. Ed. by Robin Wilson and John J. Watkins. Oxford, pp. 3–37.

Leibniz, Gottfried Wilhelm (1666). *Dissertatio de arte combinatoria*. Leipzig.

---

[4] Brief remarks on van Schooten are found in Knuth (2013, pp. 18f.). [1] For a brief summary of Leibniz' terminology, see Knobloch (1974, p. 412, 1973, esp. pp. 24, 72). [2] But used in the definition of *combinatio*. [3] See, for example, the detailed discussion of the Mersenne's terminology by Morris (2021, pp. 193–195).

Maclean, Ian (2005). "Thomas Harriot on Combinations". *Revue d'histoire des mathématiques* 11, pp. 57–88.
Mersenne, Marin (1625). *La vérité des sciences*. Paris.
Mersenne, Marin (1635). *Harmonicorum libri*. Paris.
Morris, Seán (2021). "Self-Translation of Mathematical Texts in Seventeenth-Century France: The Cases of Pascal, Mersenne and Hérigone". PhD thesis. University of Exeter.
van Schooten, Frans (1657). *Exercitationum Mathematicorum libri quinque*. Liber primus. Leiden.
Strode, Thomas (1678). *A Short Treatise of the Combinations, Elections, Permutations and Composition of Quantities*. London.
Tropfke, Johannes (1924). *Geschichte der Elementar-Mathematik*. Sechster Band. Analysis. Analytische Geometrie. Berlin and Leipzig.
Wallis, John (1685). *A Discourse of Combinations, Alternations, and Aliquot Parts*. In: *A Treatise of Algebra, both Historical and Practical*. London, 107–152 (appendix).

combinatorial topology

Topology was originally considered as combinatorial in character; this view was expressed by Dehn and Heegaard (1907, p. 170), for example, who regarded topology (called by them *analysis situs*) as a "part of combinatorics, characterized by its intuitive meaning".[1] Explicitly, *kombinatorische Analysis situs* was used by Tietze (1908, p. 3).

In the 1920s, Tietze developed a three-fold distinction of topology. The first field was *general topology* (*allgemeine Topologie*), the study of invariant properties under topological mappings between topological spaces (Tietze 1923a, p. 290). The second field was *n-dimensional topology* (*n-dimensionale Topologie*), which deals with n-dimensional manifolds (Tietze 1923b, p. 46). The third field was combinatorial topology.

The distinction was also upheld in the survey article by Tietze and Vietoris, entitled "Relations between different branches of topology" (Tietze and Vietoris 1928). Their article contains various terminological remarks. In particular, they state (p. 208) that the name *combinatorial topology* is most appropriate for the subject matter described by Dehn and Heegaard (1907), while at the same time the scope of the term has recently been extended to cover subject matter which falls into n-dimensional topology (Tietze and Vietoris 1928, p. 208). Tietze and Vietoris do not explicitly give any reference, but they do mention the English term *combinatorial analysis situs*, which suggests that they refer to papers like Alexander (1926).

Combinatorial topology also began to be opposed to set-theoretic topology in the mid 1920s; the corresponding German term was used by Alexandroff (1925, p. 296) and Klein (1926, p. 347) and the English one by Wilder (1932, p. 650): "there is only one branch of mathematics to be known as *analysis situs*, or *topology*, and that what are known as *set-theoretic topology* and *combinatorial topology* are really two methods within topology having the same aim".

See also *algebraic topology*.

References

Alexander, James W. (1926). "Combinatorial analysis situs". *Transactions of the American Mathematical Society* 28, pp. 301–329.

[1] "[...] ist die Analysis situs [...] ein durch seine anschauliche Bedeutung ausgezeichneter Teil der Kombinatorik". Translation from Epple (1995, pp. 392f.).

Alexandroff, Pavel S. (1925). "Zur Begründung der n-dimensionalen mengentheoretischen Topologie". *Mathematische Annalen* 94, pp. 296–308.
Dehn, Max and Poul Heegaard (1907). "Analysis situs". *Encyklopädie der mathematischen Wissenschaften mit Einschluss ihrer Anwendungen* III AB3, pp. 153–220.
Epple, Moritz (1995). "Branch Points of Algebraic Functions and the Beginnings of Modern Knot Theory". *Historia Mathematica* 22, pp. 371–401.
Klein, Felix (1926). *Vorlesungen über höhere Geometrie*. Dritte Auflage. Bearbeitet und herausgegeben von W. Blaschke. Berlin.
Tietze, Heinrich (1908). "Über die topologischen Invarianten mehrdimensionaler Mannigfaltigkeiten". *Monatshefte für Mathematik und Physik* 19, pp. 1–118.
Tietze, Heinrich (1923a). "Beiträge zur allgemeinen Topologie. I. Axiome für verschiedene Fassungen des Umgebungsbegriffs". *Mathematische Annalen* 88, pp. 290–311.
Tietze, Heinrich (1923b). "Über Analysis situs". *Abhandlungen aus dem Mathematischen Seminar der Universität Hamburg* 2, pp. 37–68.
Tietze, Heinrich and Leopold Vietoris (1928). "Beziehungen zwischen den verschiedenen Zweigen der Topologie". *Encyklopädie der mathematischen Wissenschaften mit Einschluss ihrer Anwendungen* III AB 13, pp. 141–237.
Wilder, Raymond Louis (1932). "Point sets in three and higher dimensions and their investigation by means of a unified analysis situs". *Bulletin of the American Mathematical Society* 38, pp. 649–692.

combinatorics

While combinatorial practices have been found in various ancient cultures, the terminology evolved in Western Europe. In the 13th century, the philosopher Ramon Llull used what might be called a combinatorial approach to philosophical and scientific questions. He was most influential, and his book *Ars magna* was, as summarized by Knobloch (2013, pp. 123f.), "a systematic summary of all the branches of knowledge of his time, based on the combinatorial art. In order to know something it was not necessary to analyse the objects by means of experience, but rather to understand the fundamental terms, the so-called 'principles', and to combine them with each other. Thus, combinatorics became the basic tool for exploring all that was known at that time". In the context of Llull's works, the term *ars combinatoria* began to be used. It is found, in particular, in the posthumously edited *Opera* (Llull 1598), where *combinatoria ars* appears in the index. In the 17th century, the term is found in various treatises, notably in Kircher (1669, p. 155), where the combinatorial art is described as consisting of two parts, the study of combinations and the study of permutations.[1] An important usage is Leibniz' *Dissertation on the Art of Combinations* (*Dissertatio de arte combinatoria*) (Leibniz 1666), but, as in the previous examples, he also did not use it in today's meaning: "Leibniz himself spoke of the 'ars combinatoria', a philosophical term to which he attached different meanings in the course of his life, but which always embraced more than what we call combinatorics today and was not even restricted to mathematical problems" (Knobloch 1974, p. 410).[2]

The term was also used by Bernoulli (1713, p. 73), who defined it as follows: "the Art called Combinatorics [*ars combinatoria*] [...] teaches us how to enumerate all

[1] The term is also found in various earlier books by Kircher, for example in Kircher (1645, ad lectorem) and Kircher (1663). [2] Knobloch (ibid.) also points out Llull's influence on Leibniz at that time.

possible ways in which several things can be combined, transposed, or joined with each other".[3]

The word *combinatorics* itself is of German origin; it appears in the early 19th century. At that time, the usual German term was *Combinationslehre* 'doctrine of combinations', used, for example, by Stahl (1800).[4] An early usage of *Combinatorik* is Hindenburg (1803, p. xix), where it is applied to a particular mathematical theory in which combinatorial techniques are used, namely Arbogast's calculus of derivations (Arbogast 1800).[1] Around 1820, *Combinatorik* began to be appear as a synonym of *Combinationslehre*; Egen (1820), for example, used *Combinatorik* in the table of contents but *Combinationslehre* in the actual text (p. 287).

In a review of Netto (1901), Cajori (1902, p. 469) translated German *Combinatorik* by English *combinatoric*, but this usage found no followers; MacMahon's (1915) book, for example, was called *Combinatory Analysis*.[2] It is not until around 1940 that the word *combinatorics* comes up. It was used in lectures given by Levi in February 1940. Levi (1942, p. 1) declared: "Combinatorial Mathematics is the counterpart to Analysis. Whereas Combinatorics deals with distinct objects, Analysis plays in a fluid entity; the Analyst measures, but the Combinator counts". In print, the term is first found (OED) in an article by Levi (1940) entitled "On a method of finite combinatorics which applies to the theory of infinite groups".

Note. For this history of combinatorics, see the various contributions in Wilson and Watkins (2013).

References

[anon.] (1669). "An Accompt of Some Books". *Philosophical Transactions of the Royal Society of London* 4, pp. 1086–1097.

Arbogast, Louis François Antoine (1800). *Du calcul des dérivations*. Strasbourg.

Bernoulli, Jakob (1713). *Ars coniectandi*. Basel.

Bernoulli, Jakob (2006). *The Art of Conjecturing, together with Letter to a Friend on Sets in Court Tennis*. Translated with an introduction and notes by Edith Dudley Sylla. Baltimore.

Cajori, Florian (1902). "Lehrbuch der Combinatorik [review]". *Science* 16, pp. 469–470.

Egen, Peter Nikolaus Caspar (1820). *Handbuch der allgemeinen Arithmetik*. Theil I. Die Buchstabenrechnung. Berlin.

Grattan-Guinness, Ivor (1990). *Convolutions in French Mathematics, 1800–1840*. From the Calculus and Mechanics to Mathematical Analysis and Mathematical Physics. Basel.

Hindenburg, Carl Friedrich (1795). "Allgemeine Darstellung des Polynomialtheorems nach de Moivre und Boscovich, nebst verschiedenen Bemerkungen über die dabey zum Grunde liegende lexikographischen Involutionen". *Archiv der reinen und angewandten Mathematik* 1, pp. 385–423.

---

[3] "Ars, *Combinatoria* dicta, [...] docetque sic enumerare modos omnes possibiles, secundum quos res plures permisceri, transponi vel conjungi invicem possunt". Translation from Bernoulli (2006, p. 193). [4] When the combinatorial art was applied to the calculus, it was also called *combinatorial analysis* (*combinatorische Analyse*) (Hindenburg 1795, p. 423). [1] According to Koppelman (1971, p. 161), the calculus of derivations is "essentially a method for determining the coefficient of x^n in the expansion of $\varphi(a + bx + cx^2 + ...)$ for any function φ. In this respect it is highly combinatorial and recalls the work of the German school". See also Grattan-Guinness (1990, pp. 211–216).

[2] The use of *combinatory* in this sense goes back to the 17th century. In a review of Kircher (1669), it says: "his [Kircher]'s *Combinatory* Art, as he cals [sic] it, consists chiefly in this, that any subjects being given or proposed, it shews, how often and how many ways they may be combined together, or exchanged, or transposed among themselves" ([anon.] 1669, p. 1093).

Hindenburg, Carl Friedrich (1803). *Über combinatorische Analysis und Derivations-Calcul.* Einige Fragmente gesammelt und zum Druck befördert. Leipzig.
Kircher, Athanasius (1645). *Ars magna lucis et vmbrae in decem libros digesta.* Roma.
Kircher, Athanasius (1663). *Polygraphia nova et universalis ex combinatoria arte detecta.* Roma.
Kircher, Athanasius (1669). *Ars magna sciendi, in XII libros digesta.* Amsterdam.
Knobloch, Eberhard (1974). "The Mathematical Studies of G.W. Leibniz on Combinatorics". *Historia Mathematica* 1, pp. 409–430.
Knobloch, Eberhard (2013). "Renaissance combinatorics". In: *Combinatorics.* Ancient and Modern. Ed. by Robin Wilson and John J. Watkins. Oxford, pp. 123–145.
Koppelman, Elaine (1971). "The Calculus of Operations and the Rise of Abstract Algebra". *Archive for History of Exact Sciences* 8, pp. 155–242.
Leibniz, Gottfried Wilhelm (1666). *Dissertatio de arte combinatoria.* Leipzig.
Levi, Friedrich Wilhelm (1940). "On a method of finite combinatorics which applies to the theory of infinite groups". *Bulletin of the Calcutta Mathematical Society* 32, pp. 65–68.
Levi, Friedrich Wilhelm (1942). *Finite geometrical systems.* Six public lectures delivered in February 1940 at the University of Calcutta. Calcutta.
Llull, Ramon (1598). *Opera.* Ed. by Lazarus Zetzner. Strasbourg.
MacMahon, Percy Alexander (1915). *Combinatory Analysis.* Volume I. Cambridge.
Netto, Eugen (1901). *Lehrbuch der Combinatorik.* Leipzig.
Stahl, Conrad Dietrich Martin (1800). *Grundriss der Combinationslehre.* Jena and Leipzig.
Wilson, Robin and John J. Watkins, eds. (2013). *Combinatorics.* Ancient and Modern. Oxford.

commensurable **and** ***incommensurable***

Commensurability and incommensurability were defined by Euclid (*Elements*, book X, def. 1); he used the words *σύμμετρος* and *ἀσύμμετρος*: "Those magnitudes are said to be *commensurable* [*σύμμετρα*] which are measured by the same measure, and those *incommensurable* [*ἀσύμμετρα*] which cannot have any common measure".[1] According to the first scholium to book X, the terminology is due to the Pythagoreans (Heiberg 1888, p. 416). In extant texts, it was used before Euclid by Plato and Aristotle, for example (see Mugler (1958-59, pp. 389f.)).

Martianus Capella (*De nuptiis* 6, 719) took over these terms as *symmetrus* and *ametrus* (Willis 1983, p. 256).[2] The pair *commensurabilis* and *incommensurabilis*, literal translations of the Greek terms, was first used by Boethius in his introduction to arithmetic (Boethius 1867, p. 37). His terminology eventually became standard in Latin texts, but alternatives came up in the 12th and 13th centuries under Arabic influence. For example, Gerard of Cremona, translating from that language, used the pair *communicans* and *incommunicans* (Busard 1984, p. 233); a translation made directly from the Greek used *commensurabilis* and *incommensurabilis* (Busard 1987, p. 211), on the other hand. The terminology was not fixed at that time; for example, Campanus used all of these words (Busard 2005, p. 306).

English *commensurable* and *incommensurable* were apparently first used by Recorde (1557, fol. Bi^{r}); the French terms were earlier used by Oresme: "It is a fact that things are commensurable when the proportion of one thing to another can be found in rational numbers, and, when the proportion cannot be thus stated, the things are incommensurable" (Oresme 1968, p. 196).[3]

---

1 Translation from Euclid (1956, p. 10). 2 For these two terms, see Guillaumin (2020, pp. 33f., 287). 3 "Et doit l'en savoir que les choses sont conmensurables quant telle proporcion comme une a <a> l'autre puet estre trouvée en nombres. Et quant elle ne puet, les choses sont inconmensurables". Translation from Oresme (1968, p. 197). The terms also appear at about the same time in de Conty (n.d., fol. 202^{r}).

Note. For the Greek terminology, see Mugler (1958-59, pp. 85f., 389f.); for the Latin one, see Ayuso García (2008, pp. 1283–1286) and Guillaumin (2020, pp. 33f., 66, 155, 287). Tropfke (1980, pp. 131–139) gives a historical overview of the concept and the terminology. Rommevaux (2001) discusses rationality in medieval (Latin and Arabic) versions of the *Elements*. For the much discussed discovery of incommensurability see, e. g., Fowler (1994).

References

Ayuso García, Manuel (2008). "La terminología latina de la geometría en Marciano Capela". PhD thesis. Madrid: Universidad Nacional de Educación a Distancia.

Boethius, Anicius Manlius Severinus (1867). *De institutione arithmetica libri duo, De institutione musica libri quinque*. Accedit Geometria quae fertur Boetii. Ed. by Gottfried Friedlein. Leipzig.

Busard, H.L.L. (1984). *The Latin Translation of the Arabic Version of Euclid's* Elements, *commonly ascribed to Gerard of Cremona*. Leiden.

Busard, H.L.L. (1987). *The Mediaeval Latin Translation of Euclid's Elements Made Directly from the Greek*. Stuttgart.

Busard, H.L.L. (2005). *Campanus of Novara and Euclid's* Elements. Volume I. Stuttgart.

de Conty, Évrard (n.d.). "Le Livre des Probleumes d'Aristote, c'est à dire des fortes questions [manuscript]". Bibliothèque nationale de France. Département des manuscrits. Français 210. URL: https://gallica.bnf.fr/ark:/12148/btv1b10547045k.

Euclid (1956). *The Thirteen Books of Euclid's Elements*. Volume III. Books X-XIII and Appendix. Ed. by Thomas L. Heath. New York.

Fowler, David (1994). "The story of the discovery of incommensurability, revisited". In: *Trends in the Historiography of Science*. Ed. by Kostas Gavroglu, Jean Christianidis, and Efthymios Nicolaidis, pp. 221–235.

Guillaumin, Jean-Yves (2020). *Dictionnaire de la terminologie latine ancienne de l'arithmétique et de la géométrie*. Paris.

Heiberg, J.L., ed. (1888). *Euclidis Elementa*. Vol. V. Leipzig.

Mugler, Charles (1958-59). *Dictionnaire historique de la terminologie géométrique des Grecs*. Paris.

Oresme, Nicole (1968). *Le livre du ciel et du monde*. Edited by Albert D. Menut and Alexander J. Denomy, translated with an introduction by Albert D. Menut. Madison.

Recorde, Robert (1557). *The Whetstone of Witte*. London.

Rommevaux, Sabine (2001). "Rationalité, exprimabilité: une relecture médiévale du livre X des Éléments d'Euclide". *Revue d'histoire des mathématiques* 7, pp. 91–119.

Tropfke, Johannes (1980). *Geschichte der Elementarmathematik*. Band 1: Arithmetik und Algebra. Vollständig neu bearbeitet von Kurt Vogel, Karin Reich, Helmuth Gericke. Berlin and New York.

Willis, James, ed. (1983). *Martianus Capella*. Leipzig.

commutative

The term was coined as French *commutative* by François-Joseph Servois in 1814. He applied the term to "functions" (*fonctions*, often to be understood as operators)[1] rather than algebraic operations: two functions were "commutative between each other" (*commutatives entres elles*) if they yielded the same result, no matter in what order they were applied (Servois 1814-15, p. 98). In other words, the order can be changed without affecting the result, which explains the choice of the term: Latin *commutare* means 'to change, exchange'.

---

[1] Bradley and Petrilli (2010, Functions and Operators) observe that "Servois does not make a distinction between 'function' and 'operator' as we do today. Sometimes he uses the word 'function' in the same way that students of calculus do (i. e., to describe a relation between independent and dependent variables), but he also uses it to describe operators, functions of functions such as the varied state, and difference operators".

The term was soon picked up by French mathematicians and appears in Lacroix (1819, p. 727), for example. In England, the term was introduced in what was called the "calculus of operations", a field in which certain analogies between different "operations" such as differentiation and exponentiation were pursued.[2] The term was used by Jarrett (1831, p. 40); De Morgan (1836, p. 163), who does not seem to have been aware of Servois' work at that time, at first called such operations *convertible*. Gregory (1840, p. 211), discussing commutative and distributive functions, states that it is "advisable to comprehend these functions under a common name" and points out that "Servois, in a paper which does not seem to have received the attention it deserves, has called them [...] *distributive functions*, and [...] *commutative functions*". De Morgan (1840, p. 444) then also begins using the term *commutative*, as does Hamilton (1844, p. 429) in connection with quaternions.

In Germany, Graßmann (1844, pp. 3–4) paraphrased the notions of associativity and commutativity, but used the term *einfach* 'simple' for compositions that have both properties. The terms *associative*, *commutative*, and *distributive* were popularized by Hankel (1867, p. 3), who explained that "[t]hese names have been adopted universally in England since 1840 and hence I have not hesitated to transplant them to German soil".[1] While Hankel's phrasing suggests that he was the first to use German *kommutativ*, the translation had already been used by Schlömilch (1856, p. 29), for example.[2]

With respect to permutations (called *substitutions*), Cauchy (1844, p. 154) used the word *permutable*.[3] The term was in common use in this sense in the 19th century.

The frequently used *Abelian group* (German *Abel'sche Gruppe*) seems to have been first used by Weber (1882, p. 304); the term also appears in his widely read textbook on algebra (Weber 1896, p. 6). The English term appears in the mid-1890s, for example in Burnside (1894, p. 83), who later, referring to Abel (1829), also explains the name: "The connection of Abel's name with groups of permutable operations is due to his having been the first to investigate, with complete generality, the application of such groups to the theory of equations" (Burnside 1897, p. 46).[4] The term is a natural extension of the term *Abelian equation*, introduced by Jordan (1870, p. 287): "We thus call *abelian equations* all of those whose group only contains substitutions that are exchangeable among each other".[5]

Note that Jordan (1870, pp. 171f.) used *groupe abélien* in a different way, namely for the symplectic group. The term *groupe eulérien*, suggested by Perott (1889, pp. 99, 101), found no appeal.

---

[2] A classic example is the analogy between the raising of a sum to a power and the differential of a product. See Koppelman (1971). [1] "Diese Namen sind in England seit 1840 vollkommen eingebürgert und ich habe daher nicht Anstand genommen, sie auch auf deutschen Boden zu verpflanzen". Translation from Crowe (1985, p. 16). Hankel also named Servois as the originator of *commutative* and *distributive*, and added it was "probably" Hamilton who introduced *associative*.

[2] In fact, the German terms *assoziativ* and *distributiv* are also found in German texts before Hankel. See the respective entries. [3] For a brief discussion, see Wußing (1984, p. 88). [4] Weber (1895, p. 536) explained the term in the same way. [5] "Nous appellerons donc *équations abéliennes* toutes celles dont le groupe ne contient que des substitutions échangeables entre elles". Translation from Cox (2011, p. 9). As Jordan points out, this term is itself a generalization of Kronecker (1853, p. 368), who used *Abelsche Gleichungen* for equations that have a cyclic group.

Note. The context in which Servois introduced the terminology is explained in Koppelman (1971) and Allaire and Bradley (2002), for example. Crowe (1985, pp. 15f.) gives a concise overview of the history of the terminology. For the term *Abelian group*, see Miller (1941, pp. 489f.) and Sørensen (2013).

References

Abel, Niels Henrik (1829). "Mémoire sur une classe particulière d'équations résolubles algébriquement". *Journal für die reine und angewandte Mathematik* 4, pp. 131–156.

Allaire, Patricia R. and Robert E. Bradley (2002). "Symbolical Algebra as a Foundation for Calculus: D.F. Gregory's Contribution". *Historia Mathematica* 29, pp. 395–426.

Bradley, Robert E. and Salvatore J. Petrilli (2010). "Servois' 1814 Essay on a New Method of Exposition of the Principles of Differential Calculus, with an English Translation". *Convergence* (November 2010). URL: `https://old.maa.org/press/periodicals/convergence/servois-1814-essay-on-a-new-method-of-exposition-of-the-principles-of-differential-calculus-with-an-3`.

Burnside, William Snow (1894). "On certain composite groups". *The Messenger of Mathematics* 24, pp. 82–96.

Burnside, William Snow (1897). *Theory of Groups of Finite Order*. Cambridge.

Cauchy, Augustin-Louis (1844). *Exercises d'analyse et de physique mathématique*. Tome troisième. Paris.

Cox, David A. (2011). "Why Eisenstein Proved the Eisenstein Criterion and Why Schonemann Discovered It First". *The American Mathematical Monthly* 118, pp. 3–21.

Crowe, Michael J. (1985). *A History of Vector Analysis*. The Evolution of the Idea of a Vectorial System. New York.

Graßmann, Hermann (1844). *Die Wissenschaft der extensiven Grösse, oder die Ausdehnungslehre, eine neue mathematische Disziplin*. Erster Theil, die lineale Ausdehnungslehre enthaltend. Leipzig.

Gregory, Duncan Farquharson (1840). "On the Real Nature of Symbolical Algebra". *Transactions of the Royal Society of Edinburgh* 14, pp. 208–216.

Hamilton, William Rowan (1844). "On a new Species of Imaginary Quantities connected with a theory of Quaternions". *Proceedings of the Royal Irish Academy* 2, pp. 424–434.

Hankel, Hermann (1867). *Vorlesungen über die complexen Zahlen und ihre Functionen*. In zwei Theilen. I. Theil: Theorie der complexen Zahlsysteme. Leipzig.

Jarrett, Thomas (1831). *An Essay on Algebraic Development*. Cambridge.

Jordan, Camille (1870). *Traité des substitutions et des équations algébriques*. Paris.

Koppelman, Elaine (1971). "The Calculus of Operations and the Rise of Abstract Algebra". *Archive for History of Exact Sciences* 8, pp. 155–242.

Kronecker, Leopold (1853). "Über die algebraisch auflösbaren Gleichungen". *Bericht über die zur Bekanntmachung geeigneten Verhandlungen der Königl. Preuss. Akademie der Wissenschaften zu Berlin* 1853, pp. 365–374.

Lacroix, Sylvestre François (1819). *Traité du calcul différentiel et du calcul intégral*. Seconde édition, revue et augmentée. Tome troisième. Paris.

Miller, George Abram (1941). "Maximal Invariant Proper Subgroups of a Finite Group". *Proceedings of the National Academy of Sciences* 27, pp. 587–590.

De Morgan, Augustus (1836). *The Differential and Integral Calculus*. London.

De Morgan, Augustus (1840). "Operation". *The Penny Cyclopædia* 16, pp. 442–446.

Perott, Joseph (1889). "Remarque au sujet du théorème d'Euclide sur l'infinité du nombre des nombres premiers". *American Journal of Mathematics* 11, pp. 99–138.

Schlömilch, Oskar (1856). "A treatise on the calculus of operations [review]". *Literaturzeitung der Zeitschrift für Mathematik und Physik* 1, pp. 28–32.

Servois, François-Joseph (1814-15). "Essai sur un nouveau mode d'exposition des principes du calcul différentiel". *Annales des mathématiques pures et appliquées* 5, pp. 93–140.

Sørensen, Henrik Kragh (2013). "What's Abelian about abelian groups?" *Journal of the British Society for the History of Mathematics* 23, pp. 37–51.
Weber, Heinrich (1882). "Beweis des Satzes, dass jede eigentlich primitive quadratische Form unendlich viele Primzahlen darzustellen fähig ist". *Mathematische Annalen* 20, pp. 301–329.
Weber, Heinrich (1895). *Lehrbuch der Algebra*. Erster Band. Braunschweig.
Weber, Heinrich (1896). *Lehrbuch der Algebra*. Zweiter Band. Braunschweig.
Wußing, Hans (1984). *The Genesis of the Abstract Group Concept*. A Contribution to the History of the Origin of Abstract Group Theory. Tr. Abe Shenitzer. Cambridge, Mass.

commutative algebra

At the end of the 1920s, the expressions *commutative algebra* and *non-commutative algebra* came up in the sense of, broadly speaking, 'the study of commutative and non-commutative rings'.[1] Emmy Noether is known to have lectured on the latter subject in 1928 (Koreuber 2015, p. 49), and the term may be due to her. An early usage in print is Brauer (1932, p. 44), who used quotation marks, indicating the novelty of the terminology. It became generally known through a speech by Noether at the International Congress of Mathematicians in 1932.[2] The talk was entitled "Hypercomplex Systems and their Relations to Commutative Algebra and Number Theory" and began as follows:

> During the past few years the theory of hypercomplex systems, of algebras, has made good progress; but it is only very recently that the significance of this theory for commutative problems has become evident. Today I should like to comment on this significance of the noncommutative for the commutative: and indeed, I want to do this particularly in view of two classical problems originating from Gauss, the principal genus theorem and the closely related norm principle. The formulation of these problems has undergone continuous change: with Gauss they appear as the culmination of his theory of quadratic forms; then they play an essential role in the characterization of the relative cyclic and abelian number fields via class field theory; and finally they manifest themselves as theorems on automorphisms and on the decomposition of algebras, and at the same time this last formulation allows the theorems to be extended to arbitrary relative Galois number fields (Noether 1932, p. 167).[3]

The terminology also featured prominently in one of Noether's next papers, which was entitled "Non-commutative algebra" (Noether 1933).

---

[1] Note that there are earlier uses of English *(non-)commutative algebra*, but not in this specific sense. The term may refer to an algebraic structure, as it does in Peirce (1881, p. 219), or it generally refers to an approach to algebraic structures in which the commutative law is or is not presupposed, as it seems to be the case in Macfarlane (1899, p. 354): "It is a common belief that quaternions has only a remote connection with algebra; that it is only one of several systems of non-commutative algebra, and that the mathematician can get on very well without it". [2] Curtis (2007) discusses the talk in detail. [3] "Die Theorie der hyperkomplexen Systeme, der Algebren, hat in den letzten Jahren einen starken Aufschwung genommen; aber erst in allerneuester Zeit ist die Bedeutung dieser Theorie für kommutative Fragestellungen klar geworden. Über diese Bedeutung des Nichtkommutativen für das Kommutative möchte ich heute berichten: und zwar will ich das im einzelnen verfolgen an zwei klassischen, auf Gauss zurückgehenden Fragestellungen, dem Hauptgeschlechtssatz und dem eng damit verbundenen Normensatz. Diese Fragestellungen haben sich im Laufe der Zeit in ihrer Formulierung immer wieder gewandelt: bei Gauss treten sie auf als Abschluss seiner Theorie der quadratischen Formen; dann spielen sie eine wesentliche Rolle in der Charakterisierung der relativ zyklischen und abelschen Zahlkörper durch die Klassenkörpertheorie, und schliesslich lassen sie sich aussprechen als Sätze über Automorphismen und über das Zerfallen von Algebren, und diese letztere Formulierung gibt dann zugleich eine Übertragung der Sätze auf beliebige relativ galoissche Zahlkörper". Translation from Noether (1981, p. 167).

References

Brauer, Richard (1932). "Über die Konstruktion der Schiefkörper, die von endlichem Rang in bezug auf ein gegebenes Zentrum sind". *Journal für die reine und angewandte Mathematik* 168, pp. 44–64.

Curtis, Charles W. (2007). "Emmy Noether's 1932 ICM Lecture on Noncommutative Methods in Algebraic Number Theory". In: *Episodes in the History of Modern Algebra (1800–1950)*. Ed. by Jeremy J. Gray and Karen Hunger Parshall. Providence, R.I., pp. 199–220.

Koreuber, Mechthild (2015). *Emmy Noether, die Noether-Schule und die moderne Algebra.* Zur Geschichte einer kulturellen Bewegung. Berlin and Heidelberg.

Macfarlane, Alexander (1899). "The Fundamental Principles of Algebra". *Science* 10, pp. 345–364.

Noether, Emmy (1932). "Hyperkomplexe Systeme in ihren Beziehungen zur kommutativen Algebra und Zahlentheorie". In: *Verhandlungen des Internationalen Mathematiker-Kongresses Zürich 1932*. I. Band. Ed. by Walter Saxer. Zürich and Leipzig, pp. 189–194.

Noether, Emmy (1933). "Nichtkommutative Algebra". *Mathematische Zeitschrift* 37, pp. 514–541.

Noether, Emmy (1981). "Hypercomplex Systems and Their Relations to Commutative Algebra and Number Theory". In: *Emmy Noether*. A Tribute to Her Life and Work. Ed. by James W. Brewer and Martha K. Smith. New York and Basel, pp. 167–173.

Peirce, Benjamin (1881). "Linear Associative Algebra". *American Journal of Mathematics* 4, pp. 97–229.

commutative diagram

An early example of a commutative diagram is found in Hausdorff's notes from March 1933, as pointed out by Scholz (2008, p. 884), who explains that Hausdorff used the diagram for homomorphisms between the terms of two sequences of groups. Purkert and Wußing (1994, p. 757) remark that the concept was also used in lectures by Hurewicz from 1941.

In print, some such diagrams appear in Eilenberg and Mac Lane (1942, e. g., p. 777); they did not use the term but did say that "the application of η 'commutes' with the application of any homomorphism T to the free group F and its subgroup R".

Several years later, Spanier (1949, p. 108, 1949, p. 116) on various occasions said that "commutativity holds in the diagram". Explicitly, French *diagramme commutatif* is found in some contributions to the Séminaire Henri Cartan (1950-51), for example in Eilenberg (1950-51, p. 2) and Cartan (1950-51, p. 2).[1] The term also appears in Serre (1951, p. 452) and, in English, in Whitehead (1952, p. 427).

References

Cartan, Henri (1950-51). "Théorie axiomatique de la cohomologie". *Séminaire Henri Cartan* 3, exposé 16, pp. 1–9.

Eilenberg, Samuel (1950-51). "Homologie des groupes. I: axiomes et unicité". *Séminaire Henri Cartan* 3, exposé 1, pp. 1–5.

Eilenberg, Samuel and Saunders Mac Lane (1942). "Group Extensions and Homology". *Annals of Mathematics* 43, pp. 757–831.

Purkert, Walter and Hans Wußing (1994). "Fundamental concepts of abstract algebra". In: *Companion Encyclopedia of the History and Philosophy of the Mathematical Sciences*. Volume 1. Ed. by Ivor Grattan-Guinness. Baltimore and London, pp. 741–760.

Scholz, Erhard (2008). "Hausdorffs Blick auf die entstehende algebraische Topologie". In: *Felix Hausdorff. Gesammelte Werke*. Band III. Mengenlehre (1927, 1935). Deskriptive Mengenlehre und Topologie. Ed. by Ulrich Felgner et al. Berlin and Heidelberg, pp. 865–892.

---

[1] Eilenberg's paper was read on Nov. 13, 1950, Cartan's on April 23, 1951.

Serre, Jean-Pierre (1951). "Homologie singulière des espaces fibrés". *Annals of Mathematics* 54, pp. 425–505.
Spanier, Edwin H. (1949a). "Borsuk's Cohomotopy Groups". *Annals of Mathematics* 50, pp. 203–245.
Spanier, Edwin H. (1949b). "The Mayer homology theory". *Bulletin of the American Mathematical Society* 55, pp. 102–112.
Whitehead, George W. (1952). "Fiber Spaces and the Eilenberg Homology Groups". *Proceedings of the National Academy of Sciences* 38, pp. 426–430.

commutator

The term was coined as German *Commutator*[1] by Dedekind (1897, p. 553). He introduces the element $\varepsilon = (\psi^{-1}\varphi^{-1}\psi)\varphi$ and calls it, "for brevity", "the *Commutator of the elements* φ and ψ".[2] The term had earlier been made public by Frobenius (1896, p. 1348): "Following Dedekind, I call the element F, which is obtained from A and B with the help of the equation $BA = ABF$, the *commutator* of A and B".[3] Frobenius knew this term from correspondence with Dedekind, in which the latter also considered groups of commutators (see Dedekind 1931, p. 422).

English *commutator* appears almost immediately afterwards;[4] it was used by Miller (1898, p. 135), who also considers the *commutator subgroup* and points out that *first derived subgroup* is an "equivalent" term.[5] For the latter term, Miller refers to Lie, who had used German *derivirte Gruppe* in Lie (1888, p. 19), for example. This terminology is actually due to Betti (1852), however.

In this paper, Betti (1852, p. 51) defines a substitution (*sostituzione*) as "the operation by means of which one passes from one permutation of several quantities to another", i. e. he uses *substitution* for 'permutation'.[6] He then defines the "substitution derived from θ by means of ψ"[7] as $\theta_1 = \psi^{-1}\theta\psi$ (Betti 1852, p. 55); a derived group (*gruppo derivato*) G is defined as $\psi^{-1}G\psi$ (Betti 1852, p. 61). Betti's term was in common use for some time; it is found, for example, in Serret (1866, p. 255) and still named as a synonym of *commutator* in de Séguier (1904, p. 60).

Note. On the history of the concept and term, see Kappe and Morse (2007); note, however, that the concept of derived group is ignored in that article. A concise terminological overview (excluding Betti) is given by Magnus (1939, p. 17). On Betti's use of *derived substitution* and *derived group*, see Wußing (1984, pp. 125f.) and Kiernan (1971, p. 104).

References

Betti, Enrico (1852). "Sulla risoluzione delle equazioni algebriche". *Annali di scienze matematiche e fisiche* 3, pp. 49–115.
Burnside, William Snow (1911). *Theory of Groups of Finite Order*. Second edition. Cambridge.

---

[1] Modern spelling: *Kommutator*. [2] "welches [Element] wir der Kürze halber den *Commutator der Elemente* φ and ψ nennen wollen". [3] "Das Element F, das sich mittelst der Gleichung $BA = ABF$ aus A und B ergiebt, nenne ich nach Dedekind den *Commutator* von A und B". Translation from Kappe and Morse (2007, p. 688). [4] French *commutateur* is found in de Séguier (1904, p. 8). [5] A similar statement was given by Burnside (1911, p. 55): "The group generated by the commutators of a group G is called the *commutator sub-group* or the *derived group*". [6] "Una sostituzione è la operazione mediante la quale si passa da una ad un altra permutazione di più quantità". Translation from Kiernan (1971, p. 104). For the term *substitution* in this sense, see *permutation*. [7] "sostituzione derivata di θ per mezzo di ψ".

Dedekind, Richard (1897). "Ueber Gruppen, deren sämtliche Teiler Normalteiler sind". *Mathematische Annalen* 48, pp. 548–561.
Dedekind, Richard (1931). *Gesammelte mathematische Abhandlungen.* Zweiter Band. Ed. by Robert Fricke, Emmy Noether, and Öystein Ore. Braunschweig.
Frobenius, Georg (1896). "Über die Primfaktoren der Gruppendeterminante". *Sitzungsberichte der Königlich Preußischen Akademie der Wissenschaften zu Berlin* (1896), pp. 1343–1382.
Kappe, Luise-Charlotte and Robert Fitzgerald Morse (2007). "On commutators in groups". In: *Groups St. Andrews 2005.* Volume 2. Ed. by C.M. Campbell et al., pp. 531–558.
Kiernan, B. Melvin (1971). "The Development of Galois Theory from Lagrange to Artin". *Archive for History of Exact Sciences* 8, pp. 40–154.
Lie, Sophus (1888). "Beiträge zur allgemeinen Transformationstheorie". *Berichte über die Verhandlungen der Königlich Sächsischen Gesellschaft der Wissenschaften zu Leipzig (Mathematisch-Physische Klasse)* 40, pp. 14–21.
Magnus, Wilhelm (1939). "Allgemeine Gruppentheorie". *Enzyklopädie der mathematischen Wissenschaften mit Einschluss ihrer Anwendungen.* Zweite, völlig neubearbeitete Auflage I 1,9, pp. 1–51.
Miller, George Abram (1898). "On the commutator groups". *Bulletin of the American Mathematical Society* 4, pp. 135–139.
de Séguier, Jean Armand (1904). *Théorie des groupes finis.* Éléments de la théorie des groupes abstraits. Paris.
Serret, Joseph-Alfred (1866). *Cours d'algèbre supérieure.* Troisième édition. Tome second. Paris.
Wußing, Hans (1984). *The Genesis of the Abstract Group Concept.* A Contribution to the History of the Origin of Abstract Group Theory. Tr. Abe Shenitzer. Cambridge, Mass.

compact

The French term *ensemble compact* was introduced by Fréchet (1904, p. 849): "We will call *compact set* any set E such that there is at least one element common to an arbitrary infinite sequence of sets E_1, E_2, ... , E_n, ... contained in E, where these (possessing at least one element each) are closed and each is contained in the previous".[1] In 1906, he gave a slightly rephrased definition which more explicitly invokes the concept of limit point: "We will say that a set is *compact* if it contains only a finite number of points or if every one of its infinite subsets gives rise to at least one limit point" (Fréchet 1906, p. 6).[2] Fréchet never gave a reason why he chose the name, and when asked much later, he said he had completely forgotten the reasons for his choice (Pier 1980, p. 440).

The term did not meet everyone's approval. In particular, Schoenflies (1908, p. 281) rejected it, suggesting the German alternatives *lückenlos* 'without gaps' and *abschließbar* 'closeable'.[3] Nevertheless, the term was soon in common use; it was adapted by Hahn (1908, p. 251) and Hedrick (1911, p. 285), for example.[4]

---

[1] "Nous appellerons *ensemble compact* tout ensemble E tel qu'il existe toujours au moins un élément commun à une suite infinie quelconque d'ensembles E_1, E_2, ... E_n, ... contenus dans E, lorsque ceux-ci (possédant au moins un élément chacun) sont fermés et chacun contenu dans le précédent". Translation from Guerraggio, Jaëck, and Mazliak (2016, p. 164). [2] "Nous dirons qu'un ensemble est *compact* lorsqu'il ne comprend qu'un nombre fini d'éléments ou lorsque toute infinité de ses éléments donne lieu à au moins un élément limite". Translation from Raman-Sundström (2015, p. 625). For Fréchet and compactness, see also Taylor (1982). [3] See Taylor (1982, p. 266).
[4] At about the same time, the term *compact* was also used in rather different ways. In particular, Russell (1903, p. 271) defined *compact series* and added: "Such series are called by Cantor *überall dicht* [everywhere dense]". Russell's use of the term is discussed in Pier (1961, pp. 172f.).

Hausdorff (1914, p. 230) used German *kompakt* for metric spaces and linked the concept to the Heine-Borel Theorem. A change in the terminology occurred when Alexandroff and Urysohn applied the concept to topological spaces. They listed three equivalent properties, among them the Heine-Borel property, and called such sets *bicompact* (French *bicompact* in Alexandroff and Urysohn (1923, p. 6), German *bikompakt* in Alexandroff and Urysohn (1924, p. 260)).[1]

Hildebrandt (1926, p. 468) repeats the three equivalent properties and explains the name:

THEOREM III. The following properties of a class E in a Hausdorff space $\mathfrak{B}$ are equivalent:

A Every infinite set of power μ has a complete limiting element in E.

B If $\mathfrak{F}$ is a well-ordered monotonic decreasing family of closed sets F, of power $\geq \mu$ such that for each α, $F_\alpha E$ contains at least one element, then the sets $F_\alpha E$ have a common element.

C If E is contained in a family $\mathfrak{G}$ of open sets G, the power of $\mathfrak{G}$ being $\geq \mu$, then E is covered by a subfamily of $\mathfrak{G}$ of power $\leq \mu$.

Hildebrandt goes on to state: "Alexandroff and Urysohn call a set E satisfying the conditions of Theorem II[2] *bicompact*, because it is a meeting place of the generalization of Theorem I[3] and of Theorem III". Alexandroff and Urysohn's terminology continued to be used throughout the 1930s, but by around 1940, another change set in.

As explained by Folland (2010, p. 665), "[t]he words 'compact' and 'bicompact' continued to be used in the senses of Hausdorff and Alexandroff-Urysohn throughout the 1930s, but in the ensuing decade the common usage gradually shifted to the modern 'countably compact' and 'compact.' This shift was initiated by André Weil and the Bourbaki group [...][4], although they imposed the further stipulation that compact spaces should be Hausdorff, and not until about 1950 did the new names become completely standard".

The shift is reflected by Lefschetz (1942, p. 17):

We emphasize at the outset the following important departures from hitherto accepted terminology: (a) with Bourbaki we shall replace the term *bicompact* of Alexandroff-Urysohn by the term *compact*; (b) what has been known hitherto as *compact* (following Fréchet who introduced the concept) shall be called *countably compact*.

The term *countably compact* had earlier been suggested by Tukey (1940, p. 42):

I suggest the following additional terminology, "sequentially compact" = every sequence contains a convergent subsequence. "countably compact" = Fréchet's "compact" = every infinite set has a non-empty derivative (the term "countably" is justified by the fact that we may clearly restrict ourselves to countably infinite sets).

---

[1] Alexandroff and Urysohn reserved the term *compact* for what would now be called *countably compact*. For comments on Alexandroff and Urysohn's definition, see Chandler and Faulkner (1998, p. 634) and Raman-Sundström (2015, pp. 627f.). [2] Theorem II states the equivalence of these properties in a Hausdorff space. [3] Theorem I lists statements equivalent to "If E is covered by a denumerable family $\mathfrak{G}$ of open sets G, then E is covered by a finite subfamily of $\mathfrak{G}$". [4] Folland refers to Weil (1937, p. 6) and to Bourbaki (1940, pp. 59–61).

Still, the terminological situation was not settled by such statements, as is illustrated by Munkres (1974, p. 178):

> In the early days of topology, [...] the open covering formulation was called 'bicompactness.' Later, the word 'compact' was shifted to apply to the open covering definition, leaving this one to search for a new name. It still has not found a name on which everyone agrees. On historical grounds, some call it 'Fréchet compactness'; others call it the 'Bolzano-Weierstrass property.' We have invented the term 'limit point compactness.' It seems as good a term as any; at least it describes what the property is about.

See also *complete continuity*.

Note. On the the history of compactness, see, for example, Pier (1961, 1980), Raman-Sundström (2015) as well as Pietsch (2007, pp. 63–65) and Hildebrandt (1926). A detailed account of the history of open sets and related concepts is found in Moore (2008).

References

Alexandroff, Pavel S. and Pavel Urysohn (1923). "Sur les espaces topologiques compacts". *Bulletin international de l'Académie polonaise des sciences et des lettres. Classe des sciences mathématiques et naturelles* (1923), pp. 5–8.

Alexandroff, Pavel S. and Pavel Urysohn (1924). "Zur Theorie der topologischen Räume". *Mathematische Annalen* 92, pp. 258–266.

Bourbaki, Nicolas (1940). *Topologie génerale*. Chapitre I. Structures topologiques. II. Structures uniformes. Paris.

Chandler, Richard E. and Gary D. Faulkner (1998). "Hausdorff Compactifications: A Retrospective". In: *Handbook of the History of General Topology*. Volume 2. Ed. by C.E. Aull and R. Lowen. Dordrecht et al., pp. 631–667.

Folland, Gerald B. (2010). "A Tale of Topology". *The American Mathematical Monthly* 117, pp. 663–672.

Fréchet, Maurice (1904). "Généralisation d'un théorème de Weierstrass". *Comptes rendus de l'Académie des sciences* 139, pp. 848–850.

Fréchet, Maurice (1906). "Sur quelques points du calcul fonctionnel". *Rendiconti del Circolo Matematico di Palermo* 22, pp. 1–74.

Guerraggio, Angelo, Frédéric Jaëck, and Laurent Mazliak (2016). "Lines on the Horizon". In: *Images of Italian Mathematics in France*. The Latin Sisters, from Risorgimento to Fascism. Ed. by Frédéric Brechenmacher et al. Basel, pp. 127–179.

Hahn, Hans (1908). "Bemerkungen zu den Untersuchungen des Herrn M. Fréchet: Sur quelques points du calcul fonctionnel". *Monatshefte für Mathematik und Physik* 19, pp. 247–257.

Hausdorff, Felix (1914). *Grundzüge der Mengenlehre*. Leipzig.

Hedrick, Earle Raymond (1911). "On Properties of a Domain for Which Any Derived Set is Closed". *Transactions of the American Mathematical Society* 12, pp. 285–294.

Hildebrandt, T.H. (1926). "The Borel theorem and its generalizations". *Bulletin of the American Mathematical Society* 32, pp. 423–474.

Lefschetz, Solomon (1942). *Algebraic Topology*. Providence, R.I.

Moore, Gregory H. (2008). "The emergence of open sets, closed sets, and limit points in analysis and topology". *Historia Mathematica* 35, pp. 220–241.

Munkres, James R. (1974). *Topology*. A First Course. Englewood Cliffs.

Pier, Jean-Paul (1961). "Genèse et évolution de l'idée de compact". *Revue d'histoire des sciences* 14, pp. 169–179.

Pier, Jean-Paul (1980). "Historique de la notion de compacité". *Historia Mathematica* 7, pp. 425–443.

Pietsch, Albrecht (2007). *History of Banach Spaces and Linear Operators*. Boston, Mass. et al.

Raman-Sundström, Manya (2015). "A Pedagogical History of Compactness". *The American Mathematical Monthly* 122, pp. 619–635.

Russell, Bertrand (1903). *The Principles of Mathematics*. Vol. I. Cambridge.
Schoenflies, Arthur (1908). *Die Entwickelung der Lehre von den Punktmannigfaltigkeiten*. Bericht erstattet der Deutschen Mathematiker-Vereinigung. Zweiter Teil. Leipzig.
Taylor, Angus E. (1982). "A Study of Maurice Fréchet: I. His Early Work on Point Set Theory and the Theory of Functionals". *Archive for History of Exact Sciences* 27, pp. 233–295.
Tukey, John W. (1940). *Convergence and Uniformity in Topology*. Princeton.
Weil, André (1937). *Sur les espaces à structure uniforme et sur la topologie générale*. Paris.

companion

The earliest mathematical term that contains the word *companion* presumably was *companion of the cycloid*. This curve turned out to be a sine curve and was first discussed by de Roberval in the 1630s. The name *companion* is based on its mode of construction. Figure 17 reproduces an illustration given by Roberval himself. The cycloid is shown in the figure by the points $A, 8, 9, \ldots 14, D$. To construct the companion, take any point on the cycloid, say 10. The corresponding point on the circle is G. Measure the distance from G to the diameter AB of the circle and move 10 to the right by this distance. In that way point 3 on the companion is found. The companion is the locus of all points constructed in this way.

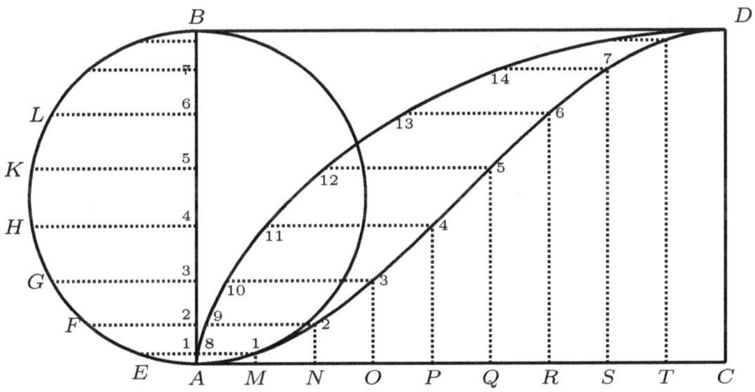

Figure 17: Roberval's companion of the cycloid. From de Roberval (1693a, p. 192), slightly simplified.

de Roberval (1693a, p. 252) used the Latin *trochoidis comitem, seu sociam*, which can be explained thus: *trochoid* was one alternative terms for *cycloid*, and the nouns *comes* and *socius* can both be translated as 'companion'. In French, de Roberval (1693b, p. 232) used *compagne*. These terms were also known to Pascal (1658, p. 5) and Wallis (1659, p. 100), for example.

The term *companion matrix* of a monic polynomial was introduced by Loewy (1917, p. 255) as German *Begleitmatrix*, of which the English term is a translation (*Begleit-* comes from a verb meaning 'to accompany'). The matrix "accompanies" a system of linear equations. The English term appears to have been first used by

MacDuffee (1933, p. 20): "A. LOEWY called B (or its negative) a 'Begleitmatrix'. We shall take the liberty of calling it the *companion matrix* of the equation $f(\lambda) = 0$".[1]

In knot theory, the *companion knot* was introduced by Schubert (1953, p. 245) as German *Begleitknoten.*

References

Hawkins, Thomas (2013). *The Mathematics of Frobenius in Context.* A Journey Through 18th to 20th Century Mathematics. New York.

Loewy, Alfred (1917). "Die Begleitmatrix eines linearen homogenen Differentialausdruckes". *Nachrichten von der Königl. Gesellschaft der Wissenschaften zu Göttingen (mathematisch-physikalische Klasse)* (1917), pp. 255–263.

MacDuffee, Cyrus Colton (1933). *The Theory of Matrices.* Berlin.

Pascal, Blaise (1658). *Histoire de la roulette, appellée autrement la trochoïde, ou la cycloïde.* Paris.

de Roberval, Gilles Personne (1693a). *De trochoide ejusque spatio.* In: *Divers ouvrages de mathématique et de physique.* Paris, pp. 246–278.

de Roberval, Gilles Personne (1693b). *Traité des indivisibles.* In: *Divers ouvrages de mathématique et de physique.* Paris, pp. 190–245.

Schubert, Horst (1953). "Knoten und Vollringe". *Acta Mathematica* 90, pp. 131–286.

Wallis, John (1659). *Tractatus Duo.* Prior, De Cycloide et corporibus inde genitis. Posterior, epistolaris in qua agitur De Cissoide, et Corporibus inde genitis: et De Curvarum, tum Linearum *Εὐθύνσει*, tum Superficierum *Πλατυσμῷ*. Oxford.

complement **(parallelogram)**

The Greek term *παραπλήρωμα*, from a verb meaning 'to fill up', was introduced by Euclid (*Elements*, book I, prop. 43): "In any parallelogram the complements of the parallelograms about the diameter are equal to one another" (see fig. 18).[1] Proclus later explained the absence of a definition: "The term 'complements' was derived by the author of the Elements from the thing itself, since complements fill the whole of the area outside the two parallelograms. This is why he does not regard it as deserving of special mention in the Definitions" (Proclus 1873, p. 418).[2]

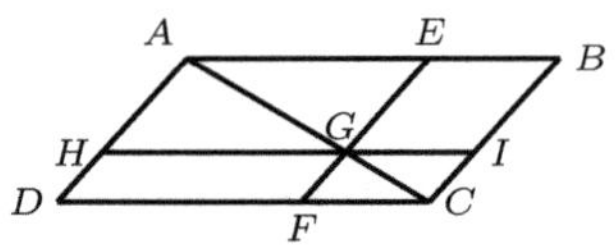

Figure 18: Complements. From Euclid (1574, fol. 68v). The complements of the parallelograms $AEGH$ and $CFGI$ are $DFGH$ and $EBIG$.

In the first translations of the *Elements* into Latin, *supplementum* was usually used to translate the Greek term. The term was used, for example, in the pseudo-Boethian *Geometria* (Folkerts 1970, pp. 126, 199), and later also in various 12th-century translations such as the one made in Sicily on the basis of a Greek original (Busard 1987, p. 50).

The word *complementum* in this sense does not appear to come up until the 16th century, when it was used, for example, by Ramus (1569, Geometria, p. 78) and by

[1] For Loewy's work, see also Hawkins (2013, pp. 587–592). [1] Translation from Euclid (1956, p. 340). [2] Translation from Proclus (1970, p. 331).

Clavius (Euclid 1574, fol. 68^{v}). Billingsley used *supplement* in the text but also said that *complement* can be used in the same sense (Euclid 1570, fol. 53^{v}).[3]

Xylander literally translated the term by German *Erfüllung* 'fulfilment, completion' (Euclid 1562, p. 32); *Komplement* was later used by Marius (Euclid 1610, p. 34). The first French translation of the *Elements* uses *supplément* (Euclid 1564, fol. 35^{r}); in Henrion's translation, this word is also used in prop. 43 (Euclid 1615, p. 49), but later Henrion slips into *complément* (Euclid 1615, p. 199).

Note. For the Greek term, see Mugler (1958-59, p. 332) and the comments by Heath in Euclid (1956, p. 341).

References

Busard, H.L.L. (1987). *The Mediaeval Latin Translation of Euclid's Elements Made Directly from the Greek*. Stuttgart.

Euclid (1557). *In Euclidis Elementa Geometrica Demonstrationum Libri sex*. Ed. by Jacques Peletier. Lyon.

Euclid (1562). *Die Sechs Erste Bücher Euclidis*. Auß Griechischer sprach in die Teütsch gebracht ... Durch Wilhelm Holtzmann genannt Xylander. Basel.

Euclid (1564). *Les Six Premiers Livres des Elements*. Traduicts et commentez par Pierre Forcadel de Besies. Paris.

Euclid (1570). *The Elements of Geometrie of the most auncient Philosopher Euclide of Megara*. Faithfully (now first) translated into the Englishe toung, by H. Billingsley. London.

Euclid (1574). *Elementorum libri XV*. Ed. by Christopher Clavius. Roma.

Euclid (1610). *Die Ersten Sechs Bücher Elementorum Evclidis*. Mit sonderm Fleiß und Mühe auß Griechischer in unsere Hohe deutsche Sprach übergesetzet ... und aigentlichest erkläret ... Durch Simonem Marium. Ansbach.

Euclid (1615). *Les quinze livres des Elements d'Euclide*. Traduicts de Latin en François ... par D. Henrion. Paris.

Euclid (1956). *The Thirteen Books of Euclid's Elements*. Volume I. Introduction and Books I, II. Ed. by Thomas L. Heath. New York.

Folkerts, Menso (1970). *‚Boethius' Geometrie II*. Ein mathematisches Lehrbuch des Mittelalters. Wiesbaden.

Mugler, Charles (1958-59). *Dictionnaire historique de la terminologie géométrique des Grecs*. Paris.

Proclus (1873). *In primum Euclidis elementorum librum commentarii*. Ed. by Gottfried Friedlein. Leipzig.

Proclus (1970). *Commentary on the First Book of Euclid's Elements*. Translated with Introduction and Notes by Glenn R. Morrow. Princeton.

Ramus, Petrus (1569). *Arithmeticae libri duo. Geometriae septem et viginti*. Basel.

complement (trigonometry)

The complement of an angle is the difference between the angle and a right angle, and, analogously, the complement of an arc is the difference between that arc and a quadrant. The notion is found in Greek texts on trigonometry. Ptolemy, in particular, used a paraphrase that can be literally translated as "the remainder to the quadrant" (Toomer 1984, p. 17).[1]

Trigonometry was developed in India, where, in particular, the sine and its complement, the cosine, were first systematically studied.[2] One term used for the comple-

---

[3] Billingsley (ibid.) also points out that Peletier differentiated between *supplement* and *complement*: "The parallelogrammes about the diameter he [Peletier] calleth Complementes, the other two figures he calleth Supplementes". Peletier's distinction is found in Euclid (1557, pp. 40f.). [1] See *remainder*. [2] The original definitions of these concepts are illustrated in the article *sine*.

ment was *koṭi*, which Datta and Singh (1983, p. 41) explain as follows: "*koṭi* means, amongst others 'the curved end of a bow' or 'the end or extremity in general'; hence in trigonometry it came to denote 'the complement of an arc to 90°'".[3]

In Arabic texts, the Sanskrit term appears as *tamām*, which literally means 'completion' and 'perfection, fulfillment'. This word was used, for example, by al-Battānī and by al-Bīrūnī.[1] This terminology seems to have been based on the Sanskrit terms (Bossong 1978, p. 172).[2]

When medieval translators were looking for a Latin equivalent of the Arabic terms, they came up with different solutions.

One of these can be illustrated by Plato of Tivoli (12th century), who translated one of al-Battānī's works, his Book of Astronomical Tables, under the title *De motu stellarum*. He frequently chose the paraphrase "which lacks to the completion (perfection) of 90 [degrees]", using the verb *perficere* 'to perfect' or the noun *perfectio*.[3] To refer to the cosine, the complement of the sine, Plato of Tivoli also used the word *residuum* 'remainder'; he thus used *chorda residui* 'chord[4] of the remainder' for the cosine (al-Battānī 1537, fol. 13^{r}). Gerard of Cremona, in his translation of the "Toledan tables" by al-Zarqālī, chose *sinus residui* 'sine of the remainder' (Curtze 1900, p. 342). In the *Almagesti minor*, a summary of the first six books of Ptolemy's *Almagest*, the alternative *sinus perfectionis* is found (Zepeda 2018, e. g., p. 164).

The word *complementum* in the sense under discussion also comes up in the 12th century. In particular, it was used by Gerard of Cremona in his translation of Jābir ibn Aflaḥ's "Correction of the Almagest"; in the text, *complementum arcus* and *complementum anguli* are also defined at the beginning of the text (ibn Aflaḥ 1534, p. 3).[5] The corresponding Spanish term *complimiento* is already found in the second half of 13th century (Bossong 1978, p. 172).[6]

In the 15th century, Regiomontanus used Latin *complementum* regularly; he gave a definition of the complement of an arc in his *On Triangles*: "the *complement* of any *arc* is the difference between [the arc] itself and a quadrant" (Regiomontanus 1533, p. 7).[7] At the time when Regiomontanus' work was posthumously published, *complementum* was already a standard term, used, for example, by Apian (1533, def. 10) and Fine (1532, fol. 56^{v}).[8] Apian (1532) also uses the German term *Complement* throughout his description of Halley's comet; the French term is found in Bassantin (1557, p. 8),[9] and Borough (1581, e. g., fol. C iij^{r}) and Blagrave (1585, p. 75) use the English term.

Note. The word *complement* was occasionally also used to refer to the supplement of an angle or arc. For example, Magini (1609, fol. 2^{r}) distinguished between the

---

[3] On *koṭi*, see also Burgess (1858, pp. 203, 413). [1] For al-Battānī, see Nallino (1907, p. 325) and Bossong (1978, p. 180), for al-Bīrūnī, see al-Bīrūnī (1934, p. 5). [2] In terms of the bow metaphor (described in the article *chord*), *koṭi* is still part of the metaphor, while *tamām* is not. [3] For example: "quod ad perficiendum 90 deficit" (al-Battānī 1537, fol. 16^{v}); "usque ad perfectionem 90" (al-Battānī 1537, fol. 7^{r}). [4] Plato of Tivoli used *chorda* to refer to the sine. See *sine*. [5] "complementum arcus, est superfluitas quae est inter ipsum & quartam circuli, sive sit arcus minor quarta circuli sive maior. Et similiter complementum anguli, est superfluitas quae est inter ipsum & angulum rectum, sive angulus sit minor recto sive maior". [6] For more Spanish attestations from about the same time, see Chabás and Goldstein (2003, p. 107). [7] "Complementum arcus cuiuslibet dicitur, quae sibi & quadranti interest differentia". Translation from Hughes (1967, p. 31). [8] Stevin (1608, p. 331) later suggested, without much success, the Dutch word *verschil* 'difference' as an alternative. [9] Bassantin writes *compliment*.

complementum anguli ad semicirculum 'supplement' and the *complementum anguli ad quadrantem* 'complement'. See *supplement*.

Note. Brief overviews of the terminology are found in Tropfke (1940, pp. 70f.) and Tropfke (1923, pp. 33f.). Bossong (1978, pp. 171–173) discusses the Spanish terms and its Arabic and Sanskrit sources.

References

ibn Aflaḥ, Jābir (1534). *Gebri Filii Affla Hispalensis De astronomia libri IX* [Translated by Gerard of Cremona]. In: Apian, Peter. *Instrumentum primi mobilis*. Nürnberg.

Apian, Peter (1532). *Ein kurtzer bericht d'Observation unnd urtels des Jüngst erschinnen Cometen*. Ingolstadt.

Apian, Peter (1533). *Introductio geographica*. Ingolstadt.

Bassantin, James (1557). *Astronomique discours*. Lyon.

al-Battānī, Muḥammad ibn Jābir (1537). *De motu stellarum*. Nürnberg.

al-Bīrūnī, Abū Rayḥān Muḥammad ibn Aḥmad (1934). *The Book of Instruction in the Elements of the Art of Astrology*. Written in Ghaznah, 1029 A. D. Reproduced from Brit. Mus. MS. Or. 8349. The Translation facing the Text by R. Ramsay Wright. London.

Blagrave, John (1585). *The mathematical ievvel*. London.

Borough, William (1581). *A Discours of the Variation of the Cumpas, or Magneticall Needle*. London.

Bossong, Georg, ed. (1978). *Los Canones de Albateni*. Herausgegeben sowie mit Einleitung, Anmerkungen und Glossar versehen. Tübingen.

Burgess, Ebenezer (1858). "Translation of the Surya-Siddhanta, a text-book of Hindu Astronomy". *Journal of the American Oriental Society* 6, pp. 141–498.

Chabás, José and Bernard R. Goldstein (2003). *The Alfonsine Tables of Toledo*. Dordrecht.

Curtze, Maximilian (1900). "Urkunden zur Geschichte der Trigonometrie im christlichen Mittelalter". *Bibliotheca Mathematica* 3, pp. 321–416.

Datta, Bibhutibhusan and Avadhesh Narayan Singh (1983). "Hindu Trigonometry (revised by Kripa Shankar Shukla)". *Indian Journal of History of Science* 18, pp. 39–108.

Fine, Oronce (1532). *Protomathesis*. Tradotte da Cosimo Bartoli. Paris.

Hughes, Barnabas (1967). *Regiomontanus: On Triangles. De triangulis omnimodis*, by Johann Müller, otherwise known as Regiomontanus. Translated, with an introduction and notes. Madison, Wisconsin.

Magini, Giovanni Antonio (1609). *Primum mobile duodecim libris contentum*. Bologna.

Nallino, Carlo Alfonso, ed. (1907). *Al-Battānī sive Albatenii Opus astronomicum*. Pars secunda. Milano.

Regiomontanus, Johannes (1533). *De triangulis omnimodis*. Ed. by Johannes Schöner. Nürnberg.

Stevin, Simon (1608). *Wisconstighe Ghedachtenissen*. Eerste deel des Weereltschrifts van den Driehouckhandel. Leiden.

Toomer, G.J. (1984). *Ptolemy's Almagest*. London.

Tropfke, Johannes (1923). *Geschichte der Elementar-Mathematik*. Fünfter Band. I. Ebene Trigonometrie. II. Sphärik und sphärische Trigonometrie. Berlin and Leipzig.

Tropfke, Johannes (1940). *Geschichte der Elementar-Mathematik*. Vierter Band. Ebene Geometrie. Dritte, verbesserte und vermehrte Auflage besorgt von Dr. Kurt Vogel. Berlin and Leipzig.

Zepeda, Henry, ed. (2018). *The First Latin Treatise on Ptolemy's Astronomy: The* Almagesti minor *(c. 1200)*. Turnhout.

complement **(set)**

The concept of the complement of a set is found in Cantor (1883, p. 51): "if Q is a set contained in P and R is that set that remains when Q is removed from P".[1] Cantor wrote $R \equiv P - Q$, but did not use any specific name for the set R. The terminology

[1] "wenn Q eine in P enthaltene Menge und R diejenige Menge ist, welche übrig bleibt, wenn man Q von P entfernt".

complementary set seems to be due to Jordan (1892, p. 72),[2] who used *ensemble complémentaire*: "If the set E does not contain all possible points, the points which do not belong to it form a *complementary* set E_1".[1] Jordan (1893, p. 20) gave the same definition in his well-known *Calcul différentiel*.

Some years after Jordan introduced the term, Schoenflies (1900, p. 63) introduced the German word *Complementärmenge*: "If the set P decomposes into the subsets P_1 and P_2, they are called *complementary* sets".[2] In English, Osgood (1897, p. 173) used *complementary set* and Pierpont (1905, p. 149) just used *complement*. Pierpont named *difference* as a synonym in the case that one of the sets is a subset of the other, as did Hausdorff (1914, p. 4).[3]

Note. The history of this term does not seem to have attracted much attention, and very few writers pointed to the first use of the term. The only reference that can be provided here is Janiszewski (1912, p. 99), who refers to Jordan (1893, p. 20) for the term.

References

Cantor, Georg (1883). "Ueber unendliche, lineare Punktmannichfaltigkeiten. 4". *Mathematische Annalen* 21, pp. 51–58.
Cantor, Georg (1885). "Über verschiedene Theoreme aus der Theorie der Punctmengen in einem n-fach ausgedehnten stetigen Raume G_n. Zweite Mittheilung". *Acta Mathematica* 7, pp. 105–124.
Hausdorff, Felix (1914). *Grundzüge der Mengenlehre*. Leipzig.
Janiszewski, Zygmunt (1912). "Sur les continus irréductibles entre deux points". *Journal de l'École polytechnique* 16, pp. 79–170.
Jordan, Camille (1892). "Remarques sur les intégrales définies". *Journal de mathématiques pures et appliquées* 8, pp. 69–99.
Jordan, Camille (1893). *Cours d'analyse*. Tome premier. Calcul différentiel. Paris.
Osgood, William F. (1897). "Non-Uniform Convergence and the Integration of Series Term by Term". *American Journal of Mathematics* 19, pp. 155–190.
Pierpont, James (1905). *Lectures on the Theory of Functions of Real Variables*. Volume I. Boston.
Schoenflies, Arthur (1900). "Die Entwickelung der Lehre von den Punktmannigfaltigkeiten". *Jahresbericht der Deutschen Mathematiker-Vereinigung* 8, pp. 1–250.
Schröder, Ernst (1877). *Der Operationskreis des Logikkalkuls*. Leipzig.

complete **(axioms)**

There are several types of completeness and in early discussions of the concept, it is often difficult to determine which of these types may be meant, or if completeness was meant at all when the term *complete* was used. This applies, in particular, to various publications by Hilbert, who began using the term *vollständig* in the late 1890s and resumed it on various occasions.[4] His book on the foundations of geometry begins as follows:

> The following investigation is a new attempt to choose for geometry a simple and complete set [*vollständiges System*] of independent axioms and to deduce from these the most important

---

[2] Schröder (1877, p. 7) had earlier uses German *komplementär* in connection with classes (*Klassen*).

[1] "Si l'ensemble E ne contient pas tous les points possibles, les points qui ne lui appartiennent pas forment un ensemble *complémentaire* E_1". [2] "Zerfällt die Menge P in die Teilmengen P_1 und P_2, so heissen sie *Complementärmengen* bezüglich P". [3] Cantor used *Differenz* occasionally at best, for example in Cantor (1885, p. 108). [4] The following paragraphs on Hilbert are based on Awodey and Reck (2002).

geometrical theorems in such a manner as to bring out as clearly as possible the significance of the different groups of axioms and the scope of the conclusions to be derived from the individual axioms (Hilbert 1899, Einleitung).[1]

Awodey and Reck (2002, p. 9) remark that "[t]hroughout *Grundlagen*, Hilbert does not elaborate much on what he means by 'completeness' in passages such as these. It is clear [...], however, that he takes to be one of his primary goals what we have called relative completeness (in the informal sense), namely with respect to 'the most important geometric theorems' recognized by the mathematicians of his time".

A second example is found in Hilbert's article "On the concept of number", where a definition is given: "The necessary task then arises of showing the *consistency* and the *completeness* of these axioms, i. e., it must be proved that the application of the given axioms can never lead to a contradiction and, further, that the system of axioms is adequate to prove all geometrical propositions" (Hilbert 1900, p. 181).[2] Awodey and Reck (2002, p. 11) interpret this usage as a case of "logical completeness: the (informal) provability of all truths of geometry from his axioms" and remark: "Overall it seems fair to say, however, that Hilbert is just not entirely clear on the notion of 'completeness' at the time of writing *Grundlagen* and '*Über den Zahlbegriff*'".

A final example[3] from Hilbert is from lectures he gave in 1917; Zach (1999, p. 340) cites it as "the first time that completeness is formulated as a precise mathematical question to be answered for a system of axioms": "Let us now turn to the question of *completeness*. We want to call the system of axioms under consideration complete if we always obtain an inconsistent system of axioms by adding a formula which is so far not derivable to the system of basic formulas (Ewald and Sieg 2013, p. 157).[4]

The term *complete* was used in a different way by Huntington. In a paper entitled "A Complete Set of Postulates for the Theory of Absolute Continuous Magnitude", he presented six postulates that "form a *complete set*; that is, they are (I) *consistent*, (II) *sufficient*[5], (III) *independent* (or *irreducible*). By these three terms we mean: (I) there is at least one assemblage in which the chosen rule of combination satisfies all the six requirements; (II) there is essentially only one such assemblage possible; (III) none of the six postulates is a consequence of the other five".[6] Scanlan (1991, p. 982) comments on Huntington: "The use of 'complete' here is an idiosyncratic usage of Huntington's, which he drops in later papers".

---

[1] "Die vorliegende Untersuchung ist ein neuer Versuch, für die Geometrie ein einfaches und vollständiges System von einander unabhängiger Axiome aufzustellen und aus denselben die wichtigsten geometrischen Sätze in der Weise abzuleiten, dass dabei die Bedeutung der verschiedenen Axiomgruppen und die Tragweite der aus den einzelnen Axiomen zu ziehenden Folgerungen möglichst klar zu Tage tritt". Translation from Hilbert (1902, Introduction). [2] "Es entsteht dann die Aufgabe, die *Widerspruchslosigkeit* und *Vollständigkeit* dieser Axiome zu zeigen, d. h. es muss bewiesen werden, daß die Anwendung der aufgestellten Axiome nie zu Widersprüchen führen kann, und ferner, daß das System der Axiome zum Nachweis aller geometrischen Sätze ausreicht". Translation from Ewald (1996, pp. 1092f.). [3] For Hilbert and the *axiom of completeness*, not discussed here any further, see Awodey and Reck (2002, pp. 11–13). [4] "Wenden wir uns nun zu der Frage der *Vollständigkeit*. Wir wollen vorgelegte Axiomen-System vollständig nennen, falls durch die Hinzufügung einer bisher nicht ableitbaren Formel zu dem System der Grundformeln stets ein widerspruchsvolles Axiomen-System entsteht". Translation from Zach (1999, p. 340). [5] *sufficient* was Huntington's term for *categorical*. [6] For some comments on Huntington'a paper, see Awodey and Reck (2002, pp. 15f.) and Scanlan (1991).

In the 1920s, Fraenkel published two revised editions of his book on set theory; these contain discussions of completeness (1923, pp. 226–228, 1928, pp. 347–354). He remarked that "this property [completeness] has been studied much less [than independence] so far and, when studied at all, has was not always been understood in the same sense" (Fraenkel 1928, p. 347).[1] The three kinds of "completeness" (*Vollständigkeit*) he discussed are categoricity, decidability, and non-forkability.[2]

Further terminological differentiations are of later origin. For example, *semantically complete* was used by Tarski in a lecture from 1940 entitled "On the completeness and categoricity of deductive systems".[3] In this lecture, which was first published in 2010, Tarski defined: "a system of sentences of a given deductive theory is called *semantically complete* if every sentence which can be formulated in the given theory is such that either it or its negation is a logical consequence of the considered set of sentences" (Mancosu 2010, p. 490).

In print, the term appeared in Myhill (1950, pp. 188f.):

> [...] there are three senses in which a system may be called complete.
>
> (1) It may be called complete if the addition to the axioms of any undemonstrable formula would render every formula demonstrable [...]
>
> (2) It may be called complete if either p or (p) is provable for every formula p [...]
>
> (3) It may be called complete if every formula which is true (with reference to a specified interpretation) is provable. We shall refer to this as semantical completeness.

See also *consistent* (axioms), *categorical*, and *independent* (axioms).

Note. The development of the terminology is much more complex than has been presented here. For further details, see Awodey and Reck (2002), Zach (1999), Mancosu, Zach, and Badesa (2009), and Corcoran (1981).

References

Awodey, Steve and Erich H. Reck (2002). "Completeness and Categoricity, Part I: Nineteenth-century Axiomatics to Twentieth-century Metalogic". *History and Philosophy of Logic* 23, pp. 1–30.

Corcoran, John (1981). "From Categoricity to Completeness". *History and Philosophy of Logic* 3, pp. 113–119.

Ewald, William (1996). *From Kant to Hilbert*. A Source Book in the Foundations of Mathematics. Volume II. Oxford.

Ewald, William and Wilfried Sieg, eds. (2013). *David Hilbert's Lectures on the Foundations of Arithmetic and Logic 1917-1933*. Berlin and Heidelberg.

Fraenkel, Adolf (1923). *Einleitung in die Mengenlehre*. Zweite erweiterte Auflage. Berlin.

Fraenkel, Adolf (1928). *Einleitung in die Mengenlehre*. Dritte umgearbeitete und stark erweiterte Auflage. Berlin.

Hilbert, David (1899). *Grundlagen der Geometrie*. Leipzig.

Hilbert, David (1900). "Über den Zahlbegriff". *Jahresbericht der Deutschen Mathematiker-Vereinigung* 8, pp. 180–183.

Hilbert, David (1902). *The Foundations of Geometry*. Authorised translation by E.J. Townsend. Chicago.

---

[1] "Diese Eigenschaft ist bisher viel weniger behandelt und, soweit es überhaupt geschah, nicht immer im gleichen Sinn verstanden worden". Translation from Awodey and Reck (2002, p. 22).

[2] For accounts of Fraenkel's discussion, see Mancosu (2010, pp. 472–474) and Awodey and Reck (2002, pp. 21–26).

[3] The German terms *semantisch vollständig* und *syntaktisch vollständig* are found in Scholz (1942-44, p. 221).

Mancosu, Paolo (2010). *The Adventure of Reason.* Interplay Between Philosophy of Mathematics and Mathematical Logic, 1900–1940. Oxford.
Mancosu, Paolo, Richard Zach, and Calixto Badesa (2009). "The Development of Mathematical Logic from Russell to Tarski, 1900–1935". In: *The Development of Modern Logic*. Ed. by Leila Haaparanta. Oxford, pp. 318–470.
Myhill, John R. (1950). "A Complete Theory of Natural, Rational, and Real Numbers". *The Journal of Symbolic Logic* 15, pp. 185–196.
Scanlan, Michael (1991). "Who were the American Postulate Theorists?" *Journal of Symbolic Logic* 56, pp. 981–1002.
Scholz, Heinrich (1942-44). "Was will die formalisierte Grundlagenforschung?" *Deutsche Mathematik* 7, pp. 206–248.
Zach, Richard (1999). "Completeness Before Post: Bernays, Hilbert, and the Development of Propositional Logic". *Bulletin of Symbolic Logic* 5, pp. 331–366.

complete continuity

The German term *vollstetig* (*voll* 'full(y)', *stetig* 'continuous') is due to Hilbert (1906, p. 200). He used it for functions of infinitely many variables, which he wrote as $F(x_1, x_2, \dots)$, with the property that the values $F(x_1 + \varepsilon, x_2 + \varepsilon, \dots)$ converge towards $F(x_1, x_2, \dots)$ if the ε tend to zero.

The term was soon adopted by others, for example by Hilb (1908, p. 9), who referred to Hilbert. English *completely continuous* was used by Hart (1916, p. 310). Riesz (1913, p. 96) applied the term *complètement continue* to operators; in Riesz (1918, p. 74), he defined a linear transformation to be completely continuous if it transforms every bounded sequence into a compact one.

Hille (1948, p. 14) later suggested a change in the terminology. After having given a definition of *compact transformation*, he remarked: "Here the usual term is 'completely continuous transformation.' This is a misnomer, however, since a compact transformation need not be continuous anywhere".

See also *strong* and *weak* (convergence, topology) and *compact*.

Note. Historical remarks on the concept and the terminology can be found in Pietsch (2007, pp. 45, 49f.).

References

Hart, William L. (1916). "Differential Equations and Implicit Functions in Infinitely Many Variables". *Proceedings of the National Academy of Sciences* 2, pp. 309–313.
Hilb, Emil (1908). "Über Integraldarstellungen willkürlicher Funktionen". *Mathematische Annalen* 66, pp. 1–66.
Hilbert, David (1906). "Grundzüge einer allgemeinen Theorie der linearen Integralgleichungen. (Vierte Mitteilung)". *Nachrichten von der Königl. Gesellschaft der Wissenschaften zu Göttingen (mathematisch-physikalische Klasse)* (1906), pp. 157–228.
Hille, Einar (1948). *Functional Analysis and Semi-Groups*. New York.
Pietsch, Albrecht (2007). *History of Banach Spaces and Linear Operators*. Boston, Mass. et al.
Riesz, Frigyes (1913). *Les systèmes d'équations linéaires à une infinité d'inconnues*. Paris.
Riesz, Frigyes (1918). "Über lineare Funktionalgleichungen". *Acta Mathematica* 41, pp. 71–98.

complete graph

According to König (1936, p. 25), the term was introduced as French *réseau complet* by Sainte-Laguë (1923, p. 28), who called a graph "*complete* if it has edges that join all vertices pairwise".[1] König (1936, p. 25) used the literal German translation *vollständiger Graph*, and *complete graph* is found in Erdős and Szekeres (1935, p. 466).

References

Erdős, Paul and George Szekeres (1935). "A combinatorial problem in geometry". *Compositio Mathematica* 2, pp. 463–470.

Golumbic, Martin Charles and André Sainte-Laguë (2021). *The Zeroth Book of Graph Theory*. An Annotated Translation of Les Réseaux (ou Graphes)–André Sainte-Laguë (1926). Cham.

König, Dénes (1936). *Theorie der endlichen und unendlichen Graphen*. Leipzig.

Sainte-Laguë, André (1923). "Les réseaux". *Annales de la Faculté des sciences de Toulouse* 15, pp. 27–86.

complete quadrangle* and *complete quadrilateral

The term *complete quadrilateral* was coined by Carnot (1801, p. 122) as French *quadrilatère complet*.[1] Carnot proceeded from four straight lines in a plane, no three of which pass through a same point. This produces six points of intersection A, B, C, D, E, F and three quadrilaterals $ABDC$, $AFDG$, $BFCG$ (see fig. 19). Each of these quadrilaterals has two diagonals, which, in the given configuration, reduce to three (AD, BC, FG); for example, the quadrilaterals $ABCD$ and $AFDG$ share the diagonal AD. The "assemblage" (*assemblage*) of these three quadrilaterals was called *quadrilatère complet* by Carnot.

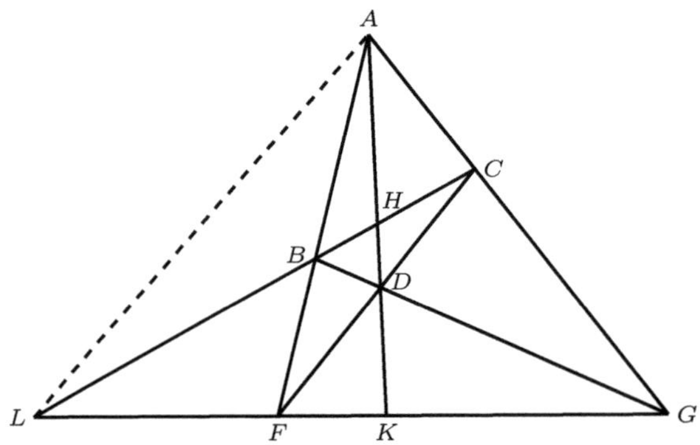

Figure 19: A complete quadrilateral. From Carnot (1801, Pl. II).

The term was soon adopted by Brianchon (1806, p. 298) and others. In Carnot (1810, p. 208), it was translated by German *vollständige vierseitige Figur* 'complete four-sided figure'. English *complete quadrilateral* seems to have been first used by Davies (1835, p. 125).

About 30 years after Carnot introduced the term, Steiner (1832, p. 72) remarked that Carnot's term is frequently used, but he also criticized that "two essentially

[1] "[Un réseau] est *complet* s'il contient tous les chemins joignant deux à deux ses sommets". Translation from Golumbic and Sainte-Laguë (2021, p. 10). [1] The origin of the terminology in the work of Carnot and Steiner (see below) has frequently been pointed out, for example by Tropfke (1940, p. 128).

different figures have not been distinguished". Accordingly, he differentiates between a complete quadrilateral (German *vollständiges Vierseit*) and its dual, a complete quadrangle (*vollständiges Viereck*).

Townsend (1863, p. 144) coined the neologisms *polystigm* (Greek *στίγμα* 'point') and *polygram* (*γράμμα* 'line'): "Every system of points or lines, whatever be their number and position, in which the several lines of connection or points of intersection of each point or line with all the others are taken into account without exception, is said to form a complete figure, which in the absence of any as yet generally recognized nomenclature may be termed a *polystigm* in the former case and a *polygram* in the latter". For the special case of four points or lines, he correspondingly used *tetrastigm* and *tetragram* (Townsend 1863, p. 145).

The notion is considerably older than the term. Complete spherical quadrilaterals were used by Menelaus, Ptolemy, and later by various Arabic astronomers and mathematicians.[1] In Arabic texts, the usual term was *al-shakl al-qaṭṭāʿ*, used by Thābit ibn Qurra in the 9th century (see Rashed 2017, pp. 415–417). The Arabic term is commonly translated by *sector figure*; the corresponding Latin term was used in the 12th century in a translation by Gerard of Cremona of Thābit's treatise on the sector figure (Bjørnbo 1924, p. 6).

References

Bjørnbo, Axel Anthon (1924). *Thabits Werk über den Transversalensatz (liber de figura sectore)*. Mit Bemerkungen von H. Suter. Ed. by H. Bürger and K. Kohl. Erlangen.

Brianchon, Charles Julien (1806). "Sur les surfaces courbes du second degré". *Journal de l'École polytechnique* 6, cahier 13, pp. 297–311.

Carnot, L.N.M. (1801). *De la corrélation des figures de géométrie*. Paris.

Carnot, L.N.M. (1810). *Geometrie der Stellung*. Übersetzt von H.C. Schumacher. Erster Theil. Altona.

Davies, Thomas Stephens (1835). "Question 524". *New Series of The Mathematical Repository* 6, pp. 124–125.

Rashed, Roshdi, ed. (2017). *Lexique historique de la langue scientifique arabe*. Hildesheim.

Sidoli, Nathan (2006). "The Sector Theorem Attributed to Menelaus". *Sciamvs: Sources and Commentaries in Exact Sciences* 7, pp. 43–79.

Steiner, Jacob (1832). *Systematische Entwickelung der Abhängigkeit geometrischer Gestalten von einander*. Erster Theil. Berlin.

Townsend, Richard (1863). *Chapters on the Modern Geometry of the Point, Line, and Circle*. Vol. I. Dublin.

Tropfke, Johannes (1940). *Geschichte der Elementar-Mathematik*. Vierter Band. Ebene Geometrie. Dritte, verbesserte und vermehrte Auflage besorgt von Dr. Kurt Vogel. Berlin and Leipzig.

complete space

The notion is due to Fréchet (1906, p. 23), who did not use the term *complete*, however. Instead, he said that such a space "allows a generalisation of the theorem of Cauchy".[2] Hausdorff (1914, p. 315) introduced German *vollständig* 'complete', explicitly referring to Fréchet for the notion; he defined: "We shall call a metric space *complete if every fundamental sequence converges*".[3] The English term is found in Blumberg (1920, p. 127).

---

[1] For a detailed historical account, see Sidoli (2006). [2] "admet une généralisation du théorème de Cauchy". [3] "Wir wollen einen metrischen Raum *vollständig* nennen, *wenn jede Fundamentalfolge konvergiert*".

Note. For the history of the concept, see Dugac (1984); see also Taylor (1982, p. 256) on Fréchet's later use of the same word in a different meaning.

References

Blumberg, Henry (1920). "Hausdorff's Grundzüge der Mengenlehre [review]". *Bulletin of the American Mathematical Society* 27, pp. 116–129.
Dugac, Pierre (1984). "Histoire des espaces complets". *Revue d'histoire des sciences* 37, pp. 3–28.
Fréchet, Maurice (1906). "Sur quelques points du calcul fonctionnel". *Rendiconti del Circolo Matematico di Palermo* 22, pp. 1–74.
Hausdorff, Felix (1914). *Grundzüge der Mengenlehre*. Leipzig.
Taylor, Angus E. (1982). "A Study of Maurice Fréchet: I. His Early Work on Point Set Theory and the Theory of Functionals". *Archive for History of Exact Sciences* 27, pp. 233–295.

complex number

This term was first used by Gauß (1831, p. 628), who introduced German *complexe ganze Zahl* 'complex whole number':

> For the true grounding of the theory of biquadratic residues, one must extend the field of higher arithmetic, which has hitherto been confined to the real integers, into the imaginary integers, and must concede to the latter the same legitimacy as the former [...] The author calls every quantity $a+bi$, where a and b denote real quantities and where i is an abbreviation for $\surd -1$, a complex integer, provided that a and b are integers. So the complex quantities are not opposed to the real, but contain them as a special case where $b = 0$.[1]

In the actual article, written in Latin, Gauß (1832, § 30) used *numeros integros complexos*. The corresponding French term *nombre entier complexe* was already used by Lejeune Dirichlet (1832, p. 380); according to the OED, the English term is first attested in H. Smith (1860, § 25).

Gauß (1831, p. 634) also discusses the earlier term *impossible numbers*: "only the imaginary quantities – which are contrasted with the real ones, and which were formerly, and are still occasionally (although improperly) called *impossible* – are still merely tolerated rather than fully accepted".[2]

The word *impossible* is found early on. Cardano (1545, ch. 37, fol. 66^{r}) discusses the equation $x(10-x) = 40$, an "impossible case". He nevertheless proceeds and gives $5+\sqrt{-15}$ and $5-\sqrt{-15}$ as solutions, calling $\sqrt{-15}$ a "sophistic quantity" (*quantitas sophistica*).[3] French *solutions impossibles* was used by Girard (1629, fol. F 1^{r}).

As a curiosity, it may be worth pointing out one terminological suggestion that does not appear to have had any followers at all: Matzka (1850, p. 51) preferred the

---

1 "Es ist dieses nichts anders, als daß für die wahre Begründung der Theorie der biquadratischen Reste das Feld der höhern Arithmetik, welches man sonst nur auf die reellen ganzen Zahlen ausdehnte, auch über die imaginären erstreckt werden, und diesen das völlig gleiche Bürgerrecht mit jenen eingeräumt werden muß [...] Der Verf. nennt jede Größe $a+bi$, wo a und b reelle Größen bedeuten, und i der Kürze wegen anstatt $\surd -1$ geschrieben ist, eine complexe ganze Zahl, wenn zugleich a und b ganze Zahlen sind. Die complexen Größen stehen also nicht den reellen entgegen, sondern enthalten diese". Translation from Gauß (1996, pp. 308f.). The word *legitimacy* here translates German *Bürgerrecht*, which lit. means '(right of) citizenship'. 2 "Allein die den reellen Größen gegenübergestellten imaginären – ehemals, und hin und wieder noch jetzt, obwohl unschicklich, unmögliche genannt – sind noch immer weniger eingebürgert als nur geduldet". Translation from Gauß (1996, p. 311). 3 Remmert (1995, p. 57) suggests translating this by "formal number".

German neologisms *Bifarial* or *Bifarielle*, "from [Latin] *bis* and *fari*, to mean two things, or from *bifariam* 'in two directions'".

See also *imaginary*.

Note. Of the vast amount of literature on complex numbers, see, in particular, Tropfke (1980, pp. 151–158), D. E. Smith (1958, pp. 261–267), and Remmert (1995, § 1).

References

Cardano, Gerolamo (1545). *Artis magnae sive de regulis algebraicis liber unus*. Nürnberg.

Gauß, Carl Friedrich (1831). "Theoria residuorum biquadraticorum, commentatio secunda [announcement]". *Göttingische gelehrte Anzeigen* (1831), pp. 625–638.

Gauß, Carl Friedrich (1832). "Theoria residorum biquadraticorum. Commentatio secunda". *Commentationes Societatis Regiae Scientiarum Gottingensis recentiores* 7, pp. 89–148.

Gauß, Carl Friedrich (1996). "Notice on the Theory of Biquadratic Residues". In: *From Kant to Hilbert*. A Source Book in the Foundations of Mathematics. Volume I. Ed. by William Ewald. Oxford, pp. 306–313.

Girard, Albert (1629). *Invention nouvelle en algèbre*. Amsterdam.

Lejeune Dirichlet, Gustav (1832). "Démonstration d'une propriété analogue à la loi de réciprocité qui existe entre deux nombres premiers quelconques". *Journal für die reine und angewandte Mathematik* 9, pp. 379–389.

Matzka, Wilhelm (1850). *Versuch einer richtigen Lehre von der Realitaet der vorgeblich imaginären Grössen der Algebra*. Prag.

Remmert, Reinhold (1995). "Complex Numbers". In: Ebbinghaus, Heinz-Dieter et al. *Numbers*. With an Introduction by K. Lamotke. Translated by H.L.S. Orde. Edited by J.H. Ewing. New York, pp. 55–96.

Smith, David Eugene (1958). *History of Mathematics*. Volume II. Special Topics of Elementary Mathematics. New York.

Smith, H.J.S. (1860). "Report on the Theory of Numbers. Part I". *Report of the British Association for the Advancement of Science* (1859), pp. 228–267.

Tropfke, Johannes (1980). *Geschichte der Elementarmathematik*. Band 1: Arithmetik und Algebra. Vollständig neu bearbeitet von Kurt Vogel, Karin Reich, Helmuth Gericke. Berlin and New York.

***complex* (topology)**

By a *complex*, Listing (1862, p. 181) understood "any aggregate of points, lines, surfaces, in- or excluding the solid spaces bounded by them".[1] He chose the German name *Complex* because complexes are made up of simpler objects[2], which he called *constituents* (*Constituenten*). The word first appeared in Listing (1861, p. 354); detailed discussions are found in Listing (1862, §§ 1, 42). Tait (1884, p. 46) used English *complex* in Listing's sense.

The terminology spread in the 20th century.[3] It was used by Dehn and Heegaard in their article on "analysis situs" (Dehn and Heegaard 1907); in their study on the history of complexes, Burde and Zieschang (1999, p. 105) remark that Dehn and Heegaard gave "the first abstract definition of a complex". After pointing out that the term was due to Listing, Dehn and Heegaard (1907, p. 156) proceeded from "points" and called their totality a *point complex* (*Punktkomplex*).[4] From these points, new

---

[1] "jedes Aggregat von Punkten, Linien und Flächen ein- oder ausschließlich der dadurch abgegrenzten körperlichen Räume". [2] An explicit explanation of the term along these lines was given by Alexandroff and Hopf (1935, p. 1). [3] Poincaré did not use the term in his papers on analysis situs.

[4] The definitions as sketched in the following are found in Dehn and Heegaard (1907, pp. 156–163). For an account of their definition, see Bollinger (1972, pp. 144–146); see also Vanden Eynde (1992, pp. 165–168).

elements called *line segments* (*Strecken oder Linienstücke*) could be built, resulting in a *Streckenkomplex*, which they also called *Linienkomplex* or *one-dimensional complex* (*eindimensionaler Komplex*). Using the concepts of "circle" (*Kreis*) and "sphere" (*Kugel*), they went on to define higher-dimensional complexes.

With varying definitions, the terminology is found in numerous texts, for example in Veblen and Alexander (1912-13, p. 165). For various kinds of complexes, often just referred to as a *complex*, see *simplicial*, *cell complex*, *CW-complex*, and *chain complex*. The terms *geometric complex* and *algebraic complex* seem to be due to Alexandroff (1932, pp. 6, 16)

Note. Historical remarks on complexes are found in Burde and Zieschang (1999) and Klette (2000).

References

Alexandroff, Pavel S. (1932). *Einfachste Grundbegriffe der Topologie*. Berlin.

Alexandroff, Pavel S. and Heinz Hopf (1935). *Topologie*. Erster Band. Berlin.

Bollinger, Maja (1972). "Geschichtliche Entwicklung des Homologiebegriffs". *Archive for History of Exact Sciences* 9, pp. 94–170.

Burde, Gerhard and Heiner Zieschang (1999). "Development of the Concept of a Complex". In: *History of Topology*. Ed. by Ioan Mackenzie James. Amsterdam, pp. 103–110.

Dehn, Max and Poul Heegaard (1907). "Analysis situs". *Encyklopädie der mathematischen Wissenschaften mit Einschluss ihrer Anwendungen* III AB3, pp. 153–220.

Vanden Eynde, Ria (1992). "Historical Evolution of the Concept of Homotopic Paths". *Archive for History of Exact Sciences* 45, pp. 127–188.

Klette, Reinhard (2000). "Cell Complexes through Time". *Proceedings of the SPIE* 4117, pp. 134–145.

Listing, Johann Benedict (1861). "Der Census räumlicher Complexe, oder Verallgemeinerung des Euler'schen Satzes von den Polyëdern". *Nachrichten von der Georg-Augusts-Universität und der Königl. Gesellschaft der Wissenschaften zu Göttingen* (1861), pp. 352–358.

Listing, Johann Benedict (1862). "Der Census räumlicher Complexe, oder Verallgemeinerung des Euler'schen Satzes von den Polyëdern". *Abhandlungen der Königlichen Gesellschaft der Wissenschaften zu Göttingen* 10, pp. 97–182.

Tait, Peter Guthrie (1884). "Listing's *Topologie*". *The London, Edinburgh and Dublin Philosophical Magazine* 17, pp. 30–46.

Veblen, Oswald and James W. Alexander (1912-13). "Manifolds of N Dimensions". *Annals of Mathematics* 14, pp. 163–178.

composite number

See *prime number*.

composition series

The term *composition series* can be ascribed to Jordan and Netto. In his book on the theory of substitutions he introduced German *Reihe der Zusammensetzung*, lit. 'series of composition', Netto defined:

> If G is a compound group, and if the series of groups
>
> $$G, G_1, G_2, \ldots, G_\mu, 1$$

> is so taken that every G_λ is a maximal self-conjugate subgroup of the preceding one, then this series is called *the series belonging to the compound group G*, or *the series of composition of G*, or, still more briefly, *the series of G* (Netto 1882, p. 87).[1]

While Netto explicitly used *series of composition*, Jordan (1869, pp. 152f., 1870, pp. 41f.) already used *suite* and *facteurs de composition* in this context. At that time, the composition factors were actual numbers rather than quotient groups. The corresponding transition was made by Hölder (1889, p. 30): "But this theory of the factors of composition must then be deepened, so that the factors are interpreted as *groups*".[2]

English *series of composition* and *factors of composition* were used in Jordan's sense by Bolza (1890-91, p. 97).

Note. Historical background can be found in Corry (2004, ch. 1.1), for example.

References

Barnett, Janet Heine (2018). "Otto Hölder's Formal Christening of the Quotient Group Concept". *Abstract Algebra* 3. URL: https://digitalcommons.ursinus.edu/triumphs_abstract/3/ (retrieved June 30, 2020).

Bolza, Oskar (1890-91). "On the Theory of Substitution-Groups and its Applications to Algebraic Equations". *American Journal of Mathematics* 13, pp. 59–144.

Corry, Leo (2004). *Modern Algebra and the Rise of Mathematical Structures*. Second revised edition. Basel.

Hölder, Otto (1889). "Zurückführung einer beliebigen algebraischen Gleichung auf eine Kette von Gleichungen". *Mathematische Annalen* 34, pp. 26–56.

Jordan, Camille (1869). "Commentaire sur Galois". *Mathematische Annalen* 1, pp. 141–160.

Jordan, Camille (1870). *Traité des substitutions et des équations algébriques*. Paris.

Netto, Eugen (1882). *Substitutionentheorie und ihre Anwendungen auf die Algebra*. Leipzig.

Netto, Eugen (1892). *The Theory of Substitutions and its Applications to Algebra*. Revised by the author and translated with his permission by F.N. Cole. Ann Arbor, Mich.

computable

The French term *calculable* was first defined by Borel (1912, p. 162) for numbers: "We say that a number α is computable if, given a natural number n, we can obtain a rational number that differs from α by at most $\frac{1}{n}$".[3] Borel (1912, p. 166) then went on to define *fonction calculable*:

> We say that a function is computable if its value is computable for any computable value of the variable. In other words, if α is a computable number, one has to know how to compute the value of $f(\alpha)$ with precision $\frac{1}{n}$ for any n. One should not forget that, by definition, to be given a computable number α just means to be given a method to obtain an arbitrary approximation to α".[4]

---

[1] "Ist G zusammengesetzt und die Reihe der Gruppen $G, G_1, G_2, \ldots, G_\mu$, 1 so beschaffen, dass jede folgende Gruppe eine ausgezeichnete Maximaluntergruppe der vorhergehenden ist, so heisst diese Reihe *die zur zusammengesetzten Gruppe G gehörige Reihe*, oder auch *die Reihe der Zusammensetzung von G*, respektive kurz *die Reihe von G*". Translation from Netto (1892, p. 96).

[2] "Diese Theorie von den Factoren der Zusammensetzung muss aber dahin vertieft werden, dass die Factoren als *Gruppen* aufgefasst werden". Translation from Barnett (2018, p. 8).

[3] "Nous dirons qu'un nombre α est calculable lorsque, étant donné un nombre entier quelconque n, on sait obtenir un nombre rationnel qui diffère de α de moins de $\frac{1}{n}$". Translation from Avigad and Brattka (2014, p. 5).

[4] "Nous dirons qu'une fonction est calculable, lorsque sa valeur est calculable pour toute valeur calculable de la variable. En d'autres termes, si α est un nombre calculable, on doit savoir calculer la valeur $f(\alpha)$ à $\frac{1}{n}$ près, quel que soit n [...]". Translation from Avigad and Brattka (2014, p. 5).

Computable numbers were so called by Turing (1936, p. 230):

> The 'computable' numbers may be described briefly as the real numbers whose expressions as a decimal are calculable by finite means [...] According to my definition, a number is computable if its decimal can be written down by a machine.

In the same paper, Turing (1936, p. 254) also defined *computable function.*

Note. For the history of the concept, see Adams (2011) and Soare (1999), for example.

References

Adams, Rod (2011). *An Early History of Recursive Functions and Computability from Gödel to Turing*. Boston.

Avigad, Jeremy and Vasco Brattka (2014). "Computability and analysis: the legacy of Alan Turing". In: *Turing's Legacy*. Developments from Turing's Ideas in Logic. Ed. by Rod Downey. Cambridge, pp. 1–47.

Borel, Émile (1912). "Le calcul des intégrales définies". *Journal de mathématiques pures et appliquées* 8, pp. 159–210.

Soare, Richard I. (1999). "The History and Concept of Computability". In: *Handbook of Computability Theory*. Ed. by Edward R. Griffor. Amsterdam et al., pp. 3–36.

Turing, Alan (1936). "On Computable Numbers, with an Application to the Entscheidungsproblem". *Proceedings of the London Mathematical Society* 42, pp. 230–265.

concave

See *convex*.

conchoid

In book 4 of his *Collection*, Pappus of Alexandria (1876, p. 244)[1] attributes the term *conchoid* to the Greek geometer Nicomedes, about whom little is known but who may have lived in the second century. Pappus used the term *cochleoid* (Greek *κοχλοειδής*); cp. Greek *κόχλος* 'shell-fish with a spiral shell, land snail' (LSJ). According to Heath (1921, p. 238) and Mugler (1958-59, p. 253), this seems to have been the original term; later writers such as Proclus (1873, p. 111) used the term *κογχοειδής* (*κόγχη* 'mussel', LSJ). As is very common for curve names, the term is based on the shape of the curve.

The Latin term is found in the late medieval ages. It was used by William of Moerbeke (13th century) in a translation of Eutocius' commentary on Archimedes' *On the Sphere and Cylinder* (Clagett 1976, p. 249).[2] Moerbeke used *linea conchoydalis* (other spellings occur). Early printed attestations are found in Valla (1501, book XIV, ch. 2) and Werner (1522, fol. f ii^{v}). Occasionally, the term is also re-latinized as *linea conchilis*, from *concha* 'shell'. This term is found in Eutocius (1544, p. 25) and Clavius (1604, p. 301), for example.

In modern languages, the term is found in the 17th century. Thus, as for French, Mersenne (1625, p. 862) uses *conchoide*; similarly, Schwenter (1636, p. 146), writing in German, uses *Conchoide*. In English, the term appears in its Latin form *Conchois* in Ramus (1636, p. 17), where it is explained as "Cockleshell-like". In correspondence from 1666, the term *conchoid line* was used by John Collins (Rigaud 1842, p. 469).[3]

---

[1] For an English translation, see Pappus of Alexandria (2010, p. 127). [2] For the passage in Eutocios, see Archimedes (2004, p. 301). [3] The original spelling was modernized in Rigaud's edition.

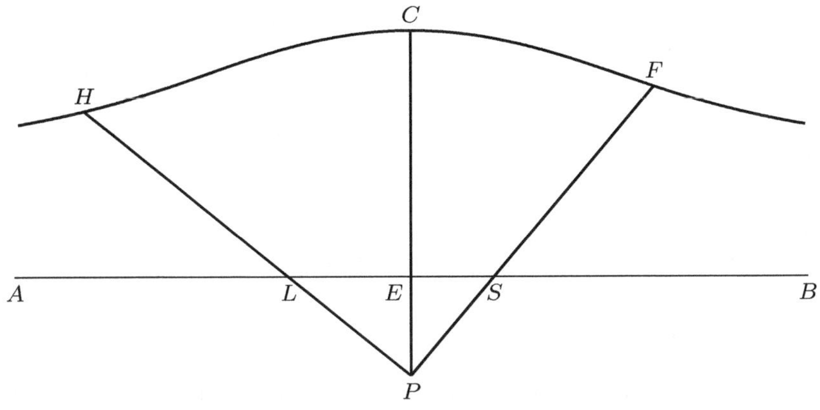

Figure 20: The conchoid of Nicomedes. Based on Clavius (1604, p. 301). The conchoid is constructed from a given pole P, a directrix AB, and a distance $d = CE$. All lines through P (except the parallel to AB) intersect AB, as PS intersects AB at S, for example. Extend the line in the direction of S. The points on the conchoid are, by definition, those for which $FS = HL = CE$ etc.

Hobbes (1656, p. 233) has "the Conchoeides", and the noun *Conchoid* appears in [anon.] (1670, p. 907).[1]

Note. Dürer's "conchoid" is not a conchoid as discussed above. He called his curve *Muschellini*, from *Muschel* 'mussel' in Dürer (1525, fol. C 6^{v}). On Dürer's conchoid, see Beech (1990) and Loria (1910, pp. 228–231). – On de Sluse's conchoid, see Loria (1910, pp. 74–77) and Teixeira (1908, pp. 26–30). It was first mentioned in a letter by de Sluse in 1662 (Huygens 1891, p. 247).

Note. For the Greek terms, see Mugler (1958-59, pp. 253, 257f.). For the history of the curve, see Heath (1921, pp. 238–240), who focuses on Greek authors, and Loria (1910, pp. 136–142).

References

[anon.] (1629). *Recreations Mathematiqves*. Rouen.

[anon.] (1670). "An Account of Two Books". *Philosophical Transactions of the Royal Society of London* 4, pp. 903–912.

Archimedes (2004). *The Works of Archimedes*. Translation and Commentary. Volume 1: The Two Books *On the Sphere and the Cylinder*. Edited and translated by Reviel Netz. Cambridge.

Beech, Martin (1990). "Albrecht Dürer and his conchoid". *Bulletin of the Institute of Mathematics and its Applications* 26, pp. 38–40.

Clagett, Marshall (1976). *Archimedes in the Middle Ages*. Volume II. The Translations from the Greek by William of Moerbeke. Philadelphia.

Clavius, Christopher (1604). *Geometria practica*. Roma.

Dürer, Albrecht (1525). *Underweysung der Messung, mit dem Zirckel und Richtscheyt*. Nürnberg.

Van Etten, Henry (1633). *Mathematicall Recreations*. London.

---

[1] The French term *Conchoïde* is mistranslated as *Conoyde Line* in Van Etten (1633, p. 236). The book is a translation of [anon.] (1629); on its authorship, see Heeffer (2006).

Eutocius (1544). *Commentarii Evtocii Ascalonitae*. In: *Archimedis syracusani philosophi ac geometriae excellentissimi Opera*. Ed. by Thomas Gechauff. Basel.
Heath, Thomas L. (1921). *A History of Greek Mathematics*. Volume I. From Thales to Euclid. Oxford.
Heeffer, Albrecht (2006). "*Récréations Mathématiques*. A Study on Its Authorship, Sources and Influence". *Gibecière* 1, pp. 77–167.
Hobbes, Thomas (1656). *Elements of Philosophy*. London.
Huygens, Christiaan (1891). *Œuvres complètes*. Tome quatrième. Correspondance 1662–1663. La Haye.
Loria, Gino (1910). *Spezielle algebraische und transzendente ebene Kurven*. Theorie und Geschichte. Autorisierte, nach dem italienischen Manuskript bearbeitete deutsche Ausgabe von Fritz Schütte. Zweite Auflage. Erster Band. Leipzig.
Mersenne, Marin (1625). *La vérité des sciences*. Paris.
Mugler, Charles (1958-59). *Dictionnaire historique de la terminologie géométrique des Grecs*. Paris.
Pappus of Alexandria (1876). *Pappi Alexandrini Collectionis*. Volumen I. Ed. by Friedrich Hultsch. Berlin.
Pappus of Alexandria (2010). *Book 4 of the* Collection. Edited With Translation and Commentary by Heike Sefrin-Weis. London.
Proclus (1873). *In primum Euclidis elementorum librum commentarii*. Ed. by Gottfried Friedlein. Leipzig.
Ramus, Petrus (1636). *Via regia ad geometriam. The Vvay to Geometry*. Written in Latine by Peter Ramus, and now Translated and much enlarged by the Learned Mr. William Bedwell. London.
Rigaud, Stephen Peter, ed. (1842). *Correspondence of Scientific Men of the Seventeenth Century*. Vol. II. Oxford.
Schwenter, Daniel (1636). *Deliciae Physico-Mathematicae*. Nürnberg.
Teixeira, Francisco Gomes (1908). *Traité des courbes spéciales remarquables plans et gauches*. Tome I. Coïmbra.
Valla, Giorgio (1501). *De expetendis, et fugiendis rebus opus*. Venezia.
Werner, Johannes (1522). *Libellus super viginti duobus elementis conicis*. Nürnberg.

conditional probability

Kolmogorov's book on the foundations of probability was a major step in the axiomatization of probability. He defined conditional probability (*bedingte Wahrscheinlichkeit*) in the following terms: "If $\mathrm{P}(A) > 0$, then the quotient $\mathrm{P}_A(B) = \frac{\mathrm{P}(AB)}{\mathrm{P}(A)}$ is called the *conditional probability* of the event B under the condition A" (Kolmogorov 1933, p. 6).[1]

Shafer and Vovk (2006, p. 89), who describe Kolmogorov's treatment of conditional probability as "novel",[2] also remark that "[b]efore the *Grundbegriffe*, this term [*conditional probability*] was less common than 'relative probability'". The importance of this concept was stressed by Hausdorff (1901, p. 152), who also used the term *relative Wahrscheinlichkeit*.[3] The terminology was adopted by Broggi (1907, p. 27) and Czuber (1908, p. 45), who both referred to Hausdorff; Markov (1912, p. 15), on the other hand, used no special term. In the 1920s, the term *bedingte Wahrscheinlichkeit* began to be used by some authors such as Tschuprow (1925, p. 39) and Baur (1928, p. 9). Piaggio (1931, p. 405), who presented some postulates

[1] "Wenn $\mathrm{P}(A) > 0$ ist, so nennt man den Quotienten $\mathrm{P}_A(B) = \frac{\mathrm{P}(AB)}{\mathrm{P}(A)}$ die *bedingte Wahrscheinlichkeit* des Ereignisses B unter der Bedingung A". Translation from Kolmogorov (1950, p. 6) (slightly modified). [2] The novelty was pointed out by Kolmogorov (1933, p. iii) himself. [3] He used the symbol $p_F(E)$ for the relative probability of E "posito F". The concept (not the terminology) also appears in Bohlmann (1901, p. 860). For Bohlmann's work, see Krengel (2011, p. 6).

for probability, stated that "the probability of two independent events both occurring is the product of their separate probabilities, with a similar law for dependent events, except that the second factor is now the *conditional* probability of the second event when the first is certain".

The germs of the concept can be found in the 18th century. de Moivre (1738, p. 7), for example, wrote: "the Probability of the happening of two Events dependent, is the product of the Probability of the happening of one of them, by the Probability which the other will have of happening, when the first shall have been consider'd as having happen'd".[1]

In conclusion, the terminology cannot be attributed to *one* person, but it can be said that the uses by Hausdorff and Kolmogorov were influential and that after the publication of Kolmogorov's book, the term *conditional*, as opposed to *relative probability*, gradually became standard. It was used, for example, by Uspensky (1937, p. 31) and by Lévy (1937, p. 68).

Note. Some remarks on the history of the concept are found in Purkert (2006, pp. 564–567); Shafer and Vovk (2006) focus on Kolmogorov. See also the articles on conditional and posterior probability by Aldrich in Miller (n.d.).

References

Baur, Franz (1928). *Korrelationsrechnung*. Leipzig and Berlin.

Bohlmann, Georg (1901). "Lebensversicherungs-Mathematik". *Encyklopädie der mathematischen Wissenschaften mit Einschluss ihrer Anwendungen* I D 4b, pp. 852–917.

Broggi, Ugo (1907). *Die Axiome der Wahrscheinlichkeitsrechnung*. Göttingen.

Czuber, Emanuel (1908). *Wahrscheinlichkeitsrechnung und ihre Anwendung auf Fehlerausgleichung, Statistik und Lebensversicherung*. Erster Band. Leipzig and Berlin.

Hausdorff, Felix (1901). "Beiträge zur Wahrscheinlichkeitsrechnung". *Berichte über die Verhandlungen der Königlich Sächsischen Gesellschaft der Wissenschaften zu Leipzig (Mathematisch-Physische Klasse)* 53, pp. 152–178.

Kolmogorov, Andrey Nikolaevich (1933). *Grundbegriffe der Wahrscheinlichkeitsrechnung*. Berlin.

Kolmogorov, Andrey Nikolaevich (1950). *Foundations of the Theory of Probability*. Translation edited by Nathan Morrison. New York.

Krengel, Ulrich (2011). "On the contributions of Bohlmann to probability theory". *Journ@l électronique d'histoire des probabilités et de la statistique* 7.

Lévy, Paul (1937). *Théorie de l'addition des variables aléatoires*. Paris.

Markov, Andrey A. (1912). *Wahrscheinlichkeitsrechnung*. Nach der zweiten Auflage des russischen Werkes übersetzt von Heinrich Liebmann. Leipzig and Berlin.

Miller, Jeff (n.d.). "Earliest Known Uses of Some of the Words of Mathematics". URL: `https://mathshistory.st-andrews.ac.uk/Miller/mathword/`.

de Moivre, Abraham (1738). *The Doctrine of Chances: or, A Method of Calculating the Probability of Events in Play*. The Second Edition. London.

Piaggio, H. T. H. (1931). "Probability and Its Applications". *The Mathematical Gazette* 15, pp. 404–411.

Purkert, Walter (2006). "Kommentar zu [H 1901a]". In: *Felix Hausdorff. Gesammelte Werke*. Band V. Astronomie, Optik und Wahrscheinlichkeitstheorie. Herausgegeben von J. Bemelmans, Ch. Binder, S.D. Chatterji, S. Hildebrand, W. Purkert, F. Schmeidler und E. Scholz. Berlin, pp. 556–590.

Schneider, Ivo (1968). "Der Mathematiker Abraham de Moivre (1667–1754)". *Archive for History of Exact Sciences* 5, pp. 177–317.

Shafer, Glenn and Vladimir Vovk (2006). "The Sources of Kolmogorov's *Grundbegriffe*". *Statistical Science* 21, pp. 70–98.

---

[1] On this passage, see Schneider (1968, pp. 279f.).

Tschuprow, A.A. (1925). *Grundbegriffe und Grundprobleme der Korrelationstheorie*. Leipzig.
Uspensky, J.V. (1937). *Introduction to Mathematical Probability*. New York and London.

conductor

The English term *conductor* was suggested by Dickson et al. (1923, p. 74) as a translation of German *Führer* (from *führen* 'to lead, conduct'): "Let us denote by G the ring of all integers of a field $k(\alpha)$ and by R any sub-ring. [...] There exists in G a set of integers K such that the product of any integer of G by one of k is an integer of R. This set is an ideal which is the greatest common divisor of all ideals of G which belong to R. This ideal is called the conductor (Führer) of R". Dickson et al. (1923) also refer to Dedekind (1882) in this passage, where the term is found on p. 28; it first appeared in Dedekind (1877, p. 22). The term was adopted by Hilbert, for example, who used it in his *Zahlbericht* (Hilbert 1897, p. 238).

Other translations such as *bearer* (Harkin 1929, p. 216) or *leader* (Sheffer 1931, p. 655) do not seem to have found much appeal.

Note. Some historical remarks can be found in Narkiewicz (2018, pp. 43f.).

References

Dedekind, Richard (1877). *Über die Anzahl der Ideal-Classen in den verschiedenen Ordnungen eines endlichen Körpers*. Braunschweig.
Dedekind, Richard (1882). "Über die Diskriminanten endlicher Körper". *Abhandlungen der Königlichen Gesellschaft der Wissenschaften zu Göttingen* 29, pp. 1–56.
Dickson, Leonard Eugene et al. (1923). *Algebraic Numbers*. Report of the Committee on Algebraic Numbers, National Research Council. Washington, D.C.
Harkin, Duncan (1929). "The Abstract Identity of Modular Systems and Ideals". *American Journal of Mathematics* 51, pp. 205–216.
Hilbert, David (1897). "Die Theorie der algebraischen Zahlkörper". *Jahresbericht der Deutschen Mathematiker-Vereinigung* 4, pp. 175–546.
Narkiewicz, Władysław (2018). *The Story of Algebraic Numbers in the First Half of the 20th Century*. From Hilbert to Tate. Cham.
Sheffer, I.M. (1931). "Vorlesungen über die Singulären Moduln und die Komplexe Multiplikation der Elliptischen Funktionen. By R. Fueter [review]". *Bulletin of the American Mathematical Society* 37, pp. 654–655.

cone

Greek *κῶνος* was used to refer to the pine cone and the pine tree (LSJ); the mathematical term is due to the similarity to the pine cone. In the *Elements* (book XI, def. 18), Euclid gave a genetic definition of a right cone: "When, one side of those about the right angle in a right-angled triangle remaining fixed, the triangle is carried round and restored again to the same position from which it began to be moved, the figure so comprehended is a *cone*".[1] The term was in use before Euclid (Mugler 1958-59, p. 288). A more general definition of a cone as a double oblique cone (generated by a straight line fixed at a point and moving around a circle) was later given by Apollonius in his *Conics* (Apollonius of Perga 1891, p. 6).[2]

---

[1] Translation from Euclid (1956, p. 261). [2] For discussion of the differences between Euclid's and Apollonius' definitions, see Barrow-Green, Gray, and Wilson (2019, p. 155) and Fried and Unguru (2001, pp. 74–79). See also Netz' comments on the question of who first conceived of cones in the Apollonian style (Archimedes 2004, pp. 61f.).

In classical Latin, *conus* is thinly attested in clear mathematical senses (Ayuso García 2008, p. 1329). The term was mentioned by Martianus Capella in *The Marriage of Philology and Mercury* (sections 721 and 722; see Willis 1983, pp. 257f.), and a rare instance of a definition was given by Isidorus of Sevilla in his *Etymologies*: "A cone (*conon*) is a figure that from a wide base narrows at the top" (Isidore of Seville 1911, book III, xii).[1]

In the 12th century, Latin *conus* is used in a translation, based on a Greek text, of Euclid's *Elements* (Busard 1987, p. 310). In translations based on Arabic texts, however, a cone is called a 'round pyramid' (*pyramis rotunda*), for example by Robert of Chester (Busard and Folkerts 1992, p. 265).[2] This overlap between the words *pyramid* and *cone* appears to be based on Arabic. As Hogendijk (1994, p. 207) points out, "[t]he Arabic word is *makhrūt* (literally: 'cone-shaped'), which can be used for a cone as well as for a (not necessarily regular) pyramid". For example, al-Bīrūnī (10th/11th c.) defined the term *al-makhrūt* as follows: "A cone is a solid whose base is a circle or other figure, from which it diminishes to a point" (al-Bīrūnī 1934, p. 18).

In English, the term seems to occur first in its geometrical meaning; it is first found (MED, s. v. cōnǒun) in John Trevisa's 14th-century translation of *De proprietatibus rerum*: "Conoun is a figure broode at oon ende and sharp at þat oþer" (Seymour 1975, p. 1368). English *cone* is then used in Billingley's translation of the *Elements* (Euclid 1570, fol. 316^{v}), where it is also pointed out that the cone is sometimes called "a round Pyramis, which is not so aptly. For a Pyramis, and a Cone, are farre distant, & of sundry natures. [...] Therefore can not a Cone by any iust reason beare the name of a Pyramis" (Euclid 1570, fol. 317^{r}).

The French term dates from the 16th century (Möhren 2006, p. 105) and is first attested in Rabelais (1552, ch. 33); in a purely mathematical text, it is first found in Errard (1594, definitions).[3] In earlier texts such as de Bouelles (1511, fol. xxx^{r}), *pyramide ronde* 'round pyramid' is used instead. The German term for a cone is *Kegel*, lit. 'peg, pole, stake'. It is found in the 16th century (Dürer 1525, fol. G i).[4]

In the French *Encylopédie*, a new meaning is recorded: "The name *cone* is also given to solids other than those whose surfaces are produced by the movement of a line around the circumference of a circle; it extends to all types of solids that can be formed in the same way, taking any curve as the circumference of the base" (d'Alembert 1753, pp. 845f.).[5]

---

[1] "Conon, figura quae ab amplo in angustum finit". Translation from Isidore of Seville (2006, p. 94). [2] In some texts, Latin *pinealis* (from *pinea* 'pine cone') is also used for the cone. This translation is found, for example, in a translation by Gerard of Cremona of a text called *De Scientiis* by the philosopher al-Fārābī (9th/10th c.). See al-Fārābī (2005, p. 68). As the editor of the translation points out (ibib., p. 207), in the original Arabic the word *ṣanawbara* 'pine' is used. [3] Möhren (2006, p. 105) also points out a potential earlier attestation from the 13th century (Victor 1979, p. 526). The adjective *conique* was used by Évrard de Conty in the 14th century (de Conty n.d., fol. 214^{r}); a conical figure is there described as "pyramidal all round" ("une figure pyramidale toute ronde"), "thick" ("grosse") at one end and "pointed" ("ague") at the other. [4] Around 1400, von Megenberg had used *virekter kegel*, lit. 'quadrangular (four-cornered) cone', in the sense of 'pyramid' (von Megenberg 1980, p. 26). [5] "Le nom de *cone* se donne encore à d'autres solides qu'à ceux dont les surfaces sont produites par le mouvement d'une ligne autour de la circonférence d'un cercle; il s'étend à toutes les especes de corps que l'on peut former de la même maniere, en prenant une courbe quelconque pour circonférence de la base".

This extended meaning may be related to Euler (1732, pp. 188f.), who, while he did not explicitly apply the extended meaning to *cone*, did define a "conoidical solid" (*corpus conoidicus*) as "a solid bounded by straight lines drawn through the points of any curve and a fixed point taken outside of the plane of the curve".[1] Euler (1732, p. 117) also gave a similar definition for a cylindrical solid, which is "any solid of which sections perpendicular to an axis are similar and equal to each other".[2] There are also earlier usages of *elliptic*, *hyperbolic*, and *parabolic cone* and *cylinder*, for example in Parent (1711, p. 118); Tacquet (1651, p. 1) applied these adjectives to cylinders only.[3]

The extended terminology still caused some discussion at the end of the 18th century. When Monge lectured at the École normale, objections were raised in an ensuing debate against his use of *cylindre à base circulaire* on the grounds that *à base circulaire* seemed superfluous and would imply that there could be other kinds of cylinders (Monge 1795, p. 66). Monge replied that indeed circular cylinders used to be the cylinders originally considered, but that in the construction of vaults, curves other than circles were used but the name *cylinder* was kept. Therefore, it became necessary to specify what kind of cylinder was meant (Monge 1795, pp. 66f.).

Note. For overviews of the terminology and definitions, see Tropfke (1924, pp. 30–32), and Heath (1956, p. 270). Ayuso García (2008, pp. 1327–1330) focuses on the Latin term; for the Greek, see Mugler (1958-59, pp. 286–288), for the Arabic, see Souissi (1968, pp. 153f., 221) and Rashed (2017, pp. 232, 418–422).

References

d'Alembert, Jean Le Rond (1753). "Cone". *Encyclopédie, ou Dictionnaire raisonné des sciences, des arts et des métiers* 3, pp. 845–846.

Apollonius of Perga (1891). *Apollonii Pergaei quae graece exstant cum commentariis antiquis*. Vol. I. Ed. by J.L. Heiberg. Leipzig.

Archimedes (2004). *The Works of Archimedes*. Translation and Commentary. Volume 1: The Two Books *On the Sphere and the Cylinder*. Edited and translated by Reviel Netz. Cambridge.

Ayuso García, Manuel (2008). "La terminología latina de la geometría en Marciano Capela". PhD thesis. Madrid: Universidad Nacional de Educación a Distancia.

Barrow-Green, June, Jeremy Gray, and Robin Wilson (2019). *The History of Mathematics: A Source-Based Approach*. Volume 1. Providence, R.I.

al-Bīrūnī, Abū Rayḥān Muḥammad ibn Aḥmad (1934). *The Book of Instruction in the Elements of the Art of Astrology*. Written in Ghaznah, 1029 A. D. Reproduced from Brit. Mus. MS. Or. 8349. The Translation facing the Text by R. Ramsay Wright. London.

de Bouelles, Charles (1511). *Geometrie en francoys*. Paris.

Boyer, Carl B. (1956). *History of Analytic Geometry*. New York.

Busard, H.L.L. (1987). *The Mediaeval Latin Translation of Euclid's Elements Made Directly from the Greek*. Stuttgart.

Busard, H.L.L. and Menso Folkerts, eds. (1992). *Robert of Chester's (?) Redaction of Euclid's* Elements, *the so-called Adelard II Version*. Basel.

de Conty, Évrard (n.d.). "Le Livre des Probleumes d'Aristote, c'est à dire des fortes questions [manuscript]". Bibliothèque nationale de France. Département des manuscrits. Français 210. URL: `https://gallica.bnf.fr/ark:/12148/btv1b10547045k`.

[1] "Corpora conoidica hic mihi denotant silida lineis rectis ex curvae cuiuslibet singulis punctis ad punctum fixum extra planum curvae assumtum ductis terminata". Translation from Boyer (1956, p. 166). [2] "omnia corpora, quorum sectiones ad axem perpendiculares sunt inter se aequales et similes". Translation from Boyer (1956, p. 165). [3] See also the definition by al-Bīrūnī cited above.

Dürer, Albrecht (1525). *Underweysung der Messung, mit dem Zirckel und Richtscheyt*. Nürnberg.
Errard, Jean (1594). *La Geometrie et Practique generalle d'icelle*. Paris.
Euclid (1570). *The Elements of Geometrie of the most auncient Philosopher Euclide of Megara*. Faithfully (now first) translated into the Englishe toung, by H. Billingsley. London.
Euclid (1956). *The Thirteen Books of Euclid's Elements*. Volume III. Books X-XIII and Appendix. Ed. by Thomas L. Heath. New York.
Euler, Leonhard (1732). "De linea brevissima in superficie quacunque duo quaelibet puncta jungente". *Commentarii academiae scientiarum imperialis Petropolitanae* 3, pp. 110–124.
al-Fārābī (2005). *Über die Wissenschaften. De scientiis*. Lateinisch-deutsch. Übersetzt, mit einer Einleitung und Anmerkungen herausgegeben von Franz Schupp. Hamburg.
Fried, Michael N. and Sabetai Unguru (2001). *Apollonius of Perga's* Conica. Text, Context, Subtext. Leiden et al.
Heath, Thomas L. (1956). *The Thirteen Books of Euclid's Elements*. Volume I. Introduction and Books I, II. New York.
Hogendijk, Jan P. (1994). "An Arabic text on the comparison of the five regular polyhedra: Book XV of the revision of the Elements by Muhyī al-Dīn al-Maghribī". *Zeitschrift für Geschichte der arabisch-islamischen Wissenschaften* 8, pp. 133–233.
Isidore of Seville (1911). *Etymologiarum sive Originum libri XX*. Tomus I. Ed. by Wallace M. Lindsay. Oxford.
Isidore of Seville (2006). *The* Etymologies *of Isidore of Seville*. Ed. by Stephen A. Barney et al. Cambridge.
von Megenberg, Konrad (1980). *Die deutsche Sphaera*. Ed. by Francis B. Brévart. Tübingen.
Möhren, Frankwalt (2006). "Les débuts de l'écriture française de la géométrie au XIIIe siècle". In: *L'écriture du texte scientifique*. Des origines de la langue française au XVIIIe siècle. Ed. by Claude Thomasset. Paris, pp. 93–116.
Monge, Gaspard (1795). "Géométrie descriptive [deuxième débat]". In: *Séances des écoles normales*. Seconde partie. Débats. Tome premier. Paris, pp. 63–74.
Mugler, Charles (1958-59). *Dictionnaire historique de la terminologie géométrique des Grecs*. Paris.
Parent, Antoine (1711). "Problème géométrique". *Mémoires de l'Académie royale des sciences* (1709), pp. 118–131.
Rabelais, François (1552). *Le Quart Livre des faicts et dicts Heroiques du bon Pantagruel*. Paris.
Rashed, Roshdi, ed. (2017). *Lexique historique de la langue scientifique arabe*. Hildesheim.
Seymour, Maurice Charles, ed. (1975). *On the Properties of Things*. John Trevisa's Translation of Bartholomaeus Anglicus' *De Proprietatibus Rerum*. A Critical Text. Volume II. Oxford.
Souissi, Mohammed (1968). *La langue des mathématiques en arabe*. Tunis.
Tacquet, André (1651). *Cylindricorum et annularium libri IV*. Antwerpen.
Tropfke, Johannes (1924). *Geschichte der Elementar-Mathematik*. Siebenter Band. Stereometrie. Verzeichnisse. Berlin and Leipzig.
Victor, Stephen K. (1979). *Practical Geometry in the High Middle Ages*. "Artis cuiuslibet consummatio" and the "Pratike de geometrie". Philadelphia.
Willis, James, ed. (1983). *Martianus Capella*. Leipzig.

confidence interval

The term *confidence interval* was introduced by Neyman (1934, p. 562):

> The form of this solution consists in determining certain intervals, which I propose to call the confidence intervals [...], in which we may assume are contained the values of the estimated characters of the population, the probability of an error is a statement of this sort being equal to or less than $1 - \varepsilon$, where ε is any number $0 < \varepsilon < 1$, chosen in advance. The number ε I call the confidence coefficient.

Clopper and Pearson (1934, p. 406) immediately adopted the term and attributed it to Neyman. Neyman (1941, p. 128) later reported that he started his work on confidence intervals in 1930 and that he originally used the Polish term *przedział ufności*.

Note. For the history of the concept, see David and Edwards (2001, pp. 187–193) and Miller (n.d.).

References

Clopper, C.J. and E.S. Pearson (1934). "The Use of Confidence or Fiducial Limits Illustrated in the Case of the Binomial". *Biometrika* 26, pp. 404–413.

David, H.A. and A.W.F. Edwards (2001). *Annotated Readings in the History of Statistics*. New York.

Miller, Jeff (n.d.). "Earliest Known Uses of Some of the Words of Mathematics". URL: `https://mathshistory.st-andrews.ac.uk/Miller/mathword/`.

Neyman, Jerzy (1934). "On the Two Different Aspects of the Representative Method". *Journal of the Royal Statistical Society* 97, pp. 558–625.

Neyman, Jerzy (1941). "Fiducial Argument and the Theory of Confidence Intervals". *Biometrika* 32, pp. 128–150.

configuration

Configurations are certain arrangements of points and lines and certain types of them, for example Desargues' configuration, have been known for a long time. The formal introduction of the notion and terminology is due to Theodor Reye. He mentioned the German term *Configuration* in passing in his book on the geometry of position (Reye 1877, p. 4) and gave a definition in Reye (1882, p. 94): "A configuration n_i in the plane consists of n points and n lines in such a position that each of the n lines contains i of the n points and that through the each of the n points pass i of the n lines".[1] This definition had earlier been published by Kantor (1880, p. 717), who referred to private communication with Reye.[2]

The English term *configuration* is found in Moore (1895, p. 36), who refers to Reye; it also appears in Reye (1898, p. 5).

The term is now used in various meanings; for example, Grannell and Griggs (2007, p. 161) remark: "The term 'configuration' is nowadays used rather loosely; it has come to refer to any fixed small number of blocks which form part of a design".[1]

Note. The history of configurations has been studied in various articles by Gropp. See, in particular, Gropp (1990) and Gropp (2004). See also Grünbaum (2009, pp. 8–13) for a brief overview and Steinitz (1910) for a summary of early work on configurations.

References

Grannell, Mike and Terry Griggs (2007). "Designs and topology". In: *Surveys in Combinatorics 2007*. Ed. by A. Hilton and J. Talbot. Cambridge, pp. 121–174.

Gropp, Harald (1990). "On the History of Configurations". In: *Structures in Mathematical Theories*. Reports of the San Sebastian International Symposium, September 25-29, 1990. Ed. by A. Díez, J. Echeverría, and A. Ibarra. Bilbao, pp. 263–268.

Gropp, Harald (2004). "Configurations between geometry and combinatorics". *Discrete Applied Mathematics* 138, pp. 79–88.

Grünbaum, Branko (2009). *Configurations of Points and Lines*. Providence, R.I.

---

[1] "Eine Configuration n_i in der Ebene besteht aus n Punkten und n Geraden in solcher Lage, dass jede der n Geraden i von den n Punkten enthält und durch jeden der n Punkte i von den n Geraden gehen". [2] As pointed out by Gropp (1990, pp. 264f.), Martinetti (1887) already studied configurations from a purely combinatorial point of view. [1] For further terminological remarks, see Gropp (1990, p. 267).

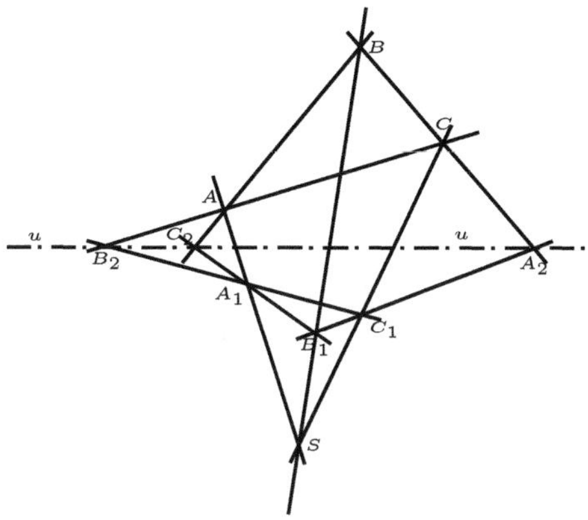

Figure 21: A Desargues configuration. From Reye (1877, Tafel I). If two triangles ABC and $A_1B_1C_1$ are situated in such a way that $\overline{AA_1}$, $\overline{BB_1}$, and $\overline{CC_1}$ meet in a point S, then the intersection points of the sides $\overline{AB}$ and $\overline{A_1B_1}$, $\overline{BC}$ and $\overline{B_1C_1}$, and $\overline{CA}$ and $\overline{C_1A_1}$ lie on a straight line u, and vice versa.

Kantor, Seligmann (1880). "Ueber eine Gattung von Configurationen in der Ebene und im Raume". *Sitzungsberichte der Kaiserlichen Akademie der Wissenschaften (mathematisch-naturwissenschaftliche Klasse)* 80, pp. 715–723.

Martinetti, Vittorio (1887). "Sulle configurazioni piani μ_3". *Annali di Matematica Pura ed Applicata* 15, pp. 1–26.

Moore, E.H. (1895). "Concerning Jordan's linear groups". *Bulletin of the American Mathematical Society* 2, pp. 33–43.

Reye, Theodor (1877). *Die Geometrie der Lage*. Zweite vermehrte Auflage. Erste Abtheilung. Hannover.

Reye, Theodor (1882). "Das Problem der Configurationen". *Acta Mathematica* 1, pp. 92–96.

Reye, Theodor (1898). *Lectures on the Geometry of Position*. Part I. Translated and edited by Thomas F. Holgate. New York.

Steinitz, Ernst (1910). "Konfigurationen der projektiven Geometrie". *Encyklopädie der mathematischen Wissenschaften mit Einschluss ihrer Anwendungen* III AB 5a, pp. 481–516.

confocal

The notion of confocality seems to occur first in connection with conic sections.[2] For example, confocal parabolas have been dealt with by Tschirnhaus (1695, p. 492), and confocal ellipses were treated by Maclaurin (1742, § 648), who says that they have "the same focus", which is what *confocal* literally means. The term itself seems to be a 19th-century invention.

---

[2] For the following, see Staude (1910, p. 951).

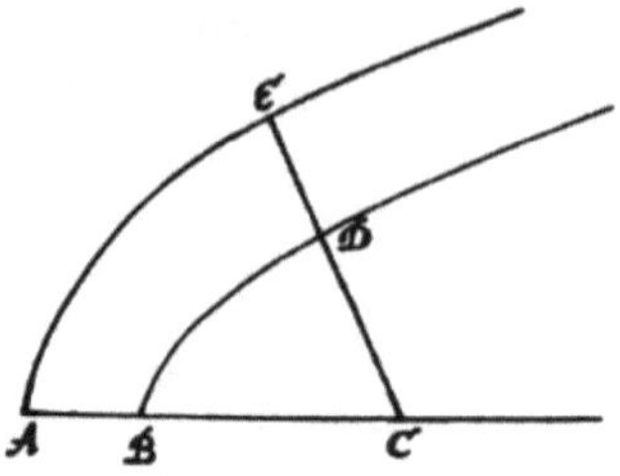

Figure 22: Confocal parabolas. From Tschirnhaus (1695).

It appears to be due to Poncelet (1826-27, p. 270), who applied it to "lines of the second order or surfaces of revolution of the same order, which have a *common focus* [*foyer commun*] or are *confocal* [*confocales*]".[1] In Poncelet (1829, p. 69), he gives *homofocal* (French *homofocale*) as a synomym. Magnus (1833, p. 204) applied German *confocal* to conic sections and later, in Magnus (1837, p. 353), to surfaces of revolution. Jacobi (1834, p. 137) used the term for second-order surfaces whose principal sections have the same focus.

See also *focus*.

References

Jacobi, Carl Gustav Jacob (1834). "Aus einem Schreiben des Herrn Prof. Dr. C.G.J. Jacobi zu Königsberg i. Pr., an den Herrn Prof. Dr. J. Steiner in Berlin". *Journal für die reine und angewandte Mathematik* 12, pp. 137–140.

Maclaurin, Colin (1742). *A Treatise of Fluxions*. In Two Books. Edinburgh.

Magnus, Ludwig Immanuel (1833). *Sammlung von Aufgaben und Lehrsätzen aus der analytischen Geometrie*. Berlin.

Magnus, Ludwig Immanuel (1837). *Sammlung von Aufgaben und Lehrsätzen aus der analytischen Geometrie des Raumes*. Erste Abtheilung. Berlin.

Poncelet, Jean Victor (1829). "Mémoire sur la théorie générale des polaires réciproques". *Journal für die reine und angewandte Mathematik* 4, pp. 1–71.

Poncelet, Jean Victor (1826-27). "Analyse d'un mémoire présenté à l'Académie royale des sciences". *Annales de mathématiques pures et appliquées* 17, pp. 265–272.

Staude, Otto (1910). *Analytische Geometrie des Punktepaares, des Kegelschnittes und der Fläche zweiter Ordnung*. Zweiter Teilband. Leipzig and Berlin.

Tschirnhaus, Ehrenfried Walter (1695). "Nova et singularis Geometriae promotio, circa dimensionem quantitatum curvarum". *Acta Eruditorum* (Nov. 1695), pp. 489–493.

conformal

Functions that preserve angles have been known since antiquity, but were not systematically studied until the 18th century, when Lambert, Euler and Lagrange began to publish on the subject.[2] These authors paraphrased the notion. Euler (1778a, p. 128), for example, described conformality thus: "all very small shapes on the Sphere are represented by similar shapes on the plane".[3] Lagrange (1781, p. 163) used a very

---

[1] "lignes du second ordre ou des surfaces de révolution du même ordre, qui ont un *foyer common* ou sont *confocales*." [2] See the references in the note at the end of this article. [3] "omnes figurae valde exiguae in Sphaera sumtae per similes figuras in plano exprimantur". Translation from Euler (1778b, p. 31).

similar wording: "any infinitely small portion of that surface keeps the same shape in the projection and is alterated only in size".[1] These paraphrases were also used by Gauß (1825) in his prize-winning paper "General solution of the problem: to represent the parts of a given surface on another given surface, so that the smallest parts of the representation shall be similar to the corresponding Parts of the Surface represented" and they continued to be used throughout the 19th century.

The term *conformal* dates from the end of the 18th century. Schubert (1789, p. 131) used Latin *projectio conformis* for the special case of the mapping of the terrestrial spheroid on the plane.[2] The term means 'of the same shape' and its choice can be explained thus: "From the circumstance that the shapes of two corresponding elements in the two planes are the same, the process of passing from one plane to the other is known as *conformal representation*" (Jeans 1908, p. 260).[3]

Schubert's introduction of the term does not seem to have had an immediate terminological effect. Instead, it was through Gauß that the term became known. When he resumed the subject in 1844, Gauß said that the "need is felt for a special denomination", so he decided to "call these [maps] *conformal* maps or transfers by giving this otherwise vague epithet a sharply delineated meaning" (Gauß 1845, p. 4).[4] After Gauß, the term was first adopted by other German writers and then transferred to other languages. For example, Amstein (1877, p. 178) uses French *conforme*, and Peirce (1879, p. 394) uses English *conforme*, while later writers such as Cole (1890, p. 159) preferred *conformal*.

Various alternatives have been suggested for *conformal*.

The term *orthomorphic*, for example, suggests correct shapes (Greek *ὀρθός* 'right', *μορφή* 'shape, form'.). It is usually attributed to Germain (1866, p. 62),[5] who used French *orthomorphe*, but, as Germain points out himself, he took over the term from d'Azevac (1863, p. 144). The term gained some currency as it was adopted by Cayley (1891, p. 262), for example.

Tissot (1879, p. 75) preferred French *autogonal*, a term made up of Greek *αὐτός* 'self' and *γόνος* 'angle'; the term *isogonal* (Greek *ἴσος* 'same') was suggested by Siebeck (1858, p. 228).

The German term *winkeltreu* (from *Winkel* 'angle' and *treu* 'faithful') was suggested by Breusing in 1882, according to Voss (1903, p. 365).[6] It is not until the end of the 19th century that English *conformal* can be said to be the preferred term.

Note. Concise historical overviews of conformal mappings are given by Ivanov and Trubetskov (1995, pp. 1–4) and Voss (1903, pp. 364–371); Papadopoulos (2018) is more detailed. Kastrup (2008) makes various useful historical and terminological comments. See also Lee (1944) for a general terminological discussion concerning map projections.

---

[1] "une portion quelconque infiniment petite de cette surface conserve la même figure dans la projection et n'est altérée que dans la grandeur" [2] Schubert's use of the term has been pointed out by Stäckel (1917, p. 111) and Kastrup (2008, p. 641), for example. [3] The same explanation was given by Hilbert and Cohn-Vossen (1952, p. 262). [4] Translation based on Kastrup (2008, p. 676). For "maps or transfers", Gauß uses "Abbildungen oder Übertragungen". See *mapping*. [5] For example, by Lee (1944, p. 191). [6] Voss says that Breusing used the term at the second congress of German geographers. In the proceedings of that congress, the term is not found. Nevertheless, various contemporary writers such as Debes (1883, p. 20) attribute the term to Breusing.

References

Amstein, Hermann (1877). "Un exemple de représentation conforme". *Bulletin de la Société Vaudoise des Sciences Naturelles* 15, pp. 175–198.

d'Azevac, Armand (1863). *Coup d'oeil historique sur la projection des cartes de géographie*. Paris.

Cayley, Arthur (1891). "On some problems of orthomorphosis". *Journal für die reine und angewandte Mathematik* 107, pp. 262–277.

Cole, Frank Nelson (1890). "The Linear Functions of a Complex Variable". *Annals of Mathematics* 5, pp. 121–176.

Debes, Ernst (1883). "Dr. Nells modifizierte Globular-Projektion". *Mitteilungen des Vereins für Erdkunde* (1882), pp. 19–34.

Euler, Leonhard (1778a). "De repraesentatione superficiei sphaericae super plano". *Acta Academiae Scientarum Imperialis Petropolitanae* (1777), pp. 107–132.

Euler, Leonhard (1778b). "On the mapping of Spherical Surfaces onto the Plane". Tr. George Heine. URL: `https://scholarlycommons.pacific.edu/euler-works/490/` (retrieved May 25, 2018).

Gauß, Carl Friedrich (1825). "Allgemeine Auflösung der Aufgabe: Die Theile einer gegebnen Fläche auf einer andern gegebnen Fläche so abzubilden, daß die Abbildung dem Abgebildeten in den kleinsten Theilen ähnlich wird". *Astronomische Abhandlungen* 3, pp. 1–30.

Gauß, Carl Friedrich (1828). "General solution of the problem: to represent the parts of a given surface on another given surface, so that the smallest parts of the representation shall be similar to the corresponding Parts of the Surface represented". *The Philosophical Magazine* 4, pp. 104–113, 206–215.

Gauß, Carl Friedrich (1845). "Untersuchungen über Gegenstände der höhern Geodaesie". *Abhandlungen der Königlichen Gesellschaft der Wissenschaften zu Göttingen* 2, pp. 3–45.

Germain, Adrien (1866). *Traité des projections des cartes géographiques*. Paris.

Hilbert, David and Stefan Cohn-Vossen (1952). *Geometry and the Imagination*. Translated by P. Nemenyi. New York.

Ivanov, Valentin I. and Michael K. Trubetskov (1995). *Handbook of Conformal Mapping with Computer-Aided Visualization*. Boca Raton.

Jeans, James (1908). *The Mathematical Theory of Electricity and Magnetism*. Cambridge.

Kastrup, H.A. (2008). "On the advancements of conformal transformations and their associated symmetries in geometry and theoretical physics". *Annalen der Physik* 17, pp. 631–690.

Lagrange, Joseph-Louis (1781). "Sur la Construction des Cartes Géographiques". *Nouveaux Mémoires de l'Académie Royale des Sciences et Belles-Lettres de Berlin* (1779), pp. 161–210.

Lee, I.P. (1944). "The nomenclature and classification of map projections". *Empire Survey Review* 51, pp. 190–200.

Papadopoulos, Athanase (2018). "Quasiconformal mappings, from Ptolemy's geography to the work of Teichmüller". In: *Uniformization, Riemann-Hilbert Correspondence, Calabi-Yau Manifolds, & Picard-Fuchs Equations*. Ed. by Lizhen Ji and Shing-Tung Yau. Somerville, pp. 237–314.

Peirce, Charles Sanders (1879). "A Quincuncial Projection of the Sphere". *American Journal of Mathematics* 2, pp. 394–397.

Schubert, Friedrich Theodor (1789). "De proiectione sphaeroidis ellipticae geographica. Dissertatio prima". *Nova Acta Academiae Scientiarum Imperialis Petropolitanae* 5, pp. 130–146.

Siebeck, Paul (1858). "Ueber die graphische Darstellung imaginärer Functionen". *Journal für die reine und angewandte Mathematik* 55, pp. 221–253.

Stäckel, Paul (1917). "C.F. Gauß als Geometer". *Nachrichten von der Königl. Gesellschaft der Wissenschaften zu Göttingen (mathematisch-physikalische Klasse)* (1917. Beiheft), pp. 25–142.

Tissot, Auguste (1879). "Mémoire sur la représentation des surfaces et les projections des cartes géographiques". *Nouvelles annales de mathématiques* 18, pp. 532–548.

Voss, Aurel (1903). "Abbildung und Abwickelung zweier Flächen aufeinander". *Encyklopädie der mathematischen Wissenschaften mit Einschluss ihrer Anwendungen* III D 6a, pp. 355–440.

congruent **(arithmetic)**

The term *congruent* was defined by Gauß at the beginning of his *Disquisitiones arithmeticae*:[1]

> If the number a measures the difference of the numbers b, c, then b and c are said to be *congruent according to* a; if not, *incongruent*; this a we call the *modulus*. Each of the numbers b, c are called a *residue* of the other in the first case, a *nonresidue* in the second (Gauß 1801, § 1).[2]

While Gauß was most influential in spreading this term, he was not the first to use it. In fact, Goldbach had used the term much earlier in 1732: In a discussion of the equation $x^e = dp + r$, he calls the "residual number" r, "for brevity's sake", a *congruum* (Goldbach 1738, p. 101).[3] Goldbach's coinage remained largely unnoticed in the 18th century, however, and there is no reason to assume that Gauß borrowed the term from Goldbach (Juškevič and Kopelevič 1994, p. 113).

After Gauß, it did not take long for the term to be transferred to other languages. French *congrus* appears in a translation of Gauß' book (Gauß 1807, § 1), and in an English review of Gauß (1807), the word *congruous* is used ([anon.] 1808, p. 529) in the definition of congruence, but *congruent* also appears on the same page. In 19th-century texts, *congruous* remained popular; it features in Smith's "Report on the Theory of Numbers", for example (Smith 1860, p. 230).

One alternative was suggested by Cauchy (1829, p. 206). While Cauchy does cite Gauß' term, he defined: "Two positive or negative integers u and v are *equivalent* with reference to the modulus n, if the difference $u - v$ or $v - u$ is divisible by n".[4] On Gauß's term and the alternative *equivalent*, Kronecker (1901, p. 86) commented: "Instead of the word *congruent*, which already had a definite meaning in geometry, where it characterizes a relation which carries the character of the identity, Gauß could very well have chosen the exceptionally distinctive *equivalent*, or *equal-valued* [...] But the Gaussian denomination has already so generally permeated science that we need to wean ourselves off the idea of equality which adheres to the word *congruent* in geometry".[5]

In some 19th-century German texts, the word *gleichrestig*, from *gleich* 'equal' and *Rest* 'remainder', is found. It was used, for example, by Brennecke (1855, p. 6) and also by Lejeune Dirichlet (1863, p. 34).[6]

---

[1] Gauß already used this terminology in 1797. See Bullynck (2009, p. 64). [2] "Si numerus a numerorum b, c differentiam metitur, b et c *secundum* a *congrui* dicuntur, sin minus, *incongrui*: ipsum a *modulum* appellamus. Uterque numerorum b, c, priori in casu alterius *residuum*, in posteriori vero *nonresiduum* vocatur". Translation from Goldstein and Schappacher (2007, p. 6). [3] "$x^e = dp + r$, hunc numerum residuum r, breuitatis causa *congruum*". – On this passage, see Cantor (1901, p. 611), Miller (1946, p. 487), and Juškevič and Kopelevič (1994, p. 113). [4] "les nombres entiers positifs ou négatifs u et v sont *équivalens* suivant le module n lorsque la différence $u - v$ ou $v - u$ sera divisible par n". [5] "*Gauss* hätte an Stelle des Wortes 'kongruent', das in der Geometrie bereits eine bestimmte Bedeutung hatte, und dort eine Beziehung charakterisiert, die den Charakter der Identität trägt, sehr wohl das außerordentlich bezeichnende 'äquivalent', oder 'gleichwertig' nehmen können [...] Doch ist die Gaussische Ausdrucksweise schon so allgemein in die Wissenschaft eingedrungen, daß wir uns eben von der Vorstellung der Gleichheit, die dem Worte 'kongruent' in der Geometrie anhaftet, entwöhnen müssen". [6] The term was already mentioned in a letter to Gauß dated Dec. 30, 1809 (Peters 1860, p. 24).

The terminology was later generalized; Dedekind (1871, p. 452) applied it to rings in general[1]: "We further call two numbers ω, ω' contained in $\mathfrak{o}$, and whose difference is divisible by $\mathfrak{a}$, *congruent with respect to* $\mathfrak{a}$ [...], and denote this with the congruence $\omega \equiv \omega' \pmod{\mathfrak{a}}$".[2]

Note. For the history of the notion, see Bullynck (2009) and Boucard and Verdier (2015); in Weil (2001), many remarks on congruences can be found. For Gauß' *Disquisitiones arithmeticae*, see also Goldstein and Schappacher (2007).

References

[anon.] (1808). "Gauss's *Arithmetical Researches*". *The Monthly Review* 55, pp. 527–544.

Avigad, Jeremy (2004). "Dedekind's 1871 version of the theory of ideals". URL: `https://kilthub.cmu.edu/articles/Dedekind_s_1871_Version_of_the_Theory_of_Ideals/6491342/1` (retrieved July 25, 2007).

Boucard, Jenny and Norbert Verdier (2015). "Circulations mathématiques et congruences dans les périodiques de la première moitié du XIXe siècle". *Philosophia Scientiae* 19, pp. 57–78.

Brennecke, Wilhelm (1855). *Einige Sätze aus den Anfangsgründen der Zahlenlehre*. Posen.

Bullynck, Maarten (2009). "Modular arithmetic before C.F. Gauss: Systematizations and discussions on remainder problems in 18th-century Germany". *Historia Mathematica* 36, pp. 48–72.

Cantor, Moritz (1901). *Vorlesungen über Geschichte der Mathematik*. Dritter Band. Von 1668 bis 1758. Leipzig.

Cauchy, Augustin-Louis (1829). "Mémoire sur la théorie des nombres". *Bulletin des sciences mathématiques, physiques et chimiques* 12, pp. 205–221.

Dedekind, Richard (1871). "Ueber die Composition der binären quadratischen Formen". In: *Vorlesungen über Zahlentheorie von P.G. Lejeune Dirichlet*. Herausgegeben und mit Zusätzen versehen von R. Dedekind. Zweite, umgearbeitete und vermehrte Auflage. Braunschweig, pp. 380–497.

Gauß, Carl Friedrich (1801). *Disquisitiones arithmeticae*. Leipzig.

Gauß, Carl Friedrich (1807). *Recherches arithmétiques*. Traduites par A.-C.-M. Poullet-Delisle. Paris.

Goldbach, Christian (1738). "Criteria quaedam aequationum, quarum nulla radix rationalis est". *Commentarii academiae scientiarum Petropolitanae* 6, pp. 98–106.

Goldstein, Catherine and Norbert Schappacher (2007). "A Book in Search of a Discipline (1801-1860)". In: *The Shaping of Arithmetic after C. F. Gauss's Disquisitiones Arithmeticae*. Ed. by Catherine Goldstein, Norbert Schappacher, and Joachim Schwermer. Berlin and Heidelberg, pp. 3–65.

Juškevič, Adolf P. and Judith Kh. Kopelevič (1994). *Christian Goldbach, 1690-1764*. Aus dem Russischen übersetzt von Annerose und Walter Purkert. Basel et al.

Kronecker, Leopold (1901). *Vorlesungen über Zahlentheorie*. Erster Band. Ed. by Kurt Hensel. Leipzig.

Lejeune Dirichlet, Gustav (1863). *Vorlesungen über Zahlentheorie*. Ed. by Richard Dedekind. Braunschweig.

Miller, George Abram (1946). "Notes in Tribune Publique". In: *The Collected Works of George Abram Miller*. Volume III. Urbana, Ill., pp. 472–491.

Peters, C.A.F., ed. (1860). *Briefwechsel zwischen C.F. Gauss und H.C. Schumacher*. Erster Band. Altona.

Smith, H.J.S. (1860). "Report on the Theory of Numbers. Part I". *Report of the British Association for the Advancement of Science* (1859), pp. 228–267.

Weil, André (2001). *Number Theory*. An approach through history from Hammurapi to Legendre. Boston.

---

[1] In the definition, $\mathfrak{o}$ is a ring, $\mathfrak{a}$ an ideal. The definition is not fully general, however, since Dedekind deals with complex number fields. [2] "Wir nennen ferner zwei in $\mathfrak{o}$ enthaltene Zahlen ω, ω', deren Differenz durch $\mathfrak{a}$ theilbar ist, congruent nach $\mathfrak{a}$ [...], und bezeichnen dies durch die Congruenz $\omega \equiv \omega' \pmod{\mathfrak{a}}$". Translation based on Avigad (2004, p. 38).

***congruent* (geometry)**

According to Fowler (1999, p. 11), "congruence does not seem to be a basic concept in the *Elements*, since Euclid has no single word for it; the idea of coincidence in Common Notion 4 might seem a candidate, but this piece of text may be a later interpolation[1] [...] and Euclid does not make systematic use of this terminology (it appears elsewhere only in I 4 & 8 and III 24). Another way of describing congruence is found, for the first time, in XI Definition 10".[2]

Fowler is referring to the following:

- In common notion 4, it says: "Things which coincide with one another are equal one another" (Euclid 1956a, p. 155). Coincidence (or superposition) is here expressed by the verb *ἐφαρμόζειν*, which means 'to fit exactly, to coincide with'.[3]
- Common notion 4 is used in the proofs of some congruence theorems (not in the theorems themselves), e. g. in book I, prop. 4, which states: "If two triangles have two sides equal to two sides respectively, and have the angles contained by the equal straight lines equal, then they also have the base equal to the base, the triangle equals the triangle, and the remaining angles equal the remaining angles respectively, namely those opposite the equal sides" (Euclid 1956a, p. 247). This is the side-angle-side congruence theorem for triangles.
- In Book XI, def. 10, Euclid defines: "Equal and similar solid figures are those contained by similar planes equal in multitude and in magnitude" (Euclid 1956b, p. 261). For "equal and similar", Euclid uses *ἴσος καί ὅμοιος*. See *equal* and *similar* (geometry).

A comparison of the first translations of the *Elements* into Latin, French, English, German, and Italian and of some related texts shows when the term *congruent* was first used in these contexts:

- In the Pseudo-Boethian *Geometria II*, *convenire* is used in common notion 4 (Boethius 1867, p. 378).
- In Adelard's translation of the *Elements* into Latin, *superponere* and *superpositio* are used in common notion 4 and propositions 4 and 8 (Busard 1983, pp. 33, 36). These words also appear in other versions, for example in Campanus' translation (Busard 2005, pp. 58, 62).
- The verb *congruere* first appears in Zamberti's edition of the *Elements* (Euclid 1505, e. g., fol. A iii^r), but only in the propositions; in the common notions, Zamberti uses *convenire* (Euclid 1505, fol. A ii^r).
- In the common notions, *congruere* is then used by Barozzi in a translation of Proclus' commentary on the *Elements* (Proclus 1560, p. 111); in Commandinus' influential edition of the *Elements*, *congruere* is also used in both the common notions and the propositions (Euclid 1572, fol. 6^v, 9^v).

---

[1] Fowler refers to Euclid (1956a, pp. 224f.). [2] No definition of congruence is given in the *Elements*. [3] More precisely, the word "has a different meaning according as it is used in the active or in the passive. In the passive, [...] it means 'to be *applied* to' without any implication that the applied figure will exactly fit, or coincide with, the figure to which it is applied; on the other hand the active [...] is used intransitively and means 'to fit exactly,' 'to coincide with'" (Euclid 1956a, pp. 224f.). – The numbering of the common notions varies.

- Alternatives in modern languages include French *accomoder* (Euclid 1564, fol. 8^{r}) and *convenir* (Euclid 1598, fol. a $\mathrm{iiij}^{\mathrm{v}}$), English *agree* (Euclid 1570, fol. 7^{r}, 13^{r}); in German and Italian, terms corresponding to *superpose* are used at first (Euclid 1562, p. 5, Euclid 1543, fol. x $\mathrm{iiii}^{\mathrm{r}}$).

Definitions of the term *congruent* do not seem to appear until the 16th century. The first ones found are due to Ramus. In the anonymously published *Geometria*, he wrote: "Magnitudines congruae sunt, quarum partes applicatae partibus conveniunt" (Ramus 1567, p. 4). Two years later, he repeated the definition with the modification that *conveniunt* was replaced by *aequalum locum occupant* (Ramus 1569, p. 6). The latter definition was translated to English soon afterwards: "Congruall or agreeable magnitudes are those, whose parts being applied one to another, fill an equall place" (Ramus 1590, p. 1). Ramus' use of the adjective found some followers in the next century; Pardies (1671, p. 14) used French *congrus*, Hobbes (1656, p. 132) English *congruous*, while *congruent* sporadically appeared in Barrow's translation of the *Elements* (Euclid 1660, e. g., p. 246). Such examples are not particularly frequent, however; in French, for example, *égale* 'equal' long remained in common use: Legendre (1794, pp. 57, 280), for example, used *équivalent* for figures that have the same area and *égale* for congruent objects.

In the 18th century, the expression *equal and similar* was frequently used for 'congruent'. This usage is reflected in Wolff's mathematical dictionary, in which Latin *congruentia* is defined as "the union of equality and similarity in a figure or line" (Wolff 1716, p. 418).[1] Tropfke (1940, pp. 95f.), who cites numerous works, observes that in books that discuss the congruence theorems *congruent* is not used in the theorems and that a change sets in during the second half of the 18th century. An example of the transition is found in Clemm (1764, p. 183): "If in two triangles two sides with an included angle are equal, then the triangles are equal and similar, that is congruent, and cover each other".[2]

The first person who equated congruence with equality and similarity appears to have been Leibniz. He wrote on congruence on a number of occasions, for example in a manuscript on "analysis situs" in 1693: "a true geometric analysis ought not only consider equalities and proportions which are truly reducible to equalities but also similarities and, arising from the combination of equality and similarity, congruences" (Gerhardt 1858, p. 179).[3] Similar statements by Leibniz are found from the late 1670s onwards; for instance, in 1678/79 he noted: "All things that are similar and equal are congruent" (Leibniz 1999, p. 1987).[4]

Later authors noted that congruence is not to be equated with similarity and equality. For example, Kant remarked that "[objects] equal and similar are not congruent unless

---

[1] "die Vereinigung der Gleichheit und Ähnlichkeit in einer Figur oder Linie". This definition was also given in the *Encyclopédie*, for example (Formey 1753, p. 869). [2] "Wenn also in zween Dreyecken zwo Seiten mit dem eingeschlossenen Winkel einander gleich sind, so sind die ganze Dreyecke gleich und ähnlich, das ist congruent, und decken einander". [3] "Itaque Analysis vere geometrica non tantum aequalitates spectat et proportionalitates , quae revera ad aequalitates reducuntur, sed similitudines etiam, et ex aequalitate ac similitudine conjunctis natas congruentias adhibere debet". Translation from Loemker (1989, p. 255). [4] "Omnia similia et aequalia sunt congrua". Translation from Loemker (1989, p. 278).

they lie in a plane" (Kant 1968, p. 15).[1] Kant's statement did not come out of nowhere; see *chiral*.

Note. A historical overview of the terminology is found in Tropfke (1940, pp. 94–96). For the Greek terminology, see also Mugler (1958-59, pp. 208, 231), Heath in Euclid (1956a, pp. 224–231), von Fritz (1959) as well as Fried (2009). For the history of superposition, see Murdoch (1964) and Palmieri (2009).

References

Boethius, Anicius Manlius Severinus (1867). *De institutione arithmetica libri duo, De institutione musica libri quinque*. Accedit Geometria quae fertur Boetii. Ed. by Gottfried Friedlein. Leipzig.

Busard, H.L.L. (1983). *The First Latin Translation of Euclid's Elements Commonly Ascribed to Adelard of Bath*. Books I-VIII and books X.36-XV.2. Toronto.

Busard, H.L.L. (2005). *Campanus of Novara and Euclid's* Elements. Volume I. Stuttgart.

Clemm, Heinrich Wilhelm (1764). *Mathematisches Lehrbuch*. Stuttgart.

Euclid (1505). *Elementorum libri XIII ... cum expositione Theonis*. Ed. by Bartolomeo Zamberti. Venezia.

Euclid (1543). *Euclide Megarense philosopho*. Diligentemente reassettato, et alla integrita ridotto per ... Nicolo Tartalea. Venezia.

Euclid (1562). *Die Sechs Erste Bücher Euclidis*. Auß Griechischer sprach in die Teütsch gebracht ... Durch Wilhelm Holtzmann genannt Xylander. Basel.

Euclid (1564). *Les Six Premiers Livres des Elements*. Traduicts et commentez par Pierre Forcadel de Besies. Paris.

Euclid (1570). *The Elements of Geometrie of the most auncient Philosopher Euclide of Megara*. Faithfully (now first) translated into the Englishe toung, by H. Billingsley. London.

Euclid (1572). *Euclidis Elementorum libri XV*. A Federico Commandino Urbinate nuper in latinum conversi, commentariisque quibusdam illustrati. Pesaro.

Euclid (1598). *Les six premiers livres des elemens d'Euclide*. Traduicts & commentez par I. Errard de Bar-Le-Duc. Paris.

Euclid (1660). *Euclide's Elements*. The whole fifteen Books compendiously demonstrated by Isaac Barrow. London.

Euclid (1956a). *The Thirteen Books of Euclid's Elements*. Volume I. Introduction and Books I, II. Ed. by Thomas L. Heath. New York.

Euclid (1956b). *The Thirteen Books of Euclid's Elements*. Volume III. Books X-XIII and Appendix. Ed. by Thomas L. Heath. New York.

Formey, Jean Henri Samuel (1753). "Congruence". *Encyclopédie, ou Dictionnaire raisonné des sciences, des arts et des métiers* 3, pp. 869–870.

Fowler, David (1999). *The Mathematics of Plato's Academy*. A New Reconstruction. Oxford.

Fried, Michael N. (2009). "Similarity and Equality in Greek Mathematics: Semiotics, History of Mathematics, and Mathematics Education". *For the Learning of Mathematics* 29, pp. 2–7.

von Fritz, Kurt (1959). "Gleichheit, Kongruenz und Ähnlichkeit in der antiken Mathematik bis auf Euklid". *Archiv für Begriffsgeschichte* 4, pp. 7–81.

Gerhardt, C.I., ed. (1858). *Leibnizens mathematische Schriften*. Zweite Abtheilung. Band I. Halle.

Hobbes, Thomas (1656). *Elements of Philosophy*. London.

Kant, Immanuel (1968). *Gesammelte Schriften*. Band XXVIII. Ed. by Gerhard Lehmann. Berlin.

Legendre, Adrien-Marie (1794). *Éléments de géométrie*. Avec des notes. Paris.

Leibniz, Gottfried Wilhelm (1999). *Sämtliche Schriften und Briefe*. Philosophische Schriften. Vierter Band. 1677-Juni 1690. Teil A. Berlin.

Loemker, Leroy E., ed. (1989). *Gottfried Wilhelm Leibniz. Philosophical Papers and Letters*. Second edition. Dordrecht et al.

Mugler, Charles (1958-59). *Dictionnaire historique de la terminologie géométrique des Grecs*. Paris.

[1] "Aequalia et similia congruunt non nisi in *plano*". Translation from Rusnock and George (1995, p. 263).

Murdoch, John E. (1964). "Superposition, Congruence and Continuity in the Middle Ages". In: *Mélanges Alexandre Koyré. I.* Paris, pp. 416–441.
Palmieri, Paolo (2009). "Superposition: on Cavalieri's practice of mathematics". *Archive for History of Exact Sciences* 63, pp. 471–495.
Pardies, Ignace Gaston (1671). *Elemens de geometrie*. Paris.
Proclus (1560). *In primum Euclidis elementorum librum commentariorum libri quattuor*. Ed. by Francesco Barozzi. Padua.
Ramus, Petrus (1567). *Geometria*. Paris.
Ramus, Petrus (1569). *Arithmeticae libri duo. Geometriae septem et viginti*. Basel.
Ramus, Petrus (1590). *Elementes of geometrie*. Faithfully translated by Tho. Hood. London.
Rusnock, Paul and Rolf George (1995). "A Last Shot at Kant and Incongruent Counterparts". *Kant-Studien* 86, pp. 257–277.
Tropfke, Johannes (1940). *Geschichte der Elementar-Mathematik*. Vierter Band. Ebene Geometrie. Dritte, verbesserte und vermehrte Auflage besorgt von Dr. Kurt Vogel. Berlin and Leipzig.
Wolff, Christian (1716). *Mathematisches Lexicon*. Leipzig.

conic section, *parabola*, *ellipse*, and *hyperbola*

This article deals with the term *conic section* and the three non-degenerate types of conic sections. In the following, the original Greek terminology will be explained first. There are essentially two stages: before and after Apollonius. Subsequently, conic sections in Latin texts will be discussed, followed by an overview of the terminology in German, French, and English.

The discovery of conic sections has been attributed to Menaechmus, as pointed out by Proclus (1873, p. 111): "For the helix on a sphere or a cone is around a solid, but a conic or spiric section arises from such-and-such a section of a solid. Some of these sections, in particular the conic, were discovered by Menaechmus".[1]

Apollonius, who contributed substantially to their study, referred to a conic section simply as a *section* (*τομή*, from a verb meaning 'to cut'). This word could generally refer to the intersection of various geometrical objects including conics and also to the result of the intersection, for example a point (Mugler 1958-59, pp. 425f.). Apollonius also used the more specific *section of a cone* (*κώνου τομή*). The explicit *conic section* (*κωνική τομή*) is later found in Proclus (1873, pp. 111f.) and Pappus of Alexandria (1986, p. 107).

In book VII of his *Collection*, Pappus discussed the traditional names of the three types of conic sections: "Aristaeus [...] and Apollonius's (other) predecessors, named the first of the three conic curves 'section of an acute-angled cone', the second 'of a right-angled', the third 'of an obtuse-angled'" (Pappus of Alexandria 1877, p. 672).[2]

To understand Pappus' remarks, it is useful to first consider a passage in Euclid's *Elements*, in which he gave a genetic definition of a right cone and added a distinction between different kinds of cones (book XI, def. 18): "When a right triangle with one side of those about the right angle remains fixed is carried round and restored again to the same position from which it began to be moved, the figure so comprehended is a *cone*. And, if the straight line which remains fixed equals the remaining side about the right angle which is carried round, the cone will be *right-angled*; if less, *obtuse-angled*; and if greater, *acute-angled*".[3] As explained by Heath, "[t]his distinction is

[1] Translation from Proclus (1970, p. 91). [2] Translation from Pappus of Alexandria (1986, pp. 114, 116). [3] Translation from Euclid (1956b, p. 261). The words for *right-angled*, *obtuse-angled*, and *acute-angled* are *ὀρθογώνιος*, *ἀμβλυγώνιος*, and *ὀξυγώνιος*.

quite unnecessary for Euclid's purpose and is not used by him in Book XII; it is no doubt a relic of the method, still in use in Euclid's time, by which the earlier Greek geometers produced conic sections, namely, by cutting right cones only by sections always perpendicular to an edge" (Euclid 1956b, p. 270).

For each of these three kinds of cones, a particular kind of conic section arises, as is also explained by Eutocius (see fig. 23):

> in the right-angled cone it is that which we now call the parabola, in the obtuse-angled cone [that which we now call] the hyperbola, and in the acute-angled cone [that which we now call] the ellipse. [...] And that which is called the section of a right-angled cone they (the Ancients) sought only in a right-angled cone when it was cut by a plane perpendicular to a side [i. e. generator] of the cone, while they were accustomed to demonstrate that the section of an obtuse-angled cone arises [only] in an obtuse-angled cone and the section of an acute-angled cone [only] in an acute-angled cone, when in each case the plane perpendicular to a side of the cone is drawn. This is what these ancient names of those lines indicate (Apollonius of Perga 1893, pp. 168, 170).[1]

These were the classical names of the conic sections; as seen above, Pappus attributed them to Aristaeus, whose work on conic sections is lost. The classical terms were used by Archimedes in his treatise *On Conoids and Spheroids* (Archimedes 1880), for example.[2]

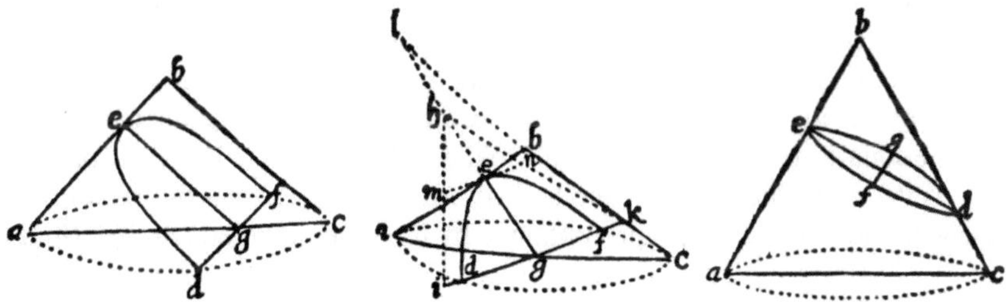

Figure 23: Conic sections (pre-Apollonian style). From Sturm (1670, pp. 316, 317, 319).

Pappus also indicated a change in the terminology suggested by Apollonius:

> But since the three curves occur in each of these three cones, when cut variously, Apollonius was apparently at a loss to know why on earth his predecessors selectively named the one 'section of an acute-angled cone' when it can also be (a section) of a right-angled and obtuse-angled one, the second (cone), and the third 'of an obtuse-angled' when it can be of an acute-angled and a right-angled (cone), so, replacing the names, he called the (section) of 'an acute-angled (cone) 'ellipse', that of a right-angled 'parabola', and that of an obtuse-angled 'hyperbola', each from a certain property of its own. For a certain area applied to a certain line, in the section of an acute-angled cone, falls short by a square, in that of an obtuse-angled

[1] Translation and insertions from Clagett (1980, p. 76). [2] For the term *section of a right-angled cone*, see Archimedes (1880, p. 274); the terms *section of an obtuse-angled cone* and *section of an acute-angled cone* are in Archimedes (1880, pp. 276, 280).

(cone) exceeds by a square, but in that of a right-angled (cone) neither falls short nor exceeds (Pappus of Alexandria 1877, pp. 672, 674).[1]

Pappus here alludes to the so-called *application of areas*.[2] This method was explained and linked to conic sections by Proclus:

These things, says Eudemus [...], are ancient and are discoveries of the Muse of the Pythagoreans, I mean the *application of areas* (*παραβολὴ τῶν χόριων*), their *exceeding* (*ὑπερβολή*) and their *falling-short* (*ἔλλειψις*). It was from the Pythagoreans that later geometers [...] took the names, which they again transferred to the so-called *conic* lines, designating one of these a *parabola* (application), another a *hyperbola* (exceeding) and another an *ellipse* (falling-short), whereas those godlike men of old saw the things signified by these names in the construction, in a plane, of areas upon a finite straight line. For, when you have a straight line set out and lay the given area exactly alongside the whole of the straight line, then they say that you *apply* the said area; when however you make the length of the area greater than the straight line itself, it is said to *exceed*, and when you make it less, in which case, after the area has been drawn, there is some part of the straight line extending beyond it, it is said to *fall short* (Proclus 1873, pp. 419f.).[3]

As explained by Heath, in the case of the parabola the basic property is that "the square on the ordinate at any point is equal to a rectangle applied to p [the parameter] as base with altitude equal to the corresponding abscissa" (Euclid 1956a, p. 345). With respect to fig. 24, this means – in modern terms – that the relation $K\Lambda^2 = \Theta Z \cdot Z\Lambda$ holds ($K\Lambda$ ordinate, ΘZ parameter, $Z\Lambda$ abscissa), which corresponds to the Cartesian equation $y^2 = px$. For the hyperbola, the basic property is that "the square on the ordinate is equal to the rectangle applied to p [the parameter] having as its width the abscissa and *exceeding* [...] by a figure similar and similarly situated to the rectangle contained by the given diameter and p" (ibid.) and for the ellipse, the basic property is that "the square on the ordinate is equal to the rectangle applied to p [the parameter] having as its width the abscissa and [...] *falling-short* [...] by a figure similar and similarly situated to the rectangle contained by the given diameter and p" (ibid.).

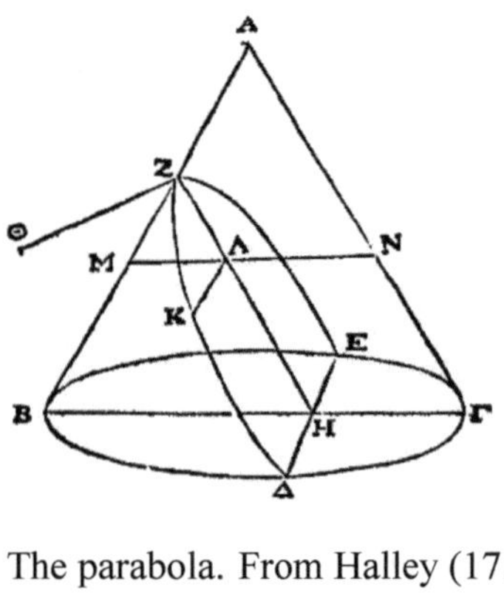

Figure 24: The parabola. From Halley (1710, p. 31).

[1] Translation from Pappus of Alexandria (1986, pp. 114, 116). [2] For explanations of the application of areas and its relation to conic sections, see, e. g., Heath in Euclid (1956a, pp. 343–345) and Grootendorst (2000, pp. 263–284). [3] Translation and insertions from Heath in Euclid (1956a, p. 343).

The terms *parabola*, *hyperbola*, and *ellipse* are, as indicated by Pappus, first attested in the writings of Apollonius;[1] they are found in his *Conics*, in particular. Apollonius considers *oblique* cones rather than just right ones,[2] and thus only needs *one* cone to generate all three kinds of conic sections (see fig. 25). This is his definition of a parabola:[3]

> If a cone is cut by a plane through its axis, and also cut by another plane cutting the base of the cone in a straight line perpendicular to the base of the axial triangle, and if further the diameter of the section is parallel to one side of the axial triangle, then any straight line which is drawn from the section of the cone to its diameter parallel to the common section of the cutting plane and of the cone's base, will equal in square the rectangle contained by the straight line cut off by it on the diameter beginning from the section's vertex and by another straight line which has the ratio to the straight line between the angle of the cone and the vertex of the section that the square on the base of the axial triangle has to the rectangle contained by the remaining two sides of the triangle. And let such a section be called a *parabola* (*παραβολή*) (Apollonius of Perga 1891, p. 38).[4]

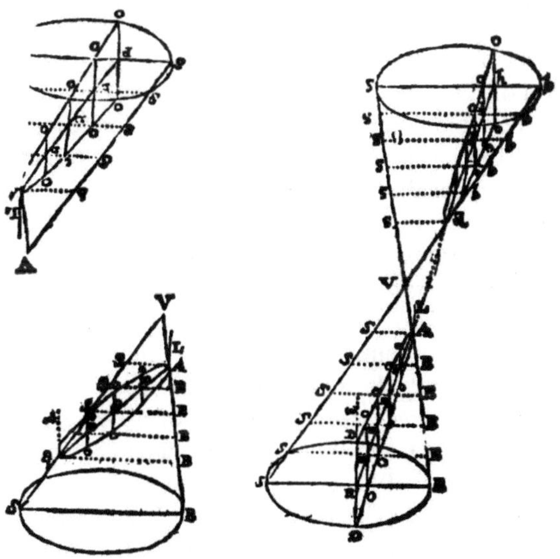

Figure 25: Conic sections (Apollonian style). From Wallis (1655, p. 16).

In Latin, terminology involving conic sections does not appear until the 13th century since "[b]efore the twelfth century the knowledge of conic sections in the Latin West was non-existent" (Clagett 1980, p. 3). This began to change when, in the

[1] The uses of *parabola* and *ellipse* in the Archimedean corpus are believed to be later interpolations (Fried and Unguru 2001, p. 81, Toomer 1976, p. 4). [2] For discussion of the differences between Euclid's and Apollonius' definitions, see Barrow-Green, Gray, and Wilson (2019, p. 155) and Fried and Unguru (2001, pp. 74–79). [3] This definition is from Apollonius' *Conics* (book I, prop. 11). The definition of the hyperbola is found in the following prop. 12, that of the ellipse in prop. 13.
[4] Translation from Apollonius of Perga (2013, p. 19) (emphasis added).

12th century, Gerard of Cremona translated some passages from Apollonius on the basis of Arabic sources; other early Latin texts in which conic sections occur are the *Perspectiva* by the 13th-century philosopher Witelo and, also dating from the 13th century, the various translations of many of the works of Archimedes by William of Moerbeke.

In the relevant passages, Gerard used *sectio* for 'conic section' (Clagett 1980, pp. 7–10), while Witelo used the full *sectio conica* (Unguru 1977, p. 273). Early attestations in print are Valla (1501, bk. XIII, ch. 3), where the relevant chapter is misprinted as *De comica* [sic] *sectione* but otherwise *conica* is used, and Werner (1522, fol. a iiij^{r}), who gave this definition: "If a plane cutting a cone does not pass through its apex and is not parallel to the base, the section common to this same plane and the conical surface is a certain curved line which the mathematicians call a conic section [...] Of conic sections, one is a parabola, the second is a hyperbola, and the third is an ellipse".[1]

As for the names of the three kinds of conic sections, in the texts by Gerard, Witelo, and Moerbeke both the pre-Apollonian and the Apollonian terminology is found. Gerard used *sectio mukefi* (Clagett 1980, p. 9) for the parabola, a term based on Arabic *mukāfi*, which is itself a literal translation of the Greek *παραβολή* (Clagett 1980, p. 10). In his translation of *On conoids*, Moerbeke used *sectio rectangula* (Clagett 1976, e. g., p. 288). At the same time, Moerbeke also transliterated the Apollonian term as *parabola* (Clagett 1976, e. g., p. 142). Witelo was aware of both the Greek and the Arabic terms: "the section EFG will be called a right-angled section, or a parabola, and is [the same as] that which the Arabs call *Mukefi*" (Unguru 1977, p. 273).[2]

For the ellipse, Gerard used *sectio diminuta* (Clagett 1980, p. 9) (Clagett 1980, p. 10). Moerbeke used *sectio oxygonia* (Clagett 1976, e. g., p. 289), again a literal translation; he also used *ellipsis* (Clagett 1976, e. g., p. 268). Witelo called this section "acute-angled, or ellipse, or diminished *Mukefi*" (Unguru 1977, p. 274).[3]

The names for the hyperbola are analogous: Gerard wrote *sectio addita* (Clagett 1980, p. 9), Moerbeke *sectio amblygonia* (Clagett 1976, e. g., p. 288) or *yperbola* (Clagett 1976, e. g., p. 268), and Witelo "obtuse-angled [section], or hyperbole, or increased *Mukefi*" (Unguru 1977, p. 273).[4]

To conclude this article, a look at German, French, and English shall now be taken. In all of these languages, the terminology came up in the 16th century. For example, Dürer (1525, fol. C iiii^{v}) used *Schnydt durch ein Kegel* 'cut through a cone', and Kepler in his *Messekunst* from 1616 preferred the compound *Kegelschnitt* still used today (Kepler 1616, p. 19).[5] The English term is first found (OED) in 1570: "By Wheles, may a straight line be described: Likewise, a Spirall line in plaine, Conicall

---

[1] "Si planum secans conum per eius verticem non venerit neque basi parallelum extiterit, communis sectio eiusdem plani et conicae superficiei inflexa quaedam est linea quam Mathematici conicam vocant sectionem [...] Conicarum sectionum alia parabole, alia hyperbole, alia Ellipsis". Translation from Clagett (1980, p. 291). [2] "sectio EFG dicetur sectio rectangula, vel parabola, et est ilia quam Arabes dicunt *mukefi*." Translation from Unguru (1977, p. 116). [3] "sectio EFG dicetur oxigonia, vel elipsis, vel *mukefi* diminuta". Translation from Unguru (1977, p. 116). [4] "sectio EFG dicetur ampligonia, vel yperbole, vel *mukefi* addita". Translation from Unguru (1977, p. 116).
[5] The term is due to him (Kothmann 1998, p. 21).

Section lines, and other Irregular lines, at pleasure, may be drawen" (Dee 1570, fol. ciiij^{v}). An early usage of French *section conique* is L'Hoste (1619, p. 40).

For the three kinds of conic sections, Dürer (1525, fol. C iij^{v}) coined some new words: "I know of no German words for these three sections, but I want to give them names by which they can be called. The ellipse I call *eyer linie* ["egg line"] because it looks like an egg. The parabola I call *brenn linie* ["burn line"] because if a mirror is shaped according to this line, it will set things on fire. And the hyperbola I shall call *gabellinie* ["fork line"]".[1] Dürer's suggestions did find some followers,[2] but *Parabel*, *Ellipse*, and *Hyperbel* eventually became standard.

In French, *parabole* first appears (TLF) in Gessner (1555, p. 70), and is then found in Cardano (1556, fol. 299^{v}), where also *hyperbole* and *ellipsis* are used; *ellipse* first appears in Stevin (1608, second volume, p. 17).[3]

English *parabola* first appears (OED) in Gessner (1559, p. 71). An early use of English *ellipse* is Blagrave (1585, p. 20); L. Digges and T. Digges (1579, p. 187) earlier wrote *Eleipsis*; *hyperbola* appears on the following page.[4]

Note. For the Greek terminology, see Mugler (1958-59, pp. 171, 268, 325f., 436f.), for the Arabic, see Rashed (2017, pp. 236–239, 652–663). Tropfke (1924, pp. 128–164) discusses the history of conic sections with an emphasis on the terminology. For a recent overview of conic sections in Greek antiquity, see Fried (2019).

References

Apollonius of Perga (1891). *Apollonii Pergaei quae graece exstant cum commentariis antiquis*. Vol. I. Ed. by J.L. Heiberg. Leipzig.

Apollonius of Perga (1893). *Apollonii Pergaei quae graece exstant cum commentariis antiquis*. Vol. II. Ed. by J.L. Heiberg. Leipzig.

Apollonius of Perga (2013). *Conics*. Books I-IV. Books I-III translated by R. Catesby Taliaferro. Book IV translated by Michael N. Fried. Santa Fe.

Archimedes (1880). *Opera omnia*. Cum commentariis Eutocii. Volumen I. Ed. by J.L. Heiberg. Leipzig.

Barrow-Green, June, Jeremy Gray, and Robin Wilson (2019). *The History of Mathematics: A Source-Based Approach*. Volume 1. Providence, R.I.

Blagrave, John (1585). *The mathematical ievvel*. London.

Cardano, Gerolamo (1556). *Les livres de Hierome Cardanus medecin milannois intitulés de la Subtilité, & subtiles inventions, ensemble les causes occultes, & raisons d'icelles*. Traduis de Latin en Francois, par Richard le Blanc. Paris.

Clagett, Marshall (1976). *Archimedes in the Middle Ages*. Volume II. The Translations from the Greek by William of Moerbeke. Philadelphia.

Clagett, Marshall (1980). *Archimedes in the Middle Ages*. Volume IV. A Supplement on the Medieval Latin Traditions of Conic Sections (1150-1566). Philadelphia.

Dee, John (1570). "Mathematicall Praeface". In: *The Elements of Geometrie of the most auncient Philosopher Euclide of Megara*. Ed. by Henry Billingsley. London.

---

[1] "diser dreyer schnydt namen weis jch auf deutzsch nit zuosagenn / wir woellenn jn aber namen geben / dabey man sie kennen mueg / Die Elipsis will jch ein eyer lini nennen / darumb das sie schyer einem ey gleich ist / Die Parabola sey genennt eyn brenn lini / darumb so man auß jr ein spiegel macht so zuendt sie an / Aber die Hiperbole will jch einn gabellini nennen". Translation from Dürer (1977, p. 95). [2] For example, Kepler (1616, p. 5) used *Ey lini* and also gave *ablenger Circkel* 'oblong circle' as an alternative. [3] Steinfeld and Pescarini (2013, p. 244) cite Henrion (1613, p. 164) as a first use. [4] The spelling used there is *hyperbole*. In L. Digges and T. Digges (1571, fol. T.j.^{r}), the adjectives *parabollical*, *hyperbollical* and *ellepseycal* [sic] are found.

Digges, Leonard and Thomas Digges (1571). *A geometrical practise, named Pantometria diuided into three bookes, Longimetra, Planimetra, and Stereometria*. London.
Digges, Leonard and Thomas Digges (1579). *An Arithmeticall Militare Treatise, named Stratioticos*. Augmented, digested, and lately finished, by Thomas Digges. London.
Dürer, Albrecht (1525). *Underweysung der Messung, mit dem Zirckel und Richtscheyt*. Nürnberg.
Dürer, Albrecht (1977). *The Painter's Manual*. A Manual of Measurement of Lines, Areas, and Solids by Means of Compass and Ruler Assembled by Albrecht Dürer for the Use of All Lovers of Art with Appropriate Illustrations Arranged to be Printed in the Year MDXXV. New York.
Euclid (1956a). *The Thirteen Books of Euclid's Elements*. Volume I. Introduction and Books I, II. Ed. by Thomas L. Heath. New York.
Euclid (1956b). *The Thirteen Books of Euclid's Elements*. Volume III. Books X-XIII and Appendix. Ed. by Thomas L. Heath. New York.
Fried, Michael N. (2019). "Conic sections". In: *Oxford Classical Dictionary*. Oxford. URL: `https://oxfordre.com/classics/display/10.1093/acrefore/9780199381135.001.0001/acrefore-9780199381135-e-8161`.
Fried, Michael N. and Sabetai Unguru (2001). *Apollonius of Perga's* Conica. Text, Context, Subtext. Leiden et al.
Gessner, Conrad (1555). *Tresor Evonyme Philiatre des Remedes secretz*. Lyon.
Gessner, Conrad (1559). *The treasure of Euonymus*. Translated (with great diligence, [et] laboure) out of Latin, by Peter Morvvying. London.
Grootendorst, Albert W. (2000). *Jan de Witt's* Elementa Curvarum Linearum, Liber Primus. Text, Translation, Introduction, and Commentary. New York.
Halley, Edmund (1710). *Apollonii Pergaei Conicorum libri octo, et Sereni Antissensis De sectione cylindri & coni libri duo*. Oxford.
Henrion, D. (1613). *Memoires mathematiques*. Premier volume. Paris.
L'Hoste, Jean (1619). *Epipolimetrie ov Art de mesvrer tovtes svperficies*. Saint-Mihiel.
Kepler, Johannes (1616). *Außzug auß der Uralten MesseKunst Archimedis*. Linz.
Kothmann, Hella (1998). "Vom Kunst-Wort zur Wissenschaftssprache. Johannes Keplers Beitrag zur deutschen Fachsprache". In: *Johannes Keplers Beitrag zur Deutschen Fachsprache*. Ed. by Ulrich Grigull and Volker Bialas. München, pp. 7–49.
Mugler, Charles (1958-59). *Dictionnaire historique de la terminologie géométrique des Grecs*. Paris.
Pappus of Alexandria (1877). *Pappi Alexandrini Collectionis*. Volumen II. Ed. by Friedrich Hultsch. Berlin.
Pappus of Alexandria (1986). *Book 7 of the* Collection. Part 1. Introduction, Text, and Translation. Edited with Translation and Commentary by Alexander Jones. London.
Proclus (1873). *In primum Euclidis elementorum librum commentarii*. Ed. by Gottfried Friedlein. Leipzig.
Proclus (1970). *Commentary on the First Book of Euclid's Elements*. Translated with Introduction and Notes by Glenn R. Morrow. Princeton.
Rashed, Roshdi, ed. (2017). *Lexique historique de la langue scientifique arabe*. Hildesheim.
Steinfeld, Nadine and Sandrine Pescarini (2013). "Pleins feux sur l'ellipse en étymologie: un fait linguistique et un outil métalinguistique. Premier volet: étude historique et épistémologique de l'ellipse". *Neuphilologische Mitteilungen* 114, pp. 237–268.
Stevin, Simon (1608). *Memoires mathematiques*. Translaté en François par Jean Tuning. Leiden.
Sturm, Johann Christoph (1670). *Des unvergleichlichen Archimedis Kunst-Bücher*. Nürnberg.
Toomer, G.J., ed. (1976). *Diocles. On Burning Mirrors*. The Arabic Translation of the Lost Greek Original. Berlin et al.
Tropfke, Johannes (1924). *Geschichte der Elementar-Mathematik*. Sechster Band. Analysis. Analytische Geometrie. Berlin and Leipzig.
Unguru, Sabetai, ed. (1977). *Witelonis Perspectivae liber primus*. Book I of Witelo's *Perspectiva*. Wrocław.
Valla, Giorgio (1501). *De expetendis, et fugiendis rebus opus*. Venezia.
Wallis, John (1655). *De sectionibus conicis*. Oxford.
Werner, Johannes (1522). *Libellus super viginti duobus elementis conicis*. Nürnberg.

***conjugate* (overview)**

This adjective is used in various combinations with terms from a wide range of mathematical fields. An early use already occurs in antiquity; see *conjugate axis*. Much later, Newton used the word to refer to an *isolated point*. Various new usages are found in the 19th century; see *conjugate* (algebra), *conjugate* (complex numbers).

For further uses, see also *normal subgroup*, *group*, *dual polyhedron*, *isogonal*, and *syzygy*.

conjugate axis* and *diameter

The terms *conjugate axis* and *conjugate diameter* were first defined by Apollonius:

> The two straight lines, each of which, being a diameter, bisects the straight lines parallel to the other, I call the *conjugate diameters* of a curved line and of two curved lines [...] And I call those straight lines the *conjugate axes* of a curved line and of two curved lines which, being conjugate diameters, cut the straight lines parallel to each other at right angles (Apollonius of Perga 1891, p. 8).[1]

Apollonius used the word *συζυγής*, which is a variant of *σύζυγος* 'yoked together, paired' (LSJ). The term *conjugate diameter* (of an ellipse) was also used once by Archimedes in his *On Conoids and Spheroids* (Archimedes 1880, p. 324).[2]

The corresponding term is not found in classical Latin in this sense, since conic sections remained unknown in the Latin West until the 12th century (Clagett 1980, p. 3). At that time, Gerard of Cremona translated a book by Ibn al-Haiṭam on parabolic burning mirrors, to which he prefaced a fragment of the beginning of Apollonius' *Conics*. He translated from the Arabic and left the Arabic word for 'conjugate' essentially unchanged; he thus wrote *diametros muzdaguageni* and *axes muzdaguageni* (Clagett 1980, pp. 6f.). The Arabic word comes from a root meaning 'pair'.

The 13th-century translator William of Moerbeke then once used Latin *coniuga* in connection with diameters in his translation of Archimedes' *On conoids and spheroids* (Clagett 1976, p. 300). Valla (1501, book XIII, ch. 3), on the other hand, took over the Greek term and wrote *syzyges*. In the first Latin translation of Apollonius' works, *conjunctus* is used (Apollonius of Perga 1537, fol. 2^{v}), but in Commandinus' influential edition, *conjugatus* is preferred (Apollonius of Perga 1566, fol. 6^{v}).

In French, the term *diamètres conjugués* appears, apparently first, in Desargues (1639, p. 29);[3] in the *advertissement* (notice) at the end, Desargues also uses *coniugale*. English *conjugate diameter* seems to be due to Foster (1654, p. 184). In German, Sturm (1670, p. 208) used *creutzender Durchmesser* (*kreuzen* means 'to cross'), while Wolff (1716, p. 516) used *Neben-Diameter* (*neben* 'beside, next to'). Neither term found much appeal; the use of German *konjugiert* in this sense is of later origin and is found in de La Chapelle (1771, p. 305).

Note. The history of the term is discussed by Tropfke (1924, pp. 144f.). For the Greek term, see Mugler (1958-59, p. 386); for the Arabic, see Rashed (2017, p. 650).

---

[1] Translation from Apollonius of Perga (2013, p. 4) (emphasis added). [2] See also Heath in Archimedes (1897, p. clxx). [3] Desargues wrote *diametres coniugez*. Pascal (1640) uses *conjugué* as well.

References

Apollonius of Perga (1537). *Opera*. Per ... Ioannem Baptistam Memum ... de Graeco in Latinum traducta & noviter impressa. Venezia.

Apollonius of Perga (1566). *Conicorum libri quattuor*. Ed. by Federico Commandino. Bologna.

Apollonius of Perga (1891). *Apollonii Pergaei quae graece exstant cum commentariis antiquis*. Vol. I. Ed. by J.L. Heiberg. Leipzig.

Apollonius of Perga (2013). *Conics*. Books I-IV. Books I-III translated by R. Catesby Taliaferro. Book IV translated by Michael N. Fried. Santa Fe.

Archimedes (1880). *Opera omnia*. Cum commentariis Eutocii. Volumen I. Ed. by J.L. Heiberg. Leipzig.

Archimedes (1897). *The Works of Archimedes*. Edited in Modern Notation with Introductory Chapters by T. L. Heath. Cambridge.

de La Chapelle, Jean-Baptiste (1771). *Abhandlung von den Kegelschnitten*. Uebersetzt und mit Anmerkungen versehen von Johann Lorenz Böckmann. Karlsruhe.

Clagett, Marshall (1976). *Archimedes in the Middle Ages*. Volume II. The Translations from the Greek by William of Moerbeke. Philadelphia.

Clagett, Marshall (1980). *Archimedes in the Middle Ages*. Volume IV. A Supplement on the Medieval Latin Traditions of Conic Sections (1150-1566). Philadelphia.

Desargues, Girard (1639). *Brouillon project d'une atteinte aux evenemens des rencontres du Cone avec un Plan*. Paris. URL: http://gallica.bnf.fr/ark:/12148/bpt6k105071b.

Foster, Samuel (1654). *Elliptical, or azimuthal horologiography*. London.

Mugler, Charles (1958-59). *Dictionnaire historique de la terminologie géométrique des Grecs*. Paris.

Pascal, Blaise (1640). *Essay pour les coniques*. Paris. URL: http://gallica.bnf.fr/ark:/12148/btv1b86262279/.

Rashed, Roshdi, ed. (2017). *Lexique historique de la langue scientifique arabe*. Hildesheim.

Sturm, Johann Christoph (1670). *Des unvergleichlichen Archimedis Kunst-Bücher*. Nürnberg.

Tropfke, Johannes (1924). *Geschichte der Elementar-Mathematik*. Sechster Band. Analysis. Analytische Geometrie. Berlin and Leipzig.

Valla, Giorgio (1501). *De expetendis, et fugiendis rebus opus*. Venezia.

Wolff, Christian (1716). *Mathematisches Lexicon*. Leipzig.

conjugate (algebra)

Conjugate elements and conjugate groups appear in the 19th century. Galois uses French *groupes conjugués* in his second memoir (Galois 2011); he used the term "for groups of permutations contained in a larger group and such that their groups of substitutions are all equal and form a normal subgroup of the group of substitutions of the large one" (Neumann 2011, p. 24). He does not apply the term to the permutations themselves and appears to use the word *conjugate* in its sense 'grouped together' (Dahan 1980, p. 295).

Conjugate elements (in connection with permutation groups) are found in Cauchy (1844, pp. 165–170), but he does not give the concept a special name.[1] Shortly afterwards, Betti (1852, p. 55) defines the "permutation derived from θ by means of ψ"[2] as $\theta_1 = \psi^{-1}\theta\psi$, and a *derived group* (*gruppo derivato*) G as $\psi^{-1}G\psi$ (Betti 1852, p. 61).

Jordan (1869, pp. 141f.) referred to the permutation $b^{-1}ab$ as the *transformed* (French *transformé*) of the permutation a by the permutation b, and correspondingly called a conjugate subgroup a *groupe transformé*. He kept this terminology in Jordan

[1] Cauchy otherwise uses *conjuguer* in the sense 'to join'; see *group*. [2] "sostituzione derivata di θ per mezzo di ψ".

(1870, pp. 23f.) and was followed by Sylow (1872, p. 585) and Netto (1878, p. 242), for example.

Klein (1879, pp. 430f.), on the other hand, called conjugate elements *gleichberechtigt*, which literally means 'having equal rights'; he also used the term in connection with subgroups.[1] The term found some appeal, and was also used by Gierster (1881, p. 331) and Dyck (1883, p. 75), for example. Bolza (1890-91, p. 89) translated this as *conjugate subgroup*; using *conjugate* here made sense because Cole (1886, p. 268) had earlier used *self-conjugate sub-group* for a normal subgroup. In a translation of Netto's *Theory of Substitutions*, where *transformed* (German *transformiert*) was used (Netto 1882, p. 47), *conjugate* was also favoured. The term found additional acceptance when Weber (1895, p. 506) adopted it in his widely read book on algebra.

See also *transposition* (matrix).

Note. For the history of concept, see the various remarks in Dahan (1980), Kiernan (1971), and Wußing (1984, pp. 94f., 118–144); see also Burkhardt and Vogt (1909, p. 565) for a brief terminological overview.

References

Betti, Enrico (1852). "Sulla risoluzione delle equazioni algebriche". *Annali di scienze matematiche e fisiche* 3, pp. 49–115.

Bolza, Oskar (1890-91). "On the Theory of Substitution-Groups and its Applications to Algebraic Equations". *American Journal of Mathematics* 13, pp. 59–144.

Burkhardt, Heinrich and Henri Vogt (1909). "Sur les groupes finis discontinus". *Encyclopédie des sciences mathématiques pures et appliquées* I.1, pp. 532–616.

Cauchy, Augustin-Louis (1844). *Exercises d'analyse et de physique mathématique*. Tome troisième. Paris.

Cole, Frank Nelson (1886). "A Contribution to the Theory of the General Equation of the Sixth Degree". *American Journal of Mathematics* 8, pp. 265–286.

Dahan, Amy (1980). "Les travaux de Cauchy sur les substitutions. Étude de son approche du concept de groupe". *Archive for History of Exact Sciences* 23, pp. 279–319.

Dyck, Walther (1883). "Gruppentheoretische Studien. II. Ueber die Zusammensetzung einer Gruppe discreter Operationen, über ihre Primitivität und Transitivität". *Mathematische Annalen* 22, pp. 70–108.

Galois, Évariste (2011). "The Second Memoir". In: *The Mathematical Writings of Évariste Galois*. Ed. by Peter M. Neumann. Zürich, pp. 169–191.

Gierster, Joseph (1881). "Die Untergruppen der Galois'schen Gruppe der Modulargleichungen für den Fall eines primzahligen Transformationsgrades". *Mathematische Annalen* 18, pp. 319–365.

Jordan, Camille (1869). "Commentaire sur Galois". *Mathematische Annalen* 1, pp. 141–160.

Jordan, Camille (1870). *Traité des substitutions et des équations algébriques*. Paris.

Kiernan, B. Melvin (1971). "The Development of Galois Theory from Lagrange to Artin". *Archive for History of Exact Sciences* 8, pp. 40–154.

Klein, Felix (1879). "Ueber die Transformation siebenter Ordnung der elliptischen Functionen". *Mathematische Annalen* 14, pp. 428–471.

Lie, Sophus (1874). "Ueber Gruppen von Transformationen". *Nachrichten von der Königl. Gesellschaft der Wissenschaften und der Georg-Augusts-Universität zu Göttingen* 22, pp. 529–542.

Netto, Eugen (1878). "Einleitung in die Theorie der Substitutionen und ihre Anwendungen". *Archiv der Mathematik und Physik* 62, pp. 225–258.

Netto, Eugen (1882). *Substitutionentheorie und ihre Anwendungen auf die Algebra*. Leipzig.

[1] Klein may have chosen the term on the basis of a remark by Lie (1874, p. 531).

Neumann, Peter M. (2011). *The Mathematical Writings of Évariste Galois*. Zürich.
Sylow, Ludvig (1872). "Théorèmes sur les groupes de substitutions". *Mathematische Annalen* 5, pp. 584–594.
Weber, Heinrich (1895). *Lehrbuch der Algebra*. Erster Band. Braunschweig.
Wußing, Hans (1984). *The Genesis of the Abstract Group Concept*. A Contribution to the History of the Origin of Abstract Group Theory. Tr. Abe Shenitzer. Cambridge, Mass.

conjugate **(complex number)**

The conjugate of a complex number was introduced as French *conjugué* by Cauchy (1821, p. 180): "We say that two imaginary expressions are *conjugate* to each other when the two expressions differ only in the signs of the coefficient of $\sqrt{-1}$".[1] The term *complex number* had not yet been coined at that time, which is why Cauchy speaks of "imaginary expressions".

Cauchy's term was soon adopted, for example by Gauß (1832, § 31): "we shall call the number that arises from a complex number by the exchange of i and $-i$ its *conjugate* number".[2] English *conjugate* in this sense appears in Young (1835, p. 19).

Note. The origin of the term is briefly mentioned by Tropfke (1980, p. 158) and Smith (1958, p. 267).

References

Bradley, Robert E. and C. Edward Sandifer (2009). *Cauchy's* Cours d'analyse. An Annotated Translation. Dordrecht et al.
Cauchy, Augustin-Louis (1821). *Cours d'analyse de l'École Royale Polytechnique*. I.re partie. Analyse algébrique. Paris.
Gauß, Carl Friedrich (1832). "Theoria residorum biquadraticorum. Commentatio secunda". *Commentationes Societatis Regiae Scientiarum Gottingensis recentiores* 7, pp. 89–148.
Smith, David Eugene (1958). *History of Mathematics*. Volume II. Special Topics of Elementary Mathematics. New York.
Tropfke, Johannes (1980). *Geschichte der Elementarmathematik*. Band 1: Arithmetik und Algebra. Vollständig neu bearbeitet von Kurt Vogel, Karin Reich, Helmuth Gericke. Berlin and New York.
Young, John Radford (1835). *On the Theory and Solution of Algebraical Equations*. London.

connected

An early attempt at defining connectedness was made by Bolzano (1817, pp. 20f.): "A spatial object of which every part which can be viewed as a line according to the definition just given, has at least *one point* in common with the remaining part which then likewise must be viewed as a line, is called an *absolutely connected line*".[3] In the following pages, he also defined connectedness for surfaces and solids.[4]

Path-connectedness can be found in Riemann's thesis: "We regard two parts of a surface as connected, or belonging to a single piece, if from any point of one part to any point of the other, a line can be drawn interior to the surface. We regard two parts

---

[1] "On dit que deux expressions imaginaires sont *conjuguées* l'une à l'autre, lorsque ces deux expressions ne diffèrent entre elles que par le signe du coefficient de $\sqrt{-1}$". Translation from Bradley and Sandifer (2009, p. 121). [2] "Contra numero complexo *coniunctum* vocamus eum, qui per permutationem ipsius i cum $-i$ inde oritur". [3] "Ein Raumding, dessen jeder Theil, der sich nach eben gegebener Erklärung als Linie ansehen heisst, mit dem noch übrigen Theile, der sich dann gleichfalls als Linie muss ansehen lassen, wenigstens *einen Punct* gemein hat, heisst eine *durchaus zusammenhängende Linie*". Translation from Russ (2004, p. 302). [4] A discussion of his definitions is found in Johnson (1977, pp. 275–280).

as separate, when this procedure is not possible" (Riemann 1851, p. 6).[1] Riemann simply used the word *zusammenhängend* 'connected'.

A very different concept of connectedness was defined by Cantor (1883, pp. 575f.), who defined a *connected set* (*zusammenhängende Menge*) in the following terms: "We call T a *connected* point set, if for any two of its points t and t', and ε being an arbitrarily small number, a *finite* number of points $t_1, t_2, \ldots\ t_\nu$ of T exist such that the distances $\overline{tt_1}, \overline{t_2t_3}, \ldots, \overline{t_\nu t'}$ are all less than ε".[2] Ferreirós (2007, p. 209) remarks that "Cantor's notion of connectedness is somewhat strange, since according to it the set of rational numbers is connected".

A different definition was suggested by Jordan (1892, p. 75), who also used another name: "We say that a perfect and bounded set is *of one piece* if it cannot be decomposed into several perfect separated sets".[3] According to Moore (2008, p. 226), this means that a set E in $\mathbb{R}^n$ is connected "if and only if E cannot be partitioned into two closed and 'separated' sets", where "two sets A and B are 'separated' if the greatest lower bound of the distance ab, where the point a belongs to A and b to B, is not zero". Jordan's definition also appears in Schoenflies (1904, p. 209) under the name *connected* (*zusammenhängend*).

General definitions of connectedness were given independently by Lennes (1906, p. 284, 1911, p. 303), Riesz (1905a, p. 650, 1905b, p. 413, 1909, p. 321), and slightly later by Hausdorff (1914, p. 244). Lennes' definition, given at a talk in December 1905, reads: "A set of points is connected if in every pair of complementary subsets at least one subset contains a limit point of points in the other set". Riesz's first definition dates from October 1905 but his contributions in this respect remained almost completely unnoticed;[4] in the first publication cited, he used Jordan's *d'un seul tenant*, afterwards he used *zusammenhängend*.

In connection with graphs, *connected* essentially means 'path-connected'. A corresponding definition of French *réseaux connexe* was first given by Brunel (1895, p. 165): "connected graphs, that means one can, by following the edges of the graph, go from any arbitrary vertex of the graph to another equally arbitrary vertex".[5] The equivalent German term *zusammenhängend* was used by König (1916, p. 453). Jordan (1869, p. 185) earlier used *continu* in this sense.

See also *locally connected*, *simply connected*, and *arcwise connected*.

Note. For the history of connectedness, see Wilder (1978, 1980), upon which this article is largely based. See also Herrlich, Hušek, and Preuß (2002), Zitarelli (2009), and the editorial notes in Brouwer (1976, pp. 486f.).

---

[1] "Wir betrachten zwei Flächentheile als zusammenhängend oder Einem Stücke angehörig, wenn sich von einem Punkte des einen durch das Innere der Fläche eine Linie nach einem Punkte des andern ziehen lässt, als getrennt, wenn diese Möglichkeit nicht Statt findet". Translation from Riemann (2004, p. 6). The passage was pointed out by Remmert (1991, p. 39). [2] "Wir nennen T eine *zusammenhängende* Punktmenge, wenn für je zwei Punkte t und t' derselben, bei vorgegebener beliebig kleiner Zahl ε immer eine *endliche* Anzahl Punkte $t_1, t_2, \ldots\ t_\nu$ von T auf mehrfache Art vorhanden sind, so dass die Entfernungen $\overline{tt_1}, \overline{t_2t_3}, \ldots, \overline{t_\nu t'}$ sämmtlich kleiner sind als ε". [3] "Nous disons qu'un ensemble parfait et borné est *d'un seul tenant* lorsqu'il ne peut être décomposé en plusieurs ensembles parfaits séparés". [4] Tietze and Vietoris (1928, p. 157) belong to the few who mention Riesz in this respect. [5] "réseaux connexes, c'est-à-dire tels que l'on peut, en suivant les arêtes du réseau, aller d'un sommet quelconque du réseau à un autre sommet également quelconque".

References

Bolzano, Bernhard (1817). *Die drey Probleme der Rectification, der Complanation und der Cubirung*. Leipzig.

Brouwer, L.E.J. (1976). *Collected Works*. 2: Geometry, Analysis, Topology and Mechanics. Ed. by Hans Freudenthal. Amsterdam and Oxford.

Brunel, Georges (1895). “Analysis situs. Recherches sur les réseaux”. *Mémoires de la Société des sciences physiques et naturelles de Bordeaux* 5, pp. 165–215.

Cantor, Georg (1883). “Ueber unendliche, lineare Punktmannichfaltigkeiten. 5”. *Mathematische Annalen* 21, pp. 545–586.

Ferreirós, José (2007). *Labyrinth of Thought*. A History of Set Theory and Its Role in Modern Mathematics. Second revised edition. Basel.

Hausdorff, Felix (1914). *Grundzüge der Mengenlehre*. Leipzig.

Herrlich, Horst, Mirek Hušek, and Gerhard Preuß (2002). “Zusammenhang”. In: *Felix Hausdorff. Gesammelte Werke*. Band II. Grundzüge der Mengenlehre. Ed. by Egbert Brieskorn et al. Berlin and Heidelberg, pp. 752–756.

Johnson, Dale M. (1977). “Prelude to Dimension Theory: The Geometrical Investigations of Bernard Bolzano”. *Archive for History of Exact Sciences* 17, pp. 261–295.

Jordan, Camille (1869). “Sur les assemblages de lignes”. *Journal für die reine und angewandte Mathematik* 70, pp. 185–190.

Jordan, Camille (1892). “Remarques sur les intégrales définies”. *Journal de mathématiques pures et appliquées* 8, pp. 69–99.

König, Dénes (1916). “Über Graphen und ihre Anwendung auf Determinantentheorie und Mengenlehre”. *Mathematische Annalen* 77, pp. 453–465.

Lennes, Nels Johann (1906). “Curves in non-metrical analysis situs”. *Bulletin of the American Mathematical Society* 12, pp. 280, 284–285.

Lennes, Nels Johann (1911). “Curves in Non-Metrical Analysis Situs with an Application in the Calculus of Variations”. *American Journal of Mathematics* 33, pp. 287–326.

Moore, Gregory H. (2008). “The emergence of open sets, closed sets, and limit points in analysis and topology”. *Historia Mathematica* 35, pp. 220–241.

Remmert, Reinhold (1991). *Theory of Complex Functions*. Translated by Robert B. Burckel. New York.

Riemann, Bernhard (1851). “Grundlagen für eine allgemeine Theorie der Functionen einer veränderlichen complexen Grösse”. PhD thesis. Georg-August-Universität Göttingen.

Riemann, Bernhard (2004). *Collected Papers*. Translated from the 1892 edition by Roger Baker, Charles Christenson and Henry Orde. Heber City.

Riesz, Frigyes (1905a). “Sur les ensembles discontinus”. *Comptes rendus de l'Académie des sciences* 141, pp. 650–653.

Riesz, Frigyes (1905b). “Über mehrfache Ordnungstypen I”. *Mathematische Annalen* 61, pp. 406–421.

Riesz, Frigyes (1909). “Die Genesis des Raumbegriffs”. *Mathematische und naturwissenschaftliche Berichte aus Ungarn* 24, pp. 309–353.

Russ, Steve (2004). *The Mathematical Works of Bernard Bolzano*. Oxford.

Schoenflies, Arthur (1904). “Beiträge zur Theorie der Punktmengen”. *Mathematische Annalen* 58, pp. 195–234.

Tietze, Heinrich and Leopold Vietoris (1928). “Beziehungen zwischen den verschiedenen Zweigen der Topologie”. *Encyklopädie der mathematischen Wissenschaften mit Einschluss ihrer Anwendungen* III AB 13, pp. 141–237.

Wilder, Raymond Louis (1978). “Evolution of the Topological Concept of ‘Connected’”. *The American Mathematical Monthly* 85, pp. 720–726.

Wilder, Raymond Louis (1980). “Correction and Addendum to ‘Evolution of the Topological Concept of ‘Connected’”. *The American Mathematical Monthly* 87, pp. 31–32.

Zitarelli, David E. (2009). “Connected Sets and the AMS, 1901–1921”. *Notices of the American Mathematical Society* 56, pp. 450–458.

connection

Right after the publication of Levi-Civita's paper on parallelism (Levi-Civita 1917), Weyl (1918, pp. 389f.) coined the term *connection* (*Zusammenhang*) in his article on "Purely Infinitesimal Geometry":

> If P' is infinitely close to the fixed point P, then P' is *affinely connected with* P, if for each vector at P it is determined into which vector at P' it will transform under *parallel displacement* from P to P'. The parallel displacement of all vectors at P from there to P' must evidently satisfy the following requirement.
>
> **A.** *The transfer of the totality of vectors from P to the infinitely close point P' by means of parallel displacement produces an affine transformation of the vectors at P to the vectors at P'.* [...] A second requirement is still to be imposed on the concept of parallel displacement, that of *commutativity.*
>
> **B.** *If P_1, P_2 are two points infinitely close to P and if the infinitesimal vector $\overrightarrow{PP_1}$ becomes $\overrightarrow{P_2P_{21}}$ under parallel displacement from P to P_2, and $\overrightarrow{PP_2}$ becomes $\overrightarrow{P_1P_{12}}$ under parallel displacement to P_1, then the points P_{12} and P_{21} coincide.*[1]

The term also appears in Weyl (1919, pp. 100f.), where also (p. 109) *metric connection* is found. French *connexion* appears in Dienes (1922, p. 1168) and in various articles by Cartan, for example in Cartan (1923, p. 325); English *connection* was used in Weyl (1921, p. 800).

Note. For the history of the concept, see Reich (1992), Bourguignon (1992), and Scholz (1999).

References

Bourguignon, Jean-Pierre (1992). "Transport parallèle et connexions en géométrie et en physique". In: *1830–1930: A Century of Geometry*. Epistemology, History and Mathematics. Ed. by Luciano Boi, Dominique Flament, and Jean-Michel Salanskis. Berlin, pp. 150–166.

Cartan, Élie (1923). "Sur les variétés à connexion affine et la théorie de la relativité généralisée (première partie)". *Annales scientifiques de l'École Normale Supérieure* 40, pp. 325–412.

Dienes, Paul (1922). "Sur la connexion du champ tensoriel". *Comptes rendus de l'Académie des Sciences de Paris* 174, pp. 1167–1169.

Levi-Civita, Tullio (1917). "Nozione di parallelismo in una varietà qualunque e conseguente specificazione geometrica della curvatura riemanniana". *Rendiconti del Circolo matematico di Palermo* 42, pp. 172–205.

Reich, Karin (1992). "Levi-Civitasche Parallelverschiebung, affiner Zusammenhang, Übertragungsprinzip: 1916/17 – 1922/23". *Archive for History of Exact Sciences* 44, pp. 77–105.

Scholz, Erhard (1999). "Weyl and the theory of connections". In: *The Symbolic Universe*. Geometry and Physics 1890–1930. Ed. by Jeremy Gray, pp. 260–284.

Weyl, Hermann (1918). "Reine Infinitesimalgeometrie". *Mathematische Zeitschrift* 2, pp. 384–411.

[1] "Ist P' ein zu dem festen Punkt P unendlich benachbarter, so *hängt* P' mit P *affin zusammen*, wenn von jedem Vektor in P feststeht, in welchen Vektor in P' er durch *Parallelverschiebung* von P nach P' übergeht. Die Parallelverschiebung der sämtlichen Vektoren in P von dort nach P' muß dabei selbstverständlich der folgenden Forderung genügen. **A.** *Die Verpflanzung der Gesamtheit der Vektoren von P nach dem unendlich benachbarten Punkte P' durch Parallelverschiebung liefert eine affine Abbildung der Vektoren in P auf die Vektoren in P'.* [...] An den Begriff der Parallelverschiebung ist noch eine zweite Forderung, die der *Kommutativität*, zu stellen. **B.** Sind P_1, P_2 zwei zu P unendlich benachbarte Punkte und geht der infinitesimale Vektor $\overrightarrow{PP_1}$ durch Parallelverschiebung von P nach $\overrightarrow{P_2P_{21}}$ über, $\overrightarrow{PP_2}$ aber durch Parallelverschiebung nach P_1 in $\overrightarrow{P_1P_{12}}$, so fallen P_{12} und P_{21} zusammen". Translation from Weyl (2007, p. 1094).

Weyl, Hermann (1919). *Raum · Zeit · Materie*. Vorlesungen über allgemeine Relativitätstheorie. Dritte, umgearbeitete Auflage. Berlin.
Weyl, Hermann (1921). "Electricity and gravitation [Translated by Dr. Robert W. Lawson]". *Nature* 106, pp. 800–802.
Weyl, Hermann (2007). "Purely Infinitesimal Geometry (Excerpt)". In: *The Genesis of General Relativity*. Volume 4. Gravitation in the Twilight of Classical Physics: The Promise of Mathematics. Ed. by Jürgen Renn. Dordrecht, pp. 1089–1105.

conoid

The term *conoid* was introduced by Archimedes; he defined the concept in his *On Conoids and Spheroids* and distinguished two types:

> Concerning the *right-angled conoid* [paraboloid of revolution] it was laid down that, if a section of a right-angled cone [a parabola] be made to revolve about the diameter [axis] which remains fixed and return to the position from which it started, the figure comprehended by the section of the right-angled cone is called a *right-angled conoid* [...] The figure comprehended by the section of the obtuse-angled cone [hyperbola] is called an *obtuse-angled conoid* [hyperboloid of revolution] (Archimedes 1880, pp. 274, 276).[1]

Archimedes used the word κωνοειδής, an adjective (referring to σχῆμα 'figure') often used as a noun and meaning 'cone-like'. Heath considers the Greek term "perhaps more satisfactory than *paraboloid* [and] *hyperboloid*" but also remarks that "when κωνοειδής is qualified by the adjective *right-angled ὀρθογώνιον* to denote the paraboloid of revolution, and by ἀμβλύγώνιον *obtuse-angled* to denote the hyperboloid of revolution, the expressions are less logical, as the solids do not resemble right-angled and obtuse-angled cones respectively; in fact, since the angle between the asymptotes of the generating hyperbola may be acute, a hyperboloid of revolution would in that case more resemble an acute-angled cone. The terms *right-angled* and *obtuse-angled* were merely transferred to the conoids from the names for the respective conics without any more thought of their meaning" (Archimedes 1897, p. clxix).

The corresponding Latin term does not appear in classical Latin, and is indeed a much later creation. In the 13th century, William of Moerbeke translated various works by Archimedes and entitled his translation of the above-mentioned treatise *De conoydalibus et speroydalibus* (Clagett 1976, p. 288). The usual form *conoides*, however, is of later origin and found in Valla (1501, book 10, ch. vi). This word was also used by Commandinus in his edition of Archimedes' works, where Archimedes' book is entitled *De conoidibus et de sphaeroidibus* (Archimedes 1558, fol. 27^{r}).[2]

As for modern languages, French *conoïde* is found in Cardano (1556, fol. 303^{v}); the English term is used by Dee (1570, fol. c iiij^{v}) but does not become commonly used until the second half of the 17th century, as for example by Anderson (1668, e. g., p. 8). In German, the term did not become naturalized until the 18th century; in earlier German texts, the word usually appears in its Latin form. Sturm (1670, Vorbericht) suggested *After-Kegel*, where *after* is used in its obsolete meaning 'false'.

---

[1] Translation and notes from Archimedes (1897, p. 99). – For the terms *section of a right-angled cone* and *section of an obtuse-angled cone*, see *conic section*. [2] In the commentary, Commandinus suggested the new terms *conoides parabolicum* and *conoides hyperbolicum* as a substitute for the Archimedean names for the two kinds of conoids (Archimedes 1558, commentary, fol. 26^{v}).

In the 18th century, a change in the terminology can be perceived. On the one hand, Archimedes' classification of conoids gradually fell out of use and was eventually replaced by the terms developed in France, namely *paraboloid* and *hyperboloid* (see the respective entries).

On the other hand, in France, the term *conoid* received a new meaning. As pointed out by Boyer (1956, p. 207), the new meaning appears to go back to Tinseau (1780, p. 626), who used the term for a particular kind of ruled surface. Tinseau constructs it like this (see fig. 26): He considers a plane ACQ, a straight line AB (called *axis*), and a curve DEQ (called *directrix*); then, he imagines a "mobile" line AQ that always remains parallel to the horizon and successively passes through the corresponding points on the axis and on the directrix, thus generating the conoid.

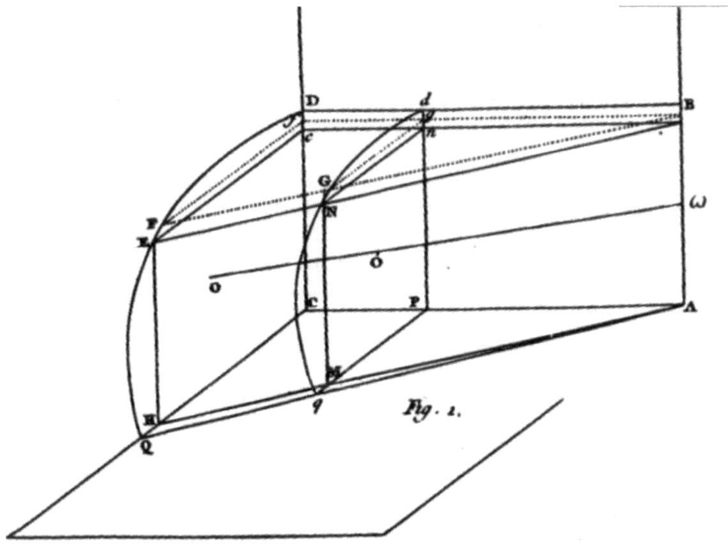

Figure 26: The term *conoid* given a new meaning. From Tinseau (1780).

Tinseau (1780, p. 626) does not claim the term is new but points out that this kind of skew surface is known in architecture under this name.[1] The term is also found, without definition, in Monge (1780, p. 385), and spread through various publications such as Monge (1807, seconde partie, p. 22). The English term appears in this sense in Davies (1826, pp. 192f.), who also explains that it is based on "the analogy existing between them [conoids], and the surfaces of cones", that is: "If all the points in which the elements touch the rectilinear directrix were brought together into one point, the elements still passing through the points in which they touch the second directrix, the surface becomes a conic surface".

For the term *Plücker's conoid*, see *cylindroid*.

[1] He also distinguished between a *right* and an *oblique* conoid.

Note. For the Greek term, see Mugler (1958-59, pp. 265f.) and Heath in Archimedes (1897, pp. clxix–clxx).

References

Anderson, Robert (1668). *Stereometrical Propositions*. London.
Archimedes (1558). *Opera*. Ed. by Federico Commandino. Venezia.
Archimedes (1880). *Opera omnia*. Cum commentariis Eutocii. Volumen I. Ed. by J.L. Heiberg. Leipzig.
Archimedes (1897). *The Works of Archimedes*. Edited in Modern Notation with Introductory Chapters by T. L. Heath. Cambridge.
Boyer, Carl B. (1956). *History of Analytic Geometry*. New York.
Cardano, Gerolamo (1556). *Les livres de Hierome Cardanus medecin milannois intitulés de la Subtilité, & subtiles inventions, ensemble les causes occultes, & raisons d'icelles*. Traduis de Latin en Francois, par Richard le Blanc. Paris.
Clagett, Marshall (1976). *Archimedes in the Middle Ages*. Volume II. The Translations from the Greek by William of Moerbeke. Philadelphia.
Davies, Charles (1826). *Elements of Descriptive Geometry*. Philadelphia.
Dee, John (1570). "Mathematicall Praeface". In: *The Elements of Geometrie of the most auncient Philosopher Euclide of Megara*. Ed. by Henry Billingsley. London.
Monge, Gaspard (1780). "Mémoire sur les Propriétés de plusieurs genres de Surfaces courbes, particulièrement sur celles des Surfaces développables, avec une Application à la Théorie des Ombres et des Pénombres". *Mémoires de mathématique et de physique, presentés à l'Académie royale des sciences, par divers sçavans* 9, pp. 382–440.
Monge, Gaspard (1807). *Application de l'analyse à la géométrie*. Paris.
Mugler, Charles (1958-59). *Dictionnaire historique de la terminologie géométrique des Grecs*. Paris.
Sturm, Johann Christoph (1670). *Des unvergleichlichen Archimedis Kunst-Bücher*. Nürnberg.
Tinseau, Charles (1780). "Sur quelques propriétés des solides renfermés par des surfaces composées de lignes droites". *Mémoires de mathématique et de physique, presentés à l'Académie royale des sciences, par divers sçavans* 9, pp. 625–642.
Valla, Giorgio (1501). *De expetendis, et fugiendis rebus opus*. Venezia.

consistent **(axioms)**

Consistency in the sense of freedom from contradiction has always been a central concern in mathematics and logic. In the 19th century, the terminology was also applied to systems of axioms. For example, Frischauf (1872, Vorwort), discussing the new geometries developed by Lobachevsky and Bolyai, remarked that "the question of a proof for this axiom [the parallel postulate] was finally settled when, by the construction of a consistent [*widerspruchsfreie*] geometry based on the opposite assumption, the unprovability of this axiom became obvious".[1] For *consistent*, Frischauf used *widerspruchsfrei*, which means 'free from contradiction'.

The term *widerspruchslos*, of the same meaning, was used by Hilbert (1899, p. 19) in his *Foundations of Geometry* and various other texts. In the studies on postulates from the beginning of the 20th century, *consistency* was used in this sense. Thus, Huntington (1902, p. 264) defined *consistent* in the following terms: "there is at least one assemblage in which the chosen rule of combination satisfies all the six requirements".[2] In a talk given at the ICM in 1900, Padoa (1902, p. 249) used

---

[1] "Die Frage nach einem Beweis dieses Axioms wurde erst dann vollständig erledigt, als durch Aufstellung einer in sich widerspruchsfreien Geometrie, die sich auf die entgegengesetzte Voraussetzung stützt, die Unbeweisbarkeit dieses Axioms ersichtlich war". [2] Huntington speaks of six requirements because he presents six postulates for "absolute continuous magnitude".

French *compatible* in connection with sets of axioms; he defined postulates to be compatible when they do not contradict each other and remarked that incompatibility can be proved by finding an "interpretation of the non-defined symbols that satisfies simultaneously all the postulates".[1]

An early statement of relative consistency, pointed out by Webb (1995, p. 13), is from Russell (1897, p. 38): "This correspondence removes, once for all, the possibility of a lurking contradiction in Metageometry, since, to a proposition in the one, corresponds one and only one proposition in the other, and contradictory results in one system, therefore, would correspond to contradictory results in the other. Hence Metageometry cannot lead to contradictions, unless Euclidean Geometry, at the same moment, leads to corresponding contradictions".

Concerning early uses of the term *consistent*, a general caveat is in order. Scanlan (1991, p. 985), who discusses the early American postulate theorists like Huntington and Veblen in the period from about 1900-1930, remarks: "It is clear that Huntington, Veblen, and other authors in this period are not observing the post-1950 terminological convention by which 'consistency' is reserved for a syntactic notion [...] in the period under discussion terminology was evolving and not yet standardized. In particular, what authors have in mind when using the term 'consistent' must be gathered from the context of usage".

See also *independent* (axioms), *categorical*, and *complete* (axioms).

References

Frischauf, Johannes (1872). *Absolute Geometrie nach Johann Bolyai*. Leipzig.
Hilbert, David (1899). *Grundlagen der Geometrie*. Leipzig.
Huntington, Edward V. (1902). "A Complete Set of Postulates for the Theory of Absolute Continuous Magnitude". *Transactions of the American Mathematical Society* 3, pp. 264–279.
Padoa, Alessandro (1902). "Un nouveau système irréductible de postulats pour l'algèbre". In: *Compte rendu du deuxième Congrès international des mathématiciens*. Ed. by E. Duporcq. Paris, pp. 249–256.
Russell, Bertrand (1897). *An Essay on the Foundations of Geometry*. Cambridge.
Scanlan, Michael (1991). "Who were the American Postulate Theorists?" *Journal of Symbolic Logic* 56, pp. 981–1002.
Webb, Judson (1995). "Tracking Contradictions in Geometry: The Idea of a Model from Kant to Hilbert". In: *From Dedekind to Gödel*. Essays on the Development of the Foundations of Mathematics. Ed. by Jaakko Hintikka. Dordrecht, pp. 1–20.

constant

See *variable*.

constructible (set theory, algebraic geometry)

According to Kanamori (2007, pp. 159f.), "[s]et theory was launched on an independent course as a distinctive field of mathematics by Gödel's formulation of the class L of constructible sets through which he established the relative consistency of AC [Axiom of Choice] in mid-1935 and CH [Continuum Hypothesis] in mid-1937". In his first communication on this topic, Gödel (1938, p. 556) explained:

> the term "constructible" is to be understood in the semiintuitionistic sense which excludes impredicative procedures. This means "constructible" sets are defined to be those sets which

[1] "il faut trouver *une interprétation des symboles non définies, qui vérifie simultanément tous les postulates*".

> can be obtained by Russell's ramified hierarchy of types, if extended to include transfinite orders. The extension to transfinite orders has the consequence that the model satisfies the impredicative axioms of set theory, because an axiom of reducibility can be proved for sufficiently high orders.

The term is also used in algebraic geometry. In what seems to be the first use in English, Hironaka (1960, p. 824) defined: "in general, a subset of a topological space is said to be *constructible* if it is a finite union of locally closed subsets, i. e., intersections of closed subsets and open subsets". In this sense, French *constructible* was introduced by Chevalley; see Chevalley (1955, p. 3) and Chevalley and Cartan (1955-56, p. 8).

Note. For historical comments on Gödel's constructible universe, see Kanamori (2007) and Ferreirós (2007, pp. 382–386).

References

Chevalley, Claude (1955). "Sur la théorie des variétés algébriques". *Nagoya Mathematical Journal* 7, pp. 1–43.

Chevalley, Claude and Henri Cartan (1955-56). "Schémas normaux; morphismes; ensembles constructibles". *Séminaire Henri Cartan* 8, exposé 7, pp. 1–10.

Ferreirós, José (2007). *Labyrinth of Thought.* A History of Set Theory and Its Role in Modern Mathematics. Second revised edition. Basel.

Gödel, Kurt (1938). "The Consistency of the Axiom of Choice and of the Generalized Continuum-Hypothesis". *Proceedings of the National Academy of Sciences* 24, pp. 556–557.

Hironaka, Heisuke (1960). "A Generalized Theorem of Krull-Seidenberg on Parameterized Algebras of Finite Type". *American Journal of Mathematics* 82, pp. 831–850.

Kanamori, Akihiro (2007). "Gödel and Set Theory". *The Bulletin of Symbolic Logic* 13, pp. 153–188.

contact angle

See *angle of contingence*.

contact transformation

The term *contact transformation* is due to Lie. Having dealt with the concept in his PhD thesis (Lie 1872b), he introduced the term in two self-announcements of the thesis: In Lie (1874, p. 420) he used German *Berührungstransformation* (from the verb *berühren* 'to touch'), and in Lie (1872a, p. 365), he used French *transformation de contact.* The German term is also found in Lie (1872c, p. 476).

The term was explained by Klein (1872, p. 33):

> By a contact-transformation is to be understood, analytically speaking, any substitution which expresses the values of the variables x, y, z and their partial derivatives $\frac{dz}{dx} = p$, $\frac{dz}{dy} = q$ in terms of the new variables x', y', z', p', q'. It is evident that such substitutions, in general, convert surfaces that are in contact into surfaces in contact, and this accounts for the name.[1]

The above translation of Klein's Erlangen Program from 1893 also seems to be the first use of the English term; the earlier *tangential transformation* (Forsyth 1890, p. 232) did not catch on. Lovett (1897, p. 321) explicitly regarded the former as "a

---

[1] "Unter einer Berührungstransformation hat man, analytisch zu reden, jede Substitution zu verstehen, welche die Variabel-Werthe x, y, z und ihre partiellen Differentialquotienten $\frac{dz}{dx} = p$, $\frac{dz}{dy} = q$ durch neue x', y', z', p', q' ausdrückt. Dabei gehen, wie ersichtlich, sich berührende Flächen im Allgemeinen wieder in sich berührende Flächen über, was den Namen Berührungstransformation begründet". Translation from Klein (1893, p. 233).

more precise designation, since by such a transformation not only is the property of tangency, but also in general the order of contact preserved".

Note. For the history of the notion, see Geiges (2001), Hawkins (2000, passim), and Demidov (1982). Lie (1896, ch. 1) discusses the prehistory of the notion himself.

References

Demidov, Sergej S. (1982). "The Study of Partial Differential Equations of First Order in the 18th and 19th Centuries". *Archive for History of Exact Sciences* 25, pp. 325–350.

Forsyth, Andrew Russell (1890). *Theory of Differential Equations*. Part I. Cambridge.

Geiges, Hansjörg (2001). "A brief history of contact geometry and topology". *Expositiones Mathematicae* 19, pp. 25–53.

Hawkins, Thomas (2000). *Emergence of the Theory of Lie Groups*. An Essay in the History of Mathematics 1869-1926. New York.

Klein, Felix (1872). *Vergleichende Betrachtungen über neuere geometrische Forschungen*. Programm zum Eintritt in die philosophische Facultät und den Senat der k. Friedrich-Alexanders-Universität zu Erlangen. Erlangen.

Klein, Felix (1893). "A comparative review of recent researches in geometry". *Bulletin of the New York Mathematical Society* 2, pp. 215–249.

Lie, Sophus (1872a). "Lie (S.). – Over en Classe geometriske Transformationer [self-announcement]". *Bulletin des sciences mathématiques et astronomiques* 3, pp. 365–367.

Lie, Sophus (1872b). "Over en Classe geometriske Transformationer". *Forhandlinger i Videnskabsselskabet i Christiania* (1871), pp. 67–109.

Lie, Sophus (1872c). "Zur Theorie partieller Differentialgleichungen erster Ordnung, insbesondere über eine Classification derselben". *Nachrichten von der Königl. Gesellschaft der Wissenschaften und der Georg-Augusts-Universität zu Göttingen* (1872), pp. 473–489.

Lie, Sophus (1874). "Lie (S.). – Ueber eine Classe geometrischer Transformationen [self-announcement]". *Jahrbuch über die Fortschritte der Mathematik* 3, pp. 420–422.

Lie, Sophus (1896). *Geometrie der Berührungstransformationen*. Dargestellt von Sophus Lie und Georg Scheffers. Leipzig.

Lovett, Edgar Odell (1897). "Geometrie der Berührungstransformationen [review]". *Bulletin of the American Mathematical Society* 3, pp. 321–350.

continued fraction

This term may be attributed to Wallis (1656, p. 182), who, in proposition 191 of his *Arithmetica infinitorum*, published an expansion of $\frac{4}{\pi}$ into continued fractions by Lord Brouncker. Wallis there described it in terms of "a fraction that has a denominator continually broken".[1] In his *Treatise of Algebra*, Wallis (1685, p. 318) similarly speaks of "a fraction still fracted continually".[2]

A general definition of *continued fraction* (Latin *fractio continua*) was given by Euler (1744, p. 99). He explains: "In order that what I mean by the name continued fraction will be more clearly understood, I exhibit first of all the most general example"[3] and displays a continued fraction as seen in fig. 27.

The term spread through Euler's *Introduction to Analysis of the Infinite* (Euler 1748, ch. 18).

---

[1] "fractio, quae denominatorem habeat continue fractum". Translation from Stedall (2004, p. 167).

[2] The notion is found in earlier texts. See Brezinski (1980, ch. 2). In particular, Brezinski (1980, p. 65) calls Pietro Antonio Cataldi the "real discoverer of the theory of continued fractions". He refers, in particular, to Cataldi (1613).

[3] "Quo igitur, quid nomine fractionum continuarum intelligam, clarius percipiatur, amplissimum earum exemplum ante omnia exhibeo". Translation from Euler (1985, p. 296).

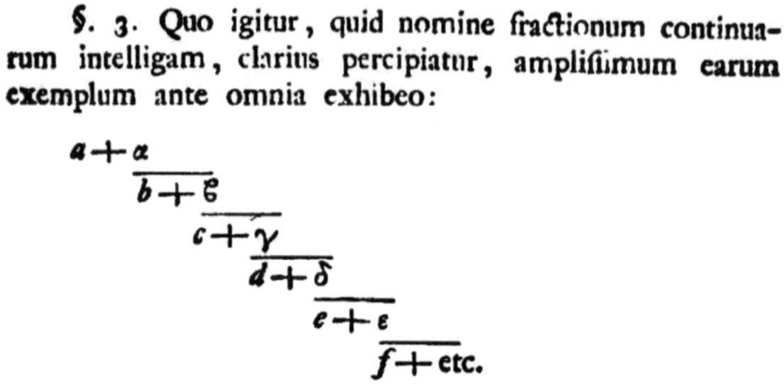

§. 3. Quo igitur, quid nomine fractionum continuarum intelligam, clarius percipiatur, amplissimum earum exemplum ante omnia exhibeo:

a + α
b + β
c + γ
d + δ
e + ε
f + etc.

Figure 27: A continued fraction. From Euler (1744, p. 99).

In German, the standard designation for continued fractions is *Kettenbruch*, lit. 'chain fraction'. This term was used by Bürja (1786, p. 168) and soon afterwards appeared in a mathematical encyclopaedia (Rosenthal 1794, p. 415).

The convergents of a continued fraction

$$x = b_0 + \cfrac{a_1}{b_1 + \cfrac{a_2}{b_2 + \cfrac{a_3}{b_3 + \cdots}}}$$

are b_0, $\frac{b_1 b_0 + a_1}{b_1}$, $\frac{b_2(b_1 b_0 + a_1) + a_2 b_0}{b_2 b_1 + a_2}$, ..., which stand in a recurrence relation. For these convergents, Legendre (1797, p. 23) suggested the name *fractions convergentes* because "these fractions must approach the total value of the continued fraction more and more".[1] This term was quickly adopted, for example by Garnier in his edition of Clairaut's *Algebra* (Clairaut 1801, p. 300) and by Lacroix (1801, p. 217). English *converging fraction* appeared in Lagrange (1806, p. 40). A shortened version was suggested by Parker (1827, p. 113): "they are called converging fractions, or more shortly convergents". German writers preferred the term *Näherungsbruch* (from *Näherung* 'approximation'), which seems to have been first used by Grüson (1799, p. 114).

Note. For the history of continued fractions, see Brezinski (1980, ch. 2), Cretney (2014), and Tropfke (1924, pp. 74–84). The early history of the notion is discussed under the name *anthyphairetic ratio* (reciprocal subtraction). For this concept, see, for example, Fowler (1979) and Hogendijk (2002).

References

Brezinski, Claude (1980). *History of Continued Fractions and Padé Approximants*. Berlin and Heidelberg.

Bürja, Abel (1786). *Der selbstlernende Algebrist*. Erster Theil. Berlin and Libau.

[1] "Ces valeurs doivent approcher de plus en plus de la valeur totale de la fraction continue, c'est pourquoi nous les appelons *fractions convergentes*".

Cataldi, Pietro Antonio (1613). *Trattato del modo brevissimo di trouare la Radice quadra delli numeri*. Bologna.
Clairaut, Alexis Claude (1801). *Elémens d'algebre*. Sixième édition. Tome second. Ed. by Jean Guillaume Garnier. Paris.
Cretney, Rosanna (2014). "The origins of Euler's early work on continued fractions". *Historia Mathematica* 41, pp. 139–156.
Euler, Leonhard (1744). "De fractionibus continuis dissertatio". *Commentarii academiae scientiarum imperialis Petropolitanae* 9, pp. 98–137.
Euler, Leonhard (1748). *Introductio in analysin infinitorum*. Tomus primus. Lausanne.
Euler, Leonhard (1985). "An Essay on Continued Fractions. Translated by Myra F. Wyman and Bostwick F. Wyman". *Mathematical Systems Theory* 18, pp. 295–328.
Fowler, David (1979). "Ratio in early Greek mathematics". *Bulletin of the American Mathematical Society* 1, pp. 807–846.
Grüson, Johann Philipp (1799). *Grundriss der reinen und angewandten Mathematik*. Erster Theil. Die Arithmetik. Halle.
Hogendijk, Jan (2002). "Anthyphairetic Ratio Theory in Medieval Islamic Mathematics". In: *From China to Paris*. Ed. by Yvonne Dold-Samplonius et al. Stuttgart, pp. 187–202.
Lacroix, Sylvestre François (1801). *Complément des Élémens d'algèbre*. Deuxième édition. Paris.
Lagrange, Joseph-Louis (1806). "An Essay on Numerical Analysis, on the Transformation of Fractions". *New Series of The Mathematical Repository* 1, part III, pp. 24–40.
Legendre, Adrien-Marie (1797). *Essai sur la théorie des nombres*. Paris.
Parker, James (1827). *Arithmetic and Algebra*. Part I (Library of Useful Knowledge). London.
Rosenthal, Gottfried Erich, ed. (1794). *Encyklopädie der reinen Mathematik und praktischen Geometrie*. Erster Band. A und B. Gotha.
Stedall, Jacqueline A. (2004). *The Arithmetic of Infinitesimals*. John Wallis 1656. Translated from Latin to English with an Introduction. New York.
Tropfke, Johannes (1924). *Geschichte der Elementar-Mathematik*. Sechster Band. Analysis. Analytische Geometrie. Berlin and Leipzig.
Wallis, John (1656). *Arithmetica infinitorum*. In: *Operum Mathematicorum*. Pars altera. Oxford, pp. 1–198.
Wallis, John (1685). *A Treatise of Algebra, both Historical and Practical*. London.

continuous

The Greek word *συνεχής*, which literally means 'holding together', used to be in common use in philosophy and was explained by Aristotle (*Physics* 5, 227a10-14), for example: "things in virtue of which they touch each other become one and the same thing, and (as the very name indicates) are 'held together,' which can only be if the two limits do not remain two but become one and the same".[1] While this definition is relational, there are also non-relational uses: "To quantity let us turn next. This is either discrete or continuous [...] Of quantities that are discrete we may here instance number and speech, of quantities that are continuous line, superficies and solid, to which time and place may be added" (Aristotle, *Categories* 6, 4b20-25).[2] Aristotle used the word *διορισμένον* (from a verb meaning 'to divide, separate') for 'discrete'.

Such examples just illustrate Greek usage; an overview of the meanings is found in Sattler (2021), for example. On the whole, continuity in Greek antiquity is a philosophical rather than a mathematical concept; De Risi (2019, p. 116) remarks: "the notion of continuity was never thematized in any ancient mathematical text, and neither Euclid, Archimedes, nor Apollonius ever mentioned this notion in their works

---

[1] Translation from Aristotle (1934, p. 39). [2] Translation from Aristotle (1962, pp. 35, 37).

(or defined geometrical objects as 'continuous')". Nevertheless, the term was kept (i. e., its Latin counterpart *continuus*) when continuity was used in the calculus (see *continuous* (calculus)) and in topology (see *continuum*).

As a mathematical term, *συνεχής* appears in the Greek term for continuous proportion (in which the consequent of one ratio is the antecedent of the next). In this sense, *συνεχής* was used by Aristotle (*Nicomachean Ethics* 1131a32). Boethius used the corresponding *continua proportionalitas* (Boethius 1867, p. 138).[1]

French *continue* was first used (DMF) by Oresme in the 14th century (Oresme 1968, p. 44); English *continual*, later (in the 17th century) replaced by *continuous*, was first used (MED, OED) at about the same time by John Trevisa: "oon is bigynnyng of all þinge þat is continual and discrete" (Seymour 1975, p. 1355).

References

Aristotle (1934). *The Physics*. With an English translation by Philip H. Wicksteed and Francis M. Cornford. In two volumes. II. London.

Aristotle (1962). *The Categories. On Interpretation. Prior Analytics*. Ed. by Harold P. Cook and Hugh Tredennick. London.

Boethius, Anicius Manlius Severinus (1867). *De institutione arithmetica libri duo, De institutione musica libri quinque*. Accedit Geometria quae fertur Boetii. Ed. by Gottfried Friedlein. Leipzig.

Guillaumin, Jean-Yves (2020). *Dictionnaire de la terminologie latine ancienne de l'arithmétique et de la géométrie*. Paris.

Heath, Thomas L. (1956). *The Thirteen Books of Euclid's Elements*. Volume II. Books III–IX. New York.

Nicomachus of Gerasa (1866). *Nicomachi Geraseni Pythagorei Introductionis Arithmeticae Libri II*. Ed. by Richard Hoche. Leipzig.

Oresme, Nicole (1968). *Le livre du ciel et du monde*. Edited by Albert D. Menut and Alexander J. Denomy, translated with an introduction by Albert D. Menut. Madison.

De Risi, Vincenzo (2019). "Leibniz on the Continuity of Space". In: *Leibniz and the Structure of Sciences*. Modern Perspectives on the History of Logic, Mathematics, Epistemology. Ed. by Vincenzo De Risi. Cham, pp. 111–169.

Sattler, Barbara M. (2021). "Divisibility or Indivisibility. The Notion of Continuity from the Presocratics to Aristotle". In: *The History of Continua*. Philosophical and Mathematical Perspectives. Ed. by Stewart Shapiro and Geoffrey Hellman. Oxford, pp. 6–26.

Seymour, Maurice Charles, ed. (1975). *On the Properties of Things*. John Trevisa's Translation of Bartholomaeus Anglicus' *De Proprietatibus Rerum*. A Critical Text. Volume II. Oxford.

***continuous* (calculus)**

The first definition[2] of a *continuous curve* (*curva continua*) appears to be due to Euler (1748, § 9):

---

[1] Boethius used Nicomachus' *Introduction to Arithmetic* (II, 21) where, however, *συνημμένη* 'connected' rather than *συνεχής* appears (Nicomachus of Gerasa 1866, p. 121). For further remarks on the term, see Heath (1956, p. 293) and Guillaumin (2020, pp. 84f.). Latin *discretus* is discussed by Guillaumin (2020, pp. 109–111). [2] Stedall (2008, p. 301) points out the first use of the concept: "The concept of continuity arises most naturally from the consideration of curves: intuitively a continuous curve is one that can be drawn without the pen leaving the paper. This idea was first made explicit by Wallis in the *Arithmetica infinitorum* of 1656, where he defined a smooth curve (*aequabilis curva*) as one that does not jump about (*non subsultans*)". For the original text, see Wallis (1656, p. 193). Stedall (2000, p. 238) also cites Pell's critical comment on Wallis' definition: "no man will be able to find what he [Wallis] means by *aequabilis curva*". The word *aequabilis* means 'similar, equal, uniform'.

> A *continuous* curve is one such that its nature can be expressed by a single function of x. If a curve is of such a nature that for its various parts BM, MD, DM, &c different functions of x are required for its expression [...] then we call such a curve *discontinuous* or *mixed* and *irregular*. This is because such a curve cannot be expressed by one constant law, but is formed from several continuous parts.[1]

His definition is usually interpreted in the following terms: "Euler's (and generally 18th-century's) 'continuous' functions broadly correspond to modern analytic functions, while most 'discontinuous' functions would now be called piecewise analytic" (Domingues 2008, p. 197). A function was continuous in Euler's sense "if its formula remained unchanged – hence, if the function was describable by just one calculational expression. Thus, the hyperbola with its two branches belonged to the continuous functions" (Schubring 2006, pp. 21f.).[2]

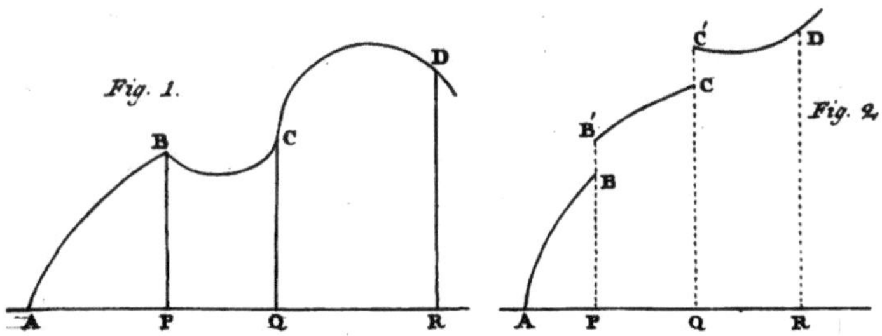

Figure 28: Continuous and discontinuous functions. From Arbogast (1791, Planche I).

Bolzano criticized older definitions[3] and presented a new one which is quite similar to the modern one:[4]

> According to a *correct definition*, the expression *that a function fx varies according to the law of continuity for all values of x inside or outside certain limits* means only that, *if x is any such value the difference $f(x+\omega) - fx$ can be made smaller than any given quantity, provided ω can be taken as small as we please* (Bolzano 1817, p. 11).[5]

[1] "Linea scilicet curva *continua* ita est comparata, ut ejus natura per unam ipsius x Functionem definitam exprimatur. Quod si autem linea curva ita sit comparata, ut variae ejus portiones BM, MD, DM, &c., per varias ipsius x Functiones exprimantur; [...] hujusmodi lineas curvas *discontinuas* seu *mixtas* & *irregulares* appellamus: propterea quod non secundum unam legem constantem formantur, atque ex portionibus variarum curvarum continuarum componuntur". Translation from Euler (1990, § 9). [2] The hyperbola was explicitly given as an example of a continuous function in Euler (1767, § 2). [3] Bolzano (1817, p. 5) refers to various other works by Kästner, Clairaut, Lacroix, and Lagrange, for example. See also Bolzano (1930, pp. 15f.). [4] For discussions of Bolzano's definition, see Rusnock and Kerr-Lawson (2005) and Smoryński (2017, pp. 54f.), for example. [5] "Nach einer *richtigen Erklärung* nähmlich versteht man unter der Redensart, *daß eine Function fx für alle Werthe von x, die inner- oder außerhalb gewisser Grenzen liegen, nach dem Gesetze der Stetigkeit sich ändre*, nur so viel, *daß, wenn x irgend ein solcher Werth ist, der Unterschied $f(x+\omega) - fx$ kleiner als jede gegebene Größe gemacht werden könne, wenn man ω so klein, als man nur immer will, annehmen kann*". Translation from Russ (2004, p. 256).

Further milestones in the definition of continuous functions include Cauchy (1821, pp. 34f. 1823, pp. 7f.),[1] and the ε-δ-definition first given by Weierstraß in a lecture in 1861 (see Dugac 1973, pp. 119f.).[2]

Note. For the history of the notion, see, for example, Smoryński (2017, esp. pp. 45-79), Grabiner (2005, pp. 87–97), Ferraro (2000), Dhombres (1988), Papadopoulos (2017), and Schubring (2006). Harper (2016) discusses various definitions of continuity given in the 19th and 20th centuries. See also the references cited in the article *function*.

References

Arbogast, Louis François Antoine (1791). *Mémoire sur la nature des fonctions arbitraires qui entrent dans les intégrales des équations aux différentielles partielles*. St. Petersburg.

Bolzano, Bernhard (1817). *Rein analytischer Beweis des Lehrsatzes, daß zwischen je zwei Werthen, die ein entgegengesetztes Resultat gewähren, wenigstens eine reelle Wurzel der Gleichung liege*. Prag.

Bolzano, Bernhard (1930). *Bernard Bolzano's Schriften*. Band 1. Functionenlehre. Herausgegeben und mit Anmerkungen versehen von Dr. Karel Rychlík. Prag.

Bradley, Robert E. and C. Edward Sandifer (2009). *Cauchy's* Cours d'analyse. An Annotated Translation. Dordrecht et al.

Cates, Dennis M. (2019). *Cauchy's* Calcul Infinitésimal. An Annotated English Translation. Cham.

Cauchy, Augustin-Louis (1821). *Cours d'analyse de l'École Royale Polytechnique*. I.re partie. Analyse algébrique. Paris.

Cauchy, Augustin-Louis (1823). *Résumé des leçons données à l'École Royale Polytechnique sur le calcul infinitésimal*. Tome premier. Paris.

Dhombres, Jean (1988). "Un texte d'Euler sur les fonctions continues et les fonctions discontinues, véritable programme d'organisation de l'analyse au 18e siècle". *Cahiers du séminaire d'histoire des mathématiques* 9, pp. 23–68.

Domingues, João Caramalho (2008). *Lacroix and the Calculus*. Basel.

Dugac, Pierre (1973). "Eléments d'analyse de Karl Weierstrass". *Archive for History of Exact Sciences* 10, pp. 41–176.

Euler, Leonhard (1748). *Introductio in analysin infinitorum*. Tomus secundus. Lausanne.

Euler, Leonhard (1767). "De usu functionum discontinuarum in analysi". *Novi Commentarii academiae scientiarum Petropolitanae* 11, pp. 67–102.

Euler, Leonhard (1990). *Introduction to Analysis of the Infinite*. Book II. Translated by John D. Blanton. New York.

Ferraro, Giovanni (2000). "Functions, Functional Relations, and the Laws of Continuity in Euler". *Historia Mathematica* 27, pp. 107–132.

Grabiner, Judith V. (2005). *The Origins of Cauchy's Rigorous Calculus*. Mineola, N.Y.

Harper, J.F. (2016). "Defining continuity of real functions of real variables". *BSHM Bulletin* 31, pp. 189–204.

Papadopoulos, Athanase (2017). "Looking Backward: From Euler to Riemann". In: *From Riemann to Differential Geometry and Relativity*. Ed. by Lizhen Ji, Athanase Papadopoulos, and Sumio Yamada. Cham, pp. 1–93.

Rusnock, Paul and Angus Kerr-Lawson (2005). "Bolzano and uniform continuity". *Historia Mathematica* 32, pp. 303–311.

Russ, Steve (2004). *The Mathematical Works of Bernard Bolzano*. Oxford.

Schubring, Gert (2006). *Conflicts Between Generalization, Rigor, and Intuition*. Number Concepts Underlying the Development of Analysis in 17th-19th Century France and Germany. New York.

[1] English translations of Cauchy's definitions can be found in Bradley and Sandifer (2009, p. 26) and Cates (2019, 9, fn. 4). [2] An English translation of Weierstraß' definition is found in Smoryński (2017, pp. 56f.). See also Sinkevich (2016, p. 197), who points out that "[i]t is in these notes that the definition of continuous function in the language of epsilontics appears for the first time".

Sinkevich, Galina I. (2016). "On History of Epsilontics". *Antiquitates Mathematicae* 10, pp. 183–204.
Smoryński, Craig (2017). *MVT: A Most Valuable Theorem*. Cham.
Stedall, Jacqueline A. (2000). "A large discourse concerning algebra: John Wallis's 1685 *Treatise of algebra*". PhD thesis. Open University.
Stedall, Jacqueline A. (2008). *Mathematics Emerging*. A Sourcebook 1540 – 1900. Oxford.
Wallis, John (1656). *Arithmetica infinitorum*. In: *Operum Mathematicorum*. Pars altera. Oxford, pp. 1–198.

continuum

Cantor (1883, p. 32) defined a *continuum* as a perfect and connected set; in a note at the end of his article, he remarked:

> Observe that this definition of a continuum is free from every reference to that which is called the *dimension* of a continuous structure; the definition includes also continua that are composed of connected pieces of different dimensions, such as lines, surfaces, solids, etc. On a later occasion I shall show how it is possible to proceed in an orderly fashion from this general continuum to the special continua with definite dimension. I know very well that the word "continuum" has previously *not* had a precise meaning in mathematics; so my definition will be judged by some as too *narrow*, by others as too *broad*. I trust that I have succeeded in finding a proper *mean* between the two. In my opinion, a *continuum* can only be a *perfect* and *connected* structure.[1]

Cantor (1883, p. 32) explicitly mentions that he regards Bolzano's definition as wrong because "it expresses one-sidedly just *one* property of the continuum";[2] he refers to Bolzano (1851, § 38), where the following definition was given: "If we attempt to gain a clear awareness of the concept which we designate by the expression a *continuous extension* or a *continuum* then we cannot help defining a continuum to exist where, and only where, a collection of simple objects (of points in time or space or even of substances) occurs which are so arranged that every single one of them has at least one neighbour in this collection at every distance however small".[3] Cantor also rejected the definition given by Dedekind (1872, p. 18).[4]

[1] "Man beachte, dass diese Definition eines Continuums frei ist von jedem Hinweis auf das, was man die Dimension eines stetigen Gebildes nennt; die Definition umfasst nämlich auch solche Continua, die aus zusammenhängenden Stücken verschiedener Dimension, wie Linien, Flächen, Körper etc. bestehen. Bei einer späteren Gelegenheit will ich zeigen, wie man von diesem allgemeinen Continuum zu den spezielleren Continuis mit bestimmter Dimension ordnungsmässig hinkommt. Ich weiss sehr wohl, dass das Wort ‚*Continuum*' in der Mathematik eine *feste* Bedeutung bisher *nicht* angenommen hat; es wird daher meine Definition desselben von Einigen als zu *eng*, von Anderen als zu *weit* beurtheilt werden; hoffentlich ist es mir gelungen, dabei die *richtige Mitte* zu finden. Nach meiner Auffassung kann unter einem *Continuum nur* ein *perfectes* und *zusammenhängendes* Gebilde verstanden werden". Translation from Ewald (1996, p. 919). [2] "sie drückt einseitig bloss *eine* Eigenschaft des Continuums aus". Translation from Ewald (1996, p. 906). [3] "Versuchen wir nämlich, uns den Begriff, den wir mit den Benennungen ‚eine *stetige Ausdehnung* oder ein *Kontinuum*' bezeichnen, zu einem deutlichen Bewußtsein zu bringen: so können wir nicht umhin zu erklären, dort, aber auch nur dort sei ein Kontinuum vorhanden, wo sich ein Inbegriff von einfachen Gegenständen (von Punkten in der Zeit oder im Räume oder auch von Substanzen) befindet, die so gelegen sind, daß jeder einzelne derselben für jede auch noch so kleine Entfernung wenigstens einen Nachbar in diesem Inbegriffe habe". Translation from Russ (2004, p. 642). – For critical comments on Cantor's criticism, see Johnson (1977, pp. 283f.). [4] Dedekind there defined *Stetigkeit* (continuity).

Moore (2008, p. 223) remarks that Cantor's "essentially topological definition would be modified around 1930 to a compact, connected set"; such definitions are found in Urysohn (1928, p. 296) and Menger (1928, pp. 200f.), for example.

At the time Cantor published his definition of *continuum*, the term was also used in (at least) one other sense: Weierstraß used the word for open and connected sets in n-dimensional Euclidean space, a concept later designated as a *domain* (German *Gebiet*). For discussion and references, see Moore (2008, pp. 237f.). The term was also used for the real numbers or intervals of real numbers (Cantor 1874, p. 259).

Note. For the history of the concept, see, for example, Moore (2008), Noble and Fernández de Castro (2017), and Charatonik (1998). Various historical definitions are referred to by Herrlich, Hušek, and Preuß (2008, pp. 799f.).

References

Bolzano, Bernhard (1851). *Paradoxien des Unendlichen.* Herausgegeben aus dem schriftlichen Nachlasse des Verfassers von Dr. Fr. Přihonsky. Leipzig.

Cantor, Georg (1874). "Ueber eine Eigenschaft des Inbegriffs aller reellen algebraischen Zahlen". *Journal für die reine und angewandte Mathematik* 77, pp. 258–262.

Cantor, Georg (1883). *Grundlagen einer allgemeinen Mannichfaltigkeitslehre.* Ein mathematisch-philosophischer Versuch in der Lehre des Unendlichen. Leipzig.

Charatonik, J.J. (1998). "History of Continuum Theory". In: *Handbook of the History of General Topology.* Volume 2. Ed. by C.E. Aull and R. Lowen. Dordrecht et al., pp. 703–786.

Dedekind, Richard (1872). *Stetigkeit und irrationale Zahlen.* Braunschweig.

Ewald, William (1996). *From Kant to Hilbert.* A Source Book in the Foundations of Mathematics. Volume II. Oxford.

Herrlich, Horst, Mirek Hušek, and Gerhard Preuß (2008). "Hausdorffs Studien über Kurven, Bögen und Peano-Kontinua". In: *Felix Hausdorff. Gesammelte Werke.* Band III. Mengenlehre (1927, 1935). Deskriptive Mengenlehre und Topologie. Ed. by Ulrich Felgner et al. Berlin and Heidelberg, pp. 798–839.

Johnson, Dale M. (1977). "Prelude to Dimension Theory: The Geometrical Investigations of Bernard Bolzano". *Archive for History of Exact Sciences* 17, pp. 261–295.

Menger, Karl (1928). *Dimensionstheorie.* Leipzig.

Moore, Gregory H. (2008). "The emergence of open sets, closed sets, and limit points in analysis and topology". *Historia Mathematica* 35, pp. 220–241.

Noble, Eduardo and Max Fernández de Castro (2017). "The Ancient *versus* the Modern Continuum". *Revista Portuguesa de Filosofia* 73, pp. 1343–1380.

Russ, Steve (2004). *The Mathematical Works of Bernard Bolzano.* Oxford.

Urysohn, Pavel (1928). "Über im kleinen zusammenhängende Kontinua. (Aus dem Nachlaß herausgegeben von Paul Alexandroff)". *Mathematische Annalen* 98, pp. 296–308.

contour

This term was introduced into mathematics by Cauchy. He did not formally define the term, but spoke of "a system of straight or curved lines, which form a contour, which starts at the point (x_0, y_0) and ends at the point (X, Y)" (Cauchy 1825, p. 26).[1] It did not take long for the term to be adopted; for example, it appears in Sturm and Liouville (1836, p. 279).

The explicit term *contour-integral* was used by Smith (1874, p. 162).

See also *path* and *loop* (complex function).

---

[1] "un système de lignes droites ou courbes, formant un contour qui partirait du point (x_0, y_0) pour aboutir au point (X, Y)".

Note. For a discussion of Cauchy's 1825 memoir and the introduction of the term, see Smithies (1997, ch. 4, esp. p. 100).

References

Cauchy, Augustin-Louis (1825). *Mémoire sur les intégrales définies, prises entre des limites imaginaires*. Paris.

Smith, H.J.S. (1874). "On the Higher Singularities of Plane Curves". *Proceedings of the London Mathematical Society* 6, pp. 153–182.

Smithies, Frank (1997). *Cauchy and the Creation of Complex Function Theory*. Cambridge.

Sturm, Charles-François and Joseph Liouville (1836). "Démonstration d'un théorème de M. Cauchy, relatif aux racines imaginaires des équations". *Journal de mathématiques pures et appliquées* 1, pp. 278–289.

contractible space

As explained by Lefschetz (1935, p. 16), "[a]ccording to Borsuk to whom this notion is due, a topological space is called locally contractible whenever for every open set U containing any point x there is another, $V \supset U$, which is homotopic to a point on U". In this sense the terminology was introduced in Borsuk (1932a, p. 952), where French *contractibilité locale* and *localement contractile* were used, and then in Borsuk (1932b, p. 236), where the corresponding German terminology *lokal zusammenziehbar* was introduced.

The explicit distinction between *contractible* and *locally contractible* appeared in Čech (1935, p. 14).

References

Borsuk, Karol (1932a). "Sur la notion de contractibilité locale des ensembles". *Comptes rendus de l'Académie des sciences* 194, pp. 951–953.

Borsuk, Karol (1932b). "Über eine Klasse von lokal zusammenhängenden Räumen". *Fundamenta Mathematicae* 19, pp. 220–242.

Čech, Eduard (1935). "Sur la connexité locale d'ordre supérieur". *Compositio Mathematica* 2, pp. 1–25.

Lefschetz, Solomon (1935). "Chain-deformations in topology". *Duke Mathematical Journal* 1, pp. 1–18.

convergence in measure

The French term *convergence en mesure* was coined by Riesz (1909, p. 1303):

> Let $f_1(x), f_2(x), \ldots, f(x)$ be measurable functions defined on a set E; if ε is any positive number, we will designate by $m(n, \varepsilon)$ the measure of the set
>
> $$[|f(x) - f_n(x)| > \varepsilon];$$
>
> we shall then say that the sequence $[f_n(x)]$ converges in measure towards the function $f(x)$, if, no matter how small the number ε, we have
>
> $$\lim_{n=\infty} m(n, \varepsilon) = 0.$$ [1]

[1] Soient $f_1(x), f_2(x), \ldots, f(x)$ des fonctions mesurables, définies sur l'ensemble E; ε étant une quantité positive quelconque, nous désignérons par $m(n, \varepsilon)$ la mesure de l'ensemble $[|f(x) - f_n(x)| > \varepsilon]$; alors nous dirons que la suite $[f_n(x)]$ tend en mesure vers la fonction $f(x)$, si, quelque petit que soit la quantité ε, on a $\lim_{n=\infty} m(n, \varepsilon) = 0$.

Shortly afterwards, Riesz (1910, p. 450) himself used the corresponding German term *Konvergenz dem Maße nach.* Riesz (1912, p. 641) also pointed out that Borel used the alternative *asymptotic convergence*, which is found in Borel (1912b, p. 415) and Borel (1912a, p. 192).[1]

Note. For the history of the concept and the terminology, see Medvedev (1991, pp. 120–128).

References

Borel, Émile (1912a). "Le calcul des intégrales définies". *Journal de mathématiques pures et appliquées* 8, pp. 159–210.

Borel, Émile (1912b). "Sur les théorèmes fondamentaux de la théorie des fonctions de variables réelles". *Comptes rendus de l'Académie des sciences* 154, pp. 413–415.

Medvedev, Fyodor A. (1991). *Scenes from the History of Real Functions.* Translated from the Russian by Roger Cooke. Basel.

Riesz, Frigyes (1909). "Sur les suites de fonctions mesurables". *Comptes rendus de l'Académie des sciences* 148, pp. 1303–1305.

Riesz, Frigyes (1910). "Untersuchungen über Systeme integrierbarer Funktionen". *Mathematische Annalen* 69, pp. 449–497.

Riesz, Frigyes (1912). "Sur quelques points de la théorie des fonctions sommables". *Comptes rendus de l'Académie des sciences* 154, pp. 641–643.

convergent

Before this term came into mathematics, it was used in optics: "I use the words 'converge', 'diverge', and so on, for the sake of brevity, of straight lines which, when produced in all directions, all converge at the same point" (Kepler 1604, p. 444).[2]

There may have been a loose analogy when the mathematical term was coined by Gregory (1667, p. 10).[3] Whether or not this is the case, the terminology is influenced by Gregory's method of calculating the sine of an arc: he establishes a sequence $a_1, ...a_n$ (larger than the sine) and a sequence $b_1, ..., b_n$ (smaller than the sine) which fulfil $a_1 - b_1 > a_2 - b_2 > a_3 - b_3 > ....$ This pair of sequences is then called *convergent* since the sequences tend or verge towards the same value (Reiff 1889, p. 16). Thus, "[Gregory's] definition is rather different from the modern one, even if it contains the basic idea of quantities approaching a limit" (Ferraro 2008, p. 20).

The term was immediately adopted by Huygens (1668, p. 53) as French *convergent*; the English term *converging* appears in [anon.] (1668, p. 641). Leibniz also used the Latin term in various manuscripts (see, e. g., Leibniz 2003, p. xxiii).

In various letters, Leibniz also used *advergent* (Latin *advergens*). He began using this word in 1705 (see Gerhardt 1859, p. 272) and gave a definition along modern lines in 1713 (see Gerhardt 1855-56, pp. 985f.), arguing that *advergent* is more correct since Gregory used *convergent* for two sequences rather than one. Leibniz's terminological

---

[1] Literally, Borel defined the *asymptotic equivalence* (French *équivalence asymptotique*) of a function to polynomials but he also used *convergence asymptotique*. [2] "vocibus conuergere, diuergere, utor breuitatis causâ, de rectis, quae si producantur altrobique, in idem omnes punctum concurrerent". Translation from Kepler (2000, p. 214). Another definition is found in Kepler (1611, p. 8). [3] Gregory was aware of Kepler's terminology and had used it himself before (Gregory 1663, p. 1). See also Yuschkevitch (1976, p. 58), according to whom Gregory "possibly transplant[ed] it [the term *convergent*] into mathematics from optics, with which he occupied himself a good deal". The term *divergent* was not used by Gregory.

suggestion did not find much approval, though, and *convergent* became the standard term.

Schubring (2006, p. 28) notes that "[t]he understanding of convergence of series prevailing in the eighteenth century was that a series converges if its terms (seen absolutely) become ever smaller, approximating the value 0". Such a definition was given by d'Alembert (1754, p. 165), for instance.[1]

Schubring (2006, p. 30) also explains that Cauchy "assigns a central function to the concept of convergence. Only at that time did the concept of convergence attain the status of a fundamental idea in analysis". A definition is found in Cauchy (1821, p. 123):

> We call a *series* an indefinite sequence of quantities,
>
> $$u_0, u_1, u_2, u_3, \ldots,$$
>
> which follow from one to another according to a determined law. These quantities themselves are the various *terms* of the series under consideration. Let
>
> $$s_n = u_0 + u_1 + u_2 + \ldots + u_{n-1}$$
>
> be the sum of the first n terms, where n denotes any integer number. If, for ever increasing values of n, the sum s_n indefinitely approaches a certain limit s, the series is said to be *convergent*, and the limit in question is called the *sum* of the series. On the contrary, if the sum s_n does not approach any fixed limit as n increases indefinitely, the series is *divergent*, and does not have a sum.[2]

See also the entries on various types of convergence in this book. For convergents as used in connection with continued fractions, see *continued fraction.*

Note. Many useful comments on the history of the notion and terminology are found in Dugac (2003), Ferraro (2008), and Schubring (2006). For Gregory, see also Scriba (1983), Heinrich (1901), and Hischer (2016). For Leibniz's terminology, see Leibniz (2003, p. xxiii) and Ferraro (2008, p. 20).

---

[1] Domingues (2008, p. 84) notes a change in d'Alemberts understanding of convergence: "By the article 'Série', d'Alembert's ideas seem to have changed a little: a series was then convergent if it approached more and more a finite quantity and, continued to infinity, it would finally become equal to that quantity. That its terms would be decreasing was by then a consequence, not the definition". This remark refers to d'Alembert (1765, p. 93). For some discussion of the progress concerning convergence in the time between d'Alembert and Cauchy, see Schubring (2006, pp. 27–30). [2] "On appelle *série* une suite indéfinie de quantités u_0, u_1, u_2, u_3, &c. ... qui dérivent les unes des autres suivant une loi déterminée. Ces quantités elles-mêmes sont les différents *termes* de la série que l'on considère. Soit $s_n = u_0 + u_1 + u_2 + \ldots + u_{n-1}$ la somme des n premiers termes, n désignant un nombre entier quelconque. Si, pour des valeurs de n toujours croissantes, la somme s_n s'approche indéfiniment d'une certaine limite s, la série sera dite *convergente*, et la limite en question s'appellera la *somme* de la série. Au contraire, si, tandis que n croit indéfiniment, la somme s ne s'approche d'aucune limite fixe, la série sera *divergente* et n'aura plus de somme". Translation from Bradley and Sandifer (2009, p. 85). – Bolzano (1817, § 5) also gave a definition of convergence in modern terms but, as is well known, his text exerted little influence at that time. On his definition, Russ and Trlifajová (2016, p. 40) comment that "he [Bolzano] articulates before Cauchy a general criterion for the convergence of an infinite sequence. And he does so more clearly and concisely than Cauchy and without reference to the infinitely large or the infinitely small". See also Spalt (2022, pp. 105–107).

References

[anon.] (1668). "Vera circuli et hyperbolae quadratura [review]". *Philosophical Transactions of the Royal Society of London* 3, pp. 640–641.

d'Alembert, Jean Le Rond (1754). "Convergent". *Encyclopédie, ou Dictionnaire raisonné des sciences, des arts et des métiers* 4, p. 165.

d'Alembert, Jean Le Rond (1765). "Série ou suite". *Encyclopédie, ou Dictionnaire raisonné des sciences, des arts et des métiers* 15, pp. 93–95.

Bolzano, Bernhard (1817). *Rein analytischer Beweis des Lehrsatzes, daß zwischen je zwei Werthen, die ein entgegengesetztes Resultat gewähren, wenigstens eine reelle Wurzel der Gleichung liege*. Prag.

Bradley, Robert E. and C. Edward Sandifer (2009). *Cauchy's* Cours d'analyse. An Annotated Translation. Dordrecht et al.

Cauchy, Augustin-Louis (1821). *Cours d'analyse de l'École Royale Polytechnique*. I.re partie. Analyse algébrique. Paris.

Domingues, João Caramalho (2008). *Lacroix and the Calculus*. Basel.

Dugac, Pierre (2003). *Histoire de l'analyse*. Autour de la notion de limite et de ses voisinages. Ed. by Bernard Bru and Roger Laurent. Paris.

Ferraro, Giovanni (2008). *The Rise and Development of the Theory of Series up to the Early 1820s*. New York.

Gerhardt, C.I., ed. (1859). *Leibnizens mathematische Schriften*. Erste Abtheilung. Band IV. Halle.

Gerhardt, C.I., ed. (1855-56). *Leibnizens mathematische Schriften*. Erste Abtheilung. Band III. Halle.

Gregory, James (1663). *Optica promota*. London.

Gregory, James (1667). *Vera circuli et hyperbolae quadratura*. Padova.

Heinrich, Georg (1901). "James Gregorys 'Vera circuli et hyperbolae quadratura'". *Bibliotheca Mathematica* 2, pp. 77–85.

Hischer, Horst (2016). "James Gregory und ‚Konvergenz' – auf den Spuren zu seinem Algorithmus". Universität des Saarlandes Preprint Nr. 379. URL: `https://www.math.uni-sb.de/PREPRINTS/` (retrieved Aug. 23, 2017).

Huygens, Christiaan (1668). "Vera Circuli et Hyperboles Quadratura, in propriâ suâ proportionis specie inventa & demonstrata à Jacobo Gregorio Scoto". *Journal des sçavans* (July 1668), pp. 52–56.

Kepler, Johannes (1604). *Ad Vitellionem paralipomena, quibus astronomiae pars optica traditur*. Frankfurt.

Kepler, Johannes (1611). *Dioptrice seu Demonstratio eorum quae visui & visibilibus propter conspicilla non ita pridem inventa accidunt*. Augsburg.

Kepler, Johannes (2000). *Optics*. Paralipomena to Witelo & Optical Part of Astronomy. Translated by William H. Donahue. Santa Fe.

Leibniz, Gottfried Wilhelm (2003). *Sämtliche Schriften und Briefe*. Siebente Reihe. Mathematische Schriften. Dritter Band. 1672-1676. Differenzen, Folgen, Reihen. Hannover.

Reiff, Richard August (1889). *Geschichte der unendlichen Reihen*. Tübingen.

Russ, Steve and Kateřina Trlifajová (2016). "Bolzano's measurable numbers: are they real?" In: *Research in History and Philosophy of Mathematics*. The CSHPM 2015 Annual Meeting in Washington, D.C. Ed. by Maria Zack and Elaine Landry. Cham, pp. 39–56.

Schubring, Gert (2006). *Conflicts Between Generalization, Rigor, and Intuition*. Number Concepts Underlying the Development of Analysis in 17th-19th Century France and Germany. New York.

Scriba, Christoph J. (1983). "Gregory's converging double sequence: A new look at the controversy between Huygens and Gregory over the 'analytical' quadrature of the circle". *Historia Mathematica* 10, pp. 274–285.

Spalt, Detlef D. (2022). *A Brief History of Analysis*. With Emphasis on Philosophy, Concepts, and Numbers, Including Weierstraß' Real Numbers. Cham.

Yuschkevitch, Adolph P. (1976). "The Concept of Function up to the Middle of the 19th Century". *Archive for History of Exact Sciences* 16, pp. 37–85.

convex* and *concave

The close connection between convexity and concavity was already stated by Aristotle in his *On the Heavens* (270b-271a): "Concave and convex are, it would seem, contraries not only of each other but also of the straight line, when they are considered together and taken as a unity".[1] For *concave*, Aristotle uses *κοίλος*, lit. 'hollow'; for *convex*, he uses *κυρτός*, lit. 'bulging, swelling'. The first definition goes back to Archimedes, who defined *concave in the same direction*: "I apply the term *concave in the same direction* to a line such that, if any two points on it are taken, either all the straight lines connecting the points fall on the same side of the line, or some fall on one and the same side while others fall on the line itself, but none on the other side" (Archimedes 1880, p. 6).[2]

This definition requires some explanation: In fig. 29 (taken from an 18th-century book on the calculus)[3], the curve on the left is concave in the same direction because if any two of the points A to F are joined by a straight line segment, the segments are all on the same side of the curve, while the curve on the right is not concave in the same direction because, for example, AB and EF are on different sides of the curve. The second part of the definition ("while others fall on the line itself") takes into consideration that curves may contain line segments.[4] For surfaces, Archimedes (1897, p. 3) gave an analogous definition. The terms are also found, without definition, in Euclid's *Elements* (book III, prop. 8).

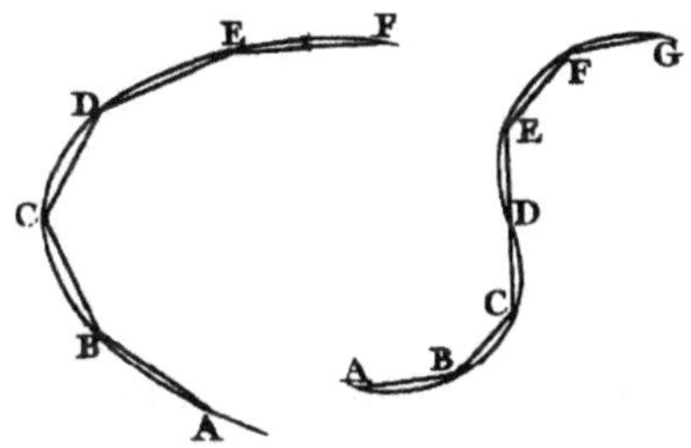

Figure 29: Convacity and convexity. From Deidier (1740, Livre II, Planche 6).

In Latin texts, *κοίλος* was commonly translated by *concavus* or *cavus*, which also lit. mean 'hollow', while *κυρτός* was translated by *convexus* 'vaulted, arched'. This terminology is found, for example, in some 13th-century translations by William of Moerbeke (see Clagett 1964, pp. 163, 307). However, *convexus* could be used to translate *κοίλος* as well (Clagett 1964, p. 167).

[1] Translation from Aristotle (1939, p. 27). [2] Translation from Archimedes (1897, p. 2). [3] The term discussed there is *toujours concave ou toujours convexe du même côté* 'always concave or always convex of the same side' (Deidier 1740, p. 313). [4] Archimedes' understanding of *curve* is pointed out by his commentator Eutocius: "it should be known that he [Archimedes] calls 'curved lines' not simply the circular or the conical or those which have continuity without breaking, but he terms 'curved' any line in a plane, without qualification, which is other than straight; any single line in a plane, compounded in whatever way, so that even if it is composed of straight lines" (Archimedes 2004, p. 244).

In the vernacular the terminology often appears virtually unchanged in comparison to the Latin. This is the case, for example, for the earliest attestation in French (DMF), dating from the 14th century: "Concave and convex are related opposites and not true contraries; for a circular body is said to be concave with respect to that which is inside it, and it is called convex with respect to what is outside it, but it is nonetheless the same thing" (Oresme 1968, p. 88).[1] The first English example (acc. to the OED) refers to surfaces: "Manyfolde superficies, playne Conuex and Concaue" (L. Digges and T. Digges 1571, fol. Aij^{r}).

Sometimes, however, literal translations were given. For example, Dürer (1525, fol. A iij^{r}) used German *hohl* 'hollow' in connection with surfaces, and in the first English translation of Euclid's *Elements* the expression "concauitie or hollownes" is used (Euclid 1570, fol. 87^{r}).

In the 17th century, the term is used in connection with inflection points. In Fermat's manuscript "Methodus ad disquirendam maximam et minimam" (method for the study of maxima and minima), which was finished in the 1630s but not published until 1679, Fermat discussed the conchoid of Nicomedes and suggested that "to be able to draw the curve well", it would be suitable "to study the points of inflection [*puncta inflexionum*], at which the curvature changes from convex to concave, and conversely" (de Fermat 1679, p. 73).[2]

The first definition in connection with polygons appears to be due to Legendre (1794, p. 27): "we will only consider polygons with salient angles, which one can otherwise call *convex polygons*. A convex polygon is so [constructed] that a straight line, drawn any way you wish, cannot meet the periphery of this polygon in more than two points".[3] The term *salient angle* (*angle saillant*) in this definition means that its vertex is directed outward; its counterpart is a *re-entering angle* (*angle rentrant*), where the vertex is directed inward (see fig. 30).[4] Legendre (1794, p. 168) also defined convex polyhedra analogously.

Legendre's definition was applied to more general figures by Brunn (1887, p. 1), who defined a convex set under the German name *Oval* as "a closed curve that has two and only two points in common with every intersecting straight line in its plane".[5] Brunn called a corresponding figure in space *Eifläche* 'egg-surface'.

Such 3-dimensional objects were then considered by Minkowski (1897, p. 198), who began his paper thus: "A *convex body* is completely characterized by the properties that it is a closed set of points, has inner points, and that every straight line that takes up some of its inner points always has two points in common with its

[1] "concave et convex ou courve ne sont pas seullement opposite<s> un a l'autre, mais touz .ii. semblent opposites a ce que est droit". Translation from Oresme (1968, p. 89). [2] "ut perfecte curva possit delineari, investiganda sunt ex arte puncta inflexionum, in quibus curvatura ex convexa fit concava vel contra". Translation from Mahoney (1994, p. 201). [3] "nous ne considérerons [...] que les polygones à angles saillants, qu'on peut appeler autrement *polygones convexes*. Tout polygone convexe est tel qu'une ligne droite menée comme on voudra ne peut rencontrer le contour de ce polygone qu'en deux points". Translation from Hon and Goldstein (2008, pp. 227f., fn. 29) (slightly adapted). [4] The term *salient* in this sense appears to go back to de Bouelles (1542, fol. 22^{v}), who used it in connection with star polygons. See *star polygon*. [5] "eine [...] geschlossene Kurve, die mit jeder sie durchsetzenden Geraden zwei und nur zwei Punkte gemein hat, heisse ein Oval". Translation from Kjeldsen (2008, p. 82).

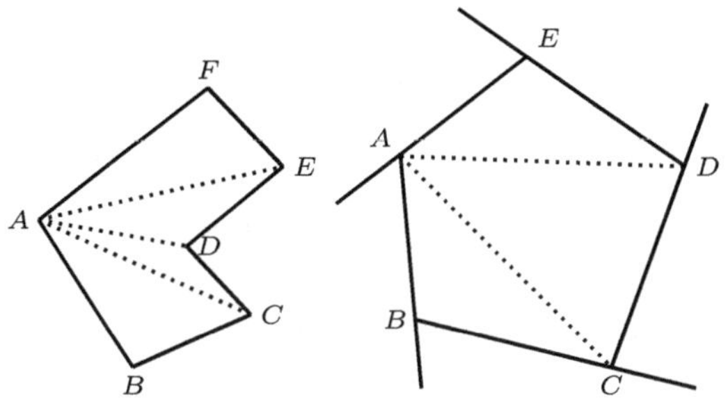

Figure 30: Re-entering and salient angles. From Bézout (1770, Géom., Pl. II). In the figure on the left, the angle CDE is re-entering; in the figure on the right, all angles are salient.

boundary".[1] Minkowski (1897, p. 201) also defined plane convex sets: "I want to call *convex domain* each closed set of points such that for each two points the whole line segment joining the two points always belongs to the set of points".[2]

The modern term *convex function* goes back to the Danish mathematician Johan Jensen. He introduced the terminology in Danish in Jensen (1905, p. 96), but his article from 1906, written in French, reached a wider audience: "If a function $\varphi(x)$ of the real variable x, which is real, finite, and uniform, satisfies in a certain interval the inequality $\varphi(x) + \varphi(y) \geq 2\varphi(\frac{x+y}{2})$, we say that $\varphi(x)$ is a convex function in that interval" (Jensen 1906, p. 176).[3]

The term *convex hull* was introduced by Blaschke (1916, p. 54) as German *konvexe Hülle*. In his definition, Blaschke proceeded from a bounded and closed point set $\mathfrak{M}$ and stated that there are convex point sets that contain $\mathfrak{M}$. He then defined: "*Among all these convex bodies over $\mathfrak{M}$ there is a smallest one $\mathfrak{K}$, i. e., a body that is contained in all other convex bodies over $\mathfrak{M}$.* We shall call $\mathfrak{K}$ the *convex hull* of $\mathfrak{M}$".[4] The term was adopted by Pál (1920, p. 25), who also attributed it to Blaschke. The English term appears to have been first used by Birkhoff (1935, p. 359).[5]

Note. For the general history of convexity, see Balestro, Martini, and Teixeira (2024, esp. ch. 12), Fenchel (1983), Gruber (1990, 1993), Dwilewicz (2009), Papadopoulos (2019), and Kjeldsen (2008). For the history of convex functions, see Roberts and Varberg (1973, p. 8) and Guerraggio and Molho (2004).

---

[1] "Ein convexer Körper ist vollständig dadurch gekennzeichnet, daß er eine abgeschlossene Punktmenge ist, innere Punkte besitzt, und dass jede gerade Linie, die innere Punkte von ihm aufnimmt, mit seiner Begrenzung stets zwei Punkte gemein hat". Translation from Kjeldsen (2008, p. 80).
[2] "Als *convexen Bereich* will ich überhaupt jede abgeschlossene Punktmenge bezeichnen, zu der mit irgend zwei Punkten stets auch die ganze sie verbindende Strecke gehört". [3] "Lorsqu'une fonction $\varphi(x)$, réelle, finie et uniforme, de la variable réelle x, satisfait dans un certain intervalle à l'inégalité $\varphi(x)+\varphi(y) \geq 2\varphi(\frac{x+y}{2})$, on dit que $\varphi(x)$ est une fonction convexe dans cet intervalle". – The definition in Jensen (1905, p. 96) is completely analogous. [4] "*Unter allen diesen Körpern über $\mathfrak{M}$ gibt es aber einen kleinsten $\mathfrak{K}$, d. h. einen Körper, der in allen anderen konvexen Körpern über $\mathfrak{M}$ enthalten ist.* Wir wollen $\mathfrak{K}$ die *konvexe Hülle* von $\mathfrak{M}$ nennen". [5] A slightly earlier use in print is in an abstract of Birkhoff's paper (Birkhoff 1934, p. 664).

References

Archimedes (1880). *Opera omnia.* Cum commentariis Eutocii. Volumen I. Ed. by J.L. Heiberg. Leipzig.

Archimedes (1897). *The Works of Archimedes.* Edited in Modern Notation with Introductory Chapters by T. L. Heath. Cambridge.

Archimedes (2004). *The Works of Archimedes.* Translation and Commentary. Volume 1: The Two Books *On the Sphere and the Cylinder.* Edited and translated by Reviel Netz. Cambridge.

Aristotle (1939). *On the Heavens.* Translated by W.K.C. Guthrie. Cambridge, Mass.

Balestro, Vitor, Horst Martini, and Ralph Teixeira (2024). *Convexity from the Geometric Point of View.* Cham.

Bézout, Étienne (1770). *Cours de mathématiques.* Tome premier, contenant l'Arithmétique, la Géométrie & la Trigonométrie rectiligne. Paris.

Birkhoff, Garrett (1934). "Integration of functions with values in a Banach Space". *Bulletin of the American Mathematical Society* 40, p. 664.

Birkhoff, Garrett (1935). "Integration of Functions with Values in a Banach Space". *Transactions of the American Mathematical Society* 38, pp. 357–378.

Blaschke, Wilhelm (1916). *Kreis und Kugel.* Leipzig.

de Bouelles, Charles (1542). *Livre singulier & utile, touchant l'art et practique de geometrie.* Paris.

Brunn, Hermann (1887). *Ueber Ovale und Eiflächen.* München.

Clagett, Marshall (1964). *Archimedes in the Middle Ages.* Volume I. The Arabo-Latin Tradition. Madison.

Deidier, Antoine (1740). *Le calcul différentiel et le calcul intégral, expliqués et appliqués à la geometrie.* Paris.

Digges, Leonard and Thomas Digges (1571). *A geometrical practise, named Pantometria diuided into three bookes, Longimetra, Planimetra, and Stereometria.* London.

Dürer, Albrecht (1525). *Underweysung der Messung, mit dem Zirckel und Richtscheyt.* Nürnberg.

Dwilewicz, Roman J. (2009). "A short history of Convexity". *Differential Geometry – Dynamical Systems* 11, pp. 112–129.

Euclid (1570). *The Elements of Geometrie of the most auncient Philosopher Euclide of Megara.* Faithfully (now first) translated into the Englishe toung, by H. Billingsley. London.

Fenchel, Werner (1983). "Convexity Through the Ages". In: *Convexity and Its Applications.* Ed. by Peter M. Gruber and Jörg M. Wills. Basel.

de Fermat, Pierre (1679). "Methodus ad disquirendam maximam & minimam". In: *Varia opera mathematica.* Toulouse, pp. 63–73.

Gruber, Peter M. (1990). "Zur Geschichte der Konvexgeometrie und der Geometrie der Zahlen". In: *Ein Jahrhundert Mathematik 1890 - 1990.* Festschrift zum Jubiläum der DMV. Ed. by Gerd Fischer et al. Braunschweig and Wiesbaden, pp. 421–455.

Gruber, Peter M. (1993). "History of convexity". In: *Handbook of Convex Geometry.* Volume A. Ed. by Peter M. Gruber and Jörg M. Wills. Amsterdam, pp. 1–15.

Guerraggio, Angelo and Elena Molho (2004). "The Origins of quasi-concavity: a Development between Mathematics and Economics". *Historia Mathematica* 31, pp. 62–75.

Hon, Giora and Bernard R. Goldstein (2008). *From Summetria to Symmetry.* The Making of a Revolutionary Scientific Concept. New York.

Jensen, Johan (1905). "Om konvekse Funktioner og Uligheder imellem Middelværdier". *Nyt tidsskrift for matematik B* 16, pp. 49–68.

Jensen, Johan (1906). "Sur les fonctions convexes et les inégalites entre les valeurs moyennes". *Acta Mathematica* 30, pp. 175–193.

Kjeldsen, Tinne Hoff (2008). "From measuring tool to geometrical object: Minkowski's development of the concept of convex bodies". *Archive for History of Exact Sciences* 62, pp. 59–89.

Legendre, Adrien-Marie (1794). *Éléments de géométrie.* Avec des notes. Paris.

Mahoney, Michael S. (1994). *The Mathematical Career of Pierre de Fermat, 1601–1665.* Second Edition. Princeton.

Minkowski, Hermann (1897). “Allgemeine Lehrsätze über die convexen Polyeder”. *Nachrichten von der Königl. Gesellschaft der Wissenschaften zu Göttingen (mathematisch-physikalische Klasse)* (1897), pp. 198–220.
Oresme, Nicole (1968). *Le livre du ciel et du monde*. Edited by Albert D. Menut and Alexander J. Denomy, translated with an introduction by Albert D. Menut. Madison.
Pál, Julius (1920). “Über ein elementares Variationsproblem”. *Danske Videnskabernes Selskab. Matematisk-fysiske Meddelelser* III, 2, pp. 1–35.
Papadopoulos, Athanase (2019). “Convexity in Greek Antiquity”. In: *Geometry in History*. Ed. by S.G. Dani and Athanase Papadopoulos. Cham, pp. 139–152.
Roberts, A. Wayne and Dale E. Varberg (1973). *Convex Functions*. New York.

convolution

This term was originally introduced in the context of invariant theory as German *Faltung*, from the verb *falten* ‘to fold’. Gordan (1875, p. 3) used the word for a symbolical product in invariant theory. In this sense it was quickly picked up, and while it remained untranslated in MacKinnon (1894-95, p. 104), Grace and Young (1903, p. 409) suggested the English translation *convolution.*

Nowadays, however, the term is mainly known in connection with bilinear forms and integrals. A germ of the notion is already found in Frobenius (1878, pp. 2–4), who defined a symbolical product of bilinear forms. Hilbert (1906, pp. 159, 179) extended Frobenius’ definition[1] and introduced the German term *Faltung*. Schur (1911, p. 4) suggested German *Komposition* ‘composition’ in addition to Hilbert’s *Faltung*.

Volterra used Italian *operazione di composizione* (1910, p. 169) and also French *opération de composition* ‘operation of composition’ (1913, p. 69) with respect to integrals. The term *Faltung* first appears[2] in Bernstein (1921, p. 817), where also the following notation is introduced:

$$\xi(t) * \eta(t) = \int_0^t \xi(\tau)\eta(t-\tau)d\tau = \int_0^t \xi(t-\tau)\eta(\tau)d\tau$$

Bernstein also used the new term in a joint paper with Doetsch (Doetsch and Bernstein 1922).

In English texts, the term was often left untranslated until the 1930s. Haviland (1934, p. 626) uses *symbolical product* and mentions *Faltung* and *convolution* in a footnote, while Wiener (1933, p. 45) stuck to *Faltung*, saying “there is no good English word”. Wintner (1932, p. 251) first uses *folding expression*, and then switches to *convolution* (Wintner 1934, p. 660).

The reason for choosing the term was later discussed by Doetsch. He considers two converging infinite series $\sum_{i=0}^{\infty} a_i z^i$ and $\sum_{i=0}^{\infty} b_i z^i$, and multiplies them using the Cauchy product, which yields $\sum_{i=0}^{\infty}\sum_{j=0}^{\infty} a_j b_{i-j} z^i = \sum_{n=0}^{\infty} c_n z^n$, where $c_n = a_0 b_0 + a_1 b_1 + ... + a_n b_n$. Doetsch (1937, p. 156) then explains: “If we write the indices 0, 1, ..., n on a strip of paper and fold it in the middle, one obtains the covering

[1] See Pietsch (2007, pp. 269f.). [2] See Friedman (2018, p. 351).

of indexes, whose corresponding elements are to be multiplied together. This is why we call such an operation 'Faltung'".[1]

Note. Comments on convolutions as used by Frobenius and Hilbert can be found in Pietsch (2007, pp. 269f.) and Dieudonné (1981, pp. 113f.). On convolutions with respect to integrals, see Domínguez (2015) and Godard (2017). Above all, see Friedman (2018, pp. 340–354), who discusses all of the meanings in detail.

References

Bernstein, Felix (1921). "Die Integralgleichung der elliptischen Thetanullfunktion. Zweite Note". *Koninklijke Akademie van Wetenschappen te Amsterdam. Proceedings of the Section of Sciences* 23, pp. 817–823.

Dieudonné, Jean (1981). *History of Functional Analysis*. Amsterdam.

Doetsch, Gustav (1937). *Theorie und Anwendung der Laplace-Transformation*. Berlin.

Doetsch, Gustav and Felix Bernstein (1922). "Über die Integralgleichung der elliptischen Theta-funktion". *Jahresbericht der Deutschen Mathematiker-Vereinigung* 31, pp. 148–153.

Domínguez, Alejandro (2015). "A History of the Convolution Operation". *IEEE Pulse* 6, pp. 38–49.

Friedman, Michael (2018). *A History of Folding in Mathematics*. Mathematizing the Margins. Cham.

Frobenius, Georg (1878). "Ueber lineare Substitutionen und bilineare Formen". *Journal für die reine und angewandte Mathematik* 84, pp. 1–63.

Godard, Roger (2017). "The Convolution as a Mathematical Object". In: *Research in History and Philosophy of Mathematics*. The CSHPM 2016 Annual Meeting in Calgary, Alberta. Ed. by Maria Zack and Dirk Schlimm. Cham, pp. 199–212.

Gordan, Paul (1875). *Ueber das Formensystem binärer Formen*. Leipzig.

Grace, John Hilton and Alfred Young (1903). *The Algebra of Invariants*. Cambridge.

Haviland, E. K. (1934). "On the Theory of Absolutely Additive Distribution Functions". *American Journal of Mathematics* 56, pp. 625–658.

Hilbert, David (1906). "Grundzüge einer allgemeinen Theorie der linearen Integralgleichungen. (Vierte Mitteilung)". *Nachrichten von der Königl. Gesellschaft der Wissenschaften zu Göttingen (mathematisch-physikalische Klasse)* (1906), pp. 157–228.

MacKinnon, Annie L. (1894-95). "Concomitant Binary Forms in Terms of the Roots". *Annals of Mathematics* 9, pp. 95–157.

Pietsch, Albrecht (2007). *History of Banach Spaces and Linear Operators*. Boston, Mass. et al.

Schur, Issai (1911). "Bemerkungen zur Theorie der beschränkten Linearformen mit unendlich vielen Veränderlichen". *Journal für die reine und angewandte Mathematik* 140, pp. 1–28.

Volterra, Vito (1910). "Questioni generali sulle equazioni integrali ed integro-differenziali". *Atti della Reale Accademia dei Lincei. Rendiconti* 19, pp. 169–180.

Volterra, Vito (1913). *Leçons sur les fonctions de lignes*. Paris.

Wiener, Norbert (1933). *The Fourier Integral and Certain of its Applications*. Cambridge.

Wintner, Aurel (1932). "Remarks on the Ergodic Theorem of Birkhoff". *Proceedings of the National Academy of Sciences* 18, pp. 248–251.

Wintner, Aurel (1934). "On Analytic Convolutions of Bernoulli Distributions". *American Journal of Mathematics* 56, pp. 659–663.

coordinates

The term *coordinates* (Latin *coordinatas*) was coined by Leibniz, who presented it to the public in 1692. He explicitly grouped the older terms *ordinate* and *abscissa* together: "both the *ordinate* and the *abscissa*, which are usually designated by x and

---

[1] "Schreibt man die Indizes 0, 1, ..., n auf einen Streifen Papier und faltet ihn in der Mitte, so kommen gerade diejenigen Indizes zur Deckung, deren Glieder miteinander zu multiplizieren sind. Daher nennen wir eine derartige Bildung eine 'Faltung'". Translation from Friedman (2018, p. 353).

y (which I am in the habit of calling *coordinates*)" (Leibniz 1692, p. 170).[1] In this paper, Leibniz also described a quadrant of a system of coordinates.

The new term was soon adopted and appeared in various early textbooks on the calculus such as Guisnée (1705, p. 18) and Reyneau (1708, p. 541);[2] the English term was used by Hayes (1704, p. 313), the German one by Wolff (1710, p. 333).

An earlier attempt at unifying the terminology is found in Pascal (1658, p. 18), who suggested *ordonnée à l'axe* 'ordinate to the axis' (abscissa) and *ordonnée à la base* 'ordinate to the base' (ordinate).

The epithet *Cartesian* in connection with coordinates came up in the 19th century. For example, Chasles (1837, p. 635) used French *système de coordonnées de Descartes*, and Hamilton (1847, p. 8), in a talk held in late 1844, *Cartesian method of coordinates*. Explicitly, French *coordonnées cartésiennes* was used by Terquem (1846, p. 487). Presumably, the terminology began to be used at this time to distinguish Cartesian from various new types of coordinates. By the end of the century, the term was in common use.

Note. A historical survey of the terminology is found in Tropfke (1924, pp. 116–118). For the history of coordinates, see Müller (1910), Loria (1948), and Amadeo (2018).

References

Amadeo, Marcello (2018). "Textbooks revealing the development of a concept—the case of the number line in the analytic geometry (1708–1829)". *ZDM* 50, pp. 907–920.

Chasles, Michel (1837). *Aperçu historique sur l'origine et le développement des méthodes en géométrie*. Bruxelles.

Guisnée, Nicolas (1705). *Application de l'algebre a la geometrie*. Paris.

Hamilton, William Rowan (1847). "Theory of Quaternions". *Proceedings of the Royal Irish Academy* 3, pp. 1–16.

Hayes, Charles (1704). *A Treatise of Fluxions*. London.

de l'Hôpital, Guillaume François Antoine (1696). *Analyse des infiniments petits, pour l'intelligence des lignes courbes*. Paris.

Leibniz, Gottfried Wilhelm (1692). "De linea ex lineis numero infinitis ordinatim ductis inter se concurrentibus formata [...]" *Acta Eruditorum* (Apr. 1692), pp. 168–171.

Loria, Gino (1948). "Perfectionnements, évolution, métamorphoses du concept de «coordonnées»: Contribution à l'histoire de la géométrie analytique". *Osiris* 8, pp. 218–288.

Müller, Emil (1910). "Die verschiedenen Koordinatensysteme". *Encyklopädie der mathematischen Wissenschaften mit Einschluss ihrer Anwendungen* III AB 7, pp. 596–770.

Pascal, Blaise (1658). *Lettre de A. Dettonville à Monsieur de Carcavy*. Paris.

Reyneau, Charles-René (1708). *Usage de l'analyse*. Tome II. Paris.

Terquem, Olry (1846). "*Gnomonique graphique et analytique*, par Born. Paris 1846 [review]". *Nouvelles annales de mathématiques* 5, pp. 483–487.

Tropfke, Johannes (1924). *Geschichte der Elementar-Mathematik*. Sechster Band. Analysis. Analytische Geometrie. Berlin and Leipzig.

Wolff, Christian (1710). *Der Anfangs-Gründe aller Mathematischen Wissenschaften Letzter Theil*. Halle.

---

[1] "Verum tam *ordinata* quam *abscissa*, quas per x & y designari mos est (quas & *coordinatas* appellare soleo". Note that Leibniz called x the *ordinate* and y the *abscissa*, which is unusual.

[2] de l'Hôpital (1696, p. 1) stuck to the older terms.

coplanar

This elementary term appears to be due to Hamilton (1846, p. 31), who introduced it in an instalment of articles on quaternions: "to express that three lines are in or parallel to one common plane, we may write the *equation of coplanarity* $S.\alpha\alpha'\alpha'' = 0$".

It took some time until the term was also transferred to other languages: French *coplanaire* was used by Hoüel (1867, p. 347), while the German term appears in Gravelius (1889, p. 608) and Schröder (1891, p. 575)

References

Gravelius, Harry, ed. (1889). *Theoretische Mechanik starrer Systeme*. Auf Grund der Arbeiten und mit einem Vorworte von Sir Robert S. Ball. Berlin.

Hamilton, William Rowan (1846). "On Quaternions; or on a new System of Imaginaries in Algebra". *The London, Edinburgh and Dublin Philosophical Magazine* 29, pp. 26–31.

Hoüel, Jules (1867). "Théorie élémentaire des quantités complexes. Première partie. Algèbre des quantités complexes". *Mémoires de la Société des sciences physiques et naturelles de Bordeaux* 5, pp. 1–64.

Schröder, Ernst (1891). *Vorlesungen über die Algebra der Logik* (Exakte Logik). Zweiter Band. Erste Abteilung. Leipzig.

coprime

See *relatively prime*.

correlation (statistics)

The statistical term *correlation* is due to Galton (1888, p. 135):

> Two variable organs are said to be co-related when the variation of the one is accompanied on the average by more or less variation of the other, and in the same direction [...] It is easy to see that co-relation must be the consequence of the variations of the two organs being partly due to common causes. If they were wholly due to common causes, the co-relation would be perfect, as is approximately the case with the symmetrically disposed parts of the body. If they were in no respect due to common causes, the co-relation would be *nil*.

The term as such was not a new creation; Galton himself indicated where it came from:

> "Co-relation or correlation of structure" is a phrase much used in biology, and not least in that branch of it which refers to heredity, and the idea is even more frequently present than the phrase; but I am not aware of any previous attempt to define it clearly, to trace its mode of action in detail, or to show how to measure its degree (Galton 1888, p. 135).

What Galton alludes to here as the "correlation of structure" goes back to Cuvier (1812, p. 58), who discussed what he called the "principle of correlation" ("principe de corrélation"): "Every organized individual forms an entire system of its own, all the parts of which naturally correspond, and concur to produce a certain definite purpose, by reciprocal reaction, or by combining towards the same end".[1]

Galton's term was soon adopted, for example by Edgeworth (1892, p. 429).

Note. For the history of the concept, see Stigler (1989). Aldrich's article in Miller (n.d.) contains more information on some related terms.

---

[1] "Tout être organisé forme un ensemble, un système unique et clos, dont les parties se correspondent mutuellement et concourent à la même action définitive par une action réciproque. Aucune de ces parties ne peut changer sans que les autres changent aussi, et, par conséquent, chacune d'elle prise séparément indique et donne toutes les autres". Translation from Cuvier (1815, p. 90).

References

Cuvier, Georges (1812). *Recherches sur les ossemens fossiles de quadrupèdes*. Tome premier. Paris.
Cuvier, Georges (1815). *Essay on the Theory of the Earth*. Translated by Robert Kerr. Edinburgh.
Edgeworth, F.Y. (1892). "The law of error and correlated averages". *The London, Edinburgh and Dublin Philosophical Magazine*, pp. 429–438, 518–526.
Galton, Francis (1888). "Co-relations and their Measurement, chiefly from Anthropometric Data". *Proceedings of the Royal Society of London* 45, pp. 135–145.
Miller, Jeff (n.d.). "Earliest Known Uses of Some of the Words of Mathematics". URL: `https://mathshistory.st-andrews.ac.uk/Miller/mathword/`.
Stigler, Stephen M. (1989). "Francis Galton's Account of the Invention of Correlation". *Statistical Science* 4, pp. 73–79.

correlation (geometry)

The French adjective *corrélatif* was used by Chasles (1837, p. 255) in his discussion of the principle of duality: "we shall say that two figures that have between each other the dependencies prescribed by the laws of this principle are *correlative*".[1]

The corresponding English terminology was used by Hirst (1874, p. 40): "A correlation is said to be established between two planes, when their points and right lines are so associated, that to each point in one of the planes, and to each line passing through that point, respectively correspond, in the other plane, *one* line and *one* point in that line"; *correlative* had earlier been used by Townsend (1865, p. 339).

Other terms have been in use for the concept; for instance, Veblen and Young (1910, p. 262) remarked: "The terms *reciprocity* and *duality* are sometimes used in place of *correlation*".

Note. Some notes on Chasles and on some of the terms he introduced are found in Laptev and Rosenfeld (1996, pp. 38–40).

References

Chasles, Michel (1837). *Aperçu historique sur l'origine et le développement des méthodes en géométrie*. Bruxelles.
Hirst, T.A. (1874). "On the Correlation of two Planes". *Proceedings of the London Mathematical Society* 5, pp. 40–70.
Laptev, B.L. and Boris A. Rosenfeld (1996). "Geometry". In: *Mathematics of the 19th Century. Geometry. Analytic Function Theory*. Ed. by Andrey Nikolaevich Kolmogorov and Adolph Andrei Pavlovich Yushkevich. Basel, pp. 1–117.
Townsend, Richard (1865). *Chapters on the Modern Geometry of the Point, Line, and Circle*. Vol. II. Dublin.
Veblen, Oswald and John Wesley Young (1910). *Projective Geometry*. Volume I. Boston.

cosecant

When Thomas Fincke introduced the trigonometric term *tangent* and *secant*, he also considered their complements, using Latin *secans complementi* for the complement of the secant (Fincke 1583, p. 76). Fincke's model was followed by various writers such as Clavius in his edition of Theodosius' *Sphaerica* (Theodosius 1586, p. 191), Pitiscus (1600, p. 119), and Snel (1627, pp. 25f.).[2]

While Edmund Gunter (1620, descriptio canonis) did use the abbreviation *cosine* and *cotangens*, he does not appear to have used *cosecant* as well. The English term is

[1] "nous dirons que deux figures qui auront entre elles les dépendances voulues par les lois de ce principe, sont *corrélatives*". [2] Magini (1592, pp. 14f.) was aware of the term, but preferred *secans secundus*.

found in Urquhart (1645, Lexicidion): "Anti-secant, Anti-sine, and Anti-tangent, are the complements of the Secant, Sine, and Tangent, and are called sometime Co-secant, Co-sine, and Co-tangent: they have anti prefixed, because they are not in the same colume, and co, because they are in the next to it". Urquhart's phrasing suggests that the term was already in use when he wrote his text. As with the other *co*-terms, continental writers were reluctant to adopt *cosecant*, but its use increased after it was included in Wolff's mathematical dictionary (Wolff 1716, pp. 442f.).

Note. As Miller (n.d.), citing Rouse Ball (1908, p. 201) and Smith (1958, p. 622), points out, it is sometimes stated that Rheticus already used the term in his posthumously published *Opus Palatinum de triangulis*. Rheticus did not use this term, though, nor did he use *sinus*. Instead, he preferred the terminology used in triangles, i. e. *basis*, *hypotenusa*, and *perpendiculum*. Thus, in *De triquetris rectarum linearum in planitie*, which is part of the *Opus Palatinum de triangulis*, he explicitly calls the secant *hypotenusa* and also disapprovingly says that *sinus* is a "barbaric word" ("barbara vocabula") (Rheticus 1596, p. 90). Also, in the tables in the *Opus Palatinum* we find *perpendiculum* for the sine, *basis* for the cosine, and *hypotenusa* for both the secant and cosecant. See von Braunmühl (1900, pp. 146f.), Tropfke (1923, pp. 19f., 36), and Van Brummelen (2009, pp. 275f.).

Note. The terms are briefly discussed by Tropfke (1923, pp. 19f., 36) and von Braunmühl (1900, p. 146). Various remarks on the history of the concept can be found in the works cited in the note in the entry *trigonometry*. The cosecant as used by Arabic-Islamic mathematicians is briefly discussed in the article *cotangent*.

References

von Braunmühl, Anton (1900). *Vorlesungen über die Geschichte der Trigonometrie*. Erster Teil. Von den ältesten Zeiten bis zur Erfindung der Logarithmen. Leipzig.

Van Brummelen, Glen (2009). *The Mathematics of the Heavens and the Earth*. The Early History of Trigonometry. Princeton and Oxford.

Fincke, Thomas (1583). *Geometriae rotundi libri XIII*. Basel.

Gunter, Edmund (1620). *Canon triangulorum, sive Tabulae Sinuum et Tangentium artificialium*. London.

Magini, Giovanni Antonio (1592). *De planis triangulis*. Venice.

Miller, Jeff (n.d.). "Earliest Known Uses of Some of the Words of Mathematics". URL: `https://mathshistory.st-andrews.ac.uk/Miller/mathword/`.

Pitiscus, Bartholomaeus (1600). *Trigonometriae, siue, De dimensione triangulor[um] libri quinque*. Augsburg.

Rheticus, Georg Joachim (1596). *Opus Palatinum de triangulis*. Ed. by Valentin Otho. Neustadt.

Rouse Ball, Walter William (1908). *A Short Account of the History of Mathematics*. London.

Smith, David Eugene (1958). *History of Mathematics*. Volume II. Special Topics of Elementary Mathematics. New York.

Snel, Willebrord (1627). *Doctrinae triangulorum canonicae libri quattuor*. Leiden.

Theodosius (1586). *Sphaericorum Libri III*. Ed. by Christopher Clavius. Roma.

Tropfke, Johannes (1923). *Geschichte der Elementar-Mathematik*. Fünfter Band. I. Ebene Trigonometrie. II. Sphärik und sphärische Trigonometrie. Berlin and Leipzig.

Urquhart, Thomas (1645). *The Trissotetras*. Or, a most exquisite table for resolving all manner of triangles, whether plaine or sphericall, rectangular or obliquangular. London.

Wolff, Christian (1716). *Mathematisches Lexicon*. Leipzig.

coset

The concept of *coset* appears in the writings of Galois, who, according to Miller, Blichfeldt, and Dickson (1916, p. 24), "did not use a special name for it". That is to say, Galois used French *groupe* not just for "real" groups, but also for any set of permutations, which includes cosets (Wußing 1984, p. 111). Also, Galois used the word *groupe partiel* in the sense of both coset and subgroup (Neumann 2011, p. 26).[1]

The terminology becomes fixed towards the end of the 19th century. Weber (1895, p. 502) suggested the German term *Nebengruppe* (*neben* lit. means 'beside, next to'). This seems an unhappy choice because it might suggest that a coset is also a group, and that is the reason why Netto (1900, p. 306) proposed the alternative *conjuger Complex*, which can be translated as *conjugate complex*; Netto's term, however, did not catch on.

The English term *co-set* was introduced by Miller (1910a, p. 382, 1910b, pp. 307f.);[2] in a paper published one year later, Miller (1911, p. 326) points out Weber's use of *Nebengruppe*. While Miller's term soon became standard for English, the situation in Germany remained unsettled for a while. This is illustrated by van der Waerden (1930, p. 26), who uses *Nebenklasse* and mentions *Nebengruppe*, *Nebenkomplex*, and *Restklasse* as synonyms.

Note. Miller (1911, pp. 326–328) contains some historical information on the notion.

References

Kiernan, B. Melvin (1971). "The Development of Galois Theory from Lagrange to Artin". *Archive for History of Exact Sciences* 8, pp. 40–154.
Miller, George Abram (1910a). "On a method due to Galois". *The Quarterly Journal of Pure and Applied Mathematics* 41, pp. 382–384.
Miller, George Abram (1910b). "Some relations between substitution group properties and abstract groups". *Proceedings of the American Philosophical Society held at Philadelphia for promoting useful knowledge* 49, pp. 307–314.
Miller, George Abram (1911). "On the Use of the Co-Sets of a Group". *Transactions of the American Mathematical Society* 12, pp. 326–334.
Miller, George Abram, Hans Frederick Blichfeldt, and Leonard Eugene Dickson (1916). *Theory and Application of Finite Groups*. New York.
Netto, Eugen (1900). *Vorlesungen über Algebra*. In zwei Bänden. Zweiter Band. Leipzig.
Neumann, Peter M. (2011). *The Mathematical Writings of Évariste Galois*. Zürich.
van der Waerden, Bartel Leendert (1930). *Moderne Algebra*. Erster Teil. Berlin.
Weber, Heinrich (1895). *Lehrbuch der Algebra*. Erster Band. Braunschweig.
Wußing, Hans (1984). *The Genesis of the Abstract Group Concept*. A Contribution to the History of the Origin of Abstract Group Theory. Tr. Abe Shenitzer. Cambridge, Mass.

cosine

Like the sine, the cosine was first systematically used in India.[3] One term used for the cosine was *koṭi-jyā* (*jyā* is the chord or sine). Datta and Singh (1983, p. 41) explain the terminology as follows: "*koṭi* means, amongst others 'the curved end of a bow' or 'the end or extremity in general'; hence in Trigonometry it came to denote 'the

[1] For the germ of the concept in Cauchy's work, see Kiernan (1971, p. 66). [2] The former reference is cited by Miller, Blichfeldt, and Dickson (1916, p. 24) as the likely first use of this term; it is also given in the OED as the first attestation. [3] The original definition of these concepts is illustrated in the article *sine*.

complement of an arc to 90°'. So the radical significance of the term *koṭi-jyā* is 'the *jyā* of the complementary arc'".

In Arabic texts, this term appears as *jayb al-tamām* 'sine of completion'.[1] The term was used, for example, by Abū al-Wafā Būzhjānī in his *Almagest* (see Rashed 2017, pp. 162, 164) and al-Bīrūnī in his *Book of Instruction in the Elements of the Art of Astrology* (al-Bīrūnī 1934, p. 5). This terminology is based on the Sanskrit terms (Bossong 1978, p. 172).

In the 12th century, numerous Arabic texts were translated into Latin, and the term *tamām* required a Latin counterpart. Different solutions were found. Thus, in the *Almagesti minor*, a summary of the first six books of Ptolemy's *Almagest*, *sinus perfectionis* was regularly used (Zepeda 2018, e. g., p. 164). Plato of Tivoli, who translated one of al-Battānī's works under the title *De motu stellarum*, paraphrases the complement as that "which lacks to the completion of 90° (al-Battānī 1537, fol. 16ᵛ)[2] and also used *chorda residui* 'chord[3] of the remainder' (al-Battānī 1537, p. 13). Gerard of Cremona, in his translation of the "Toledan tables" by al-Zarqālī, chose *sinus residui* 'sine of the remainder' (Curtze 1900, p. 342).

The term *sinus complementi* 'sine of the complement' also dates from the 12th century. It was used, in particular, by Gerard of Cremona in his translation of Jābir ibn Aflaḥ's "Correction of the Almagest" (ibn Aflaḥ 1534, e. g., p. 13),[4] and also in his translation of the "Tabulae Jahen" by al-Jayyānī (al-Jayyānī 1549, fol. O iijᵛ).[5]

In the 15th century, *sinus complementi* was regularly used by Regiomontanus, for example in his book *On Triangles* (Regiomontanus 1533, e. g., p. 21). At the time his book was posthumously published, the term had already been used in various other publications, for example in Fine (1532, fol. 56ᵛ).

The synonymous *sinus rectus secundus* has been attributed by various writers such as Tropfke (1923, p. 34) to Apian, who used it, for example, in Apian (1533, fol. A iijʳ). The term is actually older; it was already used by Regiomontanus (1496, fol. n 2ᵛ), where it was explicitly introduced as a shorter alternative to "remainder of the first sine".[6] The term remained in use until the 18th century.[7]

The actual term *cosinus* was introduced by Edmund Gunter: "the sine of the complement [...] which for brevity's sake we call *cosinus*" (Gunter 1620, descriptio canonis).[8] Shortly afterwards, he also used the English term *cosine* (Gunter 1623, The first booke of the crosse-staffe, p. 53).[9] During the 17th century, the term was almost exclusively used by English writers, while continental writers stuck to *sinus complementi* and its equivalents in other languages. That does not mean that the term remained unnoticed on the continent, though. For example, Kepler (1627, preface,

---

[1] For *tamām*, see *complement* (trigonometry). [2] "quod ad perficiendum 90 deficit" [3] Plato of Tivoli used *chorda* to refer to the sine. See *sine*. [4] In this text, definitions of the complement of an arc and of an angle are given (ibn Aflaḥ 1534, p. 3). See *complement* (trigonometry). [5] In the latter text, a definition of the cosine is given. [6] "Residuique sinum primum, ut brevius dicam, sinum quoque secundum sive sinum complementi eius elicias". [7] An idiosyncratic use may be noted here. In several publications, Rheticus, considered trigonometric functions independently of circles, used *basis* for the cosine, and *perpendiculum* for the sine (Rheticus 1551, fol. Cijʳ, Rheticus 1596, pp. 86, 89). See Tropfke (1923, pp. 19f., 34). [8] "sinus complementi [...] quem brevitatis causa *cosinũ* dicimus". [9] It is not infrequently claimed that Gunter wrote *co-sinus* with a hyphen. In the two works of Gunter just cited, the hyphen is *not* used, however.

p. 18) points out that the term was due to Gunter, and Kircher (1641, p. 208) mentions it as a recent addition to the mathematical vocabulary. Caramuel y Lobkowitz (1670, p. 783) also mentions the term but finds *antisinus* more appropriate.

Note. Brief overviews of the terminology are given by Tropfke (1923, pp. 33–35) and Smith (1958, p. 619). For the Arabic term, see Rashed (2017, pp. 161–165). Various remarks on the history of the concept can be found in the works cited in the note in the entry *trigonometry*.

References

ibn Aflaḥ, Jābir (1534). *Gebri Filii Affla Hispalensis De astronomia libri IX* [Translated by Gerard of Cremona]. In: Apian, Peter. *Instrumentum primi mobilis*. Nürnberg.

Apian, Peter (1533). *Introductio geographica*. Ingolstadt.

al-Battānī, Muḥammad ibn Jābir (1537). *De motu stellarum*. Nürnberg.

al-Bīrūnī, Abū Rayḥān Muḥammad ibn Aḥmad (1934). *The Book of Instruction in the Elements of the Art of Astrology*. Written in Ghaznah, 1029 A. D. Reproduced from Brit. Mus. MS. Or. 8349. The Translation facing the Text by R. Ramsay Wright. London.

Bossong, Georg, ed. (1978). *Los Canones de Albateni*. Herausgegeben sowie mit Einleitung, Anmerkungen und Glossar versehen. Tübingen.

Caramuel y Lobkowitz, Juan (1670). *Mathesis biceps, vetus et nova*. Campania.

Curtze, Maximilian (1900). "Urkunden zur Geschichte der Trigonometrie im christlichen Mittelalter". *Bibliotheca Mathematica* 3, pp. 321–416.

Datta, Bibhutibhusan and Avadhesh Narayan Singh (1983). "Hindu Trigonometry (revised by Kripa Shankar Shukla)". *Indian Journal of History of Science* 18, pp. 39–108.

Fine, Oronce (1532). *Protomathesis*. Tradotte da Cosimo Bartoli. Paris.

Gunter, Edmund (1620). *Canon triangulorum, sive Tabulae Sinuum et Tangentium artificialium*. London.

Gunter, Edmund (1623). *De sectore & radio. The description and use of the Sector*. London.

al-Jayyānī, Abū'Abd Allāh Muḥammad Ibn Mu'ādh (1549). *Scriptum antiquum saraceni cuiusdam, de diversarum gentium Eris, annis ac mensibus, et de reliquis Astronomiae principiis*. Nürnberg.

Kepler, Johannes (1627). *Tabulae Rudolphinae*. Ulm.

Kircher, Athanasius (1641). *Magnes sive arte magnetica*. Roma.

Rashed, Roshdi, ed. (2017). *Lexique historique de la langue scientifique arabe*. Hildesheim.

Regiomontanus, Johannes (1496). *Epytoma Joannis de Monte Regio in Almagestum Ptolomei*. Venezia.

Regiomontanus, Johannes (1533). *De triangulis omnimodis*. Ed. by Johannes Schöner. Nürnberg.

Rheticus, Georg Joachim (1551). *Canon doctrinae triangvlorvm*. Leipzig.

Rheticus, Georg Joachim (1596). *Opus Palatinum de triangulis*. Ed. by Valentin Otho. Neustadt.

Smith, David Eugene (1958). *History of Mathematics*. Volume II. Special Topics of Elementary Mathematics. New York.

Tropfke, Johannes (1923). *Geschichte der Elementar-Mathematik*. Fünfter Band. I. Ebene Trigonometrie. II. Sphärik und sphärische Trigonometrie. Berlin and Leipzig.

Zepeda, Henry, ed. (2018). *The First Latin Treatise on Ptolemy's Astronomy: The* Almagesti minor *(c. 1200)*. Turnhout.

cotangent

When Thomas Fincke introduced the trigonometrical term *tangent*, he also considered its complement (Fincke 1583, pp. 73–75), using Latin *tangens complementi*. As in the case of *tangent*, Fincke's model was followed by Lansberg (1591, p. 9) and Clavius in his edition of Theodosius' *Sphaerica* (Theodosius 1586, p. 191). In English, the corresponding expression *tangent of the complement* was already used by Blundeville (1594, fol. 56^r).

As with *cosine*, the abbreviation to Latin *cotangens* is due to Edmund Gunter (1620, descriptio canonis); as in the case of *cosine*, Gunter (1623, The first booke of the crosse-staffe, p. 55) also contains the English word *cotangent*. According to Urquhart (1645, Lexicidion), the complement of the tangent is "called sometime" *Co-tangent*. In the 17th century, the term was mainly used by English writers and continental writers were reluctant to adopt it. For example, the first attestation for German *cotangens* seems to be Wolff (1710, p. 130).

Before Fincke, it was customary to use Latin *umbra recta* 'direct shadow' or *umbra extensa* 'extended shadow'.[1] This expression was widely used by medieval writers and goes back to Arabic terminology. A so-called *gnomon* was an instrument (a stick) used to determine the length of shadows and also the time of day. The terminology is shown in figure 31.[2]

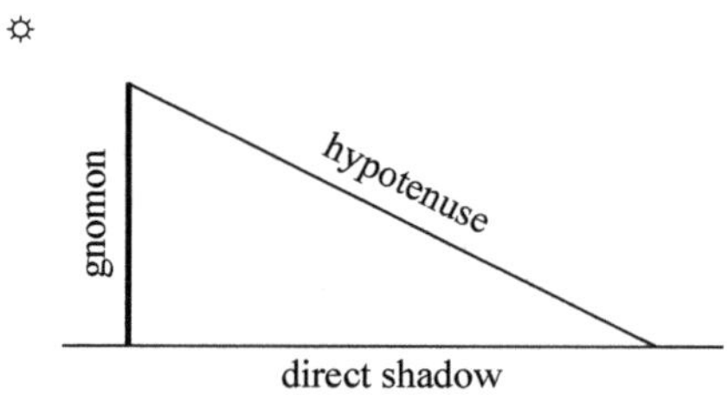

Figure 31: Direct shadow

For example, al-Bīrūnī (1976, p. 64) defined the cotangent in this way and his term *al-ẓill al-mustawī*, meaning 'direct shadow', is based on that fact that the gnonom is normal to the horizon.[3] He goes on to point out that this shadow is also called "extended [...] because its extension is along the face of the earth which has neither protrusions nor concavities" (ibid.). The term *extended shadow* was also used by al-Battānī (Nallino 1903, p. 182), for example. Nallino (ibid.) points out that *second shadow* (*ẓill thānī*) was also in use.

The direct shadow is the basis for Latin *umbra recta*, while the extended shadow is the basis for Latin *umbra extensa*. Both terms are found in the 12th century: The former term was used by Gerard of Cremona (Curtze 1900, p. 342), for example, while the latter was used by Plato of Tivoli (al-Battānī 1537, fol. 14^r).

In fig. 31, the "hypotenuse of the direct shadow" (*quṭr al-ẓill al-mustawī*) corresponds to the cosecant. The term also appears in al-Bīrūnī (1976, p. 64) but had

[1] For the following, see Van Brummelen (2009, pp. 149–156) and Tropfke (1923, pp. 24–26).
[2] The figure is based on Van Brummelen (2009, p. 151) and al-Bīrūnī (1976, p. 65). [3] The term was already used by al-Khwārizmī. See Suter (1914, p. 77). A brief overview of the Arabic terminology for trigonometric functions is given by Kennedy (1956, pp. 17f.).

been used earlier, for example by Abū al-Wafā᾽ (Carra de Vaux 1892, p. 420) and by Ḥabash al-Ḥāsib (Nallino 1903, p. 182).[1]

Note. An account of the history of the concept and term is given by Tropfke (1923, pp. 23-28, 36). For the medieval Latin term, see also Kunitzsch (1982, p. 562). The trigonometric functions in the Arabic-Islamic world are explained in Moussa (2010); for the Arabic terminology, see also Rashed (2017, pp. 474–476). Various remarks on the history of the concept can be found in the works cited in the note in the entry *trigonometry*.

References

al-Battānī, Muḥammad ibn Jābir (1537). *De motu stellarum*. Nürnberg.

al-Bīrūnī, Abū Rayḥān Muḥammad ibn Aḥmad (1976). *The Exhaustive Treatise on Shadows*. Translation & Commentary by E. S. Kennedy. Volume I. Aleppo.

Blundeville, Thomas (1594). *M. Blundevile His Exercises*. London.

Van Brummelen, Glen (2009). *The Mathematics of the Heavens and the Earth*. The Early History of Trigonometry. Princeton and Oxford.

Van Brummelen, Glen and James Steven Byrne (2021). “Maurolico, Rheticus, and the Birth of the Secant Function”. *Journal for the History of Astronomy* 52, pp. 189–211.

Carra de Vaux, Bernard (1892). “L’almageste de Abulwéfa Albuzdjani”. *Journal asiatique* 19, pp. 408–471.

Curtze, Maximilian (1900). “Urkunden zur Geschichte der Trigonometrie im christlichen Mittelalter”. *Bibliotheca Mathematica* 3, pp. 321–416.

Fincke, Thomas (1583). *Geometriae rotundi libri XIII*. Basel.

Gunter, Edmund (1620). *Canon triangulorum, sive Tabulae Sinuum et Tangentium artificialium*. London.

Gunter, Edmund (1623). *De sectore & radio. The description and use of the Sector*. London.

Kennedy, Edward Stewart (1956). *A Survey of Islamic Astronomical Tables*. Philadelphia.

Kunitzsch, Paul (1982). “Glossar der arabischen Fachausdrücke in der mittelalterlichen europäischen Astrolabliteratur”. *Nachrichten der Akademie der Wissenschaften in Göttingen (Philologisch-historische Klasse)* 11, pp. 459–571.

Lansberg, Johan Philip (1591). *Triangulorum geometriae libri quatuor*. Lyon.

Moussa, Ali (2010). “The Trigonometric Functions, as they were in the Arabic-Islamic Civilization”. *Arabic Sciences and Philosophy* 20, pp. 93–104.

Nallino, Carlo Alfonso, ed. (1903). *Al-Battānī sive Albatenii Opus astronomicum*. Pars prima. Milano.

Rashed, Roshdi, ed. (2017). *Lexique historique de la langue scientifique arabe*. Hildesheim.

Rheticus, Georg Joachim (1551). *Canon doctrinae triangvlorvm*. Leipzig.

Suter, Heinrich, ed. (1914). *Die astronomischen Tafeln des Muhammed ibn Musa al-Khwarizmi in der Bearbeitung des Maslama ibn Ahmed al-Madjriti und der latein. Übersetzung des Athelhard von Bath auf Grund der Vorarbeiten von A. Björnbo und R. Besthorn*. København.

Theodosius (1586). *Sphaericorum Libri III*. Ed. by Christopher Clavius. Roma.

Tropfke, Johannes (1923). *Geschichte der Elementar-Mathematik*. Fünfter Band. I. Ebene Trigonometrie. II. Sphärik und sphärische Trigonometrie. Berlin and Leipzig.

Urquhart, Thomas (1645). *The Trissotetras*. Or, a most exquisite table for resolving all manner of triangles, whether plaine or sphericall, rectangular or obliquangular. London.

Wolff, Christian (1710). *Der Anfangs-Gründe aller Mathematischen Wissenschaften Dritter Theil*. Halle.

---

[1] Corresponding medieval Latin terms do not seems to exist. This is not surprising because the cosecant was rarely considered. According to Van Brummelen and Byrne (2021, p. 190), the cosecant as a trigonometric function first appeared in print in Rheticus (1551). In this work, Rheticus expresses the trigonometric functions by the terminology used in triangles, that is, *basis*, *hypotenusa*, and *perpendiculum*. See *cosecant*.

countable* and *uncountable

In his first article on set theory, Cantor (1874) proved that the set of real numbers is uncountably infinite, without using this terminology. Cantor (1879, p. 4) then introduces the new term, mentioning "the class of point sets enumerable to infinity" by which he means "those point sets that have the same power as the natural series of numbers".[1] For what has here been translated by *enumerable*, Cantor uses German *abzählbar*, from *abzählen* 'to count'. *abzählbar unendlich* 'countably infinite' is found in the same place in Cantor's article. Three years later, he suggested *countable set* (*abzählbare Menge*) as a shorter and simpler mode of expression (Cantor 1882, p. 116).[2]

Various English translations have been suggested for the concept. The *Century Dictionary*, published from 1889 to 1891, discusses it in an entry for *enumerable*; C. S. Peirce was responsible for the mathematical entries. In one of his articles, he used *numerable*, however (Peirce 1892, p. 541); later he switched to *denumerable* (Peirce 1897, p. 214), a term also used by Russell (1903, p. 296) and judged by Huntington (1905, p. 217) to be "preferable to the other terms because it avoids a suggestion of finitude". Finally, *countable* was used by Young (1902, p. 285) and earlier in a talk by Bolza (1894, p. 97). The French term *dénombrable* was settled upon in the various translations of some of Cantor's papers in the *Acta Mathematica* from 1883; see, e. g., Cantor (1883a, p. 365).

Uncountability was originally expressed, and still is, simply by saying that a set is not countable. Weber (1896, p. 748) also used *unzählbar*,[3] while Hausdorff (1914, p. 56) has *unabzählbar*; *überabzählbar* is of later origin and appears to have been first used by Frankel (1923, p. 188, 1927, e. g., p. 57). English translations include *innumerable* (Peirce 1892, p. 541), *abnumeral* (Peirce 1897, p. 215),[4] and *unenumerable* (Weber 1897, p. 177). Bolza (1894, p. 97) preferred *non-countable*, while *uncountable* is used in Slaught (1908, p. 67).

Note. An early overview of the terminology was given by Huntington (1905, p. 217). For Cantor's introduction of countability, see Ferreirós (2007, ch. VI).

References

Bolza, Oskar (1894). "Cantor's proof of the existence of transcendental numbers [abstract]". *Quarterly Calendar* 3, p. 97.

Cantor, Georg (1874). "Ueber eine Eigenschaft des Inbegriffs aller reellen algebraischen Zahlen". *Journal für die reine und angewandte Mathematik* 77, pp. 258–262.

Cantor, Georg (1879). "Ueber unendliche, lineare Punktmannichfaltigkeiten". *Mathematische Annalen* 15, pp. 1–7.

Cantor, Georg (1882). "Ueber unendliche, lineare Punktmannichfaltigkeiten. 3". *Mathematische Annalen* 20, pp. 113–121.

Cantor, Georg (1883a). "Sur les ensembles infinis et linéaires de points". *Acta Mathematica* 2, pp. 349–380.

Cantor, Georg (1883b). "Ueber unendliche, lineare Punktmannichfaltigkeiten. 5". *Mathematische Annalen* 21, pp. 545–586.

Dauben, Joseph W. (1982). "Peirce's place in mathematics". *Historia Mathematica* 9, pp. 311–325.

---

[1] "die Classe der *in's Unendliche abzählbaren* Punktmengen [...], d. h., diejenigen Punktmengen, welche mit der natürlichen Zahlenreihe: $1, 2, 3, \ldots, \nu, \ldots$ gleiche Mächtigkeit haben". [2] See also Cantor (1883b, p. 549), where a revised definition is given. [3] Weber only uses the word in a section title, but not in the text. [4] On Peirce's terminology, see Dauben (1982).

Ferreirós, José (2007). *Labyrinth of Thought*. A History of Set Theory and Its Role in Modern Mathematics. Second revised edition. Basel.
Fraenkel, Adolf (1923). *Einleitung in die Mengenlehre*. Zweite erweiterte Auflage. Berlin.
Fraenkel, Adolf (1927). *Zehn Vorlesungen über die Grundlegung der Mengenlehre*. Leipzig.
Hausdorff, Felix (1914). *Grundzüge der Mengenlehre*. Leipzig.
Huntington, Edward V. (1905). "A Set of Postulates for Ordinary Complex Algebra". *Transactions of the American Mathematical Society* 6, pp. 209–229.
Peirce, Charles Sanders (1892). "The law of mind". *The Monist* 2, pp. 533–559.
Peirce, Charles Sanders (1897). "The logic of relatives". *The Monist* 7, pp. 161–217.
Russell, Bertrand (1903). *The Principles of Mathematics*. Vol. I. Cambridge.
Slaught, H.E. (1908). "The fifteenth summer meeting of the American Mathematical Society". *Bulletin of the American Mathematical Society* 15, pp. 57–71.
Weber, Heinrich (1896). *Lehrbuch der Algebra*. Zweiter Band. Braunschweig.
Weber, Heinrich (1897). "Transcendental numbers. An authorized translation by Professor W. W. Beman of Chapter XXV. of Vol. II., of the *Lehrbuch der Algebra*". *Bulletin of the American Mathematical Society* 3, pp. 174–195.
Young, William Henry (1902). "On the Density of Linear Sets of Points". *Proceedings of the London Mathematical Society* 34, pp. 285–290.

covariance

The term *covariance* seems to be due to Fisher (1930, p. 198), who defined it as "[t]he mean product of the deviations from the means".[1] It also appears in Hotelling (1930, p. 848).

Note. See also Miller (n.d.) for notes on the term and also for *analysis of covariance*.

References

David, H.A. (1998). "First (?) Occurrence of Common Terms in Mathematical Statistics – A Second List, with Corrections". *The American Statistician* 52, pp. 36–40.
Fisher, Ronald A. (1930). *The Genetical Theory of Natural Selection*. Oxford.
Hotelling, Harold (1930). "The Consistency and Ultimate Distribution of Optimum Statistics". *Transactions of the American Mathematical Society* 32, pp. 847–859.
Miller, Jeff (n.d.). "Earliest Known Uses of Some of the Words of Mathematics". URL: `https://mathshistory.st-andrews.ac.uk/Miller/mathword/`.

covariant and *contravariant*

Like various other terms from invariant theory, these terms are due to Sylvester. He defined *covariant* as "[a] function which stands in the same relation to the primitive function from which it is derived as any of its linear transforms do to a similarly derived transform of its primitive", and *contravariant* as "[a] function which stands in the same relation to the primitive function from which it is derived as any of its linear transforms do to an inversely derived transform of its primitive" (Sylvester 1853, p. 544). Sylvester had originally introduced the terms in print in a paper published in 1851 (Sylvester 1851, p. 290). The French term appears in Cayley (1851, p. 368): "In making use of Sylvester's new terms, I call a *Covariant* of a given function any function which does not change form as a result of a linear transformation of the

[1] The origin of the term was pointed out by David (1998, p. 37), for example.

variables, and an *Invariant* any function of the coefficients alone with that property".[1] The term was soon in common use; it was adopted, for example, by Aronhold (1858, p. 98) and Hermite (1856, p. 3).

The term *covariant differentiation* goes back to Ricci (1887, p. 17):

> By means of the theorem demonstrated above, we can thus construct successively expressions with $2, 3, \ldots, p$ indices, such that those with p indices are the coefficients of forms covariant with respect to φ^2 and include the derivatives of U up to the order p. Further it will be easily seen that these [expressions] will all be linear relative to the derivatives themselves and that each one will include just one derivative of order p. We shall call them *covariant derivatives of order* p in the manifold which is intrinsically defined by the expression φ^2 of the square of its line element.[2]

The English term *covariantive differentiation* was used by Maschke (1903, p. 457): "For the formation of invariants and covariants, which involve explicitly derivatives higher than the first, it turns out to be of the greatest advantage to use instead of ordinary differentiation another process which might be called *covariantive differentiation*". Maschke refers to Ricci (1892, p. 175) for the term.

The terms *covariant tensor* and *contravariant tensor* are due to Einstein and Grossmann (1913, p. 25); the terms are related there to *covariant* and *contravariant systems* of Ricci and Levi-Civita (1901, p. 131). The terminology spread quickly; an early use of the English terms is in Fokker (1915, p. 90).

For *covariant functor*, see *functor*.

Note. For the history of covariant differentiation, see Dell'Aglio (1996) and Reich (1994, pp. 59–65).

References

Aronhold, Siegfried (1858). "Theorie der homogenen Functionen dritten Grades von drei Veränderlichen". *Journal für die reine und angewandte Mathematik* 55, pp. 97–191.

Cayley, Arthur (1851). "Note sur la théorie des Hyperdéterminants". *Journal für die reine und angewandte Mathematik* 42, pp. 368–371.

Dell'Aglio, Luca (1996). "On the genesis of the concept of covariant differentiation". *Revue d'histoire des mathématiques* 2, pp. 215–264.

Einstein, Albert and Marcel Grossmann (1913). *Entwurf einer verallgemeinerten Relativitätstheorie und einer Theorie der Gravitation*. Leipzig.

Fokker, A.D. (1915). "A Summary of Einstein and Grossmann's Theory of Gravitation". *The London, Edinburgh, and Dublin Philosophical Magazine and Journal of Science* 29, pp. 77–96.

Hermite, Charles (1856). "Sur la théorie des fonctions homogènes à deux indéterminées. Premier mémoire". *Journal für die reine und angewandte Mathematik* 52, pp. 1–17.

Maschke, Heinrich (1903). "A Symbolic Treatment of the Theory of Invariants of Quadratic Differential Quantics of n Variables". *Transactions of the American Mathematical Society* 4, pp. 445–469.

---

[1] "en me servant des nouveaux termes de M. Sylvester, je nomme '*Covariant*' d'une fonction donnée, toute fonction qui ne change pas de forme en faisant subir aux variables des transformations linéaires quelconques, et '*Invariant*' toute fonction des seuls coefficients qui a la propriété mentionnée". Translation from Wußing (1984, p. 169). [2] "Mediante il teorema sopra dimostrato possiamo dunque costruire successivamente delle espressioni con $2, 3, \ldots, p$ indici, per guisa che quelle con p indici siano coefficienti di forme covarianti a φ^2 e contengano le derivate di U fino all'ordine p. Si vede di più facilmente che esse saranno tutte lineari rispetto alle derivate stesse, e che contengono ciascuna una sola derivata di ordine p. Noi le chiameremo *derivate covarianti di ordine* p nella varietà, che è definita in sè dalla espressione φ^2 del quadrato del suo elemento lineare". Translation from Dell'Aglio (1996, p. 254).

Reich, Karin (1994). *Die Entwicklung des Tensorkalküls*. Vom absoluten Differentialkalkül zur Relativitätsgeometrie. Basel.
Ricci, Gregorio (1887). "Sulla derivazione covariante ad una forma quadratica differenziale". *Atti della Reale Accademia dei Lincei. Rendiconti* 3, pp. 15–18.
Ricci, Gregorio (1892). "Résumé de quelques travaux sur les systèmes variables de fonctions associés à une forme différentielle quadratique". *Bulletin des Sciences Mathématiques* 16, pp. 167–189.
Ricci, Gregorio and Tullio Levi-Civita (1901). "Méthodes de calcul différentiel absolu et leurs applications". *Mathematische Annalen* 54, pp. 125–201.
Sylvester, James Joseph (1851). "On the general theory of associated algebraical forms". *The Cambridge and Dublin Mathematical Journal* 6, pp. 289–293.
Sylvester, James Joseph (1853). "On a Theory of the Syzygetic relations of two rational integral functions, comprising an application to the Theory of Sturm's Functions, and that of the greatest Algebraical Common Measure". *Philosophical Transactions of the Royal Society of London* 143, pp. 407–548.
Wußing, Hans (1984). *The Genesis of the Abstract Group Concept*. A Contribution to the History of the Origin of Abstract Group Theory. Tr. Abe Shenitzer. Cambridge, Mass.

cover

In 1895, Émile Borel stated what was soon afterwards often referred to as the *Heine-Borel Theorem*:[1]

> Here is the theorem: *If one has on a straight line an infinite number of partial intervals, such that any point on the line is interior to at least one of the intervals, one can effectively determine a* LIMITED NUMBER *of intervals chosen among the given intervals and having the same property* (*any point on the line is interior to at least one of them*) (Borel 1895, p. 51).[2]

The theorem was much referred to in the following years. In 1904, Lebesgue presented a proof of the theorem for uncountable covers, which he stated in the following terms: "If one has a family of intervals Δ such that any point of an interval (a, b), including a and b, is *interior* to at least one of Δ, there exists a family formed of a *finite* number of intervals Δ and that has the same property [any point of (a, b) is interior to one of them]" (Lebesgue 1904, pp. 104f.).[3]

In his proof, Lebesgue used French *être couvert* and thus introduced the term *cover* (as a verb, not as a noun):[4]

> Let (a, b) be one of the intervals Δ containing a, the property to demonstrate is evident for the interval (a, x), if x is contained between α and β; I want to say that this interval may be

---

[1] Heine (1872, p. 188) made use of the theorem, without proving it. [2] "Voici ce théorème: *Si l'on a sur une droite une infinité d'intervalles partiels, tels que tout point de la droite soit intérieur à l'un au moins des intervalles, on peut déterminer effectivement un* NOMBRE LIMITÉ *d'intervalles choisis parmi les intervalles donnés et ayant la même propriété* (*tout point de la droite est intérieur à au moins l'un d'eux*)". Translation from Andre, Engdahl, and Parker (2013), who also note that *straight line* means 'closed and bounded interval', *partial interval* 'open interval', and *limited number* 'finite number'. [3] "*Si l'on a une famille d'intervalles Δ tels que tout point d'un intervalle (a, b), y compris a et b, soit* intérieur *à l'un au moins des Δ, il existe une famille formée d'un nombre fini des intervalles Δ et qui jouit de la même propriété* [*tout point de (a, b) est intérieur à l'un d'eux*]". Translation from Andre, Engdahl, and Parker (2013). Note that (a, b) refers to a closed and bounded interval. [4] In his proof of the Heine-Borel Theorem, Schoenflies (1900, p. 52) informally used the German verb *bedecken*, which also means 'to cover'.

> covered with the help of a finite number of intervals Δ, which I express in saying that the point x is reached" (Lebesgue 1904, p. 105).[1]

Soon afterwards, Veblen and Lennes (1907, p. 33) used the verb *to cover* in the same sense. The noun is of later origin. It came up in the 1920s and appears to be due to Urysohn, who used French *couverture* (Urysohn 1923, p. 14) as well as German *Überdeckung* (Urysohn 1924, p. 276).

See also *compact*.

Note. The history of the Heine-Borel Theorem is discussed in detail by Maurey and Tacchi (2005) and Andre, Engdahl, and Parker (2013).

References

Andre, Nicole R., Susannah M. Engdahl, and Adam E. Parker (2013). "An Analysis of the First Proofs of the Heine-Borel Theorem". *Convergence* (August 2013). URL: `https://old.maa.org/press/periodicals/convergence/an-analysis-of-the-first-proofs-of-the-heine-borel-theorem-introduction`.

Borel, Émile (1895). "Sur quelques points de la théorie des fonctions". *Annales scientifiques de l'École Normale Supérieure* 12, pp. 9–55.

Heine, Eduard (1872). "Die Elemente der Functionenlehre". *Journal für die reine und angewandte Mathematik* 74, pp. 172–188.

Lebesgue, Henri (1904). *Leçons sur l'intégration et la recherche des fonctions primitives*. Paris.

Maurey, Bernard and Jean-Pierre Tacchi (2005). "La genèse du théorème de recouvrement de Borel". *Revue d'histoire des mathématiques* 11, pp. 163–204.

Schoenflies, Arthur (1900). "Die Entwickelung der Lehre von den Punktmannigfaltigkeiten". *Jahresbericht der Deutschen Mathematiker-Vereinigung* 8, pp. 1–250.

Urysohn, Pavel (1923). "Sur la métrisation des espaces topologiques". *Bulletin international de l'Académie polonaise des sciences et des lettres. Classe des sciences mathématiques et naturelles* (1923), pp. 13–16.

Urysohn, Pavel (1924). "Über die Metrisation der kompakten topologischen Räume". *Mathematische Annalen* 92, pp. 275–293.

Veblen, Oswald and Nels Johann Lennes (1907). *Introduction to Infinitesimal Analysis*. Functions of a real variable. New York.

critical graph

In graph theory, the term *critical* was first suggested by Dirac (1951, p. 347), who defined:[2]

> If a is a node of the graph Γ and $K(\Gamma - a) = K(\Gamma) - 1$, then a will be called a *critical* node of Γ. It is easy to see that *if a is critical, the degree of a is at least* $K(\Gamma) - 1$. (The degree of a is the number of edges incident with a.) If every node of a graph is critical, the graph will be called a *critical graph*.

The terminology was quickly adopted and appears in Ore's *Theory of Graphs* (Ore 1962, p. 229), for example.

---

[1] "Soit (α, β) l'un des intervalles Δ contenant a, la propriété à démontrer est évidente pour l'intervalle (a, x), si x est compris entre α et β; je veux dire que cet intervalle peut être couvert à l'aide d'un nombre fini d'intervalles Δ, ce que j'exprime en disant que le point x est atteint". Translation from Andre, Engdahl, and Parker (2013).

[2] In the definition, $K(\Gamma)$ denotes the chromatic number of the graph Γ; $K(\Gamma - a)$ is "the graph obtained from Γ by deleting a and all edges incident in a, leaving all other nodes and edges unchanged" (Dirac 1951, p. 347).

References

Dirac, Gabriel Andrew (1951). "Note on the colouring of graphs". *Mathematische Zeitschrift* 54, pp. 347–353.
Ore, Øystein (1962). *Theory of Graphs*. Providence, R.I.

critical point

In the 19th century, *critical point* began to be used in various mathematical senses. An early definition was given by Price (1848, p. 163): "Thus far [...] we have considered what geometrical properties belong to the first and second differential coefficients, when they have determinate values; but suppose at any point in a curve, say $(x = a, y = b)$, which we shall call a *critical point*, $\frac{dy}{dx}$ assumes the indeterminate form $\frac{0}{0}$, a question arises, what geometrical meaning belongs to it".[1]

The word also appeared in complex analysis. Forsyth (1893, p. 15), after defining *branch point*, explained: "Also [called] *critical point*, which, however, is sometimes used to include all special points of a function". This use for *branch points* is found in Briot and Bouquet (1875, p. 12).

In the 20th century, the term began to be defined in terms of vanishing derivatives, for example by Morse (1925, p. 347): "A point of R at which all of the first partial derivatives of f vanish will be called a *critical point* of f". The term was not infrequently used in the same meaning as *stationary point* (Morrill 1956, p. 78); however, the meaning of the two terms began to be distinguished later on, for example by Lynch (1964, p. 237): "The *critical points* of a function are those points of its domain (a) at which the derivative is zero (stationary points), or (b) at which the derivative does not exist, or (c) that are endpoints of intervals of the domain".

See also *stationary point*.

References

Briot, Charles and Jean Claude Bouquet (1875). *Théorie des fonctions elliptiques*. Deuxième édition. Paris.
Forsyth, Andrew Russell (1893). *Theory of Functions of a Complex Variable*. Cambridge.
Lynch, Ransom U. (1964). *A First Course in Calculus*. Including Analytic Geometry. Boston.
Morrill, William K. (1956). *Calculus*. Princeton.
Morse, Marston (1925). "Relations Between the Critical Points of a Real Function of n Independent Variables". *Transactions of the American Mathematical Society* 27, pp. 345–396.
Peacock, George (1834). "Report on the Recent Progress and Present State of certain Branches of Analysis". *Report of the British Association for the Advancement of Science* (1833), pp. 185–352.
Price, Bartholomew (1848). *A Treatise on Infinitesimal Calculus*. London.

cross-cap

As described by Kerékjártó (1923, p. 151), "[a] *cross-cap* arises in the following way: take a half-sphere, cut it open along an arc acb and identify, on the newly created boundary, the arc ac' with bc and ac with bc, such that a coincides with b and c with c'" (see fig. 32).[2] This construction also justifies the name.

Kerékjártó used the German name *Kreuzhaube*, of which the English term is a literal translation. The term is not due to him, however; the earliest use in print is

---

[1] Peacock (1834, p. 247) earlier used the term as an alternative to *singular point*. [2] "Eine *Kreuzhaube* entsteht folgendermaßen: man nimmt eine Halbkugel, schneidet längs eines Kreisbogens acb auf [...] und identifiziert auf dem entstehenden Rand den Bogen ac' mit bc und den Bogen ac mit bc', so daß a mit b und c mit c' zusammenfällt".

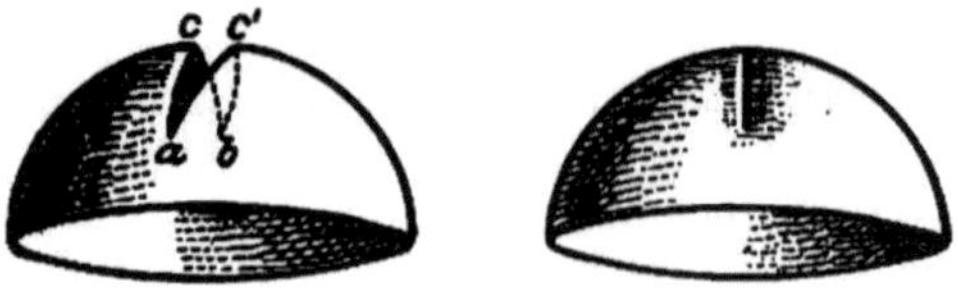

Figure 32: A cross-cap. From Kerékjártó (1923, p. 151).

Dehn and Heegaard (1907, p. 198). The concept is slightly older; earlier illustrations can be found in Dyck (1888, fig. 10) and Boy (1903, p. 172).[1]

The term was translated into English as *one-sided cap* by Brahana and Coble (1926, p. 18); *cross-cap* came up in the 1930s and was used, for example, by Franklin (1935, p. 182).

See also *handle*.

Note. Brief comments on the history of the concept can be found in Volkert (2002, pp. 56f., 63).

References

Boy, Werner (1903). "Über die Curvatura integra und die Topologie geschlossener Flächen". *Mathematische Annalen* 57, pp. 151–184.
Brahana, H.R. and A.B. Coble (1926). "Maps of Twelve Countries with Five Sides with a Group of Order 120 Containing an Ikosahedral Subgroup". *American Journal of Mathematics* 48, pp. 1–20.
Dehn, Max and Poul Heegaard (1907). "Analysis situs". *Encyklopädie der mathematischen Wissenschaften mit Einschluss ihrer Anwendungen* III AB3, pp. 153–220.
Dyck, Walther (1888). "Beiträge zur Analysis situs. I. Aufsatz. Ein- und zweidimensionale Mannigfaltigkeiten". *Mathematische Annalen* 32, pp. 457–512.
Franklin, Philip (1935). "On Minkowski's Definition of Length and Area". *Journal of Mathematics and Physics* 14, pp. 179–185.
Hilbert, David and Stefan Cohn-Vossen (1932). *Anschauliche Geometrie*. Berlin and Heidelberg.
Kerékjártó, Béla (1923). *Vorlesungen über Topologie*. I. Flächentopologie. Berlin.
Volkert, Klaus (2002). *Das Homöomorphismusproblem insbesondere der 3-Mannigfaltigkeiten, in der Topologie 1892-1935*. Paris.

cross-ratio

Traces of this concept can already be found in antiquity. In his *Collections* (book 7, prop. 129), Pappus proved the following theorem (in modern terminology): When four lines all issue from the same point, the cross-ratio is the same for every transversal intersecting these lines.[2]

The terminology, however, is a 19th-century invention. The concept was called *Doppelschnittsverhältniss* 'ratio of bisection' by Möbius (1827, p. 244). Steiner (1832, p. 7) shortened this to *Doppelverhältniss* 'double ratio', a custom which Möbius (1853, p. 15) himself and other authors quickly followed. The term is still standard in German.[3]

Chasles (1837, p. 34) introduced the alternative term *rapport anharmonique*: "This ratio being called *harmonic* in the special case where it equals unity, we will

[1] See also Hilbert and Cohn-Vossen (1932, pp. 279) for a later but frequently cited illustration.
[2] This formulation follows Chasles (1837, p. 302). [3] In the now usual spelling *Doppelverhältnis*.

call it, in the general case, *anharmonic ratio* or *function*".[1] von Staudt (1856, p. 15) introduced German *Wurf* 'throw', a term he chose because he generally avoided all metric concepts (such as ratio). Finally, the term *cross-ratio* was introduced by Clifford (1878, p. 42): "The ratio *ab.cd* : *ac.bd* is called a *cross-ratio* of the four points *abcd*" (see fig. 33).[2]

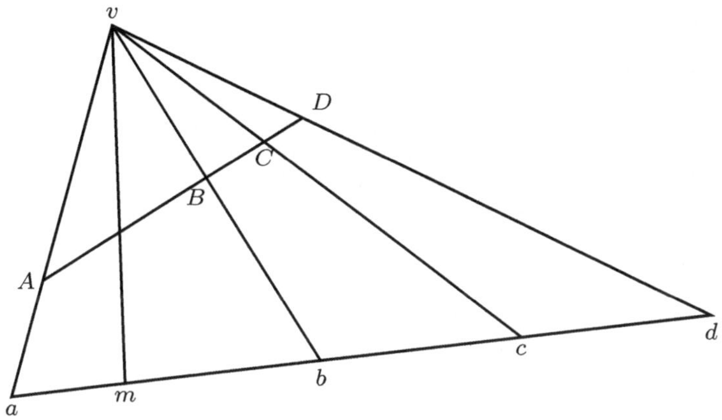

Figure 33: The cross-ratio. From Clifford (1878, p. 42).

Chasles' term was originally very successful, as Casey (1885, p. 46) witnesses: "Chasles' nomenclatrue [sic], although perhaps the least appropriate, is almost universally adopted". By 1911, Milne (1911, p. 2) was able to state that *cross-ratio* "is now generally adopted".

In sum, "Chasles was the first to use extensively the concept of cross ratio as a projective invariant.[3] [...] The concept of cross ratio was already known; it is present in some theorems due to Pappus and was utilized in geometry before Chasles (e. g. by Poncelet). However, Möbius, Chasles and Steiner were the first to define the cross ratio and to consider it as one of the main objects of projective geometry [...] The history of the cross ratio begins with these three mathematicians. Before them, one can only speak of a protohistory" (Bussotti 2024, p. 34).

Note. The history of the concept is discussed by Coolidge (1940, book I, ch. 5) and Gray (2011, pp. 70–73), for example. For some modern generalizations, see the brief overview in Labourie (2008).

References

Bussotti, Paolo (2024). *Chasles and the Projective Geometry.* The Birth of a Global Foundational Programme for Mathematics, Mechanics and Philosophy. Cham.

[1] "Ce rapport étant dit *harmonique* dans la cas particulier où il est égal à l'unité, nous l'appellerons dans le cas général *rapport* ou *fonction anharmonique*". [2] The name is explained by Labourie (2008, p. 1234): "In projective coordinates, the cross ratio is computed as a ratio involving four terms with some 'crossing symmetries', hence its name". [3] That is, it is preserved by the projective transformations of a projective line.

Casey, John (1885). *A Treatise on the Analytical Geometry of the Point, Line, Circle, and Conic Sections*. Dublin.
Chasles, Michel (1837). *Aperçu historique sur l'origine et le développement des méthodes en géométrie*. Bruxelles.
Clifford, William Kingdon (1878). *Elements of Dynamic*. Part I. Kinematic. London.
Coolidge, Julian Lowell (1940). *A History of Geometrical Methods*. Oxford.
Gray, Jeremy (2011). *Worlds Out of Nothing*. A Course in the History of Geometry in the 19th Century. London et al.
Labourie, François (2008). "What is a cross ratio?" *Notices of the American Mathematical Society* 55, pp. 1234–1235.
Milne, John J. (1911). *An Elementary Treatise on Cross-Ratio Geometry*. Cambridge.
Möbius, August Ferdinand (1827). *Der barycentrische Calcul*. Leipzig.
Möbius, August Ferdinand (1853). "Ueber eine neue Verwandtschaft zwischen ebenen Figuren". *Berichte über die Verhandlungen der Königlich Sächsischen Gesellschaft der Wissenschaften zu Leipzig (Mathematisch-Physische Klasse)* 5, pp. 14–24.
von Staudt, Karl Georg Christian (1856). *Beiträge zur Geometrie der Lage*. Erstes Heft. Nürnberg.
Steiner, Jacob (1832). *Systematische Entwickelung der Abhängigkeit geometrischer Gestalten von einander*. Erster Theil. Berlin.

crossed product and *crossed module*

The notion of crossed product and the term go back to Emmy Noether. As Hasse (1932, p. 180) reports, she used them in a lecture in 1929 in Göttingen as German *verschränktes Produkt*, where *verschränkt* means 'crossed, folded'. The translation *crossed product* first appears in Hasse's paper.[1] He gave the following definition: "A *crossed product* of Z (by G) is defined as an algebra A of the following type: A possesses a Z-basis u_s, uniquely corresponding to the n elements S of G, for which the relations $zu_S = u_S z^S$, $u_S u_T = u_{ST} a_{S,T}$, where $a_{S,T} \neq 0$ in Z, hold" (Hasse 1932, p. 181).

Noether wrote to Hasse after having read his paper and asked: "Are the 'crossed products' your English invention? This word is good" (Lemmermeyer and Roquette 2006, p. 109).[2] Roquette (2005, p. 21) comments: "We do not know whether Hasse himself invented this terminology, or perhaps it was H. T. Engstrom, the American mathematician who helped Hasse to translate his manuscript from German into English. In any case, in the English language the terminology 'crossed product' has been in use since then".

Soon afterwards, the word *crossed* is applied to other structures. The term *crossed module*, for example, was introduced by Whitehead (1949, p. 453).[3]

Note. For the introduction of the crossed product, see Roquette (2005, pp. 19–24) and Frei (2003, p. 164).[4]

References

Brauer, Richard, Helmut Hasse, and Emmy Noether (1932). "Beweis eines Hauptsatzes in der Theorie der Algebren". *Journal für die reine und angewandte Mathematik* 167, pp. 399–404.
Frei, Günther (2003). "Zur Geschichte der Arithmetik der Algebren (1843-1932)". *Elemente der Mathematik* 58, pp. 156–168.

[1] The German term was also mentioned in Brauer, Hasse, and Noether (1932, p. 399). [2] "Sind die ‚crossed products' Ihre englische Erfindung? Das Wort ist gut". Translation from Roquette (2005, p. 21). [3] The notion is earlier found in Whitehead (1946). [4] See also Frei and Roquette (2008).

Frei, Günther and Peter Roquette, eds. (2008). *Emil Artin und Helmut Hasse*. Die Korrespondenz 1923–1934. Unter Mitwirkung von Franz Lemmermeyer. Göttingen.
Hasse, Helmut (1932). "Theory of cyclic algebras over an algebraic number field". *Transactions of the American Mathematical Society* 34, pp. 171–214.
Lemmermeyer, Franz and Peter Roquette, eds. (2006). *Helmut Hasse und Emmy Noether*. Die Korrespondenz 1925–1935. Göttingen.
Roquette, Peter (2005). *The Brauer-Hasse-Noether Theorem in Historical Perspective*. Berlin et al.
Whitehead, J.H.C. (1946). "Note on a Previous Paper Entitled 'On Adding Relations to Homotopy Groups'". *Annals of Mathematics* 47, pp. 806–810.
Whitehead, J.H.C. (1949). "Combinatorial homotopy. II". *Bulletin of the American Mathematical Society* 55, pp. 453–496.

crossing number

This graph-theoretical term seems to be due to Harary and Hill (1963, p. 334):[1]

> For any graph G, we say that the *crossing number* $c(G)$ is the minimum number of crossings with which it is possible to draw G in the plane. We note that the edges of G need not be straight line segments, and also that the result is the same whether G is drawn in the plane or on the surface of a sphere. Another invariant of G is the *rectilinear crossing number*, $\bar{c}(G)$, which is the minimum number of crossings when G is drawn in the plane in such a way that every edge is a straight line segment.

The concept became known in mathematics through a problem posed by Turán in the 1940s, the so-called "brick factory problem".[2] Turán (1977, p. 8) later recalled:

> We worked near Budapest, in a brick factory. There were some kilns where the bricks were made and some open storage yards where the bricks were stored. All the kilns were connected by rail with all the storage yards. The bricks were carried on small wheeled trucks to the storage yeards. All we had to do was to put the bricks on the trucks at the kilns, push the trucks to the storage yards, and unload them there. We had a reasonable piece rate for the trucks, and the work itself was not difficult; the trouble was only at the crossings. The trucks generally jumped the rails there, and the bricks fell out of them; in short this caused a lot of trouble and loss of time [...] but *nolens-volens* the idea occurred to me that this loss of time could have been minimized if the number of crossings of the rails had been minimized. But what is the minimum number of crossings?

Note. For the history of the concept, see Beineke and Wilson (2010) and Schaefer (2022).

References

Beineke, Lowell W. and Robin J. Wilson (2010). "The Early History of the Brick Factory Problem". *The Mathematical Intelligencer* 32, pp. 41–48.
Harary, Frank and Anthony Hill (1963). "On the Number of Crossings in a Complete Graph". *Proceedings of the Edinburgh Mathematical Society* 13, pp. 333–338.
Schaefer, Marcus (2022). "The Graph Crossing Number and its Variants: A Survey". *The Electronic Journal of Combinatorics (Dynamic Surveys)* DS21, pp. 1–154.
Turán, Pál (1977). "A note of welcome". *Journal of Graph Theory* 1, pp. 7–9.

---

[1] On this paper, Schaefer (2022, p. 44) remarks that it is presumably "the first paper defining the crossing number for arbitrary graphs". [2] Schaefer (2022, pp. 7–9) points out that the problem is found before Turán in sociological contexts.

cryptography **and** ***cryptology***

The terms *cryptography* and *cryptology* both contain Greek *κρυπτός* 'hidden, secret', combined with *γραφία* (from *γράφειν* 'to write') and *λογία* (from the highly polysemous *λόγος*).[1] The terminology features prominently in the book *Cryptomenytices et cryptographiae libri IX* (Selenus 1624), which was frequently referred to in the 17th century.[2] Selenus (1624, p. 6) defines:

> Cryptography is the art of revealing safely and without risk, by means of hidden writing, and without the knowledge of any third person, some fact, to some person to whom we wish that fact to be known.[3]

The word *cryptography* is here defined in the narrow sense; Selenus (1624, pp. 7, 10) then also decides to use it in a broader sense, including any means of secret communication, not just writing. He also discusses various related terms and indicates who used them before:

> As regards the Name, this appears in a variety of forms in different authors; as, Cryptography and Cryptology, Steganology and Steganography, and Polygraphy, in Trithemius[4]; the Art of writing by Notes and the Art of Cipher-writing, in Cardano[5], Scaliger (Exerc., 327)[6], Porta[7] and De Vigenère[8]; Cryptomenytics, in the writing of the brief notes to Porta[9]; and, finally, Scotography, in Abram Colorno[10], the Jew of Mantua. All these terms are commonly used synonymously; but examination shows that they differ not a little from one another. For some are general, some are specific, and some combine both characteristics.

The most important references are Trithemius (1606)[11] and Trithemius (1518). It is in the latter book, at the beginning of book 4, that *cryptologia* is first found, without further explanation.

The word *cryptographia* does not seem to have been used by Trithemius. However, French *cryptographie* does appear on several occasions in Trithemius (1561) – the book is a translation of the *Polygraphia*, but contains various redactions and additions by its translator Gabriel de Collange. The word is found in a section written by de Collange – see, e. g., Trithemius (1561, fol. 282^{v}), where the term is explained by "occulte escriture" (secret writing). de Collange's text was known to Selenus, who referred to it on several occasions (see, e. g., Selenus 1624, p. 5.).

---

[1] As will be seen in the following, when *cryptology* was originally coined, the relevant meaning of *λόγος* was 'speech' or 'verbal expression'. [2] The name of the author is a pseudonym: *Gustavus* is an anagram of *Augustus* (with *u* = *v* in Latin), and *Selenus* refers to the lunar goddess Selene, which is a pun on *Lüneburg* (lunar castle), the place where the book was published (Kahn 1967, p. 154). The author actually is August the Younger, Duke of Brunswick-Luneburg. [3] "Cryptographia est Ars, aliquid alicui, cuirem exploratum volumus, clàm omnibus aliis, tutò & sine periculo manifestandi, per occultam Scriptionem". Translation from Selenus (c1900, p. 42). [4] This refers to Trithemius' books *Steganographia* and *Polygraphia*, which are discussed below. As for the words, Trithemius coined *steganographia* from *στεγανός* 'covered', and used *polygraphia* "because of the multiplicity of ways of writing that it included (Kahn 1967, p. 133). [5] This refers to Cardano (1550, p. 319), who uses *ars notarum occultandi*. [6] Scaliger (1557, p. 445) has *ars zipherarum* but denies its status of an art. [7] della Porta (1563, p. 12) has *furtivarum notarum ars* (*furtivus* 'secret, hidden'). [8] de Vigenère (1586, fol. 10^{v}): "nostre subiect des occultes manières d'escrire, communément appellees Chiffres". [9] See the explanations below. [10] Colorni (1593) used *Scotographia* (Greek *σκότος* 'darkness') as the title of his book. [11] The book already circulated in manuscript form in the early 16th century.

Another major reference by Selenus is to della Porta (1593). More precisely, in the above passage he refers to an appendix to this book, written by an anonymous scholar. On the second page, Selenus' *cryptomenytice* is used in the Greek form *κρυπτομηνυτικῆς* (della Porta 1593, Animadversiones, p. 2); the second part is based on the verb *μηνύειν* 'to disclose'. It is also mentioned there that *cryptologia* was used by Trithemius – an early reference to its likely originator.

Selenus goes on to critically comment on all of these terms.[1] For example, *art of writing by notes*, already used by Suetonius in his description of the Caesar cipher[2], "is far too specific to embrace all branches of the art", while *polygraphy* is too general "and embraces a broader field than is covered by each and every method of secret writing" even though Trithemius "restricts the use of the word to hidden writing in general". Likewise, Selenus regards *steganography*, *steganology*, and *scotography* as too general. The word *cryptology*, on the other hand, "would seem to answer our purpose. But if we take the word in the general sense, of the theory of all that is hidden, it is too broad. If, on the other hand, we restrict its application to the voice and to writing, it is too narrow to include every branch of the art. But if even Cryptology is too narrow a term, most certainly Cryptography, which is included under Cryptology, can denote nothing more than a subdivision of the art". He concludes:

> There remains, therefore, the word Cryptomenytics, which embraces every conceivable method of secret communication, whether it be by the voice, by writing, or by signs. Be, then, Cryptomenytics the most general term, embracing, as subdivisions in nearest line, Cryptology, which is of the voice, and Cryptography, which is of writing, and a third division which has to do with signs, and follows in order the other two divisions. This we may call Synthemology [*synthemologia*], or Semaeology [*semaeologia*].[3]

A summary of the terminology used by Selenus was given shortly afterwards by Wilkins (1641, p. 14):

> The art of secret information in the generall, as it includes all significatory signes, may be stiled *Cryptomenysis*, or private Intimations. The particular ways of discoursing, were before intimated to bee threefold. 1. By Speaking. 2. By Writing. 3. By signes or gestures. According to which variety, there also different ways of Secrecy. 1. Cryptologia. 2. Cryptographia. 3. Semaeologia.

The only modification here is that Wilkins used *cryptomenysis* (*μήνυσις* 'laying of information', also from *μηνύειν*) instead of *cryptomenytice*. Wilkins does not use the English terms *cryptology* and *cryptography*. These do appear in print shortly afterwards, however: Howell (1645, Section III, p. 119) has "*Cryptology*, or Epistolizing in a Clandestin way". Howell's book is a collection of letters, and in the second edition (Howell 1650, Section III, p. 99), the letter in which the cited passage occurs is dated March 15, 1623. English *cryptography* was then used by Browne (1646, p. 77). Howell and Browne are cited by the OED as first uses for the respective terms.

---

[1] The following quotations are all from Selenus (1624, pp. 6f.); the translations are from Selenus (c1900, pp. 42f.). [2] Suetonius (1979, book I, 56). For this criticism, Selenus refers to Hugo (1617, ch. 27). [3] For a person involved in the subject, Selenus (1624, p. 438) also coins the terms *cryptomenyticus* and *cryptologus*.

New definitions of the terminology as well as several additions came up in the 20th century. In particular, Friedman (1918, p. 9) redefined: “Cryptology is that branch of knowledge which deals with the origin, development, and methods of all forms of secret communication. Cryptography is that branch of cryptology which deals with the methods of secret writing”. Shortly afterwards, he introduced some new terms: “Cryptanalytics is the name recently applied to the science which embraces all the principles, methods, and means employed in the ANALYSIS of cryptograms[1] [...] Cryptanalysis is the name applied to the steps performed in the application of the principles of cryptanalytics to cryptograms” (Friedman 1923, p. 2). Even *steganography* was revived and given a new, specialized meaning: “The term ‘steganography’ is now obsolete but it would have been a useful designation for that form of secret writing in which not only is the secret meaning kept unintelligible without the key, but the very fact that it exists is hidden” (McCracken 1948, p. 216). This suggestion was taken up, in particular, by Kahn (1967, p. 975). Kahn (1967, p. xvi) also explicitly used *cryptology* in a now commonly found way: “Cryptology is the science that embraces cryptography and cryptanalysis”.

Note. There are numerous works on the history of cryptography and cryptology. In the preparation of this article, Kahn (1967), Strasser (2007), and Ernst (1996) have been very useful.

References

Browne, Thomas (1646). *Pseudodoxia epidemica*. London.

Cardano, Gerolamo (1550). *De subtilitate libri XXI*. Nürnberg.

Colorni, Abramo (1593). *Scotographia, ouero scienza di scrivere oscuro*. Prag.

Ernst, Thomas (1996). “Schwarzweiße Magie. Der Schlüssel zum dritten Buch der *Steganographia* des Trithemius”. *Daphnis* 25, pp. 1–205.

Friedman, William F. (1918). *An Introduction to Methods for the Solution of Ciphers*. Geneva, Ill.

Friedman, William F. (1923). *Elements of Cryptanalysis*. Washington, D.C.

Howell, James (1645). *Epistolae Ho-Elianae*. Familiar Letters Domestic and Forren. London.

Howell, James (1650). *Epistolae Ho-Elianae*. Familiar Letters Domestic and Forren. The second edition. London.

Hugo, Herman (1617). *De prima scribendi origine*. Antwerpen.

Kahn, David (1967). *The Codebreakers*. New York.

McCracken, George E. (1948). “Athanasius Kircher’s Universal Polygraphy”. *Isis* 39, pp. 215–228.

della Porta, Giambattista (1563). *De furtivis literarum notis vulgo de ziferis*. Napoli.

della Porta, Giambattista (1593). *De occultis literarum notis*. Editio secunda. Montbeliard.

Scaliger, Julius Caesar (1557). *Exotericarum exercitationum liber quintus decimus, de subtilitate*. Paris.

Selenus, Gustavus (1624). *Cryptomenytices et cryptographiae libri IX*. Lüneburg.

Selenus, Gustavus (c1900). *The Cryptomenytics and Cryptography*. Trans. by John William Henry Walden. [unpublished].

Strasser, Gerhard F. (2007). “The Rise of Cryptology in the European Renaissance”. In: *The History of Information Security*. A Comprehensive Handbook. Ed. by Karl de Leeuw and Jan Bergstra. Amsterdam, pp. 277–325.

Suetonius (1979). *Lives of the Caesars*. With an English Translation by J.C. Rolfe. Volume I. Cambridge, Mass.

Trithemius, Johannes (1518). *Polygraphiae libri VI*. Basel.

Trithemius, Johannes (1561). *Polygraphie et Universelle escriture Cabalistique*. Trans. by Gabriel de Collange. Paris.

[1] The word refers to encrypted text.

Trithemius, Johannes (1606). *Steganographia*. Frankfurt.
de Vigenère, Blaise (1586). *Traicté des chiffres ou secretes manieres d'escrire*. Paris.
Wilkins, John (1641). *Mercury, or the Secret and Swift Messenger*. London.

cube

Greek *κύβος* means 'cube' (esp. 'cubical die') and 'anything of cubic shape'. The etymology of *κύβος* is unknown, but it is assumed to be a loanword (Chantraine 1968-80, p. 595). In his *Elements* (book XI, def. 25), Euclid gave the following definition: "A cube is a solid figure contained by six equal squares".[1] The term was used before Euclid (Mugler 1958-59, p. 259). Euclid also used the term in connection with numbers: "And a cube [number] is equal multiplied by equal and again by equal, or a number which is contained by three equal numbers" (book VII, def. 19).[2] Pappus of Alexandria used the synonym *ἑξάεδρον* for 'cube' in the geometrical sense (Pappus of Alexandria 1876, p. 352); the term is also found in the Pseudo-Heronian *Definitions* (Heron of Alexandria 1912, p. 64).[3]

In Latin, *cubus* (also *cybus*) was used to translate *κύβος*. The first definition of the term in this language is due, according to Ayuso García (2008, p. 1306), to Vitruvius: "a cube is a body with all its sides squared and their surfaces equal".[4]

The 2nd-century writer Aulus Gellius discussed the word at some length:

> the *κύβος* is a figure which is square on all its sides, "like the dice," says Marcus Varro, "with which we play on a gaming-board, for which reason the dice themselves are called *κύβοι*." Similarly in numbers too the term *κύβος* is used [...] Pythagoras declared that the cube of the number three controls the course of the moon, since the moon passes through its orbit in twenty-seven days, and the ternio, or "triad," which the Greeks call *τριάς*, when cubed makes twenty-seven (Gellius 1946, p. 92).[5]

The French term *cube* dates from the 14th century (DMF); it was used by Oresme: "the second regular body has six faces, each one a square quadrangle, and it is shaped like a die and is called cube or hexahedron" (Oresme 1968, p. 640).[6] The German term *Würfel* is first found (DWB) in this sense in Schreiber (1518, fol. G iij^{r}). The English term comes up in the 16th century.[7] Recorde (1551, fol. A iii^{r}) first calls it a *dye* but also uses *cube* (Recorde 1556, p. 156); the term was also used in Billingsley's translation of the *Elements* (Euclid 1570, fol. 318^{r}).

Note. For a concise terminological overview, see Tropfke (1924, pp. 16–18). For the Greek term, see Mugler (1958-59, pp. 258f.), for Greek *ἑξάεδρον*, see Mugler

[1] Translation from Euclid (1956b, p. 262). [2] Translation from Euclid (1956a, p. 278). [3] Latin *hexahedrum* is extremely rare before the late Middle Ages and only known to have been used by the 4th-century philosopher Calcidius (TLL, Ayuso García 2008, p. 1307); see Calcidius (2016, p. 206). [4] "cybus autem est corpus ex lateribus aequali latitudine planitiarum perquadratus". Translation from Vitruvius (1931, p. 253). [5] "*κύβος* enim est figura ex omni latere quadrata, 'quales sunt,' inquit M. Varro, 'tesserae quibus in alveolo luditur, ex quo ipsae quoque appellatae *κύβοι*.' In numeris etiam similiter *κύβος* dicitur [...] Huius numeri cubum Pythagoras vim habere lunaris circuli dixit, quod et luna orbem suum lustret septem et viginti diebus et numerus ternio, qui *τριὰς* Graece dicitur, tantundem efficiat in cubo". Translation from Gellius (1946, p. 93). [6] "Le secont [corps regulier] a .vi. faces dont chascune est quadrangle quarré et est comme un dé et est appellé cubus ou exacedron". Translation from Oresme (1968, p. 640). Note that Oresme here uses the word in its Latin form, but he does use *cube* otherwise (e. g., Oresme 1968, p. 642). [7] John Trevisa (14th century) left the term in its Latin form *cubus*. The adjective *cubic* (number) is earlier found in *The Art of Nombryng* ([anon.] 1922, p. 47).

(1958-59, p. 180). For Latin *cubus*, see Ayuso García (2008, pp. 1305–1310), and for the Arabic term, see Rashed (2017, pp. 734f.).

References

[anon.] (1922). "The Art of Nombryng". In: *The Earliest Arithmetics in English*. Ed. by Robert Steele. London, pp. 33–51.
Ayuso García, Manuel (2008). "La terminología latina de la geometría en Marciano Capela". PhD thesis. Madrid: Universidad Nacional de Educación a Distancia.
Calcidius (2016). *On Plato's* Timaeus. Edited and translated by John Magee. Cambridge, Mass.
Chantraine, Pierre (1968-80). *Dictionnaire étymologique de la langue grecque*. Histoire des mots. Paris.
Euclid (1570). *The Elements of Geometrie of the most auncient Philosopher Euclide of Megara*. Faithfully (now first) translated into the Englishe toung, by H. Billingsley. London.
Euclid (1956a). *The Thirteen Books of Euclid's Elements*. Volume II. Books III-IX. Ed. by Thomas L. Heath. New York.
Euclid (1956b). *The Thirteen Books of Euclid's Elements*. Volume III. Books X-XIII and Appendix. Ed. by Thomas L. Heath. New York.
Gellius, Aulus (1946). *Attic Nights*. Volume I. Translated by J.C. Rolfe. Cambridge, Mass.
Heron of Alexandria (1912). *Heronis Alexandrini opera quae supersunt omnia*. Volumen IV. Heronis definitiones cum variis collectionibus. Heronis quae feruntur geometrica. Ed. by J.L. Heiberg. Leipzig.
Mugler, Charles (1958-59). *Dictionnaire historique de la terminologie géométrique des Grecs*. Paris.
Oresme, Nicole (1968). *Le livre du ciel et du monde*. Edited by Albert D. Menut and Alexander J. Denomy, translated with an introduction by Albert D. Menut. Madison.
Pappus of Alexandria (1876). *Pappi Alexandrini Collectionis*. Volumen I. Ed. by Friedrich Hultsch. Berlin.
Rashed, Roshdi, ed. (2017). *Lexique historique de la langue scientifique arabe*. Hildesheim.
Recorde, Robert (1551). *The patheway to knowledg*. London.
Recorde, Robert (1556). *The Castle of Knowledge*. London.
Schreiber, Heinrich (1518). *Ayn new kunstlich Buech*. Nürnberg.
Tropfke, Johannes (1924). *Geschichte der Elementar-Mathematik*. Siebenter Band. Stereometrie. Verzeichnisse. Berlin and Leipzig.
Vitruvius (1931). *On Architecture*. In two volumes. I. Edited from the Harleian manuscript 2767 and translated into English by Frank Granger. London and New York.

cuboid

In the mistaken belief that he was coining a new term, Hayward (1890, p. 78) suggested: "The need of some short word in the place of the polysyllabic 'rectangular parallelepiped' has been long felt. I have coined the word 'cuboid' in the hope that it may obtain currency among mathematicians". The fact is, however, that the same suggestion had already been made by Cayley (1874, p. 27): "the rectangular parallelepiped, or, say for shortness, the 'cuboid'".

Hayward (1890, p. 78) also pointed out that De Morgan had earlier suggested *right solid*, a term that did not catch on. It was proposed in De Morgan (1849, p. 106); on *rectangular parallelepiped*, he commented: "The length of this phrase is intolerable".

In German texts, the word *Quader* is in common use. The older *Quaderstück* was already used by Kepler (1616, e. g., p. 23), who took it over from the language of architecture, where it meant 'brick'. The word is made up of *Quader* (from medieval Latin *quadrus* 'ashlar block') and *Stück* 'piece'.[1] Kepler's term did not immediately

[1] See Götze (1919, p. 135) for come comments on Kepler' term.

catch on, however; it did not become popular until the second half on the 19th century, when *Quader* was re-introduced by Kommerell (1865, p. v).[1]

References

Cayley, Arthur (1874). "On the Potentials of Polygons and Polyhedra". *Proceedings of the London Mathematical Society* 6, pp. 20–34.

Götze, Alfred (1919). *Anfänge einer mathematischen Fachsprache in Keplers Deutsch.* Berlin.

Hayward, Robert Baldwin (1890). *The Elements of Solid Geometry.* London.

Kepler, Johannes (1616). *Außzug auß der Uralten MesseKunst Archimedis.* Linz.

Kommerell, Ferdinand (1865). *Lehrbuch der Stereometrie.* Tübingen.

De Morgan, Augustus (1849). *Trigonometry and Double Algebra.* London.

Schmid, K.A. (1873). *Encyklopädie des gesammten Erziehungs- und Unterrichtswesens.* Neunter Band. Spanien–Vives. Gotha.

curl

When he introduced this term, Maxwell (1871, p. 231) was very explicit about his choice:

> I propose, but with great diffidence, to call this vector the *Curl* or *Version* of the original vector function. It represents the direction and magnitude of the rotation of the subject matter carried by the vector σ. I have sought for a word which shall neither, like Rotation, Whirl, or Twirl, connote motion, nor, like Twist, indicate a helical or screw structure which is not of the nature of a vector at all.

Some of Maxwell's earlier thoughts on the terminology are found in a letter to Tait from 1870: "the vector part I would call the Twist of the vector function. Here the word twist has nothing to do with a screw or helix. If the word *turn* or *version* would do they would be better than twist, for twist suggests a screw. Twirl is free from the screw notion and is sufficiently racy. Perhaps it is too dynamical for pure mathematicians, so for Cayley's sake I might say Curl (after the fashion of Scroll[2])" (Knott 1911, p. 143).

In the first edition of his *Treatise on Electricity and Magnetism*, Maxwell also uses *curl*, and names *version* as an alternative (Maxwell 1873, p. 28); in the second edition, however, the word *rotation* is suggested instead, again "with great diffidence" (Maxwell 1881, p. 29). This variety is reflected in translations of his book. Thus, in a German translation, *Rotation* is used (Maxwell 1883, p. 29), while in a French translation, *tournoiment* (from *tournoyer* 'to wheel around, whirl') and *version* are found, which recalls the terms used in the first English edition.

The term first spread in English publications; for example, Gibbs (1881-84, p. 16) and Heaviside (1883, p. 8) adopted it. There were also critical voices: Thomson (1890, p. 22) notes that *curl* "has, I grieve to say, been adopted by Mr. Oliver Heaviside. [...] I object to 'curl,' because 'rotation' or 'spin' is good, and 'curl' is bad". In spite of such criticism, Wilson (1901, p. 155) was able to state: "To the operator $\nabla\times$ Maxwell gave the name *curl*. This nomenclature has become widely accepted".

The translations of Maxwell's book have already been pointed out. However, there were, around the year 1900, numerous other publications in which further terms

---

[1] Kommerell's authorship of the term was pointed out by Schmid (1873, p. 222), for example. Kommerell does not appear to have been aware of Kepler's usage. [2] Cayley's *scroll* refers to skew surfaces; see *skew surface*.

and translations were suggested. For example, Föppl (1894, p. 41) suggested German *Quirl* 'whisk', which he regarded as etymologically related to *curl*; the term was also used Wiechert (1896, p. 288) and various others. Jaumann (1905, p. 148) also names *Rotor* and *Wirbeldichte* ('vortex density') as already known synonyms.[1] A French alternative was *tourbillon*, lit. 'vortex' (Ricci and Levi-Civita 1901, p. 194).

Note. The origin of the terminology is discussed in Crowe (1985, pp. 131–135, 163–167).

References

Clifford, William Kingdon (1873). "Preliminary Sketch of Biquaternions". *Proceedings of the London Mathematical Society* 4, pp. 381–395.

Crowe, Michael J. (1985). *A History of Vector Analysis*. The Evolution of the Idea of a Vectorial System. New York.

Föppl, August (1894). *Einführung in die Maxwell'sche Theorie der Elektricität*. Leipzig.

Gibbs, Josiah Willard (1881-84). *Elements of Vector Analysis*. New Haven.

Heaviside, Oliver (1883). "The Relations between Magnetic Force and Electric Current". *The Electrician* 10, pp. 6–8.

Jaumann, Gustav (1905). *Die Grundlagen der Bewegungslehre*. Leipzig.

Knott, Cargill Gilston (1911). *Life and Scientific Work of Peter Guthrie Tait*. Cambridge.

Maxwell, James Clerk (1871). "Remarks on the Mathematical Classification of Physical Quantities". *Proceedings of the London Mathematical Society* 3, pp. 224–232.

Maxwell, James Clerk (1873). *A Treatise on Electricity and Magnetism*. Vol. I. Oxford.

Maxwell, James Clerk (1881). *A Treatise on Electricity and Magnetism*. Vol. I. Second Edition. Oxford.

Maxwell, James Clerk (1883). *Lehrbuch der Electricität und des Magnetismus*. Autorisierte deutsche Übersetzung von B. Weinstein. Erster Band. Berlin.

Ricci, Gregorio and Tullio Levi-Civita (1901). "Méthodes de calcul différentiel absolu et leurs applications". *Mathematische Annalen* 54, pp. 125–201.

Thomson, William (1890). "Ether, electricity, and ponderable matter". *Journal of the Institution of Electrical Engineers* 48, pp. 4–35.

Wiechert, Emil (1896). "Ueber die Grundlagen der Electrodynamik". *Annalen der Physik und Chemie* 59, pp. 283–323.

Wilson, Edwin B. (1901). *Vector Analysis*. New York and London.

Zermelo, Ernst (1902). "Hydrodynamische Untersuchungen über die Wirbelbewegungen in einer Kugelfäche". *Zeitschrift für Mathematik und Physik* 47, pp. 201–237.

current

This term for a particular kind of functional is due to Georges de Rham. As explained by Lützen (1982, p. 146), "inspired by the similarities between forms and chains, de Rham in [1936] defined the concept of a current. He defined an elementary current as a pair (c^{p+k}, ω^k) consisting of a $(p+k)$-chain and a k-form; a current is then a linear combination of elementary currents".[2] In this sense, *current* (French *courant*) was introduced in de Rham (1936, p. 221). de Rham explained the term on several occasions, for example in de Rham (1984, p. 34): "The choice of the term 'current'

[1] Clifford (1873, p. 381) suggested *rotor* in the following sense: "I propose to use the name *rotor* (short for *rotator*) to mean a quantity having magnitude, direction, and position, of which the simplest type is a velocity of rotation about a certain axis". *Wirbeldichte* was proposed by Zermelo (1902, p. 205). [2] Lützen's explanation is almost a literal translation of de Rham's definition. de Rham revised the definition after the publication of Schwarz' work on distributions.

is motivated by the fact that in ordinary 3-space, '1-dimensional currents' can be interpreted as electrical currents".

de Rham (1937, p. 324) also used German *Strömung*, a literal translation of the term. The English term does not seem to have been used, at least in print, until around 1950; it appears in de Rham and Kodaira (1950, p. 22).

Note. See Lützen (1982, pp. 144–147) for a discussion of de Rham's work on currents.

References

Lützen, Jesper (1982). *The Prehistory of the Theory of Distributions*. Berlin et al.

de Rham, Georges (1936). "Relations entre la topologie et la théorie des intégrales multiples". *L'Enseignement Mathématique* 35, pp. 213–228.

de Rham, Georges (1937). "Über mehrfache Integrale". *Abhandlungen aus dem Mathematischen Seminar der Universität Hamburg* 12, pp. 313–339.

de Rham, Georges and Kunihiko Kodaira (1950). *Harmonic Integrals*. Princeton.

de Rham, Georges (1984). *Differentiable Manifolds*. Forms, Currents, Harmonic Forms. Translated from the French by F.R. Smith. Berlin.

curvature

In a study on the history of curvature, Gericke (1982, p. 3) found only one place where curvature is explicitly mentioned in Greek antiquity, namely in Plutarch's *Platonic Questions*: "while a straight line, whatever its length, keeps the same straightness throughout, we see that the circumferences of circles are more curved, that is are more highly concentrated in their convexity, if they are smaller, and more relaxed, if they are larger" (Plutarch 1976, p. 59). Plutarch here used the word *καμπυλωτέρας*, a comparative form of *καμπύλος* 'bent' (see *curve*).

This passage was later echoed by Wallis (1685b, p. 90), who asked "How comes it to pass, that in lesser Circles, the Curve doth more flye off, than in greater?" and answered "because the lesser Circumference is more crooked. For it hath as much Curvity, in a shorter length".

In the 14th century, curvature was discussed by Nicole Oresme under the name *curvitas*:

> [W]e do not know with what, or with regard to what, the intensity of curvature is measured. For now it appears to me that there are only two [possible] ways [to speak of the measure of curvature]. The first is that the increase in curvature is a function of its departure from straightness, i. e., of its distance from straightness. This is [to be measured] by the quantity of the angle constituted of a straight line and a curve, e. g., an angle of contingence or perhaps another angle also constructed from a straight line and a curve (Clagett 1968, p. 214).[1]

With regard to circles, Oresme states that "the intensity of curvature is a direct function of the smallness of the radius" (Clagett 1968, p. 222).[2]

[1] "ignotum enim est penes quid vel circa quid attenditur intensio curvitatis. Et pro nunc non apparet mihi nisi alter duorum modorum: unus est quod maioritas curvitatis attenditur penes recessum ipsius a rectitudine et distantiam ab ipsa rectitudine, que est secundum quantitatem anguli constituti ex recta et curva sicut est angulus contingentie aut forsan unus alter etiam constitutus ex recta et curva". Translation from Clagett (1968, p. 215). – Oresme's notion of curvature is discussed by Serrano and Suceavă (2015). [2] "intensio curvitatis attenditur penes semidyametri parvitatem". Translation from Clagett (1968, p. 223).

In the early 17th century, Kepler (1604, pp. 75f.) discussed the circle of curvature, at least in rudimentary form (see fig. 34). He wanted to find the image of a certain point reflected in a mirror, and remarked: "a truer reason directs you to find the circle which contains the pattern of curvature that the section has at the point of reflection" (Kepler 1604, pp. 75f.).[1]

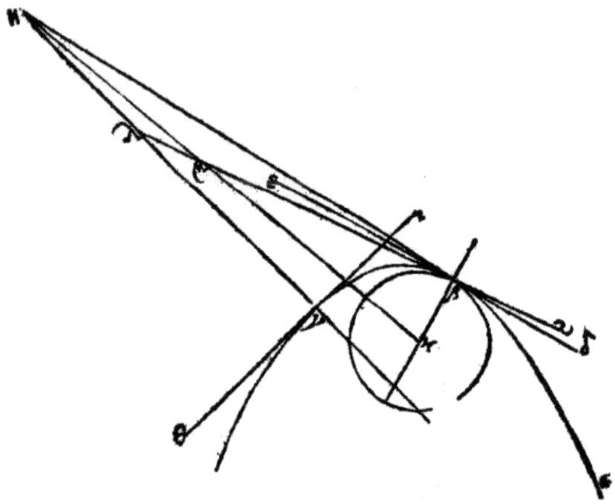

Figure 34: The circle of curvature. From Kepler (1604, p. 75).

The term *curvature* began to be more frequently used in the 17th century, particularly in England. Hobbes (1666, p. 16), for example, defined the contact angle with reference to curvature: "The angle of contact is the quantity of curvature that is in the arc of the circle made by continual and uniform bending of a right line".[2] In Wallis (1685a, p. 99), the following statement is found: "That what he [Clavius] calls an Angle of Contact, is but what I call the Degree of Curvity".

Newton also discussed the notion in his *Method of series and fluxions*, written around 1671 but not published until 1736. In this treatise, he used the terms *centre of curvature* (Latin *centrum curvaminis*) and *radius of curvature* (*radius curvaminis*) (Newton 1969, p. 150).[3] Newton here dealt with the circle of curvature, which was used to measure the curvature of a plane curve. Leibniz (1686, p. 289) called this circle the *osculating circle*, and stated that "a circle is the most suitable curve for determining

[1] "verior ratio iubet invenire circulum, qui contineat rationem curvitatis quam habet sectio in beta puncto repercussus". Translation from Kepler (2000, pp. 90f.). – On this passage, see Coolidge (1952, pp. 376f.) and Gericke (1982, p. 4). Eneström (1912-13, pp. 73f.) is skeptical and argues that Kepler only had a vague idea of the osculating circle. Kepler did not give a definition nor did he show how the circle can be constructed. [2] "Angulus contactus est quantitas curvitatis que est in arcu circuli facta a continua & uniformi flexione lineae rectae". Translation from Jesseph (1999, p. 100). [3] In notes from the 1660s written in English, Newton used *crookednesse* or *curvity* (Newton 1967, e. g., pp. 248, 424). Latin *curvitas* also appears (Newton 1967, p. 422).

curvature, because the curvature of the one circle is the same everywhere".[4] Leibniz here used the Latin noun *flexura*, lit. 'bending', for 'curvature'. See *osculating*.

In the 18th century, the term began to be used in connection with objects other than plane curves. In particular, Pitot (1726, p. 113) and Clairaut (1731) in his *Recherches sur les courbes à double courbure* use the term in connection with space curves (see *space curve*).

Several years later, the term is also found with regard to surfaces. In particular, Euler (1767) published an important paper in which he studied the curvature at a point on a surface. As explained by Papadopoulos (2017, p. 61), "Euler's idea [...] was to introduce a notion of curvature at a point of a differentiable surface based on the curvature of curves that pass through that point. His intuition was that to understand curvature at a point of a surface, it suffices to study the curvature of curves that are intersections of that surface with Euclidean planes". Euler began his paper with these words:

> In order to know the curvature of curved lines, the determination of the radius of the osculating circle offers the proper method, which for each point on the curve provides us with a circle whose curvature is the same. But when one asks for the curvature of a surface, the question is rather equivocal, and not at all subject to a definitive answer, as in the previous case. [...] Thus the question about curvature of surfaces is not amenable to a simple answer (Euler 1767, p. 119).[1]

A new approach to the curvature of surfaces was presented by Gauß (1828b) in his ground-breaking *General Investigations of Curved Surfaces*, in which he introduced the terms *total* or *integral curvature* (*curvatura totalis* or *integra*) as well as *measure of curvature* (*mensura curvaturae*), later referred to as *Gaussian curvature*:

> Thus, to each part of a curved surface inclosed within definite limits we assign a *total* or *integral curvature*, which is represented by the area of the figure on the sphere corresponding to it. From this integral curvature must be distinguished the somewhat more specific curvature which we shall call the *measure of curvature*. The latter refers to a point of the surface, and shall denote the quotient obtained when the integral curvature of the surface element about a point is divided by the area of the element itself; and hence it denotes the ratio of the infinitely small areas which correspond to one another on the curved surface and on the sphere (Gauß 1828b, § 6).[2]

The corresponding German terms *ganze Krümmung*, *Totalkrümmung* and *Krümmungsmaß* appear slightly earlier in an account of his book (Gauß 1827, p. 1763);

---

[4] "circulus aptissimus ad determinandam flexuram, quia ubique eadem unius circuli est flexura". Translation from Leibniz (2014, p. 1). [1] "Pour connoitre la courbure des lignes courbes, la détermination du rayon osculateur en fournit la plus juste mesure, en nous présentant pour chaque point de la courbure un cercle, dont la courbure est précisément la même. Mais quand on demande la courbure d'une surface, la question est fort équivoque et point du tout susceptible d'une réponse absolue, comme dans le cas précédent [...] Donc la question sur la courbure des surfaces n'est pas susceptible d'une réponse simple". Translation from Euler (n. d., p. 1). [2] "Scilicet cuilibet parti superficiei curvae limitibus determinatis cinctae *curvaturam totalem* seu *integram* adscribimus, quae per aream figurae illi in superficie sphaerica respondentem exprimetur. Ab hac curvatura integra probe distinguenda est curuatura quasi specifica, quam nos mensuram curvaturae vocabimus: haec posterior ad *punctum* superficiei refertur, et denotabit quotientem qui oritur, dum curvatura integra elementi superficialis puncto adiacentis per aream ipsius elementi dividitur, et proin indicat rationem arearum infinite parvarum in superficie curva et in superficie sphaerica sibi mutuo respondentium". Translation from Gauß (1902, § 6).

he first used *Krümmungsmaß* in notes from 1822 (Gauß 1900, p. 381). The English terms *entire* and *total curvature* as well as *measure of curvature* appear in Gauß (1828a, p. 332); the corresponding French terms were used in [anon.] (1828, p. 5).

Gauß continued with a terminological remark:

> As for the terminology, we have thought it especially desirable that all ambiguity be avoided. For this reason we have not thought it advantageous to follow strictly the analogy of the terminology commonly adopted (though not approved by all) in the theory of plane curves, according to which the measure of curvature should be called simply curvature, but the total curvature, the amplitude (Gauß 1828b, § 6).[1]

The concept is commonly designated as *Gaussian curvature*. This term was used by Schnuse (1848, p. 1496) (*Gauß'sches Krümmungsmaß*); Bonnet (1866, p. 71) used French *courbure de Gauss*, while *Gaussian measure of curvature* appears in Craig (1879, p. 5).

The concept of curvature was further generalized to manifolds of n dimensions by Riemann in 1854 in his famous "On the hypotheses which lie at the bases of geometry", which was published posthumously (Riemann 1868).[2]

See also *torsion* (differential geometry) and *osculating*.

Note. The history of the notion is discussed by Coolidge (1952) and Gericke (1982). See also A'Campo-Neuen and Papadopoulos (2019), Knoebel et al. (2007, pp. 159–227), and Friedelmeyer (1993). For Gauß' paper, see the detailed account by Dombrowski (1977) or the English translation Dombrowski (1979); see also Merker (2004).

References

[anon.] (1828). "Disquisitiones generales circa superficies curvas; par M. Gauss". *Bulletin des sciences mathématiques, physiques et chimiques* 10, pp. 4–11.

A'Campo-Neuen, Annette and Athanase Papadopoulos (2019). "A Path in History, from Curvature to Convexity". In: *Geometry in History*. Ed. by S.G. Dani and Athanase Papadopoulos. Cham, pp. 305–353.

Bonnet, Ossian (1866). "Note sur les lignes d'ombre et d'ombre portée". *Nouvelles annales de mathématiques* 5, pp. 71–72.

Clagett, Marshall, ed. (1968). *Nicole Oresme and the Medieval Geometry of Qualities and Motions*. A Treatise on the Uniformity and Difformity of Intensities Known as *Tractatus de configurationibus qualitatum et motuum*. Madison, Milwaukee.

Clairaut, Alexis Claude (1731). *Recherches sur les courbes à double courbure*. Paris.

Coolidge, Julian Lowell (1952). "The Unsatisfactory Story of Curvature". *The American Mathematical Monthly* 59, pp. 375–379.

Craig, Thomas (1879). *A General Differential Equation in the Theory of the Deformation of Surfaces*. Philadelphia.

Dombrowski, Peter (1977). "Differentialgeometrie – 150 Jahre nach den ‚Disquisitiones generales circa superficies curvas' von Carl Friedrich Gauß". *Abhandlungen der Braunschweigischen Wissenschaftlichen Gesellschaft* 27, pp. 63–102.

Dombrowski, Peter (1979). "Differential Geometry – 150 years after Carl Friedrich Gauss' Disquisitiones Generales Circa Superficies Curvas". *Astérisque* 62, pp. 97–153.

---

[1] "Quod vero attinet ad terminologiam, imprimis prospiciendum esse duximus, ut omnis ambiguitas arceatur, quapropter haud congruum putavimus, analogiam terminologiae in doctrina de lineis curvis planis vulgo receptam (etsi non omnibus probatam) stricte sequi, secunddum quam mensura curvaturae simpliciter audire debuisset curvatura, curvatura integra autem amplitudo". Translation from Gauß (1902, § 6). [2] For discussion of this paper and further references, see Gray (2005).

Eneström, Gustaf (1912-13). "Kleine Bemerkungen zur letzten Auflage von Cantors ‚Vorlesungen über Geschichte der Mathematik'". *Bibliotheca Mathematica* 13, pp. 65–82.

Euler, Leonhard (1767). "Recherches sur la courbure des surfaces". *Mémoires de l'Académie Royale des Sciences et Belles-Lettres de Berlin* (1760), pp. 119–143.

Euler, Leonhard (n.d.). "Researches on the curvature of surfaces". URL: `https://scholarlycommons.pacific.edu/euler-works/333/` (retrieved Nov. 23, 2021).

Friedelmeyer, Jean-Pierre (1993). "De l'angle de contingence au rayon de courbure: comment penser, comparer, mesurer le courbe". In: *La figure et l'espace*. Actes du huitième Inter IREM Epistémologie et histoire des mathématiques, Lyon 31 Mai et 1er Juin 1991. Ed. by IREM de Lyon. Villeurbanne, pp. 365–396.

Gauß, Carl Friedrich (1827). "Disquisitiones circa superficies curvas". *Göttingische gelehrte Anzeigen* (1827), pp. 1761–1768.

Gauß, Carl Friedrich (1828a). "Account of a Paper by Prof. Gauss, intitled Disquisitiones generales circa superficies curvas". *The Philosophical Magazine* 3, pp. 331–336.

Gauß, Carl Friedrich (1828b). *Disquisitiones generales circa superficies curvas*. Göttingen.

Gauß, Carl Friedrich (1900). "Stand meiner Untersuchung über die Umformung der Flächen". In: *Werke*. Achter Band. Göttingen, pp. 374–385.

Gauß, Carl Friedrich (1902). *General Investigations of Curved Surfaces*. Translated with notes and a bibliography by James Caddall Morehead and Adam Miller Hiltebeitel. Princeton.

Gericke, Helmuth (1982). "Zur Vorgeschichte und Entwicklung des Krümmungsbegriffs". *Archive for History of Exact Sciences* 27, pp. 1–21.

Gray, Jeremy (2005). "Bernhard Riemann, posthumous thesis 'On the hypotheses which lie at the foundation of geometry'". In: *Landmark Writings in Western Mathematics 1640-1940*. Ed. by Ivor Grattan-Guinness. Amsterdam, pp. 506–520.

Hobbes, Thomas (1666). *De principiis & ratiocinatione geometrarum*. London.

Jesseph, Douglas M. (1999). *Squaring the Circle*. The War Between Hobbes and Wallis. Chicago and London.

Kepler, Johannes (1604). *Ad Vitellionem paralipomena, quibus astronomiae pars optica traditur*. Frankfurt.

Kepler, Johannes (2000). *Optics*. Paralipomena to Witelo & Optical Part of Astronomy. Translated by William H. Donahue. Santa Fe.

Knoebel, Arthur et al. (2007). *Mathematical Masterpieces*. Further Chronicles by the Explorers. New York.

Leibniz, Gottfried Wilhelm (1686). "Meditatio nova de natura anguli contactus et osculi". *Acta Eruditorum* (June 1686), pp. 289–292.

Leibniz, Gottfried Wilhelm (2014). "New Considerations regarding the nature of the Angle of contact and of osculation". Transl. with notes by Ian Bruce. URL: `http://www.17centurymaths.com/contents/leibnizcontents.html` (retrieved Aug. 3, 2017).

Merker, Claude (2004). *Le mémoire de Gauss sur les surfaces courbes et la naissance de la géométrie différentielle intrinsèque*. Besançon.

Newton, Isaac (1967). *The Mathematical Papers of Isaac Newton*. Volume I. 1664-1666. Ed. by D.T. Whiteside. Cambridge.

Newton, Isaac (1969). "De methodis serierum et fluxionum". In: *The Mathematical Papers of Isaac Newton*. Volume III. 1670-1673. Ed. by D.T. Whiteside. Cambridge, pp. 32–353.

Papadopoulos, Athanase (2017). "Looking Backward: From Euler to Riemann". In: *From Riemann to Differential Geometry and Relativity*. Ed. by Lizhen Ji, Athanase Papadopoulos, and Sumio Yamada. Cham, pp. 1–93.

Pitot, Henri (1726). "Quadrature de la moitié d'une Courbe des Arcs, appellée la Compagne de la Cycloïde". *Mémoires de l'Académie royale des sciences* (1724), pp. 107–113.

Plutarch (1976). "Platonic Questions". In: *Moralia*. Volume XIII. Part I. Ed. by Harold Cherniss. Cambridge, Mass.

Riemann, Bernhard (1868). "Ueber die Hypothesen, welche der Geometrie zu Grunde liegen". *Abhandlungen der Königlichen Gesellschaft der Wissenschaften zu Göttingen* 13, pp. 133–152.

Schnuse, C.H. (1848). "Navier: Lehrbuch der Differential- und Integralrechnung [review]". *Göttingische gelehrte Anzeigen* 149, pp. 1490–1496.
Serrano, Isabel M. and Bogdan D. Suceavă (2015). "A Medieval Mystery: Nicole Oresme's Concept of Curvitas". *Notices of the American Mathematical Society* 62, pp. 1030–1034.
Wallis, John (1685a). *A Defense of the Treatise of the Angle of Contact*. In: *A Treatise of Algebra, both Historical and Practical*. London, 69–105 (appendix).
Wallis, John (1685b). *A Treatise of Angular Sections*. In: *A Treatise of Algebra, both Historical and Practical*. London, 1–68 (appendix).

curve

Something that is crooked or bent was called *καμπύλος* in Greek (from a verb meaning 'to bend'). According to Mugler (1958-59, p. 237), this word was used in mathematics in opposition to straight lines and applied to circles, conics, spirals, etc. Euclid did not use the term in his *Elements*, but it is found in Archimedes' *On the Sphere and Cylinder*, for example: "There are in a plane some limited curved lines, which are either wholly on the same side as the straight [lines] joining their limits or have nothing on the other side" (Archimedes 1880, p. 6).[1] The word *καμπύλος* is also found in the Pseudo-Heronian *Definitions*, in which a classification of lines was given; lines are there divided into *straight* and *not straight*, with the latter being further subdivided into *circular circumferences*, *spiral-shaped* and *curved* (Heron of Alexandria 1912, pp. 16, 18).[2]

In Latin texts, the word *curvus*, which has the same meaning as *καμπύλος*, was used as *linea curva*, which is found, for example, in the encyclopaedic *The Marriage of Philology and Mercury* by Martianus Capella, who flourished in the 5th century: "Next comes the second kind of figures, those that are formed by curved lines; the Greek name for these figures is *καμπυλόγραμμον*" (Willis 1983, p. 253).[3] Earlier, the Roman land surveyor Balbus had used the word *flexuosus* 'crooked, winding': "There are three kinds of line: straight, circular, curving [...] A curving line is one of many shapes, as in the case of fields or mountain ridges or rivers" (Blume, Lachmann, and Rudorff 1848, p. 99).[4]

It was the Latin form *linea curva*, later shortened to *curva*, which gave rise to *curve* used in modern languages. In French, for example, it is found as *ligne courbe*, which was, according to the DMF, first used in the 14th century by Oresme (1968, p. 378).

In early English attestations attempts at using a native word can be found. For example, Recorde (1551, fol. A^{v}) used *croked line*: "A Straight lyne, is the shortest that maye be drawenne betweene two prickes. And all other lines, that go not right forth from prick to prick, but boweth any waye such are called *Croked lynes*".[5] The term was also used by L. Digges and T. Digges (1571, fol. B j^{r}), in whose book also *curve line* can be found (L. Digges and T. Digges 1571, fol. N iij^{v}).[6] The German

[1] Translation from Archimedes (2004, p. 35). [2] See Heath (1956, p. 159). [3] "Sequitur secundum schematum genus, quod curvis lineis informatur, quod *καμπυλόγραμμον* appellatur, cuius species duae sunt". Translation from Stahl (1977, p. 267). [4] "Linearum genera sunt trea, rectum, circum ferens, flexuosum [...] Flexuosa linea est multiformis, uelut aruorum aut iugorum aut fluminum". Translation from Campbell (2000, p. 209). [5] *croked lyne* was earlier used by Benese (c1537, fol. a iv^{r}). [6] The English adjectives *curve* and *curved* are earlier found in non-mathematical meanings in the 16th century (OED), but the example in Digges appears to be the earliest purely mathematical use in English. The noun is dated to 1658 by the OED.

term *krumme Linie*, first attested in the late 14th century (Mendthal 1886, p. 65) and later replaced by *Kurve*, is similar to *curved* or *crooked line* in that *krumm* means 'bent'.

In the 19th century and afterwards, the concept was much discussed and new definitions came up. For example, Jordan (1887, p. 587) gave his famous definition of a curve by the equations $x = f(t), y = \varphi(t)$, with f and φ being functions of the independent variable t.[1] The various remarkable properties of curves led Klein (1925, p. 187) to a much-quoted remark: "Every person believes that he knows what a curve is until he has learned so much mathematics that the countless possible abnormalities confuse him".[2]

See also *line* and *space curve*.

Note. For the early terminology, see Tropfke (1940, pp. 53–56) and Heath (1956, pp. 158–165); see also Ayuso García (2008, pp. 985f.), Guillaumin (2020, p. 93), and Mugler (1958-59, p. 237). For the Arabic terminology, see Souissi (1968, pp. 145, 293).[3] Historical summaries on various definitions for *curve* can be found in von Mangoldt (1921), Menger (1932, pp. 77f.), Boyer (1945), and Herrlich, Hušek, and Preuß (2008).

References

Archimedes (1880). *Opera omnia*. Cum commentariis Eutocii. Volumen I. Ed. by J.L. Heiberg. Leipzig.

Archimedes (2004). *The Works of Archimedes*. Translation and Commentary. Volume 1: The Two Books *On the Sphere and the Cylinder*. Edited and translated by Reviel Netz. Cambridge.

Ayuso García, Manuel (2008). "La terminología latina de la geometría en Marciano Capela". PhD thesis. Madrid: Universidad Nacional de Educación a Distancia.

Benese, Richard (c1537). *This boke sheweth the maner of measurynge of all maner of lande*. London.

Blume, Friedrich, Karl Lachmann, and Adolf Rudorff, eds. (1848). *Die Schriften der römischen Feldmesser*. Erster Band. Texte und Zeichnungen. Berlin.

Boyer, Carl B. (1945). "Historical Stages in the Definition of Curves". *National Mathematics Magazine* 19, pp. 294–310.

Campbell, Brian (2000). *The Writings of the Roman Land Surveyors*. Introduction, Text, Translation and Commentary. London.

Digges, Leonard and Thomas Digges (1571). *A geometrical practise, named Pantometria diuided into three bookes, Longimetra, Planimetra, and Stereometria*. London.

Guillaumin, Jean-Yves (2020). *Dictionnaire de la terminologie latine ancienne de l'arithmétique et de la géométrie*. Paris.

Heath, Thomas L. (1956). *The Thirteen Books of Euclid's Elements*. Volume I. Introduction and Books I, II. New York.

Heron of Alexandria (1912). *Heronis Alexandrini opera quae supersunt omnia*. Volumen IV. Heronis definitiones cum variis collectionibus. Heronis quae feruntur geometrica. Ed. by J.L. Heiberg. Leipzig.

Herrlich, Horst, Mirek Hušek, and Gerhard Preuß (2008). "Hausdorffs Studien über Kurven, Bögen und Peano-Kontinua". In: *Felix Hausdorff. Gesammelte Werke*. Band III. Mengenlehre (1927, 1935). Deskriptive Mengenlehre und Topologie. Ed. by Ulrich Felgner et al. Berlin and Heidelberg, pp. 798–839.

[1] For references to further definitions, see the note at the end of this article. [2] "Was eine Kurve ist, glaubt jeder Mensch zu wissen, bis er so viel Mathematik gelernt hat, daß ihn die unzähligen möglichen Abnormitäten verwirrt gemacht haben". Translation from Klein (2016, p. 199). [3] Rashed (2017) does not have a corresponding entry but does deal with the Arabic word in some entries; see his index under *ligne courbe*.

Jordan, Camille (1887). *Cours d'analyse*. Tome troisième. Calcul intégral. Équations différentielles. Paris.
Klein, Felix (1925). *Elementarmathematik vom höheren Standpunkte aus*. Zweiter Band. Ausgearbeitet von E. Hellinger. Berlin.
Klein, Felix (2016). *Elementary Mathematics from a Higher Standpoint*. Volume II: Geometry. Translated by Gert Schubring. Berlin and Heidelberg.
von Mangoldt, Hans (1921). "Die Begriffe ‚Linie' und ‚Fläche'". *Encyklopädie der mathematischen Wissenschaften mit Einschluss ihrer Anwendungen* III A B 2, pp. 130–152.
Mendthal, Hans (1886). *Geometria culmensis*. Ein agronomischer Tractat aus der Zeit des Hochmeisters Conrad von Jungingen (1393-1407). Leipzig.
Menger, Karl (1932). *Kurventheorie*. Herausgegeben unter Mitarbeit von Georg Nöbeling. Berlin and Leipzig.
Mugler, Charles (1958-59). *Dictionnaire historique de la terminologie géométrique des Grecs*. Paris.
Oresme, Nicole (1968). *Le livre du ciel et du monde*. Edited by Albert D. Menut and Alexander J. Denomy, translated with an introduction by Albert D. Menut. Madison.
Rashed, Roshdi, ed. (2017). *Lexique historique de la langue scientifique arabe*. Hildesheim.
Recorde, Robert (1551). *The patheway to knowledg*. London.
Souissi, Mohammed (1968). *La langue des mathématiques en arabe*. Tunis.
Stahl, William Harris (1977). *Martianus Capella and the Seven Liberal Arts*. Volume II. The Marriage of Philology and Mercury. New York.
Tropfke, Johannes (1940). *Geschichte der Elementar-Mathematik*. Vierter Band. Ebene Geometrie. Dritte, verbesserte und vermehrte Auflage besorgt von Dr. Kurt Vogel. Berlin and Leipzig.
Willis, James, ed. (1983). *Martianus Capella*. Leipzig.

curvilinear angle

In Euclid's *Elements* (book I, def. 9), the notion of a rectilinear angle (formed by two straight lines) is explicitly defined, but its counterpart, a curvilinear angle (formed by curved lines), is not, even though an example of such angles, the angle of contingence (between the circumference of a circle and its tangent) *is* discussed (see *angle of contingence*).

A Greek term sometimes used in this connection is *περιφερόγραμμος* (cf. *περιφέρεια* 'circumference', *γράμμα* 'line') 'bounded by a curved line' (LSJ). Proclus uses this word for "the angle bounded by circular lines" (Proclus 1873, p. 104).[1] It should also be pointed out that the word *καμπυλόγραμμος* would correspond in structure to *curvilinear*, but it is not documented in ancient Greek. However, this Greek word is cited by Martianus Capella (Ayuso García 2008, pp. 967f.).

The Latin words *curvilineus* and *curvilinearis* do not seem to appear until the 12th century. Robert of Chester, for example, uses *figura curvilineus* without definition (Busard and Folkerts 1992, p. 307). Gerard of Cremona calls a curvilinear angle *circumflexus* (from *circumflectere* 'to bend, turn about') (Tummers 1994, p. 13).

In the 13th century, *curvilineus* is used by Roger Bacon: "A rectilinear angle is that which is contained by straight lines. A curvilinear is that which is included by curved lines" (Molland 1993, p. 280).[2]

In modern languages, the terminology is found from the late 15th century on. Thus, Pacioli (1494, Tractatus Geometrie, fol. 1^r) uses Italian *angolo curvilineo* in his treatise on geometry. French *angle curvilinéaire* is found (as *angle curvilineare*) in

[1] This paraphrase is from Proclus (1970, p. 84). – The word is earlier found in Aristotle's *de Caelo* (286b). See Mugler (1958-59, p. 346). [2] "Angulus rectilineus est qui ex rectis lineis continetur. Curvilineus est qui curvis lineis includitur". Translation from Molland (1993, p. 281).

Piccolomini (1550, fol. B viij^{v}). In Billingsley's translation of the *Elements*, the term *crooked lyned angle* is used (Euclid 1570, fol. 2^{v}); according to the OED, *curvilineal angle* is first found in Hobbes (1656, p. 24).

Note. For the Latin and Greek terms, see Ayuso García (2008, pp. 967f.).

References

Ayuso García, Manuel (2008). "La terminología latina de la geometría en Marciano Capela". PhD thesis. Madrid: Universidad Nacional de Educación a Distancia.

Busard, H.L.L. and Menso Folkerts, eds. (1992). *Robert of Chester's (?) Redaction of Euclid's* Elements, *the so-called Adelard II Version*. Basel.

Euclid (1570). *The Elements of Geometrie of the most auncient Philosopher Euclide of Megara*. Faithfully (now first) translated into the Englishe toung, by H. Billingsley. London.

Hobbes, Thomas (1656). *Six Lessons to the Professors of the Mathematiques, One of Geometry, the Other of Astronomy*. London.

Molland, George (1993). "Roger Bacon's *Geometria Speculativa*". In: *Vestigia Mathematica*. Studies in medieval and early modern mathematics in honour of H.L.L. Busard. Ed. by Menso Folkerts and Jan P. Hogendijk. Amsterdam, pp. 265–303.

Mugler, Charles (1958-59). *Dictionnaire historique de la terminologie géométrique des Grecs*. Paris.

Pacioli, Luca (1494). *Summa de Arithmetica Geometria Proportioni et Proportionalita*. Venezia.

Piccolomini, Alessandro (1550). *La sphere du monde*. Traduitte du Tuscan en François par Iacques Goupyl. Paris.

Proclus (1873). *In primum Euclidis elementorum librum commentarii*. Ed. by Gottfried Friedlein. Leipzig.

Proclus (1970). *Commentary on the First Book of Euclid's Elements*. Translated with Introduction and Notes by Glenn R. Morrow. Princeton.

Tummers, Paul M. J. E. (1994). *Anaritius' Commentary on Euclid*. The Latin Translation, I-IV. Nijmegen.

curvilinear coordinates

The term *curvilinear coordinates* was coined in the 19th century. The general notion was used by Lamé (1833), who later introduced the term *elliptic coordinates* (French *coordonnées elliptiques*) (Lamé 1837, p. 156) and, one year later, presented a "mémoire sur les coordonnées curvilignes" (Lamé 1838).

The German term *krummlinige Coordinaten* was then used by Böklen (1859, p. 26); the English term appeared in Gregory and Walton (1845, p. 124).

Note. For the history of the notion, see Loria (1948, pp. 275–286), Kline (1972, pp. 687–689), and Boyer (1956, pp. 260–264). Guitart (2009) focuses on Lamé.

References

Böklen, Otto (1859). "Ueber krummlinige Coordinaten". *Archiv der Mathematik und Physik* 33, pp. 26–32.

Boyer, Carl B. (1956). *History of Analytic Geometry*. New York.

Gregory, Duncan Farquharson and William Walton (1845). *A Treatise on the Application of Analysis to Solid Geometry*. Cambridge.

Guitart, René (2009). "Les coordonnées curvilignes de Gabriel Lamé - Représentation des situations physiques et nouveaux objets mathématiques". *Bulletin de la SABIX* 44, pp. 119–129.

Kline, Morris (1972). *Mathematical Thought From Ancient To Modern Times*. Volume 2. New York and Oxford.

Lamé, Gabriel (1833). "Mémoire sur la propagation de la chaleur dans les Polyèdres, et principalement dans le Prisme triangulaire régulier". *Journal de l'École polytechnique* 14, cahier 22, pp. 194–251.

Lamé, Gabriel (1837). “Sur les surfaces isothermes dans les corps solides homogènes en équilibre de température”. *Journal de mathématiques pures et appliquées* 2, pp. 147–183.
Lamé, Gabriel (1838). “Mémoire sur les coordonnées curvilignes”. *Comptes rendus de l'Académie des sciences* 6, pp. 43–45.
Loria, Gino (1948). “Perfectionnements, évolution, métamorphoses du concept de «coordonnées»: Contribution à l'histoire de la géométrie analytique”. *Osiris* 8, pp. 218–288.

cusp

For this notion, a relatively large number of terms have been used. A small part of the variety is indicated by de Bragelongne (1732, p. 167), who states “that we call *point de rebroussement* what Mr. Newton and the other English Geometers have called *cuspis*”.[1] de Bragelonge here mentions the French term, which literally means ‘point of turning back’, and he also indicates that *cuspis*, literally ‘pointed end’, was first used by Newton. But more terms have been used for this notion; these include *point of reflection, point of regression* or *point of retrogression* as well as *spinode* and *stationary point*.

The French term *point de rebroussement* contains the noun *rebroussement*, which comes from the verb *rebrousser* ‘to turn back’; a *point de rebroussement* is thus a ‘point of turning back’. There is also a French phrase *rebrousser chemin* ‘to turn around’, and this phrase was used when de l'Hôpital (1696, p. 59) published the first printed definition of *point de rebroussement*:

> When a curved line AFK is in part concave and in part convex towards a straight line AB or towards a fixed point B, the point F, that separates the concave part from the convex part and, consequently, is the end of one and the beginning of the other, is called an *inflection point* [*point d'infléxion*] when the curve, having arrived at F, continues along its path in the same direction, and a *cusp* [*point de rebroussement*] when it turns back [*rebrousse chemin*] on its path on the side of its origin.[2]

The term is not due to de l'Hospital, though; it had been coined by Johann Bernoulli, who used it in a latter to de l'Hospital from April 1694: “there are curves that I call *bicornes*, in which the greatest or least ordinates pass through a point which I call *de rebroussement*” (see Bernoulli 1955, p. 207).[3]

In German, the literal translation *Wiederkehr-Punct* or *Wiederkehrungs-Punct* was suggested by Wolff (1710, p. 340); Kästner (1761, p. 469) preferred *Rückkehrpunct*, which was more commonly used than Wolff's term. In his mathematical dictionary, Wolff (1716, p. 1125) also included a Latin literal variant, *punctum regressus*; Hayes (1704, p. 153) used English *Point of Retrogression*.

Bernoulli also called a cusp *punctum reflexus*, for example in a letter from 1696 (see Leibniz 2011, p. 122).[4] This term also appears in other languages, for example

[1] “[q]ue nous nommons *point de rebroussement* ce que M. Newton & les autres Géometres Anglois one nommé *cuspis*”. [2] “Lors qu'une ligne courbe AFK est en partie concave & en partie convexe vers une ligne droite AB ou vers und point fixe B; le point F qui sépare la partie concave de la convexe, & qui par-conséquent est la fin de l'une & le commencement de l'autre, est appellé point d'*infléxion*, lorsque la courbe étant parvenue en F continue son chemin vers le même côté: & point de *rebroussement* lors qu'elle rebrousse chemin du côté de son origine”. Translation from Bradley, Petrilli, and Sandifer (2015, pp. 66f.). [3] “il y a des courbes que j'appelle *bicornes*, dont les plus grandes ou plus petites ordonnées passent par le point que j'appelle *de rebroussement*”. Translation from Bradley, Petrilli, and Sandifer (2015, p. 251). [4] Bernoulli also used *punctum reversionis* (see Bernoulli 1988, p. 364).

as French *point de Reflexion* (de la Hire 1707, pp. 341f.) and as English *point of Reflexion* (Maclaurin 1742, p. 215).

The term *cusp* itself is derived from Latin *cuspis* 'point, spear, pointed end' and was used (as Latin *cuspis*) by Newton in the 1660s: "A cusp is where a curve is terminated in a sharp vertex" (Newton 1968, p. 96).[1] In English texts, *cusp* is at first left in the Latinate form. For example, Harris (1710, s. v. curves) writes that "the *Nodus* will be changed into a *Cuspis*", but *cusp* begins to be used soon afterwards, for instance in the second edition of Stone's *Mathematical Dictionary*: "If these Roots are equal, the Nodus will be changed into a very acute Cusp or Point" (Stone 1743, s. v. deficient hyperbola). The German term *Spitze* has the same literal meaning as *cusp* and was apparently first used by Kästner (1761, p. 469).

In the 19th century, Cayley (1852, p. 166) proposed the term *spinode*, a contraction of *spine* and *node*: "for symmetry of nomenclature and in order to avoid ambiguities I shall [...] use the term 'node' as synonymous with double point, and the term 'spinode' as synonymous with cusp".

The distinction between two kinds of cusps, those of the first and second kind, is due to de l'Hôpital (1696, p. 102). As can be seen in fig. 35, the two branches of the curve have a common tangent in the cusp. If the branches are on opposite sides of the tangent, they are of the first kind, and if they are on the same side, they are of the second kind.[2] de l'Hôpital called the two kinds of cusps *points de rebroussement de la première sorte* and *points de rebroussement de la seconde sorte*.

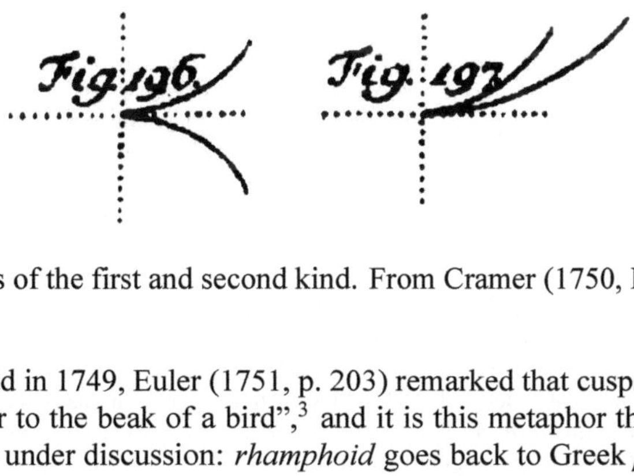

Figure 35: Cusps of the first and second kind. From Cramer (1750, Planche XXV).

In a paper read in 1749, Euler (1751, p. 203) remarked that cusps of the second kind "are similar to the beak of a bird",[3] and it is this metaphor that gave rise to one of the terms under discussion: *rhamphoid* goes back to Greek *ῥάμφος*, which designates the beak of a bird of prey. The term was first used by Francœur (1809, p. 315), who says that "we have preferred the denominations of *keratoid* [*Ceratoïde*]

[1] "Cuspis est ubi curva acutâ vertice terminatur". Translation from Newton (1968, p. 97). – Newton also called such a point a *punctum flexionis infinitae* 'point of infinite inflection'. See Newton (1969, p. 181). [2] de l'Hôpital gave a geometrical construction of cusps of the second kind, but could not find any explicit equation. de Gua de Malves (1740, pp. 69f.) then published a proof claiming that algebraic curves can never have such cusps, while d'Alembert and Euler soon found counterexamples to this claim (see Bradley 2006). [3] "semblables à un bec d'oiseau". The similarity of a cusp of the second kind to a beak was already mentioned by Euler in 1744. See his letter to Cramer in Joffredo (2017, p. 138). A corresponding statement is also found in Cramer (1750, p. 572).

and *rhamphoid* [*Ramphoïde*] to those of cusp of the first and second kind" and gives the above-mentioned etymology of *rhamphoid*; he also explains that *keratoid* is based on *κέρας* 'horn'.[1] The terms immediately made their way into Barlow's dictionary (1814, s. v. inflection), where they appear as nouns. It took until the middle of the century for the terms to be more widely used. They appear in some textbooks such as Walton (1846, p. 179). In the second half of the century the more correct spelling *rhamphoid* becomes more common. The German terms *Schnabel* or *Schnabelspitze* (*Schnabel* 'beak', *Spitze* 'cusp') are based on the same metaphor; *Schnabel* was used by Vega (1784, p. 432).

The alternative term *node-cusp* for a cusp of the second kind goes back to Cayley (1866, p. 215) and was chosen because "the cusp of the second kind was equivalent to a double point [which Cayley also called a *node*] and cusp" (ibid.).

A disapproving terminological comment was made by De Morgan (1856, p. 611): "I have long ago ceased trying to keep in memory which is the cusp of the *first* kind, and which of the *second* kind. Nor do I admit the uncouth terms *ceratoid* and *ramphoid*; for the cusp called ramphoid is as much like the point of a horn as that of a beak, and the cusp called ceratoid is not the point of a horn at all, but that of a spear". For the keratoid cusp, other writers perceived different similarities: Fiedler (1875, p. 302) found it looked like a thorn (German *Dorn*), while von Lilienthal (1908, p. 5) found *Hellebardenspitze* (*Hellebarde* 'halberd') more expressive than *cusp of the first kind*.

See also *stationary point*.

Note. For the history of the concept, see Bradley (2006, 2010).

References

Barlow, Peter (1814). *A New Mathematical and Philosophical Dictionary*. London.

Bernoulli, Johann (1955). *Der Briefwechsel von Johann I. Bernoulli*. Band 1: Der Briefwechsel mit Jacob Bernoulli, dem Marquis de l'Hôpital u.a. Ed. by Otto Spiess. Basel.

Bernoulli, Johann (1988). *Der Briefwechsel von Johann I. Bernoulli*. Band 2: Der Briefwechsel mit Pierre Varignon. Erster Teil: 1692-1702. Ed. by Pierre Costabel and Jeanne Peiffer. Basel.

Bradley, Robert E. (2006). "The Curious Case of the Bird's Beak". *International Journal of Applied Mathematics and Computer Science* 1, pp. 243–268.

Bradley, Robert E. (2010). "Cusps: Horns and Beaks". In: *Mathematical Time Capsules*. Ed. by Dick Jardine and Amy Shell-Gellasch. Washington, D.C., pp. 89–99.

Bradley, Robert E., Salvatore J. Petrilli, and C. Edward Sandifer (2015). *L'Hôpital's Analyse des infiniments petits*. An Annotated Translation with Source Material by Johann Bernoulli. Cham.

de Bragelongne, Christophe-Bernard (1732). "Examen des lignes du quatriéme ordre ou courbes du troisiéme ordre". *Mémoires de l'Académie royale des sciences* (1730), pp. 158–216.

Cayley, Arthur (1852). "On the singularities of surfaces". *The Cambridge and Dublin Mathematical Journal* 7, pp. 166–171.

Cayley, Arthur (1866). "On the higher singularities of a plane curve". *The Quarterly Journal of Pure and Applied Mathematics* 7, pp. 212–222.

Cramer, Gabriel (1750). *Introduction à l'analyse des lignes courbes algébriques*. Genève.

Euler, Leonhard (1751). "Sur le point de rebroussement de la seconde espèce de M. le Marquis de l'Hôpital". *Histoire de l'Académie Royale des Sciences et des Belles-Lettres de Berlin* (1749), pp. 203–221.

---

[1] "Nous avons préféré les dénominations de *Cératoïde* et *Ramphoïde* à celles de rebroussement de la 1re, et la 2e espèce sous lesquelles ces points sont connus". Translation from Francœur (1830, p. 314).

Fiedler, Wilhelm (1875). *Die darstellende Geometrie in organischer Verbindung mit der Geometrie der Lage*. Leipzig.
Francœur, Louis-Benjamin (1809). *Cours complet de mathématiques pures*. Tome second. Paris.
Francœur, Louis-Benjamin (1830). *A Complete Course of Pure Mathematics*. Vol. II. Translated from the French by R. Blakelock. Cambridge and London.
de Gua de Malves, Jean Paul (1740). *Usages de l'analyse de Descartes pour découvrir, sans le secours du calcul différentiel, les propriétés des lignes géométriques de tous les ordres*. Paris.
Harris, John (1710). *Lexicon Technicum*. Vol. II. London.
Hayes, Charles (1704). *A Treatise of Fluxions*. London.
de la Hire, Philippe (1707). "Traité des roulettes". *Mémoires de l'Académie royale des sciences* (1706), pp. 340–379.
de l'Hôpital, Guillaume François Antoine (1696). *Analyse des infiniments petits, pour l'intelligence des lignes courbes*. Paris.
Joffredo, Thierry (2017). "Approches biographiques de l'*Introduction à l'analyse des lignes courbes algébriques* de Gabriel Cramer". PhD thesis. Université de Lorraine.
Kästner, Abraham Gotthelf (1761). *Anfangsgründe der Analysis des Unendlichen*. Göttingen.
Leibniz, Gottfried Wilhelm (2011). *Sämtliche Schriften und Briefe*. Dritte Reihe. Mathematischer, naturwissenschaftlicher und technischer Briefwechsel. Siebenter Band. Juli 1696-Dezember 1698. Hannover.
Maclaurin, Colin (1742). *A Treatise of Fluxions*. In Two Books. Edinburgh.
De Morgan, Augustus (1856). "On the Singular Points of Curves, and on Newton's Method of Co-ordinated Exponents". *Transactions of the Cambridge Philosophical Society* 9, pp. 608–627.
Newton, Isaac (1968). *The Mathematical Papers of Isaac Newton*. Volume II. 1667-1670. Ed. by D.T. Whiteside. Cambridge.
Newton, Isaac (1969). *The Mathematical Papers of Isaac Newton*. Volume III. 1670-1673. Ed. by D.T. Whiteside. Cambridge.
Stone, Edmund (1743). *A New Mathematical Dictionary*. London.
Vega, Georg (1784). *Vorlesungen über die Mathematik*. Zweyter Band. Wien.
von Lilienthal, Reinhold (1908). *Vorlesungen über Differentialgeometrie*. Erster Band: Kurventheorie. Leipzig.
Walton, William (1846). *A Treatise on the Differential Calculus*. Cambridge.
Wolff, Christian (1710). *Der Anfangs-Gründe aller Mathematischen Wissenschaften Letzter Theil*. Halle.
Wolff, Christian (1716). *Mathematisches Lexicon*. Leipzig.

cut

The German term *Schnitt* (from *schneiden* 'to cut') was introduced by Dedekind in his famous *Continuity and Irrational Numbers*:

> every rational number a effects a separation of the system R into two classes such that every number a_1 of the first class A_1 is less than every number a_2 of the second class A_2; the number a is either the greatest number of the class A_1 or the least number of the class A_2. If now any separation of the system R into two classes A_1, A_2 is given which possesses only *this* characteristic property that every number a_1 in A_1 is less than every number a_2 in A_2, then for brevity we shall call such a separation a *cut* [*Schnitt*] and designate it by (A_1, A_2) (Dedekind 1872, p. 19).[1]

---

1 "[...] daß jede rationale Zahl a eine Zerlegung des Systems R in zwei Classen A_1, A_2 von der Art hervorbringt, daß jede Zahl a_1 der ersten Classe A_1 kleiner ist als jede Zahl a_2 der zweiten Classe A_2; die Zahl a ist entweder die größte Zahl der Classe A_1 oder die kleinste Zahl der Klasse A_2. Ist nun irgendeine Eintheilung des Systems R in zwei Classen A_1, A_2 gegeben, welche nur *die* charakteristische Eigenschaft besitzt, daß jede Zahl a_1 in A_1 kleiner ist als jede Zahl a_2 in A_2, so wollen wir der Kürze halber eine solche Eintheilung einen *Schnitt* nennen und mit (A_1, A_2) bezeichnen". Translation from Dedekind (1909, pp. 12f.).

The French translation *section* was used by Dedekind (1877, p. 7); Couturat (1896, p. 60) preferred *coupure*, which became the standard term. English *cut* appeared in Smith (1896, p. 514); Pierpont (1905, p. 82) preferred *partition.*

References

Couturat, Louis (1896). *De l'infini mathématique*. Paris.
Dedekind, Richard (1872). *Stetigkeit und irrationale Zahlen*. Braunschweig.
Dedekind, Richard (1877). *Sur la théorie des nombres entiers algébriques*. Paris.
Dedekind, Richard (1909). *Essays on the Theory of Numbers*. I. Continuity and Irrational Numbers, II. The Nature and Meaning of Numbers. Authorized translation by Wooster Woodruff Beman. Chicago.
Pierpont, James (1905). *Lectures on the Theory of Functions of Real Variables*. Volume I. Boston.
Smith, David Eugene (1896). "History of Modern Mathematics". In: *Higher Mathematics*. A Text-book for Classical and Engineering Colleges. Ed. by Mansfield Merriman and Robert S. Woodward. New York, pp. 508–570.

CW-complex

CW-complexes are generalizations of simplicial complexes. The historical context in which the concept was developed is explained by Dieudonné (2009, p. 221):

> In his study of the homology of grassmannians [...][1] begun in 1933, Ehresmann was led to consider a partition of a space X into "cells" of a more general type than those in use up to that time. For each dimension p a cell of dimension p is still homeomorphic to an open ball in R^p, but is boundary in X is not necessarily homeomorphic to S_{p-1} as for usual cells. This idea was developed in more general contexts by J.H.C. Whitehead in 1941 [...][2] under the name of "membrane complex," which he changed in 1950 to CW-*complex*.

The term first appeared in Whitehead (1949, p. 223):

> We shall describe a complex, K, as closure finite if, and only if, if $K(e)$ is a finite subcomplex, for every cell $e \in K$. [...] We shall say that K has the weak topology [...] if, and only if, a subset $X \subset K$ is closed (open) provided $X \cap e$ is closed (relatively open) for each cell $e \in K$ [...] By a CW-*complex* we mean one which is closure finite and has the weak topology.[3]

Note. Historical comments are found in Dieudonné (2009, pp. 220–223) and Burde and Zieschang (1999, pp. 107f.).

References

Burde, Gerhard and Heiner Zieschang (1999). "Development of the Concept of a Complex". In: *History of Topology*. Ed. by Ioan Mackenzie James. Amsterdam, pp. 103–110.
Dieudonné, Jean (2009). *A History of Algebraic and Differential Topology, 1900–1960*. Boston.
Ehresmann, Charles (1934). "Sur la topologie de certains espaces homogènes". *Annals of Mathematics* 35, pp. 396–443.
Milnor, John W. and James D. Stasheff (1974). *Characteristic Classes*. Princeton.
Whitehead, J.H.C. (1941). "On Incidence Matrices, Nuclei and Homotopy Types". *Annals of Mathematics* 42, pp. 1197–1239.
Whitehead, J.H.C. (1949). "Combinatorial homotopy. I". *Bulletin of the American Mathematical Society* 55, pp. 213–245.

[1] See Ehresmann (1934). [2] Dieudonné refers to Whitehead (1941, p. 1211). [3] Dieudonné (2009, p. 222) cites a comment on the name *weak topology* by Milnor and Stasheff (1974, p. 63): "It is customary in algebraic topology to call this the 'weak topology,' a weak topology being one with many open sets. This usage is unfortunate since analysts use the term weak topology with precisely the opposite meaning. On the other hand the terms 'fine topology' or 'large topology' or 'Whitehead topology' are certainly acceptable".

***cycle* (topology)**

In topology, the terminology goes back to Listing. In his "Census of spatial complexes" (Listing 1862), he introduced what he called *cyclosis* (German *Cyklose*), which meant, roughly, the existence of cycles (*Cyklen*), which may be understood as closed curves (Listing 1862, pp. 109f.).

While Listing's paper at first remained largely unnoticed in Germany, it did come to the attention of Maxwell.[1] He adopted some of Listing's terms, and in his *Treatise on Electricity and Magnetism* he defined:

> Let there be p points in space, and let l lines of any form be drawn joining these points so that no two lines intersect each other, and no point is left isolated. We shall call a figure composed of lines in this way a Diagram. Of these lines, $p-1$ are sufficient to join the p points so as to form a connected system. Every new line completes a loop or closed path, or, as we shall call it, a Cycle. [...] The existence of cycles is called Cyclosis, and the number of cycles in a diagram is called its Cyclomatic number (Maxwell 1873, p. 16).[2]

Poincaré used French *cycle* in his famous series of papers on topology; he gave a definition right after one of his definitions of *homology*:

> Given a closed manifold V of p dimensions, I consider other manifolds contained in it, closed or not, of fewer dimensions. I denote by W_q such a manifold of q dimensions.
>
> If $\sum W_q$ is a set of q-dimensional manifolds and $\sum W_{q-1}$ is a set of $(q-1)$-dimensional manifolds, the congruence
>
> $$\sum W_q \equiv \sum W_{q-1}$$
>
> signifies (by definition) that $\sum W_{q-1}$ is the complete boundary of the set of manifolds $\sum W_q$. I express the same fact without showing $\sum W_q$ by writing the relation
>
> $$\sum W_{q-1} \sim 0,$$
>
> which I call a *homology*.
>
> Then the congruence
>
> $$\sum W_q \equiv 0$$
>
> signifies that the manifold $\sum W_q$ is closed.
>
> If we have $\sum W_q \equiv 0$ without having $\sum W_q \sim 0$ (or $n \sum W_q \sim 0$, for some integer n), I say that the manifold $\sum W_q$ is a cycle of q dimensions (Poincaré 1902, pp. 169f.).[3]

Closed chains appear under this name in Alexander (1926, p. 313): "A chain will be said to be *closed* (as distinguished from *open*, or *bounded*) if its boundary

---

[1] According to Epple (1998, p. 346), Maxwell got to know the paper in late 1868. [2] On Maxwells' definition, see Epple (1998, pp. 339f.). [3] "Je rappelle qu'étant donné une variété V fermée à p dimensions, je trace sur cette variété d'autres variétés, fermées ou non, d'un moins grand nombre de dimensions; je désigne par W_q une variété à q dimensions tracée de la sorte sur V. Si $\sum W_q$ est un ensemble de variétés à q dimensions et $\sum W_{q-1}$ un ensemble de variétés à $q-1$ dimensions, la congruence $\sum W_q \equiv \sum W_{q-1}$ signifie (par définition) que $\sum W_{q-1}$ forme la frontière complète de l'ensemble de variétés $\sum W_q$. J'exprime le même fait sans mettre en évidence $\sum W_q$ en écrivant la relation $\sum W_{q-1} \sim 0$ que j'appelle *une homologie*. Alors la congruence $\sum W_q \equiv 0$ signifiera que la variété $\sum W_q$ est fermée. Si l'on a $\sum W_q \equiv 0$ sans avoir $\sum W_q \sim 0$ (ou $n \sum W_q \sim 0$, n étant entier) je dirai que la variété $\sum W_q$ constitue un cycle à q dimensions". Translation from Poincaré (2010, p. 151).

vanishes". At the same time, Vietoris (1926, p. 1008) gave the following definition: "A homogeneous n-dimensional complex without boundary shall be called an *n-dimensional cycle*".[1] The term was also preferred by Brown (1930, p. 251): "We shall use the nomenclature of analysis situs as defined by Alexander, substituting the terms *cycle* and *Betti number* for *closed chain* and *connectivity number* respectively".

With the generalization of homology theories during the 20th century, the concept of cycle was concomitantly generalized to designate any closed chain (see *chain*), i. e. one that is mapped to zero by a generalized "boundary operator" of the respective theory.

References

Alexander, James W. (1926). "Combinatorial analysis situs". *Transactions of the American Mathematical Society* 28, pp. 301–329.

Brown, Arthur B. (1930). "Relations Between the Critical Points of a Real Analytic Functions of *N* Independent Variables". *American Journal of Mathematics* 52, pp. 251–270.

Epple, Moritz (1998). "Topology, Matter, and Space, I: Topological Notions in 19th-Century Natural Philosophy". *Archive for History of Exact Sciences* 52, pp. 297–392.

Listing, Johann Benedict (1862). "Der Census räumlicher Complexe, oder Verallgemeinerung des Euler'schen Satzes von den Polyëdern". *Abhandlungen der Königlichen Gesellschaft der Wissenschaften zu Göttingen* 10, pp. 97–182.

Maxwell, James Clerk (1873). *A Treatise on Electricity and Magnetism*. Vol. I. Oxford.

Poincaré, Henri (1902). "Sur les cycles des surfaces algébriques. Quatrième complément à l'analysis situs". *Journal de mathématiques pures et appliquées* 8, pp. 169–214.

Poincaré, Henri (2010). *Papers on Topology. Analysis Situs* and Its Five Supplements. Translated by John Stillwell. Providence, R.I.

Vietoris, Leopold (1926). "Ueber den höheren Zusammenhang von kompakten Räumen und eine Klasse von Abbildungen, welche ihn ungeändert lässt". *Koninklijke Akademie van Wetenschappen te Amsterdam. Proceedings of the Section of Sciences* 29, pp. 1008–1013.

cyclic (algebra)

In 1815, Cauchy published two papers which are regarded as the beginning of a systematic treatment of permutations (see, e. g., Wußing 1984, ch. 2). He called a cyclic permutation a *substitution circulaire* (Cauchy 1815, p. 17).[2] This terminology remained standard in the 19th century, although the adjective *cyclic* began to be more commonly used in the second half of the 19th century, for example by Sylvester (1861, p. 380) and by Petersen (1878, p. 252), who used German *cyklisch*.

But even before Cauchy, the adjective is occasionally found in this sense. In particular, Hirsch (1809, p. 305) defined what he called (in German) a *cyklische Versetzung* 'cyclic permutation':[3]

> The characteristic feature of such permutations is that each root moves one place forward or backward, and in the first case assumes the place of the first, but in the second case assumes the place of the last, so that a kind of circular movement takes place [...] The permutations

---

[1] "Ein homogen n-dimensionaler Komplex ohne Rand heisse ein *n-dimensionaler Zyklus*". [2] See *permutation* for the term *substitution*. [3] The word *Versetzung* literally means 'displacement' or 'move, transfer'; it was commonly used as an alternative to *permutation*.

shall also be called cyclic if only some of the roots move in the indicated way, while the other remain at their place.[1]

The concept of cyclic group can be found in various publications from around 1800. It occurs in Ruffini (1799, pp. 247f.) (see Wußing 1984, p. 82), and in Gauß (1801, § 54, § 339) (see Wußing 1984, pp. 52f.), for example.[2] The term, however, is of later origin.

Frobenius and Stickelberger (1879, p. 219) called cyclic groups *elementary* (German *elementar*): "If a group has a basis of r elements and no basis of less than r elements, r is called the rank (*Rang*) of the group. A group is called *elementary*, if it is of the first rank, i. e. if all its elements can be represented as powers of one if its elements".[3]

The explicit term *cyclic group* appears to come up around 1880. It is found in Gierster (1881, p. 319) as German *cyklische Gruppe*; Gierster does not give a definition but refers to Serret (1866, pp. 363ff.), where the notion, but not the term is found.

Netto (1882, p. 133), in his book on the theory of permutations, then gave a definition of cyclic groups (German *cyklische Gruppen*): "One type can be established at once, in its full generality. This type is formed by the powers of a circular substitution. A group Ω of this type is called a *cyclical group*, and every function of n elements which belongs to a cyclical group is called a *cyclical function*".[4] This definition also shows the relation to cyclic functions.[5] The term *fonction cyclique* in this sense appears to be due to Serret (1854, p. 563). It was soon in common usage and appears, for example, in Mathieu (1861, p. 320): "A function is called *cyclic* if it remains invariable by a certain circular permutation containing all its variables".[6] This link to Cauchy's "circular substitutions" (cyclic permutations) is the reason why the term was chosen, as Brioschi (1878, p. 148) explicitly states.

The English term *cyclic group* appears in Klein (1888, p. 6); French *groupe cyclique* is in Molk (1883, p. 64).

In sum, the term *cyclic permutation*, the oldest of the terms discussed here, was the reason why also cyclic functions were so called, and because of the relation between cyclic functions and cyclic groups, the term was eventually also transferred to groups.

---

[1] "Das karakteristische Merkmal solcher Versetzungen bestehet darin, dass jede Wurzel um eine Stelle vor- oder rückwärts rückt, und im ersten Falle die letzte die Stelle der ersten, im zweyten Falle aber, die erste die Stelle der letzten einnimmt, so dass eine Art von Kreisbewegung stattfindet [...] Die Versetzungen sollen auch alsdann noch cyklisch heißen, wenn bloß einige der Wurzeln sich auf die angegebene Art bewegen, die übrigen aber an ihrer Stelle bleiben". [2] On Galois' use of the concept and his terminology, see Neumann (2011, p. 26). [3] "Besitzt eine Gruppe eine Basis von r Elementen und keine Basis von weniger als r Elementen, so heisst r der *Rang* der Gruppe. Eine Gruppe heisst *elementar*, wenn sie vom ersten Range ist, wenn sich also alle ihre Elemente als Potenzen von einem derselben darstellen lassen". [4] "Einen Typus können wir ganz allgemein aufstellen. Er wird durch die Potenzen einer Cirkularsubstitution gebildet. Wir nennen eine solche Gruppe Ω ein[e] *cyklische Gruppe* und jede Funktion von n Veränderlichen, welche unter dem Einflüsse einer cyklischen Gruppe des Grades n ungeändert bleibt, eine *cyklische Funktion*". Translation from Netto (1892, p. 145). [5] Terminological transfers between certain types of functions or equations to groups or vice versa are quite common. See, for example, *solvable group* or *transitive group*. [6] "On appelle fonction cyclique une fonction qui reste invariable par une certaine substitution circulaire contenant toutes ses variables".

Transfers to other algebraic structures are of later origin. Hilbert (1896, p. 29), for example, applied the term to fields.

Note. For Cauchy's theory of permutations, see Wußing (1984, ch. 2) and Dahan (1980). Remarks on the history of cyclic groups can be found in Wußing (1984, pp. 52f., 82) and Kiernan (1971).

References

Brioschi, Francesco (1878). "Ueber die Auflösung der Gleichungen vom fünften Grade". *Mathematische Annalen* 13, pp. 109–160.

Cauchy, Augustin-Louis (1815). "Sur le nombre des valeurs qu'une fonction peut acquérir, lorsqu'on y permute de toutes les manières possibles les quantités qu'elle renferme". *Journal de l'École polytechnique* 10, cahier 17, pp. 1–28.

Dahan, Amy (1980). "Les travaux de Cauchy sur les substitutions. Étude de son approche du concept de groupe". *Archive for History of Exact Sciences* 23, pp. 279–319.

Frobenius, Georg and Ludwig Stickelberger (1879). "Ueber Gruppen von vertauschbaren Elementen". *Journal für die reine und angewandte Mathematik* 86, pp. 217–262.

Gauß, Carl Friedrich (1801). *Disquisitiones arithmeticae*. Leipzig.

Gierster, Joseph (1881). "Die Untergruppen der Galois'schen Gruppe der Modulargleichungen für den Fall eines primzahligen Transformationsgrades". *Mathematische Annalen* 18, pp. 319–365.

Hilbert, David (1896). "Ein neuer Beweis des Kroneckerschen Fundamentalsatzes über Abelsche Zahlkörper". *Nachrichten von der Königl. Gesellschaft der Wissenschaften zu Göttingen (mathematisch-physikalische Klasse)* (1896), pp. 29–39.

Hirsch, Meier (1809). *Sammlung von Aufgaben aus der Theorie der algebraischen Gleichungen*. Erster Theil. Berlin.

Kiernan, B. Melvin (1971). "The Development of Galois Theory from Lagrange to Artin". *Archive for History of Exact Sciences* 8, pp. 40–154.

Klein, Felix (1888). *Lectures on the Ikosahedron and the Solution of Equations of the Fifth Degree*. Translated by George Gavin Morrice. London.

Mathieu, Émile (1861). "Mémoire sur l'étude des fonctions de plusieurs quantités, sur la manière de les former et sur les substitutions qui les laissent invariables". *Journal de mathématiques pures et appliquées* 6, pp. 241–323.

Molk, Jules (1883). "Netto, E. Substitutionentheorie und ihre Anwendung auf die Algebra [review]". *Bulletin des sciences mathématiques et astronomiques* 7, pp. 57–72.

Netto, Eugen (1882). *Substitutionentheorie und ihre Anwendungen auf die Algebra*. Leipzig.

Netto, Eugen (1892). *The Theory of Substitutions and its Applications to Algebra*. Revised by the author and translated with his permission by F.N. Cole. Ann Arbor, Mich.

Neumann, Peter M. (2011). "Introduction". In: *The Mathematical Writings of Évariste Galois*. Ed. by Peter M. Neumann. Zürich, pp. 1–32.

Petersen, Julius (1878). *Theorie der algebraischen Gleichungen*. Kopenhagen.

Ruffini, Paolo (1799). *Teoria generale delle equazioni, in cui si dimostra impossibile la soluzione algebraica delle equazioni generali di grado superiore al quarto*. Bologna.

Serret, Joseph-Alfred (1854). *Cours d'algèbre supérieure*. Deuxième édition. Paris.

Serret, Joseph-Alfred (1866). *Cours d'algèbre supérieure*. Troisième édition. Tome second. Paris.

Sylvester, James Joseph (1861). "On a Generalization of a Theorem of Cauchy on Arrangements". *The London, Edinburgh, and Dublin Philosophical Magazine and Journal of Science* 22, pp. 378–382.

Wußing, Hans (1984). *The Genesis of the Abstract Group Concept*. A Contribution to the History of the Origin of Abstract Group Theory. Tr. Abe Shenitzer. Cambridge, Mass.

cyclide

According to Charles Dupin's (1822, p. 200) original definition, surfaces are called "cyclide surfaces" (French *surfaces cyclides*) if all their lines of curvature are circles.[1] Dupin (1822, p. 201) points out himself that the term is based on this characteristic property; the word *cyclide* consists of a Greek element meaning 'circle' (see *circle*). Dupin coined the term by 1816: Dupin (1822) is based on a talk he gave on Jan. 22, 1816. In print, *surfaces cyclides* first appears in a report on a paper read by Dupin in 1817 (Arago, Ampère, and Cauchy 1818, p. lj).

The term was soon accepted and the concept extended by Maxwell (1868) and Cayley (1873), in particular.

Note. For the history of the concept, see Chandru, Dutta, and Hoffmann (1989).

References

Arago, François, André-Marie Ampère, and Augustin-Louis Cauchy (1818). "Mémoire sur les routes suivies par la lumière dans les phénomènes de réflexion, par M. Dupin". *Mémoires de l'Académie royale des sciences de l'Institut de France* 2, l–lij.

Cayley, Arthur (1873). "On the cyclide". *The Quarterly Journal of Pure and Applied Mathematics* 12, pp. 148–163.

Chandru, V., D. Dutta, and C. M. Hoffmann (1989). "On the geometry of Dupin cyclides". *The Visual Computer* 5, pp. 277–290.

Dupin, Charles (1822). "Sur les routes suivies par la lumière et par les corps élastiques, en général, dans les phénomènes de la réflexion et de la réfraction". In: *Applications de géométrie et de méchanique*. Paris, pp. 187–245.

Maxwell, James Clerk (1868). "On the Cyclide". *The Quarterly Journal of Pure and Applied Mathematics* 9, pp. 111–126.

cycloid

The invention of this curve has been attributed to various mathematicians.[2] It is Charles de Bouelles who is usually cited as having been the first who made use of it. He discusses the cycloid, without using a name, in a treatise on the quadrature of the circle (de Bouelles 1503, fol. lxxxvi^{r}); see fig. 36.[3]

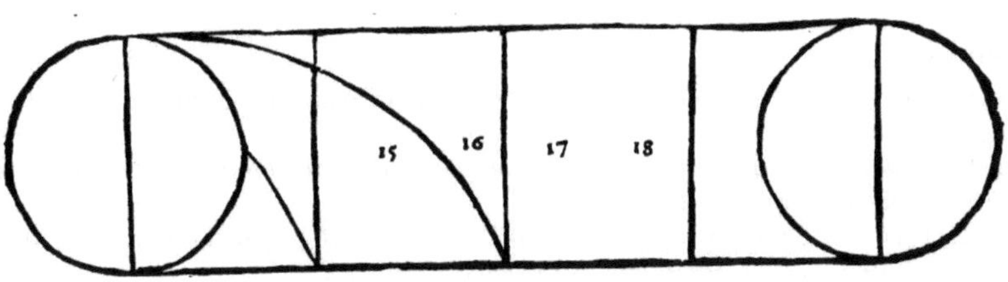

Figure 36: de Bouelles' illustration of the cycloid. From de Bouelles (1503, fol. lxxxvi^{r}).

[1] "surfaces dont toutes les lignes de courbure sont des cercles". – Because Dupin originated the term, such a surface is also called *Dupin's cyclide*. [2] Tannery (1883, p. 284), citing Iamblichus as evidence, argued that the curve was already known in antiquity. See also Boyer (1956, p. 33) and Loria (1911, p. 74). [3] de Bouelles' use of the cycloid was already pointed out by Wallis (1697). See Günther (1887, p. 13), who concludes that de Bouelles gave "the first genetic construction of the cycloid", and also Klibansky (1980, p. 359). The construction is also found in de Bouelles (1542, fol. 34^{v}).

In print, the Latin term *cyclois* is first found in Torricelli (1644, p. 85), who also stated, naming Galileo Galilei, that the curve "was so named by our predecessors 45 years ago".[1] Also, in a letter from 1643 to Roberval, he said that the term was coined by Galilei (see Torricelli 1693, p. 283).[2] Since Galilei's text is lost (Favaro 1912-13), Torricelli is the only witness for this attribution. What does remain is a letter by Galilei, dated Feb. 24, 1640, in which, however, the term does *not* occur: "More than 50 years ago I had it in mind to describe that curved line [*linea arcuata*, i. e. the cycloid] [...] so as to adapt it to the arches of a bridge" (Galilei 1856, p. 366).[3]

In 1658 Pascal published an account of the history of the cycloid, written under the pseudonym Dettonville. He states there that Mersenne had studied the curve in 1615 and called it a *roulette* (Pascal 1658, p. 1); the word contains French *roue* 'wheel'. In print, it is first found in Mersenne (1637, p. 24).[4] Mersenne first used Latin *cyclois* in January 1640 in a letter to John Pell (de Waard 1965, p. 51), and French *cycloide* in May 1640 (de Waard 1965, p. 306).[5]

Pascal (1658, p. 2) further mentions the term *trochoid* (*trochoïde*), preferred by Roberval and found in a posthumously published treatise (de Roberval 1693, p. 246). The word is an analogue to *roulette* based on the Greek word *τροχός* 'wheel'. Wolff (1710, p. 155) used the corresponding German *Rad-Linie* (from *Rad* 'wheel').

Finally, Pascal (1658, p. 3) states that Jean de Beaugrand in 1638 changed these terms, *roulette* and *trochoid*, to *cycloid* (*cycloïde*).

The curve had been the topic of a fierce debate about questions of priority of discovery and plagiarism. That is why Montucla (1798, pp. 52, 55) also described the cycloid as the "apple of discord" (*pomme de discorde*) and the "Helen of Geometers" (*Hélène des Géomètres*).

Note. The history of the cycloid has been much discussed. For overviews, see Whitman (1943), Martin (2010), Loria (1911, pp. 73–75), and Freguglia and Giaquinta (2016, pp. 27–37). For details concerning the terminology, see also the editorial notes by Rochot in de Waard (1965, pp. 51, 126f., 584–586, 607f.). Bold (1996, ch. 8) discusses some of the terminological choices Pascal made when writing his history of the cycloid.

References

Bold, Stephen C. (1996). *Pascal Geometer*. Discovery and Invention in Seventeenth-century France. Genève.

de Bouelles, Charles (1503). *Liber de circuli quadratura*. In: *In hoc libro contenta*. Ed. by Jacques Lefèvre d'Étaples. Paris, fol. 85–87.

de Bouelles, Charles (1542). *Livre singulier & utile, touchant l'art et practique de geometrie*. Paris.

Boyer, Carl B. (1956). *History of Analytic Geometry*. New York.

Favaro, Antonio (1912-13). "Sulla invenzione del nome 'Cicloide'". *Bibliotheca Mathematica* 13, pp. 351–352.

[1] "Vocata est à praedecessoribus nostris. Praecipue à Galileo iam supra 45". [2] On this letter, see Itard (1975, p. 114). [3] "Quella linea arcuata sono più di cinquant'anni che mi venne in mente di descriverla [...] per adattarla agli archi di un ponte". Translation from Freguglia and Giaquinta (2016, p. 34). [4] To be precise, the word appears in a supplement to the *Harmonie universelle* entitled "Nouvelles observations physiques et mathématiques", which was printed in 1638 (de Waard 1965, p. 584). Hara (1971, p. 245, fn. 28) has collected the places where Mersenne discusses the cycloid in his publications. [5] The English term appeared in Wallis (1662, p. 124).

Freguglia, Paolo and Mariano Giaquinta (2016). *The Early Period of the Calculus of Variations*. Basel.

Galilei, Galileo (1856). *Le opere di Galileo Galilei*. Supplemento. Firenze.

Günther, Siegmund (1887). "War die Zykloide bereits im XVI. Jahrhundert bekannt?" *Bibliotheca Mathematica* 1, pp. 8–14.

Hara, Kokiti (1971). "Pascal et Wallis au sujet de la cycloïde (II)". *Gallia* 10-11, pp. 231–249.

Itard, Jean (1975). "La lettre de Torricelli à Roberval d'octobre 1643". *Revue d'histoire des sciences* 28, pp. 113–124.

Klibansky, Raymond (1980). "Nicolas de Cues, Charles de Bovelles et la cycloïde". In: *Correspondance du P. Marin Mersenne*. XIV. Ed. by Cornelis de Waard and Armand Beaulieu. Paris, pp. 358–362.

Loria, Gino (1911). *Spezielle algebraische und transzendente ebene Kurven*. Theorie und Geschichte. Autorisierte, nach dem italienischen Manuskript bearbeitete deutsche Ausgabe von Fritz Schütte. Zweite Auflage. Zweiter Band. Leipzig.

Martin, John (2010). "The Helen of Geometry". *The College Mathematics Journal* 41, pp. 17–27.

Mersenne, Marin (1637). *Nouvelles observations physiques et mathematiques*. In: *Harmonie universelle, contenant la theorie et la pratique de la musique*. Paris.

Montucla, Jean-Étienne (1798). *Histoire des mathématiques*. Nouvelle édition, considérablement augmentée. Tome second. Paris.

Pascal, Blaise (1658). *Histoire de la roulette, appellée autrement la trochoïde, ou la cycloïde*. Paris.

de Roberval, Gilles Personne (1693). *De trochoide ejusque spatio*. In: *Divers ouvrages de mathématique et de physique*. Paris, pp. 246–278.

Tannery, Paul (1883). "Notes pour l'histoire des lignes et surfaces courbes dans l'antiquité (1)". *Bulletin des sciences mathématiques et astronomiques* 7, pp. 278–291.

Torricelli, Evangelista (1644). *Appendix de dimensione cycloidis*. In: *Opera geometrica*. Firenze, pp. 85–92.

Torricelli, Evangelista (1693). *Clarissimo viro Robervallio*. In: *Divers ouvrages de mathématique et de physique par Messieurs de l'Académie Royale des Sciences*. Paris, pp. 283–302.

de Waard, Cornelius, ed. (1965). *Correspondance du P. Marin Mersenne, Religieux Minime*. IX. Paris.

Wallis, John (1662). *Hobbius Heauton-timorumenos, or, A consideration of Mr. Hobbes his dialogves in an epistolary discourse, addressed, to the Honourable Robert Boyle, Esq.* Oxford.

Wallis, John (1697). "An extract of a letter from Dr. Wallis, of May 4. 1697, concerning the cycloeid known to Cardinal Cusanus, about the Year 1450; and to Carolus Bovillus about the Year 1500". *Philosophical Transactions of the Royal Society of London* 19, pp. 561–566.

Whitman, E.A. (1943). "Some Historical Notes on the Cycloid". *The American Mathematical Monthly* 50, pp. 309–315.

Wolff, Christian (1710). *Der Anfangs-Gründe aller Mathematischen Wissenschaften Letzter Theil*. Halle.

cyclomatic

In graph theory, the cyclomatic number indicates the number of independent cycles in a given graph. The germs of this notion can already be found in Kirchhoff (1847), but the term is due to Listing (1861, p. 355, 1862, p. 130), who used German *cyklomatische Ordnungszahl* 'cyclomatic order number'. The term was chosen because a cycle in a graph was called *Cyklus* by Listing (see *cycle*).

The term was picked up early, for example by Maxwell (1873, p. 16): "the number of cycles in a diagram is called its Cyclomatic number"; the French term *nombre cyclomatique* is found in a translation of Maxwell's treatise (Maxwell 1885, p. 17).

Alternative terms include French *degré de continuité* 'degree of continuity' for the number 1 greater than the cyclomatic number (Jordan 1869, p. 186). Whitney (1931, p. 125) used *nullity* because the number is equal to the nullity of the oriented incidence matrix associated with the graph. König (1936, p. 53) used German *Zusammenhangszahl* 'connectivity number' and also *Index*, terms that did not catch on since they already had other mathematical meanings. The terms *cycle rank* and *circuit rank* appear to be of later origin; the latter may be due to Ore (1962, p. 67), while the former was used by Harary (1967, p. 10), whose text is based on lectures held in 1962-63.

Note. The history of the notion and terminology is briefly summarized in Dehn and Heegaard (1907, pp. 171f.), König (1990, p. 135, fn. 5), and Biggs, Lloyd, and Wilson (1998, p. 136).

References

Biggs, Norman L., E. Keith Lloyd, and Robin J. Wilson (1998). *Graph Theory, 1736-1936*. Oxford.

Dehn, Max and Poul Heegaard (1907). "Analysis situs". *Encyklopädie der mathematischen Wissenschaften mit Einschluss ihrer Anwendungen* III AB3, pp. 153–220.

Harary, Frank (1967). "Some Theorems and Concepts of Graph Theory". In: *A Seminar on Graph Theory*. Ed. by Frank Harary and Lowell Beineke. New York et al., pp. 1–12.

Jordan, Camille (1869). "Sur les assemblages de lignes". *Journal für die reine und angewandte Mathematik* 70, pp. 185–190.

Kirchhoff, Gustav Robert (1847). "Ueber die Auflösung der Gleichungen, auf welche man bei der Untersuchung der linearen Vertheilung galvanischer Ströme geführt wird". *Annalen der Physik und Chemie* 12, pp. 497–508.

König, Dénes (1936). *Theorie der endlichen und unendlichen Graphen*. Leipzig.

König, Dénes (1990). *Theory of Finite and Infinite Graphs*. Translated by Richard McCoart. Boston.

Listing, Johann Benedict (1861). "Der Census räumlicher Complexe, oder Verallgemeinerung des Euler'schen Satzes von den Polyëdern". *Nachrichten von der Georg-Augusts-Universität und der Königl. Gesellschaft der Wissenschaften zu Göttingen* (1861), pp. 352–358.

Listing, Johann Benedict (1862). "Der Census räumlicher Complexe, oder Verallgemeinerung des Euler'schen Satzes von den Polyëdern". *Abhandlungen der Königlichen Gesellschaft der Wissenschaften zu Göttingen* 10, pp. 97–182.

Maxwell, James Clerk (1873). *A Treatise on Electricity and Magnetism*. Vol. I. Oxford.

Maxwell, James Clerk (1885). *Traité d'électricité et de magnétisme*. Traduit de l'anglais sur la deuxième édition, par G. Séligman-Lui. Tome I. Paris.

Ore, Øystein (1962). *Theory of Graphs*. Providence, R.I.

Whitney, Hassler (1931). "Non-Separable and Planar Graphs". *Proceedings of the National Academy of Sciences* 17, pp. 125–127.

cyclotomic

This adjective is related to the division of the circle. Greek *κύκλος* means 'circle', and the ending is related to the verb *τέμνειν* 'to cut'. While the division of the circle has, of course, been known for a long time, the adjective was not introduced in an algebraic context until 1879.

It was Sylvester (1879, p. 357) who used *cyclotomic* with respect to functions (polynomials) as a translation of German *Kreistheilungsfunction*[1], a word previously coined by Klein (1877, p. 168). *Kreisteilung* literally means 'division of the circle'. Sylvester (1879, p. 362) defined: "let $p = mk + 1$. Then $x^k - 1$ is a factor of $x^{p-1} - 1$. Let $\chi_k x$ be the factor of $x^k - 1$, which contains all its primitive roots; this is what I term a *cyclotomic function of the first species* to the index k", and the

[1] The current spelling is *Kreisteilungsfunktion*.

term thus referred to cyclotomic polynomials.[1] Explicitly, *cyclotomic polynomial* (German *Kreisteilungspolynom*) is found in Netto (1882).[2]

The term *cyclotomic equation* (German *Kreistheilungsgleichung*) was defined by Bachmann (1872, p. 18): "The equation [$\frac{x^p-1}{x-1} = x^{p-1} + x^{p-2} + \cdots + x + 1 = 0$] shall henceforth be called *cyclotomic equation*".[3] At the time, this term had already been in common use. It appears, for example, in Kummer (1850, p. 160), where it is used without definition since its meaning must have been clear. In earlier texts, the notion was often paraphrased, for example by Legendre (1797, p. 143), who spoke of "the equations relative to the division of the circle";[4] see also Gauß (1801, p. xii).

The term *cyclotomic field* appears shortly after Sylvester (1879) as French *corps cyclotomique* in Dedekind (1880, pp. 1206f.). A definition was given by Weber (1886, p. 202), who used German *Kreiskörper*, from *Kreis* 'circle' and *Körper*, Dedekind's term for 'field': "By a cyclotomic field one understands each number field composed of rational numbers and roots of unity".[5] Weber (1896, p. 68) later used the longer *Kreistheilungskörper*, which is still the standard term in German.

Note. For the history of the notion, see, e. g., Neumann (2007) and Weil (1975).

References

Bachmann, Paul (1872). *Die Lehre von der Kreistheilung und ihre Beziehungen zur Zahlentheorie*. Leipzig.

Dedekind, Richard (1880). "Sur la théorie des nombres complexes idéaux". *Comptes rendus de l'Académie des sciences* 90, pp. 1205–1207.

Gauß, Carl Friedrich (1801). *Disquisitiones arithmeticae*. Leipzig.

Klein, Felix (1877). "Ueber lineare Differentialgleichungen". *Mathematische Annalen* 12, pp. 167–179.

Kummer, Ernst Eduard (1850). "Allgemeine Reciprocitätsgesetze für beliebig hohe Potenzreste". *Monatsberichte der Königlich Preußischen Akademie der Wissenschaften zu Berlin* (1850), pp. 154–165.

Legendre, Adrien-Marie (1797). *Essai sur la théorie des nombres*. Paris.

Netto, Eugen (1882). *Substitutionentheorie und ihre Anwendungen auf die Algebra*. Leipzig.

Neumann, Olaf (2007). "Cyclotomy: From Euler through Vandermonde to Gauss". In: *Leonhard Euler*. Life, Work and Legacy. Ed. by Robert E. Bradley and C. Edward Sandifer. Amsterdam, pp. 323–362.

Sylvester, James Joseph (1879). "On Certain Ternary Cubic-Form Equations". *American Journal of Mathematics* 2, pp. 357–393.

Weber, Heinrich (1886). "Theorie der Abel'schen Zahlkörper". *Acta Mathematica* 8, pp. 193–263.

Weber, Heinrich (1896). *Lehrbuch der Algebra*. Zweiter Band. Braunschweig.

Weil, André (1975). "La cyclotomie jadis et naguère". *Séminaire Bourbaki* 16, exposé 452, pp. 318–338.

cylinder

The Greek term *κύλινδρος* 'cylinder, roll' is derived from the verb *κυλίνδειν* 'to roll'. Euclid (*Elements*, book 11, def. 21) gave the following definition: "When a rectangular parallelogram with one side of those about the right angle remains fixed

[1] Cyclotomic polynomials were defined by Gauß (1801, § 339). [2] The term only appears in the table of contents, not in the text. [3] "Die Gleichung [$\frac{x^p-1}{x-1} = x^{p-1} + x^{p-2} + \cdots + x + 1 = 0$] soll hinfort als *Kreistheilungsgleichung* bezeichnet werden". [4] "les équations relatives à la division du cercle". [5] "Unter einem Kreiskörper versteht man jeden aus rationalen Zahlen und Einheitswurzeln zusammengesetzten Zahlkörper".

is carried round and restored again to the same position from which it began to be moved, the figure so comprehended is a *cylinder*”.[1]

The Greek term was taken over into Latin as *cylindrus*, a word attested since Cicero (LS). In Arabic texts, the Greek term was translated by the word *usṭuwānā*, which literally means ‘column’ (Rashed 2017, pp. 368–372). The Arabic term is the basis of the alternative Latin term *columna*, which is found in the 12th century, for instance in a translation of an Arabic text (Busard 1968, p. 123). The term was also used by Fibonacci: “A *column* is a figure erected perpendicularly upon a circular base and ending in a circle equal to its base, or it is formed by rotating a rectangle around one fixed edge back to its starting position” (Fibonacci 1862, p. 159).[2] Since a column does not necessarily have a circular base, the term *round column* (*rotunda columna* was also in common use; it appears, for example, in a 12th-century translation by Plato of Tivoli (Curtze 1902, p. 164).

The French term is found in the 14th century (DMF), where in a manuscript *chilindre* and the synonymous *coulombe* are found (de Conty n.d., fol. 214ʳ). The explicit *colonne ronde* was also used by Chuquet in the late 15th century (Chuquet 1979, p. 153); similarly, de Bouelles (1511, fol. xxxᵛ) defined a *colonne* in general terms as a solid with any base that proceeds upward “without any diminution”,[3] and he used a suitable adjective such as *ronde* to specify the base. By the second half of the 16th century, however, *cylindre* has become the customary term.

In German, the earliest uses are based on Latin *columna*; the corresponding term *Säule* is found in 1477 (Curtze 1899, p. 56);[4] the explicit *runde Säule* was used by Dürer (1525, fol. Gʳ) and others. Schmid (1539, p. 25) writes *runde columna* and points out that it is called *chilindrus* by the learned; in Jacob (1565, fol. 317ᵛ) the German word is fully naturalized. Kepler (1616, p. 22) later suggested, without much success, some alternative terms: when a cylinder was flat, he called it *Täller* (modern *Teller*) ‘plate’ or *Rad* ‘wheel’; if high and long, and lying flat, a *Walger* ‘rolling log’, *Waltzen* ‘roller, barrel’, or *Wellen* ‘bundle of brushwood’.[5]

In English, *cylinder* is first found (OED) in Euclid (1570, fol. 317ᵛ).

For the use of the term with an arbitrary base, see also *cone*.

Note. A brief historical overview if the terminology is found in Tropfke (1924, p. 16, pp. 29f.). For the Greek term, see Mugler (1958-59, p. 263); for the Latin, see Ayuso García (2008, pp. 1335–1338); for the Arabic, see Rashed (2017, pp. 368–372).

References

Ayuso García, Manuel (2008). “La terminología latina de la geometría en Marciano Capela”. PhD thesis. Madrid: Universidad Nacional de Educación a Distancia.

de Bouelles, Charles (1511). *Geometrie en francoys*. Paris.

Busard, H.L.L. (1968). “L’algèbre au Moyen Âge: le «Liber mensurationum» d’Abû Bekr”. *Journal des savants* (1968), pp. 65–124.

[1] Translation from Euclid (1956, p. 262). [2] “colvmna uero est figura orthogonaliter eleuata super basem circularem terminata in circulo, qui equalis est sue basi; uel est illud quod ambit paralilogramum rectiangulum circa latus unum immobiliter fixum ad idem punctmi reductum; que etiam uocatur figura teres”. Translation from Hughes (2008, pp. 277f.). [3] “La coulonne est ung corps irregulier procedant de sa base en haut egalement sans diminution nulle”. [4] The figure drawn there shows an object with a rectangular base. [5] For Kepler’s terms, see Götze (1919, pp. 140, 186, 202–204, 208–210).

Chuquet, Nicolas (1979). *La Géométrie*. Première géométrie algébrique en langue française (1484). Introduction, texte et notes par Hervé l'Huillier. Paris.
de Conty, Évrard (n.d.). "Le Livre des Probleumes d'Aristote, c'est à dire des fortes questions [manuscript]". Bibliothèque nationale de France. Département des manuscrits. Français 210. URL: `https://gallica.bnf.fr/ark:/12148/btv1b10547045k`.
Curtze, Maximilian (1899). "Der Tractatus Quadrantis des Robertus Anglicus in deutscher Übersetzung aus dem Jahr 1477". *Abhandlungen zur Geschichte der Mathematik* 9, pp. 41–63.
Curtze, Maximilian, ed. (1902). *Urkunden zur Geschichte der Mathematik im Mittelalter und der Renaissance*. Erster Theil. Leipzig.
Dürer, Albrecht (1525). *Underweysung der Messung, mit dem Zirckel und Richtscheyt*. Nürnberg.
Euclid (1570). *The Elements of Geometrie of the most auncient Philosopher Euclide of Megara*. Faithfully (now first) translated into the Englishe toung, by H. Billingsley. London.
Euclid (1956). *The Thirteen Books of Euclid's Elements*. Volume III. Books X-XIII and Appendix. Ed. by Thomas L. Heath. New York.
Fibonacci (1862). *Scritti di Leonardo Pisano*. Pubblicati da Baldassare Boncompagni. Volume II. Roma.
Götze, Alfred (1919). *Anfänge einer mathematischen Fachsprache in Keplers Deutsch*. Berlin.
Hughes, Barnabas (2008). *Fibonacci's* De Practica Geometrie. Berlin.
Jacob, Simon (1565). *Ein New und Wolgegründt Rechenbuch*. Frankfurt.
Kepler, Johannes (1616). *Außzug auß der Uralten MesseKunst Archimedis*. Linz.
Mugler, Charles (1958-59). *Dictionnaire historique de la terminologie géométrique des Grecs*. Paris.
Rashed, Roshdi, ed. (2017). *Lexique historique de la langue scientifique arabe*. Hildesheim.
Schmid, Wolffgang (1539). *Das erst buch der Geometria*. Nürnberg.
Tropfke, Johannes (1924). *Geschichte der Elementar-Mathematik*. Siebenter Band. Stereometrie. Verzeichnisse. Berlin and Leipzig.

cylindroid

The word *cylindroid* came up in the second half of the 17th century. It was used by Michael Dary in correspondence from April 1663: "I call them cylindroids (by which I mean) a solid contained under three surfaces, viz. two planes parallel called the bases, and these bases are either both circles or both ellipses, or else one a circle and the other an ellipsis, and the third superficies is convex begirting the whole body, which I call the zone, in which zone a right line may be everywhere applied directly from base to base" (Rigaud 1841, pp. 99f.).[1]

Dary (ibid.) also wrote that he earlier considered the names *tri-surface* and *tessera-hedrum*. In print, *cylindroid* appears in Dary (1669, pp. 35f.). It is only occasionally found in this sense, for example in Wallis (1670, p. 556) or in Harris' *Lexicon technicum*, where it is defined as follows: "Cylindroid, is a Solid Figure with Elliptical Bases, parallel, and alike situated" (Harris 1704). As can be seen from the quotation, the term's meaning was soon restricted to what was otherwise called an *elliptic cylinder*, a term that appears to be due to Tacquet (1651, p. 2).

At about the same time when Dary coined the term, Wren (1669, p. 962) used *hyperbolic cylindroid* in a different sense, namely for what was later called a *hyperboloid of one sheet* (see *hyperboloid*). d'Alembert (1753, p. 898) later explained that, while solids generated by the revolution of a curve around an axis are called *conoids*, the term *cylindroid* was chosen because the solid is more similar to a cylinder than to

[1] The 17th-century spelling was modernized in Rigaud (1841). The passage is cited by the OED as the first use in English.

a cone.[1] In the 18th century, the term is more commonly found in Wren's sense than in Dary's.

Another definition of *cylindroid* was given in the 19th century. In the 1860s, Plücker introduced a new surface, a particular kind of right conoid (Plücker 1865, p. 756, Plücker 1868, p. 97). He did not give it a special name, though. In print, the term *cylindroid* for this surface first appears in Ball (1871, p. 182), who attributes it to Cayley. In a later publication, Ball presents more details on the term:

> The name cylindroid was suggested by Professor Cayley in 1871 in reply to a request of the writer. The word originated in the following construction, which was then communicated by Professor Cayley. Cut the cylinder $x^2 + y^2 = (p_\beta - p_\alpha)^2$ in an ellipse by the plane $z = x$, and consider the line $x = 0$, $y = p_\beta - p_\alpha$. If any plane $z = c$ cuts the ellipse in the points A, B and the line in C, then CA, CB are two generating lines of the surface (Ball 1876, p. 16).

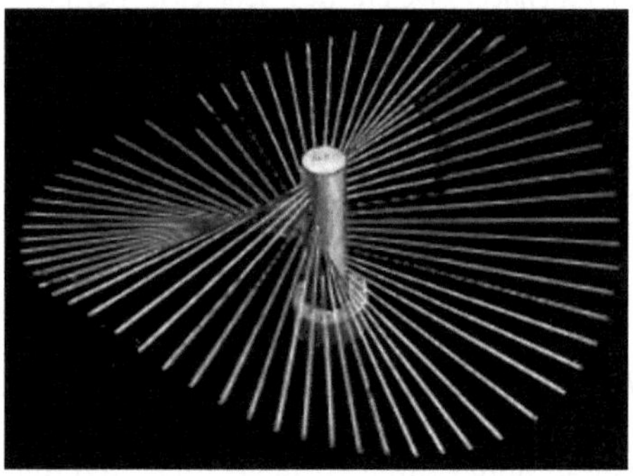

Figure 37: Plücker's conoid (cylindroid). From Willson (1892, p. 125).

Rittershaus (1876, p. 564) suggested German *cubisches Conoid* 'cubical conoid' because the surface is of the third degree. However, this term only partly caught on and the figure is still usually referred to as *Plücker's conoid* or *Plücker's cylindroid.*

Note. For Plücker's cylindroid, see Fiedler (1876, p. 204) and Obenrauch (1897, pp. 321–323).

References

d'Alembert, Jean Le Rond (1753). "Conoïde". *Encyclopédie, ou Dictionnaire raisonné des sciences, des arts et des métiers* 3, p. 898.

d'Alembert, Jean Le Rond (1754). "Cylindroïde". *Encyclopédie, ou Dictionnaire raisonné des sciences, des arts et des métiers* 4, p. 593.

Ball, Robert Stawell (1871). "Description of a model of a conoidal cubic surface called the 'Cylindroid', which is presented in the theory of the geometrical freedom of a rigid body". *The London, Edinburgh, and Dublin Philosophical Magazine and Journal of Science* 42, pp. 181–183.

Ball, Robert Stawell (1876). *The Theory of Screws*. A Study in the Dynamics of a Rigid Body. Dublin.

[1] "Ce *conoïde* s'appelle *cylindroïde*, à cause qu'il ressemble plus à un cylindre qu'à un cone, ne se terminant pas en pointe comme les autres conoïdes". See also d'Alembert (1754).

Dary, Michael (1669). *Dary's Miscellanies: being, for the most part, a brief collection of mathematical theorems*. London.

Fiedler, Wilhelm (1876). "Geometrie und Geomechanik. Eine Uebersicht zur Kennzeichnung ihres Zusammenhangs nach seiner gegenwärtigen Entwickelung". *Vierteljahrsschrift der Naturforschenden Gesellschaft in Zürich* 21, pp. 186–228.

Harris, John (1704). *Lexicon Technicum*. Vol. I. London.

Obenrauch, Ferdinand (1897). *Geschichte der darstellenden und projectiven Geometrie*. Brünn.

Plücker, Julius (1865). "On a New Geometry of Space". *Philosophical Transactions of the Royal Society of London* 155, pp. 725–791.

Plücker, Julius (1868). *Neue Geometrie des Raumes*. Erste Abtheilung. Leipzig.

Rigaud, Stephen Peter, ed. (1841). *Correspondence of Scientific Men of the Seventeenth Century*. Vol. I. Oxford.

Rittershaus, Trajan (1876). "Die kinematische Kette. Ihre Beweglichkeit und Zwangsläufigkeit". *Der Civilingenieur* 22, pp. 559–572.

Tacquet, André (1651). *Cylindricorum et annularium libri IV*. Antwerpen.

Wallis, John (1670). *Mechanica*. Sive, De Motu, Tractatus Geometricus. London.

Willson, Frederick Newton (1892). *Theoretical and Practical Graphics*. Princeton.

Wren, Christopher (1669). "Generatio corporis cylindroidis hyperbolici". *Philosophical Transactions of the Royal Society of London* 4, pp. 961–962.

D

decimal

Decimal arithmetic was popularized by Stevin (1585a) in his book *De thiende* ('the tenth'); the French translation, made by Stevin himself, was correspondingly entitled *La disme* (Stevin 1585c, p. 132), while Latin *decarithmia* appeared in Stevin (1585b).[1] An English translation of *De thiende*, entitled *Disme: The Art of Tenths, or Decimall Arithmetike* (Stevin 1608), seems to be the first time that *decimal arithmetic* was used.[2] Latin *decimalis arithmetica* is then found in Napier (1617, p. 21), French *arithmétique décimale* in Mersenne (1625, p. 526); both Napier and Mersenne referred to Stevin. Beyer (1619) used *logistica decimalis*, and also German *Decimal-Rechnung* (p. 8) as well as *Decimal-Bruch* 'decimal fraction' (p. 7). For the latter, Beyer (1603, p. 1) had earlier used *zehentheyliger Bruch*, from the words for 'ten' and 'part'.[3]

Decimal numbers were called *thiendetalen* 'tenth numbers' by Stevin (1585a, p. 12); in French, he used *nombres de dismes* (Stevin 1585c, p. 141) and in the English translation, *disme numbers* was used (Stevin 1608, fol. C 2^{v}); explicitly, *decimal number* was used by Johnson (1623, p. 227).

Note. For the history of decimal fractions, see, for example, Tropfke (1980, pp. 114–118) and Smith (1958, pp. 235–250). For the history of the decimal system, see Folkerts (2001) and the literature cited there.

References

Beyer, Johann Hartmann (1603). *Ein newe und schöne Art der Vollkommenen Visier-Kunst*. Frankfurt.

Beyer, Johann Hartmann (1619). *Logistica Decimalis*. Frankfurt.

Folkerts, Menso (2001). "Early Texts on Hindu-Arabic Calculation". *Science in Context* 14, pp. 13–38.

Johnson, John (1623). *Iohnsons Arithmatick in 2. Bookes*. London.

Mersenne, Marin (1625). *La vérité des sciences*. Paris.

Napier, John (1617). *Rabdologiae seu numerationis per virgulas libri duo*. Edinburgh.

[1] The word appears in the front matter ("De opere operibusque autoris sermo"). [2] This place is cited by the OED as the first use of *decimal* in English. Latin *decimalis*, from *decimus* 'tenth', is of post-classical origin. [3] Note that Beyer's *Logistica decimalis* is sometimes erroneously dated 1603; see Sarton (1935, pp. 178f.) for discussion.

H. Becker, *A Historical Dictionary of Mathematical Terms*, Science Networks. Historical Studies 66, https://doi.org/10.1007/978-3-032-02873-0_4

Sarton, George (1935). "The First Explanation of Decimal Fractions and Measures (1585). Together with a History of the Decimal Idea and a Facsimile (No. XVII) of Stevin's Disme". *Isis* 23, pp. 153–244.
Smith, David Eugene (1958). *History of Mathematics*. Volume II. Special Topics of Elementary Mathematics. New York.
Stevin, Simon (1585a). *De thiende*. Leiden.
Stevin, Simon (1585b). *L'Arithmetique*. Leiden.
Stevin, Simon (1585c). *La pratique d'arithmetique*. Leiden.
Stevin, Simon (1608). *Disme: The Art of Tenths, or Decimall Arithmetike*. Published in English with some additions by Robert Norton, Gent. London.
Tropfke, Johannes (1980). *Geschichte der Elementarmathematik*. Band 1: Arithmetik und Algebra. Vollständig neu bearbeitet von Kurt Vogel, Karin Reich, Helmuth Gericke. Berlin and New York.

Dedekind domain **and** ***Dedekind ring***

The terms *Dedekind domain* and *Dedekind ring* have been attributed to various mathematicians. In a study on the origin of Dedekind rings, for example, Ullrich (1999, pp. 131f.) remarked that *Dedekind domain* was due to Cohen (1950). In a review of Ullrich's paper,[1] Lemmermeyer referred to Dieudonné (1947, p. 621) for *Dedekind ring*, and Roquette (2018, p. 18) suspected: "Perhaps 'Bourbaki' can be regarded as the one who created this terminology [*Dedekind ring*]?".

The latter assumption appears indeed to be true, as is confirmed by Dieudonné (1982, p. 621), who stated that Bourbaki introduced *Noetherian ring*, *Artinian ring*, *Dedekind ring*, and *factorial ring*.[2] French *anneau de Dedekind* appeared in Bourbaki (1942, p. 59), where it was defined as a "ring of integrity having a unit element, and in which every ideal $\neq (0)$ can be decomposed in a unique manner into a product of prime ideals".[3]

The terminology is more commonly found in the 1950s; for example, Kaplansky (1952, p. 330) defined: "A Dedekind ring is an integral domain in which classical ideal theory holds: that is, every ideal is uniquely a product of prime ideals or alternatively, every ideal is invertible". Emmy Noether is known to have referred to Dedekind rings as "5-Axiome-Ringe" (Lemmermeyer and Roquette 2006, p. 98); the five axioms are found at the beginning of Noether (1927).

Note that *Dedekind ring* was earlier in occasional use in a different sense. In an article (finished in 1936) written for the second edition of the *Enzyklopädie der mathematischen Wissenschaften* but not published at the time, Brauer wrote: "A ring is called semisimple (or Dedekind), if its radical only consists of 0" (Brauer 1980,

---

[1] See `https://zbmath.org/?q=an%3A0943.11002`. [2] Dieudonné extensively commented on the reason for introducing these terms, namely "that in many fields of recent origin, a number of mathematicians, mainly of Anglo-Saxon or German origin, were guilty of particularly acute carelessness in showing a total lack of imagination and a complete contempt for their languages; a typical sin was their abuse of abbreviations". Thus, *Dedekind ring* was meant to replace *ZPI-Ring*, where *ZPI* stands for "decomposition into prime ideals". Such abbreviations came up in the late 1920s; see Krull (1939, pp. 27f., fn. 45) for some references. [3] "un anneau d'intégrité, ayant un élément unité, et dans lequel tout idéal $\neq (0)$ se décompose d'une manière unique en un produit d'idéaux premiers". – For the term Bourbaki refers to chapter five of his book *Algèbre*, which was not published until 1950 (Bourbaki 1950): However, in this book the term was not used, presumably because by that time the arrangement of the book on algebra had been changed.

p. 166).[1] This terminology came up in the late 1920s, when Brandt (1928, p. 1) coined *Dedekindsche Algebra*, which he defined as an "algebra without radical".[2]

Note. For the history of the concept, see Ullrich (1999, esp. pp. 131f.).

References

Bourbaki, Nicolas (1942). *Topologie génerale*. Chapitre III. Groupes topologiques. Chapitre IV. Nombres réels. Paris.

Bourbaki, Nicolas (1950). *Algèbre*. Chapitre IV. Polynômes et fractions rationnelles. Chapitre V. Corps commutatifs. Paris.

Brandt, Heinrich (1928). "Idealtheorie in Quaternionenalgebren". *Mathematische Annalen* 99, pp. 1–29.

Brauer, Richard (1980). *Collected Papers*. Volume I. Theory of Algebras, and Finite Groups. Ed. by Paul Fong and Warren J. Wong. Cambridge, Mass.

Cohen, I.S. (1950). "Commutative rings with restricted minimum conditions". *Duke Mathematical Journal* 17, pp. 27–41.

Dieudonné, Jean (1947). "Sur les produits tensoriels". *Annales scientifiques de l'École Normale Supérieure* 64, pp. 101–117.

Dieudonné, Jean (1982). "The Work of Bourbaki During the Last Thirty Years". *Notices of the American Mathematical Society* 29, pp. 618–623.

Frobenius, Georg (1903). "Theorie der hyperkomplexen Größen". *Sitzungsberichte der Königlich Preußischen Akademie der Wissenschaften zu Berlin* (1903), pp. 504–537.

Hawkins, Thomas (2013). *The Mathematics of Frobenius in Context*. A Journey Through 18th to 20th Century Mathematics. New York.

Kaplansky, Irving (1952). "Modules over Dedekind rings and Valuation Rings". *Transactions of the American Mathematical Society* 72, pp. 327–340.

Krull, Wolfgang (1939). "Allgemeine Modul-, Ring und Idealtheorie". *Enzyklopädie der mathematischen Wissenschaften mit Einschluss ihrer Anwendungen*. Zweite, völlig neubearbeitete Auflage I 1, 11, pp. 1–54.

Lemmermeyer, Franz and Peter Roquette, eds. (2006). *Helmut Hasse und Emmy Noether*. Die Korrespondenz 1925–1935. Göttingen.

Noether, Emmy (1927). "Abstrakter Aufbau der Idealtheorie in algebraischen Zahl- und Funktionenkörpern". *Mathematische Annalen* 96, pp. 26–61.

Roquette, Peter (2018). *The Riemann Hypothesis in Characteristic p in Historical Perspective*. Cham.

Ullrich, Peter (1999). "Die Entdeckung der Analogie zwischen Zahl- und Funktionenkörpern: der Ursprung der 'Dedekind-Ringe'". *Jahresbericht der Deutschen Mathematiker-Vereinigung* 101, pp. 116–134.

deficient number

See *perfect number*.

definite

This term goes back to Latin *definitus* 'defined, bounded'. Its etymology is closely related to that of *define*, both words ultimately stemming from *finis* 'limit, boundary'. English *definite* and French *définie* both appear in the 16th century with the general meaning 'having exact limits, clearly defined'. In the following, three common mathematical uses of *definite* will be discussed.

---

[1] "Ein Ring heißt *halbeinfach* (oder ein *Dedekindscher Ring*), wenn sein Radikal nur aus 0 besteht".

[2] Brandt remarks that he chose this term in analogy to Frobenius' *Dedekindsche Gruppe* (Frobenius 1903, p. 519). For some comments on the introduction of this term, see Hawkins (2013, p. 530).

definite integral

While definite integrals have been considered from the 17th century on, with the creation of the calculus "the concept of a definite integral, which had earlier been the basic object of study, fades into the background, into the area of applications, and not only was there no special symbol for it for quite some time, there wasn't even any special name for it" (Medvedev 1991, p. 179).

The name indeed did not appear until the second half of the 18th century. In a paper read in 1779 (published in 1782), Laplace defined French *intégrale définie*: "I call *definite integral* [*intégrale définie*] an integral taken from a determined value of the variable to another determined value" (Laplace 1782, p. 209).[1] This appears to be the first definition of the term (Domingues 2008, p. 148),[2] and the usage is clearly derived from the general meaning. Laplace did not define *indefinite integral* but did use the term (Laplace 1782, p. 275).

The term was soon adopted, for example, by Lacroix (1798, p. 142), and also by Poisson (1820, p. 319), who used the notation $\int fxdx = Fb - Fa$. Cauchy presented his influential definition in his *Résumé des leçons données à l'École polytechnique sur le calcul infinitésimal* (Cauchy 1823, p. 83), for example.[3] The English term is found in [anon.] (1801, p. 499).

See also *limit* (integral) and *integral*.

Note. For the history of definite integrals, see, for example, Schubring (2006, pp. 30–32), Domingues (2008, p. 148), and Medvedev (1991, ch. 4).

References

[anon.] (1801). "La Croix on *Differences and Series*". *The Monthly Review* 36, pp. 498–501.

Cauchy, Augustin-Louis (1823). *Résumé des leçons données à l'École Royale Polytechnique sur le calcul infinitésimal*. Tome premier. Paris.

Cauchy, Augustin-Louis (1827). "Mémoire sur les intégrales définies". *Mémoires de l'Académie royale des sciences de l'Institut de France* 1, pp. 601–799.

Domingues, João Caramalho (2008). *Lacroix and the Calculus*. Basel.

Euler, Leonhard (1766). "Observationes circa integralia formularum $\int x^{p-1}dx(1-x^n)^{\frac{q}{n}-1}$ posito post integrationem $x = 1$". *Mélanges de Philosophie et de Mathématique de la Société Royale de Turin* 3, pp. 156–177.

Lacroix, Sylvestre François (1798). *Traité du calcul différentiel et du calcul intégral*. Tome second. Paris.

Laplace, Pierre-Simon (1782). "Mémoire sur les suites". *Mémoires de l'Académie royale des sciences* (1779), pp. 207–309.

Laplace, Pierre-Simon (2010). "Mémoire sur les suites". Translated by Richard J. Pulskamp. URL: `https://www.probabilityandfinance.com/pulskamp/Laplace/` (retrieved July 19, 2024).

Medvedev, Fyodor A. (1991). *Scenes from the History of Real Functions*. Translated from the Russian by Roger Cooke. Basel.

Poisson, Siméon Denis (1820). "Suite du Mémoire sur les intégrales définies, inseré dans les deux précédens volumes de ce Journal". *Journal de l'École polytechnique* 11, cahier 18, pp. 295–341.

Schubring, Gert (2006). *Conflicts Between Generalization, Rigor, and Intuition*. Number Concepts Underlying the Development of Analysis in 17th-19th Century France and Germany. New York.

---

[1] "je nomme *intégrale définie*, une intégrale prise depuis une valeur déterminée de la variable jusqu'à une autre valeur déterminée". Translation from Laplace (2010, p. 2) (slightly adapted).

[2] The term *indefinite integral* is found, without definition, in Euler (1766, p. 156).

[3] Cauchy uses the term on numerous occasions, for example in Cauchy (1827, p. 614), a paper written in 1814.

definite form

Shortly after Laplace introduced the term *definite integral*, *definite* was introduced into number theory by Gauß (1801, § 271) in connection with quadratic forms:

> Certain ternary forms are so constructed that they can represent positive and negative numbers without distinction, e. g. the form $x^2 + y^2 - z^2$. We will call them *indefinite forms*. On the other hand there are forms that cannot represent negative numbers but (except for zero which is obtained by making each unknown $= 0$) only positive numbers, e. g. $x^2 + y^2 + z^2$. We will call these *positive forms*. Finally there are others that cannot represent positive numbers, e. g. $-x^2 - y^2 - z^2$. We will call these *negative forms*. Positive and negative forms will be called *definite forms*.[1]

An *indefinite form* (*forma indefinita*) can represent both positive and negative numbers and is in this sense not clearly delimited. For definite forms the opposite holds.

The corresponding French term *forme définie* is found in Gauß (1807, § 271), the English term in Smith (1867, pp. 277f.).

References

Gauß, Carl Friedrich (1801). *Disquisitiones arithmeticae*. Leipzig.
Gauß, Carl Friedrich (1807). *Recherches arithmétiques*. Traduites par A.-C.-M. Poullet-Delisle. Paris.
Gauß, Carl Friedrich (1986). *Disquisitiones Arithmeticae*. English Edition. Translated by A.A. Clarke. Revised by W.C. Waterhouse. New York et al.
Smith, H.J.S. (1867). "On the Orders and Genera of Ternary Quadratic Forms". *Philosophical Transactions of the Royal Society of London* 157, pp. 255–298.

definite (set theory)

Another use of *definite* is found in the context of set theory and logic. Zermelo (1908, p. 263) introduced *definite assertions* (German *definite Aussagen*) and defined:

> A question or assertion $\mathfrak{E}$ is said to be "*definite*" if the fundamental relations of the domain, by means of the axioms and the universally valid laws of logic, determine without arbitrariness whether it holds or not. Likewise a "propositional function" ($\mathfrak{E}(x)$), in which the variable term x ranges over all individuals of a class $\mathfrak{K}$, is said to be "definite" if it is definite for *each single* individual x of the class $\mathfrak{K}$. Thus the question whether $a \in b$ or not is always definite, as is the question whether $M \subseteq N$ or not.[2]

Note. For Zermelo's introduction of the term, see van Heijenoort (2002, pp. 199f.) and Felgner (2010).

---

[1] "Quaedam formae ternariae ita sunt comparatae, ut per ipsas sine discrimine repraesentari possint numeri positivi et negativi, *e. g.* forma $xx + yy - zz$, quam obrem *formae indefinitae* vocabuntur. Contra per alias numeri negativi repraesentari nequeunt, sed (praeter cifram quae prodit, ponendo singulas indeterminatas $= 0$) positivi tantum, ut $xx + yy + zz$, quare *formae positivae* dicentur; denique per alias numeri positivi repraesentari nequeunt, ut $-xx - yy - zz$, unde appellabuntur *formae negativae*; formae positivae et negativae nomine communi *formae definitae* dicentur". Translation from Gauß (1986, § 271). [2] "Eine Frage oder Aussage $\mathfrak{E}$, über deren Gültigkeit oder Ungültigkeit die Grundbeziehungen des Bereiches vermöge der Axiome und der allgemeingültigen logischen Gesetze ohne Willkür entscheiden, heißt ‚*definit*'. Ebenso wird auch eine „Klassenaussage" ($\mathfrak{E}(x)$), in welcher der variable Term x alle Individuen einer Klasse $\mathfrak{K}$ durchlaufen kann, als ‚definit' bezeichnet, wenn sie für *jedes einzelne* Individuum x der Klasse $\mathfrak{K}$ definit ist. So ist die Frage, ob $a \in b$ oder nicht ist, immer definit, ebenso die Frage, ob $M \subseteq N$ oder nicht". Translation from Zermelo (2009, p. 193).

References

Felgner, Ulrich (2010). "Introductory note to 1908b". In: *Ernst Zermelo. Collected Works. Gesammelte Werke*. Volume I. Ed. by Heinz-Dieter Ebbinghaus and Akihiro Kanamori. Heidelberg et al., pp. 160–189.

van Heijenoort, Jean (2002). *From Frege to Gödel*. A Source Book in Mathematical Logic, 1879–1931. Cambridge, Mass.

Zermelo, Ernst (1908). "Untersuchungen über die Grundlagen der Mengenlehre". *Mathematische Annalen* 65, pp. 261–281.

Zermelo, Ernst (2009). *Collected Works. Gesammelte Werke*. Volume I. Band I. Set Theory, Miscellanea. Mengenlehre, Varia. Edited by, Herausgegeben von Heinz-Dieter Ebbinghaus, Akihiro Kanamori. Heidelberg et al.

degenerate

The use of the verb *degenerate* in mathematical contexts appears to originate in the 17th century. It was used, for example, by Wallis (1685, p. 272): "ofttimes what was presumed to be a Circle, may be found to be an Ellipse, or Hyperbola; and again may degenerate into a Parabola, or a Triangle, or perhaps a Streight Line". When exactly the terminology came up is difficult to determine, but Kepler (1615, fol. H^v) already used *degenerare* for a conic section that degenerated into two straight lines.

In the second half of the 19th century, the term began to be used in algebraical contexts as well. It was often applied when a determinant was zero. For example, Study (1889, p. 6) used the German adjective *ausgeartet* in connection with transformations: "A transformation of determinant zero is called a *degenerate* transformation".[1] At the same time, Scheffers (1889, p. 441) applied the term to hypercomplex number systems that have no identity element.[2]

Weyl (1918, p. 23) later applied the term to bilinear forms:

> We shall call the form "*non-degenerate*" if it vanishes identically in $\mathfrak{y}$ only when the vector $\mathfrak{x} = 0$. This happens if, and only if, the homogeneous equations $\sum_{i=1}^{n} a_{ik} = 0$ have a single solution $\xi_i = 0$ or when the determinant $|a_{ik}| \neq 0$".[3]

References

Kepler, Johannes (1615). *Nova Stereometria Doliorum Vinariorum*. Linz.

Scheffers, Georg (1889). "Ueber die Berechnung von Zahlensystemen". *Berichte über die Verhandlungen der Königlich Sächsischen Gesellschaft der Wissenschaften zu Leipzig (Mathematisch-Physische Klasse)* 41, pp. 400–457.

Study, Eduard (1889). *Methoden zur Theorie der ternären Formen*. Leipzig.

Wallis, John (1685). *A Treatise of Algebra, both Historical and Practical*. London.

Weyl, Hermann (1918). *Raum · Zeit · Materie*. Vorlesungen über allgemeine Relativitätstheorie. Berlin.

Weyl, Hermann (1922). *Space–Time–Matter*. Translated from the German by Henry L. Rose. London.

degree (overview)

This term has a large number of mathematical meanings and overlaps to a considerable extent with the terms *order* and *rank*. The general word came into English via Old

---

[1] "Eine Transformation von der Determinante Null heisst eine ‚*ausgeartete Transformation*'".

[2] Scheffers expressed this criterion in terms of determinants. [3] "Wir wollen die Form „*nicht-ausgeartet*" nennen, wenn sie für einen Vektor $\mathfrak{x}$ identisch in $\mathfrak{y}$ nur dann verschwindet, falls $\mathfrak{x} = 0$ ist. Das ist dann und nur dann der Fall, wenn die homogenen Gleichungen $\sum_{i=1}^{n} a_{ik} = 0$ die einzige Lösung $\xi_i = 0$ besitzen oder wenn die Determinante $|a_{ik}| \neq 0$ ist". Translation adapted from Weyl (1922, p. 26).

French *degre* (mod. *degré*), which comes from supposed (unattested) Latin *degradus*, from *de-* 'down' and *gradus* 'step; that upon which one steps; stage, degree'.[1] As will be seen in the discussion of the degrees of a circle, however, this etymological outline does not yet give the whole development, as Arabic translations play a major role in the history of the term.

The following entries discuss some selected meanings of the term: the degree of a circle, of equations and curves, of graphs as well as the topological concept.

For the origin of the degree of a field extension, see *extension*. See also *class* (curve).

degree (circle)

The subdivision of the circle into 360 degrees is of Babylonian origin and was adopted in Greek astronomy.[2] It is first explicitly found in an inscription, which Neugebauer (1975, p. 699) tentatively dates "around 100 B.C.". The inscription contains the sentence "[a] circle comprises 360 degrees" (Jones 2006, p. 13) and uses Greek *μοῖρα*, lit. 'part', for 'degree'. The same word was also used by the astronomer Geminus, who lived in the 1st century (Geminus 1898, p. 4).

In the *Almagest*, however, Ptolemy uses *τμήμα*, lit. 'section'. In book I, 10, which is about the size of chords, he states: "For the user's convenience, then, we shall subsequently set out a table of their amounts, dividing the circumference into 360 parts, and tabulating the chords subtended by the arcs at intervals of half a degree, expressing each as a number of parts in a system where the diameter is divided into 120 parts" (Ptolemy 1898, p. 31).[3] At the same time, Ptolemy does use *μοῖρα* (Ptolemy 1898, e. g., pp. 242, 246) with reference to arcs (and implicitly, to angles).[4]

There were, thus, two Greek words for 'degree', and consequently there were also two words in other languages. In Arabic, *τμήμα* was translated as as *juz'*, which also literally means 'part'; the term was used, for example, by al-Ḥajjāj in his translation of the *Almagest* (Rashed 2017, p. 252). al-Bīrūnī, on the other hand, used Arabic *daraja* (al-Bīrūnī 1934, pp. 24, 57), which originally means 'step' and thus differs from *μοῖρα*; why this term was chosen remains an open question.[5]

The Latin term *gradus* (also meaning 'step') is not used in classical Latin in the sense under discussion (TLL), and it does not appear to be used as a term for 'degree' until the 12th century, when various Arabic texts were translated into Latin (Tropfke 1930, p. 59). For example, Plato of Tivoli used this word in a translation of an astronomical treatise by al-Battānī (al-Battānī 1537, fol. 4^{v}); in other texts of

---

[1] This word gave rise to German *Grad*. [2] According to Neugebauer (1975, p. 305), "we are not sure whether it was Hipparchus who first introduced the division of *c* [circumference of the earth] into 360 degrees or Eratosthenes (who perhaps remained at a strictly sexagesimal division of the circle)". [3] Translation from Toomer (1984, p. 48). [4] Tropfke (1930, p. 59) states that Ptolemy thus sharply distinguished between *μοῖρα* and *τμήμα*. [5] Kunitzsch (1974, p. 156) regards *daraga* as a calque of Greek *βαθμός* 'step' (the word is also used in astronomy; see Neugebauer 1975, p. 699), a view that Toomer (1977, p. 206) rejects. al-Bīrūnī (1934, p. 57) himself states: "Astronomers have agreed to divide the circumference of circles, whether great or small, into 360 equal parts [...] Those of the ecliptic are named degrees, *darajah*, because the sun on the two sides of its course ascends and descends among them as by the steps of a ladder; in other circles they are simply called divisions".

the 12th century, *pars* is used (see Kunitzsch 1974, pp. 156–158 for discussion and references).

Spanish *grado* is attested in the 13th century (Bossong 1978, pp. 233f.), and so is French *degré* (Dörr 1998, p. 122). The English term is found in the late 14th century (OED), and so is German *Grad* (Tropfke 1930, p. 59, DWB).

Note. Brief historical notes on the term can be found in Tropfke (1930, p. 59) and Smith (1958, p. 232). For usages in the 12th century, see Kunitzsch (1974, pp. 156–158). For the Arabic terminology, see also Rashed (2017, pp. 251–253); for the Greek, see Thomas (1957, pp. 48f.).

References

al-Battānī, Muḥammad ibn Jābir (1537). *De motu stellarum*. Nürnberg.

al-Bīrūnī, Abū Rayḥān Muḥammad ibn Aḥmad (1934). *The Book of Instruction in the Elements of the Art of Astrology*. Written in Ghaznah, 1029 A. D. Reproduced from Brit. Mus. MS. Or. 8349. The Translation facing the Text by R. Ramsay Wright. London.

Bossong, Georg, ed. (1978). *Los Canones de Albateni*. Herausgegeben sowie mit Einleitung, Anmerkungen und Glossar versehen. Tübingen.

Dörr, Stephen (1998). *Der älteste Astronomietraktat in französischer Sprache: L'Introductoire d'astronomie*. Edition und lexikalische Analyse. Tübingen.

Geminus (1898). *Elementa Astronomiae*. Ed. by Karl Manitius. Leipzig.

Jones, Alexander (2006). "The Keskintos Astronomical Inscription: Text and Interpretations". *Sciamvs: Sources and Commentaries in Exact Sciences* 7, pp. 3–41.

Kunitzsch, Paul (1974). *Der Almagest*. Die Syntaxis Mathematica des Claudius Ptolemäus in arabisch-lateinischer Überlieferung. Wiesbaden.

Neugebauer, Otto (1975). *A History of Ancient Mathematical Astronomy*. New York et al.

Ptolemy, Claudius (1898). *Syntaxis Mathematica*. Volumen I. Ed. by J.L. Heiberg. Leipzig.

Rashed, Roshdi, ed. (2017). *Lexique historique de la langue scientifique arabe*. Hildesheim.

Smith, David Eugene (1958). *History of Mathematics*. Volume II. Special Topics of Elementary Mathematics. New York.

Thomas, Ivor (1957). *Selections Illustrating the History of Greek Mathematics*. I. From Thales to Euclid. London and Cambridge, Mass.

Toomer, G.J. (1977). "Ptolemaic Astronomy in Islam". *Journal for the History of Astronomy* 8, pp. 204–210.

Toomer, G.J. (1984). *Ptolemy's Almagest*. London.

Tropfke, Johannes (1930). *Geschichte der Elementar-Mathematik*. Erster Band. Rechnen. Dritte, verbesserte und vermehrte Auflage. Berlin and Leipzig.

***degree* (equations and curves)**

The systematic use of the term *degree* may be said to begin with Descartes; before him, the term is only found sporadically. For example, Viète (1615, p. 15) uses Latin *gradus* for the degree of an equation, without defining the notion.

In his famous *Géométrie*, Descartes (1637, p. 372) classified equations according to their *dimension*: "Every equation can have as many distinct roots (values of the unknown quantity) as the number of dimensions of the unknown quantity in the equation".[1] The corresponding curves, however, were classified in terms of their *degree* (*degré*):

---

[1] "Scachés donc qu'en chasque Equation, autant que quantité inconnue a de dimensions, autant peut il y auoir de diuerses racines, c'est a dire de valeurs de cete quantité". Translation from Descartes (1925, p. 159).

> The ancients were familiar with the fact that the problems of geometry may be divided into three classes, namely, plane, solid, and linear problems.[1] This is equivalent to saying that some problems require only circles and straight lines for their construction, while others require a conic section and still others require more complex curves. I am surprised, however, that they did not go further, and distinguish between different degrees of these more complex curves (Descartes 1637, p. 315).[2]

The term *degré* was also used by Prestet (1675, p. 357) in connection with the geometric problems mentioned by Descartes, but Prestet (ibid.) also applied it to equations (*égalités*). He may be the first who used *degree* in this sense (Tropfke 1937, p. 30).

Newton, on the other hand, used Latin *ordo* in his famous "enumeration" of curves, first published in 1704 but written more than 30 years earlier:

> Geometrical lines are best divided into orders, according to the dimensions of the equation expressing the relation between abscissa and ordinate, or, which is the same thing, according to the number of points in which they can be cut by a straight line. So that a line of the first order will be a straight line; those of the second or quadratic order will be conic sections and the circle; and those of the third or cubic order will be the cubic Parabola, the Neilian Parabola, the Cissoid of the ancients, and others we are about to describe (Newton 1704, p. 139).[3]

In the first English translation of Newton's text, Harris (1710, s. v. curve) translated *ordo* by the three words *class*, *gender*, and *order*. Wolff (1710, pp. 122f.) used German *Geschlecht* 'genus' in connection with curves but preferred *Dimension* in connection with equations; occasionally, he also used *Grad* 'degree' with reference to equations (pp. 123f.), a custom that is now standard.

Thus, in the 17th century, the term *dimension* for the degree of an equation is preferred by Descartes and Newton (and others), while Prestet suggests *degree*; for the degree of a curve, Descartes suggested *degree*, while Newton preferred *order*. At the end of the 17th century, the term *degree* is also used in connection with differential equations (Bernoulli 1694, p. 435).

Note. For a brief historical overview of the terminology, see Tropfke (1937, p. 30).

References

Bernoulli, Johann (1694). "Modus generalis construendi omnes aequationes differentiales primi gradus". *Acta Eruditorum* (Nov. 1694), pp. 435–437.

Descartes, René (1637). *La Géométrie*. In: *Discours de la méthode*. Leiden, pp. 295–413.

---

[1] This classification of geometric problems follows Pappus (*Collection*, book III, prop. 5; see Pappus of Alexandria 1876, p. 54 and Descartes 1925, p. 40 for a translation). [2] "Les anciens ont fort bien remarqué, qu'entre les Problesmes de Geometrie, les vns sont plans, les autres solides, & les autres lineaires, c'est a dire, que les vns peuuent estre construits, en ne traçant que des lignes droites, & des cercles; au lieu que les autres ne le peuuent estre, qu'on n'y employe pour le moins quelque section conique; qu'on n'y employe quelque autre ligne plus composée. Mais ie m'estonne de ce qu'ils n'ont point outre cela distingué diuers degrés entre ces lignes plus composées". Translation from Descartes (1925, p. 40). [3] "Lineae Geometricae secundum numerum dimensionum aequationis qua relatio inter Ordinatas & Abscissas definitur, vel (quod perinde est) secundum numerum punctorum in quibus a linea recta secaeri possunt, optimè distinguuntur in Ordines. Qua ratione linea primi Ordinis erit Recta sola, eae secundi sive quadratici ordinis erunt sectiones Conicae & Circulus, & eae tertii sive cubici Ordinis Parabola Cubica, parabola Neiliana, Cissois veterum & reliquae quas hic enumerare suscepimus". Translation from Newton (1860, p. 7), slightly adapted.

Descartes, René (1925). *The Geometry of René Descartes*. Translated from the French and Latin by David Eugene Smith and Marcia L. Latham. Chicago and London.
Harris, John (1710). *Lexicon Technicum*. Vol. II. London.
Newton, Isaac (1704). "Enumeratio linearum tertii ordinis". In: *Opticks*. London, pp. 138–162.
Newton, Isaac (1860). *Enumeration of Lines of the Third Order*. Trans. by C.M.R. Talbot. London.
Pappus of Alexandria (1876). *Pappi Alexandrini Collectionis*. Volumen I. Ed. by Friedrich Hultsch. Berlin.
Prestet, Jean (1675). *Elemens des mathematiques*. Paris.
Tropfke, Johannes (1937). *Geschichte der Elementar-Mathematik*. Dritter Band. Proportionen, Gleichungen. Dritte, verbesserte und vermehrte Auflage. Berlin and Leipzig.
Viète, François (1615). *De aequationum recognitione et emendatione tractatus duo*. Ed. by Alexander Anderson. Paris.
Wolff, Christian (1710). *Der Anfangs-Gründe aller Mathematischen Wissenschaften Letzter Theil*. Halle.

degree (graph)

The term *degree* in connection with graphs was introduced by Petersen (1891, p. 194), who used German *Grad* in connection with regular graphs and defined it as the number of edges through each vertex. The French *degré* in this sense was used by Petersen (1898, p. 225) himself, while the English term appears in Brahana (1917, p. 59).

The degree of a vertex of a graph was defined by Sainte-Laguë (1923, p. 28) as the number of edges that are incident with the vertex. In this sense, *Grad* was also defined by König (1936, p. 3).

References

Brahana, H.R. (1917). "A Proof of Petersen's Theorem". *Annals of Mathematics* 19, pp. 59–63.
König, Dénes (1936). *Theorie der endlichen und unendlichen Graphen*. Leipzig.
Petersen, Julius (1891). "Die Theorie der regulären Graphs". *Acta Mathematica* 15, pp. 193–220.
Petersen, Julius (1898). "Sur la théorème de Tait". *L'Intermédiaire des Mathématiciens* 5, pp. 225–227.
Sainte-Laguë, André (1923). "Les réseaux". *Annales de la Faculté des sciences de Toulouse* 15, pp. 27–86.

degree (topology)

The mapping degree, today more commonly called *Brouwer degree*, was introduced by Brouwer in 1911. The term *Grad* 'degree' in this sense appears in a letter to Hilbert from 1910:

> In order to define this degree, we introduce homogeneous co-ordinates x, y, z on κ and homogeneous co-ordinates ξ, η, ζ on λ, and consider first the univalent mapping, which is determined for each domain by a correspondence
>
> $$\xi : \eta : \zeta = f_1(x, y, z) : f_2(x, y, z) : f_3(x, y, z),$$
>
> where f_1, f_2, f_3 are polynomials.
>
> If we fix a positive orientation on both spheres, and choose in each point of κ this positive orientation, then each generic point of λ occurs p times with positive orientation and q times

with negative orientation. One can show that $p - q$ is a constant for all generic points, which we call the degree of the mapping (Brouwer 1976, pp. 421f.).[1]

In print, the term first appears in Brouwer (1912, p. 105); occasionally, Brouwer also *Abbildungsgrad* 'mapping degree' (e. g., p. 106). The term was quickly adopted, for example by Nielsen (1913, p. 40) and Tietze (1914, p. 248).

For the complex prehistory of the concept, see also *index* (functions), *characteristic* (topology), and *winding number*.

Note. For the history of the concept, see Siegberg (1981) and Johnson (1981); in Dinca and Mawhin (2021), numerous further references are found.

References

Brouwer, L.E.J. (1912). "Über Abbildungen von Mannigfaltigkeiten". *Mathematische Annalen* 71, pp. 97–115.

Brouwer, L.E.J. (1976). *Collected Works.* 2: Geometry, Analysis, Topology and Mechanics. Ed. by Hans Freudenthal. Amsterdam and Oxford.

van Dalen, Dirk (2013). *L.E.J. Brouwer – Topologist, Intuitionist, Philosopher.* How Mathematics Is Rooted in Life. London.

Dinca, George and Jean Mawhin (2021). *Brouwer Degree.* The Core of Nonlinear Analysis. Cham.

Johnson, Dale M. (1981). "The Problem of the Invariance of Dimension in the Growth of Modern Topology, Part II". *Archive for History of Exact Sciences* 25, pp. 85–266.

Nielsen, Niels (1913). "Kurvennetze auf Flächen". PhD thesis. Christian-Albrechts-Universität zu Kiel.

Siegberg, Hans Willi (1981). "Some Historical Remarks Concerning Degree Theory". *The American Mathematical Monthly* 88, pp. 125–139.

Tietze, Heinrich (1914). "Über stetige Abbildungen einer Quadratfläche auf sich selbst". *Rendiconti del Circolo Matematico di Palermo* 38, pp. 247–304.

degrees of freedom

According to Walker (1940, p. 253), "[i]n the work of modern statisticians, the concept of degrees of freedom is not found before 'Student's' paper of 1908, it was first made explicit by the writings of R. A. Fisher, beginning with his paper of 1915 on the distribution of the correlation coefficient, and has only within the decade or so received general recognition. Nevertheless the concept was familiar to Gauss and his astronomical associates".[2] The statistical terminology came up in the 1920s, when it was used by Fisher (1922): "We shall show that Elderton's Tables of Goodness of Fit [..] may still be applied, but that the value of n' with which the table should be entered is not now equal to the number of cells, but to *one more than the number of degrees of freedom in the distribution*".[3]

---

[1] "Um diesen Grad zu definieren, führen wir auf κ homogene Koördinaten x, y, z ein, und auf λ homogene Koördinaten ξ, η, ζ, und betrachten zunächst die eindeutige Abbildung, welche gebietsweise bestimmt ist durch eine Beziehung $\xi : \eta, \zeta = f_1(x, y, z) : f_2(x, y, z) : f_3(x, y, z)$, wo f_1, f_2, f_3 Polynome sind. Nehmen wir sodann auf beiden Kugeln einen positiven Umlaufssinn an, und wählen wir in jedem Punkte von κ diesen positiven Umlaufssinn, so tritt jeder Punkt allgemeiner Lage von λ p Male auf mit positivem Umlaufssinn und q Male mit negativem Umlaufssinn. Man kann dann zeigen, dass $p - q$ für alle Punkte allgemeiner Lage von λ eine Konstante ist, welche wir den Grad der Abbildung nennen". Translation from van Dalen (2013, pp. 153f.). [2] Works referred to by Walker are Student (1908), Fisher (1915), and Gauß (1823). [3] This usage is pointed out by the OED as the first; see also Aldrich in Miller (n.d.), who remarks that Fisher used the term in connection with the χ^2-test.

The term was originally used in mechanics, where it was introduced by Thomson and Tait (1867, p. 130):

> A free point has *three* degrees of freedom, inasmuch as the most general displacement which it can take is resolvable into three, parallel respectively to any three directions, and independent of each other [...] If the point be constrained to remain always on a given surface, *one* degree of constraint is introduced, or there are left but *two* degrees of freedom.

A definition was given by Maxwell (1873, p. 185): "The number of degrees of freedom of a system is the number of data which must be given in order completely to determine its position".

Note. For the history of the concept, see Rodgers (2019).

References

Fisher, Ronald A. (1915). "Frequency distribution of the values of the correlation coefficient in samples from an indefinitely large population". *Biometrika* 10, pp. 507–521.

Fisher, Ronald A. (1922). "On the Interpretation of χ^2 from Contingency Tables, and the Calculation of *P*". *Journal of the Royal Statistical Society* 85, pp. 87–94.

Gauß, Carl Friedrich (1823). *Theoria combinationis observationum erroribus minimis obnoxiae*. Göttingen.

Maxwell, James Clerk (1873). *A Treatise on Electricity and Magnetism*. Vol. I. Oxford.

Miller, Jeff (n.d.). "Earliest Known Uses of Some of the Words of Mathematics". URL: `https://mathshistory.st-andrews.ac.uk/Miller/mathword/`.

Rodgers, Joseph Lee (2019). "Degrees of Freedom at the Start of the Second 100 Years: A Pedagogical Treatise". *Advances in Methods and Practices in Psychological Science* 2, pp. 396–405.

Student (1908). "The probable error of a mean". *Biometrika* 6, pp. 1–25.

Thomson, William and Peter Guthrie Tait (1867). *Treatise on Natural Philosophy*. Vol. I. Oxford.

Walker, H.M. (1940). "Degrees of Freedom". *Journal of Educational Psychology* 31, pp. 253–269.

dendrite

In Greek, *δενδρίτες* means 'pertaining to a tree' (from *δένδρον* 'tree'). In mathematics, this term is not found until the 1920s, when it began to be used for locally connected continua that do not contain any simple closed curve. Menger (1926, p. 146) points out that the notion is first found in Mazurkiewicz (1921, p. 130), where, however, no explicit term is used. It was coined soon afterwards, though, and appears to be due to Ważewski (1923, pp. 9, 33), who used French *dendrite*.

Menger (1926, p. 146) translated this by the German *Baumkurve*, lit. 'tree curve'. It is not surprising that Menger chose this particular term, since in graph theory a *tree* had already been defined as a connected graph without cycles. Menger explicitly states the he uses the term as a generalization of the word *tree* as used in graph theory.[1]

A new suggestion was made by Gehman (1927, p. 553): "we propose to use the word *acyclic* in place of the phrase *containing no simple closed curve*. That is, an acyclic curve is a continuous curve containing no simple closed curve". This term met with approval and was adopted by Kline (1927, p. 162), for example, who also states that "Menger and a number of foreign writers call the same sets *Baumkurven*", a statement that also shows that *dendrite* was not used at the time in the English-

[1] Menger also points out that the concept was dealt with by Scherrer (1926) under the name *ungeschlossene stetige Kurve* (unclosed continuous curve).

speaking world. The word is found in Aitchison (1932, p. 800) and Whyburn (1933, p. 131).

Note. Some historical notes on the concept are found in Charatonik (1998).

References

Aitchison, Beatrice (1932). "Concerning the mapping of locally connected continua onto simple arcs". *Bulletin of the American Mathematical Society* 38, p. 800.

Charatonik, J.J. (1998). "History of Continuum Theory". In: *Handbook of the History of General Topology*. Volume 2. Ed. by C.E. Aull and R. Lowen. Dordrecht et al., pp. 703–786.

Gehman, H.M. (1927). "Concerning acyclic continuous curves". *Transactions of the American Mathematical Society* 29, pp. 553–568.

Kline, J.R. (1927). "Separation theorems and their relation to recent developments in analysis situs". *Bulletin of the American Mathematical Society* 34, pp. 155–192.

Mazurkiewicz, Stefan (1921). "Un théorème sur les lignes de Jordan". *Fundamenta Mathematicae* 2, pp. 119–130.

Menger, Karl (1926). "Bericht über die Dimensionstheorie". *Jahresbericht der Deutschen Mathematiker-Vereinigung* 35, pp. 113–150.

Scherrer, W. (1926). "Über ungeschlossene stetige Kurven". *Mathematische Zeitschrift* 24, pp. 125–130.

Ważewski, Tadeusz (1923). "Sur les courbes de Jordan ne renfermant aucune courbe simple fermée de Jordan". PhD thesis. Université de Paris.

Whyburn, Gordon T. (1933). "Characterizations of Certain Curves by Continuous Functions Defined upon them". *American Journal of Mathematics* 55, pp. 131–134.

denominator* and *numerator

The terminology for numerators and denominators was introduced in the 12th and 13th centuries. For the denominator, the terms included[1] *denominatio*, found in writings of the 12th century (see, for example, Curtze 1898, p. 23). Fibonacci used *numerus denominatus* for the denominator and *numerus denominans* for the numerator: "If over any number will be made a fraction line, and over the same line will be written another number, the upper number means the number of parts determined by the lower number; the lower is called the denominator (*denominatus*) and the upper is called the numerator (*denominans*)" (Fibonacci 1857, p. 24).[2]

The actual words *denominator* and *numerator* seem, according to Busard (1998, p. 80), to have been first used in the 13th century in a treatise with the incipit "Minutiarum vulgarium scribes superius numeratorem"; in the 14th century, they are found in the widely distributed *Algorismus de minuciis* (*minutiae* 'fractions') by Johannes de Lineriis (Busard 1968, p. 21), for example.

The modern terminology is attested in the 15th century. Thus, in a manuscript dating from around 1430 and written in Occitan, the terms *denominador* and *nombrador* are found (Sesiano 2018, p. 256);[3] Chuquet used French *dénominateur* and

[1] For more terms and references, see Tropfke (1980, p. 119). [2] "Cum super quemlibet numerum quedam uirgula protracta fuerit, et super ipsam quilibet alius numerus descriptus fuerit, superior numerus partent uel partes inferioris numeri affirmat; nam inferior denominatus, et superior denominans appellatur". Translation from Sigler (2002, p. 49). – From today's perspective, it would be expected that *denominatus* and *denominans* were used the other way round. According to Tropfke (1980, p. 119), the words were erroneously interchanged in Boncampagni's edition. This opinion is not shared by Busard (1998, p. 75) and Lüneburg (1993, p. 62). [3] These terms are also found in print in Pellos (1492, fol. 19^{r}). This text dates from around 1460.

nominateur (Chuquet 1881, p. 52), terms earlier found alongside *nombre denominant* and *nombre nombrant* in a manuscript from 1449 ([anon.] c1449, fol. 104^r). Italian *denominatore* and *numeratore* are first found in print in the *Treviso Arithmetic* ([anon.] 1478, fol. 32^r).[1]

German *Nenner* 'denominator' (from *nennen* 'to name'), still used today, is found in the *Algorismus Ratisbonensis*, dating from around 1450 (Vogel 1954, p. 196); German *Zähler* (from *zählen* 'to count') is found (as *czeler*) in the *Geometria culmensis* from around 1400 (Mendthal 1886, p. 45). In print, both terms are found in the *Bamberger Rechenbuch* (Wagner 1483, ch. 6).[2] The same terminology is found in Dutch: in the oldest known arithmetic printed in that language ([anon.] 1508), *noemer* (lit. 'namer') and *teller* (from *tellen* 'to count') were used.[3]

The first English attestation is found in the oldest known printed arithmetic in English that is fully extant, entitled *An Introduction for to lerne to rekyn with the pen*, two early editions of which are from 1536/37 and 1539 ([anon.] 1537, [anon.] 1539).[4] This text is influenced by Dutch and French sources, as is also shown in the vocabulary it uses.[5] In the 1539 edition, the following definition is given: "In fraccyons there be two maners of numbers, whereof the fyrste is called the numerator for he sheweth the number of the denominator, that standeth vnder hym. The other is called the denominator, for he sheweth euer howe much the parte is, and standeth under the numerator" ([anon.] 1539, fol. fi^v). As pointed out by Bockstaele (1960, p. 319), however, in the following the terms *numerator* and *denominator* are not used anymore; instead, *teller* and *number* are preferred ([anon.] 1539, fol. fiii^r), which are analogous to the Dutch terminology mentioned above.[6] Nevertheless, *numerator* and *denominator* soon became standard. They were also used by Recorde (1543, fol. 117^v, fol. 118^r) and other 16th-century authors. Exceptions do occur: for example, in Ramus (1592, p. 29), *numberer* and *namer* are found: "the upper is called the numberer, the under is called the namer".

Note. Overviews of the terminology are found in Tropfke (1980, p. 119) and Smith (1958, p. 220). For Latin *denominare* and *denominatio*, see also Guillaumin (2020, pp. 100f.) and Allard (1990, pp. 157f.). Rashed (2017, p. 793) briefly discusses the Arabic terminology. For various remarks on the terminology in the early 16th century, see also Williams (2012, p. 181).

References

[anon.] (c1449). "[Traitté d'arithmetique]". Manuscript, BNF Français 7287. URL: `https://gallica.bnf.fr/ark:/12148/btv1b525140414`.

[anon.] (1478). *Larte de labbacho*. Treviso.

[anon.] (1508). *Die maniere om te leeren cyffren na die rechte consten Algorismi*. Brussel.

[anon.] (1537). *An introduccion for to lerne to rekyn with the pen*. St. Albans.

[anon.] (1539). *An introduction for to lerne to recken with the pen*. London.

Allard, André (1990). "La formation du vocabulaire latin de l'arithmétique médiévale". In: *Méthodes et instruments du travail intellectuel au moyen âge*. Études sur le vocabulaire. Ed. by Olga Weijers. Turnhout, pp. 137–181.

[1] For the Italian terminology, see Ricci (1994, p. 33). [2] For the German terminology, see also Schirmer (1912, pp. 47, 78) and Reiner (1961, pp. 88f., 117). [3] The book has not been consulted and is cited on the basis of Bockstaele (1959, pp. 61, 69) and Kool (1992, p. 97). [4] [anon.] (1537) has not been consulted. [5] See Bockstaele (1960) and Williams (2012) for discussion. [6] These terms are extremely rare in English.

Bockstaele, Paul (1959). "Het oudste gedrukte Nederlandse Rekenboekje". *Scientiarum Historia* 1, pp. 53–71.
Bockstaele, Paul (1960). "Notes on the First Arithmetics Printed in Dutch and English". *Isis* 51, pp. 315–321.
Busard, H.L.L. (1968). *Het rekenen met breuken in de middeleeuwen, in het bijzonder bij Johannes de Lineriis*. Brussel.
Busard, H.L.L. (1998). "Über eine Vorlage des Traktats Algorismus de minutiis von Johannes de Lineriis". *Sudhoffs Archiv* 82, pp. 74–91.
Chuquet, Nicolas (1881). *Le Triparty en la Science des Nombres*. Ed. by Aristide Marre. Rome.
Curtze, Maximilian (1898). "Über eine Algorismus-Schrift des XII. Jahrhunderts". *Abhandlungen zur Geschichte der Mathematik* 8, pp. 1–27.
Fibonacci (1857). *Scritti di Leonardo Pisano*. Pubblicati da Baldassare Boncompagni. Volume I. Roma.
Guillaumin, Jean-Yves (2020). *Dictionnaire de la terminologie latine ancienne de l'arithmétique et de la géométrie*. Paris.
Kool, Marjolein (1992). "De rekenkundige termen van Simon Stevin". *Scientiarum Historia* 18, pp. 91–107.
Lüneburg, Heinz (1993). *Leonardi Pisani Liber Abbaci oder Lesevergnügen eines Mathematikers*. 2., überarbeitete und erweiterte Auflage. Mannheim.
Mendthal, Hans (1886). *Geometria culmensis*. Ein agronomischer Tractat aus der Zeit des Hochmeisters Conrad von Jungingen (1393-1407). Leipzig.
Pellos, Francés (1492). *Sen segue de la art de arithmeticha*. Torino.
Ramus, Petrus (1592). *The Art of Arithmeticke in Whole Numbers and Fractions*. Translated into English by William Kempe. London.
Rashed, Roshdi, ed. (2017). *Lexique historique de la langue scientifique arabe*. Hildesheim.
Recorde, Robert (1543). *The Ground of Artes*. London.
Reiner, Karl (1961). *Die Terminologie der ältesten mathematischen Werke in deutscher Sprache nach den Beständen der Bayerischen Staatsbibliothek*. München.
Ricci, Laura (1994). "Il lessico matematico della Summa di Luca Pacioli". *Studi di lessicografia italiana* 12, pp. 5–71.
Schirmer, Alfred (1912). *Der Wortschatz der Mathematik nach Alter und Herkunft untersucht*. Straßburg.
Sesiano, Jacques (2018). *L'arithmétique de Pamiers*. Traité mathématique en langue d'oc du XVe siècle. Lausanne.
Sigler, Laurence (2002). *Fibonacci's Liber Abaci*. A Translation into Modern English of Leonardo Pisano's Book of Calculation. New York.
Smith, David Eugene (1958). *History of Mathematics*. Volume II. Special Topics of Elementary Mathematics. New York.
Tropfke, Johannes (1980). *Geschichte der Elementarmathematik*. Band 1: Arithmetik und Algebra. Vollständig neu bearbeitet von Kurt Vogel, Karin Reich, Helmuth Gericke. Berlin and New York.
Vogel, Kurt, ed. (1954). *Die Practica des Algorismus Ratisbonensis*. Ein Rechenbuch des Benediktinerklosters St. Emmeram aus der Mitte des 15. Jahrhunderts nach den Handschriften der Münchner Staatsbibliothek und der Stiftsbibliothek St. Florian. München.
Wagner, Ulrich (1483). *[Bamberger Rechenbuch]*. Bamberg.
Williams, Travis D. (2012). "The Earliest English Printed Arithmetic Books". *The Library* 13, pp. 164–184.

dense set

The notion of a dense set was developed in the 19th century. Early traces of the concept can be found in Lejeune Dirichlet (1829, p. 169):

> it is necessary that the function $\varphi(x)$ be such that, if a and b designate any two arbitrary quantities within $-\pi$ and π, one may always locate between a and b two other quantities r and s that are close enough, so that the function remains continuous within the interval from

> r to s. [...] It will now be apparent that the integral of a function does not mean anything unless the function satisfies the condition we have enunciated.[1]

Ferreirós (2007, p. 150) remarks on this condition: "The condition is purely topological in character; to put it in modern terminology, it is necessary and sufficient that the set of discontinuities of the function be nowhere dense in $\mathbb{R}$".

Hankel later drew upon the work of Dirichlet, studying functions with infinitely many discontinuities in a finite interval. He seems have been the first author, according to Ferreirós (2007, p. 156), "who explicitly introduced notions of point-set theory – dense and nowhere dense –, although these ideas had been suggested by Dirichlet". Hankel defined the terms *fill the segment* (*die Strecke erfüllen*) and *lie scattered* (*zerstreut liegen*):

> Whenever within a segment there lies a multitude of points possessing a certain property, I say that these points
>
> *fill the segment* if no interval, however small, can be given within the segment in which one does not find at least one point of that multitude;
>
> that on the contrary this multitude of points does *not* fill the segment, but that the points lie *scattered* on it, if between any two arbitrarily close points of the segment one can always give an interval, in which no point of the multitude lies (Hankel 1870, pp. 25f.).[2]

Several years later, Cantor (1879, p. 2) then introduced the term *überall-dicht*, lit. 'everywhere dense', with reference to intervals: "If P lies partially or entirely in the $(\alpha \cdots \beta)$, then the remarkable case can occur that any arbitrarily small interval $(\gamma \cdots \delta)$ in $(\alpha \cdots \beta)$ contains some points of P. In such a case we will say that P is everywhere-dense in the interval $(\alpha \cdots \beta)$".[3]

At the same time, du Bois-Reymond (1879, p. 287) used the neologism *pantachisch*, from Greek *πανταχῆ* 'everywhere'; the term was only in occasional use, however: Cantor remarked in a letter from 1888 that he did not adopt it nor, apparently, did anyone else (Meschkowski and Nilson 1991, p. 316). On that occasion, Cantor also pointed out that he first used the concept in Cantor (1874, p. 258).

Cantor's term was soon in common use and translated into other languages. For example, English *everywhere dense* appears in Harnack (1891, p. 246). French *partout dense* was used by Harnack (1882, p. 245); originally, however, *ensemble condensé* was the usual French term, introduced in Cantor (1883, p. 351). Borel

---

[1] "Il est nécessaire qu'alors la fonction $\varphi(x)$ soit telle que, si l'on désigne par a et b deux quantités quelconques comprises entre $-\pi$ et π, on puisse toujours placer entre a et b d'autres quantités r et s assez rapprochées pour que la fonction reste continue dans l'intervalle de r à s. [...] On verra alors que l'intégrale d'une fonction ne signifie quelque chose qu'autant que la fonction satisfait à la condition précédemment énoncée". Translation from Ferreirós (2007, p. 150). [2] "Wenn auf einer Strecke eine Schaar von Punkten liegt, denen eine gewisse Eigenschaft zukommt, so sage ich, dass diese Punkte *die Strecke erfüllen*, wenn in der Strecke kein noch so kleines Intervall angegeben werden kann, in dem nicht wenigstens Ein Punkt jener Schaar lage; dass dagegen diese Schaar von Punkten die Strecke *nicht* erfüllen, sondern die Punkte *zerstreut* auf ihr liegen, wenn zwischen je zwei beliebig nahen Punkten der Strecke immer ein Intervall angegeben werden kann, in dem kein Punkt jener Schaar liegt". Translation from Ferreirós (2007, p. 155). [3] "Liegt P theilweise oder ganz im Intervalle $(\alpha \cdots \beta)$, so kann der bemerkenswerthe Fall eintreten, dass *jedes noch so kleine* in $(\alpha \cdots \beta)$ enthaltene Intervall $(\gamma \cdots \delta)$ Punkte von P enthält. In einem solchen Falle wollen wir sagen, dass P *im Intervalle* $(\alpha \cdots \beta)$, *überall-dicht* sei". Translation from Dauben (1990, p. 78).

(1898, p. 38) rejected *ensemble condensé*; one reason he gave that was the Germans do not say *verdichtet* (condensed) either. Afterwards, *ensemble dense* became the usual term in French. Additionally, the epithet *everywhere* was often left out at that time, for example by Hausdorff (1907, p. 541).

The term *dense in itself* (German *insichdicht*) is also due to Cantor (1884, p. 471); he defined a set P as *dense in itself* if all of its elements are also limit points of P.

Note. For the history of the concept and the terminology, see Hawkins (1979, chs. 3.2, 3.3) and Ferreirós (2007, esp. chs. v, vi).

References

du Bois-Reymond, Paul (1879). "Erläuterungen zu den Anfangsgründen der Variationsrechnung". *Mathematische Annalen* 15, pp. 283–314.

Borel, Émile (1898). *Leçons sur la théorie des fonctions*. Paris.

Cantor, Georg (1874). "Ueber eine Eigenschaft des Inbegriffs aller reellen algebraischen Zahlen". *Journal für die reine und angewandte Mathematik* 77, pp. 258–262.

Cantor, Georg (1879). "Ueber unendliche, lineare Punktmannichfaltigkeiten". *Mathematische Annalen* 15, pp. 1–7.

Cantor, Georg (1883). "Sur les ensembles infinis et linéaires de points". *Acta Mathematica* 2, pp. 349–380.

Cantor, Georg (1884). "Ueber unendliche, lineare Punktmannichfaltigkeiten. 6". *Mathematische Annalen* 23, pp. 453–488.

Dauben, Joseph W. (1990). *Georg Cantor*. His Mathematics and Philosophy of the Infinite. Princeton.

Ferreirós, José (2007). *Labyrinth of Thought*. A History of Set Theory and Its Role in Modern Mathematics. Second revised edition. Basel.

Hankel, Hermann (1870). *Untersuchungen über die unendlich oft oscillirenden und unstetigen Functionen*. Tübingen.

Harnack, Axel (1882). "Théorie de la série de Fourier". *Bulletin des sciences mathématiques et astronomiques* 6, pp. 242–260.

Harnack, Axel (1891). *An Introduction to the Study of the Elements of the Differential and Integral Calculus*. Trans. by George L. Cathcart. London and Edinburgh.

Hausdorff, Felix (1907). "Über dichte Ordnungstypen". *Jahresbericht der Deutschen Mathematiker-Vereinigung* 16, pp. 541–546.

Hawkins, Thomas (1979). *Lebesgue's Theory of Integration*. Its Origins and Development. Second Edition. Providence, R.I.

Lejeune Dirichlet, Gustav (1829). "Sur la convergence des séries trigonométriques qui servent à répresenter une fonction arbitraire entre des limites données". *Journal für die reine und angewandte Mathematik* 4, pp. 157–169.

Meschkowski, Herbert and Wilfried Nilson, eds. (1991). *Georg Cantor*. Briefe. Berlin.

denumerable

See *countable*.

derive* and *derivative

Latin *derivare* can be traced to *rivus* 'brook' and originally meant 'to divert a stream of water'; it was also used as a grammatical term to express that one word is "derived" from another. In mathematics, it was originally used to express that one mathematical entity is derived from one another by certain rules.[1] The first person to use the term may have been Stevin (1585, p. 22), who distinguished "primitive" polynomials (*quantitez primitiues*) such as $ax + b$ from "derivative" ones (*quantitez derivatiues*)

[1] This is how Ozanam (1702, p. 212) explains the name.

such as $ax^2 + b$, where the latter "can be obtained from primitive polynomials by replacing x by a power of x" (Struik 1969, p. 83).

As for the calculus, Leibniz uses *derivare* in a letter from 1677 to Henry Oldenburg, in which he gives a definition of "differential equation" and says that such an equation is "derived" (*derivata*) from another one (Leibniz 1987, p. 172). This definition has been regarded by various writers as the origin of the term under discussion,[1] but it is simply an example of *derivare* used in the meaning described above.[2] It is not terminological yet, and Leibniz made no systematic use of it, nor did other writers at the time.

The first more systematic links between verbs meaning 'to derive' and differentiation do not seem to appear until about 1760. In particular, Euler (1766, p. 258), in a paper read in 1759, wrote that "the characters Φ and Ψ indicate some regular or irregular functions, from which by differentiation one will have the derived functions Φ' and Ψ'".[3] And in a textbook on the calculus, we find $p = \frac{dy}{dx}$ explicitly referred to as the *dérivée*, the derivative (Leseur and Jacquier 1768, p. 467). No definitions of the terms are given in these texts.

In spite of such attestations, it still seems correct to say that, as Grabiner (1983, p. 203) put it, "Lagrange's term 'derived function' (*fonction dérivée*) is the origin of our term 'derivative'". Lagrange systematically used the term from the 1770s onwards. For example, it appeared in Lagrange (1774, pp. 186f.)[4] and was explicitly defined in Lagrange (1797, p. 14): "We will call the function fx *primitive function* with respect to the functions $f'x$, $f''x$, &c. which are derived from it, and we will call the latter *derived functions* with respect to the former".[5] Thus, while Lagrange may not have been the first who used the term, it is through his consistent use in its current meaning that the term became known. After the publication of Lagrange's *Théorie des fonctions analytiques*, it frequently appeared in textbooks and articles. Cauchy (1823, p. 9) also adopted it and explained it thus: "The form of the new function which serves as the limit of the ratio $\frac{f(x+i)-f(x)}{i}$ will depend on the form of the proposed function $y = f(x)$. In order to indicate this dependence, one gives the new function the name of *derived function*".[6]

In English, the term *derived function* is found in an account of Lagrange's *Théorie* ([anon.] 1799, p. 494), while the noun *derivative* is found in an account of Arbogast's *Du calcul des dérivations* ([anon.] 1801, p. 525). The German term *abgeleitete Function*, which is a literal translation of the French, is found in Lagrange (1798, p. 23).

---

[1] E. g., Cantor (1901, p. 189). [2] This interpretation was also hinted at by John Conway in Miller (n.d.). [3] "les caracteres Φ & Ψ marquent des fonctions quelconques régulieres ou irrégulieres, d'où par la différentiation on aura les fonctions dérivées Φ' & Ψ'". Translation from Euler (2013, p. 18). [4] The paper was read in 1772. [5] "Nous appellerons la fonction fx, *fonction primitive*, par rapport aux fonctions $f'x$, $f''x$, &c. qui en dérivent, et nous appelerons celles-ci, *fonctions dérivées*, par rapport à celle-là". Translation from Hairer and Wanner (2008, p. 81). Lagrange goes on to discuss the distinction between the first and higher derivatives. [6] "la forme de la fonction nouvelle qui servira de limite au rapport $\frac{f(x+i)-f(x)}{i}$ dépendra de la forme de la fonction proposée $y = f(x)$. Pour indiquer cette dépendance, on donne á la nouvelle fonction le nom de *fonction dérivée*". Translation from Edwards (1979, p. 313).

The synonymous term *differential coefficient* (French *coefficient différentielle*) first appears in print in Lacroix (1797, p. 98).[7] Domingues (2008, p. 73) points out that "Lacroix had already used it in 1785, but in a memoir that remained unpublished,[1] and only for partial derivatives" and that the "name was probably 'on the air'" because it also appears, for partial derivatives, in Bossut (1798, p. 351).[2] That the term was already known when Lacroix formally introduced it is also confirmed by an earlier use by Monge (1770-73, p. 85): "Let now φ' be the coefficient of the differential [*coefficient de la différentielle*] of φ, so that ω being any variable, one has $d\varphi(\omega) = \varphi'(\omega)d\omega$".[3] The last equation also explains the term, as was explicitly stated by Lacroix in a later work: "the quantity which it [the differential quotient] expresses is nothing else than the multiplier of the differential dx, in the expression of du" (Lacroix 1802, p. 5).[4]

The term was quickly adopted, for example by Arbogast (1800, p. 307). The corresponding German term appears in a translation of Lacroix's book (Lacroix 1799, p. 206); the English term is found in an account of the same book ([anon.] 1800, pp. 503f.) and also in Woodhouse (1803, p. xxvi), who called it a "fluxionary or differential coefficient".

The alternative term *differential quotient* also came up in the late 18th century. It originated in the German combinatorial school (Voss and Molk 1909, p. 278); *Differentialquotient* appears, for example, in Klügel (1789, p. 13). The term is mainly found in German texts. Pasquich (1791, p. 9) preferred *Verschwindungs-Quotient*, from *verschwinden* 'to vanish', but this did not catch on.

See also *differential* and *derivative* (algebra).

Note. For the history of the notion, see Bos (1974) and Grabiner (1983). Averbukh and Smolyanov (1968) focus on the history of the concept in the late 19th and in the 20th century. A brief terminological overview is given by Voss and Molk (1909, p. 278).

References

[anon.] (1799). "La Grange's *Théorie des fonctions analytiques*". *The Monthly Review* 28, pp. 481–499.

[anon.] (1800). "La Croix *on the Differential and Integral Calculus*". *The Monthly Review* 31, pp. 493–505.

[anon.] (1801). "Arbogast *on the Calculus of Derivations*". *The Monthly Review* 36, pp. 524–532.

Arbogast, Louis François Antoine (1800). *Du calcul des dérivations*. Strasbourg.

Averbukh, V.I. and O.G. Smolyanov (1968). "The various definitions of the derivative in linear topological spaces". *Russian Mathematical Surveys* 23, pp. 67–113.

Bos, Henk J.M. (1974). "Differentials, Higher-Order Differentials and the Derivative in the Leibnizian Calculus". *Archive for History of Exact Sciences* 14, pp. 1–90.

Bossut, Charles (1798). *Traités de Calcul Différentiel et de Calcul Intégral*. Tome second. Paris.

Cantor, Moritz (1901). *Vorlesungen über Geschichte der Mathematik*. Dritter Band. Von 1668 bis 1758. Leipzig.

---

[7] See, e. g., Domingues (2008, p. 73). [1] The memoir is reproduced in Domingues (2008, pp. 355–366). [2] Bossut (1798, ibid.) also gives *formule différentielle* as an alternative. [3] "Soit actuellement φ' le coefficient de la différentielle de φ, de manière que ω étant une variable quelconque, on ait $d\varphi\omega = \varphi'\omega d\omega$". [4] "parce que la quantité qu'elle représente n'est autre chose que le multiplicateur de la différentielle". Translation from Lacroix (1816, p. 4).

Cauchy, Augustin-Louis (1823). *Résumé des leçons données à l'École Royale Polytechnique sur le calcul infinitésimal*. Tome premier. Paris.
Domingues, João Caramalho (2008). *Lacroix and the Calculus*. Basel.
Edwards, C.H. (1979). *The Historical Development of the Calculus*. New York et al.
Euler, Leonhard (1766). "Continuation des recherches sur la propagation du son". *Histoire de l'Académie Royale des Sciences et des Belles-Lettres de Berlin* (1759), pp. 241–264.
Euler, Leonhard (2013). "Continued investigations into the propagation of sound". Translated by Ian Bruce. URL: `http://www.17centurymaths.com/contents/eulercontents.html` (retrieved Mar. 3, 2017).
Grabiner, Judith V. (1983). "The changing concept of change: the derivative from Fermat to Weierstrass". *Mathematics Magazine* 56, pp. 195–206.
Hairer, Ernst and Gerhard Wanner (2008). *Analysis by Its History*. New York.
Klügel, Georg Simon (1789). "Grundsätze der reinen Mechanik (Beschluß)". *Philosophisches Magazin* 2, pp. 1–28.
Lacroix, Sylvestre François (1797). *Traité du calcul différentiel et du calcul intégral*. Tome premier. Paris.
Lacroix, Sylvestre François (1799). *Lehrbegriff des Differential- und Integralcalculs*. Erster Theil. Berlin.
Lacroix, Sylvestre François (1802). *Traité élémentaire de calcul différentiel et de calcul intégral*. Paris.
Lacroix, Sylvestre François (1816). *An Elementary Treatise on the Differential and Integral Calculus*. Translated from the French. Cambridge.
Lagrange, Joseph-Louis (1774). "Sur une nouvelle espèce de calcul relatif à la différentiation et à l'intégration des quantités variables". *Nouveaux Mémoires de l'Académie Royale des Sciences et Belles-Lettres de Berlin* (1772), pp. 185–206.
Lagrange, Joseph-Louis (1797). *Théorie des fonctions analytiques*. Paris.
Lagrange, Joseph-Louis (1798). *Theorie der analytischen Funktionen*. Aus dem Französischen übersetzt von Johann Philipp Grueson. Berlin.
Leibniz, Gottfried Wilhelm (1987). *Sämtliche Schriften und Briefe*. Dritte Reihe. Mathematischer, naturwissenschaftlicher und technischer Briefwechsel. Zweiter Band. 1676-1679. Hannover.
Leseur, Thomas and François Jacquier (1768). *Elemens du calcul integral*. Seconde partie. Parma.
Miller, Jeff (n.d.). "Earliest Known Uses of Some of the Words of Mathematics". URL: `https://mathshistory.st-andrews.ac.uk/Miller/mathword/`.
Monge, Gaspard (1770-73). "Second mémoire sur le calcul intégral de quelques équations aux différences partielles". *Mélanges de Philosophie et de Mathématique de la Societé Royale de Turin* 5, pp. 79–122.
Ozanam, Jacques (1702). *Nouveaux elemens d'algebre*. Premiere partie. Amsterdam.
Pasquich, Johann (1791). *Unterricht in der Mathematischen Analysis und Maschinen-Lehre*. Zweyter Band, enthaltend die Differential- und Integralrechnung. Leipzig.
Stevin, Simon (1585). *L'Arithmetique*. Leiden.
Struik, Dirk Jan, ed. (1969). *A Source Book in Mathematics, 1200-1800*. Harvard and London.
Voss, Aurel and Jules Molk (1909). "Calcul différentiel". *Encyclopédie des sciences mathématiques pures et appliquées* II.1, pp. 242–336.
Woodhouse, Robert (1803). *The Principles of Analytical Calculation*. Cambridge.

derived (further uses)

While the verb *derive* is especially known from the calculus, it is also found in various other areas of mathematics.[1] Examples are *derived set*, *derived functor*, and *derived group*.

---

[1] Early attestations involving the verb *derive* and its general etymology are discussed under *derive* and *derivative*.

In his famous 1872 paper, Cantor introduced some elementary topological notions. One of them was that of a *derived set* (*abgeleitete Menge*[2]), which he defined as the set of all limit points of a point set (Cantor 1872, p. 129). French *ensemble dérivé* was used in Cantor (1883, p. 350), and English *derived set* in Harnack (1891, p. 245).

The term *derived functor* as used in category theory is due to Cartan and Eilenberg (1956, p. 83). Because the printing of the book was delayed, the term is earlier found in Buchsbaum (1955, p. 1), who referred to Cartan's and Eilenberg's forthcoming publication.

For *derived group*, see *commutator*.

Note. For the history of derived sets, see Ferreirós (2007, esp. pp. 141-144).

References

Buchsbaum, David A. (1955). "Exact Categories and Duality". *Transactions of the American Mathematical Society* 80, pp. 1–34.
Cantor, Georg (1872). "Ueber die Ausdehnung eines Satzes aus der Theorie der trigonometrischen Reihen". *Mathematische Annalen* 5, pp. 123–132.
Cantor, Georg (1883). "Sur les ensembles infinis et linéaires de points". *Acta Mathematica* 2, pp. 349–380.
Cartan, Henri and Samuel Eilenberg (1956). *Homological Algebra*. Princeton.
Ferreirós, José (2007). *Labyrinth of Thought*. A History of Set Theory and Its Role in Modern Mathematics. Second revised edition. Basel.
Harnack, Axel (1891). *An Introduction to the Study of the Elements of the Differential and Integral Calculus*. Trans. by George L. Cathcart. London and Edinburgh.

descriptive geometry

The term *descriptive geometry* was coined by Monge (1794, p. 1): "descriptive geometry is the art of representing on drawing paper, which has only two dimensions, objects that have three and that can be rigorously defined".[1] Monge then specified the purposes of descriptive geometry: "The purpose of this art is two-fold. First it allows one to represent three-dimensional objects susceptible of being rigorously defined on a two-dimensional drawing. [...] Second [...] by taking the description of such objects to its logical conclusion, we can deduce something about their shape and relative positioning" (Monge 1794, p. 1).[2]

He points out that only two planes are required in order to plot spatial objects as long as one introduces the notion of projection as opposed to that of distance as in analytical geometry. This is where he presents the basic principle of descriptive geometry: given two orthogonal planes, each point in space can be defined in terms of its projections onto these planes.

In what appears to be the earliest attestation for the English term *descriptive geometry*, it is remarked that "[d]escriptive geometry [...] is divided into three

---

[2] In German, *ableiten* is generally used in the calculus sense and other meanings where English prefers *derive*. [1] "La géométrie descriptive est l'art de représenter sur des feuilles de dessin qui n'ont que deux dimensions, les objets qui en ont trois, et qui sont susceptibles d'une définition rigoureuse". Translation from Middleton and Baudouin-Matuszek (2007, p. 267). [2] "Cet art a deux objets principaux: Le premier est de représenter avec exactitude, sur des dessins qui n'ont que deux dimensions, les objets qui en ont trois, et qui sont susceptibles de définition rigoureuse [...] Le second [...] est de déduire de la description exacte des corps tout ce qui suit nécessairement de leurs formes et de leurs positions respectives. Dans ce sens, c'est un moyen de rechercher la vérité". Translation from Sakarovitch (2005, p. 226).

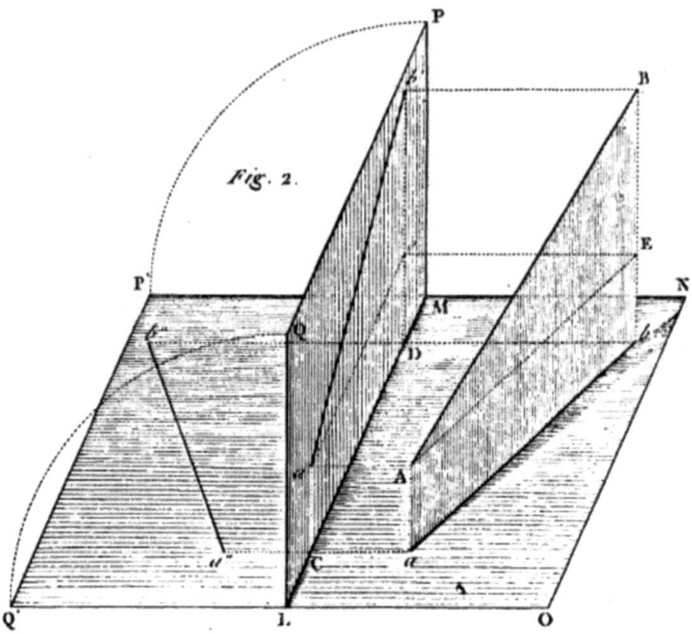

Figure 38: Projections of points and lines onto the plane and the elevation. From Monge (1799, Planche I).

principal branches—stereotomy, civil works, and military architecture" ([anon.] 1799, pp. 114f.).[3]

Note. For the history of the concept, see, for example, Loria (1921) and Andersen and Grattan-Guinness (1994). See also Sakarovitch (2005).

References

[anon.] (1799). "Account of the schools for the public service, in the French republic". *The Monthly Magazine* 7, pp. 114–118.

Andersen, Kirsti and Ivor Grattan-Guinness (1994). "Descriptive Geometry". In: *Companion Encyclopedia of the History and Philosophy of the Mathematical Sciences*. Volume 2. Ed. by Ivor Grattan-Guinness. Baltimore and London, pp. 887–896.

Loria, Gino (1921). *Storia della geometria descrittiva dalle origini sino ai giorni nostri*. Milano.

Middleton, Robin and Marie-Noëlle Baudouin-Matuszek (2007). *Jean Rondelet*. The Architect as Technician. New Haven and London.

Monge, Gaspard (1794). "Stéréotomie". *Journal de l'École polytechnique* 1, cahier 1, pp. 1–14.

Monge, Gaspard (1799). *Géométrie descriptive*. Paris.

Sakarovitch, Joël (2005). "Gaspard Monge, *Géométrie descriptive*, first edition (1795)". In: *Landmark Writings in Western Mathematics 1640-1940*. Ed. by Ivor Grattan-Guinness. Amsterdam, pp. 225–241.

---

3 The OED gives a reference from 1800 as a first use.

descriptive set theory

The term *descriptive set theory* was not in common use before the 1950s. A relatively early example is found in Alexandroff and Hopf (1935, p. 19): "To topology also belongs the '*descriptive theory of point sets*' (Théorie descriptive des ensembles), i. e. essentially the theory of Borel sets, of A-sets [analytic sets], and of projective sets".[1] Alexandroff and Hopf go on to give a brief historical account and cite various works, for example by Lebesgue, Suslin, and Luzin.

The shorter *deskriptive Mengenlehre* had been in occasional use several years before; in print, it appeared in Alexandroff (1928, p. 36) and in Rademacher (1924-25, p. 177). In French texts, the subject was originally referred to as the *théorie descriptive des fonctions*, for example by Luzin (1930, p. xiii), who named Borel, Baire, and Lebesgue as the founders of the theory. The adjective *descriptif* in this sense was earlier used by Denjoy (1915, p. 707) and de la Vallée Poussin (1916, p. vii), and English *descriptive* by Hobson (1921, ch. ii).

Kanovei and Koepke (2002, p. 776) note that ultimately the term *descriptive* in this sense goes back to a discussion between Baire, Borel, Hadamard, and Lebesgue in the early 20th century. In this exchange of letters, published by Hadamard (1905),[1] they discussed Zermelo's proof of the well-ordering theorem, in which the axiom of choice was used. In this context, Baire, Borel, and Lebesgue demanded that it must be possible to define or at least describe (*décrire*) mathematical objects by certain agreed-upon procedures and, although *describe* was not precisely specified, certain arbitrary sets of arbitrarily large cardinality, and Zermelo's choice function were not regarded as describable (Kanovei and Purkert 2008, p. 776).[2]

Note. For the history of the concept, see Kanamori (1995).

References

Alexandroff, Pavel S. (1928). "Darstellung der Grundzüge der Urysohnschen Dimensionstheorie". *Mathematische Annalen* 98, pp. 31–63.

Alexandroff, Pavel S. and Heinz Hopf (1935). *Topologie*. Erster Band. Berlin.

Denjoy, Arnaud (1915). "Sur la théorie descriptive des nombres dérivés d'une fonction continue". *Comptes rendus de l'Académie des sciences* 160, pp. 707–709.

Hadamard, Jacques (1905). "Cinq lettres sur la théorie des ensembles". *Bulletin de la Société Mathématique de France* 33, pp. 261–273.

Hobson, Ernest William (1921). *The Theory of Functions of a Real Variable and the Theory of Fourier's Series*. Second edition. Volume I. Cambridge.

Kanamori, Akihiro (1995). "The Emergence of Descriptive Set Theory". In: *From Dedekind to Gödel*. Essays on the Development of the Foundations of Mathematics. Ed. by Jaakko Hintikka. Dordrecht, pp. 241–262.

Kanovei, Vladimir and Koepke (2002). "Deskriptive Mengenlehre in Hausdorffs *Grundzügen der Mengenlehre*". In: *Felix Hausdorff. Gesammelte Werke*. Band II. Grundzüge der Mengenlehre. Ed. by Egbert Brieskorn et al. Berlin, pp. 773–787.

Kanovei, Vladimir and Walter Purkert (2008). "Mengenlehre – Historische Einführung". In: *Felix Hausdorff. Gesammelte Werke*. Band III. Mengenlehre (1927, 1935). Deskriptive Mengenlehre und Topologie. Ed. by Ulrich Felgner et al. Berlin and Heidelberg, pp. 1–40.

---

[1] "Zur Topologie gehört auch die '*deskriptive Punktmengenlehre*' (Théorie descriptive des ensembles), d. h. im wesentlichen die Theorie der Borelschen Mengen, der A-Mengen und der projektiven Mengen". [1] For a detailed discussion of the correspondence, see Moore (1982, pp. 92–98).

[2] Hadamard referred to Tannery (1897) for the above distinction. See also Moore (1982, p. 94).

Luzin, Nikolai (1930). *Leçons sur les ensembles analytiques et leurs applications*. Paris.
Moore, Gregory H. (1982). *Zermelo's Axiom of Choice*. Its Origins, Development, & Influence. New York.
Rademacher, Hans (1924-25). "K. Yoneyama. On continuous set of points. II. [review]". *Jahrbuch über die Fortschritte der Mathematik* 47, p. 177.
Tannery, Jules (1897). "De l'infini mathématique". *Revue générale des sciences pures et appliquées* 8, pp. 129–140.
de la Vallée Poussin, Charles Jean (1916). *Intégrales de Lebesgue. Fonctions d'ensemble. Classes de Baire*. Leçons professées au Collège de France. Paris.

design

As defined by Gropp (1992, p. 316), "[a] design (or more exactly) a $t - (v, k, \lambda)$ design is a set of v elements together with a family of subsets of k elements each such that every t elements occur together in the same subset exactly λ times. The number of subsets is denoted by b". One frequently cited starting point of design theory is a problem posed by W.S.B. Woolhouse in 1844 in the *Lady's and Gentleman's Diary* (p. 84): "Determine the number of combinations that can be made of n symbols, p symbols in each; with this limitation, that no combination of q symbols which may appear in any one of them shall be repeated in any other". One of those who addressed the problem was Kirkman (1847); on this paper, Gropp (1992, p. 317) remarks that "[t]his was the first time that a general theorem on such combinatorial structures was proved which are called designs today".

The name came up in the 1930s in a completely different context, namely the design of experiments. Gropp (1992, p. 319) explains that "[t]his method tries to find out how much an experiment depends on certain parameters. It looks for a possibility to get the correct result with a high statistical certainty without too many expensive and long trials".[1] The method was discussed by Yates (1936, p. 121): "Most biological workers are probably by now familiar with the methods of experimental design known as randomized blocks and the Latin square. These were originally developed by Prof. R. A. Fisher, when Chief Statistician at Rothamsted Experimental Station, for use in agricultural field trials".

He went on to explain:

> The present paper describes another possible modification of the randomized block type of arrangement, analogous to confounding, which enables us to dispense with the restriction that the number of experimental units in a block shall equal the number of treatments. In this modified type of arrangement the number of experimental units per block is fixed, being less than the number of treatments, and the treatments are so allotted to the blocks that every two treatments occur together in a block equally frequently. With six treatments a, b, c, d, e, f, and blocks of three experimental units, for example, the following grouping of treatments into ten blocks satisfies the above conditions:

1 $a\ b\ c$	6 $b\ c\ f$
2 $a\ b\ d$	7 $b\ d\ e$
3 $a\ c\ e$	8 $b\ e\ f$
4 $a\ d\ f$	9 $c\ d\ e$
5 $a\ e\ f$	10 $c\ d\ f$

[1] For a historical account of the design of experiments, see Verdooren (2020).

> In this set of groupings every two treatments occur together in a block twice, there being five replications of each treatment.
>
> It is proposed to call this type of arrangement a *symmetrical incomplete randomized block arrangement*, or more briefly, when the symmetry and randomization are understood, an *incomplete block arrangement* (Yates 1936, p. 122).[2]

In a classic paper published shortly afterwards, Bose (1939, p. 354) stated that "[t]he object of [his] paper is to study the combinatorial problem involved in the construction of a certain type of design, first introduced in experimental studies by F. Yates (1936) and called *Balanced Incomplete Block Design*". The term spread via Bose's widely read article, which is of a purely combinatorial nature; it also appeared in Weiss and Cox (1939, p. 294). In a paper published shortly before Bose (1939), Bose and Nair (1939, p. 338) state: "These new designs came to be called 'balanced incomplete block designs'".

Note. For the history of designs, see Wilson (2003), Gropp (1992), and Anderson et al. (2007).

References

Anderson, Ian et al. (2007). "Design Theory: Antiquity to 1950". In: *Handbook of Combinatorial Designs*. Second Edition. Ed. by Charles J. Colbourn and Dinitz Jeffrey H. Boca Raton et al., pp. 11–22.

Bose, R.C. (1939). "On the construction of balanced incomplete block designs". *Annals of Eugenics* 9, pp. 353–399.

Bose, R.C. and K.R. Nair (1939). "Partially Balanced Incomplete Block Designs". *Sankhyā: The Indian Journal of Statistics* 4, pp. 337–372.

Gropp, Harald (1992). "The Birth of a Mathematical Theory in British India". *Colloquia Mathematica Societatis János Bolyai* 60, pp. 315–327.

Kirkman, Thomas P. (1847). "On a problem in combinations". *The Cambridge and Dublin Mathematical Journal* 2, pp. 191–204.

Verdooren, L. Rob (2020). "History of the Statistical Design of Agricultural Experiments". *Journal of Agricultural, Biological, and Environmental Statistics* 25, pp. 457–486.

Weiss, Martin G. and Gertrude M. Cox (1939). "Balanced incomplete block and lattice square designs for testing yield differences among large numbers of soybean varieties". *Iowa Agriculture and Home Economics Experiment Station Research Bulletin* 23, pp. 291–316.

Wilson, Robin (2003). "The early history of block designs". *Rendiconti del Seminario Matematico di Messina* 9, pp. 267–276.

Yates, Frank (1936). "The formation of Latin squares for use in field experiments". *Empire Journal of Experimental Agriculture* 1, pp. 235–244.

determinant

The term *determinant* was originally introduced by Gauß with respect to binary forms. He defined a particular kind of determinant that was later called the *discriminant*: "We shall see in what follows that the properties of the form (a, b, c) depend in a special way on the nature of this number $bb - ac$. We shall call it the *determinant* of this form" (Gauß 1801, § 154).[1] While this definition already implies the reason for chosing this name, Gauß later is slightly more explicit: "The properties of a binary

[2] Yates (1936, p. 127) also introduced a notation similar to the one used nowadays.

[1] "Numerum $bb - ac$, a cuius indole proprietates formae (a, b, c) imprimis pendere in sequentibus docebimus, *determinantem* huius formae vocabimus". Translation from Gauß (1986, § 154). – On Gauß's paper, see Muir (1960a, pp. 63–66).

form $axx + 2bxy + cyy$ primarily depend on the number $bb - ac$, which is therefore called the determinant of that form" (Gauß 1831, p. 1067).[2] The term was adopted by Jacobi (1827, p. 354), for example, a step that Muir (1960a, p. 178) described as "decisive of the fate of the word".

In its current meaning the term goes back to Cauchy (1815).[3] Cauchy regarded determinants as special cases of alternating symmetric functions and explained that Gauß "has used them to his advantage [...] He called these functions *determinants*. I shall retain this name, which gives me a simple means of stating the results. I shall only note that one sometimes calls the functions of which we are speaking the *resultants*" (Cauchy 1815, p. 51).[1] The word *resultant* mentioned by Cauchy was suggested by Laplace (1776, pp. 296f.).[2] For some time, it remained in use as a synonym of *determinant*; Liouville (1841, p. 348) remarked: "Instead of the word *resultant*, the mathematicians often use the word *determinant*".[3] Even Cauchy used *resultant* in many publications after his 1815 article, for example in Cauchy (1841, p. 161).

The English term appears to go back to Cayley (1849, p. 75); his paper was read in 1843.

See also *resultant* and *discriminant*.

Note. For the history of the concept, see Muir (1960a, 1960b). On Leibniz and determinants, see Knobloch (1980, 2001, 2013). Tropfke (1980, p. 406) gives a concise terminological overview.

References

Cauchy, Augustin-Louis (1815). "Sur les fonctions qui ne peuvent obtenir que deux valeurs égales et de signes contraires par suite des transpositions opérées entre les variables qu'elles renferment". *Journal de l'École polytechnique* 10, cahier 17, pp. 29–112.

Cauchy, Augustin-Louis (1841). *Exercises d'analyse et de physique mathématique*. Tome deuxième. Paris.

Cayley, Arthur (1849). "On the Theory of Determinants". *Transactions of the Cambridge Philosophical Society* 8, pp. 75–88.

Gauß, Carl Friedrich (1801). *Disquisitiones arithmeticae*. Leipzig.

Gauß, Carl Friedrich (1831). "Untersuchungen über die Eigenschaften der positiven ternären quadratischen Formen von Ludwig August Seeber [review]". *Göttingische gelehrte Anzeigen* (1831, 2), pp. 1065–1077.

Gauß, Carl Friedrich (1986). *Disquisitiones Arithmeticae*. English Edition. Translated by A.A. Clarke. Revised by W.C. Waterhouse. New York et al.

Jacobi, Carl Gustav Jacob (1827). "Ueber die Pfaffsche Methode, eine gewöhnliche lineare Differentialgleichung zwischen $2n$Variabeln durch ein System von n Gleichungen zu integriren". *Journal für die reine und angewandte Mathematik* 2, pp. 347–357.

Knobloch, Eberhard (1980). *Der Beginn der Determinantentheorie*. Leibnizens nachgelassene Studien zum Determinantenkalkül. Textband. Hildesheim.

---

[2] "Die Eigenschaften einer binären Form $axx + 2bxy + cyy$ hängen vornehmlich von der Zahl $bb - ac$ ab, welche daher der Determinant jener Form heißt". [3] Cauchy's paper is discussed in detail by Muir (1960a, pp. 92–131). [1] "M. Gauss s'en est servi avec avantage [...] et il a désigné ces mêmes fonctions sous le nom de *déterminants*. Je conserverai cette dénomination qui fournit un moyen facile d'énoncer les résultats; j'observerai seulement qu'on donne aussi quelquefois aux fonctions dont il s'agit le nom de *résultantes*". Translation from Wußing (1984, p. 267). [2] On earlier uses of the term *resulting equation*, for example by Leibniz and Newton, see *resultant*. [3] "Au lieu du mot *résultante*, les géomètres emploient souvent le mot *déterminant*".

Knobloch, Eberhard (2001). "Déterminants et élimination chez Leibniz". *Revue d'histoire des sciences* 54, pp. 143–164.
Knobloch, Eberhard (2013). "Leibniz's Theory of Elimination and Determinants". In: *Seki, Founder of Modern Mathematics in Japan*. A Commemoration on His Tercentenary. Ed. by Eberhard Knobloch, Hikosaburo Komatsu, and Dun Liu. Tokyo et al., pp. 229–244.
Laplace, Pierre-Simon (1776). "Recherches sur le calcul intégral et sur le système du monde". *Mémoires de l'Académie royale des sciences* (1772, seconde partie), pp. 267–376.
Liouville, Joseph (1841). "Mémoire sur quelques propositions générales de géométrie et sur la théorie de l'élimination dans les équations algébriques". *Journal de mathématiques pures et appliquées* 6, pp. 345–411.
Muir, Thomas (1960a). *The Theory of Determinants in the Historical Order of Development*. Volume I. General and Special Determinants up to 1841. New York.
Muir, Thomas (1960b). *The Theory of Determinants in the Historical Order of Development*. Volume II. The Period 1841 to 1860. New York.
Tropfke, Johannes (1980). *Geschichte der Elementarmathematik*. Band 1: Arithmetik und Algebra. Vollständig neu bearbeitet von Kurt Vogel, Karin Reich, Helmuth Gericke. Berlin and New York.
Wußing, Hans (1984). *The Genesis of the Abstract Group Concept*. A Contribution to the History of the Origin of Abstract Group Theory. Tr. Abe Shenitzer. Cambridge, Mass.

developable surface

According to Gaspard Monge, the founder of descriptive geometry, "[a] surface is developable whenever, by supposing it to be flexible and inextensible, one may conceive of mapping it onto a plane, like cones and cylinders can be, so that the way in which it rests on the plane is without duplication or disruption of continuity" (Monge 1780, p. 383).[1] Monge used the French term *surface développable*, and his terminological innovation consists in the creation of the adjective *développable*, derived from the verb *développer* 'to unfold, develop'.

Before Monge, the concept was dealt with in an influential paper by Euler, in which he points out that it is generally known that the cylinder and the cone can be "developed" onto a plane (Euler 1772, p. 3). Euler uses the Latin verb *explicare*, which means 'to spread out, unfold' (and also, 'to explain') and thus overlaps with *développer*. As Reich (1973, p. 296) points out, Euler had earlier described his results in two letters to Lagrange. In these letters, he used French *expliquer*, which corresponds to Latin *explicare* (see Lagrange 1892, p. 214).

The concept and terminology occur even before Euler, in particular in some works by the French engineer Frézier, who, in a theoretical chapter of his book on stereotomy[2], defined the verb *développer*: "To develop a solid, that is to stretch out on a plane surface all those surfaces in which it is enveloped" (Frézier 1737, p. 321).[3] As Stäckel (1893, p. 453) points out, however, Frézier's understanding of development differs from the modern one since Frézier (1737, pp. 338f.) also considers the "development" of the sphere onto a plane. In a later treatise, Frézier

---

[1] "Une surface est développable, lorsqu'en la supposant flexible et inextensible, on peut concevoir appliquée sur un plan, comme celles des cônes et des cylindres, de maniere qu'elle le touche sans duplicature ni solution de continuité". Translation from Reich (2007, p. 490). The paper was read in January 1775. Monge had already used the term in a paper read in 1771, but that paper was not published until 1785 (Monge 1785, e. g., p. 515). [2] Sakarovitch (2003, p. 69) defines stereotomy as "the art of drawing the shapes to be given to stones (and bricks) for future assembly".
[3] "Développer un corps, c'est étendre sur une surface plane toutes celles, dont il est enveloppé".

(1760, p. 120) mentions some solids such as the torus which "cannot be developed, i. e. spread out in all length and width on a plane".[4]

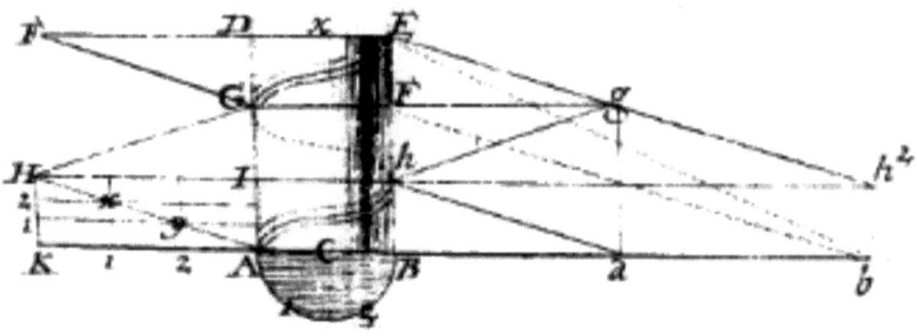

Figure 39: A developable surface. From Frézier (1737, Planche 23). Frézier uses this illustration to show that when a cylinder is developed onto a plane, a helix on the cylinder is turned into a straight line.

The English term *developable* does not seem to occur until 1814, when, in a summary of Monge (1806), Leybourn (1814, p. 103) points out that a "developable surface is that which may be expanded upon a plane, with tearing, or duplicature". In German, the notion was paraphrased as "surfaces that can be developed"[5] in [anon.] (1782, p. 28); Klügel (1808, p. 339) used *abwickelbare Fläche*, which corresponds directly to *developable surface*.

Note. The history of the concept and terminology is discussed by Reich (1973, pp. 295–302); see also Stäckel (1893, pp. 453–456), Lawrence (2011), Cajori (1929), Calvo-López (2011), and Friedman (2018, pp. 157–165).

References

[anon.] (1782). "Mathematische Aufsätze". *Zugabe zu den Göttingischen Anzeigen von gelehrten Sachen* (1782), pp. 25–29.

Cajori, Florian (1929). "Generalizations in Geometry as Seen in the History of Developable Surfaces". *The American Mathematical Monthly* 36, pp. 431–437.

Calvo-López (2011). "From Mediaeval Stonecutting to Projective Geometry". *Nexus Network Journal* 13, pp. 503–533.

Euler, Leonhard (1772). "De solidis quorum superficiem in planum explicare licet". *Novi Commentarii academiae scientiarum Petropolitanae* 16, pp. 3–34.

Frézier, Amédée François (1737). *La théorie et la pratique de la coupe des pierres et des bois*. Tome premier. Strasbourg and Paris.

Frézier, Amédée François (1760). *Elémens de stéréotomie*. A l'usage de l'architecture pour la coupe des pierres. Tome second. Paris.

Friedman, Michael (2018). *A History of Folding in Mathematics*. Mathematizing the Margins. Cham.

Klügel, Georg Simon (1808). *Mathematisches Wörterbuch*. Erste Abtheilung: Die reine Mathematik. Dritter Theil von K bis P. Leipzig.

Lagrange, Joseph-Louis (1892). *Œuvres*. Tome quatorzième et dernier. Ed. by J.-A. Serret and Gaston Darboux. Paris.

[4] "ne peuvent être développées, c'est-à-dire étendues de toute leur longeur & largeur sur un plan". See also Frézier (1760, pp. 160f.).
[5] "Flächen, die sich abwickeln lassen".

Lawrence, Snežana (2011). "Developable Surfaces: Their History and Application". *Nexus Network Journal* 13, pp. 701–714.
Leybourn, Thomas (1814). "Brief Notices of the Principal Papers relating to Mathematics, in the Memoirs of the French National Institute". *New series of the Mathematical repository* 3, part I, pp. 49–51.
Monge, Gaspard (1780). "Mémoire sur les Propriétés de plusieurs genres de Surfaces courbes, particulièrement sur celles des Surfaces développables, avec une Application à la Théorie des Ombres et des Pénombres". *Mémoires de mathématique et de physique, presentés à l'Académie royale des sciences, par divers sçavans* 9, pp. 382–440.
Monge, Gaspard (1785). "Mémoire sur les développées, les rayons de courbure, et les différens genres d'inflexions des courbes à double courbure". *Mémoires de mathématique et de physique, presentés à l'Académie royale des sciences, par divers sçavans* 10, pp. 511–550.
Monge, Gaspard (1806). "Sur la Surface courbe, dont toutes les Normales sont tangentes à une même Surface développable quelconque". *Journal de l'École polytechnique* 6, cahier 13, pp. 1–59.
Reich, Karin (1973). "Die Geschichte der Differentialgeometrie von Gauss bis Riemann (1828-1868)". *Archive for History of Exact Sciences* 11, pp. 273–382.
Reich, Karin (2007). "Euler's Contribution to Differential Geometry and its Reception". In: *Leonhard Euler*. Life, Work and Legacy. Ed. by Robert E. Bradley and C. Edward Sandifer. Amsterdam, pp. 479–502.
Sakarovitch, Joël (2003). "Stereotomy, a multifaceted technique". In: *Proceedings of the First International Congress on Construction History, Madrid, 20th-24th January 2003*. Volume III. Ed. by Santiago Huerta. Reverte, pp. 69–79.
Stäckel, Paul (1893). "Bemerkungen zur Geschichte der geodätischen Linien". *Berichte über die Verhandlungen der Königlich Sächsischen Gesellschaft der Wissenschaften zu Leipzig (Mathematisch-Physische Klasse)* 45, pp. 444–467.

diagonal

In Greek, *διαγώνιος* literally means 'from angle to angle'. The term is very rare with Euclid,[1] who, following a usage common in Plato's time (*Menon* 85B), preferred *διάμετρος* (which gave rise to *diameter*), for example, in book I, prop. 34 of the *Elements*: "In parallelogrammic areas the opposite sides and angles are equal to one another, and a diameter bisects the areas".[2] The term is not used at all by writers such as Archimedes and Apollonius, and it is not found before Euclid (Mugler 1958-59, p. 127). The preference of *διάμετρος*, as can be seen in the article *diameter*, is reflected in much later texts, where *diameter* is used where, from today's point of view, *diagonal* would be expected.[3] Later, Hero of Alexandria used *διαγώνιος* regularly, but without definition, in his book *Metrica*. In the pseudo-Heronian *Geometrica*, it is defined as "the straight line drawn, in a fourgon and similar figures, from vertex to vertex" (Bruins 1964, p. 2).[4]

The word was taken over to Latin virtually unchanged; *(linea) diagonios* was used used by Vitruvius, but he also created the extended *(linea) diagonalis*.[5] The term is later also found with Roman land surveyors (Blume, Lachmann, and Rudorff

[1] In his *Elements*, it is only found in book XI, prop. 28. [2] Translation from Euclid (1956, p. 323).
[3] For example, Bradwardine declares: "a diagonal line is that which is drawn from angle to opposite angle, and, if it is in a square, it is called a diameter" (Molland 1989, p. 52) ("est linea diagonalis que ducitur ab angulo in angulum oppositum, et, si est in quadrato, vocatur dyameter"). Translation from Molland (1989, p. 53). [4] A definition is also given in the pseudo-Heronian *Definitions* (Heron of Alexandria 1912, p. 46). [5] See OLD, LS, TLL.

1848, p. 196).[6] The term *linea diagonalis* remains rare, however,[7] until about the 12th century, when it is found in some texts such as Robert of Chester's translation of Euclid's *Elements* (Busard and Folkerts 1992, p. 299).

The French term appears in second the 13th century (DEAF) in a treatise called *Pratike de geometrie* (Victor 1979, e. g., p. 474). In German, the term first appears in the mixed form *lini diagonalis* (Widmann 1489, fol. E 1^{v}).[8] Various alternative terms such as *Ecklini* 'corner line' (Schmid 1539, p. 6) and *Scheidlini* 'separating line' (Schottel 1663, p. 481) ultimately remained unsuccessful. The English term appears in the 16th century (OED); as the first attestation in a purely mathematical context, it is found in Billingley's translation of the *Elements*: "A diagonall line is a right line which in angular figures is drawne from one angle and extended to his contrary angle" (Euclid 1570, fol. 340^{v}).

Note. For the Greek term, see Mugler (1958-59, pp. 127f.). A brief terminological overview is found in Tropfke (1940, pp. 121f.).

References

Blume, Friedrich, Karl Lachmann, and Adolf Rudorff, eds. (1848). *Die Schriften der römischen Feldmesser*. Erster Band. Texte und Zeichnungen. Berlin.

Bruins, E.M., ed. (1964). *Codex Constantinopolitanus*. Palatii Veteris No. 1. Part Three. Translation and Commentary. Leiden.

Bubnov, Nicolaus, ed. (1899). *Gerberti postea Silvestri II papae Opera Mathematica*. Berlin.

Busard, H.L.L. and Menso Folkerts, eds. (1992). *Robert of Chester's (?) Redaction of Euclid's* Elements, *the so-called Adelard II Version*. Basel.

Euclid (1570). *The Elements of Geometrie of the most auncient Philosopher Euclide of Megara*. Faithfully (now first) translated into the Englishe toung, by H. Billingsley. London.

Euclid (1956). *The Thirteen Books of Euclid's Elements*. Volume I. Introduction and Books I, II. Ed. by Thomas L. Heath. New York.

Folkerts, Menso (2014). "Die Mathematik der Agrimensoren – Quellen und Nachwirkung". In: *In den Gefilden der römischen Feldmesser*. Juristische, wissenschaftsgeschichtliche, historische und sprachliche Aspekte. Ed. by Eberhard Knobloch and Cosima Möller. Berlin and Boston, pp. 131–148.

Heron of Alexandria (1912). *Heronis Alexandrini opera quae supersunt omnia*. Volumen IV. Heronis definitiones cum variis collectionibus. Heronis quae feruntur geometrica. Ed. by J.L. Heiberg. Leipzig.

Molland, George, ed. (1989). *Thomas Bradwardine, Geometria Speculativa*. Latin Text and English Translation with an Introduction and a Commentary. Stuttgart and Wiesbaden.

Mugler, Charles (1958-59). *Dictionnaire historique de la terminologie géométrique des Grecs*. Paris.

Schmid, Wolffgang (1539). *Das erst buch der Geometria*. Nürnberg.

Schottel, Justus Georg (1663). *Ausführliche Arbeit Von der Teutschen HaubtSprache*. Braunschweig.

Strauß, Gerhard et al., eds. (1999). *Deutsches Fremdwörterbuch*. 2. Auflage. Band 4: da capo – Dynastie. Berlin and New York.

Tropfke, Johannes (1940). *Geschichte der Elementar-Mathematik*. Vierter Band. Ebene Geometrie. Dritte, verbesserte und vermehrte Auflage besorgt von Dr. Kurt Vogel. Berlin and Leipzig.

Victor, Stephen K. (1979). *Practical Geometry in the High Middle Ages*. "Artis cuiuslibet consummatio" and the "Pratike de geometrie". Philadelphia.

---

[6] In a treatise on land surveying that appears to have been originally composed not later than the second century (Folkerts 2014, p. 132), *diagonus* is found (Bubnov 1899, p. 530). [7] It is not found, for example, in Martianus Capella's *De nuptiis* or Isidore of Seville's *Etymologies*. [8] See Strauß et al. (1999, pp. 484f.). The word *lini* (mod. *Linie*) is the Germanized version of Latin *linea*; *diagonalis* remains unchanged (the word was thus not naturalized yet). In the same text, the purely Latin form *linea diagonalis* is also found.

Widmann, Johannes (1489). *Behende vnd hubsche Rechenung auff allen kauffmanschafft*. Leipzig.

diagonal (linear algebra) and *diagonal matrix*

The term *diagonal* of a matrix was coined in analogy to the common geometrical meaning. French *diagonale* in this sense appears to go back to Cauchy (1815, p. 54), who used it in a definition of the diagonal elements of a matrix (he called these elements *principal terms* (*termes principaux*)): "they [the diagonal elements] are all situated in the system [matrix] on a diagonal of the square formed by the system".[1] A definition was given by Lebesgue (1837, p. 344), who stated "that the system [matrix] is symmetric, since the numbers that form the system are placed symmetrically with regard to the numbers with equal indices $A_{1,1}A_{2,2}\ldots A_{n,n}$, which form the diagonal of the system".[2]

The term *diagonal matrix*, for a matrix in which every entry that is not on the main diagonal is zero, goes back to Kronecker (1889, p. 360). He used German *Diagonalsystem*, a term that is also found in various texts such as Bachmann (1898, p. 297). German *Diagonalmatrix* then appears in Schur (1901, p. 11), English *diagonal matrix* in Baker (1907, p. 20).

References

Bachmann, Paul (1898). *Zahlenheorie*. Vierter Theil. Die Arithmetik der quadratischen Formen. Leipzig.

Baker, Henry Frederick (1907). *An Introduction to the Theory of Multiply Periodic Functions*. Cambridge.

Cauchy, Augustin-Louis (1815). "Sur les fonctions qui ne peuvent obtenir que deux valeurs égales et de signes contraires par suite des transpositions opérées entre les variables qu'elles renferment". *Journal de l'École polytechnique* 10, cahier 17, pp. 29–112.

Kronecker, Leopold (1889). "Über symmetrische Systeme". *Sitzungsberichte der Königlich Preußischen Akademie der Wissenschaften zu Berlin* (1889), pp. 349–362.

Lebesgue, Victor-Amédée (1837). "Thèses de mécanique et d'astronomie". *Journal de mathématiques pures et appliquées* 2, pp. 337–365.

Schur, Issai (1901). *Ueber eine Klasse von Matrizen, die sich einer gegebenen Matrix zuordnen lassen*. Berlin.

diameter

Greek *διάμετρος*, from *διά* 'across, through' and *μέτρον* 'measure', is the source of this term, which was already used by Euclid and some writers before him. The term could refer to the diameter of circles, spheres, and conic sections, but also for a line that joins the opposite vertices of a parallelogram (Mugler 1958-59, pp. 132–135). In the latter sense, it is today translated as *diagonal*, but *diameter* has long remained in use in this sense. For example, Ramus (1569a, Geometria, p. 21) gave the following definition, which he applied not just to circles, but also to triangles and squares: "A diameter is a right line inscribed in a figure through the centre".[1] On the terminology, Proclus commented:

[1] "ils sont tous situés dans le système [...] sur une diagonale du carré formé par le système".
[2] "On peut dire alors que le système est symétrique, puisque les nombres qui le forment sont placés symétriquement par rapport aux nombres à indices égaux $A_{1,1}A_{2,2}\ldots A_{n,n}$ qui forme la diagonale du système". [1] "Diameter est recta inscripta figurae per centrum". Translation from Ramus (1590, fol. 2v).

> The author of the *Elements* himself makes it clear that he is defining not every diameter, but the diameter of the circle. The square also has a diameter, and so does the parallelogram in general, and among solid figures the sphere. But in these cases such a line is also called a "diagonal," and in the case of the sphere an "axis" also, "diameter" alone being used for the circle (Proclus 1873, p. 156).[2]

In Euclid's *Elements* (book I, def. 17), a diameter of a circle is defined as "any straight line drawn through the center and terminated in both directions by the circumference of the circle, and such a straight line also bisects the circle".[3] The sixth-century commentator Simplicius is reported by the Persian mathematician al-Nairīzī to have commented on the term: "As for the diameter, it is only called a diameter in the Greek language on account of its (all of it) passing through the circle as if it were measuring it. And, lo, as for a measuring stick, it is what, all of it, passes through a thing. And, what is more, it was already called a diameter in the Greek language prior to its dividing the circle into halves" (Lo Bello 2009, p. 12).

With reference to conic sections, a generalized definition was given by Apollonius in his *Conics* (def. 4): "Of any curved line which is in one plane, I call that straight line the *diameter* which, drawn from the curved line, bisects all straight lines drawn to this curved line parallel to some straight line" (Apollonius of Perga 1891, p. 6).[1]

Latin writers usually used *diametros* rather than *diameter*.[2] The first definition in Latin was, according to Ayuso García (2008, p. 939), given by Pseudo-Censorinus: "The diameter is the straight line drawn through the center and cutting the circle in half" (Sallmann 1983, p. 69).[3] The term was also used by Martianus Capella (*De nuptiis* 711): "A diameter is any straight line, drawn through the center, which divides the circle into equal parts" (Willis 1983, p. 252).[4]

Occasionally, *linea dimetiens* is found, where *dimetiens* has an etymology analogous to the Greek term (from *dis-* and *mētior* 'to measure'). The term was used by Pliny, but was in use until well into modern times, especially by writers who preferred purely Latin word forms such as Ramus (1569b, p. 155).

In modern languages, the term often remains virtually unchanged. In the first known text on geometry in any vernacular, the French *Pratike de geometrie*, dating from the second half of the 13th century, *dyametre* (modern *diamètre*) is used (Victor 1979, p. 481).[5] English *diametre* is first found in Trevisa's translation (14th c.) of *De proprietatibus rerum* (OED). In German, *Dyameter* is found in the *Geometria Culmensis* from around 1400 (Mendthal 1886, p. 66), but alternatives such as *mittellini* 'middle line' were in common use (see, e. g., Köbel 1536, fol. F iiij^r); Stevin (1594, p. 34) uses the analogous Dutch *middellijn*. The German standard term, however, is

---

2 Translation from Proclus (1970, p. 124). Note that Euclid defines both an axis and a diameter of the sphere (book XI, deff. 15 and 17). 3 Translation from Euclid (1956, p. 185). Euclid's definition of the diameter of a sphere is in book XI, def. 17. 1 Translation from Apollonius of Perga (2013, p. 3) (emphasis added). 2 According to Ayuso García (2008, p. 938), *diameter* is not attested in classical Latin at all. 3 "diametron est recta linea per centron inmissa et in utramque partem secans circulum". 4 "diametros est directa linea quaedam per punctum supra dictum ducta, quae orbem aequalibus partibus dividit". Translation from Stahl (1977, p. 266). – In 714, Capella gives *distermina* 'separating, parting (line)' (LS) as a Latin alternative. This term is extremely rare, though. 5 See also DEAF, s. v. diametre. Italian *diametro* is first attested at about the same time, in both the sense of 'diameter' and 'diagonal' (TLIO), and so is Spanish *diámetro* (Bossong 1978, p. 195).

Durchmesser, from *durch* 'through' and *messer* 'measurer'; it spread through Sturm (1667),[6] but was already used by Goldmann (1656, e. g., pp. 8, 42).

See also *diagonal*.

Note. For the Greek term, see Mugler (1958-59, pp. 132–135) and Heath in Euclid (1956, pp. 185f.), for the Latin, see Ayuso García (2008, pp. 937–945). Terminological remarks can also be found in Tropfke (1940, p. 132) and Smith (1958, p. 278). For the Arabic term *quṭr*, see Souissi (1968, pp. 283f.) and Rashed (2017, pp. 648–651).

References

Apollonius of Perga (1891). *Apollonii Pergaei quae graece exstant cum commentariis antiquis*. Vol. I. Ed. by J.L. Heiberg. Leipzig.

Apollonius of Perga (2013). *Conics*. Books I-IV. Books I-III translated by R. Catesby Taliaferro. Book IV translated by Michael N. Fried. Santa Fe.

Ayuso García, Manuel (2008). "La terminología latina de la geometría en Marciano Capela". PhD thesis. Madrid: Universidad Nacional de Educación a Distancia.

Bossong, Georg, ed. (1978). *Los Canones de Albateni*. Herausgegeben sowie mit Einleitung, Anmerkungen und Glossar versehen. Tübingen.

Euclid (1956). *The Thirteen Books of Euclid's Elements*. Volume I. Introduction and Books I, II. Ed. by Thomas L. Heath. New York.

Goldmann, Nikolaus (1656). *Tractatus de usu proportionatorii. Eine Anleitung zum Gebrauch des Ebenpassers*. Leiden.

Köbel, Jakob (1536). *Geometrei*. Oppenheim.

Lo Bello, Anthony (2009). *The Commentary of al-Nayrizi on Books II-IV of Euclid's* Elements of Geometry. With a Translation of That Portion of Book I missing from MS Leiden Or. 399.1 but present in the Newly Discovered Qom Manuscript edited by Rüdiger Arnzen. Leiden and Boston.

Mendthal, Hans (1886). *Geometria culmensis*. Ein agronomischer Tractat aus der Zeit des Hochmeisters Conrad von Jungingen (1393-1407). Leipzig.

Mugler, Charles (1958-59). *Dictionnaire historique de la terminologie géométrique des Grecs*. Paris.

Proclus (1873). *In primum Euclidis elementorum librum commentarii*. Ed. by Gottfried Friedlein. Leipzig.

Proclus (1970). *Commentary on the First Book of Euclid's Elements*. Translated with Introduction and Notes by Glenn R. Morrow. Princeton.

Ramus, Petrus (1569a). *Arithmeticae libri duo. Geometriae septem et viginti*. Basel.

Ramus, Petrus (1569b). *Scholarum Mathematicarum, Libri Unus et Triginta*. Basel.

Ramus, Petrus (1590). *Elementes of geometrie*. Faithfully translated by Tho. Hood. London.

Rashed, Roshdi, ed. (2017). *Lexique historique de la langue scientifique arabe*. Hildesheim.

Sallmann, Klaus, ed. (1983). *Censorini de die natali liber ad Q. Caerellium*. Leipzig.

Smith, David Eugene (1958). *History of Mathematics*. Volume II. Special Topics of Elementary Mathematics. New York.

Souissi, Mohammed (1968). *La langue des mathématiques en arabe*. Tunis.

Stahl, William Harris (1977). *Martianus Capella and the Seven Liberal Arts*. Volume II. The Marriage of Philology and Mercury. New York.

Stevin, Simon (1594). *De Sterctenbouwing*. Leiden.

Sturm, Johann Christoph (1667). *Des Unvergleichlichen Archimedis Sand-Rechnung*. Nürnberg.

Tropfke, Johannes (1940). *Geschichte der Elementar-Mathematik*. Vierter Band. Ebene Geometrie. Dritte, verbesserte und vermehrte Auflage besorgt von Dr. Kurt Vogel. Berlin and Leipzig.

Victor, Stephen K. (1979). *Practical Geometry in the High Middle Ages*. "Artis cuiuslibet consummatio" and the "Pratike de geometrie". Philadelphia.

Willis, James, ed. (1983). *Martianus Capella*. Leipzig.

---

6 Sturm suggested it in a note at the end of his preface.

diametral plane

Latin *planum diametrale* was suggested by Euler (1748, appendix, § 13): "Just as when a plane curve is divided into two similar and equal parts by a straight line, that line is called a diameter, so in the solid case, that plane which divides the solid into two similar parts, is called a *diametral plane*".[1] He also used the term in his *Scientia navalis* (Euler 1749, p. 44).

English *diametral plane* seems to have been used for the first time in Rees' *Cyclopædia* (Rees 1819); the article on *geometry* contains this definition: "A diametral plane is that which divides a system of parallel chords, each into equal parts".[2] In this sense, the terminology goes back to Monge and Hachette (1802, p. 151).[3]

Note. Historical references can be found in Staude (1910, p. 936).

References

Euler, Leonhard (1748). *Introductio in analysin infinitorum*. Tomus secundus. Lausanne.
Euler, Leonhard (1749). *Scientia navalis*. Pars prior. St. Petersburg.
Euler, Leonhard (1990). *Introduction to Analysis of the Infinite*. Book II. Translated by John D. Blanton. New York.
Monge, Gaspard and Jean Nicolas Pierre Hachette (1802). "Application de l'algèbre à la géométrie". *Journal de l'École polytechnique* 4, cahier 11, pp. 143–169.
Rees, Abraham (1819). *The Cyclopædia; Or, Universal Dictionary of Arts, Sciences and Literature*. Vol. XVI. London.
Staude, Otto (1910). *Analytische Geometrie des Punktepaares, des Kegelschnittes und der Fläche zweiter Ordnung*. Zweiter Teilband. Leipzig and Berlin.

diffeomorphism

A diffeomorphism is a differentiable mapping between manifolds which has a differentiable inverse. The term was coined by Ambrose and Singer (1953, p. 428): "We use the word 'diffeo' for a 1:1 differentiable mapping of one manifold onto another whose inverse is also differentiable and call the manifolds diffeomorphic". Thus, in the article the shorter *a diffeo* is used instead of *diffeomorphism*. The term was soon adopted, for example by Milnor (1956, p. 401) and Palais (1957, p. 4); the latter attributes the term to Ambrose. Both Milnor and Palais use *diffeomorphism*.

See *homeomorphism*, where it shown that Poincaré's use of *homéomorphisme* corresponds much more to *diffeomorphism* than to *homeomorphism*.

Note. The history of the concept is discussed in detail by Moore (2007).

References

Ambrose, Warren and I.M. Singer (1953). "A Theorem on Holonomy". *Transactions of the American Mathematical Society* 75, pp. 428–443.
Milnor, John W. (1956). "On Manifolds Homeomorphic to the 7-Sphere". *Annals of Mathematics* 64, pp. 399–405.
Moore, Gregory H. (2007). "The evolution of the concept of homeomorphism". *Historia Mathematica* 34, pp. 333–343.

---

[1] "Quemadmodum ergo in Figuris planis ea Linea recta, qua Figura in duas partes similes & aequales dirimebatur, Diameter est appellata, ita in solidis id planum, quo Corpus in duas partes similes dividitur, *Diametrale* vocemus". Translation from Euler (1990, p. 383). [2] The article may have been written by Peter Barlow or by James Ivory, who are named as contributors to entries on geometry. [3] Their definition was: "On appelle [...] *plan diamétral*, celui qui divise un système de cordes parallèles entre elles, chacune en deux parties égales".

Palais, Richard S. (1957). *A global formulation of the Lie theory of transformation groups*. Providence, R.I.

difference

Latin *differentia* begins to be used in the 12th century in the meaning 'difference'. For example, Fibonacci sometimes used the word in this sense (Fibonacci 1857, e. g., p. 416). At this time, however, *differentia* was more commonly used in other senses like 'decimal place'.[1] Jordanus de Nemore, who flourished in the 13th century, used *differentia* quite regularly, for example in his *De numeris datis* (Hughes 1981, e. g., p. 57).

The OED cites the late 14th century *Equatorie of Planetis* for the English term: "The quantite of that difference þat I clepe [call] the remenaunt" (Price 1955, p. 36). In print, the term appeared in Recorde (1543, fol. 37^{v}): "The remayner is a summe lefte after dewe workyng which declareth the excesse or difference of the two other nombers".

The French term was used in a 13th-century algorism: "there is a difference between eight and ten, viz. two" (Karpinski and Waters 1928, p. 74).[1] Stevin (1585, p. 71) later used the term in his definition of subtraction: "To subtract is to find the difference of two given numbers".[2] Italian *differenza* was in common use in the 14th century (TLIO).

In the Latin form *differentia*, the word appeared in a German text in Widmann (1489, fol. i 4^{v}); the German form *Differentz* was used slightly later by Schreiber (1518, fol. A vi^{r}) and Rudolff (1525, fol. A vi^{r}).[3]

See also *remainder*, where also some of the Greek terminology is discussed. For the difference of sets, see *complement* (set).

Note. Some remarks on the terminology are found in Tropfke (1930, pp. 14, 96, 1980, p. 204), Guillaumin (2020, pp. 106f.), and Smith (1958, p. 97).

References

Allard, André (1990). "La formation du vocabulaire latin de l'arithmétique médiévale". In: *Méthodes et instruments du travail intellectuel au moyen âge*. Études sur le vocabulaire. Ed. by Olga Weijers. Turnhout, pp. 137–181.

Fibonacci (1857). *Scritti di Leonardo Pisano*. Pubblicati da Baldassare Boncompagni. Volume I. Roma.

Folkerts, Menso (1997). *Die älteste lateinische Schrift über das indische Rechnen nach al-Ḫwārizmī*. Edition, Übersetzung und Kommentar ... unter Mitarbeit von Paul Kunitzsch. München.

Guillaumin, Jean-Yves (2020). *Dictionnaire de la terminologie latine ancienne de l'arithmétique et de la géométrie*. Paris.

Hughes, Barnabas (1981). *Jordanus de Nemore*. De numeris datis. A critical edition and translation. Berkeley et al.

Karpinski, Louis Charles and E.G.R. Waters (1928). "A Thirteenth Century Algorism in French Verse". *Isis* 11, pp. 45–84.

Price, Derek J. (1955). *The Equatorie of the Planetis*. Cambridge.

---

[1] For discussion of further meanings and for further references, see Tropfke (1930, p. 96, 1980, p. 204) and Guillaumin (2020, pp. 106f.). In the sense 'place', the term is commonly found in texts on calculating with Hindu-Arabic numerals (Allard 1990, p. 159). For comments, see Folkerts (1997, p. 170). [1] "s'a differenche // Qui est entre les .viii. et .x., // Che sunt .ii.". Translation from Karpinski and Waters (1928, p. 75). [2] "Soubstraire est trouver la difference de deux nombres donnez". [3] For the German term, see Strauß et al. (1999, pp. 535–539).

Recorde, Robert (1543). *The Ground of Artes*. London.
Rudolff, Christoph (1525). *Behend vnnd Hübsch Rechnung durch die kunstreichen regeln Algebra*. Straßburg.
Schreiber, Heinrich (1518). *Ayn new kunstlich Buech*. Nürnberg.
Smith, David Eugene (1958). *History of Mathematics*. Volume II. Special Topics of Elementary Mathematics. New York.
Stevin, Simon (1585). *L'Arithmetique*. Leiden.
Strauß, Gerhard et al., eds. (1999). *Deutsches Fremdwörterbuch*. 2. Auflage. Band 4: da capo – Dynastie. Berlin and New York.
Tropfke, Johannes (1930). *Geschichte der Elementar-Mathematik*. Erster Band. Rechnen. Dritte, verbesserte und vermehrte Auflage. Berlin and Leipzig.
Tropfke, Johannes (1980). *Geschichte der Elementarmathematik*. Band 1: Arithmetik und Algebra. Vollständig neu bearbeitet von Kurt Vogel, Karin Reich, Helmuth Gericke. Berlin and New York.
Widmann, Johannes (1489). *Behende vnd hubsche Rechenung auff allen kauffmanschafft*. Leipzig.

differential

Leibniz' first publication on the calculus was his paper "A new method for finding maxima and minima" from 1684 (Leibniz 1684). In this paper, Leibniz introduces the symbol dx to the public and calls it (p. 467) the *difference* (Latin *differentia*). However, the terminology was still fluctuating at the time, and in the same article, Leibniz also speaks of *differential quantities* (*quantitates differentiales*) and uses the noun *differentiale*.[1] Additionally, the article contains the expression *differential equation* (*aequatio differentialis*), and Leibniz also refers to "this calculation which I call *differential*" (p. 469).[2] The editors of Leibniz' works point out that Leibniz used the adjective *differentialis* from 1673 on (Leibniz 2008, p. xxxii).

The terminology is found in various publications from the 1690s. In the first textbook on the calculus, however, de l'Hôpital (1696, p. 2) used *différence* for differential: "The infinitely small portion by which a variable quantity continually increases or decreases is called the *Difference*".[3] This is surprising because de l'Hôpital (1693, p. 130) had used *différentielle* before, and also because his teacher Johann Bernoulli regularly used the word.[4]

Early or earliest attestations in print for French, German, and English are Bernoulli (1692, p. 281) for the French noun *différentielle*, and Wolff (1710, p. 243), who uses German *Differential-Grössen* (differential quantities). The English noun *differential* is first found (OED) in Harris (1702, p. 117), who notes that the method of fluxions "is much more natural and shorter than [...] the *French* one with the Differential *d* Multiplied into the Flowing Quantity, to denote the Fluxion".

Euler (1755, part I, § 114), of course, also adopted the term, stating that "[w]ith Leibniz we call infinitely small differences by the name *differentials*".[5] He went on to discuss the British and the continental terminology:

---

[1] See Eneström in Voss and Molk (1909, p. 276), who also notes that Leibniz mostly used *differentia* for 'differential'. [2] "calculi hujus, quem voco differentialem". Translation from Leibniz (2014, p. 2). [3] "La portion infiniment petite dont une quantité variable augmente ou diminuë continuellement, en est appellée la *Différence*". Translation modified from Bradley, Petrilli, and Sandifer (2015, p. 2). [4] Not just in the articles he published at the time, but also in his lectures on the differential calculus (Bernoulli 1922-23, p. 3), which de l'Hôpital used for his book. [5] "Differentias igitur infinite parvas hic cum LEIBNIZIO *differentialia* vocabimus". Translation from Euler (2000, part I, § 114).

> The English mathematicians [...] use different names and symbols. They call infinitely small differences, which we call *differentials, fluxions* and sometimes *increments*. These words seem to fit better in Latin, and they signify reasonably well the things themselves. A variable quantity by continuously increasing takes on various different values, and for this reason can be thought of as being in flux, from which comes the word fluxion. This was first used by Newton for the rate of change, to designate an infinitely small increment that a quantity receives, as if, by analogy, it were flowing (Euler 1755, part I, § 114).[6]

In the 19th century, Cauchy (1823, p. 13) then gave the following definition:

> Let $y = f(x)$ always be a function of the independent variable x, i an infinitely small quantity, and h a finite quantity. If we set $i = \alpha h$, α will also be an infinitely small quantity, and we will have identically
>
> $$\frac{f(x+i)-f(x)}{i} = \frac{f(x+\alpha h)-f(x)}{\alpha h},$$
>
> where we will deduce
>
> $$\frac{f(x+\alpha h)-f(x)}{\alpha} = \frac{f(x+i)-f(x)}{i}h \tag{1}$$
>
> The limit toward which the first member of equation (1) converges, while the variable α indefinitely approaches zero and the quantity h remaining constant, is what we call the *differential* of the function $y = f(x)$. We indicate this differential by the character d as follows dy or $df(x)$.[1]

On the changing role of the differential, Stolz (1893, p. 18) later remarked: "The introduction of differentials alongside derivatives may seem superfluous. In fact, the differentials of a function of one variable are only of significance as special cases of a general concept of the complete differential of order m of a function of several variables".

According to Taylor (1974, p. 355), the modern definition of differentiability of real functions of several variables seems to be due to Stolz (1893, pp. 130–133). An early adopter[2] of this definition was Pierpont (1905, pp. 268f.):

> Let Δf be the increment that $f(x_1 \cdots x_m)$ receives when we pass from the point $a = (a_1 \cdots a_m)$ to the point $a + h = (a_1 + h_1 \cdots a_m + h_m)$. Here any of the h's may $= 0$. Let
>
> $$\Delta f = f'_{x_1}(a)h_1 + \cdots + f'_{x_m}(a)h_m + \alpha_1 h_1 + \cdots + \alpha_m h_m,$$

---

[6] "Mathematici Angli [...] aliis tam nominibus, quam signis utuntur. Differentias enim infinite parvas, quas nos differentialia vocamus, potissimum *fluxiones* nominare solent, interdum quoque *incrementa*; quae voces uti latino sermoni magis conveniunt, ita quoque res, quas denotant, satis commode exprimunt. Quantitas enim variabilis crescendo continuo alios atque alios valores recipiens tanquam fluens considerari potest hincque vox fluxionis, quae primum a NEUTONO ad celeritatem crescendi adhibebatur, ad incrementum infinite parvum, quod quantitas quasi fluendo accipit, designandum analogice est translata". Translation from Euler (2000, part I, § 115).

[1] "[...] La limite vers laquelle converge le premier membre de l'équation (1), tandis que la variable α s'approche indéfiniment de zéro, la quantité h demeurant constante, est ce qu'on appelle la *différentielle* de la fonction $y = f(x)$. On indique cette différentielle par la caractéristique d, ainsi qu'il suit: dy ou $df(x)$". Translation from Cates (2019, p. 17).

[2] Pierpont does not explicitly say that he took over the definition from Stolz, but he does point out his indebtedness to Stolz in his preface (Pierpont 1905, p. v). On this point, see also Taylor (1974, p. 373) and Taylor (1987, p. 44).

where the α_r are functions of $h_1 \cdots h_m$, such that

$$\lim_{h=0} \alpha_1 = 0, \cdots \lim_{h=0} \alpha_m = 0.$$

The function f is, in this case, said to be a *totally differentiable function at* a. We call

$$df = f'_{x_1}(a)h_1 + \cdots + f'_{x_m}(a)h_m$$

the *total differential of* f *at* a.

Stolz' definition was also adopted by Young (1909, p. 82) and by Fréchet (1912, p. 388); the latter adapted it to functionals. Both Young and Fréchet attributed the definition to Stolz.[3]

For *differential coefficient* and *differential quotient*, see *derivative*.

Note. Bos (1974) discusses differentials and related notions in the Leibnizian Calculus. For the later development of the notion, see Taylor (1974) and Averbukh and Smolyanov (1968).

References

Averbukh, V.I. and O.G. Smolyanov (1968). "The various definitions of the derivative in linear topological spaces". *Russian Mathematical Surveys* 23, pp. 67–113.

Bernoulli, Johann (1692). "Solution du problème de la courbure que fait une voile enflée par le vent". *Journal des sçavans* (Apr. 1692), pp. 280–282.

Bernoulli, Johann (1922-23). "Lectiones de calculo differentialium". *Verhandlungen der Naturforschenden Gesellschaft in Basel* 24, pp. 1–32.

Bos, Henk J.M. (1974). "Differentials, Higher-Order Differentials and the Derivative in the Leibnizian Calculus". *Archive for History of Exact Sciences* 14, pp. 1–90.

Bradley, Robert E., Salvatore J. Petrilli, and C. Edward Sandifer (2015). *L'Hôpital's Analyse des infiniments petits*. An Annotated Translation with Source Material by Johann Bernoulli. Cham.

Cates, Dennis M. (2019). *Cauchy's* Calcul Infinitésimal. An Annotated English Translation. Cham.

Cauchy, Augustin-Louis (1823). *Résumé des leçons données à l'École Royale Polytechnique sur le calcul infinitésimal*. Tome premier. Paris.

Euler, Leonhard (1755). *Institutiones calculi differentialis*. St. Petersburg.

Euler, Leonhard (2000). *Foundations of Differential Calculus*. Translated by John D. Blanton. New York.

Fréchet, Maurice (1911). "Sur la notion de différentielle". *Comptes rendus de l'Académie des sciences* 152, pp. 845–847, 1050–1051.

Fréchet, Maurice (1912). "Sur la notion de différentielle totale". *Nouvelles annales de mathématiques* 12, pp. 385–403, 433–449.

Harris, John (1702). *A New Short Treatise of Algebra*. London.

de l'Hôpital, Guillaume François Antoine (1693). "Méthode facile pour déterminer les points des caustiques par réfraction, avec une méthode nouvelle de trouver les développements". *Mémoires de mathématique et de physique tirés des registres de l'Académie Royale des Sciences* (Aug. 1693), pp. 129–133.

de l'Hôpital, Guillaume François Antoine (1696). *Analyse des infiniments petits, pour l'intelligence des lignes courbes*. Paris.

Leibniz, Gottfried Wilhelm (1684). "Nova Methodus pro maximis et minimis, itemque tangentibus, quae nec fractas nec irrationales quantitates moratur, et singulare pro illis calculi genu". *Acta Eruditorum* (Oct. 1684), pp. 467–473.

Leibniz, Gottfried Wilhelm (2008). *Sämtliche Schriften und Briefe*. Siebente Reihe. Mathematische Schriften. Fünfter Band. 1674-1676. Infinitesimalmathematik. Berlin.

[3] Fréchet had originally published his definition in 1911 (Fréchet 1911); at that time, he was not yet aware of the works by Stolz, Pierpont, and Young.

Leibniz, Gottfried Wilhelm (2014). "A new method of finding the maxima and minima, and likewise for tangents, and with a single kind of calculation for these, which is hindered neither by fractions nor irrational quantities". Transl. with notes by Ian Bruce. URL: http://www.17centurymaths.com/contents/leibnizcontents.html (retrieved Aug. 3, 2017).
Pierpont, James (1905). *Lectures on the Theory of Functions of Real Variables*. Volume I. Boston.
Stolz, Otto (1893). *Grundzüge der Differential- und Integralrechnung*. Erster Theil: Reelle Veränderliche und Functionen. Leipzig.
Taylor, Angus E. (1974). "The Differential: Nineteenth and Twentieth Century Developments". *Archive for History of Exact Sciences* 12, pp. 355–383.
Taylor, Angus E. (1987). "A Study of Maurice Fréchet: III. Fréchet as Analyst, 1909–1930". *Archive for History of Exact Sciences* 37, pp. 25–76.
Voss, Aurel and Jules Molk (1909). "Calcul différentiel". *Encyclopédie des sciences mathématiques pures et appliquées* II.1, pp. 242–336.
Wolff, Christian (1710). *Der Anfangs-Gründe aller Mathematischen Wissenschaften Letzter Theil*. Halle.
Young, William Henry (1909). "Un théorème sur les différentielles". *Comptes rendus de l'Académie des sciences* 148, pp. 82–84.

differential **and** ***derivative*** **(algebra)**

In his PhD-thesis, Blumberg (1912, p. 42) presented a system of axioms to approach the concept of derivation in a purely algebraic way. To the axioms for a field $(\mathfrak{S}, \oplus, \odot)$, he added some for an operator D (for *differentiation*):

> Axiom D_1. If a is any element of $\mathfrak{S}$, then Da is a uniquely determined element of $\mathfrak{S}$.
>
> Axiom D_2. If a and b are any two elements of $\mathfrak{S}$, then $D(a \oplus b) = Da \oplus Db$.
>
> Axiom D_3. If a and b are any two elements of $\mathfrak{S}$, then $D(a \odot b) = (a \odot Db) \oplus (b \odot Da)$.

Referring to Blumberg, Baer (1927, p. 16) called such an operator a *derivation* (German *Ableitung*). The English term was used by Jacobson (1937, p. 206), and an extension to higher derivatives was suggested by Hasse and Schmidt (1937, p. 224).

In the 1930s, *differential* was explicitly applied to algebraic structures. This step was made by Raudenbush (1933, p. 509):

> In an abstract treatment of ordinary algebraic differential equations, a system of elements, closed with respect to the three operations of addition, multiplication and differentiation, will hold a central position. Ascribing to addition and multiplication the usual properties (including the commutative property of multiplication), such a system will be a commutative ring, a domain of integrity or a commutative field. The operation of differentiation is linked with the operations of addition and multiplication by the formulas for the derivatives of a sum and of a product. Such a system will be called a *differential ring*, a *differential domain of integrity* or a *differential field*, as the case may be.

The algebraic approach to differential equations was later referred to as *differential algebra*, which name goes back to Ritt's book of the same title (Ritt 1950).[1]

References

Baer, Reinhold (1927). "Algebraische Theorie der differentiierbaren Funktionenkörper I". *Sitzungsberichte der Heidelberger Akademie der Wissenschaften* 8, pp. 15–32.
Blumberg, Henry (1912). "Über algebraische Eigenschaften von linearen homogenen Differentialausdrücken". PhD thesis. Georg-August-Universität.

[1] As Ritt indicates in the preface, the book title was suggested by his former student Ellis Kolchin.

Hasse, Helmut and Friedrich Karl Schmidt (1937). "Noch eine Begründung der Theorie der höheren Differentialquotienten in einem algebraischen Funktionenkörper einer Unbestimmten". *Journal für reine und angewandte Mathematik* 177, pp. 215–237.
Jacobson, Nathan (1937). "Abstract Derivation and Lie Algebras". *Transactions of the American Mathematical Society* 42, pp. 206–224.
Raudenbush, H.W. (1933). "Differential Fields and Ideals of Differential Forms". *Annals of Mathematics* 34, pp. 509–517.
Ritt, Joseph Fels (1950). *Differential Algebra*. Providence, R.I.

differential equation

Differential equations have been studied as soon as the calculus was founded, and the term is found in the 1670s. Newton, for example, expressed the problem of solving equations in his fluxional language: "When an equation involving the fluxions of quantities is exhibited, to determine the relation of the quantities one to another" (Newton 1969, p. 82).[2] Correspondingly, in British sources differential equations were commonly referred to as *fluxional equations*.

The first definition of Latin *aequatio differentialis* was given by Leibniz in a letter from 1677 to Henry Oldenburg: "I call a differential equation such that the value of $d\bar{x}$ is expressed, and that it is derived from another in which the value of x was expressed" (Leibniz 1987, p. 172).[1]

In print, the term first appears, without definition, in Leibniz (1684, p. 467);[2] it was then used by Craig (1685, p. 28), who credited Leibniz for the term. French *équation différentielle* was used by Bernoulli (1692, p. 280), and the English term first appears (OED) in Hayes (1704, pp. 100, 151). The term also quickly found its way into various dictionaries; for example, Wolff (1716, p. 20) has an entry "aequatio differentialis, eine differential-Gleichung", where it is defined as an equation containing differentials.

Note. For the history of differential equations, see Archibald (2003), Archibald, Fraser, and Grattan-Guinness (2004), Demidov (1992), and Gray (2021).

References

Archibald, Tom (2003). "Differential Equations: A Historical Overview to circa 1900". In: *A History of Analysis*. Ed. by Hans Niels Jahnke. Providence, R.I., pp. 325–354.
Archibald, Tom, Craig Fraser, and Ivor Grattan-Guinness (2004). "The History of Differential Equations, 1670–1950". *Oberwolfach Reports* 1, pp. 2729–2794.
Bernoulli, Johann (1692). "Solution du problème de la courbure que fait une voile enflée par le vent". *Journal des sçavans* (Apr. 1692), pp. 280–282.
Craig, John (1685). *Methodus Figurarum Lineis Rectis & Curvis Comprehensarum Quadraturas Determinandi*. London.
Demidov, Sergej S. (1992). "La théorie des équations différentielles à la limite des XVIIIe - XIXe siècles". In: *Amphora*. Festschrift for Hans Wussing on the Occasion of his 65th Birthday. Ed. by Sergej S. Demidov et al. Basel, pp. 157–169.
Gray, Jeremy (2021). *Change and Variations*. A History of Differential Equations to 1900. Cham.
Hayes, Charles (1704). *A Treatise of Fluxions*. London.

---

[2] "Exposita aequatione fluxiones quantitatum involvente, invenire relationem quantitatum inter se". Translation from Newton (1969, p. 83). [1] "Aequationem differentialem voco talem qua valor ipsius $d\bar{x}$ exprimitur, quaeque ex alia derivata est, qua valor ipsius x exprimebatur". Translation from Turnbull (1960, p. 221). [2] The translators of Leibniz' papers carefully remark that the impression that a fixed terminology already existed at this time should be avoided. See Leibniz (2011, p. 51).

Leibniz, Gottfried Wilhelm (1684). "Nova Methodus pro maximis et minimis, itemque tangentibus, quae nec fractas nec irrationales quantitates moratur, et singulare pro illis calculi genu". *Acta Eruditorum* (Oct. 1684), pp. 467–473.
Leibniz, Gottfried Wilhelm (1987). *Sämtliche Schriften und Briefe*. Dritte Reihe. Mathematischer, naturwissenschaftlicher und technischer Briefwechsel. Zweiter Band. 1676-1679. Hannover.
Leibniz, Gottfried Wilhelm (2011). *Die mathematischen Zeitschriftenartikel*. Übersetzt und kommentiert von Heinz-Jürgen Heß und Malte-Ludolf Babin. Hildesheim et al.
Newton, Isaac (1969). "De methodis serierum et fluxionum". In: *The Mathematical Papers of Isaac Newton*. Volume III. 1670-1673. Ed. by D.T. Whiteside. Cambridge, pp. 32–353.
Turnbull, H. W., ed. (1960). *The Correspondence of Isaac Newton*. Volume II. 1676–1687. Cambridge.
Wolff, Christian (1716). *Mathematisches Lexicon*. Leipzig.

differential form

While differential forms were already dealt with in the 18th century, the first (symbolical) definition is attributed, for example by Katz (1985, p. 321), to Cartan (1899, p. 244):

> Given n variables $x_1, x_2, \dots, x_n$, let us consider purely symbolical expressions ω, obtainable from the n differentials $dx_1, dx_2, \dots, dx_n$ and from certain coefficients which are functions of $x_1, x_2, \dots, x_n$ using a finite number of *symbols* of addition or multiplication; those expressions being, in the ordinary sense of the word, *homogeneous* in $dx_1, dx_2, \dots, dx_n$. As they are purely symbolical, we will constrain ourselves, each time one meets an addition or multiplication symbol, not to change the order of terms or of factors united by this symbol.[1]

Cartan here used *differential expression* (French *expression différentielle*), and this remained the usual designation until about 1945, as asserted by Katz (1985, p. 334). Exceptions did occur; for example, Kähler (1932, p. 1) used Italian *forma differenziale*.

A commonly used definition of a *differential form* was given by Chevalley (1946, ch. 5, § 3); it was paraphrased by Katz (1985, p. 334): "A differential form of order r is a rule of assignment to each point p of a subset A of a manifold V a homogeneous element of order r of the 'Cartan differential algebra at p'; i. e., of the algebra of alternate multilinear functions with Grassmann multiplication associated to the tangent space to V at p". This definition is a variant of one given by Weyl in lectures (Weyl 1934-35).

Note. For the history of differential forms, see Katz (1981, 1985) and Samelson (2001).

References

Cartan, Élie (1899). "Sur certaines expressions différentielles et le problème de Pfaff". *Annales scientifiques de l'École Normale Supérieure* 16, pp. 239–332.
Chevalley, Claude (1946). *Theory of Lie Groups I*. Princeton.

[1] "Étant données n variables $x_1, x_2, \dots, x_n$, considérons des expressions ω, purement symboliques, se déduisant, au moyen d'un nombre fini de *signes* d'addition ou de multiplication, des n différentielles $dx_1, dx_2, \dots, dx_n$ et de certains coefficients fonctions de $x_1, x_2, \dots, x_n$; ces expressions étant, dans le sens ordinaire du mot, *homogènes* en $dx_1, dx_2, \dots, dx_n$. Comme elles sont purement symboliques, nous nous astreindrons, toutes les fois qu'il y aura un signe d'addition ou de multiplication, à ne pas changer l'ordre des termes ou des facteurs réunis par ce signe". Translation from Popescu-Pampu (2016, pp. 126f.).

Kähler, Erich (1932). "Forme differenziali e funzioni algebriche". *Memorie dell'Accademia d'Italia* 3, pp. 1–19.
Katz, Victor J. (1981). "The history of differential forms from Clairaut to Poincaré". *Historia Mathematica* 8, pp. 161–188.
Katz, Victor J. (1985). "Differential Forms: Cartan to De Rham". *Archive for History of Exact Sciences* 33, pp. 321–336.
Popescu-Pampu, Patrick (2016). *What is the Genus?* Cham.
Samelson, Hans (2001). "Differential Forms, the Early Days; or the Stories of Deahna's Theorem and of Volterra's Theorem". *The American Mathematical Monthly* 108, pp. 522–530.
Weyl, Hermann (1934-35). *The structure and representation of continuous groups*. (Notes by Richard Brauer). Princeton.

differential geometry

This term is usually attributed to Bianchi (1886),[2] whose book is a lithographed edition of lectures entitled *Lezioni di geometria differenziale* ("lectures on differential geometry").[3]

However, the term was in use well before Bianchi. In particular, Auguste Comte, known as the founder of positivism, used the French term *géométrie différentielle* once in his book on analytic geometry (Comte 1843, p. 596), and in a later book he includes a whole chapter so entitled (Comte 1856, ch. 5).[1] While Comte does not give a definition of the subject, the contents he discusses clearly show that he is talking about differential geometry in the modern sense of the term.

There were other mathematicians who used the term. In 1879, Pierre Laffitte lectured on *géométrie différentielle*; an outline of the course was also published in print (Laffitte 1879). A definition of German *Differentialgeometrie* was given by Baltzer (1882, p. 76): "Those parts of analytic geometry which are based on geometric infinitesimal considerations can be subsumed under the title *differential geometry*".[2] While this seems to be the first time Baltzer used the term in print, he lectured on *Differentialgeometrie* before, e. g. in 1873.[3]

Note. The history of classical differential geometry is discussed by Struik (1933a, 1933b), Reich (1973), and Libermann (1978). For an overview of the 20th-century development, see Libermann (2005).

References

Baltzer, Richard (1882). *Analytische Geometrie*. Leipzig.
Bianchi, Luigi (1886). *Lezioni di geometria differenziale*. Pisa.
Bianchi, Luigi (1894). *Lezioni di geometria differenziale*. Pisa.
Comte, Auguste (1843). *Traité élémentaire de geómétrie analytique à deux et à trois dimensions*. Paris.
Comte, Auguste (1856). *Synthèse subjective* ou, Système universel des conceptions propres à l'état normal de l'humanité. Tome premier. Paris.
Laffitte, Pierre (1879). "Cours de géométrie différentielle". *La Revue occidentale philosophique, sociale et politique* 2, pp. 173–185.

[2] See, e. g., Reich (2007, p. 479) and Lawrence (2011, p. 703). [3] A printed version appeared in 1894 (Bianchi 1894). [1] As pointed out in the OED, the English term *differential geometry* is found in 1877 in an English translation of this book. [2] "Die auf geometrische Infinitesimalbetrachung gegründeten Abschnitte der analytischen Geometrie können unter dem Titel *Differentialgeometrie* zusammengefaßt werden". [3] *Verzeichniß der Vorlesungen, welche auf der Großherzoglich Hessischen Ludewigs-Universität zu Gießen im Sommersemester 1873 gehalten [...] werden*. Gießen, p. 5. Available online: `http://dx.doi.org/10.22029/jlupub-1491`.

Lawrence, Snežana (2011). "Developable Surfaces: Their History and Application". *Nexus Network Journal* 13, pp. 701–714.
Libermann, Paulette (1978). "Géométrie différentielle". In: *Abrégé d'histoire des mathématiques 1700-1900*. II. Ed. by Jean Dieudonné. Paris, pp. 177–210.
Libermann, Paulette (2005). "La géométrie différentielle d'Élie Cartan à Charles Ehresmann et André Lichnerowicz". In: *Géométrie au 20e siècle, 1930-2000*. Histoire et horizons. Ed. by Joseph Kouneiher et al. Paris, pp. 191–208.
Reich, Karin (1973). "Die Geschichte der Differentialgeometrie von Gauss bis Riemann (1828-1868)". *Archive for History of Exact Sciences* 11, pp. 273–382.
Reich, Karin (2007). "Euler's Contribution to Differential Geometry and its Reception". In: *Leonhard Euler*. Life, Work and Legacy. Ed. by Robert E. Bradley and C. Edward Sandifer. Amsterdam, pp. 479–502.
Struik, Dirk Jan (1933a). "Outline of a History of Differential Geometry (I)". *Isis* 19, pp. 92–120.
Struik, Dirk Jan (1933b). "Outline of a History of Differential Geometry (II)". *Isis* 20, pp. 161–191.

digit

In a 14th-century algorism, in which *digit* is first found (OED), the traditional terminology was defined:

> Somme is callyd nombre of digitys for alle ben digitys that ben withine ten as ix, viii, vii, vi, v, iv, iii, ii, i. Articules ben alle thei that mow be devyded into nombrys of ten as xx, xxx, xl, and suche other. Composittys be alle nombrys that ben componyd of a digyt and of an articule as fourtene fyftene thrittene and suche other (Halliwell 1839, p. 30).

The Latin terms *digitus* (lit. 'finger'), *articulus* (lit. "joint'), and *compositus* came up in this sense in the Middle Ages. The first two terms were used by Gerbert in the late 10th century (Bubnov 1899, pp. 7f.); in a commentary on Gerbert of slightly later origin, the third term is found (Bubnov 1899, p. 254). The distinction was also made in the Pseudo-Boethian *Geometria II* (Boethius 1867, p. 395), also dating from the 11th century; the terminology spread in the 12th and 13th centuries.

The terminology also appears in early vernacular algorisms, for example in the earliest French one (Mortet 1908/09, pp. 58, 61), where the term is spelt *degit*, and in a 13th century algorism in verse (Karpinski and Waters 1928, p. 62), where *doi* was used. In German texts, the literal translation *Fingerzahl* 'finger number' was sometimes used; an early usage is Apian (1527, fol. A v^{v}).

Note. For overviews of the terminology, see Tropfke (1980, pp. 18–21) and Smith (1958, pp. 12–15).

References

Apian, Peter (1527). *Eyn Newe unnd wolgegründte underweysung aller Kauffmanß-Rechnung*. Ingolstadt.
Boethius, Anicius Manlius Severinus (1867). *De institutione arithmetica libri duo, De institutione musica libri quinque*. Accedit Geometria quae fertur Boetii. Ed. by Gottfried Friedlein. Leipzig.
Bubnov, Nicolaus, ed. (1899). *Gerberti postea Silvestri II papae Opera Mathematica*. Berlin.
Halliwell, James Orchard, ed. (1839). *Rara Mathematica*. Or, A collection of treatises on the mathematics and subjects connected with them. London.
Karpinski, Louis Charles and E.G.R. Waters (1928). "A Thirteenth Century Algorism in French Verse". *Isis* 11, pp. 45–84.
Mortet, Victor (1908/09). "Le plus ancien traité français d'algorisme". *Bibliotheca Mathematica* 9, pp. 55–64.
Smith, David Eugene (1958). *History of Mathematics*. Volume II. Special Topics of Elementary Mathematics. New York.

Tropfke, Johannes (1980). *Geschichte der Elementarmathematik.* Band 1: Arithmetik und Algebra. Vollständig neu bearbeitet von Kurt Vogel, Karin Reich, Helmuth Gericke. Berlin and New York.

dihedral angle

This term *dihedral* seems to have been used in crystallographic contexts first. It was coined in analogy to *polyhedron* and appears as French *dièdre* in Romé de L'Isle (1772, p. xxix), where it is defined as "that which has two faces" ("qui a 2 faces"). The meaning is illustrated by fig. 40, which Romé de L'Isle (1772, p. 222) describes as a prism with a dihedral summit that consists of two unequal planes, the larger plane being an irregular pentagon, the smaller one a trapezium.

Figure 40: Dihedral summit. From Romé de L'Isle (1772, Planche III, fig. 5).

From crystallography the word soon found its way into geometry. After Legendre (1794, p. 152) had used French *coin* 'corner, wedge', it was Lacroix (1799, p. 132) who suggested *angle dièdre*: "From now on we will call *dihedral angle*, that is *angle of two faces*, the inclination of two planes that meet".[1] He goes on to point out that the concept is usually called *plane angle* (*angle plan*), which term Lacroix regards as "vicious" (*très-vicieuse*) because it only conveys the idea of a two-dimensional concept, an angle formed by two intersecting lines.[1] Finally, Lacroix explains that the term consists of two Greek words, one meaning 'two', the other meaning 'face' or 'base'.

The term was quickly picked up, for example by Kramp (1806, p. 162). In English, the term is found in Whewell (1825, p. 88) in a crystallographic context; in Francœur (1830, p. 251), it occurs in a purely geometrical context. In a German translation of Lacroix's book, the term *Flächenwinkel* (from *Fläche* 'plane' and *Winkel* 'angle') is used (Lacroix 1806, p. 183).

[1] "Nous nommerons désormais *angle dièdre*, c'est-à-dire *angle à deux faces*, l'inclinaison de deux plans qui se rencontrent". [1] In the *Elements* (book I, def. 8), Euclid defined: "A *plane angle* is the inclination to one another of two lines in a plane which meet one another and do not lie in a straight line". Translation from Euclid (1956, p. 153).

Note. Brief notes on the history of dihedral angles are found in Tropfke (1924, p. 12).

References

Euclid (1956). *The Thirteen Books of Euclid's Elements*. Volume I. Introduction and Books I, II. Ed. by Thomas L. Heath. New York.
Francœur, Louis-Benjamin (1830). *A Complete Course of Pure Mathematics*. Vol. II. Translated from the French by R. Blakelock. Cambridge and London.
Kramp, Christian (1806). *Élémens de géométrie*. Cologne.
Lacroix, Sylvestre François (1799). *Elémens de géométrie*. Paris.
Lacroix, Sylvestre François (1806). *Anfangsgründe der Geometrie*. Aus dem Französischen übersetzt von E.M. Hahn. Berlin.
Legendre, Adrien-Marie (1794). *Éléments de géométrie*. Avec des notes. Paris.
Romé de L'Isle, Jean-Baptiste (1772). *Essai de cristallographie*. Paris.
Tropfke, Johannes (1924). *Geschichte der Elementar-Mathematik*. Siebenter Band. Stereometrie. Verzeichnisse. Berlin and Leipzig.
Whewell, William (1825). "A General Method of Calculating the Angles made by any Planes of Crystals, and the Laws according to which they are formed". *Philosophical Transactions of the Royal Society of London* 115, pp. 87–130.

dihedron **and** ***dihedral group***

The term *dihedron* (German *Dieder*) was introduced by Klein in his *Lectures on the Ikosahedron*. After listing the five Platonic solids, he adds: "To these we will then add a sixth configuration, which corresponds to the *plane regular n-gon*. In fact, we can denote this latter by considering the portion of the plane limited by the sides of the n-gon to be doubled, as a regular solid—a dihedron, as we will say: only that this solid, contrary to the elementary notion of such, encloses no space" (Klein 1884, pp. 3f.).[1]

For the group of symmetries of a regular polygon, Klein (1884, p. 10) also introduced the term *dihedral group* (*Diedergruppe*). The English terminology is first found in Klein (1888, pp. 3f., 11), a translation of Klein's book.

Note. For Klein's introduction of these terms, see Wußing (1984, pp. 200–205).

References

Klein, Felix (1884). *Vorlesungen über das Ikosaeder*. Leipzig.
Klein, Felix (1888). *Lectures on the Ikosahedron and the Solution of Equations of the Fifth Degree*. Translated by George Gavin Morrice. London.
Wußing, Hans (1984). *The Genesis of the Abstract Group Concept*. A Contribution to the History of the Origin of Abstract Group Theory. Tr. Abe Shenitzer. Cambridge, Mass.

dilatation

As explained by Emch (1905, p. 60), a dilatation "may physically be illustrated by stretching a piece of rubber first in the direction of the x-axis and then in the direction of the y-axis. Equal distances along one of the axes are stretched by the same amount". Emch referred to Lie for the term, who used it, for example, in Lie

[1] "Ihnen werden wir dann noch eine sechste Configuration hinzurechnen, welche dem *ebenen regulären n Eck* correspondirt. In der That können wir letzteres, indem wir uns den von den Seiten des regulären n Ecks begrenzten Ebenentheil als doppelt denken, als einen regulären Körper, als *Dieder*, wie wir sagen wollen, bezeichnen: nur dass dieser Körper, der elementaren Vorstellung eines solchen zuwider, keinen Raum einschliesst". Translation from Klein (1888, pp. 3f.).

(1896, p. 14): "We assign to the arbitrarily chosen curve c its parallel curve of distance n. This assignment is an operation, and we shall call it, in particular, a *dilatation*" (see fig. 41).[2]

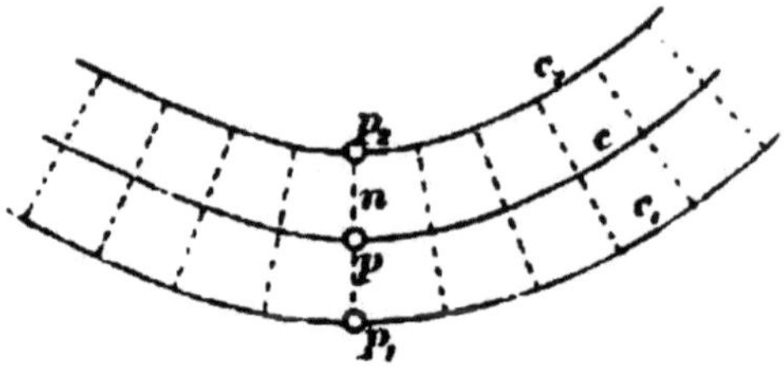

Figure 41: A dilatation. From Lie (1896, p. 14).

Lie had used this term from the early 1870s on, for example in Lie (1871, p. 552). He there used the word *parallel transformation*, which he explained by "Bonnet's dilatation". A parallel transformation is, according to Lie (1872, p. 180), the "transition from a surface to a parallel surface".[1] French *dilatation* was indeed used by Bonnet in various papers, for example in Bonnet (1853, pp. 254f.); the term expresses the idea of stretching (Latin *dilatare* 'to enlarge, expand').

A new definition in terms of coordinate geometry was given by Artin (1940, p. 15): "A mapping σ which associates with every point P a point $P' = \sigma(P)$ of the same plane is called a *dilatation*, if for any two points P, Q and their images P', Q' we can find two parallel lines l and l' so that l contains P, Q and l' contains P', Q'".

Generally, caution is necessary in connection with terms like *dilatation.* As Martin (1982, p. 136) warns his readers: "the language of similarity theory is not standardized! In general, it is impossible to tell what any of the following words means without examining the context in which that word is used: stretch, dilation, dilatation, homothety, enlargement, contraction, central similarity, radial transformation, or size transformation".

References

Artin, Emil (1940). "Coordinates in affine geometry". *Reports of a Mathematical Colloquium* 2, pp. 15–20.

Bonnet, Ossian (1853). "Mémoire sur les surfaces dont les lignes de courbure sont planes ou sphériques". *Journal de l'École polytechnique* 20, cahier 35, pp. 117–308.

Emch, Arnold (1905). *An Introduction to Projective Geometry and its Applications.* New York.

Lie, Sophus (1871). "Zur Theorie eines Raumes von n Dimensionen". *Nachrichten von der Königl. Gesellschaft der Wissenschaften und der Georg-Augusts-Universität zu Göttingen* 22, pp. 535–557.

Lie, Sophus (1872). "Ueber Complexe, insbesondere Linien- und Kugel-Complexe, mit Anwendung auf die Theorie partieller Differentialgleichungen". *Mathematische Annalen* 5, pp. 145–208.

[2] "Wir ordnen der beliebig gewählten Curve c ihre Parallelcurve im Abstande n zu. Diese Zuordnung ist eine Operation, und wir werden sie insbesondere eine *Dilatation* nennen". [1] "Uebergang von einer Fläche zu einer Parallelfäche".

Lie, Sophus (1896). *Geometrie der Berührungstransformationen*. Dargestellt von Sophus Lie und Georg Scheffers. Leipzig.
Martin, George E. (1982). *Transformation Geometry*. An Introduction to Symmetry. New York.

dimension

Greek mathematicians used various terms for the notion of dimension. Perhaps the most commonly used is *διάστασις* 'parting, separation; interval, extension, dimension' (LSJ). Use of this word originates with Aristotle (Mugler 1958-59, p. 135), who used it, for example, in his *Topics* (142b), where a body is defined as that "which has three dimensions".[2] Elsewhere, Aristotle also uses *διάστημα* 'interval, distance' (LSJ), for example in his *Physics* (209a), while Plato (*Republic* 528b) prefers *αὔξη* 'growth, increase' (LSJ).

In Latin texts, *intervallum* 'space between, interval, distance', from *vallum* 'the line of palisades about an intrenchment' (LS) is first used to express dimension. The term was used by Boethius (1867, p. 88): "Three intervals are longitude, latitude, depth, that is the line, the plane and the solid".[3] In the 10th century, Gerbert of Aurillac (1899, p. 52) also uses *intervallum* but gives *dimensio*, lit. 'a measuring' (LS), as a synonym: "A solid body is whatever is set within three boundaries or dimensions. It is whatever is held by length, width, and height".[1] Tropfke (1940, p. 48) cites this as the first use of *dimensio*, but the term was already used by Macrobius in the 5th century: "it would be well to keep in mind that all bodies have three dimensions: longitude, latitude, and altitude" (Macrobius 1963, p. 16).[2]

In modern languages, the term is attested in the 13th century. In French, for example, it is found in the *Mirror of the World* (second half of the 13th century), acc. to the DEAF. It was also regularly used by Oresme (1968, p. 46), who also commented on the possible number of dimensions: "Three dimensions or measures are: 1) length, 2) width, 3) depth or thickness; accordingly, a body is long, wide, and thick. There can be no more dimensions than these".[3] In the first printed text on geometry in French, de Bouelles (1511, fol. xxxi^{v}) recalled the Latin terminology, using *intervalles et dimensions*. The English term is first found (MED) in John Trevisa's 14th-century translation of the encyclopaedic *De proprietatibus rerum* (Seymour 1975, p. 1367). German *Dimension* is of later origin; it occurs in Ryff (1547, e. g., fol. aa ij^{v}), for example.

In the 18th century, the possibility of a fourth dimension is occasionally considered, for example by d'Alembert (1754, p. 1010): "A clever acquaintance of mine believes that one might nevertheless consider timespan as a fourth dimension, and that the product of time with volume would in a certain manner be a product of four dimensions; this idea may be contested, but it has, it would seem to me, some merit,

[2] Translation from Aristotle (1960, p. 585). [3] "Tria vero intervalla sunt: longitudo, latitudo, altitudo, id est linea, superficies atque soliditas". Translation from Masi (2006, p. 129). [1] "Solidum corpus est quidquid tribus intervallis seu dimensionibus porrigitur, id est, quidquid longitudine, latitudine altitudineque distenditur". Translation from Katz et al. (2016, p. 45). [2] "hoc loco admonendi sumus quod omne corpus longitudinis, latitudinis et altitudinis dimensionibus constat". Translation from Macrobius (1990, p. 97). [3] "Troys dimensions ou mesures sont longitude et latitude et spissitude ou parfondesce, et selonc ce, un corps est lonc et lé et espés. Et ne pueent estre plusseurs dimensions". Translation from Oresme (1968, p. 47).

if only because of its novelty".[4] Lagrange (1797, p. 223) stated that "mechanics can be regarded as a geometry of four dimensions".[5]

In the 19th century, the restriction to three or four dimensions was gradually dropped. In a letter to Gauß from 1816, Wachter spoke of "a space of arbitrarily many dimensions" (Stäckel 1901, p. 65).[6] In his "Theory of Extension", Graßmann (1844, pp. ix-x) emphasized the advantage of his general theory in these terms:

> The essential advantages of this approach are, in regard to form, that all principles expressing perceptions of space are now entirely omitted, and consequently the origins of this science are as immediate as those of arithmetic; and in regard to content, that the limitation to three dimensions is absent. Only thus do the laws come to light in their full clarity and generality, and are their essential interrelationships revealed; and many regularities, which for three dimensions either do not appear at all, or at best obscurely, present themselves perfectly clearly with this generalization.[1]

Shortly afterwards, Cayley (1845) published a paper on the "analytical geometry of n dimensions"; the term just appears in the title, while the paper is of a purely algebraic character.[2] From 1850 to 1852, Schläfli wrote a long paper on the "theory of multiple continuity", which was not published until 1901 (Schläfli 1901). In this paper Schläfli considered multidimensional objects such as hyperspheres (see *hyperplane*).[3] And in 1854, Riemann presented his famous "On the Hypotheses which lie at the Bases of Geometry", in which he introduced "n-fold extended magnitudes"[4], which he also referred to as "manifolds of n dimensions"[5] (Riemann 1868, p. 138).[6]

In the third quarter of the 19th century, the term was also applied to various algebraic structures. The usage comes up in connection with vector spaces, for example in Dedekind and Weber (1882, p. 200) (German *Dimension*) and in Peano (1888, p. 143) (Italian *dimensione*).[7]

In the 20th century, the notion of dimension was systematically studied from various perspectives. Important early contributions from topology include Poincaré's

---

[4] "Un homme d'esprit de ma connoissance croit qu'on pourroit cependant regarder la durée comme une quatrieme *dimension*, & que le produit du tems par la solidité seroit en quelque maniere un produit de quatre dimensions; cette idée peut être contestée, mais elle a, ce me semble, quelque mérite, quand ce ne seroit que celui de la nouveauté". Translation from Van Oss (1983, p. 456). – It is a matter of some dispute who d'Alembert refers to by "a clever acquaintance". Debever (1975) suggests he means Diderot, Van Oss (1983, p. 456) thinks it was d'Alembert himself. [5] "on peut regarder la mécanique comme une géométrie à quatre dimensions". [6] "ein Raum von beliebig vielen Dimensionen". [1] "Der wesentliche Vortheil, welcher durch diese Auffassung erreicht wurde, war der Form nach der, dass nun alle Grundsätze, welche Raumesanschauungen ausdrückten, gänzlich wegfielen, und somit der Anfange in eben so unmittelbarer wurde, wie der der Arithmetik, dem Inhalte nach aber der, dass die Beschränkung auf drei Dimensionen wegfiel. Erst hierdurch traten die Gesetze in ihrer Unmittelbarkeit und Allgemeinheit ans Licht und stellten sich in ihrem wesentlichen Zusammenhange dar, und manche Gesetzmässigkeit, die bei drei Dimensionen entweder noch gar nicht, oder nur verdeckt vorhanden war, entfaltete sich nun bei dieser Verallgemeinerung in ihrer ganzen Klarheit". Translation from Graßmann (1995, pp. 11f.). [2] On Cayley's paper, see Laptev and Rosenfeld (1996, pp. 76f.). [3] On Schläfli's paper, see Laptev and Rosenfeld (1996, pp. 78–81). [4] "n-fach ausgedehnte Größen". [5] "Mannigfaltigkeiten von n Dimensionen". [6] On Riemann's approach to dimension, see, e. g., Ohshika (2017, pp. 300f.) and Laptev and Rosenfeld (1996, p. 83). [7] The concept (but not the word) is earlier found in Dedekind (1871), where it is applied, in particular, to fields (§ 159) and modules (§ 161); it is also found in Graßmann (1844) and Graßmann (1862). For the history of the concept in linear algebra, see Dorier (1996); see also *basis*.

definition: "A continuum has n dimensions when it is possible to divide it into several parts by means of one or more cuts which are themselves continua of $n-1$ dimensions. The continuum of n dimensions is thus defined by the continuum of $n-1$ dimensions; this is a definition by recursion" (Poincaré 1912, p. 488).[8] This definition was the starting point for Brouwer (1913, pp. 146f.), who defined what was later called *large inductive dimension*. The small inductive dimension was approached by Urysohn (1922, p. 441) and Menger (1923, pp. 157f.), while the covering dimension was defined by Čech (1933, p. 280), based on earlier work by Lebesgue (1911).

In algebraic geometry, the dimension can be defined in purely algebraic terms. Such a definition is found in Lang (1958, p. 28), for example: "We define the *dimension* of a point (x) over k to be the transcendence degree of $k(x)$ over k. We write $\dim_k(x)$, or, briefly, $\dim(x)$ when the reference to k is clear. If V is a variety and (x) a generic point of V, we let $\dim V = \dim(x)$. The dimension of V thus defined is clearly independent of the generic point selected. To express the fact that a variety has dimension r, we sometimes write V^r". This definition seems to go back to van der Waerden (1926, p. 196);[1] it also appeared in his book *Moderne Algebra* (van der Waerden 1931, pp. 61f.).[2]

Note. The term *dimension* is also found in 15th- and 16th-century algebra. Michael Stifel, in his edition of Rudolff's *Coss*, considered "going beyond the cube just as if there were more than three dimensions, which is against nature" (Rudolff 1553, fol. 9r).[3] Descartes used the term in his *Géométrie* from 1637 for algebraic expressions (Descartes 1637, p. 299) and also for the degree of an equation: "Every equation can have as many distinct roots as the number of dimensions of the unknown quantity in the equation" (Descartes 1637, p. 372).[4]

Note. For the Greek terms, see Mugler (1948, pp. 8–17, 1958-59, pp. 88f., 135–137). Tropfke (1940) gives an overview of the terminology. The history of the fourth dimension is discussed in Wieleitner (1925), Cajori (1926), Archibald (1914), Manning (1914, pp. 1–22), and Weitzenböck (1956, pp. 11–18). For higher dimensions, see Segre (1921, pp. 772–788) and Volkert (2018). On the history of topological dimension theory, see Katětov and Simon (1997) and Crilly and Johnson (1999); see also Herrlich, Hušek, and Preuß (2008), Johnson (1979), Johnson (1977), and Engelking (1978). For broken dimensions, see also *fractal*.

References

d'Alembert, Jean Le Rond (1754). "Dimension". *Encyclopédie, ou Dictionnaire raisonné des sciences, des arts et des métiers* 4, pp. 1009–1010.

---

[8] "Un continu a n dimensions quand on peut le décomposer en plusieurs parties en y pratiquant une ou plusieurs coupures qui soient elles-mêmes des continus à $n-1$ dimensions. Le continu à n dimensions se trouve ainsi défini par le continu à $n-1$ dimensions; c'est une définition par récurrence". Translation from van Dalen (2013, p. 400). [1] See Dieudonné (1985, pp. 70f.). [2] An algebraic approach to the concept of dimension is already implicit in Kronecker (1882); see Bourbaki (1994, pp. 105f.). Noether (1923, pp. 248–251) gave an arithmetical definition. [3] "vnd vber den cubum hinauß faren / gleych als weren mehr den drey dimensiones / welchs doch auch wider die natur ist". Translation from Manning (1914, p. 3). [4] "Scachés donc qu'en chasque Equation, autant que la quantité inconnue a de dimensions, autant peut il y auoir de diuerses racines, c'est a dire de valeurs de cete quantité". Translation from Descartes (1925, p. 159).

Archibald, Raymond Clare (1914). "Time as a fourth dimension". *Bulletin of the American Mathematical Society* 20, pp. 409–412.
Aristotle (1960). *Posterior Analytics. Topica.* Translated by Hugh Tredennick, E. S. Forster. Cambridge, Mass.
Boethius, Anicius Manlius Severinus (1867). *De institutione arithmetica libri duo, De institutione musica libri quinque*. Accedit Geometria quae fertur Boetii. Ed. by Gottfried Friedlein. Leipzig.
de Bouelles, Charles (1511). *Geometrie en francoys*. Paris.
Bourbaki, Nicolas (1994). *Elements of the History of Mathematics*. Translated from the French by John Meldrum. Berlin.
Brouwer, L.E.J. (1913). "Über den natürlichen Dimensionsbegriff". *Journal für die reine und angewandte Mathematik* 142, pp. 146–152.
Cajori, Florian (1926). "Origins of Fourth Dimension Concepts". *The American Mathematical Monthly* 33, pp. 397–406.
Cayley, Arthur (1845). "Chapters in the analytical geometry of (n) dimensions". *The Cambridge Mathematical Journal* 4, pp. 119–127.
Čech, Eduard (1933). "Příspěvek k teorii dimense". *Časopis pro pěstování matematiky a fysiky* 62, pp. 277–291.
Crilly, Tony and Dale M. Johnson (1999). "The Emergence of Topological Dimension Theory". In: *History of Topology*. Ed. by Ioan Mackenzie James. Amsterdam, pp. 1–24.
van Dalen, Dirk (2013). *L.E.J. Brouwer – Topologist, Intuitionist, Philosopher*. How Mathematics Is Rooted in Life. London.
Debever, Robert (1975). "Un dialogue d'Alembert-Diderot? Un trait d'esprit á propos de la quatrième dimension". In: *Études sur le XVIIIe siècle*. Ed. by Roland Mortier and Hervé Hasquin. Bruxelles, pp. 129–133.
Dedekind, Richard (1871). "Ueber die Composition der binären quadratischen Formen". In: *Vorlesungen über Zahlentheorie von P.G. Lejeune Dirichlet*. Herausgegeben und mit Zusätzen versehen von R. Dedekind. Zweite, umgearbeitete und vermehrte Auflage. Braunschweig, pp. 380–497.
Dedekind, Richard and Heinrich Weber (1882). "Theorie der algebraischen Functionen einer Veränderlichen". *Journal für die reine und angewandte Mathematik* 92, pp. 181–290.
Descartes, René (1637). *La Géométrie*. In: *Discours de la méthode*. Leiden, pp. 295–413.
Descartes, René (1925). *The Geometry of René Descartes*. Translated from the French and Latin by David Eugene Smith and Marcia L. Latham. Chicago and London.
Dieudonné, Jean (1985). *History of Algebraic Geometry*. An Outline of the History and Development of Algebraic Geometry. Translated by Judith D. Sally. Monterey, California.
Dorier, Jean-Luc (1996). "Basis and dimension – from Grassmann to van der Waerden". In: *Hermann Günther Graßmann (1809-1877)*. Visionary Mathematician, Scientist and Neohumanist Scholar. Ed. by Gert Schubring. Dordrecht, pp. 175–196.
Engelking, Ryszard (1978). *Dimension Theory*. Amsterdam.
Gerbert of Aurillac (1899). "Geometria". In: *Opera mathematica*. Ed. by Nicolaus Bubnov. Berlin, pp. 48–97.
Graßmann, Hermann (1844). *Die Wissenschaft der extensiven Grösse, oder die Ausdehnungslehre, eine neue mathematische Disziplin*. Erster Theil, die lineale Ausdehnungslehre enthaltend. Leipzig.
Graßmann, Hermann (1862). *Die Ausdehnungslehre*. Vollständig und in strenger Form bearbeitet. Berlin.
Graßmann, Hermann (1995). *A New Branch of Mathematics*. The *Ausdehnungslehre* of 1844 and Other Works. Translated by Lloyd C. Kannenberg. Chicago.
Herrlich, Horst, Mirek Hušek, and Gerhard Preuß (2008). "Hausdorffs Studien zur Dimensionstheorie". In: *Felix Hausdorff. Gesammelte Werke*. Band III. Mengenlehre (1927, 1935). Deskriptive Mengenlehre und Topologie. Ed. by Ulrich Felgner et al. Berlin and Heidelberg, pp. 840–864.
Johnson, Dale M. (1977). "Prelude to Dimension Theory: The Geometrical Investigations of Bernard Bolzano". *Archive for History of Exact Sciences* 17, pp. 261–295.
Johnson, Dale M. (1979). "The Problem of the Invariance of Dimension in the Growth of Modern Topology, Part I". *Archive for History of Exact Sciences* 20, pp. 97–188.

Katětov, Miroslav and Petre Simon (1997). "Origins of Dimension Theory". In: *Handbook of the History of General Topology*. Volume 1. Ed. by C.E. Aull and R. Lowen. Dordrecht, pp. 113–134.

Katz, Victor J. et al., eds. (2016). *Sourcebook in the Mathematics of Medieval Europe and North Africa*. Princeton.

Kronecker, Leopold (1882). *Grundzüge einer arithmetischen Theorie der algebraischen Grössen*. Berlin.

Lagrange, Joseph-Louis (1797). *Théorie des fonctions analytiques*. Paris.

Lang, Serge (1958). *Introduction to Algebraic Geometry*. New York.

Laptev, B.L. and Boris A. Rosenfeld (1996). "Geometry". In: *Mathematics of the 19th Century*. Geometry. Analytic Function Theory. Ed. by Andrey Nikolaevich Kolmogorov and Adolph Andrei Pavlovich Yushkevich. Basel, pp. 1–117.

Lebesgue, Henri (1911). "Sur la non-applicabilité de deux domaines appartenant respectivement à des espaces, à n et $n + p$ dimensions". *Mathematische Annalen* 70, pp. 166–168.

Macrobius (1963). *Commentarius in somnium Scipionis*. Ed. by James A. Willis. Leipzig.

Macrobius (1990). *Commentary on the Dream of Scipio*. Translated with an introduction and notes by William Harris Stahl. New York.

Manning, Henry Parker (1914). *Geometry of Four Dimensions*. New York.

Masi, Michael (2006). *Boethian Number Theory*. Translation of the *De Institutione Arithmetica*. Amsterdam and New York.

Menger, Karl (1923). "Über die Dimensionalität von Punktmengen. Erster Teil". *Monatshefte für Mathematik und Physik* 33, pp. 148–160.

Mugler, Charles (1948). *Platon et la recherche mathématique de son époque*. Strasbourg and Zürich.

Mugler, Charles (1958-59). *Dictionnaire historique de la terminologie géométrique des Grecs*. Paris.

Noether, Emmy (1923). "Eliminationstheorie und allgemeine Idealtheorie". *Mathematische Annalen* 90, pp. 229–261.

Ohshika, Ken'ichi (2017). "The Origin of the Notion of Manifold: From Riemann's Habilitationsvortrag Onward". In: *From Riemann to Differential Geometry and Relativity*. Ed. by Lizhen Ji, Athanase Papadopoulos, and Sumio Yamada. Cham, pp. 295–309.

Oresme, Nicole (1968). *Le livre du ciel et du monde*. Edited by Albert D. Menut and Alexander J. Denomy, translated with an introduction by Albert D. Menut. Madison.

Van Oss, Rosine G. (1983). "D'Alembert and the fourth dimension". *Archive for History of Exact Sciences* 10, pp. 455–457.

Peano, Guiseppe (1888). *Calcolo geometrico secondo l'Ausdehnungslehre di H. Grassmann, preceduto dalle operazioni della logica deduttiva*. Torino.

Plato (1942). *The Republic*. With an English translation by Paul Shorey. In two volumes. II. Books VI–X. Cambridge, Mass.

Poincaré, Henri (1912). "Pourquoi l'espace a trois dimensions". *Revue de métaphysique et de morale* 20, pp. 483–504.

Riemann, Bernhard (1868). "Ueber die Hypothesen, welche der Geometrie zu Grunde liegen". *Abhandlungen der Königlichen Gesellschaft der Wissenschaften zu Göttingen* 13, pp. 133–152.

Rudolff, Christoph (1553). *Die Coss Christoffs Rudolffs*. Ed. by Michael Stifel. Königsberg.

Ryff, Walther Hermann (1547). *Der furnembsten, notwendigsten, der gantzen Architectur angehörigen mathematischen und mechanischen Künst eygentlicher Bericht*. Nürnberg.

Schläfli, Ludwig (1901). *Theorie der vielfachen Kontinuität*. Ed. by J.H. Graf. Basel.

Segre, Corrado (1921). "Mehrdimensionale Räume". *Encyklopädie der mathematischen Wissenschaften mit Einschluss ihrer Anwendungen* III C 7, pp. 769–972.

Seymour, Maurice Charles, ed. (1975). *On the Properties of Things*. John Trevisa's Translation of Bartholomaeus Anglicus' *De Proprietatibus Rerum*. A Critical Text. Volume II. Oxford.

Stäckel, Paul (1901). "Friedrich Ludwig Wachter, ein Beitrag zur Geschichte der nichteuklidischen Geometrie". *Mathematische Annalen* 54, pp. 49–75.

Tropfke, Johannes (1940). *Geschichte der Elementar-Mathematik*. Vierter Band. Ebene Geometrie. Dritte, verbesserte und vermehrte Auflage besorgt von Dr. Kurt Vogel. Berlin and Leipzig.

Urysohn, Pavel (1922). "Les multiplicités Cantoriennes". *Comptes rendus de l'Académie des sciences* 175, pp. 440–442.

Volkert, Klaus (2018). *In höheren Räumen*. Der Weg der Geometrie in die vierte Dimension. Berlin.
van der Waerden, Bartel Leendert (1926). "Zur Nullstellentheorie der Polynomideale". *Mathematische Annalen* 96, pp. 183–208.
van der Waerden, Bartel Leendert (1931). *Moderne Algebra*. Zweiter Teil. Berlin.
Weitzenböck, Roland W. (1956). *Der vierdimensionale Raum*. Basel.
Wieleitner, Heinrich (1925). "Zur Frühgeschichte der Räume von mehr als drei Dimensionen". *Isis* 7, pp. 486–489.

direct product **and** ***direct sum***

Historically, groups were the first objects for which a *direct product* was defined. The term was coined by Hölder (1889, p. 36) as German *directes Product*. Hölder defined the notion by considering a certain equivalence relation; his definition amounted to saying that $G \cdot H$ is the direct product of G and H if G and H have only the identity element in common and every element of G is permutable with every element of H. He chose the term, as he suggests himself, because in his opinion this product is the "simplest conceivable product" of groups; he also notes that it is uniquely determined.

The term was soon adopted by other writers, for example by Burkhardt (1899, p. 219) and Miller (1900, p. 66), where Hölder is also named as the originator of the term.[1] The notation $\mathfrak{A} \times \mathfrak{B}$ was suggested by Remak (1911, p. 293).

At about the same time when Hölder introduced the direct product, the notion of a direct sum was used by Scheffers (1891, p. 323):

> The composition of several systems $A, B, C, \ldots$ to just one system S is very similar to the addition in elementary mathematics, since here as there the commutative and the associative law hold [...] This entitles us to call the composition of several systems $A, B, C, \ldots$ to just one system S the *addition*: $S = A + B + C + \ldots$.[2]

The explicit term *direct sum* appears to be due to Wedderburn (1908, p. 84), who wrote: "If an algebra A is expressible as the sum of two algebras A_1 and A_2, which are such that $A_1 A_2 = 0 = A_2 A_1$, A is said to be *reducible*, and to be the *direct sum* of A_1 and A_2. It was in this sense that the word sum was first used by Scheffers".[3] The notation $A = B \oplus C$ is found in Dickson (1923, p. 33).

The concept has also been extended to infinite sums and products, for example by Prüfer (1924, p. 176), who considered infinite direct products of Abelian groups under the name *improper direct product* (*uneigentliches direktes Produkt*). For a module A and an arbitrary family of modules $\{A_\alpha\}$ (all over the same ring), Cartan and Eilenberg (1956, p. 4) defined the direct sum and the direct product and remarked: "If the family $\{A_\alpha\}$ is finite, the notions of direct sum and direct product coincide". In the infinite case, this assertion does not hold. The terminological situation became somewhat confusing afterwards. Rotman (1995, p. 308), for example, names *complete direct sum* or *strong direct sum* as alternatives of *direct product*, and *weak direct sum* as an alternative of *direct sum*.[4]

---

[1] The English term may have been first used by Burnside (1896, p. 360). [2] "Die Zusammensetzung mehrerer Systeme $A, B, C, \ldots$ zu einem einzigen S hat grosse Aehnlichkeit mit der Addition der elementaren Mathematik. Denn wie dort gilt hier das commutative und das associative Gesetz [...] Dies berechtigt uns, die Zusammensetzung mehrerer Systeme $A, B, C, \ldots$ zu einem einzigen S symbolisch als *Addition* zu bezeichnen: $S = A + B + C + \ldots$". [3] Note that while the use of *Summe* may be said to be implicit in Scheffers' article, it not explicitly there. [4] A comment along very similar lines is found in Halmos (1963, p. 121).

The term *direct product* was also used in vector analysis; see *scalar product*. It has also been used for *tensor products*.[5]

Note. For the history of the notions, see van der Waerden (1985, esp. ch. 11), Hawkins (2013, esp. pp. 306-317), Kleiner (1998), and Parshall (1985).

References

Burkhardt, Heinrich (1899). "Endliche discrete Gruppen". *Encyklopädie der mathematischen Wissenschaften mit Einschluss ihrer Anwendungen* I A 6, pp. 208–226.

Burnside, William Snow (1896). "On the Isomorphism of a Group with Itself". *Proceedings of the London Mathematical Society* 27, pp. 354–367.

Cartan, Henri and Samuel Eilenberg (1956). *Homological Algebra*. Princeton.

Dickson, Leonard Eugene (1923). *Algebras and their Arithmetics*. Chicago.

Halmos, Paul R. (1963). *Lectures on Boolean Algebras*. Princeton.

Hawkins, Thomas (2013). *The Mathematics of Frobenius in Context*. A Journey Through 18th to 20th Century Mathematics. New York.

Henderson, Harold V., Friedrich Pukelsheim, and Shayle R. Searle (1983). "On the History of the Kronecker Product". *Linear and Multilinear Algebra* 14, pp. 113–120.

Hölder, Otto (1889). "Zurückführung einer beliebigen algebraischen Gleichung auf eine Kette von Gleichungen". *Mathematische Annalen* 34, pp. 26–56.

Kleiner, Israel (1998). "From numbers to rings: The early history of ring theory". *Elemente der Mathematik* 53, pp. 18–35.

MacDuffee, Cyrus Colton (1933). *The Theory of Matrices*. Berlin.

Miller, George Abram (1900). "On the groups which are the direct products of two subgroups". *Transactions of the American Mathematical Society* 1, pp. 66–71.

Parshall, Karen Hunger (1985). "Joseph H. M. Wedderburn and the Structure Theory of Algebras". *Archive for History of Exact Sciences* 32, pp. 223–349.

Prüfer, Heinz (1924). "Theorie der Abelschen Gruppen. I. Grundeigenschaften". *Mathematische Zeitschrift* 20, pp. 165–187.

Remak, Robert (1911). "Über die Zerlegung der endlichen Gruppen in direkte unzerlegbare Faktoren". *Journal für die reine und angewandte Mathematik* 139, pp. 293–308.

Rotman, Joseph J. (1995). *An Introduction to the Theory of Groups*. Fourth Edition. New York.

Scheffers, Georg (1891). "Zurückführung complexer Zahlensysteme auf typische Formen". *Mathematische Annalen* 39, pp. 293–390.

van der Waerden, Bartel Leendert (1985). *A History of Algebra*. From al-Khwārizmī to Emmy Noether. Berlin.

Wedderburn, J.H.M. (1908). "On Hypercomplex Numbers". *Proceedings of the London Mathematical Society* 6, pp. 77–118.

directed graph

The term *directed graph* seems to have been coined by König (1936, p. 4) as German *gerichteter Graph* (the English term is a literal translation of this).

The English coinage *digraph*, short for *directed graph*, seems to be first found in print in Harary (1954, p. 143). The term is not due to Harary, however; Harary (1989, p. 751) himself later cites from a letter by George Pólya, in which Pólya asked Harary a favour: "From now on say 'digraph' instead of 'directed graph.' It is clear and short and it will catch on".

---

[5] Yet another usage is in matrix theory for what is now commonly called the *Kronecker product*. In this sense the term seems to go back to MacDuffee (1933, p. 81); he apparently used it because, as the says on the following page, "[t]he concept of direct product of matrices arises naturally from the concept of direct product in group theory". For the history of the Kronecker product and its rather unjustified name, see Henderson, Pukelsheim, and Searle (1983).

References

Harary, Frank (1954). "On the notion of balance of a signed graph". *Michigan Mathematical Journal* 2, pp. 143–146.
Harary, Frank (1989). "The Polya Picture Album–Encounters of a Mathematician [review]". *The American Mathematical Monthly* 96, pp. 750–753.
König, Dénes (1936). *Theorie der endlichen und unendlichen Graphen*. Leipzig.

directed set

The term *directed set* was introduced by Birkhoff (1937, p. 40):

> By a *directed set* of points x_α of $\mathfrak{X}$ [a topological space] is meant an ordering of points, one to each subscript of a class A, where (D1) given $\alpha \in A$, $\beta \in A$, either α follows β (in symbols, $\alpha > \beta$) or α does not follow β (in symbols, $\alpha \not> \beta$), (D2) if $\alpha > \beta$ and $\beta > \gamma$, then $\alpha > \gamma$, (D3) given $\alpha \in A$, $\beta \in A$, there exists $\gamma \in A$ satisfying $\gamma > \alpha$ and $\gamma > \beta$.

Birkhoff (ibid.) additionally remarks that "[i]t is primarily condition (D3) which was due to Moore and Smith, and which distinguishes directed sets from other 'partially ordered sets'". He means Moore and Smith (1922, p. 103), where condition D3 is called *composition property*. Later, in his *Lattice Theory*, Birkhoff (1940, p. 31) attributed the concept to Moore (1915), where no special name was used.[1]

As pointed out by Reitberger (1997, pp. 33f.), the concept was also discussed by Vietoris (1921, p. 184) under the name *oriented set* (German *orientierte Menge*).

Note. For historical remarks on the concept, see Reitberger (1997) and Pietsch (2007, pp. 60–62) as well as the entry *net*.

References

Birkhoff, Garrett (1937). "Moore-Smith Convergence in General Topology". *Acta Mathematica* 38, pp. 39–56.
Birkhoff, Garrett (1940). *Lattice Theory*. New York.
Birkhoff, Garrett (1948). *Lattice Theory*. Revised Edition. New York.
Cartan, Henri (1937a). "Théorie des filtres". *Comptes rendus de l'Académie des sciences* 205, pp. 595–598.
Cartan, Henri (1937b). "Filtres et ultrafiltres". *Comptes rendus de l'Académie des sciences* 205, pp. 777–779.
Moore, E.H. (1915). "Definition of Limit in General Integral Analysis". *Proceedings of the National Academy of Sciences* 1, pp. 628–632.
Moore, E.H. and H.L. Smith (1922). "A General Theory of Limits". *American Journal of Mathematics* 44, pp. 102–121.
Pietsch, Albrecht (2007). *History of Banach Spaces and Linear Operators*. Boston, Mass. et al.
Reitberger, Heinrich (1997). "The Contributions of L. Vietoris and H. Tietze to the Foundations of General Topology". In: *Handbook of the History of General Topology*. Volume 1. Ed. by C.E. Aull and R. Lowen. Dordrecht, pp. 31–40.
Tukey, John W. (1940). *Convergence and Uniformity in Topology*. Princeton.
Vietoris, Leopold (1921). "Stetige Mengen". *Monatshefte für Mathematik und Physik* 31, pp. 173–204.

direction field

An explicit definition of this concept was given by von Sanden (1914, p. 157) under the German name *Richtungsfeld*, of which *direction field* is a literal translation:

[1] In the second edition, Birkhoff (1948, p. xi) also refers to Cartan (1937a, 1937b), in which papers Cartan introduced the term *filter*, as well as to Tukey (1940, p. 30).

If a differential equation of the first order be written in the form

$$\frac{dy}{dx} = f(x, y),$$

a geometrical interpretation of the equation is immediately apparent. Taking x and y to represent rectangular co-ordinates, then at each point in the $x - y$ plane—apart from certain so-called singular points—the function $f(x, y)$ has a definite value. If $f(x, y)$ is a many-valued function of x and y, then we will confine attention to one branch of the function. Geometrically speaking, the problem may be stated as the determination of a system of curves in the plane of $x - y$ such that the tangent at each point makes an angle y with the x axis given by $\tan\varphi = f(x, y)$.

The function $f(x, y)$ defines a "field of direction" in the $x - y$ plane, with each point of which is associated a particular direction defined by φ, along which the "integral curve" through the point must pass.[1]

While Mehmke (1917, p. 117) referred to von Sanden for the term, it had actually been used earlier by Büchel (1904, e. g., pp. 154, 167) and Kasner (1910, p. 131), neither of whom gave an explanation. Thus, it remains unclear who actually coined the term.

Note. The concept is much older and was used by Johann Bernoulli in 1694; for references, see *isocline*.

References

Büchel, Wilhelm (1904). "Zur Topologie der durch eine gewöhnliche Differentialgleichung erster Ordnung und ersten Grades definierten Kurvenschar". *Mitteilungen der Mathematischen Gesellschaft in Hamburg* 4, pp. 133–168.

Kasner, Edward (1910). "The Theorem of Thomson and Tait and Natural Families of Trajectories". *Transactions of the American Mathematical Society* 11, pp. 121–140.

Mehmke, Rudolf (1917). *Leitfaden zum graphischen Rechnen*. Leipzig and Berlin.

von Sanden, Horst (1914). *Praktische Analysis*. Berlin and Leipzig.

von Sanden, Horst (1923). *Practical Mathematical Analysis*. With examples by the translator H. Levy. London.

directrix

The directrix is a fixed line used in the construction of conic sections (and other curves) and was already known to Pappus, who mentioned it in book 7, prop. 238 of his *Collection* (Pappus of Alexandria 1877, p. 1012). The term, however, is a 17th-century invention and was first used by Jan de Witt as Latin *directrix*. It is the feminine form of Latin *director* 'one who directs'; the feminine form was used because *directrix* is a short form of *linea directrix*, where *linea* 'line' is feminine. de

[1] "Schreibt man eine Differentialgleichung erster Ordnung in der Form $\frac{dy}{dx} = f(x, y)$, so liegt eine geometrische Deutung dieser Gleichung sehr nahe. Deutet man x, y als rechtwinklige Koordinaten, so hat die Funktion $f(x, y)$ in jedem Punkte der xy-Ebene, von einzelnen sogenannten singulären Stellen abgesehen, einen bestimmten Wert. Ist $f(x, y)$ mehrdeutig, so beschränken wir uns auf einen Zweig der Funktion. Die Aufgabe ist, geometrisch gesprochen, die, in der xy-Ebene Kurven so zu bestimmen, daß ihre Tangenten an jeder Stelle einen Winkel φ mit der x-Achse bilden, für den $\tan\varphi = f(x, y)$ wird. Die Funktion $f(x, y)$ definiert ein ‚Richtungsfeld', indem sie jedem Punkte der xy-Ebene durch den Winkel eine Richtung zuweist, in der er von einer jener ‚Integralkurven' passiert werden muß". Translation from von Sanden (1923, p. 167).

Witt defines the terms *generator* [*describens*] and *directrix* at the beginning of his book:

> I. When along a fixed straight line another straight line moves by means of a fixed point on it, such that this other straight line is always parallel to itself, or coincides with itself, and when during this same motion an angle rotates about a fixed point (coinciding with its vertex) such that one arm of this angle always passes through the aforementioned moving point and at the same time the point of intersection of the other arm with the aforementioned moving straight line describes a curve, then the straight line, that – as is said –always moves parallel to itself or coincides with itself, will be called the *generator*.
>
> II. The other straight line, however, that remains fixed, will be called the *directrix* (de Witt 1661, p. 159).[1]

The term is soon found in other publications such as Barrow (1670, p. 14). de la Hire (1673, p. 75) transferred the term to French as *directrice*; in Ozanam's mathematical dictionary, the term is not used in connection with conic sections but it *is* used with reference to the conchoid (Ozanam 1691, p. 107). The earliest English use appears to be in Sturm (1700, p. 163), also with reference to the conchoid rather than to conic sections. The term is also included in Wolff's mathematical dictionary, where it is defined in general terms as any line that determines the motion of another line or surface such that a surface or a solid is generated (Wolff 1716, p. 543).

Note. For brief notes on the history of the term, see Tropfke (1924, p. 147). For the origin of the focus-directrix definition of conics (which is difficult to establish), see Toomer (1976, pp. 16f.) and Acerbi (2011).

References

Acerbi, Fabio (2011). "The geometry of burning mirrors in Greek antiquity. Analysis, heuristic, projections, lemmatic fragmentation". *Archive for History of Exact Sciences* 65, pp. 471–497.

Barrow, Isaac (1670). *Lectiones geometricae. in quibus opticorum phaenomenon geminae rationes inivestigantur, ac exponantur*. London.

Grootendorst, Albert W. (2000). *Jan de Witt's* Elementa Curvarum Linearum, Liber Primus. Text, Translation, Introduction, and Commentary. New York.

de la Hire, Philippe (1673). *Nouvelle méthode en géométrie pour les sections des superficies coniques, et cylindriques*. Paris.

Ozanam, Jacques (1691). *Dictionnaire mathematique*. Paris.

Pappus of Alexandria (1877). *Pappi Alexandrini Collectionis*. Volumen II. Ed. by Friedrich Hultsch. Berlin.

Sturm, Johann Christoph (1700). *Mathesis enucleata, or, The elements of the mathematicks*. Made English by J.R.A.M. and R.S.S. London.

Toomer, G.J., ed. (1976). *Diocles. On Burning Mirrors*. The Arabic Translation of the Lost Greek Original. Berlin et al.

Tropfke, Johannes (1924). *Geschichte der Elementar-Mathematik*. Sechster Band. Analysis. Analytische Geometrie. Berlin and Leipzig.

de Witt, Johannes (1661). *Elementa curvarum linearum*. In: *Geometria, à Renato des Cartes*. Pars secunda. Ed. by Frans van Schooten. Amsterdam, pp. 153–340.

[1] "I. Si per rectam lineam immotam altera recta certo sui puncto sibi semper parallela moveatur aut incedat, eodemque illo motu anguli cujusdam rectilinei, circa punctum fixum (quod idem sit cum ejus vertice) circulariter mobilis, crus unum semper per praedictum mobile punctum transiens secum ducat, atque uta simul cruris alterius, & dictae linea incedentis intersectione curva describatur linea; recta, quae, uti praedictum est, sibi semper parallela movetur aut incedit, *Describens* dicetur. II. Altera verò recta, immota manens, *Directrix* vocat". Translation from Grootendorst (2000, p. 48) (emphasis added).

Wolff, Christian (1716). *Mathematisches Lexicon*. Leipzig.

discontinuous

See *continuous* (calculus) and *jump*.

discrete

See *continuous*.

discriminant

The discriminant of a polynomial is made up of the coefficients of such a polynomial and determines certain properties of the roots. For example, the discriminant of a quadratic polynomial $ax^2 + bx + c$ is $\Delta = b^2 - 4ac$, and, considering the roots of the polynomial $x_{1,2} = \frac{-b \pm \sqrt{\Delta}}{2a}$, the nature of the roots depends on the discriminant. In particular, the discriminant of a polynomial is zero if and only if the polynomial has a multiple root.

This notion was defined by Gauß (1801, § 154) under the name *determinant*,[1] a designation that was later used by Cauchy in the sense in which the term is known today (see *determinant*). Thus, *determinant* was in use in two senses.

In view of this situation, the term *discriminant* was introduced by Sylvester (1851, p. 406), who wanted "to avoid the obscurity and confusion arising from employing the same word [*determinant*] in two different senses" and explained that he chose the new word "because it affords the *discrimen*[2] or test for ascertaining whether or not equal factors enter into a function of two variables, or more generally of the existence or otherwise of multiple points in the locus represented or characterized by any algebraical function, the most obvious and first observed species of singularity of such function or locus".[3]

The term was almost immediately adopted by Brioschi (1854, p. 19); the corresponding German and French terms are found in translations of this book (Brioschi 1856a, p. 16, Brioschi 1856b, p. 22). Baltzer (1864, p. 115) also adopted it in the second edition of his textbook on determinants.

For the discriminant of a field, see *integral basis*.

Note. Various remarks on the history of the term and the concept can be found in Muir (1960a, 1960b).

References

Baltzer, Richard (1864). *Theorie und Anwendung der Determinanten*. Zweite vermehrte Auflage. Leipzig.

Brioschi, Francesco (1854). *La teorica dei determinanti e le sue principali applicazioni*. Pavia.

Brioschi, Francesco (1856a). *Theorie der Determinanten und ihre hauptsächlichen Anwendungen*. Berlin.

Brioschi, Francesco (1856b). *Théorie des déterminants et leurs principales applications*. Traduit de l'Italien par Edouard Combescure. Paris.

Gauß, Carl Friedrich (1801). *Disquisitiones arithmeticae*. Leipzig.

Lagrange, Joseph-Louis (1775). "Recherches d'arithmétique". *Nouveaux Mémoires de l'Académie Royale des Sciences et Belles-Lettres de Berlin* (1773), pp. 265–312.

[1] The concept is earlier found in Lagrange (1775, p. 285); see Muir (1960a, pp. 40f.). [2] Latin *discrimen* means 'distinction, difference'. [3] On this passage, see Muir (1960b, p. 63).

Muir, Thomas (1960a). *The Theory of Determinants in the Historical Order of Development*. Volume I. General and Special Determinants up to 1841. New York.
Muir, Thomas (1960b). *The Theory of Determinants in the Historical Order of Development*. Volume II. The Period 1841 to 1860. New York.
Sylvester, James Joseph (1851). "On a remarkable Discovery in the Theory of Canonical Forms and of Hyperdeterminants". *The London, Edinburgh, and Dublin Philosophical Magazine and Journal of Science* 2, pp. 391–410.

disjoint

The notion of disjoint sets appears towards the end of the 1870s. Cantor (1878, p. 248) said that two sets that have no element in common have "no connection" (*keinen Zusammenhang*) or, to use a less visual translation, "no relation".[1] He also used a slightly modified terminology, saying that disjoints sets "are without connection" (see, e. g., Cantor 1880, p. 355). This terminology did not catch on, perhaps because *Zusammenhang* was also used in topology (see *connected*).

At first, the notion is not frequently mentioned in set-theoretical texts. In the 1890s, the mathematician Ernst Schröder suggests some alternatives in his "Lectures on the algebra of logic". He uses the German term *disjunkt* in connection with "domains" (*Gebiete*), which can be thought of as subsets,[2] and says that two domains are disjoint if they have no elements in common (Schröder 1890, p. 212). Schröder (1898, pp. 318, 359) later points out that his *disjunkt* corresponds to Cantor's *ohne Zusammenhang* and that the term is generally used in logic with reference to classes. The usage referred to by Schröder was indeed established long before the term was used in connection with sets and can be exemplified by Kiesewetter (1795, p. 257), who described disjoint notions (German *disjunkte Begriffe*) as opposed to and excluding each other. Kiesewetter gives birds and amphibians as examples.[3]

In the second volume of his book, Schröder (1891, p. 104) adds a synonymous term, *gebietefremd*, from the words *Gebiet* 'domain' and *fremd* 'alien, foreign'. In a later paper, Schröder (1898, p. 318) suggests yet another term, *elementefremd*, a term that stressed that the *elements* of the sets in question are "alien". Unlike *gebietefremd*, this term is still found today; it was used again by Zermelo (1908) (minimally modified as *elementenfremd*).

The French term *disjoint* appears in Couturat (1900, p. 31) with reference to classes, and the English term *disjoint* was apparently first used by Keyser (1909, p. 956), as pointed out by Miller (n.d.).

References

Cantor, Georg (1878). "Ein Beitrag zur Mannigfaltigkeitslehre". *Journal für die reine und angewandte Mathematik* 84, pp. 242–258.
Cantor, Georg (1880). "Ueber unendliche, lineare Punktmannichfaltigkeiten. 2". *Mathematische Annalen* 17, pp. 355–358.
Couturat, Louis (1900). "Sur une définition logique du nombre". *Revue de métaphysique et de morale* 8, pp. 23–36.
Keyser, Cassius Jackson (1909). "The Thesis of Modern Logistic". *Science* 30, pp. 949–963.
Kiesewetter, J.G.C.C. (1795). *Grundriß einer reinen allgemeinen Logik nach Kantischen Grundsätzen*. Berlin.

[1] Cantor also said that two sets are connected if their intersection is not empty. [2] See Schröder (1890, p. 157) for a definition of *Gebiet*. [3] Schröder (1877, p. 7) himself had used the term *disjunkte Klassen*.

Miller, Jeff (n.d.). "Earliest Known Uses of Some of the Words of Mathematics". URL: `https://mathshistory.st-andrews.ac.uk/Miller/mathword/`.
Schröder, Ernst (1877). *Der Operationskreis des Logikkalkuls*. Leipzig.
Schröder, Ernst (1890). *Vorlesungen über die Algebra der Logik* (Exakte Logik). Erster Band. Leipzig.
Schröder, Ernst (1891). *Vorlesungen über die Algebra der Logik* (Exakte Logik). Zweiter Band. Erste Abteilung. Leipzig.
Schröder, Ernst (1898). "Ueber zwei Definitionen der Endlichkeit und G. Cantor'sche Sätze". *Nova Acta Leopoldina. Abhandlungen der Kaiserlich Leopoldinisch-Carolinisch Deutschen Akademie der Naturforscher* 71, pp. 301–362.
Zermelo, Ernst (1908). "Untersuchungen über die Grundlagen der Mengenlehre". *Mathematische Annalen* 65, pp. 261–281.

disk

See *ball*.

distribution (probability)

The term was introduced by von Mises (1919), who presented an axiomatic approach to probability based on the concept of *Kollektiv*, which can be thought of as random sequences. He designated such a sequence by $(e) = e_1, e_2, e_3, \ldots$ and called the e_i *elements* (*Elemente*). Then he explained that "[to] each element corresponds as a 'label' [*Merkmal*] a system of values of the k real variables $x_1, x_2, \ldots x_k$, i. e., a point of the k-dimensional 'label space' [*Merkmalraum*] and we assume that not all elements and even not all except finitely many elements have the same label" (von Mises 1919, p. 55).[1] After introducing his axioms, von Mises then defined the term *Verteilungsfunktion*, for which he also used the shorter *Verteilung* 'distribution':

> It is much more convenient to have to deal with an ordinary *point* function instead of with a *set* function. We therefore aim at representing the distribution within the Kollektiv, i. e., the totality of all probability values, by a function $W(x_1, x_2, \ldots x_k)$ of the k components of the label, or, written more briefly, by $W(\bar{x})$. [...] $W(\bar{x})$ shall denote the value of W_A for the point set A, whose points have the coordinates $\xi_1, \xi_2, \xi_3, \ldots \xi_k$ with $\xi_k \leq x_\kappa$ $(\kappa = 1, 2, \ldots, k)$. [...] We call $W(\bar{x})$ the *distribution function* or briefly the distribution (von Mises 1919, pp. 66f.).[2]

Siegmund-Schultze (2010, p. 213) remarks that "[t]he only difference between this definition of a distribution function and our modern notion is that von Mises assumes his notion of a 'Kollektiv' as standing behind the set function".

As to the success of his concept, von Mises (1933, p. 761) later remarked that after he introduced the term in 1919, many authors have adopted it (for example,

---

[1] "Jedem Element sei als ‚Merkmal' ein bestimmtes Wertsystem der k reellen Veränderlichen $x_1, x_2, \ldots, x_n$ oder ein Punkt des k-dimensionalen ‚Merkmalraumes' zugeordnet, wobei nicht alle Elemente und auch nicht alle bis auf endlich viele dasselbe Merkmal aufweisen sollen". Translation from Siegmund-Schultze (2010, p. 208).

[2] "Es ist wesentlich bequemer, statt mit einer *Mengen*funktion mit einer gewöhnlichen *Punkt*funktion zu tun zu haben. Wir wollen daher die Verteilung innerhalb eines Kollektivs, d. i. die Gesamtheit aller Wahrscheinlichkeitswerte, durch eine Funktion $W(x_1, x_2, \ldots x_k)$ der k Merkmalkomponenten oder kürzer geschrieben durch $W(\bar{x})$ [...] darstellen [...] Es soll $W(\bar{x})$ den Wert von W_A bedeuten für jene Punktmenge A, deren Punkte die Koordinaten $\xi_1, \xi_2, \xi_3, \ldots \xi_k$ besitzen, für die $\xi_k \leq x_\kappa$ $(\kappa = 1, 2, \ldots, k)$ [...] Wir nennen $W(\bar{x})$ die *Verteilungsfunktion* oder auch kurz die Verteilung". Translation from Siegmund-Schultze (2010, p. 213).

Coolidge (1922, p. 135) used English *distribution function*). In the now usual form, distributions are found in Kolmogorov (1933, p. 17).

Note. Detailed accounts of von Mises' approach are given by Siegmund-Schultze (2010) and Verburgt (2015, ch. 11).

References

Coolidge, Julian Lowell (1922). "The Gaussian Law of Error for any Number of Variables". *Transactions of the American Mathematical Society* 24, pp. 135–143.
Kolmogorov, Andrey Nikolaevich (1933). *Grundbegriffe der Wahrscheinlichkeitsrechnung*. Berlin.
von Mises, Richard (1919). "Grundlagen der Wahrscheinlichkeitsrechnung". *Mathematische Zeitschrift* 5, pp. 52–99.
von Mises, Richard (1933). "Über Zahlenfolgen, die ein kollektiv-ähnliches Verhalten zeigen". *Mathematische Annalen* 108, pp. 757–772.
Siegmund-Schultze, Reinhard (2010). "Sets versus trial sequences, Hausdorff versus von Mises: 'Pure' mathematics prevails in the foundations of probability around 1920". *Historia Mathematica* 37, pp. 204–241.
Verburgt, Lukas M. (2015). "'A terrible piece of bad metaphysics'? Towards a history of abstraction in nineteenth- and early twentieth-century probability theory, mathematics and logic". PhD thesis. Universiteit van Amsterdam.

distribution (generalized function)

The French term *distribution* was introduced by Schwartz (1945, p. 60), where distributions are defined as continuous linear functionals in certain linear spaces; as the title of his paper makes clear, he conceived of a distribution as a generalization of the notion of function and differentiation.

In his autobiography, Schwartz later explained the term: "Why did we choose the name distribution? Because, if μ is a measure, i. e. a particular kind of distribution, it can be considered as a distribution of electric charges in the universe. Distributions give more general types of electric charges, for example dipoles and magnetic distributions" (Schwartz 2001, p. 238).

Schwartz' theory, for which he received a Fields Medal in 1950, was an instant success and his terminology spread quickly; Harald Bohr lectured on the subject in 1948 (Barany 2016, pp. 134–136).

Lützen (1982, p. 159) concludes his study of the origins of the concept by answering the question of who invented the concept:

> If one asks about the first people to use distributions in mathematics, the answer is Fourier 1822, Kirchhoff 1882 and Heaviside 1898. If one asks for a rigorous theory, which possibly only implicitly used distributions, the answer is Bochner 1932. If one wants to know who first defined distributions rigorously as functionals, the answer is Sobolev 1935 and finally, if one wants to point to the person who saw the far-reaching applications of distributions and created a broad theory of these objects, Schwartz is the one to cite with 1945-1950 as his years of publication.

Note. For the history of the concept, see Lützen (1982) and Synowiec (1983).

References

Barany, Michael J. (2016). "Distributions in Postwar Mathematics". PhD thesis. Princeton University.
Lützen, Jesper (1982). *The Prehistory of the Theory of Distributions*. Berlin et al.
Schwartz, Laurent (1945). "Généralisation de la notion de fonction, de dérivation, de transformation de Fourier et applications mathématiques et physiques". *Annales de l'université de Grenoble* 21, pp. 57–74.

Schwartz, Laurent (2001). *A Mathematician Grappling with His Century*. Translated from the French by Leila Schneps. Basel.
Synowiec, John (1983). "Distributions: The evolution of a mathematical theory". *Historia Mathematica* 10, pp. 149–183.

distributive

This term, designating an important property of operations, became common in algebraic works in the 1840s. It was introduced as French *distributif* by Servois (1814-15, p. 98), who used the adjective with reference to "functions" (*fonctions*, often to be understood as operators)[1]:

> Let $\varphi(x + y + ...) = \varphi x + \varphi y + ...$ The functions which, like φ, are such that the function of the (algebraic) *sum* of any number of quantities is equal to the sum of the same function of each of these quantities, are called *distributive*.[2]

The reason for choosing this term seems to have been that the function of the sum is "distributed" over the sum of the functions.

The term soon found its way to the English language: Murphy (1830, p. 354), referring to Servois, dealt with "functions of operation of the distributive kind, (that is, such whose action on the whole, is the sum of the actions on the parts)". The term was subsequently used by Jarrett (1831, p. 40) and De Morgan (1836, p. 163), for example. The German term *distributiv* appears in the mid-1850s in some announcements of English publications (Schnuse 1855b, p. 1588, Schnuse 1855a, p. 2006). A first change became noticeable when Murphy (1837, p. 181) started to speak of *linear* operations (operators) instead of *distributive* functions or operations (see *operator*). Nevertheless, *distributive* as applied to functions or operators remained in use throughout the 19th century.

The wide range of the word *operation* can be seen in a geometric example that Gregory (1840, p. 212) gave:

> there is a geometrical operation which is subject to the laws of distributive and permutative[3] operations, and therefore may be represented by the same symbols. This is transference to a distance measured in a straight line. Thus if x represent a point, line, or any geometrical figure, $a(x)$ will represent the transference of this point or line; and it will be seen at once that $a(x) + a(y) = a(x + y)$; or the operation a is distributive.

Shortly afterwards, Hamilton (1844, pp. 429f.) stated that the "distributive character of the multiplication of ordinary algebraic quantities (real or imaginary) extends to the operation of the same name in the theory of quaternions". The term thus also appeared in the context of hypercomplex number systems; in this sense, the terms

[1] Bradley and Petrilli (2010, Functions and Operators) observe that "Servois does not make a distinction between 'function' and 'operator' as we do today. Sometimes he uses the word 'function' in the same way that students of calculus do (i. e., to describe a relation between independent and dependent variables), but he also uses it to describe operators, functions of functions such as the varied state, and difference operators". [2] "Soit $\varphi(x + y + ...) = \varphi x + \varphi y + ...$ Les fonctions qui, comme φ, sont telles que la fonction de la *somme* (algébrique) d'un nombre quelconque est égale à la somme des fonction pareilles de chacune de ces quantités, seront appelées *distributives*". Translation from Bradley and Petrilli (2010, p. 5), slightly modified. [3] Gregory occasionally uses *permutative* instead of *commutative*.

spread, in Germany, through Hankel (1867, p. 163), who also pointed out that the term is due to Servois.

Note. The context in which Servois introduced the terminology is explained in Koppelman (1971) and Allaire and Bradley (2002), for example. Crowe (1985, pp. 15f.) gives a concise overview of the history of the terminology.

References

Allaire, Patricia R. and Robert E. Bradley (2002). "Symbolical Algebra as a Foundation for Calculus: D.F. Gregory's Contribution". *Historia Mathematica* 29, pp. 395–426.

Bradley, Robert E. and Salvatore J. Petrilli (2010). "Servois' 1814 Essay on a New Method of Exposition of the Principles of Differential Calculus, with an English Translation". *Convergence* (November 2010). URL: `https://old.maa.org/press/periodicals/convergence/servois-1814-essay-on-a-new-method-of-exposition-of-the-principles-of-differential-calculus-with-an-3`.

Crowe, Michael J. (1985). *A History of Vector Analysis*. The Evolution of the Idea of a Vectorial System. New York.

Gregory, Duncan Farquharson (1840). "On the Real Nature of Symbolical Algebra". *Transactions of the Royal Society of Edinburgh* 14, pp. 208–216.

Hamilton, William Rowan (1844). "On a new Species of Imaginary Quantities connected with a theory of Quaternions". *Proceedings of the Royal Irish Academy* 2, pp. 424–434.

Hankel, Hermann (1867). *Vorlesungen über die complexen Zahlen und ihre Functionen*. In zwei Theilen. I. Theil: Theorie der complexen Zahlsysteme. Leipzig.

Jarrett, Thomas (1831). *An Essay on Algebraic Development*. Cambridge.

Koppelman, Elaine (1971). "The Calculus of Operations and the Rise of Abstract Algebra". *Archive for History of Exact Sciences* 8, pp. 155–242.

De Morgan, Augustus (1836). *The Differential and Integral Calculus*. London.

Murphy, Robert (1830). "On the Inverse Method of Definite Integrals, with Physical Applications". *Transactions of the Cambridge Philosophical Society* 4, pp. 353–405.

Murphy, Robert (1837). "First Memoir on the Theory of Analytical Operations". *Philosophical Transactions of the Royal Society of London* 127, pp. 179–210.

Schnuse, C.H. (1855a). "A Treatise on the Calculus of Operations [announcement]". *Göttingische gelehrte Anzeigen* 117, pp. 2002–2012.

Schnuse, C.H. (1855b). "A Treatise on the Infinitesimal Calculus [announcement]". *Göttingische gelehrte Anzeigen* 117, pp. 1586–1593.

Servois, François-Joseph (1814-15). "Essai sur un nouveau mode d'exposition des principes du calcul différentiel". *Annales des mathématiques pures et appliquées* 5, pp. 93–140.

divergence and *convergence* (vector calculus)

The *convergence* of a vector function was introduced by Maxwell (1871, p. 231):[1]

> If σ represents a vector function, $\nabla\sigma$ may contain both a scalar and a vector part, which may be written $S\nabla\sigma$ and $V\nabla\sigma$. I propose to call the scalar part the *Convergence* of σ because, if a closed surface be described about any point, the surface integral of σ, which expresses the effect of the vector σ considered as an inward flux through the surface, is equal to the volume integral of $S\nabla\sigma$ throughout the enclosed space. I think, therefore, that the convergence of a vector function is a very good name for the effect of that vector function in carrying its subject inwards towards a point.

Maxwell also used the term in his *Treatise on Electricity and Magnetism* (Maxwell 1873, p. 28).

[1] Maxwell already used the term in a letter to Tait from 1870 (Knott 1911, p. 143).

Crowe (1985, p. 132) notes that "according to quaternion multiplication, if $\sigma = it + ju + kv$, then $\nabla\sigma = \left(\frac{dt}{dx} + \frac{du}{dy} + \frac{dv}{dz}\right) + i\left(\frac{dv}{dx} - \frac{du}{dz}\right) + j\left(\frac{dt}{dz} - \frac{dv}{dx}\right) + k\left(\frac{du}{dx} - \frac{dt}{dy}\right)$. Maxwell called the scalar part of this product, represented in quaternion language, as $S\nabla\sigma$, the 'Convergence' of σ" and adds: "since $S\nabla\sigma$ (in quaternion language) is equal to the modern $S - \nabla \cdot \gamma$, Maxwell's name 'convergence' required alteration. The modern name 'divergence' for $\nabla \cdot \sigma$ was given by William Kingdon Clifford" (Crowe 1985, p. 148), namely in his *Elements of Dynamic*: "Prof. Clerk Maxwell calls the quantity $-E$ the *convergence* of σ. We might perhaps therefore call E itself the *divergence* of σ" (Clifford 1878, p. 209).

The terminology was also adopted by Heaviside (1883, p. 510):

> Let R denote the electrostatic force, and p the volume density of electricity then we may say $4\pi\rho = -conv.R$, or $4\pi\rho = -div.R$, where we use conv and div as abbreviations to be understood as follows: In terms of the components X, Y, Z of the force, we have
>
> $$4\pi\rho = \left(\frac{dX}{dx} + \frac{dY}{dy} + \frac{dZ}{dz}\right)$$
>
> [...] The expression on the right hand side of this equation (with the $-$ sign prefixed) Maxwell called the "convergence" of the force; it is really the integral amount of force, taken algebraically, entering the unit volume; but since $+$ convergence indicates negative electrification, we may as well use the term "divergence" for the same quantity with $+$ sign prefixed.

Again, Crowe's comments are instructive: "Heaviside here used the term 'divergence' for the first time in his writings. Though Clifford had introduced this term five years earlier, it by no means necessarily follows that Heaviside had read Clifford, since Heaviside did not mention Clifford and since divergence is an obvious name for the negative of the convergence. The physical representation in terms of lines of force also suggests this name" (Crowe 1985, p. 165).

The terminology quickly spread through the various books that began to use vector calculus such as Föppl (1894, pp. 41, 44), who also remarks that he uses *divergence* for what Maxwell called *convergence*.

Note. For the introduction of the terminology, see Crowe (1985, esp. pp. 131f., 142, 148).

References

Clifford, William Kingdon (1878). *Elements of Dynamic*. Part I. Kinematic. London.

Crowe, Michael J. (1985). *A History of Vector Analysis*. The Evolution of the Idea of a Vectorial System. New York.

Föppl, August (1894). *Einführung in die Maxwell'sche Theorie der Elektricität*. Leipzig.

Heaviside, Oliver (1883). "Some Electrostatic & Magnetic Relations". *The Electrician* 10, pp. 510–511, 558–559, 582–583.

Knott, Cargill Gilston (1911). *Life and Scientific Work of Peter Guthrie Tait*. Cambridge.

Maxwell, James Clerk (1871). "Remarks on the Mathematical Classification of Physical Quantities". *Proceedings of the London Mathematical Society* 3, pp. 224–232.

Maxwell, James Clerk (1873). *A Treatise on Electricity and Magnetism*. Vol. I. Oxford.

divergent

The term *divergent series* has frequently been attributed[1] to Nikolaus Bernoulli, who used Latin *series divergentes* in a letter from 1713 (see Gerhardt 1855-56, p. 982). However, the term had already been used earlier by writers such as Leibniz, Newton, and de Moivre.

For instance, *divergent* was used by Leibniz in an unpublished note written in the latter half of 1672 (see Leibniz 2003, pp. 64f., 75). He used it in analogy to *convergent*, which, at the time, was applied to two sequences rather than one (see *convergent*). Newton (1971, p. 610) used the same term in an unpublished note from 1684 on the computation of series. In print, the term is found in de Moivre (1704, preface), who used *divergent* in opposition to *convergent*.

The term gradually began to be used more frequently in the 1730s, when also the first explicit definitions are found (for example, in Wolff 1730, p. 310 and Chelucci 1738, p. 34). The English terminology *diverging series* is found in Colson (1736, p. 152); in a French translation of an English review of that book, this is rendered "the series that diverge"[1] (Eames 1738, p. 195). In a paper read in 1753, Euler (1760, p. 20) defined a divergent series thus: "series are called divergent, whose terms do not tend to zero but never decrease below a certain amount or even increase to infinity".[2]

See also *convergent* for a potential analogy of the term as it was previously used in optics and for further aspects of the conceptual development.

Note. For the terminology, see Leibniz (2003, p. xxiii) and Cajori (1923).

References

Barbeau, E.J. and P. J. Leah (1976). "Euler's 1760 paper on divergent series". *Historia Mathematica* 3, pp. 141–160.

Cajori, Florian (1923). "The name divergent series". *Bulletin of the American Mathematical Society* 29, p. 55.

Cantor, Moritz (1901). *Vorlesungen über Geschichte der Mathematik*. Dritter Band. Von 1668 bis 1758. Leipzig.

Chelucci, Domenico (1738). *Institutiones analyticae earumque usus in geometria*. Roma.

Colson, John (1736). "The Method of Fluxions and Infinite Series; or, A Perpetual Comment upon the foregoing Treatise". In: *The Method of Fluxions and Infinite Series*. London, pp. 141–339.

Eames, John (1738). "Méthode des Fluxions & des Séries Infinies avec son application à la Géométrie des Courbes". *Transactions philosophiques de la société royale de Londres* (1735), pp. 193–200.

Euler, Leonhard (1755). *Institutiones calculi differentialis*. St. Petersburg.

Euler, Leonhard (1760). "De seriebus divergentibus". *Novi Commentarii academiae scientiarum Petropolitanae* 5, pp. 19–23, 205–237.

Gerhardt, C.I., ed. (1855-56). *Leibnizens mathematische Schriften*. Erste Abtheilung. Band III. Halle.

Leibniz, Gottfried Wilhelm (2003). *Sämtliche Schriften und Briefe*. Siebente Reihe. Mathematische Schriften. Dritter Band. 1672-1676. Differenzen, Folgen, Reihen. Hannover.

de Moivre, Abraham (1704). *Animadversiones in Georgii Cheynaei Tractatum de Fluxionum Methodo Inversa*. London.

Newton, Isaac (1971). *The Mathematical Papers of Isaac Newton*. Volume IV. 1674-1684. Ed. by D.T. Whiteside. Cambridge.

Wolff, Christian (1730). *Elementa matheseos universae*. Tomus I. Halle.

---

[1] For example, by Cantor (1901, p. 370) and Cajori (1923, p. 55). [1] "les séries qui divergent".

[2] "Divergentes autem series dicuntur, quarum termini non ad nihilum tendunt, sed vel infra certum limitem nunquam decrescunt, vel adeo ad infinitum excrescunt". Translation from Barbeau and Leah (1976, p. 143). – A modified definition is found in Euler (1755, § 109).

division

In classical Greek mathematics, the notion of division can be expressed in various ways; according to Riley (2000, p. 25), *μερίζειν* (from *μέρος* 'part') is "perhaps the most common word for 'to divide'". The verb was used by Nicomachus and Diophantus, for example (see Tropfke 1980, pp. 242f.).[3]

In Latin, terms for 'divide' are hardly used before the 12th century. An agrimensorial tract, the *Podismus*, uses *partiri*: "I divide 70 by the base, i. e. by 14, which yields 5" (Blume, Lachmann, and Rudorff 1848, p. 299).[4] Latin *dividere* often means 'to divide' in general; Boethius uses it in this way: "if one takes an even number, it can be divided into two even parts, as ten is divided into fives. It can also be divided into unequal parts, as when the same ten is divided into 3 and 7" (Boethius 1867, p. 14).[1]

In the 12th and 13th centuries, there were several popular texts in which *dividere* and *divisio* were used. The terminology occurs, for example, in Fibonacci's *Liber abaci*: "And it is noted that the number which is divided is called the dividend, and the number which divides it is called the divisor" (Fibonacci 1857, p. 28).[2] In Sacrobosco's highly popular *De arte numerandi*, division is defined in the following terms:

> When two numbers have been proposed, division is the distribution of the greater by the smaller into as many parts as there are units in the smaller [number]. In division, therefore, it must be noted that three numbers are necessary–the number to be divided [the dividend], the dividing number, or divisor, and the number denoting the quotient, or number that comes forth (Curtze 1897, p. 11).[3]

The English terms *to divide* and the noun *division* are attested in *The Craft of Nombrynge* from around 1425: "þer ben 7 spices [species] or partes of þis craft [...] The 5. is called multiplicacion. The 6. is called diuision" ([anon.] 1922b, p. 7). In the *The Art of Nombryng*, the definition is this: "For to dyvyde oo nombre by a-nother, it is of .2. nombres proposede, It is forto depart the moder nombre into as many partis as ben of vnytees in the lasse nombre" ([anon.] 1922a, p. 43).

French *division* and *diviser* are found in the 13th century (Mortet 1908/09, p. 61, Karpinski and Waters 1928, p. 82); Italian *dividere* was used by Jacopo da Firenze in the early 14th century (Høyrup 2007, p. 195). German *teilen* appears in the *Geometria culmensis* from around 1400 (Mendthal 1886, p. 67).

A commonly found alternative term is *partition*. It is found, for example, in the *Introduction for to lerne to recken with the pen* ([anon.] 1539, fol. d iii^r). The French

---

[3] See the note at the end of this article for alternative terms. [4] "LXX. hoc partior ad basem, id est ad XIIII, et fit V". [1] "si enim ponatur par numerus, potest in duo aequaiia dividi, ut denarius dividitur in quinos, porro autem et per inaequalia, ut idem denarius in III et VII". Translation from Masi (2006, p. 78). [2] "Et notandum quod numerus qui diuiditur uocatur diuisus uel diuidendus; et numerus qui diuidit uocatur diuidens, uel diuisor". Translation from Sigler (2002, p. 53). [3] "Divisio est numeri per numerum duobus numeris propositis maioris per minorem in tot partes distributio, quot sunt unitates in minori. Notandum igitur, quod in divisione tres numeri sunt necessarii, scilicet numerus dividendus, et numerus dividens sive divisor, et numerus denotans sive numerus exiens". Translation from Katz et al. (2016, p. 90). – Another popular text in which *dividere* was used is the "Carmen de Algorismo" from about 1220 (de Villa Dei 1922, p. 73); for the terms in some 12th- and 13th-century texts on Hindu reckoning, see Allard (1990, p. 160).

verb *partir* was used by Brunetto Latini (13th century) in his *Trésor* (Latini 1948, p. 20).

Note. For historical overviews of the terminology, see Tropfke (1930, pp. 70–72, 110–113), Tropfke (1980, pp. 242–248), and Smith (1958, pp. 128–132). For the Greek terminology, see also Riley (2000, pp. 21–29);[4] for the Latin one, see also Guillaumin (2020, pp. 116, 211) and Schmidt (1914, p. 55). An overview of the Latin arithmetical terminology from the 12th and 13th centuries is found in Allard (1990, pp. 152–180). The Arabic terminology is discussed by Rashed (2017, pp. 642–644).

References

[anon.] (1539). *An introduction for to lerne to recken with the pen*. London.

[anon.] (1922a). "The Art of Nombryng". In: *The Earliest Arithmetics in English*. Ed. by Robert Steele. London, pp. 33–51.

[anon.] (1922b). "The Craft of Nombrynge". In: *The Earliest Arithmetics in English*. Ed. by Robert Steele. London, pp. 3–32.

Allard, André (1990). "La formation du vocabulaire latin de l'arithmétique médiévale". In: *Méthodes et instruments du travail intellectuel au moyen âge*. Études sur le vocabulaire. Ed. by Olga Weijers. Turnhout, pp. 137–181.

Blume, Friedrich, Karl Lachmann, and Adolf Rudorff, eds. (1848). *Die Schriften der römischen Feldmesser*. Erster Band. Texte und Zeichnungen. Berlin.

Boethius, Anicius Manlius Severinus (1867). *De institutione arithmetica libri duo, De institutione musica libri quinque*. Accedit Geometria quae fertur Boetii. Ed. by Gottfried Friedlein. Leipzig.

Christianidis, Jean and Jeffrey A. Oaks (2013). "Practicing algebra in late antiquity: The problem-solving of Diophantus of Alexandria". *Historia Mathematica* 40, pp. 127–163.

Curtze, Maximilian (1897). *Petri Philomeni de Dacia in Algorismum vulgarem Johannis de Sacrobosco commentarius*. Kopenhagen.

Fibonacci (1857). *Scritti di Leonardo Pisano*. Pubblicati da Baldassare Boncompagni. Volume I. Roma.

Guillaumin, Jean-Yves (2020). *Dictionnaire de la terminologie latine ancienne de l'arithmétique et de la géométrie*. Paris.

Høyrup, Jens (2007). *Jacopo da Firenze's Tractatus Algorismi and Early Italian Abbacus Culture*. Basel.

Karpinski, Louis Charles and E.G.R. Waters (1928). "A Thirteenth Century Algorism in French Verse". *Isis* 11, pp. 45–84.

Katz, Victor J. et al., eds. (2016). *Sourcebook in the Mathematics of Medieval Europe and North Africa*. Princeton.

Latini, Brunetto (1948). *Li Livres dou Trésor*. Édition critique par Francis Carmody. Berkeley.

Masi, Michael (2006). *Boethian Number Theory*. Translation of the *De Institutione Arithmetica*. Amsterdam and New York.

Mendthal, Hans (1886). *Geometria culmensis*. Ein agronomischer Tractat aus der Zeit des Hochmeisters Conrad von Jungingen (1393-1407). Leipzig.

Mortet, Victor (1908/09). "Le plus ancien traité français d'algorisme". *Bibliotheca Mathematica* 9, pp. 55–64.

Rashed, Roshdi, ed. (2017). *Lexique historique de la langue scientifique arabe*. Hildesheim.

Riley, Marc (2000). "Terminology for Arithmetic Operations in Ancient Greek". URL: `http://www.csus.edu/indiv/r/rileymt/` (retrieved Sept. 22, 2017).

Schmidt, Max Carl Paul (1914). *Zur Entstehung und Terminologie der elementaren Mathematik*. Zweite verbesserte und stark vermehrte Auflage. Leipzig.

Sigler, Laurence (2002). *Fibonacci's Liber Abaci*. A Translation into Modern English of Leonardo Pisano's Book of Calculation. New York.

[4] Christianidis and Oaks (2013, pp. 140f.) discuss words for dividing in Diophantus' works.

Smith, David Eugene (1958). *History of Mathematics*. Volume II. Special Topics of Elementary Mathematics. New York.
Tropfke, Johannes (1930). *Geschichte der Elementar-Mathematik*. Erster Band. Rechnen. Dritte, verbesserte und vermehrte Auflage. Berlin and Leipzig.
Tropfke, Johannes (1980). *Geschichte der Elementarmathematik*. Band 1: Arithmetik und Algebra. Vollständig neu bearbeitet von Kurt Vogel, Karin Reich, Helmuth Gericke. Berlin and New York.
de Villa Dei, Alexander (1922). "Carmen de Algorismo". In: *The Earliest Arithmetics in English*. Ed. by Robert Steele. London, pp. 72–80.

division algebra and *division ring*

While division algebras had already been examined in the second half of the 19th century, the term is due to Dickson (1914, p. 66): "By a *division algebra* shall be meant one in which both right-hand and left-hand division, except by zero, are possible and unique". Earlier it was customary to talk about "algebras in which division is uniquely possible" (Dickson 1906), while Wedderburn (1908, p. 91) preferred *primitive algebra*.[1] Dickson's suggestion was soon accepted and used in various treatises on the subject; this also includes Wedderburn, for example in his article "On division algebras" (Wedderburn 1921).

The term *division ring* came up in the 1930s and may go back to Albert (1937, p. 23): "A ring $\mathfrak{A}$ with a unity element is called a division ring (or quasi-field) if every non-zero element of $\mathfrak{A}$ is a regular element".[2] Weiss (1939, p. 641) explains the choice of the term and contrasts it with the synonymous *quasi-field*[3]: "The term, division ring, emphasizes the fact that such a ring is one in which division is possible, whereas the terminology quasi emphasizes the fact that multiplication is non-commutative. Unfortunately, there seems to be no uniform terminology, the terminology varying with the author".

See also *skew* (algebra), *hypercomplex*, and *quasifield*.

Note. For the history of the concept, see Frei (2003, 2007) as well as the literature cited in the entry *hypercomplex*.

References

Albert, A.A. (1937). *Modern Higher Algebra*. Chicago.
Dickson, Leonard Eugene (1906). "On Commutative Linear Algebras in Which Division is Always Uniquely Possible". *Transactions of the American Mathematical Society* 7, pp. 514–522.
Dickson, Leonard Eugene (1914). *Linear Algebras*. Cambridge.
Frei, Günther (2003). "Zur Geschichte der Arithmetik der Algebren (1843-1932)". *Elemente der Mathematik* 58, pp. 156–168.
Frei, Günther (2007). "Developments in the Theory of Algebras over Number Fields: A New Foundation for the Hasse Norm Residue Symbol and New Approaches to Both the Artin Reciprocity Law and Class Field Theory". In: *Episodes in the History of Modern Algebra (1800–1950)*. Ed. by Jeremy J. Gray and Karen Hunger Parshall. Providence, R.I., pp. 117–151.
Hazlett, O.C. (1916). "On the Classification and Invariantive Characterization of Nilpotent Algebras". *American Journal of Mathematics* 38, pp. 109–138.
Kline, Morris (1972). *Mathematical Thought From Ancient To Modern Times*. Volume 3. New York and Oxford.
Wedderburn, J.H.M. (1908). "On Hypercomplex Numbers". *Proceedings of the London Mathematical Society* 6, pp. 77–118.

[1] On Wedderburn's use of this term, see Hazlett (1916, p. 111) and Kline (1972, p. 1152). [2] Regular elements in a ring have the property $ab = ba = 1$. [3] See *skew* (algebra).

Wedderburn, J.H.M. (1921). "On Division Algebras". *Transactions of the American Mathematical Society* 22, pp. 129–135.
Weiss, Marie J. (1939). "Algebra for the Undergraduate". *The American Mathematical Monthly* 46, pp. 635–642.

divisor

This term has several mathematical meanings, the oldest of which being that of the divisor of a number. The general history is discussed under *division.*

In classical Greek, words for 'dividend' and 'divisor' do not occur at all,[4] and the terminology instead goes back to Latin origins. The Latin term *divisor* was used in the 10th century by Gerbert of Aurillac (Gerbert of Aurillac 1899, p. 15) and can be said to be standard in the 12th century, when it was used by Fibonacci, for example: "And it is noted that the number which is divided is called the dividend, and the number which divides it is called the divisor".[1] It was also used in the *Algorismus vulgaris* by Johannes de Sacrobosco (Curtze 1897, p. 11),[2] which was an extremely popular and widely spread text.

The English term is found around 1450 in *The Art of Nombrynge*, where the concept is first paraphrased: "And note wele that in makynge of dyvysioun ther ben .3. nombres necessary: that is to sey, the nombre to be dyvydede; the nombre dyvydyng and the nombre exeant [resulting number], other how oft, or quocient", and then *dyvyser* is used ([anon.] 1922, p. 43).[3] An early usage in print is [anon.] (1539, fol. d iii^r).

French *diviseur* is found (DMF) in Chuquet's *Triparty* (1484): "In division only two numbers are needed, that is to say, the divider [*partiteur*[4]] or divisor [*diviseur*] and the number to be divided" (Chuquet 1881, p. 47).[5] In a 14th-century algorism written in Anglo-Norman, the term is found earlier (Karpinski and Staubach 1935, p. 142).[6] In print, the term appears as *nombre diviseur* in [anon.] (1501, fol. b^v), where also the noun *diviseur* is found (fol. b ii^r).

In Italian, the situation is similar: in a late 14th century text, *partitore* is found (Ambrosetti 2011, p. 154), a term also used in the first printed book on arithmetic ([anon.] 1478, fol. 22^v). Pacioli (1494, fol. 31^v) names both *partitore* and *divisore.*

The German term *Teiler* (from *teilen* 'to divide') is found in a treatise from about 1460 (Curtze 1895, p. 52); the term also appears in early printed texts such as Widmann (1489, fol. c 8^v) and is still in regular use. Apian (1527, fol. B vii^r) names *Divisor* as an alternative to *Teiler.*[7]

See also *factor*.

---

[4] See Tropfke (1980, p. 242); in Riley (2000, pp. 21–29), where the Greek arithmetical terminology is discussed, no term for 'divisor' is listed. [1] "Et notandum quod numerus qui dividitur vocatur divisus vel dividendus; et numerus qui dividit vocatur dividens, vel divisor". Translation from Sigler (2002, p. 53). [2] For a translation, see Katz et al. (2016, p. 90). [3] This text is cited by OED and MED as the first usage of the term in English. [4] From *partir*, in its then common sense 'to divide'. For earlier attestations of French *partiteur* and some of its variants, see Benoit (1985, p. 82) and Spiesser (2003, pp. 230f.). In Occitan, *partidor* was used in the so-called *Arithmétique de Pamiers* from around 1430 (Sesiano 2018, p. 233). [5] "En diuision ne sont requiz que deuc nombres cestas [dire] le diuiseur ou partiteur et le nombre a partir". Translation from Flegg, Hay, and Moss (1985, p. 37), slightly adapted. [6] In a 13th-century algorism, *devisant* is used (Karpinski and Waters 1928, p. 82) instead. [7] For the German terms, see Schirmer (1912, pp. 16, 71f.).

Note. For historical overviews of the terminology, see Tropfke (1930, pp. 70–72, 110–113), Tropfke (1980, pp. 242–248), and Smith (1958, p. 131).

References

[anon.] (1478). *Larte de labbacho*. Treviso.

[anon.] (1501). *Livre de Chiffres et de getz*. Lyon.

[anon.] (1539). *An introduction for to lerne to recken with the pen*. London.

[anon.] (1922). "The Art of Nombryng". In: *The Earliest Arithmetics in English*. Ed. by Robert Steele. London, pp. 33–51.

Ambrosetti, Nadia (2011). "Una traduzione dell'algebra di al-Khwarizmi nella Firenze del XIV secolo". *Bollettino di storia delle scienze matematiche* 31, pp. 137–166.

Apian, Peter (1527). *Eyn Newe unnd wolgegründte underweysung aller Kauffmanß-Rechnung*. Ingolstadt.

Benoit, Paul (1985). "Recherches sur le vocabulaire des opérations élémentaires dans les arithmétiques en langue française de la fin du Moyen Âge". *Documents pour l'histoire du vocabulaire scientifique* 7, pp. 77–95.

Chuquet, Nicolas (1881). *Le Triparty en la Science des Nombres*. Ed. by Aristide Marre. Rome.

Curtze, Maximilian (1895). "Ein Beitrag zur Geschichte der Algebra in Deutschland im 15. Jahrhundert". *Abhandlungen zur Geschichte der Mathematik* 7, pp. 31–74.

Curtze, Maximilian (1897). *Petri Philomeni de Dacia in Algorismum vulgarem Johannis de Sacrobosco commentarius*. Kopenhagen.

Flegg, Graham, Cynthia Hay, and Barbara Moss, eds. (1985). *Nicolas Chuquet, Renaissance mathematician*. A study with extensive translation of Chuquet's mathematical manuscript completed in 1484. Dordrecht.

Gerbert of Aurillac (1899). "Geometria". In: *Opera mathematica*. Ed. by Nicolaus Bubnov. Berlin, pp. 48–97.

Karpinski, Louis Charles and Charles N. Staubach (1935). "An Anglo-Norman Algorism of the Fourteenth Century". *Isis* 23, pp. 121–152.

Karpinski, Louis Charles and E.G.R. Waters (1928). "A Thirteenth Century Algorism in French Verse". *Isis* 11, pp. 45–84.

Katz, Victor J. et al., eds. (2016). *Sourcebook in the Mathematics of Medieval Europe and North Africa*. Princeton.

Pacioli, Luca (1494). *Summa de Arithmetica Geometria Proportioni et Proportionalita*. Venezia.

Riley, Marc (2000). "Terminology for Arithmetic Operations in Ancient Greek". URL: `http://www.csus.edu/indiv/r/rileymt/` (retrieved Sept. 22, 2017).

Schirmer, Alfred (1912). *Der Wortschatz der Mathematik nach Alter und Herkunft untersucht*. Straßburg.

Sesiano, Jacques (2018). *L'arithmétique de Pamiers*. Traité mathématique en langue d'oc du XVe siècle. Lausanne.

Sigler, Laurence (2002). *Fibonacci's Liber Abaci*. A Translation into Modern English of Leonardo Pisano's Book of Calculation. New York.

Smith, David Eugene (1958). *History of Mathematics*. Volume II. Special Topics of Elementary Mathematics. New York.

Spiesser, Maryvonne, ed. (2003). *Une arithmétique commerciale du XVe siècle*. Le *Compendy de la praticque des nombres* de Barthélemy de Romans. Turnhout.

Tropfke, Johannes (1930). *Geschichte der Elementar-Mathematik*. Erster Band. Rechnen. Dritte, verbesserte und vermehrte Auflage. Berlin and Leipzig.

Tropfke, Johannes (1980). *Geschichte der Elementarmathematik*. Band 1: Arithmetik und Algebra. Vollständig neu bearbeitet von Kurt Vogel, Karin Reich, Helmuth Gericke. Berlin and New York.

Widmann, Johannes (1489). *Behende vnd hubsche Rechenung auff allen kauffmanschafft*. Leipzig.

***divisor* (further uses)**

The term *divisor* is also used to denote the subgroup of a group. In this sense it goes back to French *diviseur*, a term used by Évariste Galois, who had in mind an analogy with numbers. In one of his manuscripts, he wrote: "If a group is contained in another, the latter will be the sum of a certain number of groups similar to the first, which will be said to be a *divisor* of it" (P. M. Neumann 2011, p. 280).[1] The term was in common use for some time; see *subgroup*.

Dedekind also applied the term to fields (and other structures): "We call a field A a *divisor* of field M, and the latter a *multiple* of the former, when all the numbers in A are also found in M" (Dedekind 1871, p. 424).[1]

A different sense of the term is found in the writings of Kronecker. He presented his theory of divisors in his *Elements of an Arithmetical Theory of Algebraic Quantities* (Kronecker 1882). His divisors (German *Divisoren*) can be said to correspond to Dedekind's *ideals*, but his theory was very different from Dedekind's, as explained by Edwards (1980, p. 353): "Instead of unique factorization into primes, Kronecker emphasizes the notion of *greatest common divisor*. It is probably for this reason that the basic objects of his theory, corresponding to Dedekind's 'ideals', are called 'divisors'. The main points of his theory are the definition of divisors, the notion of divisibility of one divisor by another, the concomitant notion of absolute equivalence of divisors (two divisors are absolutely equivalent if each divides the other), and the interpretation of an algebraic integer as a divisor".

See also *ideal* (algebra).

Note. For divisor theory and its history, see Edwards (1980) and O. Neumann (2002)

References

Avigad, Jeremy (2004). "Dedekind's 1871 version of the theory of ideals". URL: `https://kilthub.cmu.edu/articles/Dedekind_s_1871_Version_of_the_Theory_of_Ideals/6491342/1` (retrieved July 25, 2007).

Corry, Leo (2004). *Modern Algebra and the Rise of Mathematical Structures*. Second revised edition. Basel.

Dedekind, Richard (1871). "Ueber die Composition der binären quadratischen Formen". In: *Vorlesungen über Zahlentheorie von P.G. Lejeune Dirichlet*. Herausgegeben und mit Zusätzen versehen von R. Dedekind. Zweite, umgearbeitete und vermehrte Auflage. Braunschweig, pp. 380–497.

Edwards, Harold M. (1980). "The Genesis of Ideal Theory". *Archive for History of Exact Sciences* 23, pp. 321–378.

Kronecker, Leopold (1882). *Grundzüge einer arithmetischen Theorie der algebraischen Grössen*. Berlin.

---

[1] "Si un groupe est contenu dans un autre, celui-ci sera la somme d'un certain nombre de groupes semblables au premier, qui en sera dit un *diviseur*". Translation from P. M. Neumann (2011, p. 281).

[1] "Wir nennen einen Körper A einen *Divisor* des Körpers M, diesen ein *Multiplum* von jenem, wenn alle in A enthaltenen Zahlen sich auch in M vorfinden". Translation from Avigad (2004, p. 12). Note that Dedekind used *divisor* and *multiple* in an opposite way in connection with modules and ideals: "If all the numbers of a module $\mathfrak{a}$ are also numbers of a module $\mathfrak{d}$, then $\mathfrak{a}$ is called a *multiple* of $\mathfrak{d}$, and $\mathfrak{d}$ a *divisor* of $\mathfrak{a}$" ("Wenn alle Zahlen eines Moduls $\mathfrak{a}$ auch Zahlen eines Moduls $\mathfrak{d}$ sind, so heisse $\mathfrak{a}$ ein *Vielfaches* von $\mathfrak{d}$, und b ein Theiler von $\mathfrak{a}$") (Dedekind 1871, p. 443). Translation from Avigad (2004, p. 30). For his analogous definition for ideals, see Dedekind (1871, p. 453). Krull (1935, p. 149) later commented that this usage is "historically clear, but impossible from the point of view of set-theory". See Corry (2004, pp. 228f.).

Krull, Wolfgang (1935). *Idealtheorie*. Berlin.
Neumann, Olaf (2002). "Was sollen und was sind Divisoren?" *Mathematische Semesterberichte* 48, pp. 139–192.
Neumann, Peter M. (2011). *The Mathematical Writings of Évariste Galois*. Zürich.

domain (function)

This terminology does not seem to appear until around 1850. In particular, the German word *Gebiet* 'domain' (or *Größengebiet*, where *Größe* means 'quantity' or 'magnitude') was used by Riemann, without definition, in his PhD thesis, where he wrote, for example: "we consider the extent of the conditions needed to determine such a function in a given domain" (Riemann 1851, § 19).[1] The terminology cannot be said to established at that time, however, and it should also be noted that *Gebiet* is commonly used with some additional topological meanings (such as 'connectedness') which do not play a role in the definition of *domain* in connection with functions in general. Lipschitz (1876, p. 150), for example, uses French *champ* in such a way: "The functions f_j are given on a continuously connected set of values of the variables x, y_1, y_2, ..., y_n; this set of values will be called the *domain G*".[2]

In the early 1880s, the terminology is found in some textbooks. For example, Harnack (1881, p. 89) uses German *Gebiet*, while Stolz (1885, p. 149) prefers *Bereich*; the latter word can also be translated by *domain* or *area*. English *domain* is found as a translation of *Gebiet* in Harnack (1891, p. 86); French *domaine* appears in Borel (1895, p. 9) and Borel (1898, p. 56).

The German term *Definitionsbereich* (a word still in common use), lit. 'domain of definition', came up towards the end of the 19th century; it was used (without definition) by Burkhardt (1897, p. 76), for example. The translation *domain of definition* was suggested by Pierpont (1905, p. 120): "The domain of the independent variable x which enters in the law defining a function $f(x)$ is also called the *domain of definition of the function*".

For the term *source*, see *range*.

References

Birkhoff, Garrett, ed. (1973). *A Source Book in Classical Analysis*. Cambridge, Mass.
Borel, Émile (1895). "Sur quelques points de la théorie des fonctions". *Annales scientifiques de l'École Normale Supérieure* 12, pp. 9–55.
Borel, Émile (1898). *Leçons sur la théorie des fonctions*. Paris.
Burkhardt, Heinrich (1897). *Funktionentheoretische Vorlesungen*. Erster Teil. Leipzig.
Harnack, Axel (1881). *Die Elemente der Differential- und Integralrechnung*. Zur Einführung in das Studium. Leipzig.
Harnack, Axel (1891). *An Introduction to the Study of the Elements of the Differential and Integral Calculus*. Trans. by George L. Cathcart. London and Edinburgh.
Lipschitz, Rudolf (1868). "Erörterung der Möglichkeit, ein gegebenes System gewöhnlicher Differenzialgleichungen vollständig zu integriren". In: *Abhandlungen aus dem Gebiete der Naturwissenschaften, Mathematik und Medicin*. Bonn, pp. 45–58.

---

[1] "Zur Orientirung auf diesem Felde wird ein Ueberschlag über den Umfang der zur Bestimmung einer solchen Function innerhalb eines gegebenen Grössengebiets erforderlichen Bedingungen dienen". Translation from Riemann (2004, p. 32). [2] "Les fonctions f^α sont données pour un ensemble de valeurs des variables x, y^1, y^2, ... , y^n, ayant entre elles une connexion continue: cet ensemble de valeurs en sera dit le champ G". Translation based on Birkhoff (1973, p. 246). In the corresponding German article, Lipschitz (1868, p. 47) uses *Größengebiet*.

Lipschitz, Rudolf (1876). "Sur la possibilité d'intégrer complètement un système donné d'équations différentielles". *Bulletin des sciences mathématiques et astronomiques* 10, pp. 149–159.

Pierpont, James (1905). *Lectures on the Theory of Functions of Real Variables*. Volume I. Boston.

Riemann, Bernhard (1851). "Grundlagen für eine allgemeine Theorie der Functionen einer veränderlichen complexen Grösse". PhD thesis. Georg-August-Universität Göttingen.

Riemann, Bernhard (2004). *Collected Papers*. Translated from the 1892 edition by Roger Baker, Charles Christenson and Henry Orde. Heber City.

Stolz, Otto (1885). *Vorlesungen über allgemeine Arithmetik*. Erster Theil: Allgemeines und Arithmetik der reellen Zahlen. Leipzig.

dot product

See *scalar product*.

dual (geometry)

The principle of duality was first stated in general form by Gergonne in his paper "Philosophical considerations on the elements of the science of the extended", in which he explained:[1]

> a very striking feature of this part of geometry that does not depend at all on metrical relationships between parts of the figures, is that, except for some theorems which are symmetrical of themselves, [...] all the theorems there are double, that is to say, in plane geometry to each theorem always corresponds necessarily another one which can be deduced from it by merely exchanging the two words *points* and *lines* with one another, whereas in space geometry *points* and *planes* are the words that have to be exchanged with one another to pass from one theorem to its correlate (Gergonne 1825-26, p. 210).[2]

Gergonne (ibid.) called this kind of correspondence between theorems *duality* (French *dualité*) and also used a two-column format to represent this correspondence. He began to use the term *principe de dualité* shortly afterwards, for example in Gergonne (1827-28, p. 152).

The terminology is soon found in various publications; for example, Steiner (1832, p. vii) used German *Prinzip der Dualität* and Booth (1840, p. 1) English *principle of duality*. Plücker (1831, p. v), on the other hand, preferred to use the "free translation" (as he called it) *principle of reciprocity* (*Princip der Reciprocität*).

See also *pole*, *polar*, and *reciprocal polar*.

Note. For accounts of the history of duality, see Chemla (1989), Specker (1958), Etwein, Voelke, and Volkert (2019) as well the various contributions in Krömer and Haffner (2024).

References

Booth, James (1840). *On the Application of a New Analytic Method to the Theory of Curves and Curved Surfaces*. Dublin.

---

[1] There was a fierce controversy between Gergonne and Poncelet on the question of priority; for discussion, see Pedoe (1975) and Lorenat (2015). [2] "un caractère extrêmement frappant de cette partie de la géométrie qui ne dépend aucunement des relations métriques entre les parties des figures; c'est qu'à l'exception de quelques théorèmes symétriques d'eux-mêmes [...] tous les théorèmes y sont doubles; c'est-à-dire que, dans la géométrie plane, à chacune théorème il en répond toujours nécessairement un autre qui s'en déduit en y échangeant simplement entre eux les deux mots *points* et *droites*; tandis que, dans la géométrie de l'espace, ce sont les mots *points* et *plans* qu'il faut échanger entre eux pour passer d'un théorème à son corrélatif". Translation from Chemla (1989, p. 332).

Chemla, Karine (1989). "The Background to Gergonne's Treatment of Duality: Spherical Trigonometry in the late 18th Century". In: *The History of Modern Mathematics*. Volume I: Ideas and Their Reception. Ed. by David E. Rowe and John McCleary. Boston, pp. 331–359.
Etwein, Frank, Jean Daniel Voelke, and Klaus Volkert (2019). *Dualität als Archetypus mathematischen Denkens*. Klassische Geometrie und Polyedertheorie. Göttingen.
Gergonne, Joseph Diez (1825-26). "Considérations philosophiques sur les élémens de la science de l'étendue". *Annales de mathématiques pures et appliquées* 16, pp. 209–231.
Gergonne, Joseph Diez (1827-28). "Rectification de quelques théorèmes énoncés dans les Annales". *Annales de mathématiques pures et appliquées* 18, pp. 149–154.
Krömer, Ralf and Emmylou Haffner, eds. (2024). *Duality in 19th and 20th Century Mathematical Thinking*. Basel.
Lorenat, Jemma (2015). "Polemics in Public: Poncelet, Gergonne, Plücker, and the Duality Controversy". *Science in Context* 28, pp. 545–585.
Pedoe, Dan (1975). "Notes on the History of Geometrical Ideas II. The Principle of Duality". *Mathematics Magazine* 48, pp. 274–277.
Plücker, Julius (1831). *Analytisch-geometrische Entwicklungen*. Zweiter Band. Essen.
Specker, Ernst (1958). "Dualität". *Dialectica* 12, pp. 451–465.
Steiner, Jacob (1832). *Systematische Entwickelung der Abhängigkeit geometrischer Gestalten von einander*. Erster Theil. Berlin.

dual basis

In linear algebra, the term is found in the 1940s.[1] It was used by Halmos (1942, p. 22): "If V is an n-dimensional vector space and $X = (x_1, \dots, x_n)$ is a basis in V, then there is a uniquely determined basis X' in V', $X' = (y_1, \dots, y_n)$, called the *dual basis* of X, with the property that $[x_i, y_i] = \delta_{ij}$".[2] French *base duale* was then used by Bourbaki (1947, p. 46), German *duale Basis* by Pickert (1953, p. 16), who referred to Bourbaki.

References

Bourbaki, Nicolas (1947). *Algèbre*. Chapitre II. Algèbre linéaire. Paris.
Halmos, Paul R. (1942). *Finite Dimensional Vector Spaces*. Princeton.
Hopf, Heinz (1930). "Zur Algebra der Abbildungen von Mannigfaltigkeiten". *Journal für die reine und angewandte Mathematik* 163, pp. 71–88.
Pickert, Günter (1953). "Lineare Algebra". *Enzyklopädie der mathematischen Wissenschaften mit Einschluss ihrer Anwendungen*. Zweite, völlig neubearbeitete Auflage I 1, 6, pp. 1–43.
Steenrod, Norman (1936). "Universal Homology Groups". *American Journal of Mathematics* 58, pp. 661–701.

dual graph

The term *dual graph* was coined by Whitney (1931, p. 127):[3]

> If H_1 is a subgraph of G, that subgraph H_2 of G containing those arcs not in H_1 is called the complement of H_1 in G. Let R, R', r, r', etc., stand for the ranks of G, G', H, H', etc., respectively, with similar definitions for nullity. Suppose there is a 1-1 correspondence between the arcs of G and G' such that if H is any subgraph of G and H' is the complement of the corresponding subgraph of G, then $r' = R' - n$. We say then G' is a *dual* of G.

Reidemeister (1932, p. 136) used the German term *dualer Streckenkomplex*, where *Streckenkomplex*, lit. 'line segment complex', was used for graphs; *dual graph* was also used by König (1936, p. 128), who referred to Whitney.

---

[1] Before the term appeared in linear algebra, it was also used in different meanings in topology, for example by Hopf (1930, p. 75) and Steenrod (1936, p. 680). [2] Halmos used *conjugate space* rather than *dual space*. [3] See also Whitney (1932, p. 351).

Note. For remarks on the history of the concept, see Soifer (2009, p. 176); for Whitney's approach to duality, see also Biggs, Lloyd, and Wilson (1998, pp. 148–150).

References

Biggs, Norman L., E. Keith Lloyd, and Robin J. Wilson (1998). *Graph Theory, 1736-1936*. Oxford.
König, Dénes (1936). *Theorie der endlichen und unendlichen Graphen*. Leipzig.
Reidemeister, Kurt (1932). *Einführung in die kombinatorische Topologie*. Braunschweig.
Soifer, Alexander (2009). *The Mathematical Coloring Book*. Mathematics of Coloring and the Colorful Life of Its Creators. New York.
Whitney, Hassler (1931). "Non-Separable and Planar Graphs". *Proceedings of the National Academy of Sciences* 17, pp. 125–127.
Whitney, Hassler (1932). "Non-separable and planar graphs". *Transactions of the American Mathematical Society* 34, pp. 339–363.

dual polyhedron

The duality of the cube and the octahedron and of the dodecahedron and the icosahedron appear to have been first alluded to by Maurolycus (1575, p. 103), who used the word *correlativa* in this respect, without defining it.[1] Kepler (1619, book IV, ch. 1, p. 181) expressed the same duality in rather metaphorical terms:

> there are two notable marriages, so to speak, of these figures, by combination from the two classes: the males, the cube and the dodecahedron from the primaries, the females, the octahedron and the icosahedron from the secondaries. In addition to these there is one which is, so to speak, celibate or hermaphrodite, the tetrahedron, because it is inscribed in itself, just as the feminine ones are inscribed in, and so to speak subject to, the males, and have the female tokens of their sex opposite to the masculine ones, or in other words the angles to the plane faces.[2]

An explicit definition of dual polyhedra was given by Meister (1785, p. 39), who called them *reciprocal solids* (*corpora reciproca*) and defined them by the condition that one of the polyhedra has as many solid angles (vertices) as the other has sides (faces), and vice versa.[3]

In an anonymously published article, the French term *polyèdres conjugués* is found in this sense ([anon.] 1818-19, p. 324). At a time when the term *dual* was already

[1] The passage by Maurolycus was pointed out by Müller (1860, pp. 1f.) and by Günther (1876, p. 52). [2] "Sunt autem notabilia duo veluti conjugia harum figurarum, ex diversis combinata classibus: Mares, Cubus et Dodecaëdron ex primarijs; foeminae, Octoëdron et Icosiëdron ex secundarijs; quibus accedit una veluti coelebs aut Androgynos, Tetraëdron; quia sibi ipsi inscribitur, ut illae foemellae maribus inscribuntur et veluti subjiciuntur, et signa sexus foeminina masculinis opposita habent, angulos scilicet planiciebus". Translation from Kepler (1997, pp. 396f.), where also the following comment by the translators is made in a footnote: "These sexual properties of the polyhedra are introduced here by Kepler for the first time. Evidently, the vertices are the masculine tokens of sex and the faces the feminine ones. It may be noted that the male polyhedra have more vertices than faces. The female polyhedra, on the other hand, have more faces than vertices, while the hermaphrodite has the same number of each. In the case of each marriage, the male member has the same number of vertices as the female member has faces, so that, when the female figure is inscribed in the male, the male and female tokens face each other. Also, it may be noted that, in the case of each related pair of polyhedra, the ratio of the radii of the circumscribed and inscribed spheres is the same for both". [3] "Corpora reciproca voco, quorum utrumcunque tot habet angulos solidos, quot alterum latera; hinc et tot latera, quot alterum angulos solidos".

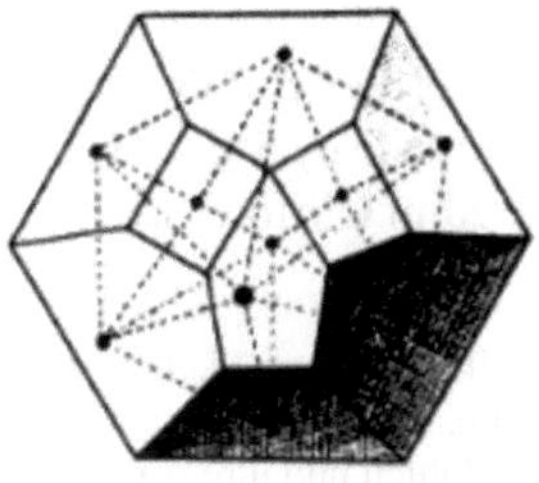

Figure 42: Dual polyhedra. From Brückner (1894, Taf. I).

established in mathematics, Möbius (1866, p. 59) explained that such polyhedra "are called *reciprocal* to each other, or also *being dually related to each other*".[4]

Note. Some historical remarks on the terminology can be found in Günther (1876, p. 52) and Brückner (1900, p. 74). For a detailed account of the history of the concept, see Etwein, Voelke, and Volkert (2019, esp. chs. 1.1, 4).

References

[anon.] (1818-19). "Recherches sur les polyèdres, renfermant en particulier un commencement de solution du problème proposé à la page 256 du VII.e volume des Annales". *Annales de mathématiques pures et appliquées* 9, pp. 321–344.

Brückner, Max (1894). "Die Elemente der vierdimensionalen Geometrie mit besonderer Berücksichtigung der Polytope". *Jahresbericht des Vereins für Naturkunde zu Zwickau in Sachsen* (1893), pp. 1–61.

Brückner, Max (1900). *Vielecke und Vielflache.* Theorie und Geschichte. Leipzig.

Etwein, Frank, Jean Daniel Voelke, and Klaus Volkert (2019). *Dualität als Archetypus mathematischen Denkens.* Klassische Geometrie und Polyedertheorie. Göttingen.

Günther, Siegmund (1876). *Vermischte Untersuchungen zur Geschichte der mathematischen Wissenschaften.* Leipzig.

Kepler, Johannes (1619). *Harmonices mundi libri V.* Linz.

Kepler, Johannes (1997). *The Harmony of the World.* Translated into English with an Introduction and Notes by E.J. Aiton, A.M. Duncan, J.V. Field. Philadelphia, PA.

Maurolycus, Franciscus (1575). *Opuscula mathematica.* Venezia.

Meister, Albrecht Ludwig Friedrich (1785). "Commentatio de solidis geometricis pro cognoscenda eorum indole in certos ordines et versus disponendis". *Commentationes Societatis Regiae Scientiarum Gottingensis* 7, pp. 3–74.

Möbius, August Ferdinand (1866). "Ueber die Bestimmung der Inhalte eines Polyeders". *Berichte über die Verhandlungen der Königlich Sächsischen Gesellschaft der Wissenschaften zu Leipzig (Mathematisch-Physische Klasse)* 17, pp. 31–68.

Müller, Johann Heinrich Traugott (1860). "Zur Geschichte des Dualismus in der Geometrie". *Archiv der Mathematik und Physik* 34, pp. 1–6.

dual numbers

Dual numbers are of the form $u + v\epsilon$, with u and v complex numbers and $\epsilon^2 = 0$. In this sense, the term *dual* was introduced by Study (1903, p. 196); he called such numbers *duale Größen* (*dual magnitudes*), while *duale Zahlen* was used by Grünwald

[4] "[...] nennt man einander *reciprok*, oder auch *zu einander in dualer Beziehung stehend*".

(1906, p. 81), who referred to Study for the term. English and French translations soon appeared in Miller and Swift (1906, p. 238) and Study and Cartan (1908, p. 455).

Note. Historical remarks were made by Study (1903, pp. 207f.) himself; he pointed out that the concept first appeared in Clifford (1873).

References

Clifford, William Kingdon (1873). "Preliminary Sketch of Biquaternions". *Proceedings of the London Mathematical Society* 4, pp. 381–395.
Grünwald, Josef (1906). "Über duale Zahlen und ihre Anwendung in der Geometrie". *Monatshefte für Mathematik* 17, pp. 81–136.
Miller, E.A. and Elijah Swift (1906). "The Meran meeting of the Deutsche Mathematiker-Vereinigung". *Bulletin of the American Mathematical Society* 12, pp. 237–240.
Study, Eduard (1903). *Geometrie der Dynamen*. Leipzig.
Study, Eduard and Élie Cartan (1908). "Nombres complexes". *Encyclopédie des sciences mathématiques pures et appliquées* I.5, pp. 329–468.

dual space

The German term *dualer Raum* appears to be due to Weyl (1928, pp. 12f.), where it is defined as the vector space consisting of the linear forms of a given vector space. The English term *dual space* is used in a translation of this book (Weyl 1931, p. 12); the term was also adopted by van der Waerden (1935, p. 5), Weil (1936-37, p. 7), and Julia (1936, pp. 116f.); in the latter text, *espace dualistique* is given as an alternative. The terminology features prominently in Dieudonné (1942, p. 107).

Other suggestions include *adjoint space* (Hildebrand 1931, p. 199)[1], and *conjugate* or *transposed space* (*konjugierter* oder *transponierter Raum*) (Schauder 1930, p. 184), which is found as *espace conjugué* in Banach (1932, p. 188).

Earlier, Hahn (1927, pp. 218f.) used German *polarer Raum* 'polar space'. For the word *polar*, Hahn refers to Helly (1921, p. 62); Helly there defined a concept of duality. He did not use the word *polar space*, but defined a norm which he called *polare Abstandsfunktion* 'polar distance function'. In this context, *polarer Raum* already appears in Hahn (1922, p. 5).[2]

Note. A detailed account of the history of the concept is given by Waldvogel (2024); see also Heuser (2001) and Bernkopf (1966-67, pp. 78–90).

References

Banach, Stefan (1932). *Théorie des opérations linéaires*. Warszawa.
Bernkopf, Michael (1966-67). "The Development of Function Spaces with Particular Reference to Their Origins in Integral Equation Theory". *Archive for History of Exact Sciences* 3, pp. 1–96.
Dieudonné, Jean (1942). "La dualité dans les espaces vectoriels topologiques". *Annales scientifiques de l'École Normale Supérieure* 59, pp. 107–139.
Hahn, Hans (1922). "Über Folgen linearer Operationen". *Monatshefte für Mathematik und Physik* 32, pp. 3–88.
Hahn, Hans (1927). "Über lineare Gleichungssysteme in linearen Räumen". *Journal für die reine und angewandte Mathematik* 157, pp. 214–229.
Helly, Eduard (1921). "Über Systeme linearer Gleichungen mit unendlich vielen Unbekannten". *Monatshefte für Mathematik und Physik* 31, pp. 60–91.
Heuser, Harro (2001). "Hahns Weg zum Satz von Hahn-Banach". *Internationale Mathematische Nachrichten* 183, pp. 1–20.

---

[1] Hildebrandt refers to Hahn (1927, pp. 218f.); see the remarks below. [2] For discussion, see Narici and Beckenstein (1997, pp. 198–200).

Hildebrand, T.H. (1931). "Linear functional transformations in general spaces". *Bulletin of the American Mathematical Society* 37, pp. 185–212.
Julia, Gaston (1936). *Introduction mathématique aux théories quantiques*. Leçons rédigées par J. Dufresnoy. Première partie. Paris.
Narici, Lawrence and Edward Beckenstein (1997). "The Hahn-Banach theorem: the life and times". *Topology and its Application* 77, pp. 193–211.
Schauder, Juliusz (1930). "Über lineare, vollstetige Funktionaloperationen". *Studia Mathematica* 2, pp. 183–196.
van der Waerden, Bartel Leendert (1935). *Gruppen von linearen Transformationen*. Berlin.
Waldvogel, Alessa (2024). "The development of dual spaces in functional analysis". In: *Duality in 19th and 20th Century Mathematical Thinking*. Ed. by Ralf Krömer and Emmylou Haffner. Basel, pp. 473–556.
Weil, André (1936-37). "Formes différentielles extérieures". *Séminaire de Mathématiques dit de Julia* 4, exposé 1, pp. 1–15.
Weyl, Hermann (1928). *Gruppentheorie und Quantenmechanik*. Leipzig.
Weyl, Hermann (1931). *The Theory of Groups and Quantum Mechanics*. Translated from the second (revised) German edition by H.P. Robertson. London.

dynamical system

The term *dynamical system* (*système dynamique*) was formally introduced by Birkhoff (1912, p. 306):

> A dynamical system in a very broad sense may be considered as being defined by any system of differential equations of the first order:
>
> $$\frac{dx_1}{X_1} = \frac{dx_2}{X_2} = \cdots = \frac{dx_n}{X_n} = dt$$
>
> where $X_1, \ldots, X_n$ are given functions, real and uniform depending on $x_1, \ldots, x_n$, analytical with respect to these variables, and where t is the independent variable. The variables are the *coordinates* of the motion and t indicates the *time*.[1]

These equations were earlier considered by Poincaré (1886, p. 168). The terminology spread via various works by Birkhoff, in particular via his book *Dynamical*

[1] "Un système dynamique, dans une très large acception, peut être considéré comme défini par un système quelconque d'équations différentielles ordinaires du premier ordre: $\frac{dx_1}{X_1} = \frac{dx_2}{X_2} = \cdots = \frac{dx_n}{X_n} = dt$, où $X_1, \ldots, X_n$ sont des fonctions données, réelles et uniformes de $x_1, \ldots, x_n$, analytiques par rapport à ces variables, et où t est la variable indépendante. Les variables sont les *coordonnées* du movement, et t indique le *temps*". Translation from Ginoux (2009, p. 6), slightly adapted.

Systems (Birkhoff 1927).[2] It was also soon applied to discrete-time systems, for example by Nemytskii (1949, pp. 96f.).[3]

Note. For the history of the concept, see Aubin and Dalmedico (2002) and Roque (2011), for example. Various generalizations of the concept are discussed by Bhatia and Szegö (1970, pp. 10f.).

References

Anosov, D.V. (1963). "Ergodic properties of geodesic flows on closed Riemannian manifolds of negative curvature [in Russian]". *Doklady Akademii Nauk SSSR* 151, pp. 1250–1252.

Anosov, D.V. et al. (1988). "Smooth Dynamical Systems. Translated from the Russian by E.R. Dawson and D. O'Shea". In: *Dynamical Systems I*. Ordinary Differential Equations and Smooth Dynamical Systems. Ed. by D.V. Anosov and V.I. Arnold. Berlin et al., pp. 149–233.

Aubin, David (2005). "George David Birkhoff, *Dynamical Systems* (1927)". In: *Landmark Writings in Western Mathematics 1640-1940*. Ed. by Ivor Grattan-Guinness. Amsterdam, pp. 412–881.

Aubin, David and Amy Dahan Dalmedico (2002). "Writing the History of Dynamical Systems and Chaos: *Longue Durée* and Revolution, Disciplines and Cultures". *Historia Mathematica* 29, pp. 273–339.

Bhatia, Nam Parshad and Giorgio P. Szegö (1970). *Stability Theory of Dynamical Systems*. New York et al.

Birkhoff, George D. (1912). "Quelques théorèmes sur le mouvement des systèmes dynamiques". *Bulletin de la Société Mathématique de France* 40, pp. 305–323.

Birkhoff, George D. (1927). *Dynamical Systems*. Providence, R.I.

Ginoux, Jean-Marc (2009). *Differential Geometry Applied to Dynamical Systems*. New Jersey.

Gottschalk, Walter Helbig and Gustav Arnold Hedlund (1949). "The Dynamics of Transformation Groups". *Transactions of the American Mathematical Society* 65, pp. 348–359.

Hedlund, Gustav Arnold (1939). "The dynamics of geodesic flows". *Bulletin of the American Mathematical Society* 45, pp. 241–260.

Nemytskii, Viktor (1949). "Topological problems of the theory of dynamical systems [in Russian]". *Uspekhi Matematicheskikh Nauk* 4, pp. 91–153.

Poincaré, Henri (1886). "Sur les courbes définies par les équations différentielles (quatrième partie)". *Journal de mathématiques pures et appliquées* 2, pp. 151–218.

Roque, Tatjana (2011). "The Qualitative Analysis of Differential Equations and the Development of Dynamical Systems Theory". In: *Brazilian Studies in the Philosophy and History of Science*. An Account of Recent Works. Ed. by Décio Krause and Antonio Videira. Dordrecht, pp. 313–324.

---

[2] For an account of this book, see Aubin (2005). [3] For this usage, Nemytskii (1949, p. 102) referred to an unpublished manuscript by Pontryagin and Schnirelman. – For discrete-time systems, Anosov (1963, p. 1250) suggested the term *cascade*. He explained that this term is based on the opposition to *flow*, a term "often used" for dynamical system with continuous time. The link is explained by Gottschalk and Hedlund (1949, p. 348): "In the study of classical dynamical systems defined by systems of ordinary differential equations, it has proved useful to regard such a system as a one-parameter transformation group acting on a topological space". In this sense, *flow* was introduced by Hedlund (1939, p. 243). Anosov et al. (1988, p. 154) explain the analogy as follows: "For illustrative purposes, an analogy is often drawn with the steady flow of a liquid or gas, where a similar family of transformations arises. After time t a fluid particle moves from a point x to the point $g^t x$. This explains the use of the word 'flow'. But it should be noted that the 'hydrodynamical' analogy is quite superficial; the imaginary 'phase fluid', 'flowing' in the phase space, differs from a real continuous medium in that there is no interaction between neighbouring particles".

E

edge

For an edge of a polyhedron, Euler introduced Latin *acies* 'sharp edge or point'. The term is found in a letter to Goldbach from 1750, in which Euler explains that he uses *acies* "for lack of an accepted word" (see Euler 2015, p. 479).[1] Euler then introduces the term in print in a paper published in 1758, but read several years earlier (Euler 1758, § 6).[2]

The French term *arête*, which corresponds in meaning to *acies*, was introduced by Legendre in 1794. Although Legendre does not say so, he seems to have chosen his term as a translation of Euler's term and defines: "The common intersection of two adjacent faces of a polyhedron is called *side* [*côté*] or *edge* [*arête*]" (Legendre 1794, p. 165).[3] In the notes to the book, Legendre (1794, p. 284) also mentions French *faîte* 'ridge (e. g., of a roof)' as a translation of *acies*, but this never caught on.

The term seems to have spread via Legendre's book, which appeared in numerous editions and was translated various times. In Legendre (1819, p. 123), *edge* is used to translate *arête*, a term picked up by De Morgan (1838, p. 323). The German term *Kante*, lit. 'edge', is, according to Tropfke (1924, p. 13), only sporadically found before Klügel (1808, p. 1) used it in his dictionary; an early usage is Lorenz (1793, p. 317).

While Euler was the first to use *acies* in the sense of 'edge', this does not mean that before him, there were no terms to refer to edges. There were words literally meaning 'side' that were more commonly used to refer to edges, but there were also some sporadically used words that are remarkably similar to Euler's term.

The use of a word meaning 'side' for the edges of a solid is already found in Greek. Pappus (*Collection*, book V, 19) makes use of Greek *πλευρά* (otherwise used for the sides of plane figures) in his list of semiregular solids, for example (see Pappus of Alexandria 1876, p. 354). This model is followed in other languages. For

---

[1] "aus Mangel eines recipirten Wortes". Translation from Biggs, Lloyd, and Wilson (1998, p. 76).

[2] In the *Göttingische Gelehrte Anzeigen* 1, 1761, p. 418, the term is translated into German as *Seitenlinien* 'side lines' or *Schärfen*, from *scharf* 'sharp'.

[3] "L'intersection commune de deux faces adjacentes d'un polyedre s'appelle *côté* ou *arête* du polyedre". Translation from Hon and Goldstein (2005, p. 117).

H. Becker, *A Historical Dictionary of Mathematical Terms*, Science Networks. Historical Studies 66, https://doi.org/10.1007/978-3-032-02873-0_5

example, Fibonacci used Latin *latus*, lit. 'side', for the edges of a pyramid (and also for the sides of a triangle); see Hughes (2008, p. 293). Similarly, Pacioli (1509, part I, ch. xxv) uses Italian *lato*. Dürer (1525, fol. M iii^v) calls the edges *scharpfe seyten* 'sharp sides'. In a translation of Ramus (1569), English *side* is used to translate the original *latus*: "The sides of a tetraedron are sixe" (Ramus 1590, fol. 20^r), but in Ramus (1636, p. 254), it says that "[t]he edges of a tetraedrum are sixe". There are also analogous attestations for French *arête*: Nicéron (1638, p. 39) explains that the icosahedron has "thirty sides or edges".[1]

Note. Euler's term has been commented upon by Federico (1982, pp. 66f.) and Lakatos (1976, p. 6). See also Tropfke (1924, p. 13), who briefly discusses the terminology, and points out Dürer's use cited above.

References

Biggs, Norman L., E. Keith Lloyd, and Robin J. Wilson (1998). *Graph Theory, 1736-1936*. Oxford.

Dürer, Albrecht (1525). *Underweysung der Messung, mit dem Zirckel und Richtscheyt*. Nürnberg.

Euler, Leonhard (1758). "Elementa doctrinae solidorum". *Novi Commentarii academiae scientiarum Petropolitanae* 4, pp. 109–140.

Euler, Leonhard (2015). *Correspondence of Leonhard Euler with Christian Goldbach*. Ed. by Franz Lemmermeyer and Martin Mattmüller. Basel.

Federico, Pasquale Joseph (1982). *Descartes on Polyhedra*. A Study of the De Solidorum Elementis. New York and Berlin.

Hon, Giora and Bernard R. Goldstein (Jan. 2005). "Legendre's Revolution (1794): The Definition of Symmetry in Solid Geometry". *Archive for History of Exact Sciences* 59, p. 107.

Hughes, Barnabas (2008). *Fibonacci's* De Practica Geometrie. Berlin.

Klügel, Georg Simon (1808). *Mathematisches Wörterbuch*. Erste Abtheilung: Die reine Mathematik. Dritter Theil von K bis P. Leipzig.

Lakatos, Imre (1976). *Proofs and Refutations*. The Logic of Mathematical Discovery. Cambridge.

Legendre, Adrien-Marie (1794). *Éléments de géométrie*. Avec des notes. Paris.

Legendre, Adrien-Marie (1819). *Elements of Geometry*. Translated from the French. Cambridge, N.E.

Lorenz, Johann Friedrich (1793). *Die Elemente der Mathematik*. Erster Theil. Die reine Mathematik. Zweyte gänzlich umgearbeitete Auflage. Leipzig.

De Morgan, Augustus (1838). "On the Relation between the Number of Faces, Edges, and Corners in a Solid Polyhedron". *The London and Edinburgh Philosophical Magazine and Journal of Science* 12, pp. 323–324.

Nicéron, Jean-François (1638). *La perspective curieuse*. Paris.

Pacioli, Luca (1509). *Divina Proportione*. Venezia.

Pappus of Alexandria (1876). *Pappi Alexandrini Collectionis*. Volumen I. Ed. by Friedrich Hultsch. Berlin.

Ramus, Petrus (1569). *Arithmeticae libri duo. Geometriae septem et viginti*. Basel.

Ramus, Petrus (1590). *Elementes of geometrie*. Faithfully translated by Tho. Hood. London.

Ramus, Petrus (1636). *Via regia ad geometriam. The Vvay to Geometry*. Written in Latine by Peter Ramus, and now Translated and much enlarged by the Learned Mr. William Bedwell. London.

Tropfke, Johannes (1924). *Geschichte der Elementar-Mathematik*. Siebenter Band. Stereometrie. Verzeichnisse. Berlin and Leipzig.

edge of regression

The edge of regression of a developable surface was introduced in a paper by Monge, which was presented in 1771 but not published until 1785. Monge (1785, p. 524) used the French term *arrête de rebroussement*; *arrête*, later spelt *arête*, here means

[1] "trente costez ou arestes".

'edge' and *rebroussement* literally means 'going back or in the opposite direction' (see *cusp*). Just as a cusp separates a curve into two parts, an edge of regression separates a developable surface into two sheets (French *nappes*). This is the motivation for this terminology, as has been pointed out, for example, by Schreiber (1828, p. 59) and Young (1833, p. 215). Because of the delay between the presention of his paper and its publication, Monge's term is already found in print in Monge (1780, p. 403), for example.

The English term *cuspidal edge* was suggested as a translation by Gregory and Walton (1845, p. 159) because, as they say, it "expresses better the meaning of the French words". Some authors, e. g. Hamilton (1853, p. 580), shortened this to *cusp-edge*. An unsuccessful suggestion was made by De Morgan (1836, p. 403): *connecting curve of characteristics*. The actual term *edge of regression* was suggested by Hymers (1830, p. 115).

The German term *Rückkehrkante*, which is a literal translation of the French term, came up in the 1820s; it is found in Brandes (1824, p. 208) and Schreiber (1828, p. 59).

References

Brandes, Heinrich Wilhelm (1824). *Lehrbuch der höhern Geometrie, in analytischer Darstellung.* Zweiter Theil. Leipzig.

Gregory, Duncan Farquharson and William Walton (1845). *A Treatise on the Application of Analysis to Solid Geometry.* Cambridge.

Hamilton, William Rowan (1853). *Lectures on Quaternions*. Dublin.

Hymers, John (1830). *A Treatise on Analytical Geometry of Three Dimensions*. Cambridge.

Monge, Gaspard (1780). "Mémoire sur les Propriétés de plusieurs genres de Surfaces courbes, particulièrement sur celles des Surfaces développables, avec une Application à la Théorie des Ombres et des Pénombres". *Mémoires de mathématique et de physique, presentés à l'Académie royale des sciences, par divers sçavans* 9, pp. 382–440.

Monge, Gaspard (1785). "Mémoire sur les développées, les rayons de courbure, et les différens genres d'inflexions des courbes à double courbure". *Mémoires de mathématique et de physique, presentés à l'Académie royale des sciences, par divers sçavans* 10, pp. 511–550.

De Morgan, Augustus (1836). *The Differential and Integral Calculus*. London.

Schreiber, Guido (1828). *Lehrbuch der darstellenden Geometrie nach Monge's Géométrie descriptive*. Erste Lieferung. Karlsruhe and Freiburg.

Young, John Radford (1833). *The Elements of the Differential Calculus*. Revised and corrected, by Michael O'Shanessey. New York.

eigenvalue*, *eigenfunction*, *eigenvector*, and *eigenspace

These terms were originally coined in German as *Eigenwert*, *Eigenfunktion*, *Eigenvektor*, and *Eigenraum*; the English terms are a rare combination of a German and an English word. German *eigen* means 'own, proper, inherent'. While in the meaning 'own' the word is quite frequent, it is otherwise mainly found in compounds such as *Eigenschaft* 'property (quality)' or *Eigenart* 'peculiarity'.[1] In both of these examples the meaning 'characteristic' is implied.

The hybrid character of the English terms provoked a good deal of controversy, as illustrated by the following statement:

> In publications dealing with wave mechanics there is a growing practice of rendering the words *Eigenfunktion* and *Eigenwert* into English by *eigen-function* and *eigen-value* re-

[1] *-schaft* is a suffix to make nouns, *Art* means 'kind'.

> spectively. The English expressions *proper function* and *proper value* are equally concise, correspond exactly in etymology to the originals, and have the sanction of Prof. Schrödinger himself in the English lectures which he gave at the Royal Institution. Is it not desirable, therefore, that we should discontinue the use of the mixed forms, which is a barbarism without sense or convenience, and equally repellent to German and to English ears? (Hinshelwood 1930, p. 604).

Such criticism is what Stewart (2001, p. 23) refers to when he points out that "[t]he day when purists and pedants could legitimately object to 'eigenvalue' as a hybrid of German and English has long since passed".

The mathematical term *Eigenwert* (*Wert* 'value') of a form was coined by Hilbert (1904, p. 64): "The zeros of $\delta(\lambda)$ will be called *eigenvalues belonging to the kernel* $K(s,t)$".[1] Hilbert also used some other *Eigen*-terms such as *Eigenfunktion* (Hilbert 1904, p. 66). Physical terms like *Eigenschwingung* (*Schwingung* meaning 'oscillation' or 'vibration'), which was in common use in the last quarter of the 19th century[2] for systems on which no external forces act, or *Eigenton* 'proper tone', used before by Helmholtz (1863, p. 69), may have been a model upon which Hilbert created the term.[3] Hilbert himself never explained why he chose the name.

The struggle with the terminology can be seen in early English attempts to translate the terms. For example, Bateman (1906, p. 235) used *fundamental function* for *eigenfunction*; then he switched to *Eigenfunction* (Bateman 1907, p. 544), and then switched again to *characteristic function* (Bateman 1908, p. 448). Similarly, Bôcher (1909, p. 57) called eigenvalues *characteristic numbers*, but also *characteristic parameter values* (Bôcher 1913, p. 170). On the other hand, Press (1928, p. 45) suggested *parametral value* or simply *parametral* for 'eigenvalue'.[4]

The term *proper value* appears in French first; an early attestation for *valeur propre* (and *fonction propre*) is Tannery (1907, p. 85). English *proper value* and *proper function* come up in the 1920s; they were used in the translation of Schrödinger's papers on wave mechanics (Schrödinger 1928, publisher's note).

There are more alternative terms; see *latent* and *characteristic* (linear algebra). Sometimes, also terms such as *secular equation* are found. It was used by H. Weber (1895, p. 276) for the characteristic equation: "we consider an equation that we will

---

[1] "die Nullstellen von $\delta(\lambda)$ mögen die zum Kern $K(s,t)$ gehörigen Eigenwerte heißen". Translation from Stewart (2011, p. 69). – Bernkopf (1966-67, p. 9) explains Hilbert's definition: "It should be noted that 'eigenvalue' will mean what today would be called reciprocal eigenvalue by most authors; *i. e.*, for HILBERT λ will turn out to be an eigenvalue for the operator K if there is a non-zero x satisfying $x - \lambda K(x) = 0$". The now usual definition is found in various texts from the 1920s, for example in Wintner (1929, p. 15) and von Neumann (1930, p. 50). [2] It was used, for example, by H. K. Müller (1884, p. 2), and by von Helmholtz in lectures in the 1880s (von Helmholtz 1898, p. 19). The term is already found in W. Weber (1827, p. 314) and appears to be a contraction of *eigenthümliche Schwingung*, used by Chladni (1802, p. 217). A link between eigenvalues, eigenfunctions, and "eigen-vibrations" is discussed by Hellinger and Toeplitz (1927, esp. pp. 1358-1361); see also Lamb (1907, pp. 220–227). – I wish to thank François Ziegler for drawing my attention to this term, for providing various references, and for pointing out some potential improvements to this entry. His comments on *Eigenschwingung* and also on *secular equation* led to substantial revisions of this entry. [3] To be precise, Helmholtz (1863, p. 69) used *eigener Ton* but started switching to *Eigenton* (p. 71). The word had already been used by J. Müller (1840, p. 438). [4] He repeated this suggestion in Press (1930, p. 883).

call, for the sake of brevity, the *secular equation*, because the investigation of the secular perturbations of the planets first led to it".[1]

Finally, the term *eigenvector* was introduced by Hahn (1911, p. 111). He considered a real symmetric matrix $K = (k_{pq})_{(p,q=1,2,\ldots,n)}$ (eigenvalue λ) and the equation $\sum_{q=1}^{n}(e_{pq} - \lambda k_{pq})\varphi_q = 0$ ($p = 1, 2, \ldots, n$) and remarked: "We interpret the numbers $(\varphi_1, \varphi_1, \ldots, \varphi_n)$ as a vector and call it an *eigenvector* of the matrix K belonging to this eigenvalue".[2] German *Eigenraum* (eigenspace) seems to be due to Weyl (1927, p. 6); in Weyl (1928, p. 20), this definition is found: "the vectors $\mathfrak{x}$ which satisfy the equation $A\mathfrak{x} = \alpha \cdot \mathfrak{x}$ constitute a linear subspace $\mathfrak{R}(\alpha)$, the *eigenspace* belonging to α".[3]

Note. For the early history of eigenvalues, see Hawkins (1975) and Tou (2018).

References

Bateman, Harry (1906). "A class of Integral Equations". *Transactions of the Cambridge Philosophical Society* 20, pp. 233–252.

Bateman, Harry (1907). "The inversion of a definite integral". *Mathematische Annalen* 63, pp. 525–548.

Bateman, Harry (1908). "On Essentially Positive Double Integrals and the Part which they play in the Theory of Integral Equations". *Report of the British Association for the Advancement of Science* (1907), pp. 447–449.

Bernkopf, Michael (1966-67). "The Development of Function Spaces with Particular Reference to Their Origins in Integral Equation Theory". *Archive for History of Exact Sciences* 3, pp. 1–96.

Bôcher, Maxime (1909). *Introduction to the Study of Integral Equations*. London.

Bôcher, Maxime (1913). "Boundary problems in one dimension". In: *Proceedings of the Fifth International Congress of Mathematicians*. Vol. I. Ed. by E.W. Hobson and A.E.H. Love. Cambridge, pp. 163–195.

Brechenmacher, Frédéric (2014). "Lagrange and the secular equation". *Lettera Matematica* 2, pp. 79–91.

Chladni, Ernst (1802). *Die Akustik*. Leipzig.

Hahn, Hans (1911). "Bericht über die Theorie der linearen Integralgleichungen". *Jahresbericht der Deutschen Mathematiker-Vereinigung* 20, pp. 69–111.

Hawkins, Thomas (1975). "Cauchy and the spectral theory of matrices". *Historia Mathematica* 2, pp. 1–29.

[1] "Als zweites Beispiel betrachten wir eine Gleichung, die wir der Kürze wegen die *Säcular-Gleichung* nennen wollen, weil die Untersuchung der säcularen Störungen der Planeten zuerst auf sie geführt hat". – Note that the term *secular equation* as used in astronomy is much older but does not refer to an equation at all. The term is explained by Laplace, who first remarks that "Halley was the first who perceived an acceleration of the mean motion of the moon" and that Richard Dunthorne and Tobias Mayer then reexamined this important point of lunar theory, trying to "represent it, by adding, to the mean longitudes of this satellite, a quantity proportional to the square of the number of centuries after the year 1700. This correction, which supposes the moon's motion to be accelerated in proportion to the times, is called the *secular equation*" ("[...] Cette correction, qui suppose que le mouvement de la Lune s'accélère en raison des temps, est ce que l'on nomme *équation séculaire*"). In this sense, the term, which is derived from Latin *saeculum* 'century', seems to go back to Kepler; see Steele (2012, p. 2) and Shook (2020, esp. pp. 7-9). Note that some authors claim that *secular equation* as used in the 18th century is the same as a characteristic equation; see Jammer (1966, p. 215) and Brechenmacher (2014, p. 79).

[2] "Die Zahlen $(\varphi_1, \varphi_1, \ldots, \varphi_n)$ deuten wir wieder als einen Vektor und nennen ihn einen zu diesem Eigenwert gehörigen Eigenvektor der Matrix K".

[3] "diejenigen Vektoren $\mathfrak{x}$, welche der Gleichung $A\mathfrak{x} = \alpha \cdot \mathfrak{x}$ genügen, bilden einen linearen Teilraum $\mathfrak{R}(\alpha)$, den zu α gehörigen *Eigenraum*". Translation based on Weyl (1931, p. 2), where *characteristic space* was used.

Hellinger, Ernst and Otto Toeplitz (1927). “Integralgleichungen und Gleichungen mit unendlichvielen Unbekannten”. *Encyklopädie der mathematischen Wissenschaften mit Einschluss ihrer Anwendungen* II C 13, pp. 1335–1616.
von Helmholtz, Hermann (1898). *Vorlesungen über theoretische Physik*. Band III. Die mathematischen Principien der Akustik. Ed. by Arthur König and Carl Runge. Leipzig.
Helmholtz, Hermann (1863). *Die Lehre von den Tonempfindungen als physiologische Grundlage für die Theorie der Musik*. Braunschweig.
Hilbert, David (1904). “Grundzüge einer allgemeinen Theorie der linearen Integralgleichungen. (Erste Mitteilung)”. *Nachrichten von der Königl. Gesellschaft der Wissenschaften zu Göttingen (mathematisch-physikalische Klasse)* (1904), pp. 49–91.
Hinshelwood, C.N. (1930). “English Equivalents of *Eigenfunktion* and *Eigenwert*”. *Nature* 126, p. 604.
Jammer, Max (1966). *The Conceptual Development of Quantum Mechanics*. New York et al.
Lamb, Horace (1907). “Schwingungen elastischer Körper, insbesondere Akustik”. *Encyklopädie der mathematischen Wissenschaften mit Einschluss ihrer Anwendungen* IV 26, pp. 215–310.
Müller, Heinrich Karl (1884). *Über Resonanzschwingungen gespannter Saiten*. Fulda.
Müller, Johannes (1840). *Handbuch der Physiologie des Menschen*. Zweiter Band. Koblenz.
von Neumann, John (1930). “Allgemeine Eigenwerttheorie Hermitescher Funktionaloperatoren”. *Mathematische Annalen* 102, pp. 49–131.
Press, A. (1928). “The Classical Reasonableness of the Quantum Theory and Simple Operative Solutions of Schrödinger’s equation”. *The London, Edinburgh and Dublin Philosophical Magazine and Journal of Science* 6, pp. 33–49.
Press, A. (1930). “English Equivalents of *Eigenfunktion* and *Eigenwert*”. *Nature* 126, p. 883.
Schrödinger, Erwin (1928). *Collected Papers on Wave Mechanics*. Translated from the second German edition. London and Glasgow.
Shook, John (2020). “The Meaning of ‘Secular’ as a Scientific Concept”. *Secularism and Nonreligion* 9, pp. 1–11.
Steele, John M. (2012). *Ancient Astronomical Observations and the Study of the Moon’s Motion (1691–1757)*. New York et al.
Stewart, G.W. (2001). *Matrix Algorithms*. Volume II: Eigensystems. Philadelphia.
Stewart, G.W. (2011). “Fredholm, Hilbert, Schmidt. Three Fundamental Papers on Integral Equations”. Translated with commentary by G.W. Stewart. URL: `http://www.umiacs.umd.edu/~stewart/FHS.pdf` (retrieved Nov. 3, 2020).
Tannery, Jules (1907). “Lebedeff (W.). – Die Theorie der Integralgleichungen in Anwendungen auf einige Reihenentwickelungen [review]”. *Bulletin des sciences mathématiques* 31, pp. 84–90.
Tou, Erik R. (2018). “Eigenvectors and Eigenvalues”. *Convergence* (February 2018). URL: `https://old.maa.org/press/periodicals/convergence/math-origins-eigenvectors-and-eigenvalues`.
Weber, Heinrich (1895). *Lehrbuch der Algebra*. Erster Band. Braunschweig.
Weber, Wilhelm (1827). “Auszug aus den die Theorie des Schalles und Klanges betreffenden Aufsätzen von Felix Savart, mit einigen Bemerkungen”. *Journal für Chemie und Physik* 51, pp. 291–329.
Weyl, Hermann (1927). “Quantenmechanik und Gruppentheorie”. *Zeitschrift für Physik* 46, pp. 1–46.
Weyl, Hermann (1928). *Gruppentheorie und Quantenmechanik*. Leipzig.
Weyl, Hermann (1931). *The Theory of Groups and Quantum Mechanics*. Translated from the second (revised) German edition by H.P. Robertson. London.
Wintner, Aurel (1929). *Spektraltheorie der unendlichen Matrizen*. Einführung in den analytischen Apparat der Quantenmechanik. Leipzig.

elastic

The Latin term *elasticus* dates from the 17th century only and seems to have been invented by Pecquet (1651, p. 54) as *vis elastica* ‘elastic force’. It was coined on the analogy of Greek ἐλαστικός 'that drives, propulsive, impulsive' and quickly entered mathematical terminology.[1]

Bernoulli (1691, p. 289) posed a special case of the elastica problem, giving the solution in encrypted form:

> If an elastic lamina shall be free from weight AB [...], with a uniform width everywhere and of a length, with the lower end A made firm somewhere and a weight B may be attached to the top, as much as it suffices for the lamina at that point to be curved, so that the line of the direction of the weight BC shall be perpendicular to the curved lamina at B, the curvature of the lamina will be of the following nature: Qrzumu bapt dxqopddbbp poyl fy bbqnfqbfp lty ge mutds udthh tuhs tmixy yxdksdhxp gqsrkfgudl bg ipqandtt tcpgkbp aqdbkzs. [The proportion between the applied line of the axis and the tangent is to the tangent itself thus as the square of the applied line to a certain constant area].[2]

As explained by Levien (2008, p. 4), "[t]he deciphered form of the anagram is hardly less cryptic than the original, but digging through his 1694 explanation, it is possible to extract the fundamental idea: at every point along the curve, the product of the radius of curvature and the distance from the line BC is a constant, i. e. the two quantities are inversely proportional". Bernoulli (1694, p. 263) then used the term *curva elastica* 'elastic curve'.

The term is soon found in other languages. Bernoulli (1706, pp. 177f.), for example, used French *courbe élastique*, and Wolff (1716, p. 576) German *elastische Linie*.

Note. On the history of the notion, see Levien (2008), Truesdell (1983, 1960), and Loria (1911, pp. 212–217).

References

Bernoulli, Jakob (1691). "Specimen alterum calculi differentialis in dimetienda Spirali Logarithmica Loxodromiis Nautarum". *Acta Eruditorum* (June 1691), pp. 282–290.

Bernoulli, Jakob (1694). "Curvatura laminae elasticae". *Acta Eruditorum* (June 1694), pp. 262–276.

Bernoulli, Jakob (1706). "Véritable hypothèse de la résistance des solides, avec la démonstration de la courbure des corps qui font ressort". *Mémoires de l'Académie royale des science* (1705), pp. 176–186.

Leibniz, Gottfried Wilhelm et al. (2014). "The curve formed by a heavy flexible cord due to its own weight". Transl. with notes by Ian Bruce. URL: `http://www.17centurymaths.com/contents/leibnizcontents.html` (retrieved Mar. 12, 2018).

Levien, Raph (2008). *The elastica: a mathematical history*. Tech. rep. EECS Department, University of California, Berkeley. URL: `http://www2.eecs.berkeley.edu/Pubs/TechRpts/2008/EECS-2008-103.html` (retrieved Sept. 11, 2008).

Loria, Gino (1911). *Spezielle algebraische und transzendente ebene Kurven*. Theorie und Geschichte. Autorisierte, nach dem italienischen Manuskript bearbeitete deutsche Ausgabe von Fritz Schütte. Zweite Auflage. Zweiter Band. Leipzig.

Pecquet, Jean (1651). *Experimenta nova anatomica*. Paris.

Truesdell, Charles (1960). *The Rational Mechanics of Flexible or Elastic Bodies, 1638-1788*. Zürich.

[1] On Pecquet's introduction of the term, see Webster (1965, esp. pp. 453f.). [2] "Si lamina elastica gravitatis expers AB (fig. 8), uniformis ubique crassitiei et latitudinis, inferiore extremitate A alicubi firmetur et superiori B pondus appendatur, quantum sufficit ad laminam eousque incurvandam, ut linea directionis ponderis BC curvatae laminae in B sit perpendicularis, erit curvatura laminae sequentis naturae: Qrzumu bapt dxqopddbbp poyl fy bbqnfqbfp lty ge mutds udthh tuhs tmixy yxdksdhxp gqsrkfgudl bg ipqandtt tcpgkbp aqdbkzs [Portio axis applicatam inter et tangentem est ad ipsam tangentem sicut quadratum applicatae ad constans quoddam.]". Translation from Leibniz et al. (2014, p. 12). The decoded sentence was published in Bernoulli (1694, p. 272).

Truesdell, Charles (1983). "The influence of elasticity on analysis: The classic heritage". *Bulletin of the American Mathematical Society* 9, pp. 293–310.
Webster, Charles (1965). "The Discovery of Boyle's Law, and the Concept of the Elasticity of Air in the Seventeenth Century". *Archive for History of Exact Sciences* 2, pp. 441–502.
Wolff, Christian (1716). *Mathematisches Lexicon*. Leipzig.

elation

As used by Emch (1895, p. 207), an *elation* is "a perspective collineation in which the center lies in the axis of collineation". Emch remarks that the term is due to Lie, who introduced it in his book on "continuous groups", in which he stated that the concept was occasionally called *degenerate perspective transformation* (*ausgeartete perspective Transformation*) and that he needed a shorter designation (Lie 1893, pp. 261f.).

As an alternative, Enriques (1904, p. 170) used *special homology* (*omologia speciale*); Poncelet generally called every perspective collination a *homologie* (see *homology*).

Note. Some historical comments on the terminology are found in Coxeter (1993, p. 57).

References

Coxeter, H.S.M. (1993). *The Real Projective Plane*. New York.
Emch, Arnold (1895). "Involutoric Transformation of the Straight Line". *Kansas University Quarterly* 4, pp. 111–116.
Enriques, Federigo (1904). *Lezioni di geometria proiettiva*. Seconda edizione aumentata. Bologna.
Lie, Sophus (1893). *Vorlesungen über continuierliche Gruppen*. Ed. by Georg Scheffers. Leipzig.

element

The word *element* goes back to Latin *elementum* 'one of the four substances (earth, air, fire, water) from which all matter was supposed to be made up', 'component or ingredient', also 'letter of the alphabet' (OLD).[1] The word translated Greek *στοιχεῖον*, which was also used to refer to the classical elements and which also appeared as the title of Euclid's *Elements*.[2]

In the 17th century, the word began to be used in the calculus. Leibniz (1710, p. 159) gave the following definition:

> Here dx signifies an element, that is, a (momentary) increment or decrement of the (continuously) increasing quantity x. It is also called a *difference*, namely that between two minimally (or inassignably) different proximal [values of] x, where one arises from the other that is (momentarily) increasing or decreasing [...] Furthermore, ddx is an element of an element, or *a difference of differences*, for the quantity dx itself is not always constant, but often in turn (continuously) increases or decreases. And similarly, one can proceed to $dddx$ or d^3x, and so on.[3]

---

[1] The general etymology of Latin *elementum* 'rudiment, first principle' is unclear. See, for example, Walde and Hofmann (1938, pp. 397f.) and Liberman (2007). A frequently referred to study is Diels (1899). [2] On the various meanings of *στοιχεῖον*, see Burkert (1959) and Crowley (2005). [3] "Hic dx significat elementum, id est incrementum vel decrementum (momentaneum) ipsius quantitatis x (continue) crescentis. Vocatur et *differentia*, nempe inter duas proximas x elementariter (seu inassignabiliter) differentes, dum una fit ex altera (momentanee) crescente vel decrescente [...] Porro ddx est elementum elementi seu *differentia differentiarum*, nam ipsa quantitas dx non semper constans est, sed plerumque rursus (continue) crescit aut decrescit. Et similiter procedi potest ad $dddx$ seu d^3x, et ita porro". Translation from Rutherford (2008, p. 257).

The term was in use much earlier. Leibniz, to whom the term appears to be due, began using it in the 1670s (see, for example, Leibniz 2003, pp. 183, 401). In print, it appears in Leibniz (1693, p. 42); an early usage of the English term is Hayes (1704, p. 4).

Another meaning is found in combinatorics, where the objects to be combined in some way were called *combinatorial elements* (Weingärtner 1800, p. 1) or simply *elements* (Stahl 1800, p. 1). In projective geometry, von Staudt (1847, p. 1) defined a range of points (set of points incident with a line). He called a range of points a “linear formation” (*lineares Gebilde*), and used the word *element* in the definition: “the set of all points (elements), which fulfil a line (unbounded straight line)”.[1]

In connection with sets, terms for ‘element’ appear in the early 1870s. In particular, Cantor (1872, p. 128) used *Punkt* ‘point’ in connection with point sets (subsets of the real numbers), and Dedekind (1872, p. 13) referred to the elements of a number field as *individuals* (*Individuen*). Later, however, Cantor generally used *Element*, for example in his definition of a subset: “By a component part [subset] of a manifold [set] M we understand any other manifold M' of which the elements are also elements of M” (Cantor 1878, p. 242).[2] Cantor’s term is soon found in other languages; English *element* was used by Royce (1900, p. 510), for example.

References

Burkert, Walter (1959). “*ΣΤΟΙΧΕΙΟΝ*. Eine semasiologische Studie”. *Philologus* 103, pp. 167–197.

Cantor, Georg (1872). “Ueber die Ausdehnung eines Satzes aus der Theorie der trigonometrischen Reihen”. *Mathematische Annalen* 5, pp. 123–132.

Cantor, Georg (1878). “Ein Beitrag zur Mannigfaltigkeitslehre”. *Journal für die reine und angewandte Mathematik* 84, pp. 242–258.

Crowley, Timothy J. (2005). “On the use of *stoicheion* in the sense of ‘element’”. *Oxford Studies in Ancient Philosophy* 29, pp. 367–394.

Dedekind, Richard (1872). *Stetigkeit und irrationale Zahlen*. Braunschweig.

Diels, Hermann (1899). *Elementum: eine Vorarbeit zum griechischen und lateinischen Thesaurus*. Leipzig.

Hayes, Charles (1704). *A Treatise of Fluxions*. London.

Leibniz, Gottfried Wilhelm (1693). “Additio G.G.L. ad solutionem problematis in Actis A. 1692, pag. 274 propositi”. *Acta Eruditorum* (Jan. 1693), p. 42.

Leibniz, Gottfried Wilhelm (1710). “Monitum de characteribus algebraicis”. *Miscellanea Berolinensia* (1710), pp. 155–160.

Leibniz, Gottfried Wilhelm (2003). *Sämtliche Schriften und Briefe*. Dritte Reihe. Mathematischer, naturwissenschaftlicher und technischer Briefwechsel. Fünfter Band. 1691-1693. Hannover.

Liberman, Anatoly (2007). “Monthly Gleanings”. URL: `https://blog.oup.com/2007/10/element_hocus_pocus/` (retrieved Mar. 15, 2020).

Royce, Josiah (1900). *The World and the Individual*. New York.

Rutherford, Donald (2008). “Leibniz on Infinitesimals and the Reality of Force”. In: *Infinitesimal Differences. Controversies between Leibniz and his Contemporaries*. Ed. by Ursula Goldenbaum and Douglas Jesseph. Berlin and New York, pp. 255–280.

Stahl, Conrad Dietrich Martin (1800). *Grundriss der Combinationslehre*. Jena and Leipzig.

von Staudt, Karl Georg Christian (1847). *Geometrie der Lage*. Nürnberg.

Walde, Alois and Johann Baptist Hofmann (1938). *Lateinisches etymologisches Wörterbuch*. Heidelberg.

[1] “der Inbegriff von allen Punkten (Elementen), welche eine Gerade (unbegrenzte gerade Linie) erfüllen”. [2] “Unter einem *Bestandtheil* einer Mannigfaltigkeit M verstehen wir jede andere Mannigfaltigkeit M', deren Elemente zugleich Elemente von M sind”.

Weingärtner, Johann Christoph (1800). *Lehrbuch der combinatorischen Analysis nach der Theorie des Herrn Professor Hindenburg*. Erster Theil. Leipzig.

elementary divisor

Elementary divisors, as used in matrix theory, were so named by Weierstraß (1869, p. 311); he used German *Elementar-Theiler*. It did not take long for translations to appear; French *diviseur élémentaire* was used by Casorati (1881, p. 177), and the English term by Young (1893, p. 170).

Bromwich (1906, p. 27) later suggested *invariant factor*; in a review of Bromwich's book, Bôcher (1908, p. 195) commented: "The author replaces Weierstrass's term *elementary divisor*, which is now in universal use, by the term *invariant factor*. This is rather a high-handed proceeding and would seem to require justification. Weierstrass's term is a suitable one and admits of satisfactory translation into English; so that it is hard to see what the author's purpose can have been in making the change".

Note. The history of the concept is discussed by Hawkins (1977).

References

Bôcher, Maxime (1908). "Quadratic Forms and their Classification by Means of Invariant-Factors [review]". *Bulletin of the American Mathematical Society* 14, pp. 194–195.
Bromwich, T. J. I'A. (1906). *Quadratic Forms and their Classification by Means of Invariant-Factors*. Cambridge.
Casorati, Felice (1881). "Sur la distinction des intégrales des équations différentielles linéaires en sous-groupes". *Comptes rendus de l'Académie des Sciences de Paris* 92, pp. 175–178.
Hawkins, Thomas (1977). "Weierstrass and the Theory of Matrices". *Archive for History of Exact Sciences* 17, pp. 119–163.
Weierstraß, Karl (1869). "Zur Theorie der bilinearen und quadratischen Formen". *Monatsberichte der Königlich Preußischen Akademie der Wissenschaften zu Berlin* (1868), pp. 310–338.
Young, J.W.A. (1893). "On the Determination of Groups Whose Order is a Power of a Prime". *American Journal of Mathematics* 15, pp. 124–178.

elementary function

Elementary functions are closely associated with Liouville's work on "integration in finite terms", which consists, in the words of Ritt (1923, p. 211), in "the determination of the circumstances under which the integral of an elementary function is itself elementary. The elementary functions are understood here to be those which are obtained by performing algebraic operations and taking exponentials and logarithms".[1]

Liouville (1834, p. 40) indeed extensively discussed the concept under the name *finite explicit function* (*fonction finie explicite*);[2] a classification of these functions is given in Liouville (1837). He seems to have taken over this expression from Laplace (1814, p. 6), who did not explain it any further.[3] As stated by Hardy (1905, p. vi), Liouville's work on this topic was largely ignored for some time.

The first definition of *elementary function* under this name and in the context of integration in finite terms seems to be due to Hardy (1905, p. 3):

---

[1] In his later monograph, Ritt (1948, pp. 3–12) gave a precise definition that essentially ran over 10 pages. [2] Regarding the term *explicit*, Lützen (1990, p. 351) explains: "One must be careful not to be mislead [sic] by Liouville's use of the word 'explicit,' for it is only the transcendental operations log and exp that are restricted to act explicitly, whereas the algebraic functions can be defined by polynomial equations that are not solvable by radicals". [3] Liouville (1833, pp. 143f.) explicitly refers to Laplace.

An *elementary function* is a member of the class of functions which comprises

(i) rational functions[1],
(ii) algebraical functions, explicit or implicit,
(iii) the exponential function e^x,
(iv) the logarithmic function $\log x$,
(v) all functions which can be defined by means of any finite combination of the symbols proper to the preceding four classes of functions.

When the context of integration in finite terms is disregarded, *elementary* is earlier found with reference to functions. For example, both Servois (1814-15, pp. 121f.) and Wallace (1815, pp. 399f.) used the word for the functions $x^m, a^x, \log x, \sin x, \cos x$.[2] Serret (1868, p. 542) later included under that name "functions which result from one of the operations of algebra" as well as exponential, logarithmic and circular (i. e., trigonometric) functions.

Note. A detailed historical account of integration in finite terms is given by Lützen (1990, pp. 351–422); see also Singer (2022).

References

Hardy, Godfrey Harold (1905). *The Integration of Functions of a Single Variable*. Cambridge.
Laplace, Pierre-Simon (1814). *Théorie analytique des probabilités*. Seconde édition. Paris.
Liouville, Joseph (1833). "Premier mémoire sur la détermination des intégrales dont la valeur est algébrique". *Journal de l'École polytechnique* 14, cahier 22, pp. 124–148.
Liouville, Joseph (1834). "Sur les Transcendantes Elliptiques de première et de seconde espèce, considérées comme fonctions de leur amplitude". *Journal de l'École polytechnique* 14, cahier 23, pp. 37–83.
Liouville, Joseph (1837). "Sur la classification des transcendantes et sur l'impossibilité d'exprimer les racines de certaines équations en fonction finie explicite des coefficients". *Journal de mathématiques pures et appliquées* 2, pp. 56–104.
Lützen, Jesper (1990). *Joseph Liouville 1809-1882*. Master of Pure and Applied Mathematics. New York.
Ritt, Joseph Fels (1923). "On the Integrals of Elementary Functions". *Transactions of the American Mathematical Society* 25, pp. 211–222.
Ritt, Joseph Fels (1948). *Integration in Finite Terms*. Liouville's Theory of Elementary Methods. New York.
Serret, Joseph-Alfred (1868). *Cours de calcul différentiel et intégral*. Tome premier. Calcul différentiel. Paris.
Servois, François-Joseph (1814-15). "Essai sur un nouveau mode d'exposition des principes du calcul différentiel". *Annales des mathématiques pures et appliquées* 5, pp. 93–140.
Singer, Michael F. (2022). "Comments on Rosenlicht's *Integration in Finite Terms*". In: *Integration in Finite Terms: Fundamental Sources*. Ed. by Clemens G. Raab and Michael F. Singer. Cham, pp. 11–30.
Wallace, William (1815). "Fluxions". *The Edinburgh Encyclopaedia* 9, pp. 382–467.

elementary matrix

The elementary matrix operations are row (or column) switching, row multiplying, and row addition; elementary matrices are obtained by performing one of these

[1] For Hardy (ibid.), a rational function is "a function defined by means of any finite combination of the elementary operations of addition, multiplication, and division, operating on the variable x".
[2] Servois also considered composite elementary functions such as $a^{\varphi x}$ or $\sin \varphi x$.

operations on the unit matrix. The German term *elementar* in this sense was used by Kronecker (1867, p. 608), who used it with reference to "systems", i. e. matrices. He also used the term *elementare Transformation* (p. 610). His paper was read in 1866, and the approach was also presented in Clebsch and Gordan (1866, p. 306), who remarked that as far as they knew, the concept was due to Kronecker.

Like many others at the time, Kronecker did not use matrix language; explicitly, *elementary matrix* began to be used in the early 20th century, for example by Schlesinger (1905, p. 285).

Instead of *elementary transformation*, the term *elementary operation* later became more common; an early usage is MacDuffee (1933, p. 32).

Note. Some historical background is given by Hawkins (2013, pp. 353–357). Wedderburn (1934, p. 169) referred to Graßmann (1862, pp. 241ff.) for the concept; MacDuffee (1933, p. 32) to Kronecker.

References

Clebsch, Alfred and Paul Gordan (1866). *Theorie der Abelschen Functionen.* Leipzig.

Graßmann, Hermann (1862). *Die Ausdehnungslehre.* Vollständig und in strenger Form bearbeitet. Berlin.

Hawkins, Thomas (2013). *The Mathematics of Frobenius in Context.* A Journey Through 18th to 20th Century Mathematics. New York.

Kronecker, Leopold (1867). "Ueber bilineare Formen". *Monatsberichte der Königlich Preußischen Akademie der Wissenschaften zu Berlin* (1866), pp. 597–612.

MacDuffee, Cyrus Colton (1933). *The Theory of Matrices.* Berlin.

Schlesinger, Ludwig (1905). "Beiträge zur Theorie der Systeme linearer homogener Differentialgleichungen". *Journal für die reine und angewandte Mathematik* 130, pp. 263–297.

Wedderburn, J.H.M. (1934). *Lectures on Matrices.* New York.

ellipse

See *conic section.*

ellipsoid

The word *ellipsoid* is of much later origin than *spheroid*; it gradually begins to be used with some regularity at around 1700; its exact origin is unclear, but the term was already used by Riccioli (1651, p. 95).[1] Both these terms are often used interchangeably, but differences can also be seen. In addition, both words varied in meaning, depending on various factors such as preferences by certain writers, language, and the topic under discussion (there are differences, for example, between texts on geometry and those on the figure of the earth). The history of the terms *ellipsoid* and *spheroid* thus remains somewhat unsatisfactory.

The term is not included in Ozanam's and Vitalis' mathematical dictionaries (Ozanam 1691, Vitalis 1690) and does not appear to have acquired any currency until about 1700. It was used, without definition, by Parent (1700, p. 171); on a later occasion, however, Parent (1718, p. 284) explicitly used *ellipsoïde* as short for *sphéroïde elliptique*, which he defined as a solid generated by the movement of an ellipse around one of its axes. In other words, Parent used the term for what was

[1] Wallis (1655, p. 108) used the word *elliptoid* [sic] for certain curves, that is for generalizations of ellipses. This usage is not particularly frequent, but is recorded in some references works such as Wolff (1716, pp. 583f.) and Chambers (1728, pp. 294f.).

usually called a *spheroid*. In this sense the term was also used by Reyneau (1708, p. 637), for example, and in various other textbooks such as Deidier (1739, pp. 550f.) and de La Chapelle (1750, p. 166).

In various contributions to the *Encyclopédie*, d'Alembert explained the terminology: "*ellipsoid* is the name given by some geometers to the solid of revolution that the ellipse forms by turning around one of its axes" (d'Alembert 1755a, p. 519).[1] He refers to the entry for *spheroid*, where practically the same definition is found (d'Alembert 1765, p. 459); in the entry for *elliptique*, he also points out: "*Conoid* or *elliptical spheroid*, the same thing as ellipsoid (d'Alembert 1755b, p. 520).[2] Earlier, d'Alembert (1747, p. 152) had also noted another shade of meaning, when he described "a kind of spheroid known by the geometers under the name of *ellipsoid*, which is not [...] a solid of revolution".[3] He appears to be referring to a triaxial ellipsoid (with three unequal axes). In this sense, French *ellipsoide* was used by Bouguer (1746, p. 206) in connection with ships[4] and explained as a solid whose horizontal and vertical sections are elliptic.[5]

The term *elliptic spheroid* came up in the second half of the 17th century. It is found, without definition, as Latin *sphaerois elliptica* in Guarini (1671, p. 563); Varignon (1692, p. 44) defined a *sphéroïde elliptique* as a solid generated by the rotation of a half-ellipse around one of its axes.[6]

In studies on the figure of the earth, the term was understood as an ellipsoid of revolution differing not much from a sphere. This sense was recorded in the *Dictionnaire de Trévoux* (1752, t. III, s. v. ellipsoïde), where de Maupertuis is cited as having used the term in this sense.[7] The term was indeed used by de Maupertuis and others in this meaning; see, for example, de Maupertuis (1738, p. 147).

Finally, it is perhaps worth pointing out that the term has been criticized many times, without much effect. The reason for the criticism is that, while a spheroid is like a sphere, an ellipsoid is not like an ellipse (see, e. g., Klügel and Mollweide 1823, p. 375 and Müller 1860, p. 37).

See also *spheroid*.

References

d'Alembert, Jean Le Rond (1747). *Réflexions sur la cause générale des vents*. Paris.

d'Alembert, Jean Le Rond (1755a). "Ellipsoïde". *Encyclopédie, ou Dictionnaire raisonné des sciences, des arts et des métiers* 5, p. 519.

d'Alembert, Jean Le Rond (1755b). "Elliptique". *Encyclopédie, ou Dictionnaire raisonné des sciences, des arts et des métiers* 5, p. 520.

[1] "est le nom que quelques géometres ont donné au solide de révolution que forme l'ellipse en tournant autour de l'un ou de l'autre de ses axes". [2] "Conoïde ou sphéroïde elliptique, c'est la même chose qu'ellipsoïde". – See also Hutton (1795, p. 481), who states that "the solid generated by the revolution is properly called an *ellipsoid*, as distinguished from the Spheroid, which is generated from the revolution of any oval whatever, whether it be an ellipse or not. But generally speaking, the term Spheroid is used for an ellipsoid". [3] "un espece de Sphéroide connu par les Geométres sous le nom d'*Ellipsoide*, & qui ne sera point [...] un solide de révolution, mais dont toutes les coupes par l'axe *OC* seront des Ellipses". [4] The idea to model a vessel as a (semi-)ellipsoid to calculate certain volumes was suggested by Varignon (1723, p. 45). [5] "l'ellipsoïde [...] corps dont toutes les tranches horisontales [sic] sont elliptiques, de même que les verticales". [6] "le sphéroide elliptique formé par le mouvement de la demi-ellipse autour de *AB*". [7] "L'ellipsoïde de la terre ne diffère pas beaucoup du globe. Maupertuis".

d'Alembert, Jean Le Rond (1765). "Sphéroïde". *Encyclopédie, ou Dictionnaire raisonné des sciences, des arts et des métiers* 15, p. 459.
Bouguer, Pierre (1746). *Traité du navire*. Paris.
Chambers, Ephraim (1728). *Cyclopaedia, or, An universal dictionary of arts and sciences*. Volume the First. London.
de La Chapelle, Jean-Baptiste (1750). *Traité des sections coniques, et autres courbes anciennes*. Paris.
Deidier, Antoine (1739). *La science des géométres ou la théorie de la pratique de la géométrie*. Paris.
Guarini, Guarino (1671). *Euclides adauctus et methodicus mathematicaque universalis*. Torino.
Hutton, Charles (1795). *A Philosophical and Mathematical Dictionary*. In two volumes. Vol. II. London.
Klügel, Georg Simon and Carl Brandan Mollweide (1823). *Mathematisches Wörterbuch*. Erste Abtheilung: Die reine Mathematik. Vierter Theil von Q bis S. Leipzig.
de Maupertuis, Pierre Louis Moreau (1738). *La figure de la terre*. Amsterdam.
Müller, Johann Heinrich Traugott (1860). *Beiträge zur Terminologie der griechischen Mathematiker*. Leipzig.
Ozanam, Jacques (1691). *Dictionnaire mathematique*. Paris.
Parent, Antoine (1700). *Élémens de méchanique et de physique*. Paris.
Parent, Antoine (1718). *Geometrie theori-pratique*. In: *Suite des elemens de mathematique*. Paris.
Reyneau, Charles-René (1708). *Usage de l'analyse*. Tome II. Paris.
Riccioli, Giovanni Battista (1651). *Almagestum novum astronomiam veterem novamque complectens*. Bologna.
Varignon, Pierre (1692). "Démonstration commune à la Sphère, & aux sphéroïdes elliptiques, tant allongez qu'applatis". In: *Mémoires de mathématique et de physique, tirez des registres de l'Academie royale des sciences*. Paris, pp. 42–48.
Varignon, Pierre (1723). "Jaugeage d'un navire ellipsoïde". *Mémoires de l'Académie royale des science* (1721), pp. 44–55.
Vitalis, Hieronymus (1690). *Lexicon mathematicum, astronomicum et geometricum*. Paris.
Wallis, John (1655). *De sectionibus conicis*. Oxford.
Wolff, Christian (1716). *Mathematisches Lexicon*. Leipzig.

elliptic

Derived from *ellipse*, this term was originally used for certain geometrical objects that are somehow related to the ellipse; later, the term was transferred to other domains such as differential equations.

In the following, the terms *elliptic conoid*, *elliptic cylinder*, *elliptic coordinates*, and *elliptic differential equation* are discussed in brief. The terms *elliptic function* and *elliptic integral* are discussed in a separate entry. For *elliptic geometry*, see *hyperbolic geometry*.

While Federico Commandino suggested the term *hyperbolic conoid* and *parabolid conoid* in his translation of the works of Archimedes, he did not suggest the analogous *elliptic conoid* for a spheroid. Instead, *conoides ellipticum* seems to be due to Wallis (1655, p. 36); it was also used in correspondence from 1659 between Huygens and Mylon (Huygens 1889, p. 373) and appears in Ozanam's dictionary (1691, p. 121), where an "Elliptic conoid, or simply spheroid" is defined as a solid "generated by the movement of an ellipse around one if its two axes".[1] See also *spheroid*.

[1] "Conoïde Elliptique, ou simplement Spheroide, quand il est produit par le mouvement achevé d'une Ellipse autour de l'un de ses deux axes".

The term *elliptic cylinder* (Latin *cylindrus ellipticus*) appears to go back to Tacquet (1651, pp. 1f.). Tacquet also used the terms *circular*, *hyperbolic*, and *parabolic cylinder*.

The term *elliptic coordinates* (French *coordonnées elliptiques*) was suggested by Lamé (1833, p. 199); it became better known through Lamé (1837, p. 156).

The terminological distinction between hyperbolic, parabolic, and elliptic differential equations was suggested by du Bois-Reymond (1889, p. 265).

Note. For du Bois-Reymond's classification of differential equations, see Gray (2021, pp. 289–293). For Lamé and the introduction of elliptic coordinates, see Reich (1973, pp. 327f.) and Boyer (1956, pp. 262–264).

References

du Bois-Reymond, Paul (1889). "Ueber lineare partielle Differentialgleichungen zweiter Ordnung". *Journal für die reine und angewandte Mathematik* 104, pp. 204–229.

Boyer, Carl B. (1956). *History of Analytic Geometry*. New York.

Gray, Jeremy (2021). *Change and Variations*. A History of Differential Equations to 1900. Cham.

Huygens, Christiaan (1889). *Œuvres complètes*. Tome deuxième. Correspondance 1657–1659. La Haye.

Lamé, Gabriel (1833). "Mémoire sur les surfaces isothermes dans les corps solides en équilibre de température". *Annales de chimie et de physique* 53, pp. 190–204.

Lamé, Gabriel (1837). "Sur les surfaces isothermes dans les corps solides homogènes en équilibre de température". *Journal de mathématiques pures et appliquées* 2, pp. 147–183.

Ozanam, Jacques (1691). *Dictionnaire mathematique*. Paris.

Reich, Karin (1973). "Die Geschichte der Differentialgeometrie von Gauss bis Riemann (1828-1868)". *Archive for History of Exact Sciences* 11, pp. 273–382.

Tacquet, André (1651). *Cylindricorum et annularium libri IV*. Antwerpen.

Wallis, John (1655). *De sectionibus conicis*. Oxford.

elliptic curve

As explained by Lemmermeyer (2011, p. 9), Clebsch (1865) showed that "curves of genus 1 can be parametrized by elliptic functions [...] This result later gave elliptic curves their name". The term was coined a few years after the publication by Clebsch:[1] Brioschi (1869, p. 559) used Italian *curva ellitica*, defining it as a plane curve of order n with $\frac{n(n-3)}{2}$ double points or cusps. Brioschi referred to Clebsch.[2]

Brioschi's article was reviewed by Jung (1873, p. 473), who used German *elliptische Curve*; in another review published in the *Bulletin des sciences mathématiques et astronomiques* 1 (1870), p. 188, French *courbe elliptique* is found. The first English usage is in an abstract in *Nature* 1 (1870), p. 297. The term was then adopted by Klein (1880, p. 536), for example, who defined it as a curve of genus 1.

In the 20th century, the theory of elliptic curves was considerably developed, and connections with other fields such as number theory and cryptography were established. Lang (1978, p. v) was already able to say that "[i]t is possible to write endlessly on elliptic curves".

---

1 While the origin of the term has already been pointed out by Müller (1900, p. 28), for example, D'Antonio (2005, p. 27) finds it "fairly obscure" and Rice and Brown (2012, p. 168) date it to the early twentieth century.

2 It is worth pointing out that the expression *elliptic curve* had been around in different senses before Brioschi. For example, Serret (1845, p. 421) used it for algebraic curves whose arc lengths are expressed by elliptic integrals of the first kind.

References

D'Antonio, Lawrence (2005). "Teaching Elliptic Curves Using Original Sources". In: *From Calculus to Computers*. Using the Last 200 Years of Mathematics History in the Classroom. Ed. by Amy Shell-Gellasch and Dick Jardine. Washington, D.C., pp. 25–44.

Brioschi, Francesco (1869). "Sulla equazione che dà i punti di flesso delle curve ellittiche". *Reale Istituto lombardo di scienze e lettere. Rendiconti* 2, pp. 559–565.

Clebsch, Alfred (1865). "Ueber diejenigen Curven, deren Coordinaten sich als elliptische Functionen eines Parameters darstellen lassen". *Journal für die reine und angewandte Mathematik* 64, pp. 210–270.

Jung, Giuseppe (1873). "J. Brioschi. Nota sulla equazione che dà i punti di flesso delle curve ellittiche [review]". *Jahrbuch über die Fortschritte der Mathematik* 2, pp. 473–474.

Klein, Felix (1880). "Ueber unendlich viele Normalformen des elliptischen Integral's erster Ordnung". *Sitzungsberichte der mathematisch-physikalischen Classe der königlich bayerischen Akademie der Wissenschaften* 10, pp. 533–541.

Lang, Serge (1978). *Elliptic Curves*. Diophantine Analysis. Berlin et al.

Lemmermeyer, Franz (2011). "Parametrizing Algebraic Curves". URL: `https://arxiv.org/abs/1108.6219` (retrieved July 1, 2024).

Müller, Felix (1900). *Mathematisches Vokabularium*. Französisch-Deutsch und Deutsch-Französisch. Enthaltend die Kunstausdrücke aus der reinen und angewandten Mathematik. Leipzig.

Rice, Adrian and Ezra Brown (2012). "Why Ellipses Are Not Elliptic Curves". *Mathematics Magazine* 85, pp. 163–176.

Serret, Joseph-Alfred (1845). "Note sur les courbes elliptiques de la première classe". *Journal de mathématiques pures et appliquées* 10, pp. 421–429.

elliptic function and *elliptic integral*

The origins of these two terms are closely intertwined. The term *elliptic* arose in connection with attempts to measure the arc lengths of certain curves, particularly ellipses, hyperbolas and lemniscates. It has been much criticized, perhaps most clearly by Pringsheim (1930, p. 130): "the denomination in question [*elliptic function*] seems to me one of the most unsuitable ones which can be found in the vocabulary of analysis".[1]

The French term *fonction elliptique* was created by Legendre (1793, p. 17):

> Since the transcendentals we have considered are always reducible to these two forms $\int (A+B\sin^2\varphi)\frac{d\varphi}{\Delta}$, $\int \frac{d\varphi}{1+n\sin^2\varphi\Delta}$ it is evident that they are comprehended in the general formula
>
> $$H = \int \frac{A+B\sin^2\varphi}{1+n\sin^2\varphi}\cdot\frac{d\varphi}{\Delta}.$$
>
> We shall in future call the integrals contained in this formula *elliptic functions* or *elliptic transcendentals*.[2]

The terms *amplitude*, *modulus* and *complement of the modulus* are introduced in the same place.[3] Legendre (1793, p. 14) goes on to discuss the terminology and finds some arguments to call it "improper", but he also has reasons to keep it: "It

[1] "erscheint mir die fragliche Bezeichnung [*elliptische Funktion*] als eine der unpassendsten, welche sich in dem Wortschatz der Analysis vorfindet". [2] "Puisque les transcendantes que nous avons considérées se réduisent toujours à ces deux formes $\int (A+B\sin^2\varphi)\frac{d\varphi}{\Delta}$, $\int \frac{d\varphi}{1+n\sin^2\varphi\Delta}$ il est clair qu'elles sont comprises dans la formule générale $H = \int \frac{A+B\sin^2\varphi}{1+n\sin^2\varphi}\cdot\frac{d\varphi}{\Delta}$. Nous appellerons désormais, *fonctions* ou *transcendantes elliptiques*, les intégrales comprise dans cette formule". Translation from Legendre (1809, p. 13). [3] Of these, *amplitude* was earlier used in Legendre (1788, p. 617).

does not appear that the function H, taken in all its generality, can be reduced to elliptic arcs [...] Thus, the term *elliptic function* is in some respects improper; we retain it however because of the great analogy which is found to subsist between the properties of this function, and those of elliptic arcs".[1]

In sum, Legendre introduced the basic terminology for elliptic integrals but called them *elliptic functions*. Arguments against this terminology were expressed, for example, by Gudermann (1844, pp. iii–iv): "to keep this denomination might lead to the mistaken opinion that the whole theory is only there to rectify the ellipse".[2] Gudermann (1844, p. iii) instead proposed the term *modular function* (German *Modular-Function*) because, as he explicitly stated, these "functions" also depend on a quantity called the *module* (*Modulus*). His suggestion did not catch on, however.

In the 1830s, the study of elliptic functions was also pursed by Abel and Jacobi, who also considered the inverse of an elliptic "function". In this context, a change in the terminology was proposed. This change was made most explicit in Jacobi's fundamental work on elliptic functions, which contains a chapter on "new notation for elliptic functions" (Jacobi 1829, § 17). Jacobi designates the inverse functions as *sin am* (sine of the amplitude), etc., and suggests that "these expressions [the inverse functions] and generally the trigonometric functions of the amplitude will in the following be designated by the name *elliptic functions*, such that we give this name a certain other meaning than has hitherto been done by analysts".[3]

This step amounts to a reversal of the terminology, and Legendre opposed this change in a letter to Jacobi from 1829, arguing that the term was generally adopted and asking Jacobi for the reason for the change, which Legendre regarded as "schismatizing" (see Borchardt 1875, pp. 269f.). Jacobi soon replied to Legendre and explained that he absolutely needed a designation for his inverse functions, whose properties completely corresponded to the trigonometric functions: he added that if Legendre found this displeasing, he would be happy to accept any other terms. Since Legendre died soon afterwards, this discussion did not lead to any further changes and Jacobi's terminology stuck even though it continued to be criticized. For example, Enneper (1890, p. 2) stated:

> The denomination 'elliptic functions' has been transferred on the basis of an analogy with circular functions to a species of functions which stands in close connection with the elliptic integrals. The terminology concerning the word 'elliptic' unfortunately suffers from an imperfection which has its origin in the seeming generalization of an isolated problem.

---

[1] "Il ne paroît pas que la fonction H prise dans toute l'généralité, puisse se réduire à des arcs d'ellipse [...] Ainsi la dénomination fonction elliptique est impropre à quelques égards; nous l'adoptons néansmois à cause de la grande analogie qu'on trouvera entre les propriétés de cette fonction et celles des arcs d'ellipse". Translation from Legendre (1809, p. 14). [2] "Die Belassung dieser Benennung könnte die irrige Meinung erwecken [...], als wäre diese ganze Theorie nur dazu da, um die Ellipse rectificiren zu können". This statement was repeated in similar terms by Weierstraß (1915, p. 2). [3] "Ipsas expressiones [...] ac generaliter functiones trigonometricas amplitudinis in sequentibus functionum ellipticarum nomine insignire convenit; ita ut ei nomini aliam quandam tribuamus notionem atque hactenus factum est ab Analystis".

For the occuring integrals and functions the epithet ‘elliptic’ is neither geometrically nor analytically characteristic in any way, but of a purely conventional nature.[1]

Note. For the history of elliptic functions and integrals, see Houzel (1978), Cooke (1994), Cooke (2005), Rice (2008), and Roy (2017). Ayoub (1984) discusses the notion in Fagnano.

References

Ayoub, Raymond (1984). “The Lemniscate and Fagnano’s Contributions to Elliptic Integrals”. *Archive for History of Exact Sciences* 29, pp. 131–149.
Borchardt, C.W. (1875). “Correspondance mathématique entre Legendre et Jacobi”. *Journal für die reine und angewandte Mathematik* 80, pp. 205–279.
Cooke, Roger (1994). “Elliptic integrals and functions”. In: *Companion Encyclopedia of the History and Philosophy of the Mathematical Sciences*. Volume 1. Ed. by Ivor Grattan-Guinness. Baltimore and London, pp. 529–539.
Cooke, Roger (2005). “C.G.J. Jacobi, book on elliptic functions (1829)”. In: *Landmark Writings in Western Mathematics 1640-1940*. Ed. by Ivor Grattan-Guinness. Amsterdam, pp. 412–430.
Enneper, Alfred (1890). *Elliptische Functionen*. Theorie und Geschichte. Zweite Auflage. Ed. by Felix Müller. Halle.
Gudermann, Christoph (1844). *Theorie der Modular-Functionen und der Modular-Integrale*. Berlin.
Houzel, Christian (1978). “Fonctions elliptiques et intégrales abéliennes”. In: *Abrégé d’histoire des mathématiques 1700-1900*. II. Ed. by Jean Dieudonné. Paris, pp. 1–113.
Jacobi, Carl Gustav Jacob (1829). *Fundamenta nova theoriae functionum ellipticarum*. Königsberg.
Legendre, Adrien-Marie (1788). “Mémoire sur les intégrations par d’arcs d’ellipse”. *Mémoires de l’Académie royale des sciences* (1786), pp. 616–643.
Legendre, Adrien-Marie (1793). *Mémoire sur les transcendantes elliptiques*. Paris.
Legendre, Adrien-Marie (1809). “A Memoir on Elliptic Transcendentals”. *New Series of The Mathematical Repository* 2, part III, pp. 1–34.
Pringsheim, Alfred (1930). “Kritisch-historische Anmerkungen zur Funktionentheorie. IV. Über die Bezeichnung ‚Elliptische Funktionen’ und die Umkehrung der Weierstraßschen *Pe*-Funktion”. *Sitzungsberichte der mathematisch-naturwissenschaftlichen Abteilung der Bayerischen Akademie der Wissenschaften* (1930), pp. 129–164.
Rice, Adrian (2008). “In Search of the ‘Birthday’ of Elliptic Functions”. *The Mathematical Intelligencer* 30, pp. 48–56.
Roy, Ranjan (2017). *Elliptic and Modular Functions from Gauss to Dedekind to Hecke*. Cambridge.
Weierstraß, Karl (1915). *Vorlesungen über die Theorie der elliptischen Functionen*. Bearbeitet von J. Knoblauch. Berlin.

embedding* and *immersion

The term *embedding* is found in various mathematical fields; its definitions vary accordingly. In the following, mostly examples from topology will be considered.

An instance of the topological concept is found in Klein (1876, p. 478), who made a distinction between absolute and relative topological properties:

I call *absolute* those properties that belong to the manifold in question independent of the surrounding space, in which they can be taken to lie. *Relative* properties depend on the

[1] “Die Bezeichnung ‚elliptische Functionen’ ist nach Analogie der Kreisfunctionen auf eine Gattung von Functionen übertragen worden, welche mit den elliptischen Integralen in enger Verbindung stehen, wobei die Terminologie hinsichtlich des Wortes ‚elliptisch’ leider an jener Unvollkommenheit leidet, die ihren Ursprung in der scheinbaren Verallgemeinerung eines isolierten Problems hat. Für die auftretenden Integrale und Functionen ist das Epitheton ‚elliptisch’ weder geometrisch noch analytisch irgendwie characteristisch, sondern rein conventioneller Natur”.

surrounding space; they are invariant in the case of distortions of the manifold that take place within the space in question, but they are not invariant under arbitrary distortions.[1]

When exactly *embedding* began to be used is difficult to determine. What can be said is that Tietze (1908, p. 10, fn. 3), when he discussed these absolute and relative properties, did not use the term. However, in a later discussion of the same topic, he did use the verb *einbetten* (Tietze 1923, p. 296). The same word was also used by Segre (1921, p. 811), who also remarks that del Pezzo used Italian *immersa* in his works (for example, in del Pezzo 1887, p. 241). The Italian is thus found much earlier than German *Einbettung*.

In the 1920s, the term is also found in algebra. For example, van der Waerden used the term in lectures on ideal theory from 1927-28: "If a ring $\mathfrak{O}$ can be embedded in a field Ω, then $\mathfrak{O}$ certainly has no zero divisors" (Bečvář and Bečvářová 2020, p. 124).[2]

Definitions of the term are scarce in the 1920s. One was given by Menger (1928, p. 115) in connection with metric spaces. Alexandroff and Hopf (1935, pp. 89f.) later used the term in their book on topology: "We shall say that the topological space R is contained in the topological space R' (or, that R can be *topologically embedded* in R'), if R is homeomorphic to a point set of R'".[3] Seifert and Threlfall (1934, e. g., pp. 2, 9) used both *Einbettung* and *Einlagerung*.

The distinction between *embeddings* and *immersions* seems to go back to Whitney. It appears in Whitney (1941, p. 127):

> It may be shown that any M_n may be *imbedded* in E^{2n+1}, that is, mapped in a (1-1) differentiable manner, so that the matrix of partial derivatives is everywhere of maximum rank. Also, it may be *immersed* in E^{2n}; this is the same as imbedding, except that we allow self-intersections.[4]

The term was adopted by Nash (1954, p. 383), for example: "An immersion is defined, as by Whitney [...], to be a non-singular imbedding except that self-intersections are allowed". Another adopter was Hirsch (1959, p. 242): "Let M and N be differentiable manifolds of dimensions k and n respectively, $k < n$. A differentiable map $f : M \to N$ is called an *immersion* if f is of class C^1 and the Jacobian matrix of f has rank k at each point of M. Such a map is also called *regular*". The term *regular map* mentioned by Hirsch appears to go back to Whitney (1935, p. 462).[5]

References

Alexandroff, Pavel S. and Heinz Hopf (1935). *Topologie*. Erster Band. Berlin.

Bečvář, Jindřich and Martina Bečvářová (2020). *Jarník's Notes of the Lecture Course Allgemeine Idealtheorie by B.L. van der Waerden (Göttingen 1927/1928)*. Praha.

[1] "*Absolut* nenne ich diejenigen Eigenschaften, welche der betr. Mannigfaltigkeit unabhängig vom dem umfassenden Raume zukommen, in welchem gelegen man sie voraussetzen mag. *Relative* Eigenschaften hängen von dem umgebenden Raume ab; sie sind invariant bei Verzerrungen der Mannigfaltigkeit, die innerhalb des betr. Raumes stattfinden, nicht aber bei beliebigen Verzerrungen". Translation from Tobies (2021, p. 141). [2] "Wenn sich ein Ring $\mathfrak{O}$ in einen Körper Ω einbetten lässt, so hat $\mathfrak{O}$ sicher keine Nullteiler". [3] "Wir sagen, daß der topologische Raum R vom topologischen Raum R' *umfaßt* wird (oder, daß R sich in R' *topologisch einbetten* läßt), wenn R einer Punktmenge von R' homöomorph ist". [4] A similar statement appears in Whitney (1944, p. 221). [5] See also Dieudonné (2009, p. 60).

Dieudonné, Jean (2009). *A History of Algebraic and Differential Topology, 1900–1960*. Boston.
Hirsch, Morris W. (1959). "Immersions of Manifolds". *Transactions of the American Mathematical Society* 93, pp. 242–276.
Klein, Felix (1876). "Ueber den Zusammenhang der Flächen". *Mathematische Annalen* 9, pp. 476–482.
Menger, Karl (1928). "Untersuchungen über allgemeine Metrik". *Mathematische Annalen* 100, pp. 75–163.
Nash, John (1954). "C^1 Isometric Imbeddings". *Annals of Mathematics* 60, pp. 383–396.
del Pezzo, Pasquale (1887). "Sulle superficie dell' n^{mo} ordine immerse nello spazio di n dimensioni". *Rendiconti del Circolo Matematico di Palermo* 1, pp. 241–271.
Segre, Corrado (1921). "Mehrdimensionale Räume". *Encyklopädie der mathematischen Wissenschaften mit Einschluss ihrer Anwendungen* III C 7, pp. 769–972.
Seifert, Herbert and William Threlfall (1934). *Lehrbuch der Topologie*. Leipzig.
Tietze, Heinrich (1908). "Über die topologischen Invarianten mehrdimensionaler Mannigfaltigkeiten". *Monatshefte für Mathematik und Physik* 19, pp. 1–118.
Tietze, Heinrich (1923). "Beiträge zur allgemeinen Topologie. I. Axiome für verschiedene Fassungen des Umgebungsbegriffs". *Mathematische Annalen* 88, pp. 290–311.
Tobies, Renate (2021). *Felix Klein*. Visions for Mathematics, Applications, and Education. Revised by the Author and Translated by Valentine A. Pakis. Cham.
Whitney, Hassler (1935). "Differentiable manifolds in Euclidean space". *Proceedings of the National Academy of Sciences* 21, pp. 462–464.
Whitney, Hassler (1941). "On the topology of differentiable manifolds". In: *Lectures in Topology*. Ed. by Raymond Louis Wilder and W.L. Ayres. Ann Arbor, pp. 101–141.
Whitney, Hassler (1944). "The Self-Intersections of a Smooth n-Manifold in $2n$-Space". *Annals of Mathematics* 45, pp. 220–246.

empty set

Calling a set *empty* means to think of sets metaphorically in terms of some kind of container. This metaphorical thinking is found in texts on sets and logic in the second half of the 19th century. For example, Schröder (1877, p. 11) used the term *empty class* (German *leere Klasse*) and designated it by the symbol 0.[1] Cantor (1880, p. 356) defined an empty "set",[2] using the paraphrase "the set contains not a single point" and introduced the letter *O* for it. Dedekind (1888, p. 2) used the term *empty system* (German *leeres System*), *system* being his term for 'set'. In English, *empty system* is found in Dedekind (1909, p. 45); slightly earlier, Royce (1905, p. 364) used *empty collection*. The French term *ensemble vide* was used by Couturat (1900, p. 30).

Hessenberg (1909, p. 86) summarizes the terminology: "There is one and only one 'empty' or 'improper' set, the 'zero set' or 'zero', 0, which contains no element".[3] The designation "improper" (*uneigentlich*), which Hessenberg uses here, is also found in Zermelo (1908, p. 263); according to Felgner (2010, p. 176), "this cautious mode of expression has its cause in Cantor's definition of the concept of set. In fact, it is not clear from Cantor's definition whether the empty class really is a set".

At least until the 1920s, however, it was more common to speak of a *null set* or *zero set* (German *Nullmenge*). This terminology is found in Schröder (1890, p. 212) and Schröder (1891, p. 2), where he defines a *null domain* (*Nullgebiet*). Similarly,

---

[1] In using this symbol, Schröder followed Boole (1847, p. 21); Kanamori (2003, p. 274) designates Boole as the inventor of the empty set. [2] Cantor says that "strictly speaking" this "set" cannot be regarded as a set. [3] "Es gibt eine und nur eine 'leere' oder 'uneigentliche' Menge, die 'Nullmenge' oder 'Null', 0, welche kein Element enthält".

Hausdorff (1908, p. 3) used *Nullmenge* 'null set', and so did Zermelo (1908, p. 263) and Fraenkel (1919, p. 236), for instance.

In English, similar expressions are found from around 1900. For example, Russell (1903, p. 73) defines a *null class*, and Huntington (1905, p. 153), who uses English *class* for Cantor's *Menge*, suggests "*null* or *empty* class"; Royce (1905, p. 364) uses *empty* or *zero collection*, W. H. Young and G. C. Young (1906, p. 146) have *null-set*. As *null set* was also used for sets of measure zero, this term eventually dropped out of use.

Note. Some historical information on the empty set is found in Kanamori (2003); see also the editorial notes in Hausdorff (2002, pp. 581–583).

References

Boole, George (1847). *The Mathematical Analysis of Logic*. Cambridge.

Cantor, Georg (1880). "Ueber unendliche, lineare Punktmannichfaltigkeiten. 2". *Mathematische Annalen* 17, pp. 355–358.

Couturat, Louis (1900). "Sur une définition logique du nombre". *Revue de métaphysique et de morale* 8, pp. 23–36.

Dedekind, Richard (1888). *Was sind und was sollen die Zahlen?* Braunschweig.

Dedekind, Richard (1909). *Essays on the Theory of Numbers*. I. Continuity and Irrational Numbers, II. The Nature and Meaning of Numbers. Authorized translation by Wooster Woodruff Beman. Chicago.

Felgner, Ulrich (2010). "Introductory note to 1908b". In: *Ernst Zermelo. Collected Works. Gesammelte Werke*. Volume I. Ed. by Heinz-Dieter Ebbinghaus and Akihiro Kanamori. Heidelberg et al., pp. 160–189.

Fraenkel, Adolf (1919). *Einleitung in die Mengenlehre*. Berlin.

Hausdorff, Felix (1908). "Grundzüge einer Theorie der geordneten Mengen". *Mathematische Annalen* 65, pp. 435–505.

Hausdorff, Felix (2002). *Gesammelte Werke*. Band II. Grundzüge der Mengenlehre. Ed. by Egbert Brieskorn et al. Berlin and Heidelberg.

Hessenberg, Gerhard (1909). "Kettentheorie und Wohlordnung". *Journal für die reine und angewandte Mathematik* 135, pp. 81–133.

Huntington, Edward V. (1905). "The Continuum as a Type of Order. An Exposition of the Modern Theory". *Annals of Mathematics* 7, pp. 15–43.

Kanamori, Akihiro (2003). "The Empty Set, the Singleton, and the Ordered Pair". *Bulletin of Symbolic Logic* 9, pp. 273–298.

Royce, Josiah (1905). "The Relation of the Principles of Logic to the Foundations of Geometry". *Transactions of the American Mathematical Society* 6, pp. 353–415.

Russell, Bertrand (1903). *The Principles of Mathematics*. Vol. I. Cambridge.

Schröder, Ernst (1877). *Der Operationskreis des Logikkalkuls*. Leipzig.

Schröder, Ernst (1890). *Vorlesungen über die Algebra der Logik* (Exakte Logik). Erster Band. Leipzig.

Schröder, Ernst (1891). *Vorlesungen über die Algebra der Logik* (Exakte Logik). Zweiter Band. Erste Abteilung. Leipzig.

Young, William Henry and Grace Chisholm Young (1906). *The Theory of Sets of Points*. Cambridge.

Zermelo, Ernst (1908). "Untersuchungen über die Grundlagen der Mengenlehre". *Mathematische Annalen* 65, pp. 261–281.

endomorphism

The term *endomorphism* is a comparatively late addition to the algebraic vocabulary. It appears in various publications from 1937. Thus, German *Endomorphismus* is found in Fitting (1937, p. 357), who says he had earlier used *Automorphismus* for the notion, but, following a suggestion of van der Waerden, he wants to use *Endomorphismus* from now on because *Automorphismus* is already in use in another special meaning. Fitting does not give a reference to a publication by van der Waerden. Zassenhaus (1937, p. 38) uses the term *operator* (German *Operator*) but points out that "some authors" say *endomorphism.*

The earliest occasion on which van der Waerden uses the term in print seems to be the second edition of his classic book *Modern Algebra*, where he defines a homomorphism and then states that if "the homomorphism [is] a mapping of $\mathfrak{M}$ into itself, we call it an *endomorphism*" (van der Waerden 1937, p. 31).[1] The term incorporates Greek *ἔνδον*, which means 'within, in' (LSJ).

At the same time, the English term is also found in Neumann (1937, p. 508), where it is attributed to Philip Hall.[2] Neumann cites two papers by Hall, namely Hall (1933) and Hall (1936). However, in these papers the term is *not* used, nor is it found in other publications by Hall before 1940.[3] It is still possible that Hall coined the term independently of van der Waerden and originally used it in private communication only.

References

Chandler, Bruce and Wilhelm Magnus (1982). *The History of Combinatorial Group Theory*. A Case Study in the History of Ideas. New York.

Fitting, Hans (1937). "Die Gruppe der zentralen Automorphismen einer Gruppe mit Hauptreihe". *Mathematische Annalen* 114, pp. 355–372.

Hall, Philip (1933). "A Contribution to the Theory of Groups of Prime-Power Order". *Proceedings of the London Mathematical Society* 36, pp. 29–95.

Hall, Philip (1936). "The Eulerian functions of a group". *The Quarterly Journal of Mathematics* 7, pp. 134–151.

Hall, Philip (1940). "The classification of prime-power groups". *Journal für die reine und angewandte Mathematik* 182, pp. 130–141.

Magnus, Wilhelm (1939). "Allgemeine Gruppentheorie". *Enzyklopädie der mathematischen Wissenschaften mit Einschluss ihrer Anwendungen.* Zweite, völlig neubearbeitete Auflage I 1,9, pp. 1–51.

Neumann, B.H. (1937). "Identical relations in groups. I". *Mathematische Annalen* 114, pp. 506–525.

van der Waerden, Bartel Leendert (1937). *Moderne Algebra*. Erster Teil. Zweite verbesserte Auflage. Berlin.

Zassenhaus, Hans (1937). *Lehrbuch der Gruppentheorie*. Erster Band. Leipzig and Berlin.

entire function

The term *entire function* (German *ganze Funktion*) goes back to Weierstraß. One of the last definitions he gave was this:

> Among the functions in question – now including all rational ones – the simplest are those for which there is only *one* (essential or non-essential) singular point in the whole domain of the independent variable. If this point is at ∞, then, as was noted above, such a function can always be represented by a series of the form
>
> $$A_0 + A_1x + A_2x^2 + \cdots$$

---

[1] "[ist] der Homomorphismus eine Abbildung von $\mathfrak{M}$ in sich, so nennen wir ihn einen *Endomorphismus*". [2] Chandler and Magnus (1982, p. 158) also attribute the term to Hall, without reference. Magnus (1939, p. 13) earlier referred to Fitting for the term. [3] The term appears in Hall (1940, p. 140) without definition.

> in which x denotes the argument of the function, while the coefficients A_0, A_1, A_2, ..., are determined values independent of x. On the other hand, every series of this form that converges for every finite value of x represents a single-valued function of x with the one singular point ∞. Such a function I will call an *entire* single-valued function of x – or simply, when no ambiguity arises, an entire function (Weierstraß 1886, pp. 7f.).[1]

An earlier version of this definition, phrased in slightly different terms, is found in Weierstraß (1877, p. 16). Essentially, the concept goes back to Weierstraß (1856, p. 332), where he spoke of functions "that essentially have the character of an *entire* rational function".[2] On the differences, Turner (2013, p. 40) remarks: "To Weierstrass, a function of the independent (complex) variable x has, for a finite value a, the *character of an entire function* if it can be expressed by a power series of only whole and positive powers of $(x - a)$. If the function has, for every finite value of the variable x, the character of an entire function, we say that it is a function of *entire character*. [...] in present-day mathematics, we do not differentiate between entire functions and functions of entire character; the term 'entire' is used to describe functions which are analytic everywhere. However, when Weierstrass and Mittag-Leffler[3] worked with these definitions, an entire function was strictly a polynomial, where a function of entire character contained infinitely many non-zero terms in its power series expansion".

The terminology spread quickly; for example, Betti (1860, p. 66) already used the Italian term *funzioni intere*. As Bottazzini and Gray (2013, p. 325) explain, he used the word for "the functions whose values can be expressed by convergent power series for any real or complex value of z".

Note that *entire function* (or *integral function*) in the sense of a polynomial is due to Euler (1748, § 9), where he explained that "the general formula of these integral functions will be $a + bz + cz^2 + dz^3 + ez^4 + fz^5 +$ & c.".[4]

See also the entries for *holomorphic*, *regular*, and *analytic* functions.

Note. For various remarks on the history of the concept, see Turner (2013) and Iurato (2014).

References

Betti, Enrico (1860). "La teorica delle funzioni ellittiche". *Annali di Matematica Pura ed Applicata* 3, pp. 65–159, 298–310.

Birkhoff, Garrett, ed. (1973). *A Source Book in Classical Analysis*. Cambridge, Mass.

---

[1] "Unter den Functionen, um welche es sich handelt – die rationalen jetzt eingeschlossen – sind die einfachsten diejenigen, für welche es im ganzen Gebiete der unabhängigen Veränderlichen nur eine (wesentliche oder ausserwesentliche) singuläre Stelle giebt. Liegt diese Stelle im Unendlichen, so kann, wie oben bemerkt, eine solche Function stets dargestellt werden durch eine Reihe von der Form $A_0 + A_1x + A_2x^2 + \cdots$ in der x das Argument der Function, die Coefficienten A_0, A_1, A_2, ..., aber von x unabhängige, bestimmte Grössen bedeuten; so wie anderseits jede Reihe von dieser Form, wenn sie für jeden endlichen Werth von x convergirt, der Ausdruck einer eindeutigen Function von x mit der einen singulären Stelle ∞ ist. Eine solche Function will ich eine *ganze* eindeutige Function von x nennen – oder auch bloss, wo keine Zweideutigkeit dadurch entsteht, ganze Function". Translation based on Birkhoff (1973, p. 84). [2] "im Wesentlichen den Charakter der ganzen rationalen Functionen besitzen". [3] See Mittag-Leffler (1876, pp. 3f.); Turner (2007, p. 7) has a translation of Mittag-Leffer's definition. [4] "Functionum integrarum haec ergo erit formula generalis $a + bz + cz^2 + dz^3 + ez^4 + fz^5 +$ & c.". Translation from Euler (2013, § 9).

Bottazzini, Umberto and Jeremy Gray (2013). *Hidden Harmony – Geometric Fantasies*. The Rise of Complex Function Theory. New York.
Euler, Leonhard (1748). *Introductio in analysin infinitorum*. Tomus primus. Lausanne.
Euler, Leonhard (2013). "Introductio in analysin infinitorum". Translated and annotated by Ian Bruce. URL: `http://www.17centurymaths.com/contents/introductiontoanalysisvol1.htm` (retrieved Feb. 25, 2018).
Iurato, Giuseppe (2014). "On some historical aspects of the theory of Riemann zeta function *[Preprint]*". URL: `http://philsci-archive.pitt.edu/11241/` (retrieved July 4, 2021).
Mittag-Leffler, Gösta (1876). "En metod att analytiskt framställa en funktion af rational karakter, hvilken blir oändlig alltid och endast uti vissa föreskrifna oändlighetspunkter, hvilkas konstanter äro påförhand angifna". *Öfversigt af kongl. vetenskaps-akademiens förhandlingar* 33, pp. 3–16.
Turner, Laura E. (2007). "The Mittag-Leffler Theorem: The Origin, Evolution, and Reception of a Mathematical Result, 1876–1884". MA thesis. Simon Fraser University.
Turner, Laura E. (2013). "The Mittag-Leffler Theorem: The origin, evolution, and reception of a mathematical result, 1876–1884". *Historia Mathematica* 40, pp. 36–83.
Weierstraß, Karl (1856). "Theorie der *Abel*'schen Functionen". *Journal für die reine und angewandte Mathematik* 52, pp. 285–379.
Weierstraß, Karl (1877). "Zur Theorie der eindeutigen analytischen Functionen". *Abhandlungen der Königlichen Akademie der Wissenschaften zu Berlin* (1876), pp. 11–60.
Weierstraß, Karl (1886). "Zur Theorie der eindeutigen analytischen Functionen". In: *Abhandlungen aus der Functionenlehre*. Berlin, pp. 1–52.

entropy

In its mathematical meaning, the term was introduced by Shannon (1948, pp. 393f.):

> Quantities of the form $H = -\sum p_i \log p_i$ (the constant K merely amounts to a choice of a unit of measure) play a central role in information theory as measures of information, choice and uncertainty. The form of H will be recognized as that of entropy as defined in certain formulations of statistical mechanics where p_i is the probability of a system being in cell i of its phase space. H is then, for example, the H in Boltzmann's famous H theorem.[1] We shall call $H = \sum p_i \log p_i$ the entropy of the set of probabilities $p_1 \ldots p_n$. If x is a chance variable we will write $H(x)$ for its entropy; thus x is not an argument of a function but a label for a number, to differentiate it from $H(y)$ say, the entropy of the chance variable y.

Shannon's paper was an immediate success and the term was adopted by Weaver (1949, p. 12), for example.

Tribus (1964, p. 354) recalled asking Shannon in 1961 about the origin of the term:

> He [Shannon] considered naming it "information" but felt that this word had unfortunate popular interpretations that would interfere with his intended uses of it in his new theory. He was inclined towards naming it "uncertainty" and discussed the matter with the late John Von Neumann. Von Neumann suggested that the function ought to be called "entropy" since it was already in use in some treatises on statistical thermodynamics [...]. Von Neumann, Shannon reports, suggested that there were two good reasons for calling the function "entropy." "It is already in use under that name," he is reported to have said, "and besides, it will give you a great edge in debates because nobody really knows what entropy is anyway."

The terminology was originally suggested in thermodynamics by Clausius (1865b, p. 390):

> We obtain the equation $\int \frac{dQ}{T} = S - S_0$ which, while somewhat differently arranged, is the same as that which was formerly used to determine S.[2] If we wish to designate S by a

---

[1] For references and discussion of this theorem, see Badino (2011) and the literature cited there.
[2] Q quantity of heat, T absolute temperature.

proper name we can say of it that it is the *transformation content* of the body, in the same way that we say of the quantity U that it is the *heat and work content* of the body. However, since I think it is better to take the names of such quantities as these, which are important for science, from the ancient languages, so that they can be introduced without change into all the modern languages, I proposed to name the magnitude S the *entropy* of the body, from the Greek word ἡ τροπὴ, a transformation. I have intentionally formed the word *entropy* so as to be as similar as possible to the word *energy*, since both these quantities, which are to be known by these names, are so nearly related to each other in their physical significance that a certain similarity in their names seemed to me advantageous.[1]

Clausius' paper was quickly translated, and the earliest uses in English and French appear to be Clausius (1867, p. 357) and Clausius (1865a, p. 391), respectively.[2] The term was also adopted but redefined by Tait (1868, p. 29): "We shall [...] use the excellent term Entropy in the opposite sense to that in which Clausius has employed it,–viz., so that the *Entropy of the Universe tends to zero*".[3]

Note. A historical account of Shannon's theory is Verdú (1998). For the concept in physics, see Müller (2007, pp. 47–126), for example; Baierlein (1992) briefly discusses the origin of the term. Bawden and Robinson (2015) present an excellent overview of the various meanings of *entropy*.

References

Badino, Massimiliano (2011). "Mechanistic slumber vs. statistical insomnia: the early history of Boltzmann's H-theorem (1868–1877)". *The European Physical Journal H* 36, pp. 353–378.

Baierlein, Ralph (1992). "How entropy got its name". *American Journal of Physics* 60, p. 1151.

Bawden, David and Lyn Robinson (2015). "'A few exciting words': Information and entropy revisited". *Journal of the Association for Information Science and Technology* 66, pp. 1965–1987.

Clausius, Rudolf (1865a). "Sur diverses formes facilement applicables qu'on peut donner aux équations fondamentales de la théorie mécanique de la chaleur". *Journal de mathématiques pures et appliquées* 10, pp. 361–400.

Clausius, Rudolf (1865b). "Ueber verschiedene für die Anwendung bequeme Formen der Hauptgleichungen der mechanischen Wärmetheorie". *Annalen der Physik und Chemie* 125, pp. 353–400.

Clausius, Rudolf (1867). *The Mechanical Theory of Heat, with Its Applications to the Steam-engine, and to the Physical Properties of Bodies*. Edited by T. Archer Hirst. London.

Klein, Martin J. (1969). "Gibbs on Clausius". *Historical Studies in the Physical Sciences* 1, pp. 127–149.

Magie, Willliam Francis, ed. (1963). *A Source Book In Physics*. Cambridge, Mass.

---

[1] "[...] Sucht man für S einen bezeichnenden Namen, so könnte man, ähnlich wie von der Grösse U gesagt ist, sie sey der *Wärme- und Werkinhalt* des Körpers, von der Grösse S sagen, sie sey der *Verwandlungsinhalt* des Körpers. Da ich es aber für besser halte, die Namen derartiger für die Wissenschaft wichtigen Grössen aus den alten Sprachen zu entnehmen, damit sie unverändert in allen neuen Sprachen angewandt werden können, so schlage ich vor, die Grösse S nach dem griechischen Worte ἡ τροπὴ, die Verwandlung, die *Entropie* des Körpers zu nennen. Das Wort *Entropie* habe ich absichtlich dem Worte *Energie* möglichst ähnlich gebildet, denn die beiden Größen, welche durch diese Worte benannt werden sollen, sind ihren physikalischen Bedeutungen nach einander so nahe verwandt, dass eine gewisse Gleichartigkeit in der Benennung mir zweckmässig zu sein scheint". Translation from Magie (1963, p. 234). [2] Clausius (1867, p. 357) is cited by the OED as the first use in English. [3] On Tait's misuse of the term, see Klein (1969, pp. 133f.): "Tait had misused the term 'entropy' in the first edition of his book: he had made a serious error in stating Clausius' definition, and then compounded it by arbitrarily deciding to use the same word 'entropy' to mean the negative of the quantity he incorrectly attributed to Clausius".

Müller, Ingo (2007). *A History of Thermodynamics*. The Doctrine of Energy and Entropy. Berlin and Heidelberg.
Shannon, C.E. (1948). "A Mathematical Theory of Communication". *The Bell System Technical Journal* 27, pp. 379–423.
Tait, Peter Guthrie (1868). *Sketch of Thermodynamics*. Edinburgh.
Tribus, Myron (1964). "Information theory and thermodynamics". In: *Heat Transfer, Thermodynamics, and Education*. Ed. by H.A. Johnson. New York, pp. 348–368.
Verdú, Sergio (1998). "Fifty Years of Shannon Theory". *IEEE Transactions on Information Theory* 44, pp. 2057–.
Weaver, Warren (1949). "The Mathematics of Communication". *Scientific American* 181, pp. 11–15.

enumerative geometry

In his book on enumerative geometry, Schubert (1879, p. 1) begins by stating that the "target of enumerative geometry is to answer all questions of the following form: '*How many* geometrical objects of given definition satisfy certain given *conditions*?'".[1] This characterization explains the choice of the term *abzählende Geometrie* (*abzählen* 'to count (off)'), which Schubert had already used slightly earlier in a series of papers entitled "Contributions to enumerative geometry", beginning with Schubert (1876). A letter by Schubert from 1876 shows that he was quite certain that *abzählende Geometrie* had not been in use earlier and that he coined the term in analogy to *darstellende Geometrie* 'descriptive geometry' (see Michel 2020, pp. 293, 431).[2]

The French translation *géométrie énumérative* appears soon afterwards, for example in Zeuthen (1881, p. 39). The English term appears at around 1890; for example, the *Century Dictionary* (1889-1891), in the entry for *geometry* (p. 2495), defines: "Enumerative or denumerative geometry, the theory of the number of solutions of geometrical problems, and of the number of incidences and coincidences in a diagram drawn under given conditions".[3]

Note. An early historical account of enumerative geometry is Loria (1888); see also Coolidge (1940, pp. 180–194), Kleiman (1976), and Michel (2020).

References

Coolidge, Julian Lowell (1940). *A History of Geometrical Methods*. Oxford.
Kleiman, Steven L. (1976). "Chasles's enumerative theory of conics: a historical introduction". In: *Studies in Algebraic Geometry*. Ed. by Abraham Seidenberg. Washington, D.C., pp. 117–138.
Loria, Gino (1888). "Notizie storiche sulla Geometria numerativa". *Bibliotheca mathematica* 2, pp. 39–48, 67–80.
Michel, Nicolas (2020). "Of Words and Numbers. The writing of generality in the emergence of enumerative geometry (1852-1893)". PhD thesis. Université de Paris.
Schubert, Hermann (1876). "Beiträge zur abzählenden Geometrie". *Mathematische Annalen* 10, pp. 1–116.
Schubert, Hermann (1879). *Kalkül der abzählenden Geometrie*. Leipzig.

---

[1] "Das äusserliche Ziel der abzählenden Geometrie ist die Beantwortung aller Fragen von folgender Form: '*Wieviel* geometrische Gebilde von bestimmter Definition erfüllen gewisse gegebene *Bedingungen*?'" (original emphasis changed from bold to italics). [2] The letter also shows that Schubert thought the only possible translation into French would be *géométrie numérique*, and that he had used *Geometrie der Anzahl* before (*Anzahl* means 'number' as in *the number of elements of set*. Schubert also notes that *Abzählung* as such had been in use in this context before. [3] In the *Encyclopædia Britannica* (9th ed. (1883), vol. xv, p. 630), Schubert's term is translated by *arithmic geometry*, which did not catch on, and neither did *geometrical totics* (Story 1880, p. 382); the term is explainable by Story's definition: "A *totient* is the number of things which satisfy certain conditions" (ibid.).

Story, William E. (1880). "On the Theory of Rational Derivation on a Cubic Curve". *American Journal of Mathematics* 3, pp. 356–387.
Zeuthen, H.G. (1881). "Théorie des figures projectives sur une surface du second ordre". *Mathematische Annalen* 18, pp. 33–68.

envelope

Envelopes of families of curves began to be studied in the 17th century. In this context, Torricelli's (1644, pp. 183f.) studies on ballistic curves are usually referred to: "Torricelli had demonstrated that all ballistic parabolas which are the trajectories of cannon-balls shot at different angles of elevation (initial velocity and vertical plane remaining) touch one fixed parabola. This parabola is called the 'safety parabola' since it defines the range of the cannon. It is the envelope of the ballistic parabolas" (Engelsman 1984, p. 22).

The concept is then found at the end of the 17th century in various studies on the calculus. In particular, Leibniz discussed envelopes in two articles (1692, 1694). He did not explicitly suggest a term for the concept, but did state the general problem: "to find the line that touches infinitely many lines given by position in ordered sequence" (Leibniz 1694, p. 311).[1]

The term *envelope* was not coined until the end of the 18th century. As has been pointed out by various historians (for example, by Scarpello and Scimone 2005, pp. 245f.), the term goes back to Monge. He introduced French *enveloppe*, which is derived from the verb *envelopper*, meaning 'envelop, enwrap', with reference to surfaces rather than lines in Monge (1786, p. 22); he also used the term in his *Feuilles d'analyse* (Monge 1801, no. 7).[2]

The English term *envelope* appears in Dupin (1827, p. 172), while the German translation *einhüllende Curve* appears to go back to Littrow (1823, p. 185).

Note. For this history of the concept, see Engelsman (1984), Scarpello and Scimone (2005), and Rothenberg (1908, pp. 358–362). Blåsjö (2012) discusses Leibniz' article from 1694 in detail.

References

Blåsjö, Viktor (2012). "The rectification of quadratures as a central foundational problem for the early Leibnizian calculus". *Historia Mathematica* 39, pp. 405–431.
Dupin, Charles (1827). *Mathematics Practically Applied to the Useful and Fine Arts*. Adapted to the State of the Arts in England by George Birkbeck. London.
Engelsman, Steven B. (1984). *Families of Curves and the Origins of Partial Differentiation*. Amsterdam and New York.
Leibniz, Gottfried Wilhelm (1692). "De linea ex lineis numero infinitis ordinatim ductis inter se concurrentibus formata [...]" *Acta Eruditorum* (Apr. 1692), pp. 168–171.
Leibniz, Gottfried Wilhelm (1694). "Nova calculi differentialis applicatio & usus ad multiplicem linearum constructionem ex data tangentium conditione". *Acta Eruditorum* (Apr. 1694), pp. 311–316.
Littrow, Joseph Johann (1823). *Analytische Geometrie*. Wien.
Monge, Gaspard (1786). "Réflexions sur quelques séries". *Mémoires de l'Académie royale des sciences de Turin* (1774-75, première partie), pp. 19–33.

---

[1] "invenire lineam, quae infinitas lineas ordinatim positione datas tangit". – By lines given by position in ordered sequence, Leibniz meant a family of curves. See *family*. [2] The term is also found in Monge (1787, e. g., p. 174).

Monge, Gaspard (1787). "Mémoire sur le calcul intégral des équations aux différences partielles". *Mémoires de l'Académie royale des sciences* (1784), pp. 118–192.
Monge, Gaspard (1801). *Feuilles d'analyse appliquée à la géométrie*. Paris.
Rothenberg, Siegfried (1908). *Geschichtliche Darstellung der Entwicklung der Theorie der singulären Lösungen totaler Differentialgleichungen von erster Ordnung mit zwei variablen Grössen*. Leipzig.
Scarpello, Giovanni Mingari and Aldo Scimone (2005). "The Work of Tschirnhaus, La Hire and Leibniz on Catacaustics and the Birth of the Envelopes of Lines in the 17th Century". *Archive for History of Exact Sciences* 59, pp. 223–250.
Torricelli, Evangelista (1644). *De motu gravium naturaliter descendentium et proiectorum libri duo*. In: *Opera geometrica*. Firenze, pp. 95–243.

epicycloid and *hypocycloid*

The first major treatise on epicycloids was published by de la Hire under the title *Traité des épicycloides* (1694).[1] At the very beginning of his treatise, de la Hire gave the following definition (see fig. 43): "If the circle ABD rolls on the arc *Abd* of some circle, the point D which is the extremity D of the diameter AD, describes in its movement the line D14, 15 16 17 d, which is called *epicycloid*" (de la Hire 1694, p. 1).[2]

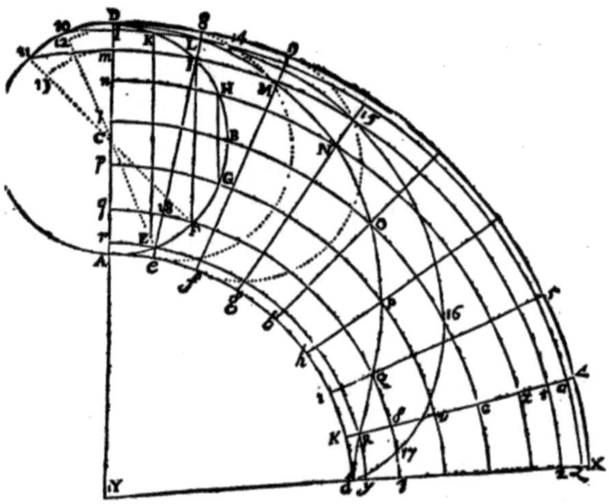

Figure 43: The epicycloid. From de la Hire (1694, p. 6).

He also distinguished between exterior (*extérieure*) and interior (*intérieure*) epicycloids, depending on whether the generating circle (*cercle générateur*) rolls outside or inside of the given circle; for the latter concept, the term *hypocycloid* was later introduced.

[1] In the preface, de la Hire states that he began to work on this treatise "about twenty years ago" and attributes the notion to Desargues. [2] "Si le cercle ABD roule sur l'arc *Abd* de quelque cercle, le point D qui est l'extrémité D du diamétre AD, décrira dans son mouvement la ligne D 14, 15 16 17 d, qu'on appelle *Epicycloide*".

As the wording of de la Hire's definition indicates, the term had already been in use at the time. It was used, for example, in notes by Huygens from 1678 which are based on communication with the Danish astronomer Ole Rømer (see Huygens 1934, p. 610); also, in 1678, Huygens gave a talk entitled "La mesure des lignes Epicycloides" (see Huygens 1934, p. 400).[1] The term was thus already in use in the 1670s, and indeed found earlier in Renaldini (1668, p. 124) and, in particular, in Wallis (1659, p. 74), which appears to be the earliest usage. Wallis described the curve as generated by the rolling movement of a circle around another circle, and explained the term on the basis of the term *epicycle* customary in ancient astronomy, where an epicyle (Greek *ἐπίκυκλος*, lit. 'on the circle') is essentially a small circle that moves around another circle.[2]

In conclusion, *epicycloid* appears to be due to Wallis, but it only gained wider use after the publication of de la Hire's treatise. In English, it was first (OED) used in Wallis (1696, p. 113).

The supplementary term *hypocycloid* is of later and unclear origin; it was used by Bernoulli (1722, p. 404) and later adopted by Euler, for example in his *Introductio in analysin infinitorum* (Euler 1748, § 523). The English term seems to have been first used in [anon.] (1807, p. 364), a review of Camus (1806), where *internal epicycloid* was used instead (Camus 1806, p. 22).

As seen above, the invention of the epicycloid has been attributed to various writers such as Rømer and Desargues. In this respect, it should also be mentioned that Dürer (1525, fol. D^v) constructed a particular epicycloid, calling it *spinnen lini*, lit. 'spider line', because, as he says, when it is drawn, it is similar to a spider.[3]

Note. For the history of the concept, see, e. g., Loria (1911, pp. 92–118), Teixeira (1909, pp. 155–171), and Woodbury (1958).

References

[anon.] (1807). "Camus *on the Teeth of Wheels*". *The Monthly Review* 53, pp. 363–366.

Bernoulli, Johann (1722). "Variae Solutiones & Constructiones Problematis". *Acta Eruditorum* (Aug. 1698), pp. 396–413.

Camus, Charles Étienne Louis (1806). *A Treatise on the Teeth of Wheels*. Translated from the French. London.

Dürer, Albrecht (1525). *Underweysung der Messung, mit dem Zirckel und Richtscheyt*. Nürnberg.

Dürer, Albrecht (1605). *Institutionum Geometricarum Libri Quatuor*. Arnheim.

Euler, Leonhard (1748). *Introductio in analysin infinitorum*. Tomus secundus. Lausanne.

de la Hire, Philippe (1694). "Traité des épicycloides". In: *Mémoires de mathématique et physique*. Paris, pp. 1–78.

Huygens, Christiaan (1934). *Œuvres complètes*. Tome dix-huitième. L'horloge à pendule ou à balancier de 1666 à 1695. Anecdota. La Haye.

[1] Some authors attribute the notion to Ole Rømer, a Danish astronomer. In particular, Leibniz supported this view, pointing out that Rømer invented the concept in 1674 (see, for example Leibniz 1692, p. 444 and Leibniz 2011, p. 729) . Leibniz does not indicate whether Rømer also used the *term*, though, and publications by Rømer in which epicycloids occur are not extant (see the editorial comments in Huygens 1934, p. 602). [2] "Non inutilis est contemplatio Cycloidum quae generantur ex circulo super Basim circularem revoluto, hujus generis vocentur Epicycloides, tales enim curvas Planetae in Epicyclis vecti describunt". – The usage appears at the very end of Wallis' book, which may be the reason why it seems to have been generally overlooked. [3] According to Loria (1911, p. 93), this is the oldest consideration of an epicycloid. The corresponding Latin term *linea aranei* (*aranea* 'spider') is found in Dürer (1605, p. 37).

Leibniz, Gottfried Wilhelm (1692). "Generalia de natura linearum". *Acta Eruditorum* (Sept. 1692), pp. 440–446.
Leibniz, Gottfried Wilhelm (2011). *Sämtliche Schriften und Briefe*. Dritte Reihe. Mathematischer, naturwissenschaftlicher und technischer Briefwechsel. Siebenter Band. Juli 1696-Dezember 1698. Hannover.
Loria, Gino (1911). *Spezielle algebraische und transzendente ebene Kurven*. Theorie und Geschichte. Autorisierte, nach dem italienischen Manuskript bearbeitete deutsche Ausgabe von Fritz Schütte. Zweite Auflage. Zweiter Band. Leipzig.
Renaldini, Carlo (1668). *De resolutione, & compositione mathematica*. Libri dvo. Padua.
Teixeira, Francisco Gomes (1909). *Traité des courbes spéciales remarquables plans et gauches*. Tome II. Coïmbra.
Wallis, John (1659). *Tractatus Duo*. Prior, De Cycloide et corporibus inde genitis. Posterior, epistolaris in qua agitur De Cissoide, et Corporibus inde genitis: et De Curvarum, tum Linearum *Εὐθύνσει*, tum Superficierum *Πλατυσμῷ*. Oxford.
Wallis, John (1696). "An extract of a letter from the Reverend Dr. John Wallis to Richard Waller". *Philosophical Transactions of the Royal Society of London* 19, pp. 111–113.
Woodbury, Robert S. (1958). "The First Epicycloidal Gear Teeth". *Isis* 49, pp. 375–377.

epimorphism

The term *epimorphism* came up in the 1950s. Early uses in print include Cartan (1953, p. 44), where, in a talk given in March 1953, French *épimorphisme* is defined for Abelian groups or modules as a homomorphism that is onto (Greek *ἐπί* means just that); the same definition was given by Schwartz (1954, p. 346) in a talk from December 1953. The English term was used by Eilenberg and Mac Lane (1954, pp. 83f.), who defined: "an *epimorphism* $f : A \to B$ is a homomorphism with $f(A) = B$. Thus 'epimorphism' means 'homomorphism onto'". The term was not used in Bourbaki's publications up to that time; it did appear, however, in a redaction (by Dieudonné, dated April 1953) of their book on set theory (Bourbaki 1953, p. 14).

References

Bourbaki, Nicolas (1953). "Rédaction n°180. Ensembles. Chapitre IV. Structures (état 7 ?)" URL: `http://archives-bourbaki.ahp-numerique.fr/items/show/594` (retrieved Aug. 19, 2022).
Cartan, Henri (1953). "Variétés analytiques complexes et cohomologie". In: *Colloque sur les fonctions de plusieurs variables tenu à Bruxelles du 11 au 14 mars 1953*. Liège, pp. 41–55.
Eilenberg, Samuel and Saunders Mac Lane (1954). "On the Groups $H(\Pi, n)$, II: Methods of Computation". *Annals of Mathematics* 60, pp. 49–139.
Schwartz, Laurent (1954). "Solution élémentaire d'une équation aux dérivées partielles à coefficients constants". *Séminaire Bourbaki* 2, exposé 87, pp. 345–350.

equal

The concept 'equal' is outlined in the common notions at the beginning of the *Elements* (but never defined in the text):

1. Things which are equal to the same thing are also equal to one another.
2. If equals be added to equals, the wholes are equal.
3. If equals be subtracted from equals, the remainders are equal.
4. Things which coincide with one another are equal to one another.
5. The whole is greater than the part.[1]

[1] Translation from Euclid (1956, p. 155).

Fried (2009, p. 3) describes the fourth common notion as "the basic criterion, the basic test, for equality: if one figure can be fitted exactly on top of another, then the figures are equal".

Euclid used *ἴσος*[1] in the sense of common notion 4 at first; then, in book I, prop. 35, it appears in the meaning 'equal in area': "Parallelograms which are on the same base and in the same parallels equal one another". The semantics of *ἴσος* is controversially discussed in the literature. According to Heath, "we are in this proposition introduced for the first time to a new conception of equality between figures"; Fried (2009, p. 3) regards this view as "misleading" and argues that "the *Elements* presents a broad notion of equality whose full extent lies in the possibilities allowed by the five common notions. It is not restricted to the case where one figure fits on another, which we call congruence; but, it is rooted in this very geometrical idea. This is a crucial point: proposition I.35 does not introduce a new idea of equality, equality through equality of area; rather, the areas are equal because the figures are equal, and the figures are equal because they are generated by shapes that fit on another".[2]

Euclid (e. g., book XI, prop. 29) also used *ἴσος* for figures that have the same volume. The word was in use and applied to various mathematical objects before Euclid; see Mugler as cited in the note at the end of this entry.

In Latin texts, the words *aequalis* and *aequus* were commonly used for *ἴσος*. Boethius gave this definition: "Something is equal [*aequale*] when that which is compared does not fall short by virtue of its smaller sum nor exceed by virtue of a larger sum" (Boethius 1867, p. 45).[3]

In modern languages, the word is usually first found in non-mathematical contexts. For example, the OED cites Chaucer's *Treatise on the astrolabe* as a first use. The term is, however, found in various early mathematical texts in the vernacular; for example, Italian *iguali* was used in the first Italian text on algebra, dating from 1328 (Van Egmond 1978, p. 166). The equal sign was first used in print by Recorde (1557, fol. Ffi^{v}): "And to auoide the tediouse repetition of these woordes: is equalle to: I will sette as I doe often in woorke vse, a paire of paralleles, or Gemowe[4] lines of one lengthe, thus: ═══, bicause noe .2. thynges, can be moare equalle".

See also *similar* (geometry) and *congruent* (geometry).

Note. For the Greek term, see Mugler (1958-59, pp. 227-231), Fried (2009), and von Fritz (1959). The Latin terms are discussed by Guillaumin (2020, pp. 25f., 28), the Arabic ones by Oaks (2010, pp. 266f.) and Rashed (2017, pp. 390f., 521–523, 712f., 750–753).

References

Boethius, Anicius Manlius Severinus (1867). *De institutione arithmetica libri duo, De institutione musica libri quinque*. Accedit Geometria quae fertur Boetii. Ed. by Gottfried Friedlein. Leipzig.

Van Egmond, Warren (1978). "The earliest vernacular treatment of algebra: The Libro di Ragioni of Paolo Gerardi (1328)". *Physis* 20, pp. 155–189.

---

[1] This was not the only word used for 'equal'. Proclus reports that Thales used *ὅμοιος* for equal angles. See *similar*. [2] In the succinct words of Fried and Unguru (2001, p. 230): "things are equal if they coincide or if they are the result of adding or subtracting coinciding things from coinciding things". [3] "aequale quidem est, quod ad aliquid comparatum neque minore summa infra est, neque maiore transgreditur". Translation from Masi (2006, p. 100). [4] From French *gémeau* 'twin', a variant form of *jumeau*.

Euclid (1956). *The Thirteen Books of Euclid's Elements*. Volume I. Introduction and Books I, II. Ed. by Thomas L. Heath. New York.
Fried, Michael N. (2009). "Similarity and Equality in Greek Mathematics: Semiotics, History of Mathematics, and Mathematics Education". *For the Learning of Mathematics* 29, pp. 2–7.
Fried, Michael N. and Sabetai Unguru (2001). *Apollonius of Perga's* Conica. Text, Context, Subtext. Leiden et al.
von Fritz, Kurt (1959). "Gleichheit, Kongruenz und Ähnlichkeit in der antiken Mathematik bis auf Euklid". *Archiv für Begriffsgeschichte* 4, pp. 7–81.
Guillaumin, Jean-Yves (2020). *Dictionnaire de la terminologie latine ancienne de l'arithmétique et de la géométrie*. Paris.
Masi, Michael (2006). *Boethian Number Theory*. Translation of the *De Institutione Arithmetica*. Amsterdam and New York.
Mugler, Charles (1958-59). *Dictionnaire historique de la terminologie géométrique des Grecs*. Paris.
Oaks, Jeffrey A. (2010). "Equations and Equating in Arabic Mathematics". *Archives Internationales d'Histoire des Sciences* 60, pp. 265–298.
Rashed, Roshdi, ed. (2017). *Lexique historique de la langue scientifique arabe*. Hildesheim.
Recorde, Robert (1557). *The Whetstone of Witte*. London.

equation

When Descartes published his *Géométrie*, the term *equation* was already in common use,[1] but Descartes certainly contributed significantly to its spread. He explained:

> [...] some general statements must be made concerning the nature of equations. An equation consists of several terms, some known and some unknown, some of which are together equal to the rest; or rather, all of which taken together are equal to nothing; for this is often the best form to consider. Every equation can have as many distinct roots (values of the unknown quantity) as the number of dimensions of the unknown quantity in the equation (Descartes 1637, pp. 371f.).[2]

When the term was actually used first is a difficult matter to decide. In early uses of *aequatio*, the word usually has a different meaning: "As with the terms *aequatio* and *aequationis* in early Latin works on algebra, we should be careful interpreting these terms in the modern way. They basically mean the act of making even–an essential operation in the algebraic solution of problems. They do not necessarily mean an equation in the sense of symbolic algebra" (Heeffer 2007, p. 96).[3] This statement also appears to apply to the use of the word by Fibonacci, who used it once in the *Liber abaci* (Fibonacci 1857, p. 407),[4] which Hughes (1996, p. 216) cites as "the certain earliest use of the word *equation*".[5] Høyrup (2019, p. 419), on the other

[1] That does not mean that the term was universally adopted. Prestet (1675), for example, used *égalité* instead of *équation*. [2] "il faut que ie die quelque chose en general de la nature des Equations; c'est a dire des sommes composées de plusieurs termes partie connus, & partie inconnus, dont les vns sont esgaux aux autres, ou plutost qui considerés tous ensemble sont esgaux a rien, car ce sera souuent le meilleur de les considerer en cete sorte. Scachés donc qu'en chasque Equation, autant que la quantité inconnue a de dimensions, autant peut il y auoir de diuerses racines, c'est a dire de valeurs de cete quantité". Translation from Descartes (1925, pp. 156, 159). [3] Another sense is discussed by Hughes (1996, pp. 217f.): "during the Middle Ages which we are discussing – the twelfth through fourteenth century, the commonly accepted meaning of equation was the method of dividing the celestial sphere into twelve houses of equal size [...] Such a preponderant astronomical meaning of *equation*, therefore, may well account for the absence of the word in mathematical, specifically algebraic, treatises". [4] The word is also once found in the *Practica Geometrie* (Fibonacci 1862, p. 210). [5] Referring to the same use, Tropfke (1937, p. 28) even regarded the word as already "fully established". In view of the rarity of the term, this seems a premature conclusion.

hand, rejects the view that Fibonacci used *equation* in the usual sense: "Fibonacci only has the action of equating – the isolated appearance of *equatio* [...] is to be understood as a corresponding verbal noun".

Another difficulty is based on the question of what actually constitutes an equation. As Heeffer (2007, p. 98) understands the concept, "[t]he observation that two polynomials are numerically identical does not in itself constitute an equation. An operation on an equation, however, would be". The first instance of such an operation that Heeffer found is Cardano (1539, fol. HH^r).

Viète (1591, fol. 4^r) later attempted to explain the word: "In analysis the word 'equation,' standing by itself, means an equality properly constructed in accordance with [the rules of] zetetics[1]. 2. Thus an equation is a comparison of an unknown magnitude and a known magnitude".[2]

The Arabic term *muʿādala* is derived from a verb meaning 'to make equal'; according to Rashed (2017, p. 534), it was first used by Qusta ibn Luqa in his translation of Diophantus' works on arithmetic. Diophantus himself used *ἰσότης* and *ἴσωσις*, which are also derived from a verb meaning 'to make equal' (for references and discussion, see Christianidis and Oaks 2013, pp. 141–143).

Note. For the Greek terminology, see Christianidis and Oaks (2013, pp. 141–143) and Thomaidis (2011). For the Arabic terminology, see Rashed (2017, pp. 534f.) and Oaks (2010). The early Latin terminology is discussed by Hughes (1996, pp. 215–219).

References

Cardano, Gerolamo (1539). *Practica arithmetice*. Milano.

Christianidis, Jean and Jeffrey A. Oaks (2013). "Practicing algebra in late antiquity: The problem-solving of Diophantus of Alexandria". *Historia Mathematica* 40, pp. 127–163.

Descartes, René (1637). *La Géométrie*. In: *Discours de la méthode*. Leiden, pp. 295–413.

Descartes, René (1925). *The Geometry of René Descartes*. Translated from the French and Latin by David Eugene Smith and Marcia L. Latham. Chicago and London.

Fibonacci (1857). *Scritti di Leonardo Pisano*. Pubblicati da Baldassare Boncompagni. Volume I. Roma.

Fibonacci (1862). *Scritti di Leonardo Pisano*. Pubblicati da Baldassare Boncompagni. Volume II. Roma.

Heeffer, Albrecht (2007). "Learning concepts through the history of mathematics". In: *Philosophical Dimensions in Mathematics Education*. Ed. by Karen François and Jean Paul Van Bendegem. New York, pp. 83–103.

Høyrup, Jens (2019). *Selected Essays on Pre- and Early Modern Mathematical Practice*. Cham.

Hughes, Barnabas (1996). "Arabic Algebra. Victim of Religious and Intellectual Animus". In: *Mathematische Probleme im Mittelalter*. Der lateinische und arabische Sprachbereich. Ed. by Menso Folkerts. Wiesbaden, pp. 197–220.

Klein, Jacob (1992). *Greek Mathematical Thought and the Origin of Algebra*. Translated by Eva Brann. New York.

Oaks, Jeffrey A. (2010). "Equations and Equating in Arabic Mathematics". *Archives Internationales d'Histoire des Sciences* 60, pp. 265–298.

Prestet, Jean (1675). *Elemens des mathematiques*. Paris.

[1] Zetetics is defined by Viète (1591, fol. 4^r) as the art "by which is found the equation or proportion between the magnitude that is being sought and those that are given". Translation from Klein (1992, p. 321). [2] "Aequationis vox simpliciter prolata in Analyticis de æqualitate per Zetesim rite ordinata accipitur. 2. Aequatio est magnitudinis incertae cum certa comparatio". Translation from Viète (1983, p. 29).

Rashed, Roshdi, ed. (2017). *Lexique historique de la langue scientifique arabe*. Hildesheim.
Thomaidis, Yannis (2011). "Some remarks on the meaning of equality in Diophantos's *Arithmetica*". *Historia Mathematica* 38, pp. 28–41.
Tropfke, Johannes (1937). *Geschichte der Elementar-Mathematik*. Dritter Band. Proportionen, Gleichungen. Dritte, verbesserte und vermehrte Auflage. Berlin and Leipzig.
Viète, François (1591). *In artem analyticam isagoge*. Tours.
Viète, François (1983). *The Analytic Art*. Nine Studies in Algebra, Geometry and Trigonometry from the *Opus Restitutae Mathematicae Analyseos, seu Algebrâ Novâ*. Translated by T. Richard Witmer. New York.

equiangular

The word ισογώνιος, from the words for 'equal' and 'angle', was used in Greek mathematics in essentially two senses, namely when the angles of two or more figures are equal one by one and when the angles of a polygon are all equal (Mugler 1958-59, pp. 222f.). Euclid used the term in both senses in the *Elements*, for example in book IV, prop. 2: "In a given circle to inscribe a triangle equiangular with a given triangle",[1] and in book IV, prop. 11: "In a given circle to inscribe an equilateral and equiangular pentagon".[2]

The Latin term *aequiangulus* is only attested once before the 12th century (Ayuso García 2008, p. 1517), namely in the Pseudo-Boethian *Geometria I* (Blume, Lachmann, and Rudorff 1848, p. 392). Afterwards, it is found in some of the first translations of the *Elements* into Latin, for example in the first translation made directly from the Greek (Busard 1987, p. 15).

The German term *gleichwinklig*, which is still in use, is a direct translation of *aequiangulus* and early found in the *Geometria culmensis* from around 1400 (Mendthal 1886, p. 20)[3]; in French, *équiangle* appears in Cardano (1556, fol. 303^r) and *équiangulaire* in Jacquinot (1558, fol. 96^v). English *equiangle* appears in the 16th century in L. Digges and T. Digges (1571, fol. B iij^v) and Euclid (1570, fol. 111^r); *equiangular* is of a later date and is, apparently first, found in Ramus (1636, p. 142).[4]

For the term *equiangular spiral*, see *logarithmic spiral*. See also *isogonal*.

References

Ayuso García, Manuel (2008). "La terminología latina de la geometría en Marciano Capela". PhD thesis. Madrid: Universidad Nacional de Educación a Distancia.
Blume, Friedrich, Karl Lachmann, and Adolf Rudorff, eds. (1848). *Die Schriften der römischen Feldmesser*. Erster Band. Texte und Zeichnungen. Berlin.
Busard, H.L.L. (1987). *The Mediaeval Latin Translation of Euclid's Elements Made Directly from the Greek*. Stuttgart.
Cardano, Gerolamo (1556). *Les livres de Hierome Cardanus medecin milannois intitulés de la Subtilité, & subtiles inventions, ensemble les causes occultes, & raisons d'icelles*. Traduis de Latin en Francois, par Richard le Blanc. Paris.
Digges, Leonard and Thomas Digges (1571). *A geometrical practise, named Pantometria diuided into three bookes, Longimetra, Planimetra, and Stereometria*. London.
Euclid (1570). *The Elements of Geometrie of the most auncient Philosopher Euclide of Megara*. Faithfully (now first) translated into the Englishe toung, by H. Billingsley. London.
Euclid (1956). *The Thirteen Books of Euclid's Elements*. Volume II. Books III-IX. Ed. by Thomas L. Heath. New York.

[1] Translation from Euclid (1956, p. 81). [2] Translation from Euclid (1956, p. 100). [3] The spelling there is *gleychwinkelyck*. [4] The word is used only a few times in the text; usually, *equiangled* is preferred (e. g., Ramus 1636, p. 42).

Jacquinot, Dominique (1558). *L'usage de l'astrolabe, avec un petit traicté de la Sphere*. Seconde Edition. Paris.
Mendthal, Hans (1886). *Geometria culmensis*. Ein agronomischer Tractat aus der Zeit des Hochmeisters Conrad von Jungingen (1393-1407). Leipzig.
Mugler, Charles (1958-59). *Dictionnaire historique de la terminologie géométrique des Grecs*. Paris.
Ramus, Petrus (1636). *Via regia ad geometriam. The Vvay to Geometry*. Written in Latine by Peter Ramus, and now Translated and much enlarged by the Learned Mr. William Bedwell. London.

equicontinuity

In an article translated from Italian, Arzelà (1904, p. 52) presented a new English term:

> A set of functions taken from the set a is said to be *equably continuous* in T if, given an arbitrarily small positive number σ, a positive number δ can always be found such that in every part of T whose maximum chord does not exceed δ the oscillation of each of the functions in question is less than σ.

The terminology goes back to Ascoli (1884, p. 545), who used Italian *egualmente continua*; the term was adopted by Arzelà (1889, p. 344), for example.

English *equicontinuous* appears to have been first used by Moore (1921, p. 42), who referred to Ascoli, as did Hahn (1921, p. 300), when he proposed the German term *gleichgradig stetig* (*gleichgradig* is made up of the words for 'equal' and 'degree'). French *égale continuité* was used by Fréchet (1906, p. 11).

References

Arzelà, Cesare (1889). "Funzioni di linee". *Atti della Reale Accademia dei Lincei. Rendiconti* 5, pp. 342–348.
Arzelà, Cesare (1904). "Note on Series of Analytic Functions". *Annals of Mathematics* 5, pp. 51–63.
Ascoli, Guilio (1884). "Le curve limite di una varietà data di curve". *Atti della Reale Accademia dei Lincei. Memorie* 18, pp. 521–586.
Fréchet, Maurice (1906). "Sur quelques points du calcul fonctionnel". *Rendiconti del Circolo Matematico di Palermo* 22, pp. 1–74.
Hahn, Hans (1921). *Theorie der reellen Funktionen*. Berlin.
Moore, Robert Lee (1921). "Concerning Certain Equicontinuous Systems of Curves". *Transactions of the American Mathematical Society* 22, pp. 41–55.

equilateral

In Euclid's *Elements* (book I, def. 20), equilateral triangles are defined as follows: "Of trilateral figures, an *equilateral triangle* is that which has its three sides equal".[1] Euclid used the word *ισόπλευρος*, which is composed of the words for 'equal' and 'side'. The term was in use before Euclid; see Mugler (1958-59, pp. 224f.).[2]

Latin *aequilateralis* is first attested, according to Ayuso García (2008, p. 717), with the 3rd-century writer Censorinus, who used it with reference to hexagons (Sallmann 1983, p. 14).[3] Ausonius (Evelyn-White 1919, p. 374) and Martianus Capella (Willis 1983, p. 252) name the Greek term *ισόπλευρος* and point out that *aequilaterus* is its Latin counterpart.

---

[1] Translation from Euclid (1956, p. 154). [2] The term was also used in connection with numbers; in the *Theaetetus* (147), Plato defined: "Any number which is the product of a number multiplied by itself we likened to the square figure, and we called such a number 'square' or 'equilateral'" (translation from Cornford 1957, p. 23). [3] In the fragment *De figuris* by Pseudo-Censorinus, *aequilaterus* is used with reference to triangles (Sallmann 1983, p. 69).

As for the vernacular languages, the French term *triangle équilatère* is first found, according to Pötters and Schreiber (2000, pp. 283f.), in a 13th-century text on practical geometry (Victor 1979, p. 476). In the first printed French book on geometry, de Bouelles (1511, fol. viij^{v}) uses *triangle ysopleure* in his definition but also occasionally uses *équilatère* (de Bouelles 1511, fol. x^{v}). The corresponding Italian terminology is found in the 14th century (Feola 2008, p. 180).

The notion is discussed by Recorde (1551, fol. B iii^{r}) in the following terms: "There is also an other distinction of the names of triangles, according to their sides, whiche other be all equal [...] and that the Greekes doo call *Isopleuron*, and Latine men *æquilaterum*: and in english it may be called a *threlike triangle*" (*threlike* is to be explained as *three-like*).[1] In the first English translation of the *Elements*, *equilater* (also *equilatre*) is used (Euclid 1570, fol. 4^{r}). The word *equilateral* is found, in a non-mathematical text, in Kett (1585, fol. N 3^{r}); Rathborne (1616, p. 9) uses it in a mathematical setting.

In the *Geometria culmensis* from around 1400, the word *glychwendig* is used (Mendthal 1886, p. 42); the word is made up of *gleich* 'equal' and *Wand*, which literally means 'wall' but is sometimes used in that treatise for 'side'. The standard term is *gleichseitig*; according to the DWB, this word first appears in Schmid (1539, p. 16).[2] The word is also found in the first German translation of books I-VI of the *Elements* (Euclid 1562, p. 3).

See also *side*.

Note. A concise overview of the terminology is found in Tropfke (1940, p. 81). For the Greek term, see Mugler (1958-59, pp. 224f.); for the Latin, see Ayuso García (2008, pp. 717f.) and Guillaumin (2020, p. 28).

References

Ayuso García, Manuel (2008). "La terminología latina de la geometría en Marciano Capela". PhD thesis. Madrid: Universidad Nacional de Educación a Distancia.

de Bouelles, Charles (1511). *Geometrie en francoys*. Paris.

Cornford, Francis Macdonald (1957). *Plato's Theory of Knowledge*. The *Theaetetus* and the *Sophist* of Plato translated with a running commentary. New York.

Euclid (1562). *Die Sechs Erste Bücher Euclidis*. Auß Griechischer sprach in die Teütsch gebracht ... Durch Wilhelm Holtzmann genannt Xylander. Basel.

Euclid (1570). *The Elements of Geometrie of the most auncient Philosopher Euclide of Megara*. Faithfully (now first) translated into the Englishe toung, by H. Billingsley. London.

Euclid (1956). *The Thirteen Books of Euclid's Elements*. Volume I. Introduction and Books I, II. Ed. by Thomas L. Heath. New York.

Evelyn-White, Hugh G., ed. (1919). *Ausonius*. In two volumes. I. London and New York.

Feola, Francesco (2008). *Gli esordi della geometria in volgare*. Un volgarizzamento trecentesco della *Practica Geometriae* di Leonardo Pisano. Firenze.

Guillaumin, Jean-Yves (2020). *Dictionnaire de la terminologie latine ancienne de l'arithmétique et de la géométrie*. Paris.

Kett, Francis (1585). *The glorious and beautifull garland of mans glorification*. London.

---

[1] Recorde correspondingly uses *tweyleke* (*two-like*) for isosceles triangles, and *noveleke* (*non-like*) for scalene triangles. [2] Schmid does not give a definition of *gleichseitig*. However, he *does* define an equilateral triangle (Schmid 1539, p. 13), but there he simply says it is "called *Triangulus equilaterus* or *Isopleurus* by the learned" ("bey den gelerten Triangulus equilaterus oder Isopleurus genannt").

Mendthal, Hans (1886). *Geometria culmensis*. Ein agronomischer Tractat aus der Zeit des Hochmeisters Conrad von Jungingen (1393-1407). Leipzig.
Mugler, Charles (1958-59). *Dictionnaire historique de la terminologie géométrique des Grecs*. Paris.
Pötters, Wilhelm and Gerhard Schreiber (2000). "Le pratike de geometrie (1276/77) – Sprach- und kulturhistorische Betrachtungen zu einem mathematischen Fachtext des Mittelalters". In: *Romania una et diversa*. Philologische Studien für Theodor Berchem zum 65. Geburtstag. Ed. by Martine Guille and Reinhard Kiesler. Tübingen, pp. 267–292.
Rathborne, Aaron (1616). *The surueyor*. London.
Recorde, Robert (1551). *The patheway to knowledg*. London.
Sallmann, Klaus, ed. (1983). *Censorini de die natali liber ad Q. Caerellium*. Leipzig.
Schmid, Wolffgang (1539). *Das erst buch der Geometria*. Nürnberg.
Tropfke, Johannes (1940). *Geschichte der Elementar-Mathematik*. Vierter Band. Ebene Geometrie. Dritte, verbesserte und vermehrte Auflage besorgt von Dr. Kurt Vogel. Berlin and Leipzig.
Victor, Stephen K. (1979). *Practical Geometry in the High Middle Ages*. "Artis cuiuslibet consummatio" and the "Pratike de geometrie". Philadelphia.
Willis, James, ed. (1983). *Martianus Capella*. Leipzig.

equivalent and *equivalence relation*

The names for reflexive, symmetric, and transitive relations are all 19th-century creations (see the respective entries). The general notion of equivalence relation is found in the 1890s, when reflexive, symmetric, and transitive relations are grouped together, without using the term *equivalence relation,* which only comes up in the 20th century.

The term *equivalent,* however, had been used earlier for various relations which *are* equivalence relations in the modern sense.[1] For example, Legendre (1794, p. 57) defined equivalence with respect to geometrical figures: "I will call *equivalent figures* those whose surfaces are equal".[2] Another example is found in Gauß' *Disquisitiones arithmeticae*, where *equivalent forms* are defined: "if F' is contained in F and F is contained in F', the determinants of the forms will be equal, and $(\alpha\delta - \beta\gamma)^2 = 1$. In this case we will call the forms *equivalent*" (Gauß 1801, § 157).[3] The corresponding French term *formes équivalentes* appears in Gauß (1807, § 157); the English term is of later origin and seems to appear first in Smith (1862, pp. 80f.). For this notion, Eisenstein (1844, p. 91) introduced the symbol $\sim$.[4]

A last example is the use of *equivalent* in set theory, which goes back to Georg Cantor. In a paper from 1878, he defined what it meant for two sets to have the same power (the existence of a bijection): "If two well-defined manifolds M and N can be coordinated with each other univocally and completely, element by element (which,

[1] The Latin word *aequivalentia* is made up of words meaning 'equal' and 'value'. The word comes up in medieval Latin. No attempt has been made here to trace its exact origins; suffice it to say that by the 14th century, it was in use in the sense of logical equivalence. For example, John Buridan used the term in this way: "equipollence [*aequipollentia*] is the equivalence [*aequiualentia*] of two propositions in truth or falsity so that either of them follows from the other by a formal consequence, and so that it is necessary that if they are formed at the same time, then, if the one is true, then so is the other, and if the one is false, the other is false too" (Buridan 2001, p. 42).
[2] "J'appellerai *figures équivalentes* celles dont les surfaces sont égaux". By saying "the surfaces are equal" Legendre means that the surfaces are equal in area. [3] "si tum F' sub F, tum F sub F' contenta est, formarum determinantes erunt aequales atque $(\alpha\delta - \beta\gamma)^2 = 1$". Translation from Gauß (1986, § 157). – The notion is found earlier in Lagrange (1775, p. 285), which Gauß (1801, § 222) duly acknowledges. See Cox (1989, p. 24). [4] For the symbol, see Cajori (1929, p. 122).

if possible in one way, can always happen in many others), we shall employ in the sequel the expression, that those manifolds have the *same power* or, also, that they are *equivalent*" (Cantor 1878, p. 242).[1]

In the 1890s, the relations that are reflexive, transitive, and symmetric begin to be regularly classed together. For example, Peano (1894, pp. 45f.) states that such a relation "has the properties of equality".[2] It remained common practice, however, to simply list these three criteria, without using a generic name. One such example is found in Krull (1923, p. 103): "The property of two elements of being congruent modulo $\mathfrak{p}*$ is symmetric, reflexive, transitive".[3]

An early suggestion for a generic term was made by Jourdain (1913, p. 492) at the very beginning of his paper: "I call a relation which is reflexive, symmetrical, and transitive an *isoid relation*".[4] The term, which is based on Greek *ἴσος* 'equal', stresses the similarity of equivalence relations to equality.

While Jourdain's term did not find much appeal, soon afterwards Loewy (1915, p. 21) defined an *equivalence* (German *Äquivalenz*) in the following way: "For a system of elements, a relation is called an *equivalence* [...] if it is *a determined, reflexive, symmetric and transitive relation*".[5] The German compound *Äquivalenzrelation* spread through Hasse (1926, p. 17) and van der Waerden (1930, p. 14); the corresponding English term appears in the 1930s, for example in Clifford (1933, p. 869): "Conjugacy as thus defined is an equivalence relation, that is, it is reflexive, symmetric, and transitive".[6]

Note. For the history of the concept and terminology, see Asghari (2019); many of the examples discussed in this article (although not the important one by Loewy) have also been pointed out there.

References

Asghari, Amir (2019). "Equivalence: an attempt at a history of the idea". *Synthese* 196, pp. 4657–4677.

---

[1] "Wenn zwei wohldefinirte Mannigfaltigkeiten M und N sich eindeutig und vollständig, Element für Element, einander zuordnen lassen (was, wenn es auf eine Art möglich ist, immer auch noch auf viele andere Weisen geschehen kann), so möge für das Folgende die Ausdrucksweise gestattet sein, dass diese Mannigfaltigkeiten *gleiche Mächtigkeit* haben, oder auch, dass sie *äquivalent* sind". Translation from Ferreirós (2007, p. 188). [2] "a les propriétés de l'égalité". – According to Felgner (2020, p. 117), Peano here introduces the term *equivalence relation*. This is not the case. [3] "Die Eigenschaft zweier Elemente, modulo $\mathfrak{p}*$ kongruent zu sein, ist symmetrisch, reflexiv, transitiv". – Numerous further examples are easily found, for example in Russell (1903, p. 114) and Burali-Forti (1899, p. 247). [4] Jourdain refers to Peano and his school for the notion. Jourdain's usage is pointed out by Asghari (2019, p. 4659), who also states that "Jourdain was one of the first who suggested a decontextualized term for what we now know as 'equivalence relation'. However, he was not successful at popularizing the term". [5] "Eine Relation heißt für ein System von Elementen eine *Äquivalenz* [...] wenn sie *eine determinierte, reflexive, symmetrische und transitive Relation* ist". For Loewy (1915, p. 20), *determined* means that for any two elements, the relation either holds or it does not. [6] In view of these examples, it is surprising that Tarski (1937, p. 60) stated: "Until now, no term has been introduced to designate the totality of relations that are reflexive, symmetric and transitive at the same time. Sometimes relations of this kind are called *equalities* or *equivalences*" ("Man hat bisher keinen besonderen Terminus zur Bezeichnung der Gesamtheit von zugleich reflexiven, symmetrischen und transitiven Beziehungen eingeführt. Manchmal werden Beziehungen dieser Art allgemein *Gleichheiten* oder *Äquivalenzen* genannt").

Burali-Forti, Cesare (1899). "Sur l'égalite, et sur l'introduction des éléments". *L'Enseignement mathématique* 1, pp. 246–261.
Buridan, John (2001). *Summulae de Dialectica*. An annotated translation, with a philosophical introduction by Gyula Klima. New Haven and London.
Cajori, Florian (1929). *A History of Mathematical Notations*. Volume II. Notations mainly in higher mathematics. Chicago.
Cantor, Georg (1878). "Ein Beitrag zur Mannigfaltigkeitslehre". *Journal für die reine und angewandte Mathematik* 84, pp. 242–258.
Clifford, A.H. (1933). "A System Arising from a Weakened Set of Group Postulates". *Annals of Mathematics* 34, pp. 865–871.
Cox, David A. (1989). *Primes of the Form* $x^2 + ny^2$. Fermat, Class Field Theory, and Complex Multiplication. New York.
Eisenstein, Gotthold (1844). "Untersuchungen über die cubischen Formen mit zwei Variabeln". *Journal für die reine und angewandte Mathematik* 27, pp. 89–104.
Felgner, Ulrich (2020). "Die Begriffe der Äquivalenz, der Gleichheit und der Identität". *Jahresbericht der Deutschen Mathematiker-Vereinigung* 122, pp. 109–129.
Ferreirós, José (2007). *Labyrinth of Thought*. A History of Set Theory and Its Role in Modern Mathematics. Second revised edition. Basel.
Gauß, Carl Friedrich (1801). *Disquisitiones arithmeticae*. Leipzig.
Gauß, Carl Friedrich (1807). *Recherches arithmétiques*. Traduites par A.-C.-M. Poullet-Delisle. Paris.
Gauß, Carl Friedrich (1986). *Disquisitiones Arithmeticae*. English Edition. Translated by A.A. Clarke. Revised by W.C. Waterhouse. New York et al.
Hasse, Helmut (1926). *Höhere Algebra*. I. Lineare Gleichungen. Berlin and Leipzig.
Jourdain, Philip E.B. (1913). "On isoid relations and theories of irrational number". In: *Proceedings of the 5th International Congress of Mathematicians*. Vol. II. Ed. by E.B. Hobson and A.E.H. Love. Cambridge, pp. 492–496.
Krull, Wolfgang (1923). "Algebraische Theorie der Ringe. I". *Mathematische Annalen* 88, pp. 80–122.
Lagrange, Joseph-Louis (1775). "Recherches d'arithmétique". *Nouveaux Mémoires de l'Académie Royale des Sciences et Belles-Lettres de Berlin* (1773), pp. 265–312.
Legendre, Adrien-Marie (1794). *Éléments de géométrie*. Avec des notes. Paris.
Loewy, Alfred (1915). *Lehrbuch der Algebra*. Erster Teil. Grundlagen der Arithmetik. Leipzig.
Peano, Guiseppe (1894). *Notations de logique mathématique*. Turin.
Russell, Bertrand (1903). *The Principles of Mathematics*. Vol. I. Cambridge.
Smith, H.J.S. (1862). "Report on the Theory of Numbers. Part III". *Report of the British Association for the Advancement of Science* (1861), pp. 292–340.
Tarski, Alfred (1937). *Einführung in die Mathematische Logik und in die Methodologie der Mathematik*. Wien.
van der Waerden, Bartel Leendert (1930). *Moderne Algebra*. Erster Teil. Berlin.

ergodic

The first mathematical publications in which ergodic theorems were proved are Birkhoff (1931) and von Neumann (1932).[1] As related by Birkhoff and Koopman (1932, pp. 280f.), these theorems are:

> Under the above hypotheses, if $Z_{\alpha,\beta}(P; M)$ denotes the relative sojourn of the moving point $P_t(PO = P)$ in M in the time interval $\alpha \leq t \leq \beta$ (i. e., it is the length of this sojourn divided by $\beta - \alpha$), then, as $\beta - \alpha \longrightarrow +\infty$, M remaining fixed, $Z_{\alpha,\beta}(P; M)$ *converges in the mean to a limit* $Z(P; M)$ (a limiting process involving neighboring motions) ["The Mean Ergodic Theorem"].

[1] While Birkhoff's paper was published first, von Neumann proved his theorem before Birkhoff. See Zund (2002).

Under the same conditions as before, $Z_{\alpha,\beta}(P; M) \longrightarrow Z(P; M)$ as $\beta - \alpha \longrightarrow +\infty$, for all P of Q except for a set of measure zero ["The Ergodic Theorem"].

In Birkhoff and Koopman (1932), the term *ergodic theory* seems to have been used for the first time; it became more frequently used after the publication of the first monograph on the subject, Hopf's *Ergodentheorie* (Hopf 1937). Ergodic theorems quickly also appeared in stochastics, where they were phrased in different terms. For example, Doob (1934, p. 761) remarked: "The law of large numbers will be seen to correspond to the ergodic theorem of Birkhoff".[1]

While the mathematical ergodic theory may be said to begin with the papers from the 1930s just cited, the origin of the theory was well known at the time to be found in thermodynamics; papers usually cited (for example by von Neumann 1932, p. 81) in this connection are Boltzmann (1871) and Maxwell (1879). As expressed by Birkhoff and Koopman (1932, p. 279), the ergodic hypothesis "states that, in the systems considered in the kinetic theory, each particular motion will pass, sooner or later, through every state of motion, or phase (combination of position and velocity), consistent with its energy; and (implicitly) that the length of time during which it exhibits a given set of phases is proportional to the relative abundance of the latter". The name *ergodic hypothesis* appears to be due to P. Ehrenfest and T. Ehrenfest (1911, p. 31), who used German *Ergodenhypothese*; because ergodic systems were regarded as unlikely to exist,[2] for systems with certain weaker requirements the name *quasi-ergodic system* (*quasi-ergodisches System*) was suggested in the same article (P. Ehrenfest and T. Ehrenfest 1911, p. 32).

The German terms *Ergode* and *ergodisches System* are due to Boltzmann.[3] The former first appeared in Boltzmann (1885a, p. 242) and Boltzmann (1885b, p. 79), the latter once appeared in Boltzmann (1885b, p. 86).[4] In these papers (the latter one is an extended version of the former), Boltzmann indeed developed a whole series of terms, for example *monodisches System* or *Monode* (Boltzmann 1885b, p. 75), *Holode* (p. 77), *Ergode* (p. 79), and also *isodisch* (pp. 83, 90).[5] Since Boltzmann's *isodisch* corresponds to *ergodic* as used later on, Mathieu (1988, p. 376) concludes: "with the 'Ergodic Hypothesis' in mind, Ergodic Theory ought to have been called 'Isodic Theory'".

The actual etymology of *Ergode* is debated, and Boltzmann himself never explained his choice of terms. The classical etymology was given by the Ehrenfests, who explained it as being made up of Greek *ἔργον* 'work' and *ὁδός* 'path' (P. Ehrenfest and T. Ehrenfest 1911, p. 30). They also erroneously remarked that the term was first used in Boltzmann (1887).[6] A different etymology has been suggested by Mathieu

---

[1] For the link to the law of large numbers, Doob referred to Khintchine (1933) and Hopf (1934).
[2] No examples of ergodic systems in mechanics were known at the time; indeed Plancherel (1913) and Rosenthal (1913) proved that such systems cannot exist. See Brush (1971). [3] English *ergodic system* was, perhaps first, used by van Vleck (1915, p. 339). [4] That these are indeed first uses has variously been pointed out, for example by Brush in Boltzmann (1964, p. 297) and von Plato (1991, p. 72). [5] Cercignani (1998, p. 134) compares the terminology of Boltzmann to that of Gibbs (1902): *Monode - statistical ensemble, Holode - canonical ensemble, Ergode - microcanonical ensemble.* See also Darrigol (2018, pp. 280–282). [6] In that paper, *Ergode* does occur once (Boltzmann 1887, p. 208); the other terms mentioned above do not occur at all.

(1988, pp. 375f.) and in various publications by Gallavotti (e. g., Gallavotti 1995 and 2016), according to whom the term is derived from ἔργον and εἶδος 'form, shape'.

Boltzmann (1885b, p. 79) himself defined:

> We will call those Monodes which have only the equation of motion as a constraint Ergodes. Those Ergodes which are constrained by other equations are called sub-Ergodes. [...] So for an Ergode there is only one value of φ, which is the same for all systems, and which during the evolution of each system is constant and equal to its energy $\chi + \psi = \frac{\Phi + L}{N}$.[1]

Discussing this definition, Mathieu (1988, pp. 375f.) argues that "Boltzmann thought of systems which were described by equations involving only the energy, which were 'work-like'. Probably he knew the Greek adjective ἐργ-ώδησ which means 'laborious' and is composed from ἔργον and εἶδος, 'work-full' [...] However, we feel that much more than the etymological origin Boltzmann's own definition speaks in favour of our alternative explanation". This is essentially also the explanation Gallavotti gives, who also extends it to the previously mentioned related terms; for example, he explains *Monode* as composed of μόνος and εἶδος (Gallavotti 2016, p. 252). Whether these etymologies are also what Boltzmann had in mind remains an open question.

Note. For the etymology of the term, see Mathieu (1988) and Gallavotti (1995, 2016); see also Uffink (2022, note 8) and Cercignani (1998, p. 141). For the history of the ergodic hypothesis, see, e. g., von Plato (1991).

References

Birkhoff, George D. (1931). "Proof of the Ergodic Theorem". *Proceedings of the National Academy of Sciences* 17, pp. 656–660.

Birkhoff, George D. and B.O. Koopman (1932). "Recent Contributions to the Ergodic Theory". *Proceedings of the National Academy of Sciences* 18, pp. 279–282.

Boltzmann, Ludwig (1871). "Einige allgemeine Sätze über Wärmegleichgewicht". *Sitzungsberichte der Kaiserlichen Akademie der Wissenschaften (mathematisch-naturwissenschaftliche Klasse)* 63, pp. 679–711.

Boltzmann, Ludwig (1885a). "Ueber die Eigenschaften monocyklischer und anderer damit verwandter Systeme". *Sitzungsberichte der Kaiserlichen Akademie der Wissenschaften (mathematisch-naturwissenschaftliche Klasse)* 90, pp. 231–245.

Boltzmann, Ludwig (1885b). "Ueber die Eigenschaften monocyklischer und anderer damit verwandter Systeme". *Journal für die reine und angewandte Mathematik* 98, pp. 68–94.

Boltzmann, Ludwig (1887). "Ueber die mechanischen Analogien des zweiten Hauptsatzes der Thermodynamik". *Journal für die reine und angewandte Mathematik* 100, pp. 201–212.

Boltzmann, Ludwig (1964). *Lectures on Gas Theory*. Translated by Stephen G. Brush. Berkeley.

Boltzmann, Ludwig (2019). "On the Properties of Monocyclic Systems, and others related to them". Translated and with comments by Kim A. Sharp and Franz Matschinsky. URL: `https://arxiv.org/abs/1906.09221` (retrieved Nov. 4, 2022).

Brush, Stephen G. (1971). "Proof of the impossibility of ergodic systems: The 1913 papers of Rosenthal and Plancherel". *Transport Theory and Statistical Physics* 1, pp. 287–311.

Cercignani, Carlo (1998). *Ludwig Boltzmann*. The Man Who Trusted Atoms. Oxford.

Darrigol, Olivier (2018). *Atoms, Mechanics, and Probability*. Ludwig Boltzmann's Statistico-Mechanical Writings. Oxford.

---

[1] "Monoden, welche nur durch die Gleichung der lebendigen Kraft beschränkt sind, will ich als Ergoden, solche, welche ausser der Gleichung auch noch durch andere beschränkt sind, als Subergoden bezeichnen. [...] Für Ergoden existirt also nur ein φ, welches gleich der für alle Systeme gleichen und während der Bewegung jedes Systems constanten Energie eines Systems $\chi + \psi = \frac{\Phi + L}{N}$ ist". Translation from Boltzmann (2019, p. 10).

Doob, Joseph L. (1934). "Probability and Statistics". *Transactions of the American Mathematical Society* 36, pp. 759–775.
Ehrenfest, Paul and Tatiana Ehrenfest (1911). "Begriffliche Grundlagen der statistischen Auffassung in der Mechanik". *Encyklopädie der mathematischen Wissenschaften mit Einschluss ihrer Anwendungen* IV 32, pp. 3–90.
Gallavotti, Giovanni (1995). "Ergodicity, ensembles, irreversibility in Boltzmann and beyond". *Journal of Statistical Physics* 78, pp. 1571–1589.
Gallavotti, Giovanni (2016). "Ergodicity: a historical perspective. Equilibrium and Nonequilibrium". *The European Physical Journal H* 41, pp. 181–259.
Gibbs, Josiah Willard (1902). *Elementary Principles in Statistical Mechanics.* Developed with especial Reference to the Rational Foundations of Thermodynamics. New York.
Hopf, Eberhard (1934). "On Causality, Statistics and Probability". *Journal of Mathematics and Physics* 13, pp. 51–102.
Hopf, Eberhard (1937). *Ergodentheorie*. Berlin et al.
Khintchine, Aleksandr (1933). "Zur mathematischen Begründung der statistischen Mechanik". *Zeitschrift für angewandte Mathematik und Mechanik* 13, pp. 101–103.
Mathieu, Martin (1988). "On the origin of the notion 'Ergodic Theory'". *Expositiones Mathematicae* 6, pp. 373–377.
Maxwell, James Clerk (1879). "On Boltzmann's Theorem on the average distribution of energy in a system of material points". *Transactions of the Cambridge Philosophical Society* 12, pp. 547–570.
von Neumann, John (1932). "Proof of the Quasi-Ergodic Hypothesis". *Proceedings of the National Academy of Sciences* 18, pp. 70–82.
Plancherel, Michel (1913). "Beweis der Unmöglichkeit ergodischer mechanischer Systeme". *Annalen der Physik* 42, pp. 1061–1063.
von Plato, Jan (1991). "Boltzmann's ergodic hypothesis". *Archive for History of Exact Sciences* 42, pp. 71–89.
Rosenthal, Arthur (1913). "Beweis der Unmöglichkeit ergodischer Gassysteme". *Annalen der Physik* 42, pp. 796–806.
Uffink, Jos (2022). "Boltzmann's Work in Statistical Physics". In: *The Stanford Encyclopedia of Philosophy*. Ed. by Edward N. Zalta. Stanford. URL: `https://plato.stanford.edu/archives/sum2022/entries/statphys-Boltzmann/` (retrieved Sept. 6, 2022).
van Vleck, E.B. (1915). "The rôle of the point-set theory in geometry and dynamics". *Bulletin of the American Mathematical Society* 21, pp. 321–341.
Zund, Joseph D. (2002). "George David Birkhoff and John von Neumann: A Question of Priority and the Ergodic Theorems, 1931–1932". *Historia Mathematica* 29, pp. 138–156.

error theory

The beginnings of the systematic study of errors is commonly attributed to Lambert, for example by Swijtink (1994, p. 1379): "The Alsatian natural philosopher Johann Heinrich Lambert published [...] the first systematic discussion of the 'theory of errors', as he called it".[1] He expounded his theory in Lambert (1760) as well as in Lambert (1765a) and Lambert (1765b). In the latter works, he used German *Theorie der Fehler* and explained that the theory should deal with the relation between errors, their consequences, the circumstances of measurement and the quality of the instruments (Lambert 1765a, § 321).[2] In this connection, he refers to Marinoni (1751) on several occasions (Lambert 1765a, e. g., pp. 8, 228).[3]

[1] Similar statements can be found in Gray and Tilling (1978, p. 23) and Knobloch (2006, p. 103).
[2] For discussion, see Sheynin (2021, ch. 2.3). (As of February 2024, Sheynin's unpaginated manuscript is available on Sheynin's website and contained in a file entitled "Studies in the history of statistics and probability, vol. 25".) [3] For a discussion of Marinoni's work, see Knobloch (2006, pp. 106f.).

Marinoni was also referred to by Kästner (1756, p. 131) when he briefly discussed the theory of errors:

> If one looks for an unknown length by measuring lines or angles, an error can easily occur that causes an uncertainty concerning the actual magnitude of the length sought. And since lines can be measured much more accurately than angles, the error based on the latter is mainly to be considered. About this theory of errors (*Theoria aberrationum*), something is found in Herr Bouguer's *Figure de la terre*[1], in Herr Marinoni's work *de re ichnographica.*[2]

While this seems to be the earliest use of *theory or errors*, remarks on errors in observations can be found in earlier times. Galilei (1632, p. 283), for example, about whom Sheynin (1971, p. 221) says that he "had formulated the main propositions of the probabilistic error theory",[3] noted:

> [T]here is some error in every combination of these observations. This I believe to be absolutely unavoidable, for the observations used in every investigation being four in number [...] made by different observers in different places and with different instruments), anybody who knows anything about matters will say that it cannot be that no error will have fallen in among the four. Especially when we know that in taking a single polar elevation with the same instrument in the same place and by the same observer (who may have made it many times), there will be a variance of a minute or so; even of many minutes, as may be seen in plenty of places in this same book.[4]

Statements on observational errors can be found, however, in much earlier times; for examples from antiquity, see, e. g., Sheynin (1993).

To come back to the term *Theorie der Fehler*, it was first adopted in some German publications such as Mayer (1777, p. iv) and Späth (1789, p. iii). The corresponding English *theory of errors* seems to have been first used by A. De Morgan (1845, pp. 449f.),[5] and French *théorie des erreurs* by Liagre (1852).

Note. For the history of error theory, see the works by Knobloch and Sheynin cited in this article.

References

Bouguer, Pierre (1749). *La figure de la terre*. Paris.
Galilei, Galileo (1632). *Dialogo sopra i due massimi sistemi del mondo*. Firenze.

---

[1] This refers to Bouguer (1749). See Knobloch (2006, pp. 105f.) for discussion. [2] "Wenn man eine unbekannte Länge durch Abmessung der Linien oder der Winkel sucht, so kann man bey der Abmessung leicht einen Fehler begehen, der eine Ungewißheit wegen der eigentlichen Größe der gesuchten Länge verursachet. Und da man die Linien viel genauer messen kann, als die Winkel, so hat man vornehmlich den Fehler, der von dem letzteren ausgeht, in Betrachtung zu ziehen. Von dieser Theorie der Fehler (*Theoria aberrationum*) findet sich etwas in Herrn Bouguer *Figure de la terre*, und Herrn Marinoni Werke *de re ichnographica* [...]". [3] For details, see Knobloch (2006, pp. 97f.). [4] "in ciascuna combinazione delle osservazioni sia qualche errore: il che credo che assolutamente sia necessario; perché, sendo 4 in numero le osservazioni che servono per una indagine [...] fatte da diversi osservatori, in diversi luoghi e con diversi strumenti, chiunque abbia qualche cognizione di tal pratica dirà non potere essere che tra tutte 4 non sia caduto qualche errore, e massime mentre che noi veggiamo che nel prender una sola altezza di polo, co 'l medesimo strumento, nel medesimo luogo e dal medesimo osservatore, che l'ha potuta far mille volte, tuttavia si va titubando di qualche minuto, e spesso anco di molti, come in questo medesimo libro potete vedere in diversi luoghi". Translation from Galilei (1967, pp. 289f.). [5] According to S. E. De Morgan (1882, p. 92), the text was written between 1836 and 1837.

Galilei, Galileo (1967). *Dialogue Concerning the Two Chief World Systems*. Translated, wuth revised notes, by Stillman Drake. Berkeley and Los Angeles.
Gray, Jeremy and Laura Tilling (1978). "Johann Heinrich Lambert, mathematician and scientist, 1728 – 1777". *Historia Mathematica* 5, pp. 13–41.
Kästner, Abraham Gotthelf (1756). "Von Fehlern beym Feldmessen". *Der Königl. Schwedischen Akademie der Wissenschaften Abhandlungen* 15, pp. 131–140.
Knobloch, Eberhard (2006). "On the Origin of Error Theory". *Historia de la probabilidad y la estadística* 3, pp. 95–116.
Lambert, Johann Heinrich (1760). *Photometria*. Augsburg.
Lambert, Johann Heinrich (1765a). "Anmerkungen und Zusätze zur practischen Geometrie". In: *Beyträge zum Gebrauch der Mathematik und deren Anwendung*. Berlin, pp. 1–313.
Lambert, Johann Heinrich (1765b). "Theorie der Zuverlässigkeit der Beobachtungen und Versuche". In: *Beyträge zum Gebrauch der Mathematik und deren Anwendung*. Berlin, pp. 424–488.
Liagre, Jean Baptiste Joseph (1852). *Calcul des probabilités et théorie des erreurs*. Bruxelles.
Marinoni, Giovanni Giacomo (1751). *De re ichnographica*. Vienna.
Mayer, Johann Tobias (1777). *Gründlicher und ausführlicher Unterricht zur praktischen Geometrie*. Erster Theil. Göttingen.
De Morgan, Augustus (1845). "Theory of probabilities". *Encyclopædia Metropolitana* 2, pp. 393–490.
De Morgan, Sophia Elizabeth (1882). *Memoir of Augustus De Morgan*. London.
Sheynin, Oscar (1971). "Newton and the Classical Theory of Probability". *Archive for History of Exact Sciences* 7, pp. 217–243.
Sheynin, Oscar (1993). "The Treatment of Observations in Early Astronomy". *Archive for History of Exact Sciences* 46, pp. 153–192.
Sheynin, Oscar (2021). *History of the theory of errors*. Second edition, essentially revised. Berlin. URL: `http://www.sheynin.de`.
Späth, Johann Leonhard (1789). *Analytische Untersuchungen über die Zuverläßigkeit, mit welcher ein Landmesser vermittelst verschiedener Geometer-Werkzeuge Winkel und Linien abmessen kann*. Altdorf and Nürnberg.
Swijtink, Zeno G. (1994). "Probability and statistics in mechanics". In: *Companion Encyclopedia of the History and Philosophy of the Mathematical Sciences*. Volume 2. Ed. by Ivor Grattan-Guinness. Baltimore and London, pp. 1377–1385.

error function

The name for this function was introduced by Glaisher (1871, p. 296), who considered the integral $\int_x^\infty e^{-x^2}\,dx$, which he says is "well known for its use in physics", and defined:

> As it is necessary that the function should have a name, and as I do not know that any has been suggested, I propose to call it the Error-function, on account of its earliest and still most important use being in the connexion with the theory of Probability, and to write $\int_x^\infty e^{-x^2}\,dx = \operatorname{Erf} x$.

Glaisher (1871, p. 421) also defined the *Error-function-complement* as $\int_0^x e^{-u^2}\,du = \operatorname{Erfc} x$ and towards the end of the article considered some applications, naming refraction and the theory of the conduction of heat (pp. 427-431). He singles out the theory of errors of observation, in which "the function is of paramount importance; and this fact is the justification of the name and symbol by which it has been here denoted" (Glaisher 1871, p. 429).

In England, the term was immediately accepted, for example by Pendlebury (1871, p. 439). In other countries, the terminology is not frequently found in the 19th century; for example, Opitz (1903, p. 42) remarked that it spread only little in Germany.

Note. The history of the concept is discussed in Nahin (2023).

References

Glaisher, J.W.L. (1871). "On a Class of Definite Integrals". *The London, Edinburgh and Dublin Philosophical Magazine and Journal of Science* 42, pp. 294–302, 421–436.

Nahin, Paul J. (2023). *The Probability Integral.* Its Origin, Its Importance, and Its Calculation. Cham.

Opitz, Hans R.G. (1903). "Über die Auflösung der transcendenten Gleichung $\int_0^x e^{-x^2} dx = \sum_{\lambda=0}^{\infty} \frac{(-1)^\lambda x^{2\lambda+1}}{\lambda!(2\lambda+1)} = \frac{\pi}{4}$". *Archiv der Mathematik und Physik* 5, pp. 42–46.

Pendlebury, Richard (1871). "Note on some Definite Integrals". *The London, Edinburgh and Dublin Philosophical Magazine and Journal of Science* 42, pp. 437–440.

Euclidean field and *Pythagorean field*

Both of these terms were originally introduced in Weyl's book *The Classical Groups*. In this text, Weyl (1939, p. 13) first defined a *Pythagorean field* by the condition that "[t]he Pythagorean equation $\alpha^2 + \beta^2 = \gamma^2$ has a solution γ for any two given numbers α and β". Later on in the text, Weyl (1939, p. 273) additionally defined: "The constructions in Euclidean geometry with ruler and compass are algebraically equivalent to the four species and the extraction of square roots. A field in which every quadratic equation $x^2 - \rho = 0$ is solvable may therefore be called a *Euclidean field*".

It took several years for these terms to spread; around 1950, they were used by Baer (1950, pp. 439, 449), Dieudonné (1951, pp. 89, 201), and Ellis (1951, p. 22). In none of these papers is Weyl mentioned as their originator. The concept is already found in Hilbert (1899); for discussion, see Hallett and Majer (2004, esp. pp. 202f.).

Note that *Euclidean field* is also used for a field in which the Euclidean algorithm holds; as remarked by Brauer (1940, p. 697), in this sense the term is due to Hardy and Wright (1938, p. 212). See also *Euclidean ring*, which term has apparently served as a model.

Note also that Podehl and Reidemeister (1934, p. 255) used German *Euklidischer Körper* for what Weyl called *Pythagorean field*; Bachmann (1940, p. 206) confirmed that he and Reidemeister had earlier used the term in this sense and also suggested *Hilbert field* (*Hilbertscher Körper*) instead.[1] This suggestion was as unsuccessful as the term *Platonic field* (*Platonischer Körper*) used by Schreiber (1975, p. 73).

References

Bachmann, Friedrich (1940). "Stufen der absoluten Geometrie. Die Frage nach der Unabhängigkeit der Anordnung". *Mathematische Annalen* 117, pp. 197–234.

Baer, Reinhold (1950). "Free Mobility and Orthogonality". *Transactions of the American Mathematical Society* 68, pp. 439–460.

Brauer, Alfred (1940). "On the Non-Existence of the Euclidean Algorithm in Certain Quadratic Number Fields". *American Journal of Mathematics* 62, pp. 697–716.

Dieudonné, Jean (1951). "On the Orthogonal Groups Over the Rational Field". *Annals of Mathematics* 54, pp. 85–93.

Ellis, David (1951). "Geometry in abstract distance spaces". *Publicationes Mathematicae* 2, pp. 1–25.

Hallett, Michael and Ulrich Majer, eds. (2004). *David Hilbert's* Lectures on the Foundations of Geometry *1891–1902*. Berlin et al.

[1] Reidemeister (1957, p. 128) later used *Euklidischer Körper* in Weyl's sense.

Hardy, Godfrey Harold and E.M. Wright (1938). *An Introduction to the Theory of Numbers*. Oxford.
Hilbert, David (1899). *Grundlagen der Geometrie*. Leipzig.
Podehl, Erich and Kurt Reidemeister (1934). "Eine Begründung der ebenen elliptischen Geometrie". *Abhandlungen aus dem Mathematischen Seminar der Universität Hamburg* 10, pp. 231–255.
Reidemeister, Kurt (1957). *Raum und Zahl*. Berlin et al.
Schreiber, Peter (1975). *Theorie der geometrischen Konstruktionen*. Berlin.
Weyl, Hermann (1939). *The Classical Groups*. Princeton.

Euclidean and *non-Euclidean geometry*

In the 18th century, a distinction was made between elementary and higher geometry. The distinction is found in Wolff (1716, pp. 667, 671) and was also adopted by Chambers in his entry *geometry*: "Elementary, or Common Geometry, Is that employed in the consideration of Right Lines, and plain Surfaces, and Solids generated therefrom. [...] The Higher, or Sublimer Geometry is that employ'd in the consideration of Curve Lines, Conic Sections, and Bodies form'd thereof" (Chambers 1728, p. 141).

Euclidean geometry was used in the title of Hodgson (1723); a definition is not given there, however. Sulzer (1745, p. 65) was more explicit: "Geometry is of two kinds, ordinary or Euclidean geometry, and higher geometry".[1] The definitions he gave correspond to those given by Wolff and Chambers.

Non-Euclidean geometry did not come up in opposition to *Euclidean geometry* in the above sense. Instead, it arose in connection with attempts to prove or disprove the well-known parallel postulate as it appeared in Euclid's *Elements*. The new term as well as some alternatives are found in the correspondence of Gauß. In particular, Gauß used German *Nicht-Euklidische Geometrie* in a letter dated November 8, 1824: "All of my efforts to find a contradiction, an inconsistency in this non-Euclidean geometry have been fruitless" (Gauß 1900, p. 187).[2] As reported by Sartorius von Waltershausen (1856, p. 81), Gauß also used *anti-euklidisch*; this term is found in Gauß' correspondence in 1816 (Gauß 1900, p. 175).

Another alternative was suggested in a letter from 1818 by Schweikart to Gauß: "There are two kinds of geometry: a geometry in the strict sense–the Euclidean; and an astral geometry".[3] As explained by Martin (1998, p. 307), "Schweikart's *astral geometry* is hyperbolic geometry and gets its name from the fact that a segment having a critical angle of measure $\pi/4$ would have to be astronomically long". The term did not catch on, but it was occasionally mentioned as a synonym of *non-Euclidean geometry*, for example by Baltzer (1874, p. 13), who also listed *pangeometry* and *imaginary geometry* as further alternatives.

Both of these terms are due to Lobachevsky, who used them in various publications, for example in Lobachevsky (1855, p. 5):

> Thus, there are only two possible assumptions: either the sum of the three angles of any rectilinear triangle is equal to two right angles, and this assumption leads to the known geometry – or in any rectilinear triangle this sum is less than two right angles and this

[1] "Die Geometrie ist von zweyerley Art, die gemeine oder Euclydische Geometrie und die höhere Geometrie". [2] "Alle meine Bemühungen, einen Widerspruch, eine Inconsequenz in dieser Nicht-Euklidischen Geometrie zu finden, sind fruchtlos gewesen". Translation from Abardia, Reventós, and Rodríguez (2012, p. 309). [3] "Es gibt eine zweifache Geometrie,–eine Geometrie im engern Sinn–die Euklidische; und eine astralische Grössenlehre" (Gauß 1900, p. 180). Translation from Gray (2011, p. 95).

> assumption serves as a basis for another geometry, to which I gave the name *Imaginary geometry*, but which it may be more convenient to call *Pangeometry*, because this name designates a general geometric theory which includes ordinary geometry as a particular case.[1]

Another term worth mentioning in this respect is *absolute geometry*, which may be said to go back to Bolyai's "Appendix Explaining the Absolutely True Science of Space" (Bolyai 1832). Bolyai did not explicitly use *absolute geometry* in this text, even though Frischauf (1872, p. 16) claimed that he did. Instead, the term seems to have been first used by Frischauf. It referred to geometry in which the parallel postulate is dropped. Prenowitz and Jordan (1965, p. 37) later suggested that this field be called *neutral geometry*.

By the end of the 19th century, *non-Euclidean geometry* was in common use, even though Felix Klein criticized the terminology. He argued that a geometry can only properly be called *non-Euclidean* if the parallel postulate is negated, pointing out that this is the way Gauß used the term. However, after Gauß the term was often used when the parallel postulate was not assumed and, therefore "the expression *non-Euclidean* or *anti-Euclidean* is not well chosen; the word *pangeometry* is better" (Klein 1892, pp. 169f.).[2]

See also the entry for *hyperbolic, elliptic,* and *parabolic geometry*.

References

Abardia, Judit, Agustí Reventós, and Carlos J. Rodríguez (2012). "What did Gauss read in the Appendix?" *Historia Mathematica* 39, pp. 292–323.

Baltzer, Richard (1874). *Die Elemente der Mathematik*. Zweiter Band. Planimetrie, Stereometrie, Trigonometrie. Vierte verbesserte Auflage. Leipzig.

Bolyai, János (1832). "Appendix. Scientiam Spatii absolute veram exhibens". In: Bolyai, Farkas. *Tentamen juventutem studiosam in elementa matheseos purae, elementaris ac sublimioris, methodo intuitiva, evidentiaque huic propria, introducendi*. Tomus primus. Maros Vásárhely.

Chambers, Ephraim (1728). *Cyclopaedia, or, An universal dictionary of arts and sciences*. Volume the First. London.

Frischauf, Johannes (1872). *Absolute Geometrie nach Johann Bolyai*. Leipzig.

Gauß, Carl Friedrich (1900). *Werke*. Achter Band. Leipzig.

Gray, Jeremy (2011). *Worlds Out of Nothing*. A Course in the History of Geometry in the 19th Century. London et al.

Hodgson, James (1723). *A System of the Mathematics*. Vol. I. London.

Klein, Felix (1892). *Nicht-Euklidische Geometrie*. I. Vorlesung. Göttingen.

Lobachevsky, Nikolai (1855). *Pangéométrie ou précis de géométrie fondée sur une théorie générale et rigoureuse des parallèles*. Kazan.

Lobachevsky, Nikolai (2010). *Pangeometry*. Edited and translated by Athanase Papadopoulos. Zürich.

Martin, George E. (1998). *The Foundations of Geometry and the Non-Euclidean Plane*. New York.

Prenowitz, Walter and Meyer Jordan (1965). *Basic Concepts of Geometry*. New York et al.

---

[1] "Ainsi il n'y a que deux suppositions possibles: ou la somme des trois angles de tout triangle rectiligne est égale à deux angles droits, cette supposition donne la géométrie connue – ou dans tout triangle rectiligne cette somme est moindre que deux angles droits et cette supposition sert de base à une autre géométrie, à laquelle j'avais donné le nom de géométrie imaginaire, mais qu'il est peut être plus convenable de nommer Pangéométrie parce que ce nom désigne une théorie géométrique générale, qui comprend la géométrie ordinaire comme cas particulier". Translation from Lobachevsky (2010, pp. 5f.). See also pp. 229f. of the same volume for possible explanations of why Lobachevsky chose this term. [2] "Daher ist der Ausdruck nicht-euklidisch oder antieuklidisch nicht gut gewählt, besser ist das Wort Pangeometrie".

Sartorius von Waltershausen, Wolfgang (1856). *Gauss zum Gedächtniss*. Leipzig.
Sulzer, Johann Georg (1745). *Kurzer Begriff aller Wissenschaften und andern Theil der Gelehrsamkeit*. Frankfurt and Leipzig.
Wolff, Christian (1716). *Mathematisches Lexicon*. Leipzig.

Euclidean ring

The term *Euclidean ring* appears to be due to van der Waerden; for example, van Dantzig (1935, p. 222) attributed it to him. German *euklidischer Ring* was indeed used by van der Waerden in lectures on ideal theory in 1927/28, where it was defined thus: "A Euclidean ring is a ring without zero divisors and with a unit element, where each ideal is a principal ideal" (Bečvář and Bečvářová 2020, p. 135).[1] van der Waerden there also showed that in a Euclidean ring a generalized Euclidean algorithm exists and unique factorial decomposition is possible (Bečvář and Bečvářová 2020, pp. 136f.).

The term is not found in the first edition of van der Waerden's classic *Moderne Algebra* (van der Waerden 1930),[2] and was included only in the second edition (van der Waerden 1937, p. 60), which contained a new chapter on Euclidean rings and principal ideal rings. The term must have spread before, though, as the above cited van Dantzig shows; the term was also known to MacDuffee (1933, p. 565): "The term *euclidean ring* has been applied to a principal ideal ring with a Euclid algorithm".

Note. Some historical notes and a comparison of definitions are found in Ağargün and Fletcher (1995).

References

Ağargün, A.G. and C.R. Fletcher (1995). "Euclidean rings". *Turkish Journal of Mathematics* 19, pp. 291–299.
Bečvář, Jindřich and Martina Bečvářová (2020). *Jarník's Notes of the Lecture Course Allgemeine Idealtheorie by B.L. van der Waerden (Göttingen 1927/1928)*. Praha.
van Dantzig, David (1935). "Zur topologischen Algebra. II. Abstrakte $\mathfrak{b}_{\mathfrak{v}}$-adische Ringe". *Compositio Mathematica* 2, pp. 201–223.
MacDuffee, Cyrus Colton (1933). "Matrices with elements in a principal ideal ring". *Bulletin of the American Mathematical Society* 39, pp. 564–584.
van der Waerden, Bartel Leendert (1930). *Moderne Algebra*. Erster Teil. Berlin.
van der Waerden, Bartel Leendert (1937). *Moderne Algebra*. Erster Teil. Zweite verbesserte Auflage. Berlin.

Eulerian graph

This terminology was introduced by König (1936, p. 19) in his famous book on graph theory:

> Some questions, which relate to certain mathematical games and to which graph theory and especially topology owe their first interest, lead to the question of when a graph can be represented as a closed trail. If a closed trail contains all the edges of the graph, then it shall be called an *Euler trail* of the graph. In the above mentioned question the so-called

---

[1] "Ein euklidischer Ring ist ein Ring ohne Nullteiler und mit dem Einselement, wo jedes Ideal Hauptideal ist". [2] The concept is there; see van der Waerden (1930, § 16).

Euler graphs play an important role. A graph is called an *Euler graph* if it is finite and all its vertices are of even degree.[1]

König also remarks that Veblen and Alexander (1912-13, p. 166) called such graphs *closed* but that "[t]his designation was also used sometimes for graphs without endpoints". König chose the term because Euler (1741) used the concept in his well-known solution of the Königsberg bridge problem.

The corresponding English terminology does not seem to have come up until the 1950s; *Euler graph* is found in Ore (1951, p. 49), *Eulerian graph* in Fejes Tóth (1957, p. 124).

References

Euler, Leonhard (1741). "Solutio problematis ad geometriam situs pertinentis". *Commentarii academiae scientiarum imperialis Petropolitanae* 8, pp. 128–140.

Fejes Tóth, László (1957). "Filling of a domain by isoperimetric discs". *Publicationes Mathematicae* 5, pp. 119–127.

König, Dénes (1936). *Theorie der endlichen und unendlichen Graphen*. Leipzig.

König, Dénes (1990). *Theory of Finite and Infinite Graphs*. Translated by Richard McCoart. Boston.

Ore, Øystein (1951). "A problem regarding the tracing of graphs". *Elemente der Mathematik* 6, pp. 49–53.

Veblen, Oswald and James W. Alexander (1912-13). "Manifolds of *N* Dimensions". *Annals of Mathematics* 14, pp. 163–178.

even **and** ***odd*** **(numbers and functions)**

The origins of these terms are found in ancient Greece. An early definition of even numbers was given by Plato (*Laws* 895e): "We have instances of a thing divisible into two halves, both in arithmetic and elsewhere; in arithmetic the name of this is 'the even,' and the definition is 'a number divisible into two equal parts'" (Plato 1936, p. 337). In the *Elements*, Euclid gave the same definition (book 7, def. 6) and also defined an odd number as "that which is not divisible into two equal parts, or that which differs by a unit from an even number" (def. 7).[2]

For 'even', Plato and Euclid used the term *ἄρτιος*, which literally means 'complete, perfect, suitable', while for 'odd', Euclid used *περισσός*, the meanings of which include 'out of the common, extraordinary, strange' and 'more than sufficient' (LSJ).

In Latin, the Greek terms were translated by *par* 'equal, similar, matching' and its counterpart *impar*; both words were used by Cicero with respect to numbers (OLD). These terms were also used, for example, by Boethius in his book on arithmetic (Boethius 1867, p. 13).[3]

---

[1] "Manche Fragestellungen, die sich auf gewisse mathematische Spiele beziehen und denen die Graphentheorie und überhaupt die Analysis Situs ihre ersten Untersuchungen verdankt, führen auf die Frage, wann ein Graph als ein geschlossener Kantenzug aufgefaßt (gezeichnet) werden kann. Enthält ein geschlossener Kantenzug sämtliche Kanten des Graphen, so soll er als eine *Eulersche Linie* des Graphen bezeichnet werden. Bei der erwähnten Frage spielen die sog. *Eulerschen Graphen* eine ausgezeichnete Rolle; ein Graph wird so bezeichnet, falls er endlich ist und sämtliche Knoten von geradem Grade sind". Translation from König (1990, p. 25) (slightly modified). [2] Translation from Euclid (1956, p. 277). For a discussion of different ancient definitions, see Nicomachus of Gerasa (1926, pp. 190f.) and Heath's comments in Euclid (1956, p. 281), according to whom the distinction between odd and even numbers goes back to Pythagoras. [3] They are also found in Martianus Capella (Willis 1983, p. 271), Isidore of Seville (2006, p. 90), and also in the 12th-century translations of the *Elements* (see, for example, Busard 1983, p. 196).

The French terms *pair* and *impair* in this sense are found as *per* and *nonper* in a 13th-century algorism (Karpinski and Waters 1928, p. 72); they were also used by Oresme (1968, p. 384). The English terms *even* and *odd* date from the 14th century (OED); they were first used by John Trevisa in his translation of Bartholomaeus Anglicus' *De proprietatibus rerum*: "þe euene nombre may be departed euene atwynne, as tweyne, foure, sixe, and eighte. Some nombre [i]s odde þat may nouȝt be departed into euene parties, for þe oone partye is more, and þe oþer lasse" (Seymour 1975, p. 1360). In a German arithmetic from the middle of the 15th century (Unger 1888, p. 127), the words *effen* and *oneffen* (which etymologically correspond to English *even* and *uneven*) are found; as Tropfke (1930, pp. 122f.) notes, however, the terms *gerade* and *ungerade* have been standard since the late 15th century.

The distinction between even and odd functions was first made by Euler (1729, p. 98): "In the first place functions are to be noted that I call *even*, of which this is the property: that they remain unchanged if $-x$ is put in place of x [...] In the second place, I take heed of *odd* functions, which in short produce the negative of these, if x is changed to $-x$".[1] He used the Latin terms *par* and *impar*; the terms are also in his *Introduction to Analysis of the Infinite* (Euler 1748, §§ 19, 21).[2]

Note. For overviews of the terminology, see Tropfke (1930, pp. 122f.) and Smith (1958, pp. 16–18). For the Arabic terms, which literally mean 'couple, pair' and 'sole, singular', respectively, see Souissi (1968, pp. 84f.) and Rashed (2017, pp. 489–494). For the Latin terms, see Guillaumin (2020, pp. 150f., 206).

References

Boethius, Anicius Manlius Severinus (1867). *De institutione arithmetica libri duo, De institutione musica libri quinque*. Accedit Geometria quae fertur Boetii. Ed. by Gottfried Friedlein. Leipzig.

Busard, H.L.L. (1983). *The First Latin Translation of Euclid's Elements Commonly Ascribed to Adelard of Bath*. Books I-VIII and books X.36-XV.2. Toronto.

Euclid (1956). *The Thirteen Books of Euclid's Elements*. Volume II. Books III-IX. Ed. by Thomas L. Heath. New York.

Euler, Leonhard (1729). "Problematis traiectoriarum reciprocarum solutio". *Commentarii academiae scientiarum imperialis Petropolitanae* 2, pp. 90–111.

Euler, Leonhard (1748). *Introductio in analysin infinitorum*. Tomus primus. Lausanne.

Euler, Leonhard (2006). "The Solution of the Problem of Reciprocal Trajectories". Translated by Ian Bruce. URL: `http://www.17centurymaths.com/contents/eulercontents.html` (retrieved Aug. 17, 2019).

Guillaumin, Jean-Yves (2020). *Dictionnaire de la terminologie latine ancienne de l'arithmétique et de la géométrie*. Paris.

Isidore of Seville (2006). *The* Etymologies *of Isidore of Seville*. Ed. by Stephen A. Barney et al. Cambridge.

Karpinski, Louis Charles and E.G.R. Waters (1928). "A Thirteenth Century Algorism in French Verse". *Isis* 11, pp. 45–84.

Nicomachus of Gerasa (1926). *Introduction to Arithmetic*. Translated by Martin Luther D'Ooge, with studies in Greek Arithmetic by Frank Egleston Robbins and Louis Charles Karpinski. New York.

Oresme, Nicole (1968). *Le livre du ciel et du monde*. Edited by Albert D. Menut and Alexander J. Denomy, translated with an introduction by Albert D. Menut. Madison.

---

[1] "Primo loco notandae sunt functiones, quas pares appello, quarum haic est proprietas, ut immutatae maneant, etsi loco x ponantur $-x$ [...] Secundo functiones impares observo, quae prorsus sui negativas producunt, si x abit in $-x$". Translation from Euler (2006, p. 20) (emphasis added).

[2] For Euler's introduction of the term, see Sandifer (2007, pp. 20f.).

Plato (1936). *Laws*. Translated by R.G. Bury. In two volumes. II. London.
Rashed, Roshdi, ed. (2017). *Lexique historique de la langue scientifique arabe*. Hildesheim.
Sandifer, C. Edward (2007). *The Early Mathematics of Leonhard Euler*. Washington, D.C.
Seymour, Maurice Charles, ed. (1975). *On the Properties of Things*. John Trevisa's Translation of Bartholomaeus Anglicus' *De Proprietatibus Rerum*. A Critical Text. Volume II. Oxford.
Smith, David Eugene (1958). *History of Mathematics*. Volume II. Special Topics of Elementary Mathematics. New York.
Souissi, Mohammed (1968). *La langue des mathématiques en arabe*. Tunis.
Tropfke, Johannes (1930). *Geschichte der Elementar-Mathematik*. Erster Band. Rechnen. Dritte, verbesserte und vermehrte Auflage. Berlin and Leipzig.
Unger, Friedrich (1888). "Das älteste deutsche Rechenbuch". *Zeitschrift für Mathematik und Physik (hist.-lit. Abtheilung)* 33, pp. 125–145.
Willis, James, ed. (1983). *Martianus Capella*. Leipzig.

event

The word *event* (Latin *eventus*) appears in the earliest definitions of *probability*, for example in Bernoulli's *Ars conjectandi*:

> *Probability*, indeed, is degree of certainty, and differs from the latter as a part differs from the whole. If, for example, the whole or absolute certainty – which we designate by the letter a or by the unity symbol 1 – is supposed to consist of five probabilities or parts, three of which stand for the existence or future existence of some event, the remaining two standing against its existence or future existence, this event is said to have $\frac{3}{5}a$ or $\frac{3}{5}$ certainty (J. Bernoulli 1713, p. 211).[1]

English *event* was then used by de Moivre:

> The Probability of an Event is greater, or less, according to the number of Chances by which it may Happen, compar'd with the whole number of all the Chances, by which it may either Happen or Fail. Thus, If an Event has 3 Chances to Happen, and 2 to Fail; the Probability of its Happening may be estimated to be $\frac{3}{5}$, and the Probability of its Failing is $\frac{2}{5}$ (de Moivre 1718, p. 1).[2]

The term is regularly found in treatises on probability and was also used by Kolmogorov (1933, p. 2) in his famous axiomatisation: "Let E be a collection of elements $\xi, \eta, \zeta, \ldots$, which we shall call *elementary events*, and $\mathfrak{F}$ a set of subsets of E; the elements of the set $\mathfrak{F}$ will be called *random events*".[3]

References

Bernoulli, Jakob (1713). *Ars coniectandi*. Basel.
Bernoulli, Jakob (2006). *The Art of Conjecturing, together with Letter to a Friend on Sets in Court Tennis*. Translated with an introduction and notes by Edith Dudley Sylla. Baltimore.
Bernoulli, Nicolaus (1709). *De usu artis conjectandi in jure*. Basel.

---

[1] "Probabilitas enim est gradus certitudinis et ab hac differt ut pars a toto. Nimirum si certitudo integra & absoluta, quam litera a vel unitate 1 designamus, quinque verb. gr. probabilitatibus ceu partibus constare supponantur, quarum tres militant pro existential aut futuritione alicujus eventus, reqlique contra; eventus ille dicetur habere $\frac{3}{5}a$, seu $\frac{3}{5}$, certitudinis". Translation from J. Bernoulli (2006, p. 315). For a recent account of Bernoulli's book, see Sylla (2014). – In print, the same definition appears earlier in N. Bernoulli (1709, p. 5), who took it over from his uncle. [2] The term is also found in de Moivre (1711, p. 215). [3] "Es sei E eine Menge von Elementen $\xi, \eta, \zeta, \ldots$, welche man *elementare Ereignisse* nennt, und $\mathfrak{F}$ eine Menge von Teilmengen aus E; die Elemente der Menge $\mathfrak{F}$ werden weiter *zufällige Ereignisse* genannt". Translation from Kolmogorov (1950, p. 2).

Kolmogorov, Andrey Nikolaevich (1933). *Grundbegriffe der Wahrscheinlichkeitsrechnung*. Berlin.
Kolmogorov, Andrey Nikolaevich (1950). *Foundations of the Theory of Probability*. Translation edited by Nathan Morrison. New York.
de Moivre, Abraham (1711). "De mensura sortis, seu, de probabilitate eventuum in ludis a casu fortuito pendentibus". *Philosophical Transactions of the Royal Society of London* 27, pp. 213–264.
de Moivre, Abraham (1718). *The Doctrine of Chances: or, A Method of Calculating the Probability of Events in Play*. London.
Sylla, Edith Dudley (2014). "Tercentenary of *Ars Conjectandi* (1713): Jacob Bernoulli and the Founding of Mathematical Probability". *International Statistical Review* 82, pp. 27–45.

evolute

The naming of the term is explained by Mahoney (2005, p. 229, fn. 8): "The term 'evolute' derived from the technique of wrapping a string (or imagining a string wrapped) around a curve and then tracing the locus of its endpoint as the curve is unwrapped (evoluta), that is, as the string is pulled away while being held tangent to the curve".

The name *evolute* was introduced by Huygens (1673, p. 60); he used Latin *evoluta*, which is ultimately derived from Latin *evolvere* 'to unroll'. Huygens defined the term as follows (see fig. 44):

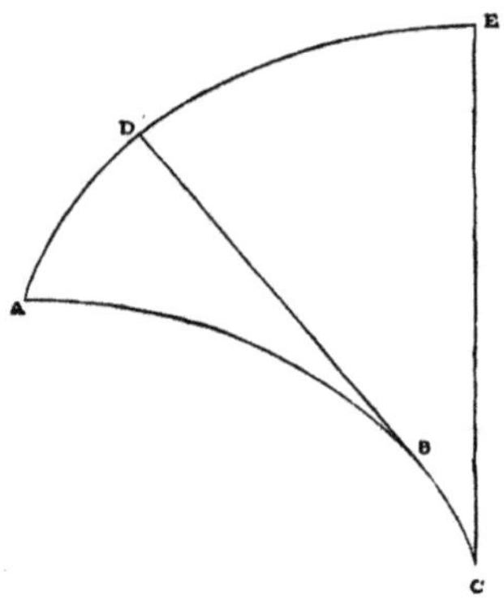

Figure 44: The evolute. From Huygens (1673, p. 59).

> Truly, the curve on which the thread is initially wound around is called the Evolute. In the above figure, ABC is the evolute, and ADE is described by the evolution from ABC; for indeed when the end of the thread from A reaches D, the extended part of the thread is the right line DB, and the rest of the thread BC is thus still in contact with the curve ABC. Moreover it is clear that DB is a tangent to the evolute at B.[1]

The term was a success and was soon found in various writings on the calculus, for example in Bernoulli (1692, p. 207). English *evolute* is first found (OED) in

[1] "Illa vero cui filum circumplicatum erat, dicatur Evoluta. In figura superiori, ABC est evoluta, ADE descripta ex evolutione ABC, ut nempe cum extremitas fili ex A venit in D, pars fili extensa sit DB recta, reliqua parte BC adhuc applicata curvae ABC. Manifestum est autem DB tangere evolutam in B". Translation from Huygens (2013, p. 61).

Hayes (1704, p. 109). The French translated the term by the word *développée*, from *développer* 'to develop'. It appears in the widely read *Analyse des infiniments petits* (de l'Hôpital 1696, p. 71).[1]

See also *involute*.

Note. Huygens construction of the evolute is discussed by Yoder (1988, passim).

References

Bernoulli, Jakob (1692). "Lineae Cycloidales, Evolutae, Anti-Evolutae, Causticae, Anti-Causticae, Peri-Causticae". *Acta Eruditorum* (May 1692), pp. 207–213.

Hayes, Charles (1704). *A Treatise of Fluxions*. London.

de l'Hôpital, Guillaume François Antoine (1696). *Analyse des infiniments petits, pour l'intelligence des lignes courbes*. Paris.

Huygens, Christiaan (1673). *Horologium oscillatorium*. Paris.

Huygens, Christiaan (2013). "Horologium oscillatorium". Translated and annotated by Ian Bruce. URL: `http://www.17centurymaths.com/contents/huygenscontents.html` (retrieved Oct. 12, 2019).

Mahoney, Michael S. (2005). "Charting the Globe and Tracking the Heavens: Navigation and the Sciences in the Early Modern Era". In: *The Heirs of Archimedes*. Technology, Science and the Art of Warfare through the Age of Enlightenment. Ed. by Brett D. Steele and Tamera Dorland. Cambridge, Mass., pp. 221–230.

Yoder, Joella G. (1988). *Unrolling Time*. Christiaan Huygens and the Mathematization of Nature. Cambridge.

exact (category theory)

According to Dieudonné (2009, p. 87), the term *exact sequence* first appears in an article by Kelley and Pitcher (1947, p. 687).[2] They defined: "A homomorphism sequence is said to be *exact* if the kernel of the homomorphism on each group is identical with the image of the preceding group". Kelley and Pitcher (1947, p. 682) also point out that the notion was earlier used by Hurewicz (1941).[3]

On the basis of private communication with Kelley, Sarason (1990, p. 18) reports on the origin of the term; Kelley attributed the term to Samuel Eilenberg and Norman Steenrod:

> Before Kelley and Pitcher published their results, they communicated them at two American Mathematical Society meetings in late 1945 and early 1946 [...] At one of these meetings they encountered Eilenberg, who at the time, in collaboration with Steenrod, was writing the now famous book *Foundations of Algebraic Topology* [...] Eilenberg informed Kelley and Pitcher that he and Steenrod had settled upon the term 'exact sequence' to use in their book. The term seemed so marvelously fitting that Kelley and Pitcher adopted it.

In the meetings alluded to, Kelley and Pitcher (1946, p. 74) used *natural homomorphism sequence*, a usage based on Hurewicz (1941, p. 562), who spoke of *natural homomorphisms*.

The term *exact functor* appears as French *foncteur exact* in Cartan (1950-51, p. 2); Buchsbaum (1955, p. 2) introduced the term *exact category* for what was later called an *abelian category*.

---

[1] The spelling *Dévelopée* was used there. [2] Marquis (2009, p. 80) claims that the term was already used in Eilenberg and Steenrod (1945); this is not the case, however. [3] In that note, the notion is paraphrased: "the kernel of each of these homomorphisms is the image of the preceding homomorphism".

Note. For the history of exact sequences, see Dieudonné (2009, pp. 85–90), Krömer (2007, esp. pp. 71f., 97–100), and Sarason (1990). For exact categories, see Krömer (2007, pp. 100f.).

References

Buchsbaum, David A. (1955). "Exact Categories and Duality". *Transactions of the American Mathematical Society* 80, pp. 1–34.

Cartan, Henri (1950-51). "Théorie de la cohomologie des espaces". *Séminaire Henri Cartan* 3, exposé 17, pp. 1–11.

Dieudonné, Jean (2009). *A History of Algebraic and Differential Topology, 1900–1960*. Boston.

Eilenberg, Samuel and Norman Steenrod (1945). "Axiomatic Approach to Homology Theory". *Proceedings of the National Academy of Sciences* 31, pp. 117–120.

Hurewicz, Witold (1941). "On duality theorems". *Bulletin of the American Mathematical Society* 47, pp. 562–563.

Kelley, John L. and Everett Pitcher (1946). "Applications of natural homomorphism sequences". *Bulletin of the American Mathematical Society* 52, p. 74.

Kelley, John L. and Everett Pitcher (1947). "Exact Homomorphism Sequences in Homology Theory". *Annals of Mathematics* 48, pp. 682–709.

Krömer, Ralf (2007). *Tool and Object*. A History and Philosophy of Category Theory. Basel.

Marquis, Jean-Pierre (2009). *From a Geometrical Point of View*. A Study of the History and Philosophy of Category Theory. Dordrecht.

Sarason, Donald (1990). "The exact answer to a question of Shields". *Mathematical Intelligencer* 12, pp. 18–19.

exact differential

The concept of exact differential was defined by Clairaut (1743, p. 37) under the name *differentielle complette*: "By a complete differential I understand a quantity whose integral is a function of x and y".[1] The terminology varied, however. For example, in an earlier publication, Clairaut (1742, p. 297) also defined a *differentielle complette*, but alternatively used *differentielle exacte* in the same sense (p. 304).[2]

While *exact* seems to be due to Clairaut, *complete* had earlier been used by Bernoulli (1719, pp. 196f.) with respect to differential equations: "I call such a differential equation of the curves which are to be intersected *complete* which expresses the relation which the differentials have to each other, not only those of the coordinates of these curves, but also of the variable parameter";[3] he used the designation $dx = pdy + qda$.

Lacroix (1819, p. 690) later mentioned that in his *Traité du calcul différentiel et du calcul intégral*, he sometimes used *complète* instead of *exacte* and remarked that the latter is not only more commonly used but also more appropriate since *complète* is often opposed to *partielle* and synonymous with *totale*.

The first use of English *exact differential* seems to be [anon.] (1800, p. 469).

Note. For the history of the concept, see de Oliveira (2020) and Greenberg (1995).

References

[anon.] (1800). "La Place *on Celestial Mechanics*". *The Monthly Review* 31, pp. 464–479.

---

[1] "J'entends par differentielle complette, une quantité qui a pour intégrale une fonction de x & et de y". Translation based on Guicciardini (1989, p. 80). [2] For comments on Clairaut's terminology, see Greenberg (1995, esp. pp. 351, 370). [3] "Voco autem aequationem differentialem Curvarum secandarum *completam* illam, quae exprimit relationem, quam habent inter se differentialia". Translation from Engelsman (1984, p. 98).

Bernoulli, Nicolaus (1719). "Tentamen solutio generalis Problematis de construenda Curva, quae alias ordinatim positione datas ad angulos rectos secat". *Acta Eruditorum* (June 1719), pp. 295–304.

Clairaut, Alexis Claude (1742). "Sur l'intégration ou la construction des équations différentielles du premier ordre". *Mémoires de l'Académie royale des sciences* (1740), pp. 293–323.

Clairaut, Alexis Claude (1743). *Théorie de la figure de la terre*. Paris.

Engelsman, Steven B. (1984). *Families of Curves and the Origins of Partial Differentiation*. Amsterdam and New York.

Greenberg, John L. (1995). *The Problem of the Earth's Shape from Newton to Clairaut*. The Rise of Mathematical Science in Eighteenth-Century Paris and the Fall of 'Normal' Science. Cambridge and New York.

Guicciardini, Niccolò (1989). *The Development of Newtonian Calculus in Britain, 1700–1800*. Cambridge.

Lacroix, Sylvestre François (1819). *Traité du calcul différentiel et du calcul intégral*. Seconde édition, revue et augmentée. Tome troisième. Paris.

de Oliveira, Mário J. (2020). "Exact and inexact differentials in the early development of mechanics and thermodynamics". *Revista Brasileira de Ensino de Física* 42, pp. 1–11.

expectation

In probability theory, the mathematical expectation is the average amount "expected" to be gained. The term originates with a phrase by Huygens. In his short treatise "Van Rekeningh in Spelen van Geluck", he wrote:

> I take it as a fundament [...] that in gambling the chance that somebody has toward something is worth as much as that [with] which, having it, he can arrive at the same chance by an equitable game, that is, where no loss is offered to anybody. For instance: if somebody without my knowledge hides 3 shillings in one hand and 7 shillings in the other and lets me choose which of either I want to get, I say this is worth the same to me as if I had 5 shillings for sure. Because, if I have 5 shillings, I can again arrive at having the same chance to get 3 or 7 shillings, and this by an equitable game; which will be shown afterward (Huygens 1660, pp. 489f.).[1]

Huygens expressed the expectation in terms of "being worth", as in proposition 1, which follows right after the text cited above: "If I have the same chance to get a or b it is worth as much to me as $(a+b)/2$" (Huygens 1660, p. 490).[2]

The actual word *expectation* was then used by van Schooten in his translation of Huygens' treatise (Huygens 1657). His usage is explained by Freudenthal (1980, p. 114):

> The Latin equivalents which Van Schooten preferred when translating Huygens' terminology have been a source of misunderstanding and confusion ever since. Historically it was Van Schooten who introduced the term *expectatio*, though not in the sense of our "expectation," in probability. [...] in Van Schooten's version, *expectatio* means the payoff table of a

[1] "Ick neeme tot beyder fondament, dat in het speelen de kansse, die yemant ergens toe heeft, even soo veel weerdt is als het geen, het welck hebbende hy weder tot deselfde kansse kan geraecken met rechtmatigh spel, dat is, daer in niemandt verlies geboden werdt. By exempel. So yemandt sonder mijn weeten in d'eene handt 3 schellingen verbergt, en in d'ander 7 schellingen, ende my te kiesen geeft welck van beyde ick begeere te hebben, ick segge dit my even soo veel weerdt te zijn, als of ick 5 schellingen seecker hadde. Om dat, als ik 5 schellingen hebbe, ick wederom daer toe kann geraecken, dat ick gelijcke kans sal hebben, om 3 of 7 schellingen te krijgen, en dat met rechtmatigh spel: gelijck hier naer sal betoont werden". Translation from Freudenthal (1980, pp. 114f.). [2] "Als ick gelijke kans hebbe om a of b te hebben, dit is my so veel weerdt als $\frac{a+b}{w}$". Translation from Freudenthal (1980, p. 115).

> hazard game, whereas our "expectation" is covered by such terms as *aestimandam esse* (valuating) and *valor expectationis* (value of expectation).[1] In other words, *expectatio* means the catalogue of what the player may expect when playing the game, whereas the cash value of this *expectatio* corresponds to what nowadays is called expectation.

In particular, the use of *valor expectationis* can be regarded as the origin of expectation in the now usual sense.

The term is soon found in other works, for example in Arbuthnot (1692, p. 3), who translated Huygens' text: "In both Cases I shall make use of this Principle, *Ones Hazard or Expectation to gain any Thing is worth so much, as, if he had it, he could purchase the like Hazard or Expectation again in a just and equal Game*". This appears to be the first printed use of *expectation* in English.[2]

J. Bernoulli (1713, p. 5) also adopted the term in his *Ars conjectandi* but warned the reader:

> It can be seen from what we have said that we are not using the word *expectation* in its ordinary sense, according to which we are commonly said to expect or to hope for what is best of all, though worse things can happen to us. Here account is taken of the extent to which our hope of getting the best is tempered and diminished by fear of getting something worse. So by its "value" we always mean something intermediate between the best we hope for and the worst we fear. This should be understood here and throughout the following.[3]

Bernoulli and others often used *sors* 'lot' instead of *expectatio*; for a detailed discussion, see J. Bernoulli (2006, esp. 71f., 116–121). de Moivre (1718, p. 2) defined the expectation in the following terms: "The Expectation of obtaining any Thing, is estimated by the Value of that Thing multiplied by the Probability of obtaining it".

In 1728, Cramer used French *espérance mathématique* 'mathematical expectation' in the now usual sense; the term is found in a letter to Nicolaus Bernoulli, which was first published by D. Bernoulli (1738, p. 191); Cramer also used *espérance morale* 'moral expectation' (p. 192).[4] The terminology was later adopted by Laplace, who used it and gave definitions on several occasions, starting in 1773 (Laplace 1776, p. 116) and ending in 1825 (Laplace 1825, p. 24) (these definitions differed only in minor details):

> The probability of events may be useful in determining the hope or fear of people affected by their occurrence. The word *expectation* has various interpretations: in general it expresses the advantage to anyone who expects any benefit whatsoever, under assumptions that are only probable. This advantage, in the theory of chances, is the product of the sum expected and the probability of getting it; it is the partial sum that ought to be paid out when one does not want to run the risks of the event, supposing that the total sum is to be shared out in proportion to the probabilities. This allocation is the only fair one when all irrelevant circumstances are set aside, because an equal degree of probability gives an equal claim to

---

[1] The latter term is found in Huygens (1657, p. 523). [2] Newton used it in notes that may go back to 1664 (Newton 1967, p. 60). The OED dates the term to 1718. [3] "Ex dictis colligi potest, vocabulum *Expectationis* non sumi hic sensu vulgari, quo communiter expectare vel sperare dicimur quod omnium optimum est, licet nobis pejus accidere possit; sed quatenus spes nostra impetrandi optimum temperata & imminuta est metu consequendi pejus: adeo ut per valorem ejus semper significetur intermedii quidpiam inter optimum quod speramus, & pessimum quod metuimus; quod hic & in sequentibus ubique est intelligendum". Translation from J. Bernoulli (2006, p. 134).
[4] The moral expectation was later called *utility*. For some comments on Cramer, see Faccarello (2016).

the sum expected. We shall call this advantage *mathematical expectation* (Laplace 1995, p. 24).[1]

English *mathematical expectation* was used, apparently for the first time, by Etrington (1802, p. 414); Cramer's term was first translated into German as *mathematische Hoffnung*, lit. 'mathematical hope' (D. Bernoulli 1738, p. 88). Haberl (1875, p. 199) later used *Hoffnungswert* (*Wert* 'value') and named the three synonyms *wahrscheinlicher Wert* (probable value), *Erwartungswert* (*Erwartung* 'expectation'), and *mathematische Hoffnung*. Lexis (1886, p. 438) settled on *Erwartungswert*; he was later referred to by Fisher (1915, p. 97): "Russian and Scandinavian writers and the followers of the Lexis statistical school of Germany have preferred to make another quantity known as the 'probable' or 'expected value,' the nucleus of their investigations". Fisher is cited by the OED for the first use of *expected value*.

Note. Freudenthal (1980) discusses the notion and terminology in Huygens' treatise. See also Dale (2003, pp. 298–334), J. Bernoulli (2006, pp. 71f., 116–121), and Brönnimann (2001).

References

Arbuthnot, John (1692). *Laws of Chance, or, A Method of Calculation of the Hazards of Game*. London.

Bernoulli, Daniel (1738). "Specimen theoriae novae de mensura sortis". *Commentarii academiae scientiarum imperialis Petropolitanae* 5, pp. 175–192.

Bernoulli, Jakob (1713). *Ars coniectandi*. Basel.

Bernoulli, Jakob (2006). *The Art of Conjecturing, together with Letter to a Friend on Sets in Court Tennis*. Translated with an introduction and notes by Edith Dudley Sylla. Baltimore.

Brönnimann, Daniel (2001). "Die Entwicklung des Wahrscheinlichkeitsbegriffs von 1654 bis 1718". MA thesis. ETH Zürich. URL: `ftp://stat.ethz.ch/Masters-Theses/Daniel_Broennimann-Wahrschkeit.pdf` (retrieved Mar. 17, 2020).

Dale, Andrew I. (2003). *Most Honourable Remembrance*. The Life and Work of Thomas Bayes. New York.

Etrington, Thomas (1802). "Letter from the Rev. Doctor Etrington". *The Anti-Jacobin Review and Magazine* 11, pp. 410–414.

Faccarello, Gilbert (2016). "Daniel Bernoulli (1700–1782)". In: *Handbook on the History of Economic Analysis*. Volume I: Great Economists since Petty and Boisguilbert. Ed. by Gilbert Faccarello and Heinz D. Kurz. Cheltenham, pp. 40–47.

Fisher, Arne (1915). *The Mathematical Theory of Probabilities and its Application to Frequency Curves and Statistical Methods*. Volume I. Trans. by F.W. Frankland. New York.

Freudenthal, Hans (1980). "Huygens' foundation of probability". *Historia Mathematica* 7, pp. 113–117.

Haberl, Josef (1875). *Lehrbuch der politischen Arithmetik*. Wien.

Huygens, Christiaan (1657). "De ratiociniis in ludo aleae". In: van Schooten, Frans. *Exercitionum Mathematicorum libri quinque*. Leiden, pp. 521–534.

[1] "La probabilité des événements sert à déterminer l'espérance ou la crainte des personnes intéressées à leur existence. Le mot *espérance* a diverses acceptions: il exprime généralement l'avantage de celui qui attend un bien quelconque, dans des suppositions qui ne sont que probables. Cet avantage, dans la théorie des hasards, est le produit de la somme espérée, par la probabilité de l'obtenir: c'èst la somme partielle qui doit revenir lorsqu'on ne veut pas courir les risques de l'évènement, en supposant que la répartition se fasse proportionnellement aux probabilités. Cette répartition est la seule équitable, lorsqu'on fait abstraction de toutes circonstances étrangères; parce qu'un égal degré de probabilité, donne un droit égal sur la somme espérée. Nous nommerons cet avantage, *espérance mathématique*". Translation from Laplace (1995, p. 11).

Huygens, Christiaan (1660). "Van Rekeningh in Spelen van Geluck". In: van Schooten, Frans. *Mathematische Oeffeningen*. Amsterdam, pp. 485–500.
Laplace, Pierre-Simon (1776). "Recherches sur l'intégration des Équations différentielles aux différences finies, & sur leur usage dans la théorie des hasards". *Mémoires de mathématique et de physique, presentés à l'Académie royale des sciences, par divers sçavans* 7, pp. 37–232.
Laplace, Pierre-Simon (1825). *Essai philosophique sur les probabilités*. Cinquième édition. Paris.
Laplace, Pierre-Simon (1995). *Philosophical Essay on Probabilities*. Translated from the fifth French edition of 1825 by Andrew I. Dale. New York.
Lexis, Wilhelm (1886). "Über die Wahrscheinlichkeitsrechnung und deren Anwendung auf die Statistik". *Jahrbücher für Nationalökonomie und Statistik* 47, pp. 433–450.
de Moivre, Abraham (1718). *The Doctrine of Chances: or, A Method of Calculating the Probability of Events in Play*. London.
Newton, Isaac (1967). *The Mathematical Papers of Isaac Newton*. Volume I. 1664-1666. Ed. by D.T. Whiteside. Cambridge.

exponent

The term *exponent* goes back to the *Arithmetica integra* by Michael Stifel: "the progression of numbers which progress naturally represents [*exponat*] the terms of the geometric progressions [...] each term of a cossic progression has its exponent [*exponentem*] in its order" (Stifel 1544, fol. 235ᵛ).[1] The term goes back to the Latin verb *exponere*, which means 'to set forth, to explain, to expose' (LS). Castel (1728, p. 223) later explains the term to his readers, giving a^2 as an example: "2 is called *exponent* because it exposes, it makes visible the degree of power to which a is exalted or elevated".[2]

Stifel's term was soon adapted to French by Peletier (1554, p. 8), who used *exposant*, while Ramus (1560, fol. 3ʳ) preferred Latin *index*.[3] The English term *exponent* appears to have been first used by Oughtred (1647, p. 1), who also names *index* as a synonym: "the upper numbers are Indices or Exponents of terms continually proportional from an unit both wayes". German *Exponent* is due to Wolff (1710, p. 21).

Note. On the origin of the term, see Tropfke (1933, p. 151) and Smith (1958, p. 521). See also Rosenfeld and Černova (1967) for the early history of exponents and Wieleitner (1924) for the history of fractional exponents.

References

Castel, Louis (1728). *Mathématique universelle abrégée*. Paris.
Oughtred, William (1647). *The Key of the Mathematicks*. London.
Peletier, Jacques (1554). *L'algebre de Jaques Peletier du Mans, departie an deus livres*. Lyon.
Ramus, Petrus (1560). *Algebra*. Paris.
Rosenfeld, Boris A. and M.L. Černova (1967). "Algebraic exponents and their geometric interpretation". *Organon* 4, pp. 109–112.
Smith, David Eugene (1958). *History of Mathematics*. Volume II. Special Topics of Elementary Mathematics. New York.
Stifel, Michael (1544). *Arithmetica integra*. Nürnberg.
Tropfke, Johannes (1933). *Geschichte der Elementar-Mathematik*. Zweiter Band. Allgemeine Arithmetik. Dritte, verbesserte und vermehrte Auflage. Berlin and Leipzig.

---

[1] "ut progressio numerorum naturaliter progredientium, exponat terminos progressionum Geometricarum [...] quemlibet terminum progressionis cossicae habere exponentem in suo ordine". [2] "2 s'appelle un *Exposant*, parce qu'il expose, il fait voir le degré de Puissance auquel a est exalté ou élevé". [3] For a brief discussion of Ramus' usage, see Verdonk (1966, pp. 210f.).

Verdonk, Johannes Jacobus (1966). *Petrus Ramus en de wiskunde*. Assen.
Wieleitner, Heinrich (1924). "Zur Geschichte der gebrochenen Exponenten". *Isis* 6, pp. 509–520.
Wolff, Christian (1710). *Der Anfangs-Gründe aller Mathematischen Wissenschaften Letzter Theil*. Halle.

exponential curve

According to Bos (1996, p. 7), variable exponents were first considered by Leibniz in September 1679, when he wrote a letter to Huygens in which he considered the equation $x^x - x = 24$ (Leibniz 1987, p. 844). Huygens was sceptical about their usefulness, as illustrated by a letter he sent to Leibniz in 1690: "I must confess that the nature of that sort of supertranscendental lines, in which the unknowns enter the exponent, seems to me so obscure that I would not think about introducing them into geometry unless you could indicate some notable usefulness of them" (Huygens 1901, p. 537).[1]

In his reply from November 1690, Leibniz assumed that Huygens' word *supertranscendental* was meant jokingly, and he used *expressions exponentiales* himself (Huygens 1901, p. 548). The term appears to be his creation (Cantor 1901, p. 232).

In a letter to Leibniz from 1694, Johann Bernoulli suggested the Latin term *percurrens*, which literally means 'through-running', a designation that is explained in the following quotation:

> I call in-between or percurrent a curve whose equation is of indeterminable degree, that is, in which the variables and constants go up to indeterminable dimensions and hence runs through all possible dimensions. These equations differ from your transcendentals in that the number of dimensions in those is inconstant and undetermined, whereas in these it is determined but unknown (Leibniz 2004, pp. 90f.).[2]

In print, the term *exponential curve* is then found in Nieuwentijt (1694, p. 34), who found it "suitable", and in Bernoulli (1697, p. 126), who by this time favoured the term over his own *percurrent*. The English term *exponential curve* is first found (OED) in Hayes (1704, p. 306); Wolff (1710, p. 336) uses German *Exponential-Linie* and also *Exponential-Aequation*.

In the 17th century and also later exponential curves have also been called *logarithmic*; see also *logarithmic curve*.

Note. For the history of the notion, see Cajori (1913), Curtis (1978), Bos (1996), and Blåsjö (2017, pp. 140–151).

References

Bernoulli, Johann (1697). "Principia calculi exponentialium, seu percurrentium". *Acta Eruditorum* (Mar. 1697), pp. 125–133.
Blåsjö, Viktor (2017). *Transcendental Curves in the Leibnizian Calculus*. Cambridge, Mass.

---

[1] "Pour moy j'avoue que la nature de ces lignes supertranscendantes, où les inconnues entrent dans l'Exposant, me parait si obscure, que je ne serais pas d'avis de les introduire dans la geometrie, à moins que vous n'y remarquiez quelque notable utilité". Translation from Bos (1997, pp. 23f.).

[2] "Medias autem vel percurrentes appello quarum aequatio est indeterminati gradus, id est, in qua literae indeterminatae et constantes ascendunt ad dimensionem indeterminatam et proinde omnes possibiles dimensiones percurrunt: Hae aequationes a Tuis transcendentalibus in eo differunt quod numerus dimensionum in illis est vagus et indeterminatus, in his vero determinatus sed incognitus". Translation from Blåsjö (2017, p. 146).

Bos, Henk J.M. (1996). "Johann Bernoulli on exponential curves, ca. 1695. Innovation and habituation in the transition from explicit constructions to implicit functions". *Nieuw Archief voor Wiskunde* 14, pp. 1–19.
Bos, Henk J.M. (1997). *Lectures in the History of Mathematics*. Providence, R.I.
Cajori, Florian (1913). "History of the Exponential and Logarithmic Concepts". *The American Mathematical Monthly* 20, pp. 5–14, 35–47, 75–84, 107–117, 148–151, 173–182, 205–210.
Cantor, Moritz (1901). *Vorlesungen über Geschichte der Mathematik*. Dritter Band. Von 1668 bis 1758. Leipzig.
Curtis, Lorenzo J. (1978). "Concept of the exponential law prior to 1900". *American Journal of Physics* 46, pp. 896–906.
Hayes, Charles (1704). *A Treatise of Fluxions*. London.
Huygens, Christiaan (1901). *Œuvres complètes*. Tome neuvième. Correspondance 1685–1690. La Haye.
Leibniz, Gottfried Wilhelm (1987). *Sämtliche Schriften und Briefe*. Dritte Reihe. Mathematischer, naturwissenschaftlicher und technischer Briefwechsel. Zweiter Band. 1676-1679. Hannover.
Leibniz, Gottfried Wilhelm (2004). *Sämtliche Schriften und Briefe*. Dritte Reihe. Mathematischer, naturwissenschaftlicher und technischer Briefwechsel. Sechster Band. 1694-Juni 1696. Hannover.
Nieuwentijt, Bernard (1694). *Considerationes Circa Analyseos ad Quantitates infinitè parvas applicatae Principia*. Amsterdam.
Wolff, Christian (1710). *Der Anfangs-Gründe aller Mathematischen Wissenschaften Letzter Theil*. Halle.

extension

The term *extension* (German *Erweiterung*) was suggested by Steinitz. He noted that a "partial system" K' of a field K can also be a field with respect to the operations inherited from K and stated the necessary and sufficient conditions. If these conditions hold, "we say that *the field K comprises the field K', K' is contained in K*, and call K' a *subfield* of K, K *extension field* or *extension* of K'" (Steinitz 1910, pp. 174f.).[1] The English term *extension* is found in Gokhale (1922, p. 299) and Dickson et al. (1923, p. 76). Fraenkel (1916, pp. 44f.) used the term in connection with rings, in which context English *extension* was used by Wahlin (1918, p. 495).

In his paper, Steinitz also introduced various related terms. These include *intermediate field* (p. 176), *equivalent extensions* and *simple extension* (p. 182), *algebraic extension* and *algebraically closed* (p. 198), *degree* of an extension (p. 199), *normal extension* (p. 207), *transcendental extension* (p. 288) and *transcendence degree* (p. 299). See also *separable* (algebra).

For the extension of a function, see *restriction* and *extension* (functions).

Note. For an account of Steinitz' work, see Corry (2004, pp. 192–196).

References

Corry, Leo (2004). *Modern Algebra and the Rise of Mathematical Structures*. Second revised edition. Basel.
Dickson, Leonard Eugene et al. (1923). *Algebraic Numbers*. Report of the Committee on Algebraic Numbers, National Research Council. Washington, D.C.
Fraenkel, Adolf (1916). *Über gewisse Teilbereiche und Erweiterungen von Ringen*. Leipzig.
Gokhale, Vishnu Dattatreya (1922). "Concerning Compact Kürschak Fields". *American Journal of Mathematics* 44, pp. 297–316.

---

[1] "Wir sagen dann: *der Körper K umfaßt den Körper K', K' ist in K enthalten*, und nennen K' *Teilkörper* von K, K *Erweiterungskörper* oder *Erweiterung* von K'". – The term *subfield* was in use before Steinitz; see *subgroup*.

Steinitz, Ernst (1910). "Algebraische Theorie der Körper". *Journal für die reine und angewandte Mathematik* 137, pp. 167–309.
Wahlin, G.E. (1918). "Ueber gewisse Teilbereiche und Erweiterungen von Ringen [review]". *Bulletin of the American Mathematical Society* 24, pp. 493–496.

exterior algebra and *wedge product*

The *exterior*-terminology essentially goes back to the work of Graßmann, who used it, in particular, in the two versions of his *Ausdehnungslehre* (Graßmann 1844, 1862). See *vector product*.

Graßmann's ideas spread slowly; one line of reception is the work of Élie Cartan, who, from around 1900 on, developed his calculus of differential forms, in which he used Graßmann's exterior product. He coined the terms *exterior differential form* (*forme différentielle extérieure*) as well as *exterior derivative* (*dérivé extérieure*), and said he chose this terminology because exterior differential forms "obey the rules of Graßmann's exterior multiplication" (É. Cartan 1922, pp. ix, x).[1]

The terms *exterior algebra* (*algèbre extérieure*) and the synonym *Grassmann algebra* (*algèbre de Grassmann*) were then coined by Weil (1936-37, p. 1):

> A hypercomplex system over k, $G = G(k^n)$ with the following properties shall be called *exterior algebra* or *Grassmann algebra* over k^n:
>
> 1^0–G contains a unit element (designated by 1) and all the elements of k^n; and G is generated by these elements.
>
> 2^0–If x, y are two elements of k^n, their product in G obeys the law: $yx = -xy$. (If k is not of characteristic 2, it follows:
>
> 3^0–for all x in k^n, $xx = 0$; if k is of characteristic 2, we add 3^0 as an axiom).
>
> 4^0–The elements of G do not satisfy other relations than those that can be deduced from 2^0 and 3^0.[2]

The terminology spread after Bourbaki (1948, pp. 69, 71) used *produit extérieure* and *algèbre extérieure* in his volume on multilinear algebra.

When Weil introduced the term *exterior algebra*, he also remarked that the multiplication on such an algebra is called *exterior multiplication*; he designated this multiplication by the symbol $\wedge$. Weil's paper was presented at the Séminaire de Julia, and others attending this seminar also used the symbol (for example, H. Cartan 1936-37, p. 11).[3]

Explicitly, *wedge product* is found in later works, for example in Chern (1952, p. 229): "We shall, following Bourbaki, use wedge product to denote exterior multiplication". Bourbaki (1948, p. 69) did use the symbol but not the term.

---

[1] "parce qu'elles obéissent aux règles de la multiplication extérieur de H. Grassmann". [2] "On appellera *algèbre extérieure* ou *algèbre de Grassmann* sur k^n, un système hypercomplexe sur k, $G = G(k^n)$ avec les propriétés suivantes: 1^0– G contient un élément unité (désigné par 1) et tous les éléments de k^n; et G est engendré par ces éléments. 2^0– Si x, y, sont deux éléments de k^n, leur produit dans G obéit à la loi: $yx = -xy$. (Si k n'est pas de caractéristique 2, il en résulte que: 3^0– quel que soit x dans k^n, $xx = 0$; si k est de caractéristique 2, on ajoute 3^0 comme axiome). 4^0– Les éléments de G ne satisfont pas à d'autres relations qu'à celles qui se déduisent de 2^0 and 3^0".
[3] The symbol appeared earlier in Burali-Forti and Marcolongo (1907, p. 78) and Burali-Forti and Marcolongo (1909, p. 31), where it is used for the cross or vector product.

The concept was earlier used by Graßmann, whose term was *progressive product* (Graßmann 1862, pp. 60f.).[1]

Note. Explanations and historical comments on Graßmann's exterior multiplication can be found in Dieudonné (1979) as well as in Zaddach (1996) and Brini and Teolis (1996). For Cartan's theory, see Katz (1985) and Pier (2001, pp. 283–287).

References

Beutelspacher, Albrecht (1996). "A Survey of Grassmann's *Lineale Ausdehnungslehre*". In: *Hermann Günther Graßmann (1809-1877)*. Visionary Mathematician, Scientist and Neohumanist Scholar. Ed. by Gert Schubring. Dordrecht, pp. 3–6.

Bourbaki, Nicolas (1948). *Algèbre*. Chapitre III. Algèbre multilinéaire. Paris.

Brini, Andrea and Antonio G.B. Teolis (1996). "Grassmann progressive and regressive products and CG-Algebras". In: *Hermann Günther Graßmann (1809-1877)*. Visionary Mathematician, Scientist and Neohumanist Scholar. Ed. by Gert Schubring. Dordrecht, pp. 231–242.

Burali-Forti, Cesare and Roberto Marcolongo (1907). "Per l'unificazione delle notazioni vettoriali. Nota II". *Rendiconti del Circolo Matematico di Palermo* 24, pp. 65–80.

Burali-Forti, Cesare and Roberto Marcolongo (1909). *Elementi di calcolo vettoriale con numerose applicazioni*. Bologna.

Cartan, Élie (1922). *Leçons sur les invariants intégraux*. Paris.

Cartan, Henri (1936-37). "Systèmes de Pfaff". *Séminaire de Mathématiques dit de Julia* 4, exposé 2, pp. 1–33.

Chern, Shiing-Shen (1952). "On the Kinematic Formula in the Euclidean Space of *N* Dimensions". *American Journal of Mathematics* 74, pp. 227–236.

Dieudonné, Jean (1979). "The Tragedy of Grassmann". *Séminaire de Philosophie et Mathématiques* 1979, pp. 1–14.

Graßmann, Hermann (1844). *Die Wissenschaft der extensiven Grösse, oder die Ausdehnungslehre, eine neue mathematische Disziplin*. Erster Theil, die lineale Ausdehnungslehre enthaltend. Leipzig.

Graßmann, Hermann (1862). *Die Ausdehnungslehre*. Vollständig und in strenger Form bearbeitet. Berlin.

Katz, Victor J. (1985). "Differential Forms: Cartan to De Rham". *Archive for History of Exact Sciences* 33, pp. 321–336.

Pier, Jean-Paul (2001). *Mathematical Analysis During the 20th Century*. Oxford.

Weil, André (1936-37). "Formes différentielles extérieures". *Séminaire de Mathématiques dit de Julia* 4, exposé 1, pp. 1–15.

Zaddach, Arno (1996). "Regressive products and Bourbaki". In: *Hermann Günther Graßmann (1809-1877)*. Visionary Mathematician, Scientist and Neohumanist Scholar. Ed. by Gert Schubring. Dordrecht, pp. 285–295.

extremal graph

The work of Turán is often seen as the beginning of extremal graph theory (e. g., by Simonovits 1977, p. 112). In particular, in an article published in 1941, he posed an extremal problem and gave a bound for the number of edges of a graph that does not contain a complete subgraph (Turán 1941). Explicitly, the term *extremal graph* appears in Turán (1954, p. 21). It is more generally used from the 1960s onwards, for example by Sauve (1961, p. 110) and Lorden (1962, p. 114).

References

Lorden, Gary (1962). "Blue-Empty Chromatic Graphs". *The American Mathematical Monthly* 69, pp. 114–120.

[1] For a brief account of Graßmann's work, see Beutelspacher (1996).

Sauve, Leopold (1961). "On Chromatic Graphs". *The American Mathematical Monthly* 68, pp. 107–111.
Simonovits (1977). "On Paul Turán's influence on graph theory". *Journal of Graph Theory* 1, pp. 102–116.
Turán, Pál (1941). "Egy gráfelméleti szélsőérték-feladatról". *Matematikai és Fizikai Lapok* 49, pp. 436–452.
Turán, Pál (1954). "On the theory of graphs". *Colloquium Mathematicae* 3, pp. 19–30.

extremum

The term combines the notions of maximum and minimum and was suggested as German *Extremum* by Paul du Bois-Reymond (1879, p. 564): "On this occasion, I take the liberty of introducing a new term. It is so often necessary to speak of the existence of a maximum *or* minimum that this cumbersome alternative eventually becomes tedious. I suggest that these two cases be designated with the *one* word *extremum*".[1]

Much earlier, however, the notions of maximum and minimum had already been combined and expressed with a compound meaning 'extreme value'. For example, Nieuwentijt (1695, p. 267) used Latin *valores extremum*, Stone (1726, see Maximis and Minimis) *extreme value*, Baltzer (1872, p. 217) German *extremer Werth*, and du Bourguet (1810, p. 105) suggested French *extrêmes grandeurs* 'extreme magnitudes'.

The English term *extremum* appears in Hedrick (1903, p. 141).

References

Baltzer, Richard (1872). *Die Elemente der Mathematik*. Erster Band. Gemeine Arithmetik, Allgemeine Arithmetik, Algebra. Vierte verbesserte Auflage. Leipzig.
du Bois-Reymond, Paul (1879). "Fortsetzung der Erläuterungen zu den Anfangsgründen der Variationsrechnung". *Mathematische Annalen* 15, pp. 564–577.
du Bourguet, J.-B.-E. (1810). *Traités élémentaires de calcul différentiel et de calcul intégral*. Tome premier. Paris.
Dorofeeva, A.V. (1998). "The Calculus of Variations". In: *Mathematics of the 19th Century*. Function Theory According to Chebyshev. Ordinary Differential Equations. Calculus of Variations. Theory of Finite Differences. Ed. by Andrey Nikolaevich Kolmogorov and Adolph Andrei Pavlovich Yushkevich. Basel, pp. 197–260.
Hedrick, Earle Raymond (1903). "On the Characteristics of Differential Equations". *Annals of Mathematics* 4, pp. 121–144.
Nieuwentijt, Bernard (1695). *Analysis infinitorum seu curvilineorum proprietates ex polygonorum natura deductae*. Amsterdam.
Stone, Edmund (1726). *A New Mathematical Dictionary*. London.

[1] "Bei dieser Gelegenheit erlaube ich mir eine neue Namensgebung in Vorschlag zu bringen. Es ist so oft nöthig, von dem Vorhandensein eines Maximums *oder* Minimums und damit Zusammenhängendem zu reden, dass diese schleppende Alternative schliesslich lästig wird. Ich schlage vor, diese beiden Fälle mit dem *einen* Wort *Extremum* zu bezeichnen [...]". Translation partially based on Dorofeeva (1998, p. 233).

F

face

The faces of a polyhedron are the surfaces at its sides. In classical Greek, one way to refer to the faces was to use the term *βάσις*, which Plato used in his *Timaeus* (55b), where he says the the icosahedron has "twenty equilateral triangular bases".[1] The corresponding Latin translation *basis* was used by Bacon in a discussion of the Platonic solids (Bridges 1900, p. 160) and earlier by Fibonacci, who referred to polyhedra as *solida multarum basium* 'solids with many bases' (Fibonacci 1862, p. 159).

Another term was *ἕδρα*, which means 'seat, basis'. According to Mugler (1958-59, p. 159), who cites Hero of Alexandria and Pappus as evidence, the term is rare and of late origin; in its Latin form *hedra*, it was still used by Kepler (1616, Erclärung) and, in the 18th century, by Euler (1758, § 6), for example.

The mathematical origin of the term *face*, however, is unclear. Federico (1982, p. 51) points out that French *face* was consistently used in Descartes' manuscript *De Solidorum Elementis*, which dates from around 1637; he remarks that "this usage is unusual and perhaps even unique" (ibid.). There are, however, various texts in which *face* was used well before Descartes.

For example, Marsilio Ficino, in his commentary on the *Timaeus*, used Latin *facies* in connection with various kinds of polyhedra (Ficino 1496, ch. xli). Similarly, French *face* appears in de Boissière (1561, p. 51) in connection with the cube, and in Italian Barbaro (1558, p. 53) states that the icosahedron has "twenty triangular faces"[2]. Both the French and the Italian term can be found earlier, however. The Italian term was already used in the 14th century (see Simi 1999, p. 84 for references), and the French term appears in Oresme's *Livre du ciel et du monde*, for example in his definition of regular solids: "a body is called regular whose faces or plane surfaces and angles are similar and equal" (Oresme 1968, p. 640).[3]

---

[1] Translation from Plato (1929, p. 133). [2] "uenti faccie triangolari". [3] "un corps est dit regulier qui a faces ou superfices semblables et equales et angles aussi". Translation from Oresme (1968, p. 640), slightly adapted. The term appears in other places of that book; see pp. 596, 640, 642.

H. Becker, *A Historical Dictionary of Mathematical Terms*, Science Networks. Historical Studies 66, https://doi.org/10.1007/978-3-032-02873-0_6

The term was therefore in use at least since the 14th century and well established in the middle of the sixteenth century. What remains an open question is whether Oresme's use is really the first one.

Note. See Mugler (1958-59, p. 95) for *βάσις* and Mugler (1958-59, p. 159) for *ἕδρα*. Mugler (p. 193) also notes the rare use of *ἐπίπεδον* in this sense.

References

Barbaro, Daniele (1558). *La pratica della perspettiva*. Venezia.
de Boissière, Claude (1561). *L'art d'arithmétique contenant toute dimension*. Reueu & augmentée par Lucas Tremblay. Paris.
Bridges, John Henry, ed. (1900). *The 'Opus Majus' of Roger Bacon*. Volume I. London et al.
Euler, Leonhard (1758). "Elementa doctrinae solidorum". *Novi Commentarii academiae scientiarum Petropolitanae* 4, pp. 109–140.
Federico, Pasquale Joseph (1982). *Descartes on Polyhedra*. A Study of the De Solidorum Elementis. New York and Berlin.
Fibonacci (1862). *Scritti di Leonardo Pisano*. Pubblicati da Baldassare Boncompagni. Volume II. Roma.
Ficino, Marsilio (1496). *Commentaria in Platonem*. Firenze.
Kepler, Johannes (1616). *Außzug auß der Uralten MesseKunst Archimedis*. Linz.
Mugler, Charles (1958-59). *Dictionnaire historique de la terminologie géométrique des Grecs*. Paris.
Oresme, Nicole (1968). *Le livre du ciel et du monde*. Edited by Albert D. Menut and Alexander J. Denomy, translated with an introduction by Albert D. Menut. Madison.
Plato (1929). *Timaeus. Critias. Cleitophon. Menexenus. Epistles*. Translated by R.G. Bury. London.
Simi, Annalisa (1999). "La geometria nel Rinascimento: il Codice Ottoboniano latino 3307 della Biblioteca Apostolica Vaticana". *Contributi di filologia dell'Italia mediana* 13, pp. 41–109.

factor (arithmetic)

Many of the terms in Greek arithmetic were originally used in geometry. For example, a factor was called *πλευρά* 'side', the side of a rectangle being the geometrical equivalent of the factors in a product. In Euclid' *Elements*, the concept is found in book VII, def. 16, for example: "And, when two numbers having multiplied one another make some number, the number so produced is called *plane*, and its *sides* are the numbers which have multiplied one another".[1]

The term *factor* is derived from the Latin verb *facere* 'to make, to do'; factors are so called, in Harris' (1704, s. v. Factors) words, "because they do *Facere Productum*, make or constitute the Product". While *factor* dates from the 16th century, the verb *facere* was already used by Roman land surveyors in the context of multiplication (see *multiplication*).

It appears that the first who used *factor* in the sense under discussion was Ramus in 1569. His terminology evolved over time. Originally, he expressed division in terms of measuring; for example, in Ramus (1557, p. 15), he stated: "Qui metitur mensorem, metitur mensum" – "[a number] that measures the factor [lit., 'the measurer'], also measures the product [lit., 'the measured']". In later publications, however, Ramus points out that the concept of measuring is a geometric metaphor which has, in his opinion, no place in arithmetic (Ramus 1569b, p. 126); consequently, he then uses a different terminology, stating, for example: "qui multos factores habet, habet

---

[1] Translation from Euclid (1956, p. 278). – The term was in use before Euclid; see LSJ, Heath in Euclid (1956, p. 288), and Szabó (1978, p. 42).

etiam multos diuisores" – "[a number] that has many factors, also has many divisors" (Ramus 1569a, p. 8).[1]

The term is then found in Geyger (1609, p. 23) and Beyer (1619, p. 14); Beyer also gives a definition: "*factores numeri*, two numbers whose multiplication produces the third number".[2] Oughtred (1631, p. 5) also adopts the term.

The English term dates from the 17th century and seems to be due to Oughtred (1633, p. 6), who states the following rule: "*In Multiplication, As an Vnite is to one of the factores* (or numbers to be multiplied:) *so is the other of the factores, to the product*". In his *Key of the Mathematicks*, he defines: "The number found out by Multiplication, is called the Factus, the Product, the Rectangle, or the Plain [...] And the numbers so multiplied are called the Factores and the Sides" (Oughtred 1647, p. 7).[3]

The German term is of later origin; in early 17th-century texts, it appears in Latin term and is typographically marked as a foreign word, for example in Geyger (1609, p. 7).[4] Even in Wolff (1710, p. 39), it not yet completely naturalized. In France, the term seems to have become known through English texts; referring to Harris, the *Dictionnaire de Trévoux* (1721, vol. 2, p. 1618) remarks: "the English Arithmeticians call the multiplicand and the multiplier the *factors*".[5]

Note. An overview of the terminology is found in Tropfke (1980, pp. 224–227).

References

Atwell, George (1658). *The faithfull surveyour discovering divers errours in land measuring*. Cambridge?

Beyer, Johann Hartmann (1619). *Logistica Decimalis*. Frankfurt.

Euclid (1956). *The Thirteen Books of Euclid's Elements*. Volume II. Books III-IX. Ed. by Thomas L. Heath. New York.

Geyger, Philipp (1609). *Grundtliche und ordentliche Erklerung deß newen und kunstreychen Rechentisches*. Zürich.

Harris, John (1704). *Lexicon Technicum*. London.

Oughtred, William (1631). *Arithmeticae in numeris et speciebus institutio*. London.

Oughtred, William (1633). *The circles of proportion and the horizontal instrument*. Translated into English, and set out for the public benefit, by William Forster. London.

Oughtred, William (1647). *The Key of the Mathematicks*. London.

Ramus, Petrus (1557). *Arithmeticae libri tres*. Paris.

Ramus, Petrus (1569a). *Arithmeticae libri duo. Geometriae septem et viginti*. Basel.

Ramus, Petrus (1569b). *Scholarum Mathematicarum, Libri Unus et Triginta*. Basel.

Szabó, Árpád (1978). *The Beginnings of Greek Mathematics*. Dordrecht.

Tropfke, Johannes (1980). *Geschichte der Elementarmathematik*. Band 1: Arithmetik und Algebra. Vollständig neu bearbeitet von Kurt Vogel, Karin Reich, Helmuth Gericke. Berlin and New York.

Verdonk, Johannes Jacobus (1966). *Petrus Ramus en de wiskunde*. Assen.

Wolff, Christian (1710). *Der Anfangs-Gründe aller Mathematischen Wissenschaften Erster Theil*. Halle.

[1] For a discussion of Ramus' terminology, see Verdonk (1966, pp. 155–158). [2] "*factores numeri*, zwo Zahlen/ deren Multiplication die dritte Zahl produzirt". Geyger explains *factores* by *latus* and the corresponding German word *Seite*. [3] The OED refers to Atwell (1658, p. 112) as the first usage in English. [4] Geyger writes: "Such den einen *Factorem* oder die ein zal welche *multipliciert* [...]" ("Find the one factor or the number which multiplies"). [5] "Les Arithméticiens Anglois appellent *facteurs*, le multiplicand & le multiplicateur".

factor (algebra)

The term has been transferred to various mathematical fields. In particular, the transfer to algebra had already been prepared by Galois, who called a subgroup a *divisor* (French *diviseur*); see *subgroup*.

Jordan (1869, p. 152) introduced a composition series of a group G as a series of subgroups G_0, G_1, ..., G_n, with G_0 being the trivial subgroup and G_{i+1} being a maximal subgroup of G_i. The *composition factors* (*facteurs de composition*) are the numbers $|G_i|/|G_{i+1}|$.[1]

Hölder (1889, p. 30) later remarked that "[t]his theory of composition factors must however be deepened, in the sense that the factors are to be understood as *groups*".[2] He correspondingly called the groups G_i/G_{i+1} *factor groups* (*Factorgruppen*), a notion which he described as "until now, not sufficiently appreciated" (Hölder 1889, p. 28). The English term appeared soon afterwards in Netto (1892, p. 96).

Note. The history of the notion is discussed in Nicholson (1993) and Schlimm (2008).

References

Hölder, Otto (1889). "Zurückführung einer beliebigen algebraischen Gleichung auf eine Kette von Gleichungen". *Mathematische Annalen* 34, pp. 26–56.
Jordan, Camille (1869). "Commentaire sur Galois". *Mathematische Annalen* 1, pp. 141–160.
Jordan, Camille (1870). *Traité des substitutions et des équations algébriques*. Paris.
Netto, Eugen (1892). *The Theory of Substitutions and its Applications to Algebra*. Revised by the author and translated with his permission by F.N. Cole. Ann Arbor, Mich.
Nicholson, Julia (1993). "The Development and Understanding of the Concept of Quotient Group". *Historia Mathematica* 20, pp. 68–88.
Schlimm, Dirk (2008). "On Abstraction and the Importance of Asking the Right Research Questions: Could Jordan have Proved the Jordan-Hölder Theorem?" *Erkenntnis* 68, pp. 409–420.

factor (graph theory)

Another extension of the term is found in graph theory, where the *factor* (German *Factor*) of a graph was introduced by Petersen (1891, p. 194):

> By the *order* of a graph I mean the number of its vertices [...] and by its *degree* the number of edges incident with each vertex [...] I shall denote by G_α^n, or simply G_α, a graph of order n and degree α. Such a graph can be decomposed into factors when other graphs of the same order but lesser degree can be found, such that superimposing them results in the given graph. A graph which cannot be split in this way is said to be *primitive*.[3]

In English, the terms *factor* and *factorable* were used by Brahana (1917, p. 59).

Note. For the history of the concept, see Biggs, Lloyd, and Wilson (1998, ch. 10).

---

[1] See Schlimm (2008, pp. 411f.) for discussion. Jordan also presented the concept in his *Traité* (Jordan 1870, pp. 41f.). [2] "Diese Theorie von den Factoren der Zusammensetzung muss aber dahin vertieft werden, dass die Factoren als *Gruppen* aufgefasst werden". Translation from Schlimm (2008, p. 412). [3] "Durch die *Ordnung* eines *graphs* werde ich die Anzahl der Punkte (die Ordnung der binären Grundform) verstehen, durch den *Grad* die Anzahl der in jedem Punkte zusammenlaufenden Linien [...] Durch G_α^n oder einfach G_α werde ich einen *graph* von der Ordnung n und vom Grade α verstehen. Ein solcher lässt sich zerlegen oder in Factoren auflösen, wenn man andere *graphs* von derselben Ordnung aber von niedrigerem Grade finden kann, die durch Überlagerung den gegebenen *graph* herstellen. Ein *graph*, der sich nicht in solcher Weise auflösen lässt, heisst *primitiv*". Translation from Biggs, Lloyd, and Wilson (1998, p. 191).

References

Biggs, Norman L., E. Keith Lloyd, and Robin J. Wilson (1998). *Graph Theory, 1736-1936*. Oxford.
Brahana, H.R. (1917). "A Proof of Petersen's Theorem". *Annals of Mathematics* 19, pp. 59–63.
Petersen, Julius (1891). "Die Theorie der regulären Graphs". *Acta Mathematica* 15, pp. 193–220.

factorial

This term was originally introduced by Kramp, who used German *Facultät* in a letter from 1796 (Kramp 1798, p. 109), Latin *facultas* in a paper read in 1797 (Kramp 1799b, p. 257), and French *faculté* in Kramp (1799a, p. 46), where he defined: "I will refer to products whose factors constitute an arithmetic progression by the name of *faculties*, such as $a(a+r)(a+2r)\ldots(a+mr-r)$. This class of analytic functions was never given a name: we will see that it deserved to be given one".[1] One year later, Arbogast (1800, p. 364) suggested "*factorial quantity*, or simply *factorial*" as an alternative.[2] Kramp (1808, pp. xi, xii) then adopted Arbogast's term: "I gave them the name *faculties* [*facultées*]. Arbogast replaced this by the more suitable and more French term *factorial* [*factorielle*]; I recognized the advantage of this denomination and on adopting his idea, I am happy to pay hommage to the memory of my friend".[3]

The German term *Fakultät* was adopted by other German writers (for example, by Weingärtner 1800, p. 227) and is still in use in the mathematical sense. Klügel (1805, p. 175) remarked that it was coined in analogy to the terms *power* (*Potenz*) and *dignity* (*Dignität*, an obsolete synonym of *Potenz*). In English, *factorial* appears in an anonymous text presumably written by Herschel and Babbage (1813, p. xiv); the term *faculty*, found in Wallace (1824, p. 285), is rare and never caught on.

Note. For a brief historical discussion of the concept, see Noble (2022, pp. 218–222).

References

[Herschel, J.F.W. and Babbage, Charles] (1813). "Preface". *Memoirs of the Analytical Society* (1813), pp. i–xxii.
Arbogast, Louis François Antoine (1800). *Du calcul des dérivations*. Strasbourg.
Klügel, Georg Simon (1805). *Mathematisches Wörterbuch*. Erste Abtheilung: Die reine Mathematik. Zweyter Theil von E bis I. Leipzig.
Kramp, Christian (1798). "Aus einem Schreiben Herrn D. Kramp's vom 30 May 1796, dessen weitere Fortschritte in der combinatorischen Analysis betreffend". *Archiv der reinen und angewandten Mathematik* 2, Heft 5, pp. 107–112.
Kramp, Christian (1799a). *Analyse des réfractions astronomiques et terrestres*. Strasbourg.
Kramp, Christian (1799b). "Fractionum Wallisianarum analysis". *Abhandlungen der Kurfürstlich-Mainzischen Akademie nützlicher Wissenschaften zu Erfurt* 1, pp. 257–292.
Kramp, Christian (1808). *Éléments d'arithmétique universelle*. Köln.
Noble, Eduardo (2022). *The Rise and Fall of the German Combinatorial Analysis*. Cham.
Wallace, J. (1824). "New algebraic series". *American Journal of Science and Arts* 7, pp. 278–286.

---

[1] "Je désignerai par le nom des *facultés*, les produits dont les facteurs constituent une progression arithmétique; tel que $a(a+r)(a+2r)\ldots(a+mr-r)$. Cette classe de fonctions analytiques n'a jamais eu de dénomination particulière; nous allons voir qu'elle méritoit d'en avoir une". Translation from Noble (2022, p. 220). [2] "quantité factorielle, ou simplement *factorielle*". [3] "[...] je leur avais donné le nom de *facultés*. Arbogast lui avait substitué la nomination plus nette et plus française de *factorielles*; j'ai reconnu l'avantage de cette nouvelle dénomination; et en adoptant son idée, je me suis félicité de pouvoir rendre hommage à la mémoire de mon ami".

Weingärtner, Johann Christoph (1800). *Lehrbuch der combinatorischen Analysis nach der Theorie des Herrn Professor Hindenburg*. Erster Theil. Leipzig.

factorial ring

See *unique factorization domain.*

faithful functor

This term is presumably due to Eilenberg.[1] He presented it at a conference in 1956 in a joint paper with Henri Cartan and gave the following definition (Cartan and Eilenberg 1958, p. 18):

> A functor $T : \mathscr{A} \to \mathscr{B}$ is called *faithful* if the following axioms hold

(Fid. 1) If $f, g \in Hom_{\mathscr{A}}(A, A')$ and $T(f) = T(g)$ then $f = g$

(Fid. 2) If $f : A \to A'$ is an isomorphism in $\mathscr{A}$ and $T(f) = i_{T(A)}$ then $A = A'$

The attribution to Eilenberg rests on the fact that he earlier used the term in lectures from 1957, where exactly the same definition was given (Eilenberg 1957, § 4). The term was soon in common use and appears, for example, in Mitchell's *Theory of Categories* (Mitchell 1965, pp. 51f.), who also remarks that a functor is "faithful if it preserves distinctness of morphisms", which explains the name.

References

Cartan, Henri and Samuel Eilenberg (1958). "Foundations of fiber bundles". In: *Symposium Internacional de Topología Algebraica*. Ciudad de México, pp. 16–23.
Eilenberg, Samuel (1957). *Foundations of Fiber Bundles*. Chicago.
Kurosch, A.G., A.Ch. Liwschitz, and E.G. Schulgeifer (1963). "Grundzüge der Theorie der Kategorien". In: Kurosch, A.G. et al. *Zur Theorie der Kategorien*. Trans. by Helmut Boseck and Hans-Jürgen Hoehnke. Berlin, pp. 7–70.
Mitchell, Barry (1965). *Theory of Categories*. New York.

family

Families of curves began to be investigated at the end of the 17th century. Engelsman (1984, p. 8) discusses the early terminology used by Leibniz: "Leibniz provided the technical term for such a family of curves, the term that remained in use through the half of the 18th century: 'infinitae curvae ordinatim positione datae' ('infinitely many curves, given by position in ordered sequence')".

The terminology first appears in print in Leibniz (1692, p. 168), where Leibniz proceeds from the usual meaning of *ordinate* (certain parallel lines used in connection with conic sections) and points out that its meaning was generalized by Desargues, who also used the term for straight lines that "converge toward a common point, or diverge from it".[2] Leibniz then generalizes further: "But I accept as 'drawn in ordered sequence' not only straight lines but also any curves, provided the law is known according to which in a given point of some curve (the ordinatrix) the [curved] line corresponding to that point can be drawn, which [line] will be one of those that

[1] This attribution is suggested in Kurosch, Liwschitz, and Schulgeifer (1963, p. 21); it is due to the translators of that paper. [2] "sub Ordinatim applicatis etiam comprehendit rectas convergentes ad unum punctum commune, aut ab eo divergentes". – This generalization is found in Desargues (1639, p. 15).

are to be drawn in ordered sequence, or that are given by position".[1] An example Leibniz gives are light rays that are reflected in a mirror.

The first writer who used the term *family* regularly seems to have been Christian Wolff.[2] He first uses German *Familie* in Wolff (1710, p. 123) and then gives a more general definition of Latin *familia* in Wolff (1713, p. 365): "*Familia curvarum* vocatur plurium curvarum diversi generis congeries, quæ omnes, per eandem æquationem indeterminati gradus, sed pro diversitate generis diversimode explicandi definiuntur". This definition appears in various 18th-century dictionaries[3] and was translated by Chambers (1728) as follows: "*Family of Curves*, a Congeries of several Curves of different Orders, or Kinds; all which are defin'd by the same indeterminate Equation, but in a different Manner, according to their different Orders". As an example, Wolff gives $a^{m-1}x = y^m$ (parameter m).

In the 1780s, the French term *famille* begins to be applied to surfaces as well. Monge (1787, p. 86) appears to have been the first writer to do so (Baltzer 1882, p. 441). This extension of the meaning spreads through Monge and other writers of the *École Normale*. The English term is found in this meaning in [anon.] (1802, p. 465).

Note. On families of curves, see the thorough account by Engelsman (1984) and also the editorial notes in Leibniz (2011, pp. xli–xlvi).

References

[anon.] (1802). "The Sittings of the Normal Schools". *The Monthly Review* 37, pp. 459–467.

d'Alembert, Jean Le Rond et al. (1784). *Encyclopédie méthodique*. Mathématiques. Tome premier. Paris.

Baltzer, Richard (1882). *Analytische Geometrie*. Leipzig.

Chambers, Ephraim (1728). *Cyclopaedia, or, An universal dictionary of arts and sciences*. Volume the First. London.

Cramer, Gabriel (1750). *Introduction à l'analyse des lignes courbes algébriques*. Genève.

Desargues, Girard (1639). *Brouillon project d'une atteinte aux evenemens des rencontres du Cone avec un Plan*. Paris. URL: `http://gallica.bnf.fr/ark:/12148/bpt6k105071b`.

Eneström, Gustaf (1909-10). "Über das angebliche Vorkommen krummliniger Koordinaten bei Leibniz". *Bibliotheca Mathematica* 10, pp. 43–47.

Engelsman, Steven B. (1984). *Families of Curves and the Origins of Partial Differentiation*. Amsterdam and New York.

Halley, Edmond (1695). "Propositio Generalis Arearum dimensionem exhibens in universo illo Curvarum Genere, quae revolutione aequabili Circuli super Basin quamuis, vel Rectilineam vel Circularem describi possint". *Philosophical Transactions of the Royal Society of London* 19, pp. 125–128.

Halley, Edmond (1706). "A General Proposition shewing the Dimension of the Areas in all those kinds of Curves which are describ'd by the Equable Revolution of a Circle upon any Basis, either a Rectilineal or a Circular one". *Miscellanea Curiosa* 2, pp. 177–182.

---

[1] "Ego sub ordinatim ductis non tantum rectas, sed et curvas lineas qualescunque accipio, modo lex habeatur, secundum quam dato lineae cujusdam datae (tanquam ordinatricis) puncto, respondens ei puncto linea duci possit, quae una erit ex ordinatim ducendis, seu ordinatim positione datis". Translation from Engelsman (1984, p. 24), who also explains that the concept of parameter is still lacking in this passage but that it *is* present in Leibniz (1694). See also Eneström (1909-10). [2] The terms *familia* and *family* were also used by Halley (1695, p. 128, 1706, p. 182). [3] The term also appears in d'Alembert et al. (1784, p. 462) and Hutton (1795, p. 353), who also uses *tribe* as an alternative. One of the scarce non-dictionary uses in the 18th century is Cramer (1750, p. 223).

Hutton, Charles (1795). *A Philosophical and Mathematical Dictionary*. In two volumes. Vol. I. London.
Leibniz, Gottfried Wilhelm (1692). "De linea ex lineis numero infinitis ordinatim ductis inter se concurrentibus formata [...]" *Acta Eruditorum* (Apr. 1692), pp. 168–171.
Leibniz, Gottfried Wilhelm (1694). "Nova calculi differentialis applicatio & usus ad multiplicem linearum constructionem ex data tangentium conditione". *Acta Eruditorum* (Apr. 1694), pp. 311–316.
Leibniz, Gottfried Wilhelm (2011). *Sämtliche Schriften und Briefe*. Dritte Reihe. Mathematischer, naturwissenschaftlicher und technischer Briefwechsel. Siebenter Band. Juli 1696-Dezember 1698. Hannover.
Monge, Gaspard (1787). "Mémoire sur l'expression analytique de la génération des surfaces courbes". *Mémoires de l'Académie royale des sciences* (1784), pp. 85–117.
Wolff, Christian (1710). *Der Anfangs-Gründe aller Mathematischen Wissenschaften Letzter Theil*. Halle.
Wolff, Christian (1713). *Elementa matheseos universae*. Tomus I. Halle.

fibre, fibre bundle, fibre space

This topological terminology was introduced by Seifert as German *Faser* 'fibre'. He first used it in publications from 1932 onwards and also introduced the term *gefaserter Raum*, lit. 'fibered space'. He provided this definition:

> One could say a manifold is fibered if it is decomposed into curves, called *fibers*, such that each point lies on exactly one fiber and a neighborhood of each point can be mapped homeomorphically onto a neighborhood of a point in a Euclidean space in such a way that fibers are mapped to line segments of a bundle of parallel lines. This requirement is a local one. But even if we postulated this for all points of the manifold, we would still find this definition of a fibered manifold to be too general.
>
> In the present paper we consider only those fibered manifolds which satisfy in addition to the four manifold axioms the three following axioms[1] which relate to properties of the fibering in the large. We call these manifolds *fibered spaces* (Seifert 1932, p. 150).[2]

The term *fibre-bundle* (as well as *sphere-bundle*) was coined by Whitney (1940, p. 148, 1941, pp. 117, 119), while *fibre space* appeared in Hurewicz and Steenrod (1941, p. 60).

From these works the terminology spread, with modified definitions. Massey (1955, pp. 327f.) remarked: "There are many different definitions current for the

---

[1] The three axioms are: "The manifold can be decomposed into fibers, where each fiber is a simple closed curve. [...] Each point lies on exactly one fiber. [...] For each fiber H there exists a *fiber neighborhood*, that is, a subset consisting of fibers and containing H, which can be mapped under a *fiber preserving* map onto a *fibered solid torus*, where H is mapped onto the 'middle fiber'". Translation from Seifert and Threlfall (1980, p. 361). [2] "Eine Mannigfaltigkeit wird man als gefasert bezeichnen, wenn ihre Punkte sich auf eine Menge von Kurven, *Fasern*, so verteilen, dass durch jeden Punkt genau eine Faser geht und eine Umgebung jeden Punktes sich auf die Umgebung eines Punktes eines euklidischen Raumes so topologisch abbilden lässt, dass die Fasern in die Geraden-Stücke eines Parallelgeradenbündels übergehen. Diese Forderung bezieht sich nur auf die Umgebung eines Punktes. Aber auch wenn wir sie an alle Punkte der Mannigfaltigkeit ohne Ausnahme stellten, so wäre doch dieser Begriff der gefaserten Mannigfaltigkeit uns noch zu allgemein. In der vorliegenden Arbeit betrachten wir nur solche gefaserte Mannigfaltigkeiten, die ausser den vier Mannigfaltigkeitsaxiomen noch den drei folgenden Axiomen genügen, welche Eigenschaften der Faserung im Grossen fordern. Solche Mannigfaltigkeiten nennen wir kurz *gefaserte Räume*". Translation from Seifert and Threlfall (1980, p. 361). – For comments on Seifert's definition, see Zisman (1999, pp. 607f.).

terms 'fibre space' and 'fibre bundle'. Each of these definitions involves a continuous mapping onto, $p : E \to B$, called a 'projection'; the image space B is always called the 'base space', and the inverse images of the points of B are called 'fibres'". In the following, Massey discussed five different definitions.

Note. For the history of the concept, see McCleary (2004), Eckmann (1999), and Zisman (1999).

References

Eckmann, Beno (1999). "Birth of Fibre Spaces, and Homotopy". *Expositiones Mathematicae* 17, pp. 23–34.

Hurewicz, Witold and Norman Steenrod (1941). "Homotopy Relations in Fibre Spaces". *Proceedings of the National Academy of Sciences* 27, pp. 60–64.

Massey, William S. (1955). "Some Problems in Algebraic Topology and the Theory of Fibre Bundles". *Annals of Mathematics* 62, pp. 327–359.

McCleary, John (2004). "A History of Manifolds and Fibre Spaces: Tortoises and Hares". *Rendiconti del Circolo Matematico di Palermo* (Serie II, Suppl. 72), pp. 9–29.

Seifert, Herbert (1932). "Topologie dreidimensionaler geschlossener Räume". *Acta Mathematica* 60, pp. 147–238.

Seifert, Herbert and William Threlfall (1980). *Seifert and Threlfall: A Textbook of Topology and Seifert: Topology of 3-dimensional fibered spaces*. Ed. by Joan S. Birman and Julian Eisner. New York et al.

Whitney, Hassler (1940). "On the Theory of Sphere-Bundles". *Proceedings of the National Academy of Sciences* 26, pp. 148–153.

Whitney, Hassler (1941). "On the topology of differentiable manifolds". In: *Lectures in Topology*. Ed. by Raymond Louis Wilder and W.L. Ayres. Ann Arbor, pp. 101–141.

Zisman, Michel (1999). "Fibre Bundles, Fibre Maps". In: *History of Topology*. Ed. by Ioan Mackenzie James. Amsterdam, pp. 605–629.

field

Field-theoretic ideas developed in connection with the study of polynomial equations. It is especially in the works of Abel and Galois that such ideas can be found.[1] The latter, in particular, proceeds from polynomial equations and remarks: "When we say that a function is rationally known, we wish to say that its numerical value is expressible as a rational function of the coefficients of the equation and of some adjoined quantities" (Galois 2011, p. 112).[2] What Galois essentially does (as interpreted from today's perspective), in the words of van der Waerden (1985, p. 105), is this: "a certain 'ground field' is presupposed, which may be extended by adjunctions in the course of the investigations".[3] While Galois does not use a term corresponding to *ground field*, he does use the verb *adjoindre*.

The first explicit terms appear in the 1850s: Dedekind used *rationales Gebiet* 'rational domain' in early lectures (see Dedekind 1981, p. 83), while Kronecker (1880, p. 214) used *Rationalitäts-Bezirk* 'district of rationality' first, and later changed this to *Rationalitäts-Bereich* 'domain of rationality': "The 'domain of rationality' ($\mathfrak{R}'$,

[1] For work on polynomial equations before Galois, see, for example, van der Waerden (1985, pp. 76–102) and Wußing (1984, pp. 96–101). [2] "Quand nous disons qu'une fonction est rationnellement connue, nous voulons dire que sa valeur numérique est exprimable en fonction rationnelle des coëfficients de l'équation et des quantités adjointes". Translation from Galois (2011, p. 112). – As Kiernan (1971, p. 47) notes, the word *known* (*connu*) recalls Lagrange's use of the word. [3] For further discussion, see, for example, Wußing (1984, pp. 102–117) and Edwards (2012).

$\mathfrak{R}''$, $\mathfrak{R}'''$, ...) contains, as one can already see in the denotation, all these magnitudes which are rational functions of the magnitudes $\mathfrak{R}'$, $\mathfrak{R}''$, $\mathfrak{R}'''$, ... with integral numbers as coefficients" (Kronecker 1882, p. 4).[1] Kronecker also says that in his opinion it is "appropriate to avoid the expressions with a decidedly spatial character"[2], which is why he prefers *Bereich* to Dedekind's *Körper*, a term which Kronecker apparently primarily connected with its geometrical meaning 'solid'. As will be seen immediately, however, this spatial meaning is not what Dedekind had in mind.[3]

The term that stuck in the end is Dedekind's *Körper*, lit. 'body'. He defined it thus:

> By a *field* [*Körper*] we mean an infinite system of real or complex numbers, which is closed and complete in itself, so that the addition, subtraction, multiplication, and division of any two of these numbers always yields yet another number of the same system (Dedekind 1871, p. 424).[4]

Dedekind wrote in the context of algebraic number theory and originally limited *Körper* to complex number fields, as can be seen from the quotation. He later commented on this new term and said that he had coined it in analogy to uses of the word in various disciplines, where the term commonly conveys the idea, as Dedekind puts it, of an "organic whole":

> Similarly as in the sciences, in geometry and in the life of human society, this name [*Körper*] is to denote also in the present context a system which exhibits a certain completeness, perfection, self-containedness through which it appears as an organic whole, a natural unity (Dedekind 1894, p. 452).[5]

In his widely read book on algebra, Weber (1895, p. 452) used Dedekind's *Körper* and remarked that Kronecker's *Rationalitätsbereich* is sometimes useful, that is, when in a particular investigation the elements of the field are to be regarded as known or rational.[6] This suggestion was accepted by Moore (1897, p. 50), for example, who said "[t]hat this is surely a desirable usage".

The English term *field* (or, more precisely, *field of order s* and *Galois-field*) was first used by Moore (1893, p. 75), and was a translation Weber's (1893, pp. 526f.) *endlicher Körper* 'finite field', as Moore says (Moore 1896, p. 208).[7] His choice of

---

[1] "Der Rationalitäts-Bereich ($\mathfrak{R}'$, $\mathfrak{R}''$, $\mathfrak{R}'''$, ...) enthält, wie schon die Bezeichnung deutlich erkennen lässt, alle diejenigen Größen, welche rationale Functionen der Grössen $\mathfrak{R}'$, $\mathfrak{R}''$, $\mathfrak{R}'''$, ... mit ganzzahligen Coefficienten sind". Translation from Haffner (2014, p. 307). [2] "halte ich es für angemessen, die Ausdrücke mit entschieden räumlichem Gepräge zu vermeiden". [3] Right after the publication of Kronecker (1882), Dedekind wrote down some notes in which he explicitly rejected Kronecker's criticism as incorrect (Edwards, Neumann, and Purkert 1982, p. 54). [4] "Unter einem *Körper* wollen wir jedes System von unendlich vielen reellen oder komplexen Zahlen verstehen, welches in sich so abgeschlossen und vollständig ist, daß die Addition, Subtraktion, Multiplikation und Division von je zwei Zahlen immer wieder eine Zahl desselben Systems hervorbringt". Translation from Avigad (2004, p. 12). [5] "Dieser Name [*Körper*] soll, ähnlich wie in den Naturwissenschaften, in der Geometrie und im Leben der menschlichen Gesellschaft, auch hier ein System bezeichnen, das eine gewisse Vollständigkeit, Vollkommenheit, Abgeschlossenheit besitzt, wodurch es als ein organisches Ganzes, als eine natürliche Einheit erscheint". Translation from Schappacher (2010, p. 3268). [6] "Es ist bisweilen nützlich, den Ausdruck Rationalitätsbereich für einen Körper dann zu brauchen, wenn damit ausgedrückt werden soll, dass bei einer bestimmten Aufgabe die Grössen dieses Körpers als bekannt oder rational betrachtet werden sollen". [7] Note that when Dedekind speaks of a *finite field*, he means a finitely generated field.

terms may have been influenced by *field* in physics, but whether this is really the case is an open question. But just like Dedekind's term for German, Moore's translation is still used today, although, as Schappacher (2010, p. 3268) points out, "[w]hat is totally lost in the translation *field* is Dedekind's motivation for choosing the term *Körper*".

Moore's translation was not immediately accepted, however. For example, in a translation of two of Dedekind's papers (Dedekind 1909, p. 5), the term *body*, which is a literal translation of the German, was used following Pierpont (1900, p. 25). Even earlier, Hathaway (1887, p. 167) suggested *Universe*, which, like *Körper*, emphasizes the idea of a whole. Mathews (1893, p. 320) and W. H. Young and G. C. Young (1906, p. 6) preferred *corpus*, and Reid (1910, p. vi) discussed the terminology as follows: "The author has adopted the term 'realm' as the equivalent of körper, corpus, campus, body, domain and field, as it has the advantage, he believes, of not having been used in any other branch of mathematics".[1] The French term *corps*, finally, already appears in Dedekind (1877, p. 65).

In most of the sources cited so far the term is used in Dedekind's or Kronecker's sense. The articles by Moore are an exception, because Moore refers to Weber's paper from 1893. This makes an important difference, because Weber (1893, pp. 526f.) there presented the first abstract axiomatic approach to fields. As seen from today's perspective, however, his definition does not quite match the concept as used today because Weber omitted associativity of multiplication, for example. He singled out the rational numbers as the "most important" example but also stressed that fields can be finite, giving $\mathbb{Z}/p\mathbb{Z}$ (p prime) as examples (Weber 1893, pp. 526f.).

A major landmark in the study of fields is Steinitz' article entitled "Algebraic Theory of Fields" (Steinitz 1910). In this long paper, Steinitz discussed fields as objects in their own right.[2] Apart from the obvious closure, he defined a field on the basis of the following criteria (Steinitz 1910, pp. 172f.):

1. associativity of addition,
2. commutativity of addition,
3. associativity of multiplication,
4. commutativity of multiplication,
5. distributivity,
6. "unlimited and unique subtraction", meaning that if a and b are in the field, then there is a unique element x for which $a + x = b$,
7. "unlimited and unique division", meaning that if a and b are in the field and $a \neq 0$, then there is a unique element x for which $a \cdot x = b$.

Apart from giving this definition, Steinitz introduced a large number of field-theoretic concepts still used today and proved many important theorems.[3]

Note. For a historical discussion of the concept, see, for example, Purkert (1971, 1973) and Kleiner (2007, pp. 63–78). Lüneburg (2001), Gow (2013), and Brechen-

[1] König (1903, p. 8) suggested, with little success, German *orthoider Bereich* 'orthoid domain'.
[2] See the last section in the entry *algebra*. [3] For further discussions, see, for example, Lüneburg (2008, pp. 193–242) and Corry (2004, pp. 192–196).

macher (2016) focus on finite fields. Becker (2005, ch. 10) discusses the terminology from a linguistic perspective.

References

Avigad, Jeremy (2004). "Dedekind's 1871 version of the theory of ideals". URL: `https://kilthub.cmu.edu/articles/Dedekind_s_1871_Version_of_the_Theory_of_Ideals/6491342/1` (retrieved July 25, 2007).

Becker, Holger (2005). "Semantische und lexikalische Aspekte der mathematischen Fachsprache des 19. Jahrhunderts". PhD thesis. Carl von Ossietzky Universität Oldenburg. URL: `http://oops.uni-oldenburg.de/59/`.

Brechenmacher, Frédéric (2016). "A history of Galois fields". *Khronos* 3, pp. 181–260.

Corry, Leo (2004). *Modern Algebra and the Rise of Mathematical Structures*. Second revised edition. Basel.

Dedekind, Richard (1871). "Ueber die Composition der binären quadratischen Formen". In: *Vorlesungen über Zahlentheorie von P.G. Lejeune Dirichlet*. Herausgegeben und mit Zusätzen versehen von R. Dedekind. Zweite, umgearbeitete und vermehrte Auflage. Braunschweig, pp. 380–497.

Dedekind, Richard (1877). *Sur la théorie des nombres entiers algébriques*. Paris.

Dedekind, Richard (1894). "Ueber die Theorie der ganzen algebraischen Zahlen". In: *Vorlesungen über Zahlentheorie von P.G. Lejeune Dirichlet*. Herausgegeben und mit Zusätzen versehen von R. Dedekind. Vierte, umgearbeitete und vermehrte Auflage. Braunschweig, pp. 434–657.

Dedekind, Richard (1909). *Essays on the Theory of Numbers*. I. Continuity and Irrational Numbers, II. The Nature and Meaning of Numbers. Authorized translation by Wooster Woodruff Beman. Chicago.

Dedekind, Richard (1981). "Eine Vorlesung über Algebra". In: *Richard Dedekind 1831-1981. Eine Würdigung zu seinem 150. Geburtstag*. Ed. by Winfried Scharlau. Braunschweig and Wiesbaden, pp. 59–100.

Edwards, Harold M. (2012). "Galois for 21st-Century Readers". *Notices of the American Mathematical Society* 59, pp. 912–923.

Edwards, Harold M., Olaf Neumann, and Walter Purkert (1982). "Dedekind's 'Bunte Bemerkungen' zu Kroneckers 'Grundzüge'". *Archive for History of Exact Sciences* 27, pp. 49–85.

Galois, Évariste (2011). "The First Memoir". In: *The Mathematical Writings of Évariste Galois*. Ed. by Peter M. Neumann. Zürich, pp. 105–168.

Gow, Roderick (2013). "History of finite fields". In: *Handbook of Finite Fields*. Ed. by Gary L. Mullen and Daniel Panario. Boca Raton, pp. 3–11.

Haffner, Emmylou (2014). "The 'Science of Numbers' in action in Richard Dedekind's works: between mathematical explorations and foundational investigations". PhD thesis. Université Paris Diderot.

Hathaway, Arthur S. (1887). "A Memoir in the Theory of Numbers". *American Journal of Mathematics* 9, pp. 162–179.

Kiernan, B. Melvin (1971). "The Development of Galois Theory from Lagrange to Artin". *Archive for History of Exact Sciences* 8, pp. 40–154.

Kleiner, Israel (2007). *A History of Abstract Algebra*. Basel.

König, Julius (1903). *Einleitung in die allgemeine Theorie der algebraischen Gröszen*. Aus dem Ungarischen übertragen vom Verfasser. Leipzig.

Kronecker, Leopold (1880). "Einige Entwickelungen aus der Theorie der algebraischen Gleichungen". *Monatsberichte der Königlich Preußischen Akademie der Wissenschaften zu Berlin* (1879), pp. 205–229.

Kronecker, Leopold (1882). *Grundzüge einer arithmetischen Theorie der algebraischen Grössen*. Berlin.

Lüneburg, Heinz (2001). "On the Early History of Galois Fields". In: *Finite Fields and Applications*. Ed. by Dieter Jungnickel and Harald Niederreiter. Berlin and Heidelberg, pp. 341–355.

Lüneburg, Heinz (2008). *Von Zahlen und Größen*. Dritthalbtausend Jahre Theorie und Praxis. Band II. Basel.

Mathews, George Ballard (1893). "On the Complex Integers connected with the Equation $\theta^3 - 2 = 0$". *Proceedings of the London Mathematical Society* 24, pp. 319–327.
Moore, E.H. (1893). "A doubly-infinite system of simple groups". *Bulletin of the New York Mathematical Society* 3, pp. 73–78.
Moore, E.H. (1896). "A doubly-infinite system of simple groups". In: *Mathematical Papers read at the International Mathematical Congress (Chicago, 1893)*. New York, pp. 208–243.
Moore, E.H. (1897). "Concerning transcendentally transcendental functions". *Mathematische Annalen* 48, pp. 49–74.
Pierpont, James (1900). "Galois' Theory of Algebraic Equations. Part II. Irrational Resolvents". *Annals of Mathematics* 2, pp. 22–56.
Purkert, Walter (1971). "Zur Genesis des abstrakten Körperbegriffs. 1. Teil". *NTM-Schriftenreihe für Geschichte der Naturwissenschaften, Technik und Medizin* 8, pp. 23–37.
Purkert, Walter (1973). "Zur Genesis des abstrakten Körperbegriffs. 2. Teil". *NTM-Schriftenreihe für Geschichte der Naturwissenschaften, Technik und Medizin* 10, pp. 8–20.
Reid, Legh Wilber (1910). *The Elements of the Theory of Algebraic Numbers*. With an Introduction by David Hilbert. New York.
Schappacher, Norbert (2010). "Rewriting Points". In: *Proceedings of the International Congress of Mathematicians*. Hyderabad 2010. Ed. by Rajendra Bhatia. Vol. 4. New Delhi, pp. 3258–3291.
Steinitz, Ernst (1910). "Algebraische Theorie der Körper". *Journal für die reine und angewandte Mathematik* 137, pp. 167–309.
van der Waerden, Bartel Leendert (1985). *A History of Algebra*. From al-Khwārizmī to Emmy Noether. Berlin.
Weber, Heinrich (1893). "Die allgemeinen Grundlagen der Galois'schen Gleichungstheorie". *Mathematische Annalen* 43, pp. 521–549.
Weber, Heinrich (1895). *Lehrbuch der Algebra*. Erster Band. Braunschweig.
Wußing, Hans (1984). *The Genesis of the Abstract Group Concept*. A Contribution to the History of the Origin of Abstract Group Theory. Tr. Abe Shenitzer. Cambridge, Mass.
Young, William Henry and Grace Chisholm Young (1906). *The Theory of Sets of Points*. Cambridge.

figure

The Greek term for a 2- or 3-dimensional bounded figure is *σχῆμα*. It was defined by Euclid in the *Elements* (book I, def. 14): "A *figure* is that which is contained by any boundary or boundaries".[1] Before Euclid, the term had already been used by Plato and Aristotle, for example (Mugler 1958-59, pp. 408f.).

In Latin texts, the term was sometimes transliterated as *schema*; other terms were *forma* and *figura*.[2] The agrimensor Balbus, for example, defined: "A figure [*forma*] is an area contained within one or several boundaries" (Blume, Lachmann, and Rudorff 1848, p. 104).[3] The Pseudo-Boethian *Geometria I* has almost the same definition for *figura* (Blume, Lachmann, and Rudorff 1848, p. 377), which term also appears in the various translations of the *Elements* in the 12th and 13th centuries.

English *figure* is found in the 14th century (MED, OED);[4] it was used, for example, by John Trevisa in his translation of *De propretatibus rerum*: "Playne figures ben conteyned in lengþe and in brede [...] And solid figures ben conteyned in lengþe, brede, and depnesse" (Seymour 1975, p. 1367). French *figure* was used in a French translation, dating from 1372, of the same text (Bartholomaeus Anglicus 1482, book 19, ch. Cxxvii); for Italian *figura*, TLIO cites Dante (*Convivio* II, xiii, 26). Recorde (1551, fol. A iii^r) offered some alternatives: "*A forme, figure, or shape*, is that thyng that

1 Translation from Euclid (1956, p. 153). 2 For references, see the literature cited in the note.
3 "Forma est quae sub aliquo aut aliquibus finibus continetur". Translation from Campbell (2000, p. 211). 4 The OED cites a text from 1340 as the earliest use.

is inclosed within one bond or manie bondes". The German term is found in the *Geometria culmensis* from around 1400 (Mendthal 1886, p. 52).[5]

Note. A concise overview of the terminology was given by Tropfke (1940, pp. 58f.). For the Greek term, see also Mugler (1958-59, pp. 408f.) and Heath's comments in Euclid (1956, pp. 182f.). The Latin terminology is discussed by Ayuso García (2008, pp. 287–290, 305–308, 325–329) and by Guillaumin (2020, pp. 133, 136, 258).

References

Ayuso García, Manuel (2008). "La terminología latina de la geometría en Marciano Capela". PhD thesis. Madrid: Universidad Nacional de Educación a Distancia.

Bartholomaeus Anglicus (1482). *Le Proprietaire des choses*. Ed. by Pierre Farget. Trans. by Jean Corbichon. Lyon.

Blume, Friedrich, Karl Lachmann, and Adolf Rudorff, eds. (1848). *Die Schriften der römischen Feldmesser*. Erster Band. Texte und Zeichnungen. Berlin.

Campbell, Brian (2000). *The Writings of the Roman Land Surveyors*. Introduction, Text, Translation and Commentary. London.

Euclid (1956). *The Thirteen Books of Euclid's Elements*. Volume I. Introduction and Books I, II. Ed. by Thomas L. Heath. New York.

Guillaumin, Jean-Yves (2020). *Dictionnaire de la terminologie latine ancienne de l'arithmétique et de la géométrie*. Paris.

Mendthal, Hans (1886). *Geometria culmensis*. Ein agronomischer Tractat aus der Zeit des Hochmeisters Conrad von Jungingen (1393-1407). Leipzig.

Mugler, Charles (1958-59). *Dictionnaire historique de la terminologie géométrique des Grecs*. Paris.

Recorde, Robert (1551). *The patheway to knowledg*. London.

Schirmer, Alfred (1912). *Der Wortschatz der Mathematik nach Alter und Herkunft untersucht*. Straßburg.

Seymour, Maurice Charles, ed. (1975). *On the Properties of Things*. John Trevisa's Translation of Bartholomaeus Anglicus' *De Proprietatibus Rerum*. A Critical Text. Volume II. Oxford.

Tropfke, Johannes (1940). *Geschichte der Elementar-Mathematik*. Vierter Band. Ebene Geometrie. Dritte, verbesserte und vermehrte Auflage besorgt von Dr. Kurt Vogel. Berlin and Leipzig.

filter

This topological term was first used as French *filtre* by Cartan (1937, p. 595). Choosing a set $\mathcal{E}$, he defined:

> A family $\boldsymbol{F}$ of subsets of $\mathcal{E}$ is called *filter* (constructed on $\mathcal{E}$) if it fulfils the following three conditions:
>
> F-I: the family $\boldsymbol{F}$ is not empty and does not contain the empty set;
>
> F-II: the intersection of two sets of $\boldsymbol{F}$ belongs to $\boldsymbol{F}$;
>
> F-III: each set that contains a set of $\boldsymbol{F}$ belongs to $\boldsymbol{F}$.[1]

His terminology was quickly adopted; *filter* appears, for example, in Weil (1937, p. 4), Dieudonné (1939, p. 277), Tukey (1940, p. 86), and Bourbaki (1940, p. 20).

One important precursor is Vietoris (1921, p. 185), who introduced the notion of "filter base", which he called *Kranz*, lit. 'wreath'. In his theory, Vietoris eliminated

---

[5] For the German term, see Schirmer (1912, pp. 23f.).

[1] "Une famille $\boldsymbol{F}$ de sous-ensembles de $\mathcal{E}$ prend le nom de *filtre* (construit sur $\mathcal{E}$) si elle remplit les trois conditions suivantes: F-I: la famille $\boldsymbol{F}$ n'est pas vide et ne contient pas le sous-ensemble vide; F-II: l'intersection de deux ensembles de $\boldsymbol{F}$ appartient à $\boldsymbol{F}$; F-III: tout ensemble qui contient un ensemble de $\boldsymbol{F}$ appartient à $\boldsymbol{F}$".

countability from topology in much the same way Cartan did about 20 years later (Reitberger 1997, 2002).

See also *net* for a closely related notion.

Note. For the history of the concept, see the remarks by Pietsch (2007, pp. 60–62), Reitberger (1997, pp. 33f., 2002, pp. 2–5), and Dudley (2002, p. 76).

References

Bourbaki, Nicolas (1940). *Topologie génerale*. Chapitre I. Structures topologiques. II. Structures uniformes. Paris.

Cartan, Henri (1937). "Théorie des filtres". *Comptes rendus de l'Académie des sciences* 205, pp. 595–598.

Dieudonné, Jean (1939). "Sur les espaces uniformes complets". *Annales scientifiques de l'École normale supérieure* 56, pp. 277–291.

Dudley, R.M. (2002). *Real Analysis and Probability*. Cambridge.

Pietsch, Albrecht (2007). *History of Banach Spaces and Linear Operators*. Boston, Mass. et al.

Reitberger, Heinrich (1997). "The Contributions of L. Vietoris and H. Tietze to the Foundations of General Topology". In: *Handbook of the History of General Topology*. Volume 1. Ed. by C.E. Aull and R. Lowen. Dordrecht, pp. 31–40.

Reitberger, Heinrich (2002). "Leopold Vietoris zum Gedenken (4.6.1891 - 9.4.2002)". *Internationale Mathematische Nachrichten* 56, pp. 1–16.

Tukey, John W. (1940). *Convergence and Uniformity in Topology*. Princeton.

Vietoris, Leopold (1921). "Stetige Mengen". *Monatshefte für Mathematik und Physik* 31, pp. 173–204.

Weil, André (1937). *Sur les espaces à structure uniforme et sur la topologie générale*. Paris.

finite

See *infinite*.

finite difference

While Newton has been called "the single most significant contributor to the theory of finite differences" (Roy 2011, p. 160), he did not publish much about this topic during his lifetime, the major exception being his "Methodus differentialis" (The method of differences) (Newton 1711, pp. 93–101).[1] The first book-length publication on the topic was Taylor's *Methodus Incrementorum* (Method of Increments), in which Taylor wrote at the very beginning:

> Indeterminate quantities I consider here as continually augmented by increments or diminished by decrements. The integral indeterminates I designate by the letters z, x, v, &c., while their increments, or the parts then added, I mark by the same letters dotted below, $\underset{\cdot}{z}$, $\underset{\cdot}{x}$, $\underset{\cdot}{v}$ &c. I represent the increments of the increments, or the second increments of the integrals themselves, by the same letters twice dotted, $\underset{\cdot\cdot}{z}$, $\underset{\cdot\cdot}{x}$, $\underset{\cdot\cdot}{v}$, &c. I designate the third increments by the letters triply dotted, $\underset{\cdot\cdot\cdot}{z}$, $\underset{\cdot\cdot\cdot}{x}$, $\underset{\cdot\cdot\cdot}{v}$, and so on (Taylor 1715, p. 1).[2]

---

[1] An edition and translation is found in Newton (1981, pp. 244–255); for Newton's manuscripts on finite differences, see Newton (1971, pp. 13–73). [2] "Quantitates indeterminatas in his considero ut Incrementis perpetuò auctas, vel Decrementis diminutas. Indeterminatas ipsas Integrales designo literis z, x, v, &c. earumque Incrementa, seu partes mox addendas designo iisdem literis a parte inferiori punctatis $\underset{\cdot}{z}$, $\underset{\cdot}{x}$, $\underset{\cdot}{v}$ &c. Quorum Incrementorum Incrementa, seu Integralium ipsarum Incrementa secunda designo iisdem literis bis punctatis $\underset{\cdot\cdot}{z}$, $\underset{\cdot\cdot}{x}$, $\underset{\cdot\cdot}{v}$, &c. Incrementa tertia designo literis ter punctatis $\underset{\cdot\cdot\cdot}{z}$, $\underset{\cdot\cdot\cdot}{x}$, $\underset{\cdot\cdot\cdot}{v}$ & sic porro". Translation from Cajori (1929, p. 263). – A detailed account of Taylor's method of increments is given by Feigenbaum (1985).

The explicit term *finite difference* seems to originate with Nicole (1719), who wrote a paper entitled "Traité du calcul des différences finies". The concept was not very frequently studied in the following years; Emerson (1763, p. iv) judged that because Taylor's "way of writing is so very short and abstracted, as not to be obvious to common readers; it has happened, that this elegant method of computation has lain ever since in obscurity, has been read by few, and improved by none".[3] Euler devoted the first chapter of his book on integral calculus on the subject and defined:

> We express the increment by Δy, which the function y undergoes when we substitute $x + \omega$ for x, where ω takes any value we wish. In the discussion of series it is usual to take $\omega = 1$, but here it is preferable to leave the value general, so that it can be arbitrarily increased or decreased. We usually call this increment Δy of the function y its *difference* (Euler 1755, p. 5).[1]

For a detailed discussion of the further history of the concept, see Lubet (2015).[2]

References

Cajori, Florian (1929). *A History of Mathematical Notations*. Volume II. Notations mainly in higher mathematics. Chicago.

Chabert, Jean-Luc, ed. (1999). *A History of Algorithms*. From the Pebble to the Microchip. Berlin and Heidelberg.

Emerson, William (1763). *The Method of Increments*. London.

Euler, Leonhard (1755). *Institutiones calculi differentialis*. St. Petersburg.

Euler, Leonhard (2000). *Foundations of Differential Calculus*. Translated by John D. Blanton. New York.

Feigenbaum, Lenore (1985). "Brook Taylor and the Method of Increments". *Archive for History of Exact Sciences* 34, pp. 1–140.

Goldstine, Herman H. (1977). *A History of Numerical Analysis from the 16th through the 19th Century*. New York.

Lubet, Jean-Pierre (2015). "Le calcul aux différences finies, une nouvelle branche de l'analyse". In: *Sciences mathématiques 1750-1850*. Continuités et ruptures. Ed. by Christian Gilain and Alexandre Guilbaud. Paris, pp. 443–473.

Newton, Isaac (1711). *Analysis per quantitatum series, fluxiones, ac differentias*. Cum enumeratione linearum tertii ordinis. Ed. by William Jones. London.

Newton, Isaac (1971). *The Mathematical Papers of Isaac Newton*. Volume IV. 1674-1684. Ed. by D.T. Whiteside. Cambridge.

Newton, Isaac (1981). *The Mathematical Papers of Isaac Newton*. Volume VIII. 1687-1722. Ed. by D.T. Whiteside. Cambridge.

Nicole, François (1719). "Traité du calcul des différences finies". *Mémoires de l'Académie royale des sciences* (1717), pp. 7–21.

Roy, Ranjan (2011). *Sources in the Development of Mathematics*. Infinite Series and Products from the Fifteenth to the Twenty-first Century. Cambridge.

Taylor, Brook (1715). *Methodus Incrementorum Directa & Inversa*. London.

---

[3] Especially the very last assertion is certainly an overstatement. [1] "Exprimet ergo Δy incrementum, quod functio y capit, si in ea loco x ponatur $x + \omega$, denotante ω numerum quemcunque pro lubitu assumtum. In doctrina quidem serierum sumi solet $\omega = 1$; verum hic ad nostrum institutum expedit, valore generali vti, qui pro arbitrio augeri diminuiue queat. Vocari quoque solet hoc incrementum Δy functionis y eius DIFFERENTIA [...]". Translation from Euler (2000, p. 2) (emphasis added). [2] Many historical comments can also be found in Goldstine (1977), Chabert (1999), and Roy (2011).

fixed point

In the sense of a point that is not changed by a given transformation, function, etc., the term does not appear until the end of the 19th century. It is found in Klein and Fricke (1890, pp. 163f.), who considered a "general linear substitution"[3] $z' = \frac{az+b}{cz+d}$ (z complex) and, setting $z' = z$, called such "points that correspond to themselves" *fixed points* (*Fixpunkte*).[4] Their result was that the fixed points of the above substitution are $z_1, z_2 = \frac{a-d\pm\sqrt{(a-d)^2+4bc}}{2c}$.

Forsyth (1893, p. 514), who frequently refers to Klein and Fricke's book, seems to have adopted the term from them: "If we assume that w and z are two points in the same plane, then there will in general be two different points which are unaltered by the transformation; they are called the *fixed* (or *double*) points of the transformation".[1]

The terminology gradually spread and became more frequent after Brouwer (1912, p. 115) used it in the formulation of his fixed point theorem and Weyl (1913, p. 150) in his *Concept of a Riemann Surface*.

References

Brouwer, L.E.J. (1912). "Über Abbildungen von Mannigfaltigkeiten". *Mathematische Annalen* 71, pp. 97–115.
Cremona, Luigi (1858). "*Beiträge zur Geometrie der Lage*, von Dr. Karl Georg Christian von Staudt [review]". *Annali di Matematica Pura e Applicata* 1, pp. 125–128.
Forsyth, Andrew Russell (1893). *Theory of Functions of a Complex Variable*. Cambridge.
Gretschel, Heinrich (1868). *Lehrbuch zur Einführung in die organische Geometrie*. Leipzig.
Klein, Felix and Robert Fricke (1890). *Vorlesungen über die Theorie der elliptischen Modulfunctionen*. Erster Band. Grundlegung der Theorie. Leipzig.
Weyl, Hermann (1913). *Die Idee der Riemannschen Fläche*. Leipzig und Berlin.

flat module

The term *flat module* (French *module plat*) was introduced by Serre (1956, p. 34): "Let B be an A-module. One says that B is A-flat (or flat) if, for every exact sequence of A-modules $F \to F \to G$, the sequence $E \otimes_A B \to F \otimes_A B \to G \otimes_A B$ is exact".[2]

Serre's term was quickly adopted, for example by Cartan and Eilenberg (1956, p. 122), who used English *flat module*.

Note. Brief historical remarks can be found in Maury (1968).

References

Cartan, Henri and Samuel Eilenberg (1956). *Homological Algebra*. Princeton.
Maury, G. (1968). "Modules projectifs, Modules injectifs. Généralisations". *L'enseignement mathématique* 13, pp. 257–262.
Serre, Jean-Pierre (1956). "Géométrie algébrique et géométrie analytique". *Annales de l'Institut Fourier* 6, pp. 1–42.

---

[3] "allgemeine lineare Substitution". [4] "Nennen wir diese sich selbst entsprechenden Punkte Fixpunkte". [1] This was already pointed out by OED2 as the first use in English. – As for the term *double point*, Forsyth seems to mean its use in projective geometry, where it is used for points that correspond to themselves (see, e. g., Gretschel 1868, p. 34). Ultimately, it may go back to Cremona (1858, p. 126), who used *elementi doppi* in connection with involutions. [2] "Soit B un A-module. On dit que B est A-plat (ou plat) si, pout toute suite exacte de A-modules $F \to F \to G$, la suite $E \otimes_A B \to F \otimes_A B \to G \otimes_A B$ est exacte".

floor function and *ceiling function*

The terms *floor* and *ceiling* were introduced by Iverson (1962, p. 12) in a book on programming:

> A number represented in positional notation [...] must, in practice, employ only a finite number of digits. It is therefore often desirable to approximate a number x by an integer. For this purpose two functions are defined:
>
> 1. the *floor of* x (or integral part of x) denoted by $\lfloor x \rfloor$ and defined as the largest integer not exceeding x,
>
> 2. the *ceiling of* x denoted by $\lceil x \rceil$ and defined as the smallest integer not exceeded by x.

The concept is much older. It had been used by Gauß (1808, p. 71), who also introduced the notation $[x]$ for the floor. At the same time, Legendre (1808, p. 8) used $E(\frac{n}{a})$ for "the largest integer contained in the fraction $\frac{n}{a}$".[1] The E stands for *entier*. Neither Gauß nor Legendre explicitly said that $[x]$ or $E(\frac{n}{a})$ were considered as functions; statements to this effect came up later, for example in Bouniakowski (1869, p. 26). This was the basis for the alternative *greatest integer function*, used by Dickson (1920, p. v), for instance. van Dantzig (1942, p. 369) preferred *entier-function*.

References

Bouniakowski, Viktor (1869). "Sur quelques formules qui résultent de la combinaison des résidus quadratiques et non quadratiques des nombres premiers". *Bulletin de l'Académie impériale des sciences de St.-Pétersbourg* 13, pp. 25–32.

van Dantzig, David (1942). "On the affirmative content of Peano's theorem on differential equations". *Koninklijke Akademie van Wetenschappen te Amsterdam. Proceedings of the Section of Sciences* 45, pp. 367–373.

Dickson, Leonard Eugene (1920). *History of the Theory of Numbers*. Volume II. Diophantine analysis. Washington, D.C.

Gauß, Carl Friedrich (1808). "Theorematis arithmetici demonstratio nova". *Commentationes Societatis Regiae Scientiarum Gottingensis* 16, pp. 69–74.

Iverson, Kenneth E. (1962). *A Programming Language*. New York and London.

Legendre, Adrien-Marie (1808). *Essai sur la théorie des nombres*. Seconde édition. Paris.

fluxion and *fluent*

Fluxions and fluents are basic elements in Newton's calculus. According to Panza (2012, p. 219), Newton first used the terms in his *De methodis serierum et fluxionum*, written in 1670/71; he gave the following definition:

> But to distinguish the quantities which I consider as just perceptibly but indefinitely growing from others which in any equations are to be looked on as known and determined and are designated by the initial letters a, b, c, &c., I will hereafter call them 'fluents' and designate them by the final letters v, x, y and z. And the speeds with which they each flow and are increased by their generating motion (which I might more readily call 'fluxions' or simply 'speeds') I will designate by the letters l, m, n, and r (Newton 1969, p. 72).[2]

---

[1] "l'entier le plus grand contenu dans la fraction $E(\frac{n}{a})$". [2] "Quantitates autem quas ut sensim crescentes indefinitè considero, quo distinguam ab alijs quantitatibus quae in aequationibus quibuscunque pro determinatis et cognitis habendæ sunt ac initialibus literis a, b, c, &c designantur, posthac denominabo fluentes, ac designabo finalibus literis v, x, y, et z. Et celeritates quibus singulae a motu generante fluunt et augentur (quas possim fluxiones vel simpliciter celeritates vocitare) designabo literis l, m, n et r". Translation from Panza (2012, p. 251).

Even though *De methodis* was not published until 1736 (Newton 1736), Newton's notes were circulated and the term gradually became known in England and on the continent. In print, it was mentioned by Newton (1687, pp. 251, 254) himself, but also by Wallis (1693, p. 391) and Nieuwentijt (1695, p. 279), for example.

On the further development of the terminology, Cajori (1919, p. 55) remarked: "A hopeless confusion arose in the use of the term 'fluxion.' Newton always took it to be a velocity, but many writers [...] simply said 'fluxion' instead of 'differential'". For a corresponding statement by Euler, see *differential*.[3]

Already in the 18th century, Newton's terminology was mainly used by British writers, and in the 19th century, it eventually fell out of use completely.

Note. A detailed account of Newton's creation of fluents and fluxions is Panza (2012).

References

Cajori, Florian (1919). *A History of the Conceptions of Limits and Fluxions in Great Britain from Newton to Woodhouse*. Chicago and London.

Guicciardini, Niccolò (1989). *The Development of Newtonian Calculus in Britain, 1700–1800*. Cambridge.

Newton, Isaac (1687). *Philosophiæ Naturalis Principia Mathematica*. London.

Newton, Isaac (1736). *The Method of Fluxions and Infinite Series*. Trans. by John Colson. London.

Newton, Isaac (1969). *The Mathematical Papers of Isaac Newton*. Volume III. 1670-1673. Ed. by D.T. Whiteside. Cambridge.

Nieuwentijt, Bernard (1695). *Analysis infinitorum seu curvilineorum proprietates ex polygonorum natura deductae*. Amsterdam.

Panza, Marco (2012). "From velocities to fluxions". In: *Interpreting Newton*. Critical Essays. Ed. by Andrew Janiak and Eric Schliesser. Cambridge, pp. 219–254.

Wallis, John (1693). *De Algebra tractatus*. In: *Opera Mathematica*. Volumen alterum. Oxford, pp. 1–482.

focus

The focus of conic sections was already discussed by Apollonius. However, he only dealt with foci in connection with ellipses and hyperbolas, not with parabolas.[1] When Apollonius spoke about foci, he called them "points arising from the application" (*τὰ ἐκ τῆς παραβολῆς γενηθέντα σημεῖα*); the expression is found in book 3 of his *Conics*:

> If in an hyperbola or ellipse or circumference of a circle or opposite sections straight lines are drawn from the vertex of the axis at right angles, and a rectangle equal to the fourth part of the figure is applied to the axis on each side and exceeding by a square figure in the case of the hyperbola and opposite sections, but deficient in the case of the ellipse, and some straight line is drawn tangent to the section and meeting the perpendicular straight lines, then the straight lines drawn from the points of meeting to the points produced by application make right angles at the aforesaid points (Apollonius of Perga 1891, p. 424).[2]

These points are determined (fig. 45) by the condition $BH \cdot HA = BZ \cdot ZA = \frac{1}{4}$ of "the figure", which is the rectangle formed by the major axis and the parameter. In terms of the application of areas (see *conic section* for an explanation of that concept), this means that "a rectangle is applied to AB as base equal to one-fourth part of the

---

[3] For additional examples, see Guicciardini (1989, pp. 41f.). [1] That does not necessarily mean that Apollonius was unaware of the focus of the parabola. See Toomer (1976, p. 15). [2] Translation from Apollonius of Perga (2013, p. 249).

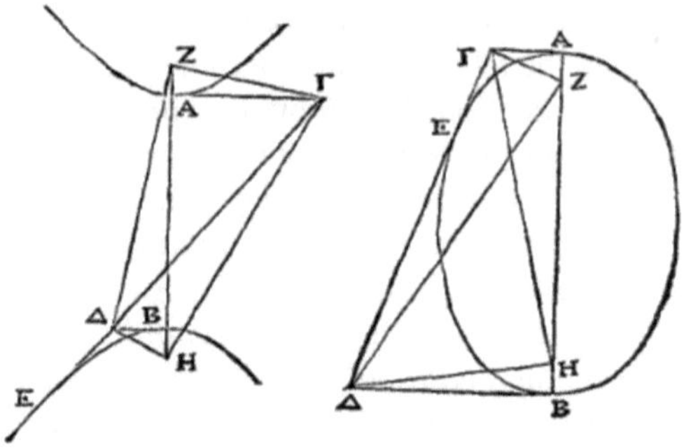

Figure 45: The "points arising from the application". From Halley (1710, p. 206).

'figure', and in the case of the hyperbola exceeding, but in the case of the ellipse falling short, by a square figure" (Heath 1921, p. 156).[1]

The corresponding Latin expression for "points arising from the application" is *punctum ex comparatione factum*, used by Commandino in his edition of Apollonius' works (Apollonius of Perga 1566, fol. 95[v]), and later mentioned in various works on conic sections such as Guldin (1640, p. 73) and de Saint-Vincent (1647, p. 243).

The term still used today, *focus*, was suggested by Kepler (1604, p. 93), who explicitly points out that there is "no name" for foci yet and explains: "For the sake of light, and with an eye turned towards mechanics, we shall call these points 'foci'. We might have called them 'centers', since they are on the axes of the sections, but in the hyperbola and the ellipse the writers on conic sections called another point the center".[2] The term, from Latin *focus* 'fire-place, hearth' (LS), is based on physical properties: as explained by Donahue in Kepler (2000, pp. 108, fn. 14), the word "refers to the properties of conoidal mirrors of focussing light rays at one of these points when the rays originate at the other point (in the parabola, the originating 'focus' is infinitely distant, and the rays coming from it are parallel)".

Not long afterwards, Mydorge (1631, p. 8) suggested another term: Latin *umbilicus*, lit. 'navel' but also 'centre' (see *umbilic*). Another alternative was *pole*, proposed by de Saint-Vincent (1647, p. 243), who defined the *poles or foci* of an ellipse (Latin *poli seu foci*). Kepler's and Mydorge's terms were in common use in the 17th century, but Mydorge's eventually fell out of use; *pole* was not used very frequently.[3]

An early usage of *focus* in French is Mersenne (1624, p. 169);[4] the word was soon changed to *foyer*, found in Hérigone (1637, p. 552), for example, where the term is also attributed to Kepler. Descartes (1637a, p. 91) used the literal *points bruslans*

[1] Heath's designation for the axis has been adapted to fig. 45. [2] "Nos lucis causa & oculis in Mechanicam intentis ea puncta Focos appellabimus. Centra dixissemus, quia sunt in axibus sectionum, nisi in hyperbola et ellipsi conici auctores aliud punctum centri nomine appellarent". Translation from Kepler (2000, pp. 107f.). [3] Occasionally, *centrum reflexionis* is also used for the focus, for example by Bramer (1634, p. 4), who points out (p. 5) that the alternative *focus* is due to Kepler, and by Guldin (1640, p. 74), who points out the use of *centrum reflexionis* by Bramer.
[4] Mersenne (ibid.) also suggests the focus could be called *point originaire*.

'burning points'.[5] Desargues (1639, p. 28) lists the terms known at the time: "the two points called *Navels*, *burning points* or *foci* of the figure".[6] The English term is, according to the OED, due to Hobbes (1656, p. 49): "the *Burning Points*, which writers call the *Foci* and *Umbilici* of the Section".[7] The German term *Brennpunkt*, lit. 'burning point', does not seem to appear in connection with conic sections until the 18th century; it is found in Gaupp (1708, p. 219) and is also included in Wolff's mathematical dictionary (Wolff 1716, p. 641).[8]

Note. Tropfke (1924, p. 147) makes some concise historical comments on the terminology.

References

Apollonius of Perga (1566). *Conicorum libri quattuor*. Ed. by Federico Commandino. Bologna.

Apollonius of Perga (1891). *Apollonii Pergaei quae graece exstant cum commentariis antiquis*. Vol. I. Ed. by J.L. Heiberg. Leipzig.

Apollonius of Perga (2013). *Conics*. Books I-IV. Books I-III translated by R. Catesby Taliaferro. Book IV translated by Michael N. Fried. Santa Fe.

Bramer, Benjamin (1634). *Apollonius Cattus, Oder Geometrischer Wegweiser*. Kassel.

Desargues, Girard (1639). *Brouillon project d'une atteinte aux evenemens des rencontres du Cone avec un Plan*. Paris. URL: `http://gallica.bnf.fr/ark:/12148/bpt6k105071b`.

Descartes, René (1637a). *La Dioptrique*. In: *Discours de la méthode*. Leiden, pp. 1–153.

Descartes, René (1637b). *La Géométrie*. In: *Discours de la méthode*. Leiden, pp. 295–413.

Field, Judith V. and Jeremy Gray (1986). *The Geometrical Work of Girard Desargues*. New York.

Gaupp, Johannes (1708). *Gnomonica mechanica oder mechanische Sonnen-Uhr-Kunst*. Lindau.

Guldin, Paul (1640). *De centro gravitatis*. Liber secundus. Vienna.

Halley, Edmund (1710). *Apollonii Pergaei Conicorum libri octo, et Sereni Antissensis De sectione cylindri & coni libri duo*. Oxford.

Heath, Thomas L. (1921). *A History of Greek Mathematics*. Volume II. From Aristarchus to Diophantus. Oxford.

Hérigone, Pierre (1637). *Cursus mathematicus*. Tomus quintus. Paris.

Hobbes, Thomas (1656). *Six Lessons to the Professors of the Mathematiques, One of Geometry, the Other of Astronomy*. London.

Kepler, Johannes (1604). *Ad Vitellionem paralipomena, quibus astronomiae pars optica traditur*. Frankfurt.

Kepler, Johannes (2000). *Optics*. Paralipomena to Witelo & Optical Part of Astronomy. Translated by William H. Donahue. Santa Fe.

Mersenne, Marin (1624). *L'Impiété des Deistes*. Seconde partie. Paris.

Moore, Jonas (1660). *A new Contemplation Geometrical upon the Ellipsis*. In: *Moor's Arithmetick. In two books*. London.

Mydorge, Claude (1631). *Prodromi catoptricorum et dioptricorum*. Libri primus et secundus. Paris.

de Saint-Vincent, Grégoire (1647). *Opus Geometricum quadraturae circuli et sectionum coni*. Antwerpen.

Schwenter, Daniel (1636). *Deliciae Physico-Mathematicae*. Nürnberg.

Toomer, G.J., ed. (1976). *Diocles. On Burning Mirrors*. The Arabic Translation of the Lost Greek Original. Berlin et al.

Tropfke, Johannes (1924). *Geschichte der Elementar-Mathematik*. Sechster Band. Analysis. Analytische Geometrie. Berlin and Leipzig.

Wolff, Christian (1716). *Mathematisches Lexicon*. Leipzig.

---

[5] Descartes (1637b, p. 359) also used this term in the *Géométrie*. [6] "les deux poincts qu'on nomme *nombrils*, *bruslans*, autrements *foyers* de la figure". Translation from Field and Gray (1986, pp. 128f.). [7] See also Moore (1660, p. 9): "the Umbilici or Navels, as *Mydorgius* calls them, or the Foci or Focusses or burning Points, as others terme them". [8] In connection with mirrors, however, the term is already found in Schwenter (1636, p. 301).

foliation

As informally described by Plante (1988, p. 92), "[a] foliation is simply a decomposition of a manifold into a disjoint union of immersed submanifolds (called leaves) of constant dimension such that the decomposition is locally homeomorphic to the decomposition of $R^n = R^k \times R^{n-k}$ into the parallel submanifolds $R^k \times \{\text{point}\}$".

In this sense, the term goes back to Reeb (1947, p. 1613), who introduced the French term *variété feuilleté.* Chern (1953, p. 262) translated it by *leaved manifold,* Hirzebruch (1954, p. 221) by *laminated manifold.* In French texts, the term *feuilletage* was soon used instead of *variété feuilleté* (e. g., Ehresmann 1951, p. 64); in English, this appeared as *foliation* in lectures from 1953-54 given by Ehresmann.[1]

Reeb (1978, p. 104) later remarked that, as far as the term is concerned, one should not so much think of a botanical metaphor: "In lieu of a botanist, the assembly may admit a pastry-cook member. Puff-pastry – I have good reasons to believe – gives a good idea of a foliation (of codimension 1 in $\mathbb{R}^3$) for which it draws the leaves well and suggests properties".[1]

Note. For the history of the concept and the terminology, see Haefliger (2005).

References

Chern, Shiing-Shen (1953). "*Sur les espaces fibrés et les variétés feuilletées.* By W.T. Wu and G. Reeb [review]". *Bulletin of the American Mathematical Society* 59, pp. 258–263.
Ehresmann, Charles (1951). "Sur la théorie des variétés feuilletées". *Rendiconti di matematica e delle sur applicazioni* 10, pp. 64–82.
Ehresmann, Charles (1982). *Œuvres complètes et commentées.* Partie II.2. Catégories ordonnées. Applications des ordres en topologie. Éditée et commentée par Andrée Charles Ehresmann. Amiens.
Haefliger, André (2005). "Naissance des feuilletages, d'Ehresmann-Reeb à Novikov". In: *Géométrie au 20e siècle, 1930-2000.* Histoires et horizons. Ed. by Joseph Kouneiher et al. Paris, pp. 257–271.
Hirzebruch, Friedrich (1954). "Some Problems on Differentiable and Complex Manifolds". *Annals of Mathematics* 60, pp. 213–236.
Pier, Jean-Paul (2001). *Mathematical Analysis During the 20th Century.* Oxford.
Plante, Joseph F. (1988). "*Geometric theory of foliations*, by César Camacho and Alcides Lins Neto [review]". *Bulletin of the American Mathematical Society* 92, pp. 92–97.
Reeb, Georges (1947). "Variétés feuilletées, feuilles voisines". *Comptes rendus de l'Académie des sciences* 224, pp. 1613–1614.
Reeb, Georges (1978). "Structures feuilletées". In: *Differential Topology, Foliations and Gelfand-Fuks Cohomology.* Ed. by Paul A. Schweitzer. Berlin, New York, pp. 104–113.

folium

The word *folium* means 'leaf'. Descartes first mentioned his folium, a curve defined by the equation $x^3 + y^3 = axy$, in a letter to Mersenne, dated 18 January, 1638; at the time, he did not give the curve a name and had no or a wrong understanding of what the curve looked like (Loria 1910, p. 54).[2]

---

[1] See the beginning of Ehresmann (1982). [1] "À défaut du botaniste l'assemblée comptera peut-être un pâtissier. La pâte feuilletée – j'ai de bonnes raisons de le croire – donne une bonne idée d'un feuilletage (de codimension 1 dans $\mathbb{R}^3$) dont elle dessine bien les feuilles et en suggère des propriétés". Translation from Pier (2001, p. 303). [2] His sketch of the curve is reproduced in Descartes (1897-1908, vol. I, p. 490).

In another letter, dated 18 August, 1638, Descartes (1897-1908, vol. II, p. 313) incorrectly stated that the curve had several leaves (*feuilles*).[3] This time he gave a correct representation of the curve in the first quadrant, and the statement that the curve has several leaves is based on the wrong assumption that the curve has the same shape in the other quadrants as well (Gray 1994, pp. 847f.). In the same letter, he also pointed out that Roberval called the curve *galand* (knot of ribbons) or *fleur de jasmin* (jasmine flower). In further texts of the 17th century, the curve is frequently not named at all or Roberval's terms are preferred.[4]

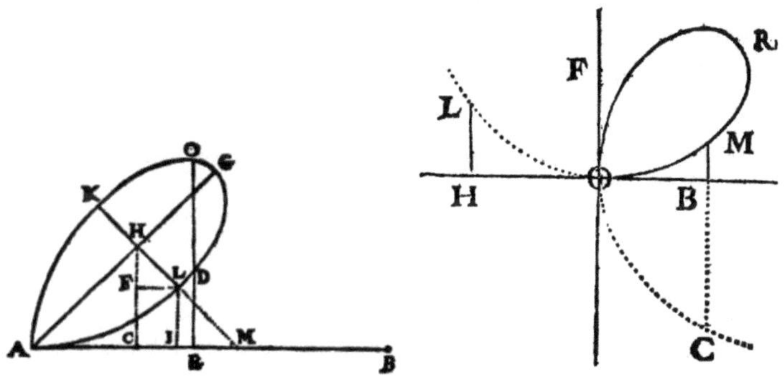

Figure 46: Illustrations from van Schooten (1657, p. 498) and Saurin (1703, p. 524).

In a letter from 1693, Huygens referred to the curve as *la Feuille de M. des Cartes ou de Roberval* (Huygens 1905, p. 510);[5] in print, the term *feuille de M. Descartes* appears in Saurin (1703, p. 841). de Moivre (1714) called the curve the *Foliate*.

Note. For the history of the curve, see Loria (1910, pp. 52–59), Teixeira (1908, pp. 85–91), and Hofmann (1954).

References

Descartes, René (1897-1908). *Œuvres*. Ed. by Charles Adam and Paul Tannery. 12 vols. Paris.

Gray, Jeremy (1994). “Algebraic and analytic geometry”. In: *Companion Encyclopedia of the History and Philosophy of the Mathematical Sciences*. Volume 2. Ed. by Ivor Grattan-Guinness. Baltimore and London, pp. 847–865.

Hofmann, Joseph Ehrenfried (1954). “Altes und Neues von der Quadratur des Descartesschen Blattes”. *Centaurus* 3, pp. 279–295.

Huygens, Christiaan (1905). *Œuvres complètes*. Tome dixième. Correspondance 1691-1695. La Haye.

Loria, Gino (1910). *Spezielle algebraische und transzendente ebene Kurven*. Theorie und Geschichte. Autorisierte, nach dem italienischen Manuskript bearbeitete deutsche Ausgabe von Fritz Schütte. Zweite Auflage. Erster Band. Leipzig.

[3] According to Loria (1910, p. 54), this is the first time that the term *leaf* was used in connection with this curve. [4] For references, see Loria (1910, p. 55). [5] Huygens also gave the first correct “complete” drawing of the folium (Huygens 1905, p. 352).

de Moivre, Abraham (1714). "A ready Description and Quadrature of a Curve of the Third Order, resembling that commonly called the Foliate". *Philosophical Transactions of the Royal Society of London* 29, pp. 329–331.
Saurin, Joseph (1703). "Réponse à l'écrit de M. Rolle de l'Académie Royale des Sciences inseré dans le journal du 13 avril 1702". *Journal des sçavans* (Aug 1702), pp. 519–534.
van Schooten, Frans (1657). *Exercitationum Mathematicorum libri quinque*. Liber primus. Leiden.
Teixeira, Francisco Gomes (1908). *Traité des courbes spéciales remarquables plans et gauches*. Tome I. Coïmbra.

forest

In graph theory, a forest is a collection of *trees*. Whitney defined the term in 1932: "A *forest* is a graph containing no circuits. A *tree* is a connected forest" (Whitney 1932, p. 340).[1] In Whitney's terminology, the *components* of a forest are trees, which explains why the term was chosen.

References

Whitney, Hassler (1931). "Non-Separable and Planar Graphs". *Proceedings of the National Academy of Sciences* 17, pp. 125–127.
Whitney, Hassler (1932). "Non-separable and planar graphs". *Transactions of the American Mathematical Society* 34, pp. 339–363.

forgetful functor

The term *forgetful functor* appears to date from the early 1960s, while in the 1950s occasionally phrases such as these can be found: "There is a covariant functor taking the category $d\mathcal{G}$ into the category $\mathcal{G}$, namely that functor which forgets about the differential operator" (Gugenheim and Moore 1957, p. 270).

Explicitly, the term was used by Freyd (1962, p. 22, 1964, p. 62); Felscher (1965, p. 8) referred to Freyd for the term but translated it by *projection functor* (*Projektionsfunktor*), which did not catch on. Pareigis (1969, p. 12) noted that *forgetful functor* was in frequent use, as did Mac Lane (1971, p. 14), who added that *underlying functor* was also in use.

References

Felscher, Walter (1965). *Adjungierte Funktoren und primitive Klassen*. Heidelberg.
Freyd, Peter (1962). *Functor categories and their application to relative homology*. New York.
Freyd, Peter (1964). *Abelian Categories*. New York.
Gugenheim, V.K.A.M. and J.C. Moore (1957). "Acyclic Models and Fibre Spaces". *Transactions of the American Mathematical Society* 85, pp. 265–306.
Mac Lane, Saunders (1971). *Categories for the Working Mathematician*. New York et al.
Pareigis, Bodo (1969). *Kategorien und Funktoren*. Stuttgart.

form

This number-theoretical term was coined by Gauß. He first used Latin *forma* in his diary in 1796 (Klein 1903, p. 8); the term first appears in print in his *Disquisitiones arithmeticae* (§ 153) from 1801: "In this section we shall treat particularly of functions

[1] The term is not yet found in Whitney (1931), an earlier short account of Whitney (1932).

in two unknowns x, y of the form $ax^2 + 2bxy + cy^2$ where a, b, c are given integers. We will call these functions *forms of the second degree* or simply *forms*".[1]

Earlier writers like Lagrange (1775, p. 265) called an expression like $Bt^2 + Ctu + Du^2$ a *formula* (*formule*); Legendre (1788, p. 532) also used French *expression* in this context. Lagrange (1775, p. 265) also said that a number could be *of the form* $Bt^2 + Ctu + Du^2$, for example. Very occasionally, Legendre (1797, p. 263) also referred to an expression like $19y^2 + 2yz - 2z^2$ as a *quadratic form* (*forme quadratique*).

The term shows a relatively common change of meaning: before Gauß, it was common to say that something *has* a certain form; Gauß changed this to say that something *is* a form. The words *structure* and *function*, for example, underwent a similar development (Becker 2005, pp. 199f.).

French *forme* in Gauß' sense appears in Gauß (1807, § 153); German *Form* begins to be used around 1830, for example by Seeber (1831). The English term comes up in the 1850s: in particular, Sylvester (1852) used it in his article "On the principles of the calculus of forms".

Note. For the history of forms, see Wußing (1984, ch. 3) and Dickson (1952); for the terminology, see Reich (1994, p. 43) and Bachmann (1911, p. 466).

References

Bachmann, Paul (1911). "Über Gauß' zahlentheoretische Arbeiten". *Nachrichten von der Königl. Gesellschaft der Wissenschaften zu Göttingen (mathematisch-physikalische Klasse)* (1911), pp. 455–508.

Becker, Holger (2005). "Semantische und lexikalische Aspekte der mathematischen Fachsprache des 19. Jahrhunderts". PhD thesis. Carl von Ossietzky Universität Oldenburg. URL: `http://oops.uni-oldenburg.de/59/`.

Dickson, Leonard Eugene (1952). *History of the Theory of Numbers*. Volume III. Quadratic and higher forms. New York.

Gauß, Carl Friedrich (1801). *Disquisitiones arithmeticae*. Leipzig.

Gauß, Carl Friedrich (1807). *Recherches arithmétiques*. Traduites par A.-C.-M. Poullet-Delisle. Paris.

Gauß, Carl Friedrich (1986). *Disquisitiones Arithmeticae*. English Edition. Translated by A.A. Clarke. Revised by W.C. Waterhouse. New York et al.

Klein, Felix (1903). "Gauß' wissenschaftliches Tagebuch 1796-1814". *Mathematische Annalen* 57, pp. 1–34.

Lagrange, Joseph-Louis (1775). "Recherches d'arithmétique". *Nouveaux Mémoires de l'Académie Royale des Sciences et Belles-Lettres de Berlin* (1773), pp. 265–312.

Legendre, Adrien-Marie (1788). "Recherches d'analyse indéterminée". *Mémoires de l'Académie royale des Sciences* (1785), pp. 465–559.

Legendre, Adrien-Marie (1797). *Essai sur la théorie des nombres*. Paris.

Reich, Karin (1994). *Die Entwicklung des Tensorkalküls*. Vom absoluten Differentialkalkül zur Relativitätsgeometrie. Basel.

Seeber, Ludwig August (1831). *Untersuchungen über die Eigenschaften der positiven ternären quadratischen Formen*. Freiburg.

Sylvester, James Joseph (1852). "On the principles of the calculus of forms". *The Cambridge and Dublin Mathematical Journal* 7, pp. 52–97, 179–214.

---

[1] "In hac sectione imprimis de functionibus duarum indeterminatarum x, y, huius formae, $axx + 2bxy + cyy$, ubi a, b, c sunt integri dati tractabimus, quas *formas secundi gradus*, sive simpliciter formas *dicemus*". Translation from Gauß (1986, § 153).

Wußing, Hans (1984). *The Genesis of the Abstract Group Concept*. A Contribution to the History of the Origin of Abstract Group Theory. Tr. Abe Shenitzer. Cambridge, Mass.

Fourier series

As explained by Gray (2015, p. 13), "Fourier series are trigonometric series used to represent a function, and they are widely used throughout pure and applied mathematics. Fourier was not the first to use them, but his name is rightly attached to them because he was the first to use them in the study of heat diffusion, to display their use in the solution of a partial differential equation, and to argue successfully for their generality". Fourier used these series in his classic *Théorie analytique de la chaleur* (Fourier 1822) and in other publications.

The origin of the name is quite obscure. The evidence suggests that it was used independently by various writers, and in various languages, at the end of the 1830s. Thus, Kelland (1837, p. 64) used *Fourier's series*, and Ostrogradsky (1841, p. 310), in a talk held in 1839, French *série de Fourier*. German *Reihe des Fourier* is found in Ohm (1837, p. 490) and *Fouriersche Reihe* in Ohm (1838, p. 528). In none of these texts is there any indication that the respective author is coining a new term.

Note. For a historical discussion of Fourier's work and of the concept in the 18th century, see Langer (1947).

References

Fourier, Joseph (1822). *Théorie analytique de la chaleur*. Paris.
Gray, Jeremy (2015). *The Real and the Complex*. A History of Analysis in the 19th Century. Cham et al.
Kelland, Philip (1837). *Theory of Heat*. Cambridge and London.
Langer, Rudolph E. (1947). "Fourier's Series: The Genesis and Evolution of a Theory". *The American Mathematical Monthly* 54, pp. 1–86.
Ohm, Martin (1837). *Lehrbuch der Mechanik*. Zweiter Band. Statik fester Körper. Berlin.
Ohm, Martin (1838). *Lehrbuch der Mechanik*. Dritter Band. Dynamik fester Körper. Berlin.
Ostrogradsky, Mikhail (1841). "Mémoire sur les quadratures définies". *Mémoires de l'Académie Impériale des Sciences de Saint-Petersbourg* 1, pp. 309–336.

Fourier transform

According to Domínguez (2016, p. 57), the "definitive and modern meaning" of *Fourier transform* is due to Wiener and Paley. In 1934, they presented the following theorem[1] including a definition:

> Let $f(x)$ belong to L_2 over $(-\infty, \infty)$. Then there exists a function $g(u)$ belonging to L_2 over $(-\infty, \infty)$, and such that
>
> $$\lim_{A\to\infty} \int_{-\infty}^{\infty} \left| g(u) - (2\pi)^{-\frac{1}{2}} \int_{-A}^{A} f(x)e^{iux}dx \right|^2 du = 0.$$
>
> Furthermore
>
> $$\int_{-\infty}^{\infty} |g(u)|^2 du = \int_{-\infty}^{\infty} |f(x)|^2 dx,$$
>
> and

[1] The theorem is due to Plancherel (1910), which is why Wiener and Paley called it *Plancherel's theorem*.

$$\lim_{A\to\infty}\int_{-\infty}^{\infty}\left|f(x)-(2\pi)^{-\frac{1}{2}}\int_{-A}^{A}g(u)^{-iux}du\right|^{2}dx=0.$$

The function $g(u)$ is called the Fourier Transform of $f(x)$. It is determined except over a set of points of zero measure (Paley and Wiener 1934, p. 2).

The terminology originally goes back to Plancherel, who used both German *Fouriersche Transformierte* (Plancherel 1913, p. 574) and French *transformée de Fourier* (Plancherel 1915, p. 316). As explained by Domínguez (2016, p. 57), "the term used by Plancherel does not correspond to its modern use because it applies only to those cases in which f is produced from F by the same integral operator that gives F from f". Titchmarsh (1923, p. 463) used the English term first (for the cosine Fourier transform).

Note. For the history of the concept, see Domínguez (2016).

References

Domínguez, Alejandro (2016). "Highlights in the History of the Fourier Transform". *IEEE Pulse* 7, pp. 53–61.

Paley, Raymond E.A.C. and Norbert Wiener (1934). *Fourier Transforms in the Complex Domain.* New York.

Plancherel, Michel (1910). "Contribution à l'étude de la représentation d'une fonction arbitraire par des intégrales définies". *Rendiconti del Circolo Matematico di Palermo* 30, pp. 289–335.

Plancherel, Michel (1913). "Zur Konvergenztheorie der Integrale $\lim\limits_{z=\infty}\int_a^z f(x)\cos xydx$". *Mathematische Annalen* 74, pp. 573–578.

Plancherel, Michel (1915). "Sur la convergence et sur la sommation par les moyennes de Cesàro de $\lim\limits_{z=\infty}\int_a^z f(x)\cos xydx$". *Mathematische Annalen* 76, pp. 315–326.

Titchmarsh, E.C. (1923). "Hankel transforms". *Proceedings of the Cambridge Philosophical Society* 21, pp. 463–473.

fractal

The term first appeared in print in 1975, when Mandelbrot published his book *Les objets fractals* (Mandelbrot 1975). He there explained that his neologism *objet fractal* is derived from Latin *fractus*, from *frangere* 'to break' (Mandelbrot 1975, p. 1). Mandelbrot commented on the terminology on various occasions; in Mandelbrot (1977, p. 294) he explicitly stated that he coined the term in 1975. The reason for choosing the word is clear from Mandelbrot's introductory explanation:

> Many important spatial patterns of Nature are either irregular or fragmented to such an extreme degree that Euclid – a term used in this Essay to denote all of classical geometry – is hardly of any help in describing their form. The coastline of a typical oceanic island, to take an example, is neither straight, nor circular, nor elliptic, and no other classical curve can serve without undue artificiality in the presentation and the organization of empirical measurements and in the search for explanations. Similarly, no surface in Euclid represents adequately the boundaries of clouds or of rough turbulent wakes. More generally, many patterns of Nature that had already attracted the attention of primitive Man involve, in comparison to Euclid, not only a higher degree but an altogether different level of complexity.
>
> In the present Essay I hope to show that it is possible in many cases to remedy this absence of geometric representation by using a family of shapes I propose to call *fractals* – or *fractal sets* (Mandelbrot 1977, p. 1).

Mandelbrot was certainly partly inspired by Hausdorff's (1919, p. 157) use of German *gebrochene Dimension* 'broken dimension'; *fractional dimension* was used by Besicovitch (1929, p. 161). As pointed out by Russell (1897, p. 162), broken dimensions – "Geometries with m/n dimensions" – were mentioned, without explanation or reference, by Delbœuf (1893, p. 450).

Note. On the prehistory of fractals, see Jones (1991) and Chabert (1994).

References

Besicovitch, A.S. (1929). "On linear sets of points of fractional dimensions". *Mathematische Annalen* 101, pp. 161–193.
Chabert, Jean-Luc (1994). "The early history of fractals, 1870–1920". In: *Companion Encyclopedia of the History and Philosophy of the Mathematical Sciences*. Volume 1. Ed. by Ivor Grattan-Guinness. Baltimore and London, pp. 367–374.
Delbœuf, Joseph (1893). "L'ancienne et les nouvelles géométries. Première étude". *Revue philosophique de la France et de l'étranger* 36, pp. 449–484.
Hausdorff, Felix (1919). "Dimension und äußeres Maß". *Mathematische Annalen* 79, pp. 157–179.
Jones, Huw (1991). "Fractals before Mandelbrot: a selective history". In: *Fractals and Chaos*. Ed. by A.J. Crilly, Rae A. Earnshow, and Huw Jones. New York, pp. 7–33.
Mandelbrot, Benoît (1975). *Les objets fractals*. Forme, hasard, et dimension. Paris.
Mandelbrot, Benoît (1977). *Fractals*. Form, chance and dimension. San Francisco.
Russell, Bertrand (1897). *An Essay on the Foundations of Geometry*. Cambridge.

fraction

According to Tropfke (1980, p. 118), the names for the notion of fraction are generally connected to the notions of dividing, breaking off, or making smaller. This applies to the term *fraction* as well: Latin *frangere* means 'to break'.

An implicit definition of fractions is found in definitions 3 and 4 in book VII of Euclid's *Elements*: "A number is a part of a number, the less of the greater, when it measures the greater; but parts when it does not measure it".[1] For *part* (submultiple), Euclid uses *μέρος* 'share, part', and *parts* (proper fractions) are expressed by the plural *μέρη*.[2] The corresponding Latin term *pars* is later found in the 12th century in a translation of the Elements made directly from the Greek (Busard 1987, p. 155).

The notion of breaking is linked to fractions in Arabic (Tropfke 1930, p. 164), where *kasr*, from a root meaning 'to break', is the standard term.[3] It is found in the earliest mathematical texts; in particular, al-Khwārizmī used it in his treatise on algebra (Rashed 2017, pp. 724f.).

The Latin term *fractio* is found in the 12th century, for example in a translation of al-Khwārizmī's treatise on Indian arithmetic (see al-Khwārizmī 1963, p. 47 and Folkerts 1997, p. 197). It is also in the *Liber mahameleth*, which is mainly based on Arabic sources: "You should know that a number is sometimes considered by itself, without reference to another, in which case it is called an integer; sometimes in relation to another, that is, as a part or parts, in which case it is called a fraction" (Sesiano 2014, p. 28).[4] Fibonacci uses various terms such as *numerus fractus* and

[1] Translation from Euclid (1956, p. 277). [2] For further Greek terms, see Tropfke (1980, pp. 101–103). [3] It is worth noting that division is expressed in Akkadian by a verb meaning 'to split' (Tropfke 1980, p. 242). [4] "Sciendum est autem quod numerus aliquando consideratur per se, sine respectu alterius, et dicitur integer, aliquando in comparatione alterius, scilicet ut pars vel partes, et dicitur fractio". Translation from Sesiano (2014, p. 605). – For references to further attestations, see Tropfke (1980, p. 118).

fractio (Fibonacci 1857, p. 24), but also (p. 27) *numerus ruptus* 'broken number' (from *rumpere* 'to break') and (p. 23) *minutum* (the verb *minuo* means 'to make smaller, diminish').[5] There are thus at least four Latin terms: *minuta*, *numerus fractus*, *fractio*, and *numerus ruptus*.

All of these terms reappear in modern languages.[6] For example, the French *Introductoire d'astronomie*, dating from around 1270, has *fraccïon* (mod. *fraction*) (Dörr 1998, p. 141); Spanish *fraction* (mod. *fracción*) is attested at about the same time (Bossong 1978, p. 230).[7] The corresponding German term *Bruch* dates from the middle of the 15th century; it is found (as *pruch*) in the so-called *Algorismus Ratisbonensis* (Vogel 1954, p. 136).

Terms meaning 'broken number' are found from the late 13th century onwards. Italian *rotto* was used by Maestro Umbro (Arrighi 1991, e. g., p. 21), whose text dates from about 1290.[8] Slightly later, Jacopo da Firenze also used Italian *numero rocto* or *rotto* (Høyrup 2007, e. g., p. 235). A manuscript from around 1430 written in Occitan has *nombre rot* (Sesiano 2018, p. 214); French *nombre rout* was used in a mid 15th-century manuscript ([anon.] c1449, fol. 104v), and *nombre rompu* appeared in another manuscript slightly afterwards (Benoit 1985, p. 93).[9]

Various alternatives have long remained in use, as is illustrated by Jeake (1696, p. 41): "*Fractions*, sometime called *Fragments*, sometime *Parts*, sometime *Broken Numbers*".

Note. For the history of fractions and extensive comments on the terminology, see Tropfke (1980, pp. 93–121). See also Smith (1958, pp. 208–250). For the history of the notion, see the various contributions in Benoit, Chemla, and Ritter (1992). On the Arabic terms, see Souissi (1968, p. 303), Souissi (1997), Saidan (1978, pp. 415–417), and Rashed (2017, pp. 724–726).

References

[anon.] (c1449). "[Traitté d'arithmetique]". Manuscript, BNF Français 7287. URL: `https://gallica.bnf.fr/ark:/12148/btv1b525140414`.

Arrighi, Gino (1991). "Maestro umbro (sec. XIII). Amastramento de l'arte de la geometria (Cod. 2404 dalla Biblioteca Riccardiana di Firenze)". *Bollettino della deputazione di Storia patria per l'Umbria* 87, pp. 5–31.

Benoit, Paul (1985). "Recherches sur le vocabulaire des opérations élémentaires dans les arithmétiques en langue française de la fin du Moyen Âge". *Documents pour l'histoire du vocabulaire scientifique* 7, pp. 77–95.

Benoit, Paul, Karine Chemla, and Jim Ritter, eds. (1992). *Historie de fractions, fractions d'histoire*. Basel et al.

Bossong, Georg, ed. (1978). *Los Canones de Albateni*. Herausgegeben sowie mit Einleitung, Anmerkungen und Glossar versehen. Tübingen.

Busard, H.L.L. (1987). *The Mediaeval Latin Translation of Euclid's Elements Made Directly from the Greek*. Stuttgart.

Dörr, Stephen (1998). *Der älteste Astronomietraktat in französischer Sprache: L'Introductoire d'astronomie*. Edition und lexikalische Analyse. Tübingen.

[5] The word *minutum* often refers to sixtieths with Fibonacci. For a discussion of Fibonacci's terms for fractions, see Moyon and Spiesser (2015, p. 404). [6] In the following, the focus will be on the terms *fraction* and *broken number*. [7] English *fraction* is first attested in Chaucer (OED). [8] This usage was pointed out by Feola (2008, p. 172). [9] See also Spiesser (2004, pp. 291f.).

Euclid (1956). *The Thirteen Books of Euclid's Elements*. Volume II. Books III-IX. Ed. by Thomas L. Heath. New York.
Feola, Francesco (2008). *Gli esordi della geometria in volgare*. Un volgarizzamento trecentesco della *Practica Geometriae* di Leonardo Pisano. Firenze.
Fibonacci (1857). *Scritti di Leonardo Pisano*. Pubblicati da Baldassare Boncompagni. Volume I. Roma.
Folkerts, Menso (1997). *Die älteste lateinische Schrift über das indische Rechnen nach al-Ḫwārizmī*. Edition, Übersetzung und Kommentar ... unter Mitarbeit von Paul Kunitzsch. München.
Høyrup, Jens (2007). *Jacopo da Firenze's Tractatus Algorismi and Early Italian Abbacus Culture*. Basel.
Jeake, Samuel (1696). Λογιστολογία, *or Arithmetick Surveighed and Reviewed*. London.
al-Khwārizmī, Muḥammad ibn Mūsā (1963). *Mohammed ibn Musa Alchwarizmi's Algorismus*. Das früheste Lehrbuch zum Rechnen mit indischen Ziffern. Ed. by Kurt Vogel. Aalen.
Moyon, Marc and Maryvonne Spiesser (2015). "L'arithmétique des fractions dans l'œuvre de Fibonacci: fondements & usages". *Archive for History of Exact Sciences* 69, pp. 391–427.
Rashed, Roshdi, ed. (2017). *Lexique historique de la langue scientifique arabe*. Hildesheim.
Saidan, A.S. (1978). *The Arithmetic of Al-Uqlīdisī*. The Story of Hindu-Arabic Arithmetic as told in Kitāb al-Fuṣūl fī al-Ḥisāb al-Hindī. Dordrecht.
Sesiano, Jacques, ed. (2014). *Liber Mahameleth*. Cham.
Sesiano, Jacques (2018). *L'arithmétique de Pamiers*. Traité mathématique en langue d'oc du XVe siècle. Lausanne.
Smith, David Eugene (1958). *History of Mathematics*. Volume II. Special Topics of Elementary Mathematics. New York.
Souissi, Mohammed (1968). *La langue des mathématiques en arabe*. Tunis.
Souissi, Mohammed (1997). "Kasr". *Encyclopaedia of Islam* 4, pp. 725–726.
Spiesser, Maryvonne (2004). "Nombre et pratiques calculatoires dans la tradition commerciale du xve siècle: l'exemple des «arithmétiques» du sud de la France". *Sciences et Techniques en Perspective* 8, pp. 285–301.
Tropfke, Johannes (1930). *Geschichte der Elementar-Mathematik*. Erster Band. Rechnen. Dritte, verbesserte und vermehrte Auflage. Berlin and Leipzig.
Tropfke, Johannes (1980). *Geschichte der Elementarmathematik*. Band 1: Arithmetik und Algebra. Vollständig neu bearbeitet von Kurt Vogel, Karin Reich, Helmuth Gericke. Berlin and New York.
Vogel, Kurt, ed. (1954). *Die Practica des Algorismus Ratisbonensis*. Ein Rechenbuch des Benediktinerklosters St. Emmeram aus der Mitte des 15. Jahrhunderts nach den Handschriften der Münchner Staatsbibliothek und der Stiftsbibliothek St. Florian. München.

free

Free groups are first found in Dyck (1882), who proceeded on the basis of "certain *generating operations* A_1, A_2, A_3, ... about whose *particular* nature no assumptions are made. *Then it is possible to specify every group that can be built by means of iteration and combination of these operations through the knowledge of certain relations which occur in connection with the composition of the original operations*" (Dyck 1882, p. 5).[1]

Dyck worked in a function-theoretic setting, and as objects in their own right, free groups were first studied by Nielsen (Stillwell 1993, p. 90). He first introduced the term in a paper written in Danish, where *frie Gruppe* is used: "we will call it the 'free group G_n' generated by the generators a_1, ... a_n and write it $[a_1, a_2, ...a_n]$" (Nielsen

[1] "Wir gehen [...] von gewissen *erzeugenden Operationen* A_1, A_2, A_3, ... aus, über deren *speciellen* Charakter keinerlei Annahmen gemacht werden. *Dann kann man jede Gruppe, welche durch Iteration und Combination dieser Operationen sich bilden lässt, individualisiren, durch die Kenntnis gewisser Relationen, die bei der Zusammensetzung dieser ursprünglichen Operationen auftreten*". Translation from Wußing (1984, p. 241).

1921, p. 86). The term became better known through his 1924 paper, which he wrote in German: "A group shall be called *free* if it can be generated by an independent system of *generating* elements" (Nielsen 1924, p. 169).[2]

The word *free* (German *frei*) had earlier been used by Dehn in a related sense, namely for a generator not appearing in any defining relation (Chandler and Magnus 1982, p. 20): "we arrive at the following normal form of the group: among the generators are w generators which appear in no relation ('free, arbitrary generators')" (Dehn 1912, p. 137).[3] Chandler and Magnus (1982, p. 83) claim that Dehn also coined the term *free group*; they do not cite any evidence, however.[4]

The term *free product* first appears in Klein (1926, p. 361), in a section written by Artin:[5]

> Consider a finite or infinite set of groups $\mathfrak{g}_1, \mathfrak{g}_2, \ldots$. We admit the occurrence of isomorphic groups in this set but we postulate that there is no rule of composition for any two elements from distinct groups (whether they are isomorphic or not). We define now the free product $\mathfrak{G}$ of our groups as the set of the following symbols:
>
> 1. The unit 1.
> 2. All symbols of the form $A = \alpha_1\alpha_2 \ldots \alpha_n$, where α_i denotes any element of our groups $\mathfrak{g}_1, \mathfrak{g}_2 \ldots$[1]

Artin adds the conditions that no α_i may be equal to 1 and that adjacent elements α_i and α_{i+1} must belong to different groups.

The term is soon found in other contexts; Birkhoff (1933, p. 463) uses *free lattice*, and *free algebra* appears in Birkhoff (1935, p. 440). In connection with modules, the term was introduced by Bourbaki (1947, pp. 10f.). In geometry, free projective planes were introduced by Hall (1943, p. 229) and so named because they are "comparable in certain ways to free groups".

Note. For the history of free groups, see Chandler and Magnus (1982); for the history of free products, see Chandler and Magnus (1982, esp. pp. 11, 102–112).

References

Birkhoff, Garrett (1933). "On the combination of subalgebras". *Proceedings of the Cambridge Philosophical Society* 29, pp. 441–464.

[2] "Eine Gruppe heiße *frei*, wenn sie durch ein unabhängiges System *erzeugender* Elemente hervorgebracht werden kann". [3] "[gelangen wir] zu folgender *Normalform der Gruppe*: *Unter den Erzeugenden sind* w *Erzeugende, die in keiner Relation vorkommen* (*'freie, willkürliche Erzeugende'*)". Translation from Dehn (1987, p. 166), slightly modified. [4] The likely source is a talk "On general group theory and topology" ("Über allgemeine Gruppentheorie und Topologie") which Dehn gave in 1922; the talk is mentioned in the *Jahresbericht der Deutschen Mathematiker-Vereinigung* 31 (1922), part 2, p. 38, for example, but was never published. On this talk, Levi (1929, p. 296) wrote that it probably gave "the first exact foundation of free groups". A copy of the manuscript, not consulted in the making of this dictionary, is in the Dolph Briscoe Center for American History, where the Max Dehn Papers are archived. [5] The section is credited to Artin in the preface of the book. [1] "Es sei eine beliebige endliche oder unendliche Menge von Gruppen $\mathfrak{g}_1, \mathfrak{g}_2, \ldots$ vorgelegt. Unter ihnen dürfen auch isomorphe Gruppen vorkommen, es sei aber zwischen zwei Elementen aus verschiedenen Gruppen (seien sie isomorph oder nicht) keinerlei Kompositionsvorschrift gegeben. Unter dem freien Produkt unserer Gruppen verstehen wir nun die Menge $\mathfrak{G}$ folgender Symbole: 1. Die Einheit 1. 2. Alle Symbole der Form $A = \alpha_1\alpha_2 \ldots \alpha_n$, wobei α_i irgendein Element einer unserer Gruppen $\mathfrak{g}_1, \mathfrak{g}_2, \ldots$ ist". Translation from Chandler and Magnus (1982, p. 103), slightly modified.

Birkhoff, Garrett (1935). "On the structure of abstract algebras". *Proceedings of the Cambridge Philosophical Society* 31, pp. 433–454.
Bourbaki, Nicolas (1947). *Algèbre*. Chapitre II. Algèbre linéaire. Paris.
Chandler, Bruce and Wilhelm Magnus (1982). *The History of Combinatorial Group Theory*. A Case Study in the History of Ideas. New York.
Dehn, Max (1912). "Über unendliche diskontinuierliche Gruppen". *Mathematische Annalen* 71, pp. 116–144.
Dehn, Max (1987). *Papers on Group Theory and Topology*. Translated and Introduced by John Stillwell. New York.
Dyck, Walther (1882). "Gruppentheoretische Studien". *Mathematische Annalen* 20, pp. 1–44.
Hall, Marshall (1943). "Projective Planes". *Transactions of the American Mathematical Society* 54, pp. 229–277.
Klein, Felix (1926). *Vorlesungen über höhere Geometrie*. Dritte Auflage. Bearbeitet und herausgegeben von W. Blaschke. Berlin.
Levi, Friedrich (1929). *Geometrische Konfigurationen*. Mit einer Einführung in die kombinatorische Flächentopologie. Leipzig.
Nielsen, Jakob (1921). "Om Regning med ikke kommutative Faktoren og dens Anvendelse i Grupteorien". *Matematisk Tidsskrift. B*, pp. 77–94.
Nielsen, Jakob (1924). "Die Isomorphismengruppe der freien Gruppen". *Mathematische Annalen* 91, pp. 169–209.
Stillwell, John (1993). *Classical Topology and Combinatorial Group Theory*. New York.
Wußing, Hans (1984). *The Genesis of the Abstract Group Concept*. A Contribution to the History of the Origin of Abstract Group Theory. Tr. Abe Shenitzer. Cambridge, Mass.

frustum

See *truncated*.

function

According to Yuschkevitch (1976, p. 58), what is presumably the first attempt to define a function as an analytic expression is found in Gregory (1667, p. 9): "We say a quantity is composed of quantities when it derives from the addition, subtraction, multiplication, division or extraction of roots of quantities, or from any other imaginable operation".[1] As can be seen, Gregory did not use the term *function*, which was introduced shortly afterwards by Leibniz.

Latin *functio* means 'performance, execution', and it was in this very literal meaning that the term was originally used. Leibniz began using it in 1673 in some manuscripts.[2] In particular, in "Methodus tangentium inversa, seu de functionibus", he says that "other kinds of lines which, in a given figure, perform some function" (Leibniz 2008, p. 665).[3] As explained by Mahnke (1926, p. 47), 'to perform a function in a figure' for Leibniz means, for example, to touch the curve or to be perpendicular to it, taking into account only a limited segment of the line under consideration.

In print, the term is first found in Leibniz (1692, p. 170): "the tangent and some other functions depending on it, such as perpendiculars from the axis conducted to

[1] "Quantitatem dicimus a quantitatibus esse compositam, cum a quantitatum additione, subductione, multiplicatione, divisione, radicum extractione, vel quacumque alia imaginabili operatione, sit alia quantitas". Translation from Crippa (2019, p. 49). [2] For details, see the editorial remarks in Leibniz (2008, p. xviii). It is also pointed out there that in these early stages, the terminology was not yet fixed: Leibniz also used *facere officium* instead of *facere functionem*. [3] "ex aliis linearum in figura data functiones facientum generibus assumptis". Translation from Yuschkevitch (1976, p. 56).

the tangent";[4] a similar passage is found in Leibniz (1694, p. 316): "I call *function* that part of a straight line that is cut off of straight lines that are drawn using only a fixed point and a point of the curve given with its curvature".[5] Leibniz gave various examples such as the abscissa, the ordinate, the normal, the subtangent, and others. The term was adopted by Jakob Bernoulli (1694, p. 391), who referred to Leibniz. His brother Johann also used the term in 1698 (Gerhardt 1855-56, p. 507); as explained by Bos (1974, p. 9), he "wrote about the powers of a variable 'or any function in general of a variable. Leibniz agreed[6] to this use of the term, which thus lost its initial geometric connotations and became a concept connected with formulas rather than with figures".

In 1718, Johann Bernoulli presented a definition of *function* as an analytic expression: "Here one calls a *function* of a variable magnitude, a quantity composed in any manner possible from this variable magnitude and constants" (Johann Bernoulli 1719, p. 106).[7] At this time, the term was not yet well known; it is not found, as pointed out by Yuschkevitch (1976, pp. 59f.), in Wolff's mathematical dictionary (Wolff 1716).[8]

The term was also adopted by Euler, who gave it a prominent place at the beginning of his *Introduction to Analysis of the Infinite*, in which he gave the following definition: "A function of a variable quantity is an analytic expression composed in any way whatsoever of the variable quantity and numbers or constant quantities" (Euler 1748, § 4).[1]

In English, the term was apparently first used by Maclaurin (1742, p. 510) and Simpson (1750, p. 484).[2] The German term *Funktion* is found, at about the same time, in Wolff (1747, p. 539) and [anon.] (1749, p. 393).

In the unpublished *Traité du calcul intégral*, Condorcet based the definition of *function* on the concept of value; according to Spalt (2015, p. 170), he apparently was the first to do so. Condorcet's definition reads: "I assume that I have a certain number of quantities $x, y, z, \dots F$ and that for each determined value of x, y, z, *etc.*, F has one or more determined values which correspond to it; I say that F is a function of x, y, z" (de Condorcet 1786, p. 2).[3] Although the treatise was unpublished, it was known to Lacroix, who cited it in the preface of his *Traité* and also used the concept of value (Lacroix 1797, pp. xxx, 1).

[4] "tangens, vel aliae nonnullae functiones ab ea pendentes, verb.gr. perpendiculars ad tangentem ab axe ad curvam ductae". Translation from Struik (1969, p. 272). [5] "*Functionem* voco portionem rectae, quae ductis ope sola puncti fixi & puncti curvae cum curvedine sua dati rectis, abscinditur". [6] See Gerhardt (1855-56, p. 525). [7] "On appelle ici *Fonction* d'une grandeur variable, une quantité composée de quelque maniere que ce soit de cette grandeur variable & de constantes". Translation from Stedall (2008, p. 232). – A Latin version of this definition is found in Johann Bernoulli (1718, p. 21). [8] The term was first included in the third edition (Wolff 1747, p. 539). [1] "Functio quantitatis variabilis est expressio analytica quomodocunque composita ex illa quantitate variabili et numeris seu quantitatibus constantibus". Translation from Euler (1988, § 4). – An earlier definition by Euler is found in an unpublished manuscript on the differential calculus from around 1727-1730 (Juškevič 1983, p. 161). For a discussion of Euler's definitions of *function*, see Spalt (2011) and Domingues (2008, p. 196). [2] The OED dates the term to 1758. [3] "Je suppose que j'aie un certain nombre de quantités $x, y, z, \dots F$, & que pour chaque valeur déterminée de x, y, z, *etc.*, F ait une ou plusieurs valeurs déterminées qui y répondent; je dis que F est une fonction de x, y, z".

Various further mathematicians such as Fourier[4] and Lobachevsky[5] contributed to the development of the function concept. Lejeune Dirichlet (1829, p. 169) used *fonction* for what was later called the *Dirichlet function*, a remarkable extension because this function cannot be considered as given by an analytic expression. For Dirichlet, however, a function could, in certain cases, have more than one value: "Where an interruption of continuity takes place and the function $f(x)$ actually has two values [...]" (Lejeune Dirichlet 1837, p. 173).[6]

Riemann (1851, p. 1) later gave a definition of a real function which was just based on the condition that to each value only one value may be assigned: "Denote by z a variable that can take successively all possible real values. When there is a unique value of the variable w corresponding to each z, we say that w is a function of z".[7]

Numerous further definitions can be found in the literature.[8] A last example, dating from the 20th century, is the one given by Bourbaki:

> Let E and F be two sets, distinct or not. A relation between a variable x of E and a variable y of F is called *functional relation in* y, or *functional relation from* E *to* F *if, for all* $x \in E$ *there is one and only one element* y *of* F *that stands in the respective relation with* x.
>
> The operation which thus associates with each element $x \in E$ the element $y \in F$ which stands in the given relation with x is called *function*; y is called the *value* of the function for the element x (Bourbaki 1939, p. 7).[1].

On Bourbaki's approach, Monna (1972, p. 82) comments: "Here all the questions about what is a genuine function are removed from consideration".

Finally, it should be noted that there is considerable disagreement about the interpretation of many of the definitions of *function*. This applies, for instance, to the definitions given by Euler, Fourier, and Dirichlet. See the references cited.

See also *mapping*.

Note. The literature on the history of the notion abounds. Overviews are given by Yuschkevitch (1976), Monna (1972), Medvedev (1991, ch. 2), and Kleiner (1989), for example. For Leibniz' terminology, see also Mahnke (1926) and Herrera Castillo (2013).

References

[anon.] (1749). "Introductio in analysin infinitorum [review]". *Zuverlaessige Nachrichten von dem gegenwaertigen Zustande, Veraenderung und Wachsthum der Wissenschaften* 109, pp. 391–417.

---

[4] For Fourier, see Hawkins (1979, pp. 5–8) and Medvedev (1991, pp. 50–55). [5] For Lobachevsky, see Yuschkevitch (1976, pp. 77f.) and Medvedev (1991, pp. 57–60). [6] "Wo eine Unterbrechnung der Stetigkeit eintritt und also die Funktion $\varphi(x)$ eigentlich zwei Werthe hat [...]". See also Lejeune Dirichlet (1837, p. 170) and, for discussion, Spalt (2019, pp. 173–175). [7] "Denkt man sich unter z eine veränderliche Grösse, welche nach und nach alle möglichen reellen Werthe annehmen kann, so wird, wenn jedem ihrer Werthe ein einziger Werth der unbestimmten Grösse w entspricht, w eine Function von z genannt". Translation from Riemann (2004, p. 1). [8] For a brief historical discussion of definitions of functions in terms of relations, see Felgner (2002, pp. 626f.). [1] "Soient E et F deux ensembles, distincts ou non. Une relation entre une variable x de E et une variable y de F est dite *relation fonctionnelle en* y, ou *relation fonctionnelle de* E *vers* F, *si, quelque soit* $x \in E$, *il existe un élément* y *de* F, *et un seul, qui soit dans la relation considérée avec* x. On donne le nom de *fonction* à l'opération qui associe ainsi à tout élément $x \in E$ l'élément $y \in F$ qui se trouve dans la relation donnée avec x; on dit que y est la *valeur* de la fonction pour l'élément x [...]".

Bernoulli, Jakob (1694). “De methodo tangentium inversa”. *Acta Eruditorum* (Oct. 1694), pp. 391–394.
Bernoulli, Johann (1718). “De Solutionibus quae extant Problematum isoperimetricorum”. *Acta Eruditorum* (Jan. 1718), pp. 15–31.
Bernoulli, Johann (1719). “Remarques Sur ce qu’on a donné jusqu’ici de solutions des Problêmes sur les Isoperimetres”. *Mémoires de l’Académie royale des sciences* (1718), pp. 100–138.
Bos, Henk J.M. (1974). “Differentials, Higher-Order Differentials and the Derivative in the Leibnizian Calculus”. *Archive for History of Exact Sciences* 14, pp. 1–90.
Bourbaki, Nicolas (1939). *Théorie des ensembles*. Fascicule de résultats. Paris.
de Condorcet, Jean-Antoine-Nicolas (1786). *Traité du calcul intégral*. Paris. URL: `http://gallica.bnf.fr/ark:/12148/btv1b86262724`. [unpublished].
Crippa, Davide (2019). *The Impossibility of Squaring the Circle in the 17th Century*. A Debate Among Gregory, Huygens and Leibniz. Basel.
Domingues, João Caramalho (2008). *Lacroix and the Calculus*. Basel.
Euler, Leonhard (1748). *Introductio in analysin infinitorum*. Tomus primus. Lausanne.
Euler, Leonhard (1988). *Introduction to Analysis of the Infinite*. Book I. Translated by John D. Blanton. New York.
Felgner, Ulrich (2002). “Der Begriff der Funktion”. In: *Felix Hausdorff. Gesammelte Werke*. Band II. Grundzüge der Mengenlehre. Ed. by Egbert Brieskorn et al. Berlin and Heidelberg, pp. 621–633.
Gerhardt, C.I., ed. (1855-56). *Leibnizens mathematische Schriften*. Erste Abtheilung. Band III. Halle.
Gregory, James (1667). *Vera circuli et hyperbolae quadratura*. Padova.
Hawkins, Thomas (1979). *Lebesgue’s Theory of Integration*. Its Origins and Development. Second Edition. Providence, R.I.
Herrera Castillo, Laura E. (2013). “El concepto leibniziano matemático de función en 1673”. *Cultura* 32, pp. 127–144.
Juškevič, Adolf P. (1983). “L. Euler’s unpublished manuscript *Calculus Differentialis*”. In: *Leonhard Euler 1707–1783*. Beiträge zu Leben und Werk. Ed. by J.J. Burckhardt, E.A. Fellmann, and W. Habicht. Basel, pp. 161–168.
Kleiner, Israel (1989). “Evolution of the Function Concept: A Brief Survey”. *The College Mathematics Journal* 20, pp. 282–300.
Lacroix, Sylvestre François (1797). *Traité du calcul différentiel et du calcul intégral*. Tome premier. Paris.
Leibniz, Gottfried Wilhelm (1692). “De linea ex lineis numero infinitis ordinatim ductis inter se concurrentibus formata [...]” *Acta Eruditorum* (Apr. 1692), pp. 168–171.
Leibniz, Gottfried Wilhelm (1694). “Nova calculi differentialis applicatio & usus ad multiplicem linearum constructionem ex data tangentium conditione”. *Acta Eruditorum* (Apr. 1694), pp. 311–316.
Leibniz, Gottfried Wilhelm (2008). *Sämtliche Schriften und Briefe*. Siebente Reihe. Mathematische Schriften. Vierter Band. 1670-1673. Infinitesimalmathematik. Berlin.
Lejeune Dirichlet, Gustav (1829). “Sur la convergence des séries trigonométriques qui servent à répresenter une fonction arbitraire entre des limites données”. *Journal für die reine und angewandte Mathematik* 4, pp. 157–169.
Lejeune Dirichlet, Gustav (1837). “Ueber die Darstellung ganz willkürlicher Functionen durch Sinus- und Cosinusreihen”. In: *Repertorium der Physik*. Herausgegeben von Heinrich Wilhelm Dove und Ludwig Moser. I. Band. Berlin, pp. 152–174.
Maclaurin, Colin (1742). *A Treatise of Fluxions*. In Two Books. Edinburgh.
Mahnke, Dietrich (1926). *Neue Einblicke in die Entdeckungsgeschichte der höheren Analysis*. Berlin.
Medvedev, Fyodor A. (1991). *Scenes from the History of Real Functions*. Translated from the Russian by Roger Cooke. Basel.
Monna, A.F. (1972). “The Concept of Function in the 19th and 20th Centuries, in Particular with Regard to the Discussions between Baire, Borel and Lebesgue”. *Archive for History of Exact Sciences* 9, pp. 57–84.
Riemann, Bernhard (1851). “Grundlagen für eine allgemeine Theorie der Functionen einer veränderlichen complexen Grösse”. PhD thesis. Georg-August-Universität Göttingen.

Riemann, Bernhard (2004). *Collected Papers*. Translated from the 1892 edition by Roger Baker, Charles Christenson and Henry Orde. Heber City.
Simpson, Thomas (1750). *The Doctrine and Application of Fluxions*. Part I. London.
Spalt, Detlef D. (2011). "Welche Funktionsbegriffe gab Leonhard Euler?" *Historia Mathematica* 38, pp. 485–505.
Spalt, Detlef D. (2015). "Die Einführung des Wertbegriffs in die Analysis oder: die Konstituierung der Differenzial- und Integralrechnung als mathematische Lehre". *Archives Internationales d'Histoire des Sciences* 65, pp. 117–193.
Spalt, Detlef D. (2019). *Eine kurze Geschichte der Analysis für Mathematiker und Philosophen*. Berlin.
Stedall, Jacqueline A. (2008). *Mathematics Emerging*. A Sourcebook 1540 – 1900. Oxford.
Struik, Dirk Jan, ed. (1969). *A Source Book in Mathematics, 1200-1800*. Harvard and London.
Wolff, Christian (1716). *Mathematisches Lexicon*. Leipzig.
Wolff, Christian (1747). *Vollständiges Mathematisches Lexicon*. Erster Theil. Leipzig.
Yuschkevitch, Adolph P. (1976). "The Concept of Function up to the Middle of the 19th Century". *Archive for History of Exact Sciences* 16, pp. 37–85.

function space

A function space can be broadly characterized as "a topological space, the 'points' of which are functions" (Birkhoff and Kreyszig 1984, p. 259). The terminology came up in the late 19th century when, in an important paper on the "functional calculus", Pincherle (1897a, p. 85) used Italian *spazio funzionale*[1] for the set of all power series and explicitly remarked that these series can be regarded as points in this space:

> In some of my recently published papers I introduced a notion that can be conveniently regarded in many investigations of analysis as the totality of analytic functions of one variable or, for greater definiteness, as the totality of series of positive integer powers of $x - a$ totality, or space, of which every series is an element. Such a manifold, which obviously has an infinite number of dimensions, may be called a function space. Every power series in x is a point of this space and the coefficients of the series can be regarded as the coordinates of this point".[1]

The terminology was not yet in common use in the first decade of the 20th century; it gained momentum, however, after Schmidt (1908, p. 56) included a section "geometry in a function space" in his paper on linear equations with infinitely many unknowns, in which he used German *Functionenraum* 'function space' with reference to the space later designated as l^2.[2] He remarked that he "owes the concepts and theorems developed in this chapter to Kowalewski", who, however, does not

[1] Pincherle (1897b, pp. 331f.) also used French *espace fonctionnel* at the same time. [1] "In alcuni lavori recentemente pubblicati ho espresso il concetto che, per varie ricerche d'Analisi, è opportuno considerare la totalità delle funzioni analitiche di una variabile x o – per meglio fissare le idee – la totalità delle serie di potenze intere positive di x, come una varietà o spazio di cui ogni singola serie costituisce un elemento. Ad una tale varietà, evidentemente ad un numero infinito di dimensioni, si può dare il nome di *spazio funzionale*; ogni serie di potenze di x sarà un *punto* di questo spazio ed i coefficienti della serie si potranno riguardare come le coordinate del punto". Translation from Rosenfeld (1988, p. 273). – The term has been attributed to Pincherle by Kreyszig (1986, p. 35). For a brief discussion of Pincherle's paper, see Bottazzini and Gray (2013, p. 661). [2] For a discussion of Schmidt's paper, see Bernkopf (1966-67, pp. 46–48).

appear to have published[3] on this matter before Schmidt, but he did so soon afterwards: "The functions on the interval $(0, 1)$ form a set which I here call the *functional space* R_x. Each function $f(x)$ is for us a *point* in this space" (Kowalewski 1910, p. 1338).[4]

The French term *espace des fonctions* was used by Riesz in a letter to Fréchet dating from May 1907; this letter is reproduced in Rodríguez Hernández (2006, p. 207). Riesz may have heard this term during one of his stays in Göttingen (Rodríguez Hernández 2006, pp. 13, 250).

The English term appeared soon after the publication of Schmidt's paper in Richardson and Hurwitz (1909, p. 15).

Note. For the history of the concept, see Bernkopf (1966-67) and Birkhoff and Kreyszig (1984).

References

Bernkopf, Michael (1966-67). "The Development of Function Spaces with Particular Reference to Their Origins in Integral Equation Theory". *Archive for History of Exact Sciences* 3, pp. 1–96.

Birkhoff, Garrett and Erwin Kreyszig (1984). "The establishment of functional analysis". *Historia Mathematica* 11, pp. 258–321.

Bottazzini, Umberto and Jeremy Gray (2013). *Hidden Harmony – Geometric Fantasies*. The Rise of Complex Function Theory. New York.

Kowalewski, Gerhard (1910). "Les formules de Frenet dans l'espace fonctionnelle". *Comptes rendus de l'Académie des sciences* 151, pp. 1338–1340.

Kreyszig, Erwin (1986). "Zur Entwicklung der zentralen Ideen in der Funktionalanalysis". *Elemente der Mathematik* 41, pp. 25–48.

Pincherle, Salvatore (1897a). "Cenno sulla Geometria dello spazio funzionale". *Rendiconto della R. Accademia delle Scienze dell'Istituto di Bologna* 1, pp. 85–95.

Pincherle, Salvatore (1897b). "Mémoire sur le calcul fonctionnel distributif". *Mathematische Annalen* 49, pp. 325–382.

von Renteln, Michael (2007). "Zur Entstehungsgeschichte der Funktionalanalysis". In: *"Es gibt für Könige keinen besonderen Weg zur Geometrie"*. Festschrift für Karin Reich. Ed. by Gudrun Wolfschmidt. Augsburg, pp. 139–144.

Richardson, R.G.D. and Wallie Abraham Hurwitz (1909). "Note on determinants whose terms are certain integrals". *Bulletin of the American Mathematical Society* 16, pp. 14–19.

Rodríguez Hernández, Laura Regina (2006). "Friedrich Riesz' Beiträge zur Herausbildung des modernen mathematischen Konzepts abstrakter Räume". PhD thesis. Johannes Gutenberg-Universität Mainz.

Rosenfeld, Boris A. (1988). *A History of Non-Euclidean Geometry*. Evolution of the Concept of a Geometric Space. Translated by Abe Shenitzer. New York.

Schmidt, Erhard (1908). "Über die Auflösung linearer Gleichungen mit unendlich vielen Unbekannten". *Rendiconti del Circolo Matematico di Palermo* 25, pp. 53–77.

functional

The origin of the term *functional* can be found in Italy and France. In 1886, the Italian mathematician Salvatore Pincherle introduced the term "functional operation" (*operazione funzionale*): "I call *functional operation* every operation which, when executed

---

[3] Kowalewski did, however, write a letter to Friedrich Engel on Feb. 1, 1907, in which he interpreted the theory of integral equations geometrically. In this letter, he used the term *Funktionenraum*, which he defined as "the set of all *continuous* functions that exist in a fixed interval (a, b)" ("der Inbegriff aller *stetigen* Funktionen [...], die es in einem festen Intervall (a, b) gibt"). See von Renteln (2007, pp. 140f.).

[4] "Les fonctions dans l'intervalle $(0, 1)$ forment un ensemble que j'appelle ici l'*espace fonctionnel* R_x. Chaque fonction $f(x)$ est pour nous un *point* dans cet espace".

over an analytic function, yields an analytic function as a result" (Pincherle 1886, p. 393).[1] Pincherle (1897, p. 333) also used the French term *opération fonctionnelle*, for which he gave practically the same definition.[2]

At about the same time, Volterra discussed functionals under the names "functions that depend on other functions" (Volterra 1887a, p. 98) and "functions of lines" (Volterra 1887b, pp. 225f.).[3]

Hadamard (1903, p. 351) also used the term *opération fonctionelle*. He did not, however, use the noun *functional* in this paper, as is sometimes claimed. But in a letter to Fréchet, written around 1904 or 1905, he suggests that the word *fonctionnelle* should be used as a noun: "I decided to call *functionals* the 'functions of functions' or 'functions of lines'. U is a functional of $\phi(x)$. In any case, I propose this term to you" (Taylor 1982, p. 251).[4] The noun then appears in print in Fréchet (1906, p. 38).[5]

Note. The notion of functional in Volterra, Hadamard, and Fréchet is discussed by Taylor (1982, pp. 250f.). For Pincherle, see Bottazzini (2012) and Siegmund-Schultze (2003, p. 391). The terminology was briefly discussed by Pincherle (1912, p. 1) himself.

References

Bottazzini, Umberto (2012). "Pincherle's Early Contributions to Complex Analysis". In: *Mathematicians in Bologna 1861-1960*. Ed. by Salvatore Coen. Basel, pp. 57–72.

Fréchet, Maurice (1906). "Sur quelques points du calcul fonctionnel". *Rendiconti del Circolo Matematico di Palermo* 22, pp. 1–74.

Hadamard, Jacques (1903). "Sur les opérations fonctionelles". *Comptes rendus de l'Académie des sciences* 136, pp. 351–354.

Pincherle, Salvatore (1886). "Studi sopra alcuni operazioni funzionali". *Memorie dell'Accademia di Scienze dell'Istituto di Bologna* 7, pp. 393–442.

Pincherle, Salvatore (1897). "Mémoire sur le calcul fonctionnel distributif". *Mathematische Annalen* 49, pp. 325–382.

Pincherle, Salvatore (1912). "Équations et opérations fonctionnelles". *Encyclopédie des sciences mathématiques pures et appliquées* II.26, pp. 1–81.

Siegmund-Schultze, Reinhard (2003). "The Origins of Functional Analysis". In: *A History of Analysis*. Ed. by Hans Niels Jahnke. Providence, R.I., pp. 385–404.

Taylor, Angus E. (1982). "A Study of Maurice Fréchet: I. His Early Work on Point Set Theory and the Theory of Functionals". *Archive for History of Exact Sciences* 27, pp. 233–295.

Volterra, Vito (1887a). "Sopra le funzioni che dipendono da altre funzioni". *Atti della Reale Accademia dei Lincei. Rendiconti* 3, pp. 97–105, 141–146, 153–158.

Volterra, Vito (1887b). "Sopra le funzioni dipendenti da linee". *Atti della Reale Accademia dei Lincei. Rendiconti* 3, pp. 225–230, 274–281.

functional analysis

Siegmund-Schultze (2003, p. 385) begins his article on the origins of functional analysis with these words: "The beginnings of functional analysis between 1887 and 1932

---

[1] "Chiamo *operazione funzionale* qualunque operazione che eseguita sopra una funzione analitica d'a per risultato una funzione analitica". Translation from Bottazzini (2012, p. 68). [2] "Nous appelerons *opération fonctionnelle* toute opération qui, exécutée sur une fonction, donne comme résultat une fonction". [3] The Italian expressions used by Volterra are *funzioni che dipendono da altre funzioni* and *funzioni delle linee*. [4] "Je me suis decidé d'appeler *fonctionnelle* les 'fonctions de fonctions' ou 'fonctions de lignes'. U est fonctionnelle de $\phi(x)$. Je vous propose à tout hasard cette dénomination". Translation from Siegmund-Schultze (2003, p. 391). [5] Fréchet mistakenly stated that the noun already appears in Hadamard (1903).

developed in three different schools, the Italian, the French, and the German, under the names of *Calcolo Funzionale*, *Calcul Fonctionnel*, and *Methode der unendlich vielen Variablen*, respectively".

The Italian term *Calcolo funzionale* mentioned in this quotation came up in the 1890s. It was used by Pincherle (1893, p. 27) for "that branch of the calculus in which the function is considered as an arbitrarily variable element".[1] Pincherle (1897, p. 325) also used French *calcul fonctionnel*, saying that in this field "the variable element is not a number anymore, but the function as such".[2]

The French name *calcul fonctionnelle* is also found in the title of Fréchet (1905); in Fréchet (1906, p. 1) he first gives a definition of *functional* (French *opération fonctionnelle*) and then goes on to declare: "The research of the properties of these operations constitutes the object of the *Functional Calculus*".[3]

The German term mentioned by Siegmund-Schultze goes back to the early 20th century. For example, Hellinger (1907, p. 5) began his PhD thesis by saying that recently in the calculus, "theorems have been understood as analogous to algebraic theorems, which is done by speaking of infinitely many variables instead of the finitely many ones of algebra".[4] Hellinger referred to Hilbert (1906, p. 5) as a work in which this approach is most consistently followed.

The term *analyse fonctionnelle* is due to Lévy and spread through his book *Leçons d'analyse fonctionnelle*, in which he distinguished between *algèbre fonctionnelle*, which deals with "problems in which the unknowns are ordinary functions" (but which uses methods of the functional calculus to solve these problems), and *analyse fonctionnelle*, which deals with "problems in which the unknowns are functionals" (or with problems that cannot be conceived of independently of the concept of functional) (Lévy 1922, p. 4).[1] This distinction is slightly earlier found in Lévy (1920, p. 6).

The term was quickly adopted, for example by Bernstein and Doetsch (1925, p. 285); Doetsch (1927) gave an overview of the field. The English term first appears (OED) in Hildebrand and Graves (1927, p. 136). At the ICM in 1928, Fréchet (1929, p. 267) gave a talk in which he mentioned Lévy's term as a "happy phrase" ("heureuse locution") and compared the field to Moore's "general analysis".[2] At the same congress, the term was also used by Hadamard (1929, p. 151), who mentioned the works of Fréchet, Wiener, and Banach in this respect. Siegmund-Schultze (2003, p. 404) remarks that "Hadamard's understanding of 'functional analysis' as the

---

[1] "[...] ramo della Analisi in cui la funzione si considera come elemento arbitrariamente variabile e che si potrebbe chiamare Calcolo funzionale". [2] "l'élément variable n'est plus le nombre, mais la fonction considérée en elle-même". Translation from Bottazzini and Gray (2013, p. 661). [3] "nous dirons qu'une *opération fonctionnelle* U est définie dans un ensemble E d'éléments *de nature quelconque* (nombres, courbes, etc.) lorsqu'à tout élément A de E correspond une valeur numérique déterminée de $U : U(A)$. La recherche des propriétés des ces opérations constitue l'objet du *Calcul Fonctionnel*". [4] "[...] Theoreme als Analoga algebraischer Sätze aufzufassen, indem man von unendlich vielen Variablen statt von den endlich vielen der Algebra spricht". [1] "[l'algèbre fonctionnelle] comprendra des problèmes où les inconnues sont des fonctions ordinaires, mais qui se rattachent au calcul fonctionnel par les méthodes employées pour les résoudre. L'analyse fonctionnelle [...] comprend des problèmes où les inconnues sont des fonctionnelles, ou d'une manière générale des problèmes que l'on ne peut concevoir indépendamment de la notion de fonctionnelle". [2] For this field of study, see Siegmund-Schultze (1998). The term is due to E.H. Moore (1909, 1910).

abstract study of functionals within the *Calcul fonctionnel* may perhaps be the reason why the term functional analysis was finally chosen to denote the whole field". The word *finally* is well justified, because in major publications such as Banach (1932) and Stone (1932) the term was *not* used (it was only briefly mentioned in the respective introductions). Accounts of the growing importance of functional analysis (under this name) were given by Curtiss (1937, pp. 565f.) and Birkhoff (1938, pp. 296–298).

Note. For the history of functional analysis see, for example, Siegmund-Schultze (2003) and Dieudonné (1981).

References

Banach, Stefan (1932). *Théorie des opérations linéaires*. Warszawa.

Bernstein, Felix and Gustav Doetsch (1925). "Probleme aus der Theorie der Wärmeleitung. I. Mitteilung". *Mathematische Zeitschrift* 22, pp. 285–306.

Birkhoff, George D. (1938). "Fifty Years of American Mathematics". In: *Semicentennial Addresses of the American Mathematical Society*. Volume Two. Ed. by Raymond C. Archibald. New York, pp. 270–315.

Bottazzini, Umberto and Jeremy Gray (2013). *Hidden Harmony – Geometric Fantasies*. The Rise of Complex Function Theory. New York.

Curtiss, D.R. (1937). "Fashions in Mathematics". *The American Mathematical Monthly* 44, pp. 559–566.

Dieudonné, Jean (1981). *History of Functional Analysis*. Amsterdam.

Doetsch, Gustav (1927). "Überblick über Gegenstand und Methode der Funktionalanalysis". *Jahresbericht der Deutschen Mathematiker-Vereinigung* 36, pp. 1–30.

Fréchet, Maurice (1905). "La notion d'écart dans le calcul fonctionnel". *Comptes rendus de l'Académie des sciences* 140, pp. 772–774.

Fréchet, Maurice (1906). "Sur quelques points du calcul fonctionnel". *Rendiconti del Circolo Matematico di Palermo* 22, pp. 1–74.

Fréchet, Maurice (1929). "L'analyse générale et les espaces abstraits". In: *Atti dei Congresso Internazionale dei Matematici*. Tomo I. Bologna, pp. 267–274.

Hadamard, Jacques (1929). "Le développement et le rôle scientifique du calcul fonctionnel". In: *Atti dei Congresso Internazionale dei Matematici*. Tomo I. Bologna, pp. 143–161.

Hellinger, Ernst (1907). "Die Orthogonalinvarianten quadratischer Formen von unendlichvielen Variabelen". PhD thesis. Georg-Augusts-Universität zu Göttingen.

Hilbert, David (1906). "Grundzüge einer allgemeinen Theorie der linearen Integralgleichungen. (Vierte Mitteilung)". *Nachrichten von der Königl. Gesellschaft der Wissenschaften zu Göttingen (mathematisch-physikalische Klasse)* (1906), pp. 157–228.

Hildebrand, T.H. and Lawrence M. Graves (1927). "Implicit Functions and Their Differentials in General Analysis". *Transactions of the American Mathematical Society* 29, pp. 127–153.

Lévy, Paul (1920). *Notice sur les titres et travaux scientifiques*. Paris.

Lévy, Paul (1922). *Leçons de l'analyse fonctionnelle*. Paris.

Moore, E.H. (1909). "On a form of general analysis with applications to linear differential and integral equations". In: *Atti del IV Congresso internazionale dei matematici*. Vol. II. Ed. by Guido Castelnuovo. Roma, pp. 98–114.

Moore, E.H. (1910). "Introduction to a Form of General Analysis". In: *The New Haven Mathematical Colloquium*. New Haven, pp. 1–150.

Pincherle, Salvatore (1893). "L'algebra delle forme lineari alle differenze". *Rendiconto delle sessioni della R. Accademia delle scienze dell'Istituto di Bologna* (1892-93), pp. 25–28.

Pincherle, Salvatore (1897). "Mémoire sur le calcul fonctionnel distributif". *Mathematische Annalen* 49, pp. 325–382.

Siegmund-Schultze, Reinhard (1998). "Eliakim Hastings Moore's 'General Analysis'". *Archive for History of Exact Sciences* 52, pp. 51–89.

Siegmund-Schultze, Reinhard (2003). "The Origins of Functional Analysis". In: *A History of Analysis*. Ed. by Hans Niels Jahnke. Providence, R.I., pp. 385–404.

Stone, Marshall Harvey (1932). *Linear Transformations in Hilbert Space and Their Applications to Analysis*. New York.

functional determinant

This term was coined by Jacobi (1841a, p. 319) as Latin *determinantium functionalium*. He first refers to a previous paper (Jacobi 1841b) in which he had dealt with the concept, and points out the important role these determinants play in the calculus and concludes: "It should therefore be convenient to give such determinants a special name and call them *functional determinants*".[1]

The English term appears in Sylvester (1853, p. 476); in the same paper, he introduced, "for greater brevity" (p. 413), the term *Jacobian*, which he defined as follows: "The Jacobian to n homogeneous functions of n variables is the determinant represented by the symmetrical collocation in a square of the n differential coefficients of each of the n functions" (p. 546).

Note. For the history of the concept, see Muir (1960a, pp. 346–394) and Muir (1960b, pp. 230–253).

References

Jacobi, Carl Gustav Jacob (1841a). "De determinantibus functionalibus". *Journal für die reine und angewandte Mathematik* 22, pp. 319–352.
Jacobi, Carl Gustav Jacob (1841b). "De formatione et proprietatibus Determinatium". *Journal für die reine und angewandte Mathematik* 22, pp. 285–318.
Muir, Thomas (1960a). *The Theory of Determinants in the Historical Order of Development*. Volume I. General and Special Determinants up to 1841. New York.
Muir, Thomas (1960b). *The Theory of Determinants in the Historical Order of Development*. Volume II. The Period 1841 to 1860. New York.
Sylvester, James Joseph (1853). "On a Theory of the Syzygetic relations of two rational integral functions, comprising an application to the Theory of Sturm's Functions, and that of the greatest Algebraical Common Measure". *Philosophical Transactions of the Royal Society of London* 143, pp. 407–548.

functional equation

The term *aequationum functionalium* was first used by Pfaff (1788, p. 4, 1797, p. 48). He did not give a formal definition, but it is clear from his texts that he was using the term in the usual sense. Pfaff considered, for example, the logarithm, stating the equation $\log xu = \log x + \log y$ (Pfaff 1788, p. 6); he also remarked that $\varphi(x+y) = \varphi(x) + \varphi(y)$ might be regarded as the simplest functional equation (p. 8). Pfaff's terminology did not catch on at that time, however.

Instead, the usage of this term was established, independently of Pfaff, in England in the first quarter of the 19th century. In 1813, Babbage, citing various writers who worked on the subject before (but not Pfaff), contributed a paper entitled "On functional equations" to the *Memoirs of the Analytical Society* (Babbage 1813), and in 1815, he wrote:

> If an unknown quantity as x, be given by means of an equation, it becomes a question how to determine its value; similarly if an unknown function as ψ, be given by means of any functional equation, it is required to assign its form. In the first case, it is quantity which is to be determined; in the second, it is the form assumed by quantity, that becomes the subject

[1] "Qua de re fortasse convenit ea Determinantia propria appellatione Determinantium functionalium insignire".

> of investigation. In the one case, the various powers of the unknown quantity enter into the equation; in the other, the different orders of the function are concerned (Babbage 1815, p. 96).

Gergonne regarded Babbage's work as important and translated parts of one of his papers into French under the title "Des équations fonctionnelles" (Babbage 1821-22). The work was also noticed in Germany; in the *Göttingische Anzeigen von gelehrten Sachen* (1820, vol. 1, p. 338), Babbage's *Essay towards the Calculus of Functions* is briefly discussed and the German term *Functionalgleichung* is used.[1]

Note. For the history of functional equations, see Dhombres (1984), Dhombres (1992), and Aczél (1984). Dhombres (1993) contains a translation of Pfaff (1788).

References

Aczél, János (1984). "On history, applications and the theory of functional equations". In: *Functional Equations*. History, Applications and Theory. Ed. by János Aczél. Dordrecht et al., pp. 3–12.

Babbage, Charles (1813). "On functional equations". *Memoirs of the Analytical Society* (1813), pp. 96–114.

Babbage, Charles (1815). "An Essay towards the Calculus of Functions". *Philosophical Transactions of the Royal Society of London* 105, pp. 389–423.

Babbage, Charles (1821-22). "Des équations fonctionelles. (Extrait) Par M. Gergonne". *Annales de mathématiques pures et appliquées* 12, pp. 73–103.

Dhombres, Jean (1984). "On the historical role of functional equations". In: *Functional Equations*. History, Applications and Theory. Ed. by János Aczél. Dordrecht et al., pp. 17–31.

Dhombres, Jean (1992). "Le rôle des équations fonctionnelles dans l'*Analyse algébrique* de Cauchy". *Revue d'histoire des sciences* 45, pp. 25–50.

Dhombres, Jean (1993). "La méthode fonctionnelle chez J.F. Pfaff: une filiation leibnizienne". *Sciences et techniques en perspective* 26, pp. 97–147.

Pfaff, Johann Friedrich (1788). *Programma inaugurale in quo peculiarem differentialia investigandi rationem ex theoria functionum deducit*. Helmstedt.

Pfaff, Johann Friedrich (1797). "Observationes analyticae ad L. Euler Institutiones Calculi Integralis, vol. IV. Supplementum II. et IV." *Nova Acta Academiae Scientiarum Imperialis Petropolitanae* 11, pp. 37–57.

functor

The term *functor* was introduced in mathematics by Eilenberg and Mac Lane (1942, p. 538):

> we shall regard the character group $Ch(G)$ of G as a function of a variable group G, *together* with a prescription which assigns to any homomorphism γ of G into a second group G',
>
> $$\gamma : G \to G',$$
>
> the induced homomorphism [...]
>
> $$Ch(\gamma) : Ch(G') \to Ch(G).$$
>
> The functions $Ch(G)$ and $Ch(\gamma)$ jointly form what we shall call a 'functor'; in this case, a 'contravariant' one, because the mapping $Ch(\gamma)$ works in a direction opposite to that of γ.[1]

---

[1] *Functional-Gleichung* is indeed earlier found in a brief discussion of Pfaff (1788) in the *Allgemeine Literatur-Zeitung vom Jahre 1789*, no. 173, pp. 575f., published on June 10, 1789. [1] In the opposite case, they called a functor *covariant*.

They borrowed the term from Carnap (1934, p. 13), who used it in a linguistic context. Mac Lane (1971, pp. 29f.) himself points out "the pleasure of purloining words from the philosophers: 'Category' from Aristotle and Kant, 'Functor' from Carnap". There appears to be no direct relation between these two uses, as Mac Lane (1996, p. 131) later remarked: "Carnap [...] had talked of functors in a different sense and made some corresponding mistakes. It seemed in order to take over that word for a better and less philosophical purpose".[2]

Dieudonné (2009, p. 96) suggests the following explanation of the term: "Perhaps the custom they [Eilenberg and Mac Lane] had adopted of systematically using notations such as $Hom(G, H)$, $Ext(G, H)$, ..., for the various groups they defined in their 1942 paper, suggested to them that they were defining each time a kind of 'function' which assigned a commutative group to an arbitrary commutative group (or to a pair of such groups) according to a fixed rule. Perhaps to avoid speaking of the 'paradoxical' 'set of all commutative groups', they coined the word 'functor' for this kind of correspondence".

Note. See Krömer (2007, pp. 69f.) for a discussion of the origin of the term.

References

Carnap, Rudolf (1934). *Logische Syntax der Sprache*. Wien.

Dieudonné, Jean (2009). *A History of Algebraic and Differential Topology, 1900–1960*. Boston.

Eilenberg, Samuel and Saunders Mac Lane (1942). "Natural Isomorphisms in Group Theory". *Proceedings of the National Academy of Sciences* 28, pp. 537–543.

Krömer, Ralf (2007). *Tool and Object*. A History and Philosophy of Category Theory. Basel.

Mac Lane, Saunders (1938). "Carnap on Logical Syntax [review]". *Bulletin of the American Mathematical Society* 44, pp. 171–176.

Mac Lane, Saunders (1971). *Categories for the Working Mathematician*. New York et al.

Mac Lane, Saunders (1996). "The development and prospects for category theory". *Applied Categorical Structures* 4, pp. 129–136.

fundamental group

As a topological concept, French *groupe fondamental* was introduced by Poincaré (1895, p. 63):[1]

> [...] we can envisage a group G satisfying the following conditions:
>
> 1^0 Each closed contour M_0BM_0 corresponds to a substitution S of the group,
>
> 2^0 The necessary and sufficient condition for S to reduce to the identity substitution is that $M_0BM_0 \equiv 0$,
>
> 3^0 If S and S' correspond to contours C and C' and if $C'' \equiv C + C'$, then the substitution corresponding to C'' will be SS'.
>
> The group G is called the *fundamental group* of the manifold V.[2]

[2] Krömer (2007, p. 69) comments that this statement "obscures the fact that Carnap's terminology has always since been widely employed in logical analysis of language". – Mac Lane (1938) himself reviewed Carnap's book. [1] In an earlier paper, Poincaré (1892, p. 635) just used *groupe*.

[2] "on peut imaginer un groupe G satisfaisant aux conditions suivantes: 1^0 A chaque contour fermé, M_0BM_0 correspondra une substitution S du groupe; 2^0 La condition nécessaire et suffisante pour que S se réduise à la substitution identique, c'est que $M_0BM_0 \equiv 0$; 3^0 Si S et S' correspondent aux contours C et C' et si $C'' \equiv C + C'$, la substitution correspondant à C' sera SS'. Le groupe G s'appellera le *groupe fondamental* de la variété V". Translation from Poincaré (2010, p. 40).

It did not take long for the terminology to spread; Tietze (1906, p. 843), for example, used German *Fundamentalgruppe*, and the English term was used by Veblen and Alexander (1912-13, p. 163).

Kneser (1929, p. 256) suggested *Wegegruppe* 'path group' as an alternative;[3] he found it more expressive than Poincaré's term and also remarked that the homology group, for example, could also be regarded as fundamental. Kneser's term was also used by Reidemeister (1932, p. 51), for example. In English, it never found much appeal; *group of paths* was mentioned by Lefschetz (1942, p. 310), who himself preferred *Poincaré group*, as he had already done in an earlier book (Lefschetz 1930, p. 82).

Note. For the history of the concept, see Chandler and Magnus (1982, esp. pp. 14-21), Vanden Eynde (1992), Krömer (2013), and Brown (2021).

References

Brown, Ronald (2021). "'Not Just an Idle Game': The Story of Higher-Dimensional Versions of the Poincaré Fundamental Group". *The Mathematical Intelligencer* 43, pp. 5–15.
Chandler, Bruce and Wilhelm Magnus (1982). *The History of Combinatorial Group Theory*. A Case Study in the History of Ideas. New York.
Vanden Eynde, Ria (1992). "Historical Evolution of the Concept of Homotopic Paths". *Archive for History of Exact Sciences* 45, pp. 127–188.
Kneser, Hellmuth (1929). "Geschlossene Flächen in dreidimensionalen Mannigfaltigkeiten". *Jahresbericht der Deutschen Mathematiker-Vereinigung* 38, pp. 248–259.
Krömer, Ralf (2013). "The set of paths in a space and its algebraic structure. A historical account". *Annales de la faculté des sciences de Toulouse* 22, pp. 915–968.
Lefschetz, Solomon (1930). *Topology*. New York.
Lefschetz, Solomon (1942). *Algebraic Topology*. Providence, R.I.
Poincaré, Henri (1892). "Sur l'*Analysis situs*". *Comptes rendus de l'Académie des sciences* 115, pp. 633–636.
Poincaré, Henri (1895). "Analysis situs". *Journal de l'École polytechnique* 1, pp. 1–121.
Poincaré, Henri (2010). *Papers on Topology*. *Analysis Situs* and Its Five Supplements. Translated by John Stillwell. Providence, R.I.
Reidemeister, Kurt (1932). *Knotentheorie*. Berlin.
Tietze, Heinrich (1906). "Zur Analysis situs mehrdimensionaler Mannigfaltigkeiten". *Sitzungsberichte der Kaiserlichen Akademie der Wissenschaften (mathematisch-naturwissenschaftliche Klasse)* 115, pp. 841–846.
Veblen, Oswald and James W. Alexander (1912-13). "Manifolds of *N* Dimensions". *Annals of Mathematics* 14, pp. 163–178.

fundamental sequence

The concept of fundamental sequence, or Cauchy sequence, is found in Cauchy (1821, pp. 124f.), where no special term is used.

The name *fundamental sequence* was introduced by Cantor, who used the German word *Fundamentalreihe*, where *Reihe* is today to be translated as *series* rather than *sequence*, but at the time a clear terminological distinction between *series* and *sequences* was not yet common.

Cantor used the term in correspondence from 1882 (Meschkowski and Nilson 1991, p. 83); in print, he used it in Cantor (1883a, p. 23) and Cantor (1883b, p. 567):

---

[3] This usage was pointed out by Krömer (2013, pp. 928f.).

I postulate that after the choice of an arbitrarily small rational number ε a finite number of members of the set can be separated off, so that those remaining have pairwise a difference which in absolute terms is smaller than ε. Every such set (a_ν) which can also be characterized by the postulate $\lim_{\nu=\infty}(a_{\nu+\mu} - a_\nu) = 0$ (for arbitrary μ) I call a *fundamental series*.[1]

The corresponding English terms date from the early 20th century. Thus, Russell (1903, p. 195) used *fundamental series*, while Veblen (1905, p. 171) translated Cantor's term by *fundamental sequence*.

With Miller (n.d.), it can be said that the term *Cauchy sequence* indirectly goes back to Fréchet (1906, p. 23), who spoke of "a sequence of elements [...] satisfying the conditions of Cauchy".[2] Soon afterwards, *Cauchy sequence* appears as *Cauchysche Folge* in Kowalewski (1909, p. 219). It became more common from around 1920 on, with a use in English by Lamson (1920, p. 245) and a first textbook occurrence in Hahn (1921, p. 99).

Note. For the term as introduced by Cantor, see Ferreirós (2007, pp. 127–131). For Cauchy's use, see Lützen (2003, pp. 167–169). See also Stedall (2008, esp. 495-524).

References

Cantor, Georg (1883a). *Grundlagen einer allgemeinen Mannichfaltigkeitslehre*. Ein mathematisch-philosophischer Versuch in der Lehre des Unendlichen. Leipzig.

Cantor, Georg (1883b). "Ueber unendliche, lineare Punktmannichfaltigkeiten. 5". *Mathematische Annalen* 21, pp. 545–586.

Cantor, Georg (1976). "Foundations of a General Theory of Manifolds. Translated by Uwe Parpart". *The Campaigner* 9, pp. 69–96.

Cauchy, Augustin-Louis (1821). *Cours d'analyse de l'École Royale Polytechnique*. I.re partie. Analyse algébrique. Paris.

Ferreirós, José (2007). *Labyrinth of Thought*. A History of Set Theory and Its Role in Modern Mathematics. Second revised edition. Basel.

Fréchet, Maurice (1906). "Sur quelques points du calcul fonctionnel". *Rendiconti del Circolo Matematico di Palermo* 22, pp. 1–74.

Hahn, Hans (1921). *Theorie der reellen Funktionen*. Berlin.

Kowalewski, Gerhard (1909). *Die klassischen Probleme der Analysis des Unendlichen*. Leipzig.

Lamson, K.W. (1920). "A General Implicit Function Theorem with an Application to Problems of Relative Minima". *American Journal of Mathematics* 42, pp. 243–256.

Lützen, Jesper (2003). "The Foundation of Analysis in the 19th century". In: *A History of Analysis*. Ed. by Hans Niels Jahnke. Providence, R.I., pp. 155–195.

Meschkowski, Herbert and Wilfried Nilson, eds. (1991). *Georg Cantor*. Briefe. Berlin.

Miller, Jeff (n.d.). "Earliest Known Uses of Some of the Words of Mathematics". URL: `https://mathshistory.st-andrews.ac.uk/Miller/mathword/`.

Russell, Bertrand (1903). "Recent work on the philosophy of Leibniz". *Mind* 12, pp. 177–201.

Stedall, Jacqueline A. (2008). *Mathematics Emerging*. A Sourcebook 1540 – 1900. Oxford.

Veblen, Oswald (1905). "Definition in Terms of Order Alone in the Linear Continuum and in Well-Ordered Sets". *Transactions of the American Mathematical Society* 6, pp. 165–171.

[1] "ich fordre, daß nach Annahme einer beliebig kleinen rationalen Zahl ε eine endliche Anzahl von Gliedern der Menge abgeschieden werden kann, sodaß die übrig bleibenden paarweise einen Unterschied haben, der seiner absoluten Größe nach kleiner ist als ε. Jede derartige Menge (a_ν) welche auch durch die Forderung $\lim_{\nu=\infty}(a_{\nu+\mu} - a_\nu) = 0$ (bei beliebig gelassenem μ) charakterisiert werden kann, nenne ich eine *Fundamentalreihe*". Translation from Cantor (1976, p. 81), slightly modified. [2] "une suite d'éléments [...] satisfait aux conditions de Cauchy".

fundamental system

The term *fundamental system* is used in various senses. In the theory of differential equations, it refers to linearly independent solutions. In this sense, the German term *Fundamentalsystem* was suggested by Fuchs (1865, pp. 7f., 1866, p. 126).[1] The term was soon in general use; it was adopted by Hamburger (1873, p. 113), for example, who referred to Fuchs' paper just cited. The French term appears in Floquet (1883, p. 48), who also referred to Fuchs; the English term was used by Craig (1883, p. 140) in his discussion of Floquet's paper.

Schlesinger (1909, p. 136) suggested that Fuchs took over the term from number theory. However, this statement is not borne out by the facts. German *Fundamentalsystem* was first used by Kronecker (1871, p. 885) in connection with groups for a system of elements "$\theta_1, \theta_2, \theta_3, \ldots$, which has the property that the expression $\theta_1^{h_1}\theta_2^{h_2}\theta_3^{h_3} \ldots$ $(h_i = 1, 2, 3, \ldots n_i)$ [...] represents each element θ one and only once". In different contexts, the term was also used by Dedekind (1871, p. 478) and earlier by Smith (1861, p. 123), who defined *fundamental system* in connection with "complex units": "A system of units, capable of thus representing all units whatsoever, is called a fundamental system".

See also *basis* and and *generating* (algebra).

References

Craig, Thomas (1883). "Linear differential equations". *Science* 2, pp. 139–140.

Dedekind, Richard (1871). "Ueber die Composition der binären quadratischen Formen". In: *Vorlesungen über Zahlentheorie von P.G. Lejeune Dirichlet.* Herausgegeben und mit Zusätzen versehen von R. Dedekind. Zweite, umgearbeitete und vermehrte Auflage. Braunschweig, pp. 380–497.

Demidov, Sergej S. (1983). "On the History of the Theory of Linear Differential Equations". *Archive for History of Exact Sciences* 28, pp. 369–387.

Floquet, Gaston (1883). "Sur les équations différentielles linéaires à coefficients périodiques". *Annales scientifiques de l'École Normale Supérieure* 12, pp. 47–88.

Fuchs, Lazarus (1865). "Zur Theorie der linearen Differenzialgleichungen mit veränderlichen Coefficienten". In: *Jahresbericht über die städtische Gewerbeschule*. Berlin, pp. 1–38.

Fuchs, Lazarus (1866). "Zur Theorie der linearen Differentialgleichungen mit veränderlichen Coefficienten". *Journal für die reine und angewandte Mathematik* 66, pp. 121–160.

Hamburger, J.L. (1873). "Bemerkung über die Form der Integrale der linearen Differentialgleichungen mit veränderlichen Coefficienten". *Journal für die reine und angewandte Mathematik* 76, pp. 113–125.

Kronecker, Leopold (1871). "Auseinandersetzung einiger Eigenschaften der Klassenzahl idealer complexer Zahlen". *Monatsberichte der Königlich Preußischen Akademie der Wissenschaften zu Berlin* (1870), pp. 881–889.

Schlesinger, Ludwig (1909). "Bericht über die Entwickelung der Theorie der linearen Differentialgleichungen seit 1865". *Jahresbericht der Deutschen Mathematiker-Vereinigung* 18, pp. 133–266.

Smith, H.J.S. (1861). "Report on the Theory of Numbers. Part II". *Report of the British Association for the Advancement of Science* (1860), pp. 120–169.

fuzzy

In a paper on system theory, Zadeh (1962, p. 857) stated that "we need a radically different kind of mathematics, the mathematics of fuzzy or cloudy quantities which are not describable in terms of probability distributions". While this statement just

[1] The origin of the term has been pointed out by Demidov (1983, p. 376), for example.

contains an informal use of the term, a formal definition of a fuzzy set is found soon afterwards:

> a *fuzzy set* A in a space $Q = \{x\}$ is represented by a characteristic function f which is defined on Ω and takes values in the interval $[0, 1]$, with the value of f at x, $f(x)$, representing the "grade of membership" of x in A. Thus, if A is a set in the usual sense, $f(x)$ is 1 or 0 according as x belongs or does not belong to A. When A is a fuzzy set, then the nearer the value of $f(x)$ to 0, the more tenuous is the membership of x in A, with the "degree of belonging" increasing with increase in $f(x)$ (Bellman, Kalaba, and Zadeh 1964, pp. 1f.)[1]

In an interview, Zadeh later commented on his choice: "I coined the word 'fuzzy' because I felt it most accurately described what was going on in the theory. I could have chosen another term that would have been more 'respectable' with less pejorative connotations. I had thought about 'soft,' but that really didn't describe accurately what I had in mind. Nor did 'unsharp,' 'blurred,' or 'elastic.' In the end, I couldn't think of anything more accurate so I settled on 'fuzzy'" (Blair 1994, p. 46).

Note. For the history of the notion, see Seising (2007).

References

Bellman, Richard Ernest, Robert E. Kalaba, and Lotfi A. Zadeh (1964). *Abstraction and Pattern Classification*. Santa Monica: RAND Memorandum RM-4307-PR.

Bellman, Richard Ernest, Robert E. Kalaba, and Lotfi A. Zadeh (1966). "Abstraction and Pattern Classification". *Journal of Mathematical Analysis and Applications* 13, pp. 1–7.

Blair, Betty (1994). "Interview with Lotfi Zadeh. Creator of Fuzzy Logic". *Azerbaijan International* 2, pp. 46–47, 50.

Seising, Rudolf (2007). *The Fuzzification of Systems*. The Genesis of Fuzzy Set Theory and its Initial Applications – Developments up to the 1970s. Berlin and Heidelberg.

Zadeh, Lotfi A. (1962). "From Circuit Theory to System Theory". *Proceedings of the Institute of Radio Engineers* 50, pp. 856–865.

---

[1] The same definition is found in Bellman, Kalaba, and Zadeh (1966, p. 1).

G

Galois group* and *Galois field

In what appears to be the first use in print of English *Galois group*, Cole (1887, p. 58) recalled a definition given by Klein in lectures: "The Galois group of an equation is the group of permutations of its roots which possesses the two properties–1st, that all corresponding rational functions of the roots are rationally known, and 2d [sic] that it is the equation group for which this is true".[1] The precise origin of the terminology is not entirely clear but, according to Gray (2018, p. 252), "it seems that Klein and his students in the 1880s were the first to use the term Galois group, but they applied it specifically to the groups $PSL(2,p), p = 5, 7, 11$ that derive from a part of Galois's letter to Chevalier". Similarly, Brechenmacher (2016, p. 219) remarks that "[a]t the turn of the 1870s-1880s, the expression 'Galois groups' was used by Klein and his followers for designating the groups associated to the three modular equations".

Apparently, *Galois'sche Gruppe* was first used in print by Klein (1875, p. 207); not long afterwards, it was also used by Klein's students Dyck (1879, p. 5) and Gierster (1881, p. 319), for example.[2] The definition was changed over the course of time. As witnessed by van der Waerden (1972, p. 242), the Galois group was understood by many – he names Hilbert, Hecke, Emmy Noether, Krull, Hasse, Artin, and himself – as a group of automorphisms, "as we learned it from Dedekind and Weber".[3]

The term *Galois-field* was first used by Moore (1893, p. 75), and was a translation of Weber's (1893, pp. 526f.) *endlicher Körper* 'finite field', as Moore says (Moore 1896, p. 208). Several years later, Schottenfels (1899, p. 147) reported that "[t]he expression Galois Field is perhaps not yet in general use" but this soon changed; a prominent example is Dickson's monograph *Linear Groups with an Exposition of the Galois Field Theory* (Dickson 1901).

Note. For the history of Galois groups, see Kiernan (1971) and van der Waerden (1972); for Galois fields, see Gow (2013) and Brechenmacher (2016). Ehrhardt (2012) studies the history of Galois theory in detail.

---

[1] This usage was also pointed out by Miller (n.d.). [2] As noted by Hölder (1899, p. 485) and Brechenmacher (2016, p. 250), the related *Galois resolvent* (Italian *risolvente di Galois*) is due to Betti (1852, p. 86). [3] For references and discussion, see the literature cited in the note.

H. Becker, *A Historical Dictionary of Mathematical Terms*, Science Networks. Historical Studies 66, https://doi.org/10.1007/978-3-032-02873-0_7

References

Betti, Enrico (1852). "Sulla risoluzione delle equazioni algebriche". *Annali di scienze matematiche e fisiche* 3, pp. 49–115.
Brechenmacher, Frédéric (2016). "A history of Galois fields". *Khronos* 3, pp. 181–260.
Cole, Frank Nelson (1887). "Klein's Ikosaeder". *American Journal of Mathematics* 9, pp. 45–61.
Dickson, Leonard Eugene (1901). *Linear Groups with an Exposition of the Galois Field Theory*. Leipzig.
Dyck, Walther (1879). *Ueber regulär verzweigte Riemannsche Flächen und die durch sie definirten Irrationalitäten*. München.
Ehrhardt, Caroline (2012). *Itinéraire d'un texte mathématique*. Les réélaborations d'un mémoire d'Évariste Galois au XIXe siècle. Paris.
Gierster, Joseph (1881). "Die Untergruppen der Galois'schen Gruppe der Modulargleichungen für den Fall eines primzahligen Transformationsgrades". *Mathematische Annalen* 18, pp. 319–365.
Gow, Roderick (2013). "History of finite fields". In: *Handbook of Finite Fields*. Ed. by Gary L. Mullen and Daniel Panario. Boca Raton, pp. 3–11.
Gray, Jeremy (2018). *A History of Abstract Algebra*. From Algebraic Equations to Modern Algebra. Oxford.
Hölder, Otto (1899). "Galois'sche Theorie mit Anwendungen". *Encyklopädie der mathematischen Wissenschaften mit Einschluss ihrer Anwendungen* I B 3c, d, pp. 480–520.
Kiernan, B. Melvin (1971). "The Development of Galois Theory from Lagrange to Artin". *Archive for History of Exact Sciences* 8, pp. 40–154.
Klein, Felix (1875). "Ueber binäre Formen mit linearen Transformationen in sich selbst". *Mathematische Annalen* 9, pp. 183–208.
Miller, Jeff (n.d.). "Earliest Known Uses of Some of the Words of Mathematics". URL: `https://mathshistory.st-andrews.ac.uk/Miller/mathword/`.
Moore, E.H. (1893). "A doubly-infinite system of simple groups". *Bulletin of the New York Mathematical Society* 3, pp. 73–78.
Moore, E.H. (1896). "A doubly-infinite system of simple groups". In: *Mathematical Papers read at the International Mathematical Congress (Chicago, 1893)*. New York, pp. 208–243.
Schottenfels, Ida May (1899). "Two Non-Isomorphic Simple Groups of the Same Order 20,160". *Annals of Mathematics* 1, pp. 147–152.
van der Waerden, Bartel Leendert (1972). "Die Galois-Theorie von Heinrich Weber bis Emil Artin". *Archive for History of Exact Sciences* 9, pp. 240–248.
Weber, Heinrich (1893). "Die allgemeinen Grundlagen der Galois'schen Gleichungstheorie". *Mathematische Annalen* 43, pp. 521–549.

game

In his famous "On the Theory of Games of Strategy", von Neumann (1928b, p. 296) defined a game (German *Spiel*) in the following terms:[1]

> A game of strategy consists of a certain series of events each of which may have a finite number of distinct results. In some cases, the outcome depends on chance, i.e., the probabilities with which each of the possible results will occur are known, but nobody can influence them. All other events depend on the free decision of the player S_1, S_2, ..., S_n. In other words, for each of these events it is known which player, S_m, determines its outcome and what is his state of information with respect to the results of other ("earlier") events at the time when he

[1] The definition is followed by a more rigorous one in terms of axioms.

> makes his decision. Eventually, after the outcome of all events is known, one can calculate according to a fixed rule what payments the players S_1, S_2, ..., S_n must make to each other.[1]

von Neumann (1928a) also presented some results in a brief note written in French, the title of which is "On the theory of games" ("Sur la théorie des jeux"). von Neumann's ideas and the terminology mainly spread through the seminal book *Theory of Games and Economic Behavior* (von Neumann and Morgenstern 1944).

There have been earlier works in which games were approached from a mathematical perspective. Two commonly referred to papers are Zermelo (1913), entitled "On an application of set theory to the theory of the game of chess",[2] and Borel (1921); Kalmár (1928-29) wrote "On the theory of abstract games".

The publication of von Neumann and Morgenstern (1944) sparked further research, for example the article by Nash on "Non-Cooperative Games", in which he defined:

> For us an *n-person game* will be a set of *n players*, or *positions*, each with an associated finite set of *pure strategies*; and corresponding to each player, i, a *payoff function*, p_i, which maps the set of all n-tuples of pure strategies into the real numbers. When we use the term *n-tuple* we shall always mean a set of n items, with each item associated with a different player (Nash 1951, p. 286).

Note. For the history of the concept, see, e. g., M. A. Dimand and R. W. Dimand (1996).

References

Borel, Émile (1921). "La théorie du jeu et les équations intégrales à noyau symétrique". *Comptes rendus de l'Académie des sciences* 173, pp. 1304–1308.

Dimand, Mary Ann and Robert W. Dimand (1996). *The History of Game Theory, Volume 1.* From the beginnings to 1945. London.

Ebbinghaus, Heinz-Dieter [In Cooperation with Volker Peckhaus] (2007). *Ernst Zermelo.* An Approach to His Life and Work. Berlin and Heidelberg.

Kalmár, László (1928-29). "Zur Theorie der abstrakten Spiele". *Acta scientiarum mathematicarum* 4, pp. 65–85.

Nash, John (1951). "Non-Cooperative Games". *Annals of Mathematics* 54, pp. 286–295.

von Neumann, John (1928a). "Sur la théorie des jeux". *Comptes rendus de l'Académie des sciences* 186, pp. 1689–1691.

von Neumann, John (1928b). "Zur Theorie der Gesellschaftsspiele". *Mathematische Annalen* 100, pp. 295–320.

von Neumann, John and Oskar Morgenstern (1944). *Theory of Games and Economic Behavior.* Princeton.

---

[1] "Ein Gesellschaftsspiel besteht aus einer bestimmten Reihe von Ereignissen, deren jedes auf endlich viele verschiedene Arten ausfallen kann. Bei gewissen unter diesen Ereignissen hängt der Ausfall vom Zufall ab, d. h.: es ist bekannt mit welchen Wahrscheinlichkeiten die einzelnen Resultate eintreten werden: aber niemand vermag sie zu beeinflussen. Die übrigen Ereignisse aber hängen vom Willen der einzelnen Spieler S_1, S_2, ..., S_n ab. D. h.: es ist bei jedem dieser Ereignisse bekannt, welcher Spieler S_m seinen Ausfall bestimmt, und von den Resultaten welcher anderer ('früherer') Ereignisse er im Moment seiner Entscheidung bereits Kenntnis hat. Nachdem der Ausfall aller Ereignisse bereits bekannt ist, kann nach einer festen Regel berechnet werden, welche Zahlungen die Spieler S_1, S_2, ..., S_n aneinander zu leisten haben". Translation from von Neumann (1959, p. 14). [2] Ebbinghaus (2007, p. 132) credits Zermelo "for having written the first paper that mirrors the spirit of modern game theory". For Zermelo's paper, see also Schwalbe and Walker (2001).

von Neumann, John (1959). “On the Theory of Games of Strategy [A translation by Mrs. Sonya Bargmann]”. In: *Contributions to the Theory of Games*. Volume IV. Ed. by A.W. Tucker and R.D. Luce. Princeton, pp. 13–42.

Schwalbe, Ulrich and Paul Walker (2001). “Zermelo and the Early History of Game Theory”. *Games and Economic Behavior* 34, pp. 123–137.

Zermelo, Ernst (1913). “Über eine Anwendung der Mengenlehre auf die Theorie des Schachspiels”. In: *Proceedings of the Fifth International Congress of Mathematicians*. Vol. II. Ed. by E.W. Hobson and A.E.H. Love. Cambridge, pp. 501–504.

gamma function **and** ***beta function***

The gamma function was already considered by Euler in a letter to Goldbach from 1729,[1] but the name is due to Legendre, who introduced the symbol Γ in 1809:

> it suffices to consider the integral $\int dx(\log \frac{1}{x})^{a-1}$, in which a is positive and smaller than unity. This quantity being simply function of a, we shall designate it by $\Gamma(a)$, and we set
>
> $$\Gamma(a) = \int dx(\log \frac{1}{x})^{a-1}$$
>
> (Legendre 1810, p. 477).[2]

The same statement is also found in Legendre (1811, p. 277).[3] Explicitly, *fonction gamma* was used by Poisson (1816, p. 397) and in an anonymous review of Legendre's *Exercices de calcul intégral* ([anon.] 1817, p. 5). According to the OED, the English term is first found in Peacock (1834, p. 222).

Legendre is not known to have commented on why he chose the symbol Γ. Indeed, this is why Lacroix (1819, p. 481) critically remarked that it was a “purely arbitrary convention”.[4]

The closely related beta function was also already considered by Euler in 1729 (Euler 1738), but again the name dates from the 19th century. Binet (1839, p. 131) suggested the following notation:

> The results I will discuss in this memoir are principally founded on the properties of two functions intimately related to one another, and which the mathematicians have often occupied themselves with; these are the definite integrals
>
> $$\int_0^1 x^{p-1}dx(1-x)^{q-1} = \int_0^\infty \frac{z^{p-1}dz}{(1+z)^{p+q}} = \int_0^\infty \frac{z^{q-1}dz}{(1+z)^{p+q}}$$
>
> and
>
> $$\int_0^1 (\log \frac{1}{x})^{p-1}dx = \int_0^\infty z^{p-1}de^{-z},$$

[1] For the letter as well as an English translation and comments, see Euler (2015, pp. 99–102, 583–588); in print, Euler discussed the gamma function in a paper read in 1729 (Euler 1738). [2] “il suffira de considérer l'intégrale $\int dx(\log \frac{1}{x})^{a-1}$, dans laquelle a est positif et plus petit que l'unité. Cette quantité étant simplement fonction de a, nous la désignerons par $\Gamma(a)$, et nous ferons $\Gamma(a) = \int dx(\log \frac{1}{x})^{a-1}$”. [3] In the same text, Legendre (1811, p. 221) also introduces the term *Eulerian integrals* (*intégrales Euleriennes*) *of the first and second kind*; those of the first kind correspond to the beta function, those of the second kind to the gamma function. [4] “une convention purement due au hasard”.

where e represents the hyperbolic base. I shall designate the first of these functions by $\mathrm{B}(p,q)$, and for the second I shall adopt the notation $\Gamma(p)$ suggested by Legendre.[1]

Just like Legendre, Binet never seems to have commented on the choice of the symbol.

Note. For the history of the gamma function, see Davis (1959), Gronau (2003), and Remmert (1998, ch. 2). The origin of the symbols was pointed out by Cajori (1929, pp. 271f.).

References

[anon.] (1817). "Exercices de calcul intégral sur divers ordres de transcendantes et sur les quadratures, par A.M. Le Gendre [review]". *Bibliotheque universelle des sciences, belles lettres et arts* 60, pp. 3–11.

Binet, Jacques (1839). "Sur les intégrales définies eulériennes et sur leur application à la théorie des suites ainsi qu'à l'évaluation des fonctions des grands nombres". *Journal de l'École polytechnique* 16, cahier 27, pp. 123–343.

Cajori, Florian (1929). *A History of Mathematical Notations*. Volume II. Notations mainly in higher mathematics. Chicago.

Davis, Philip J. (1959). "Leonhard Euler's Integral: A Historical Profile of the Gamma Function". *The American Mathematical Monthly* 66, pp. 849–869.

Euler, Leonhard (1738). "De progressionibus transcendentibus seu quarum termini generales algebraice dari nequeunt". *Commentarii academiae scientiarum Petropolitanae* 5, pp. 36–57.

Euler, Leonhard (2015). *Correspondence of Leonhard Euler with Christian Goldbach*. Ed. by Franz Lemmermeyer and Martin Mattmüller. Basel.

Gronau, Detlef (2003). "Why is the gamma function so as it is?" *Teaching Mathematics and Computer Science* 1, pp. 43–53.

Lacroix, Sylvestre François (1819). *Traité du calcul différentiel et du calcul intégral*. Seconde édition, revue et augmentée. Tome troisième. Paris.

Legendre, Adrien-Marie (1810). "Recherches sur diverses sortes d'intégrales définies". *Mémoires de la classe des sciences mathématiques et physiques de l'Institut de France* (1809), pp. 416–509.

Legendre, Adrien-Marie (1811). *Exercices de calcul intégral sur divers ordres de transcendantes et sur les quadratures*. Paris.

Peacock, George (1834). "Report on the Recent Progress and Present State of certain Branches of Analysis". *Report of the British Association for the Advancement of Science* (1833), pp. 185–352.

Poisson, Siméon Denis (1816). "Rapport fait à l'Institut, par M. Poisson, sur un Mémoire de M. Hachette". *Correspondance sur l'École impériale polytechnique* 3, pp. 395–401.

Remmert, Reinhold (1998). *Classical Topics in Complex Function Theory*. Translated by Leslie Kay. New York.

generating

In geometry, certain moving objects are commonly regarded as "generating" another objects. For example, a line can be thought of as generated by a point in motion. Greek geometers such as Archimedes used the verb *γεννάω* (related to *generate*) to express this (Mugler 1958-59, p. 100). Similarly, *γίγνεσθαι*, which often serves as the passive of *γεννάω*, was used when a geometric figure comes about by geometric operations or as a solution to a problem (Mugler 1958-59, p. 103). In this sense, uses of *generate* therefore already go back to antiquity.

---

1. "Les résultats que je vais exposer dans ce Mémoire sont principalement fondés sur les propriétés de deux fonctions intimement liées l'une à l'autre, et qui ont souvent occupé les géomètres: ce sont les intégrales définies $\int_0^1 x^{p-1}dx(1-x)^{q-1} = \int_0^\infty \frac{z^{p-1}dz}{(1+z)^{p+q}} = \int_0^\infty \frac{z^{q-1}dz}{(1+z)^{p+q}}$ et $\int_0^1 (\log \frac{1}{x})^{p-1}dx = \int_0^\infty z^{p-1}de^{-z}$, où e représente la base hyperbolique. Je désignerai la première de ces fonctions par $\mathrm{B}(p,q)$; et pour la seconde j'adopterai la notation $\Gamma(p)$ proposée par M. Legendre".

In Latin texts, *generare* and related forms are originally rare and not regularly used until the 13th century.[1] Guillaumin (2020, p. 138) just cites Boethius, who, in his *Institutio arithmetica* (I, 17), used *genitrix* in the statement that unity "constitutes the primary unit of all numbers which are in the natural order and is rightly recognized as the generator (*genitrix*) of the total extended plurality of number".[2] Boethius also used *generare* and other words in a chapter on the generation of perfect numbers (I, 20).

Geometrical examples are found at a later stage. Albertus Magnus, for instance, used *generare* when he stated that a surface, by its motion, generates a solid (Tummers 2014, p. 2).

In the following entries, some selected examples of "generating"-terminology will be discussed. These are *generating circle*, *generating function*, and *generating* (*generator*, *system of generators*) as used in connection with algebraic structures.

References

Boethius, Anicius Manlius Severinus (1867). *De institutione arithmetica libri duo, De institutione musica libri quinque*. Accedit Geometria quae fertur Boetii. Ed. by Gottfried Friedlein. Leipzig.

Guillaumin, Jean-Yves (2020). *Dictionnaire de la terminologie latine ancienne de l'arithmétique et de la géométrie*. Paris.

Masi, Michael (2006). *Boethian Number Theory*. Translation of the *De Institutione Arithmetica*. Amsterdam and New York.

Mugler, Charles (1958-59). *Dictionnaire historique de la terminologie géométrique des Grecs*. Paris.

Tummers, Paul M.J.E. (2014). *Alberti Magni Ordinis Fratrum Praedicatorum Super Euclidem*. Münster.

generating **(algebra)**

In the second half of the 19th century, the term gradually began to be used in connection with algebraic structures. A definition was given by Gierster (1881, p. 339), according to whom *generating substitutions* (*erzeugende Substitutionen*) are characterized by the property that "through their repeated combination all substitutions of the group can be derived".[3]

Another definition was given by Dyck (1882, p. 5): "Let $A_1, A_2, A_3, \ldots, A_m$ be m operations of any kind which can be applied to an object J (identity) which, subsequently, will always be denoted by 1. Then these A_i may always be considered as the generating operations of a group which will be obtained by applying all operations on our object J in iteration and combination".[4] Dyck (1883, p. 77) also used the term *system of generating substitutions* (*System erzeugender Substitutionen*).

The terminology was soon in common use and translated into English and other languages; Bolza (1890-91, p. 77), for example, defined what it means for a group to

---

[1] The OLD, LS, and the TLL do not list any mathematical meanings. [2] "Quare constat primam esse unitatem cunctorum, qui sunt in naturali dispositione, numerorum et eam rite totius quamvis prolixae genitricem pluralitatis agnosci". Translation from Masi (2006, p. 79). [3] "Erzeugende Substitutionen sind solche, durch deren wiederholte Combination alle Substitutionen der Gruppe abgleitet werden können". [4] "Seien A_1, A_2, A_3, ..., A_m irgendwelche m Operationen, welche auf ein Object J (Identität), das wir in der Folge stets als 1 bezeichnen, angewandt werden können, so lassen sich diese A_i stets als die ‚*erzeugenden*' Operationen einer Gruppe auffassen, die wir erhalten, wenn wir auf unser Object J alle Operationen in Iteration und Combination anwenden". Translation from Chandler and Magnus (1982, pp. 5f.).

be *generated* by substitutions. The terminology is also found in Burnside's *Theory of Groups of Finite Order* (Burnside 1897, p. 21). de Séguier (1904, p. 7) popularized the term *générateur* and *système de générateurs* in France.

See also *basis*.

Note. The history of the concept is discussed in Wußing (1984, passim); see esp. p. 89 on Cauchy and pp. 137f. on Jordan, who both use *derive* instead of *generate*.

References

Bolza, Oskar (1890-91). "On the Theory of Substitution-Groups and its Applications to Algebraic Equations". *American Journal of Mathematics* 13, pp. 59–144.

Burnside, William Snow (1897). *Theory of Groups of Finite Order*. Cambridge.

Chandler, Bruce and Wilhelm Magnus (1982). *The History of Combinatorial Group Theory*. A Case Study in the History of Ideas. New York.

Dyck, Walther (1882). "Gruppentheoretische Studien". *Mathematische Annalen* 20, pp. 1–44.

Dyck, Walther (1883). "Gruppentheoretische Studien. II. Ueber die Zusammensetzung einer Gruppe discreter Operationen, über ihre Primitivität und Transitivität". *Mathematische Annalen* 22, pp. 70–108.

Gierster, Joseph (1881). "Die Untergruppen der Galois'schen Gruppe der Modulargleichungen für den Fall eines primzahligen Transformationsgrades". *Mathematische Annalen* 18, pp. 319–365.

de Séguier, Jean Armand (1904). *Théorie des groupes finis*. Éléments de la théorie des groupes abstraits. Paris.

Wußing, Hans (1984). *The Genesis of the Abstract Group Concept*. A Contribution to the History of the Origin of Abstract Group Theory. Tr. Abe Shenitzer. Cambridge, Mass.

generating circle

The generating circle of a cycloid was named by Torricelli (1644, p. 85) the *circulus genitor*, a term that was later also used by Wallis (1659, p. 1), for example. French *cercle générateur* was already used by de Roberval (1693, p. 220) in his *Traité des indivisibles*, written around 1630.

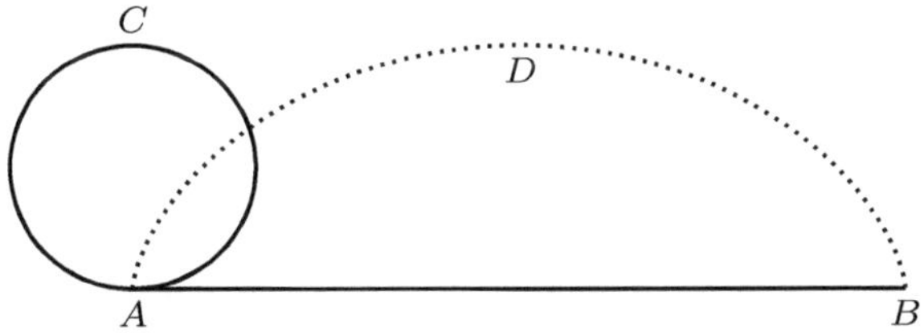

Figure 47: Generation of the cycloid. From Torricelli (1644, p. 85).

The English term is, according to the OED, first attested in [anon.] (1671, p. 2057).

References

[anon.] (1671). "An Account of Two Books". *Philosophical Transactions of the Royal Society of London* 5, pp. 2052–2062.

de Roberval, Gilles Personne (1693). *Traité des indivisibles*. In: *Divers ouvrages de mathématique et de physique*. Paris, pp. 190–245.

Torricelli, Evangelista (1644). *Appendix de dimensione cycloidis*. In: *Opera geometrica*. Firenze, pp. 85–92.

Wallis, John (1659). *Tractatus Duo*. Prior, De Cycloide et corporibus inde genitis. Posterior, epistolaris in qua agitur De Cissoide, et Corporibus inde genitis: et De Curvarum, tum Linearum *Εὐθύνσει*, tum Superficierum *Πλατυσμῷ*. Oxford.

generating function

While the use of generating functions has been traced back to de Moivre (1730, pp. 196f.),[1] the term was introduced as French *fonction génératrice* by Laplace in a paper read in 1779:

> Let y_x be any function whatever of x; if we form the infinite series $y_0 + y_1 t + y_2 t^2 + y_3 t^3 + \cdots + y_x t^x + y_{x+1} t^{x+1} + \cdots + y_\infty t^\infty$, and if we name u the sum of this series, or, which amounts to the same thing, the function of which the expansion forms this series, this function will be that which I call the *generating function* of the variable y_x (Laplace 1782, p. 207).[2]

The term quickly found acceptance and was used, for example, by Lacroix (1800, p. 301); the English term is first found, acc. to the OED, in a review of that book ([anon.] 1801, p. 498).

Note. For the history of generating functions, see Seal (1949) and Galuzzi (1995).

References

[anon.] (1801). "La Croix on *Differences and Series*". *The Monthly Review* 36, pp. 498–501.
Galuzzi, Massimo (1995). "Lagrange's essay 'Recherches sur la manière de former des tables des planètes d'après les seules observations'". *Revue d'histoire des mathématiques* 1, pp. 201–233.
Lacroix, Sylvestre François (1800). *Traité des différences et des séries*. Faisant suite du Calcul différentiel et du Calcul intégral. Paris.
Laplace, Pierre-Simon (1782). "Mémoire sur les suites". *Mémoires de l'Académie royale des sciences* (1779), pp. 207–309.
Laplace, Pierre-Simon (2010). "Mémoire sur les suites". Translated by Richard J. Pulskamp. URL: https://www.probabilityandfinance.com/pulskamp/Laplace/ (retrieved July 19, 2024).
de Moivre, Abraham (1730). *Miscellanea analytica de seriebus et quadraturis*. London.
Seal, H.L. (1949). "The Historical Development of the Use of Generating Functions in Probability Theory". *Bulletin de l'Association des actuaires suisses* 49, pp. 209–228.
Stigler, Stephen M. (1986). *The History of Statistics*. The Measurement of Uncertainty before 1900. Cambridge, Mass. and London.

generic point

In some recollections published in the early 1970s, van der Waerden (1971, p. 172) recalled a question he asked in the 1920s which he needed to answer to put algebraic geometry on solid foundations:

> What do the Italian geometers mean by a generic point on a variety (punto generico) [...]? Obviously, a generic point is supposed not to have certain undesirable properties. For instance, a generic plane is not tangent to a given curve, it does not pass through a given point, etc. I asked: is it possible to find, on a given variety U, a point ξ not having any special properties except those which hold for all points of U?

---

[1] See, e. g., Seal (1949) and Stigler (1986, pp. 91f.). [2] "Soit y_x, une function quelconque de x; si l'on forme la suite infinie $y_0 + y_1 t + y_2 t^2 + y_3 t^3 + \cdots + y_x t^x + y_{x+1} t^{x+1} + \cdots + y_\infty t^\infty$ & que l'on nomme u la somme de cette suite, ou ce qui revient au même, la fonction dont le développement forme cette suite; cette function sera ce que je nomme *function génératrice* de la variable y_x". Translation based on Laplace (2010, pp. 3f.).

The solution to this question was presented in van der Waerden (1926, p. 197):

> If M is irreducible, so that $\mathfrak{m}$ is prime, then every generic zero of the ideal $\mathfrak{m}$ is called a *generic point of the variety* M. This terminology agrees with the meaning that the words generic and special have in geometry. Indeed, by generic point of a variety, one usually means, even if this is not always clearly explained, a point which satisfies no special equation, except those equations which are met at every point. For a specific point of M, this is of course impossible to fulfil, and so one has to consider points that depend on sufficiently many parameters, that is, points that lie in a space $C_n(\Omega)$ where Ω is a transcendental extension of $\boldsymbol{P}$. But requiring of a point of $C_n(\Omega)$ that it be a zero of all those and only those polynomials of $\boldsymbol{P}[x_1, \ldots, x_n]$ that vanish at all points of the variety M yields precisely our definition of a generic point of the variety M.[1]

van der Waerden remarks that Italian geometers did not define the term. However, an attempt at a characterization was made by Enriques and Chisini (1915, p. 139):

> The notion of a *generic* "point" or "element" of a variety, that is, the distinction between properties that pertain *in general* to the points of a variety and properties that only pertain to *exceptional* points, now takes on a precise meaning for all algebraic varieties.
>
> A property is said to pertain in general to the points of a variety V_n, of dimension n, if the points of V_n not satisfying it form – inside V_n – a variety of less than n dimensions.[2]

The term *generic*, as opposed to van der Waerden's *general*, spread through Weil (1946, p. 68). When exactly *punto generico* came up is difficult to establish; it was in common use in the 1890s and appeared, for example, in Castelnuovo (1891, p. 863).

Note. For the history of the concept, see Schappacher (2010, pp. 3276–3286) and Schappacher (2007).

References

Castelnuovo, Guido (1891). "Richerche di geometria della retta nello spazio a quattro dimensioni". *Atti del Reale Istituto veneto di scienze, lettere ed arti* 49, pp. 855–901.

[1] "Ist M irreduzibel, also $\mathfrak{m}$ prim, so heißt jede allgemeine Nullstelle des Ideals $\mathfrak{m}$ *allgemeiner Punkt der Mannigfaltigkeit* M. Diese Bezeichnung ist in Übereinstimmung mit der in der Geometrie geläufigen Bedeutung der Wörter allgemein und speziell. Man versteht doch meistens, wenn es auch nicht immer deutlich gesagt wird, unter einem allgemeinen Punkt einer Mannigfaltigkeit einen solchen Punkt, der keiner einzigen speziellen Gleichung genügt, außer denjenigen Gleichungen, die in allen Punkten erfüllt sind. Diese Forderung kann natürlich ein bestimmter Punkt von M niemals erfüllen, und so ist man genötigt, Punkte zu betrachten, die von hinreichend vielen Parametern abhängen, d. h. in einem Raum $C_n(\Omega)$ liegen, wo Ω eine transzendente Erweiterung von $\boldsymbol{P}$ ist. Fordert man aber von einem Punkt von $C_n(\Omega)$, daß er Nullstelle ist für alle die und nur die Polynome von $\boldsymbol{P}[x_1, \ldots, x_n]$, die in allen Punkten der Mannigfaltigkeit M verschwinden, so kommt man gerade auf unsere Definition eines allgemeinen Punktes der Mannigfaltigkeit M". Translation from Schappacher (2007, p. 253). van der Waerden (1971, pp. 172f.) remarked that he found this definition by studying some papers by Emmy Noether.

[2] "La nozione di 'punto' o 'elemento' *generico* di una varietà, cioè la distinzione fra proprietà spettanti *in generale* ai punti d'una varietà e proprietà che spettano solo a punti *eccezionali*, acquista ora un significato preciso per tutte le varietà algebriche. Si dice che una proprietà spetta in generale ai punti d'una varietà V_n, ad n dimensioni, se i punti di V_n per cui essa non è soddisfatta formano— entro V_n—una varietà, a meno di n dimensioni". Translation from Schappacher (2007, p. 254). Concerning the differences between this characterization and van der Waerden's, Schappacher (ibid.) remarks: "Contrary to van der Waerden's notion of generic points, Enriques's 'points' are always points with complex coordinates, and genericity has to do with negligible exceptional sets, not with introducing parameters".

Enriques, Federigo and Oscar Chisini (1915). *Lezioni sulla teoria geometrica delle equazioni e delle funzioni algebriche*. Volume I. Bologna.
Schappacher, Norbert (2007). "A Historical Sketch of B.L. van der Waerden's work in algebraic geometry: 1926–1946". In: *Episodes in the History of Modern Algebra (1800–1950)*. Ed. by Jeremy Gray and Karen Hunger Parshall. Providence, R.I., pp. 245–283.
Schappacher, Norbert (2010). "Rewriting Points". In: *Proceedings of the International Congress of Mathematicians*. Hyderabad 2010. Ed. by Rajendra Bhatia. Vol. 4. New Delhi, pp. 3258–3291.
van der Waerden, Bartel Leendert (1926). "Zur Nullstellentheorie der Polynomideale". *Mathematische Annalen* 96, pp. 183–208.
van der Waerden, Bartel Leendert (1971). "The Foundation of Algebraic Geometry from Severi to André Weil". *Archive for History of Exact Sciences* 7, pp. 171–180.
Weil, André (1946). *Foundations of Algebraic Geometry*. New York.

genus

In his *Geometria Organica*, Maclaurin (1720, p. 137) proved that the maximum number of double points an irreducible curve of degree n can have is $\frac{(n-1)(n-2)}{2}$. For the difference between this number and the actual number of double points, he did not use a term, though.

It took almost 150 years for the notion to be given a name, until, in 1865, two mathematicians suggested two different terms. Cayley (1865, p. 2) called this number the *deficiency* of the curve: "[...] it will be convenient to introduce the term ‚Deficiency', viz., a curve of the order n with $\frac{1}{2}(n-1)(n-2) - D$ dps [double points], is said to have a deficiency $= D$".

In the same year, Clebsch (1865, p. 43), under the influence of Riemann (1857),[1] also defined the notion and suggested the German term *Geschlecht* 'genus': "Instead of classifying the algebraic curves by their orders and, in every class obtained in this way, by the numbers of double points and of cusps which appear on them, one can subdivide them into *Genera* by the number p; in the first Genus all those for which $p = 0$; in the second Genus all those for which $p = 1$, etc.".[2]

As Popescu-Pampu (2016, p. 60) points out, the genera are here defined as classes of curves rather than as numbers and that it is "a little strange that the n-th genus corresponds to $p = n + 1$". Clebsch (1868, p. 1238) soon changed the definition: "I suggested to call the *genus* of a curve the number $p = \frac{1}{2}(n-1)(n-2) - d$ (Mr. Cayley's *deficiency*) where n is the order of the curve and d the number of its double points and cusps. The genera of two algebraic curves have to be the same in order to be able to establish a one-to-one correspondence between their points".[3] Since Clebsch' note was written in French, he used the word *genre*, which corresponds in meaning to German *Geschlecht*.

In the same paper, Clebsch (1868, p. 1239) also used the term with respect to surfaces:

---

[1] In Riemanns's paper, the notion is found but not the term. [2] "Statt die algebraischen Curven nach Ordnungen einzutheilen, und in diesen Unterabtheilungen zu machen nach der Anzahl der Doppel- und Rückkehrpunkte, welche dieselben aufweisen, kann man dieselben in *Geschlechter* eintheilen nach der Zahl p; zu dem ersten Geschlecht also alle diejenigen für welche $p = 0$, zum zweiten diejenigen, für welche $p = 1$, u.s.w.". Translation from Popescu-Pampu (2016, p. 60).
[3] "J'ai proposé de nommer *genre* d'une courbe le nombre $p = \frac{1}{2}(n-1)(n-2) - d$ (*deficiency* de M. Cayley), n étant l'ordre de la courbe, d le nombre des ses points doubles ou de rebroussement". Translation from Popescu-Pampu (2016, p. 60).

I managed to prove the following theorem:

"Let $f = 0$ and $\varphi = 0$ define two surfaces of orders m and n respectively. Assume that one may associate to each point of $f = 0$ a single point of $\varphi = 0$ and conversely, in such a way that those surfaces are transformable one into the other in an algebraic and rational way. Assume, for more simplicity, that those surfaces have only regular singularities [...]. Let then p denote the number of arbitrary coefficients of a surface of order $n - 4$ which passes through the double curves and the cuspidal curves lying on $f = 0$, and let p' be the corresponding number relative to $\varphi = 0$. One will always have $p' = p$".

This theorem allows us to classify those surfaces with respect to their *genus* p.[1]

This concept was immediately adopted by Noether (1869, p. 299), who referred to Clebsch for the term.[2] Soon afterwards, it became clear that there are actually two notions of genus for surfaces, and these were given two names:

Let F be an algebraic surface of order n in ordinary space; assume that F does not have other singularities than a double curve of order d (≥ 0) and genus π, and a certain finite number t (≥ 0) of triple points of the surface which are also triple points of the double curve considered before. Next, consider the surfaces of order $n - 4$ which pass simply through the double curve (and therefore twice, in general, through its triple points), surfaces which we will call *adjoints* to F; *the number of such surfaces which are linearly distinct is* (according to the proof of Mr. Nöther [sic]) an invariant character (relative to the birational transformations) of the surface F; one calls it the *geometric genus* [*genre géométrique*] of F, and one denotes it by p_g. Now, if we notice that a surface of order m *big enough* must satisfy

$$md - 2t - \pi + 1$$

conditions in order to contain the double curve of F, we are led (with Cayley), to form the expression

$$p_n = \frac{(n-1)(n-3)(n-3)}{6} - (n-4)d + 2t + \pi - 1,$$

whose value is exactly p_g, if $m = n - 4$ is so big that one may apply the previous formula. Be the last condition verified or not, the *expression for* p_n *always has the invariance property*; this was shown by Zeuthen and Nöther. This expression is perfectly analogous to the one considered above when speaking about plane curves; the name of *numerical genus* [*genre numérique*] of the surface F was given to it (Castelnuovo and Enriques 1897, pp. 248f.).[3]

This distinction is earlier found in Castelnuovo (1891, p. 128), who used Italian *genere numerico* and *genere geometrico*; Noether (1883, p. 399) had already used *numerisches Flächengeschlecht*. It became necessary after Cayley (1871) showed that there are classes of surfaces for which the arithmetic genus is negative.[4] With Castelnuovo and Enriques (1901, p. 13), it became more common to say *arithmetical genus* instead of *numerical genus*. Further generalizations are due to Todd (1938) and Hirzebruch (1953).[5]

---

[1] Translation from Popescu-Pampu (2016, pp. 82f.). For discussion, see Gray (2013, pp. 500f.), who also remarks that the "geometric genus is the dimension of the space of holomorphic 2-forms or double integrals on the surface". [2] Noether (1869, p. 302) also used the compound *Flächengeschlecht*. [3] Translation from Popescu-Pampu (2016, pp. 83f.). [4] See Gray (2013, pp. 500f.). [5] Hirzebruch (1953, p. 951) also coined *Todd genus*. See also Hirzebruch (1956, pp. 1–4) for a discussion of the different definitions of the arithmetical genus known at that time. Hirzebruch's generalized definition is discussed by Hirzebruch and Kreck (2009), Popescu-Pampu (2016, pp. 159-161), and Dieudonné (2009, pp. 580–595).

The term has also been applied to Riemann surfaces; according to Lê (2020, p. 98), who discusses this aspect in detail, this seems to have been first done, in passing, by Klein (1879, p. 127).

See also *handle*.

Note. For the history of the concept and terminology, see Popescu-Pampu (2016) as well as Lê (2020) and Hirzebruch and Kreck (2009).

References

Castelnuovo, Guido (1891). "Intorno alla geometria sopra una superficie algebrica. Nota I". *Reale Istituto lombardo di scienze e lettere. Rendiconti* 24, pp. 127–137.

Castelnuovo, Guido and Federigo Enriques (1897). "Sur quelques récents résultats dans la théorie des surfaces algébriques". *Mathematische Annalen* 48, pp. 241–316.

Castelnuovo, Guido and Federigo Enriques (1901). "Sopra alcune questioni fondamentali nella teoria delle superficie algebriche". *Annali di Matematica Pura e Applicata* 6, pp. 165–225.

Cayley, Arthur (1865). "On the Transformation of Plane Curves". *Proceedings of the London Mathematical Society* 1, pp. 1–11.

Cayley, Arthur (1871). "On the deficiency of certain surfaces". *Mathematische Annalen* 3, pp. 526–529.

Clebsch, Alfred (1865). "Ueber diejenigen ebenen Curven, deren Coordinaten rationale Functionen eines Parameters sind". *Journal für die reine und angewandte Mathematik* 64, pp. 43–65.

Clebsch, Alfred (1868). "Sur les surfaces algébriques". *Comptes rendus de l'Académie des sciences* 67, pp. 1238–1239.

Dieudonné, Jean (2009). *A History of Algebraic and Differential Topology, 1900–1960*. Boston.

Gray, Jeremy (2013). *Henri Poincaré*. A Scientific Biography. Princeton.

Hirzebruch, Friedrich (1953). "On Steenrod's reduced powers, the index of inertia, and the Todd genus". *Proceedings of the National Academy of Sciences* 39, pp. 951–956.

Hirzebruch, Friedrich (1956). *Neue topologische Methoden in der algebraischen Geometrie*. Berlin and Heidelberg.

Hirzebruch, Friedrich and Matthias Kreck (2009). "On the Concept of Genus in Topology and Complex Analysis". *Notices of the American Mathematical Society* 56, pp. 713–719.

Klein, Felix (1879). "Ueber die Transformation siebenter Ordnung der elliptischen Functionen". *Mathematische Annalen* 14, pp. 428–471.

Lê, François (2020). "'Are the *genre* and the *Geschlecht* one and the same number?' An inquiry into Alfred Clebsch's *Geschlecht*". *Historia Mathematica* 53, pp. 71–107.

Maclaurin, Colin (1720). *Geometria Organica*. London.

Noether, Max (1869). "Zur Theorie der algebraischen Functionen mehrerer complexer Variablen". *Nachrichten von der Königl. Gesellschaft der Wissenschaften und der Georg-Augusts-Universität zu Göttingen* (1869), pp. 298–306.

Noether, Max (1883). "Ueber eine Fläche 6ter Ordnung vom Flächengeschlecht -1". *Mathematische Annalen* 21, pp. 399–410.

Popescu-Pampu, Patrick (2016). *What is the Genus?* Cham.

Riemann, Bernhard (1857). "Theorie der Abel'schen Functionen". *Journal für die reine und angewandte Mathematik* 54, pp. 115–155.

Todd, J.A. (1938). "The Arithmetical Invariants of Algebraic Loci". *Proceedings of the London Mathematical Society* 43, pp. 190–225.

geodesic

The Greek noun *γεωδαισία* literally means 'division of land' and was used by Aristotle and others in opposition to *γεωμετρία*: "geometry is to differ from geodesy only in this, that the latter is concerned with things we sense but the former with non-sensible things" (*Metaphysics* 997b).[1] As Heath (1949, pp. 203f.) explains, geodesy was

[1] Translation from Aristotle (1966, p. 44).

"not confined to land measuring: it covered generally the practical measurement of surfaces and volumes", the difference to geometry being that geodesy deals with concrete rather than abstract objects. The corresponding Latin term is rarely found in the Middle Ages,[1] but by the 16th century the term was generally used in the sense of 'land measuring'. John Dee's "mathematical preface" contains the explanation "Geodesie, or Land Measuring" (Dee 1570, fol. aiij^{v}).[2]

According to Stäckel (1893, p. 445), the history of geodesic lines starts with a problem published by Bernoulli (1697, p. 395):

> Given two points on a convex surface, we request a manner to describe geometrically the shortest line going from one point to the other, assuming that the surface is geometrical, as are those of the sphere, the cone, the cylinder, where the problem is very easy, no matter where the points are located; but in the conoids or spheroids the problem becomes very difficult. This is the reason why we propose, for example, the surface of the parabolic conoid, where it is asked to draw the shortest line joining two points not on the same meridian.[3]

Throughout the 18th century, geodesic lines were simply called *shortest lines* (Latin *linea brevissima*), for example by Euler (1732, p. 110). This custom is also found until well into the 19th century.[4]

In the 18th century, the French expression *opération géodésique* is used for operations of degree measurement to distinguish them from astronomical observations.[5] Since these measurements essentially concerned the length of the shortest line on the spheroid of the earth, the shortest lines were called *geodesic lines* (Stäckel 1893, p. 459). This term is first found[6] as French *ligne géodésique* in Laplace (1798, p. 112), who defines it in connection with spheroids: "the shortest line upon the surface of the earth [...] We shall call this line the *geodesic line*".[7] This term is found soon afterwards in other languages. For example, in a German translation of Laplace's treatise, *geodätische Linie* is found (Laplace 1802, p. 132), and the English term first appears (OED) in Biot (1809, p. 255).

In the first half of the 19th century, the term begins to be used with reference to a bigger variety of surfaces. For example, Bessel (1825, p. 242) defines a geodesic line of an ellipsoid of revolution, and Jacobi (1839) that of a triaxial ellipsoid. Eventually, Liouville (1844b, p. 1262) generally applies the notion to any surface: "On an

---

[1] In his 13th-century translation of Aristotle's *Metaphysics*, William of Moerbeke used *geodesia* (Vuillemin-Diem 1995, p. 54). [2] This is the first use of *geodesy* in English, according to the OED. [3] "Deux points etant donés [sic] sur une superficie convexe, on demande une maniere d'y décrire Geometriquement d'un de ces points à l'autre, la ligne la plus courte, suposé que cete surface soit Geometrique, telles que sont celles de la sphere, du cone, du cilindre, dans lesquelles le problême est fort facile, de quelque maniere que les points soient situez; mais les conoides & dans les spheroides il devient tres-difficile. C'est pourquoi on propose pour exemple la superficie du conoide parabolique, dans laquelle il faille tirer la ligne la plus courte qui joint deux points situez non pas dans le même meridien". Translation based on Freguglia and Giaquinta (2016, p. 101). [4] See Reich (1973, p. 308) for references. [5] This is explicitly confirmed by d'Alembert (1757, p. 608). [6] As pointed out by Rosenfeld (1988, p. 282) and Nabonnand (1995, p. 160). [7] "la ligne la plus courte sur la surface de la terre [...] Nous désignerons cette ligne, sous le nom de *ligne géodésique*".

arbitrary surface, the geodesic line has the property that its radius of curvature is in each point normal to the surface".[8] This model soon became established usage.

Using the term *geodesic* rather than *shortest* line also made sense because in the 1830s Jacobi had pointed out cases in which the "shortest" line actually was not the shortest one (see, e. g., Jacobi 1837, p. 76 and Jacobi 1866, p. 46).

The notion of geodesic curvature is found in Minding (1830),[1] but the term first appears in print in Bonnet (1848, pp. 43f.). It is due to Liouville, however, as is explicitly pointed out both by Bonnet (1849, p. 449) himself: "This name is due to Mr. Liouville, who has employed it in a Memoir which was written long ago but has not yet been published",[2] and by Liouville (1850, p. 568): "Let us limit ourselves to emphasizing among the relative curvatures the one of a curve drawn on a surface relative to the tangential geodesic line. I have suggested to denote it by the expressive name *geodesic curvature*, which Mr. Bonnet has kindly adopted in a remarkable memoir".[3] The term is thus motivated by the fact that the geodesic curvature "is equal to the absolute value of the infinitely small angle formed by two geodesic lines, which touch the given curve in two infinitely close points A and B, where this value has to be divided by the arc" (Reich 1973, p. 298). This motivation had already been indicated by Bonnet (1848, pp. 43f.).

Note. On the history of geodesic lines, see especially Stäckel (1893), Reich (1973, pp. 308f.), and Nabonnand (1995). Karney (2009) presents an extensive bibliography with terminological comments. See also Sager (1903) and Struik (1961, pp. 32f.). For the history of geodesic curvature, see Reich (1973, pp. 298f.), Nabonnand (1995), and Lützen (1990, pp. 748–750).

References

d'Alembert, Jean Le Rond (1757). "Géodésie". *Encyclopédie, ou Dictionnaire raisonné des sciences, des arts et des métiers* 7, pp. 606–608.

Aristotle (1966). *Metaphysics*. Translated with commentaries and glossary by Hippocrates G. Apostle. Bloomington.

Bernoulli, Johann (1697). "Problèmes à résoudre". *Journal des sçavans* (Aug. 1697), pp. 394–396.

Bessel, Friedrich Wilhelm (1825). "Ueber die Berechnung der geographischen Längen und Breiten aus geodätischen Vermessungen". *Astronomische Nachrichten* 4, pp. 241–254.

Biot, Jean-Baptiste (1809). "Analysis of the Mécanique Céleste of M. LaPlace (continued)". *The Philosophical Magazine* 34, pp. 36–43, 252–260.

Bonnet, Ossian (1848). "Mémoire sur la théorie générale des surfaces". *Journal de l'École polytechnique* 19, cahier 32, pp. 1–146.

Bonnet, Ossian (1849). "Lettre de M. O. Bonnet concernant l'emprunt qu'il a fait, à un travail inédit de M. Liouville d'une expression employée dans un travail imprimé dont il adresse un exemplaire a l'Académie". *Comptes rendus de l'Académie des sciences* 28, pp. 448–449.

---

[8] "Sur une surface quelconque, la ligne géodésique jouit de cette propriété, que son rayon de courbure est en chaque point normale à la surface". Translation from Lützen (1990, p. 717). – Almost the same statement is found in Liouville (1844a, p. 404). [1] As pointed out by Reich (1973, p. 298), for example, the notion was also known to Gauß. It appears in notes, probably taken between 1822 and 1825, in which he uses the term *Seitenkrümmung* 'lateral curvature' (Gauß 1900, p. 387). [2] Translation from Lützen (1990, p. 749). [3] "Bornons-nous à signaler parmi les courbes relatives, celle d'une courbe tracée sur une surface, par rapport à la ligne géodésique tangente. J'ai proposé pour la désigner le nom expressif de *courbure géodésique*, que M. Bonnet a bien voulu adopter dans un Mémoire remarquable". Translation from Lützen (1990, pp. 749f.).

Dee, John (1570). "Mathematicall Praeface". In: *The Elements of Geometrie of the most auncient Philosopher Euclide of Megara*. Ed. by Henry Billingsley. London.

Euler, Leonhard (1732). "De linea brevissima in superficie quacunque duo quaelibet puncta jungente". *Commentarii academiae scientiarum imperialis Petropolitanae* 3, pp. 110–124.

Freguglia, Paolo and Mariano Giaquinta (2016). *The Early Period of the Calculus of Variations*. Basel.

Gauß, Carl Friedrich (1900). "[Die Seitenkrümmung]". In: *Werke*. Achter Band. Göttingen, pp. 386–396.

Heath, Thomas L. (1949). *Mathematics in Aristotle*. Oxford.

Jacobi, Carl Gustav Jacob (1837). "Zur Theorie der Variations-Rechnung und der Differential-Gleichungen". *Journal für die reine und angewandte Mathematik* 17, pp. 68–82.

Jacobi, Carl Gustav Jacob (1839). "Lettre de M Jacobi à M Arago, concernant les lignes géodésiques tracées sur un ellipsoïde à trois axes". *Comptes rendus de l'Académie des sciences* 8, p. 284.

Jacobi, Carl Gustav Jacob (1866). *Vorlesungen über Dynamik*. Ed. by Alfred Clebsch. Berlin.

Karney, Charles (2009). "A geodesic bibliography". URL: `https://geographiclib.sourceforge.io/geodesic-papers/biblio.html` (retrieved May 6, 2018).

Laplace, Pierre-Simon (1798). *Traité de mécanique céleste*. Tome premier. Paris.

Laplace, Pierre-Simon (1802). *Mechanik des Himmels*. Aus dem Französischen übersetzt und mit erläuternden Anmerkungen versehen von J.C. Burckhardt. Zweyter Theil. Berlin.

Liouville, Joseph (1844a). "De la ligne géodésique sur un ellipsoide quelconque". *Journal de mathématiques pures et appliquées* 9, pp. 401–408.

Liouville, Joseph (1844b). "Remarques". *Comptes rendus de l'Académie des sciences* 19, pp. 1261–1263.

Liouville, Joseph (1850). "Sur les courbes à double courbure". In: Monge, Gaspard. *Applications de l'analyse à la géométrie*. Cinquième édition. Ed. by Joseph Liouville. Paris, pp. 547–568.

Lützen, Jesper (1990). *Joseph Liouville 1809-1882*. Master of Pure and Applied Mathematics. New York.

Minding, Ferdinand (1830). "Über die Curven des kürzesten Perimeters auf krummen Flächen". *Journal für die reine und angewandte Mathematik* 5, pp. 297–304.

Nabonnand, Philippe (1995). "Contribution à l'histoire de la théorie des géodésiques au XIXe siècle". *Revue d'histoire des mathématiques* 1, pp. 159–200.

Reich, Karin (1973). "Die Geschichte der Differentialgeometrie von Gauss bis Riemann (1828-1868)". *Archive for History of Exact Sciences* 11, pp. 273–382.

Rosenfeld, Boris A. (1988). *A History of Non-Euclidean Geometry*. Evolution of the Concept of a Geometric Space. Translated by Abe Shenitzer. New York.

Sager, Paul (1903). *Übersicht über die Entwickelung der Theorie der geodätischen Linien seit Gauss*. Rostock.

Stäckel, Paul (1893). "Bemerkungen zur Geschichte der geodätischen Linien". *Berichte über die Verhandlungen der Königlich Sächsischen Gesellschaft der Wissenschaften zu Leipzig (Mathematisch-Physische Klasse)* 45, pp. 444–467.

Struik, Dirk Jan (1961). *Lectures on Classical Differential Geometry*. Second Edition. Reading, Mass.

Vuillemin-Diem, Gudrun, ed. (1995). *Aristoteles Latinus*. XXV 3.2. Metaphysica. Lib. I-XIV. Recensio et Translatio Guillelmi de Moerbeka. Leiden.

geometric **(mean, proportion, progression)**

In general, the adjective *γεωμετρικός* means 'relating to geometry'; it was first used at the time of Archytas and Plato (Mugler 1958-59, p. 102). Archytas also used it in his definition of *geometric mean*[1]: "The mean is geometric, whenever they [the terms] are such that as the first is to the second so the second is to the third" (Huffman 2005, p. 162).[2]

---

[1] Archytas' terms for 'mean' are discussed by Huffman (2005, pp. 177f.). [2] Translation from Huffman (2005, p. 163).

The term *geometric proportion* (*ἀναλογία γεωμετρική*) was used by Aristotle in his *Nicomachean Ethics* (V.iii.13) and by Nicomachus in his *Introduction to Arithmetic* (II, 22) (Nicomachus of Gerasa 1866, p. 122), for example. This term appeared as *proportionalitas geometrica* in Boethius' *Institutio arithmetica* (II, 42) (Boethius 1867, p. 139); he often used *medietas geometrica* 'geometric mean' interchangeably.[3]

The first French use of the adjective *géométrique* occurred, according to the TLF, in Oresme's translation of Aristotle's *Ethics* (Oresme 1940, p. 288); he wrote *proporcionalité geometrique*. Recorde (1543, fol. 97^{v}) appears to have been the first who used *progression geometricall*.[1] The origin of the corresponding Latin *progressio geometrica* is quite obscure; in print, it appeared in von Peuerbach (c1495, s.v. De progressione). This text was published posthumously and so the term must have been known much earlier; it was also recorded in Reisch (1503, fol. g iij^{r}).

Note. For the general adjective, see Mugler (1958-59, p. 102). For the history of the terminology, see Cajori (1923, 1928).

References

Boethius, Anicius Manlius Severinus (1867). *De institutione arithmetica libri duo, De institutione musica libri quinque*. Accedit Geometria quae fertur Boetii. Ed. by Gottfried Friedlein. Leipzig.

Cajori, Florian (1923). "Origin of the names arithmetical and geometrical progression and proportion". *School Science and Mathematics* 22, pp. 734–737.

Cajori, Florian (1928). "Ciruelo on the Names 'Arithmetical' and 'Geometrical' Proportions and Progressions". *Isis* 10, pp. 363–366.

Huffman, Carl A. (2005). *Archytas of Tarentum*. Pythagorean, Philosopher and Mathematician King. Cambridge.

Mugler, Charles (1958-59). *Dictionnaire historique de la terminologie géométrique des Grecs*. Paris.

Nicomachus of Gerasa (1866). *Nicomachi Geraseni Pythagorei Introductionis Arithmeticae Libri II*. Ed. by Richard Hoche. Leipzig.

Oresme, Nicole (1940). *Le livre de éthiques d'Aristote*. Edited by Albert D. Menut. New York.

von Peuerbach, Georg (c1495). *Algorismus*. Wien.

Recorde, Robert (1543). *The Ground of Artes*. London.

Reisch, Gregor (1503). *Margarita philosophica*. Freiburg.

geometry

The Greek term *γεωμετρία* comes from the words for 'earth' and 'measure' and can be translated as 'land surveying' (geodesy), and in this sense the term was originally used. Hero of Alexandria mentions the origin of the term in his *Metrica*: "The first geometry, as the old tradition learns us, occupied itself with measurements and distributions of fields, whence it was called geometry" (Heron of Alexandria 1903, p. 2).[2] A similar account was given by Proclus: "geometry was first discovered among the Egyptians and originated in the remeasuring of their lands. This was necessary for them because the Nile overflows and obliterates the boundary lines between their properties" (Proclus 1873, p. 64).[3]

Aristotle (*Metaphysics* 997b) opposed *γεωμετρία* and *γεωδαισία*. As explained by Heath (1949, pp. 203f.), the latter literally means "'division' or 'allocation' of land. But the term was not confined to land-measuring: it covered generally the

---

[3] The adjective *geometricus* dates back to time of Cicero (LS, TLL). [1] English *geometrical* dates from the late 14th century (OED). [2] Translation from Bruins (1964, p. 182). [3] Translation from Proclus (1970, p. 52). The Egyptian origin of geometry was earlier pointed out by Herodotus (*Histories* ii, 109).

practical measurement of surfaces and volumes [...] It is different from geometry in that geometry deals with lines, planes, and solids in abstraction from matter and all other attributes [...] while mensuration deals with planes and solids not in abstraction from matter but in abstraction from any particular kind of matter".[4]

The term appears in classical Latin as *geometria* or *geometrica*; both forms were used by Cicero (TLL). Geometry was part of the quadrivium (which also included arithmetic, music, and astronomy) and the term was in frequent use. Boethius, for example, defined it as the study of "immobile magnitude [immobilis magnitudinis]" (Boethius 1867, p. 9).[1] This became a frequently found definition of geometry.[2]

In 12th-century Latin texts, a distinction between *geometria theorica* and *geometria practica* came up; according to Victor (1979, p. 3), it seems to have been first made by Hugh of St. Victor, who wrote: "we note that geometry is either theoretical (speculative) or practical (active). Theoretical geometry uses sheer intellectual reflection to study spaces and intervals of rational dimensions. But practical geometry uses instruments, and gets its results by working proportionally from one figure to another" (Baron 1956, p. 187).[3] This distinction between practical and theoretical geometry is earlier found in Arabic sources (Victor 1979, pp. 11f.).[4]

A quick look at some modern languages: The OED cites *The Seven Sages of Rome* from around 1330 as the first use in English. The TLF cites the 12th-century poem *Roman de Thèbes* for the French term; the German term also dates from the 12th century (Strauß et al. 2008, pp. 244–249). Reflecting the older uses in Greek and Latin, the term could be used for theoretical and practical geometry as well as for geodesy.

Geometry developed over the centuries, of course.[5] In the 17th century, analytic geometry begins to be developed in the works of Fermat and Descartes, and the rudiments of projective geometry are presented by Desargues.[6] It is also at his time that *elementary geometry* began to be used, for example by Schott (1661, p. 61), who equated *geometria elementaris* with the contents of the first six books of Euclid's *Elements*. The term also appears in Wolff's mathematical dictionary, where its meaning is extended as to also include solid geometry (Wolff 1716, p. 667). Wolff's dictionary also contains an entry for *higher geometry* (*geometria sublimior*), which is

[4] See also *geodesic*. It is worth pointing out that Euclid does not use the word *γεωμετρία* in his *Elements*. [1] A similar definition of geometry as studying "magnitude at rest" was already given by Proclus (Proclus 1873, p. 36) and others. A diverging usage is found in Quintilian, who, in his *Institutio Oratoria* (I, x, 35), used *geometria* for both geometry and arithmetic, saying that *geometria* deals with numbers on the one hand, and with figures on the other. [2] Because of the discrepancy between the literal meaning of *geometry* and its definition as the study of magnitudes, Dee (1570, fol. a ij^{v}) later suggested, rather unsuccessfully, a new designation: "An other name, therfore, must nedes be had, for our Mathematicall Science of Magnitudes: which regardeth neither clod, nor turff: neither hill, nor dale: neither earth nor heauen: but is absolute *Megethologia*". The term is based on the Greek *μέγεθος* 'magnitude'. [3] "His breuiter prelibatis deinceps considerandum est, quod omnis geometrica disciplina aut theorica est, id est speculatiua, aut practica, id est actiua. Theorica siquidem est, que spacia et interualla dimensionum rationabilium sola rationis speculatione uestigat; practica uero est, que quibusdam instrumentis agitur et ex aliis alia proportionaliter coniciendo diiudicat". Translation from Hugh of St. Victor (1991, p. 33). [4] For the history of practical geometry, see also Axworthy (2022) and Gautier Dalché and Querrien (2015). [5] As in the other entries for branches of mathematics, the following can only offer a brief sketch. [6] See the respective entries.

said to deal with curved lines and the solids generated by them (Wolff 1716, pp. 671f.). As examples of what belongs to higher geometry, Wolff mentions conic sections and "the new analysis" by Leibniz, who had indeed used *geometria sublimior* on various occasions, for example in Leibniz (1684, pp. 472f.), where, after having sketched the basics of differential calculus, he remarked: "And this is only the beginning of much more sublime Geometry, pertaining to even the most difficult and most beautiful problems of applied mathematics, which without our differential calculus or something similar no one could attack with any such ease".[1] A slightly different subdivision of geometry is found in d'Alembert (1757, p. 633), according to whom *géométrie élémentaire* "considers only the properties of straight lines, of circular lines, the simplest figures and solids, that is to say rectilinear or circular shapes, and the solids bounded by these shapes", while *géométrie transcendante* considers curves without using differential and integral calculus and *géométrie sublime* is the geometry of the new calculus.[2]

Another development was noted by De Risi (2016b, p. 104), who points out "the slow transformation of geometry from a discipline dealing with figures and magnitudes into a science of space: a new definition of geometry that no one endorsed at the beginning of the [18th] century (with the exception of Leibniz) but that would become widespread and almost universally accepted in the first decades of the nineteenth century".[3] Precursors of this conception are Patrizi (1587) and Pascal in his "Introduction to Geometry" (Itard 1962).[4]

The period from the end of the 18th century (when Monge's *Géométrie descriptive* was published)[5] until the early 1870s (when Klein published his *Erlanger Programm*) has been described by Bourbaki (1994, p. 131) as the "golden age of geometry".[6] Klein's programme (Klein 1872) has been succinctly summarized by Gray (2011, p. 235): "Klein shifted attention from the figures to the transformations, and argued that henceforth geometry should be about groups as well as the properties of shapes. So a geometric property was one that was invariant under all the operations of the group associated to that geometry".[7]

Some examples of further important developments in geometry are discussed in the entries *analytic geometry*, *projective geometry*, *differential geometry*, *Euclidean geometry*, *hyperbolic geometry*, *algebraic geometry*, *enumerative geometry*, and *Riemannian geometry*.

Note. Brief overviews of the terminology are given by Tropfke (1940, pp. 46f.) and Smith (1958, p. 273).[8] For the Greek term, see Mugler (1958-59, pp. 101f.), for the

---

[1] "Et haec quidem initia sunt tantum Geometriae cujusdam multo sublimioris, ad difficillima et pulcherrima quaeque etiam mistae Matheseos problemata pertingentis, quae sine calculo nostro differentiali, aut simili, non temere quisquam pari facilitale tractabit". Translation from Struik (1969, p. 279). [2] Translation from d'Alembert (2019). [3] For discussion, see the overview by De Risi (2016b, pp. 40–49) and the detailed account of geometry as the science of space in Leibniz' writings by De Risi (2007). [4] For discussions of these sources, see De Risi (2016a), Edelheit (2009), and Debuiche (2016). [5] See *descriptive geometry*. [6] Bourbaki (1994, pp. 131–138) also gives a brief overview of the major innovations in 19th-century geometry. [7] For different assessments of the influence of Klein's programme, see Hawkins (1984) and Birkhoff and Bennett (1988). [8] Note, however, that, according to Smith, Plato "always [uses γεωμετρία] to indicate surveying". That this is not correct is shown by various passages in Plato's *Republic* (book vii), for example.

Latin one, Guillaumin (2020, pp. 139f.). The Arabic term *handasa*, which is derived from Persian, is discussed by Rashed (2017, pp. 830–832) and by Souissi (2004). The Sanskrit terminology is discussed by Datta (1929) and Datta and Singh (1979, pp. 126f.). Some recent books on the history of geometry are Scriba and Schreiber (2015) and Ostermann and Wanner (2012). For the development in the 20th century, see also Berger (2010), Karzel and Kroll (1988), and the various contributions in Kouneiher et al. (2005).

References

d'Alembert, Jean Le Rond (1757). "Géométrie". *Encyclopédie, ou Dictionnaire raisonné des sciences, des arts et des métiers* 7, pp. 629–638.

d'Alembert, Jean Le Rond (2019). "Geometry". *The Encyclopedia of Diderot & d'Alembert Collaborative Translation Project*. Translated by David Fleming. URL: `http://hdl.handle.net/2027/spo.did2222.0003.663` (retrieved Sept. 15, 2024).

Axworthy, Angela (2022). "The Hybridization of Practical and Theoretical Geometry in the Sixteenth-Century Euclidean Tradition". *Journal of Interdisciplinary History of Ideas* 11, 4:1–4:104.

Baron, Roger (1956). "Hugonis de Sancto Victore *Practica Geometriae*". *Osiris* 12, pp. 176–224.

Berger, Marcel (2010). "Geometry in the 20th Century". In: *History of Mathematics*. Ed. by Vagn Lundsgaard Hansen and Jeremy Gray. Oxford, pp. 269–314.

Birkhoff, Garrett and M.K. Bennett (1988). "Felix Klein and his 'Erlanger Programm'". In: *History and Philosophy of Modern Mathematics*. Ed. by William Aspray and Philip Kitcher. Minneapolis, pp. 145–176.

Boethius, Anicius Manlius Severinus (1867). *De institutione arithmetica libri duo, De institutione musica libri quinque*. Accedit Geometria quae fertur Boetii. Ed. by Gottfried Friedlein. Leipzig.

Bourbaki, Nicolas (1994). *Elements of the History of Mathematics*. Translated from the French by John Meldrum. Berlin.

Bruins, E.M., ed. (1964). *Codex Constantinopolitanus*. Palatii Veteris No. 1. Part Three. Translation and Commentary. Leiden.

Datta, Bibhutibhusan (1929). "Origin and History of the Hindu names for Geometry". *Quellen und Studien zur Geschichte der Mathematik* 1, pp. 113–119.

Datta, Bibhutibhusan and Avadhesh Narayan Singh (1979). "Hindu Geometry (revised by Kripa Shankar Shukla)". *Indian Journal of History of Science* 15, pp. 121–188.

Debuiche, Valérie (2016). "L'invention d'une géométrie pure au 17e siècle: Pascal et son lecteur Leibniz". *Studia Leibnitiana* 48, pp. 42–67.

Dee, John (1570). "Mathematicall Praeface". In: *The Elements of Geometrie of the most auncient Philosopher Euclide of Megara*. Ed. by Henry Billingsley. London.

Edelheit, Amos (2009). "Francesco Patrizi's two books on space: geometry, mathematics, and dialectic beyond Aristotelian science". *Studies in History and Philosophy of Science* 40, pp. 243–257.

Gautier Dalché, Patrick and Armelle Querrien (2015). "Mesure du sol et géométrie au Moyen Âge". *Archives d'histoire doctrinale et littéraire du Moyen Âge* 82, pp. 97–139.

Gray, Jeremy (2011). *Worlds Out of Nothing*. A Course in the History of Geometry in the 19th Century. London et al.

Guillaumin, Jean-Yves (2020). *Dictionnaire de la terminologie latine ancienne de l'arithmétique et de la géométrie*. Paris.

Hawkins, Thomas (1984). "The *Erlanger Programm* of Felix Klein: Reflections on its place in the history of mathematics". *Historia Mathematica* 1, pp. 442–470.

Heath, Thomas L. (1949). *Mathematics in Aristotle*. Oxford.

Heron of Alexandria (1903). *Heronis Alexandrini opera quae supersunt omnia*. Volumen III. Rationes dimetiendi et Commentatio Dioptrica. Ed. by J.L. Heiberg. Leipzig.

Hugh of St. Victor (1991). *Practical Geometry (Practica Geometriae)*. Attributed to Hugh of St. Victor. Translated from the Latin with an Introduction, Notes and Appendices by Frederick A. Homann. Milwaukee.

Itard, Jean (1962). "*L'introduction à la géométrie* de Pascal". *Revue d'histoire des sciences* 15, pp. 269–286.
Karzel, Helmut and Hans-Joachim Kroll (1988). *Geschichte der Geometrie seit Hilbert*. Darmstadt.
Klein, Felix (1872). *Vergleichende Betrachtungen über neuere geometrische Forschungen*. Programm zum Eintritt in die philosophische Facultät und den Senat der k. Friedrich-Alexanders-Universität zu Erlangen. Erlangen.
Kouneiher, Joseph et al., eds. (2005). *Géométrie au 20e siècle, 1930-2000*. Histoire et horizons. Paris.
Leibniz, Gottfried Wilhelm (1684). "Nova Methodus pro maximis et minimis, itemque tangentibus, quae nec fractas nec irrationales quantitates moratur, et singulare pro illis calculi genu". *Acta Eruditorum* (Oct. 1684), pp. 467–473.
Mugler, Charles (1958-59). *Dictionnaire historique de la terminologie géométrique des Grecs*. Paris.
Ostermann, Alexander and Gerhard Wanner (2012). *Geometry by Its History*. Berlin and Heidelberg.
Patrizi, Francesco (1587). *Della nuova geometria libri XV*. Ferrara.
Proclus (1873). *In primum Euclidis elementorum librum commentarii*. Ed. by Gottfried Friedlein. Leipzig.
Proclus (1970). *Commentary on the First Book of Euclid's Elements*. Translated with Introduction and Notes by Glenn R. Morrow. Princeton.
Rashed, Roshdi, ed. (2017). *Lexique historique de la langue scientifique arabe*. Hildesheim.
De Risi, Vincenzo (2007). *Geometry and Monadology*. Leibniz's *Analysis Situs* and Philosophy of Space. Basel and Boston.
De Risi, Vincenzo (2016a). "Francesco Patrizi and the New Geometry of Space". In: *Boundaries, Extents and Circulations*. Space and Spatiality in Early Modern Natural Philosophy. Ed. by Koen Vermeir and Jonathan Regier. Dordrecht, pp. 55–106.
De Risi, Vincenzo (2016b). *Leibniz on the Parallel Postulate and the Foundations of Geometry*. The Unpublished Manuscripts. Cham.
Schott, Caspar (1661). *Cursus Mathematicus, Sive Absoluta Omnium Mathematicarum Disciplinarum Encyclopaedia, In Libros XXVIII. digesta*. Würzburg.
Scriba, Christoph J. and Peter Schreiber (2015). *5000 Years of Geometry*. Mathematics in History and Culture. Basel.
Smith, David Eugene (1958). *History of Mathematics*. Volume II. Special Topics of Elementary Mathematics. New York.
Souissi, Mohammed (2004). "'ilm al-handasa". *Encyclopaedia of Islam* 12, pp. 411–414.
Strauß, Gerhard et al., eds. (2008). *Deutsches Fremdwörterbuch*. 2. Auflage. Band 6: Gag – Gynäkologie. Berlin and New York.
Struik, Dirk Jan, ed. (1969). *A Source Book in Mathematics, 1200-1800*. Harvard and London.
Tropfke, Johannes (1940). *Geschichte der Elementar-Mathematik*. Vierter Band. Ebene Geometrie. Dritte, verbesserte und vermehrte Auflage besorgt von Dr. Kurt Vogel. Berlin and Leipzig.
Victor, Stephen K. (1979). *Practical Geometry in the High Middle Ages*. "Artis cuiuslibet consummatio" and the "Pratike de geometrie". Philadelphia.
Wolff, Christian (1716). *Mathematisches Lexicon*. Leipzig.

germ

As described by Fasanelli (1981, p. 6), "[a] germ summarizes the local behavior of a function at a point", which explains the name. According to Gray (1979, p. 15), the concept already appeared in the works of Weierstraß; the term, however, is of later origin and appears to be due to Kneser and Cartan. Kneser (1950, p. 383) defined the German term *Mengenkeim* (*germ of sets*):

> Let P be a topological space, a one of its points, f and g two functions defined in neighbourhoods of a, X and Y two subsets of P. We call f and g *equivalent in* a, if both of them are (defined and) agree in a neighbourhood of a; we call X and Y equivalent in a, if $X \cap U = Y \cap U$ with a suitable neighbourhood U of a. It is easily seen that both concepts

of equivalence are reflexive, symmetric, and transitive and therefore give rise to a partition of all functions defined on some neighbourhood of a and of all subsets of P, respectively. The classes of these partitions are called *function elements* and *germs of sets in a*, respectively.[1]

Kneser (ibid.) remarked that he defined this concept following Cartan (1950), where the term was not yet used; Cartan did, however, use French *germe*, apparently independently of Kneser, in Cartan (1951-52, p. 1). Cartan explicitly remarked that he introduced the term for what Bruhat (1951-52, p. 2) called a "variété analytique au point x", a term which also appeared in Cartan (1950, p. 33). The term spread quickly; English *germ* was already used by Kodaira and Spencer (1953, p. 643).

Kneser additionally pointed out that the term *germ* had earlier been introduced in connection with topological groups. The English term seems to have been first used in this sense by Weyl in lectures given in Princeton in 1933/34 (Weyl 1934, p. 1); it also appeared in Mayer and Thomas (1935, p. 788): "By a *group germ* will be meant a neighborhood $\mathring{\mathfrak{A}}$ of the identity element $\mathring{a}$ of $\mathring{a}$ topological group such that the inverse of any element of $\mathring{\mathfrak{A}}$ is contained in $\mathring{\mathfrak{A}}$; in the case of a coordinate group it will be understood that the group germ is a coordinate neighborhood". It is a literal translation of German *Gruppenkeim*, which was suggested by Schreier (1928, p. 114).

References

Bruhat, François (1951-52). "Sous-variétés analytiques; variétés principales". *Séminaire Henri Cartan* 4, exposé 12, pp. 1–11.

Cartan, Henri (1950). "Idéaux et modules de fonctions analytiques de variables complexes". *Bulletin de la Société Mathématique de France* 78, pp. 29–64.

Cartan, Henri (1951-52). "La notion d'espace analytique général et de fonction holomorphe sur un tel espace". *Séminaire Henri Cartan* 4, exposé 13, pp. 1–16.

Fasanelli, Florence Dowdell (1981). "The Creation of Sheaf Theory". PhD thesis. American University.

Gray, John W. (1979). "Fragments of the history of sheaf theory". In: *Applications of Sheaves*. Ed. by M.P. Fourman, C.J. Mulvey, and D.S. Scott. Berlin and Heidelberg, pp. 1–79.

Kneser, Hellmuth (1950). "Analytische Mannigfaltigkeiten im komplexen projektiven Raum". *Mathematische Nachrichten* 4, pp. 382–391.

Kodaira, Kunihiko and D.C. Spencer (1953). "On Arithmetic Genera of Algebraic Varieties". *Proceedings of the National Academy of Sciences* 39, pp. 641–649.

Mayer, Walther and T.Y. Thomas (1935). "Foundations of the Theory of Lie Groups". *Annals of Mathematics* 36, pp. 770–822.

Schreier, Otto (1928). "Über neue Untersuchungen in der Theorie der kontinuierlichen Gruppen". *Jahresbericht der Deutschen Mathematiker-Vereinigung* 37, pp. 113–122.

Weyl, Hermann (1934). *The structure and representation of continuous groups*. (Notes by Nathan Jacobson). Princeton.

---

[1] "Sei zunächst P ein topologischer Raum, a einer seiner Punkte, f und g zwei in Umgebungen von a erklärte Funktionen, X und Y zwei Teilmengen von P. Wir nennen f und g *äquivalent in* a, wenn sie in einer Umgebung von a (beide erklärt sind und) übereinstimmen; X und Y nennen wir äquivalent in a, wenn $X \cap U = Y \cap U$ mit einer geeigneten Umgebung U von a. Man stellt leicht fest, daß beide Äquivalenzbeziehungen reflexiv, symmetrisch und transitiv sind und deshalb Anlaß geben zu einer Klasseneinteilung aller in irgendwelchen Umgebungen von a erklärten Funktionen bzw. aller Teilmengen von P. Die Klassen dieser Einteilungen heißen *Funktionselemente* bzw. *Mengenkeime in a*".

glide reflection

A glide reflection is a combination of a translation and a reflection. The term seems to have been introduced by Schoenflies (1891, p. 345) as German *Gleitspiegelung*. An early use of the term in English is Hilton (1903, p. 150), where it appears as *gliding-reflexion*, a literal translation of Schoenflies' term. The choice of the first part of the term, *Gleit-*, from *gleiten* 'to glide', is due the fact that German *Gleitung* was a synonym of *translation* at the time (Schoenflies 1891, p. 255).

References

Hilton, Harold (1903). *Mathematical Crystallography and the Theory of Groups of Movements.* Oxford.

Schoenflies, Arthur (1891). *Krystallsysteme und Krystallstructur*. Leipzig.

golden section

The golden section has been known under various names since antiquity. Euclid defined it thus: "A straight line is said to have been cut in extreme and mean ratio when, as the whole line is to the greater segment, so is the greater to the less" (*Elements*, book VI, def. 3).[1] A similar wording is found in some of the earliest Latin translations of the *Elements*, for example in Gerard of Cremona's: "A straight line is said to be divided according to a proportion having a mean and two extremes, when the proportion of the whole line to its larger section is as the proportion of its larger division to its smaller section" (Busard 1984, p. 137).[2] This expression has its origin in Arabic texts (see Herz-Fischler 1998, p. 166).

The term *divine proportion* (Italian *divina proportione*) goes back to Pacioli (1509).[3] He gives an extensive discussion on why the name *divine* is justified. One of his arguments is that "just like God cannot properly be defined, nor can be understood through words, likewise this proportion of ours cannot ever be designated through intelligible numbers, nor can it be expressed through any rational quantity, but always remains occult and secret, and is called irrational by the mathematicians" (Pacioli 1509, fol. 4^{r}).[4] This term was soon adopted by other writers such as Ramus (1569, Geometria, p. 96), who attributed it to Pacioli.

Kepler knew the Euclidean term *extreme and mean ratio* and explained it thus: "this name means that whereas at other times ordinary division of the whole into parts is not concerned with proportion, or if some line [is constructed] that bears to the whole the ratio of the smaller part to the greater, then there will result four terms, two extremes and two means: in this case, on the contrary, there are only three terms, the whole and the smaller part being the two extremes; and the greater part being

---

[1] Translation from Euclid (1956, p. 188). [2] "Linea recta dicitur dividi secundum proportionem habentem medium et duo extrema, quando fuerit proportio totius linee ad maiorem eius sectionem sicut proportio maioris divisionis ipsius ad minorem eius sectionem". Translation from Herz-Fischler (1998, p. 166). [3] Campanus of Novara had already designated the golden section as "wonderful" (*mirabilis*). See Busard (2005, p. 510). [4] "La terza conuenientia e che si commo idio propriamente non se po diffinire ne per parolle a noi intendere. cosi questa nostra proportione non se po mai per numero intendibile asegnare ne per quantita alcuna rationale exprimere: ma sempre fia occulta e secreta e dali Mathematici chiamata irrationale". Translation from Herz-Fischler (1998, p. 172).

the unique mean term" (Kepler 1619, book I, p. 16).[1] He mentions another term, *proportional section* (*sectio proportionalis*), chosen for the same reason (ibid.), and also uses the terms *divine section* and *divine proportion* (Kepler 1619, pp. 29, 183).

The use of the epithet *golden* for this notion is not easy to trace. It is often claimed that the term was used in antiquity, in the Middle Ages, and also that it was coined by Leonardo da Vinci. None of this has ever been proven. Instead, it appears to be a German creation – this applies to both the German term *goldener Schnitt* and to the Latin neologism *sectio aurea*. Herz-Fischler (1998, p. 168) cited Ohm (1835, p. 268) as the first known use but also pointed out that Ohm used the term as an existing expression.[2] Indeed, since the publication of Herz-Fischler's book, various attestations before Ohm have been found (Herz-Fischler 2019)[3] and the term has been predated to Gehler (1789, p. 120), who mentions "the so-called golden section (*media et extrema ratione*, *sectione aurea* or *divina*)".[4]

Even the attestation in Gehler is not the first one, however, since, as pointed out by Becker (2019), as early as 1717, the following statement is found: "If a line is cut as discussed, it is called *media & extrema ratione secta*. The ancients called this section the golden one" (Kaschube 1717, p. 566).[5] Since the publication of Becker (2019), I have found still earlier uses: *güldene Section* was used by Sturm (1707, pt. I, p. 189); the term is also found in the second edition of the book (Sturm 1710a, pt. I, p. 175) and in Sturm (1710b, e. g., pp. 2-4, 13). Whether Sturm is really the first who used the term must remain an open question; he does not comment on it any further. It should also be noted that no further attestions between Kaschube and Gehler have been found and that, as noted above, Kaschube's statement that the ancients already used this expression cannot be confirmed by any reference. Gehler may well have known Sturm's and Kaschube's books.

The terminology eventually spread through Zeising (1854). Zeising was not sure about the reason for the name but supposed that it was based on the "exceptional benefits" of this ratio (Zeising 1854, p. 163). This explanation is presumably correct, since long before the 17th century *golden* could be used metaphorically for something valuable or fruitful (DWB), and the high esteem for the golden section is clear from Pacioli, for example, or from Kepler (1621, p. 47), who compared the golden section to a gemstome.[6] English *golden section* appeared, apparently first, in a review of Zeising's book.[7]

---

[1] "Nimirum hoc sibi vult Nomen, quòd cùm aliàs vulgaris sectio totius in partes duas, non respiciat proportionem, aut si ad totam aliqua comparetur in eà proportione, in quà est minor pars ad majorem, tunc fiant quatuor termini, duo extremi et duo Medij: hic contrà fiant tantum tres termini, tota quidem et pars minor, duo extremi; pars verò major, medius terminus unicus". Translation from Kepler (1997, p. 31). [2] Ohm states that the notion "is known under the name *golden section*".
[3] See also the letters by Lieberknecht on `http://herz-fischler.ca/new_zeising.html`.
[4] "den sogenannten güldnen Schnitt (*media et extrema ratione*, *sectione aurea s[ive] divina*)".
[5] "Wenn eine Linie AB, wie gemeldet, geschnitten ist; so heisset sie media & extrema ratione secta. Die Alten hissen diesen Schnitt den goldenen". [6] The term *golden rule* (rule of three) is a similar example (see Smith 1958, pp. 484–488 for attestations). [7] *The British Quarterly Review* 21 (1855), p. 267.

Finally, the last term mentioned here[1] is *medial section*, coined by Leslie (1809, p. 66). It is not commonly used anymore, even though Davies (1850, p. 128), who approved of Leslie's term after having compared it with others, predicted that it was likely that *medial section* "will be the ultimately prevailing one".

Note. Various remarks concerning the history of the terminology can be found, above all, in Herz-Fischler (1998). See also Nagy (1996, 1997), van der Schoot (2016), Archibald (1920, pp. 152–154), and Tropfke (1940, pp. 245–247).

References

Archibald, Raymond Clare (1920). "Notes on the logarithmic spiral, golden sections and the Fibonacci series". In: Hambidge, Jay. *Dynamic Symmetry*. The Greek Vase. New Haven, pp. 146–157.

Becker, Holger (2019). "An even earlier (1717) usage of the expression 'golden section'". *Historia Mathematica* 49, pp. 82–83.

Behr, Johann Heinrich (1677). *Der Verschantzte Turenne Oder Gründliche Alt- und Neue Kriegs-Bau-Kunst*. Frankfurt und Leipzig.

Busard, H.L.L. (1984). *The Latin Translation of the Arabic Version of Euclid's* Elements, *commonly ascribed to Gerard of Cremona*. Leiden.

Busard, H.L.L. (2005). *Campanus of Novara and Euclid's* Elements. Volume I. Stuttgart.

Davies, Thomas Stephens (1850). "On the flat knot". *The Mechanic's Magazine, Register, Journal and Gazette* 8, pp. 125–129.

Euclid (1956). *The Thirteen Books of Euclid's Elements*. Volume II. Books III-IX. Ed. by Thomas L. Heath. New York.

Gehler, Johann Samuel Traugott (1789). *Physikalisches Wörterbuch oder Versuch einer Erklärung der vornehmsten Begriffe und Kunstwörter der Naturlehre*. Zweyter Theil von Erd bis Lin. Leipzig.

Herz-Fischler, Roger (1998). *A Mathematical History of the Golden Number*. Mineola, N.Y.

Herz-Fischler, Roger (2019). "An early usage of the expression 'golden section'". *Historia Mathematica* 49, pp. 80–81.

Kaschube, Johann Wenceslaus (1717). *Cursus mathematicus, oder deutlicher Begrief der mathematischen Wissenschaften*. Jena.

Kepler, Johannes (1619). *Harmonices mundi libri V*. Linz.

Kepler, Johannes (1621). *Mysterium cosmographicum*. Frankfurt.

Kepler, Johannes (1997). *The Harmony of the World*. Translated into English with an Introduction and Notes by E.J. Aiton, A.M. Duncan, J.V. Field. Philadelphia, PA.

Leslie, John (1809). *Elements of Geometry, Geometrical Analysis, and Plane Trigonometry*. Edinburgh.

Nagy, Dénes (1996). "Golden Section(ism): From mathematics to the theory of art and musicology. Part 1". *Symmetry: Culture and Science* 7, pp. 413–442.

Nagy, Dénes (1997). "Golden Section(ism): From mathematics to the theory of art and musicology. Part 2". *Symmetry: Culture and Science* 8, pp. 74–112.

Ohm, Martin (1835). *Die reine Elementar-Mathematik*. Zweiter Band. Zweite Auflage. Berlin.

Pacioli, Luca (1509). *Divina Proportione*. Venezia.

Ramus, Petrus (1569). *Arithmeticae libri duo. Geometriae septem et viginti*. Basel.

van der Schoot, Albert (2016). *Die Geschichte des goldenen Schnitts*. Aufstieg und Fall der göttlichen Proportion. 2., aktualisierte Auflage. Aus dem Niederländischen von Stefan Häring. Stuttgart-Bad Cannstatt.

---

[1] A little known synonym is found in a few German texts, dating from around 1700, on military architecture. The term is *Solomon's wisdom*; *Sapientia Salomonis* appeared in Behr (1677, p. 179), and Stahl (1687, p. 52) also has the corresponding German *Weisheit Salomons*. What is surprising is that Behr remarks that the term was used "by many"; however, no earlier usages have been found.

Smith, David Eugene (1958). *History of Mathematics*. Volume II. Special Topics of Elementary Mathematics. New York.
Stahl, Georg Conrad (1687). *Neu-aufgeführter Europäischer Ingenieur: oder Kriegs-Bau-Kunst*. Nürnberg.
Sturm, Leonhard Christoph (1707). *Kurtzer Begriff Der gesambten Mathesis*. Frankfurt an der Oder.
Sturm, Leonhard Christoph (1710a). *Kurtzer Begriff Der gesambten Mathesis*. Anitzo zum zweythenmahl herausgegeben und an vielen Orten vermehret und gebessert. Frankfurt an der Oder.
Sturm, Leonhard Christoph (1710b). *Vier kurtze Abhandlungen*. Frankfurt an der Oder.
Tropfke, Johannes (1940). *Geschichte der Elementar-Mathematik*. Vierter Band. Ebene Geometrie. Dritte, verbesserte und vermehrte Auflage besorgt von Dr. Kurt Vogel. Berlin and Leipzig.
Zeising, Adolf (1854). *Neue Lehre von den Proportionen des menschlichen Körpers*. Leipzig.

graded

A *graded ring* (*anneau gradué*) was defined by Leray (1950, p. 10) in the following terms: "A ring $\mathcal{A}$ shall be called *graded*, if it is, as an additive group, the direct sum of subgroups $\mathcal{A}^{[p]}$ (p an integer) such that $\mathcal{A}^{[p]}\mathcal{A}^{[q]} \subset \mathcal{A}^{[p+q]}$".[1] Leray had already used it in a talk given in 1947 (Leray 1949, pp. 64f.), where the formulation "nous le nommerons" ('we shall call it') suggests that the term may be due to him.

Other early uses appear in Cartan (1948, p. 148), where *module gradué* is found, and in Serre (1948-49, p. 5), who defined *structure gradué* and *groupe gradué*.

The English term appears in the early 1950s. For example, in lectures held in 1952, Chern (1953, p. 70) defined both *graded ring* and *graded group*; *graded ring* was also used Hu (1952, p. 393) at about the same time.[2]

References

Cartan, Henri (1948). "Sur la cohomologie des espaces où opère un groupe. Notions algébriques préliminaires". *Comptes rendus de l'Académie des sciences* 226, pp. 148–150.
Chern, Shiing-Shen (1953). *Differentiable manifolds*. Spring 1952. Chicago.
Hu, Sze-Tsen (1952). "Cohomology Rings of Compact Connected Groups and Their Homogeneous Spaces". *Annals of Mathematics* 55, pp. 391–419.
Leray, Jean (1949). "L'Homologie filtrée". In: *XII Colloque international de Topologie algébrique*. Paris, 26 Juin – 2 Juillet 1947. Paris, pp. 61–82.
Leray, Jean (1950). "L'anneau spectral et l'anneau filtré d'homologie d'un espace localement compact et d'une application continue". *Journal de mathématiques pures et appliquées* 29, pp. 1–80.
Serre, Jean-Pierre (1948-49). "Groupe d'homologie d'un complexe simplicial. Généralités sur les groupes à dérivation". *Séminaire Henri Cartan* 1, exposé 2, pp. 1–9.

gradient

The word *gradient* ultimately goes back to Latin *gradientem*, a participle of *gradi* 'to go, to walk'. Before it entered the mathematical vocabulary, it had been used in two technical senses that appear to have given rise to its mathematical sense.

The first meaning appeared in connection with roads and railways for 'degree of slope'. For example, in [anon.] (1839, p. 211) the word is defined as "the inclinations of the road" and *slope* is given as a synonym; it is also stated there that "[t]he word *gradient* ha[s] always been used to express the rate of inclination, whatever that might be". Apparently, the word began to be used in English as a translation of French *pente*

---

[1] "Un anneau $\mathcal{A}$ sera dite gradué quand il est, comme groupe additive, somme directe des sous-groupes $\mathcal{A}^{[p]}$ (p entiers) tels que $\mathcal{A}^{[p]}\mathcal{A}^{[q]} \subset \mathcal{A}^{[p+q]}$". [2] Hu's paper was received on May 7, 1951; he cites Leray, but not explicitly for the term.

'inclination'; as the translator of Navier (1835) states, "I have rendered the word *pente* by *slope* in preference to *inclination, inclined plane*, or *gradient*, considering the two former, though generally used, as improper expressions, and the latter, to say the least of it, as having so little to recommend it, that I hope it will have an extremely short existence in our nomenclature" (Navier 1836, p. xiv).

The second meaning is found in metereology, where *barometric gradient* is used in connection with differences of pressure. The term was defined by Privat-Deschanel (1870, p. 168), where also the recency of the term was pointed out: "The isobaric curves correspond to differences of five millimetres [...] of pressure, and according as they are near together or far apart, the variation of pressure in passing from one to another is more or less sudden (or to use a very expressive modern phrase, the barometric gradient is more or less steep)".[1] Both of these meanings are also found in French. Van Tricht (1880, p. 229), for example, attests both of them and states that the second meaning is derived from the first one.

The gradient as a differential operator was introduced by Maxwell (1869, p. 231) under the name *slope*:

> Let us next consider the Hamiltonian operator ∇. First apply it to a scalar function P. The quantity is a vector, indicating the direction in which P decreases most rapidly, and measuring the rate of that decrease. I venture, with much diffidence, to call this the *slope* of P. Lamé calls the *magnitude* of ∇P the 'first differential operator' of P,[2] but its *direction* does not enter into his conception. We require a vector word, which shall indicate both direction and magnitude, and one not employed in another mathematical sense.

Maxwell (ibid.) also indicates the above-mentioned origin of the word: "I have taken the liberty of extending the ordinary sense of the word slope from topography, where only two independent variables are used, to space of three dimensions". The term *slope* then enters Maxwell's *Treatise on Electricity and Magnetism* (Maxwell 1873, p. 15), but in the second edition, the term *space-variation* is used instead (Maxwell 1881, p. 16). He says he is "using the phrase to indicate the direction, as well as the magnitude, of the most rapid decrease of Ψ".[3]

Since Maxwell does not appear to have used the term *gradient* himself, the question arises where it is found first. This is the case in a French translation of the second edition of Maxell's *Treatise*, where French *gradient* is found as a free translation of *space-variation* (Maxwell 1885, p. 17). The term was also adopted by Carvallo (1891, p. 233), who indicates that it is "sometimes" used. The English term is then found in Greenhill (1894, p. 487); German *Gradient* was used by Weber (1900, p. 213).[4] As Klein (1925, p. 71) confirms, the usage is based on the meteorological meaning.

The word *gradient* is also sometimes found instead of *slope*; this usage appears to go back to Lamb (1897, p. 67): "It is convenient to have a name for the property of a curve which is measured by the derived function. We shall use the term 'gradient' in this sense".

---

[1] The term is not found in the original French text (Privat-Deschanel 1869, p. 155). [2] See, for example, Lamé (1834, p. 225). In a letter from 1870, Maxwell commented on the terms: "slope is a vector word, whereas parameter has, to say the least, a scalar sound" (Knott 1911, p. 143).

[3] See also Tai (1995, p. 40). The term *space-variation* is also in the third edition of the *Treatise*.

[4] Wilson (1901, p. 138) also calls it *derivative* but gives *gradient* and *slope* as synonyms.

References

[anon.] (1839). "The Railway System in Ireland". *The Dublin Review* 6, pp. 207–269.

Carvallo, Emmanuel (1891). "Sur les systèmes linéaires, le calcul des symboles différentiels et leur application à la physique mathématique". *Monatshefte für Mathematik und Physik* 2, pp. 177–216, 225–266, 311–330.

Greenhill, Alfred George (1894). *A Treatise on Hydrostatics*. London.

Klein, Felix (1925). *Elementarmathematik vom höheren Standpunkte aus*. Zweiter Band. Ausgearbeitet von E. Hellinger. Berlin.

Knott, Cargill Gilston (1911). *Life and Scientific Work of Peter Guthrie Tait*. Cambridge.

Lamb, Horace (1897). *An Elementary Course on the Infinitesimal Calculus*. Cambridge.

Lamé, Gabriel (1834). "Mémoire sur les lois de l'équilibre du fluide éthéré". *Journal de l'École polytechnique* 15, cahier 23, pp. 191–288.

Maxwell, James Clerk (1869). "Remarks on the Mathematical Classification of Physical Quantities". *Proceedings of the London Mathematical Society* 3, pp. 224–233.

Maxwell, James Clerk (1873). *A Treatise on Electricity and Magnetism*. Vol. I. Oxford.

Maxwell, James Clerk (1881). *A Treatise on Electricity and Magnetism*. Vol. I. Second Edition. Oxford.

Maxwell, James Clerk (1885). *Traité d'électricité et de magnétisme*. Traduit de l'anglais sur la deuxième édition, par G. Séligman-Lui. Tome I. Paris.

Navier, Claude-Louis (1835). "Note sur la comparaison des avantages respectifs des diverses lignes de chemin de fer, et sur l'emploi des machines locomotives". *Annales des ponts et chaussées* 1, pp. 129–179.

Navier, Claude-Louis (1836). *On the Means of Comparing the Respective Advantages of Different Lines of Railway and on the Use of Locomotive Engines*. Translated from the French by John MacNeill. London.

Privat-Deschanel, Auguste (1869). *Traité élémentaire de physique*. Paris.

Privat-Deschanel, Auguste (1870). *Elementary Treatise on Natural Philosophy*. Translated and edited, with extensive additions, by J.D. Everett. Part I. London.

Tai, C.T. (1995). *A Historical Study of Vector Analysis* (Technical Report RL 915). Ann Arbor: The University of Michigan.

Van Tricht, R.P. (1880). "La météorologie et les stations météorologiques belges". *Annales de la Société scientifique de Bruxelles* 4, pp. 193–259.

Weber, Heinrich (1900). *Die partiellen Differential-Gleichungen der mathematischen Physik*. Nach Riemann's Vorlesungen. In vierter Auflage. Neu bearbeitet von Heinrich Weber. Erster Band. Braunschweig.

Wilson, Edwin B. (1901). *Vector Analysis*. New York and London.

graph* and *graph theory

The term originally is a shortened form of *graphic notation*, a representation of atoms and bonds used in chemistry (Brown 1864, p. 708, Frankland 1866, pp. 23–26). According to Biggs, Lloyd, and Wilson (1998, pp. 64f.), Clifford carried over the graphic notation from chemistry to invariant theory, and Sylvester conceived an analogy between chemistry and algebra he believed to have been deeper than it eventually turned out to be.[1]

It was in this context that Sylvester (1878, p. 284) introduced the term into mathematics: "Every invariant and covariant thus becomes expressible by a *graph* precisely identical with a Kekuléan diagram or chemicograph". König (1936, p. 1), who lists a number of alternative terms, commented on *graph* thus: "As a short, precise designation workable in all languages it appears to us the most suitable".[2]

---

[1] The analogy is discussed in detail by Griffiths (1964). [2] "Als kurze, prägnante und in allen Sprachen brauchbare Bezeichnung scheint sie uns die entsprechendste zu sein". Translation from König (1990, p. 77).

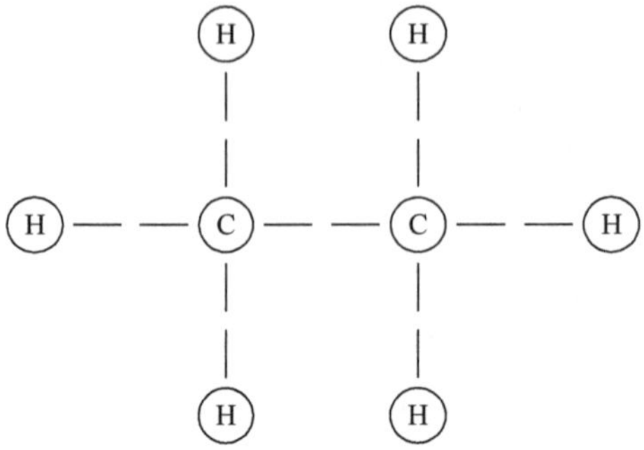

Figure 48: Graphic notation for C_2H_6. From Brown (1864, p. 708).

About 30 years before Sylvester, Listing (1847, p. 867) had already considered what he called a *Linearkomplexion* 'line complex', but this term did not catch on for long.[1] Other suggestions include French *assemblages de lignes* 'line assemblages' (Jordan 1869, p. 185); the now standard French term *réseau*, lit. 'net', was suggested by Lucas (1882, p. 46). In German, Lippich (1874, p. 92) used *Liniensystem*, lit. 'line system', while Petersen (1891, p. 194) took over the English term in unchanged form: "English authors have given the name *graph* to such a diagram; I shall retain this name".[2]

A general commment on the terminology in graph theory was made by Harary (1969, p. 8):

> Most graph theorists use personalized terminology in their books, papers, and lectures. In order to avoid quibbling at conferences on graph theory, it has been found convenient to adopt the procedure that each man state in advance the graph theoretic language he would use. Even the very word "graph" has not been sacrosanct. Some authors actually define a "graph" as a graph, but others intend such alternatives as multigraph, pseudograph, directed graph, or network. We believe that uniformity in graphical terminology will never be attained, and is not necessarily desirable.

The term *theory of graphs* appeared in the subtitle of Clifford (1881). It became more common in the 20th century and was used (as German *Graphentheorie*), for example, by Dehn (1910, p. 179) and König (1916, p. 462).

Note. For the history of graphs and graph theory, see Biggs, Lloyd, and Wilson (1998). König (1990) contains numerous valuable historical comments. See also Sachs (1989).

References

Ahrens, Wilhelm (1918). *Mathematische Unterhaltungen und Spiele*. Zweiter Band. Leipzig.
Biggs, Norman L., E. Keith Lloyd, and Robin J. Wilson (1998). *Graph Theory, 1736-1936*. Oxford.
Brown, Alexander Crum (1864). "On the Theory of Isomeric Compounds". *Transactions of the Royal Society of Edinburgh* 23, pp. 707–719.

[1] Ahrens (1918, p. 171) regarded it as undesirable because the term *linearer Komplex* had acquired a different meaning in the meantime. [2] "Englische Verfasser haben für ähnliche Figuren den Namen *graph* eingeführt; ich werde diesen Namen beibehalten". Translation from Biggs, Lloyd, and Wilson (1998, p. 191).

Clifford, William Kingdon (1881). *Mathematical Fragments*. Being Facsimiles of His Unfinished Papers Relating to the Theory of Graphs. London.
Dehn, Max (1910). "Topologie". In: *Repertorium der höheren Mathematik*. Erster Band. Geometrie. Ed. by Paul Epstein and Heinrich Emil Timerding. Leipzig and Berlin, pp. 174–194.
Frankland, Edward (1866). *Lecture Notes for Chemical Students*. Embracing Mineral and Organic Chemistry. London.
Griffiths, J. S. (1964). "Sylvester's chemico-algebraic theory. A partial anticipation of modern quantum chemistry". *Mathematical Gazette* 48, pp. 57–65.
Harary, Frank (1969). *Graph Theory*. Reading, Mass.
Jordan, Camille (1869). "Sur les assemblages de lignes". *Journal für die reine und angewandte Mathematik* 70, pp. 185–190.
König, Dénes (1916). "Über Graphen und ihre Anwendung auf Determinantentheorie und Mengenlehre". *Mathematische Annalen* 77, pp. 453–465.
König, Dénes (1936). *Theorie der endlichen und unendlichen Graphen*. Leipzig.
König, Dénes (1990). *Theory of Finite and Infinite Graphs*. Translated by Richard McCoart. Boston.
Lippich, Ferdinand (1874). "Bemerkung zu einem Satze aus Riemann's Theorie der Functionen einer veränderlichen complexen Grösse". *Sitzungsberichte der Kaiserlichen Akademie der Wissenschaften (mathematisch-naturwissenschaftliche Klasse)* 69, pp. 91–100.
Listing, Johann Benedict (1847). "Vorstudien zur Topologie". *Göttinger Studien* 1, pp. 811–875.
Lucas, Édouard (1882). *Récréations mathématiques*. Les traversées. Les ponts. Les labyrinthes. Les reines. Le solitaire. La numération. Le baguenaudier. Le taquin. Paris.
Petersen, Julius (1891). "Die Theorie der regulären Graphs". *Acta Mathematica* 15, pp. 193–220.
Sachs, Horst (1989). "Einige Gedanken zur Geschichte und zur Entwicklung der Graphentheorie". *Mitteilungen der Mathematischen Gesellschaft in Hamburg* 9, pp. 623–641.
Sylvester, James Joseph (1878). "Chemistry and Algebra". *Nature* 17 (432), p. 284.

graph (function)

The term *graph* of a function appears to be due to Chrystal (1886, p. 307): "The representative point will therefore trace out a *continuous* curve [...] This curve we may call the graph of the function". This usage was already pointed out by OED2, and earlier ones do not seem to exist, neither in English nor in other languages. Indeed, Greenhill (1891, p. 7) witnesses its recency and points out its originator: "the curve APB is now called the graph of the function fx (Chrystal, Algebra I, chap. xv.)".

The term is most likely a shortening of *graphical representation* or a similar expression. Such expressions had been in use earlier. For example, Schlömilch (1845, p. 16), in a chapter on the "geometrical representation" ("geometrische Darstellung") of functions, calls the graph the *image* (*Bild*); von Ott (1865, p. 9) named the graph a *graphical image* (*graphisches Bild*).

Chrystal's term first spread in English texts such as Wentworth (1891, p. 453). It gradually also entered other languages; for example, Osgood (1907, e. g., p. 89) used German *Graph*. It took much longer until French *graphe* was established. While the term was already used by de Possel (1932, p. 21), Bourbaki (1939, p. 16) at first preferred *ensemble représentatif*; however, Bourbaki (1952) contains a sheet of errata according to which the relevant passage in the 1939 volume should be changed to *ensemble représentatif ou graphe*. In Bourbaki (1954, p. 71), *graphe* became the regularly used term.

References

Bourbaki, Nicolas (1939). *Théorie des ensembles*. Fascicule de résultats. Paris.

Bourbaki, Nicolas (1952). *Algèbre*. Chapitre VI. Groupes et corps ordonnés. Chapitre VII. Modules sur les anneaux principaux. Paris.
Bourbaki, Nicolas (1954). *Théorie des ensembles*. Chapitre I. Déscription de la mathématique formelle. Chapitre II. Théorie des ensembles. Paris.
Chrystal, George (1886). *Algebra*. An Elementary Text Book for the Higher Classes of Secondary Schools and for Colleges. Part I. Edinburgh.
Greenhill, Alfred George (1891). *Differential and Integral Calculus With Applications*. Second edition. London.
Osgood, William F. (1907). *Lehrbuch der Funktionentheorie*. Leipzig and Berlin.
von Ott, Karl (1865). "Graphische Darstellung der Funktionen". In: *Fünftes Programm der k.k. deutschen Ober-Realschule in Prag*. Prag, pp. 1–38.
de Possel, René (1932). "Quelques problèmes de représentation conforme". *Journal de l'École polytechnique* 30, pp. 1–98.
Schlömilch, Oskar (1845). *Handbuch der algebraischen Analysis*. Jena.
Wentworth, George Alber (1891). *A Higher Algebra*. Boston.

great circle

The Greek term *ὁ μέγιστος κύκλος* literally means 'largest circle' and was already used by Autolycus of Pitane in his book *On a moving sphere* (Mogenet 1950, p. 197). This terminology was in common use in ancient Greek.

In 12th- and 13th century Latin texts and translations, the terminology varies. For example, in a translation of Ptolemy's *Almagest* that dates from the middle of the 12th century, *circulus maximus*, which corresponds to the Greek term, is used.[1] Gerard of Cremona, in his translation of Jābir ibn Aflaḥ's "Correction of the Almagest", used *circulus magnus* instead (ibn Aflaḥ 1534, p. 3).[2] In the widely diffused *Sphaera* by Johannes de Sacrobosco, a great circle is called *maior*: "For a great circle in the sphere is one which, described on the surface of the sphere about its center, divides the sphere into two equal parts, while a small circle is one which, described on the surface of the sphere, divides it not into two equal but into two unequal portions" (Thorndike 1949, pp. 85f.).[3]

The French term *grand cercle* was used in the so-called *Calendrier des bergers* ([anon.] 1493, fol. I ii^{v}), and English *great circle* appears in an English translation of that text from 1503 (Sommer 1892, fol. i ii^{r}).[4]

Note. For brief notes on the history of the term, see Tropfke (1923, p. 115); for the Greek term, see Mugler (1958-59, pp. 260f.).

References

[anon.] (1493). *[Le Kalendrier des bergers]*. [Paris].
[anon.] (1530?). *Here begynneth The Compost of Ptholomeus, Prynce of Astronomye*. Translated oute of Frenche in to Englyssh. London.
ibn Aflaḥ, Jābir (1534). *Gebri Filii Affla Hispalensis De astronomia libri IX* [Translated by Gerard of Cremona]. In: Apian, Peter. *Instrumentum primi mobilis*. Nürnberg.

---

[1] See `https://ptolemaeus.badw.de/ms/227/327/transcription/1`, section I. 8. [2] The Arabic terminology is based on a word meaning 'great'; it was used, for example, by al-Bīrūnī (1934, p. 20). [3] "Maior enim circulus in spera dicitur qui, descriptus in superficie spere super eius centrum, speram dividit in duo equalia; minor vero qui, descriptus in superficie spere, eam non dividit in duo equalia sed in portiones inequales". Translation from Thorndike (1949, p. 123).
[4] The OED cites [anon.] (1530?, fol. g iv^{v}) as a first use.

al-Bīrūnī, Abū Rayḥān Muḥammad ibn Aḥmad (1934). *The Book of Instruction in the Elements of the Art of Astrology*. Written in Ghaznah, 1029 A. D. Reproduced from Brit. Mus. MS. Or. 8349. The Translation facing the Text by R. Ramsay Wright. London.
Mogenet, Joseph (1950). *Autolycus de Pitane*. Histoire du texte suivie de l'édition critique des traités de la sphère en mouvement et des levers et couchers. Louvain.
Mugler, Charles (1958-59). *Dictionnaire historique de la terminologie géométrique des Grecs*. Paris.
Sommer, H. Oskar, ed. (1892). *The Kalender of Shepherdes*. The Edition of Paris 1503 in Photographic Facsimile, a Faithful Reprint of R. Pynson's Edition of London 1506. Vol. II. London.
Thorndike, Lynn (1949). *The Sphere of Sacrobosco and Its Commentators*. Chicago.
Tropfke, Johannes (1923). *Geschichte der Elementar-Mathematik*. Fünfter Band. I. Ebene Trigonometrie. II. Sphärik und sphärische Trigonometrie. Berlin and Leipzig.

greatest common divisor

In Greek mathematics, division was expressed by the notion of measuring. Correspondingly, the term for the greatest common divisor was called *μέγιστον κοινὸν μέτρον*, which literally means 'greatest common measure'. The term was used by Euclid (*Elements*, book VII, prop. 2): "Given two numbers not prime to one another, to find their greatest common measure".[1]

In the 12th-century translations of Euclid's *Elements*, *maximus numerus communis* is usually used, for example by Adelard (Busard 1983, p. 198); in a translation made directly from the Greek, *maxima mensura communis* is used instead (Busard 1987, p. 155), a term already known to Martianus Capella (Willis 1983, p. 294). Of later origin is *maximus communis divisor*; it was used by Ramus (1562, p. 24, 1569, p. 15) and soon became standard in Latin texts.[2]

In modern languages, corresponding terms appear to occur first in Italian. Cataneo (1546, fol. 14^{r}), for example, uses Italian *maggior commune ripiego*.[3] The term was not new at the time, however, as it can already be found in a 15th-century manuscript (Arrighi 2004, p. 174). In French, Forcadel uses *la plus grande commune mesure* (Euclid 1565, p. 16), as does Stevin (1585, p. 87); *le plus grand diviseur commun* appears in Prestet (1675, p. 44).

Similarly, Billingsley has *greatest common measure* (Euclid 1570, fol. 188^{v}), while in Ramus (1592, p. 25), *greatest common diuider* is preferred. English *greatest common divisor* is first found (OED) in Leybourn (1657, p. 104), while *highest common factor* was introduced by Peacock (1830, p. 126).

In the 19th century, *greatest common divisor* and *least common multiple* begin to be used in algebraic number theory and set theory; see *intersection* (sets) and *union*.

Note. For brief overviews of the terminology, see Tropfke (1930, p. 136, 1980, pp. 251f.) and Smith (1958, p. 222).

---

[1] Translation from Euclid (1956, p. 298). The least common multiple, which has not been traced here any further, is also found in the *Elements* (book VII, prop. 34). For historical information on the term, seen (Tropfke 1980, pp. 251f.). [2] In earlier editions of Ramus' *Arithmetica*, the Latinized Euclidean terminology is used. [3] This usage was pointed out by Smith (1958, p. 222). The verb *piegare* literally means 'to fold'. *ripiego* thus means 'refolding' and used to be a term for 'divisor'. See Ricci (1994, pp. 55f.).

References

Arrighi, Gino (2004). "Nuovi contributi per la storia della matematica in Firenze nell'età di mezzo: Il codice Palatino 573 della Biblioteca Nazionale di Firenze". In: *La matematica nell'Età di Mezzo*. Scritti scelti. A cura di Francesco Barberini, Raffaella Franci, Laura Toti Rigatelli. Pisa, pp. 159–194.

Busard, H.L.L. (1983). *The First Latin Translation of Euclid's Elements Commonly Ascribed to Adelard of Bath*. Books I-VIII and books X.36-XV.2. Toronto.

Busard, H.L.L. (1987). *The Mediaeval Latin Translation of Euclid's Elements Made Directly from the Greek*. Stuttgart.

Cataneo, Pietro (1546). *Le pratiche delle due prime mathematiche di Pietro de Catani da Siena*. Venezia.

Euclid (1565). *Les septième, huictième et neufième livres des Élémens d'Euclide*. Traduicts & commentez par Pierre Forcadel de Besies. Paris.

Euclid (1570). *The Elements of Geometrie of the most auncient Philosopher Euclide of Megara*. Faithfully (now first) translated into the Englishe toung, by H. Billingsley. London.

Euclid (1956). *The Thirteen Books of Euclid's Elements*. Volume II. Books III-IX. Ed. by Thomas L. Heath. New York.

Leybourn, William (1657). *Arithmetick, vulgar, decimal, instrumental, and algebraical*. London.

Peacock, George (1830). *A Treatise on Algebra*. Cambridge.

Prestet, Jean (1675). *Elemens des mathematiques*. Paris.

Ramus, Petrus (1562). *Arithmetica*. Paris.

Ramus, Petrus (1569). *Arithmeticae libri duo. Geometriae septem et viginti*. Basel.

Ramus, Petrus (1592). *The Art of Arithmeticke in Whole Numbers and Fractions*. Translated into English by William Kempe. London.

Ricci, Laura (1994). "Il lessico matematico della Summa di Luca Pacioli". *Studi di lessicografia italiana* 12, pp. 5–71.

Smith, David Eugene (1958). *History of Mathematics*. Volume II. Special Topics of Elementary Mathematics. New York.

Stevin, Simon (1585). *L'Arithmetique*. Leiden.

Tropfke, Johannes (1930). *Geschichte der Elementar-Mathematik*. Erster Band. Rechnen. Dritte, verbesserte und vermehrte Auflage. Berlin and Leipzig.

Tropfke, Johannes (1980). *Geschichte der Elementarmathematik*. Band 1: Arithmetik und Algebra. Vollständig neu bearbeitet von Kurt Vogel, Karin Reich, Helmuth Gericke. Berlin and New York.

Willis, James, ed. (1983). *Martianus Capella*. Leipzig.

group

The general etymology of this word is unclear, but it first appears as Italian *gruppo*/*groppo* 'heap'. The term was commonly used in the arts; according to Bosse (1649, sommaire), the word was used "when we see in a painting two, three, four or more figures or other bodies together".[1] In the 18th century, the word began to be used in other scientific contexts as well, for objects classed together on account of some scientific criterion.[2] A mathematical example of this general scientific meaning is found in Cauchy (1815, p. 21), who used French *groupe* for certains set of indices.[3]

The use of the term in algebra is based on this general scientific sense. Évariste Galois applied French *groupe* to collections of permutations which belong together because they are derived from the same equation. In particular, he did not only use

---

[1] "lorsqu'on voit dans un Tableau, deux, trois, quatre, ou plusieurs Figure ou autres Corps ensemble". Translation from Dethlefs (2018, p. 227). Dethlefs' article contains a thorough discussion of the term used in this sense. [2] The TLF gives the meaning 'set of beings or things having some properties in common which are used to classify them' and dates this meaning 1726. [3] Cauchy's use of the word is explained by Dahan (1980) and Kiernan (1971).

groupe for sets of permutations that are closed but also applied the term to cosets.[4] An example for *groupe* being used in the sense of a closed collection of objects is found in Galois' "First memoir": "if in such a group one has substitutions S and T, one is sure to have the substitution ST" (Galois 2011, p. 114).[1] Galois (2011, p. 166) also spoke of the *group of an equation* (*groupe de l'équation*).

Cayley (1854, p. 41), referring to Galois, transferred the term to English and gave an abstract definitions of *group*: "A set of symbols $1, \alpha, \beta, ...$, all of them different, and such that the product of any two of them (no matter in what order) or the product of any one of them into itself, belongs to the set, is said to be a *group*". Cayley also presented examples other than permutation groups.[2] In Germany, *Gruppe* was first used in the 1850s as well, by Dedekind in lectures (see Dedekind 1981, p. 64) and by Kronecker (1856, p. 211). Both Dedekind and Kronecker, like Cayley and (partly) Galois, used only closure as a defining criterion.

In the early 1880s, the first axiomatic definitions similar to the ones that are used today came up. Such a definition is found in Weber (1882, p. 302); Dyck (1883, p. 74), who considered free groups, explicitly demanded: "We require for our considerations that a group which contains the operation T_k must also contain its inverse T_k^{-1}".[3] Further axiomatic definitions were given by Hölder (1889, p. 29) and Weber (1893, p. 522):

> A system $\mathfrak{S}$ of finitely many or infinitely many arbitrary things (elements) forms a *group* if it satisfies the following conditions.
>
> 1) *A rule is given that assigns to a first and second element of the system a definite third element of that system* [...]
> 2) *We assume the associative law* [...]
> 3) *We assume that if* $AB = AB'$ *or if* $AB = A'B$, *then, necessarily,* $B = B'$ *or* $A = A'$. If $\mathfrak{S}$ contains a finite number of elements, then the group is called a *finite* group and the number of its elements is called its *degree*. In the case of finite groups 1), 2), and 3) imply that
> 4) *If two of three elements* A, B, C *are taken arbitrarily from* $\mathfrak{S}$ *then the third can always be uniquely determined so that* $AB = C$.[4].

[4] In other words, Galois used *groupe* not just for groups in the modern mathematical sense of the word, but also for sets which are *not* groups in this sense (see also Wußing 1984, pp. 111–114). Until well into the 19th century, the only criterion for groups was closure. [1] "si dans un pareil groupe on a les substitutions S et T, on est sûr d'avoir la substitution ST". Translation from Galois (2011, p. 115). [2] For Cayley's concept of group, see Ehrhardt (2016, pp. 105–108) and Wußing (1984, pp. 230–233). [3] "Wir setzen nun für unsere Betrachtungen noch fest, dass eine Gruppe, welche die Operation T_k enthält, stets auch ihre inverse Operation T_k^{-1} mit einbegreifen soll". Translation from Kleiner (2007, p. 33). [4] "Ein System $\mathfrak{S}$ von Dingen (Elementen) irgend welcher Art in endlicher oder unendlicher Anzahl wird zur Gruppe, wenn folgende Voraussetzungen erfüllt sind. 1) *Es ist eine Vorschrift gegeben, nach der aus einem ersten und einem zweiten Element des Systems ein ganz bestimmtes drittes Element desselben Systems abgeleitet wird* [...] 2) [wird] *das associative Gesetz vorausgesetzt* [...] 3) *Es wird vorausgesetzt, dass, wenn* $AB = AB'$ *oder* $AB = A'B$ *ist, nothwendig* $B = B'$ oder $A = A'$ *sein muss*. Wenn $\mathfrak{S}$ eine endliche Anzahl von Elementen umfasst, so heisst die Gruppe eine *endliche* und die Anzahl ihrer Elemente ihr *Grad*. Bei endlichen Gruppen ergibt sich aus 1) 2) 3) die Folgerung 4) *Wenn von den drei Elementen* A, B, C *zwei beliebig aus* $\mathfrak{S}$ *genommen werden, so kann man das dritte immer und nur auf eine Weise so bestimmen, dass* $AB = C$ *ist*". Translation from Wußing (1984, pp. 247f.)

Weber (1893, p. 523) goes on to point out that in the case of infinite groups, the last condition is a required axiom for a group, and he also points out the special case of commutative groups and gives various examples. Wußing (1984, p. 247) remarks that "[w]hat is new in Weber is the consistently abstract axiomatic definition of a group, a definition that includes infinite groups". His definition spread through his widely read textbook on algebra (Weber 1896, pp. 3f.). Soon afterwards, the first monograph on abstract group theory appeared (de Séguier 1904).

Thus, by the end of the 19th century the essential properties of groups had become clear.

To conclude this entry, some early competitors of the term shall be discussed in brief. The first word to denote the concept[1] may be said to be Italian *permutazione*. This terminology is found in Ruffini (1799, ch. 13) and Abbati (1803) but remained restricted to Italian, however.[2]

Cauchy's term *système de substitutions conjuguées* 'system of conjugate substitutions' appears frequently in 19th-century texts. He defined it thus: "I shall call *derived* substitutions all that can arise from the given substitutions by multiplying them one or more times by each other or by themselves; and the given substitutions together with all the derived substitutions form what I shall call a *system of conjugate substitutions*" (Cauchy 1844, p. 183).[3] According to Silvestri (1979, p. 317), Cauchy "did not use *conjugate* in its modern meaning; the permutations are 'conjugate' in that they 'join together' in a group structure".

Jordan (1870, p. 22), who used both Galois' and Cauchy's terms, also suggested the additional alternative *faisceau*, lit. 'bundle': "One says that a system of substitutions forms a *group* (or a *bundle*) if the product of any two of the substitutions of the system is again a member of that system".[4] Jordan's alternative remained unsuccessful.

Klein and Lie (1871, p. 56) use German *geschlossenes System* 'closed system' and add that "[t]he term 'a closed system of transformations' therefore corresponds exactly to what is called 'a group of substitutions' in the theory of substitutions".[5]

While at least some of these alternatives enjoyed some popularity, in 1890 Bolza was able to state that the term *group* "is now generally adopted" (Bolza 1890-91, p. 71).

Note. The history of the concept is examined in depth in Wußing (1984), who also discusses many terminological aspects, as do Dahan (1980) and Kiernan (1971). Axiomatic definitions of groups are discussed by Hollings (2017); see also Hofmann (1992). Becker (2005, ch. 8) discusses the history of the word *group* in detail. An

---

[1] As in the early examples cited above, in the following examples only closure is used as a defining criterion for groups. [2] On this term, see Burkhardt (1892, p. 133), Wußing (1984, p. 81 and p. 267, fn. 57), and Cassinet (1988, pp. 32f.). [3] "je nommerai substitutions *dérivées* toutes celles que l'on pourra déduire des substitutions données, multipliées une ou plusieurs fois les unes par les autres, ou par elles-mêmes, dans un ordre quelconque; et les substitutions données, jointes aux substitutions dérivées, formeront ce que j'appellerai un *système de substitutions conjuguées*". Translation from Burns (1913, p. 145). [4] "On dira qu'un système de substitutions forme un *groupe* (ou un *faisceau*) si le produit de deux substitutions quelconques du système appartient lui-même au système". Translation from Wußing (1984, p. 142). [5] "Der Ausdruck ‚ein geschlossenes System von Transformationen' entspricht also ganz dem, was man in der Theorie der Substitutionen als ‚eine Gruppe von Substitutionen' zu bezeichnen pflegt".

excellent discussion of the meanings of *group* in the 19th century is Ehrhardt (2016). For a thorough account of the spread of Galois' ideas, see also Ehrhardt (2012).

References

Abbati, Pietro (1803). "Lettera di Pietro Abbati al socio Paolo Ruffini". *Memorie di matematica e di fisica della Società italiana delle scienze* 10 (parte due), pp. 385–406.

Becker, Holger (2005). "Semantische und lexikalische Aspekte der mathematischen Fachsprache des 19. Jahrhunderts". PhD thesis. Carl von Ossietzky Universität Oldenburg. URL: `http://oops.uni-oldenburg.de/59/`.

Bolza, Oskar (1890-91). "On the Theory of Substitution-Groups and its Applications to Algebraic Equations". *American Journal of Mathematics* 13, pp. 59–144.

Bosse, Abraham (1649). *Sentimens sur la distinction des diverses manieres de peinture, dessein & graveure*. Paris.

Burkhardt, Heinrich (1892). "Die Anfänge der Gruppentheorie und Paolo Ruffini". *Abhandlungen zur Geschichte der Mathematik* 6, pp. 119–159.

Burns, Josephine E. (1913). "The Foundation Period in the History of Group Theory". *The American Mathematical Monthly* 20, pp. 141–148.

Cassinet, Jean (1988). "Paolo Ruffini (1765-1822): la résolution algébrique des équations et les groupes de permutations". *Bolletino di storia delle scienze matematiche* 8, pp. 21–69.

Cauchy, Augustin-Louis (1815). "Sur le nombre des valeurs qu'une fonction peut acquérir, lorsqu'on y permute de toutes les manières possibles les quantités qu'elle renferme". *Journal de l'École polytechnique* 10, cahier 17, pp. 1–28.

Cauchy, Augustin-Louis (1844). *Exercises d'analyse et de physique mathématique*. Tome troisième. Paris.

Cayley, Arthur (1854). "On the theory of groups, as depending on the symbolic equation $\theta^n = 1$". *The London, Edinburgh, and Dublin Philosophical Magazine and Journal of Science* 7, pp. 40–47, 408–409.

Dahan, Amy (1980). "Les travaux de Cauchy sur les substitutions. Étude de son approche du concept de groupe". *Archive for History of Exact Sciences* 23, pp. 279–319.

Dedekind, Richard (1981). "Eine Vorlesung über Algebra". In: *Richard Dedekind 1831-1981. Eine Würdigung zu seinem 150. Geburtstag*. Ed. by Winfried Scharlau. Braunschweig and Wiesbaden, pp. 59–100.

Dethlefs, Hans-Joachim (2018). "Group". In: *LexArt. Words for Painting*. (France, Germany, England, The Netherlands, 1600-1750). Ed. by Michèle-Caroline Heck. Montpellier, pp. 227–232.

Dyck, Walther (1883). "Gruppentheoretische Studien. II. Ueber die Zusammensetzung einer Gruppe discreter Operationen, über ihre Primitivität und Transitivität". *Mathematische Annalen* 22, pp. 70–108.

Ehrhardt, Caroline (2012). *Itinéraire d'un texte mathématique*. Les réélaborations d'un mémoire d'Évariste Galois au XIXe siècle. Paris.

Ehrhardt, Caroline (2016). "Notations, proof practices and the circulation of mathematical objects. The example of groups (1800-1860)". *Khronos* 3, pp. 93–121.

Galois, Évariste (2011). "The First Memoir". In: *The Mathematical Writings of Évariste Galois*. Ed. by Peter M. Neumann. Zürich, pp. 105–168.

Hofmann, Karl H. (1992). "Zur Geschichte des Halbgruppenbegriffs". *Historia Mathematica* 19, pp. 40–59.

Hölder, Otto (1889). "Zurückführung einer beliebigen algebraischen Gleichung auf eine Kette von Gleichungen". *Mathematische Annalen* 34, pp. 26–56.

Hollings, Christopher D. (2017). "'Nobody could possibly misunderstand what a group is': a study in early twentieth-century group axiomatics". *Archive for History of Exact Sciences* 71, pp. 409–481.

Jordan, Camille (1870). *Traité des substitutions et des équations algébriques*. Paris.

Kiernan, B. Melvin (1971). "The Development of Galois Theory from Lagrange to Artin". *Archive for History of Exact Sciences* 8, pp. 40–154.

Klein, Felix and Sophus Lie (1871). "Ueber diejenigen ebenen Curven, welche durch ein geschlossenes System von einfach unendlich vielen vertauschbaren Transformationen in sich übergehen". *Mathematische Annalen* 4, pp. 50–84.
Kleiner, Israel (2007). *A History of Abstract Algebra*. Basel.
Kronecker, Leopold (1856). "Über die algebraisch auflösbaren Gleichungen". *Monatsberichte der Königlich Preußischen Akademie der Wissenschaften zu Berlin* 1856, pp. 203–215.
Ruffini, Paolo (1799). *Teoria generale delle equazioni, in cui si dimostra impossibile la soluzione algebraica delle equazioni generali di grado superiore al quarto*. Bologna.
de Séguier, Jean Armand (1904). *Théorie des groupes finis*. Éléments de la théorie des groupes abstraits. Paris.
Silvestri, Richard (1979). "Simple Groups of Finite Order in the Nineteenth Century". *Archive for History of Exact Sciences* 20, pp. 313–356.
Weber, Heinrich (1882). "Beweis des Satzes, dass jede eigentlich primitive quadratische Form unendlich viele Primzahlen darzustellen fähig ist". *Mathematische Annalen* 20, pp. 301–329.
Weber, Heinrich (1893). "Die allgemeinen Grundlagen der Galois'schen Gleichungstheorie". *Mathematische Annalen* 43, pp. 521–549.
Weber, Heinrich (1896). *Lehrbuch der Algebra*. Zweiter Band. Braunschweig.
Wußing, Hans (1984). *The Genesis of the Abstract Group Concept*. A Contribution to the History of the Origin of Abstract Group Theory. Tr. Abe Shenitzer. Cambridge, Mass.

group character

See *character*.

groupoid and *magma*

The term *groupoid* was introduced by Brandt (1927, pp. 360–362).[1] He used German *Gruppoid*, which means 'group-like'. Brandt defined a groupoid as a finite set on which a product is defined (but not necessarily for all pairs of elements of the set) for which four axioms hold; the axioms concerned the uniqueness of the product (if it exists), certain associativity requirements, and requirements for units and inverse elements. The groupoid Brandt defined was later called a *Brandt groupoid*.[2] Brandt's term was quickly adopted, for example by Speiser (1927, p. 5).

A different definition was given by Birkhoff (1934, p. 351); by a groupoid he understood "any system whose elements satisfy the first three group axioms", namely the existence of the product, associativity, the the existence of a neutral element. Birkhoff referred to Specht (1933, p. 323), where German *Gruppe* was used for this concept.

Even weaker requirements were made by Hausmann and Ore (1937, p. 984): "We consider systems $\mathfrak{A}$ consisting of an arbitrary finite or infinite number of elements a_1, $a_2, \ldots$. Such a system shall be called a *groupoid* if it satisfies the PRODUCT AXIOM. *To any two elements a' and a'' in $\mathfrak{A}$ there corresponds a unique third $a = a' \cdot a''$*".[3]

The alternative term *magma* appears to have been first suggested in print by Serre (1965, p. LA 4.1): "A set M with a map $M \times M \to M$ denoted by $(x, y) \mapsto xy$ is called a *magma*". Bénabou (1968, p. 85) attributed it to Bourbaki, who used it in a revised edition of the book *Algèbre* (Bourbaki 1970, p. 1). The term may be based on the meaning 'confused mixture' ("ensemble confus d'éléments", TLF).

---

[1] That Brandt coined this term has been pointed out by various authors, for example by Borůvka (1974, p. 138) and Brown (1987, p. 113). [2] Apparently, Clifford (1942, p. 327) was the first who used this term. [3] The property here used as a definition is exactly the property that a Brandt groupoid does *not* need to have.

Note. This notion has been used in a wide variety of mathematical fields, with various definitions. For surveys and historical discussion, see Brown (1987) and Krömer (2004, pp. 193–197, 2013, pp. 939–947).

References

Bénabou, Jean (1968). "Structures algébriques dans les catégories". *Cahiers de topologie et géométrie différentielle catégoriques* 10, pp. 1–126.
Birkhoff, Garrett (1934). "Hausdorff Groupoids". *Annals of Mathematics* 35, pp. 351–360.
Borůvka, Otakar (1974). *Foundations of the Theory of Groupoids and Groups*. Berlin.
Bourbaki, Nicolas (1970). *Algèbre*. Chapitres 1 à 3. Paris.
Brandt, Heinrich (1927). "Über eine Verallgemeinerung des Gruppenbegriffes". *Mathematische Annalen* 96, pp. 360–366.
Brown, Ronald (1987). "From Groups to Groupoids: a Brief Survey". *Bulletin of the London Mathematical Society* 19, pp. 113–134.
Clifford, A.H. (1942). "Matrix Representations of Completely Simple Semigroups". *American Journal of Mathematics* 64, pp. 327–342.
Hausmann, B.A. and Øystein Ore (1937). "Theory of Quasi-Groups". *American Journal of Mathematics* 59, pp. 983–1004.
Krömer, Ralf (2004). "Die Kategorientheorie: ihre mathematischen Leistungen, ihre erkenntnistheoretischen Implikationen. Eine historische und philosophische Würdigung". PhD thesis. Université Nancy 2.
Krömer, Ralf (2013). "The set of paths in a space and its algebraic structure. A historical account". *Annales de la faculté des sciences de Toulouse* 22, pp. 915–968.
Serre, Jean-Pierre (1965). *Lie Algebras and Lie Groups*. 1964 Lectures given at Harvard University. New York and Amsterdam.
Specht, Wilhelm (1933). "Eine Verallgemeinerung der Permutationsgruppen". *Mathematische Zeitschrift* 37, pp. 321–341.
Speiser, Andreas (1927). *Die Theorie der Gruppen von endlicher Ordnung*. Zweite Auflage. Berlin and Heidelberg.

gyroid

This denomination for a particular kind of minimal surface was taken from the language of cristallographers, where *gyroidal* (Greek *γύρος* 'ring, circle, spiral') means 'spiral' or 'gyratory'. Miller (n.d.) cites a letter by Schoen from 1969 concerning the origin of the term: "The gyroid. This is my latest choice of a name for this surface, which is the only surface associate to the two intersection-free adjoint Schwarz surfaces ('P' and 'D') that is free of self-intersections. (Webster's 3d International Dictionary defines gyroidal as 'spiral or gyratory in arrangement – used esp. of the planes of crystals'.)". In print, the term first appears in Schoen (1970, p. 13).[1]

References

Miller, Jeff (n.d.). "Earliest Known Uses of Some of the Words of Mathematics". URL: `https://mathshistory.st-andrews.ac.uk/Miller/mathword/`.
Schoen, Alan H. (1970). *Infinite periodic minimal surfaces without self-intersections*. NASA Technical Note. Washington, D.C.

[1] For Schoen's illustration, see Schoen (1970, p. 14).

H

Hamiltonian cycle **and** ***Hamiltonian graph***

In his famous book on graph theory, König (1936, p. 26) raised the following question: "when is there a cycle in a graph which contains every vertex of the graph? Such a cycle — which can, of course, exist only in the case of connected graphs — shall be called a *Hamiltonian cycle*".[1] The terminology was transferred to graphs by Dirac (1952, p. 71): "A graph is said to be *Hamiltonian* if it contains a circuit which embraces all the nodes of the graph".

The study of such graphs was initiated by Kirkman and Hamilton in the second half of the 1850s; for discussion and references, see Gropp (2006) and Wilson (1988).

References

Dirac, Gabriel Andrew (1952). "Some Theorems on Abstract Graphs". *Proceedings of the London Mathematical Society* 2, pp. 69–81.

Gropp, Harald (2006). "Hamiltonian graphs from Kirkman to König". *Electronic Notes in Discrete Mathematics* 24, pp. 81–88.

König, Dénes (1936). *Theorie der endlichen und unendlichen Graphen*. Leipzig.

König, Dénes (1990). *Theory of Finite and Infinite Graphs*. Translated by Richard McCoart. Boston.

Sainte-Laguë, André (1926). *Les réseaux (ou graphes)*. Paris.

Wilson, Robin J. (1988). "A Brief History of Hamiltonian Graphs". *Annals of Discrete Mathematics* 41, pp. 487–496.

handle

Handles are essential for topological classification. As with various other topological notions, they first occur in writings about the theory of complex functions. They are found in one such work by Felix Klein, who wrote: "Let the normal surface for $p = 0$ be the sphere, for $p = 1$, the anchor-ring. For greater values of p we may imagine

[1] "wann gibt es im Graphen einen Kreis, der jeden Knotenpunkt des Graphen enthält? Ein solcher Kreis — der natürlich ebenfalls nur für zusammenhängende Graphen existieren kann — soll als eine *Hamiltonsche Linie* bezeichnet werden". Translation from König (1990, p. 33) (slightly modified). – Such cycles were called *complete* by Sainte-Laguë (1926, p. 4), who also called a Hamiltonian graph *réseau cerclée*.

H. Becker, *A Historical Dictionary of Mathematical Terms*, Science Networks. Historical Studies 66, https://doi.org/10.1007/978-3-032-02873-0_8

a sphere with p appendages (handles) as in the following figure for $p = 3$" (Klein 1882, p. 27).[1]

Figure 49: Handles. From Klein (1882, p. 27).

Klein's terminology – *Anhängsel* (*appendage*) and *Handhabe* (*handle*) – was soon changed to German *Henkel* (*handle*): For example, Lüroth (1886, p. 332) first spoke of *henkelartige Ansätze* 'handle-like appendages' and then also used *Henkel* (p. 333). Also, it did not take long for the term to be adopted in other languages. For example, in Battaglini (1882, p. 130), French *anse* 'handle' is found. De Paolis (1890, p. 69) uses Italian *sfera con un manico* 'sphere with a handle'. English *handle* is also found in Harkness and Morley (1893, p. 236).

As mentioned above, handles play an important role in the classification of surfaces, which is expressed in the "principal theorem of surface topology".[2] Seifert and Threlfall (1934, p. 141) present the theorem in the following form: "Two closed surfaces are homeomorphic if and only if they have the same Euler characteristic and orientability. The most general orientable closed surface is the 2-sphere with h handles attached ($h \geq 0$). The most general nonorientable closed surface is the 2-sphere with k cross caps attached ($k \geq 1$)".[3] The number of handles of a surface, also called its *genus*, is an important topological invariant that was already known in the 19th century.[4]

Note. On the history of the notion, brief comments are found in Volkert (2002, p. 56) and Severi (1921, p. 214). Pont (1974, p. 126) briefly comments on Klein, Popescu-Pampu (2016, pp. 53–57) on Clifford. See also *unicoherent*.

[1] "Bei höherem p mag man sich eine Kugel mit p Anhängseln (Handhaben) versehen denken". Translation from Klein (1893, p. 25). – The notion is earlier found in Tonelli (1875) and Clifford (1877). See Severi (1921, p. 60) and Volkert (2002, p. 56). [2] The name *Hauptsatz der Flächentopologie* was used by Kerékjártó (1923, p. 147). [3] "Zwei geschlossene Flächen sind dann und nur dann homöomorph, wenn sie in Charakteristik und Orientierbarkeitscharakter übereinstimmen. Die allgemeinste orientierbare geschlossene Fläche ist die Kugel mit h (≥ 0) Henkeln, die allgemeinste nichtorientierbare geschlossene Fläche die Kugel mit k (≥ 1) Kreuzhauben". Translation from Seifert and Threlfall (1980, p. 145). – For the history of this theorem, which originates in the 19th century, see Kónya (2005) and the detailed account by Gallier and Xu (2013). [4] See Popescu-Pampu (2016, pp. 51–57).

References

Battaglini, Giuseppe (1882). “Klein (F.). – Ueber Riemann’s Theorie der algebraischen Functionen und ihrer Integrale [review]”. *Bulletin des sciences mathématiques et astronomiques* 6, pp. 125–136.

Clifford, William Kingdon (1877). “On the Canonical Form and Dissection of a Riemann’s Surface”. *Proceedings of the London Mathematical Society* 2, pp. 292–304.

Gallier, Jean and Dianna Xu (2013). *A Guide to the Classification Theorem for Compact Surfaces*. Berlin and Heidelberg.

Harkness, James and Frank Morley (1893). *A Treatise on the Theory of Functions*. New York and London.

Kerékjártó, Béla (1923). *Vorlesungen über Topologie*. I. Flächentopologie. Berlin.

Klein, Felix (1882). *Ueber Riemann’s Theorie der algebraischen Functionen und ihrer Integrale*. Leipzig.

Klein, Felix (1893). *On Riemann’s Theory of Algebraic Functions and Their Integrals*. Cambridge.

Kónya, Eszter (2005). “On the fundamental theorem of compact and noncompact surfaces”. *Annales Mathematicae et Informaticae* 32, pp. 211–224.

Lüroth, Jakob (1886). “Ueber die kanonischen Perioden der Abel’schen Integrale”. *Abhandlungen der Bayerischen Akademie der Wissenschaften (Mathematisch-physikalische Klasse)* 15, pp. 329–366.

De Paolis, Riccardo (1890). “Teoria dei gruppi geometrici e delle corrispondenze che si possono stabilire tra i loro elementi”. *Memorie di matematica e di fisica della Società italiana delle Scienze* 7, pp. 1–165.

Pont, Jean-Claude (1974). *La topologie algébrique des origines à Poincaré*. Paris.

Popescu-Pampu, Patrick (2016). *What is the Genus?* Cham.

Seifert, Herbert and William Threlfall (1934). *Lehrbuch der Topologie*. Leipzig.

Seifert, Herbert and William Threlfall (1980). *Seifert and Threlfall: A Textbook of Topology and Seifert: Topology of 3-dimensional fibered spaces*. Ed. by Joan S. Birman and Julian Eisner. New York et al.

Severi, Francesco (1921). *Vorlesungen über algebraische Geometrie*. Geometrie auf einer Kurve. Riemannsche Flächen. Abelsche Integrale. Berechtigte deutsche Übersetzung von Eugen Löffler. Leipzig and Berlin.

Tonelli, Alberto (1875). “Osservazioni sulla Teoria della Connessione”. *Atti della Reale Accademia dei Lincei* 2, pp. 594–601.

Volkert, Klaus (2002). *Das Homöomorphismusproblem insbesondere der 3-Mannigfaltigkeiten, in der Topologie 1892-1935*. Paris.

harmonic

The origin of this term is Greek *ἁρμονικός*, from *ἁρμονία* ‘articulation, agreement, harmony’, from *ἁρμός* ‘joint’; the adjective *ἁρμονικός* means ‘skilled in music’. The term *harmonic mean* (*μεσότης ἁρμονική*) was already used in Greek mathematics, and later gave rise to the use of *harmonic* in projective geometry (e. g., *harmonic division, harmonic pencil*) and in *harmonic series*.[1]

As a mathematical term, *ἁρμονικός* was used by the philosopher Archytas of Tarentum[2] in a discussion of three mathematical means[3]:

> There are three means in music: one is the arithmetic, the second geometric and the third subcontrary [, which they call “harmonic”].[4] [...] The mean is subcontrary, which we call harmonic, whenever they [the terms] are such that, by which part of itself the first term

---

[1] See also *spherical harmonic*. [2] Archytas may have been born between 435 and 410 (Huffman 2005, p. 5). [3] Archytas’ terms for ‘mean’ are discussed by Huffman (2005, pp. 177f.). [4] The words “which they call ‘harmonic’” are bracketed because Huffman (2005, p. 173) regards them as a later gloss not used in Archytas’ original text.

> exceeds the second, by this part of the third the middle exceeds the third. It turns out that, in this proportion, the interval of the greater terms is greater and that of the lesser is less (Huffman 2005, p. 162).[1]

It thus appears that Archytas was responsible for the term – the *we* in "which we call harmonic" seems to refer to Archytas himself – and that the concept had been known earlier (Huffman 2005, p. 173).

Several hundred years later, Iamblichus commented on the term and concept:[2]

> In antiquity there were only three means, in the time of Pythagoras and the mathematicians around him: the arithmetic, the geometric, and third in order, the mean which was once called subcontrary, but the name of which was changed to harmonic by Archytas and Hippasus and their followers, because it manifestly embraced the ratios of what is harmonic and melodic. It was called subcontrary before, because it had characteristics that were contrary to the arithmetic mean (Iamblichus 2014, p. 173).[3]

In his discussion of the origin of the term, Huffman (2005, p. 175) concludes that "[c]ertainty about the origin of the name 'harmonic' for the third mean is impossible to reach" but that the assumption of Archytas' authorship is plausible and less problematic than other explanations.

In Latin, the term appears as *medietas armonica*[4] in Boethius' *De institutione arithmetica* (Boethius 1867, p. 152), who used it in the sense of 'harmonic proportion'.[5] In modern languages, the corresponding French term *proportion harmonique*[6] was used by Oresme in his *Book on Heavens and the World* (Oresme 1968, p. 480). The English term comes up in this sense in the 16th century (see OED, s. v. harmonical).

Note. For a detailed discussion of the Greek terminology, see especially Huffman (2005, pp. 170–181) and the literature cited there; see also Brown (1975).

References

Boethius, Anicius Manlius Severinus (1867). *De institutione arithmetica libri duo, De institutione musica libri quinque*. Accedit Geometria quae fertur Boetii. Ed. by Gottfried Friedlein. Leipzig.

Brown, Malcolm (1975). "Pappus, Plato and the Harmonic Mean". *Phronesis* 20, pp. 173–184.

Guillaumin, Jean-Yves (2020). *Dictionnaire de la terminologie latine ancienne de l'arithmétique et de la géométrie*. Paris.

Huffman, Carl A. (2005). *Archytas of Tarentum*. Pythagorean, Philosopher and Mathematician King. Cambridge.

Iamblichus (2014). *In Nicomachi arithmeticam*. Introduction, texte critique, traduction française et notes de commentaire. Ed. by Nicolas Vinel. Pisa and Roma.

Nicomachus of Gerasa (1866). *Nicomachi Geraseni Pythagorei Introductionis Arithmeticae Libri II*. Ed. by Richard Hoche. Leipzig.

Oresme, Nicole (1968). *Le livre du ciel et du monde*. Edited by Albert D. Menut and Alexander J. Denomy, translated with an introduction by Albert D. Menut. Madison.

---

[1] Translation from Huffman (2005, p. 163). The definition means that the harmonic mean is of the form $(a-b):(b-c)::a:c$. [2] It is possible that the source of Iamblichus' comment is the above passage by Archytas (Huffman 2005, p. 173). [3] Translation from Huffman (2005, p. 164). For the term *subcontrary*, see Huffman (2005, pp. 176f.). – A diverging account of the origin of the term was given by Nicomachus of Gerasa (1866, pp. 134f.); for discussion, see Huffman (2005, pp. 173f.). [4] The word *medietas* is from *medius* 'middle' and is thus constructed in analogy to the Greek terminology. [5] See Guillaumin (2020, pp. 44f.) for a brief discussion. [6] His spelling was *proporcion armonique*.

***harmonic* (projective geometry)**

In his first treatise on conic sections, Philippe de la Hire introduced the term *coupé harmoniquement* 'harmonically cut (or divided)' in the following terms: "I call the straight line AD cut in 3 parts harmonically when the rectangle contained by all AD and the middle part BC is equal to the rectangle contained by the two extreme parts AB, CD; or else if AD is to one of the two extreme parts AB, CD as the other extreme part CD or AB is to the middle part AB, which is the same thing" (de la Hire 1673, p. 1).[1] de la Hire (1673, Avant-propos) justified the term by pointing out that if one considers one of the extreme parts, the middle part, and the whole line segment, then "these three lines [line segments] will be in harmonic proportion, and that is what obliged me to give it this name [*harmonically divided*]".[2]

At exactly the same time as de la Hire, the concept is also discussed by Blondel (1673, p. 18). He points out that Desargues used the term *involution* (of four points; see *involution*), and that "others" – Blondel names Grégoire de Saint-Vincent – used the term *moyenne & extrême raison proportionnelle*, which is indeed found in Latin in de Saint-Vincent (1647, p. 2). Blondel (ibid.) goes on to add that "no one, as far as I know, neither the ancients nor the moderns, has given it its true name", which, in his opinion, was *harmonique*.[3] Thus, both de la Hire and Blondel suggested the same term at the same time. An additional suggestion was made shortly afterwards by Borelli (1679, p. 5), who used *conterminali analogia*. The word did not enjoy much success; its use was pointed out by Krafft (1753, pp. 69f., 85), for example.

In a revised and expanded version of his book, this time written in Latin, de la Hire also considered what was later called a *harmonic pencil* – four concurrent lines where any transversal intersects these four lines in such a way that the intersection points are harmonically divided – and called these lines *harmonicales* (de la Hire 1685, p. 5). The term never enjoyed wide popularity, but it was used by Milnes (1702, p. 44) and Maclaurin (1748, p. 20); the latter attributes it to de la Hire.[4] The term *harmonic pencil* was later coined by Brianchon (1817, p. 9), who used French *faisceau harmonique* (*faisceau* 'bundle'), a term also adopted by Poncelet (1822, p. 13). Steiner (1832, p. 19) used German *harmonische Strahlen* 'harmonic rays', while Plücker (1830, p. 285) preferred to translate Brianchon's term by German *Harmonikalen*. English *harmonic pencil* appears to go back to Lardner (1838, p. 329).[5]

Note. On the history of the concepts and terminology discussed here, much useful information is given by Baltus (2020). For the terminology, see also Tropfke (1940, pp. 235f.).

---

[1] "J'appelle une ligne droitte AD couppeé en 3 parties harmoniquement quand le rectangle contenu sous la toutte AD & la partie du milieu BC est égal au rectangle contenu sous les deux parties extremes AB, CD: ou bien lorsque la toutte AD est à l'une des 2 extremes AB, CD comme l'autre extreme CD ou AB est à la partie du milieu ce qui est la mesme chose". Translation based on Baltus (2020, p. 62). In a later publication written in Latin, de la Hire (1685, p. 1) used the verb *dividere* 'to divide' instead of *couper*. [2] "ces trois lignes seront en proportion harmonique, et c'est ce qui m'a obligé de luy donner ce nom". de la Hire (1673, Avant-propos) also points out Apollonius already made use of the concept. [3] These usages have all been pointed out by Brianchon (1817, p. 66). [4] The term was picked up again as German *Harmonikale* by Plücker (1830, p. 285). [5] The OED dates the term to 1840.

References

Baltus, Christopher (2020). *Collineations and Conic Sections*. An Introduction to Projective Geometry in its History. Cham.
Blondel, François (1673). *Résolution des quatre principaux problèmes d'architecture*. Paris.
Borelli, Giovanno Alfonso (1679). *Elementa conica Apollonii Paergei et Archimedis opera nova & breviori methodo demonstrata*. Roma.
Brianchon, Charles Julien (1817). *Mémoire sur les lignes du second ordre*. Paris.
de la Hire, Philippe (1673). *Nouvelle méthode en géométrie pour les sections des superficies coniques, et cylindriques*. Paris.
de la Hire, Philippe (1685). *Sectiones conicae*. Paris.
Krafft, Georg Wolfgang (1753). *Institutiones Geometricae sublimiores*. Tübingen.
Lardner, Dionysius (1838). *The First Six Books of the Elements of Euclid*. Sixth edition. London.
Maclaurin, Colin (1748). *De Linearum Geometricarum Proprietatibus Generalibus*. In: *A Treatise of Algebra*. London.
Milnes, James (1702). *Sectionum conicarum elementa nova methodo demonstrata*. Oxford.
Plücker, Julius (1830). "Über ein neues Princip der Geometrie und den Gebrauch allgemeiner Symbole und unbestimmter Coefficienten". *Journal für die reine und angewandte Mathematik* 5, pp. 268–286.
Poncelet, Jean Victor (1822). *Traité des propriétés projectives des figures*. Paris.
de Saint-Vincent, Grégoire (1647). *Opus Geometricum quadraturae circuli et sectionum coni*. Antwerpen.
Steiner, Jacob (1832). *Systematische Entwickelung der Abhängigkeit geometrischer Gestalten von einander*. Erster Theil. Berlin.
Tropfke, Johannes (1940). *Geschichte der Elementar-Mathematik*. Vierter Band. Ebene Geometrie. Dritte, verbesserte und vermehrte Auflage besorgt von Dr. Kurt Vogel. Berlin and Leipzig.

harmonic function

See *spherical harmonic*.

harmonic series

The harmonic series has been a research topic since medieval times. Around 1350, Oresme proved the divergence of this series,[1] and in 1650, Mengoli published another proof (Mengoli 1650, preface). Oresme did not call the series *harmonic*, however, and neither did Mengoli, but Mengoli did say that its terms are "harmonically arranged" ("harmonice disposita").

The English term *harmonic series* was used by Brouncker (1668, p. 646);[2] the term *progressio harmonica* was frequently used by Leibniz, for example in notes from late 1672 (Leibniz 2003, p. 88).[3] The term *series harmonica* was also used by Jakob Bernoulli in his presentation of his brother Johann's proof of the divergence of the series (Bernoulli 1689, fol. B1[v]).

The term appears to be derived from the series' relation to the harmonic mean, i. e. each term of the series is the harmonic mean of its two neighbouring terms (Stammbach 1999, p. 94), which is in line with Mengoli's description of the series cited above.

Note. An overview of the history of the concept is found in Stammbach (1999); see also Busard (1962). An additional discussion of Oresme is found, for example, in Mazet (2003); Karamata (1959) and Bell and Blåsjö (2018) discuss Mengoli.

---

[1] A translation of Oresme's proof is found in Katz et al. (2016, p. 184). [2] Brouncker referred to various propositions on the harmonic series in Wallis (1656). [3] A definition is found in Leibniz (2003, p. 715); see also the editorial comments in Leibniz (2003, pp. xxiii–xxiv).

References

Bell, Jordan and Viktor Blåsjö (2018). "Pietro Mengoli's 1650 Proof that the Harmonic Series Diverges". *Mathematics Magazine* 91, pp. 341–347.

Bernoulli, Jakob (1689). *Positiones arithmeticae de seriebus infiniti*. Basel.

Brouncker, William (1668). "The squaring of the hyperbola, by an Infinite Series of Rational Numbers". *Philosophical Transactions of the Royal Society of London* 3, pp. 645–649.

Busard, H.L.L. (1962). "Über unendliche Reihen im Mittelalter". *L'enseignement mathématique* 8, pp. 281–290.

Karamata, Jovan (1959). "Divergence de la série harmonique d'après Mengoli (1650)". *L'Enseignement Mathématique* 5, pp. 86–88.

Katz, Victor J. et al., eds. (2016). *Sourcebook in the Mathematics of Medieval Europe and North Africa*. Princeton.

Leibniz, Gottfried Wilhelm (2003). *Sämtliche Schriften und Briefe*. Siebente Reihe. Mathematische Schriften. Dritter Band. 1672-1676. Differenzen, Folgen, Reihen. Hannover.

Mazet, Edmond (2003). "La théorie des séries de Nicole Oresme dans sa perspective aristotélicienne. 'Questions 1 et 2 sur la Géométrie d'Euclide'". *Revue d'histoire des mathématiques* 9, pp. 33–80.

Mengoli, Pietro (1650). *Novae quadraturae arithmeticae, seu de additione fractionum*. Bologna.

Stammbach, Urs (1999). "Die harmonische Reihe: Historisches und Mathematisches". *Elemente der Mathematik* 54, pp. 93–106.

Wallis, John (1656). *Arithmetica infinitorum*. In: *Operum Mathematicorum*. Pars altera. Oxford, pp. 1–198.

height

In Greek geometry, *ὕψος* 'height; summit, crown' was used by Euclid, Archimedes, and other Greek writers to designate the height of various two- or three-dimensional geometrical objects (Mugler 1958-59, pp. 444f.). Euclid uses the term in the *Elements* (book VI, def. 4): "The height of any figure is the perpendicular drawn from the vertex to the base".[1] Heath comments on this definition: "The definition of 'height' is not found in Campanus and is perhaps rightly suspected, since it does not apply in terms to parallelograms, parallelepipeds, cylinders and prisms, though it is used in the *Elements* with reference to these latter figures. Aristotle does not appear to know altitude (*ὕψος*) in the mathematical sense; he uses *κάθετος* of triangles (*Meteorologica* III.3, 373a11). The term is however readily understood, and scarcely requires definition" (Euclid 1956, p. 189).

Latin *altitudo*, also meaning 'height' or 'depth' (LS), was used to translate this term. The term is found in classical Latin (TLL), and can be considered standard since. In some of the medieval translations of the *Elements*, for example in Gerard's (Busard 1984, p. 137), definitions of *altitudo* are given.[2]

The first translations into the vernacular languages already contain the later standard terms. Thus, Tartaglia has Italian *altezza* (Euclid 1543, fol. lxxviii^{v}), Xylander German *Höhe* (Euclid 1562, p. 154), Forcadel French *hauteur* (Euclid 1564, fol. 154^{r}), Zamorano Spanish *altura* (Euclid 1576, fol. 96^{r}); Billingsley defines *height* (Euclid 1570, fol. 154^{r}) but frequently also uses *altitude* (ibid.).

Note. For the Greek term, see Mugler (1958-59, pp. 444f.), for the Arabic, see Rashed (2017, pp. 573–575) and Souissi (1968, pp. 182, 257).

---

[1] Translation from Euclid (1956, p. 188). [2] *altitudo* is also commonly used to indicate the third dimension (see Guillaumin 2020, pp. 32f.). In this sense, the word corresponds to Greek *βάθος* 'depth' and *πάχος* 'thickness'. For these terms, see Mugler (1958-59, pp. 93, 334).

References

Busard, H.L.L. (1984). *The Latin Translation of the Arabic Version of Euclid's* Elements, *commonly ascribed to Gerard of Cremona*. Leiden.
Euclid (1543). *Euclide Megarense philosopho*. Diligentemente reassettato, et alla integrita ridotto per ... Nicolo Tartalea. Venezia.
Euclid (1562). *Die Sechs Erste Bücher Euclidis*. Auß Griechischer sprach in die Teütsch gebracht ... Durch Wilhelm Holtzmann genannt Xylander. Basel.
Euclid (1564). *Les Six Premiers Livres des Elements*. Traduicts et commentez par Pierre Forcadel de Besies. Paris.
Euclid (1570). *The Elements of Geometrie of the most auncient Philosopher Euclide of Megara*. Faithfully (now first) translated into the Englishe toung, by H. Billingsley. London.
Euclid (1576). *Los seis libros primeros dela geometria de Euclides*. Traduzidos en lengua Española por Rodrigio çamorano. Seville.
Euclid (1956). *The Thirteen Books of Euclid's Elements*. Volume II. Books III-IX. Ed. by Thomas L. Heath. New York.
Guillaumin, Jean-Yves (2020). *Dictionnaire de la terminologie latine ancienne de l'arithmétique et de la géométrie*. Paris.
Mugler, Charles (1958-59). *Dictionnaire historique de la terminologie géométrique des Grecs*. Paris.
Rashed, Roshdi, ed. (2017). *Lexique historique de la langue scientifique arabe*. Hildesheim.
Souissi, Mohammed (1968). *La langue des mathématiques en arabe*. Tunis.

helicoid

The general meaning of this term can be paraphrased 'in the form of a helix' or 'related to a helix'.[1] Mathematical uses include Bernoulli (1691, p. 14), who used *Parabola helicoïdes* (*helicoidal parabola*) as an alternative for *parabolic spiral*. Another example is found in Frézier (1737, p. 34), who used *Hélicoïde* for a certain solid generated by a circle or an ellipse turning around a spiral.

Today, however, the term is mainly used to refer to a particular surface that has a helix as its boundary (see fig. 50). The surface was discussed in 1776 by Meusnier, who showed that the helicoid is a minimal surface (Meusnier 1785, pp. 506f.). He did not use a particular term, though.

The term *surface hélicoïde* came up in the 19th century; an early use is Hachette (1816, p. 18). This was translated as *helicoidal surface* by De Morgan (1836, p. 397), who described the helicoid as a "winding staircase (neglecting the irregularities of the steps)". In German texts such as Scherk (1835, p. 185), *Schraubenfläche* (from *Schraube* 'screw') was used but other terms came up as well. For example, Peters (1835, p. 66) suggested *Wendeltreppe*, lit. 'spiral staircase'; while this term was not successful, the modified form *Wendelfläche* (e. g., Leroy 1840, p. 196) is still in use.

Note. Some historical information on the concept can found in Reich (1973, 2007).

References

Ayuso García, Manuel (2008). "La terminología latina de la geometría en Marciano Capela". PhD thesis. Madrid: Universidad Nacional de Educación a Distancia.
Bernoulli, Jakob (1691). "Specimen calculi differentialis in dimensione parabolae helicoidis". *Acta Eruditorum* (Jan. 1691), pp. 13–23.

---

[1] Greek ἑλικοειδής is rare and of late origin; Mugler's historical dictionary of Greek geometrical terms (Mugler 1958-59) does not have an entry for the term, although it was used by Proclus, for example (Proclus 1873, p. 104). For further discussion of Greek and Latin examples, see Ayuso García (2008, pp. 979–983).

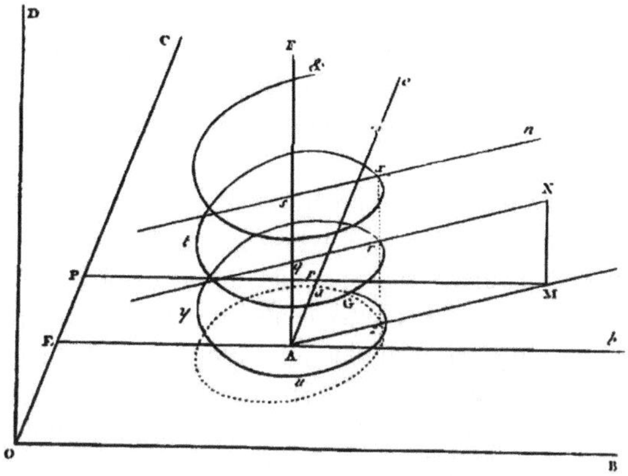

Figure 50: The right helicoid. From Meusnier (1785, Pl. X).

Frézier, Amédée François (1737). *La théorie et la pratique de la coupe des pierres et des bois*. Tome premier. Strasbourg and Paris.

Hachette, Jean Nicolas Pierre (1816). "De la courbe de contact d'un cône ou d'un cylindre, et de la surface hélicoïde des filets de la vis triangulaire". *Correspondance sur l'École royale polytechnique* 3, pp. 18–19.

Leroy, C.F.A. (1840). *Analytische Geometrie im Raum*. Uebersetzt, nach der zweiten ... Auflage von E.F. Kauffmann. Stuttgart.

Meusnier, Jean-Baptiste (1785). "Mémoire sur la courbure des surfaces". *Mémoires de mathématique et de physique, presentés à l'Académie royale des sciences, par divers sçavans* 10, pp. 477–510.

De Morgan, Augustus (1836). *The Differential and Integral Calculus*. London.

Mugler, Charles (1958-59). *Dictionnaire historique de la terminologie géométrique des Grecs*. Paris.

Peters, Adolf (1835). *Neue Curvenlehre*. Grundzüge einer Umgestaltung der höheren Geometrie durch ihre ursprüngliche analytische Methode. Dresden.

Proclus (1873). *In primum Euclidis elementorum librum commentarii*. Ed. by Gottfried Friedlein. Leipzig.

Reich, Karin (1973). "Die Geschichte der Differentialgeometrie von Gauss bis Riemann (1828-1868)". *Archive for History of Exact Sciences* 11, pp. 273–382.

Reich, Karin (2007). "Euler's Contribution to Differential Geometry and its Reception". In: *Leonhard Euler*. Life, Work and Legacy. Ed. by Robert E. Bradley and C. Edward Sandifer. Amsterdam, pp. 479–502.

Scherk, Heinrich Ferdinand (1835). "Bemerkungen über die kleinste Fläche innerhalb gegebener Grenzen". *Journal für die reine und angewandte Mathematik* 13, pp. 185–208.

helix

The Greek word *ἕλιξ* is derived from a verb meaning 'to twist' and refers to a flexible, slender willow. This etymology was already suggested in the 14th century by Oresme: "a line of uniformly difform curvature which the mathematicians call the *helix* (spiral), perhaps so called from a certain species of ivy which is similarly curved and which according to Pliny is called *helix*" (Clagett 1978, p. 139).[1] In mathematics, the

[1] "lineam curvitatis uniformiter difformis quam mathematici vocant helicen, forsan sic dictam a quadem specie hedere consimiliter tortuose, que secundum Plinium dicitur helix". Translation from Clagett (1978, p. 140). Oresme is actually referring to a spiral in this passage.

word was used to refer both to two- or three-dimensional curves, particularly the Archimedian spiral, the cylindrical helix and the spherical helix. This ambiguity had a long-lasting effect of the term *spiral*, which is not just used for plane curves but also for space curves. In the following, the focus is on three-dimensional curves (see *spiral* for plane curves).

The cylindrical helix was described by Hero of Alexandria, as mentioned by Pappus in book VIII of the *Collection*, who reports that the line generated on the surface of the cylinder is a helix and also points out the name *κοχλίας* 'screw' (Pappus of Alexandria 1878, p. 1124). This helix is also mentioned by Proclus, who says it "is traced by a point moving uniformly along a straight line that is moving around the surface of a cylinder" (Proclus 1873, p. 105).[1] In book IV, prop. 30, of the *Collection*, Pappus also discusses the spherical helix: "Just as a certain spiral is contemplated in the plane when a point travels along a straight line that describes a circle, and in solids when a point travels along one of its sides, while it describes a certain surface, so it is in fact a natural next step to contemplate a spiral described on a sphere" (Pappus of Alexandria 1876, p. 264).[2]

The Greek term was adopted in Latin as *linea (h)elica* or *(h)elix*. These terms were in common use in the 13th century.[3] For example, in his *De caelo* (I, lect. iii), Thomas Aquinas remarked: "But it must be said that a helix [*elix*], if one considers its origin, is not a simple line, but a combination of straight and circular. For a helix is produced by two imaginary motions, one of which is the motion of a line moving round a cylinder, and the other of a point moving through the line: if two such motions take place in a regular manner at the same time, a helix [*elica linea*] will be formed by the motion of the point in the moving line" (Aquinas 1886, p. 11).[4] In print, *helix* appears in Valla (1501, book X, cap. iiii), who used it for spirals and helices.

Dürer (1525, fol. B ii^{v}) called the cylindrical helix a *Schraubenlinie*[5] 'screw line'; Schmid (1539, p. 3) mentions Dürer's term but also uses *gewunden lini* (from a verb meaning 'to wind'). Schwenter (1618, p. 12) points out that *Schlangenlinie*, lit. 'snake line', is also commonly used because a helix is a curved line that 'coils around a tree like a snake".[6] Recorde (1551, fol. A iiij^{v}) refers to the same curve in the following terms: "an other fashioned line is there, which is named a twine or twist line, and it goeth as a wreyth about some other bodie". The phrase "*Helices* or Spiral lines inuented by *Archimedes*" is inserted as a commentary in Billingsley's translation of the *Elements* (Euclid 1570, fol. 19^{r}), while Digges (1576, fol. P 1^{r}) uses *Helicall line* to describe a loxodrome; Oughtred (1633, p. 37) uses *helix or spirall*

[1] Translation from Proclus (1970, p. 85). Proclus there also reports that Apollonius wrote a book on this helix (which is not extant). [2] Translation from Pappus of Alexandria (2010, p. 141). [3] See Dewender (2002, p. 82, fn. 59) for various 12th- and 13th-century discussions of a "'proportional' version the cylindrical helix" (Mancosu and Arana 2010, p. 417). – *elicis* appears (for a plane curve) in a gloss to a treatise by the agrimensor Balbus (Blume, Lachmann, and Rudorff 1848, p. 103). See Guillaumin (2020, p. 145). [4] "Sed dicendum quod elix, si quis eius originem consideret, non est linea simplex, sed mixta ex recta et circulari. Causatur enim elix ex duobus motibus imaginatis, quorum unus est lineae circumeuntis columnam, alius autem est puncti moti per lineam: si enim uterque motus simul et regulariter perficiatur, constituetur elica linea per motum puncti in linea mota". Translation from Aquinas (1963-64, book I, ch. 3). – Aquinas (ibid.) also uses *elix* for 'spiral'. [5] He wrote *schrauffen lini*. [6] "Gleich wie sich eine Schlange umb einen Baum wickelt".

line in the same meaning. “Helix about the Cylinder” is found in Salusbury (1661, p. 7).[1]

Note. For the Greek term, see Mugler (1958-59, pp. 169f.); for the Arabic one, see Rashed (2017, pp. 750, 242–245) and Hasse (2015, pp. 74f.).

References

Aquinas, Thomas (1886). *Commentaria in libros Aristotelis de caelo et mundo, De generatione et corruptione et Meteorologicorum*. Roma.

Aquinas, Thomas (1963-64). *In libros Aristotelis De caelo et mundo expositio*. Translated by Fabian R.Larcher and Pierre H. Conway. Columbus, Ohio. URL: `https://isidore.co/aquinas/english/DeCoelo.htm` (retrieved June 6, 2018).

Blume, Friedrich, Karl Lachmann, and Adolf Rudorff, eds. (1848). *Die Schriften der römischen Feldmesser*. Erster Band. Texte und Zeichnungen. Berlin.

Clagett, Marshall (1978). *Archimedes in the Middle Ages*. Volume III. The Fate of the Medieval Archimedes 1300-1565. Philadelphia.

Dewender, Thomas (2002). *Das Problem des Unendlichen im ausgehenden 14. Jahrhundert*. Eine Studie mit Textedition zum Physikkommentar des Lorenz von Lindores. Amsterdam and Philadelphia.

Digges, Leonard (1576). *A prognostication euerlasting of right good effect*. London.

Dürer, Albrecht (1525). *Underweysung der Messung, mit dem Zirckel und Richtscheyt*. Nürnberg.

Euclid (1570). *The Elements of Geometrie of the most auncient Philosopher Euclide of Megara*. Faithfully (now first) translated into the Englishe toung, by H. Billingsley. London.

Galilei, Galileo (1632). *Dialogo sopra i due massimi sistemi del mondo*. Firenze.

Guillaumin, Jean-Yves (2020). *Dictionnaire de la terminologie latine ancienne de l'arithmétique et de la géométrie*. Paris.

Hasse, Dag Nikolaus (2015). “Averroes' Critique of Ptolemy and Its Reception by John of Jandun and Agostino Nifo”. In: *Averroes' Natural Philosophy and its Reception in the Latin West*. Ed. by Paul J.J.M. Bakker. Leuven, pp. 69–88.

Mancosu, Paolo and Andrew Arana (2010). “Descartes and the cylindrical helix”. *Historia Mathematica* 37, pp. 403–427.

Mugler, Charles (1958-59). *Dictionnaire historique de la terminologie géométrique des Grecs*. Paris.

Oughtred, William (1633). *An addition vnto the vse of the instrument called the circles of proportion*. London.

Pappus of Alexandria (1876). *Pappi Alexandrini Collectionis*. Volumen I. Ed. by Friedrich Hultsch. Berlin.

Pappus of Alexandria (1878). *Pappi Alexandrini Collectionis*. Voluminis III Tomus I. Ed. by Friedrich Hultsch. Berlin.

Pappus of Alexandria (2010). *Book 4 of the* Collection. Edited With Translation and Commentary by Heike Sefrin-Weis. London.

Proclus (1873). *In primum Euclidis elementorum librum commentarii*. Ed. by Gottfried Friedlein. Leipzig.

Proclus (1970). *Commentary on the First Book of Euclid's Elements*. Translated with Introduction and Notes by Glenn R. Morrow. Princeton.

Rashed, Roshdi, ed. (2017). *Lexique historique de la langue scientifique arabe*. Hildesheim.

Recorde, Robert (1551). *The patheway to knowledg*. London.

Salusbury, Thomas (1661). *Mathematical collections and translations*. London.

Schmid, Wolffgang (1539). *Das erst buch der Geometria*. Nürnberg.

Schwenter, Daniel (1618). *Geometriae Practicae Novae*. Tractatus I. Nürnberg.

Valla, Giorgio (1501). *De expetendis, et fugiendis rebus opus*. Venezia.

---

[1] The text is a translation of Galilei (1632, p. 8), where Italian “Elica intorno al Cilindro” was used.

hereditary ring

The term *hereditary ring* apparently goes back to Cartan and Eilenberg (1956, p. 13): "A ring Λ will be called *left hereditary* if every (left) ideal of A is a projective module". They also introduced the term *semi-hereditary ring* (p. 14): "The ring Λ will be called *left semi-hereditary* if each finitely generated (left) ideal of Λ is a projective module" and remarked that "[t]he definition of right hereditary and right semi-hereditary rings is entirely similar".

Because the publication of Cartan and Eilenberg's book was delayed, the term is found earlier in print, for example in Heller (1954, p. 285).[1] Heller (p. 283) refers to the forthcoming book. The term spread quickly, and an open question on hereditary rings was soon solved by Kaplansky (1958, p. 85): "Cartan and Eilenberg [...] ask whether there exists a right hereditary ring which is not left hereditary. The answer: yes".

References

Auslander, Maurice (1953). *Notes on cohomology theory of groups*. Based on lectures given by S. Eilenberg at Columbia University (1952-1953). Chicago.

Cartan, Henri and Samuel Eilenberg (1956). *Homological Algebra*. Princeton.

Heller, Alex (1954). "Homological Resolutions of Complexes with Operators". *Annals of Mathematics* 60, pp. 283–303.

Kaplansky, Irving (1958). "On the dimension of modules and algebras. X. A right hereditary ring which is not left hereditary". *Nagoya Mathematical Journal* 13, pp. 85–88.

Hermitian form and *Hermitian matrix*

As defined by Moore (1898, p. 213), an *n-ary Hermitian form* is "a bilinear form whose two sets of variables are conjugate-imaginary, and whose corresponding coefficients are conjugate-imaginary". Moore had already used the term in a talk held in July 1896, as a printed abstract shows (Moore 1897, p. 267). Practically the same terminology was used at the same time by Loewy, who, also in 1896, used French *forme quadratique de M. Hermite* (Loewy 1896, p. 169), and, in 1898, German *Hermite'sche Form* (Loewy 1898, p. 561). Moore and Loewy seem to have coined the term independently. Such forms had been studied by Hermite (1855, 1854, p. 346).

Autonne (1902, p. 123) explicitly applied *hermitienne* to matrices.[2]

Note. A historical discussion of Hermitian forms in the works of Moore, Loewy, and others is given by Goldstein (2018).

References

Autonne, Léon (1902). "Sur les groupes linéaires, réels et orthogonaux". *Bulletin de la Société Mathématique de France* 30, pp. 121–134.

Cayley, Arthur (1866). "A Supplementary Memoir on the Theory of Matrices". *Philosophical Transactions of the Royal Society of London* 156, pp. 25–35.

Goldstein, Catherine (2018). "Long-term history and ephemeral configurations". In: *Proceedings of the International Congress of Mathematicians*. Rio de Janeiro 2018. Volume I. Ed. by Boyan Sirakov, Paulo Ney de Souza, and Marcelo Viana. Singapore, pp. 487–522.

Hermite, Charles (1854). "Sur la théorie des formes quadratiques. Second mémoire". *Journal für die reine und angewandte Mathematik* 47, pp. 343–368.

---

[1] The term is also found in some mimeographed lectures notes, based on lectures given by Eilenberg in 1952-53 (Auslander 1953, p. 15). [2] Note that Cayley (1866, p. 30) used *Hermitian matrix* in a different sense.

Hermite, Charles (1855). "Remarque sur un théorème de Cauchy". *Comptes rendus de l'Académie des sciences* 41, pp. 181–183.
Loewy, Alfred (1896). "Sur les formes quadratiques définies à indéterminées conjuguées de M. Hermite". *Comptes rendus de l'Académie des sciences* 123, pp. 168–171.
Loewy, Alfred (1898). "Ueber bilineare Formen mit conjugirt imaginären Variablen". *Mathematische Annalen* 50, pp. 557–576.
Moore, E.H. (1897). "Concerning Finite Groups of Linear Homogeneous Substitutions". *University Record* 1, p. 276.
Moore, E.H. (1898). "An universal invariant for finite groups of linear transformations: with applications in the theory of the canonical form of a linear substitution of finite period". *Mathematische Annalen* 50, pp. 213–219.

Hilbert space

As remarked by Pietsch (2007, p. 9), the closed unit ball B_{l_2} was considered by Hilbert (1906, p. 177) as a domain of linear, bilinear and quadratic forms. Therefore, Pietsch (ibid.) continues, "B_{l_2} was considered to be the decisive object for which Schoenflies coined the term '*Hilbertscher Raum*' [...]; In other words, one emphasized only the topological properties, and the significance of the linear structure was underestimated". As also pointed out by Pietsch, in this sense the term was used by Schoenflies (1908, p. 266).[1] The first English usage seems to be Hedrick (1911, p. 287).

Riesz (1913, p. 78) then used the term for the set of all square-summable sequences: "Let us consider the Hilbert space; by that we mean the set of systems (x_k) such that $\sum|x_k|^2$ converges".[2] Hausdorff (1914, p. 287) was one of many who followed this usage; in this sense the English term appeared in Wiener (1921, p. 312).

An abstract axiomatic definition was introduced by von Neumann. As paraphrased by Moore (1995, p. 286), he defined it as "a 'lincar space' (i. e., a complex vector space) with Hermitian bilinear form (i. e., a complex inner product); this made it a metric space, which was assumed to be separable, complete, and infinite-dimensional". The axioms appear in von Neumann (1928, pp. 15–17) and von Neumann (1930, pp. 63–70). In this sense the term was quickly adopted by Stone (1929, p. 198): "A complex abstract Hilbert space $\mathfrak{H}$ is a set of elements, denoted hereafter by f, g, ..., which is characterized by five properties: linearity, the existence of a unitary metric, the non-existence of a finite basis, completeness, and separability".

As pointed out by Moore (1995, pp. 285–288), the modern notion of Hilbert space appeared in Jordan and von Neumann (1935, p. 719). They dropped the conditions of infinite-dimensionality and separability, noting that "essentially new hyper-Hilbert spaces arise, but they are nevertheless similar to Hilbert space under most aspects".

Note. For the history of the concept, see Taylor (1982, pp. 281–284), Moore (1995, pp. 285–288), and Pietsch (2007, pp. 9–14)

References

Hausdorff, Felix (1914). *Grundzüge der Mengenlehre*. Leipzig.
Hedrick, Earle Raymond (1911). "On Properties of a Domain for Which Any Derived Set is Closed". *Transactions of the American Mathematical Society* 12, pp. 285–294.

[1] That this is presumably the first usage has already been stated by Taylor (1982, p. 281). [2] "Considérons l'espace hilbertien; nous y entendons l'ensemble des systèmes (x_k) tels que $\sum|x_k|^2$ converge". Translation from Pier (2001, p. 92).

Hilbert, David (1906). "Grundzüge einer allgemeinen Theorie der linearen Integralgleichungen. (Vierte Mitteilung)". *Nachrichten von der Königl. Gesellschaft der Wissenschaften zu Göttingen (mathematisch-physikalische Klasse)* (1906), pp. 157–228.
Jordan, Pascual and John von Neumann (1935). "On Inner Products in Linear, Metric Spaces". *Annals of Mathematics* 36, pp. 719–723.
Moore, Gregory H. (1995). "The axiomatization of linear algebra: 1875-1940". *Historia Mathematica* 22, pp. 262–303.
von Neumann, John (1928). "Mathematische Begründung der Quantenmechanik". *Nachrichten von der Gesellschaft der Wissenschaften zu Göttingen (mathematisch-physikalische Klasse)* (1927), pp. 1–57.
von Neumann, John (1930). "Allgemeine Eigenwerttheorie Hermitescher Funktionaloperatoren". *Mathematische Annalen* 102, pp. 49–131.
Pier, Jean-Paul (2001). *Mathematical Analysis During the 20th Century*. Oxford.
Pietsch, Albrecht (2007). *History of Banach Spaces and Linear Operators*. Boston, Mass. et al.
Riesz, Frigyes (1913). *Les systèmes d'équations linéaires à une infinité d'inconnues*. Paris.
Schoenflies, Arthur (1908). *Die Entwickelung der Lehre von den Punktmannigfaltigkeiten*. Bericht erstattet der Deutschen Mathematiker-Vereinigung. Zweiter Teil. Leipzig.
Stone, Marshall Harvey (1929). "Linear Transformations in Hilbert Space. I. Geometrical Aspects". *Proceedings of the National Academy of Sciences* 15, pp. 198–200.
Taylor, Angus E. (1982). "A Study of Maurice Fréchet: I. His Early Work on Point Set Theory and the Theory of Functionals". *Archive for History of Exact Sciences* 27, pp. 233–295.
Wiener, Norbert (1921). "On the theory of sets of points in terms of continuous transformations". In: *Comptes rendus du congrès international des mathématiciens*. (Strasbourg, 22-30 Septembre 1920). Ed. by Henri Villat. Toulouse, pp. 312–315.

holomorphic and *meromorphic*

The term *holomorphic* was introduced by Briot and Bouquet (1875, p. 14) as French *fonction holomorphe*: "If a function is continuous, monotropic[1], and has a derivative when the variable moves in a certain part of the plane, we will say that it is *holomorphic* in this part of the plane".[2] The term derives from Greek *ὅλος* 'whole' and *μορφή* 'shape, form'; as Briot and Bouquet (ibid.) explain, holomorphic functions are similar to entire functions which have the defining properties of holomorphic functions on the whole plane. Briot and Bouquet (1875, p. 15) also introduced *fonction méromorphe* (Greek *μηρός* 'part'): "When a function is holomorphic in part of the plane except for certain poles, we say that it is meromorphic in that part of the plane; that is to say, it resembles the rational fractions".[3] The terms were later also explained by Harkness and Morley (1893, p. 161) in a similar way: "The words 'holomorphic' and 'meromorphic,' introduced by Briot and Bouquet, are intended to suggest that the functions behave respectively like the integral polynomial and the rational fraction".

The term soon became known outside France. English *holomorphic* appears in Briot and Bouquet (1878, p. 161); Craig (1883, p. 197) seems to have been the first who used *meromorphic*.[4] The German terms were used in a review of Briot and Bouquet's book (Müller 1875, p. 263). Harnack (1881, p. 149) pointed out that

---

[1] Used by Briot and Bouquet (1875, p. 10) for 'single-valued'. The term was used to replace Cauchy's earlier *monodromic* (French *monodrome*) (Cauchy 1852, p. 266). See *monodromy*.
[2] "Lorsqu'une fonction est continue, monotrope, et a une dérivée, quand la variable se meut dans une certaine partie du plan, nous dirons qu'elle est *holomorphe* dans cette partie du plan".
[3] "Lorsqu'une fonction est holomorphe dans une partie du plan, excepte en certains pôles, nous dirons qu'elle est *méromorphe* dans cette partie du plan, c'est-à-dire semblable aux fractions rationnelles". Translation from Remmert (1991, p. 315).
[4] The OED dates *meromorphic* to 1890.

holomorphic was due to Briot and Bouquet; he listed other synonyms but preferred *analytic*. There were, indeed, various other terms in common use in the 19th century and later. As Remmert (1991, p. 62) explains: "Other synonyms in the older literature are 'monogenic', 'monodromic', 'analytic' and 'regular'. These and other terms originally described various properties, like having vanishing integral over every closed curve [...] or having local power series expansions or satisfying the Cauchy-Riemann equations, etc., and so were not at first recognized as synonyms. When the theory of functions reached maturity, these properties were all seen to be equivalent [...] so it is appropriate that most of these terms have now faded into oblivion".

Even more terms can be found. Briot and Bouquet (1859, p. 11) themselves earlier used *fonction synectique*, a term that was coined by Cauchy (1852, p. 267). Méray (1872, p. xvi) proposed "the epithet *olotrope*", which he chose because such functions behave like entire functions, as seen above (Greek *τρόπος* means 'turn, direction, way'). The term has been criticized because it should have been spelt *holotrope* (Greek *ὅλος*); the criticism is explained by, for example, Lovett (1900, p. 211): "Mathematics of necessity mutilates language occasionally in the construction of its technical terms, but it is to be regretted that Méray did not give heed to a correction in the form of this term suggested by the editors of the *Bulletin des Sciences mathématiques* in 1874; orthography should respect etymology and the Greek derivative is *ὅλος*". Halphen (1886, p. 454) suggested *fonction entière* for holomorphic, *fonction fractionnaire* for meromorphic functions.

See also the entries for *analytic*, *regular*, and *entire* functions.

Note. An overview of the terminology is given by Remmert (1991, pp. 61f., 315).

References

Briot, Charles and Jean Claude Bouquet (1859). *Théorie des fonctions doublement périodiques et, en particulier, des fonctions elliptiques*. Paris.

Briot, Charles and Jean Claude Bouquet (1875). *Théorie des fonctions elliptiques*. Deuxième édition. Paris.

Briot, Charles and Jean Claude Bouquet (1878). "Solution of the Equation of the Fifth Degree. Translated from the Theory of Elliptic Functions ... by Alexander Evans". *The Analyst* 5, pp. 161–166.

Cauchy, Augustin-Louis (1852). "Mémoire sur l'application du Calcul infinitésimal à la détermination des fonctions implicites". *Comptes rendus de l'Académie des sciences* 34, pp. 265–273.

Craig, Thomas (1883). "Functions of two variables". *Science* (Mar. 23, 1883), p. 197.

Halphen, Georges Henri (1886). *Traité des fonctions elliptiques et de leurs applications*. Première partie. Paris.

Harkness, James and Frank Morley (1893). *A Treatise on the Theory of Functions*. New York and London.

Harnack, Axel (1881). *Die Elemente der Differential- und Integralrechnung*. Zur Einführung in das Studium. Leipzig.

Lovett, Edgar Odell (1900). "Méray's Infinitesimal Analysis". *Bulletin of the American Mathematical Society* 6, pp. 204–212.

Méray, Charles (1872). *Nouveau précis d'analyse infinitesimale*. Paris.

Müller, Felix (1875). "Briot et Bouquet. Théorie des fonctions elliptiques [review]". *Jahrbuch über die Fortschritte der Mathematik* 7, pp. 262–268.

Remmert, Reinhold (1991). *Theory of Complex Functions*. Translated by Robert B. Burckel. New York.

holonomy group

The term *holonomy* was introduced into mathematics by Élie Cartan in connection with a classification of Riemann locally symmetric spaces. As explained by Helgason (1978, p. 197), "Cartan's method was based on the so-called holonomy group. If o is a point in a Riemannian manifold M, then the holonomy group of M is the group of all linear transformations of the tangent space M_o obtained by parallel translation along closed curves at o". Cartan coined the term *groupe d'holonomie* in the 1920s; he first used it in Cartan (1924, p. 11) and in various further publications such as Cartan (1926, p. 3). English *group of holonomy* appears in Cartan and Schouten (1926, p. 937); Vanderslice (1934, p. 163) uses *holonomy group*.

The term *holonomous* was previously coined by Hertz (1894, p. 91): "A material system between whose possible positions all conceivable continuous motions are also possible motions is called a holonomous system. The term means that such a system obeys integral (*ὅλος*) laws (*νόμος*), whereas material systems in general obey only differential conditions".[1] Cartan (1923, p. 387) himself said that he borrowed the term from mechanics.

Note. For the concept in mechanics and in the works of Cartan, see Nabonnand (2009).

References

Cartan, Élie (1923). "Sur les variétés à connexion affine et la théorie de la relativité généralisée (première partie)". *Annales scientifiques de l'École Normale Supérieure* 40, pp. 325–412.

Cartan, Élie (1924). "La théorie des groupes et les recherches récentes de géométrie différentielle". *L'enseignement mathématique* 24, pp. 5–18.

Cartan, Élie (1926). "Les groupes d'holonomie des espaces généralisés". *Acta Mathematica* 48, pp. 1–42.

Cartan, Élie and Jan Arnoldus Schouten (1926). "On RIEMANNian geometries admitting an absolute parallelism". *Koninklijke Akademie van Wetenschappen te Amsterdam. Proceedings of the Section of Sciences* 29, pp. 933–946.

Helgason, Sigurdur (1978). *Differential Geometry, Lie Groups, and Symmetric Spaces*. New York.

Hertz, Heinrich (1894). *Die Prinzipien der Mechanik in neuem Zusammenhange dargestellt*. Leipzig.

Hertz, Heinrich (1899). *The Principles of Mechanics Presented in a New Form*. Authorised English Translation by D.E. Jones and J.T. Walley. London.

Lützen, Jesper (2005). *Mechanistic Images in Geometric Form*. Heinrich Hertz's Principles of mechanics. New York.

Nabonnand, Philippe (2009). "La notion d'holonomie chez Élie Cartan". *Revue d'histoire des sciences* 62, pp. 221–245.

Vanderslice, John L. (1934). "Non-Holonomic Geometries". *American Journal of Mathematics* 56, pp. 153–193.

homaloidal

Greek *ὁμαλός* means 'flat, even' and was introduced as a mathematical term in the 19th century. The first author to use the term was Sylvester (1850, p. 368): "The Homaloidal law [...] is a corollary to a proposition which represents analytically the

[1] "Ein materielles System, zwischen dessen möglichen Lagen alle denkbaren stetigen Übergänge zugleich auch mögliche Übergänge sind, heißt ein holonomes System. Der Name soll andeuten, daß ein solches System integralen (*ὅλος*) Gesetzen (*νόμος*) gehorcht, während die materiellen Systeme im allgemeinen nur Differentialgesetzen unterworfen sind". Translation from Hertz (1899, p. 80). – For Hertz' introduction of the term, see Lützen (2005, pp. 193–195) and Nabonnand (2009, pp. 223–225).

indefinite extension of a property common to lines and surfaces to all loci (whether in ordinary or transcendental space) of the first order, all of which loci may, by an abstraction derived from the idea of levelness common to straight lines and planes, be called Homaloids". In an article published a year later, Sylvester (1851, p. 12) further explained that "[r]ight lines and planes agree in conveying an idea of levelness and uniformity; they may both be said to be homalous. I shall therefore employ the word homaloid to signify in general any theme of the first degree" (the meaning of *theme* is explained by Sylvester thus: "we may call a line a monotheme, a surface a ditheme, the species beyond a tritheme, and so on").

The term was picked up and extended by various authors. In 1871, Cremona defined certain systems of surfaces and homaloidal nets of rational curves (Cremona 1871, p. 213). In the same place, he used German *homaloidisch* also for surfaces that can be mapped onto a plane.

Note. Brief notes on the history of the term are found in Emch (1928, p. 79), for example.

References

Cremona, Luigi (1871). "Ueber die Abbildung algebraischer Flächen". *Mathematische Annalen* 4, pp. 213–230.

Emch, Arnold (1928). "Planar Cremona Transformations". In: *Selected Topics in Algebraic Geometry*. Report of the Committee on Rational Transformations. Ed. by Virgil Snyder et al. Washington, D.C., pp. 79–121.

Sylvester, James Joseph (1850). "Additions to the articles in the September Number of this Journal, 'On a new Class of Theorems,' and on *Pascal's* theorem". *The London, Edinburgh, and Dublin Philosophical Magazine and Journal of Science* 37, pp. 363–370.

Sylvester, James Joseph (1851). "On certain general properties of homogeneous functions". *The Cambridge and Dublin Mathematical Journal* 6, pp. 1–17.

homeomorphism

At the end of the 19th century, Henri Poincaré started to publish his famous series of articles on topology. After discussing certain conditions, he introduced the term *homeomorphic* (French *homéomorphe*) thus:

> It follows from these hypotheses that each point of V corresponds to exactly one point of V', and conversely; to every manifold W contained in V there corresponds a manifold W', of the same number of dimensions, contained in V'; if W is connected, finite or unbounded the same is true of W' and conversely. If all these conditions are satisfied we say that the two manifolds V and V' are equivalent from the point of view of *Analysis situs*, or simply *homeomorphic*, that is to say, of the same form" (Poincaré 1895, p. 9).[1]

Poincaré's definition was stronger than the one used today because he "worked in n-dimensional Euclidean space (rather than in an abstract space)" and "his 'homeomorphism' had strong requirements of differentiability and smoothness that have nothing directly to do with topology" (Moore 2007, p. 335).[2]

---

[1] "Il résulte de ces hypothèses qu'à tout point de V correspond un point de V' et un seul et inversement; à toute variété W contenue dans V correspondra une variété W', d'un même nombre de dimensions, contenue dans V'; si W est continue, finie ou illimitée, il en sera de même de W' et inversement. Si toutes ces conditions sont remplies, nous dirons que deux variétés V et V' sont équivalentes au point de vue de l'*Analysis situs*, ou, pour abgréger le langage, qu'elles sont *homéomorphes*, c'est-à-dire de forme pareille". Translation from Poincaré (2010, p. 10). [2] See also Bollinger (1972, p. 119).

In the cited paper, Poincaré also uses the term *homeomorphism*, although not in the sense of a mapping, but to refer to the situation that two manifolds are homeomorphic; with respect to a mapping, the term is used in Poincaré (1904, e. g., p. 66).[1]

Before Poincaré, Möbius (1863, p. 18) had discussed a related concept (see Moore 2007, p. 337), using the term *elementarily related* (German *elementar verwandt*):[2]

> Two geometric figures are said to be *elementarily related* to each other if each infinitesimal element (in all dimensions) of the one figure corresponds to a similar element of the other in such a way that if any two bounding elements of the one figure are contiguous, then the same is true of the corresponding elements in the other; or, what expresses the same thing, if a point of the one figure corresponds to a point of the other, then any two infinitely close points of the one figure correspond to infinitely close points of the other. In this regard, a curve can only be elementarily related to a curve, a surface only to a surface, and a solid body only to a solid body.[3]

Poincaré's term is soon found in German as *homöomorph* in Tietze (1906, p. 844); Dehn and Heegaard (1907, p. 159) also use the word but give Möbius' *elementar verwandt* as a "synonym", remarking that of those who dealt with the concept before, Möbius "came closest".

The English term is due to Veblen (1922, p. 3): "A $(1-1)$ continuous transformation of a complex into itself or another complex is called, following Poincaré, a *homeomorphism*. Two complexes related by a homeomorphism are said to be *homeomorphic*". Veblen's text goes back to a colloquium held in 1916. In print, the term is earlier found in a report on that colloquium (Snyder 1916, p. 87).

As stated by Moore (2007, p. 334), "[t]he evolution of the concept of 'homeomorphism' was essentially complete by 1935"; he cites the definition given by Alexandroff and Hopf (1935, p. 334): "A one-one continuous mapping f of a space X into a space Y is called a *topological mapping* or a *homeomorphism* (between X and $f(X) = Y' \subseteq Y$), if the inverse of f is a continuous mapping of Y' to X. Two spaces [...] are called *homeomorphic* if they can each be mapped topologically onto each other".[4] The alternative *topological mapping* was introduced by Brouwer

---

[1] Note that while the term was new in mathematics, it had been used in crystallography before. For example, the term was used by Breithaupt (1835, pp. 249f.) instead of *isomorphism* in its crystallographical meaning (see *isomorphism*). Breithaupt (ibid.) points out that the term was due to Naumann, who used it, for example, in Naumann (1828, p. 272). [2] Another precursor is Jordan (1866); for a brief discussion of this paper, see Popescu-Pampu (2016, pp. 51f.). [3] "Zwei geometrische Figuren sollen einander *elementar verwandt* heissen, wenn jedem nach allen Dimensionen unendlich kleinen Elemente der einen Figur ein dergleichen Element in der anderen dergestalt entspricht, dass von je zwei aneinander grenzenden Elementen der einen Figur die zwei ihnen entsprechenden Elemente der anderen ebenfalls zusammenstossen; oder, was dasselbe ausdrückt: wenn je einem Puncte der einen Figur ein Punct der anderen also entspricht, dass von je zwei einander unendlich nahen Puncten der einen auch die ihnen entsprechenden der anderen einander unendlich nahe sind. Einer Linie kann hiernach nur eine Linie, einer Fläche nur eine Fläche, und einem körperlichen Raume nur ein körperlicher Raum elementar verwandt seyn". Translation from Moore (2007, p. 337). [4] "Eine eineindeutige stetige Abbildung f des Raumes X in den Raum Y heißt eine *topologische Abbildung* oder eine *Homöomorphie* (zwischen X und $f(X) = Y' \subset Y$), wenn die Umkehrung von f eine stetige Abbildung (von Y' auf X) ist. Zwei Räume [...] heißen *homöomorph*, wenn sie aufeinander topologisch abgebildet werden können". Translation from Moore (2007, p. 334).

(1919, p. 845) under the French name *transformation topologique*; he defined them as "biunivocal and continuous".[1]

Note. The history of the concept is discussed in detail by Moore (2007).

References

Alexandroff, Pavel S. and Heinz Hopf (1935). *Topologie*. Erster Band. Berlin.

Bollinger, Maja (1972). "Geschichtliche Entwicklung des Homologiebegriffs". *Archive for History of Exact Sciences* 9, pp. 94–170.

Breithaupt, August (1835). "Ueber das Verhältniss der Formen zu den Mischungen krystallisirter Körper". *Journal für praktische Chemie* 4, pp. 249–271.

Brouwer, L.E.J. (1919). "Enumeration des groupes finis de transformations topologiques du tore". *Comptes rendus de l'Académie des sciences* 168, pp. 845–848.

Dehn, Max and Poul Heegaard (1907). "Analysis situs". *Encyklopädie der mathematischen Wissenschaften mit Einschluss ihrer Anwendungen* III AB3, pp. 153–220.

Jordan, Camille (1866). "Sur la déformation des surfaces". *Journal de mathématiques pures et appliquées* 11, pp. 105–109.

Möbius, August Ferdinand (1863). "Theorie der elementaren Verwandtschaft". *Berichte über die Verhandlungen der Königlich Sächsischen Gesellschaft der Wissenschaften zu Leipzig (Mathematisch-Physische Klasse)* 15, pp. 18–57.

Moore, Gregory H. (2007). "The evolution of the concept of homeomorphism". *Historia Mathematica* 34, pp. 333–343.

Naumann, Carl Friedrich (1828). *Lehrbuch der Mineralogie*. Berlin.

Poincaré, Henri (1895). "Analysis situs". *Journal de l'École polytechnique* 1, pp. 1–121.

Poincaré, Henri (1904). "Cinquième complément à l'analysis situs". *Rendiconti del Circolo matematico di Palermo* 18, pp. 45–110.

Poincaré, Henri (2010). *Papers on Topology. Analysis Situs* and Its Five Supplements. Translated by John Stillwell. Providence, R.I.

Popescu-Pampu, Patrick (2016). *What is the Genus?* Cham.

Snyder, Virgil (1916). "The Cambridge colloquium". *Bulletin of the American Mathematical Society* 23, pp. 81–88.

Tietze, Heinrich (1906). "Zur Analysis situs mehrdimensionaler Mannigfaltigkeiten". *Sitzungsberichte der Kaiserlichen Akademie der Wissenschaften (mathematisch-naturwissenschaftliche Klasse)* 115, pp. 841–846.

Veblen, Oswald (1922). *Analysis Situs*. New York.

homogeneous

The word *homogeneous* literally means 'of the same kind' (Greek *ὁμός* 'same', *γένος* 'kind'). The term is ancient and appears in Euclid's definition of *ratio*: "A ratio is a sort of relation in respect of size between two magnitudes of the same kind" (*Elements*, book V, def. 3). This definition should be considered in connection with def. 4: "Magnitudes are said to *have a ratio* to one another which are capable, when multiplied, of exceeding one another".[2] In his study "Homogeneity in Eudoxus's Theory of Proportion", Mueller (1970, p. 2) concludes that "[t]wo magnitudes are homogeneous if they are capable of having a ratio to one another [...] Thus triangles and squares are homogeneous; triangles and lines are not".

Latin *homogeneus* is based on the Greek term and of late (medieval) origin; the word does not appear in the 12th- and 13th-century translations and redactions of the *Elements*. The corresponding terms in modern languages were also not used in the translations of Euclid's book; Billingsley, for example, used "of one kinde"

[1] "biunivoque et continue". [2] Both translations from Euclid (1956, p. 114).

in def. 3 (Euclid 1570, fol. 126^v).[1] An early use of English *homogeneal* is found in Ramus (1590, fol. 2^r), who used it with reference to angles. In the 17th century, *homogeneous* was in common use. For example, Wallis (1662, p. 43) commented on Euclid: "But what is meant by *Homogenous*? *Euclide* having in his Definition of *λόγος*, *Ratio*, (*def.* 3 *lib.* 5) told us it was (*μεγεθῶν ὁμογενῶν*) *between Homogenous magnitudes*: That we may know what those are, he tells us (in a following Definition) that *Homogenous Magnitudes*, or (which is all one) *magnitudes which have relation to one another, are such that each of them be so multiplied as to exceed the other*".

The terminology was also used by Viète (1591, fol. A iiij^v), who influenced its further development: "The prime and perpetual law of equations or proportions which, since it deals with their homogeneity, is called the law of homogeneous terms, is this: Homogeneous terms must be compared with homogeneous terms, for, as Adrastos[2] said, it is impossible to understand how heterogeneous terms [can] affect each other".[3]

Combinations such as *homogeneous equation* are of later origin; for instance, Clairaut (1746, p. 170) defined an *equation homogene* by the condition that "all its terms are of the same dimension".[4]

References

Clairaut, Alexis Claude (1746). *Elémens d'algebre*. Paris.

Euclid (1562). *Die Sechs Erste Bücher Euclidis*. Auß Griechischer sprach in die Teütsch gebracht ... Durch Wilhelm Holtzmann genannt Xylander. Basel.

Euclid (1570). *The Elements of Geometrie of the most auncient Philosopher Euclide of Megara*. Faithfully (now first) translated into the Englishe toung, by H. Billingsley. London.

Euclid (1956). *The Thirteen Books of Euclid's Elements*. Volume II. Books III-IX. Ed. by Thomas L. Heath. New York.

Mueller, Ian (1970). "Homogeneity in Eudoxus's Theory of Proportion". *Archive for History of Exact Sciences* 7, pp. 1–6.

Ramus, Petrus (1590). *Elementes of geometrie*. Faithfully translated by Tho. Hood. London.

Theon of Smyrna (1878). *Theonis Smyrnaei philosophi Platonici expositio rerum mathematicarum ad legendum Platonem utilium*. Ed. by Eduard Hiller. Leipzig.

Theon of Smyrna (1979). *Mathematics Useful for Understanding Plato*. Translated from the 1892 Greek/French edition of J. Dupuis by Robert and Deborah Lawlor. San Diego.

Viète, François (1591). *In artem analyticam isagoge*. Tours.

Viète, François (1983). *The Analytic Art*. Nine Studies in Algebra, Geometry and Trigonometry from the *Opus Restitutae Mathematicae Analyseos, seu Algebrâ Novâ*. Translated by T. Richard Witmer. New York.

Wallis, John (1662). *Hobbius Heauton-timorumenos, or, A consideration of Mr. Hobbes his dialogves in an epistolary discourse, addressed, to the Honourable Robert Boyle, Esq*. Oxford.

---

[1] An attempt at introducing a translation was made Xylander (Euclid 1562, p. 120), who suggested German *glychförmig* (modern: *gleichförmig*), which means 'of equal form (or kind)'. [2] Viète refers to a passage in Theon of Smyrna, who quotes Adrastus: "It is impossible, says Adrastus, to find a relationship between two things which are not of the same species" (Theon of Smyrna 1878, p. 73). Translation from Theon of Smyrna (1979, p. 48). [3] "Prima & perpetua lex aequalitatum seu proportionum, quae, quoniam de homogeneis concepta est, dicitur lex homogeneorum, haec est: Homogenea homogeneis comparari. Nam quae sunt heterogenea, quomodo inter se adfecta sint, cognosci non potest, ut dicebat Adrastus". Translation from Viète (1983, p. 15). [4] "tous ses termes montent à la même dimension".

***homogeneous* (sets and spaces)**

In the first volume of the journal *Fundamenta Mathematicae* (p. 223), Knaster and Kuratowski posed the problem: "Is a bounded plane continuum that is topologically homogeneous necessarily homeomorphic to a circumference [simple closed arc]?", and gave the following definition: "A set E is called *topologically homogeneous* if for each pair of points a, b of E there is a biunivocal and bicontinuous transformation of E into itself that transforms a into b".[1]

The term was soon in common use and adopted, for example, by Fréchet (1924, p. 1513); it is not due to Knaster and Kuratowski, however, as *homogene Punktmenge* had already been used by Brouwer in correspondence from 1912 (Brouwer 1976, p. 150). In print, Brouwer (1917, p. 1194) also used English *homogeneous set of points*. Cartan (1929, p. 183) later used *espace homogène* with an analogous definition. In this sense, English *homogeneous space* was used by Montgomery and Samelson (1943, p. 455).

The term has been defined in various, often unrelated, ways. Cantor (1885, p. 110), for example, used *homogene Punctmenge* for a point set P that is dense-in-itself and, as paraphrased by Ferreirós (2007, p. 213), has the property that "sufficiently narrow neighborhoods of its elements always contain parts of P of the same power".

Note. Overview of various definitions of homogeneity in the 1920s are found in Zoretti and Rosenthal (1924, pp. 988f.) and Tietze and Vietoris (1928, p. 174).

References

Bing, R.H. (1948). "A homogeneous indecomposable plane continuum". *Duke Mathematical Journal* 15, pp. 729–742.

Brouwer, L.E.J. (1917). "On linear inner limiting sets". *Koninklijke Akademie van Wetenschappen te Amsterdam. Proceedings of the Section of Sciences* 20, pp. 1424–1426.

Brouwer, L.E.J. (1976). *Collected Works.* 2: Geometry, Analysis, Topology and Mechanics. Ed. by Hans Freudenthal. Amsterdam and Oxford.

Cantor, Georg (1885). "Über verschiedene Theoreme aus der Theorie der Punctmengen in einem n-fach ausgedehnten stetigen Raume G_n. Zweite Mittheilung". *Acta Mathematica* 7, pp. 105–124.

Cartan, Élie (1929). "Sur les invariants intégraux de certains espaces homogènes clos et les propriétés topologiques de ces espaces". *Annales de la Société Polonaise de Mathématique* 8, pp. 181–225.

Ferreirós, José (2007). *Labyrinth of Thought.* A History of Set Theory and Its Role in Modern Mathematics. Second revised edition. Basel.

Fréchet, Maurice (1924). "Sur la notion de nombre de dimensions". *Comptes rendus de l'Académie des sciences* 178, pp. 1782–1785.

Montgomery, Deane and Hans Samelson (1943). "Groups transitive on the n-dimensional torus". *Bulletin of the American Mathematical Society* 49, pp. 455–456.

Tietze, Heinrich and Leopold Vietoris (1928). "Beziehungen zwischen den verschiedenen Zweigen der Topologie". *Encyklopädie der mathematischen Wissenschaften mit Einschluss ihrer Anwendungen* III AB 13, pp. 141–237.

Zoretti, Ludovic and Arthur Rosenthal (1924). "Die Punktmengen". *Encyklopädie der mathematischen Wissenschaften mit Einschluss ihrer Anwendungen* II C 9a, pp. 852–1030.

---

[1] "Un continu (borné) plan, topologiquement homogène, est-il nécessairement homéomorphe à une circonférence? (Un ensemble E est dit *topologiquement homogène*, lorsqu'il existe pour tout couple de points a, b de E une transformation biunivoque et bicontinue de E en lui-même qui transforme a en b)". – The question was later answered in the negative by Bing (1948).

homogeneous coordinates

Homogeneous coordinates were independently introduced by various authors in the late 1820s; especially Möbius (1827) and Plücker (1830) are usually cited in this respect, as are Bobillier (1827-28) and Feuerbach (1827).[1] Neither Möbius, nor Plücker, nor any others who first used homogeneous coordinates referred to them under this name. The term does not appear to have come up before the 1850s; Graßmann (1855, p. 5) used German *homogene Coordinaten* without giving a definition.

The term was popularized by Hesse. According to Wußing (1984, p. 38), the "universal acceptance [of homogeneous coordinates] was due to Hesse. An important reason for their acceptance was the elegance of the formal apparatus associated with their use". Hesse (1861, p. 60) defined them in the following terms:

> To use the advantages of symmetry in analytic geometry, the position of a point in space is expressed by four coordinates x, y, z, p. These are four quantities, whose ratios $\frac{x}{p}$, $\frac{y}{p}$, $\frac{z}{p}$ represent the usual rectangular coordinates. From now on we will call these four quantities the coordinates, or *homogeneous* coordinates of the point, and the ratios the *rectangular* coordinates of the point.[2]

Hesse had used homogeneous coordinates much earlier and introduced some conventions such as the notation $\frac{x}{p}$, $\frac{y}{p}$, $\frac{z}{p}$ (see Loria 1948, pp. 253–257).

Note. For the history of the concept, see Loria (1948), Voelke (2010), and Boyer (1956, passim).

References

Bobillier, Étienne (1827-28). "Essai sur un nouveau mode de recherche des propriétés de l'étendue". *Journal de mathématiques pures et appliquées* 18, pp. 320–339.

Boyer, Carl B. (1956). *History of Analytic Geometry*. New York.

Feuerbach, Karl Wilhelm (1827). *Grundriss zu analytischen Untersuchungen der dreieckigen Pyramide*. Nürnberg.

Graßmann, Hermann (1855). "Allgemeiner Satz über die lineale Erzeugung aller algebraischen Oberflächen". *Journal für die reine und angewandte Mathematik* 49, pp. 1–9.

Hesse, Otto (1861). *Vorlesungen aus der analytischen Geometrie des Raumes*. Leipzig.

Loria, Gino (1948). "Perfectionnements, évolution, métamorphoses du concept de «coordonnées»: Contribution à l'histoire de la géométrie analytique". *Osiris* 8, pp. 218–288.

Möbius, August Ferdinand (1827). *Der barycentrische Calcul*. Leipzig.

Plücker, Julius (1830). "Ueber ein neues Coordinatensystem". *Journal für die reine und angewandte Mathematik* 5, pp. 1–36.

Voelke, Jean-Daniel (2010). "Le développement historique du concept d'espace projectif". In: *Eléments d'une biographie de l'Espace projectif.* Ed. by Lise Bioesmat-Martagon. Nancy, pp. 207–286.

Wußing, Hans (1984). *The Genesis of the Abstract Group Concept*. A Contribution to the History of the Origin of Abstract Group Theory. Tr. Abe Shenitzer. Cambridge, Mass.

---

[1] See the literature cited in the note. [2] "Um die Vortheile der Symmetrie in der analytischen Geometrie zu haben, drückt man den Ort eines Punktes im Räume durch vier Coordinaten x, y, z, p aus. Man versteht darunter vier Grössen, deren Verhältnisse $\frac{x}{p}$, $\frac{y}{p}$, $\frac{z}{p}$ die gebräuchlichen rechtwinkligen Coordinaten des Punktes darstellen. Wir werden fortan jene vier Grössen die Coordinaten, oder die *homogenen* Coordinaten des Punktes nennen, die angegebenen Verhältnisse dagegen die rechtwinkligen Coordinaten des Punktes".

homogeneous differential equation and *homogeneous function*

As defined by Euler (1748, § 83), "[a] homogeneous function is one in which each term has the same degree. A heterogeneous function is one in which different degrees occur".[1] The term is not due to Euler, though, and seems to go back to Bernoulli (1728, p. 175).[2]

In the same place, Bernoulli also introduced the term *homogeneous differential equation*.[3] The concept had been around for some time; Bernoulli had already used it in his lectures on the integral calculus that date from the 1690s (Bernoulli 1742, p. 422).[4]

Both terms are found in various textbooks after the publication of Euler's *Introductio*. For example, de Bougainville (1756, p. 49) has *équation homogène*[5], Leseur and Jacquier (1768, p. 109) *fonction homogène*. The English term *homogeneous function* was used by Waring (1788, p. 88).

References

Bernoulli, Johann (1728). "De integrationibus aequationum differentialium, ubi traditur methodi alicuius specimen integrandi sine praevia separatione indeterminatarum". *Commentarii academiae scientiarum imperialis Petropolitanae* 1, pp. 167–184.

Bernoulli, Johann (1742). "Lectiones mathematicae de methodo integralium". In: *Opera omnia*. Tomus tertius. Lausanne and Genf, pp. 385–558.

Bos, Henk J.M. (1974). "Differentials, Higher-Order Differentials and the Derivative in the Leibnizian Calculus". *Archive for History of Exact Sciences* 14, pp. 1–90.

de Bougainville, Louis-Antoine (1756). *Traité du calcul intégral, pour servir de suite a l'Analyse des Infiniment-Petits*. Seconde partie. Paris.

Cantor, Moritz (1901). *Vorlesungen über Geschichte der Mathematik*. Dritter Band. Von 1668 bis 1758. Leipzig.

Euler, Leonhard (1748). *Introductio in analysin infinitorum*. Tomus primus. Lausanne.

Euler, Leonhard (1988). *Introduction to Analysis of the Infinite*. Book I. Translated by John D. Blanton. New York.

Leseur, Thomas and François Jacquier (1768). *Elemens du calcul integral*. Seconde partie. Parma.

Serret, Joseph-Alfred (1911). *Lehrbuch der Differential- und Integralrechnung*. Bearbeitet von Georg Scheffers. Zweiter Band. Integralrechnung. Berlin and Leipzig.

Waring, Edward (1788). "On Centripetal Forces". *Philosophical Transactions of the Royal Society of London* 78, pp. 67–102.

homography

In his *Aperçu historique sur l'origine et le développement des méthodes en géométrie*, Chasles (1837, p. 261) introduced the French term *homographique* in connection with figures; he defined two figures as homographic "when to each point, to each line, to each plane corresponds, respectively, a point, a line, and a plane".[6] The term is made up of Greek *ὁμο-* 'same' and *γραφικός* (from *γράφειν* 'to draw, describe')

---

[1] "Functio homogenea est, per quam ubique idem regnat variabilium numerus dimensionum: Functio autem heterogenea est, in qua diversi occurrunt dimensionum numeri". Translation from Euler (1988, p. 67). [2] Cantor (1901, p. 704) cites this text for the first occurrence of the term; see also Serret (1911, p. 596). [3] Cantor (1901, p. 879) cites this text for the first occurrence of the term. [4] Bos (1974, p. 7) points out that Bernoulli already defined homogeneous differential equations in 1694 without using the term. [5] de Bougainville means differential equations. [6] "à chaque point, à chaque droite, à chaque plan de l'une correspondent respectivement un point, une droite, un plan dans l'autre". – Homography also included the requirement that the cross ratio is preserved (Chasles 1837, p. 696).

and suggests that two homographic figures are drawn in the same or similar way. Chasles (ibid.) also used *déformation homographique* (homographic transformation) in this context. Later, he pointed out that Möbius' studies on collineation are "of the same kind" ("du même genre") as his but that he got to know these studies after the publication of his *Aperçu* (Chasles 1852, pp. xxxi, footnote).[1]

The term was directly translated as German *Homographie* in Chasles (1839, p. 259); the English term is found in Gaskin (1848, p. 401) in a footnote by Thomas Stephens Davies.

Various authors have criticized the term. Poncelet (1866, p. 435) regarded it as "rather vague" ("fort vague"), and, similarly, Bôcher (1892, p. 226) remarked: "The term 'homographic transformation,' introduced by Chasles, is not as expressive as the term collineation used some years before by Möbius".

See also *collineation* for a term widely used synonymously.

Note. For a discussion of Chasles' introduction of the concept, see Bussotti (2019, esp. pp. 298-302).

References

Baltzer, Richard (1878). *Die Elemente der Mathematik*. Zweiter Band. Planimetrie, Stereometrie, Trigonometrie. Fünfte verbesserte Auflage. Leipzig.

Bôcher, Maxime (1892). "Collineation as a mode of motion". *Bulletin of the American Mathematical Society* 1, pp. 225–231.

Bussotti, Paolo (2019). "Michel Chasles' foundational programme for geometry until the publication of his Aperçu historique". *Archive for History of Exact Sciences* 73, pp. 261–308.

Chasles, Michel (1837). *Aperçu historique sur l'origine et le développement des méthodes en géométrie*. Bruxelles.

Chasles, Michel (1839). *Geschichte der Geometrie, hauptsächlich mit Bezug auf die neueren Methoden*. Aus dem Französischen übertragen von Dr. L.A. Sohncke. Halle.

Chasles, Michel (1852). *Traité de géométrie supérieure*. Paris.

Gaskin, Thomas (1848). "On the inscription of polygons in the conic sections, the sides of which pass through given points". *The Mechanic's Magazine, Museum, Register, Journal & Gazette* 48, pp. 396–401.

Poncelet, Jean Victor (1866). *Traité des propriétés projectives des figures*. Tome second. Paris.

homology **(arithmetic and geometry)**

The term goes back to Greek *ὁμόλογος*, from *ὁμός* 'same' and the highly polysemous word *λόγος*, one of whose meanings is 'ratio'. The Greek word also means 'agreeing' and 'corresponding', and this appears to be the basic meaning in Euclid's definition (*Elements*, book V, def. 11): "The term *corresponding magnitudes* is used of antecedents in relation to antecedents, and of consequents in relation to consequents".[2]

Mugler (1958-59, p. 306) points out some passages in the *Elements* in which a change of meaning can be observed: "In equiangular triangles the sides about the equal angles are proportional, and those are corresponding [*ὁμόλογοι*] sides which subtend the equal angles" (*Elements*, book VI, prop. 4).[3] The word thus begins to be applied to geometrical elements that occupy the same place in two figures that are compared (see Mugler 1958-59, p. 306, who attributes this change to Archimedes).

---

[1] It is therefore not plausible to assume, as Baltzer (1878, p. 190) did, that *homography* is a translation of *collineation*. [2] Translation from Euclid (1956, p. 114). [3] Translation from Euclid (1956, p. 200).

The Latin term is of late origin; it appeared in a 12th-century translation of the *Elements* made directly from the Greek (Busard 1987, p. 109) and was also used by William of Moerbeke in his translation of Archimedes' *On Conoids and Spheroids*: "omologas, i. e. correspondentes" (Clagett 1976, p. 290).

In Billingsley's translation of the *Elements*, *of like proportion* was regularly used (Euclid 1570, fol. 133^r); however, *homologall sides* appears in Dee's preface (Dee 1570, fol. ciij^r). Forcadel used *de semblable raison* (Euclid 1564, fol. 128^v), which is analogous to *of like proportion*;[1] *homologue* was used by Stevin (1585, p. 66) and by Henrion (Euclid 1615, p. 136), presumably under the influence of Clavius' edition of the *Elements* (Euclid 1574, p. 160).

Note. For the Greek term, see Mugler (1958-59, pp. 306f.) and Heath in Euclid (1956, pp. 114, 238).

References

Busard, H.L.L. (1987). *The Mediaeval Latin Translation of Euclid's Elements Made Directly from the Greek*. Stuttgart.

Clagett, Marshall (1976). *Archimedes in the Middle Ages*. Volume II. The Translations from the Greek by William of Moerbeke. Philadelphia.

Dee, John (1570). "Mathematicall Praeface". In: *The Elements of Geometrie of the most auncient Philosopher Euclide of Megara*. Ed. by Henry Billingsley. London.

Euclid (1505). *Elementorum libri XIII ... cum expositione Theonis*. Ed. by Bartolomeo Zamberti. Venezia.

Euclid (1564). *Les Six Premiers Livres des Elements*. Traduicts et commentez par Pierre Forcadel de Besies. Paris.

Euclid (1570). *The Elements of Geometrie of the most auncient Philosopher Euclide of Megara*. Faithfully (now first) translated into the Englishe toung, by H. Billingsley. London.

Euclid (1574). *Elementorum libri XV*. Ed. by Christopher Clavius. Roma.

Euclid (1615). *Les quinze livres des Elements d'Euclide*. Traduicts de Latin en François ... par D. Henrion. Paris.

Euclid (1956). *The Thirteen Books of Euclid's Elements*. Volume II. Books III-IX. Ed. by Thomas L. Heath. New York.

Mugler, Charles (1958-59). *Dictionnaire historique de la terminologie géométrique des Grecs*. Paris.

Stevin, Simon (1585). *L'Arithmetique*. Leiden.

homology **(projective geometry)**

In his *Traité des propriétés projectives des figures*, Poncelet gave the term *homology* a new meaning, and he also introduced the new terms *homological figures* (*figures homologiques*), *centre of homology* (*centre d'homologie*), and *axis of homology* (*axe d'homologie*):[2]

> It is not possible, in the general case which we are considering, to keep the name centre of similitude for the point where the straight lines that contain the homologous points meet; but it can very suitably be called the *centre of homology* of the two figures, and the figures

---

[1] The expressions translate *similis rationis*, used by Zamberti (Euclid 1505, fol. Eiii^r). [2] Additionally, he called the lines through the centre of homology *rayons d'homologie*. For a discussion of Poncelet's definition, see Lorenat (2015, esp. pp. 179f.).

themselves can be called *homological*; the straight line on which the homologous lines concur shall then be the *axis of concurrence* or *of homology* of the system (Poncelet 1822, p. 160).[1]

He also remarked that while these terms had not been used before, they would be simple to grasp and, in view of the ordinary meaning of *homologous*, which is that figures are similar and similarly placed, the new terms "seem to perfectly designate the particular kind of correspondence that exists between the two systems" (ibid.).[2]

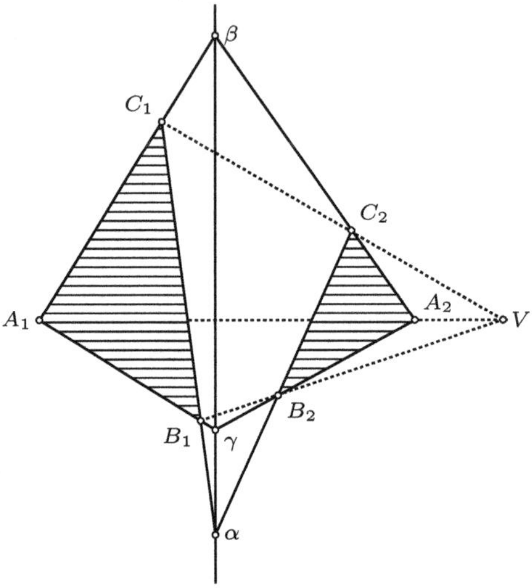

Figure 51: Homologous triangles (Emch 1905, p. 80). The lines A_1A_2, B_1B2, C_1C_2 meet in the centre of homology; the points α, β, and γ are collinear and lie on the axis of homology.

It did not take long for the terminology to spread. It was adopted by Chasles (1827-28, p. 285), who used *centre d'homologie* and attributed it to Poncelet; Plücker (1831, p. 171) translated the term by the rather cumbersome *homologous points with respect to both of the curves of the second class*,[3] which, unsurprisingly, did not catch on. In Chasles (1839, p. 80), the obvious translation *Mittelpunkt der Homologie* was used. The English terminology does not seem to have come up until the 1850s; *axis of homology*, for example, appears in Smith (1852, p. 121).

Townsend (1863, pp. 168f., 189f.), while pointing out the French origin of *homology*, spoke of *figures in perspective* as well as *centre* and *axis of perspective* instead. For the origin of this terminology, see *perspective*.

[1] "Il n'est pas possible de conserver, dans le cas général qui nous occupe, au point où concourent les droites qui renferment les points homologues, le nom de centre de similitude; mais on peut fort bien l'appeler le *centre d'homologie* des deux figures, et dire de ces figures elles-mêmes qu'elles sont *homologiques*; la droite sur laquelle concourent, deux à deux, les lignes homologues sera ainsi *l'axe de concours* ou *d'homologie* du système". [2] "elles nous paraissent parfaitement désigner l'espèce de correspondance particulière qui existe entre les deux systèmes". [3] "homologe Puncte in Beziehung auf die beiden Curven zweiter Classe".

Note. For Poncelet's introduction of the terminology, see Lorenat (2015, pp. 176–184) and Friedelmeyer (2010, pp. 120–123).

References

Chasles, Michel (1839). *Geschichte der Geometrie, hauptsächlich mit Bezug auf die neueren Methoden*. Aus dem Französischen übertragen von Dr. L.A. Sohncke. Halle.

Chasles, Michel (1827-28). "Mémoire sur les propriétés des systèmes de sections coniques, situées dans un même plan". *Annales de mathématiques pures et appliquées* 18, pp. 277–301.

Emch, Arnold (1905). *An Introduction to Projective Geometry and its Applications*. New York.

Friedelmeyer, Jean-Pierre (2010). "L'impulsion originelle de Poncelet dans l'invention de la géométrie projective". In: *Eléments d'une biographie de l'Espace projectif*. Ed. by Lise Bioesmat-Martagon. Nancy, pp. 55–158.

Lorenat, Jemma (2015). "Figures real, imagined, and missing in Poncelet, Plücker, and Gergonne". *Historia Mathematica* 42, pp. 155–192.

Plücker, Julius (1831). *Analytisch-geometrische Entwicklungen*. Zweiter Band. Essen.

Poncelet, Jean Victor (1822). *Traité des propriétés projectives des figures*. Paris.

Smith, H.J.S. (1852). "On some geometrical constructions". *The Cambridge and Dublin Mathematical Journal* 7, pp. 118–126.

Townsend, Richard (1863). *Chapters on the Modern Geometry of the Point, Line, and Circle*. Vol. I. Dublin.

homology **(topology) and** ***homology group***

In its topological sense, the term was first used by Poincaré (1895, pp. 18f.):

> Consider a manifold V of p dimensions; now let W be a manifold of q dimensions ($q \leq p$) which is a part of V. We suppose that the boundary of W is composed of λ manifolds of $q-1$ dimensions $v_1, v_2, ..., v_\lambda$. We express this fact by the notation
>
> $$v_1 + v_2 + ... + v_\lambda \backsim 0.$$
>
> More generally, the notation
>
> $$k_1 v_1 + k_2 v_2 \backsim k_3 v_3 + k_4 v_4,$$
>
> where the k are integers and the v are manifolds of $q-1$ dimensions will denote that there exists a manifold W of q dimensions forming part of V, the boundary of which is composed of k_1 manifolds similar to v_1, k_2 manifolds similar to v_2, k_3 manifolds similar to v_3 but oppositely oriented, and k_4 manifolds similar to v_4 but oppositely oriented. Relations of this form will be called *homologies*.[1]

The term was taken over by Heegaard in his PhD thesis (Heegaard 1898, p. 65);[2] the German term *homolog* appears in Wernicke (1904, p. 26) and Dehn and Heegaard (1907, p. 157). The English terms *homologous* and *homology* were used by Veblen in a colloquium held in 1916 but published only in 1922 (Veblen 1922, p. 56).

[1] "Considérons une variété V à p dimensions; soit maintenant W une variété à q dimensions ($q \leq p$) faisant partie de V. Supposons que la frontière complète de W se compose de λ variétés continues $q-1$ dimensions $v_1, v_2, \ldots, v_\lambda$. Nous exprimerons ce fait par la notation $v_1 + v_2 + \cdots + v_\lambda \backsim 0$. Plus généralement la notation $k_1 v_1 + k_2 v_2 \backsim k_3 v_3 + k_4 v_4$, où les k sont des entiers et les v des variétés à $q-1$ dimensions, signifiera qu'il existe une variété W à q dimensions faisant partie de V et dont la frontière complète se composera de k_1 variétés peu différentes de v_1, de k_2 variétés peu différentes de v_1, de k_3 variétés peu différentes de la variété opposée à v_3 et de k_4 variétés peu différentes de la variété opposée à v_4. Les relations de cette forme pourront s'appeler des *homologies*". Translation from Poincaré (2010, p. 15). [2] Heegard criticized some of Poincaré's proofs, which led Poincaré (1899) to revise his approach.

The term *homology group* was first used by Veblen (1922, p. 141): "The homologies thus correspond to the identities of the group G, which may well be called the homology group".[1] The term was also soon adopted, for example by Vietoris (1926, p. 1009, 1927, p. 457) and Mayer (1929, pp. 7f.), who used German *Homologiegruppe.*

Note. For the general history of homology, see Bollinger (1972) and McLarty (2006); for homology groups, see Basbois (2009, 2011).

References

Basbois, Nicolas (2009). "La naissance de la cohomologie des groupes". PhD thesis. Université Nice Sophia Antipolis.

Basbois, Nicolas (2011). "L'émergence de la notion de groupe d'homologie". *Gazette des Mathématiciens* 127, pp. 15–44.

Bollinger, Maja (1972). "Geschichtliche Entwicklung des Homologiebegriffs". *Archive for History of Exact Sciences* 9, pp. 94–170.

Dehn, Max and Poul Heegaard (1907). "Analysis situs". *Encyklopädie der mathematischen Wissenschaften mit Einschluss ihrer Anwendungen* III AB3, pp. 153–220.

Heegaard, Poul (1898). *Forstudier til en topologisk Teori for de algebraiske Fladers Sammenhæng.* København.

Mayer, Walther (1929). "Über abstrakte Topologie". *Monatshefte für Mathematik* 36, pp. 1–42.

McLarty, Colin (2006). "Emmy Noether's 'Set Theoretic' Topology: From Dedekind to the Rise of Functors". In: *The Architecture of Modern Mathematics.* Essays in History and Philosophy. Ed. by José Ferreirós and Jeremy Gray. Oxford, pp. 187–208.

Poincaré, Henri (1895). "Analysis situs". *Journal de l'École polytechnique* 1, pp. 1–121.

Poincaré, Henri (1899). "Complément à l'analysis situs". *Rendiconti del Circolo Matematico di Palermo* 13, pp. 285–343.

Poincaré, Henri (2010). *Papers on Topology. Analysis Situs* and Its Five Supplements. Translated by John Stillwell. Providence, R.I.

Veblen, Oswald (1922). *Analysis Situs.* New York.

Vietoris, Leopold (1926). "Ueber den höheren Zusammenhang von kompakten Räumen und eine Klasse von Abbildungen, welche ihn ungeändert lässt". *Koninklijke Akademie van Wetenschappen te Amsterdam. Proceedings of the Section of Sciences* 29, pp. 1008–1013.

Vietoris, Leopold (1927). "Über den höheren Zusammenhang kompakter Räume und eine Klasse von zusammenhangstreuen Abbildungen". *Mathematische Annalen* 97, pp. 454–472.

Wernicke, Paul (1904). *Ueber die Analysis Situs mehrdimensionaler Räume.* Göttingen.

homomorphism

See *isomorphism.*

homothetic

This term is a neologism coined from Greek *ὁμός* 'similar' and *θέσις* 'position'. It was introduced by Chasles (1827-28, p. 280) as French *homothétique*: "To avoid paraphrases, we will say in the future, of two conics traced in the same plane, that they are *homothetic*, to express that they are at the same time similar and similarly situated; by the way, we will happily abandon this adjective if we are offered another one more agreeable".[2]

---

[1] Basbois (2009, p. 119) cites this passage as likely to be the first occurrence of this term. [2] "Pour éviter les périphrases, nous dirons à l'avenir de deux coniques tracées dans un même plan, qu'elles sont *homothétiques*, pour exprimer qu'elles sont à la fois semblables et semblablement disposées; nous abandonnerons d'ailleurs très-volontiers cet adjectif, si l'on nous en offre un autre plus convenable". – The expression "similar and similarly situated" goes back to Euclid's *Elements* (book VI, prop. 18).

While Chasles introduced the term with respect to conic sections, it could easily be applied to geometric figures in general (Chasles 1837, p. 597). An illustration for the concept as applied to any systems of points A, B, C, ..., was given by Briot and Bouquet (1851, p. 204), for example (fig. 52).

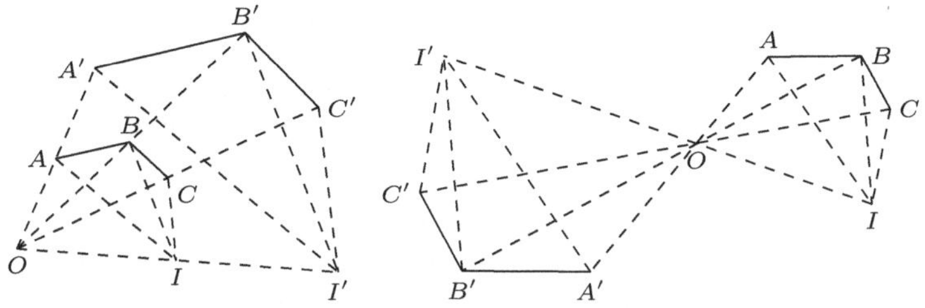

Figure 52: Illustrations from Briot and Bouquet (1851, p. 204). The points are *similarly situated* on the left (*homothétie directe*), and *inversely situated* on the right (*homothétie inverse*). The point O is called the *centre of similitude*.

The German term *homothetisch* is found in Magnus (1837, p. 364), while the English term does not appear to occur until the 1860s, when it was used by Wright (1868, p. 134). Mackay (1887, p. 69) commented on the term: "Attention should perhaps be drawn to the (unusual) use of the word 'similar' in the sense of 'similar and similarly situated,' or 'homothetic.' The word homothesis (French geometers have adopted a curious form *homothétie*) is not naturalised in English; otherwise homothesis and homothetic might have been used instead of similitude and similar".

Caution in the area of similarity is in order, as can be seen from Martin (1982, p. 136), who warns his readers: "the language of similarity theory is not standardized! In general, it is impossible to tell what any of the following words means without examining the context in which that word is used: stretch, dilation, dilatation, homothety, enlargement, contraction, central similarity, radial transformation, or size transformation".

References

Briot, Charles and Jean Claude Bouquet (1851). *Leçons nouvelles de géométrie analytique*. Deuxième édition. Paris.

Chasles, Michel (1837). *Aperçu historique sur l'origine et le développement des méthodes en géométrie*. Bruxelles.

Chasles, Michel (1827-28). "Mémoire sur les propriétés des systèmes de sections coniques, situées dans un même plan". *Annales de mathématiques pures et appliquées* 18, pp. 277–301.

Mackay, J.S. (1887). "Similitude and Inversion". *Proceedings of the Edinburgh Mathematical Society* 6, pp. 69–87.

Magnus, Ludwig Immanuel (1837). *Sammlung von Aufgaben und Lehrsätzen aus der analytischen Geometrie des Raumes*. Erste Abtheilung. Berlin.

Martin, George E. (1982). *Transformation Geometry*. An Introduction to Symmetry. New York.

Wright, Richard P. (1868). *The Elements of Plane Geometry for the Use of Schools and Colleges*. London.

homotopic* and *isotopic

According to Marquis (2006, p. 247), "Brouwer was the first mathematician to give a precise formal definition of a homotopy of paths. His definition is almost identical to the one we find in contemporary algebraic topology textbooks". Brouwer gave his definition in a footnote:

> Under a continuous modification of a univalent continuous transformation we understand in the following always the construction of a continuous series of univalent continuous transformations, i. e. a series of transformations depending in such a manner on a parameter, that the position of an arbitrary point is a continuous function of its initial position and the parameter (Brouwer 1912, p. 352).

Brouwer (1912, p. 353) also defined the homotopy class: "we shall say that two transformations *belong to the same class* if they can be transformed continuously into each other". As can be seen, the term *homotopic* was not used by Brouwer.

The term (from Greek *ὁμός* 'same' and *τόπος* 'place') had been introduced several years earlier by Dehn and Heegaard. They explained that the notion of homotopy was usually called "equivalence[1] in the sense of topology" and defined: "If two complexes transform into one another by way of a sequence of external transformations, possibly preceded by internal transformations, we call them *homotopic*" (Dehn and Heegaard 1907, p. 165).[2] They also remark (ibid.) that Jordan (1866, p. 91) called two homotopic curves on a surface *reducible*: "Any two closed contours, drawn on a given surface, are called *reducible* into one another, if one can pass from one to the other by a progressive deformation".[3] Dehn and Heegaard (1907, p. 166) also coined *isotopic* in their paper.

The terminology was quickly adopted, for example by Steinitz (1908, p. 38), who defined *homotopic paths* (*homotope Wege*). The English terms *homotopy* and *isotopy* appear in print in Snyder (1916, p. 87), a brief account of the "Cambridge colloquium", at which Veblen lectured on "analysis situs"; Veblen's definitions were given in Veblen (1922, pp. 125f.).

In short, while the words *homotopic* and *isotopic* were coined by Dehn and Heegard, their definition was not retained; instead, it was Brouwer's sense that was attached to *homotopic* (Dieudonné 2009, p. 43).

Note. For the history of the concept, see Bollinger (1972), Vanden Eynde (1992, 1999), and Krömer (2013).

References

Bollinger, Maja (1972). "Geschichtliche Entwicklung des Homologiebegriffs". *Archive for History of Exact Sciences* 9, pp. 94–170.

[1] On the ambiguous use of equivalence by Poincaré, see Vanden Eynde (1992, pp. 161f.). [2] "Gehen zwei Komplexe eventuell nach vorhergehenden internen Transformationen auseinander durch eine Reihe von externen Transformationen hervor, so nennen wir sie *homotop*". Translation from Vanden Eynde (1992, p. 166). – For a discussion of this combinatorial definition, see Vanden Eynde (1992, pp. 165–168). [3] "Deux contours fermés quelconques, tracés sur une surface donnée, seront dits *réductibles* l'un à l'autre, si l'on peut passer de l'un à l'autre par une déformation progressive". Translation from Vanden Eynde (1992, p. 147); Vanden Eynde also comments extensively on Jordan's concept.

Brouwer, L.E.J. (1912). "Continuous one-one transformations of surfaces in themselves (5th communication)". *Koninklijke Akademie van Wetenschappen te Amsterdam. Proceedings of the Section of Sciences* 15, pp. 352–361.
Dehn, Max and Poul Heegaard (1907). "Analysis situs". *Encyklopädie der mathematischen Wissenschaften mit Einschluss ihrer Anwendungen* III AB3, pp. 153–220.
Dieudonné, Jean (2009). *A History of Algebraic and Differential Topology, 1900–1960*. Boston.
Vanden Eynde, Ria (1992). "Historical Evolution of the Concept of Homotopic Paths". *Archive for History of Exact Sciences* 45, pp. 127–188.
Vanden Eynde, Ria (1999). "Development of the Concept of Homotopy". In: *History of Topology*. Ed. by Ioan Mackenzie James. Amsterdam, pp. 65–102.
Jordan, Camille (1866). "Des contours tracés sur les surfaces". *Journal de mathématiques pures et appliquées* 9, pp. 110–130.
Krömer, Ralf (2013). "The set of paths in a space and its algebraic structure. A historical account". *Annales de la faculté des sciences de Toulouse* 22, pp. 915–968.
Marquis, Jean-Pierre (2006). "A Path to the Epistemology of Mathematics: Homotopy Theory". In: *The Architecture of Modern Mathematics*. Essays in History and Philosophy. Ed. by José Ferreirós and Jeremy Gray. Oxford, pp. 239–260.
Snyder, Virgil (1916). "The Cambridge colloquium". *Bulletin of the American Mathematical Society* 23, pp. 81–88.
Steinitz, Ernst (1908). "Beiträge zur Analysis situs". *Sitzungsberichte der Berliner Mathematischen Gesellschaft* 7, pp. 29–49.
Veblen, Oswald (1922). *Analysis Situs*. New York.

homotopy group

The German term *Homotopiegruppe* was first suggested by Čech (1932, p. 203) at the ICM in 1932. It was quickly adopted, for example by Reidemeister (1934, pp. 211, 214) and Hurewicz (1935b, p. 477). The papers by Hurewicz were the ones that were most influential; Lefschetz (1957, p. 77) called Hurewicz "the founder of homotopy group theory". A definition was given in Hurewicz (1935a, p. 114):

> We now obtain two kinds of topological invariants: Depending on whether we fix the space X or the space Y, the loop groups of the spaces Y^X (and subspaces thereof) yield topological invariants of Y or X, respectively. Let us assume the former point of view. For X we choose the $(n - l)$-dimensional sphere S_{n-1} (where $n = 1, 2, \dots$). Fixing a point x_0 in S_{n-1} and a point y_0 in Y, we consider the subspace N of Y^S_{n-1} consisting of all mappings F with $F(x_0) = y_0$. (The remark at the end of the previous section implies that this space is locally contractible.) It turns out that the loop groups of N *do not depend on the components of* N. As is easily seen, the choice of points x_0 and y_0 does not matter either. Hence we obtain a group *that depends only on the space* Y *and on the number* n. We will call this group the n-th homotopy group of Y and denote it by $\pi_n(Y)$. Obviously, the group $\pi_n(Y)$ is nothing but the loop group of Y.[1]

[1] "Man erhält nun zweierlei Arten von topologischen Invarianten: Je nachdem man für X oder für Y einen festen Raum nimmt liefern die Wegegruppen der Räume Y^X (und seiner Teilmengen) topologische Invarianten von Y bzw. von X. Wir wollen uns auf den ersten Standpunkt stellen. Für X nehmen wir die $(n - l)$-dimensionale Sphäre S_{n-1} $(n = 1, 2, \dots)$. In S_{n-1} und in Y wählen wir je einen festen Punkt x_0 bzw. y_0 und betrachten die Teilmenge N von Y^S_{n-1} bestehend aus allen Abbildungen F mit $F(x_0) = y_0$. (Aus der Bemerkung am Schluss des vorigen Paragraphen folgt, dass diese Menge lokal zusammenziehbar ist). Es stellt sich heraus, dass die Wegegruppe in N von der Wahl der Komponente unabhängig ist. Da ausserdem die Wahl der Punkte x_0 und y_0, wie leicht zu sehen, keine Rolle spielt, liegt eine Gruppe vor, die nur vom Raum Y und von der Zahl $n$abhängt. Diese Gruppe wollen wir die n-te Homotopiegruppe von Y nennen und mit $\pi_n(Y)$ bezeichnen. Die Gruppe $\pi_n(Y)$ ist, wie man sofort sieht, nichts anders als die Wegegruppe in Y". Translation from Hurewicz (1995, p. 351). – For an account of Hurewicz' definition, see Fadell (1995).

After Hurewicz used the term, it spread quickly. The English term appears in Whitney (1937, p. 55).

Note. For the history of the concept, see Whitehead (1983) and Brown (2021).

References

Brown, Ronald (2021). "'Not Just an Idle Game': The Story of Higher-Dimensional Versions of the Poincaré Fundamental Group". *The Mathematical Intelligencer* 43, pp. 5–15.

Čech, Eduard (1932). "Höherdimensionale Homotopiegruppen". In: *Verhandlungen des Internationalen Mathematiker-Kongresses Zürich 1932*. II. Band / Sektions-Vorträge. Ed. by Walter Saxer. Zürich and Leipzig, p. 203.

Fadell, Edward (1995). "The contributions of Witold Hurewicz to algebraic topology". In: *Collected Works of Witold Hurewicz*. Ed. by Krystyna Kuperberg. Providence, R.I., pp. xxxiii–xl.

Hurewicz, Witold (1935a). "Beiträge zur Topologie der Deformationen (I. Höherdimensionale Homotopiegruppen)". *Koninklijke Akademie van Wetenschappen te Amsterdam. Proceedings of the Section of Sciences* 38, pp. 112–119.

Hurewicz, Witold (1935b). "Homotopie, Homologie and lokaler Zusammenhang". *Fundamenta Mathematicae* 25, pp. 467–485.

Hurewicz, Witold (1995). "Contributions to the Topology of Deformations. I. Higher-dimensional Homotopy Groups". In: *Collected Works of Witold Hurewicz*. Ed. by Krystyna Kuperberg. Providence, R.I., pp. 349–356.

Lefschetz, Solomon (1957). "Witold Hurewicz, In memoriam". *Bulletin of the American Mathematical Society* 63, pp. 77–82.

Reidemeister, Kurt (1934). "Homotopiegruppen von Komplexen". *Abhandlungen aus dem Mathematischen Seminar der Universität Hamburg* 10, pp. 211–215.

Whitehead, George W. (1983). "Fifty years of homotopy theory". *Bulletin of the American Mathematical Society* 8, pp. 1–29.

Whitney, Hassler (1937). "The maps of an n-complex into an n-sphere". *Duke Mathematical Journal* 3, pp. 51–55.

horocycle

The term *horocycle* (Greek *ὁρία* 'boundary') is due to Lobachevsky; he used it in a treatise published in German (Lobachevsky 1840, § 31):[1] "We call boundary line (oricycle) that curve lying in a plane for which all perpendiculars erected at the mid-points of chords are parallel to each other".[2] The terminology spread slowly (Lobachevsky's publications were little noticed), and first uses in French and English seem to be due to translations, that is, *courbe-limite (horicycle)* was used in Lobachevsky (1866, § 31) and English *oricycle* in Halphen's translation cited above. Beltrami (1868, p. 20) used the term in his article "Essay on the interpretation of noneuclidean geometry". In the form *horocycle*, the term appeared in the early 20th century; it was used by Coolidge (1903, p. 163).

Lobachevsky (1840, § 34) also used the related term *horosphere* (*Orisphäre*).

Note. For further comments on the origin of the concept, see the editorial comments in Lobachevsky (2010, p. 7).

References

Beltrami, Eugenio (1868). "Saggio di interpretazione della geometria non-euclidea". *Giornale di matematiche* 6, pp. 284–322.

---

[1] The Russian term for 'boundary line' is already found in "On the elements of geometry" from 1829/30 (Lobachevsky 1946, p. 195). [2] "Grenzlinie (Oricycle) nennen wir diejenige in einer Ebene liegende krumme Linie, für welche alle Perpendikel auf den Mittelpunkten der Sehnen errichtet unter sich parallel sind". Translation from Lobachevsky (1891, § 31).

Coolidge, Julian Lowell (1903). "Quadric Surfaces in Hyperbolic Space". *Transactions of the American Mathematical Society* 4, pp. 161–170.

Lobachevsky, Nikolai (1840). *Geometrische Untersuchungen zur Theorie der Parallellinien*. Berlin.

Lobachevsky, Nikolai (1866). *Études géométriques sur la théorie des parallèles*. Traduit de l'Allemand par J. Hoüel. Paris.

Lobachevsky, Nikolai (1891). *Geometrical Researches on the Theory of Parallels*. Translated from the original by George Bruce Halsted. Austin.

Lobachevsky, Nikolai (1946). *Полное собрание сочинений*. Vol. I. Ed. by V.F. Kagan et al. Moscow and Leningrad.

Lobachevsky, Nikolai (2010). *Pangeometry*. Edited and translated by Athanase Papadopoulos. Zürich.

hyperbola

See *conic section.*

hyperbolic

Derived from *hyperbola*, this term was originally used for certain geometrical objects that are somehow related to the hyperbola; later, the term was transferred to other domains such as differential equations. In the following, the terms *hyperbolic conoid, hyperbolic cylinder, acute hyperbolic solid* as well as *hyperbolic differential equation* are discussed in brief. The terms *hyperbolic spiral, hyperbolic function*, and *hyperbolic geometry* will be discussed in separate entries. For the term *hyperbolic logarithm*, see *natural logarithm*.

To begin with, the term *hyperbolic conoid* (hyperboloid of revolution) was coined by Federico Commandino in his translation of the works of Archimedes. Archimedes himself literally called a hyperboloid of revolution an 'obtuse-angled conoid' (*ἀμβλυγώνιον κωνοειδές*). In his translation, Commandino renders this by the Latin equivalent *conoides obtusiangulum* (Archimedes 1558, fol. 27^{v}); in his commentary, however, Commandino suggests *conoides hyperbolicum* as an alternative (Archimedes 1558, commentary, fol. 26^{v}). The term was very common in the 17th and 18th centuries and appears, for example, in Ozanam's dictionary (1691, p. 121): "*Conoïde Hyperbolique* [a solid] generated by the entire circumvolution of a hyperbola around its axis".[1] It is still found in various 19th century texts. See *conoid* and *hyperboloid*.

The term *hyperbolic cylinder* (Latin *cylindrus hyperbolicus*) appears to go back to Tacquet (1651, pp. 1f.). Tacquet also used the terms *circular, parabolic* and *elliptic cylinder*.

The term *acute hyperbolic solid* was used by Torricelli in his work *De solido hyperbolic acuto*; he defined it as follows: "If one rotates a hyperbola around an asymptote, as around an axis, one generates a solid infinite in length in the direction of the axis, which we call an acute hyperbolic solid" (Torricelli 1644, p. 112).[2] He proved the remarkable theorem that although the solid is infinitely long, its volume is still finite (Torricelli 1644, p. 113). The solid is called *hyperbolic* because it is generated by a hyperbola; it is called *acute* because it has a sharp end.

---

[1] "produit par la circonvolution entiere d'une Hyperbole autour de son axe". [2] "Si hyperbola circà asymptoton, tamquàm circà axem, convertatur, solidum fiet (si secundum axem consideretur) longitudine infinitum, quod quidem Acutum solidum hyperbolicum nominabimus". Translation from Mancosu (1996, p. 132).

Finally, the terminological distinction between hyperbolic, parabolic, and elliptic differential equations was suggested by du Bois-Reymond (1889, p. 265).

Note. For the origin of the term *hyperbolic conoid*, see Stäckel (1893, p. 461, fn. 8). For the hyperbolic solid, see Itard (1975), Mancosu and Vailati (1991), and Mancosu (1996, pp. 129–149). For du Bois-Reymond's classification of differential equations, see Gray (2021, pp. 289–293).

References

Archimedes (1558). *Opera*. Ed. by Federico Commandino. Venezia.

du Bois-Reymond, Paul (1889). "Ueber lineare partielle Differentialgleichungen zweiter Ordnung". *Journal für die reine und angewandte Mathematik* 104, pp. 204–229.

Gray, Jeremy (2021). *Change and Variations*. A History of Differential Equations to 1900. Cham.

Itard, Jean (1975). "La lettre de Torricelli à Roberval d'octobre 1643". *Revue d'histoire des sciences* 28, pp. 113–124.

Mancosu, Paolo (1996). *Philosophy of Mathematics and Mathematical Practice in the Seventeenth Century*. Oxford.

Mancosu, Paolo and Ezio Vailati (1991). "Torricelli's Infinitely Long Solid and Its Philosophical Reception in the Seventeenth Century". *Isis* 82, pp. 50–70.

Ozanam, Jacques (1691). *Dictionnaire mathematique*. Paris.

Stäckel, Paul (1893). "Bemerkungen zur Geschichte der geodätischen Linien". *Berichte über die Verhandlungen der Königlich Sächsischen Gesellschaft der Wissenschaften zu Leipzig (Mathematisch-Physische Klasse)* 45, pp. 444–467.

Tacquet, André (1651). *Cylindricorum et annularium libri IV*. Antwerpen.

Torricelli, Evangelista (1644). *De solido acuto hyperbolico*. In: *Opera geometrica*. Firenze, pp. 93–135.

hyperbolic function

Hyperbolic functions were introduced, in particular, by Riccati and Lambert. Indeed, Riccati coined the terms *hyperbolic sine* (*sinus hyperbolicus*) and *hyperbolic cosine* (*cosinus hyperbolicus*) and defined (see fig. 53):

> By analogy, the sine and the cosine can be carried over from the circle to the equilateral hyperbola. Let the equilateral hyperbola be described, whose semiaxis CA is equal to the radius of the circle, and from the center C to an arbitrary point F on the curve join line CF, and from the same point F let FD be dropped on the extension of axis CA. If CA is called the *total hyperbolic sine*, FD could be called the *hyperbolic sine* and CD the *hyperbolic cosine* of the double of the sector ACF divided by the total sine. Let these be the definitions of the names (Riccati 1757, p. 68).[1]

Hyperbolic functions were also considered by Lambert in 1761 (Lambert 1768), who was not aware of Riccati's work at that time and did not yet use the term *hyperbolic*. In a paper read in 1768, however, he did refer to Riccati, adopted his terminology and also introduced the term *fonction hyperbolique*, stating that it is used in analogy to *trigonometric function* (Lambert 1770, pp. 329f.).

---

[1] "Per analogiam sinus, & cosinus ex circulo traduci possunt ad hyperbolam aequilateram. Describatur hyperbola aequilatera, cujus semiaxis CA [...] radio circuli sit aequalis, & ex centro C ad punctum quodlibet F in curva positum, agatur recta CF, & ex eodem puncto F demittatur FD in axem CA productum. Si vocetur CA sinus totus hyperbolicus, FD vocari poterit sinus hyperbolicus, CD cosinus dupli sectoris ACF divisi per sinum totum. Hae sunt nominum definitiones". Translation from Van Brummelen (2021, p. 295), who also makes further comments on this definition.

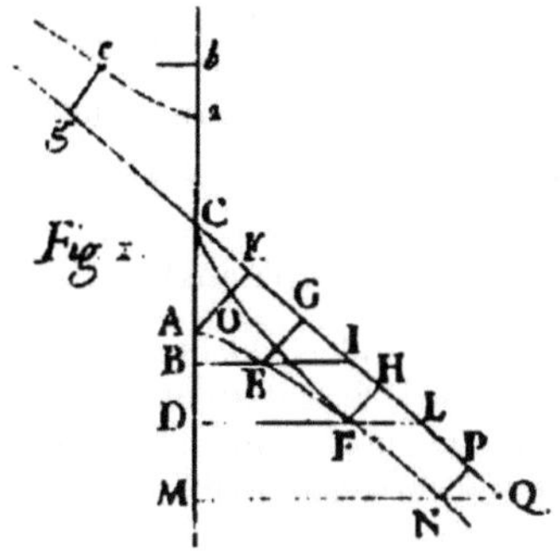

Figure 53: Hyperbolic sine and hyperbolic cosine. From Riccati (1757, following p. 94).

The terminology was soon adopted, for example by Lacroix (1798, p. 173); in English, it appears in Baily (1824, p. 415).

Note. For the history of hyperbolic functions, see Barnett (2004) and Van Brummelen (2021, pp. 294–303). The account by Günther (1881, ch. 1) is detailed and still useful.

References

Baily, Francis (1824). "On Mr. Babbage's new machine for calculating and printing mathematical and astronomical tables". *Astronomische Nachrichten* 2, pp. 409–422.

Barnett, Janet Heine (2004). "Enter, Stage Center: The Early Drama of the Hyperbolic Functions". *Mathematics Magazine* 77, pp. 15–30.

Van Brummelen, Glen (2021). *The Doctrine of Triangles*. A History of Modern Trigonometry. Princeton.

Günther, Siegmund (1881). *Die Lehre von den gewöhnlichen und verallgemeinerten Hyperbelfunktionen*. Halle.

Lacroix, Sylvestre François (1798). *Traité du calcul différentiel et du calcul intégral*. Tome second. Paris.

Lambert, Johann Heinrich (1768). "Mémoire sur quelques propriétés remarquables des quantités transcendantes, circulaires et logarithmiques". *Histoire de l'Académie Royale des Sciences et des Belles-Lettres de Berlin* (1761), pp. 265–322.

Lambert, Johann Heinrich (1770). "Observations trigonométriques". *Histoire de l'Académie Royale des Sciences et des Belles-Lettres de Berlin* (1768), pp. 327–354.

Riccati, Vincenzo (1757). *Opusculorum ad res physicas, et mathematicas pertinentium*. Tomus primus. Bologna.

hyperbolic, elliptic,* and *parabolic geometry

The terminological distinction between hyperbolic, elliptic, and parabolic geometries goes back to Klein (1871, p. 577): "Following the usual terminology in modern geometry, these three geometries will be called *hyperbolic*, *elliptic* or *parabolic* in what follows, according as the points at infinity of a line are real, imaginary or coincident".[1] In a footnote, Klein explains what he means by the "usual terminology": "E. g., one calls the points of a surface hyperbolic, elliptic or parabolic according as the principal tangents are real, imaginary or coincident. Steiner called an involution hyperbolic, elliptic or parabolic according as the double elements were real, imaginary or coincident, etc.".[1]

The terminology Klein refers to is explained by Salmon (1862, p. 182):

> A double point may be one of three different kinds according as the tangents at it are real, coincident, or imaginary. Accordingly the contact of a plane with a surface may be of three different kinds according as the tangent plane meets it in a section having a node, a cusp, or a conjugate point; or in other words according as the inflexional tangents are real, coincident, or imaginary.

Referring to Dupin (1813, p. 48), Salmon (ibid.) goes on the point out where the terms come from:

> Suppose that we confine our attention to points so near the origin that all powers of the co-ordinates above the second may be neglected, then the tangent plane (or a very near plane parallel to it) meets any surface $u_1 + u_2 + u_3 +$ &c. in the same section in which it meets the quadric $u_1 + u_2$. And according as the sections of this quadric by planes parallel to the tangent plane are ellipses, hyperbolas, or parabolas, so the section made by the tangent plane is to be considered as an infinitely small ellipse, hyperbola, or parabola. This infinitely small section Dupin calls the *indicatrix* at the point of contact, and he divides the points of the surface, according to the nature of the indicatrix into elliptic, hyperbolic, and parabolic points.

While Salmon is right in attributing the distinction to Dupin, it should be noted that Dupin does not explicitly use the term *hyperbolic point*, for example; instead, he calls the indicatrix *hyperbolic* in the respective case. The systematic use of this terminology is of later date; it is found, for example, in Young (1836, p. 232).

Steiner's terminology alluded to by Klein is found in Steiner (1867, pp. 51–53).

Klein's terminology spread quickly;[2] it was adopted by Clifford (1873, p. 387), for example:

> That geometry of three-dimensional space which assumes the Euclidian postulates has been called by Dr. Klein the *parabolic* geometry of space, to distinguish it from two other varieties, which assume uniform positive and negative curvature respectively, and which he calls the *elliptic* and *hyperbolic* geometry of space.

de Tilly (1879, p. 12) suggested, with little success, a different terminology. Instead of *parabolic* or *Euclidean geometry*, he said *common* or *commonly used geometry* (*géométrie usitée*); he also tried to replace *hyperbolic geometry* by *abstract geometry* (*géométrie abstraite*), and *elliptic geometry* by *doubly abstract geometry* (*géométrie*

---

[1] "Einem in der neueren Geometrie gewöhnlichen Sprachgebrauche folgend, sollen diese 3 Geometrieen bezüglich als *hyperbolische*, als *elliptische*) und als *parabolische* Geometrie im nachstehenden bezeichnet werden, je nachdem die beiden unendlich fernen Punkte der Geraden reell oder imaginär sind oder zusammenfallen". Translation from Stillwell (1996, p. 72). [1] "Man bezeichnet z. B. die Punkte einer Fläche als hyperbolische oder elliptische, je nachdem die Haupttangenten reell oder imaginär sind oder zusammenfallen. Steiner nennt die Involutionen hyperbolisch oder elliptisch oder parabolisch, je nachdem die Doppelelemente reell oder imaginär sind oder zusammenfallen u.s.f.". Translation from Stillwell (1996, p. 72). [2] However, *parabolic geometry* is infrequently used nowadays.

doublement abstraite). He also remarked that *hyperbolic geometry* is also called *Gaussian geometry* and *elliptic geometry Riemannian geometry*, terms which he attributed to Schering and Beltrami. de Tilly did not give a reference, but seems to mean Schering (1873, p. 14), who used *Gaussian space* (German *Gaussischer Raum*) and *Riemannian space* (*Riemannscher Raum*) but not literally *Gaussian geometry* and *Riemannian geometry*. Corresponding Italian versions of the terms were indeed used by Beltrami (1876, p. 240).

See also the entry for *Euclidean geometry* and *non-Euclidean geometry*.

Note. Useful comments on Klein's paper can be found in Stillwell (1996, pp. 63–68) and A'Campo and Papadopoulos (2015, p. 387), for example.

References

A'Campo, Norbert and Athanase Papadopoulos (2015). "On Klein's *So-called Non-Euclidean geometry*". In: *Sophus Lie and Felix Klein*. The Erlangen program and its impact in mathematics and physics. Ed. by Lizhen Ji and Athanase Papadopoulos. Zürich, pp. 91–136.

Beltrami, Eugenio (1876). "Formules fondamentales de cinématique dans les espaces de courbure constante". *Bulletin des sciences mathématiques et astronomiques* 11, pp. 233–240.

Clifford, William Kingdon (1873). "Preliminary Sketch of Biquaternions". *Proceedings of the London Mathematical Society* 4, pp. 381–395.

Dupin, Charles (1813). *Développements de géométrie*. Paris.

Klein, Felix (1871). "Ueber die sogenannte Nicht-Euklidische Geometrie". *Mathematische Annalen* 4, pp. 573–625.

Salmon, George (1862). *A Treatise on the Analytic Geometry of Three Dimensions*. Dublin.

Schering, Ernst (1873). "Linien, Flächen und höhere Gebilde in mehrfach ausgedehnten Gausschen und Riemannschen Räumen". *Nachrichten von der Königl. Gesellschaft der Wissenschaften und der Georg-Augusts-Universität zu Göttingen* (1873), pp. 13–21.

Steiner, Jacob (1867). *Vorlesungen über synthetische Geometrie*. Erster Theil: Die Theorie der Kegelschnitte in elementarer Darstellung. Leipzig.

Stillwell, John (1996). *Sources of Hyperbolic Geometry*. Providence, R.I.

de Tilly, Joseph Marie (1879). *Essai sur les principes fondamentaux de la géométrie et de la mécanique*. Bruxelles.

Young, John Radford (1836). *The Elements of the Differential Calculus*. Second edition. London.

hyperbolic logarithm

See *natural logarithm*.

hyperbolic spiral

Hyperbolic spirals were studied by in 1696 by Nicolas, who also introduced the name *spiralis hyperbolica* (Nicolas 1696, p. 23). They were also studied by Varignon in 1704, where he used French *Spirale Hyperbolique* (Varignon 1706, pp. 71f.). Varignon (ibid., p. 103) referred to Nicolas, and thus took over the term from him.

The term was adopted quickly and was also used by Johann Bernoulli (1712, p. 533) in particular, to whom the name is sometimes attributed. This wrong attribution may be due to Euler (1748, § 527), and to the fact that Bernoulli neither refers to Varignon nor to Nicolas.

The alternative name *reciprocal spiral* goes back to Cotes (1714-16, p. 26), the name being due to the fact that it can be understood as the result of a circle inversion of an Archimedean spiral. This fact had been pointed out by Bernoulli (1713, p. 128), who for this reason also gave *inverse Archimedean spiral* as an alternative.

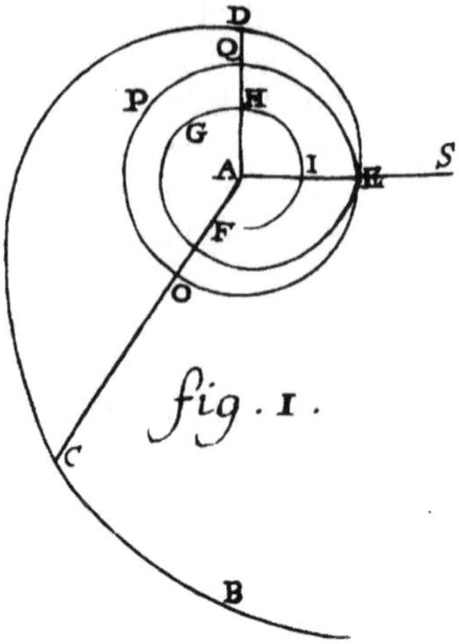

Figure 54: A hyperbolic spiral. From Nicolas (1696).

Note. For the history of the hyperbolic spiral, see Loria (1911, pp. 55–58) and Teixeira (1909, pp. 72–74); neither Loria nor Teixeira mentions Nicolas, however.

References

Bernoulli, Johann (1712). "Extrait de la Réponse de Monsieur Bernoulli à Monsieur Herman, datée de Basle le 7. Octobre 1710". *Mémoires de l'Académie royale des sciences* (1710), pp. 521–533.

Bernoulli, Johann (1713). "Continuatio Demonstrationum". *Acta Eruditorum* (Mar. 1713), pp. 115–132.

Cotes, Roger (1714-16). "Logometria Auctore Rogero Cotes". *Philosophical Transactions of the Royal Society of London* 29, pp. 5–45.

Euler, Leonhard (1748). *Introductio in analysin infinitorum*. Tomus secundus. Lausanne.

Loria, Gino (1911). *Spezielle algebraische und transzendente ebene Kurven*. Theorie und Geschichte. Autorisierte, nach dem italienischen Manuskript bearbeitete deutsche Ausgabe von Fritz Schütte. Zweite Auflage. Zweiter Band. Leipzig.

Nicolas, Pierre (1696). *De lineis logarithmicis et spiralibus hyperbolicis*. Toulouse.

Teixeira, Francisco Gomes (1909). *Traité des courbes spéciales remarquables plans et gauches*. Tome II. Coïmbra.

Varignon, Pierre (1706). "Nouvelle formation des spirales". *Mémoires de l'Académie royale des science* (1704), pp. 69–131.

hyperboloid

In his *Dictionnaire mathematique*, Ozanam (1691, p. 122) used the term *conoïde parabolique* 'parabolic conoid' (see *conoid*), which he shortened to *paraboloïde*; he did *not*, however, use the corresponding *hyperboloïde* for the hyperbolic conoid, a solid generated by the revolution of a hyperbola around its axis.[1]

The term *hyperboloid* is found soon after the publication of Ozanam's dictionary, however, for example in de Clermont (1693, p. 195), who wrote: "hiperboloïde, ou

[1] Another "missed opportunity" for using the word is found in Boyle (1671, p. 800), who mentioned "*Spheroids*, *Conoids*, *Paraboloids*, and other hard words".

conoïde hyperbolique". The shortened form was also used by Parent in 1698 (Parent 1713, p. 527), and the English term appears in Raphson (1702, p. 17).

A new classification of surfaces of second degree was developed in France around 1800, and new terminology was introduced. The results were presented, in particular, by Monge and Hachette (1802, 1805).[1] The new terms included *hyperboloïde à une nappe* 'hyperboloid of one sheet' (Monge and Hachette 1802, p. 158) and *hyperboloïde à deux nappes* 'hyperboloid of two sheets' (Monge and Hachette 1802, p. 161).[2]

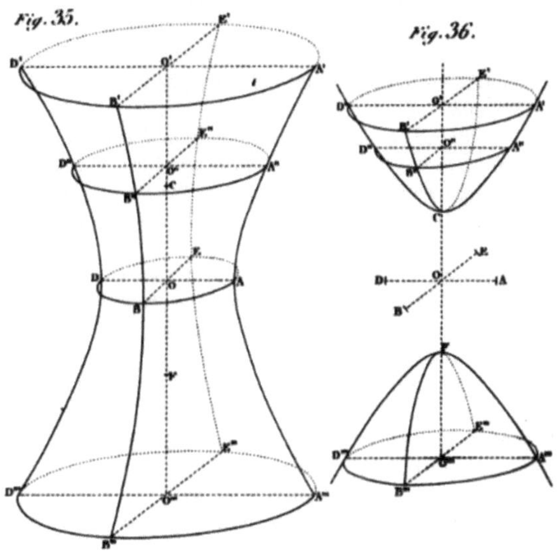

Figure 55: Hyperboloid of one sheet and two sheets. From Leroy (1834, pl. 9).

While the new terms did meet some resistance – Flauti (1815, p. 200), for example, called *hyperboloïde à une nappe* "bizarre" – they were soon in frequent use and adopted, for example, by Lacroix (1810, pp. 554, 557).[3] English *hyperboloid* was used in this sense in [anon.] (1807, p. 123).

Note that, especially in the 17th and, to a lesser extent, in the 18th century, *hyperboloid* was also used to refer to curves resembling a hyperbola; these curves are so-called "higher hyperbolas" of the kind $y^n = kx^{-m}$, for example. The term may

[1] Euler (1748, Appendix, §§ 117-125) also presented such a classification of surfaces, which was terminologically not as influential as that of Monge and Hachette, however. For example, he called the hyperboloid of one sheet *superficies elliptico-hyperbolica* (§ 120) and the hyperboloid of two sheets *superficies hyperbolico-hyperbolica* (§ 120), respectively. On Euler's classification, Boyer (1951, p. 224) commented: "The *Introductio* was the first textbook to recognize the five proper general quadric surfaces as members of a single family [...], and the names proposed were very similar to those now adopted". [2] The hyperboloid of one sheet had already been discussed by Wren (1669, p. 962), who called it a *hyperbolic cylindroid.* [3] They are not used in the first edition of Lacroix's *Traité*.

go back to Wallis (1655, p. 101), who first used *paraboloides cubicale* for the cubical parabola ($a^2y = x^3$) and then also mentioned *hyperboloid* and *elliptoid* curves.

Note. Brief remarks on the terminology can be found in Baltzer (1882, p. 491), Stäckel (1896, p. 218), and Staude (1910, p. 923). For the history of hyperbolas of higher order, see Loria (1910, pp. 316–321) and Teixeira (1909, pp. 127–130).

References

[anon.] (1807). "Extract from the Account of the Proceedings of the Society of Arts, Sciences, and Belles Lettres, of Bourdeaux [sic]". *The Philosophical Magazine* 27, pp. 121–124.
Baltzer, Richard (1882). *Analytische Geometrie*. Leipzig.
Boyer, Carl B. (1951). "The Foremost Textbook of Modern Times". *The American Mathematical Monthly* 58, pp. 223–226.
Boyle, Robert (1671). "A Continuation of Mr. Boyle's Experiments, published in the next foregoing Tract, about Fluids contiguous to other Fluids". *Philosophical Transactions of the Royal Society of London* 11, pp. 799–808.
de Clermont (1693). *La geometrie pratique de l'ingenieur, ov L'art de mesurer*. Strasbourg.
Euler, Leonhard (1748). *Introductio in analysin infinitorum*. Tomus secundus. Lausanne.
Flauti, Vincenzo (1815). *Geometria di sito sul piano, e nello spazio*. Napoli.
Lacroix, Sylvestre François (1810). *Traité du calcul différentiel et du calcul intégral*. Seconde édition, revue et augmentée. Tome premier. Paris.
Leroy, C.F.A. (1834). *Traité de géométrie descriptive*. Tome second. Paris.
Loria, Gino (1910). *Spezielle algebraische und transzendente ebene Kurven*. Theorie und Geschichte. Autorisierte, nach dem italienischen Manuskript bearbeitete deutsche Ausgabe von Fritz Schütte. Zweite Auflage. Erster Band. Leipzig.
Monge, Gaspard and Jean Nicolas Pierre Hachette (1802). "Application de l'algèbre à la géométrie". *Journal de l'École polytechnique* 4, cahier 11, pp. 143–169.
Monge, Gaspard and Jean Nicolas Pierre Hachette (1805). *Application de l'algèbre à la géométrie des surfaces du premier et second degré*. Paris.
Ozanam, Jacques (1691). *Dictionnaire mathematique*. Paris.
Parent, Antoine (1713). *Essais et recherches de mathématiques et de physique*. Troisième Volume. Paris.
Raphson, Joseph (1702). *A Mathematical Dictionary*. Abridg'd from Monsieur *Ozanam*, and Others. London.
Stäckel, Paul (1896). "Hagen, *Synopysis der höheren Mathematik* [review]". *Göttingische gelehrte Anzeigen* 158, pp. 211–227.
Staude, Otto (1910). *Analytische Geometrie des Punktepaares, des Kegelschnittes und der Fläche zweiter Ordnung*. Zweiter Teilband. Leipzig and Berlin.
Teixeira, Francisco Gomes (1909). *Traité des courbes spéciales remarquables plans et gauches*. Tome II. Coïmbra.
Wallis, John (1655). *De sectionibus conicis*. Oxford.
Wren, Christopher (1669). "Generatio corporis cylindroidis hyperbolici". *Philosophical Transactions of the Royal Society of London* 4, pp. 961–962.

hypercomplex

The investigation of hypercomplex number systems can be said to have begun in the 1840s with the works of Graßmann and Hamilton, in particular. But until well into the 19th century, it remained common to refer to hypercomplex numbers simply as *complex* numbers, a word coined by Gauß in the early 1830s. For example, in a chapter on "higher complex numbers", Hankel used the term 'complex number of the nth order'[1] for the expression $\alpha = a_1\iota_1 + \cdots + a_2\iota_2$, with a_k any complex number and ι_k being "independent, original, imaginary units" (Hankel 1867, p. 99).

[1] "complexe Zahl $n$ter Ordnung".

The term *hypercomplex* remained sporadic until at least 1900. However, a small number of authors did use the word even before the second half of the 19th century. For example, German *hypercomplex* is found in Wittstein (1846, p. 412), who used the term for a 4-dimensional number.[2] Also, Drobisch (1848, p. 174) uses the word in a talk on the geometrical construction of imaginary numbers: "Until now, however, the complex kind of number is the highest level to which arithmetical considerations have led, and a 'hypercomplex' kind of magnitude $a + ib + jc$ is, at least with respect to arithmetic, a thought devoid of content".[1]

In the 1870s, the term *hypercomplex* is used in connection with Graßmann's Extension Theory: "From n units themselves standing in no numerical relation, is formed by multiplication and summation the extensive magnitude $a = a_1e_1 + a_2e_2 + \dots a_ne_n$. The formation of a calculus of such hypercomplex [...] numbers is the fundamental idea on which Grassmann's research is based" (Sturm, Schröder, and Sohncke 1878, p. 5);[2] in the same place, *hypercomplex number* is explained as "higher complex number".[3] In the 1880s, the German word *übercomplex* (*über* literally means 'over', just like Greek *ὑπέρ*) was also occasionally used, for example by Scheffler (1880, p. 1) and Dedekind (1885, p. 142). Some time after Dedekind used this terminology, *hypercomplex* is also found in English, for example in Chapman (1892, p. 151), a review of Weierstrass' and Dedekind's work on hypercomplex numbers; the *Century Dictionary*, published between 1889 and 1891, also has an entry for *hypercomplex*.

See also *algebra* (structure) and *complex number*.

Note. For the history of hypercomplex numbers, see, for example, Kleiner (2007, ch. 3) and Stillwell (2010, ch. 20).

References

Biryukov, Boris and Zinaida Kuzicheva (2011). "On the early appraisals in Russia of H. and R. Grassmann's achievements". In: *From Past to Future*. Graßmann's Work in Context. Ed. by Hans-Joachim Petsche et al. Basel, pp. 175–183.

Chapman, C.H. (1892). "Weierstrass and Dedekind on General Complex Numbers". *Bulletin of the New York Mathematical Society* 1, pp. 150–156.

Dedekind, Richard (1885). "Zur Theorie der aus n Haupteinheiten gebildeten complexen Größen". *Nachrichten von der Königl. Gesellschaft der Wissenschaften und der Georg-Augusts-Universität zu Göttingen* (1885), pp. 141–159.

Drobisch, Moritz Wilhelm (1848). "Ueber die geometrische Construction der imaginären Grössen". *Berichte über die Verhandlungen der Königlich Sächsischen Gesellschaft der Wissenschaften zu Leipzig* 2, pp. 171–179.

Graßmann, Hermann (1862). *Die Ausdehnungslehre*. Vollständig und in strenger Form bearbeitet. Berlin.

[2] The number was of the form $x' + ix'' + j(y' + iy'')$, with $j^2 + i^2 = -1$. – Whether this is actually the first use remains an open question. [1] "Bis jetzt ist jedoch die complexe Zahlenform die höchste Stufe, auf welche arithmetische Betrachtungen geführt haben, und eine ‚hypercomplexe' Grössenform $a + ib + jc$, wenigstens in arithmetischer Beziehung, ein inhaltsleerer Gedanke". [2] "Also aus n selbst in keiner Zahlbeziehung stehenden Einheiten entsteht durch Vervielfachung und Summierung die extensive Grösse $a = a_1e_1 + a_2e_2 + \dots a_ne_n$. Die Aufstellung des *Calcüls mit solchen hypercomplexen [...] Zahlen* ist der Grundgedanke, von welchem Grassmann's Untersuchungen ausgehen". Translation from Biryukov and Kuzicheva (2011, p. 178). – *Extensive magnitude* is Graßmann's term (Graßmann 1862, p. 2). [3] "höhere complexe Zahl". This explanation is also found in Schröder (1873, e. g., pp. 139, 239).

Hankel, Hermann (1867). *Vorlesungen über die complexen Zahlen und ihre Functionen.* In zwei Theilen. I. Theil: Theorie der complexen Zahlsysteme. Leipzig.
Kleiner, Israel (2007). *A History of Abstract Algebra*. Basel.
Scheffler, Hermann (1880). *Die polydimensionalen Grössen und die vollkommenen Primzahlen.* Braunschweig.
Schröder, Ernst (1873). *Lehrbuch der Arithmetik und Algebra für Lehrer und Studirende.* Erster Band. Die sieben algebraischen Operationen. Leipzig.
Stillwell, John (2010). *Mathematics and its History.* Third Edition. New York.
Sturm, Rudolf, Ernst Schröder, and Leonhard Sohncke (1878). "Hermann Grassmann. Sein Leben und seine mathematisch-physikalischen Arbeiten". *Mathematische Annalen* 14, pp. 1–45.
Wittstein, Theodor (1846). "Ueber die geometrische Darstellung complexer Functionen". *Archiv der Mathematik und Physik* 7, pp. 411–430.

hypercube

In the January number of 1894, in *St. Nicholas*, a magazine for children, Sterns (1894, p. 250) proposed the following question: "everybody knows what a square looks like, and everybody knows just how a cube looks [...] What does the fourth power look like?". By April, two answers had arrived, as the editor reported: "Mr. Paul R. Heyl[1], of Philadelphia, incloses a model in wire open-work [see fig. 56], and writes: 'When we wish to represent the fourth power (which being of four dimensions we may call a hyper-cube), we may do so in solid perspective by placing a cube diagonally above another, and a little behind it, and joining the corresponding corners'" (Dodge 1894, p. 573).

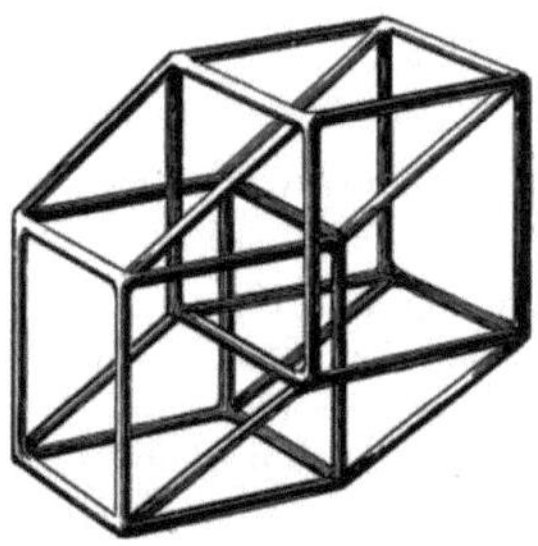

Figure 56: Hypercube. From Dodge (1894, p. 573).

While this is the first usage of the English term (OED), in French the word appeared independently at the same time in Delbœuf (1894, p. 356).

A number of alternative names have been suggested for hypercubes, some of which are still in use, and some never caught on.

The latter applies to Schläfli's German neologism *Oktaschem*, which is based on the word *Polyschem* which he used for 'polytope'; see *polytope*. The term appears in a paper written between 1850 and 1852 but not published until 1901: "The polyschemon

[1] Heyl was an American inventor and physicist.

(4, 3, 3) [...] has 16 vertices, 32 edges, 24 squares and 8 cubes, and shall therefore be called *octaschemon*" (Schläfli 1901, p. 51).[2]

Another term that was unsuccessful was suggested by Hinton (1880, p. 23): *four-square*. Several years later, he came up with a much better alternative: "We call the figure it [the cube] traces a Tessaract" (Hinton 1888, p. 118). The term is based on the Greek words *τέσσαρα* 'four' and *ἀκτίς* 'ray' and was quickly adopted, for example by Hall (1893, p. 179): "I have adopted as most convenient and complete the nomenclature in which the order of an n-fold solid is indicated by a word designating the number of its axes; as tessaract, pentact, etc.; and in which the particular solids of each order are distinguished by numeral adjectives referring to the number of $(n-1)$-fold boundaries". The spelling was later changed to *tesseract*, also by Hinton (1906, p. 182) himself.[1]

A term still found today was coined in German by Schlegel (1883, p. 434): *Achtzell* (8-cell). He gave *Oktaedroid* as an alternative; this word was introduced by Stringham (1880, p. 2). Additionally, Schröder (1890, p. 676) suggested, with little if any success, German *Achtwürfel* (8-cube) and *Oktokub*.[2]

Schoute (1905, p. 93) coined German *Maßpolytop* 'measure polytope': "In view of its applications in considerations of contents this regular polytope shall provisionally be called *measure polytope*".[3]

In the 20th century, *hypercube* also began to be used for cubes of any dimension, for example by Laurent (1904, p. 46). For this concept, *cube of n dimensions* (German *Würfel von n Dimensionen*) had already been used in the 19th century, for example by Phragmén (1885, p. 48).

References

Delbœuf, Joseph (1894). "L'ancienne et les nouvelles géométries. II". *Revue philosophique de la France et de l'étranger* 37, pp. 353–383.

Dodge, Mary Mapes (1894). "What does the fourth power look like?" *St. Nicholas: An Illustrated Magazine for Young Folks* 21, p. 573.

Hall, T. Proctor (1893). "The Projection of Fourfold Figures upon a Three-Flat". *American Journal of Mathematics* 15, pp. 179–189.

Hinton, Charles Howard (1880). "What is the fourth dimension?" *The University Magazine* 1, pp. 15–34.

Hinton, Charles Howard (1888). *A New Era of Thought*. London.

Hinton, Charles Howard (1906). "The Recognition of the Fourth Dimension". *Bulletin of the Philosophical Society of Washington* 14, pp. 179–203.

Laurent, Hermann (1904). *Théorie des nombres ordinaires et algébriques*. Paris.

Phragmén, Edvard (1885). "Über die Begrenzungen von Continua". *Acta Mathematica* 7, pp. 43–48.

Schläfli, Ludwig (1901). *Theorie der vielfachen Kontinuität*. Ed. by J.H. Graf. Basel.

Schlegel, Victor (1883). *Theorie der homogen zusammengesetzten Raumgebilde*. Halle.

Schoute, Pieter Hendrik (1905). *Mehrdimensionale Geometrie*. Zweiter Teil. Die Polytope. Leipzig.

Schröder, Ernst (1890). *Vorlesungen über die Algebra der Logik* (Exakte Logik). Erster Band. Leipzig.

---

[2] "Das Polyschem (4, 3, 3) [...] hat 16 Ecken, 32 Kanten, 24 Quadrate und 8 Würfel, und möge daher *Oktaschem* heissen". The translation *octaschemon* was chosen here because Schläfli used the analogous *polyschemon* in an English publication. [1] His paper was read in November 1901. [2] On the following pages, Schröder presents various visualizations of hypercubes. [3] "Im Hinblick auf seine Anwendung bei Inhaltsbetrachtungen möge dieses reguläre Polytop vorläufig das *Maßpolytop* heißen".

Sterns, George Warren (1894). "A serious question in mathematics". *St. Nicholas: An Illustrated Magazine for Young Folks* 21, p. 250.
Stringham, Irving (1880). "Regular Figures in n-dimensional Space". *American Journal of Mathematics* 3, pp. 1–14.

hyperelliptic

In this "Report on the Recent Progress of Analysis", Ellis (1847, p. 74) discussed the origin of the terminology:

> The first outline of a detailed theory of the higher transcendents was given by Legendre in the third supplement to his 'Traité des Fonctions Elliptiques'. He proposes to classify the transcendents comprised in the general formula
>
> $$\int \frac{f(x)dx}{(x-a)\sqrt{\varphi x}}$$
>
> according to the degree of the polynomial φx, the first class being that in which the index of this degree is three or four; the second that in which it is five or six, and so on. The first class therefore consists of elliptic integrals; all the others may be designated as *ultra-elliptic.* This epithet, however, which was proposed by Legendre, has not been so generally used as *hyper-elliptic*, which was, I believe, first used by M. Jacobi. M. Jacobi, however, has proposed to call the higher transcendents Abelian integrals.

Ellis, who was also the first to use the English terminology, was right about its origin: Legendre (1828, p. 191) defined a *fonction ultra-elliptique* in the following terms: "The simplest of these transcendents are those which have received the name *elliptic functions*; all the others, which might be given the name *ultra-elliptic functions* are divided in an infinity of classes carrying the numbers 2, 3, 4, etc.",[1] and Jacobi, in an announcement of Legendre's book (Jacobi 1832b, p. 415) and in an article published in the same year (Jacobi 1832a, p. 397), replaced the prefix *ultra-* by *hyper-*. In these papers, Jacobi also suggested *Abelian transcendent.*

The term *hyperelliptic curve* (Italian *curva iperellittica*) is due to Cremona (1869, p. 566). An abstract of the article appeared in *Nature* 1 (1870), p. 297, where also Cremona's definition was paraphrased: "hyperelliptic curve [...] those curves whose co-ordinates may be expressed rationally by means of a parameter λ, and the square root of an integral function of even order of this same parameter". The corresponding German term was used by Jung (1873, p. 475).

Clebsch (1876, p. 712) already stated that the term was common; similarly, Hancock (1898, p. 254) remarked: "[...] the more special functions in which y is connected with x by the relation $y^2 = X$ [X is a polynomial of degree > 4]; the term *hyperelliptic* is usually applied to such functions, *Abelian* being used *in general* to designate transcendents in which y is defined as any function of x through the algebraic equation $\chi(x, y) = 0$".

Note. For historical discussion and explanations of Legendre's and Jacobi's definitions, see Hancock (1898). See also *Abelian function.*

---

[1] "Les plus simples de ces transcendantes sont celles qui ont reçu le nom de fonctions elliptiques; toutes les autres, auxqelles on pourrait donner le nom de fonctions ultra-elliptiques, se divisent en une infinité de classes portant les numéros successifs 2, 3, 4, etc."

References

Clebsch, Alfred (1876). *Vorlesungen über Geometrie*. Bearbeitet und herausgegeben von Ferdinand Lindemann. Erster Band. Geometrie der Ebene. Leipzig.

Cremona, Luigi (1869). "Sulla trasformazione delle curve iperellittiche". *Reale Istituto lombardo di scienze e lettere. Rendiconti* 2, pp. 566–571.

Ellis, Robert Leslie (1847). "Report on the Recent Progress of Analysis (Theory of the Comparison of Transcendentals)". *Report of the British Association for the Advancement of Science* (1846), pp. 34–90.

Hancock, Harris (1898). "The Historical Development of Abelian Functions up to the time of Riemann". *Report of the British Association for the Advancement of Science* (1897), pp. 246–286.

Jacobi, Carl Gustav Jacob (1832a). "Considerationes generales de transcendentibus Abelianis". *Journal für die reine und angewandte Mathematik* 9, pp. 394–403.

Jacobi, Carl Gustav Jacob (1832b). "Legendre, *Théorie des fonctions elliptiques, troisième supplément* [announcement]". *Journal für die reine und angewandte Mathematik* 8, pp. 413–417.

Jung, Giuseppe (1873). "L. Cremona. Sulla trasformazione delle curve iperelliptiche [review]". *Jahrbuch über die Fortschritte der Mathematik* 2, pp. 475–477.

Legendre, Adrien-Marie (1828). *Traité des fonctions elliptiques et des intégrales eulériennes*. Tome troisième. Paris.

hypergeometric

This adjective is used in a variety of mathematical contexts. The term was coined by John Wallis, who is frequently claimed[1] to have introduced it in his *Arithmetica infinitorum*, first published in 1656. However, in this first edition the term does *not* occur.[2]

Actually, Wallis first uses the term in his *Treatise of Algebra*: "a Progression Hypergeometrical; where the Multiplier continually increaseth (or continually decreaseth;) as in 1, 2, 6, 24, 120, etc." (Wallis 1685, pp. 315f.).[3] With Wallis, *hypergeometric* refers neither to a power series nor to a function, but to a sequence, the general term of which is $\frac{1\cdot 3\cdot 5\cdots(2n+1)}{2\cdot 4\cdot 6\cdots 2n}$. As Wallis (1693b, p. 355) points out himself, he chose the term 'beyond' (*hyper*) because a hypergeometric sequence "surpasses" a geometric progression in terms of its mode of composition; this means that "in analogy with the case of geometric progressions, where each term is obtained from the preceding one by multiplying it by a constant, one defined a hypergeometric progression as a progression in which each term is obtained from the preceding one by multiplying it by a factor which increases by a unit at each step" (Papadopoulos 2017, p. 32).

The term was also used by Euler (1760, p. 213), who explicitly referred to Wallis. Euler used the term both for series and sequences. In a paper read in 1768 and published in 1769, Euler also introduced the Latin term *curva hypergeometrica* for what is today designated as a *hypergeometric function* (Euler 1769). Pfaff (1797, p. 135) was the first to use Latin *series hypergeometrica* for a power series.[4]

Kummer (1836, p. 45) considered the hypergeometric series as a function although, in contrast to the claim by Birkhoff (1973, p. 61), he did not explicitly use the term

---

[1] E. g., by Demidov, Petrova, and Simonov (1998, p. 155), Klein (1933, p. 3). [2] Dutka (1984, p. 16). The reason for the mistaken claim seems to be that the *Arithmetica infinitorum* as printed in Wallis (1693a, p. 466), where *hypergeometric* can indeed be found, is consulted. The concept *is* discussed in the first edition (Wallis 1656, pp. 173–178), in a scholium to proposition 190. [3] The term then reappears in the Latin version *De Algebra Tractatus* (Wallis 1693b, p. 353). [4] See Klein (1933, p. 3) and Dutka (1984, pp. 26–29).

hypergeometric function. Such usage came up in the second half of the 19th century, for example in Hansen (1853, p. 11) and Pochhammer (1870, p. 316). In the latter article, *hypergeometric differential equation* is also found (Pochhammer 1870, p. 336).

In probability theory and statistics, the usual definition of the hypergeometric distribution with parameters n, N, and K is given by Ethier (2010, pp. 21f.): it is "the distribution of the number of black balls in a sample of size n taken without replacement from an urn containing N balls, of which K are black and $N-K$ are white. If X is hypergeometric(n, N, K), then

$$P(X=k)=\frac{\binom{K}{n}\binom{N-K}{n-k}}{\binom{N}{h}}, k=(K+n-N)^{+},\ldots,K\wedge n;$$

here $x^{+} := \max(x,0)$ and $x \wedge y := \min(x,y)$". The hypergeometric distribution was already known in the 17th and 18th centuries (see, e. g., Hald 2003, pp. 201f., 234 and Ethier 2010, p. 68). Its probability generating function is a hypergeometric series, which, on the one hand, accounts for the name, and, on the other hand, may explain why *hypergeometric distribution* does not seem to have ever been "officially" baptized.[1] For example, *hypergeometrical distribution* was used by Tocher (1908) and by Isserlis (1917, p. 422) without explanation because it was clear what was meant. The term is more frequently found from the mid-1930s onwards.

Note. The history of the concept is discussed in detail by Dutka (1984) and Roy (2011, pp. 547–573).

References

Birkhoff, Garrett, ed. (1973). *A Source Book in Classical Analysis*. Cambridge, Mass.

Demidov, S.S., S.S. Petrova, and N.I. Simonov (1998). "Ordinary differential equations". In: *Mathematics of the 19th Century*. Function Theory According to Chebyshev. Ordinary Differential Equations. Calculus of Variations. Theory of Finite Differences. Ed. by Andrey Nikolaevich Kolmogorov and Adolph Andrei Pavlovich Yushkevich. Basel, pp. 83–196.

Dutka, Jacques (1984). "The Early History of the Hypergeometric Function". *Archive for History of Exact Sciences* 31, pp. 15–34.

Ethier, Stewart N. (2010). *The Doctrine of Chances*. Probabilistic Aspects of Gambling. Berlin.

Euler, Leonhard (1760). "De seriebus divergentibus". *Novi Commentarii academiae scientiarum Petropolitanae* 5, pp. 19–23, 205–237.

Euler, Leonhard (1769). "De curva hypergeometrica hac aequatione expressa y=1.2.3....x". *Novi Commentarii academiae scientiarum Petropolitanae* 13, pp. 3–66.

Hald, Anders (2003). *A History of Probability and Statistics and Their Applications Before 1750*. Hoboken, N.J.

Hansen, Peter Andreas (1853). "Ueber die Entwickelung des Products einer Potenz des Radius vectors mit dem Sinus oder Cosinus eines Vielfachen der wahren Anomalie in Reihen, die nach den Sinussen oder Cosinussen der Vielfachen der wahren, excentrischen oder mittleren Anomalie fortschreiten". *Berichte über die Verhandlungen der Königlich Sächsischen Gesellschaft der Wissenschaften zu Leipzig (Mathematisch-Physische Klasse)* 5, pp. 1–14.

Isserlis, Leon (1917). "On the Representation of Statistical Data". *Biometrika* 11, pp. 418–425.

[1] Analogous statements can be made in connection with other distributions.

Klein, Felix (1933). *Vorlesungen über die hypergeometrische Funktion*. Gehalten an der Universität Göttingen im Wintersemenster 1893/94. Ausgearbeitet von Ernst Ritter. Herausgegeben und mit Anmerkungen versehen von Otto Haupt. Berlin.

Kummer, Ernst Eduard (1836). "Über die hypergeometrische Reihe $1+\frac{\alpha.\beta}{1.\gamma}x+\frac{\alpha(\alpha+1)\beta(\beta+1)}{1.2.\gamma(\gamma+1)}x^2+$ [...] $+\ldots$". *Journal für die reine und angewandte Mathematik* 15, pp. 39–83, 127–172.

Papadopoulos, Athanase (2017). "Looking Backward: From Euler to Riemann". In: *From Riemann to Differential Geometry and Relativity*. Ed. by Lizhen Ji, Athanase Papadopoulos, and Sumio Yamada. Cham, pp. 1–93.

Pfaff, Johann Friedrich (1797). *Disquisitiones analyticae*. Volumen I. Helmstedt.

Pochhammer, Leo August (1870). "Ueber hypergeometrische Functionen $n$ter Ordnung". *Journal für die reine und angewandte Mathematik* 71, pp. 316–352.

Roy, Ranjan (2011). *Sources in the Development of Mathematics*. Infinite Series and Products from the Fifteenth to the Twenty-first Century. Cambridge.

Tocher, J.F. (1908). "Pigmentation Survey of School Children in Scotland". *Biometrika* 6, pp. 129–235.

Wallis, John (1656). *Arithmetica infinitorum*. In: *Operum Mathematicorum*. Pars altera. Oxford, pp. 1–198.

Wallis, John (1685). *A Treatise of Algebra, both Historical and Practical*. London.

Wallis, John (1693a). *Arithmetica Infinitorum*. In: *Opera Mathematica*. Volumen primum. Oxford, pp. 355–478.

Wallis, John (1693b). *De Algebra tractatus*. In: *Opera Mathematica*. Volumen alterum. Oxford, pp. 1–482.

hypergraph

This is a generalization of the concept of graph. In print, the term seems to first appear in Berge (1970, p. viii) as French *hypergraphe*: "A hypergraph is defined to be a family of *hyperedges* which are sets of vertices of cardinality not necessarily 2 (as for graphs)".[1] Berge (1996, pp. 8f.) points out the originator of the term: "The name of ‚hypergraph' was invented [...] by one of our students, Jean-Marie Pla, during one of our weekly meetings in 1969". Earlier terms are *graphoid* (Berge 1969, p. 50) and *set-system* (Erdős and Hajnal 1966, p. 65).

Note. Some notes on the history of the notion are found in Berge (1996).

References

Berge, Claude (1969). "The Rank of a Family of Sets and Some Applications to Graph Theory". In: *Recent Progress in Combinatorics*. Proceedings of the Third Waterloo Conference on Combinatorics, May 1968. Ed. by William Thomas Tutte. New York, pp. 49–57.

Berge, Claude (1970). *Graphes et hypergraphes*. Paris.

Berge, Claude (1973). *Graphs and Hypergraphs*. Translated by Edward Minieka. Amsterdam and London.

Berge, Claude (1996). "The history of the perfect graphs". *Southeast Asian Bulletin of Mathematics* 20, pp. 5–10.

Erdős, Paul and András Hajnal (1966). "On chromatic number of graphs and set systems". *Acta Mathematica Academiae Scientiarum Hungaricae* 17, pp. 61–99.

hyperplane, hypersurface, and *hypersphere*

In a letter to Farkas Bolyai dating from March 1832, Gauß made some terminological suggestions, among them *hypershere* (*Hypersphäre*): "One might call *hypercycle*

[1] "Un hypergraphe est défini par une famille d'«hyper-arêtes», qui sont des ensembles de sommets de cardinalité quelconque (et non plus nécesairement de cardinalité 2 comme dans les graphes)". Translation from Berge (1973, p. viii).

the collection of all points at equal distance from a straight line with which they lie in the same plane; similarly for *hypersphere*" (Gauß 1900, p. 221).[2] The term only appearing in a letter, it remained unknown at first.

Schläfli (1858, p. 280) later made a new suggestion:

> we proceed to [...] the general spheric equation $x_1^2 + x_2^2 + \cdots + x_n^2 = a^2$; the origin here assumed may be called *centre* and the constant a *radius*. For the curved continuum itself, represented by this equation, I propose the term *polysphere* or n-sphere; thus, a pair of points in a straight line will be a *monosphere*, the circle a *disphere*, and the spherical surface properly so called a *trisphere*. These terms are very harsh, but I do not know how to avoid this.

Schläfli had coined these terms several years earlier; they appear in a manuscript, written between 1850 and 1852, entitled "Theory of multiple continuity" (Schläfli 1901, p. 58), which remained unpublished at the time, however.[1]

The term *hypersphere* first appears in print in Salmon (1860, p. 285); the coinage was independent of Gauß. Salmon there also introduced the word *hyperplane*:

> It must have occurred to any person who has observed the analogy between the processes of the analytic geometry of three and of two dimensions, that as far as algebraic processes are concerned there is nothing to limit the methods employed to the case of three dimensions. If, for example, there were any beings who could conceive space of four dimensions, we could with our present faculties write as good a book of analytic geometry for their use as if we could conceive space of four dimensions ourselves. We should take four *hyper*-planes at right angle to each other, and define the position of any point by four coordinates expressing its four distances from these; and so proceed in every manner exactly as we proceed in the analytic geometry of three dimensions. [...] Now the process followed in Art. (2) may be extended word for word to space of four dimensions and gives the following relation between the distances of five points on a *hyper*-sphere.

Finally, *hypersurface* was used by Cayley (1852, p. 276) for a surface for which he gave the equation $ax^2 + by^2 + cz^2 + 2fyz + 2qzx + 2hxy \cdots + pt^2 = 0$. Craig (1882, p. 381) later gave the following definition: "This function f which satisfies say $f = const.$ sustains the same relation to the space of n dimensions as a function of the three variables x, y, z, and also of t sustains to ordinary space, so for want of a better term I shall speak of it as a hypersurface". Poincaré (1887, p. 325) also used *hypersurface*, gave a precise definition as a "manifold of three dimensions",[2] and remarked that because the "hypergeometric language" may be repugnant to "bons esprits", he will make little use of it.

Note. The uses by Schläfli (1858, p. 280) and Salmon (1860, p. 285) have also been pointed out by Miller (n.d.). For some additional historical references on hyperplanes and some synonyms, see Segre (1921, p. 789, fn. 71).

References

A'Campo, Norbert and Athanase Papadopoulos (2012). "Notes on non-Euclidean geometry". In: *Strasbourg Master Class on Geometry*. Ed. by Athanase Papadopoulos. Zürich, pp. 1–182.

---

[2] "Hypercykel könnte der Complexus aller Punkte heissen, die von einer Geraden, mit der sie in Einer Ebene liegen, gleiche Distanz haben; eben so Hypersphäre". Translation from A'Campo and Papadopoulos (2012, p. 62). [1] The terminology slightly varies there because Schläfli originally defined a hypersphere by the equation $x_1^2 + x_2^2 + \cdots + x_n^2 < a^2$. [2] "multiplicité à trois dimensions".

Cayley, Arthur (1852). "Analytical Researches Connected with Steiner's Extension of Malfatti's Problem". *Philosophical Transactions of the Royal Society of London* 142, pp. 253–278.
Craig, Thomas (1882). "General Properties of Certain Partial Differential Equations Similar to Those of Hydrodynamics". *Memoirs of the American Academy of Arts and Sciences* 10, pp. 362–388.
Gauß, Carl Friedrich (1900). *Werke*. Achter Band. Leipzig.
Miller, Jeff (n.d.). "Earliest Known Uses of Some of the Words of Mathematics". URL: `https://mathshistory.st-andrews.ac.uk/Miller/mathword/`.
Poincaré, Henri (1887). "Sur les résidus des intégrales doubles". *Acta Mathematica* 9, pp. 321–380.
Salmon, George (1860). "On the Relation which connects the Mutual Distances of Five Points in Space". *The Quarterly Journal of Pure and Applied Mathematics* 3, pp. 282–288.
Schläfli, Ludwig (1858). "On the Multiple Integral $\int^n dxdy \dots dz$, whose limits are $p_1 = a_1x + b_1y + \dots h_1z > 0, p_2 > 0, \dots p_n > 0$, and $x^2 + y^2 + \dots z^2 < 1$". *The Quarterly Journal of Pure and Applied Mathematics* 2, pp. 269–301.
Schläfli, Ludwig (1901). *Theorie der vielfachen Kontinuität*. Ed. by J.H. Graf. Basel.
Segre, Corrado (1921). "Mehrdimensionale Räume". *Encyklopädie der mathematischen Wissenschaften mit Einschluss ihrer Anwendungen* III C 7, pp. 769–972.

hyperset

Non-well-founded sets, as opposed to well-founded sets, have been investigated from the beginning of the 20th century. In particular, Mirimanov (1917, p. 42) formulated this distinction, using the term 'extraordinary set' (French *ensemble extraordinaire*):

> Let E be a set, E' one of its elements, E'' any element of E', and so on. I call a *descent* the sequence of steps from E to E', E' to E'', etc. [...] I say that a set is *ordinary* when it only gives rise to finite descents; I say that it is *extraordinary* when among its descents there are some which are infinite.[1]

The term *well-founded set* goes back to the *axiom of foundation* (an axiom introduced to avoid sets as those discussed by Mirimanov), so called by Zermelo (1930, p. 31) (German *Axiom der Fundierung*). Zermelo (1932, p. 86) also used the term *wohlfundiert* 'well-founded' in this connection.

Hyperset is a relatively recent term coined by Barwise (1984, p. 22) on the analogy of *hyperreal*[2]: "I will use the term 'set' for the cumulative conception of wellfounded set, and the term 'hyperset' for the broader notion of sets as forgetful situations. I suggest this terminology out of deference to the accepted notion, and out of similarity with the reals and hyperreals of non-standard analysis".

Note. On the history of the notion, see Ferreirós (2007, ch. XI) and Aczél (1988, pp. 103–112).

References

Aczél, Peter (1988). *Non-well-founded Sets*. Stanford.
Barwise, John (1984). "Situations, sets and the axiom of foundation". In: *Logic Colloquium '84*. Ed. by J.B. Paris, A.J. Wilkie, and G.M. Wilmers. Amsterdam, pp. 21–36.
Ferreirós, José (2007). *Labyrinth of Thought*. A History of Set Theory and Its Role in Modern Mathematics. Second revised edition. Basel.

---

[1] "Soient E un ensemble, E' un de ses élements, E'' un élement quelconque de E', et ainsi de suite. J'appelle *descente* la suite des passages de E à E', E' à E'', etc. [...] Je dirai qu'un ensemble est *ordinaire* lorsu'il ne donne lieu qu'à des descentes finies; je dirai qu'il est *extraordinaire* lorsque parmi ses descentes il y en a qui sont infinies". Translation from Aczél (1988, epigraph). [2] See the end of the article *infinitesimal* for brief notes on the introduction of this term.

Mirimanov, Dmitry (1917). "Les antinomies de Russell et de Burali-Forti et le problème fondamental de la théorie des ensembles". *L'enseignement mathématique* 19, pp. 37–52.
Zermelo, Ernst (1930). "Über Grenzzahlen und Mengenbereiche". *Fundamenta Mathematicae* 16, pp. 29–47.
Zermelo, Ernst (1932). "Über Stufen der Quantifikation und die Logik". *Jahresbericht der Deutschen Mathematiker-Vereinigung* 41, pt. 2, pp. 85–88.

hypocycloid

See *epicycloid*.

hypotenuse

The Greek verb *ὑποτείνειν* means 'to subtend, stretch under' and is the basis of the term *ὑποτείνουσα*. The terminology was not limited to right-angled triangles (a subtending line was essentially a line opposite an angle); it did become associated, however, with the Pythagorean theorem: "In right-angled triangles the square on the side subtending the right angle is equal to the squares on the sides containing the right angle" (Euclid, *Elements*, book I, prop. 47).[1] The terminology was in use before Euclid (Mugler 1958-59, p. 443).

The word may go back to the ancient technique of rope stretching: ropes are known to have been used as measuring instruments by the so-called *harpedonapts* (rope stretchers), who were surveyors in ancient Egypt and in other cultures (Gandz 1930). The rope stretchers left terminological traces. For example, in Plato's *Menon* (85B) the diagonal is described as "the line stretching from corner to corner".[2] This explanation is also suggested by Burnet (1908, p. 116): "The very name 'hypotenuse' [...] means literally 'the cord stretching over against,' and this is surely just the rope of the 'harpedonapt'".[3]

In Latin, the terminology is found in some agrimensorial texts. In particular, the literal translation *subtendens linea* was used by the land surveyor Balbus (Blume, Lachmann, and Rudorff 1848, p. 100), and *hypotenusa* appears in "De limitibus constituendis", attributed to a certain Hyginus Gromaticus, and the "podismus" (Blume, Lachmann, and Rudorff 1848, pp. 190, 298). Note that the classical names of the three sides of a right-angled triangle are *basis*, *cathetus*, and *hypotenusa*; see *cathetus*.

The German term is found in Widmann (1489, fol. C 6^{v}).[4] At about the same time, Chuquet (1979, p. 110) defined: "All triangles are constituted by three straight lines of which one is taken for the base and the two others for the two hypotenuses [sic]. The line descending perpendicularly from the angle of the hypotenuses onto the base is called the *cathetus*".[5] Chuquet's use of *hypotenuse* in the plural may seem

[1] Translation from Euclid (1956, p. 349). [2] For this and further examples such as Latin *linea*, from *linum* 'flax', also 'thread, rope', see Gandz (1930, pp. 272–275). [3] A different explanation was given by Schmidt (1914, pp. 110–120, 241–246), who argues that *ὑποτείνουσα* originally designated the chord of a harp. [4] An early use in a modern language is in a Spanish text from the second half of the 13th century, where term is *diametro* used in the sense of 'hypotenuse'; this translation is due to Arabic sources (Bossong 1978, p. 195). [5] "Tous triangles sont constituez par troys lignes droictes dont l'une est prise pour la base et les deux aultres pour les deux ypothenusses; et la ligne descendant perpendiculierement de l'angle des ypothenusses sus la base est dicte cathetuse". Translation from Flegg, Hay, and Moss (1985, p. 241), slightly modified.

surprising but becomes understandable when de la Roche's illustration is considered (see fig. 57).

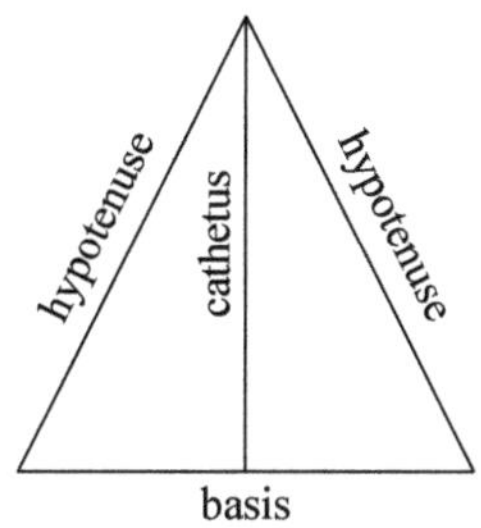

Figure 57: *hypotenuse* as defined by Chuquet. From de La Roche (1520, fol. 221^v).

The English term appears in L. Digges and T. Digges (1571); in the text, the concept is first defined under the name *subtending side* (fol. B j^v), but later on, in a discussion of proposition 47 of the *Elements*, "*Hypothenusa*, or the third side subtending the right angle" is found (fol. L iiij^v). Italian *ipotenusa* appeared (as *ypotenguza*) in the 14th-century translation of Fibonacci's *Practica Geometriae* (Feola 2008, p. 50).

Note. For an overview of the terminology, see Tropfke (1940, pp. 83–85). For the Greek terminology, see Mugler (1958-59, pp. 442f.); for the Latin one, see Guillaumin (2020, pp. 148, 277).

References

Blume, Friedrich, Karl Lachmann, and Adolf Rudorff, eds. (1848). *Die Schriften der römischen Feldmesser*. Erster Band. Texte und Zeichnungen. Berlin.

Bossong, Georg, ed. (1978). *Los Canones de Albateni*. Herausgegeben sowie mit Einleitung, Anmerkungen und Glossar versehen. Tübingen.

Burnet, John (1908). *Early Greek Philosophy*. London.

Chuquet, Nicolas (1979). *La Géométrie*. Première géométrie algébrique en langue française (1484). Introduction, texte et notes par Hervé l'Huillier. Paris.

Digges, Leonard and Thomas Digges (1571). *A geometrical practise, named Pantometria diuided into three bookes, Longimetra, Planimetra, and Stereometria*. London.

Euclid (1956). *The Thirteen Books of Euclid's Elements*. Volume I. Introduction and Books I, II. Ed. by Thomas L. Heath. New York.

Feola, Francesco (2008). *Gli esordi della geometria in volgare*. Un volgarizzamento trecentesco della *Practica Geometriae* di Leonardo Pisano. Firenze.

Flegg, Graham, Cynthia Hay, and Barbara Moss, eds. (1985). *Nicolas Chuquet, Renaissance mathematician*. A study with extensive translation of Chuquet's mathematical manuscript completed in 1484. Dordrecht.

Gandz, Solomon (1930). "Die Harpedonapten oder Seilspanner und Seilknüpfer". *Quellen und Studien zur Geschichte der Mathematik* B 1, pp. 255–277.

Guillaumin, Jean-Yves (2020). *Dictionnaire de la terminologie latine ancienne de l'arithmétique et de la géométrie*. Paris.

Mugler, Charles (1958-59). *Dictionnaire historique de la terminologie géométrique des Grecs*. Paris.

de La Roche, Estienne (1520). *Larismethique nouellement composee*. Lyon.

Schmidt, Max Carl Paul (1914). *Zur Entstehung und Terminologie der elementaren Mathematik*. Zweite verbesserte und stark vermehrte Auflage. Leipzig.

Tropfke, Johannes (1940). *Geschichte der Elementar-Mathematik.* Vierter Band. Ebene Geometrie. Dritte, verbesserte und vermehrte Auflage besorgt von Dr. Kurt Vogel. Berlin and Leipzig.
Widmann, Johannes (1489). *Behende vnd hubsche Rechenung auff allen kauffmanschafft.* Leipzig.

I

ideal (geometry)

Ideal elements in geometry were already conceived of by Kepler and Desargues, who both dealt with points at infinity. The term *ideal* itself, however, is a 19th century invention. It appeared in print in a treatise by Cauchy and Poncelet (1820-21, p. 73), where the term *corde idéale* 'ideal chord' is attributed to Poncelet. His term became better known through his *Traité des propriétés projectives des figures*, in which also *sécante idéale* 'ideal secant' and *centre idéale* 'ideal centre' are defined (Poncelet 1822, p. 27).

The German term *ideale Chorde* was used by Plücker (1828, pp. 52f.), English *ideal chord* is found in Davies (1835, p. 232). Both Plücker and Davies attribute the term to Poncelet.[1]

Note. For the history of ideal elements in geometry, see, for example, Kline (1972, pp. 288–295), Field (1997, pp. 183–185), and Gray (2011, ch. 4).

References

Cauchy, Augustin-Louis and Jean Victor Poncelet (1820-21). "Géométrie des courbes. Rapport à l'académie royale des sciences". *Annales des mathématiques pures et appliquées* 11, pp. 69–83.

Davies, Thomas Stephens (1835). "Question 555". *New Series of The Mathematical Repository* 6, pp. 229–234.

Field, Judith V. (1997). *The Invention of Infinity*. Mathematics and Art in the Renaissance. Oxford.

Gray, Jeremy (2011). *Worlds Out of Nothing*. A Course in the History of Geometry in the 19th Century. London et al.

Kline, Morris (1972). *Mathematical Thought From Ancient To Modern Times*. Volume 1. New York and Oxford.

Plücker, Julius (1828). *Analytisch-geometrische Entwicklungen*. Erster Band. Essen.

Poncelet, Jean Victor (1822). *Traité des propriétés projectives des figures*. Paris.

ideal (algebra)

After ideal elements had been established in geometry, Kummer introduced them in number-theoretical contexts. Kummer's term *ideal complex number* was later referred to by Dedekind when he introduced his term *ideal*.

---

[1] Plücker also notes that *radical axis* is more commonly used than *ideal chord*. See *radical axis*.

H. Becker, *A Historical Dictionary of Mathematical Terms*, Science Networks. Historical Studies 66, https://doi.org/10.1007/978-3-032-02873-0_9

German *ideale complexe Zahlen* is first found in Kummer (1846) and Kummer (1847c). He introduced his new findings thus:

> I have succeeded in completing and simplifying the theory of those complex numbers formed from higher roots of unity, which play an important role in cyclotomy, the theory of power residues, and of the forms of higher degrees; this was accomplished through the introduction of a special kind of imaginary divisors, which I call *ideal complex numbers* (Kummer 1847c, p. 319).[1]

Kummer called $f(\alpha) = \alpha + a_1\alpha + a_2\alpha^2 + ... + a_{\lambda-1}\alpha^{\lambda-1}$ a *complex integer* (German *ganze complexe Zahl*) and said that "those numbers $f(\alpha)$, even if they are not immediately decomposable into complex factors, still have the nature of composite numbers; but in this case the factors are not real, but *ideal complex numbers*" (ibid.).[2] To explain the introduction of this notion, Kummer made use of an extensive chemical analogy, which he explained in detail in Kummer (1847b, pp. 359–361).[3] For example, he states that the multiplication of "complex numbers" corresponds to the chemical composition, prime factors correspond to chemical elements, or more precisely, as Kummer says, to their atomic weight. In addition, "fluorine, not yet in itself representable and still counted among the elements, can be regarded as an analogon of an ideal prime factor".[4]

Kummer himself used French *nombre complexe idéale* in a letter to Liouville (Kummer 1847a, p. 136); English *ideal number* was first used (OED) by Smith (1861, p. 132).

The noun *ideal* (German *Ideal*) was later introduced by Dedekind (1871). He originally defined this algebraic structure with reference to rings of complex numbers:

> A system $\mathfrak{a}$ of infinitely many numbers contained in $\mathfrak{o}$ will be called an *ideal* if it satisfies the following pair of conditions:
>
> I. The sum and difference of any two numbers in $\mathfrak{a}$ is again a number in $\mathfrak{a}$.
>
> II. Each product of a number in $\mathfrak{a}$ and a number in $\mathfrak{o}$ is again a number in $\mathfrak{a}$ (Dedekind 1871, p. 452).[5]

Ideals can be considered as a generalization of Kummer's ideal numbers, and Dedekind explicity points out that the term was taken over from Kummer's writings (Dedekind and Weber 1882, p. 182). As Weber (1893, p. 12) put it: *Ideal* is "a name

[1] "Es ist mir gelungen, die Theorie derjenigen complexen Zahlen, welche aus höheren Wurzeln der Einheit gebildet sind und welche bekanntlich in der Kreistheilung, in der Lehre von den Potenzresten und den Formen höherer Grade eine wichtige Rolle spielen, zu vervollständigen und zu vereinfachen; und zwar durch die Einführung einer eigenthümlichen Art imaginärer Divisoren, welche ich *ideale complexe Zahlen* nenne". Translation from Lemmermeyer (2009, p. 169). [2] "Es haben vielmehr solche Zahlen $f(\alpha)$, wenn sie nicht gleich in complexe Factoren zerlegbar sind, dennoch die Natur der zusammengesetzten Zahlen; die Factoren aber sind alsdann nicht wirkliche, sondern *ideale complexe Zahlen*". [3] On Kummer's analogy, see Becker (2005, pp. 171–176); see also Edwards (1977, pp. 142f.) for critical comments on the term *ideal complex numbers*. [4] "Das Fluor, für sich bisher nicht darstellbar und noch den Elementen zugezählt, kann als Analogon eines idealen Primfactors gelten". [5] "Ein System $\mathfrak{a}$ von unendlich vielen in $\mathfrak{o}$ enthaltenen Zahlen soll ein *Ideal* heissen, wenn es den beiden Bedingungen genügt: I. Die Summe und die Differenz je zweier Zahlen in $\mathfrak{a}$ sind wieder Zahlen in $\mathfrak{a}$. II. Jedes Product aus einer Zahl in $\mathfrak{a}$ und einer Zahl in $\mathfrak{o}$ ist wieder eine Zahl in $\mathfrak{a}$". Translation from Avigad (2004, p. 38). – The citation is from § 163 of the famous "10th supplement".

which is only of historical origin, and which was formed in honour of Kummer's creation".[1] This semantic demotivation has occasionally been critically commented on: "[Dedekind] should have spoken of 'Reals' [*Reale*], since they are number aggregates that actually exist in the given domain of integrity" (Klein 1926, p. 323).[2] Dedekind was well aware that such an objection against the term might occur, but he still found it appropriate to keep it (Dedekind and Weber 1882, p. 182).

The English noun *ideal* is found in this sense in Hathaway (1887, pp. 171f.);[3] French *idéal* is apparently due to Dedekind (1877, p. 52) himself.

The concept was generalized in the 20th century, in particular by Sono (1917, p. 211) and Noether (1921, p. 30, 1927, p. 30). In these papers, the now usual definition was developed: In a commutative ring R, a subset $I \subset R$ is an ideal if it has the two properties (I) and (II) in Dedekind's definition cited above. In this form, the definition also appears in van der Waerden (1930, p. 53). With some modifications, it can also be applied to noncommutative rings.

For Kronecker's theory, see *divisor* (further uses).

Note. For Kummer's creation of ideal numbers, see Boniface (2016) and Edwards (1977, esp. ch. 4). Lemmermeyer (2009) discusses the influence of Jacobi's work on Kummer in this respect. For the history of ideals, see Edwards (1980), Soublin (1984), and Ferreirós (2007, esp. 99-107, 111–113). Mehrtens (1979) examines the creation of mathematical notions by Dedekind, and Becker (2005, pp. 236–242) discusses the metaphorical structure of Dedekind's algebraic number theory.

References

Avigad, Jeremy (2004). "Dedekind's 1871 version of the theory of ideals". URL: `https://kilthub.cmu.edu/articles/Dedekind_s_1871_Version_of_the_Theory_of_Ideals/6491342/1` (retrieved July 25, 2007).

Becker, Holger (2005). "Semantische und lexikalische Aspekte der mathematischen Fachsprache des 19. Jahrhunderts". PhD thesis. Carl von Ossietzky Universität Oldenburg. URL: `http://oops.uni-oldenburg.de/59/`.

Boniface, Jacqueline (2016). "A process of generalization: Kummer's creation of ideal numbers". In: *The Oxford Handbook of Generality in Mathematics and the Sciences*. Ed. by Karine Chemla, Renaud Chorlay, and David Rabouin. Oxford, pp. 483–500.

Dedekind, Richard (1871). "Ueber die Komposition der binären quadratischen Formen". In: *Vorlesungen über Zahlentheorie von P.G. Lejeune Dirichlet*. Herausgegeben und mit Zusätzen versehen von R. Dedekind. Zweite, umgearbeitete und vermehrte Auflage. Braunschweig, pp. 380–497.

Dedekind, Richard (1877). *Sur la théorie des nombres entiers algébriques*. Paris.

Dedekind, Richard and Heinrich Weber (1882). "Theorie der algebraischen Functionen einer Veränderlichen". *Journal für die reine und angewandte Mathematik* 92, pp. 181–290.

Edwards, Harold M. (1977). *Fermat's Last Theorem*. A Genetic Introduction to Algebraic Number Theory. New York.

Edwards, Harold M. (1980). "The Genesis of Ideal Theory". *Archive for History of Exact Sciences* 23, pp. 321–378.

Ferreirós, José (2007). *Labyrinth of Thought*. A History of Set Theory and Its Role in Modern Mathematics. Second revised edition. Basel.

[1] "ein Name, der aber nur historischen Ursprung hat, und zu Ehren der Kummerschen Schöpfung gebildet ist". [2] "[Dedekind] hätte von 'Realen' sprechen sollen. Denn es handelt sich um Zahlenaggregate, die in dem vorgelegten Integritätsbereich tatsächlich vorhanden sind". – Edwards (1977, p. 144) regards *Ideal* as "a rather bizarre adaptation of Kummer's terminology". [3] The OED dates the term to 1898.

Hathaway, Arthur S. (1887). "A Memoir in the Theory of Numbers". *American Journal of Mathematics* 9, pp. 162–179.
Klein, Felix (1926). *Vorlesungen über die Entwicklung der Mathematik im 19. Jahrhundert*. Teil I. Berlin.
Kummer, Ernst Eduard (1846). "Vervollständigung der Theorie der complexen Zahlen". *Bericht über die zur Bekanntmachung geeigneten Verhandlungen der Königlich Preussischen Akademie* (1846), pp. 87–96.
Kummer, Ernst Eduard (1847a). "Extrait d'une Lettre de M. Kummer à M. Liouville". *Journal de mathématiques pures et appliquées* 12, p. 136.
Kummer, Ernst Eduard (1847b). "Über die Zerlegung der aus Wurzeln der Einheit gebildeten complexen Zahlen in ihre Primfactoren". *Journal für die reine und angewandte Mathematik* 35, pp. 327–367.
Kummer, Ernst Eduard (1847c). "Zur Theorie der complexen Zahlen". *Journal für die reine und angewandte Mathematik* 35, pp. 319–326.
Lemmermeyer, Franz (2009). "Jacobi and Kummer's ideal numbers". *Abhandlungen aus dem Mathematischen Seminar der Universität Hamburg* 79, pp. 165–187.
Mehrtens, Herbert (1979). "Das Skelett der modernen Algebra. Zur Bildung mathematischer Begriffe bei Richard Dedekind". In: *Disciplinae Novae*. Zur Entstehung neuer Denk- und Arbeitsrichtungen in der Naturwissenschaft. Festschrift zum 90. Geburtstag von Hans Schimank. Ed. by Christoph J. Scriba. Göttingen, pp. 25–43.
Noether, Emmy (1921). "Idealtheorie in Ringbereichen". *Mathematische Annalen* 83, pp. 24–66.
Noether, Emmy (1927). "Abstrakter Aufbau der Idealtheorie in algebraischen Zahl- und Funktionenkörpern". *Mathematische Annalen* 96, pp. 26–61.
Smith, H.J.S. (1861). "Report on the Theory of Numbers. Part II". *Report of the British Association for the Advancement of Science* (1860), pp. 120–169.
Sono, Masazo (1917). "On Congruences". *Memoirs of the College of Science* 2, pp. 203–226.
Soublin, Jean-Pierre (1984). "Préhistoire des idéaux". *Cahiers du séminaire d'histoire des mathématiques* 5, pp. 13–20.
van der Waerden, Bartel Leendert (1930). *Moderne Algebra*. Erster Teil. Berlin.
Weber, Heinrich (1893). "Leopold Kronecker". *Mathematische Annalen* 43, pp. 1–25.

idèle and *adèle*

The French term *élément idéal* 'ideal element' was introduced by Chevalley (1936, p. 361).[1] He used the abbreviation *e. i.* for it. When Chevalley's paper was reviewed in some German journals, *élément idéal* was translated as *ideales Element*, and the initial letters of the German rather than the French term gave rise to the German term *Idel*, suggested in Grunwald (1937, p. 151) and then used by Hasse (1936, p. 1153), who referred to Grunwald's review for the term. Grunwald was a pupil of Hasse and it appears that it was Hasse who created the term *Idel*, as he asserted in private conversation (Roquette 2014, p. 176). Chevalley (1940, p. 394) appreciated Hasse's creation and in turn re-transferred it to French as *idèle*. Chevalley communicated with Hasse and used *idèle* already in 1937 (Frei 2005, p. 53).

The related term *adèle* is short for *additive idèle*. Its first occurrence in print appears to be in Jaffard (1954, p. 24), who defines both *adèle* and *anneau d'adèles* (ring of adèles). Some years later, Weil (1960, p. 250) points out that the terms *repartitions* and *valuation vectors* are also used for adèles. The former was used by Chevalley (1951, p. 25); Artin and Whaples (1945, p. 473) used *vector*, Iwasawa (1953) the explicit *valuation vector*. Roquette (2018, pp. 191–193) refers to a letter

[1] According to Roquette (2014, p. 176), Chevalley first used this terminology in a letter from June 1935.

by Weil for the first occurrence of the concept; whether the term is also due to Weil, as Roquette suspects, must remain an open question.

Note. For the history of the concept, see Iyanaga (2006, p. 60), Roquette (2013, pp. 43–51), and Roquette (2014). Roquette (2018, pp. 191–193) also has historical comments on adèles.

References

Artin, Emil and George Whaples (1945). "Axiomatic characterization of fields by the product formula for valuations". *Bulletin of the American Mathematical Society* 51, pp. 469–492.

Chevalley, Claude (1936). "Généralisation de la théorie du corps de classes pour les extensions infinies". *Journal de mathématiques pures et appliquées* 15, pp. 359–371.

Chevalley, Claude (1940). "La théorie du corps de classes". *Annals of Mathematics* 41, pp. 394–418.

Chevalley, Claude (1951). *Introduction to the Theory of Algebraic Functions of One Variable*. Providence, R.I.

Frei, Günther (2005). "Briefwechsel H. Hasse – C. Chevalley". URL: `https://www.mathi.uni-heidelberg.de/~roquette/Transkriptionen/hasche-051004.pdf` (retrieved Apr. 26, 2021).

Grunwald, Wilhelm (1937). "Chevalley, C.: Généralisation de la théorie du corps de classes pour les extensions infinies [review]". *Zentralblatt für Mathematik und ihre Grenzgebiete* 15, pp. 151–154.

Hasse, Helmut (1936). "Chevalley, C.: Généralisation de la théorie du corps de classes pour les extensions infinies [review]". *Jahrbuch über die Fortschritte der Mathematik* 62, p. 1153.

Iwasawa, Kenkichi (1953). "On the Rings of Valuation Vectors". *Annals of Mathematics* 57, pp. 331–356.

Iyanaga, Shokichi (2006). "Travaux de Claude Chevalley sur la théorie du corps de classes: Introduction". *Japanese Journal of Mathematics* 1, pp. 25–85.

Jaffard, Paul (1954). "Anneaux d'adèles d'après Iwasawa". *Séminaire Bourbaki* 3, exposé 103, pp. 23–33.

Roquette, Peter (2013). "What is reciprocity? On the evolution of class field theory in the 20th century". URL: `https://www.mathi.uni-heidelberg.de/~roquette/reciprocity.pdf` (retrieved Apr. 6, 2021).

Roquette, Peter (2014). "Jacques Herbrand und sein Lemma". *Mitteilungen der Mathematischen Gesellschaft in Hamburg* 34, pp. 163–194.

Roquette, Peter (2018). *The Riemann Hypothesis in Characteristic p in Historical Perspective*. Cham.

Weil, André (1960). "Adèles et groupes algébriques". *Séminaire Bourbaki* 5, exposé 186, pp. 249–257.

idempotent

The term *idempotent* was introduced by Benjamin Peirce in 1870 for an expression that, when "raised to a square or higher power, it gives itself as the result" (see Peirce 1881, p. 104). It is derived from Latin *idem* 'equal' and English *potent* 'powerful'. Frobenius (1903, p. 635) called an idempotent element a *unit* (German *Einheit*), but this usage did not catch on, presumably because of the main meaning of *unit*.

The term was later also applied to systems of elements rather than algebraic expressions; for example, Hawkes (1902, p. 314) has *idempotent system* and Taber (1904, p. 509) *idempotent algebra*.

References

Frobenius, Georg (1903). "Theorie der hyperkomplexen Größen. II". *Sitzungsberichte der Königlich Preußischen Akademie der Wissenschaften zu Berlin* (1903), pp. 634–645.

Hawkes, Herbert Edwin (1902). "On Hypercomplex Number Systems". *Transactions of the American Mathematical Society* 3, pp. 312–330.
Peirce, Benjamin (1881). "Linear Associative Algebra". *American Journal of Mathematics* 4, pp. 97–229.
Taber, Henry (1904). "On Hypercomplex Number Systems (First Paper)". *Transactions of the American Mathematical Society* 5, pp. 509–548.

identity

Identity elements of algebraic structures have been designated by at least four different terms and expressions:

1. by adjectives meaning 'identical' (esp. of permutations and transformations) and, later, by nouns meaning 'identity',
2. by nouns meaning 'unit', and, esp. in German, also by nouns containing the word *Eins* 'one',
3. by nouns meaning 'modulus' or 'module', and
4. by adjectives meaning 'neutral'.

The oldest of these different ways of expression goes back to Cauchy. In an early paper on the theory of "substitutions" (permutations), he defined an "identical permutation" and called it *substitution identique* (Cauchy 1815, p. 73).[1] This mode of expression was very common in the 19th century and was spread by textbook occurrences such as Serret (1849, p. 129). It is also found in the context of transformation groups, for example in Klein (1874, p. 120). Nouns meaning 'identity' appear from around 1880, for instance in Hurwitz (1881, p. 19) (German *Identität*) or Moore (1895), who used both English *identity* (p. 61) and *identity-element* (p. 62).

In a later paper on permutations, Cauchy (1844, e. g., p. 165) also uses the word *unité*. This was adopted by Jordan (1869, p. 141), who used it in his famous commentary on Galois for the permutation "which leaves every letter at its place".[2] Similarly, Cayley (1878, p. 50) called the identical element *unity*.[3] German *Einheit* is found, for example, in Weber (1882, p. 303), where he presented an early system of axioms for groups.

By the end of the 19th century, usage of the symbol 1 was quite common. This convention was also the basis for another term, which appears to go back to Study (1889, p. 239): he used German "die Eins des Systems" (lit., 'the one of the system') in connection with systems of hypercomplex numbers.[4] On the basis of this use, another term, *Einselement*, was introduced. It is found in Hensel (1913, p. 13), who used it in addition to *Einheitselement* 'unit element'. The term was also used by Hasse (1926, p. 11) but did not became more more popular until van der Waerden used it in his book *Moderne Algebra* (van der Waerden 1930, p. 15). The term is still in common use.

The third term *modulus* is found in connection with hypercomplex number systems and goes back to Hankel (1867, p. 23). He defined the *Modulus* of a number system but immediately shortened the term to *Modul*. Hankel's term began to be more

[1] The symbol 1 first appears in Cauchy (1844, p. 155), according to Wußing (1984, p. 94) and Burkhardt and Vogt (1909, p. 534). [2] The term also appears in Jordan (1870, p. 21). [3] In an earlier paper, Cayley (1854) just used the symbol 1. [4] The term is also in Study (1890, p. 286).

commonly used in the 1890s; its definition is illustrated by Scheffers (1891, p. 296): "In the domain of ordinary numbers there exists a number, the number one, which, when multiplied with an arbitrary number, does not change the latter. Analogously we stipulate that there be a number $\varepsilon = \varepsilon_1 e_1 + \varepsilon_2 e_2 + ... \varepsilon_n e_n$ of the kind that each $x\varepsilon = \varepsilon x = x$. This number is called the *module*".[1]

Finally, the term *neutral element* seems to originate with Fraenkel (1914, p. 8) and Loewy (1915, p. 25), who used German *neutrales Element*. The term may be due to Loewy, whose book, as yet unpublished, was cited by Fraenkel.[2]

See also *unit*.

References

Burkhardt, Heinrich and Henri Vogt (1909). "Sur les groupes finis discontinus". *Encyclopédie des sciences mathématiques pures et appliquées* I.1, pp. 532–616.

Cauchy, Augustin-Louis (1815). "Sur le nombre des valeurs qu'une fonction peut acquérir, lorsqu'on y permute de toutes les manières possibles les quantités qu'elle renferme". *Journal de l'École polytechnique* 10, cahier 17, pp. 1–28.

Cauchy, Augustin-Louis (1844). *Exercises d'analyse et de physique mathématique*. Tome troisième. Paris.

Cayley, Arthur (1854). "On the theory of groups, as depending on the symbolic equation $\theta^n = 1$". *The London, Edinburgh, and Dublin Philosophical Magazine and Journal of Science* 7, pp. 40–47, 408–409.

Cayley, Arthur (1878). "Desiderata and Suggestions: No. 1. The Theory of Groups". *American Journal of Mathematics* 1, pp. 50–52.

Fraenkel, Adolf (1914). *Über die Teiler der Null und die Zerlegung von Ringen*. Marburg.

Hankel, Hermann (1867). *Vorlesungen über die complexen Zahlen und ihre Functionen*. In zwei Theilen. I. Theil: Theorie der complexen Zahlsysteme. Leipzig.

Hasse, Helmut (1926). *Höhere Algebra*. I. Lineare Gleichungen. Berlin and Leipzig.

Hensel, Kurt (1913). *Zahlentheorie*. Berlin and Leipzig.

Hurwitz, Adolf (1881). *Grundlagen einer independenten Theorie der elliptischen Modulfunktionen und Theorie der Multiplikatorgleichungen erster Stufe*. Leipzig.

Jordan, Camille (1869). "Commentaire sur Galois". *Mathematische Annalen* 1, pp. 141–160.

Jordan, Camille (1870). *Traité des substitutions et des équations algébriques*. Paris.

Klein, Felix (1874). "Ueber die sogenannte Nicht-Euklidische Geometrie (Zweiter Aufsatz)". *Mathematische Annalen* 6, pp. 112–145.

Loewy, Alfred (1915). *Lehrbuch der Algebra*. Erster Teil. Grundlagen der Arithmetik. Leipzig.

Moore, E.H. (1895). "The group of holoedric transformation into itself of a given group". *Bulletin of the American Mathematical Society* 1, pp. 61–66.

Scheffers, Georg (1891). "Zurückführung complexer Zahlensysteme auf typische Formen". *Mathematische Annalen* 39, pp. 293–390.

Serret, Joseph-Alfred (1849). *Cours d'algèbre supérieure*. Paris.

Study, Eduard (1889). "Ueber Systeme von complexen Zahlen". *Nachrichten von der Königl. Gesellschaft der Wissenschaften und der Georg-Augusts-Universität zu Göttingen* (1889), pp. 237–268.

Study, Eduard (1890). "Ueber Systeme complexer Zahlen und ihre Anwendung in der Theorie der Transformationsgruppen". *Monatshefte für Mathematik und Physik* 1, pp. 283–355.

van der Waerden, Bartel Leendert (1930). *Moderne Algebra*. Erster Teil. Berlin.

---

[1] "Im Gebiete der gewöhnlichen Zahlen existirt eine Zahl, die Eins, die, mit einer beliebigen Zahl multiplicirt, dieselbe nicht ändert. Analog setzen wir fest, dass es eine Zahl $\varepsilon = \varepsilon_1 e_1 + \varepsilon_2 e_2 + ... \varepsilon_n e_n$ geben soll von der Beschaffenheit, dass jedes $x\varepsilon = \varepsilon x = x$ sei. Diese Zahl heisst der *Modul*". [2] Loewy was Fraenkel's married uncle, which is how Fraenkel seems to have had access to Loewy's book.

Weber, Heinrich (1882). "Beweis des Satzes, dass jede eigentlich primitive quadratische Form unendlich viele Primzahlen darzustellen fähig ist". *Mathematische Annalen* 20, pp. 301–329.
Wußing, Hans (1984). *The Genesis of the Abstract Group Concept.* A Contribution to the History of the Origin of Abstract Group Theory. Tr. Abe Shenitzer. Cambridge, Mass.

image (function)

Dedekind (1888, p. 6), proposing a general definition of mapping (*Abbildung*), introduced the word *Bild* 'image':[1]

> By a *mapping* φ of a system S we understand a law according to which, to every determinate element of S, there *belongs* a determinate thing which is called the *image* of s and which is denoted by $\varphi(s)$ [...].[2]

While the term features quite prominently in Dedekind's text and is explicitly defined, the term is not due to him, as it was already used by Riemann in his PhD thesis. In connection with conformal mappings, Riemann writes, for example: "This yields the similarity of two corresponding infinitely small triangles; and consequently, in general, similarity of the smallest parts of the plane A and their images on the plane B" (Riemann 1851, § 3).[3]

Dedekind (1888, p. 6) also used *Bild* for the set of all images (in the above sense) of the elements of a subset of S. This set was later distinguished by using the term *Bildmenge*, which was used by Weyl (1913, p. 19) and Hausdorff (1914, p. 43), for example.[4]

The term spread through various writings in set theory and the calculus such as Hausdorff (1914, p. 43) or Carathéodory (1918, p. 205). The OED cites Harris (1888, p. 65) as the first use in English, where the term is defined as follows: "If u be a function of z, and if z describe any path, u will describe another path which is called the *image* of the former with reference to the given function". Fiske (1896, p. 232) defined: "if $w = f(z)$ be such a function, to each point $x + iy$ of the independent variable will correspond a point $u + iv$ of the function. This point $u + iv$ is called the 'image'". Royce (1900, pp. 521, 524) uses the term in Dedekind's sense.[5] French *image* is found in Fouët (1902, p. 28), where no definition is given, and in Baire (1907, p. 319), for example.

In German, the term *Wertevorrat* (from the words for 'value' and 'stock, supply, reserve'), which is still in use, comes up in the second half of the 19th century. A definition was given du Bois-Reymond (1876, p. 124); according to du Bois-Reymond (1874, p. 244), the term is due to Neumann, who seems to have used it first in Neumann (1865, p. 174).

See also *range* and *codomain*.

---

[1] Dedekind (1879, p. 470) had used the word in print before; for his use of the term, which dates from the 1870s, see Ferreirós (2007, ch. III, esp. pp. 89f., 108), Benis Sinaceur (2008, pp. 11, 135), and Sieg and Schlimm (2017). [2] "Unter einer *Abbildung* φ eines Systems S wird ein Gesetz verstanden, nach welchem zu jedem bestimmten Element s von S ein bestimmtes Ding *gehört*, welches das *Bild* von s heißt und mit $\varphi(s)$ bezeichnet wird". Translation from Ewald (1996, p. 799). [3] "Es findet also zwischen zwei einander entsprechenden unendlich kleinen Dreiecken und folglich allgemein zwischen den kleinsten Theilen der Ebene A und ihres Bildes auf der Ebene B Aehnlichkeit Statt". Translation from Riemann (2004, p. 3). [4] The word was in use before; see, for example, Schoenflies (1900, p. 124). A definition is not given in that text. [5] In Dedekind (1909, p. 50), the term is translated by the noun *transform*.

References

Baire, René (1907). "Sur la non-applicabilité de deux continus à n et $n + p$ dimensions". *Comptes rendus de l'Académie des sciences* 144, pp. 318–321.

Benis Sinaceur, Hourya (2008). *La création des nombres*. Paris.

du Bois-Reymond, Paul (1874). "Ueber die sprungweisen Werthänderungen analytischer Functionen". *Mathematische Annalen* 7, pp. 241–261.

du Bois-Reymond, Paul (1876). "Beweis, dass die Coefficienten der trigonometrischen Reihe $\mathrm{f(x)} = \sum_{p=0}^{p=\infty}(\mathrm{a_p} \cos .\mathrm{px} + \mathrm{b_p} \sin .\mathrm{px})$ die Werthe $\mathrm{a_0} = \frac{1}{2\pi}\int_{-\pi}^{+\pi} \mathrm{d}\alpha \mathrm{f}(\alpha)$, $\mathrm{a_p} = \frac{1}{\pi}\int_{-\pi}^{+\pi} \mathrm{d}\alpha \mathrm{f}(\alpha) \cos .\mathrm{p}\alpha$, $\mathrm{b_p} = \frac{1}{\pi}\int_{-\pi}^{+\pi} \mathrm{d}\alpha \mathrm{f}(\alpha) \sin .\mathrm{p}\alpha$, haben, jedesmal wenn diese Integrale endlich und bestimmt sind". *Abhandlungen der Bayerischen Akademie der Wissenschaften* 12, pp. 117–166.

Carathéodory, Constantin (1918). *Vorlesungen über reelle Funktionen*. Leipzig.

Dedekind, Richard (1879). "Ueber die Theorie der ganzen algebraischen Zahlen". In: *Vorlesungen über Zahlentheorie von P.G. Lejeune Dirichlet*. Herausgegeben und mit Zusätzen versehen von R. Dedekind. Dritte, umgearbeitete und vermehrte Auflage. Braunschweig, pp. 434–627.

Dedekind, Richard (1888). *Was sind und was sollen die Zahlen?* Braunschweig.

Dedekind, Richard (1909). *Essays on the Theory of Numbers*. I. Continuity and Irrational Numbers, II. The Nature and Meaning of Numbers. Authorized translation by Wooster Woodruff Beman. Chicago.

Ewald, William (1996). *From Kant to Hilbert*. A Source Book in the Foundations of Mathematics. Volume II. Oxford.

Ferreirós, José (2007). *Labyrinth of Thought*. A History of Set Theory and Its Role in Modern Mathematics. Second revised edition. Basel.

Fiske, Thomas S. (1896). "Functions of a complex variable". In: *Higher Mathematics*. Ed. by Mansfield Merriman and Robert S. Woodward. New York, pp. 226–302.

Fouët, Edouard-André (1902). *Leçons élémentaires sur la théorie des fonctions analytiques*. Première partie. Paris.

Harris, Rollin A. (1888). "The Theory of Images in the Representation of Functions". *Annals of Mathematics* 4, pp. 65–86.

Hausdorff, Felix (1914). *Grundzüge der Mengenlehre*. Leipzig.

Neumann, Carl (1865). *Vorlesungen über Riemann's Theorie der Abel'schen Integrale*. Leipzig.

Riemann, Bernhard (1851). "Grundlagen für eine allgemeine Theorie der Functionen einer veränderlichen complexen Grösse". PhD thesis. Georg-August-Universität Göttingen.

Riemann, Bernhard (2004). *Collected Papers*. Translated from the 1892 edition by Roger Baker, Charles Christenson and Henry Orde. Heber City.

Royce, Josiah (1900). *The World and the Individual*. New York.

Schoenflies, Arthur (1900). "Die Entwickelung der Lehre von den Punktmannigfaltigkeiten". *Jahresbericht der Deutschen Mathematiker-Vereinigung* 8, pp. 1–250.

Sieg, Wilfried and Dirk Schlimm (2017). "Dedekind's Abstract Concepts: Models and Mappings". *Philosophia Mathematica* 25, pp. 292–317.

Weyl, Hermann (1913). *Die Idee der Riemannschen Fläche*. Leipzig und Berlin.

imaginary

The term originates with Descartes' book on geometry from 1637:

> Neither the true nor the false roots are always real; sometimes they are only imaginary; that is, while we can always conceive of as many roots for each equation as I have already assigned, yet there is not always a definite quantity corresponding to each root so conceived of. Thus, while we may conceive of the equation $x^3 - 6x^2 + 13x - 10 = 0$ as having three roots, yet there is only one real root, 2, while the other two, however we may increase, diminish,

> or multiply them in accordance with the rules just laid down, remain always imaginary (Descartes 1637, p. 380).[1]

As can be seen from the quotation, the expression *imaginary* is slightly dismissive in character, which is in line with other terms such as *impossible number* (see *complex number*). Also, Descartes and other writers of the 17th and 18th centuries usually spoke of imaginary *roots*, *solutions*, or *quantities* rather than of imaginary *numbers*. The actual term *imaginary number* comes up at the end of the 17th century. de Lagny (1697, p. 471), for example, remarks: "If one multiplies a real number with an imaginary number, the product is imaginary".[2]

In his *Treatise on algebra*, Wallis (1685, pp. 264, 316) already states that the term *imaginary root* or *quantity* is in common use.[3] It is also found in early mathematical dictionaries such as Ozanam (1691, p. 83) and Wolff (1716, p. 1163), where also German *eingebildete Zahl*, lit. 'imagined number', is given.

On the nature of imaginary roots, Leibniz (1702, p. 216) presented the following thoughts: "For the Nature of things, the mother of eternal varieties, or rather the Divine Mind is more jealous than she would admit to press all things into one genus. There she invented a distinguished and wonderful way out in the miracle of analysis, almost an amphibian between being and not-being, which we call imaginary root".[4]

In a paper read in 1749, Euler (1751, § 3) stated that, generally, imaginary quantities can always be written in the form $a + b\sqrt{-1}$: "Now it very often happens that not all of these roots are real quantities, and that some, or perhaps all, are imaginary quantities. We call a quantity imaginary when it is neither greater than zero, nor less than zero, nor equal to zero. This will be then something impossible, as for example $\sqrt{}-1$, or in general $a + b\sqrt{}-1$".[5]

Gauß (1831, p. 638) later critically commented on the terminology:

> We have believed that we were doing the friends of mathematics a favour by this account of the principal parts of a new theory of the so-called imaginary quantities. If one formerly contemplated this subject from a false point of view and therefore found a mysterious darkness, this is in large part attributable to clumsy terminology. Had one not called $+1$, -1,

[1] "tant les vrayes racines que les fausses ne sont pas tousiours reelles; mais quelquefois seulement imaginaires; c'est a dire qu'on peut bien toujiours en imaginer autant que aiy dit en chasque Equation; mais qu'il n'y a quelquefois aucune quantité, qui corresponde a celles qu'on imagine. comme encore qu'on en puisse imaginer trois en celle $x^3 - 6xx + 13x - 10 = 0$, il n'y en a toutefois qu'une reelle, qui est 2, & pour les deux autres, quois qu'on les augmente, ou diminue, ou multiplie en la façon que ie viens d'expliquer, on ne sçauroit les rendre autres qu'imaginaires". Translation (slightly modified) from Descartes (1925, p. 175). [2] "Si l'on multiplie nombre réel par nombre imaginaire, le produit est imaginaire". [3] The OED cites a letter from Newton from 1670 for the first English use; it was earlier used by Newton in notes from 1665 (Newton 1967, p. 494). [4] "Natura rerum, aeternarum varietatum parens, vel potius Divina Mens, quam ut omnia sub unum genus compingi patiatur. Itaque elegans & mirabile effugium reperit in illo Analyseos miraculo, idealis mundi monstro, pene inter Ens & non-Ens Amphibio, quod radicem imaginariam appellamus". Translation from Breidert (2017, p. 438). [5] "Or il arrive fort souvent que toutes ces racines ne sont pas des quantités réelles, & que quelques unes, ou peut-être toutes, sont des quantités imaginaires. On nomme quantité imaginaire, celle qui n'est ni plus grande que zero, ni plus petite que zero, ni égale à zero; ce sera donc quelque chose d'impossible, comme par example $\sqrt{}-1$, ou en général $a + b\sqrt{}-1$". Translation from Euler (2005, § 3).

$\sqrt{-1}$ positive, negative, or imaginary (or even impossible) units, but instead, say, direct, inverse, or lateral units, then there could scarcely have been talk of such darkness.[1]

See also *complex number*.

Note. For the history of the terminology, see Tropfke (1980, pp. 151–158), Smith (1958, pp. 261–267), and Remmert (1995, § 1).

References

Breidert, Wolfgang (2017). "Berkeley's Philosophy of Mathematics". In: *The Bloomsbury Companion to Berkeley*. Ed. by Richard Brook and Bertil Belfrage. London et al., pp. 435–457.

Descartes, René (1637). *La Géométrie*. In: *Discours de la méthode*. Leiden, pp. 295–413.

Descartes, René (1925). *The Geometry of René Descartes*. Translated from the French and Latin by David Eugene Smith and Marcia L. Latham. Chicago and London.

Euler, Leonhard (1751). "Recherches sur les racines imaginaires des équations". *Histoire de l'Académie Royale des Sciences et des Belles-Lettres de Berlin* 5, pp. 222–288.

Euler, Leonhard (2005). "Investigations on the imaginary roots of equations". Translated by Todd Doucet. URL: `https://scholarlycommons.pacific.edu/euler-works/170/` (retrieved Feb. 2, 2022).

Gauß, Carl Friedrich (1831). "Theoria residuorum biquadraticorum, commentatio secunda [announcement]". *Göttingische gelehrte Anzeigen* (1831), pp. 625–638.

Gauß, Carl Friedrich (1996). "Notice on the Theory of Biquadratic Residues". In: *From Kant to Hilbert*. A Source Book in the Foundations of Mathematics. Volume I. Ed. by William Ewald. Oxford, pp. 306–313.

de Lagny, Thomas Fantet (1697). *Nouveaux elemens d'arithmetique et d'algebre*. Paris.

Leibniz, Gottfried Wilhelm (1702). "Specimen Novum Analyseos pro Scientia Infiniti, circa Summas & Quadraturas". *Acta Eruditorum* (May 1692), pp. 210–219.

Newton, Isaac (1967). *The Mathematical Papers of Isaac Newton*. Volume I. 1664-1666. Ed. by D.T. Whiteside. Cambridge.

Ozanam, Jacques (1691). *Dictionnaire mathematique*. Paris.

Remmert, Reinhold (1995). "Complex Numbers". In: Ebbinghaus, Heinz-Dieter et al. *Numbers*. With an Introduction by K. Lamotke. Translated by H.L.S. Orde. Edited by J.H. Ewing. New York, pp. 55–96.

Smith, David Eugene (1958). *History of Mathematics*. Volume II. Special Topics of Elementary Mathematics. New York.

Tropfke, Johannes (1980). *Geschichte der Elementarmathematik*. Band 1: Arithmetik und Algebra. Vollständig neu bearbeitet von Kurt Vogel, Karin Reich, Helmuth Gericke. Berlin and New York.

Wallis, John (1685). *A Treatise of Algebra, both Historical and Practical*. London.

Wolff, Christian (1716). *Mathematisches Lexicon*. Leipzig.

immersion

See *embedding*.

implicit function

The terminological distinction between explicit and implicit functions dates from the 18th century and is due to Euler (1748, § 8):

---

[1] "Wir haben geglaubt, den Freunden der Mathematik durch diese kurze Darstellung der Hauptmomente einer neuen Theorie der sogenannten imaginären Größen einen Dienst zu erweisen. Hat man diesen Gegenstand bisher aus einem falschen Gesichtspunct betrachtet und eine geheimnißvolle Dunkelheit dabey gefunden, so ist dieß großentheils den wenig schicklichen Benennungen zuzuschreiben. Hätte man $+1$, -1, $\sqrt{-1}$ nicht positive, negative, imaginäre (oder gar unmögliche) Einheit, sondern etwa directe, inverse, laterale Einheit genannt, so hätte von einer solchen Dunkelheit kaum die Rede seyn können". Translation from Gauß (1996, p. 313).

> It is convenient to distinguish these [irrational functions] into explicit and implicit irrational functions. The explicit functions are those which are expressed with radical signs, as in the given examples. The implicit are those irrational functions which arise from the solution of equations.[1]

The term is not frequently found at first, and translations only appear at the end of 18th century. For instance, the Italian term is found in Franchini (1792, p. 55); de Nieuport (1794, e. g., p. 79) uses French *fonction implicite*, as does Lacroix (1797, p. 3). The English term appears in Woodhouse (1803, p. 79).

Note. On the history of the notion, see Krantz and Parks (2002) and Scarpello and Ritelli (2002).

References

Euler, Leonhard (1748). *Introductio in analysin infinitorum*. Tomus primus. Lausanne.
Euler, Leonhard (1988). *Introduction to Analysis of the Infinite*. Book I. Translated by John D. Blanton. New York.
Franchini, Pietro (1792). *Teoria dell'analisi da servire d'introduzione al metodo diretto ed inverso de'limiti*. Tomo primo. Roma.
Krantz, Steven G. and Harold R. Parks (2002). *The Implicit Function Theorem*. Basel.
Lacroix, Sylvestre François (1797). *Traité du calcul différentiel et du calcul intégral*. Tome premier. Paris.
de Nieuport, Charles (1794). *Mélanges mathématiques ou mémoires sur différens sujets de mathématiques tant pures qu'appliquées*. Premier receuil. Bruxelles.
Scarpello, Giovanni Mingari and Daniele Ritelli (2002). "A Historical Outline of the Theorem of Implicit Functions". *Divulgaciones Matemáticas* 10, pp. 171–180.
Woodhouse, Robert (1803). *The Principles of Analytical Calculation*. Cambridge.

inaccessible cardinal

Weakly inaccessible cardinals were first mentioned by Hausdorff (1907, p. 546), where they are characterized as being of "exorbitant magnitude" (see *large cardinal*). The term *inaccessible cardinal* first appears in some articles published in the journal *Fundamenta Mathematicae* in 1930: German *unerreichbare Kardinalzahl* (lit. 'unreachable cardinal') is found in Banach (1930, p. 98), and French *nombre inaccessible* in Sierpiński and Tarski (1930, p. 292), where it is attributed, without reference, to Kuratowski.[2]

The terminological distinction between *strongly inaccessible* and *weakly inaccessible* cardinals (German *schwach / stark unerreichbar*) seems to be due to Tarski (1939, p. 49); in Tarski (1938, p. 69), he called strongly inaccessible cardinals *inaccessible in the narrower sense* (*unerreichbar im engeren Sinne*).

Note. For a historical overview of set theory and inaccessible cardinals, see Kanamori (2003, pp. xi–xxii); see also Ferreirós (2007, p. 334), Plotkin (2005, pp. 173, 185, 195), and Ewald (1996, p. 1210), who also explains Hausdorff's word *exorbitant* in detail.

---

[1] "*Hae commode distinguuntur in explicitas et implicitas*. Explicitae sunt, quae per signa radicalia sunt evolutae, cuiusmodi exempla modo sunt data. Implicitae vero functiones irrationales sunt, quae ex resolutione aequationum ortum habent". Translation from Euler (1988, § 8). [2] They do refer to Kuratowski (1924), but not specifically for the term, which is not found in that paper. This attribution is repeated in Tarski (1938, p. 87), who provides the same reference.

References

Banach, Stefan (1930). "Über additive Massfunktionen in abstrakten Mengen". *Fundamenta Mathematicae* 15, pp. 97–101.
Ewald, William (1996). *From Kant to Hilbert.* A Source Book in the Foundations of Mathematics. Volume II. Oxford.
Ferreirós, José (2007). *Labyrinth of Thought.* A History of Set Theory and Its Role in Modern Mathematics. Second revised edition. Basel.
Hausdorff, Felix (1907). "Über dichte Ordnungstypen". *Jahresbericht der Deutschen Mathematiker-Vereinigung* 16, pp. 541–546.
Kanamori, Akihiro (2003). *The Higher Infinite.* Perspectives in Mathematical Logic. Berlin.
Kuratowski, Kazimierz (1924). "Sur l'état actuel de l'axiomatique de la théorie des ensembles". *Annales de la Société Polonaise de Mathématique* 3, pp. 146–147.
Plotkin, Jacob M., ed. (2005). *Hausdorff on Ordered Sets.* Providence, R.I.
Sierpiński, Wacław and Alfred Tarski (1930). "Sur une propriété caractéristique des nombres inaccessibles". *Fundamenta Mathematicae* 15, pp. 292–300.
Tarski, Alfred (1938). "Über unerreichbare Kardinalzahlen". *Fundamenta Mathematicae* 30, pp. 68–89.
Tarski, Alfred (1939). "Ideale in vollständigen Mengenkörpern. I". *Fundamenta Mathematicae* 32, pp. 45–63.

incidence

The geometrical concept of incidence was defined by Reye (1898, p. 26) in the following terms: "For the sake of brevity we shall call two straight lines 'incident' if they intersect; a straight line, or plane, and a point are 'incident' if the latter lies in the former; and finally, a ray, or point, and a plane, if the latter passes through the former". The terminology is not due to Reye, however; various 19th-century mathematicians attributed it to Graßmann, for example Sturm (1877, p. 258), Schubert (1879, p. 336), and Pasch (1882, p. 73).

Graßmann indeed used the term in his *Ausdehnungslehre* of 1862, where he defined: "If every magnitude of a domain (A) is simultaneously a magnitude of another (B) (without the converse obtaining) , then I call the two domains *incident* upon one another, and say that the first domain (A) is *subordinate* to the second, the second *superordinate* to the first" (Graßmann 1862, p. 7).[1] Later in the text (pp. 50f.), he added: "I call an elementary magnitude A *superordinate*, *subordinate*, or *incident* on another, according as this applies to the domains of these magnitudes".[2] Graßmann (1878, p. 294) later pointed out that he had earlier used the word *eingeordnet* in the same sense in Graßmann (1844, p. 202).

The terminology was also adopted by Hirst (1891, p. 92) and Klein (1893, p. 244), for example. This does not mean the term was universally accepted at that time, however; Hilbert (1899), for instance, did not use it in his *Grundlagen der Geometrie*.

---

[1] "Wenn jede Grösse eines Gebietes (A) zugleich Grösse eines andern (B) ist (ohne dass das Umgekehrte nothwendig stattfindet), so nenne ich beide Gebiete einander *incident*, und sage dann, das erste Gebiet (A) sei dem zweiten untergeordnet, das zweite dem ersten übergeordnet". Translation from Graßmann (2000, p. 8). [2] "Ich nenne endlich eine einfache Grösse A einer andern *übergeordnet*, *untergeordnet*, oder mit ihr *incident*, je nachdem dies von den Gebieten dieser Grössen gilt". Translation from Graßmann (2000, p. 44).

References

Graßmann, Hermann (1844). *Die Wissenschaft der extensiven Grösse, oder die Ausdehnungslehre, eine neue mathematische Disziplin.* Erster Theil, die lineale Ausdehnungslehre enthaltend. Leipzig.

Graßmann, Hermann (1862). *Die Ausdehnungslehre.* Vollständig und in strenger Form bearbeitet. Berlin.

Graßmann, Hermann (1878). *Die Ausdehnungslehre von 1844.* Leipzig.

Graßmann, Hermann (2000). *Extension Theory.* Translated by Lloyd C. Kannenberg. Providence, R.I.

Hilbert, David (1899). *Grundlagen der Geometrie.* Leipzig.

Hirst, T.A. (1891). "On the Correlation of two Spaces, each of Three Dimensions". *Proceedings of the London Mathematical Society* 21, pp. 92–118.

Klein, Felix (1893). *Einleitung in die höhere Geometrie, I.* Vorlesung, gehalten im Wintersemester 1892-93. Göttingen.

Pasch, Moritz (1882). *Vorlesungen über die neuere Geometrie.* Leipzig.

Reye, Theodor (1898). *Lectures on the Geometry of Position.* Part I. Translated and edited by Thomas F. Holgate. New York.

Schubert, Hermann (1879). *Kalkül der abzählenden Geometrie.* Leipzig.

Sturm, Rudolf (1877). "Ueber correlative oder reciproke Bündel". *Mathematische Annalen* 12, pp. 254–368.

incidence matrix

Incidence matrices were used by Poincaré (1900, p. 280), who simply called them *tables* (*tableaux*). The concept was picked up by Veblen (1912-13, p. 86), who referred to Poincaré but used *matrix* instead; French *matrice* was adopted in this sense by Sainte-Laguë (1926, p. 30).[1] In a later publication, Veblen (1922, p. 4) first defined the term *incident*: "A 0-cell is said to be *incident* with a 1-cell if and only if it is an end of the 1-cell; and under the same conditions the 1-cell is said to be incident with the 0-cell". He went on to explain that "[t]he incidence relations between the 0-cells and the 1-cells may be represented in a table or matrix of α_0 rows and α_1 columns".

The German term *Inzidenztafel* 'incidence table' was then used by Levi (1929, p. 3); *Inzidenzmatrix* was used in a review of Levi's book (Mayer 1930, p. 15). English *incidence matrix* appeared in Cohen (1931, p. 8).

Note. Some notes on Poincarés use of incidence matrices can be found in Basbois (2011); see also Gropp (1996).

References

Basbois, Nicolas (2011). "Die Entstehung des Begriffs der Homologiegruppe". *Internationale Mathematische Nachrichten* 217, pp. 1–37.

Cohen, L.W. (1931). "*Geometrische Konfigurationen.* By Friedrich Levi [review]". *Bulletin of the American Mathematical Society* 37, pp. 8–9.

Gropp, Harald (1996). "Poincaré and graph theory". *Philosophia Scientiae* 1, pp. 85–95.

Levi, Friedrich (1929). *Geometrische Konfigurationen.* Mit einer Einführung in die kombinatorische Flächentopologie. Leipzig.

Mayer, Walther (1930). "F. Levi, Geometrische Konfigurationen [review]". *Monatshefte für Mathematik und Physik* 37, A 15.

Poincaré, Henri (1900). "Second complément à l'analysis situs". *Proceedings of the London Mathematical Society* 32, pp. 277–308.

Sainte-Laguë, André (1926). *Les réseaux (ou graphes).* Paris.

[1] Sainte-Laguë there also gives various historical references.

Veblen, Oswald (1922). *Analysis Situs*. New York.
Veblen, Oswald (1912-13). "An Application of Modular Equations in Analysis Situs". *Annals of Mathematics* 14, pp. 86–94.

incidence structure

The concept of incidence structure was developed in the study of projective planes. A definition was given by Hall (1943, p. 233) under the name *partial plane*: "A partial plane is any system of points and lines such that: p1. There is at most one line through any two distinct points. p2. There is at most one point common to two distinct lines". The concept was picked up by Pickert (1955, pp. 2f.), who used the German term *Inzidenzstruktur* and referred to Hall for the earlier term.[1]

Note. Brief terminological notes are found in Dembowski (1968, p. 1) and André (1961, p. 87). Karzel and Kroll (1988) discuss the history of incidence structures in detail.

References

André, Johannes (1961). "Über Parallelstrukturen. Teil I: Grundbegriffe". *Mathematische Zeitschrift* 76, pp. 85–102.
Dembowski, Peter (1968). *Finite Geometries*. New York.
Hall, Marshall (1943). "Projective Planes". *Transactions of the American Mathematical Society* 54, pp. 229–277.
Karzel, Helmut and Hans-Joachim Kroll (1988). *Geschichte der Geometrie seit Hilbert*. Darmstadt.
Pickert, Günter (1955). *Projektive Ebenen*. Berlin.

incircle and some related terms

In a letter to the journal *Nature*, Hudson (1883a, p. 7) proposed a series of terms: "I beg leave to suggest the following names: *circumcircle*, *incircle*, *excircle*, and *midcircle*". In a follow-up letter, he made some more suggestions: "I propose not only to call the circle circumscribing a triangle the *circumcircle*, but also to call its centre the *circumcentre*, and in the same way to speak of the *incentre*, the three *excentres* [...], and the *midcentre*" (Hudson 1883b, p. 104).

Mackay (1883, p. 32) drew attention to these suggestions, but also remarked that some corresponding terms "have been more or less in use in Germany since 1866". He does not give a specific reference, but may mean Reuschle (1866, p. 482), who used *Umkreis* (circumcircle) and *Inkreis* (incircle). In fact, Reuschle had already suggested these and some other terms in 1853 (Reuschle 1853, p. 18). Tropfke (1940, p. 140), who refers to Reuschle, also points out that *Umkreis* is indeed considerably older and already found in Widmann (1489, fol. D 5^{v}). The term is based on the preposition *um* 'around' and thus is, like *circumcircle*, a shortening of *umschriebener Kreis* 'circumscribed circle'.

References

Hudson, William Henry Hoar (1883a). "The Circles of a Triangle". *Nature* 38, p. 7.
Hudson, William Henry Hoar (1883b). "The Centres of a Triangle". *Nature* 38, p. 104.
Mackay, J.S. (1883). "The Triangle and its Six Scribed Circles". *Proceedings of the Edinburgh Mathematical Society* 1, pp. 4–128.
Reuschle, Carl Gustav (1853). "Mathematische Abhandlung". In: *Einladungsschrift des Königlichen Gymnasiums in Stuttgart*. Stuttgart, pp. 1–39.

---

[1] *Inzidenzstruktur* was used in a different sense by Dembowski (1968, p. 1).

Reuschle, Carl Gustav (1866). "Ueber die Punkte des Dreiecks, deren Verbindungsstrecken vom Schwerpunkt gedrittelt werden". *Zeitschrift für Mathematik und Physik* 11, pp. 475–496.
Tropfke, Johannes (1940). *Geschichte der Elementar-Mathematik*. Vierter Band. Ebene Geometrie. Dritte, verbesserte und vermehrte Auflage besorgt von Dr. Kurt Vogel. Berlin and Leipzig.
Widmann, Johannes (1489). *Behende vnd hubsche Rechenung auff allen kauffmanschafft*. Leipzig.

incommensurable

See *commensurable*.

independent (axioms)

In his "Erlanger Programm", Klein (1872, p. 44) remarked that the recent investigations on the theory of parallels "show, in the first place,—and this function of theirs may be regarded as concluded once for all,—that the axiom of parallels is not a mathematical consequence of the other axioms usually assumed".[1] The word *independent* (German *unabhängig*) in this sense was used by Frischauf (1872, p. 56), for example.

It later also appears in Peano (1891, p. 93) and Hilbert (1899, e. g., p. 21). It was also common to say that a system of axioms (or postulates) was *irreducible*, as Padoa (1902, p. 250) did at a talk at the ICM in 1900. Huntington (1902, p. 264) used both *independent* and *irreducible* in this sense.

Generally, the use of *independent* in the sense 'incapable of being expressed in terms of, or of being derived or deduced from, the others' (OED) is much older. The examples cited here relate to the development of modern axiomatics.

See also *consistent* (axioms), *categorical*, and *complete* (axioms).

References

Frischauf, Johannes (1872). *Absolute Geometrie nach Johann Bolyai*. Leipzig.
Hilbert, David (1899). *Grundlagen der Geometrie*. Leipzig.
Huntington, Edward V. (1902). "A Complete Set of Postulates for the Theory of Absolute Continuous Magnitude". *Transactions of the American Mathematical Society* 3, pp. 264–279.
Klein, Felix (1872). *Vergleichende Betrachtungen über neuere geometrische Forschungen*. Programm zum Eintritt in die philosophische Facultät und den Senat der k. Friedrich-Alexanders-Universität zu Erlangen. Erlangen.
Klein, Felix (1893). "A comparative review of recent researches in geometry". *Bulletin of the New York Mathematical Society* 2, pp. 215–249.
Padoa, Alessandro (1902). "Un nouveau système irréductible de postulats pour l'algèbre". In: *Compte rendu du deuxième Congrès international des mathématiciens*. Ed. by E. Duporcq. Paris, pp. 249–256.
Peano, Guiseppe (1891). "Sul concetto di numero". *Rivista di matematica* 1, pp. 87–102.

independence (probability)

The concept of independence of events was implicitly used before the 17th century, and it was de Moivre who made it explicit.[2] He did so in *De mensura sortis*, where he used the verb *pendere* '(lit.) to hang; depend' (de Moivre 1711, p. 215). In his famous *Doctrine of Chances*, de Moivre (1718, p. 4) defined the independence of events in the following terms: "if a Fraction expresses the Probability of an Event,

---

[1] "Sie zeigen einmal – und dieses ihr Geschäft kann man als ein einmaliges, abgeschlossenes betrachten –, dass das Parallelenaxiom keine mathematische Folge der gewöhnlich vorangestellten Axiome ist [...]". Translation from Klein (1893, p. 246). [2] According to Sheynin (2017, pp. 61, 63), he was the first to do so.

and another Fraction the Probability of another Event, and those two Events are independent; the Probability that both those Events will Happen, will be the Product of those Fractions". In the second edition of the book, de Moivre (1738, pp. 6f.) revised the definition:

> Two Events are independent, when they have no connexion one with the other, and that the happening of one neither forwards nor obstructs the happening of the other.
>
> Two Events are dependent, when they are so connected together as that the Probability of either's happening is altered by the happening of the other. [...]

As pointed out by Krengel (2011, p. 1), the first to give the formal definition of independence appears to have been Bohlmann (1901, p. 859): "E_1 and E_2 are called independent if $p = p_1 p_2$".[1]

Note. For the history of the concept, see Leobacher and Prochno (2021, esp. pp. 71f.). Krengel (2011) gives an an account of Bohlmann's work.

References

Bohlmann, Georg (1901). "Lebensversicherungs-Mathematik". *Encyklopädie der mathematischen Wissenschaften mit Einschluss ihrer Anwendungen* I D 4b, pp. 852–917.

Krengel, Ulrich (2011). "On the contributions of Bohlmann to probability theory". *Journ@l électronique d'histoire des probabilités et de la statistique* 7.

Leobacher, Gunther and Joscha Prochno (2021). "Statistical independence in mathematics – the key to a Gaussian law". *Mathematische Semesterberichte* 68, pp. 69–104.

de Moivre, Abraham (1711). "De mensura sortis, seu, de probabilitate eventuum in ludis a casu fortuito pendentibus". *Philosophical Transactions of the Royal Society of London* 27, pp. 213–264.

de Moivre, Abraham (1718). *The Doctrine of Chances: or, A Method of Calculating the Probability of Events in Play*. London.

de Moivre, Abraham (1738). *The Doctrine of Chances: or, A Method of Calculating the Probability of Events in Play*. The Second Edition. London.

Sheynin, Oscar (2017). *Theory of Probability*. A Historical Essay. Revised and enlarged edition. Berlin. URL: `http://www.sheynin.de` (retrieved Jan. 2, 2018).

***index* (overview)**

Latin *index* is derived from *indicare* 'to point out'. Thus, an index literally indicates something. The term has been in wide use in various mathematical senses since the 16th century. Since that time, the term is occasionally found as an alternative to *exponent*. At the beginning of the 18th century, the term is first found in various specialized dictionaries: "INDEX, in Arithmetic, is the same what is sometimes called the *Characteristic* or Exponent of a Logarithm" (Harris 1704). In this sense, the term was suggested by Briggs; see *characteristic* (logarithm).

An example of a later use still common today occurs in modular arithmetic. In this field, the term was introduced by Gauß (1801, § 57): "We will arbitrarily choose some primitive root a as a *base* to which we will refer all numbers not divisible by p. And if $a^e \equiv b(\text{mod}.p)$, we will call e the *index* of b".[2]

Some further selected meanings of the term are discussed in the following entries.

---

[1] "Man sagt, dass E_1 und E_2 von einander unabhängig sind, wenn $p = p_1 p_2$". – p_i is the probability of the event E_i. [2] "Radicem aliquam primitivam, a, ad lubitum pro *basi* adoptamibus, ad quam omnes numeros per p non divisibiles referemus, et si fuerit $a^e \equiv b(\text{mod}.p)$, e ipsius b *indicem* vocabimus". Translation from Gauß (1986, § 57).

The term *index of inertia* is discussed in the entry *signature*.

References

Gauß, Carl Friedrich (1801). *Disquisitiones arithmeticae*. Leipzig.

Gauß, Carl Friedrich (1986). *Disquisitiones Arithmeticae*. English Edition. Translated by A.A. Clarke. Revised by W.C. Waterhouse. New York et al.

Harris, John (1704). *Lexicon Technicum*. Vol. I. London.

index (functions)

The Cauchy index is first found in various publications from 1831; he expounded the concept in Cauchy (1833), an article that was only circulated in lithographed form, however. In a summary of this article, also published in 1831, Cauchy defined:[1]

> Let x be a real variable and $f(x)$ a function that becomes infinite at $x = a$. As x increases, the function $f(x)$ in becoming infinite may pass from negative to positive, or from positive to negative, or it may not change sign. I call the *index* of the function $f(x)$ at the given value a of the variable x the quantity $+1$ in the first case, -1 in the second, and 0 in the third. The *integral index* of $f(x)$, taken between the limits $x = x_0$ and $x = X > x_0$, is the sum of the indices corresponding to the diverse values of x (Cauchy 1831, p. 122).[2]

The term *Cauchy index* appears to be due to Hurwitz (1895, p. 276).

The concept was generalized by Kronecker (1870, p. 161), who used the term *Charakteristik*. For this concept, Poincaré (1886, pp. 177f.) used French *indice*; Hadamard (1910, p. 437) called it *indice de Kronecker*.

See also *characteristic* (topology), *degree* (topology), and *winding number*.

Note. For the history of the concept, see Smithies (1997, pp. 179–183) and Johnson (1981, ch. 6).

References

Cauchy, Augustin-Louis (1831). "Formules extraites d'un mémoire présenté le 27 novembre 1831 à l'Académie des Sciences de Turin". *Bulletin des sciences mathématiques, physiques et chimiques* 16, pp. 119–128.

Cauchy, Augustin-Louis (1833). *Mémoire sur les rapports qui existent entre le Calcul des résidus et le Calcul des limites, et sur les avantages qu'offrent ces deux nouveaux calculs dans la résolution des équations algébriques ou transcendantes* [lithograph]. Turin.

Cauchy, Augustin-Louis (1837). "Calcul des indices des fonctions". *Journal de l'École polytechnique* 15, cahier 24, pp. 176–229.

Hadamard, Jacques (1910). "Sur quelques applications de l'indice de Kronecker". In: Tannery, Jules. *Introduction à la théorie des fonctions d'une variable*. Tome deuxième. Paris, pp. 437–477.

Hurwitz, Adolf (1895). "Ueber die Bedingungen, unter welchen eine Gleichung nur Wurzeln mit negativen reellen Theilen besitzt". *Mathematische Annalen* 46, pp. 273–284.

Johnson, Dale M. (1981). "The Problem of the Invariance of Dimension in the Growth of Modern Topology, Part II". *Archive for History of Exact Sciences* 25, pp. 85–266.

Kronecker, Leopold (1870). "Über Systeme von Functionen mehrerer Variabeln". *Monatsberichte der Königlich Preußischen Akademie der Wissenschaften zu Berlin* (1869), pp. 159–193.

---

[1] A full account is found in the frequently cited Cauchy (1837). [2] "Soient maintenant x une variable réelle, et $f(x)$ une fonction de cette variable, qui devienne infinie pour $x = a$. Si l'on fait croître x, la fonction $f(x)$ passera, en devenant infinie, du négatif au positif, ou du positif au négatif, ou bien elle ne changera pas de signe. La quantité $+1$, dans le premier cas, -1, dans le second, zéro, dans le troisième, est ce que je nomme l'*indice* de la function $f(x)$ pour la valeur donnée a de la variable x. L'*indice intégral* de $f(x)$, pris entre les limites $x = x_0$, et $x = X > x_0$, n'est autre chose que la somme des indices correspondants aux diverses valeurs de x".

Poincaré, Henri (1886). “Sur les courbes définies par les équations différentielles (quatrième partie)”. *Journal de mathématiques pures et appliquées* 2, pp. 151–218.
Smithies, Frank (1997). *Cauchy and the Creation of Complex Function Theory*. Cambridge.

index (group)

The group-theoretical term originally came up in Cauchy’s studies on permutations. In an article published in 1815, he studied the number of values of an algebraic function under all the permutations of its arguments and introduced the following terminology (with K being an arbitrary function of order n, and N the product $1 \cdot 2 \cdot 3 \cdots n$):

> R, the number of values of the function K, must be a divisor of N; that is to say, of the product $1 \cdot 2 \cdot 3 \cdots n$ [...][1] To abbreviate, I will henceforth call the *index of the function* K the number R which indicates how many essentially different values this function can assume [...] (Cauchy 1815, pp. 6f.).[2]

Cauchy’s theorems are not expressed in terms of group theory, but, as explained by Miller (1935, pp. 432f.), “[o]ne of the most important theorems proved therein is that there is no permutation group on n letters whose index under the symmetric group is less than the largest prime number which does not exceed n unless this index is either 2 or 1”.[3] In conclusion, Cauchy’s paper contains the germs of the concept of index and the term, used in connection with functions.

When group-theoretical language was established, however, Cauchy’s term *indice* was kept; in this sense, the word was used by Serret (1866, p. 287): “I shall call *index* of a system of conjugate substitutions[4] formed with n letters, the quotient obtained by dividing the product $N = 1 \cdot 2 \cdot 3 \cdots n$ by the number that expresses the system’s order”.[5] The corresponding German and English terms also appear in Weber (1895, p. 502) and Burnside (1897, p. 382), two influential textbooks on algebra.

Note. The origin of the term has been pointed out by Burkhardt and Vogt (1909, p. 543) and Wußing (1984, p. 95), for example.

References

Bashmakova, I.G. and E.I. Rudakov (2001). “Algebra and Algebraic Number Theory”. In: *Mathematics of the 19th Century*. Mathematical Logic. Algebra. Number Theory. Probability Theory. Second revised edition. Ed. by Andrey Nikolaevich Kolmogorov and Adolph Andrei Pavlovich Yushkevich. Basel et al., pp. 35–85.
Burkhardt, Heinrich and Henri Vogt (1909). “Sur les groupes finis discontinus”. *Encyclopédie des sciences mathématiques pures et appliquées* I.1, pp. 532–616.
Burnside, William Snow (1897). *Theory of Groups of Finite Order*. Cambridge.
Cauchy, Augustin-Louis (1815). “Sur le nombre des valeurs qu’une fonction peut acquérir, lorsqu’on y permute de toutes les manières possibles les quantités qu’elle renferme”. *Journal de l’École polytechnique* 10, cahier 17, pp. 1–28.

---

[1] n is the *order* of the function (the number of its variables). [2] “R, ou le nombre des valeurs différentes de la function K, ne peut être qu’un diviseur de N, c’est-à-dire du produit $1 \cdot 2 \cdot 3 \cdots n$ [...] Afin d’abréger, j’appellerai désormais *indice de la function* K le nombre R qui indique combien cette function peut obtenir de valeurs essentiellement différentes [...]”. Translation based on Cauchy (2014, pp. 6f.). [3] See also Dahan (1980, p. 280) and Bashmakova and Rudakov (2001, p. 65). [4] A *system of conjugate substitutions* is a group. See *group*. [5] “je nommerai *indice* d’un système de substitutions conjuguées formées avec n lettres, le quotient obtenu en divisant le produit $N = 1 \cdot 2 \cdot 3 \cdots n$ par le nombre qui exprime l’ordre du système”.

Cauchy, Augustin-Louis (2014). "Cauchy's *Mémoire sur le nombre des valeurs*". Translated by Mike Bertrand and Stephen Gaschignard. URL: `http://nonagon.org/ExLibris/cauchys-memoire-sur-le-nombre-des-valeurs` (retrieved Apr. 2, 2020).
Dahan, Amy (1980). "Les travaux de Cauchy sur les substitutions. Étude de son approche du concept de groupe". *Archive for History of Exact Sciences* 23, pp. 279–319.
Miller, George Abram (1935). "History of the theory of groups to 1900". In: *The Collected Works of George Abram Miller*. Volume I. Urbana, Ill., pp. 427–467.
Serret, Joseph-Alfred (1866). *Cours d'algèbre supérieure*. Troisième édition. Tome second. Paris.
Weber, Heinrich (1895). *Lehrbuch der Algebra*. Erster Band. Braunschweig.
Wußing, Hans (1984). *The Genesis of the Abstract Group Concept*. A Contribution to the History of the Origin of Abstract Group Theory. Tr. Abe Shenitzer. Cambridge, Mass.

index **(operator)**

In their famous paper "The index of elliptic operators on compact manifolds", Atiyah and Singer (1963, p. 423) defined:

> One of the basic properties of elliptic operators is that Ker D (i. e., the null space) and Coker $D = \Gamma(F)/D\Gamma(E)$ are both finite-dimensional. The *index* $\gamma(D)$ is defined by
>
> $$\gamma(D) = \dim \mathrm{Ker} D - \dim \mathrm{Coker} D.$$

The term was not new at the time. According to Nikolski (2020, p. 384), it first appeared in some publications from the early 1950s; he cites Gohberg (1951, p. 631), Yood (1951, p. 600), and Kreĭn and Krasnoselskii (1952, p. 219) as papers in which the term was first used (independently).[1] Nikolski also remarks that the concept already appeared in Noether (1921, p. 52) but that the word *Index* as used by Noether refers to the Cauchy index.

Note. For the history of the concept and the terminology, see Nikolski (2020, pp. 384f.), Gohberg (2004), and Dieudonné (1985).

References

Atiyah, Michael F. and Isadore M. Singer (1963). "The index of elliptic operators on compact manifolds". *Bulletin of the American Mathematical Society* 69, pp. 422–433.
Dieudonné, Jean (1985). "The index of operators in Banach spaces". *Integral Equations and Operator Theory* 8, pp. 580–589.
Gohberg, I.C. (1951). "О линейных операторах, аналитически зависящих от параметра". *Doklady Akademii Nauk SSSR* 78, pp. 629–632.
Gohberg, I.C. (2004). "The Atiyah-Singer index formula". *Nieuw Archief voor Wiskunde* 5, p. 319.
Kreĭn, M.G. and M.A. Krasnoselskii (1952). "Устойчивость индекса неограниченного оператора". *Matematicheskii Sbornik* 30, pp. 219–224.
Nikolski, Nikolaï (2020). *Toeplitz Matrices and Operators*. Translated by Danièle Gibbons, Greg Gibbons. Cambridge.
Noether, Fritz (1921). "Über eine Klasse singulärer Integralgleichungen". *Mathematische Annalen* 82, pp. 42–63.
Yood, Bertram (1951). "Properties of linear transformations preserved under addition of a completely continuous transformation". *Duke Mathematical Journal* 18, pp. 599–612.

indicatrix

There are various indicatrices in mathematics; the most well-known of these is perhaps that of Dupin. As described by Kline (1972, p. 569), Dupin constructed it as

[1] Gohberg as well as Krein and Krasnoselskii used Russian индекс.

follows: "Given the tangent plane of a surface at a point M, he [Dupin] laid off in each direction and starting from M a segment whose length is equal to the square root of the radius of curvature of the normal section of the surface in that direction. The locus of the endpoints of these segments is a conic, the indicatrix, which gives a first approximation to the form of the surface around M (because it is almost similar to a section of the surface made by a plane near M and parallel to the tangent plane at M)". Dupin introduced French *indicatrice* in this sense in his *Développements de géométrie* (Dupin 1813, p. 48);[1] he also remarks that his indicatrix indicates the directions of the two principal curvatures at a given point (Dupin 1813, p. 50).[2] An early adopter of the term was Gergonne (1813-14, p. 376).

The English term appears in Dupin (1829, p. 161); an account of Dupin's indicatrix is given in Hymers (1830, p. 100). In German, the term first appears in Leroy (1838, p. 212) as *anzeigende Ellipse* (*anzeigen* 'to indicate'); *Indicatrix* is used in Cournot (1845, p. 299).

Note. For Dupin's introduction of the term, see, e. g., A'Campo-Neuen and Papadopoulos (2019, pp. 334–337).

References

A'Campo-Neuen, Annette and Athanase Papadopoulos (2019). "A Path in History, from Curvature to Convexity". In: *Geometry in History*. Ed. by S.G. Dani and Athanase Papadopoulos. Cham, pp. 305–353.

Cournot, Antoine Augustin (1845). *Elementarlehrbuch der Theorie der Functionen oder der Infinitesimalanalysis*. Deutsch bearbeitet von C.H. Schnuse. Darmstadt.

Dupin, Charles (1813). *Développements de géométrie*. Paris.

Dupin, Charles (1829). "Remarks on the Stability of Ships". In: *Papers on Naval Architecture and Other Subjects Connected with Naval Science*. Vol. II. Ed. by William Morgan and Augustin Creuze. London, pp. 153–166.

Gergonne, Joseph Diez (1813-14). "Démonstration des principaux théorèmes de M. Dupin sur la courbure des surfaces". *Annales de mathématiques pures et appliquées* 4, pp. 368–378.

Hymers, John (1830). *A Treatise on Analytical Geometry of Three Dimensions*. Cambridge.

Kline, Morris (1972). *Mathematical Thought From Ancient To Modern Times*. Volume 2. New York and Oxford.

Leroy, C.F.A. (1838). *Die darstellende Geometrie*. Deutsch, mit Anmerkungen von E.F. Kauffmann. Stuttgart.

Poisson, Siméon Denis (1812). "*Développements de Géométrie rationnelle et analytique* [...] par M. Dupin". *Nouveau bulletin des sciences, par la Société philomathique de Paris* 3, pp. 246–248.

indicatrix (Tissot)

The so-called *Tissot indicatrix*, also called *Tissot ellipse* or *ellipse of distortion*, was developed by Tissot in the second half of the 19th century. It is a device to characterize the distortions of a map projection. As explained by Snyder (1993, p. 147), the basic idea is this: "Any very small (theoretically infinitesimal) circle on the sphere or ellipsoid representing the earth is plotted on a flat map as a circle or ellipse centered about the same point. If these very small circles are plotted at various locations on the maps, using the same radius for each circle on the globe, the distortion at any given point on the map is indicated by the shape and size or the circle or ellipse".

---

[1] The term first appeared in print in an account of Dupin's book (Poisson 1812, p. 247). [2] For this justification of the term, see also Dupin (1813, p. xvii).

Tissot (1878, pp. 51, 152) called this circle or ellipse "a kind of indicatrix"[3] or *ellipse indicatrice*.

The corresponding German and English terms are found in Zöppritz (1884, p. 104) and [anon.] (1888, p. 207). Haentzschel (1903, p. 79) suggested German *Verzerrungsellipse* 'ellipse of distortion'; he rejected *indicatrix* because the word was also used in Dupin's sense.[4]

Note. For the history of the concept, see Snyder (1993, pp. 147–149) and Papadopoulos (2017).

References

[anon.] (1888). "Tissot's Theory of the Projection of Maps". *Science* 12, p. 207.
Haentzschel, Emil (1903). *Das Erdsphäroid und seine Abbildung*. Leipzig.
Papadopoulos, Athanase (2017). "Nicolas-Auguste Tissot: a link between cartography and quasiconformal theory". *Archive for History of Exact Sciences* 71, pp. 319–336.
Snyder, John P. (1993). *Flattening the Earth*. Two Thousand Years of Map Projections. Chicago.
Tissot, Auguste (1878). "Mémoire sur la représentation des surfaces et les projections des cartes géographiques". *Nouvelles annales de mathématiques* 17, pp. 49–55, 145–163, 351–366.
Tissot, Auguste (1881). *Mémoire sur la représentation des surfaces et les projections des cartes géographiques*. Paris.
Tissot, Auguste (1887). *Die Netzentwürfe geographischer Karten*. Autorisierte deutsche Bearbeitung mit einigen Zusätzen besorgt von E. Hammer. Stuttgart.
Zöppritz, Karl (1884). *Leitfaden der Kartenentwurfslehre für Studierende der Erdkunde und deren Lehrer*. Leipzig.

***indicatrix* (topology)**

In topology, the term *indicatrix* was introduced by Klein (1876) with respect to surfaces. As described by Dieudonné (2009, p. 19), "Klein's idea was to take a small circuit [...] around a point, and to 'orient' it[1] by choosing on it a positive direction. He then conceived this circuit to be moved around continuously on the connected surface, and he distinguished two cases, depending on whether when coming back to the initial point the direction along the circuit is always preserved or may be reversed for some closed paths on the surface". Klein (1876, p. 479) called the circle an *Indicatrix*.

Klein used the concept to give a new, intrinsic, definition of one-sidedness and two-sidedness: if the direction of the indicatrix is reversed, the surface is one-sided, if the direction is preserved, the surface is two-sided. Klein (1876, p. 479) used the term *double surface* (*Doppelfläche*) in the first case,[2] and *simple surface* (*einfache Fläche*) in the second case (e. g., Klein 1874, p. 555).

This distinction was picked up by Dyck (1888, pp. 473f.), who, having generalized Klein's notion to n-dimensional manifolds, explicitly coined the terms *surface with a reversible* (German *umkehrbar*) *indicatrix* and *surface with a non-reversible indicatrix*. In 1915, the new terms *non-orientable* (reversible indicatrix) and *orientable* (non-reversible indicatrix) were suggested instead; see *orientable*.

It took some time until the term is found in other languages; for example, Mannoury (1898, pp. 133f.) uses the corresponding French term, Heegaard (1898, p. 46) the

[3] "une sorte d'indicatrice". [4] To be precise, Haentzschel did not object to Tissot's description as "a kind of indicatrix" but to the German translation of Tissot (1881), in which *a kind of* was omitted (Tissot 1887, p. 8). [1] In Klein's words, "to give it a sense (*Sinn*)". [2] The term is found earlier in Klein (1874, p. 550).

Danish one, while the English term does not seem to appear before Veblen (1922, p. 129).

Note. For the history of the concept, see Volkert (2002, pp. 42–49, 60–67).

References

Dieudonné, Jean (2009). *A History of Algebraic and Differential Topology, 1900–1960*. Boston.

Dyck, Walther (1888). "Beiträge zur Analysis situs. I. Aufsatz. Ein- und zweidimensionale Mannigfaltigkeiten". *Mathematische Annalen* 32, pp. 457–512.

Heegaard, Poul (1898). *Forstudier til en topologisk Teori for de algebraiske Fladers Sammenhæng*. København.

Klein, Felix (1874). "Bemerkungen über den Zusammenhang der Flächen". *Mathematische Annalen* 7, pp. 549–557.

Klein, Felix (1876). "Ueber den Zusammenhang der Flächen". *Mathematische Annalen* 9, pp. 476–482.

Mannoury, Gerrit (1898). "Lois cyclomatiques". *Nieuw Archief voor Wiskunde* 3, pp. 126–152.

Veblen, Oswald (1922). *Analysis Situs*. New York.

Volkert, Klaus (2002). *Das Homöomorphismusproblem insbesondere der 3-Mannigfaltigkeiten, in der Topologie 1892-1935*. Paris.

indivisible

Usage of the term *indivisible* peaked in the 17th century; it is in the work of Cavalieri that "the method of the indivisibles reached its summit of explicit elaboration and application" (Schubring 2006, p. 25). The *Geometria indivisibilibus* is expounded in Cavalieri (1635). Cavalieri did not formally define the term but on one occasion he equated *indivisibilia* with "omnes lineas figurae" (Cavalieri 1635, II, p. 21); the expression "all the lines" (*omnes lineae*) was defined at the beginning of book II:

> If through opposite tangents to a given plane figure two parallel and indefinitely produced planes are drawn either perpendicular or inclined to the plane of the given figure, and if one of the parallel planes is moved toward the other, still remaining parallel to it, until it coincides with it; then the single lines which during the motion form the intersections between the moving plane and the given figure, collected together, are called all the lines of the figure taken with one of them as *regula* (Cavalieri 1635, II, pp. 1f.).[1]

On a later occasion, Cavalieri (1647, pp. 3f.) explained the concept of indivisible by comparing, as Baron (1969, p. 135) put it, "the line indivisibles of a plane surface and the parallel threads of a woven fabric, and the surface indivisibles of a solid and the parallel leaves of a book".

The term spread through Cavalieri's publications, and various 17th-century mathematicians used it, for example Galilei: "let us remember that we are among infinites and indivisibles, the former incomprehensible to our finite understanding by reason of their largeness, and the latter by their smallness" (Galilei 1638, p. 27).[2] The exact understanding of the concept with different authors may, however, "vary considerably" (Jullien 2015a, p. 4).

---

[1] "Si per oppositas tangentes cuiuscunq; datae planae figurae ducantur duo plana invicem parallela, recta sive inclinata ad planum datae figurae, hinc inde indefinitè producta; quorum alterum moveatur versus reliquum eidem semper aequidistans donec illi congruerit: singulae rectae lineae, quae in toto motu fiunt communes sectiones plani moti, & datae figurae, simul collectae vocentur: Omnes lineae talis figurae, sumptae regula una earundem". Translation from Andersen (1985, pp. 300f.).

[2] "mà ricordiamoci, che siamo trà gl' infiniti, e gl' indivisibili, quelli incomprensibili dal nostro intelletto finito per la lor grandezza, e questi per la lor piccolezza". Translation from Galilei (1989, p. 34).

The Latin term *indivisibilis* and its Greek counterparts are, as such, much older. In particular, these words occur in philosophical texts on atomism and the nature of the continuum. To give just one of many possible examples (Aristotle, *Physics* 231a): "no continuum can be made up of indivisibles, as for instance a line out of points, granting that the line is continuous and the point indivisible".[3] However, there does not appear to be direct link between this ancient meaning and Cavalieri's: "Cavalieri's method is independent of theories concerning the composition of the continuum; it just happened that he decided to use the term indivisibles as an alternative for 'all the lines' and 'all the planes'" (Andersen 1985, p. 364).

The terminology soon disappeared because the concept was replaced by that of infinitesimal.

Note. For a detailed account of Cavalieri's indivisibles, see Andersen (1985); for general accounts of the history of indivisibles see, for example, Whiteside (1961, esp. pp. 311-330) and Sonar (2021, pp. 54–56, 155–232). See also Wallner (1903) and the various studies in Jullien (2015b).

References

Andersen, Kirsti (1985). "Cavalieri's Method of Indivisibles". *Archive for History of Exact Sciences* 31, pp. 291–367.

Aristotle (1934). *The Physics*. With an English translation by Philip H. Wicksteed and Francis M. Cornford. In two volumes. II. London.

Baron, Margaret E. (1969). *The Origins of the Infinitesimal Calculus*. Oxford.

Cavalieri, Bonaventura (1635). *Geometria indivisibilibus continuorum nova quadam ratione promota*. Bologna.

Cavalieri, Bonaventura (1647). *Exercitationes geometricae sex*. Bologna.

Galilei, Galileo (1638). *Discorsi e dimostrazioni matematiche, intorno à due nuove scienze*. Leiden.

Galilei, Galileo (1989). *Two New Sciences Including Centers of Gravity and Force of Percussion*. Translated, with a New Introduction and Notes, by Stillman Drake. Second Edition. Toronto.

Jullien, Vincent (2015a). "Explaining the Sudden Rise of Methods of Indivisibles". In: *Seventeenth-Century Indivisibles Revisited*. Ed. by Vincent Jullien. Cham and Heidelberg, pp. 1–18.

Jullien, Vincent, ed. (2015b). *Seventeenth-Century Indivisibles Revisited*. Cham and Heidelberg.

Schubring, Gert (2006). *Conflicts Between Generalization, Rigor, and Intuition*. Number Concepts Underlying the Development of Analysis in 17th-19th Century France and Germany. New York.

Sonar, Thomas (2021). *3000 Years of Analysis*. Mathematics in History and Culture. Cham.

Wallner, C.R. (1903). "Die Wandlungen des Indivisibilienbegriffs von Cavalieri bis Wallis". *Bibliotheca Mathematica* 4, pp. 28–47.

Whiteside, D.T. (1961). "Patterns of Mathematical Thought in the later Seventeenth Century". *Archive for History of Exact Sciences* 1, pp. 179–388.

inductive set

The term *inductive set* (French *ensemble inductif*) was coined by Bourbaki (1939, p. 36): "An ordered set E is called *inductive* if it verifies the following condition: *any totally ordered subset of* E *has an upper limit*".[1] The English term *inductively ordered* was used by Samuel (1948, p. 105), who referred to Bourbaki.

The term is used in other senses. One of these goes back to Russell (1919, pp. 21f.):

> A property is said to be "hereditary" in the natural-number series if, whenever it belongs to a number n, it also belongs to $n + 1$, the successor of n. Similarly a class is said to be

[3] Translation from Aristotle (1934, p. 93). [1] "Un ensemble ordonnée E est dit *inductif* s'il vérifie la condition suivante: *toute partie totalement ordonnée de* E *possède une borne supérieure*".

"hereditary" if, whenever n is a member of the class, so is $n + 1$. [...] A property is said to be "inductive" when it is a hereditary property which belongs to 0. Similarly a class is "inductive" when it is a hereditary class of which 0 is a member.[2]

References

Bourbaki, Nicolas (1939). *Théorie des ensembles*. Fascicule de résultats. Paris.
Russell, Bertrand (1913). "The Philosophical Importance of Mathematical Logic". *The Monist* 23, pp. 481–493.
Russell, Bertrand (1919). *Introduction to Mathematical Philosophy*. New York.
Samuel, Pierre (1948). "Ultrafilters and Compactification of Uniform Spaces". *Transactions of the American Mathematical Society* 64, pp. 100–132.

infimum* and *supremum

The English terms *infimum* and *supremum* came up in the 1930s; an early usage is found in Tukey (1940, p. 4): "If $a > b$ for all $b \in B$, then a is an *upper bound* of B. We define *lower bound* similarly. Any of the upper bounds of B which is a lower bound of the class of all upper bounds of B is a *supremum* (sometimes called 'least upper bound' or 'join') of B". Tukey added an analogous definition for the infimum. Both concepts are used by Tukey in general terms with respect to partially ordered sets. The terms were already used by Birkhoff (1938, p. 795).[1] As will be seen in the following, they were taken over from German sources.

The concept was already used by Bolzano (1817, p. 41): "If a property M does not apply to *all* values of a variable quantity x but does apply to *all* values *smaller* than a certain u, then there is always a quantity U which is the greatest of those of which it can be asserted that all smaller x possess the property M".[2]

Stolz (1881, p. 258) later discussed this passage and pointed out that "the quantity U is called, following Weierstrass, the *upper limit* [*obere Grenze*] of all values of x, which have the property M".[3] Weierstraß's term spread through his lectures; in 1878, he gave this definition: "g is called the upper limit of a variable quantity if there is no value of the variable that is greater than g and if in the interval $g - \delta \ldots g$, δ being an arbitrarily small quantity, there are still places of the domain of the variable" (Weierstraß 1988, p. 84).[4] Like Stolz, Pasch (1882, p. 16) also attributed the term to Weierstraß; he later suggested that the terms *untere Schranke* and *obere Schranke* be used instead (Pasch 1887, p. 133). *Schranke* overlaps in meaning with *Grenze*; it also means 'barrier' or 'gate' and is the basis for the word *beschränkt* 'bounded'. As explained by Pasch (1887, p. 133), he suggested *Schranke* because he wanted to

---

[2] *inductive property* appeared earlier in Russell (1913, p. 483). [1] They may already have been used by Bennett (1936, p. 12); it has, however, not been possible to consult this source. [2] "Wenn eine Eigenschaft M nicht *allen* Werthen einer veränderlichen Grösse x, wohl aber *allen*, die *kleiner* sind, als ein gewisser u, zukömmt: so gibt es allemahl eine Grösse U, welche die grösste derjenigen ist, von denen behauptet werden kann, dass alle kleineren x die Eigenschaft M besitzen". Translation from Russ (2004, p. 269). [3] "Die Grösse U heißt nach Weierstrass die *obere Grenze* aller Werthe von x, denen die Eigenschaft M zukommt". [4] "g heißt die obere Grenze einer veränderlichen Größe, wenn es keinen Werth der Veränderlichen giebt größer als g und wenn in dem Intervall $g - \delta ... g$, δ sei eine noch so kleine Größe, sich noch immer Stellen des Gebietes der Veränderlichen vorfinden".

avoid confusion with the limit inferior and the limit superior, which he called *untere* and *obere Grenze*.[5]

In a later publication, Pasch (1915, p. 304) added the notations *inf x* or *inf M* and *sup x* or *sup M* and explained that the abbreviations stand for *infimum* and *supremum*.[6] The terms introduced by Pasch were not immediately successful although they were adopted by Hausdorff (1924, p. 146, 1927, p. 9); they gradually gained acceptance after they were used by Hahn (1932, p. 179), Tukey (as cited above), and Birkhoff (1940, p. 16).

References

Bennett, Albert A. (1936). *Lectures introductory to the theory of linear transformations* [manuscript]. Providence, R.I.

Birkhoff, Garrett (1938). "Lattices and their applications". *Bulletin of the American Mathematical Society* 44, pp. 793–800.

Birkhoff, Garrett (1940). *Lattice Theory*. New York.

Bolzano, Bernhard (1817). *Rein analytischer Beweis des Lehrsatzes, daß zwischen je zwei Werthen, die ein entgegengesetztes Resultat gewähren, wenigstens eine reelle Wurzel der Gleichung liege*. Prag.

Hahn, Hans (1932). *Reelle Funktionen*. Erster Teil: Punktfunktionen. Leipzig.

Harkness, James and Frank Morley (1893). *A Treatise on the Theory of Functions*. New York and London.

Hausdorff, Felix (1924). "Die Mengen G_δ in vollständigen Räumen". *Fundamenta Mathematicae* 6, pp. 146–148.

Hausdorff, Felix (1927). *Mengenlehre*. Zweite, neubearbeitete Auflage. Leipzig.

Pasch, Moritz (1882). *Einleitung in die Differential- und Integralrechnung*. Leipzig.

Pasch, Moritz (1887). "Ueber einige Punkte der Functionentheorie". *Mathematische Annalen* 30, pp. 132–154.

Pasch, Moritz (1914). *Veränderliche und Funktion*. Leipzig.

Pasch, Moritz (1915). "Schranken und Grenzen". *Monatshefte für Mathematik* 26, pp. 303–308.

Pringsheim, Alfred (1898). "Irrationalzahlen und Konvergenz unendlicher Prozesse". *Encyklopädie der mathematischen Wissenschaften mit Einschluss ihrer Anwendungen* I A 3, pp. 49–146.

Russ, Steve (2004). *The Mathematical Works of Bernard Bolzano*. Oxford.

Stolz, Otto (1881). "B. Bolzano's Bedeutung in der Geschichte der Infinitesimalrechnung". *Mathematische Annalen* 18, pp. 255–279.

Tukey, John W. (1940). *Convergence and Uniformity in Topology*. Princeton.

Veblen, Oswald and Nels Johann Lennes (1907). *Introduction to Infinitesimal Analysis*. Functions of a real variable. New York.

Weierstraß, Karl (1988). *Einleitung in die Theorie der analytischen Funktionen*. Vorlesung Berlin 1878 in einer Mitschrift von Adolf Hurwitz. Ed. by Peter Ullrich. Braunschweig.

infinite

The usual term for 'infinite' in Greek mathematics and philosophy was *ἄπειρος*, derived from *πεῖρας* (*πέρας*) 'end, limit, boundary'. According to Mugler (1958-59, p. 70), the word is used with respect to figures that have no limit. In this sense, the word appears, for example, in Euclid's definition of *parallel*: "*Parallel* straight lines are straight lines which, being in the same plane and being produced indefinitely in

---

[5] In view of the terminological variation displayed in German texts, it is not surprising that in English texts there is similar variation. For example, Harkness and Morley (1893, p. 48) use *upper limit*, and attribute the term to Weierstraß. Veblen and Lennes (1907, p. 3) use *least upper* and *greatest lower bound*. The generally confusing terminological situation is well described by Pringsheim (1898, p. 72). [6] He already used the abbreviations in Pasch (1914, p. 43).

both directions, do not meet one another in either direction".[1] Mugler also points out that the term was used for the indefinite repetition of a geometric operation. Additionally, the term was in common use in philosophical texts.

The Latin word *infinitus* is an exact correspondence of the Greek (*finis* meaning 'boundary'). Boethius, for example, used the term in the sense that a procress can be carried on indefinitely (e. g. Boethius 1867, p. 35). Isidore of Seville wrote: "It is quite certain that numbers are 'without limit' (infinitus), since at whatever number you think the limit has been reached, that same number can be increased" (Isidore of Seville 1911, book III, iii).[1]

French *infini* was used by Oresme in his *Le livre du ciel et du monde*; he wrote, for example: "Just as it is impossible that a straight line or a length or a width having a limit should be infinite, just so is it impossible that any figure, for instance, a triangle, should be infinite, since it is enclosed and finished and limited in every direction" (Oresme 1968, p. 102).[2] The English term came up in mathematics in the 16th century; Billingsley used it in his translation of the *Elements*, for example in the definition of *parallel*: "Parallel or equidistant right lines are such, which being in one and the selfe same superficies, and produced infinitely on both sydes, do neuer in any part concurre" (Euclid 1570, fol. 5^{v}). The same use is found in L. Digges and T. Digges (1571, fol. B iv^{r}) and, for German *unendlich*, in Widmann (1489, fol. C 6^{v}).

Precise mathematical definitions are found in the later half of the 19th century.[3] Such a definition was given by Dedekind. In the 1870s, he wrote first drafts of his famous book on numbers, which contains, as Moore (1982, p. 24) remarks, "the first adequate definition of finite set that did not presuppose the natural numbers":

> A system S is called infinite (or: the number of the things contained in S is called infinitely great), when there is a part U of S, which is different from S, and in which S can be mapped distinctly; the system S is called finite (or: S consists of a finite number of things), when there is no part U of S different from S, in which S is distinctly mappable (Dugac 1976, p. 294).[4]

In print, this definition appears in a shortened form in Dedekind (1888, p. 17); Dedekind there notes that he communicated it to Cantor in September 1882 and, several years earlier, also to Schwarz and Weber.

At about the same time, Cantor (1879, p. 4) introduced the new term *countable* (German *abzählbar*), mentioning "the class of point sets enumerable to infinity" by

---

[1] Translation from Euclid (1956, p. 154). [1] "Numeros autem infinitos esse certissimum est, quoniam in quocumque numero finem faciendum putaveris". Translation from Isidore of Seville (2006, p. 93). [2] "Item, aussi comme c'est impossible que ligne droite ou longitude ou latitude qui a fins et termes soit infinie, semblablement est ce imposible que quelconque figure soit infinie, si comme seroit un triangle, quar il est comprins et finy et terminé de toutes pars". Translation from Oresme (1968, p. 102). [3] For a discussion of Bolzano's work on infinity, see, e. g., Bellomo (2022, ch. 5). [4] "Ein System S heisst ein unendliches (oder: die Anzahl der in S enthaltenen Dinge heisst unendlich gross), wenn es einen Theil U von S giebt, welcher von S verschieden ist, und in welchem S sich deutlich abbilden lässt; das System S heisst endliches (oder: S besteht aus einer endlichen Anzahl von Dingen), wenn es keinen von S verschiedenen Theil U von S giebt, in welchem S deutlich abbildbar ist". Translation from Ferreirós (2007, p. 109).

which he meant "those point sets that have the same power as the natural series of numbers".[5] For details, see *countable.*

For further terms involving infinity, see *point at infinity*, *sequence* and *series*, and *infinitesimal*.

Note. Concise overviews of the concept are found in Easwaran et al. (2021) and Pannenberg (2001). For the Greek terminology, see also Mugler (1958-59, pp. 69, 70), for the Latin term, see Guillaumin (2020, pp. 157f.).

References

Bellomo, Anna (2022). "Sums, Numbers and Infinity". Collections in Bolzano's Mathematics and Philosophy. PhD thesis. Universiteit van Amsterdam.

Boethius, Anicius Manlius Severinus (1867). *De institutione arithmetica libri duo, De institutione musica libri quinque*. Accedit Geometria quae fertur Boetii. Ed. by Gottfried Friedlein. Leipzig.

Cantor, Georg (1879). "Ueber unendliche, lineare Punktmannichfaltigkeiten". *Mathematische Annalen* 15, pp. 1–7.

Dedekind, Richard (1888). *Was sind und was sollen die Zahlen?* Braunschweig.

Digges, Leonard and Thomas Digges (1571). *A geometrical practise, named Pantometria diuided into three bookes, Longimetra, Planimetra, and Stereometria*. London.

Dugac, Pierre (1976). *Richard Dedekind et les fondements des mathématiques (avec de nombreux textes inédits)*. Paris.

Easwaran, Kenny et al. (2021). "Infinity". In: *The Stanford Encyclopedia of Philosophy*. Ed. by Edward N. Zalta. Winter 2021. URL: `https://plato.stanford.edu/archives/win2021/entries/infinity/` (retrieved Jan. 6, 2022).

Euclid (1570). *The Elements of Geometrie of the most auncient Philosopher Euclide of Megara*. Faithfully (now first) translated into the Englishe toung, by H. Billingsley. London.

Euclid (1956). *The Thirteen Books of Euclid's Elements*. Volume I. Introduction and Books I, II. Ed. by Thomas L. Heath. New York.

Ferreirós, José (2007). *Labyrinth of Thought*. A History of Set Theory and Its Role in Modern Mathematics. Second revised edition. Basel.

Guillaumin, Jean-Yves (2020). *Dictionnaire de la terminologie latine ancienne de l'arithmétique et de la géométrie*. Paris.

Isidore of Seville (1911). *Etymologiarum sive Originum libri XX*. Tomus I. Ed. by Wallace M. Lindsay. Oxford.

Isidore of Seville (2006). *The* Etymologies *of Isidore of Seville*. Ed. by Stephen A. Barney et al. Cambridge.

Moore, Gregory H. (1982). *Zermelo's Axiom of Choice*. Its Origins, Development, & Influence. New York.

Mugler, Charles (1958-59). *Dictionnaire historique de la terminologie géométrique des Grecs*. Paris.

Oresme, Nicole (1968). *Le livre du ciel et du monde*. Edited by Albert D. Menut and Alexander J. Denomy, translated with an introduction by Albert D. Menut. Madison.

Pannenberg, Wolfhart (2001). "Unendlichkeit". In: *Historisches Wörterbuch der Philosophie*. Band 11. Ed. by Joachim Ritter, Karlfried Gründer, and Gottfried Gabriel. Darmstadt, pp. 140–146.

Widmann, Johannes (1489). *Behende vnd hubsche Rechenung auff allen kauffmanschafft*. Leipzig.

infinitesimal

This term is made up of Latin *infinitus* 'infinite' and *-esimus*, by analogy with words such as *centesimus* 'hundredth'. A terminological overview is given by Malet (1997, p. 73):

[5] "die Classe der in's Unendliche anzählbaren Punktmengen [...], d. h., diejenigen Punktmengen, welche mit der natürlichen Zahlenreihe: $1, 2, 3, \ldots, v, \ldots$ gleiche Mächtigkeit haben".

> Although the notion of infinitesimals was clearly introduced into mainstream mathematical thought by the middle of the seventeenth century, the word "infinitesimal" is almost never to be found in seventeenth century texts. More often than not, the term used is "indivisible", although the French also used "infiniment petit". In Latin it was common to refer to parts that were "infinite parvae" or "exiguae" [infinitely small or scanty].[1]

Leibniz appears to have used the term from spring 1673 on,[2] but "[t]he terminology of Leibniz varies: in most cases he uses 'smaller than any assignable' ('minor assignabili' or 'inassignabilis') which occurs nearly one hundred times [...]; after this comes 'infinitely small' (about eighty times). In contrast, 'indivisible' and 'infinitesimal' ('infinitesima') are relatively rare, occurring about twenty-five and ten times respectively" (Probst 2008, p. 95).

Leibniz later recalled that Mercator had used this term: "for the calculus it is useful to imagine quantities infinitely small, or as Nicholas Mercator called them, infinitesimal" (Leibniz 1971, p. 63).[1] Mercator (1668a, pp. 31f.), however, uses *pars infinitissima* (*infinitissima* is the superlative form of *infinitus* 'infinite') in his book *Logarithmo-Technia*. The expression also appears in a brief "illustration" of this book (Mercator 1668b, p. 763), but at the very end of this article there is a correctional note to the effect that "if the reader offends to *infinitissima*, read *infinitesima*".[2]

At the same time, Wallis begins to use the term, for example in two letters dated "February 1667/68" (Beeley and Scriba 2005, pp. 411, 426).[3] In print, the term appears in his *Mechanica*, where he writes "$\frac{1}{\infty}$ (pars infinitesima)" (Wallis 1670, p. 110), and also in his *Arithmetica Infinitorum* (Wallis 1693, p. 367).[4]

In sum, it appears that Wallis used the term before Mercator, but only in correspondence, and that the first printed attestation is a note to Mercator's brief account of his *Logarithmo-Technia*. Afterwards, Leibniz used the term quite regularly.

In the second half of the 20th century, infinitesimals were revived by Robinson. He remarked that "the use of infinitesimals, strongly advocated by Leibnitz [sic] and unhesitatingly accepted by Euler fell into disrepute after the advent of Cauchy's methods which put Mathematical Analysis on a firm foundation" (Robinson 1961, p. 433) and presented the following definition (with R_0 the set of real numbers, and R^* any non-standard model of analysis):

> M_0, the set of all $a \in R^*$ such that $|a| < r$ for some $r \in R_0$. M_0 is a ring. The elements of M_0 will be said to be *finite*.

---

[1] That should not be taken to assume that infinitesimals and indivisibles (Cavalieri 1635) were the same thing. On the differences, see, e. g., Malet (1997, p. 73) and Schubring (2006, pp. 159f.).

[2] As stated by Probst (2008, p. 103), who also provides various attestations in Leibniz' writings.

[1] "pro calculo utile est fingere quantitates infinite parvam seu ut Nicolaus Mercator vocabat, infinitesimales". Translation from Jesseph (1998, p. 27). Leibniz' statement is from a letter to Wallis, dated 30 March 1699. [2] "ubicunque, Lector offenderit *infinitissimam*, legat *infinitesimam*". This passage is also pointed out in Leibniz (2015, p. 91) and Probst (2018, pp. 200–202). See also Probst (2008, p. 103): "[Mercator] does not use the term 'infinitesimal', but instead 'pars infinitissima' and he does so for both numbers and for lines [...] Mercator's expression signifies a minimal quantity and is therefore terminologically still close to Cavalieri's indivisibles, although he in fact employs infinitesimal quantities". [3] The latter passage, and some other attestations cited here, are also pointed out by Probst (2018, p. 201). [4] The term does not appear in the first edition of the *Arithmetica Infinitorum*, where *pars infinite parva* is used (Wallis 1656, p. 5). See Beeley (2008, p. 41) and Stedall (2004, xvii, fn. 27), who points out that Wallis' use of *infinitesimal* is "occasional".

> M_1, the set of all $a \in R^*$ such that $|a| < r$ for all positive $r \in R_0$. The elements of M_1 will be said to be *infinitesimal* (Robinson 1961, p. 434).[5]

In this context, the elements of R^* were later called *hyperreal numbers*. This terminology came up in the 1970s and spread through Keisler (1976b, p. 2). Keisler (1976a, p. 874) also remarked that *hyperreal* was due to Hewitt (1948, p. 74), who used it in an algebraic context with respect to ideals, not numbers.

Note. Detailed studies on the terminology are found in Probst (2008) and Probst (2018, pp. 200–202). For explanations of various related terms, see also Leibniz (2008, pp. xviii–xxi).

References

Beeley, Philip (2008). “Infinity, Infinitesimals, and the Reform of Cavalieri: John Wallis and his Critics”. In: *Infinitesimal Differences*. Controversies between Leibniz and his Contemporaries. Ed. by Ursula Goldenbaum and Douglas Jesseph. Berlin and New York, pp. 31–52.

Beeley, Philip and Christoph J. Scriba, eds. (2005). *The Correspondence of John Wallis (1616-1703)*. Volume II (1660 – September 1668). Oxford.

Cavalieri, Bonaventura (1635). *Geometria indivisibilibus continuorum nova quadam ratione promota*. Bologna.

Hewitt, Edwin (1948). “Rings of Real-Valued Continuous Functions. I”. *Transactions of the American Mathematical Society* 64, pp. 45–99.

Jesseph, Douglas M. (1998). “Leibniz on the Foundations of the Calculus: The Question of the Reality of Infinitesimal Magnitudes”. *Perspectives on Science* 6, pp. 6–40.

Keisler, H. Jerome (1976a). *Elementary Calculus*. Boston.

Keisler, H. Jerome (1976b). *Foundations of Infinitesimal Calculus*. Boston.

Leibniz, Gottfried Wilhelm (1971). *Mathematische Schriften*. Herausgegeben von C.I. Gerhardt. Band IV. Hildesheim and New York.

Leibniz, Gottfried Wilhelm (2008). *Sämtliche Schriften und Briefe*. Siebente Reihe. Mathematische Schriften. Vierter Band. 1670-1673. Infinitesimalmathematik. Berlin.

Leibniz, Gottfried Wilhelm (2015). *Sämtliche Schriften und Briefe*. Dritte Reihe. Mathematischer, naturwissenschaftlicher und technischer Briefwechsel. Achter Band. Januar 1699-Dezember 1701. Hannover.

Malet, Antoni (1997). “Barrow, Wallis, and the Remaking of Seventeenth Century Indivisibles”. *Centaurus* 39, pp. 67–92.

Mercator, Nicolaus (1668a). *Logarithmo-Technia, sive Methodus construendi Logarithmos Nova, accurata, & facilis*. London.

Mercator, Nicolaus (1668b). “Some illustration of the *Logarithmotechnia* of Mercator”. *Philosophical Transactions of the Royal Society of London* 3, pp. 759–764.

Probst, Siegmund (2008). “Indivisibles and Infinitesimals in Early Mathematical Texts of Leibniz”. In: *Infinitesimal Differences. Controversies between Leibniz and his Contemporaries*. Ed. by Ursula Goldenbaum and Douglas Jesseph. Berlin and New York, pp. 95–106.

Probst, Siegmund (2018). “The relation between Leibniz and Wallis: an overview from new sources and studies”. *Quaderns d'història de l'enginyeria* 16, pp. 189–208.

Robinson, Abraham (1961). “Non-standard Analysis”. *Proceedings of the Royal Academy of Sciences of Amsterdam A* 64, pp. 432–440.

Schmieden, Curt and Detlef Laugwitz (1958). “Eine Erweiterung der Infinitesimalrechnung”. *Mathematische Zeitschrift* 69, pp. 1–39.

Schubring, Gert (2006). *Conflicts Between Generalization, Rigor, and Intuition*. Number Concepts Underlying the Development of Analysis in 17th-19th Century France and Germany. New York.

Stedall, Jacqueline A. (2004). *The Arithmetic of Infinitesimals*. John Wallis 1656. Translated from Latin to English with an Introduction. New York.

---

[5] A slightly earlier redefinition of *infinitesimal* is found in Schmieden and Laugwitz (1958, p. 7).

Wallis, John (1656). *Arithmetica infinitorum*. In: *Operum Mathematicorum*. Pars altera. Oxford, pp. 1–198.
Wallis, John (1670). *Mechanica*. Sive, De Motu, Tractatus Geometricus. London.
Wallis, John (1693). *Arithmetica Infinitorum*. In: *Opera Mathematica*. Volumen primum. Oxford, pp. 355–478.

inflection point

Inflection points, points indicating a change of curvature of a curve, have been dealt with since the 17th century. The terms that were suggested at that time all contain the notion of bending or turning (the Latin verb *flectere* means 'to bend, to turn (around)' (LS)).

In his manuscript "Methodus ad disquirendam maximam et minimam" (method for the study of maxima and minima), which was finished in the 1630s but not published until 1679, Fermat, discussing the conchoid of Nicomedes, suggested that "to be able to draw the curve well", it would be suitable "to study the points of inflection [*puncta inflexionum*], at which the curvature changes from convex to concave, and conversely" (de Fermat 1679, p. 73).[1] This passage seems to be the earliest one in which inflection points are mentioned and named.[2]

A slightly different terminology is found in Huygens (1654, p. 69), who searched for "the points where the contrary flexure begins".[3] The term *flexus contrarii* which Huygens used was then adopted by de Sluse (1668, p. 117) as "point of contrary flexure" (*punctum flexus contrarii*), a model which various other mathematicians such as Leibniz (1684, p. 468) followed.

In modern languages, the French term *point d'inflexion* was used, without definition, by Ozanam (1684, p. 194); definitions *are* found in Ozanam (1691, p. 9) and in de l'Hôpital (1696, p. 59). In English, *point of flexure* is found in a review of de Sluse's book cited above ([anon.] 1670, p. 907). The term *Point of Inflexion* was used by Raphson (1702, p. 17); Hayes (1704, p. 153) calls it the *Point of contrary Flexion*. The German term *Wendungs-Punct*, lit. 'turning point', was suggested by Wolff (1710, p. 340).

Note. For a concise discussion of the term, see Berzolari (1915, p. 278). Miller (n.d.) gives additional information on the history of concept.

References

[anon.] (1670). "An Account of Two Books". *Philosophical Transactions of the Royal Society of London* 4, pp. 903–912.
Berzolari, Luigi (1915). "Théorie générale des courbes planes algébriques". *Encyclopédie des sciences mathématiques pures et appliquées* III.3, pp. 257–304.
de Fermat, Pierre (1679). "Methodus ad disquirendam maximam & minimam". In: *Varia opera mathematica*. Toulouse, pp. 63–73.
Hayes, Charles (1704). *A Treatise of Fluxions*. London.
de l'Hôpital, Guillaume François Antoine (1696). *Analyse des infiniments petits, pour l'intelligence des lignes courbes*. Paris.
Huygens, Christiaan (1654). *De circuli magnitudine inventa*. Leiden.

---

[1] "ut perfecte curva possit delineari, investiganda sunt ex arte puncta inflexionum, in quibus curvatura ex convexa fit concava vel contra". Translation from Mahoney (1994, p. 201). [2] See, e. g., Berzolari (1915, p. 278) and Miller (n.d.). [3] "quaesitum est ratione puncta ea determinari possent ubi contraria flexio initium capit".

Leibniz, Gottfried Wilhelm (1684). "Nova Methodus pro maximis et minimis, itemque tangentibus, quae nec fractas nec irrationales quantitates moratur, et singulare pro illis calculi genu". *Acta Eruditorum* (Oct. 1684), pp. 467–473.
Mahoney, Michael S. (1994). *The Mathematical Career of Pierre de Fermat, 1601–1665*. Second Edition. Princeton.
Miller, Jeff (n.d.). "Earliest Known Uses of Some of the Words of Mathematics". URL: `https://mathshistory.st-andrews.ac.uk/Miller/mathword/`.
Ozanam, Jacques (1684). *La géométrie pratique*. Paris.
Ozanam, Jacques (1691). *Dictionnaire mathematique*. Paris.
Raphson, Joseph (1702). *A Mathematical Dictionary*. Abridg'd from Monsieur *Ozanam*, and Others. London.
de Sluse, René-François (1668). *Mesolabum*. Liège.
Wolff, Christian (1710). *Der Anfangs-Gründe aller Mathematischen Wissenschaften Letzter Theil*. Halle.

injective

In his *What are numbers and what are they for*, Dedekind (1888, p. 8) defined an injective mapping very much along modern lines, using the names *ähnlich* 'similar' or *deutlich* 'distinct, clear': "A mapping φ of a system[1] S is called *similar* (or *distinct*), if to different elements a, b of the system S always correspond different images $a' = \varphi(a)$, $b' = \varphi(b)$".[2] Dedekind already used the terminology in the early 1870s; see Sieg and Schlimm (2017) and Ferreirós (2007, p. 108).

The terms *injective function* (*application injective*) and *injection* were coined by Bourbaki, who first used them in Bourbaki (1954, p. 80). The new terms spread quickly. In English, *injective* is found in Igusa (1956, p. 191); Chevalley (1956, p. 1) also uses the noun *injection*. German *injektiv* was used by Scheja (1958, p. 11).

Note that while Bourbaki also coined the terms *bijective* and *surjective*, the history of *injective* is somewhat different. The reason is that *injective* and *injection* had been in use before Bourbaki (1954) in different senses. In particular, as of 1950, *injection* was already in common use in the sense defined by Mac Lane (1950, p. 488): "If S is a subgroup of G (notation $S \subset G$), then the injection $\kappa = [G \supset S]$ of S into G is that homomorphism of S into G with $\kappa(s) = s$ for every $s \in S$". The term had already been used in this sense by Lefschetz (1942, p. 113).

References

Bourbaki, Nicolas (1954). *Théorie des ensembles*. Chapitre I. Déscription de la mathématique formelle. Chapitre II. Théorie des ensembles. Paris.
Chevalley, Claude (1956). *Fundamental Concepts of Algebra*. New York.
Dedekind, Richard (1888). *Was sind und was sollen die Zahlen?* Braunschweig.
Dedekind, Richard (1909). *Essays on the Theory of Numbers*. I. Continuity and Irrational Numbers, II. The Nature and Meaning of Numbers. Authorized translation by Wooster Woodruff Beman. Chicago.
Ferreirós, José (2007). *Labyrinth of Thought*. A History of Set Theory and Its Role in Modern Mathematics. Second revised edition. Basel.
Igusa, Jun-Ichi (1956). "Fibre Systems of Jacobian Varieties". *American Journal of Mathematics* 78, pp. 171–199.
Lefschetz, Solomon (1942). *Algebraic Topology*. Providence, R.I.

---

[1] *System* was Dedekind's term for *set*. [2] "Eine Abbildung φ eines Systems S heißt *ähnlich* (oder *deutlich*), wenn verschiedenen Elementen a, b des Systems S stets verschiedene Bilder $a' = \varphi(a)$, $b' = \varphi(b)$ entsprechen". Translation based on Dedekind (1909, p. 53).

Mac Lane, Saunders (1950). "Duality for groups". *Bulletin of the American Mathematical Society* 56, pp. 485–516.
Scheja, Günter (1958). "Verzweigte Holomorphiehüllen". *Sitzungsberichte der Bayerischen Akademie der Wissenschaften (mathematisch-naturwissenschaftliche Klasse)* (1958), pp. 8–18.
Sieg, Wilfried and Dirk Schlimm (2017). "Dedekind's Abstract Concepts: Models and Mappings". *Philosophia Mathematica* 25, pp. 292–317.

injective module and *injective resolution*

In their influential book *Homological Algebra*, Cartan and Eilenberg (1956, p. viii) defined injective modules and injective resolutions as follows: "A module C is *injective* if every homomorphism $B' \to C$ admits an extension $B \to C$ for each module B containing B' as a submodule. An injective resolution of C is an exact sequence $0 \to C \to C^0 \to C^1 \to \cdots \to C^n \to C^{n+1} \to \cdots$ with C^i injective for $i = 0, 1, 2, \ldots$".

For the concept of injective module, Cartan and Eilenberg (1956, p. 10) refer to Baer (1940). Baer did not use the term, though, which is first found as French *module injectif* in Eilenberg (1952, p. 379).[1] The term was quickly adopted, e. g. as German *injektiver Modul* in Eckmann and Schopf (1953, p. 75), who also refer to Baer for the concept.

Before the publication of *Homological Algebra*, the term *injective resolution* is found in Buchsbaum (1955, p. 33), who refers to Cartan and Eilenberg's book.

Note. Some remarks on the history of the concept can be found in Weibel (1999) and Krömer (2007, ch. 3).

References

Baer, Reinhold (1940). "Abelian groups that are direct summands of every containing abelian group". *Bulletin of the American Mathematical Society* 46, pp. 800–806.
Buchsbaum, David A. (1955). "Exact Categories and Duality". *Transactions of the American Mathematical Society* 80, pp. 1–34.
Cartan, Henri and Samuel Eilenberg (1956). *Homological Algebra*. Princeton.
Eckmann, Beno and Andreas Schopf (1953). "Über injektive Moduln". *Archiv der Mathematik* 4, pp. 75–78.
Eilenberg, Samuel (1950-51). "Homologie des groupes. I: axiomes et unicité". *Séminaire Henri Cartan* 3, exposé 1, pp. 1–5.
Eilenberg, Samuel (1952). "Foncteurs de modules et leurs satellites". *Séminaire Bourbaki* 1, exposé 46, pp. 379–381.
Krömer, Ralf (2007). *Tool and Object*. A History and Philosophy of Category Theory. Basel.
Weibel, Charles A. (1999). "History of Homological Algebra". In: *History of Topology*. Ed. by Ioan Mackenzie James. Amsterdam, pp. 797–836.

inner automorphism

An inner automorphism of a group is of the form $\phi(x) = g^{-1}xg$, for some fixed element g of the group. In this sense, the term was introduced by Frobenius (1901, p. 1324), who used German *innerer Automorphismus*. If the condition was not fulfilled, Frobenius called the automorphism *outer* (*äußerer*).

Shortly afterwards, Miller (1902, p. 120) used English *inner* and *outer automorphism*. He mentioned Frobenius' authorship of the term, and also pointed out that

[1] This usage was pointed out by Krömer (2007, p. 139). The term *groupe injectif* is earlier found in Eilenberg (1950-51, p. 1). The groups considered there are modules. Weibel (1999, p. 814) quotes this text as the earliest use of *injective* in this sense.

cogredient was an alternative used by Hölder (1893, p. 314, 1895, p. 324). Miller also commented that Frobenius' "new terms have the advantage of being shorter and perhaps more natural than the older ones".

Hölder (1893, p. 314) himself stated that he took over the terminology from Klein (1884, p. 232), who used it in narrower context. It was earlier used by Gierster (1881, p. 354), for example.[2] Hölder, Klein, and Gierster all used *contragredient* for *outer*.

References

Frobenius, Georg (1901). "Über auflösbare Gruppen. V". *Sitzungsberichte der Königlich Preußischen Akademie der Wissenschaften zu Berlin* (1901), pp. 1324–1329.

Gierster, Joseph (1881). "Die Untergruppen der Galois'schen Gruppe der Modulargleichungen für den Fall eines primzahligen Transformationsgrades". *Mathematische Annalen* 18, pp. 319–365.

Gordan, Paul (1878). "Ueber die Auflösung von Gleichungen vom fünftem Grade". *Mathematische Annalen* 13, pp. 375–404.

Hölder, Otto (1893). "Die Gruppen der Ordnungen p^3, pq^2, pqr, p^4". *Mathematische Annalen* 43, pp. 301–412.

Hölder, Otto (1895). "Bildung zusammengesetzter Gruppen". *Mathematische Annalen* 46, pp. 321–422.

Klein, Felix (1884). *Vorlesungen über das Ikosaeder*. Leipzig.

Miller, George Abram (1902). "Second report on recent progress in the theory of groups of finite order". *Bulletin of the American Mathematical Society* 9, pp. 106–123.

integer

In Greek arithmetic, *number* generally referred to natural numbers. Further specifications such as *whole* are therefore not commonly found. A late example is found in the works of Diophantus, who used *ὁλόκληρος* (lit., 'complete in all parts'): "And if you prefer to use integers only, avoiding the $\frac{1}{2}$, multiply throughout by 4" (Tannery 1893, p. 306).[1]

Arabic *ṣaḥīḥ* also literally means 'entire' or 'complete'. According to Rashed (2017, p. 488), the term *ʿadad ṣaḥīḥ* 'whole number' was used by Abū Kāmil (9th/10th c.), for example. The Arabic terminology had an important influence on the Latin one.

Latin *numerus integer* is found in this sense in the 12th century. In the *Liber Mahameleth*, a translation from the Arabic, the following definition is given: "You should know that a number is sometimes considered by itself, without reference to another, in which case it is called an integer; sometimes in relation to another, that is, as a part or parts, in which case it is called a fraction" (Sesiano 2014, p. 28).[2] This term was also used by Fibonacci in his *Liber abbaci* (Fibonacci 1857, e. g., p. 7); sometimes, he also uses *sanus* 'sound whole' (Fibonacci 1857, e. g., p. 264). The latter term was common for some time; it was used in a Spanish translation dating from the second half of the 13th century, for example (Bossong 1978, p. 307).

The Italian term appeared as *numero intero* in the 14th century in Paolo dell'Abbaco's *Trattato di aritmetica* (dell'Abbaco 1964, p. 24).[3] In a manuscript written in

[2] Gierster attributed the term to Gordan (1878), where it is not found, however. [1] Translation from Thomas (1957, p. 535). [2] "Sciendum est autem quod numerus aliquando consideratur per se, sine respectu alterius, et dicitur integer, aliquando in comparatione alterius, scilicet ut pars vel partes, et dicitur fractio". Translation from Sesiano (2014, p. 605). [3] This early usage was pointed out by Ricci (1994, p. 40), who also gives further references.

Occitan, dating from around 1430, *nombre entier* is used (Sesiano 2018, p. 215). The French term was used in the second half of the 15th century, for example by Chuquet in the 1480s (Chuquet 1881, e. g., p. 52) and, slightly earlier, in a manuscript from around 1449 ([anon.] c1449, fol. 104ᵛ).[4] The German term *ganze Zahl* (*ganz* 'entire') is also found in various 15th-century texts (Schirmer 1912, p. 25). For English *whole number*, the OED cites the 15th-century *Crafte of Nombrynge* ([anon.] 1922, p. 17); *integer* was used by Thomas Digges (L. Digges and T. Digges 1571, fol. V ijᵛ).

Clear examples that explicitly include the negative numbers are of later date; Euler (1771, § 20) made a statement to his effect: "All these numbers, whether positive or negative, have the known appellation *whole numbers*, which consequently are either greater or less than nothing. We call them *whole numbers* to distinguish them from broken ones, and from several other kinds of numbers of which we shall hereafter speak".[1]

Various extensions of the concept can be found in the 19th century. For example, Gauß (1831, p. 628) introduced *complex integers*: "The author calls every quantity $a + bi$, where a and b denote real quantities and where i is an abbreviation for $\sqrt{-1}$, a complex integer, provided that a and b are integers".[2] These complex integers were later also called *Gaussian integers*. The term may be attributed to Dedekind (1877, p. 38), who called them *nombres entiers complexes de Gauss*. The actual epithet *Gaussian* in this connection was used by Weber (1886, pp. 251f.).

For another generalisation, see *algebraic integer*.

Note. Some remarks on the history of the term can be found in Tropfke (1980, p. 128). For the Arabic, see Rashed (2017, p. 488).

References

[anon.] (c1449). "[Traitté d'arithmetique]". Manuscript, BNF Français 7287. URL: https://gallica.bnf.fr/ark:/12148/btv1b525140414.

[anon.] (1922). "The Craft of Nombrynge". In: *The Earliest Arithmetics in English*. Ed. by Robert Steele. London, pp. 3–32.

Benoit, Paul (1985). "Recherches sur le vocabulaire des opérations élémentaires dans les arithmétiques en langue française de la fin du Moyen Âge". *Documents pour l'histoire du vocabulaire scientifique* 7, pp. 77–95.

Bossong, Georg, ed. (1978). *Los Canones de Albateni*. Herausgegeben sowie mit Einleitung, Anmerkungen und Glossar versehen. Tübingen.

Chuquet, Nicolas (1881). *Le Triparty en la Science des Nombres*. Ed. by Aristide Marre. Rome.

Dedekind, Richard (1877). *Sur la théorie des nombres entiers algébriques*. Paris.

dell'Abbaco, Paolo (1964). *Trattato d'aritmetica*. Ed. by Gino Arrighi. Pisa.

Digges, Leonard and Thomas Digges (1571). *A geometrical practise, named Pantometria diuided into three bookes, Longimetra, Planimetra, and Stereometria*. London.

Euler, Leonhard (1771). *Vollständige Anleitung zur Algebra*. Erster Theil. St. Petersburg.

[4] In a manuscript from around 1460, *nombre sain* is used alternatively (Benoit 1985, p. 83). The corresponding Italian *sano* is found in the 13th century (Feola 2008, p. 173).

[1] "Alle diese Zahlen, so wohl positive als negative, führen den bekannten Namen der ganzen Zahlen, welche also entweder größer oder kleiner sind als nichts. Man nennt dieselbe ganze Zahlen, um sie von den gebrochenen, und noch vielerley anderen Zahlen, wovon unten gehandelt werden wird, zu unterscheiden". Translation based on Euler (1822, § 20) (emphasis added).

[2] "Der Verf. nennt jede Größe $a + bi$, wo a und b reelle Größen bedeuten, und i der Kürze wegen anstatt $\sqrt{-1}$, geschrieben ist, eine complexe ganze Zahl, wenn zugleich a und b ganze Zahlen sind". Translation from Gauß (1996, p. 309).

Euler, Leonhard (1822). *Elements of Algebra*. Translated from the French by John Hewlett. Third edition. London.
Feola, Francesco (2008). *Gli esordi della geometria in volgare*. Un volgarizzamento trecentesco della *Practica Geometriae* di Leonardo Pisano. Firenze.
Fibonacci (1857). *Scritti di Leonardo Pisano*. Pubblicati da Baldassare Boncompagni. Volume I. Roma.
Gauß, Carl Friedrich (1831). "Theoria residuorum biquadraticorum, commentatio secunda [announcement]". *Göttingische gelehrte Anzeigen* (1831), pp. 625–638.
Gauß, Carl Friedrich (1996). "Notice on the Theory of Biquadratic Residues". In: *From Kant to Hilbert*. A Source Book in the Foundations of Mathematics. Volume I. Ed. by William Ewald. Oxford, pp. 306–313.
Rashed, Roshdi, ed. (2017). *Lexique historique de la langue scientifique arabe*. Hildesheim.
Ricci, Laura (1994). "Il lessico matematico della Summa di Luca Pacioli". *Studi di lessicografia italiana* 12, pp. 5–71.
Schirmer, Alfred (1912). *Der Wortschatz der Mathematik nach Alter und Herkunft untersucht*. Straßburg.
Sesiano, Jacques, ed. (2014). *Liber Mahameleth*. Cham.
Sesiano, Jacques (2018). *L'arithmétique de Pamiers*. Traité mathématique en langue d'oc du XVe siècle. Lausanne.
Tannery, Paul, ed. (1893). *Diophanti Alexandrini Opera Omnia*. Volumen I. Leipzig.
Thomas, Ivor (1957). *Selections Illustrating the History of Greek Mathematics*. II. From Aristarchus to Pappus. London and Cambridge, Mass.
Tropfke, Johannes (1980). *Geschichte der Elementarmathematik*. Band 1: Arithmetik und Algebra. Vollständig neu bearbeitet von Kurt Vogel, Karin Reich, Helmuth Gericke. Berlin and New York.
Weber, Heinrich (1886). "Theorie der Abel'schen Zahlkörper". *Acta Mathematica* 8, pp. 193–263.

integral

The term *integral* was first brought to the public by Jakob Bernoulli (see, e. g., Pier 1996, p. 44). It first appeared in print in Jakob Bernoulli (1690, p. 218), where Bernoulli had solved a certain differential equation and from the equality of the differentials concluded: "Hence also their their integrals will be made equal".[1]

Later, in a letter dated February 28, 1695, Leibniz tried to convince Johann Bernoulli to adopt his term *sum*[2]:

> I leave it to your deliberation if it would not be better in the future, for the sake of uniformity and harmony, not only between ourselves but in the whole field of study, to adopt the terminology of summation instead of your integrals. Then for instance $\int ydx$ would signify the sum of all y multiplied by the corresponding dx, or the sum of all such rectangles. I ask this primarily because in that way the geometrical summations, or quadratures, correspond best with the arithmetical sums or sums of sequences (Gerhardt 1855-56, p. 168).[3]

---

[1] "Ergo et horum Integralia aequantur". Translation from Sonar (2018, p. 327), where also a discussion of Bernoulli's paper is found. [2] Leibniz started using the symbol $\int$ in October 1675 (Leibniz 2008, p. xxxiii); he used this elongated *S* because he regarded integrals as sums (Latin *summa*). Leibniz wrote: "It will be useful to write $\int$ for omn., so that $\int l =$ omn. l, or the sum of the the l's" (Leibniz 2008, p. 292). Translation from Child (1920, p. 179). In print, the symbol and the term *summa* first appear in Leibniz (1686, p. 297). [3] "Unde Tibi deliberandum relinquo, annon, pro Integralibus vestris, praestet in posterum uniformitatis et harmoniae gratia non inter nos tantum, sed in ipsa doctrina adhiberi Summatorias expressiones, ira ut, exempli gratia, $\int ydx$ significet summam omnium y in dx respondentes ductorum, seu summam omnium hujusmodi rectangulorum: praesertim cum tali ratione summationes geometricae seu quadraturae optime cum arithmeticis seu serierum summis conferantur. [...] Ego certe in totam hanc methodum me fateor, ex hac consideratione reciprocationis inter summas differentiasque, incidisse, et a Seriebus numerorum ad linearum seu ordinatarum considerationes processisse". Translation from Bos (1974, p. 21).

Bernoulli's reply from April 30, 1695, is instructive in that it contains a comment on the origin of the term:

> as regards the terminology of the sum of differentials I shall gladly use in the future your terminology of summations instead of our integrals. I would have done so already much earlier if the term integral were not so much appreciated by certain geometers[4] who acknowledge me as the inventor of the term. It would therefore be thought that I rather obscured matters, if I indicated the same thing now with one term and now with another. I confess that indeed the terminology does not aptly agree with the thing itself (the term suggested itself to me as I considered the differential as the infinitesimal part of a whole or *integral*; I did not think further about it) (Gerhardt 1855-56, p. 172).[1]

Johann Bernoulli here attributes it to himself;[2] Jakob does not appear to have objected. Historians are divided on the issue of who created the term: Cantor (1901, p. 219) attributes it to Jakob because, he argues, Jakob would have cited Johann if the latter had coined the term; Pier (1996, p. 44), for example, follows Cantor in this assessment. Others attribute the term to Johann (e. g., Domingues 2008, p. 140), while still others remain neutral (e. g., Roero 2005, p. 55).

Johann Bernoulli (1697, p. 462) did fulfil Leibniz' request and also uses *sum* in print as an alternative to *integral*, but by that time others had also started using the term (e. g., de l'Hôpital 1696, preface) and *sum* soon dropped out of use.

German *Integral* appears to be due to Wolff (1710, p. 280), while in English, *integral calculus* was used by Craig (1706, p. 261) and the noun *integral* is found in Harris (1710, s. v. helicoid).[3] The terminology was also adopted by Euler, of course; he explained it thus:

> If any differential is given, that quantity whose differential is the proposed quantity is called its *integral*. The reason for this name is as follows: Since a differential can be thought of as an infinitely small part by which a quantity increases, that quantity with respect to which this is a part can be thought of as a whole, that is, integral, and for this reason is called an integral (Euler 1755, part I, § 139).[4]

Some milestones in the further development are Cauchy's (1823, p. 83) definition of the definite integral (see *definite integral*), Riemann's definition from 1854 (Riemann 1876, pp. 225–227) and its modification by Darboux (1875, pp. 72f.), as well as the definitions given by Stieltjes (1894) and Lebesgue (1901). An important

---

[4] According to Bos (1974, p. 21), this is "a reference to French mathematicians, especially l'Hôpital, who had studied Bernoulli's *Integral Calculus*". [1] "Caeterum, quod nomenclationem differentialium summae attinet, lubentissime pro integralibus nostris Tuas in posterum adhibeo summatorias expressiones; quod diu ante fecissem, si nomen integralium non adeo invaluisset apud quosdam Geometras, qui me hujus nominis authorem agnoscunt, ut satis obscurus visus fuissem, unam eandemque rem, nunc hoc, nunc alio nomine designans. Fateor enim nomenclationem istam (quae, considerando differentialem tanquam partem infinitesimam totius vel *integri*, mihi non ulterius cogitanti, venit in mentem) rei ipsi non apte convenire". Translation from Bos (1974, pp. 21f.).
[2] He repeats this in his autobiographical notes (Wolf 1859, p. 73). Leibniz (1695, p. 185) also regarded Johann as the inventor of the term. [3] There is no separate article for the word in Harris' *Lexicon*. [4] "Si enim propositum fuerit differentiale quodcunque, eius respectu ea quantitas, cuius est differentiale, vocari solet integrale. Cuius denominationibus ratio est, quod, cum differentiale considerai possit, tanquam pars infinite parva, qua quantitas quaepiam crescit, ipsa illa quantitas respect huius partis tanquam totum seu integrum spectari potest, hancque ob causam eius vocatur integrale". Translation from Euler (2000, part I, § 139).

example of a 20th-century generalization is due to Kurzweil (1957) and Henstock (1961).[5]

See also *primitive function* and *quadrature*.

Note. For the history of the notion, see Bos (1974), Knobloch (1983), Pier (1996), Hawkins (1979), Hochkirchen (2003), Roero (2005), and Domingues (2008, ch. 5).

References

Bernoulli, Jakob (1690). "Analysis problematis ante hac propositi, de inventione linea descensus a corpore gravi percurrendae uniformiter, sic ut temporibus aequales altitudines emetiatur: & alterius cujusdam Problematis Propositio". *Acta Eruditorum* (May 1690), pp. 217–219.

Bernoulli, Johann (1697). "Lettre de M. Bernoulli ... à M. Varignon". *Journal des sçavans* (Dec. 1697), pp. 458–465.

Bos, Henk J.M. (1974). "Differentials, Higher-Order Differentials and the Derivative in the Leibnizian Calculus". *Archive for History of Exact Sciences* 14, pp. 1–90.

Cantor, Moritz (1901). *Vorlesungen über Geschichte der Mathematik*. Dritter Band. Von 1668 bis 1758. Leipzig.

Cauchy, Augustin-Louis (1823). *Résumé des leçons données à l'École Royale Polytechnique sur le calcul infinitésimal*. Tome premier. Paris.

Child, James Mark (1920). *The Early Mathematical Manuscripts of Leibniz*. Chicago and London.

Craig, John (1706). "Of the Quadratures of Geometrically irrational Figures". *Miscellanea Curiosa* 2, pp. 251–262.

Darboux, Gaston (1875). "Mémoire sur les fonctions discontinues". *Annales scientifiques de l'École Normale Supérieure* 4, pp. 57–112.

Domingues, João Caramalho (2008). *Lacroix and the Calculus*. Basel.

Euler, Leonhard (1755). *Institutiones calculi differentialis*. St. Petersburg.

Euler, Leonhard (2000). *Foundations of Differential Calculus*. Translated by John D. Blanton. New York.

Gerhardt, C.I., ed. (1855-56). *Leibnizens mathematische Schriften*. Erste Abtheilung. Band III. Halle.

Harris, John (1710). *Lexicon Technicum*. Vol. II. London.

Hawkins, Thomas (1979). *Lebesgue's Theory of Integration*. Its Origins and Development. Second Edition. Providence, R.I.

Henstock, Ralph (1961). "Definitions of Riemann Type of the Variational Integrals". *Proceedings of the London Mathematical Society* 11, pp. 402–418.

Hochkirchen, Thomas (2003). "Theory of Measure and Integration from Riemann to Lebesgue". In: *A History of Analysis*. Ed. by Hans Niels Jahnke. Providence, R.I., pp. 261–290.

de l'Hôpital, Guillaume François Antoine (1696). *Analyse des infiniments petits, pour l'intelligence des lignes courbes*. Paris.

Knobloch, Eberhard (1983). "Von Riemann zu Lebesgue – Zur Entwicklung der Integrationstheorie". *Historia Mathematica* 10, pp. 318–343.

Kurzweil, Jaroslav (1957). "Generalized ordinary differential equations and continuous dependence on a parameter". *Czechoslovak Mathematical Journal* 7, pp. 418–449.

Lebesgue, Henri (1901). "Sur une généralisation d'intégrale définie". *Comptes rendus de l'Académie des sciences* 132, pp. 1025–1027.

Leibniz, Gottfried Wilhelm (1686). "De Geometria recondita et Analysi Indivisibilium et infinitorum". *Acta Eruditorum* (June 1686), pp. 292–300.

Leibniz, Gottfried Wilhelm (1695). "Notatiuncula ad Constructiones Lineae in qua Sacoma, aequilibrium cum pondere moto faciens indedere debet". *Acta Eruditorum* (Apr. 1695), pp. 184–185.

Leibniz, Gottfried Wilhelm (2008). *Sämtliche Schriften und Briefe*. Siebente Reihe. Mathematische Schriften. Fünfter Band. 1674-1676. Infinitesimalmathematik. Berlin.

Pier, Jean-Paul (1996). *Histoire de l'intégration*. Vingt-cinq siècles de mathématiques. Paris.

[5] The history of the Henstock-Kurzweil integral is discussed by Shisha (1995) and Pier (1996, pp. 244–252).

Riemann, Bernhard (1876). "Ueber die Darstellbarkeit einer Function durch eine trigonometrische Reihe". In: *Bernhard Riemann's Gesammelte mathematische Werke*. Ed. by Heinrich Weber. Leipzig, pp. 213–250.
Roero, C. S. (2005). "Gottfried Wilhelm Leibniz, first three papers on the calculus (1684, 1686, 1693)". In: *Landmark Writings in Western Mathematics 1640-1940*. Ed. by Yvor Grattan-Guinness. Amsterdam et al., pp. 46–58.
Shisha, Oved (1995). "The Genesis of the Generalized Riemann Integral". *Computers & Mathematics with Applications* 30, pp. 207–211.
Sonar, Thomas (2018). *The History of the Priority Dispute between Newton and Leibniz*. Mathematics in History and Culture. Cham.
Stieltjes, Thomas Joannes (1894). "Recherches sur les fractions continues". *Annales de la Faculté des sciences de Toulouse* 8, J1–J122.
Wolf, Rudolph (1859). *Biographien zur Kulturgeschichte der Schweiz*. Zweiter Cyclus. Zürich.
Wolff, Christian (1710). *Der Anfangs-Gründe aller Mathematischen Wissenschaften Letzter Theil*. Halle.

integral basis **and** ***discriminant*** **(field)**

Both of these concepts were introduced by Dedekind in his famous "Tenth Supplement", in which, as Ewald (1996, p. 762) puts it, he "created, at a single hammer-blow, modern algebraic number theory". Dedekind did not quite use the current terminology; he used *Grundreihe* 'fundamental series' for an integral basis:

> Now since the discriminant of each system of n independent integers of the field Ω is a nonzero rational integer, there is one of these discriminants whose value – disregarding the sign – is *minimum*. From the previous investigation it follows directly that if a basis consists of numbers $\omega_1, \omega_2, \ldots, \omega_n$ for which the discriminant takes on this minimum value, the corresponding coordinates h_i of each integer ω of the field must necessarily be a rational integer. We will call such a basis $\omega_1, \omega_2, \ldots, \omega_n$ a *fundamental series* of the field Ω. [...]
>
> The minimal discriminant itself plays an important role, both with respect to the inner constitution of the field Ω, as well as with respect to its relationship to other fields. We will therefore call this positive or negative rational integer the *fundamental number* or the *discriminant* of the field Ω, and denote it by $\Delta(\Omega)$. Clearly it is identical to the ground number of each of the fields conjugate with Ω (Dedekind 1871, p. 447).[1]

Dedekind (1877, p. 77) himself used French *nombre fondamental* and *discriminant*; *série fondamentale* would have been a natural translation of *Grundreihe*, but Dedekind does not give the concept a name in the text. Westlund (1910, p. 388) used *fundamental number* as well as *integral basis* and seems to have the first to use the latter term; English *discriminant* appeared in Pierpont (1898, p. 228).

---

[1] "Da nun die Discriminante eines jeden Systems von n unabhängigen ganzen Zahlen des Körpers Ω eine von Null verschiedene ganze rationale Zahl ist, so giebt es unter allen diesen Discriminanten eine solche, deren Werth – abgesehen vom Vorzeichen – ein *Minimum* ist, und aus der vorstehenden Untersuchung folgt unmittelbar, dass, wenn eine Basis aus solchen ganzen Zahlen $\omega_1, \omega_2, \ldots, \omega_n$ besteht, deren Discriminante diesen Minimumwerth besitzt, die entsprechenden Coordinaten h_i, einer jeden *ganzen* Zahl ω des Körpers nothwendig *ganze* rationale Zahlen sein müssen. Eine solche Basis $\omega_1, \omega_2, \ldots, \omega_n$ wollen wir eine *Grundreihe* des Körpers Ω nennen [...] Die wichtigste Rolle spielt aber die Minimaldiscriminante selbst, sowohl hinsichtlich der inneren Constitution des Körpers Ω, als auch hinsichtlich seiner Verwandtschaft mit anderen Körpern; wir wollen daher diese positive oder negative ganze rationale Zahl die *Grundzahl* oder die *Discriminante* des Körpers Ω nennen und mit $\Delta(\Omega)$ bezeichnen". Translation from Avigad (2004, p. 34), slightly modified.

References

Avigad, Jeremy (2004). "Dedekind's 1871 version of the theory of ideals". URL: `https://kilthub.cmu.edu/articles/Dedekind_s_1871_Version_of_the_Theory_of_Ideals/6491342/1` (retrieved July 25, 2007).

Dedekind, Richard (1871). "Ueber die Composition der binären quadratischen Formen". In: *Vorlesungen über Zahlentheorie von P.G. Lejeune Dirichlet*. Herausgegeben und mit Zusätzen versehen von R. Dedekind. Zweite, umgearbeitete und vermehrte Auflage. Braunschweig, pp. 380–497.

Dedekind, Richard (1877). *Sur la théorie des nombres entiers algébriques*. Paris.

Ewald, William (1996). *From Kant to Hilbert*. A Source Book in the Foundations of Mathematics. Volume II. Oxford.

Pierpont, James (1898). "Weber's Algebra [review]". *Bulletin of the American Mathematical Society* 4, pp. 200–234.

Westlund, Jacob (1910). "On the Fundamental Number of the Algebraic Number-Field". *Transactions of the American Mathematical Society* 11, pp. 388–392.

integral domain

This term was coined as German *Integritäts-Bereich*, lit. 'domain of integrity', in Kronecker (1882). In this text, Kronecker first introduced what he called a *rational domain* or *domain of rationality* (German *Rationalitäts-Bereich)*, which he used for what corresponds to 'field':[1] "The 'domain of rationality' ($\mathfrak{R}'$, $\mathfrak{R}''$, $\mathfrak{R}'''$, ...) contains, as one can already see in the denotation, all these magnitudes which are rational functions of the magnitudes $\mathfrak{R}'$, $\mathfrak{R}''$, $\mathfrak{R}'''$, ... with integral numbers as coefficients" (Kronecker 1882, p. 4).[2]

If instead of rational functions entire functions are considered, Kronecker (1882, p. 14) speaks, "by analogy with the denomination *domain of rationality*", of a *domain of integrity* (*Integritäts-Bereich*). The prototypical example is the ring of integers.

Molk (1885, p. 20) translated the term in French as *domaine d'intégrité*. In English, the term became known partly through Hancock, who originally used *integral-realm* (Hancock 1894, p. 152), which he soon changed to *realm of integrity* (Hancock 1898, p. 149). Miller (1911, p. 220) defines a *domain of integrity* but also uses *integral domain*.

Kronecker's term was also mentioned by Hilbert in 1897 in his definition of *ring*: "Let $\theta, \mu, \ldots$ be any any algebraic numbers whose domain of rationality is a field k of degree m; then the system of all integer functions of $\theta, \mu, \ldots$ with rational integer coefficients will be called 'ring of numbers', 'ring' or 'integral domain' " (Hilbert 1897, p. 237).[3]

---

[1] More precisely, the notion of domain of rationality is more general than that of field because it relied on a very general notion of magnitude; see, for example, Haffner (2014, pp. 306–316).

[2] "Der Rationalitäts-Bereich ($\mathfrak{R}'$, $\mathfrak{R}''$, $\mathfrak{R}'''$, ...) enthält, wie schon die Bezeichnung deutlich erkennen lässt, alle diejenigen Größen, welche rationale Functionen der Grössen $\mathfrak{R}'$, $\mathfrak{R}''$, $\mathfrak{R}'''$, ... mit ganzzahligen Coefficienten sind". Translation from Haffner (2014, p. 307).

[3] "Sind θ, η, ... irgend welche ganze algebraische Zahlen, deren Rationalitätsbereich der Körper k vom m-ten Grade ist, so wird das System aller ganzen Funktionen von θ, η, deren Coefficienten ganze rationale Zahlen sind, ein *Zahlring*, *Ring* oder *Integritätsbereich* genannt". Translation adapted from Hilbert (1998, p. 67) and Corry (2004, p. 149). – From a present perspective, Hilbert's use of the term is restricted, as the term referred to sets of algebraic integers, which form a ring in the modern sense of the word.

An axiomatic definition in modern terms was given by Steinitz (1910, p. 177). In particular, he demanded that the product of any two non-zero elements is not zero; this definition was also used by Noether (1921, p. 29) and has been standard since.

Note. For Kronecker's introduction of the term, see Gray (2018, pp. 223–225) and Haffner (2014, p. 307).

References

Corry, Leo (2004). *Modern Algebra and the Rise of Mathematical Structures*. Second revised edition. Basel.

Gray, Jeremy (2018). *A History of Abstract Algebra*. From Algebraic Equations to Modern Algebra. Oxford.

Haffner, Emmylou (2014). "The 'Science of Numbers' in action in Richard Dedekind's works: between mathematical explorations and foundational investigations". PhD thesis. Université Paris Diderot.

Hancock, Harris (1894). "On the reduction of Kronecker's modular systems". *The Quarterly Journal of Pure and Applied Mathematics* 106, pp. 147–183.

Hancock, Harris (1898). "Canonical forms for the unique representation of *Kroneckers* modular systems". *Journal für die reine und angewandte Mathematik* 119, pp. 148–170.

Hilbert, David (1897). "Die Theorie der algebraischen Zahlkörper". *Jahresbericht der Deutschen Mathematiker-Vereinigung* 4, pp. 175–546.

Hilbert, David (1998). *The Theory of Algebraic Number Fields*. Translated from the German by Iain T. Adamson. With an introduction by Franz Lemmermeyer and Norbert Schappacher. Berlin.

Kronecker, Leopold (1882). *Grundzüge einer arithmetischen Theorie der algebraischen Grössen*. Berlin.

Miller, George Abram (1911). "The algebraic equation". In: *Monographs on topics of modern mathematics relevant to the elementary field*. Ed. by J.W.A. Young. London, pp. 211–260.

Molk, Jules (1885). "Sur une notion qui comprend celle de la divisibilité et sur la théorie générale de l'élimination". *Acta Mathematica* 6, pp. 1–166.

Noether, Emmy (1921). "Idealtheorie in Ringbereichen". *Mathematische Annalen* 83, pp. 24–66.

Steinitz, Ernst (1910). "Algebraische Theorie der Körper". *Journal für die reine und angewandte Mathematik* 137, pp. 167–309.

integral equation

This term is commonly attributed to Paul du Bois-Reymond in 1888 (e. g., by Dieudonné 1981, p. 97). du Bois-Reymond (1888, p. 228) indeed defined the German term *Integralgleichung* (*Gleichung* 'equation') in the modern sense of an equation in which the function to be determined is contained under an integral and gave $\int_0^S dsf(s)\varphi(s,x) = \psi(x) \cdot f(x) + \chi(x)$ as a simple example (f is the function to be determined, φ, ψ, and χ are known). The terminology spread through Hilbert (1904, p. 49), who attributed the term to du Bois-Reymond. An early English usage is Bateman (1906, p. 233).

The term had been used much earlier in a different sense. This older sense, commonly used in the 18th century by writers such as Euler and Lagrange, is given by Klügel (1805, p. 885) in his dictionary: "Integral equation is the finite equation between two or more variable quantities, which originated through the integration of a differential equation".[1]

Note. Overviews of the history of the concept are given by Ben-Menahem (2009, pp. 2622–2625) and Kline (1972, ch. 45).

[1] "Integralgleichung ist die endliche Gleichung zwischen zwey oder mehreren veränderlichen Größen, die durch die Integrierung einer Differentialgleichung entstanden ist".

References

Bateman, Harry (1906). "A class of Integral Equations". *Transactions of the Cambridge Philosophical Society* 20, pp. 233–252.

Ben-Menahem, Ari (2009). *Historical Encyclopedia of Natural and Mathematical Sciences*. Volume I. Berlin et al.

du Bois-Reymond, Paul (1888). "Bemerkungen über $\Delta z = \frac{\partial^2 z}{\partial x^2} + \frac{\partial^2 z}{\partial y^2} = 0$". *Journal für die reine und angewandte Mathematik* 103, pp. 204–229.

Dieudonné, Jean (1981). *History of Functional Analysis*. Amsterdam.

Hilbert, David (1904). "Grundzüge einer allgemeinen Theorie der linearen Integralgleichungen. (Erste Mitteilung)". *Nachrichten von der Königl. Gesellschaft der Wissenschaften zu Göttingen (mathematisch-physikalische Klasse)* (1904), pp. 49–91.

Kline, Morris (1972). *Mathematical Thought From Ancient To Modern Times*. Volume 3. New York and Oxford.

Klügel, Georg Simon (1805). *Mathematisches Wörterbuch*. Erste Abtheilung: Die reine Mathematik. Zweyter Theil von E bis I. Leipzig.

integral sine, ***integral cosine***, **and** ***exponential integral***

In English, these terms were first used by Glaisher (1870, p. 367) in the following way:

> It has for a long time been evident that the extension of the Integral Calculus would require the introduction of new functions; or, rather, that certain functions should be regarded as primary, so that forms reduced to dependence on them might be considered known. Thus, in the evaluation of Definite Integrals, the three transcendents
>
> $$\int_0^x \frac{\sin u}{u} du, \int_\infty^x \frac{\cos u}{u} du, \int_\infty^{-x} \frac{e^{-u}}{u} du$$
>
> called the sine-integral, the cosine-integral, and the exponential-integral, have become recognized elementary functions, and great use has been made of them to express the values of more complicated forms. They were introduced by Schlömilch [...] and denoted by him $\mathrm{Si}\,x$, $\mathrm{Ci}\,x$, $\mathrm{Ei}\,x$. Arndt also employed them in a similar manner about the same time.

As indicated by Glaisher, the terminology is somewhat older. In particular, the notation $\mathrm{ci}\,x$ and $\mathrm{si}\,x$ was introduced by Bretschneider (1843, p. 30).[1] He did not explain it any further, but it is very likely that it stood for *cosinus integralis* and *sinus integralis*. Shortly afterwards, Schlömilch (1846, p. 318), who seems to have been unaware of Bretschneider's earlier paper at the time, introduced the slightly different notation Si and Ci and also added Ei, but did not spell out the names yet. This was done by Arndt (1847, p. 233): "This function might be called the *integral cosine*, and Schlömilch seems to have had this is mind when he designated it by $Ci(p)$".[2] This is confirmed by Schlömilch (1848b, p. 87) himself, when he suggested the German terms *Integralexponentielle*[3], *Integralcosinus* und *Integralsinus*.[4] The corresponding French terms were used by Bierens de Haan (1855, p. 2).

[1] According to Cajori (1929, p. 280), this notation was earlier used in Bretschneider (1837). It is not found in that article, however. [2] "Man könnte diese Function wohl den *Integralcosinus* nennen, und Herr Prof. Schlömilch scheint diesen Gedanken gehabt zu haben, in dem er es durch $Ci(p)$ bezeichnete". [3] In the 20th century, *Integralexponentialfunktion* became the customary term in German. [4] In another publication from the same year, Schlömilch (1848a, pp. 148f.) justified the terminology by pointing out the "complete analogy", as he puts it, between the equations $Ei(w\sqrt{-1}) = Ci(w) + \sqrt{-1}Si(w)$ and $e^{w\sqrt{-1}} = \cos w + \sqrt{-1}\sin w$.

In the form *integral sine function* and *integral cosine function*, the English terms appear in the 20th century; they were used, for example, by Harkness (1917, p. 111) and van der Pol (1916, p. 279).

Note. For the history of the concepts, see Glaisher (1870) and Fischer (2007).

References

Arndt, Friedrich (1847). "Ueber einige bestimmte Integrale, welche sich auf die beiden Integrale $\int_\infty^p \frac{e^{-x}\partial x}{x}$, $\int_\infty^p \frac{\cos x}{x}\partial x$ zurückführen lassen". *Archiv der Mathematik und Physik* 10, pp. 233–240.

Bierens de Haan, David (1855). *Note sur une méthode pour la réduction d'intégrales définies et sur son application à quelques formules spéciales*. Amsterdam.

Bretschneider, Carl Anton (1837). "Theoriae logarithmi integralis lineamenta nova". *Journal für die reine und angewandte Mathematik* 17, pp. 257–285.

Bretschneider, Carl Anton (1843). "Berechnung der natürlichen Logarithmen sowie mehrerer anderer mit ihnen zusammenhängenden Zahlenwerte". *Archiv der Mathematik und Physik* 3, pp. 27–34.

Cajori, Florian (1929). *A History of Mathematical Notations*. Volume II. Notations mainly in higher mathematics. Chicago.

Fischer, Hans (2007). "Die Geschichte des Integrals $\int_0^\infty \frac{\sin x}{x}dx$. Eine Geschichte der Analysis in der Nussschale". *Mathematische Semesterberichte* 54, pp. 13–30.

Glaisher, J.W.L. (1870). "Tables of the Numerical Values of the Sine-Integral, Cosine-Integral, and Exponential-Integral". *Philosophical Transactions of the Royal Society of London* 160, pp. 367–388.

Harkness, James (1917). "The Algebraic basis of certain Bessel series, including more particularly the Neumann series of the second kind". *Transactions of the Royal Society of Canada* 10, section 3, pp. 97–120.

van der Pol, Balthasar (1916). "On the Wave-lengths and Radiation of Loaded Antennae". *Proceedings of the Physical Society of London* 29, pp. 269–289.

Schlömilch, Oskar (1846). "Note sur quelques intégrales définies". *Journal für die reine und angewandte Mathemati* 33, pp. 316–324.

Schlömilch, Oskar (1848a). *Analytische Studien*. Zweite Abtheilung. Leipzig.

Schlömilch, Oskar (1848b). *Handbuch der Differenzial- und Integralrechnung*. Dritte Lieferung. Integralrechnung. Erste Hälfte. Greifswald.

integrally closed

The term *integrally closed* (German *ganz-abgeschlossen*) was introduced by Noether (1927, p. 33)[1]: "The ring $\mathfrak{R}$ is called integrally closed in $\mathfrak{T}$, if each element of $\mathfrak{T}$ that is integral with respect to $\mathfrak{R}$ belongs to $\mathfrak{R}$".[2] Noether referred to Dedekind (1879, § 172) for the concept.

The term was soon in common use and adopted by Artin and van der Waerden (1927, p. 26), for example. The English term appears in Ore (1933, p. 741).

Note. Remarks on the concept in the works of Dedekind and Noether are found in Corry (2004, esp. pp. 96, 107f., 237–244).

References

Artin, Emil and Bartel Leendert van der Waerden (1927). "Die Erhaltung der Kettensätze der Idealtheorie bei beliebigen endlichen Körpererweiterungen". *Nachrichten von der Gesellschaft der Wissenschaften zu Göttingen (mathematisch-physikalische Klasse)* (1926), pp. 23–27.

[1] The paper was written in 1925. [2] "Der Ring $\mathfrak{R}$ heißt ganz-abgeschlossen in $\mathfrak{T}$, wenn jedes in bezug auf $\mathfrak{R}$ ganze Element aus $\mathfrak{T}$ zu $\mathfrak{R}$ gehört". – $\mathfrak{T}$ is a commutative ring with zero divisors, $\mathfrak{R}$ is a subring of $\mathfrak{T}$ that contains the unit element (Noether 1927, pp. 29f.).

Corry, Leo (2004). *Modern Algebra and the Rise of Mathematical Structures*. Second revised edition. Basel.
Dedekind, Richard (1879). "Ueber die Theorie der ganzen algebraischen Zahlen". In: *Vorlesungen über Zahlentheorie von P.G. Lejeune Dirichlet*. Herausgegeben und mit Zusätzen versehen von R. Dedekind. Dritte, umgearbeitete und vermehrte Auflage. Braunschweig, pp. 434–627.
Noether, Emmy (1927). "Abstrakter Aufbau der Idealtheorie in algebraischen Zahl- und Funktionenkörpern". *Mathematische Annalen* 96, pp. 26–61.
Ore, Øystein (1933). "Abstract ideal theory". *Bulletin of the American Mathematical Society* 39, pp. 728–745.

interior point

Before the basic topological concepts began to be studied systematically, the terms *interior point* and *interior* were used in an intuitive way and without definition. The expression is found, for example, in Briot and Bouquet's definition of *cercle de convergence*: "the series is convergent for all values z situated in the interior of this circle, which he shall call for this reason *circle of convergence*" (Briot and Bouquet 1856, p. 95).[1] Similarly, Cantor (1872, p. 129) used the term *das Innere* in his definition of *neighbourhood*; at that time, he did not define interior points themselves: "By a neighborhood of a point one should understand here any interval which has the point *in its interior*".[2]

Several years later, Cantor did present a definition, saying that the points of a "continuous manifold" M_μ can be split into two classes, "*interior points* and *points on the boundary*; a point p of M_μ is called *interior* point if around p as a centre a $K_{\mu-1}$ [a sphere] can be described with a radius so small that all points within and on $K_{\mu-1}$ belong to the domain M_μ; every point p of M_μ, for which such a construction is *not* possible, belongs to the *boundary* of M_μ" (Cantor 1879, p. 130).[3]

The terminology soon spread. For example, Italian *interno* was used by Dini (1878, p. 15) and also by Peano (1887, p. 158), who gave a definition very similar to Cantor's (and also defined exterior points).[4] The terminology *point intérieur* was also used by Jordan (1893, p. 20), and Harkness (1894, p. 137) translated this by *interior point*.

Note. For the history of the concept, see Moore (2008).

References

Briot, Charles and Jean Claude Bouquet (1856). "Étude des fonctions d'une variable imaginaire (Premier Mémoire)". *Journal de l'École polytechnique* 21, cahier 36, pp. 85–131.
Cantor, Georg (1872). "Ueber die Ausdehnung eines Satzes aus der Theorie der trigonometrischen Reihen". *Mathematische Annalen* 5, pp. 123–132.

---

[1] "la série est convergente pour toutes les valeurs de z situées dans l'intérieur de ce cercle, que nous appellerons pour cette raison *cercle de convergence*". [2] "Unter Umgebung eines Punktes sei aber hier ein jedes Intervall verstanden, welches den Punkt *in seinem Innern* hat". Translation from Ferreirós (2007, p. 142). [3] "Die Puncte einer stetigen M_μ zerfallen in zwei Classen, in sogenannte *innere Puncte* und in *Puncte auf der Begrenzung*; ein Punct p heißt *innerer* Punct, wenn sich um p als Mittelpunct eine $K_{\mu-1}$ von so kleinem r beschreiben läßt, daß alle Puncte innerhalb und auf $K_{\mu-1}$ zum Gebiete M_μ gehören; jeder Punct p von M_μ, bei welchem eine solche Construction *nicht* möglich ist, wird zur *Begrenzung* von M_μ gezählt". [4] Dolecki and Greco (2019, p. 57) claim that Peano was the first to define interior points, which, as can be seen from the examples, is not correct.

Cantor, Georg (1879). "Ueber einen Satz aus der Theorie der stetigen Mannigfaltigkeiten". *Nachrichten von der Königl. Gesellschaft der Wissenschaften und der Georg-Augusts-Universität zu Göttingen* (1879), pp. 127–135.
Dini, Ulisse (1878). *Fondamenti per la teorica delle funzioni di variabili reali*. Pisa.
Dolecki, Szymon and Gabriele H. Greco (2019). "The forgotten mathematical legacy of Peano". *Dissertationes Mathematicae* 537, pp. 1–77.
Ferreirós, José (2007). *Labyrinth of Thought*. A History of Set Theory and Its Role in Modern Mathematics. Second revised edition. Basel.
Harkness, James (1894). "Jordan's Cours d'Analyse". *Bulletin of the American Mathematical Society* 3, pp. 135–141.
Jordan, Camille (1893). *Cours d'analyse*. Tome premier. Calcul différentiel. Paris.
Moore, Gregory H. (2008). "The emergence of open sets, closed sets, and limit points in analysis and topology". *Historia Mathematica* 35, pp. 220–241.
Peano, Guiseppe (1887). *Applicazioni geometriche del calcolo infinitesimale*. Torino.

interpolation

Interpolation has been informally described as "the art of reading between the lines in a [numerical] table" (Thiele 1909, p. 2).[1] As to the origin and early meaning of the term, Stewart (1745, p. 353) wrote: "he [Wallis] thought upon another Method, which he calls *Interpolation* of series. By which he means a method of discovering certain intermediate Terms of a regular Series or Progression, by considering the Properties of the Progression, and the Relations of the Terms to each other".

Latin *interpolare* is derived from *inter* 'between' and *polare* 'to polish'. In classical Latin, it meant 'to polish, furbish, dress up', also 'to give a new form, shape, or appearance to any thing' (LS), and in medieval Latin it was applied to texts as well, where an interpolation is the careful addition of information. The word was introduced into mathematics by Wallis (1656, p. 132): "if it were possible to interpolate this sequence of ratios $\frac{1}{1}, \frac{1}{6}, \frac{1}{30}, \frac{1}{140}, \frac{1}{630}$, etc., the ratio that must be placed as intermediate between the first and second is that of a semicircle to the square of the diameter".[2] As Tropfke (1933, p. 252) points out, the verb can hardly be considered a fixed term with Wallis, considering that he also used the verb *intercalare* (Wallis 1656, p. 170) in the same meaning, a word mainly used in chronology (*intercalary day*).[3] In English, the noun appears in Collins (1669, p. 930), where also the verb *interpole* is used.

There were a number of early contributors to the theory of interpolation; among these, as Goldstine (1977, p. 68) puts it, "there is no single person who did so much for this field, as for so many others, as Newton". When he found a method by which certain functions may be expressed in series of powers of x, Newton wrote (in 1676): "I have another method not yet communicated [...] a convenient, rapid and general solution of this problem, *To draw a geometrical curve which shall pass through any number of given points*. [...] These things are done at once geometrically with no calculation intervening. But the above problem is of the second kind, and though at

---

[1] "*Interpolation* ist die Kunst zwischen den Zeilen einer Tafel zu lesen". Translation from Meijering (2002, p. 320). [2] "si interpolari poterit rationum series ista $\frac{1}{1}, \frac{1}{6}, \frac{1}{30}, \frac{1}{140}, \frac{1}{630}$, etc., ratio illa quae primae & secundae est intermedia, est ea quam habet Semicirculus ad Quadratum diametri". Translation from Stedall (2004, p. 126). [3] Both *interpolare* and *intercalare* were also used by Newton. For a discussion of the semantic differences, see Turnbull (1960, pp. 149f.).

first glance it looks unmanageable, yet the matter turns out otherwise. For it ranks among the most beautiful of all that I could wish to solve" (Turnbull 1960, p. 119).[4]

Over the centuries, the methods and techniques in interpolation have been refined and developed to a considerable extent; see Meijering (2002) for a detailed exposition.[5]

References

Collins, John (1669). "An Account Concerning the Resolution of Equations in Numbers". *Philosophical Transactions of the Royal Society of London* 4, pp. 929–934.

Goldstine, Herman H. (1977). *A History of Numerical Analysis from the 16th through the 19th Century*. New York.

Meijering, Erik (2002). "A Chronology of Interpolation: From Ancient Astronomy to Modern Signal and Image Processing". *Proceedings of the IEEE* 90, pp. 319–342.

Stedall, Jacqueline A. (2004). *The Arithmetic of Infinitesimals*. John Wallis 1656. Translated from Latin to English with an Introduction. New York.

Stewart, John (1745). "Sir Isaac Newton's Analysis by Equations of an infinite Number of Terms, explained". In: *Two Treatises of the Quadrature of Curves and Of Analysis of Equations of an infinite Number of Terms*. Ed. by John Stewart. London, pp. 344–479.

Thiele, Thorvald Nicolai (1909). *Interpolationsrechnung*. Leipzig.

Tropfke, Johannes (1933). *Geschichte der Elementar-Mathematik*. Zweiter Band. Allgemeine Arithmetik. Dritte, verbesserte und vermehrte Auflage. Berlin and Leipzig.

Turnbull, H. W., ed. (1960). *The Correspondence of Isaac Newton*. Volume II. 1676–1687. Cambridge.

Wallis, John (1656). *Arithmetica infinitorum*. In: *Operum Mathematicorum*. Pars altera. Oxford, pp. 1–198.

intersect* and *intersection

In Greek, there were several ways to express to concept of intersection (see the list of words in Mugler 1958-59, p. 34 and the respective entries in Mugler's book). A commonly used verb was *τέμνειν* 'to cut', which was applied when two geometrical objects intersect. It appears in Euclid's *Elements* and various texts before him (Mugler 1958-59, pp. 413–416).

In Latin, the verb *secare* (also meaning 'to cut') was used to translate *τέμνειν*. It appears, for example, in the Pseudo-Boethian *Geometria I* (Blume, Lachmann, and Rudorff 1848, p. 388). The verb *intersecare* (in mathematical senses) is of later origin and dates from late medieval times; Campanus, for example, occasionally used it in his version of the *Elements* (Busard 2005, p. 59).

For the French verb *intersecter*, the DMF and the TLL refer to the 14th-century philosopher Oresme, who used the verb in the form *intersequer* (Oresme 1968, p. 46); the noun *intersection* was used by Chuquet in the late 15th century (Chuquet 1979, e. g., p. 284).

The English terminology seems to have come up in astronomical contexts first. For example, Recorde (1556, p. 178) discusses "the pointe of the intersection or crossinge of the ecliptike line and the equinoctiall".[1] Soon after Recorde, *intersection* was used several times by Billingsley in his translation of the *Elements* (Euclid 1570, e. g., fol. 21^{v}).

[4] Translation from Turnbull (1960, p. 137). [5] Goldstine (1977) covers the history of the concept until the end of the 19th century. [1] The MED cites a comparable example from the early 15th century.

In German texts, the verb *durchschneiden* (from *durch* 'through' and *schneiden* 'to cut') is found in 16th century, for example in Dürer (1525, fol. J iiij^{v}).[2] Correspondingly, Schmid (1539, p. 37) used *punct des durchschnids* 'intersection point'. The shorter *Schnittpunkt*, now the usual term, did not come into common use until the 19th century.

Note. For the Greek terms, see Mugler (1958-59), as cited above. Latin *secare* is discussed by Guillaumin (2020, p. 259).

References

Blume, Friedrich, Karl Lachmann, and Adolf Rudorff, eds. (1848). *Die Schriften der römischen Feldmesser*. Erster Band. Texte und Zeichnungen. Berlin.

Busard, H.L.L. (2005). *Campanus of Novara and Euclid's* Elements. Volume I. Stuttgart.

Chuquet, Nicolas (1979). *La Géométrie*. Première géométrie algébrique en langue française (1484). Introduction, texte et notes par Hervé l'Huillier. Paris.

Dürer, Albrecht (1525). *Underweysung der Messung, mit dem Zirckel und Richtscheyt*. Nürnberg.

Euclid (1570). *The Elements of Geometrie of the most auncient Philosopher Euclide of Megara*. Faithfully (now first) translated into the Englishe toung, by H. Billingsley. London.

Guillaumin, Jean-Yves (2020). *Dictionnaire de la terminologie latine ancienne de l'arithmétique et de la géométrie*. Paris.

Mugler, Charles (1958-59). *Dictionnaire historique de la terminologie géométrique des Grecs*. Paris.

Oresme, Nicole (1968). *Le livre du ciel et du monde*. Edited by Albert D. Menut and Alexander J. Denomy, translated with an introduction by Albert D. Menut. Madison.

Recorde, Robert (1556). *The Castle of Knowledge*. London.

Reiner, Karl (1961). *Die Terminologie der ältesten mathematischen Werke in deutscher Sprache nach den Beständen der Bayerischen Staatsbibliothek*. München.

Schmid, Wolffgang (1539). *Das erst buch der Geometria*. Nürnberg.

***intersection* (sets)**

In algebraic number theory, the intersection of sets was usually called *greatest common divisor*. This usage is based on an analogy to arithmetic[1] and goes back to Dedekind (1871, p. 424), who defined the notion with respect to fields.[2] The term is also found in early set-theoretical papers such as Cantor (1880, p. 355),[3] who earlier also used paraphrases: With respect to two given sets, the intersection was called "the manifold of the elements common to both [given sets]" (Cantor 1878, p. 248).[4] A first, not very successful, alternative was suggested by Dedekind (1888, p. 5) himself: *Gemeinheit*, a noun derived from *gemein* 'common'.[5]

In Weber's textbook on algebra (Weber 1895, p. 510), the word *Durchschnitt* – it was commonly used for 'intersection' in its geometrical sense – is presented as an alternative to *greatest common divisor*. Weber defines the notion in connection with groups, not with sets in general. In Weber (1899, p. 10), he points out that the term has a "geometrical echo"[6] and that it was suggested by Eduard Study. Weber

---

[2] For various early alternatives, see Reiner (1961, p. 104). [1] The metaphorical structure of Dedekind's theory is discussed in Becker (2005, ch. 10.8). [2] Dedekind used the German term *größter gemeinsamer Teiler*, later also *größter gemeinsamer Divisor* (Dedekind 1894, p. 454). [3] The terminological transfer from algebraic number theory to set theory was pointed out by Ferreirós (2007, p. 204): "Cantor begins by introducing notation and terminology for some of the basic relations and operations on sets, and this terminology shows a strong resemblance to that of Dedekind's algebraic number theory". [4] "die Mannigfaltigkeit der ihnen gemeinschaftlichen Elemente". [5] The word appears as *community* in Dedekind (1909, p. 48). [6] "geometrischer Anklang".

mentioned Study's authorship on other occasions as well, for example in Weber (1897, p. 436), but never gave a precise reference. However, Study did indeed use the term in connection with hypercomplex number systems, for example in Study (1889b, p. 249). He had earlier explained the advantages of the use of expressions from geometry in other fields: certain connections between different fields of study are made evident just through the terminology and understanding is thus made easier (Study 1889a, p. 201, fn. 4). The term was quickly adopted and soon became standard, also in purely set-theoretical works such as Zermelo (1908, p. 264), where it is applied to sets in general.

The English translation *intersection* was a natural choice, not just because it had, like *Durchschnitt*, already been used in geometry, but also because *intersection* and *intersect* had already made their way into logical and other works. For example, Venn (1881, p. 81) used the verb with respect to classes, and Davis (1894, pp. 30, 42) used both the verb and the noun with respect to concepts; Whitehead (1898, p. 15) applied the verb to manifolds (in Riemann's sense).

Note. Brief overviews of the terminology are given by Zoretti and Rosenthal (1924, p. 866) and Ferreirós (2007, p. 204).

References

Becker, Holger (2005). "Semantische und lexikalische Aspekte der mathematischen Fachsprache des 19. Jahrhunderts". PhD thesis. Carl von Ossietzky Universität Oldenburg. URL: `http://oops.uni-oldenburg.de/59/`.

Cantor, Georg (1878). "Ein Beitrag zur Mannigfaltigkeitslehre". *Journal für die reine und angewandte Mathematik* 84, pp. 242–258.

Cantor, Georg (1880). "Ueber unendliche, lineare Punktmannichfaltigkeiten. 2". *Mathematische Annalen* 17, pp. 355–358.

Davis, Noah K. (1894). *Elements of Deductive Logic*. New York.

Dedekind, Richard (1871). "Ueber die Komposition der binären quadratischen Formen". In: *Vorlesungen über Zahlentheorie von P.G. Lejeune Dirichlet*. Herausgegeben und mit Zusätzen versehen von R. Dedekind. Zweite, umgearbeitete und vermehrte Auflage. Braunschweig, pp. 380–497.

Dedekind, Richard (1888). *Was sind und was sollen die Zahlen?* Braunschweig.

Dedekind, Richard (1894). "Ueber die Theorie der ganzen algebraischen Zahlen". In: *Vorlesungen über Zahlentheorie von P.G. Lejeune Dirichlet*. Herausgegeben und mit Zusätzen versehen von R. Dedekind. Vierte, umgearbeitete und vermehrte Auflage. Braunschweig, pp. 434–657.

Dedekind, Richard (1909). *Essays on the Theory of Numbers*. I. Continuity and Irrational Numbers, II. The Nature and Meaning of Numbers. Authorized translation by Wooster Woodruff Beman. Chicago.

Ferreirós, José (2007). *Labyrinth of Thought*. A History of Set Theory and Its Role in Modern Mathematics. Second revised edition. Basel.

Study, Eduard (1889a). *Methoden zur Theorie der ternären Formen*. Leipzig.

Study, Eduard (1889b). "Ueber Systeme von complexen Zahlen". *Nachrichten von der Königl. Gesellschaft der Wissenschaften und der Georg-Augusts-Universität zu Göttingen* (1889), pp. 237–268.

Venn, John (1881). *Symbolic Logic*. London.

Weber, Heinrich (1895). *Lehrbuch der Algebra*. Erster Band. Braunschweig.

Weber, Heinrich (1897). "Ueber Zahlengruppen von algebraischen Körpern". *Mathematische Annalen* 48, pp. 433–473.

Weber, Heinrich (1899). *Lehrbuch der Algebra*. Zweiter Band. Braunschweig.

Whitehead, Alfred North (1898). *A Treatise on Universal Algebra*. Volume I. Cambridge.

Zermelo, Ernst (1908). "Untersuchungen über die Grundlagen der Mengenlehre". *Mathematische Annalen* 65, pp. 261–281.

Zoretti, Ludovic and Arthur Rosenthal (1924). "Die Punktmengen". *Encyklopädie der mathematischen Wissenschaften mit Einschluss ihrer Anwendungen* II C 9a, pp. 852–1030.

interval

Like many terms in the calculus, *interval* cannot be attributed to any *one* person. The word comes from Latin *intervallum*, which lit. means 'space between two palisades', from *inter* 'between' and *vallum* 'rampart'. In mathematics, the term was sometimes used to mean 'dimension' (see *dimension*); it is also found in the 10th-century "Geometria incerti auctoris", in the sense of 'radius' (Bubnov 1899, p. 364).[1]

In mathematical texts of the 17th century, the term is commonly used, without definition, in the sense of 'space between'; often, the distance is meant. To give just one example: "Between two points given, there can be understood but one strait line; because there cannot be more then [sic] one least interval or length between the same points" (Hobbes 1656, p. 130) .[2]

In his *De arte combinatoria*, Leibniz (1666, p. 42) defines some elementary terms on the basis of combinations of simpler terms. To these simpler terms he assigned numbers, for example 2 = space (*spatium*), 3 = in between (*intersitum*), and 10 = whole (*totum*); the word *interval* is then defined: "*Intervallum* est 2.3.10". In other words, an interval is the space placed within a whole (Rossi 2006, p. 178).[3]

In the calculus, the term was not in common use until the 19th century, although it is occasionally found in the 18th century in the usual calculus sense (but definitions do not seem to appear anywhere). For example, Laplace (1776, p. 313) has the expression "the interval comprised between $t = h$, & $t = h + T$".[4] Cauchy later often used the expression "between given limits" ("entre des limites données") (Cauchy 1821, e. g., p. 34) but also *intervalle* (Cauchy 1821, e. g., p. 464), which is also found in Fourier (1822, p. 252) and Lejeune Dirichlet (1837, p. 152), for example.

As mentioned before, definitions are not given in any of these texts; these come up in the second half of the 19th century. Dedekind (1872, p. 27) gave the following one (with reference to rational numbers) in his book *Continuity and Irrational Numbers*: "Very useful in this connection is the notion of an *interval*, i. e., a system A of rational numbers possessing the following characteristic property: if a and a' are numbers of the system A, then are all rational numbers lying between a and a' contained in A".[5] A further definition, not limited to rational numbers, was given by Tannery (1886, p. 99): "Let a, b be two numbers, a being the smaller one of the two. I shall call *interval* (a, b) the set of numbers a, b and of all the numbers, rational or not, which are contained between a and b".[6]

[1] For some discussion of term, see Guillaumin (2020, pp. 161f.). *intervallum* often translated Greek *διάστημα* 'interval, space between', for which see Federspiel (2005). [2] In the original Latin, *intervallum* was used (Hobbes 1655, p. 107). [3] A similar definition is found in a manuscript from 1685/86 (Leibniz 1999, p. 36). [4] "l'intervalle compris entre $t = h$, & $t = h + T$". [5] "Sehr nützlich ist in dieser Hinsicht der Begriff eines *Intervalls*, d. h. eines Systems A von rationalen Zahlen, welches folgende charakteristische Eigenschaft besitzt: sind a und a' Zahlen des Systems A, so sind auch alle zwischen a und a' liegenden rationalen Zahlen in A enthalten". Translation from Smith (1929, p. 44). [6] "Soient a, b deux nombres quelconques, a étant le plus petit des deux, j'appellerai *intervalle* (a, b) l'ensemble des nombres a, b et de tous les nombres rationnels ou non qui sont compris entre a et b".

Occasionally, a distinction between *segment* and *interval* was made. For example, the translator in Burkhardt (1913, p. 127) remarked: "We call attention here to the usual distinction between *interval* and *segment*. A *segment* (a, b), for example, is understood to be the set of all numbers greater than a and less than b; that is, exclusive of the end-points a and b and an *interval* (a, b) is the *segment* (a, b) together with a and b".

References

Bubnov, Nicolaus, ed. (1899). *Gerberti postea Silvestri II papae Opera Mathematica*. Berlin.
Burkhardt, Heinrich (1913). *Theory of Functions of a Complex Variable*. Authorized translation from the fourth German edition by S.E. Rasor. Boston.
Cauchy, Augustin-Louis (1821). *Cours d'analyse de l'École Royale Polytechnique*. I.re partie. Analyse algébrique. Paris.
Dedekind, Richard (1872). *Stetigkeit und irrationale Zahlen*. Braunschweig.
Federspiel, Michel (2005). "Sur l'expression linguistique du rayon dans les mathématiques grecques". *Les études classiques* 73, pp. 97–108.
Fourier, Joseph (1822). *Théorie analytique de la chaleur*. Paris.
Guillaumin, Jean-Yves (2020). *Dictionnaire de la terminologie latine ancienne de l'arithmétique et de la géométrie*. Paris.
Hobbes, Thomas (1655). *Elementorum philosophiae sectio prima de corpore*. London.
Hobbes, Thomas (1656). *Elements of Philosophy*. London.
Laplace, Pierre-Simon (1776). "Recherches sur le calcul intégral et sur le système du monde". *Mémoires de l'Académie royale des sciences* (1772, seconde partie), pp. 267–376.
Leibniz, Gottfried Wilhelm (1666). *Dissertatio de arte combinatoria*. Leipzig.
Leibniz, Gottfried Wilhelm (1999). *Sämtliche Schriften und Briefe*. Philosophische Schriften. Vierter Band. 1677-Juni 1690. Teil A. Berlin.
Lejeune Dirichlet, Gustav (1837). "Ueber die Darstellung ganz willkürlicher Functionen durch Sinus- und Cosinusreihen". In: *Repertorium der Physik*. Herausgegeben von Heinrich Wilhelm Dove und Ludwig Moser. I. Band. Berlin, pp. 152–174.
Rossi, Paolo (2006). *Logic and the Art of Memory*. The Quest for a Universal Language. London.
Smith, David Eugene, ed. (1929). *A Sourcebook in Mathematics*. New York.
Tannery, Jules (1886). *Introduction à la théorie des fonctions d'une variable*. Paris.

intrinsic equation **and** ***intrinsic coordinates***

In the 18th century, some mathematicians began to search for equations for curves that do not depend on "extrinsic" properties of the curve. That means, an attempt was made "to define curves through connections between quantities inherent in the curve, such as radius of curvature" (Boyer 1956, p. 227). This approach also implies that a curve is not defined in relation to a system of coordinates.

Thus, Lacroix (1798, p. 392), discussing the intrinsic approach, states that "this manner of presenting the equation of a curve is remarkable in that it only uses quantities that are absolutely inherent in the proposed curve".[1] Some years later, Gergonne (1813-14, p. 44) used the word *intrinsic* (French *intrinsèque*) in this context. He criticizes that with the usual coordinates the equation of the curve is arbitrary and it does not, "if I may put it this way, [express] the *intrinsic* nature of the curve".[2] This word was also used by Whewell (1849, p. 659), when, in an article entitled "On the Intrinsic Equation of a Curve", he considered "how complex and intractable are

---

[1] "cette manière de présenter l'équation d'une courbe est remarquable en ce qu'elle n'emploie que des quantités absolument inhérentes à la courbe proposée". – See Boyer (1956, p. 228) and Loria (1948, p. 263). [2] "elle n'exprimera point, si je puis m'exprimer ainsi, la nature *intrinsèque* de la courbe".

generally the expressions for the lengths of curves referred to rectilinear coordinates" and "that this complexity arises in a considerable degree from the introduction into the investigation of the reference to the rectilinear coordinates (which are *extrinsic* lines)". He then considered "what would be the result if curves were expressed by means of a relation between two simple and *intrinsic* elements; the length of the curve and the angle through which it bends" (ibid.). It is this article by Whewell that is usually referred to concerning the origin of the term *intrinsic equation* (e. g., Boyer 1956, p. 229).

Especially in French and German mathematical texts, the notion was also discussed in terms of "naturalness". Thus, the term *natural coordinates* (French *coordonnées naturelles*) was already used by Ampère (1808, p. 164).[1] It was popularized[2] by Aoust (1869), who used both *coordonnées naturelles* and *équation naturelle* in the preface to his book. The general approach was also described as natural (German *natürlich*) by Peters (1835, pp. 11f.), although he did not use the terms *natural equation* or *natural coordinates*, as is sometimes claimed (e. g., [anon.] 1902, p. 22).

At the end of the century, intrinsic geometry was systematically studied by Cesàro. His book on intrinsic geometry (*geometrica intrinseca*) (Cesàro 1896) was translated into German (Cesàro 1901), where the term *natürliche Geometrie* was used. The difference in terminology was later commented upon by Kasner (1930, p. 32): "The terms *intrinsic* geometry and intrinsic equation, used by the Italian and English writers, seem more suggestive than *natural* geometry and natural equation, as used by the German school".

The German term *esoterische Koordinaten* ('esoteric coordinates'), from Greek *ἐσωτερικός* 'internal, inner', is rare and has been attributed to Reinhold Hoppe by Wölffing (1900, p. 145). Wölffing does not give a reference, and it is unclear if Hoppe actually used that term. Hoppe (1864, p. 122) otherwise named an intrinsic equation a *specific equation* (*specifische Gleichung*) (see also Lampe 1901, p. 39). The term *esoterische Geometrie* is attested, however, and found in some reviews by Lampe, e. g. in Lampe (1889, p. 819).

Note. For the history of the notion, see Wölffing (1900), Müller (1910, pp. 753–755), Coolidge (1940, pp. 343–354), Loria (1948, pp. 263–266), and Boyer (1956, pp. 227–230).

References

[anon.] (1902). "Vorlesungen über natürliche Geometrie [review]". *Monatshefte für Mathematik und Physik (Literatur-Berichte)* 13, pp. 21–23.

Adrian, Theodor (1906). "Tangential-Koordinaten". *Unterrichtsblätter für Mathematik und Naturwissenschaften* 12, pp. 138–139.

Ampère, André-Marie (1808). "Sur les avantages qu'on peut retirer dans la théorie des courbes de la considération des paraboles osculatrices, avec des réflexions sur les fonctions différentielles dont la valeur ne change pas lors de la transformation des axes". *Journal de l'École Polytechnique* 7, cahier 14, pp. 159–181.

Aoust, Barthélémy (1869). *Analyse infinitésimale des courbes tracées sur une surface quelconque*. Paris.

Boyer, Carl B. (1956). *History of Analytic Geometry*. New York.

[1] See also Poncelet (1866, p. 319). This text by Poncelet was written between 1822 and 1823.

[2] It has occasionally been attributed to Aoust, for example by Adrian (1906, p. 139).

Cesàro, Ernesto (1896). *Lezioni di geometria intrinseca*. Napoli.
Cesàro, Ernesto (1901). *Vorlesungen über natürliche Geometrie*. Autorisierte deutsche Ausgabe von Gerhard Kowalewski. Leipzig.
Coolidge, Julian Lowell (1940). *A History of Geometrical Methods*. Oxford.
Gergonne, Joseph Diez (1813-14). "Essai sur l'expression analitique des courbes, indépendamment de leur situation sur un plan". *Annales de mathématiques pures et appliquées* 4, pp. 42–55.
Hoppe, Reinhold (1864). "Darstellung der Curven durch Krümmung und Torsion". *Journal für die reine und angewandte Mathematik* 63, pp. 122–140.
Kasner, Edward (1930). "Vorlesungen über Natürliche Geometrie [review]". *Bulletin of the American Mathematical Society* 36, p. 32.
Lacroix, Sylvestre François (1798). *Traité du calcul différentiel et du calcul intégral*. Tome second. Paris.
Lampe, Emil (1889). "Cesàro, E. Sur l'analyse barycentrique des courbes [review]". *Jahrbuch über die Fortschritte der Mathematik* 21, pp. 818–819.
Lampe, Emil (1901). "Reinhold Hoppe". *Jahresbericht der Deutschen Mathematiker-Vereinigung* 9, pp. 33–58.
Loria, Gino (1948). "Perfectionnements, évolution, métamorphoses du concept de «coordonnées»: Contribution à l'histoire de la géométrie analytique". *Osiris* 8, pp. 218–288.
Müller, Emil (1910). "Die verschiedenen Koordinatensysteme". *Encyklopädie der mathematischen Wissenschaften mit Einschluss ihrer Anwendungen* III AB 7, pp. 596–770.
Peters, Adolf (1835). *Neue Curvenlehre*. Grundzüge einer Umgestaltung der höheren Geometrie durch ihre ursprüngliche analytische Methode. Dresden.
Poncelet, Jean Victor (1866). *Traité des propriétés projectives des figures*. Tome second. Paris.
Whewell, William (1849). "On the Intrinsic Equation of a Curve, and its Application". *Transactions of the Cambridge Philosophical Society* 8, pp. 659–671.
Wölffing, Ernst (1900). "Bericht über den gegenwärtigen Stand der Lehre von den natürlichen Koordinaten". *Bibliotheca Mathematica* 1, pp. 142–159.

invariant

Something that is invariant does not change (Latin *variare* 'to change'). Before the formal introduction of this term, it was used, for example, by Gauß (1828, § 12), who used the Latin adjective *invariata* in the formulation of his "remarkable theorem" (*theorema egregium*): "If a curved surface is developed upon any other surface whatever, the measure of curvature in each point remains unchanged".[1] In German, Gauß (1827, p. 1765) used *ungeändert*, which also literally means 'unchanged'.[2]

In 1846, Cayley formulates the major task of what was later called *invariant theory*, namely "to find all the derivatives of any number of functions, which have the property of preserving their form unaltered after any linear transformations of the variables" (Cayley 1846, p. 104).[3] Cayley also explains that "[b]y Derivative I understand a function deduced in any manner whatever from the given functions, and I give the name of Hyperdeterminant Derivative, or simply of Hyperdeterminant, to these derivatives which have the property just enunciated".[4]

---

[1] "Si superficies curva in quamcumque aliam superficiem explicatur, mensura curvaturae in singulis punctis invariata manet". Translation from Gauß (1902, § 12). [2] On these passages, see Reich (1994, p. 22). [3] This was pointed out by Crilly (1986, p. 244). [4] According to Cayley (1845a, p. 55), the name *hyperdeterminant* is due to "their analogy with the determinants". The term is also found in Cayley (1845b, p. 195). Klein (2016, p. 166) commented on the term thus: "In our examples, we have always reached our invariants by setting up determinants, and in this we find justification for the theory of determinants as the foundation for the theory of invariants. Because of this connection, Cayley originally used the name *hyperdeterminants* for invariants".

Sylvester (1851a, p. 396) suggested an alternative for *hyperdeterminant*: "The remaining coefficients are the two well-known hyperdeterminants, or, as I propose henceforth to call them, the two Invariants of the form $ax^4 + 4bx^3y + 6cx^2y^2 + 4dxy^3 + ey^4$". In another paper of the same year, he gives the following definition: "an invariant to a given form may be defined to be such a function of the coefficients of the form, as remains absolutely unaltered when instead of the given form any linear equivalent thereto is substituted" (Sylvester 1851b, p. 291). Cayley (1851, p. 368) quickly adopted Sylvester's term, and so did various other mathematicians.

Also, the term is soon afterwards found in other areas of mathematics. To illustrate: in his famous *Erlangen Programme*, Klein (1872, p. 7) states: "Given a manifold and a group of transformations of the same; to develop the theory of invariants relating to that group".[1] In a topological context, Dingeldey (1890, p. iv) uses German *topologisch invariant* 'topologically invariant'.

Note. For the history of invariant theory, see Wolfson (2008), Parshall (1989), and Crilly (1986).

References

Cayley, Arthur (1845a). "Note sur deux formules données par M. M. Eisenstein et Hesse". *Journal für die reine und angewandte Mathematik* 29, pp. 54–57.

Cayley, Arthur (1845b). "On the theory of linear transformations". *The Cambridge and Dublin Mathematical Journal* 4, pp. 193–209.

Cayley, Arthur (1846). "On linear transformations". *The Cambridge and Dublin Mathematical Journal* 1, pp. 104–122.

Cayley, Arthur (1851). "Note sur la théorie des Hyperdéterminants". *Journal für die reine und angewandte Mathematik* 42, pp. 368–371.

Crilly, Tony (1986). "The Rise of Cayley's Invariant Theory (1841-1862)". *Historia Mathematica* 13, pp. 241–254.

Dingeldey, Friedrich (1890). *Topologische Studien über die aus ringförmig geschlossenen Bändern durch gewisse Schnitte erzeugbaren Gebilde*. Leipzig.

Gauß, Carl Friedrich (1827). "Disquisitiones circa superficies curvas". *Göttingische gelehrte Anzeigen* (1827), pp. 1761–1768.

Gauß, Carl Friedrich (1828). *Disquisitiones generales circa superficies curvas*. Göttingen.

Gauß, Carl Friedrich (1902). *General Investigations of Curved Surfaces*. Translated with notes and a bibliography by James Caddall Morehead and Adam Miller Hiltebeitel. Princeton.

Klein, Felix (1872). *Vergleichende Betrachtungen über neuere geometrische Forschungen*. Programm zum Eintritt in die philosophische Facultät und den Senat der k. Friedrich-Alexanders-Universität zu Erlangen. Erlangen.

Klein, Felix (1893). "A comparative review of recent researches in geometry". *Bulletin of the New York Mathematical Society* 2, pp. 215–249.

Klein, Felix (2016). *Elementary Mathematics from a Higher Standpoint*. Volume II: Geometry. Translated by Gert Schubring. Berlin and Heidelberg.

Parshall, Karen Hunger (1989). "Towards a History of Nineteenth-Century Invariant Theory". In: *The History of Modern Mathematics*. Volume I: Ideas and Their Reception. Ed. by David E. Rowe and John McCleary. Boston, pp. 157–206.

Reich, Karin (1994). *Die Entwicklung des Tensorkalküls*. Vom absoluten Differentialkalkül zur Relativitätsgeometrie. Basel.

---

[1] "Es ist eine Mannigfaltigkeit und in derselben eine Transformationsgruppe gegeben. Man entwickele die auf die Gruppe bezügliche Invariantentheorie". Translation from Klein (1893, p. 219), slightly adapted.

Sylvester, James Joseph (1851a). “On a remarkable Discovery in the Theory of Canonical Forms and of Hyperdeterminants”. *The London, Edinburgh, and Dublin Philosophical Magazine and Journal of Science* 2, pp. 391–410.
Sylvester, James Joseph (1851b). “On the general theory of associated algebraical forms”. *The Cambridge and Dublin Mathematical Journal* 6, pp. 289–293.
Wolfson, Paul R. (2008). “George Boole and the origins of invariant theory”. *Historia Mathematica* 35, pp. 37–46.

inverse

In book V, def. 13 of the *Elements*, Euclid defined what was later called an *inverse ratio*: “*Inverse ratio* means taking the consequent as antecedent in relation to the antecedent as consequent”.[1] Euclid here uses the word *ανάπαλιν*, which can be translated as ‘contrariwise’ or ‘inversely’ (LSJ).[2] In the translations of the *Elements*, however, the term *inverse* does not appear until the 16th century.

For example, in the various Latin translations of the 12th century, *eversa* was used by Adelard of Bath (Busard 1983, p. 146) and by an unknown translator, who translated directly from Greek (Busard 1987, p. 109).

Latin *inversa ratio* was chosen by Ramus (1555, p. 47) and also appears in Clavius’ edition of the *Elements* (Euclid 1574, fol. 161^r).[3] French *raison inverse* was used by Forcadel (Euclid 1564, fol. 128^v). The English term *inverse ratio* was used by Barrow (Euclid 1660, p. 93); Billingsley had earlier used *converse ratio* (Euclid 1570, fol. 133^v). As for German, Xylander chose the noun *Umkehrung*,[4] from a verb lit. meaning ‘to turn around’ (Euclid 1562, p. 131). The corresponding adjective *umgekehrt* is later found where other languages use *inverse*.

The word *inverse* in connection with the composition of mathematical objects does not come up until the very end of the 18th century. Lagrange (1797, p. 101), for example, uses *fonction inverse* in passing, without giving a definition. At the same time, the term is also found in arithmetic: “the quantities A, $\frac{1}{A}$, are [...] said to be *inverse* or *reciprocal* of one another” (Vilant 1798, pp. iii f.), a use which appears to be the earliest in English.[5] Develey (1799, p. 28), with reference to the equation $--5=+5$, writes that it should be read as “*the inverse of* $-5=+5$”.[6]

The term was later applied to other objects. In particular, Cauchy (1844, p. 163) used it with respect to permutations: “it is natural to say that the two substitutions [permutations], $\binom{B}{A}$, $\binom{A}{B}$ are *inverse* to each other”.[7] Cauchy’s use appears to be the first in connection with group theory.

Cayley (1858, p. 21) applied the term to matrices and mentions *reciprocal* as a possible alternative; he writes L^{-1} for the inverse of L and remarks that “it may be termed the inverse or reciprocal matrix [...] a matrix which, compounded either as first or second component matrix with the original matrix, gives the matrix unity”.[8]

In connection with groups, the German term *entgegengesetzt*, lit. ‘opposite’, was used by Dyck (1882, p. 5), while Weber (1882, p. 303) preferred *reciprok* at the same

---

[1] Translation from Euclid (1956, p. 114). [2] As pointed out by Heath in Euclid (1956, p. 134), the terminology goes back to Aristotle. [3] The term also appears in Ramus (1569, p. 224). [4] He wrote *Umbkerung*. [5] The OED dates the term to 1813. For another early use in English, see [anon.] (1804, p. 549), where the term *inverse quantity* as used by Carnot (1803, p. xxiij) is discussed. [6] “Lisez: *l’inverse de* $-5=+5$”. [7] “[...] est naturel de dire que les deux substitutions, $\binom{B}{A}$, $\binom{A}{B}$ sont *inverses* l’une de l’autre”. [8] French *matrice inverse* was already used in Cayley (1855, p. 283).

time. These terms remained in use in the 20th century, but *inverse* gradually became more popular. It was used, in particular, by Burnside (1897, p. 12), de Séguier (1904, p. 8), and later by Speiser (1923, p. 1) and in the influential book on modern algebra by van der Waerden (1930, p. 15).

Note. Various remarks on the history of the concept can be found in Wußing (1984).

References

[anon.] (1804). "Carnot's Geometry of Position". *The Critical Review* 2, pp. 546–556.

Burnside, William Snow (1897). *Theory of Groups of Finite Order*. Cambridge.

Busard, H.L.L. (1983). *The First Latin Translation of Euclid's Elements Commonly Ascribed to Adelard of Bath*. Books I-VIII and books X.36-XV.2. Toronto.

Busard, H.L.L. (1987). *The Mediaeval Latin Translation of Euclid's Elements Made Directly from the Greek*. Stuttgart.

Carnot, L.N.M. (1803). *Géométrie de position*. Paris.

Cauchy, Augustin-Louis (1844). *Exercises d'analyse et de physique mathématique*. Tome troisième. Paris.

Cayley, Arthur (1855). "Remarques sur la notation des fonctions algébriques". *Journal für die reine und angewandte Mathematik* 50, pp. 282–285.

Cayley, Arthur (1858). "A Memoir on the Theory of Matrices". *Philosophical Transactions of the Royal Society of London* 148, pp. 17–37.

Develey, Isaac Emmanuel Louis (1799). *Introduction à l'algèbre*. Lausanne.

Dyck, Walther (1882). "Gruppentheoretische Studien". *Mathematische Annalen* 20, pp. 1–44.

Euclid (1562). *Die Sechs Erste Bücher Euclidis*. Auß Griechischer sprach in die Teütsch gebracht ... Durch Wilhelm Holtzmann genannt Xylander. Basel.

Euclid (1564). *Les Six Premiers Livres des Elements*. Traduicts et commentez par Pierre Forcadel de Besies. Paris.

Euclid (1570). *The Elements of Geometrie of the most auncient Philosopher Euclide of Megara*. Faithfully (now first) translated into the Englishe toung, by H. Billingsley. London.

Euclid (1574). *Elementorum libri XV*. Ed. by Christopher Clavius. Roma.

Euclid (1660). *Euclide's Elements*. The whole fifteen Books compendiously demonstrated by Isaac Barrow. London.

Euclid (1956). *The Thirteen Books of Euclid's Elements*. Volume II. Books III-IX. Ed. by Thomas L. Heath. New York.

Lagrange, Joseph-Louis (1797). *Théorie des fonctions analytiques*. Paris.

Ramus, Petrus (1555). *Arithmeticae libri tres*. Paris.

Ramus, Petrus (1569). *Scholarum Mathematicarum, Libri Unus et Triginta*. Basel.

de Séguier, Jean Armand (1904). *Théorie des groupes finis*. Éléments de la théorie des groupes abstraits. Paris.

Speiser, Andreas (1923). *Die Theorie der Gruppen von endlicher Ordnung*. Berlin and Heidelberg.

Vilant, Nicolas (1798). *The Elements of Mathematical Analysis, Abridged, for the Use of Students*. Edinburgh.

van der Waerden, Bartel Leendert (1930). *Moderne Algebra*. Erster Teil. Berlin.

Weber, Heinrich (1882). "Beweis des Satzes, dass jede eigentlich primitive quadratische Form unendlich viele Primzahlen darzustellen fähig ist". *Mathematische Annalen* 20, pp. 301–329.

Wußing, Hans (1984). *The Genesis of the Abstract Group Concept*. A Contribution to the History of the Origin of Abstract Group Theory. Tr. Abe Shenitzer. Cambridge, Mass.

***inversion* (combinatorics)**

In 1750, Cramer published the rule for solving a system of linear equations that was later named after him.[1] Cramer's notation differed from the modern one. He wrote a system of equations in the form

$$\begin{aligned}&A^1 = Z^1z + Y^1y + X^1x + V^1v{+}\&\text{c}.\\&A^2 = Z^2z + Y^2y + X^2x + V^2v{+}\&\text{c}.\\&\ldots\end{aligned}$$

Cramer placed the indices, which he called *exponents* (*exposants*), at the top and defined what he called a *derangement* (*dérangement*): "When an exponent is followed in the same term immediately or later on by an exponent less than it, I will call this a *derangement*" (Cramer 1750, p. 658).[1] For example, in the term $Z^1Y^2X^3$ there is no derangement, while in $Z^3Y^1X^2$ there are two.

Laplace (1776, p. 294), although referring to Cramer, used *variation* instead. The term now common was introduced by Gergonne (1813-14, p. 149) as French *inversion*. His example was followed by Cauchy (1841, p. 149) and this may have contributed to settling the terminology.

Note. For brief historical notes on the terminology, see Muir (1960, pp. 13, 25, 134, 276) and Netto and Vogt (1904, p. 68).

References

Cauchy, Augustin-Louis (1841). *Exercises d'analyse et de physique mathématique*. Tome deuxième. Paris.

Cramer, Gabriel (1750). *Introduction à l'analyse des lignes courbes algébriques*. Genève.

Gergonne, Joseph Diez (1813-14). "Développement de la théorie donnée par M. Laplace, pour l'élimination au premier degré". *Annales de mathématiques pures et appliquées* 4, pp. 148–155.

Katz, Victor J. and Karen Hunger Parshall (2014). *Taming the Unknown*. A History of Algebra from Antiquity to the Early Twentieth Century. Princeton.

Knobloch, Eberhard (2018). "Determinant Theory, Symmetric Functions, and Dyadic". In: *The Oxford Handbook of Leibniz*. Ed. by Maria Rosa Antognazza. New York, pp. 225–246.

Laplace, Pierre-Simon (1776). "Recherches sur le calcul intégral et sur le système du monde". *Mémoires de l'Académie royale des sciences* (1772, seconde partie), pp. 267–376.

Muir, Thomas (1960). *The Theory of Determinants in the Historical Order of Development*. Volume I. General and Special Determinants up to 1841. New York.

Netto, Eugen and Henri Vogt (1904). "Analyse combinatoire et théorie des déterminants". *Encyclopédie des sciences mathématiques pures et appliquées* I.2, pp. 63–132.

***inversion* (geometry)**

The origins of geometrical inversion has been found in the works of Apollonius (see, e. g., Rosenfeld 1988, pp. 114–116); it was independently rediscovered by a number of mathematicians in the 19th century. The origin of the principle of inversion during that time can be regarded as rather "complicated" because of "many scattered and mostly independent contributors" (Lützen 1990, p. 731).

There are various types of inversion; the most basic one is defined with respect to a circle (see fig. 58): "If X be a circle, O its centre, P and Q two points on any

---

[1] Knobloch (2018) shows that Leibniz already knew this rule. [1] "Quand un exposant est suivi dans le même terme, médiatement ou immédiatement, d'un exposant plus petit que lui, j'appellerai cela un *dérangement*". Translation Katz and Parshall (2014, p. 341), slightly adapted.

radius, such that the rectangle $OP \cdot OQ =$ square of the radius, then P and Q are called *inverse points* with respect to the circle" (Casey 1892, p. 95).

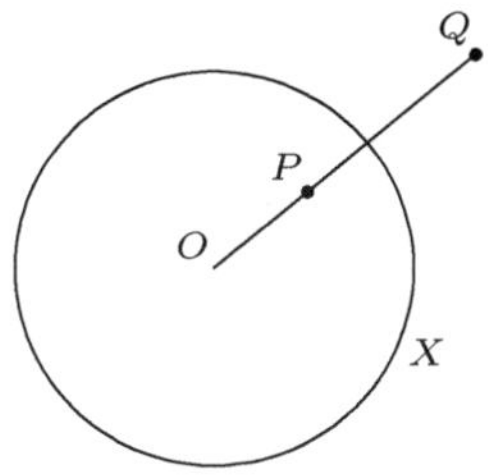

Figure 58: Circle inversion. From Casey (1892, p. 95).

It is known that in the 1820s, Steiner was already in full possession of the concept; this became obvious when Bützberger (1913, pp. 49–60) published some corresponding manuscripts from 1824.[2] In print, Steiner (1826, p. 253, 1833, p. 66) only alluded to the concept.

The French term *inverse* (of a curve) appears in Quetelet (1827, p. 105); it was Bellavitis (1836, pp. 126f.) who first used the terminology systematically, including the term *centre of inversion* (Italian *centro d'inversione*).

The English term *inverse* was, again independently, used by Ingram and Stubbs in 1843. Ingram (1843, pp. 57f.) defined: "Conceive a curve or surface B to be generated from a curve or surface A, by changing the lengths of the radii vectores of the latter to their reciprocals, without altering their direction. I propose to call B the curve or surface *inverse to* A". Shortly afterwards, Stubbs (1843, p. 145), referring to Ingram's paper, acknowledged that "Mr. Ingram has [...] noticed the generation of those surfaces whose properties I intend to develope [sic], and anticipated me in the name which I had proposed to apply to them" and additionally suggested that "the change of a curve into its inverse may be called the method of inversion".

A transformation that preserves circles was called a *Kreisverwandtschaft* by Möbius (1853, p. 20). The term is made up of the German words for 'circle' and Möbius' term *Verwandtschaft*, lit. 'relationship', which he used for 'transformation'.[1] The name *transformation by reciprocal radii* (*transformation par rayons vecteurs réciproques*) is due to Liouville (1847, p. 276).[2]

Note. A detailed discussion of the history of inversion was given by Patterson (1933); see also Zacharias (1914, pp. 1023–1036), Kötter (1901, pp. 98–106), and Lützen (1990, pp. 727–739), for example.

---

[2] In his manuscripts, Steiner used the somewhat dramatic terms *Wiedergeburt und Auferstehung* 'rebirth and resurrection' (Bützberger 1913, p. 51). [1] An extensive study is found in Möbius (1855), a work which Patterson (1933, p. 176) describes as "[t]he culmination of all these various formulations of the principle of inversion". [2] In a letter to Liouville from 1845, Thomson had called inverse points with respect to a sphere *points réciproques* (Thomson 1845, p. 364). For the term *rayon vecteur*, see *vector*.

References

Bellavitis, Giusto (1836). "Teoria delle figure inverse, e loro nella Geometria elementare". *Annali delle Scienze del Regno Lombardo-Veneto* 6, pp. 126–141.

Bützberger, Friedrich (1913). *Über bizentrische Polygone, Steinersche Kreis- und Kugelreihen und die Erfindung der Inversion*. Leipzig.

Casey, John (1892). *A Sequel to the First Six Books of the Elements of Euclid*. Sixth Edition, revised and enlarged. Ed. by Patrick A.E. Dowling. Dublin.

Ingram, John Kells (1843). "Geometrical Properties of Certain Surfaces". *Transactions of the Dublin University Philosophical Society* 1, pp. 57–63.

Kötter, Ernst (1901). "Die Entwickelung der synthetischen Geometrie von Monge bis auf Staudt (1847)". *Jahresbericht der Deutschen Mathematiker-Vereinigung* 5, pp. 1–475.

Liouville, Joseph (1847). "Note au sujet de l'article précédent". *Journal de mathématiques pures et appliquées* 12, pp. 265–290.

Lützen, Jesper (1990). *Joseph Liouville 1809-1882*. Master of Pure and Applied Mathematics. New York.

Möbius, August Ferdinand (1853). "Ueber eine neue Verwandtschaft zwischen ebenen Figuren". *Berichte über die Verhandlungen der Königlich Sächsischen Gesellschaft der Wissenschaften zu Leipzig (Mathematisch-Physische Klasse)* 5, pp. 14–24.

Möbius, August Ferdinand (1855). "Die Theorie der Kreisverwandtschaft in rein geometrischer Darstellung". *Abhandlungen der Mathematisch-Physischen Klasse der Königlich-Sächsischen Gesellschaft der Wissenschaften* 2, pp. 529–595.

Patterson, Boyd C. (1933). "The Origins of the Geometric Principle of Inversion". *Isis* 19, pp. 154–180.

Quetelet, Adolphe (1827). "Résumé d'une nouvelle théorie des caustiques, suivi de différentes applications à la théorie des projections stéréographiques". *Nouveaux mémoires de l'Académie Royale des Sciences et Belles-Lettres de Bruxelles* 4, pp. 79–113.

Rosenfeld, Boris A. (1988). *A History of Non-Euclidean Geometry*. Evolution of the Concept of a Geometric Space. Translated by Abe Shenitzer. New York.

Steiner, Jacob (1826). "Einige geometrische Betrachtungen". *Journal für die reine und angewandte Mathematik* 1, pp. 161–184, 252–288.

Steiner, Jacob (1833). *Die geometrischen Konstructionen, ausgeführt mittelst der geraden Linie und Eines festen Kreises*. Berlin.

Stubbs, John William (1843). "On the Application of a New Principle to the Geometry of Curves and Curved Surfaces". *Transactions of the Dublin University Philosophical Society* 1, pp. 145–155.

Thomson, William (1845). "Extrait d'une lettre de M. William Thomson". *Journal de mathématiques pures et appliquées* 10, pp. 364–367.

Zacharias, Max (1914). "Elementargeometrie und elementare nichteuklidische Geometrie in synthetischer Behandlung". *Encyklopädie der mathematischen Wissenschaften mit Einschluss ihrer Anwendungen* III AB 9, pp. 859–1172.

involute

The involute can be traced back to Huygens (1673, pp. 59f.), who called it a curve *descripta ex evolutione* 'described by evolution':

> If in one part of the cavity, a thread or flexible line is understood to be wound around one of the curved lines; and by keeping one end of that thread fixed to the curve, with the other end drawn away thus so that the part which is free is always kept extended; then another curve is seen to be described by the free end of this thread. Moreover this curve can be said to be described by evolution.[1]

[1] "Si lineae, in unam partem cave, filum seu linea flexilis circumplicata intelligatur, & filum seu linea flexilis circumplicata intelligatur, & manente una fili extremitate illi affixa, altera extremitas abducatur, ita ut pars ea quae soluta est semper extensa maneat; manifestum est curvam quandam aliam hac fili extremitate descibi. Vocetur autem ea, Descripta ex evolutione". Translation from Huygens (2013, p. 61).

From this relation to what Huygens called *evolution*, the term *evolvent* can be explained. The latter term was used in 18th-century English and is found, for example, in Chamber's *Cyclopaedia*, where it is explained as "a Term some Writers use for the Curve resulting from the Evolution of a Curve" (Chambers 1728, s. v. evolvent); it seems to have become less common during the 18th century. Hutton (1795, p. 452) describes it as "a term used by some writers for the Involute". It seems to have been used first in English by Gregory (1706, p. 236) as *evoluent line*.

As for *involute*, which was and still is more commonly used than *evolvent*, it appears in Nieuwentijt (1695, p. 178) as Latin *involuta*; the word appears as English *involute* in Hayes (1704, p. 109). The French term *développante* was suggested by de Fontenelle (1714, p. 64) as a counterpart to *développée* (evolute)

See also *evolute*.

Note. Huygens' construction of the involute is discussed by Yoder (1988, passim).

References

Chambers, Ephraim (1728). *Cyclopaedia, or, An universal dictionary of arts and sciences*. Volume the First. London.

de Fontenelle, Bernard (1714). "Sur le rayon de la développée". *Histoire de l'Académie royale des sciences* (1712), pp. 64–68.

Gregory, David (1706). "The Properties of the *Catenaria*". *Miscellanea Curiosa* 2, pp. 219–250.

Hayes, Charles (1704). *A Treatise of Fluxions*. London.

Hutton, Charles (1795). *A Philosophical and Mathematical Dictionary*. In two volumes. Vol. I. London.

Huygens, Christiaan (1673). *Horologium oscillatorium*. Paris.

Huygens, Christiaan (2013). "Horologium oscillatorium". Translated and annotated by Ian Bruce. URL: `http://www.17centurymaths.com/contents/huygenscontents.html` (retrieved Oct. 12, 2019).

Nieuwentijt, Bernard (1695). *Analysis infinitorum seu curvilineorum proprietates ex polygonorum natura deductae*. Amsterdam.

Yoder, Joella G. (1988). *Unrolling Time*. Christiaan Huygens and the Mathematization of Nature. Cambridge.

involution **and** ***involutorial***

In mathematics, the word has several senses, the oldest of which is found in geometry. As is well known, the term was introduced by Desargues (1639, p. 4):

> And when in a straight line AH there are three pairs of points, BH, CG, DF so disposed that the two points of each of these pairs are either both mixed with the two points of each of the other two pairs or both unmixed with them; and the related rectangles of the parts between these points are to one another as their twins, taken in the same order, are to one another: such an arrangement of these three pairs of points in a line is called here an *Involution*.[1]

Rosenfeld (1988, p. 138) explains the term and also links it to its algebraical meaning: "The term *involution*, introduced [...] by Desargues, is also[2] of botanical

[1] "Et quand en vne droicte AH, il y a comme cela trois couples de poincts BH, CG, DF ainsi conditionnées, à sçauoir que les deux poincts de chacune des couples soient de mesme, ou meslez ou demeslez, aux deux poincts de chacune des autres couples. Et que les rectangles ainsi relatifs des pieces d'entre ces poincts soient entre eux comme leurs gemeaux, pris de mesme ordre, sont entre eux vne telle disposition de ces trois couples de poincts en vne droicte, est icy nommée *Inuolution*". Translation from Field and Gray (1986, p. 77). [2] Desargues used many terms from botany in the text.

origin, and denotes, literally, the twisted state of young leaves. Since involution is a projective transformation of a line that coincides with its inverse, the term *involution* is today used to designate just such transformations. In other words, a transformation T is an involution if $T = T^{-l}$ or, equivalently, if T^2 is the identity transformation".[3]

Desargues' term was hardly used at all until the 19th century,[4] when Desargues' usage of the term was pointed out by Brianchon (1817, p. 66) and Poncelet (1822, p. 95). The German term appears in Plücker (1831, p. 29), the English one in Chasles (1841, p. 78). According to Struik (1953, p. 19), "[i]t is to Chasles that we mainly owe the theory of involution in its present form"; Struik refers to Chasles (1837, pp. 308–327).

The algebraical sense mentioned above comes up in the second half of the 19th century; Italian *trasformazione involutoria* was used by Cremona (1861, p. 90), Darboux (1870, p. 352) French *transformation involutive*. Slightly later, the term begins to be used without any connection to projective geometry. Thus, Peano (1888, p. 156) defines: "A non-numerical substitution, whose square is a number, is called an *involution*. If T is an involution, $T^2 P$ coincides with P".[1] In similar terms, Wiener (1890, p. 259) defines an involutorial transformation (*involutorische Verwandtschaft*), and *involutoric transformations* are considered by Newson (1895, p. 78) and Emch (1895, p. 259).

Note. There is an older algebraic use of the term, which is found in the 17th century in the sense 'to raise to the power'; the term may go back to Rahn (1659, p. 2), who used the German verb *involvieren* in this sense. His book was translated into English, where the definition runs as follows: "*Involution* is multiplying any quantity into its self, and that Product into the first quantity, and this last into the first, & c." (Rahn 1668, p. 2). The term is still found in the 19th century.[2]

Note. The concept as used by Desargues is discussed in Anglade and Briend (2017), Field and Gray (1986, pp. 47–59), Hogendijk (1991), and Baltus (2020); see also Struik (1953, p. 19) for a concise overview of the history of the concept.

References

Anglade, Marie and Jean-Yves Briend (2017). "La notion d'involution dans le *Brouillon Project* de Girard Desargues". *Archive for History of Exact Sciences* 71, pp. 543–588.

Baltus, Christopher (2020). "Philippe de la Hire: Was He Desargues' Schüler?" In: *Research in History and Philosophy of Mathematics*. The CSHPM 2018 Volume. Ed. by Maria Zack and Dirk Schlimm. Cham, pp. 157–172.

Blondel, François (1673). *Résolution des quatre principaux problèmes d'architecture*. Paris.

Brianchon, Charles Julien (1817). *Mémoire sur les lignes du second ordre*. Paris.

Chasles, Michel (1837). *Aperçu historique sur l'origine et le développement des méthodes en géométrie*. Bruxelles.

Chasles, Michel (1841). *Two Geometrical Memoirs on the General Properties of Cones of the Second Degree and on the Spherical Conics*. Translated from the French ... by Charles Graves. Dublin.

Cremona, Luigi (1861). "Intorno alla trasformazione geometrica di una figura piana in un'altra pur piana, sotto la condizione che ad una retta qualunque di ciascuna delle figure corrisponda

[3] See also Hogendijk (1991, pp. 20f.). [4] A rare example is Blondel (1673, p. 18). [1] "Una sostituzione non numerica, il cui quadrato sia un numero, dicesi *involuzione*. Se T è un'involuzione, $T^2 P$ coincide con P". Translation from Peano (2000, p. 132). [2] This usage was pointed out by Tropfke (1980, p. 293).

nell'altra una sola retta". *Rendiconto delle sessioni dell'accademia delle scienze dell'Istituto di Bologna* (1861-62), pp. 88–91.
Darboux, Gaston (1870). "Sur les systèmes linéaires de coniques et de surfaces du second ordre". *Bulletin des sciences mathématiques et astronomiques* 1, pp. 348–358.
Desargues, Girard (1639). *Brouillon project d'une atteinte aux evenemens des rencontres du Cone avec un Plan*. Paris. URL: `http://gallica.bnf.fr/ark:/12148/bpt6k105071b`.
Emch, Arnold (1895). "Involutoric Transformation of the Straight Line". *Kansas University Quarterly* 4, pp. 111–116.
Field, Judith V. and Jeremy Gray (1986). *The Geometrical Work of Girard Desargues*. New York.
Hogendijk, Jan P. (1991). "Desargues' *Brouillon Project* and the *Conics* of Apollonius". *Centaurus* 34, pp. 1–43.
Newson, Henry Byron (1895). "Continuous Groups of Projective Transformations Treated Synthetically". *Kansas University Quarterly* 4, pp. 71–92.
Peano, Guiseppe (1888). *Calcolo geometrico secondo l'Ausdehnungslehre di H. Grassmann, preceduto dalle operazioni della logica deduttiva*. Torino.
Peano, Guiseppe (2000). *Geometric Calculus*. According to the Ausdehnungslehre of H. Grassmann. Translated by Lloyd C. Kannenberg. Boston et al.
Plücker, Julius (1831). *Analytisch-geometrische Entwicklungen*. Zweiter Band. Essen.
Poncelet, Jean Victor (1822). *Traité des propriétés projectives des figures*. Paris.
Rahn, Johann Heinrich (1659). *Teutsche Algebra*. Zürich.
Rahn, Johann Heinrich (1668). *An Introduction to Algebra*. Translated out of the High-Dutch into English by Thomas Brancker. London.
Rosenfeld, Boris A. (1988). *A History of Non-Euclidean Geometry*. Evolution of the Concept of a Geometric Space. Translated by Abe Shenitzer. New York.
Struik, Dirk Jan (1953). *Lectures on Analytic and Projective Geometry*. Cambridge, Mass.
Tropfke, Johannes (1980). *Geschichte der Elementarmathematik*. Band 1: Arithmetik und Algebra. Vollständig neu bearbeitet von Kurt Vogel, Karin Reich, Helmuth Gericke. Berlin and New York.
Wiener, Hermann (1890). "Ueber geometrische Analysen". *Berichte über die Verhandlungen der Königlich Sächsischen Gesellschaft der Wissenschaften zu Leipzig (Mathematisch-Physische Klasse)* 42, pp. 245–267.

irrational

See *rational*.

irreducible

This adjective *irreducible* occurs in combination with various mathematical terms from different fields; its earliest usages are found in 17th-century algebra. The verb *reduce* was used at this time in connection with the factorization of equations, and an irreducible equation was one that could not be factored. In this sense, Latin *irreducibilis* was occasionally used by Hudde (1659, e. g., p. 505); mostly, Hudde spoke of equations that are *not reducible*.[1] Prestet (1675, p. 400) used French *irréductible* in the same sense.

The term *irreducible case* also came up at this time; its exact origin is unclear, but it appears to be a French creation. de Lagny (1692, p. 41) used French *cas irréductible*; he did not formally define or introduce it,[2] and the term already seems to have been in use at the time. The concept is older, of course, having been dealt with by Cardano and Bombelli, in particular, but neither of them used the term.[3]

---

[1] On Hudde's work, see Haas (1956). [2] He does give a definition in de Lagny (1697, p. 469).
[3] For the history of the concept, see Confalonieri (2013), who cites (p. 45) a text from 1781 as the earliest usage of *casus irreducibilis* known to her.

The use of *irreducible* in connection with fractions also seems to date from the second half of the 17th century; for example, Newton used Latin *irreducibilis* in a manuscript that dates from around 1676 (Turnbull 1960, p. 173).

The modern history of irreducibility may be said to begin with Gauß, who discussed cyclotomic polynomials in the final section of his *Disquisitiones Arithmeticae.* In particular, he proved the irreducibility of the cyclotomic polynomial $x^{p-1}+\cdots+x+1$ (p prime) (Gauß 1801, § 341). Gauß does not use the word *irreducible,* however.

The term *irreducible equation* goes back to Abel (1829, p. 133), who gave the following definition (in a footnote): "An equation $\varphi x = 0$, in which the coefficients are rational functions of a certain number of known quantities *a*, *b*, *c*, ..., is called *irreducible* when it is impossible to express any of its roots by an equation of lower degree, in which the coefficients are also rational functions of *a*, *b*, *c*, [...]".[1] French *réductible* was also used by Galois: "An equation is said to be reducible when it admits rational divisors; irreducible in the contrary case" (Galois 2011, p. 108),[2] and then by Wantzel (1837, p. 368). De Morgan (1836, pp. 23, 77), referring to Abel, used English *irreducible equation.*

A crucial development took place in the 19th century. In the algebraic structures originally considered, the concepts "irreducible" and "prime" are identical. However, after the discovery that unique factorization does not hold in general (see *unique factorization domain*), it also gradually became clear that the concepts generally are not the same. This distinction between irreducibility and primality was pointed out by Kummer (1847, p. 319) in connection with cyclotomic numbers $f(\alpha) = a + a_1\alpha + a_2\alpha^2 + \cdots + a_{\lambda-1}\alpha^{\lambda-1}$:[3]

> However, I have noticed that, even if $f(\alpha)$ can in no way be decomposed into complex factors, it yet does not have the true nature of a complex prime number, because it usually lacks the first and most important property of prime numbers: that is, that the product of two prime numbers is not divisible by any prime number different from them. These numbers $f(\alpha)$ thus have the nature of composite numbers even though they are not decomposable into complex factors; but the factors are then not actual, but *ideal complex numbers*.[4]

The distinction also played a central role in Dedekind's theory of algebraic numbers. In particular, Dedekind observed that[5]

---

[1] "Une équation $\varphi x = 0$, dont les coefficients sont des fonctions rationnelles d'un certain nombre de quantités connues *a*, *b*, *c*, [...] s'appelle *irréductible*, lorsqu'il est impossible d'exprimer aucune de ses racines par une équation moins élevée, dont les coefficiens soient également des fonctions rationnelles de *a*, *b*, *c*, ...". Translation from Sørensen (2010, p. 158). – Abel (1826, pp. 72, 84) had earlier used, without definition, German *irreductible Gleichung.* [2] "Une équation est dite réductible quand elle admet des diviseurs rationnels; irréductible dans le cas contraire". Translation from Galois (2011, p. 109). [3] With α being an imaginary root of the equation $\alpha^\lambda = 1$, λ prime, and a, a_1, a_2, etc., "integers". [4] "Ich habe nun aber bemerkt, daß, wenn auch $f(\alpha)$ auf keine Weise in complexe Factoren zerlegt werden kann, sie deshalb noch nicht die wahre Natur einer complexen Primzahl hat, weil sie schon gewöhnlich der ersten und wichtigsten Eigenschaft der Primzahlen ermangelt: nämlich, daß das Product zweier Primzahlen durch keine von ihnen verschiedene Primzahl theilbar ist. Es haben vielmehr solche Zahlen $f(\alpha)$, wenn gleich sie nicht in complexe Factoren zerlegbar sind, dennoch die Natur der zusammengesetzten Zahlen; die Factoren aber sind alsdann nicht wirkliche, sondern *ideale complexe Zahlen*". Translation from Goldstein and Schappacher (2007, p. 43). [5] In the quotation, $\mathfrak{o}$ refers to what Dedekind called an *Ordnung*, a structure that corresponds to a ring.

> a very peculiar property crops up for infinitely many such [algebraic number] fields Ω [...] A number in $\mathfrak{o}$ is called *decomposable* when it is the product of two numbers in $\mathfrak{o}$, neither of which is a unit; and *indecomposable* when this is not the case. Clearly every decomposable number μ is representable as a product of a *finite* number of indecomposable elements [...] But it frequently occurs that this decomposition is not totally determined, but rather there exist several *essentially different* compositions of the same number into indecomposable factors [...] This contradicts so many notions governing the character of prime numbers [...] that we therefore can no longer recognize indecomposable numbers as prime. In order to preserve the character of primes we therefore search for a criterion to characterize them that is more robust than the inadequate criterion of indecomposability (Dedekind 1871, pp. 449f.).[1]

His refined definition was: "*An integer μ, which is not a unit, will be called a* prime number *if every product $\eta\rho$ that is divisible by μ, has at least one factor η or ρ that is divisible by μ* (Dedekind 1871, p. 450).[2]

The term was later also applied to algebraic structures. For example, Emmy Noether (1921, p. 33) used *irreducible ideal* in the following way: "An ideal $\mathfrak{M}$ is called reducible if it can be expressed as the least common multiple of two proper divisors; otherwise $\mathfrak{M}$ is called irreducible".[3]

An earlier concept of irreducibility of structures is found in the late 19th century in connection with algebras or, as they were then often called, *number systems*. The terminology first appeared in Scheffers (1891, p. 317):

> A number system is called *reducible*, if its units [basis elements] may be chosen and arranged in two collections $e_1, e_2, ...$ and $\mathfrak{e}_1, \mathfrak{e}_2, ...$ so that each product of a unit e and a unit $\mathfrak{e}$, that is, each $e_i\mathfrak{e}_k$ and $\mathfrak{e}_k e_i$ is zero, and furthermore each product of two e's is expressed by the e's alone, each product of the $\mathfrak{e}$'s by the $\mathfrak{e}$'s alone. Then the whole system decomposes into two smaller ones $(e_1, e_2, \ldots)$ and $(\mathfrak{e}_1, \mathfrak{e}_2, \ldots)$, the products between which are zero.[4]

In modern terms, Scheffers defined an algebra A as reducible if it can be written as a direct sum of two proper subalgebras $A = B + C$ such that $BC = 0$. He explicitly remarked (Scheffers 1891, pp. 294, 317) that the importance of concept was pointed

[1] "Dafür tritt aber bei unendlich vielen solchen Körpern Ω ein höchst eigenthümliches Phänomen auf [...] Nennt man eine Zahl in $\mathfrak{o}$ *zerlegbar*, wenn sie das Product aus zwei Zahlen in $\mathfrak{o}$ ist, welche beide keine Einheiten sind, dagegen *unzerlegbar*, wenn dies nicht der Fall ist, so ist offenbar jede zerlegbare Zahl μ darstellbar als Product aus einer *endlichen* Anzahl von unzerlegbaren Zahlen [...]; aber es zeigt sich häufig, dass diese Zerlegung nicht eine vollkommen bestimmte ist, sondern dass mehrere *wesentlich verschiedene* Zerlegungen derselben Zahl in unzerlegbare Factoren existiren. Dies widerspricht so sehr dem in der rationalen Zahlentheorie herrschenden Begriffe des Primzahlcharakters [...], dass wir deshalb eine unzerlegbare Zahl als solche noch nicht als Primzahl anerkennen wollen; wir suchen daher für den wahren Primzahlcharakter ein kräftigeres Kriterium als diese unzulängliche Unzerlegbarkeit aufzustellen". Translation from Avigad (2004, pp. 36f.). [2] "*Eine ganze Zahl μ, welche keine Einheit ist, soll eine* Primzahl *heissen, wenn jedes durch μ theilbare Product $\eta\rho$ wenigstens einen durch μ theilbaren Factor η oder ρ besitzt*". Translation from Avigad (2004, p. 37). [3] "Ein Ideal $\mathfrak{M}$ heißt reduzibel, wenn es als kleinstes gemeinsames Vielfaches von zwei echten Teilern darstellbar ist; im entgegengesetzten Fall heißt $\mathfrak{M}$ irreduzibel". Translation from Noether (2014, p. 11). [4] "Ein Zahlensystem heisst *reducibel*, wenn seine Einheiten sich so wählen und in zwei Scharen $e_1, e_2, ...$ und $\mathfrak{e}_1, \mathfrak{e}_2, ...$ einreihen lassen, dass jedes Product einer Einheit e mit einer Einheit $\mathfrak{e}$, also jedes $e_i\mathfrak{e}_k$ und $\mathfrak{e}_k e_i$ Null ist und ferner jedes Product zweier e sich durch die e, jedes Product zweier $\mathfrak{e}$ sich durch die $\mathfrak{e}$ allein ausdrückt. Alsdann zerfällt nämlich das ganze System in zwei kleinere $(e_1, e_2, \ldots)$ und $(\mathfrak{e}_1, \mathfrak{e}_2, \ldots)$, deren Producte mit einander Null sind". Translation from Parshall (1985, p. 272) (slightly adapted). – For discussion of this passage, see Hawkins (1972, p. 253) and Parshall (1985, pp. 271–273).

out to him by Study. The terminology was quickly adopted. In English, it is found in Hawkes (1902, p. 92), who remarked that "Peirce's definitions of pure and mixed algebras[5] correspond exactly to the definitions of irreducible and reducible number systems used by Scheffers, except that the groups in a reducible system can contain no common units", and by Epsteen (1903, p. 438), who credited Scheffers for the term.

Note. For the history of the irreducibility of equations, see, e. g., Sørensen (2010), Frei (2007), and Cox (2011). For the further development, see, e. g., Edwards (1980), Neumann (2007), Avigad (2006), and Petri and Schappacher (2004). A linguistic discussion of decomposable mathematical objects is found in Becker (2005, ch. 9).

References

Abel, Niels Henrik (1826). "Beweis der Unmöglichkeit algebraische Gleichungen von höheren Graden, als dem vierten, allgemein aufzulösen". *Journal für die reine und angewandte Mathematik* 1, pp. 65–84.

Abel, Niels Henrik (1829). "Mémoire sur une classe particulière d'équations résolubles algébriquement". *Journal für die reine und angewandte Mathematik* 4, pp. 131–156.

Avigad, Jeremy (2004). "Dedekind's 1871 version of the theory of ideals". URL: `https://kilthub.cmu.edu/articles/Dedekind_s_1871_Version_of_the_Theory_of_Ideals/6491342/1` (retrieved July 25, 2007).

Avigad, Jeremy (2006). "Methodology and metaphysics in the development of Dedekind's theory of ideals". In: *The Architecture of Modern Mathematics*. Essays in History and Philosophy. Ed. by José Ferreirós and Jeremy Gray. Oxford, pp. 159–186.

Becker, Holger (2005). "Semantische und lexikalische Aspekte der mathematischen Fachsprache des 19. Jahrhunderts". PhD thesis. Carl von Ossietzky Universität Oldenburg. URL: `http://oops.uni-oldenburg.de/59/`.

Confalonieri, Sara (2013). "The telling of the unattainable attempt to avoid the *casus irreducibilis* for cubic equations: Cardano's *De Regula Aliza*". PhD thesis. Université Paris Diderot.

Cox, David A. (2011). "Why Eisenstein Proved the Eisenstein Criterion and Why Schonemann Discovered It First". *The American Mathematical Monthly* 118, pp. 3–21.

Dedekind, Richard (1871). "Ueber die Composition der binären quadratischen Formen". In: *Vorlesungen über Zahlentheorie von P.G. Lejeune Dirichlet*. Herausgegeben und mit Zusätzen versehen von R. Dedekind. Zweite, umgearbeitete und vermehrte Auflage. Braunschweig, pp. 380–497.

Edwards, Harold M. (1980). "The Genesis of Ideal Theory". *Archive for History of Exact Sciences* 23, pp. 321–378.

Epsteen, Saul (1903). "Semireducible Hypercomplex Number Systems". *Transactions of the American Mathematical Society* 4, pp. 437–444.

Frei, Günther (2007). "The Unpublished Section Eight: On the Way to Function Fields over a Finite Field". In: *The Shaping of Arithmetic after C. F. Gauss's Disquisitiones Arithmeticae*. Ed. by Catherine Goldstein, Norbert Schappacher, and Joachim Schwermer. Berlin and Heidelberg, pp. 159–198.

Galois, Évariste (2011). "The First Memoir". In: *The Mathematical Writings of Évariste Galois*. Ed. by Peter M. Neumann. Zürich, pp. 105–168.

Gauß, Carl Friedrich (1801). *Disquisitiones arithmeticae*. Leipzig.

Goldstein, Catherine and Norbert Schappacher (2007). "A Book in Search of a Discipline (1801-1860)". In: *The Shaping of Arithmetic after C. F. Gauss's Disquisitiones Arithmeticae*. Ed. by Catherine Goldstein, Norbert Schappacher, and Joachim Schwermer. Berlin and Heidelberg, pp. 3–65.

Haas, Karlheinz (1956). "Die mathematischen Arbeiten von Johann Hudde". *Centaurus* 4, pp. 235–284.

---

[5] This refers to Peirce (1881, p. 100).

Hawkes, Herbert Edwin (1902). “Estimate of Peirce’s Linear Associative Algebra”. *American Journal of Mathematics* 24, pp. 87–95.
Hawkins, Thomas (1972). “Hypercomplex Numbers, Lie Groups, and the Creation of Group Representation Theory”. *Archive for History of Exact Sciences* 8, pp. 243–287.
Hudde, Johannes (1659). “De reductione aequationum”. In: *Geometria, à Renato des Cartes*. Ed. by Frans van Schooten. Amsterdam, pp. 406–506.
Kummer, Ernst Eduard (1847). “Zur Theorie der complexen Zahlen”. *Journal für die reine und angewandte Mathematik* 35, pp. 319–326.
de Lagny, Thomas Fantet (1692). *Methodes nouvelles et abbregées pour l’extraction et l’approximation des racines*. Seconde edition. Paris.
de Lagny, Thomas Fantet (1697). *Nouveaux elemens d’arithmetique et d’algebre*. Paris.
De Morgan, Augustus (1836). *A Treatise on the Calculus of Functions*. London.
Neumann, Olaf (2007). “The *Disquisitiones Arithmeticae* and the Theory of Equations”. In: *The Shaping of Arithmetic after C. F. Gauss’s Disquisitiones Arithmeticae*. Ed. by Catherine Goldstein, Norbert Schappacher, and Joachim Schwermer. Berlin and Heidelberg, pp. 107–127.
Noether, Emmy (1921). “Idealtheorie in Ringbereichen”. *Mathematische Annalen* 83, pp. 24–66.
Noether, Emmy (2014). “Ideal Theory in Rings”. Translated by Daniel Berlyne. URL: `https://arxiv.org/abs/1401.2577` (retrieved Apr. 2, 2021).
Parshall, Karen Hunger (1985). “Joseph H. M. Wedderburn and the Structure Theory of Algebras”. *Archive for History of Exact Sciences* 32, pp. 223–349.
Peirce, Benjamin (1881). “Linear Associative Algebra”. *American Journal of Mathematics* 4, pp. 97–229.
Petri, Birgit and Norbert Schappacher (2004). “From Abel to Kronecker: Episodes from 19th Century Algebra”. In: *The Legacy of Niels Henrik Abel*. Ed. by Olav Arnfinn Laudal and Ragni Piene. Berlin and Heidelberg, pp. 227–266.
Prestet, Jean (1675). *Elemens des mathematiques*. Paris.
Scheffers, Georg (1891). “Zurückführung complexer Zahlensysteme auf typische Formen”. *Mathematische Annalen* 39, pp. 293–390.
Sørensen, Henrik Kragh (2010). *The Mathematics of Niels Henrik Abel*. Continuation and New Approaches in Mathematics During the 1820s. Aarhus.
Turnbull, H. W., ed. (1960). *The Correspondence of Isaac Newton*. Volume II. 1676–1687. Cambridge.
Wantzel, Pierre Laurent (1837). “Recherches sur les moyens de reconnaître si un Problème de Géométrie peut se résoudre avec la règle et le compas”. *Journal de mathématiques pures et appliquées* 2, pp. 366–372.

irreflexive relation

The term *reflexive relation* is due to Vailati (1891, p. 134), who used the noun *rifflessività*; he did not use a special name for irreflexive relations. At the time, numerous studies on logical relations were carried out, and new names for various kinds of relations were being suggested. In 1905, Huntington published an article entitled “The Continuum as a Type of Order”, in which he presented the following postulate: “*If* $a < b$, *then* a *and* b *are distinct*. Any relation $<$ which has this property is said to be *non-reflexive* throughout the class” (Huntington 1905, p. 158) . In a later publication, he used *irreflexive* instead (Huntington 1917, p. 10).[1]

The term *aliorelative* has also been used in this sense. It goes back to Peirce (1873, p. 369): “Simple relatives are divisible into those which contain elements of the form (A:A) and those which do not. The former express relations such as a thing may have to itself, the latter (as cousin of —, hater of —) relations which nothing can have to itself. The former may be termed *self-relatives*, the latter *alio-relatives*”. The term

[1] Huntington there mistakenly attributed the term to Peano.

was also used by Russell (1903, p. 320): "An aliorelative is a relation which no term can have to itself". Russel credited the term to Peirce, as did Schröder (1895, p. 12). Couturat (1905, p. 158) already suggested that *aliorelative* be replaced by "a word like *antireflexive*".[2]

References

Couturat, Louis (1905). *Les principes des mathématiques*. Paris.
Huntington, Edward V. (1905). "The Continuum as a Type of Order. An Exposition of the Modern Theory". *Annals of Mathematics* 7, pp. 15–43.
Huntington, Edward V. (1917). *The Continuum and other Types of Serial Order*. Second edition. Harvard.
Peirce, Charles Sanders (1873). "Description of a Notation for the Logic of Relatives, Resulting from an Amplification of the Conceptions of Boole's Calculus of Logic". *Memoirs of the American Academy of Arts and Sciences* 9, pp. 317–378.
Russell, Bertrand (1903). *The Principles of Mathematics*. Vol. I. Cambridge.
Schröder, Ernst (1895). *Vorlesungen über die Algebra der Logik* (Exakte Logik). Dritter Band. Leipzig.
Vailati, Giovanni (1891). "Le proprietà fondamentali delle operazioni della Logica deduttiva, studiate dal punto di vista d'una teoria generale delle operazioni". *Rivista di matematica* 1, pp. 127–134.

isochrone

As part of a dispute, Leibniz in 1687 challenged the Cartesians with the following problem: "To find the curve of descent along which a heavy body descends uniformly and approaches the horizontal by equal amounts in equal time intervals" (Leibniz 1687, p. 956).[1]

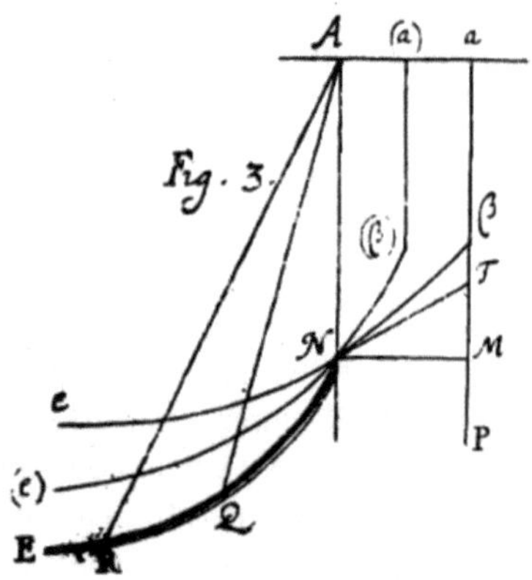

Figure 59: The isochrone. From Leibniz (1689, following p. 190).

He did not yet use the term *isochrone* (from the Greek words for 'equal' and 'time') in this publication. Huygens published his solution of the problem a month later (Huygens 1687). On this occasion, Leibniz took some notes and coined French

[2] "un mot comme *antiréflexive*". [1] "Trouver une ligne de descente, dans la quelle le corps pesant descende uniformément, & approche également de l'horison en temps égaux". Translation from Leibniz (2014, p. 1).

ligne isochrone; he recalled the problem and then stated: "This curve which one can now call *Isochronous* (after the discovery of this property) in any case is well known to the geometers, and has been the first of all the curves from ordinary geometry, *for which a rule of exact equality has been given*" (Gerhardt 1858, p. 239).[2]

In print, the corresponding Latin *linea isochrona* appears in Leibniz (1689, p. 196). In English, the term is found as *isochronal curve* in Hayes (1704, p. 145). The term *tautochrona* (*ταὐτός* 'same') was used in the same sense, for example by Bernoulli (1697, p. 207).

Note. For the history of the concept, see Hairer and Wanner (2008, pp. 134f.).

References

Bernoulli, Johann (1697). "Curvatura radii in diaphanis non uniformibus". *Acta Eruditorum* (May 1697), pp. 206–211.

Gerhardt, C.I., ed. (1858). *Leibnizens mathematische Schriften.* Zweite Abtheilung. Band I. Halle.

Hairer, Ernst and Gerhard Wanner (2008). *Analysis by Its History.* New York.

Hayes, Charles (1704). *A Treatise of Fluxions.* London.

Huygens, Christiaan (1687). "Solution du Problème proposé par M. L. dans les Nouvelles de la Republique des Lettres, du mois de Septembre 1687". *Nouvelles de la république des lettres* (Oct. 1687), pp. 1110–1111.

Leibniz, Gottfried Wilhelm (1687). "Réponse de M. L. à la Remarque de M. l'Abbé D. C. contenue dans l'Article 1. des ces Nouvelles, mois de Juin 1687, où il prétend soûtenir une Loi de la Nature avancées par M. Descartes". *Nouvelles de la république des lettres* (Sept. 1687), pp. 952–956.

Leibniz, Gottfried Wilhelm (1689). "De linea isochrona". *Acta Eruditorum* (Apr. 1689), pp. 195–198.

Leibniz, Gottfried Wilhelm (2014). "Papers relating to Isochronous curves". Transl. with notes by Ian Bruce. URL: `http://www.17centurymaths.com/contents/leibnizcontents.html` (retrieved Aug. 3, 2017).

isocline and *isoclinal*

This term originally came up in the context of magnetism, where it was used for lines drawn on a map such that the dip of the needle is the same in all points. According to Howarth (2003, pp. 63f.), these lines were first considered by Whiston (1719, p. 23), who called them *lines of equal dip*. The word *isocline*, from the Greek word for 'same' and 'slope', was suggested by Hansteen (1819, p. 225), who used German *isoklinische Linie*.

In mathematics, the term is used in connection with differential equations. It was used by Wedderburn (1903, p. 401):

> A differential equation of the first order $\phi(xyp) = 0$ besides defining the integral family, also defines a family of curves obtained by regarding p as an arbitrary constant in $\phi = 0$, and these curves have the property that all the integral curves, that intersect any particular curve of the second family, have the same direction (i. e., the same p) at the points of intersection. The latter family has been called the "Loci of Contacts of Parallel Tangents," by Hill [...][1], but, at the suggestion of Professor Chrystal, I propose to use in this paper the more convenient term "Isoclinal Family."

---

[2] "Cette ligne qu'on pourra maintenant appeller *Isochrone* (apres la decouverte de cette proprieté) est assez connue d'ailleurs aux Geometres, et *a esté la premiere de toutes les lignes courbes de la Geometrie ordinaire, à qui on ait donné une droite exactement egale*". Translation from Leibniz (2014, p. 2). [1] Wedderburn is referring to Hill (1889, p. 578).

While Wedderburn, who also spoke of *isoclinal curves* (ibid.), apparently thought that the term was new, it had indeed been suggested earlier, namely by Massau (1885, p. 697), who used French *courbes isoclines*.[2]

As pointed out by Tournès (1996, pp. 391f.), the concept itself is much earlier found in Bernoulli (1694, p. 435), who used the name *directrices*.[3]

Note. For a discussion of the history of the concept and the origin of the terminology, see Tournès (1996, pp. 389–396).

References

Bateman, Harry (1918). *Differential Equations*. London.

Bernoulli, Johann (1694). "Modus generalis construendi omnes aequationes differentiales primi gradus". *Acta Eruditorum* (Nov. 1694), pp. 435–437.

Bernoulli, Johann (1955). *Der Briefwechsel von Johann I. Bernoulli*. Band 1: Der Briefwechsel mit Jacob Bernoulli, dem Marquis de l'Hôpital u.a. Ed. by Otto Spiess. Basel.

Hansteen, Christopher (1819). *Untersuchungen über den Magnetismus der Erde*. Erster Teil. Die mechanischen Erscheinungen des Magneten. Trans. by P. Treschow Hanson. Christiania.

Hill, M.J.M. (1889). "On the c- and p-Discriminants of Ordinary Integrable Differential Equations of the First Order". *Proceedings of the London Mathematical Society* 19, pp. 561–589.

Howarth, Richard J. (2003). "Fitting Geomagnetic Fields before the Invention of Least Squares: II. William Whiston's Isoclinic Maps of Southern England (1719 and 1721)". *Annals of Science* 60, pp. 63–84.

Massau, Junius (1885). *Mémoire sur l'intégration graphique et ses applications*. Paris and Liège.

Tournès, Dominique (1996). "L'intégration approchée des équations différentielles ordinaires (1671-1914)". PhD thesis. Université Paris-Diderot.

Wedderburn, J.H.M. (1903). "On the isoclinal lines of a differential equation of the first order". *Proceedings of the Royal Society of Edinburgh* 24, pp. 400–408.

Whiston, William (1719). *The Longitude and Latitude Found by the Inclinatory or Dipping Needle*. London.

Zondadari, Enrico (1917). *Integrazione grafica e studio delle equazioni differenziali ordinarie del primo ordine coi metodi della geometria descrittiva*. Milano et al.

isogenous

The term *variétés isogènes* was introduced by Weil (1948, p. 96): "We agree to say that two Abelian varieties A and B are *isogenous* if they have the same dimension and if there is a homomorphism from A onto B".[1] The term is a neologism based on the Greek words for 'same' and 'kind'.

Weil's terminology was adopted by Rosenlicht (1953, p. 623), who later also adapted the definition to algebraic groups and pointed out that the term is due to Weil (Rosenlicht 1956, p. 417). Chevalley (1956-58, p. 1) used *isogénie* for a mapping.

Note. The origin of the term was pointed out by Weil (1979, p. 570) himself.

References

Chevalley, Claude (1956-58). "Les isogénies". *Séminaire Claude Chevalley* 2, exposé 18, pp. 1–10.

Rosenlicht, Maxwell (1953). "Simple Differentials of Second Kind on Hodge Manifolds". *American Journal of Mathematics* 75, pp. 621–626.

Rosenlicht, Maxwell (1956). "Some Basic Theorems on Algebraic Groups". *American Journal of Mathematics* 78, pp. 401–443.

---

[2] This usage was pointed out by Bateman (1918, p. 94) and Zondadari (1917, p. 5), for example.

[3] See also Bernoulli (1955, p. 248) for an illustration by Bernoulli.

[1] "Nous conviendrons de dire que deux variétés abéliennes A et B sont *isogènes* si elles ont même dimension et s'il existe un homomorphisme de A sur B".

Weil, André (1948). *Varietés abéliennes et courbes algébriques*. Paris.
Weil, André, ed. (1979). *Collected Papers*. Volume I (1926-1951). New York.

isogonal

Greek ἰσογώνιος is made up of the words for 'equal' and 'angle' and was used in the sense of *equiangular* (Mugler 1958-59, pp. 222f.). As such, it was translated by Latin *aequiangulus*; forms such as *isogonius* are rare in Latin – an example is found in a 12th-century translation of the *Elements* made directly from the Greek (Busard 1987, p. 127).[1]

The terminology was revived in the 19th century, and new special senses were given to the term. One of these senses is illustrated by Mackay (1895, p. 166): "If two angles have the same vertex and the same bisector, the sides of either angle are isogonal to each other with respect to the other angle". Mackay refers to de Longchamps (1886, p. 245) for the term, where French *droites isogonales* is found.

Mackay (1895, p. 173) further points out some closely related terminology (see fig. 60): "Points such as O, O' determined by the intersection of pairs of isogonal lines will be called *isogonal points*, or simply *isogonals*, with respect to the triangle ABC. They are sometimes called *isogonally conjugate points*, or *isogonal conjugates*, but more frequently on the continent of Europe *inverse points* with respect to the triangle ABC". For *isogonally conjugate*, Mackay refers to Neuberg (1884, p. 10), who used *points conjugués isogonaux*; *points isogonaux* is earlier found in Neuberg (1881, p. 154). Schendel (1874, p. 45) used German *isogonal entsprechende Punkte*, lit. 'points corresponding isogonally'.[2]

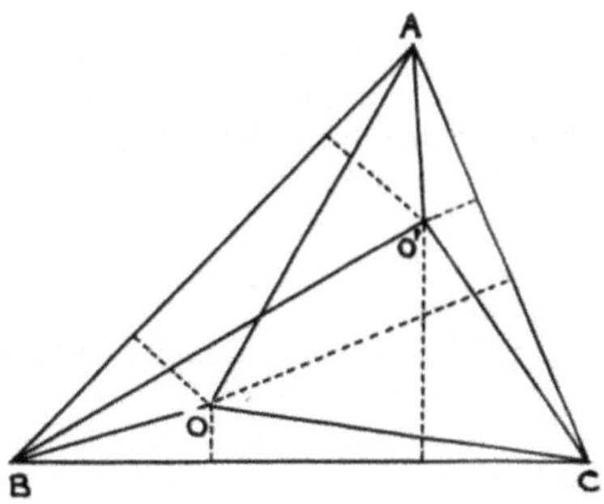

Figure 60: Isogonal points and lines. From Mackay (1895, fig. 37). BO and BO' are isogonal with respect to B, etc.

For the term *inverse points*, Mackay (1895, p. 173) refers to two virtually simultaneous publications, namely Mathieu (1865, p. 398), who has French *points*

[1] Various reference works from the 17th and 18th centuries also contain entries for the word *isagon* [sic], for example Moxon (1679, p. 72): "Isagon, A Figure of equal Angles". Moxon's entry closely follows Vitalis (1668, p. 253): "ISAGONIUS est figura Geometrica aequalibus angulis constans".
[2] The term was attributed to Schendel by Hain (1877, p. 92).

inverses, and Martin (1867, p. 37), who uses English *inverse points*.[3] Additionally, Mackay (1892, p. 100) remarks that Jacobi (1851, p. 12) used *Gegenpunkte*, lit. 'counter-points', for isogonal points.

The term was also used in other senses; for example, Siebeck (1858, p. 228) suggested *isogonal* as an alternative to *conformal*. A more common use occurs in connection with tessellations: "A plane tiling $\mathscr{T} = \{T_i | i = 1, 2, \ldots\}$ is called *isohedral* if its symmetry group $S(\mathscr{T})$ acts transitively on the tiles T_i of $\mathscr{T}$, and is called *isogonal* if $S(\mathscr{T})$ acts transitively on the vertices of $S(\mathscr{T})$" (Grünbaum and Shephard 1978, p. 335). Grünbaum and Shephard (1978, p. 337) also point out the origin: "The word 'isogons' was originally introduced by Fedorov [Fedorov 1885] to denote isogonal tilings by convex tiles". Russian изогона indeed goes back to Fedorov (1885, p. 100).

Note. In addition to the articles by Mackay, see also Berkhan and Meyer (1921, p. 1207) and Goormaghtigh (1951, pp. 30, 117f.).

References

Berkhan, Gustav and Wilhelm Franz Meyer (1921). "Neuere Dreiecksgeometrie". *Encyklopädie der mathematischen Wissenschaften mit Einschluss ihrer Anwendungen* III AB 10, pp. 1173–1276.

Busard, H.L.L. (1987). *The Mediaeval Latin Translation of Euclid's Elements Made Directly from the Greek*. Stuttgart.

Fedorov, E.S. (1885). *Начала учения о фигурах*. Moscow.

Goormaghtigh, René (1951). "Terminologie dans la géométrie du triangle et du tétraèdre". *Mathesis* 60, pp. 24–31, 116–123, 187–195, 263–274.

Grünbaum, Branko and Geoffrey Colin Shephard (1978). "The Ninety-One Types of Isogonal Tilings in the Plane". *Transactions of the American Mathematical Society* 242, pp. 335–353.

Hain, Emil (1877). "Ueber isogonal entsprechende Punkte des Dreiecks". *Archiv der Mathematik und Physik* 60, pp. 92–98.

Jacobi, Carl Friedrich Andreas (1851). *Die Entfernungsörter geradliniger Dreiecke*. Eine geometrische Abhandlung. Jena.

de Longchamps, Gaston (1886). "Généralités sur la géométrie du triangle". *Journal de mathématiques élémentaires* 5, pp. 109–114, 127–133, 154–158, 177–179, 198–206, 229–233, 243–250, 270–279.

Mackay, J.S. (1892). "Early History of the Symmedian Point". *Proceedings of the Edinburgh Mathematical Society* 11, pp. 92–103.

Mackay, J.S. (1895). "Isogonals of a Triangle". *Proceedings of the Edinburgh Mathematical Society* 13, pp. 166–178.

Martin, Hugh (1867). "A Study of Trilinear Co-ordinates: being a Consecutive Series of Seventy-two Propositions in Transversals". *Transactions of the Royal Society of Edinburgh* 24, pp. 37–52.

Mathieu, J.-J.-A. (1865). "Étude de géométrie comparée, avec applications aux sections coniques". *Nouvelles annales de mathématiques* 4, pp. 393–407, 481–493, 529–537.

Moxon, Joseph (1679). *Mathematicks made easie or, a compleat Mathematical Dictionary*. London.

Mugler, Charles (1958-59). *Dictionnaire historique de la terminologie géométrique des Grecs*. Paris.

Neuberg, Joseph (1881). "Sur le centre des médianes antiparallèles". *Mathesis* 1, pp. 153–154, 173–176, 185–190.

Neuberg, Joseph (1884). *Mémoire sur le tétraèdre*. Bruxelles.

Schendel, Leopold (1874). *Elemente der analytischen Geometrie der Ebene in trilinearen Coordinaten*. Jena.

Siebeck, Paul (1858). "Ueber die graphische Darstellung imaginärer Functionen". *Journal für die reine und angewandte Mathematik* 55, pp. 221–253.

[3] Martin's paper was read in 1865. de Longchamps (1886, p. 110) also cites Mathieu as having created this term.

Vitalis, Hieronymus (1668). *Lexicon mathematicum, astronomicum et geometricum*. Paris.

isohedron and *isohedral*

The term *isohedron* was introduced (as Russian изоэдром) by Fedorov (1885, p. 98), who defined it as a polyhedron all of whose faces are equal (congruent) or symmetric.[1] The terminology became more widely known through some papers which Fedorov published in German, such as Fedorov (1893b, pp. 579, 683). The English term seems to have been first used by Story Maskelyne et al. (1901, p. 329).

The concept had earlier been discussed by Hessel (1871, e. g., p. 18) and Hess (1875, p. 7), who used German *gleichflächige Polyeder*, where *gleichflächig* means 'equal-faced'.[1]

The terminology was later explicitly applied to tessellations: "The tiling $\mathscr{T}$ is *isohedral* if the symmetry group $S(\mathscr{T})$ (that is, the group of isometries which leave $\mathscr{T}$ invariant) is transitive on the tiles" (Grünbaum and Shephard 1977, p. 177). This involved a change of meaning; as explained by Grünbaum (2010, p. 6), "[t]he term 'isohedral'—used by Fedorov [Fedorov 1885] and Bilinski [Bilinski 1960]—nowadays indicates the more restricted class of polyhedra with the property that their symmetries act transitively on their faces. The polyhedra of Fedorov and Bilinski are not (in general) 'isohedra' by definitions that are customary today".

References

Bilinski, Stanko (1960). "Über die Rhombenisoeder". *Glasnik Matematicko-Fizicki i Astronomski* 15, pp. 251–263.

Fedorov, E.S. (1885). *Начала учения о фигурах*. Moscow.

Fedorov, E.S. (1893a). "Elemente der Gestaltenlehre". *Zeitschrift für Krystallographie und Mineralogie* 21, pp. 679–694.

Fedorov, E.S. (1893b). "Universal- (Theodolith-) Methode in der Mineralogie und Petrographie. I. Theil". *Zeitschrift für Krystallographie und Mineralogie* 21, pp. 574–678.

Fedorov, E.S. (1894). "Erwiderung auf die Bemerkungen zu E. v. Fedorow's Elemente der Gestaltenlehre von Edmund Hess". *Neues Jahrbuch für Mineralogie, Geologie und Paläontologie* (1894, II. Band), pp. 86–88.

Grünbaum, Branko (2010). "The Bilinski Dodecahedron and Assorted Parallelohedra, Zonohedra, Monohedra, Isozonohedra, and Otherhedra". *The Mathematical Intelligencer* 32, pp. 5–15.

Grünbaum, Branko and Geoffrey Colin Shephard (1977). "The eighty-one types of isohedral tilings in the plane". *Mathematical Proceedings of the Cambridge Philosophical Society* 82, pp. 177–196.

Hess, Edmund (1875). "Ueber zwei Erweiterungen des Begriffs der regelmässigen Körper der Gestaltenlehre". *Sitzungsberichte der Gesellschaft zur Beförderung der gesamten Naturwissenschaften zu Marburg* (1875), pp. 1–20.

---

[1] On the definition, Senechal and Galiulin (1984, p. 12) remark: "It seems clear from the context, though not from the definition itself, that Fedorov intends the faces of an isohedron to be equivalent with respect to its symmetry group rather than simply congruent, since Theorem 15 states that 'The points of tangency of a typical isohedron are corresponding points' and chapter 6 is a brief discussion of equifaced polyhedra which do not have this property". [1] Grünbaum (2010, p. 6) notes that *gleichflächig* "was quite established at the time of Fedorov's writing, but what it meant seems to have been more than the word implies. As explained in Edmund Hess's second note [Hess 1894] excoriating Fedorov [Fedorov 1893a, Fedorov 1894], the interpretation as 'congruent faces' (that is, monohedral) is mistaken. Indeed, by 'gleichflächig' Hess means something much more restrictive. Hess formulates it [...] very clumsily, but it amounts to symmetries acting transitively on the faces, that is, to isohedral".

Hess, Edmund (1894). “Weitere Bemerkungen zu E. v. Fedorow’s Elementen der Gestaltenlehre”. *Neues Jahrbuch für Mineralogie, Geologie und Paläontologie* (1894, II. Band), pp. 88–90.
Hessel, Johann Friedrich Christian (1871). *Uebersicht der gleicheckigen Polyeder und Hinweisung auf die Beziehungen dieser Körper zu den gleichflächigen Polyedern*. Marburg.
Senechal, Marjorie and R.V. Galiulin (1984). “An Introduction to the Theory of Figures”. *Structural Topology* 10, pp. 5–22.
Story Maskelyne, Nevil et al. (1901). “The Structure of Crystals”. *Report of the British Association for the Advancement of Science* (1901), pp. 297–337.

isolated point

With respect to curves, isolated points, also called a *conjugate points* or *acnodes*, are, metaphorically speaking, “separate points which, like islands, do not have any visible connection with a curved line, but still belong to it”.[1] Such points were considered by Newton in the 1660s. In his *Enumeration of Lines of the Third Order*, first published in 1704, he presented a classification of curves of degree 3. Under certain conditions, a curve such as a hyperbola or a parabola may have a conjugate oval, and if this oval becomes infinitely small, Newton (1704, p. 145) speaks of a *conjugate point* (Latin *punctum conjugatum*).[2] Figure 61 shows what Newton called a *parabola punctata*, a parabola with a conjugate point A.

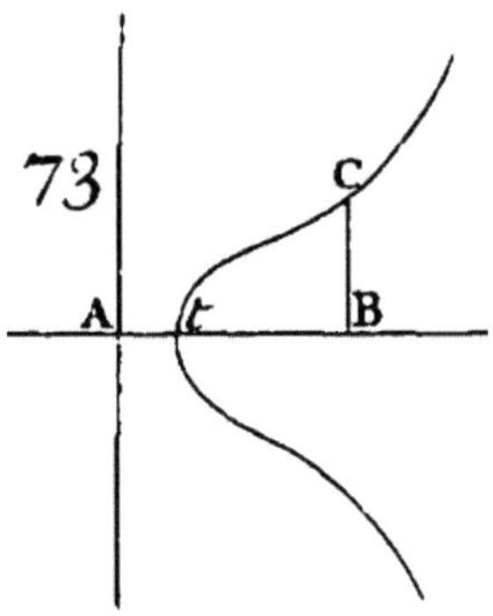

Figure 61: Newton’s *parabola punctata*. From Newton (1704, Tab. VI).

Harris (1710, s. v. curve) chose *conjugate point* in his translation of Newton’s text. The French term *point conjugué* appears in de Bragelongne (1732, p. 165), German *conjungirt*[3] is found in [anon.] (1753, pp. 803f.).

In the 1730s, such conjugate points are described in terms of ‘isolation’. For example, de Bragelongne (1733, p. 22) discusses a multiple point which is “absolutely separated and isolated”[4] from a curve but nevertheless part of the curve. The French term *point isolé* was soon afterwards used by de Gua de Malves (1740, p. 101), who

[1] “einzelne Puncte, die wie Inseln, mit einer krummen Linie keinen sichtbaren Zusammenhang haben, und doch zu ihr gehören”. See *Göttingische gelehrte Anzeigen* 2, Dec. 28, 1769, p. 1400.
[2] Newton also sometimes designated a conjugate point as a *punctum singulare* ‘singular point’. See, for example, Newton (1968, p. 150). [3] The modern Form is *konjugiert*. [4] “absolument séparé & isolé”.

gave Newton's *conjugate* as a synonym. The new term was picked up by Cramer (1750, p. 568) and Goudin and du Séjour (1756, p. 89).

English mathematicians continued to prefer Newton's term; this is the case in Lacroix (1816, p. 99), but it is mentioned in this translation that "[c]urves are sometimes accompanied by insulated points, which have the character of multiple points" (Lacroix 1816, p. 99). Peacock (1820, p. 155) then introduces *insulated* as a synonym for *conjugate.* The German term *isolirter Punct* is found in Lacroix (1800, pp. 103, 132).

The aspect of isolation was particularly emphasized by the term *hermit (point)*, apparently introduced by Ohm (1839, p. 409) as German *Einsiedler*.

The term *acnode* was introduced by Cayley (1860, p. 514): "I have elsewhere used the word 'node' to denote a double point, and I take the opportunity of suggesting the employment of the words 'crunode' (*crus*) and 'acnode' (*acus*) to denote respectively a double point with two real branches through it, and a conjugate or isolated point". Latin *acus* means 'needle', so *acnode* is probably to be understood as a point that looks like the prick of a needle.

Note. See Sauerbeck (1902, pp. 1–15) and Kline (1972, pp. 547–554) for some historical information on the concept.

References

[anon.] (1753). "Introduction à l'Analyse &c". *Zuverlässige Nachrichten von dem gegenwärtigen Zustande, Veränderung und Wachstum der Wissenschaften* 14, pp. 785–804.

de Bragelongne, Christophe-Bernard (1732). "Examen des lignes du quatriéme ordre ou courbes du troisiéme ordre". *Mémoires de l'Académie royale des sciences* (1730), pp. 158–216.

de Bragelongne, Christophe-Bernard (1733). "Examen des lignes du quatriéme ordre ou courbes du troisiéme ordre. Troisiéme Partie de la Section I". *Mémoires de l'Académie royale des sciences* (1731), pp. 10–49.

Cayley, Arthur (1860). "On a relation between two ternary cubic forms". *The London, Edinburgh and Dublin Philosophical Magazine and Journal of Science* 20, pp. 512–514.

Cramer, Gabriel (1750). *Introduction à l'analyse des lignes courbes algébriques*. Genève.

Goudin, Mathieu-Bernard and Achille Pierre Dionis du Séjour (1756). *Traité des courbes algébriques*. Paris.

de Gua de Malves, Jean Paul (1740). *Usages de l'analyse de Descartes pour découvrir, sans le secours du calcul différentiel, les propriétés des lignes géométriques de tous les ordres*. Paris.

Harris, John (1710). *Lexicon Technicum*. Vol. II. London.

Kline, Morris (1972). *Mathematical Thought From Ancient To Modern Times*. Volume 2. New York and Oxford.

Lacroix, Sylvestre François (1800). *Lehrbegriff des Differential- und Integralcalculs*. Zweiter Theil. Berlin.

Lacroix, Sylvestre François (1816). *An Elementary Treatise on the Differential and Integral Calculus*. Translated from the French. Cambridge.

Newton, Isaac (1704). "Enumeratio linearum tertii ordinis". In: *Opticks*. London, pp. 138–162.

Newton, Isaac (1968). *The Mathematical Papers of Isaac Newton*. Volume II. 1667-1670. Ed. by D.T. Whiteside. Cambridge.

Ohm, Martin (1839). *Lehrbuch der gesammten höhern Mathematik in zwei Bänden*. Erster Band. Leipzig.

Peacock, George (1820). *A Collection of Examples of the Applications of the Differential and Integral Calculus*. Cambridge.

Sauerbeck, Paul (1902). *Einleitung in die analytische Geometrie der höheren algebraischen Kurven nach den Methoden von Jean Paul de Gua de Malves*. Leipzig.

isolated point (topology)

In his series of papers on point sets, Cantor (1883, p. 51) introduced the term *isolated set* (*isolirte Punktmenge*), which he defined as a set that contains none of its limit points. Shortly afterwards, Cantor (1885, p. 110) also defined an isolated point. He considered an arbitrary point set P and said that such a set consists of two kinds of points: the former are such that "in a *sufficiently small* neighbourhood there are no other points of P; these are the so-called *isolated* points; the other points of P are also *limit points* of P".[1]

The terminology was soon adopted and appears, for example, in Jordan's *Cours d'analyse* (Jordan 1893, p. 19). The English term is found in Harkness (1894, p. 136).

Note that in the 1840s, Bolzano also developed a concept of isolated point (which did not exert much influence, however): "If the point i so lies in a spatial object that there is no distance, however small, for which it can be asserted that for this distance and for all smaller distances i possesses one or several neighbours, then I say that i stands isolated or alone in the spatial object" (Bolzano 1948, p. 143).[2] A similar definition appeared in print in Bolzano (1851, p. 73).[3]

Note. For the introduction of the term, see Johnson (1977), Dauben (1990, pp. 87–89), and Ferreirós (2007, pp. 206f.).

References

Bolzano, Bernhard (1851). *Paradoxien des Unendlichen*. Herausgegeben aus dem schriftlichen Nachlasse des Verfassers von Dr. Fr. Přihonsky. Leipzig.

Bolzano, Bernhard (1948). *Bernard Bolzano's Schriften*. Band 5. Geometrische Arbeiten. Herausgegeben und mit Anmerkungen versehen von Dr. Jan Vojtěch. Prag.

Cantor, Georg (1883). "Ueber unendliche, lineare Punktmannichfaltigkeiten. 4". *Mathematische Annalen* 21, pp. 51–58.

Cantor, Georg (1885). "Über verschiedene Theoreme aus der Theorie der Punctmengen in einem n-fach ausgedehnten stetigen Raume G_n. Zweite Mittheilung". *Acta Mathematica* 7, pp. 105–124.

Dauben, Joseph W. (1990). *Georg Cantor*. His Mathematics and Philosophy of the Infinite. Princeton.

Ferreirós, José (2007). *Labyrinth of Thought*. A History of Set Theory and Its Role in Modern Mathematics. Second revised edition. Basel.

Harkness, James (1894). "Jordan's Cours d'Analyse". *Bulletin of the American Mathematical Society* 3, pp. 135–141.

Johnson, Dale M. (1977). "Prelude to Dimension Theory: The Geometrical Investigations of Bernard Bolzano". *Archive for History of Exact Sciences* 17, pp. 261–295.

Jordan, Camille (1893). *Cours d'analyse*. Tome premier. Calcul différentiel. Paris.

isometric and *isometry*

This term is made up of the Greek words for 'equal' and 'measure'. Some early usages include Farish (1822, p. 5), who used *isometrical perspective* for a drawing technique particularly useful in engineering. Bonnet (1860, p. 174) used French *isométrique*

---

[1] "die *ersteren* liegen so, dass in *hinreichend naher* Umgebung von ihnen keine anderen Puncte von P fallen, es sind dies sogenannte *isolirte* Puncte von P; die *anderen* Puncte von P sind gleichzeitig Grenzpuncte von P". [2] "Wenn der Punct i in einem Raumdinge so liegt, daß keine auch noch so kleine Entfernung angeblich ist, von der behauptet werden könnte, *für diese und für alle kleineren Entfernungen*, die es nur überhaupt gibt, besitze i einen oder etliche Nachbarn: so sage ich, i stehe *isolirt* oder *vereinzelt* in diesem Raumdinge". Translation from Johnson (1977, p. 283). [3] For discussion of Bolzano's definition, see Johnson (1977, esp. pp. 282-284).

with respect to certain "orthogonal lines that can divide the surface into infinitely small squares" (see *isothermal*).[4] Voss (1893, p. 248) used the word *isometrisch* for a relation, namely for surfaces developable onto each other. In the following, *isometry* will only be considered in the sense of a mapping that preserves distance.

A clear and general definition was given by Hausdorff, although at first he did not use the name *isometry*. In his famous book on set theory, after having defined metric spaces, Hausdorff went on to consider what he called *Euclidean spaces*:

> In particular, we call *Euclidean n-dimensional number space* E_n the set of complexes of real numbers
>
> $$x = (x_1, x_2, ..., x_n),$$
>
> in which the distance is defined by
>
> $$\overline{xy} = \sqrt{(x_1 - y_1)^2 + (x_2 - y_2)^2 + ...(x_n - y_n)^2} \geq 0,$$
>
> and *Euclidean n-dimensional space* a metric space whose elements ξ, η, ... can be assigned to the number complexes x, y, ... just referred to by a correspondence that is one-to-one and *distance-preserving* ($\overline{\xi\eta} = \overline{xy}$) (Hausdorff 1914, p. 212).[1]

The word *entfernungstreu*, which Hausdorff uses for 'distance-preserving', is made up of *Entfernung* 'distance' and *treu* 'faithful'; it had been in use earlier and is already found in the 19th century, particularly in cartography (see, for example, H. Struve (1887, p. 53)). Hausdorff's terminology was taken over by Tietze (1920, p. 99), who added the synonym *isometrische Abbildung* 'isometric mapping', a word also used by Blaschke (1921, p. 87), who additionally used *längentreu* (*Länge* 'length') as a synonym. Urysohn (1925, p. 804) and Lindenbaum (1926, p. 213) similarly use French *transformation isométrique*. In the second edition of his book, Hausdorff (1927, p. 94) replaced *entfernungstreu* by *isometrisch*. The expression *an isometry* came up in the 1930s; Myers (1936, p. 297), for example, defined: "An isometry of M into itself is defined to be a length-preserving homeomorphism".

The term *rigid* is also used in connection with isometries. An example is found in Hathaway (1902, p. 50): "A continuous transformation is *rigid*, when it leaves invariant the distances between all pairs of points". Osgood (1901, p. 283) used *euclidean rigid motion* without definition, while Page (1888, p. 345) and Lie (1891, p. 15) preferred *Euclidean motion*. The word *rigid motion* (German *starre Bewegung*) was earlier used by some authors to define congruence, for example by Worpitzky (1873, p. 406) and K. Struve (1878, p. 5).

See also *motion*.

References

Blaschke, Wilhelm (1921). *Vorlesungen über Differentialgeometrie und geometrische Grundlagen von Einsteins Relativitätstheorie*. I. Elementare Differentialgeometrie. Berlin.

[4] "lignes orthogonales qui peuvent partager la surface en carrées infiniment petits".

[1] "Speziell bezeichnen wir als *euklidischen n-dimensionalen Zahlenraum* E_n die Menge der Komplexe reeller Zahlen $x = (x_1, x_2, ..., x_n)$, worin die Entfernung durch $\overline{xy} = \sqrt{(x_1 - y_1)^2 + (x_2 - y_2)^2 + ...(x_n - y_n)^2} \geq 0$ definiert ist, und als *euklidischen n-dimensionalen Raum* einen metrischen Raum, dessen Elemente ξ, η, ... sich den eben genannten Zahlenkomplexen x, y, ... umkehrbar eindeutig und *entfernungstreu* ($\overline{\xi\eta} = \overline{xy}$) zuordnen lassen".

Bonnet, Ossian (1860). "Mémoire sur l'emploi d'un nouveau système de variables dans l'étude des propriétés des surfaces courbes". *Journal de mathématiques pures et appliquées* 5, pp. 153–266.

Farish, William (1822). "On Isometrical Perspective". *Transactions of the Cambridge Philosophical Society* 1, pp. 1–19.

Hathaway, Arthur S. (1902). "Quaternion Space". *Transactions of the American Mathematical Society* 3, pp. 46–59.

Hausdorff, Felix (1914). *Grundzüge der Mengenlehre*. Leipzig.

Hausdorff, Felix (1927). *Mengenlehre*. Zweite, neubearbeitete Auflage. Leipzig.

Lie, Sophus (1891). *Vorlesungen über Differentialgleichungen mit bekannten infinitesimalen Transformationen*. Bearbeitet und herausgegeben von Georg Scheffers. Leipzig.

Lindenbaum, Adolf (1926). "Contributions à l'étude de l'espace métrique I". *Fundamenta Mathematicae* 8, pp. 209–222.

Myers, Sumner Byron (1936). "Isometries of 2-Dimensional Riemannian Manifolds into Themselves". *Proceedings of the National Academy of Sciences* 22, pp. 297–300.

Osgood, William F. (1901). "On a Fundamental Property of a Minimum in the Calculus of Variations and the Proof of a Theorem of Weierstrass's". *Transactions of the American Mathematical Society* 2, pp. 273–295.

Page, James M. (1888). "On the Primitive Groups of Transformations in Space of Four Dimensions". *American Journal of Mathematics* 10, pp. 293–346.

Struve, Hermann (1887). *Landkarten, ihre Herstellung und ihre Fehlergrenzen*. Berlin.

Struve, Kurt (1878). *Elemente der Mathematik*. Erster Theil: Geometrie. Berlin.

Tietze, Heinrich (1920). "Über den Richtungssinn und seine Verallgemeinerung". *Jahresbericht der Deutschen Mathematiker-Vereinigung* 29, pp. 95–112.

Urysohn, Pavel (1925). "Sur un espace métrique universel". *Comptes rendus de l'Académie des sciences* 180, pp. 803–806.

Voss, Aurel (1893). "Ueber die Fundamentalgleichungen der Flächentheorie". *Sitzungsberichte der Mathematisch-Physikalischen Classe der k.b. Akademie der Wissenschaften zu München* 22, pp. 247–278.

Worpitzky, Julius (1873). "Ueber die Grundbegriffe der Geometrie". *Archiv der Mathematik und Physik* 44, pp. 405–421.

isomorphism and *homomorphism*

The term *isomorphic* is a 19th-century coinage based on Greek ἴσος 'equal' and μορφή 'shape'. Originally it was used for chemical elements crystallizing in the same or in a related way. In the 1860s, Camille Jordan transferred the term from chemistry to algebra, and in the 1890s, *homomorphic* was introduced by Felix Klein. It took until well into the 20th century, however, that the usage as it is known today was established.

The chemical term *isomorphic* was coined by the chemist Eilhard Mitscherlich.[1] Though Mitscherlich was German, the term first appears in a paper written in Swedish: "For convenience, I have called the elements, which belong to the same group, to have a more specific term than 'of the same kind', *isomorphic*" (Mitscherlich 1821a, p. 5).[2] The term became an immediate success in crystallographical writings and appears as French *isomorphe* in Mitscherlich (1821c, p. 351) and in English, the translation *isomorphous* was originally used in Mitscherlich (1821b, p. 198).

The original mathematical meaning was coined in analogy to this by Camille Jordan (Scholz 1989b, p. 287), who originally defined a *groupe isomorphe* in Jordan (1868, p. 836), and went into more detail in Jordan (1870, p. 56):

---

1 See Authier (2013, p. 331), Schütt (1984, p. 131). 2 "För beqvämlighetens skull, har jag kallat de element, som höra till samma flock *isomorpha*, för att äga en mer bestämd technisk term än likformig".

> A group Γ is said to be *isomorphic* to a group G if it is possible to establish between their substitutions a correspondence such that: 1° to each substitution of G there corresponds a unique substitution of Γ and to each substitution of Γ one or more substitutions of G; 2° to the product of any two substitutions of G there corresponds the product of their respective substitutions.[1]

From today's perspective, Jordan here defines a surjective homomorphism from G to Γ rather than an isomorphism, but he adds the following distinction: "An isomorphism is said to be *meriedric* if many substitutions of G correspond to the same substitution of Γ, and *holoedric* in the opposite case".[2] By *meriedric* Jordan understands 'not injective', and if the opposite is the case, i. e. if the homomorphism is also injective, he calls it *holoedric*. In sum, "meriedric isomorphism" means that there is what is now called a surjective homomorphism that is non-injective, a "holoedric isomorphism" means that there is an isomorphism.[3]

The words *mériédrique* and *holoédrique* are, like *isomorphe*, taken from crystallography. They were used, in particular, by Auguste Bravais in 1849 in connection with his studies of the symmetry of crystals, in which he used group-theoretical methods.[4] The symmetry properties of crystals were studied by means of symmetry elements such as mirror planes and rotations about axes. Within a system of crystals, the symmetry properties of the crystals can be developed in different degrees. A crystal that is maximally symmetric within a system is called *holoedric* (from Greek *ὅλος* 'whole' and *ἕδρα*, here meaning 'face of a polyhedron'), since its graphical representation yields a polyhedron with a maximal number of faces. The adjective *meriedric* (Greek *μέρος* 'part') indicates partial symmetry and thus implies a reduced number of faces.

Jordan, who was well aware of Bravais' studies, took over the terms *isomorphe* to algebra on the basis of a rough analogy: crystals can have the same form, even if they are composed differently – just like groups. The adjectives *mériédrique* and *holoédrique* are then used to indicate how close the correspondence is. In connection with these terms, Jordan was influential in at least two respects: First, his terms were in current use until at least the 1930s (see below); second, homomorphisms were generally defined to be surjective. According to McLarty (1990, p. 355), the latter

---

[1] "Un groupe Γ est dit *isomorphe* à un autre groupe G, si l'on peut établir entre leurs substitutions une correspondance telle: 1° que chaque substitution de G corresponde à une seule substitution de Γ, et chaque substitution de Γ à une ou plusieurs subsitutions de G; 2° que le produit de deux substitutions quelconques corresponde au produit de leurs correspondantes respectives". Translation from Wußing (1984, p. 143). – The original definition Jordan gave in 1868 is slightly different, because it does not include the requirement of surjectivity, stated in the 1870 definition in the second part of condition 1. [2] "L'isomorphisme sera dit *mériédrique*, si plusieurs substitutions de G correspondent à une même substitution de Γ, *holoédrique* dans le cas contraire". [3] The corresponding terms are soon found in other languages. Capelli (1878, p. 33) transferred them to Italian, and the English terms are found in Cole (1887, p. 52). Netto (1882, p. 97) translated *holoédrique* by German *einstufig*, *mériédrique* by *mehrstufig*. The terms are based on the word *Stufe* 'stage, level' (the prefixes mean 'one' and 'more', respectively) and were, for a while, adopted by various authors. [4] The terms are found in reports on talks that Bravais gave before the Société philomathique. See *Bulletin de la Société philomathique de Paris* (1849), esp. pp. 51 and 85. He presented his researches in more detail in Bravais (1851), for example. The terminology may originally go back to Weiss (1818, p. 309), who suggested the German words *homoëdrisch* and *hemiëdrisch*.

condition was dropped in the 1930s (in algebraic topology rather than pure algebra);[5] he cites Seifert and Threlfall (1934, p. 297) as the first text in which this was done.[6]

The restriction of the term *isomorphic* to a one-to-one-correspondence and the introduction of the new term *homomorphic* go back to Felix Klein. This was pointed out by Ritter (1892, p. 22) and also by Fricke (1893, p. 466), according to whom Klein introduced this usage "in his more recent lectures". Both Ritter and Fricke thus restrict German *isomorph* to holoedric isomorphism and use *homomorph* and *Homomorphismus* otherwise.[1] Klein used the revised terminology in lectures from 1890/91 (Klein 1891, p. 77) and later also pointed out that the denomination *homomorphic* seems more appropriate than *isomorphic* in cases where there is not yet "identity" between groups but only "similarity" (Fricke and Klein 1897, p. 194).[2] As stated above, however, it took until well into the 20th century until Klein's suggestion was more widely adopted.[3]

In 1930, for example, van der Waerden (1930, p. 32) remarked that "a fixed usage for the words isomorphism and homomorphism does not exist",[4] a statement that he repeated in the second and later editions of his book (see, e. g., van der Waerden 1960, p. 37). He points out that Speiser (1923, p. 19) uses the words in the contrary way. There are other contemporary statements that show that in the 1930s a change is in progress. For example, Chevalley (1934, p. 373) states that "the words *homomorphism* and *isomorphism* are more convenient than the frequently used words *meriedric isomorphism* and *holoedric isomorphism*"[5], and Tschebotaröw (1931, p. 240) declares that he uses, "as is common today, the word *isomorphic* in the sense of holoedric isomorphism" and that "instead of meriedric isomorphism [he says] *homomorphic*".[6]

As of 1900, the word *isomorphic* was usually used in connection with algebraic structures, especially groups. Further generalizations are found at the beginning of the 20th century. For example, Huntington (1906, p. 26) uses the term in connection with systems of axioms: "special attention may be called to the discussion of the notion of isomorphism between two systems, and the notion of a sufficient, or categorical, set of postulates". Huntington, like Klein, demands one-to-one-correspondence. According to Mancosu, Zach, and Badesa (2009, p. 327), this is the first use of the word in this general sense of "isomorphism between arbitrary systems satisfying the same set of axioms".

---

[5] van der Waerden (1960, p. 37), for example, still required surjectivity for a homomorphism. [6] See also Krömer (2007, pp. 45f.) and *surjective*. [1] This approach was also adopted by Burnside (1895, p. 44): "Where the word 'isomorphous' is used in the present paper without qualification it is to be regarded as an abbreviation for 'holohedrically isomorphous'". [2] "Die Bezeichnung 'homomorph' erscheint treffender, als [...] 'isomorph', da es sich ja hier noch nicht um 'Gleichheit', sondern nur erst um 'Ähnlichkeit' zweier Gruppen handelt". – See also Miller (1902, p. 120) for comments on the terminology. [3] One author who did adopt the usage was de Séguier (1902, p. 257); English *homomorphism* seems to appear first in Miller (1910, p. 325). [4] "Ein fester Sprachgebrauch für die Wörter Isomorphismus und Homomorphismus existiert nicht". [5] "Ces mots d'homomorphisme et d'isomorphisme sont plus commodes que les mots isomorphisme mériédrique et isomorphisme holoédrique souvent employés". [6] "Ich gebrauche hier, wie heute üblich ist, das Wort ‚isomorph' im Sinne von ‚einstufig isomorph'. Statt *mehrstufig isomorph* sage ich ‚homomorph'".

Note. The influence of crystallography on group theory is discussed in detail in Scholz (1989b); a shorter account is found in Scholz (1989a). Li Carrillo (1981) discusses the history of the mathematical concept. See also Becker (2005, pp. 192–199), who focuses on the terminology and the transfer from crystallography to algebra.

References

Authier, André (2013). *Early Days of X-ray Crystallography*. Oxford.

Becker, Holger (2005). "Semantische und lexikalische Aspekte der mathematischen Fachsprache des 19. Jahrhunderts". PhD thesis. Carl von Ossietzky Universität Oldenburg. URL: `http://oops.uni-oldenburg.de/59/`.

Bravais, Auguste (1851). "Études cristallographiques". *Journal de l'École polytechnique* 20, cahier34, pp. 101–278.

Burnside, William Snow (1895). "On the Kinematics of non-Euclidean Space". *Proceedings of the London Mathematical Society* 26, pp. 33–56.

Capelli, Alfredo (1878). "Sopra l'isomorfismo dei gruppi di sostituzioni". *Giornale di matematiche* 16, pp. 32–87.

Chevalley, Claude (1934). "Sur la théorie du corps de classes dans les corps finis et les corps locaux". PhD thesis. Paris.

Cole, Frank Nelson (1887). "Klein's Ikosaeder". *American Journal of Mathematics* 9, pp. 45–61.

Fricke, Robert (1893). "Ueber den arithmetischen Charakter der zu den Verzweigungen (2,3,7) und (2,4,7) gehörenden Dreiecksfunctionen". *Mathematische Annalen* 41, pp. 443–468.

Fricke, Robert and Felix Klein (1897). *Vorlesungen über die Theorie der automorphen Functionen*. Erster Band: Die gruppentheoretischen Grundlagen. Leipzig.

Huntington, Edward V. (1906). "The Fundamental Laws of Addition and Multiplication in Elementary Algebra". *Annals of Mathematics* 8, pp. 1–44.

Jordan, Camille (1868). "Théorèmes généraux sur les substitutions". *Comptes rendus de l'Académie des sciences* 66, pp. 836–837.

Jordan, Camille (1870). *Traité des substitutions et des équations algébriques*. Paris.

Klein, Felix (1891). *Ausgewählte Kapitel aus der Theorie der linearen Differentialgleichungen zweiter Ordnung*. I. Vorlesung gehalten während des Wintersemesters 1890-91. Göttingen.

Krömer, Ralf (2007). *Tool and Object*. A History and Philosophy of Category Theory. Basel.

Li Carrillo, Victor (1981). "Génesis y evolución del concepto matemático de isomorfismo". *Revista Venezolana de Filosofía* 14-15, pp. 75–138.

Mancosu, Paolo, Richard Zach, and Calixto Badesa (2009). "The Development of Mathematical Logic from Russell to Tarski, 1900–1935". In: *The Development of Modern Logic*. Ed. by Leila Haaparanta. Oxford, pp. 318–470.

McLarty, Colin (1990). "The uses and abuses of the history of topos theory". *British Journal for the Philosophy of Science* 41, pp. 351–375.

Miller, George Abram (1902). "Second report on recent progress in the theory of groups of finite order". *Bulletin of the American Mathematical Society* 9, pp. 106–123.

Miller, George Abram (1910). "Historical sketch of the development of the theory of groups of finite order". *Bibliotheca Mathematica* 10, pp. 317–329.

Mitscherlich, Eilhard (1821a). "Om förhållandet emellan chemiska sammansättningen och krystallformen hos Arseniksyrade och Phosphorsyrade Salter". *Kongl. vetenskaps akademiens handlingar* (1821), pp. 4–79.

Mitscherlich, Eilhard (1821b). "On the Relation which exists between Crystalline Form and Chemical Proportions". *Quarterly Journal of Science Literature and the Arts* 14, pp. 198–207, 415–418.

Mitscherlich, Eilhard (1821c). "Sur la Relation qui existe entre la forme cristalline et les proportions chimiques. II.e Mémoire sur les arséniates et les phosphates". *Annales de chimie et de physique* 19, pp. 350–419.

Netto, Eugen (1882). *Substitutionentheorie und ihre Anwendungen auf die Algebra*. Leipzig.

Ritter, Ernst (1892). *Die eindeutigen automorphen Formen vom Geschlecht Null, eine Revision und Erweiterung der Poincaré'schen Sätze*. Leipzig.

Scholz, Erhard (1989a). "Crystallographic Symmetry Concepts and Group Theory (1850-1880)". In: *The History of Modern Mathematics*. Volume II: Institutions and Applications. Ed. by David E. Rowe and John McCleary. Boston, pp. 3–27.
Scholz, Erhard (1989b). *Symmetrie – Gruppe – Dualität*. Zur Beziehung zwischen theoretischer Mathematik und Anwendungen in Kristallographie und Baustatik des 19. Jahrhunderts. Basel.
Schütt, Hans-Werner (1984). *Die Entdeckung des Isomorphismus*. Hildesheim.
de Séguier, Jean Armand (1902). "Sur les équations de certains groupes". *Journal de mathématiques pures et appliquées* 8, pp. 253–308.
Seifert, Herbert and William Threlfall (1934). *Lehrbuch der Topologie*. Leipzig.
Speiser, Andreas (1923). *Die Theorie der Gruppen von endlicher Ordnung*. Berlin and Heidelberg.
Tschebotaröw, N. (1931). "Über ein algebraisches Problem von Herrn Hilbert. Zweite Abhandlung". *Mathematische Annalen* 105, pp. 240–255.
van der Waerden, Bartel Leendert (1930). *Moderne Algebra*. Erster Teil. Berlin.
van der Waerden, Bartel Leendert (1960). *Algebra*. Fünfte Auflage der Modernen Algebra. Erster Teil. Berlin.
Weiss, Christian Samuel (1818). "Uebersichtliche Darstellung der verschiedenen natürlichen Abtheilungen der Krystallisationssysteme". *Abhandlungen der Königlichen Preußischen Akademie der Wissenschaften zu Berlin* (1814-1815), pp. 289–336.
Wußing, Hans (1984). *The Genesis of the Abstract Group Concept*. A Contribution to the History of the Origin of Abstract Group Theory. Tr. Abe Shenitzer. Cambridge, Mass.

isoperimetric

The Greek term *ἰσοπερίμετρος* is made up of the words 'equal' and 'perimeter'. The word is found in late aniquity. For example, the Greek mathematician Theon of Alexandria (4th c.) reported in his commentary on the *Almagest* on a lost work of Zenodorus[1]: "In the same way, since the greatest of the various figures having an equal perimeter is that which has most angles, the circle is the greatest among plane figures and the sphere among solid. We shall give the proof of these propositions in a summary taken from the proofs by Zenodorus in his book *On Isoperimetric Figures*" (Rome 1936, pp. 354f.).[2] The term is also found in other sources such as Simplicius' commentary on Aristotle's *De caelo* (Simplicius 1894, p. 412).

In Latin, the term was directly taken over from the Greek. It spread through a 12th-century translation entitled *De figuris ysoperimetris* or *Liber ysoperimetrorum* (Busard 1980, p. 63); it also appears in the widely diffused *Sphere* of Sacrobosco (Thorndike 1949, p. 80).

In English, the term was used as *isoperimetrall figure* in Ramus (1590, fol. 3^{r}); French *isoperimetre* appears as an adjective in de Sacrobosco (1546, fol. B v^{v}).

Note. For the history of the isoperimetric problem, see, e. g., Blåsjö (2005) and Müller (1953).

References

Blåsjö, Viktor (2005). "The Isoperimetric Problem". *The American Mathematical Monthly* 112, pp. 526–566.
Busard, H.L.L. (1980). "Der Traktat *De isoperimetris*, der unmittelbar aus dem Griechischen ins Lateinische übersetzt worden ist". *Mediaeval Studies* 42, pp. 62–88.
Müller, Wilhelm (1953). "Das isoperimetrische Problem im Altertum". *Sudhoffs Archiv* 37, pp. 39–71.

[1] For Zenodorus, who appears to have lived in the second century B.C., see Toomer (1972).
[2] Translation from Thomas (1957, p. 387). Toomer (1972, p. 177) comments on the term in earlier editions of Theon of Alexandria's text.

Ramus, Petrus (1590). *Elementes of geometrie*. Faithfully translated by Tho. Hood. London.
Rome, Adolphe, ed. (1936). *Commentaires de Pappus et de Théon d'Alexandrie sur l'Almageste*. Tome II. Théon d'Alexandrie: *Commentaire sur les livres 1 et 2 de l'Almageste*. Città del Vaticano.
de Sacrobosco, Johannes (1546). *La sphere de Iehan de Sacrobosco*. Trans. by Martin de Perer. Paris.
Simplicius (1894). *In Aristotelis De caelo commentaria*. Ed. by J.L. Heiberg. Berlin.
Thomas, Ivor (1957). *Selections Illustrating the History of Greek Mathematics*. II. From Aristarchus to Pappus. London and Cambridge, Mass.
Thorndike, Lynn (1949). *The Sphere of Sacrobosco and Its Commentators*. Chicago.
Toomer, G.J. (1972). "The Mathematician Zenodorus". *Greek, Roman, and Byzantine Studies* 13, pp. 177–192.

isosceles

The word *isosceles* is of Greek origin; *ισοσκελής* is made of of the word *ἴσος* 'equal' and *σκέλος* 'leg' and thus literally means 'equal-legged'.[1] Euclid (*Elements*, book I, def. 20) used this word when he classified triangles on the basis of their sides: "Of trilateral figures, an *equilateral* triangle is that which has its three sides equal, an *isosceles* triangle that which has two of its sides alone equal, and a *scalene* triangle that which has its three sides unequal".[2] The term is not due to Euclid, however; it was used before him, in particular by Plato and Aristotle (Mugler 1958-59, pp. 231f.).[3]

Latin *aequicrurius* is made up in the same way as *ισοσκελής*, with *crus* meaning 'leg'. According to Ayuso García (2008, pp. 727f.), the word was first used by Martianus Capella (*De nuptiis* 712) as part of his explanation of the Greek term (Willis 1983, p. 253). In other texts, *isosceles* was used, for example by Pseudo-Censorinus (Sallmann 1983, p. 69).[4] In the Latin translations of the *Elements* made in the 12th century, the notion is usually paraphrased (in the translation of def. 20), but in the text, both *isosceles* and *aequicrurius* can be found, for example in the first translation made from the Greek (Busard 1987, p. 30) and in Robert of Chester's version (Busard and Folkerts 1992, p. 279).

In the first German translation of books I-VI of the *Elements*, the word *gleichfüssig* is used (Euclid 1562, p. 3); it is made up of *gleich* 'equal' and *Fuß*, which means 'foot' but also, in certain dialects, 'leg'.[5] The same word is found in Euclid (1610, p. 3), but in Hoffmann's *Teutscher Euclides* (Hofmann 1628, p. 57), the word *gleichschenklig*, from *Schenkel* 'thigh',[6] is suggested as an alternative.[7] This word is afterwards more commonly found and still standard.

In French, the term appears as *triangle ysochele* in de Bouelles (1511, fol. viij^{v}); in de Bouelles (1542, fol. 14^{v}), the word is spelt *isoscele*.[8] Italian *isoscele* is found as *isochiles* in the 14th century (Arrighi 1987, p. 86); in print, *ysochele* was used

---

[1] Note that while *ισοσκελής* is ancient, *σκέλος* (as a geometrical term) is not. See *leg*. [2] Translation from Euclid (1956, p. 154). This classification is earlier found in Aristotle (*Posterior Analytics* I.5); Aristotle used the same terms for 'equilateral', 'isosceles', and 'scalene' as Euclid. [3] Mugler also points out that the word was also used with respect to certain trapezia and cones. On the (rare) use of *ισοσκελής* in connection with numbers, see Allen (2013, 50, fn. 1). [4] Another term is *aequipedus*, lit. 'having equal feet'. It was used by Apuleius (*De Platone* 1,7). See Guillaumin (2020, p. 28). [5] There are earlier German texts in which *isosceles* is mentioned; for example in Widmann (1489, fol. C 5^{r}). [6] The word is etymologically related to English *shank*. [7] At the same time, the word was also used by Jungius (Elsner 2004, p. 165). [8] Chuquet earlier subsumed isosceles and scalene triangles under the name *inequilateralz* (Chuquet 1979, p. 110).

by Pacioli (1494, Tractatus Geometrie, fol. 7^v). Italian *equicrurio* appeared in the 14th-century translation of Fibonacci's *Practica Geometriae* (Feola 2008, p. 59).

In English, the concept is discussed by Recorde (1551, fol. B iij^r): "There is also an other distinction of the names of triangles, according to their sides, whiche other be all equal [...] other els two sydes bee equall and the thyrd vnequall, which the Greekes call *Isosceles*, the Latine men *æquicurio*, and in english *tweyleke* may they be called" (*tweyleke* is to be understood as *two-like*). Recorde's suggestion was not successful, and in Billingsley's translation of the *Elements*, *isosceles* is used instead (Euclid 1570, fol. 4^v).[1]

Note. A concise overview of the terminology is found in Tropfke (1940, p. 81). For the Greek term, see Mugler (1958-59, pp. 231f.); for the Latin, see Ayuso García (2008, pp. 727f., 733f.) and Guillaumin (2020, pp. 27, 163f.).

References

Allen, R.E. (2013). *Plato's "Euthyphro" and the Earlier Theory of Forms.* Volume I. London and New York.

Arrighi, Gino (1987). "Giovanni de' Danti Aretino, *Tractato de l'algorisimo*. Dal Cod. Plut. 30. 26 (sec. XIV) della Biblioteca Medicea Laurenziana di Firenze". *Atti e Memorie della Accademia Petrarca di Lettere, Arti e Scienze* 47, pp. 3–91.

Ayuso García, Manuel (2008). "La terminología latina de la geometría en Marciano Capela". PhD thesis. Madrid: Universidad Nacional de Educación a Distancia.

de Bouelles, Charles (1511). *Geometrie en francoys*. Paris.

de Bouelles, Charles (1542). *Livre singulier & utile, touchant l'art et practique de geometrie*. Paris.

Busard, H.L.L. (1987). *The Mediaeval Latin Translation of Euclid's Elements Made Directly from the Greek*. Stuttgart.

Busard, H.L.L. and Menso Folkerts, eds. (1992). *Robert of Chester's (?) Redaction of Euclid's* Elements, *the so-called Adelard II Version*. Basel.

Chuquet, Nicolas (1979). *La Géométrie*. Première géométrie algébrique en langue française (1484). Introduction, texte et notes par Hervé l'Huillier. Paris.

Elsner, Bernd, ed. (2004). *Joachim Jungius' Geometria empirica und Reiß-Kunst*. Göttingen.

Euclid (1562). *Die Sechs Erste Bücher Euclidis*. Auß Griechischer sprach in die Teütsch gebracht ... Durch Wilhelm Holtzmann genannt Xylander. Basel.

Euclid (1570). *The Elements of Geometrie of the most auncient Philosopher Euclide of Megara*. Faithfully (now first) translated into the Englishe toung, by H. Billingsley. London.

Euclid (1610). *Die Ersten Sechs Bücher Elementorum Evclidis*. Mit sonderm Fleiß und Mühe auß Griechischer in unsere Hohe deutsche Sprach übergesetzet ... und aigentlichest erkläret ... Durch Simonem Marium. Ansbach.

Euclid (1956). *The Thirteen Books of Euclid's Elements*. Volume I. Introduction and Books I, II. Ed. by Thomas L. Heath. New York.

Feola, Francesco (2008). *Gli esordi della geometria in volgare*. Un volgarizzamento trecentesco della *Practica Geometriae* di Leonardo Pisano. Firenze.

Guillaumin, Jean-Yves (2020). *Dictionnaire de la terminologie latine ancienne de l'arithmétique et de la géométrie*. Paris.

Hofmann, Heinrich (1628). *Teutscher Euclides*. Jena.

Mugler, Charles (1958-59). *Dictionnaire historique de la terminologie géométrique des Grecs*. Paris.

Pacioli, Luca (1494). *Summa de Arithmetica Geometria Proportioni et Proportionalita*. Venezia.

Recorde, Robert (1551). *The patheway to knowledg*. London.

Sallmann, Klaus, ed. (1983). *Censorini de die natali liber ad Q. Caerellium*. Leipzig.

Tropfke, Johannes (1940). *Geschichte der Elementar-Mathematik*. Vierter Band. Ebene Geometrie. Dritte, verbesserte und vermehrte Auflage besorgt von Dr. Kurt Vogel. Berlin and Leipzig.

---

[1] Billingsley uses the word both as a noun and adjective.

Widmann, Johannes (1489). *Behende vnd hubsche Rechenung auff allen kauffmanschafft*. Leipzig.
Willis, James, ed. (1983). *Martianus Capella*. Leipzig.

isothermal

Before this term was used in mathematics, it was originally introduced by the geographer Alexander von Humboldt (1817a, p. 70): "Thus we see that circles of equal annual heat, or – to use a new term – isotherms, are not parallel to the equator but – like the magnetic lines – they cut the geographic parallels under a variable angle slantwise".[1] He coined this neologism on the basis of Greek *ἴσος* 'equal' and *θέρμη* 'heat'. The corresponding French term *lignes isothermes* is found in von Humboldt (1817b, p. 466), and "isothermal lines, or lines of equal heat" is found in a translation of that paper (von Humboldt 1820, p. 3). von Humboldt (1817c, p. 111) also gave a definition of a *surface isotherme*.

The terminology *surface isotherme* was introduced into mathematics by Cauchy (1828, p. 124), who says he chose the term because on such surfaces all points are of equal heat, and adopted by Lamé (1833, p. 195).

Towards the end of the 19th century, further terminology is discussed by Darboux (1887, pp. 76f.): "Whenever the line element of a surface can be converted into the form: $ds^2 = \lambda(d\alpha^2 + d\beta^2)$, one says that the coordinate curves form an *isothermal* or *isometric* net. The former term is borrowed from the theory of heat, while the latter [...] is due to Bonnet".[2] Bonnet (1860, p. 174) indeed suggested the term *lignes isométriques* because such lines divide the surface into infinitely small squares. The terms *isometric* or *isothermal net* were not yet used by Bonnet at that time, but they appear as French *réseau isotherme ou isométrique* in Picart (1867, p. 73).

Note. See Reich (1973) and Hertrich-Jeromin (2003, pp. 103f.) for information on the history of the notion.

References

Bonnet, Ossian (1860). "Mémoire sur l'emploi d'un nouveau système de variables dans l'étude des propriétés des surfaces courbes". *Journal de mathématiques pures et appliquées* 5, pp. 153–266.
Cauchy, Augustin-Louis (1828). *Exercises de mathématiques*. Troisième année. Paris.
Darboux, Gaston (1887). *Leçons sur la théorie générale des surfaces et les applications géométriques du calcul infinitesimal*. Première partie. Généralités. Coordonnées curvilignes. Surfaces minima. Paris.
Darboux, Gaston (2019). "Lessons on the General Theory of Surfaces and the Geometrical Applications of Infinitesimal Calculus". Translated by D.H. Delphenich. URL: `http://www.neo-classical-physics.info/geometry.html` (retrieved Dec. 30, 2019).
Hertrich-Jeromin, Udo (2003). *Introduction to Möbius Differential Geometry*. Cambridge.
von Humboldt, Alexander (1817a). *De distributione geographica plantarum secundum coeli temperiem et altitudinem montium, prolegomena*. Paris.
von Humboldt, Alexander (1817b). "Des lignes isothermes et de la distribution de la chaleur sur le globe". *Mémoires de physique et de chimie, de la Société d'Arcueil* 3, pp. 462–602.

---

[1] "Ita videmus circulos aequalis caloris annui sive, ut novo vocabulo utamur, isothermos, haud aequatori parallelos esse sed, ut lineas magneticas, angulo variabili parallellos geographicos transversim intersecare". Translation from Knobloch (2007, p. 45). [2] "Toutes les fois que l'élément linéaire d'une surface peut être ramené à la forme $ds^2 = \lambda(d\alpha^2 + d\beta^2)$, on dit que les courbes coordonnées forment un réseau *isotherme* ou *isométrique*. La première dénomination est empruntée à la théorie de la chaleur; la seconde [...] est due à M. Bonnet". Translation from Darboux (2019, p. 68).

von Humboldt, Alexander (1817c). “Des lignes isothermes et de la distribution de la chaleur sur le globe”. *Annales de chimie et de physique* 5, pp. 102–111.
von Humboldt, Alexander (1820). “On Isothermal Lines, and the Distribution of Heat over the Globe”. *The Edinburgh Philosophical Journal* 3, pp. 1–20.
Knobloch, Eberhard (2007). “Alexander von Humboldt: The explorer and the scientist”. In: *The Global and the Local*. The History of Science and the Cultural Integration of Europe. Ed. by Michael Kokowski. Cracow, pp. 38–48.
Lamé, Gabriel (1833). “Mémoire sur les surfaces isothermes dans les corps solides en équilibre de température”. *Annales de chimie et de physique* 53, pp. 190–204.
Picart, Alphonse (1867). “Sur les réseaux isométriques et la déformation des surfaces de révolution”. *Bulletin de la Société philomathique de Paris* 4, pp. 73–78.
Reich, Karin (1973). “Die Geschichte der Differentialgeometrie von Gauss bis Riemann (1828-1868)”. *Archive for History of Exact Sciences* 11, pp. 273–382.

isotopic* and *isotopy

See *homotopic*.

isotropic

This word is composed of Greek *ἴσος* ‘equal’ and *τρόπος* ‘direction, turn’. Its first modern occurrence seems to be in physics. According to Müller and Timpe (1907, p. 9), the French term *isotrope* was introduced by Cauchy (1839, p. 66), who used it in the context of elasticity theory. The following definition, although more general than Cauchy’s, shows why the term was chosen: “A body in which any given quality is at all points the same in all directions is said to be *isotropic* as to that quality” (Gray 1888, p. 103).[1]

In his classic *Traité des propriétés projectives des figures*, Poncelet (1822, p. 49) discussed the notion of circular points at infinity, which are “two imaginary points situated on the line at infinity and common to any two circles” (Kline 1972, p. 845). These two points play a central role in Laguerre’s (1867, p. 51) definition of *isotropic lines* (French *droites isotropes*), a term by which he designated those lines that pass through either of these points. Klein (1926, p. 144) explains the term thus: “this name is based on the fact that under every rotation about the origin two of these lines – the lines joining the origin to the circular points of the plane orthogonal to the axis of rotation – retain their position unchanged”[2], i. e., they do not change their direction. Such an explanation had been given by Laguerre (1870, p. 165) himself, who also stated that he wanted a “short and expressive” term and that the term had the advantage of not having been used in geometry before.

A more general term is *minimal curve* (German *Minimalcurve*), introduced by Lie (1879, p. 337):[3] “To simplify the language, I take the liberty of proposing to call curves whose length is zero *minimal curves*”.[4] He remarked (ibid.) that the name “seems natural” because, among other things, minimal curves could be

---

[1] Cauchy also defined some related terms such as *isotropic function* (*fonction isotrope*). See Cauchy (1847, p. 351) and Cauchy (1848, p. 625). [2] “[die Benennung] gründet sich auf die Tatsache, daß bei jeder Drehung um den Nullpunkt zwei dieser Geraden – die Verbindungslinien mit den Kreispunkten der zur Drehachse senkrechten Ebene – unveränderte Lage behalten”. Translation from Klein (1979, p. 132). [3] The term is earlier found in Lie (1878, p. 170). [4] “Um die Sprache zu erleichtern, erlaube ich mir vorzuschlagen, Curven, deren Länge gleich Null ist, mit dem Namen *Minimalcurven* zu bezeichnen”.

understood as minimal surfaces. A special case of minimal curves are minimal lines (*Minimalgeraden*), which correspond to Laguerre's *isotropic lines*.

An overview of the *isotropic*-terminology was given by Berwald (1913, p. 149), who also included *isotroper Vektor* for a vector that lies in a minimal line; he showed that such vectors have length zero. The corresponding English term is found in Neville (1922, p. 204): "A vector whose projected square is zero is said to be *nul* or *isotropic*". In a footnote, Neville remarked that the term is due to Laguerre.

For the term *isotropic group*, see *stabilizer*.

Note. For the history of the terminology, see the brief remarks in Struik (1953, p. 101), Brieskorn (1985, pp. 393f.), and Dingeldey and Fabry (1915, p. 73).

References

Berwald, Ludwig (1913). "Über die Flächen mit einer einzigen Schar zueinander windschiefer Minimalgeraden". *Sitzungsberichte der Mathematisch-Physikalischen Classe der K.B. Akademie der Wissenschaften* (1913), pp. 143–211.

Brieskorn, Egbert (1985). *Lineare Algebra und Analytische Geometrie II*. Noten zu einer Vorlesung mit historischen Anmerkungen von Erhard Scholz. Braunschweig and Wiesbaden.

Cauchy, Augustin-Louis (1839). "Mémoire sur la réflexion et la réfraction d'un mouvement simple transmis d'un système de molécules à un autre. IV. Sur les conditions générales de la coexistence de mouvements simples, que l'on suppose propagés dans deux portions différentes d'un système moléculaire, diversement constituées et séparées l'une de l'autre par une surface plane". *Comptes rendus de l'Académie des sciences* 9, pp. 59–68.

Cauchy, Augustin-Louis (1847). "Mémoire sur les mouvements des systèmes de molécules". *Comptes rendus de l'Académie des sciences* 24, pp. 348–352.

Cauchy, Augustin-Louis (1848). "Mémoire sur les valeurs moyennes des fonctions et sur les fonctions isotropes". *Comptes rendus de l'Académie des sciences* 26, pp. 624–627.

Dingeldey, Friedrich and Eugène Fabry (1915). "Coniques". *Encyclopédie des sciences mathématiques pures et appliquées* III.17, pp. 1–162.

Gray, Andrew (1888). *The Theory and Practice of Absolute Measurements in Electricity and Magnetism*. Volume I. London.

Klein, Felix (1926). *Vorlesungen über die Entwicklung der Mathematik im 19. Jahrhundert*. Teil I. Berlin.

Klein, Felix (1979). *Development of Mathematics in the 19th Century*. Translated by M. Ackerman. Brookline, Mass.

Kline, Morris (1972). *Mathematical Thought From Ancient To Modern Times*. Volume 3. New York and Oxford.

Laguerre, Edmond (1867). "Sur les courbes résultant de l'intersection d'une sphère avec une surface". *Bulletin de la Société philomathique de Paris* 4, pp. 51–58.

Laguerre, Edmond (1870). "Sur l'emploi des imaginaires en géométrie". *Nouvelles annales de mathématiques* 9, pp. 163–175.

Lie, Sophus (1878). "Sätze über Minimalflächen". *Archiv for Mathematik og Naturvidenskab* 3, pp. 166–176, 224–233, 340–351.

Lie, Sophus (1879). "Beiträge zur Theorie der Minimalflächen. I. Projectivische Untersuchungen über algebraische Minimalflächen". *Mathematische Annalen* 14, pp. 331–416.

Müller, Conrad Heinrich and Anton Aloys Timpe (1907). "Die Grundgleichungen der mathematischen Elastizitätstheorie". *Encyklopädie der mathematischen Wissenschaften mit Einschluss ihrer Anwendungen* IV 23, pp. 1–54.

Neville, Eric Harold (1922). *Prolegomena to Analytical Geometry in Anisotropic Euclidean Space of Three Dimensions*. Cambridge.

Poncelet, Jean Victor (1822). *Traité des propriétés projectives des figures*. Paris.

Struik, Dirk Jan (1953). *Lectures on Analytic and Projective Geometry*. Cambridge, Mass.

isthmus
See *bridge*.

MIX
Papier aus verantwortungsvollen Quellen
Paper from responsible sources
FSC® C105338

If you have any concerns about our products,
you can contact us on
ProductSafety@springernature.com

In case Publisher is established outside the EU,
the EU authorized representative is:
Springer Nature Customer Service Center GmbH
Europaplatz 3, 69115 Heidelberg, Germany

Printed by Libri Plureos GmbH
in Hamburg, Germany